Heinz Bauer (31.1.1928 – 15.8.2002)

Heinz Bauer

# Selecta

Edited by
Herbert Heyer
Niels Jacob
Ivan Netuka

Walter de Gruyter · Berlin · New York 2003

## *Editors*

Herbert Heyer
Eberhard-Karls-Universität Tübingen
Mathematisches Institut
Auf der Morgenstelle 10
72076 Tübingen
Germany

Niels Jacob
University of Wales Swansea
Department of Mathematics
Singleton Park
Swansea, SA2 8PP
United Kingdom

Ivan Netuka
Faculty of Mathematics and Physics
Charles University
Sokolovska 83
186 75 PRAHA 8
Czech Republic

∞ Printed on acid-free paper which falls within the guidelines of the ANSI
to ensure permanence and durability.

*Library of Congress Cataloging-in-Publication Data*

Bauer, Heinz.
[Selections. 2003]
Selecta / Heinz Bauer ; edited by Herbert Heyer, Niels Jacob, Ivan Netuka.
p. cm.
Includes bibliographical references.
ISBN 3-11-017350-6 (cloth : acid-free paper)
1. Integrals, Generalized. 2. Measure theory. 3. Potential theory (Mathematics) 4. Convex sets. I. Heyer, Herbert. II. Jacob, Niels. III. Netuka, Ivan. IV. Title.
QA312.B3262 2003
515'.42–dc21 2003051427

ISBN 3-11-017350-6

*Bibliographic information published by Die Deutsche Bibliothek*

Die Deutsche Bibliothek lists this publication in the Deutsche Nationalbibliografie; detailed bibliographic data is available in the Internet at <http://dnb.ddb.de>.



Printed in Germany.
Printing: Werner Hildebrand, Berlin
Binding: Lüderitz & Bauer GmbH, Berlin

# Preface

In times of an overflow of written publications the intellectual mind is forced to select from this abundance and enjoys any finding that intersects his reading interest within a reduced number of pages. Selecta from the work of a highly acknowledged mathematician can be effective reading with respect to various points of view. They can serve as an overview of the interactions between the fields the author has worked in. Occasionally they contain contributions that are difficult to find in libraries. At the very best they stress in their specific layout the author's initial ideas, his perspective and the influence of his investigations on subsequent developments. And, of course, they may become an appetite stimulant for novices in the fields of research dealt with in the book, as well as for the professional interested in tracing the origin of results due to the author. In addition Selecta allow, apart from the reproduction of a representative selection of papers, the inclusion of the author's curriculum, his bibliography and detailed comments on the distinguished themes of his work.

With the "Selecta Heinz Bauer" the editors tried to provide some of these indicated features. They have collected Bauer's most important contributions, classified them within the three topics "Measure and integration", "Convexity" and "Potential theory" and had the three subcollections provided by instructive essays on the work achieved and its implications. Together with the curriculum and bibliography the resulting volume is designed to offer to the appreciative reader Bauer's main innovative ideas in a convenient presentation.

The editors' thanks go to Professors S. D. Chatterji and D. A. Edwards for having supplied profound and comprehensive essays on Bauer's work in measure and integration theory and in convexity theory respectively. Their efforts were facilitated by use of a brochure of Bauer's vita academia which colleagues of the Charles University of Prague prepared on the occasion of an honorary doctor's degree granted to him in 1992. Thanks also go to the members of the Bauer family for having offered help in collecting the necessary information for a complete bibliography and in providing a fine portrait of the author. And there is, by no means to be forgotten, the editors' sincere appreciation addressed to De Gruyter Publishers for their commitment to producing these Selecta and for giving them the traditionally attractive form.

The present volume was to be handed over to Professor Bauer on the occasion of his 75th birthday. His sudden death on August 15, 2002 which signifies a grave loss to the mathematical community converts this ceremonial idea into precious legacy.

*The Editors*

# Curriculum vitae

Geboren am 31. Januar 1928 in Nürnberg, verstorben am 15. August 2002 in Erlangen. Seit 1957 mit Irene Bauer, geb. Pöllet, verheiratet, Kinder Christian Johannes und Christina Franziska

Studium der Mathematik und Physik an den Universitäten Erlangen und Nancy (1948–1953)
Prüfung für das Lehramt an höheren Schulen (1952)
Promotion zum Dr. phil. nat. an der Universität Erlangen (1953)

Wissenschaftlicher Assistent am Mathematischen Institut der Universität Erlangen (1953–1955)
Attaché de Recherches am Centre National de la Recherche Scientifique, Paris (1956–1957)
Habilitation im Fach Mathematik an der Universität Erlangen (1956)
Privatdozent und Assistent am Mathematischen Institut der Universität Erlangen (1957–1959)
Universitätsdozent an der Universität Hamburg (1959–1961)
Lehrstuhlvertretung an der Universität München (1960/61)
Lehrstuhlvertretung an der Universität Hamburg (1961)
Ordentlicher Professor für Versicherungsmathematik und Mathematische Statistik an der Universität Hamburg (1961–1965)
Ordentlicher Professor für Mathematik an der Universität Erlangen-Nürnberg (1965–1996)
Professor Emeritus an der Universität Erlangen-Nürnberg (seit 1996)

Mitglied der Bayerischen Akademie der Wissenschaften (1975)
Ausländisches Mitglied der Finnischen Akademie der Wissenschaften (1980)
Korrespondierendes Mitglied der Östereichischen Akademie der Wissenschaften (1981)
Ausländisches Mitglied der Königlichen Dänischen Akademie der Wissenschaften (1982)
Mitglied der Deutschen Akademie der Naturforscher, Leopoldina (1986)
Mitglied der Academia Scientarium et Artium Europaea (1993)

Ehrendoktorwürde der Karls-Universität Prag (1992)
Ehrendoktorwürde der Technischen Universität Dresden (1994)

Chauvenet Prize der Mathematical Association of America (1980)
Friedensmedaille der Karls-Universität Prag (1987)
Medaille der Union Tschechoslowakischer Mathematiker und Physiker (1989)
Ehrenmitglied der Gesellschaft Carolinum (1993)
Bayerischer Verdienstorden (1994)
Bayerischer Maximiliansorden für Wissenschaft und Kunst (1998)

Mitglied des Präsidiums der Deutschen Mathematiker Vereinigung (1965–1978)
Mitglied des Wissenschaftlichen Beirats der Gesellschaft für Mathematische Forschung (Oberwolfach), (1966–1982)
Vorsitzender der Deutschen Mathematiker Vereinigung (1976–1978)
Mitglied des Senats der Deutschen Forschungsgemeinschaft (1980–1986)
Obmann für Mathematik der Deutschen Akademie der Naturforscher, Leopoldina, Mitglied des Senats der Akademie (seit 1991)
Vorsitzender des Vorstands der Gesellschaft für Mathematische Forschung (Oberwolfach), (1993–1994)
Amtsperioden als Dekan (Hamburg 1964-1965), und Senator (Erlangen-Nürnberg, 1988-1990 und 1992–1994)

Mitherausgeber der Abhandlungen aus dem Mathematischen Seminar der Universität Hamburg (1961–1966), Inventiones Mathematicae (1966–1979), Erlanger Unterreihe der Lecture Notes in Mathematics (seit 1968), Seminar über Potentialtheorie (1968), Seminar on Potential Theory II (1971), Mathematische Annalen (1970–1992); geschäftsführender Herausgeber der Mathematischen Annalen (1985–1992), de Gruyter Studies in Mathematics (1981–2001), Expositiones Mathematicae (seit 1983), Aequationes Mathematicae (seit 1988)

Im Rahmen längerer Gastaufenthalte wurden zahlreiche Universitäten in Europa, Nord- und Südamerika, Nordafrika, Neuseeland und Japan besucht.

# Curriculum vitae

| | |
|---|---|
| 1928, January 31 | Born in Nürnberg |
| 1948–1953 | Student of Mathematics and Physics, University of Erlangen and Université Nancy |
| 1953 | Dr. phil. nat. University of Erlangen |
| 1953–1955 | Assistant, University of Erlangen |
| 1955–1957 | Research Fellow (Attaché de Recherches), CNRS, Paris |
| 1956 | Habilitation at the University of Erlangen |
| 1957–1959 | Dozent, University of Erlangen |
| 1959–1961 | Dozent, University of Hamburg |
| 1960/61 | Replacements of Full Pofessorships in Hamburg and München |
| 1961–1965 | Full Professor, University of Hamburg |
| 1965–1996 | Full Professor, University Erlangen-Nürnberg |
| 1996–2002 | Professor emeritus,University of Erlangen-Nürnberg |
| 1975 | Member of the Bavarian Academy of Sciences |
| 1980 | Foreign Member of the Finnish Academy of Sciences |
| 1981 | Corresponding Member of the Austrian Academy of Sciences |
| 1982 | Foreign Member of the Danish Academy of Sciences |
| 1986 | Member of the Leopoldina, German Academy of Scientists |
| 1993 | Member of the Academia Scientarium et Artium Europaea |
| 1992 | Honorary degree from Charles University Prague |
| 1994 | Honorary degree from Technical University Dresden |
| 1980 | Chauvenet prize of the Mathematical Association of America |
| 1987 | Peace Medal from Charles University Prague |
| 1989 | Medal of the Union of Czechoslovak Mathematicians and Physicists |
| 1993 | Honorary Member of the Society Carolinum |
| 1994 | Bavarian Order of Merit |
| 1998 | Bavarian Maximilians Order for Sciences and Arts |
| 1965–1978 | Member of the Council of the German Mathematicial Society (DMV) |

| | |
|---|---|
| 1966–1982 | Member of the Council of the Mathematical Research Institute Oberwolfach |
| 1976–1978 | President of the German Mathematicial Society (DMV) |
| 1980–1986 | Member of the Senate of the German Science Foundations (DFG) |
| 1993–1994 | President of the Society of Mathematical Research (Oberwolfach) |
| 1961–1966 | Editor of Abhandlungen aus dem Mathematischen Seminar der Universität Hamburg |
| 1966–1979 | Editor of Inventiones Mathematicae |
| 1970–1992 | Editor of Mathematische Annalen, Managing Editor 1985–1992 |
| 1981–2001 | Editor of de Gruyter Studies in Mathematics |
| 1983–2002 | Editor of Expositiones Mathematicae |
| 1988–2002 | Editor of Aequationes Mathematicae |

Extended visits to several universities in Europe, North and South America, North Africa, New Zealand and Japan.

# Ph.D. students of Heinz Bauer

Guber, Siegfried
*Über die Struktur gewisser Funktionenkegel* (1961)

Heyer, Herbert
*Untersuchungen zur Theorie der Wahrscheinlichkeitsverteilungen auf lokalkompakten Gruppen* (1964)

Köhn, Johannnes
*Die Harnacksche Metrik in der Theorie der harmonischen Räume* (1965)

Hinrichsen, Diederich
*Randintegrale und nukleare Funktionenräume* (1966)

Krause, Ulrich
*Kennzeichnung und Behandlung von Markoffprozessen* (1967)

Sieveking, Malte
*Integraldarstellung superharmonischer Funktionen mit Anwendung auf parabolische Differentialgleichungen* (1967)

Hansen, Wolfhard
*Konstruktion von Halbgruppen und Markoffschen Prozessen* (1967)

Pasedach, Klaus
*Relationen und Ränder* (1968)

Sondermann, Dieter
*Maße auf lokalbeschränkten Räumen* (1968)

Bliedtner, Jürgen
*Harmonische Gruppen und Huntsche Faltungskerne* (1968)

Janßen, Klaus
*Martin-Rand und $H^p$-Theorie harmonischer Räume* (1969)

Lembcke, Jörn
*Konservative Abbildungen und Fortsetzung regulärer Maße* (1969)

Schneider, Friedrich
*Dualität und Martinrand* (1969)

Anger, Bernd
*Extremalpunktmethoden bei Koeffizientenproblemen* (1969)

van Arkel, Nicolaas
*Algebras of Holomorphic Functions of n Complex Variables* (1970)

Hommel, Gerhard
*Isotone Radon-Maße auf lokal kompakten geordneten Räumen* (1973)

Gunselmann, Winfried
*Interne Vollständigkeit* (1973)

Held, Norbert
*Über das Hochheben von stetigen Markoffprozessen auf Überlagerungsflächen* (1973)

Ritter, Gunter
*Eine Methode zur Konstruktion von Resolventen mit Hilfe von Absorptionsmengen* (1974)

Donner, Klaus
*Extremalpunktmethoden bei beschränkten, geordneten Algebren* (1974)

Bellmann, Rainern
*Normalität und verwandte Eigenschaften in lokal-konvexen Vektorräumen* (1974)

Clausing, Achim
*Koeffizientenmethoden bei der Fortsetzung invarianter Funktionale* (1975)

Schirmeier, Ursula
*Isomorphie harmonischer Räume* (1975)

Leha, Gottlieb
*Korovkinsätze für Funktionenräume* (1976)

Dembinski, Volker
*Recollement von Diffusionsprozessen und die Feller-Eigenschaft* (1976)

Schirmeier, Horst
*Geometrische Aspekte in der Theorie der nuklearen Räume* (1977)

Weippert, Georg
*Untersuchungen zur konstruktiven Begründung der Wahrscheinlichkeitstheorie* (1977)

Guggenmoss-Holzmannn, Irene
*Harmonische Funktionen auf abstrakten Wienerräumen* (1977)

Barth, Thomas
*Integraldarstellungen mehrfach superharmonischer Funktionen* (1977)

Wittmann, Rainer
*KACsche Potentialtheorie für Resolventen, Markoffsche Prozesse und harmonische Räume* (1981)

Kröger, Pawel
*Vergleichssätze für Diffusionsprozesse* (1986)

Sturm, Karl-Theodor
*Störung von Hunt-Prozessen durch signierte additive Funktionale* (1989)

Höhnle, Rainer
*Lokale Lösungen stochastischer Differentialgleichungen mit singulärer Drift* (1993)

Gallus, Cristopher
*Robustheit von Strategien zur Absicherung derivativer Finanzinstrumente* (1996)

# Contents

Entries in square brackets refer to the bibliography on pages 589–597

# The work of Heinz Bauer in measure and integration

*S. D. Chatterji*

## 1 Introduction

Bauer's mathematical researches started with investigations of certain problems in measure and integration theory; most of his research papers in this and allied areas are essentially concentrated in the period 1953–59; all the papers [R1]–[R15]* fall in this time span; [R25] (published in 1966) is a simplification and improvement of many of the results of [R14] and [R22] (published in 1964) is an isolated paper using integration theory written in the midst of a period (starting already with 1957–58) when his major interest had shifted to axiomatic potential theory. As is well-known, the latter theory has close connections with various developments in probability theory and convexity theory, subjects which Bauer's early work in measure and integration made him specially suited to study and nurture. In this survey, we describe (in 2) Bauer's original research contributions in measure and integration theory as contained in the papers mentioned above; we try to indicate also the later publications of others where Bauer's work in this area has had an influence. Finally (in 3) we shall briefly report on Bauer's expository writings in this field as incorporated in his very successful books and lecture notes.

## 2 Measure and integration

We have subdivided this central section into three subsections captioned Lattice theoretical (2.1), Integration theoretical (2.2) and Topological (2.3).

### 2.1 Lattice theoretical

The work described here is contained in the items [R1]–[R3], [R11], [R12], [S1] and [P1]–[P2]; the most important of these are [R1] and [R3] which are closely related (and form part of Bauer's 1953 Erlangen Dissertation) and the two papers [P2] and [R11] on the extension of positive linear forms.

*References containing numbers refer to the Bibliography of Heinz Bauer whereas letters in brackets are related to references placed at the end of this commentary.

Let us recall that lattice theory was very much in the air around 1950; G. Birkhoff's *"Lattice theory"* had appeared in its revised edition in 1948 and it contained a number of interesting applications of lattices to measure theory and functional analysis; Bourbaki's much awaited book on integration theory (only the first four chapters) had just appeared in 1952; this contained, besides their radically new approach to Radon measures as linear functionals, a preliminary substantial chapter on vector lattices (called Riesz spaces by Bourbaki) where F. Riesz's work (in particular, his decomposition theory) was developed in detail for use in integration theory. These two monographic references recur in Bauer's work; but, as is evident from other references, Bauer was clearly aware of different measure theoretical applications of lattice theory as contained in somewhat earlier papers of Bochner, Phillips, Kakutani and others from the 1940s. In the introduction to [R1], Bauer explicitly mentions his motivation as stemming from the proof of the Jordan decomposition of functions of bounded variation (as a difference of two monotone functions) as presented by Birkhoff via his theory of ($\mathbb{R}$-valued) valuations of lattices; Birkhoff had left the generalization of this work to valuations with values in complete lattice-ordered groups as an open problem. This induced Bauer to study in detail various decomposition theorems in the framework of complete vector lattice-valued valuations on a general lattice; the more general case of valuations with values in complete lattice-ordered groups is mentioned in both papers [R1] and [R3] but is not taken up.

Let $V$ be any lattice, $W$ a complete vector lattice; a map $f\colon V \to W$ is called a valuation if $f$ satisfies the equation

$$f(x \vee y) + f(x \wedge y) = f(x) + f(y)$$

(given in the title of [R3]) for all $x$, $y$ in $V$. If $V$ is a Boolean ring, $f$ is just a $W$-valued (finitely additive) "measure" on $V$; if $V$ is a vector lattice and $f$ is linear then $f$ is a valuation; these are the two important cases (where $V$ is a distributive lattice) which Bauer obviously has in view, although a good deal is worked out for general (not necessarily distributive) lattices $V$; some simple remarks and definitions are included even for partially ordered sets $V$ which are not necessarily lattices.

We shall now summarize the principal results of [R1] and [R3] in the set-up of a general lattice $V$ and a complete vector lattice $W$ (over the field $\mathbb{R}$). If $c$ is a fixed element of $V$, let $\Psi_c = \Psi_c(W, V)$ be the set of all valuations $f\colon V \to W$ such that $f(c) = 0$; under obvious vector operations and using a natural ordering, $\Psi_c$ becomes a partially ordered vector space. Let $\Phi_c$ denote the subset of those valuations in $\Psi_c$ which are of relatively bounded variation (a natural notion introduced by Birkhoff); an important first result is that $\Phi_c$ is a complete vector lattice (this is proved in [R1]). Let $\Phi_c^s$ be the set of continuous valuations $f$ in $\Phi_c$ (i.e. if $x_n \in V$, $n = 1, 2, \ldots, x = \sup x_n \in V$, then $f(x) = \sup f(x_n)$); then $\Phi_c^s$ is a band in $\Phi_c$ (i.e. a solid linear subspace which is also a complete vector sub-lattice). Let $\Phi_c^u$ be the set of purely discontinuous elements $f$ of $\Phi_c$ (i.e. $|g| \leq |f|$, $g$ continuous, implies $g = 0$). Then, $\Phi_c$ is a direct sum of $\Phi_c^s$ and $\Phi_c^u$, i.e. each $f \in \Phi_c$ can be written uniquely as $f = f_s + f_u$ where $f_s \in \Phi_c^s$, $f_u \in \Phi_c^u$. This decomposition theorem is one of the important results of [R1]; specialized to the case where $V$ is a Boolean ring of subsets of a basic set and $c = \emptyset$, this theorem corresponds to one proved by Hewitt and Yosida in 1952 for bounded real-valued finitely additive measures $\mu$ on a Boolean algebra of subsets according to which $\mu = \mu_c + \mu_p$ where $\mu_c$ is $\sigma$-additive and $\mu_p$ is purely finitely additive. Various other specializations of the decomposition theorem are studied in detail in [R1]; of course, the fact that $\Phi_c$ is a (complete) vector lattice automatically gives a decomposition $f = f^+ - f^-$ for all $f$ in $\Phi_c$; if $V$ is a Boolean ring and $c = o \in V$, this produces a Jordan type decomposition theorem for $W$-valued finitely additive functions of bounded variation.

The paper [R3] gives other decompositions of $W$-valued valuations of relatively bounded variation on a lattice $V$; these correspond to the well-known Lebesgue decomposition of real-valued $\sigma$-additive measures with respect to a given non-negative real-valued $\sigma$-additive measure defined on a Boolean $\sigma$-algebra of subsets of a basic set. A very general version of this is obtained as follows: let $S \subset V$; for $f \in \Phi_c = \Phi_c(W, V)$ (see definition above) with $f \geq 0$, let $f_S = \inf\{h \in \Phi_c : h \geq 0, h(x) = f(x), x \in S\}$; for any $f \in \Phi_c$, write $f_S = (f^+)_S - (f^-)_S$. Call $f \in \Phi_c$ $S$-regular if $f_S = 0$ and $S$-singular if $f_S = f$.

If $V$ is a distributive lattice and $S$ is a "saturated" sublattice of $V$ (i.e. $a, b \in S$, $a \leq x \leq b$, implies $x \in S$) then the formula

$$f = f_S + (f - f_S), \quad f \in \Phi_c$$

gives a direct sum (order preserving) decomposition of $\Phi_c$ into $S$-regular ($f - f_S = S$-regular part of $f$) and $S$-singular ($f_S = S$-singular part of $f$) elements; further, $f_S = 0$ if and only if $f(x) = 0$ for all $x \in S$. This is one of the central results of [R3]. Various specializations of $V$, $S$, $W$ are then shown to give rise to Lebesgue type decompositions; these are studied and described in detail in [R3].

The papers [R2], [S1], [P1] give further information on the above-mentioned Hewitt–Yosida decomposition for bounded real-valued finitely additive set functions defined on Boolean algebras of sets of a fixed space.

The papers [P2] and [R11] concern the possibility of extending a positive (real-valued) linear form defined originally on a vector subspace of a partially ordered vector space to the whole space (i.e. a positivity preserving Hahn–Banach theorem). As pointed out by the author in [R11], similar theorems were obtained independently by Namioka in 1957; these generalize and complete previous results of others (notably, M. Krein und Dixmier). Because of the usefulness of these results, they appear in many textbooks; a standard text giving the main results as well as exact references to Bauer and to Namioka is the book of Schaefer [Sch].

The paper [R12] gives an isolated result on ordered abelian groups; its main interest lies in the fact that it was motivated by a uniqueness question in Choquet's theory of convex cones with compact basis.

## 2.2 Integration theoretical

The two important papers here are [R9] and [R10]; [R4] is essentially a detailed résumé of [R9]; however, [R4] contains a summary of a general theory of vague convergence of positive linear functionals which was not included in [R9]. In order to describe the main contents of [R9] and [R10], we need to establish some terminology and recall briefly some historical antecedents.

That Lebesgue's and Radon's integration theories in $\mathbb{R}^n$ could be worked out without difficulty in the framework of an abstract measure space $(X, \Sigma, \mu)$ had been noticed by Fréchet in a paper of 1915 (cf. reference in Saks [S]); here, $X$ is an abstract set, $\Sigma$ is a $\sigma$-algebra (also called a Borel field) of subsets of $X$ and $\mu\colon \Sigma \to [0, \infty]$ is a $\sigma$-additive set function (called measure). Some authors take $\Sigma$ to be only a $\sigma$-ring (i.e. the whole space $X$ need not belong to $\Sigma$, although $\Sigma$ is closed under denumerable unions and differences); this is a minor point since given $\mu$ on a $\sigma$-ring, it can always be extended to a $\sigma$-algebra (non-uniquely in general and in several ways) so that the integration theory of (say real) functions on $X$ is not affected by this. An early presentation of integration theory based on $(X, \Sigma, \mu)$ is the classic book of Saks [S] (whose definitive 2nd edition in English dates from 1937); a later standard text by Halmos (published in 1950) works with $\sigma$-rings. In 1917–18, Daniell introduced another method for studying integration theory as the process of extending a basic positive linear functional $I$ defined on a vector lattice of *bounded* real-valued functions $\mathfrak{L}$ on an abstract set $X$ verifying the continuity condition $(L)$:

$$(L)\colon \text{ if } 0 \leq f_n \in \mathfrak{L},\ f_n \downarrow 0, \text{ then } I(f_n) \to 0 \text{ as } n \to \infty.$$

Daniell did not try to establish the existence of an abstract measure space $(X, \Sigma, \mu)$ whose integrable functions will correspond exactly to his summable functions; he simply said near the end of his paper that "This question requires however a separate and careful consideration" (p. 294 of his paper, cf. reference in [S]). Indeed, as shown by Stone's analysis of Daniell's theory (cf. Note II of Stone [21] referred to in Bauer's [R9]), a further condition on the vector lattice $\mathfrak{L}$ is needed to establish equivalence between integration based on an abstract measure space and that described by Daniell; a simple form of this condition is the so-called Stone condition $(S)$ on $\mathfrak{L}$:

$$(S)\colon \text{ if } f \in \mathfrak{L}, \text{ then } \min(1, f) = 1 \wedge f \in \mathfrak{L}.$$

Thus, Stone's hypotheses are that $\mathfrak{L}$ is a vector lattice of real-valued functions (not necessarily bounded) on $X$ and $I\colon \mathfrak{L} \to \mathbb{R}$ is a positive (i.e. $f \geq 0 \Rightarrow I(f) \geq 0$) linear functional satisfying $(L)$; Stone's hypothesis is not $(L)$ but, in the context concerned, equivalent to $(L)$. Assuming $(S)$, Stone shows the existence of an abstract measure space $(X, \Sigma, \mu)$ whose integrable functions are exactly those given by the Stone(–Daniell) theory; Stone also gives a detailed analysis showing what more general condition could replace $(S)$ in order to arrive at his abstract measure space. At the time of Stone's writing, Bourbaki's work on integration theory had not yet been published, although Stone seems to have had some knowledge of it; in any case, Weil's

1940 book *"L'intégration dans les groupes topologiques"* indicated explicitly that the theory was to be based on the notion of linear functionals on the space $\mathfrak{L} = \mathfrak{L}(X)$ of continuous functions with compact support in a locally compact topological space $X$; a positive linear functional $I$ on $\mathfrak{L}$ was to be called a positive Radon measure on $X$ by Bourbaki (the 1st edition of Bourbaki's book on the subject, published in 1952, is referred to in Bauer [R9] as item [8]). Happily, in the situation envisaged by Bourbaki, the vector lattice concerned verifies condition $(S)$ and that any positive linear functional on it automatically satisfies the continuity condition $(L)$; thus, Bourbaki's theory for locally compact spaces turns out to be a very special case of Stone's general theory. The object of Bauer's paper [R10] is to establish that, conversely, Stone's theory can also be deduced in a very satisfactory way from Bourbaki's; we shall elaborate on this later.

Once the theory of positive linear functionals $I$, verifying $(L)$, on Stone vector lattices $\mathfrak{L}$ (i.e. vector lattices satisfying $(S)$) had been clarified, it was natural to investigate as to what happened if $I$ did not satisfy the condition $(L)$. In the early 1950s, several authors had looked into this, some of whom (Arens, Hewitt, Loomis) are mentioned in Bauer [R9]; of these, Loomis' paper (item [14] in Bauer [R9]) is the object of Bauer's careful study [R9]. To describe the main results of [R9], it is necessary to indicate Loomis' analysis which is partially summarized in §1 of [R9]. Given a positive linear functional $I$ defined on a Stone vector lattice $\mathfrak{L}$ of (real) functions on a set $X$, Loomis extends $I$ (uniquely) to $\bar{I}$ defined on $\overline{\mathfrak{L}}$ as follows: $f \in \overline{\mathfrak{L}}$ if and only if, for any $\varepsilon > 0$, there exists $g, h$ in $\mathfrak{L}$ such that $g \leq f \leq h$ and $I(h-g) \leq \varepsilon$. Let $\mathfrak{F}$ be the family of subsets $A$ of $X$ such that $1_A \in \overline{\mathfrak{L}}$ ($1_A$ being the indicator function of the set $A$); then $\mathfrak{F}$ turns out to be a (Boolean) ring of subsets of $X$ and $\mu(A) = \bar{I}(1_A)$ defines an $\mathbb{R}_+$-valued finitely additive set function (called content) on $\mathfrak{F}$. The content $\mu$ on $\mathfrak{F}$ permits the definition of a class of ($\mu$- or $I$-)Riemann-integrable functions on $X$; $f\colon X \to \mathbb{R}$ is called Riemann-integrable if, for any $\varepsilon > 0$, there exist $g, h$ two $\mathfrak{F}$-simple functions (i.e. finite linear combinations of $1_A$, $A \in \mathfrak{F}$) such that $g \leq f \leq h$ and $\bar{I}(h-g) \leq \varepsilon$. If $\mathfrak{S}$ is the class of $\mu$-Riemann-integrable functions then $\mathfrak{S} \subset \overline{\mathfrak{L}}$ and $\bar{I}$ is a positive linear functional on each of the Stone vector lattices $\mathfrak{S}$ and $\overline{\mathfrak{L}}$. Let us call $f\colon X \to \mathbb{R}_+$ as ($I$-)measurable if the set $[f > \lambda] \in \mathfrak{F}$ for all $\lambda > 0$, except perhaps for a countable number of exceptional values of $\lambda$; here $[f > \lambda]$ denotes the set $\{x : f(x) > \lambda\}$; if $f\colon X \to \mathbb{R}$, we say that $f$ is ($I$-)measurable if $f^{\pm}$ are $I$-measurable. It turns out (Satz 1, 2, [R9]) that all functions in $\overline{\mathfrak{L}}$ are $I$-measurable and $f\colon X \to \mathbb{R}$ is $I$-Riemann integrable (i.e. $f \in \mathfrak{S}$) if and only if $f$ is bounded, $I$-measurable and $f(x) = 0$ for $x \in A^c$ for some $A \in \mathfrak{F}$; for such $f$ (i.e. $f \in \mathfrak{S}$) we write $I(f) = \int f\,d\mu$; if $f \geq 0$ is in $\mathfrak{S}$, $I(f)$ can be written as an ordinary Riemann–Stieltjes integral:

$$\bar{I}(f) = \int f\,d\mu = -\int_0^a \lambda\,d\alpha(\lambda)$$

where $0 \leq f \leq a$, $\alpha(\lambda) = \mu([f > \lambda])$.

In order to study $I$-measurable $f$'s which are either unbounded or else do not vanish on $A^c$ for any $A \in \mathfrak{F}$ (this happens if $\mathfrak{F}$ is not an algebra or if $\mu(A)$, $A \in \mathfrak{F}$, is unbounded) one has to introduce improper Riemann-integrability; by passing to $f^{\pm}$ it suffices to do so for $f \geq 0$. An $I$-measurable $f \geq 0$ is called improper Riemann-integrable (using the same symbol $\int f\, d\mu$ as for ordinary Riemann-integrability) if

$$\int f d\mu = -\int_0^{\infty} \lambda\, d\alpha(\lambda) < \infty.$$

It turns out that all $f$ in $\overline{\mathfrak{L}}$ are improper Riemann-integrable and if $f \geq 0$ is in $\overline{\mathfrak{L}}$ then

$$\int f\, d\mu \leq \bar{I}(f);$$

this is contained in Satz 3, [R9]. Further work of Loomis (as well as Bauer's) concerns the investigation of when equality holds above for all $f$ in $\mathfrak{L}$ (or $\overline{\mathfrak{L}}$). At this point, Bauer uses his decomposition theory (see 2.1 above) of writing $\mu = \mu_c + \mu_p$ and correspondingly $I = I_c + I_p$. It then turns out that

$$I_c(f) = \int f\, d\mu_c, \quad f \in \mathfrak{L}$$

and

$$D(f) = f - \int f\, d\mu, \quad f \in \mathfrak{L}$$

is purely discontinuous (Satz 5, 6, [R9]). Bauer's work in [R9] now goes in a novel direction; he first specializes the above to the case where $X$ is a locally compact space and $\mathfrak{L}$ is a Stone vector sublattice of $C_0(X)$, the space of all real continuous functions on $X$ vanishing at $\infty$ which separates the points of $X$ (conditions (21), (22) of [R9]). An important fact is that $\mathfrak{L}$ will then be dense in $C_0(X)$ in the uniform norm topology of the latter (Lemma 4, [R9]) so that any positive linear functional $I$ on $\mathfrak{L}$ can be extended uniquely to $C_0(X)$, in particular to $\mathfrak{K}(X)$, the set of continuous functions with compact support in $X$. In such a situation, we not only have the objects $\mathfrak{F}$, $\mu$, $\mathfrak{S}$ as indicated above but also a positive Radon measure written as $\varrho$ (both as a linear functional on $\mathfrak{K}(X)$ as also for its values $\varrho(A)$ at $A \subset X$, $A$ say Borelian). It turns out then that

$$\int f\, d\mu = \int f\, d\varrho, \quad f \in \mathfrak{K}(X)$$

(Satz 8, [R9]) and $\mu = \mu_c$, $\mu_\varrho = 0$ as also (Satz 10, [R9])

$$I_c(f) = \int f\, d\varrho = \int f\, d\mu, \quad f \in \mathfrak{L}.$$

The last step in Bauer's analysis in [R9] now is the most original; starting with a general Stone vector lattice $\mathfrak{L}$ of real functions on an abstract set $X$ and any positive linear functional $I$ on $\mathfrak{L}$, Bauer shows first that, without any loss of generality (in a precise sense clarified in pp. 470–471 of [R9]), one may suppose that the functions in $\mathfrak{L}$ are bounded and that they separate the points of $X$ (cf. (21), (22) of [R9]). Assuming these latter conditions, his main theorem (Satz 12, [R9]) shows how to transpose the "abstract" problem to the "concrete" one of locally compact spaces treated before. He shows that there is a locally compact space $X'$ having the following properties:

(i) $X$ is a dense subset of $X'$;

(ii) each function $g$ in $\mathfrak{L}$ can be extended (uniquely) to a function $g'$ in $C_0(X')$;

(iii) the set $\mathfrak{L}'$ consisting of the $g'$ obtained from $g \in \mathfrak{L}$ forms a Stone vector lattice verifying the original separation conditions in $X'$;

the space $X'$ is unique up to homeomorphism (see Satz 13, [R9], for exact formulation) and is independent of the positive linear functional $I$ given on $\mathfrak{L}$. How the space $X'$ is obtained from $X$ and $\mathfrak{L}$ will be indicated in 2.3. Given $I$ on $\mathfrak{L}$, it can be transferred (uniquely) to $I'$ on $\mathfrak{L}'$ in the obvious way: $I'(g') = I(g)$, $g \in \mathfrak{L}$. The analysis which precedes can now be used on $\mathfrak{L}'$, $I'$ and then transferred back to $\mathfrak{L}$, $I$ to obtain complete results, specially as regards to the equality between $I(f)$ and $\int f\,d\mu$ for all $f \in \mathfrak{L}$, see Satz 16, [R9]. As a special case of this analysis, it is shown in the Example 4.7 of [R9], the relationship between the Stone space (a compact totally disconnected topological space) of a (Boolean) algebra $\mathfrak{B}$ of subsets of a set $X$ and the associated space $X'$ given above; $X'$ now turns out to be compact; this permits a clear picture of the Hewitt–Yosida decomposition of set functions in $\mathfrak{B}$, as was done by Bauer separately in a previous paper [R2].

The idea of passing from $X$ to $X'$ is suggested at several places in Loomis' paper; the complete and precise execution of this intuitive idea is entirely the merit of Bauer. The results of [R10] flow essentially from thc work in [R9] although the exact details need elaboration; this is done with complete clarity and almost independently of the work in [R9] so that [R10] is fairly self-contained; its main result is Theorem 3, [R10] which we shall formulate as follows. Let $E$ be any set, $\mathfrak{R}$ a Stone vector lattice of functions on $\mathfrak{R}$; an abstract measure $\mu$ on $\mathfrak{R}$ (or on $E$) is a positive linear functional on $\mathfrak{R}$ satisfying the continuity condition $(L)$; then, there exists a locally compact space $E'$ such that for each (positive) abstract measure $\mu$ on $\mathfrak{R}$ there is a positive Radon measure $\mu'$ on $E'$ such that the Banach space $L^p_F(\mu')$ is isometrically isomorphic to the Banach space $L^p_F(\mu)$ for every $p$, $1 \le p < \infty$, and for every Banach space $F$ ($L^p_F$ denoting Banach spaces derived from $F$-valued $p$th power integrable functions); if $F = \mathbb{R}$, the isomorphisms are also lattice isomorphisms; the correspondence $\mu \mapsto \mu'$ is a bijection between the class of (positive) abstract measures $\mu$ on $E$ (i.e. on $\mathfrak{R}$) and the positive Radon measures $\mu'$ on $E'$; further, if $E$ is already locally compact and $\mathfrak{R}$ is $\mathfrak{K}(E)$, then $E' = E$. This theorem completely justifies the equivalence mentioned in the title of [R10]; as Bauer rightly explains in the introduction to [R10], this presentation of the equivalence is distinctly more desirable than the others known before.

The problem of representing linear functionals via integrals of various kinds has numerous ramifications, even when we restrict ourselves to real-valued objects. The monographs by Anger and Portenier [AP] and that of König [K] give references to recent work; both cite [R9]. As regards [R10], Métivier [M] has given an alternative discussion using the theory of locally convex spaces, thus generalizing Bauer's result somewhat by replacing the Stone condition on the underlying vector lattice by a more general one. Bichteler [B] has organized his whole presentation around ideas taken from [R10].

The papers [P3], [R14] and [R25] treat the following existence problem for measures: let $\varphi\colon X \to Y$ be a continuous map between locally compact spaces $X$, $Y$; $\varphi$ is called conservative if every positive Radon measure $\nu$ on $Y$, concentrated in the set $\varphi(X)$, is the image by $\varphi$ of at least one positive Radon measure $\mu$ in $X$ (i.e. $\nu = \varphi(\mu)$). The paper [P3] is a preliminary announcement of certain results concerning conservative maps and [R14] is a full development with various applications; after the publication of a decomposition result in a paper by Herz 1960, Bauer obtained a fresh proof for a more general version of Herz's decomposition result and deduced, in [R25], the main results of [P3] more easily. The decomposition result in question can be stated as follows ($\varphi$, $X$, $Y$ being as above): any positive Radon measure $\nu$ in $Y$ can be written uniquely as $\nu = \nu_1 + \nu_2$ where $\nu_1$ is $\varphi$-representable (i.e. $\nu_1 = \varphi(\mu)$ for some positive Radon measure $\mu$ in $X$) and $\nu_2$ is a positive Radon measure in $Y$ which is $\varphi$-diffuse (i.e. $\nu_2(\varphi(K)) = 0$ for every compact subset $K$ in $X$). As a simple corollary the following is obtained: $\varphi$ is conservative if $\varphi$ is locally proper (i.e. each $y \in \varphi(X)$ has a neighbourhood $V$ such that $\varphi^{-1}(V)$ is compact); in particular, $\varphi$ is conservative if $\varphi$ is proper (i.e. $\varphi^{-1}(K)$ is compact in $X$ for each compact subset $K$ of $Y$). A simple example of a non-conservative map $\varphi$ is given in [R14] (p. 405).

The problem of $\varphi$-representability has been profitably generalized by Lembcke to yield, amongst other things, various theorems concerning the existence of measures with prescribed marginals, see [$L$] for a brief overview and further references.

We end this section with [R22]; motivated by some work of Arens and Godement, this paper characterizes bilinear forms

$$\langle f, g\rangle = \int fg\,d\mu, \quad f, g \in \mathfrak{A}$$

on subalgebras $\mathfrak{A}$ of $C_0(X) \cap \mathfrak{L}^2(\mu)$ where $X$ is a locally compact space, $\mu$ is a positive Radon measure in $X$. Given $\mu$, the bilinear form has four simple and obvious properties; if now $\mathfrak{A}$ is any subalgebra of $C_0(X)$ which separates the points of $X$ and whose functions do not all vanish at any point of $X$, then any bilinear form on $\mathfrak{A}$ having these four properties is given by a unique positive Radon measure $\mu$ in the way described above.

### 2.3 Topological

Here the important papers are [R5] and [R8], both written in collaboration with Nöbeling. One of the results of [R8] is central to the work of [R9] and [R10] described above and the extension theory of topological spaces given in [R5] clarifies the construction of locally compact spaces $X'$, $E'$ indicated there.

We start with [R8]; it deals with the following general problem: if $\mathfrak{F}$ is a family of real functions defined on a set $E$, describe $\overline{\mathfrak{F}}$, the set of those functions on $E$ which can be uniformly approximated by functions in $\mathfrak{F}$.

A remarkable result is the following: if $\mathfrak{F}$ is a Stone vector lattice of *bounded* functions defined on $E$, the $\overline{\mathfrak{F}}$ is an algebra; this result is used in [R9] and [R10]. There are several other interesting results leading to some well known theorems like the Stone–Weierstrass theorem (for $E$ compact or locally compact); the considerations are entirely elementary and self-contained.

The objective of [R5] is to give a systematic overview of the various extensions of a topological space $E$. The main tool is the (Kolmogorov)–Gelfand–Shilov–Stone theory of maximal ideal spaces associated with Banach algebras. The discussion englobes the extensions which go under the names of Alexandrov, Katetov, Stone–Čech, Wallman.

Both papers [R5] and [R8] are cited in standard works devoted to spaces of continuous functions.

Finally, we mention [R7] and [R15]; [R7] discusses characterizations of Alexandrov one-point compactification and [R15] gives a generalization of a factorization theorem of Whyburn (representing a continuous map as a composition of a monotone continuous and of a right continuous map) which comes up in the work on conservative maps in §6 of [R14].

## 3 Expository writing

What obviously impresses any one reading any of the papers of Bauer is the clarity and precision of the presentation; these qualities are perceptible right from his very first publications. It is therefore not surprising that he has written several highly successful textbooks and very useful expository articles. In the first place, we must mention his basic book on probability theory and elements of measure theory published in German ([TM1]) in 1964 and which has gone through several editions both in German ([TM3]) and in English translation ([TM4]); this volume has been used by numerous readers, both to learn the essentials of modern measure and integration theory as well as to obtain a rigorous insight into the foundations of mathematical probability including conditional expectations, martingale theory, Markov processes and Brownian motion. This one volume work has now been split (and updated) into two separate volumes, one on measure theory ([TM7] in German, an English translation was just published [TM10]) and one on probability theory ([TM8] in German, [TM9] in English translation). In my opinion, these volumes still remain excellent introductions to the subjects concerned; everything essential is authoritatively presented, preparing the future researchers for the study of more advanced developments. His other introductory text on analysis known to me is the booklet on integration theory ([TM5]) written with B. Anger) which is again very readable.

A word must be said of the two sets of lecture notes ([L1], [L2]) which had considerable circulation in the 1960s amongst young researchers. [L1] (1963) concerns the modern theory of Markov processes as it was developing then under the impulse of the rapidly expanding axiomatic potential theory; [L2] (1964) concerns the convexity theory in topological vector spaces, a subject undergoing rapid expansion at that time for the same reason. In the early 60s, authoritative monographs on these topics were rather rare; the situation changed radically starting with 1964–65. However, for a certain period around 1963–64 [L1], [L2] were much appreciated and although these were not available directly from booksellers, interested people somehow managed to obtain copies of these from Germany.

Although Bauer was not "officially" a probabilist his interest and competence in probability theory was widely acknowledged; as testimony for this, we can read his invited expository articles on the subject: [S7], [S9].

Bauer's expository remarks on Radon's and Hahn's work on measure theory ([O7], [O28]) can be read with pleasure and profit. The same is no doubt true of his more numerous expository articles on potential theory, approximation theory and convexity theory as well.

## References

[AP] Anger, B., Portenier, C., *Radon integrals*, Progr. Math. 103, Birkhäuser, Basel 1992.

[B] Bichteler, K., *Integration theory*, Lecture Notes in Math. 315, Springer-Verlag, Berlin 1973.

[K] König, H., *Measure and integration*, Springer-Verlag, Berlin 1997.

[L] Lembcke, J., On simultaneous preimage measures on Hausdorff spaces, in *Measure theory* (Oberwolfach, 1981), Lecture Notes in Math. 945 (eds. D. Kölzow, D. Maharam-Stone), Springer-Verlag, Berlin 1982, 110–115.

[M] Métivier, M., Existence d'un espace de représentation de H. Bauer, *Arch. Math.* **15** (1964), 450–462.

[S] Saks, S., *Theory of the integral*, Dover Publications, New York 1964 (republication of 2nd ed., G. E. Stechert & Co., 1937).

[Sch] Schaefer, H. H., *Toplogical vector spaces*, Graduate Texts in Math. 3, Springer-Verlag, Berlin 1971.

Département de Mathématiques, École Polytechnique Federale, CH-1015 Lausanne, Switzerland
E-mail: `srishti.chatterji@epfl.ch`

# The work of Heinz Bauer in convexity theory

*D. A. Edwards*

## 1 Introduction

During the second half of the twentieth century functional analytic convexity theory underwent an intense development, which resulted from a confluence of ideas from several directions. Among these may be mentioned the Krein–Milman theorem, the Choquet theory of integral representations, boundary theory for function algebras, and minimum principles in potential theory. Over a period of almost thirty years Heinz Bauer regularly made contributions of outstanding importance to this development. The present survey is intended as a brief guide to this work. The following account is arranged thematically. The terminology here follows that of Bauer, but some of our notation differs from his.

## 2 Minimum principles ([L2], [L6], [R13], [R16], [R18], [R21], [S5], [S6])*

In [R13], [R16], [R18], [R21] Bauer obtains a number of important generalizations of the Krein–Milman theorem. These results, as well as applications of them, also appear in the other publications cited above. The key theorems of [R13], [R16] are two purely topological minimum principles, of which that in [R16] is the stronger. This concerns a non-empty compact Hausdorff space $X$ together with a set $\mathcal{T}$ of ordered pairs $(x, T)$, where $x \in X$ and $\emptyset \neq T \subseteq X$. It is assumed that for each $x \in X$ there exists at least one $T$ such that $(x, T) \in \mathcal{T}$. In this setting a point $x_0 \in X$ is said to be *$\mathcal{T}$-extreme* if $(x_0, \{x_0\})$ is the sole element of $\mathcal{T}$ whose first component is $x_0$. Now let $\mathcal{F}$ denote the set of all lower semicontinuous numerical functions $f$ on $X$ such that, whenever $(x, T) \in \mathcal{T}$ and $f(x) \leq f(y)$ for all $y \in T$, it follows that $f(x) = f(y)$ for all $y \in T$. Suppose that $\mathcal{F}$ separates the non-$\mathcal{T}$-extreme points of $X$. Under these conditions Bauer proves that each $f \in \mathcal{F}$ attains its minimum at some $\mathcal{T}$-extreme point of $X$. The proof of this purely topological result bears some resemblance to that for the Krein–Milman theorem (see [ART], [KEL]).

*References containing numbers refer to the Bibliography of Heinz Bauer whereas letters in brackets are related to references placed at the end of this commentary.

The results of [R13] can be obtained as consequences of the above theorem. Among these we may note the following. Let $K$ be a non-empty compact subset of a real Hausdorff locally convex vector space $E$, and let $f: K \to (-\infty, \infty]$ be lower semicontinuous and concave. Then $f$ attains its minimum at some extreme point of $K$. (Here an *extreme point* of $K$ is any point of $K$ not contained in an open linear segment contained in $K$, and a *concave* function on $K$ is one whose restriction to linear segments is concave in the usual sense.) The Krein–Milman theorem follows directly from this minimum principle.

There is an interesting addendum in [R21] to the above theorem. Let $K$ be a non-empty compact convex subset of $E$, and denote by $K_e$ the set of all extreme points of $K$. Let $-k_1, k_2: K \to (-\infty, \infty]$ be concave functions. Suppose that $k_1$ is upper or lower semicontinuous and that $k_2$ is upper or lower semicontinuous (all four combinations being allowed). Finally suppose that $k_1(x) \leq k_2(x)$ for all $x \in K_e$. Under these conditions it is proved that $k_1(x) \leq k_2(x)$ for all $x \in K$.

Now let $X$ be a non-empty compact Hausdorff space and $\mathcal{E}$ a set of lower semicontinuous functions that separates the points of $X$. For each $x \in X$ let $\mathcal{M}_x(\mathcal{E})$ denote the set of all Radon probability measures $\mu$ on $X$ such that $\int^* h\, d\mu \leq h(x)$ for all $h \in \mathcal{E}$. Note that equality holds here if $\mathcal{E}$ is a linear subspace of the space $C(X)$ of all real continuous functions on $X$, and that, in any case, $\varepsilon_x \in \mathcal{M}_x(\mathcal{E})$. The *Choquet boundary* of $X$ with respect to $\mathcal{E}$ is by definition the set

$$\mathrm{Ch}_{\mathcal{E}}(X) = \{x \in X : \mathcal{M}_x(\mathcal{E}) = \{\varepsilon_x\}\}.$$

(The term was first used, in a more restricted sense, in [BDL].) In this setting Bauer [R16] obtains, as a corollary of the topological minimum principle stated above, the conclusion that each function in $\mathcal{E}$ attains its minimum at some point of $\mathrm{Ch}_{\mathcal{E}}(X)$. This result of Bauer is a fundamental theorem of Choquet theory. (It is extended in [R18] to the case where $\mathcal{E}$ is not assumed to separate the points of $X$.)

Applications of Bauer's minimum principles to function algebras and classical potential theory will be described below. It should be added that these minimum principles also played a very important role in the creation of Bauer's axiomatic potential theory (see [NET]).

A theorem closely related to one of Bauer's minimum principles, though very different in formulation, is proved by Ky Fan [KYF]. For minimum theorems generalizing those of both authors, see [FUCb], [FUCc], [FH]. The topological minimum principle of [R16] is generalized in [BJ] and applied there to abstract hyperharmonic functions. Minimum principles are treated from a more geometrical standpoint in [LUMa], [LUMb]. For minimum principles for quasi-concave functions see[DUB], [BRA].

## 3 The Šilov boundary ([R17], [S2], [S5], [S6], [P4])

In [R17] Bauer offers a fresh definition of the *Šilov boundary* which considerably widens the scope of the concept. He obtains an important existence theorem for this boundary, and gives many examples. His papers [S2], [P4] are earlier and less complete accounts of this work, while [S5], [S6] include concise surveys of it.

Let $X$ be a non-empty compact Hausdorff space and $\mathcal{E}$ a non-empty set of lower semicontinuous functions $u\colon X \to (-\infty, \infty]$ that separates the points of $X$. Denote by $\mathcal{S}$ or $\mathcal{S}(\mathcal{E})$ the set of all closed subsets $S$ of $X$ with the property that each function in $\mathcal{E}$ attains its minimum at some point of $S$. The least element of $\mathcal{S}$, if such an element exists, is called by Bauer [R17] the *Šilov boundary* of $X$ relative to $\mathcal{E}$. It will be denoted here by $\check{S}_{\mathcal{E}}(X)$. A point $x \in X$ is said to be *$\mathcal{E}$-exposed* if for some $u \in \mathcal{E}$ we have $u(x) < u(y)$ for all $y \in X$, $y \neq x$. It is trivial that such points belong to $\mathrm{Ch}_{\mathcal{E}}(X)$, and also to $\check{S}_{\mathcal{E}}(X)$ if the latter exists. It is shown in [R17] that there may be points of $\mathrm{Ch}_{\mathcal{E}}(X)$ that are not *$\mathcal{E}$-exposed.* Examples are also given to show that the Šilov boundary may fail to exist, or may be a proper subset of the Choquet boundary. However, the following important existence theorem is proved in [R17]: if $\mathcal{E} + \mathcal{E} \subseteq \mathcal{E}$ then the Šilov boundary $\check{S}_{\mathcal{E}}(X)$ exists and is equal to the closure of the Choquet boundary $\mathrm{Ch}_{\mathcal{E}}(X)$. Bauer's proof requires a modification of the Hahn–Banach theorem due to Aumann (see also [EDW]). For the special case when $\mathcal{E} \subseteq C(X)$, the existence of the Šilov boundary also follows from a theorem of Arens and Singer [AS]. However, those authors do not consider the Choquet boundary. Some results of Bear [BEAa] and Siciak [SIC] can be deduced from Bauer's theorems, together with the existence theorem for the usual Šilov boundary for a function algebra. Applications of Bauer's theorems on boundaries to convexity theory, classical potential theory, and function algebras will be noted below.

Interesting alternative proofs of the existence theorem for Bauer's Šilov boundary can be found in [LUMb], [ROG], [KÖN].

## 4 The abstract Dirichlet problem ([R17], [S2], [P4])

In this section $X$ will denote a non-empty compact Hausdorff space and $\mathcal{H}$ a linear subspace of $C(X)$ that contains the constant functions and separates the points of $X$. For each $S \in \mathcal{S}(\mathcal{H})$ we denote by $\mathcal{M}_x^S$ or $\mathcal{M}_x^S(\mathcal{H})$ the set of all probability Radon measures $\mu$ on $S$ such that $h(x) = \int h_S \, d\mu$, where $h_S$ denotes the restriction of $h$ to $S$. Now let $\hat{\mathcal{H}}(S)$ be the set of all $f \in C(X)$ such that $f(x) = \int f_S \, d\mu$ whenever $x \in X$ and $\mu \in \mathcal{M}_x^S(\mathcal{H})$. Bauer [R17], [S2] shows that $\hat{\mathcal{H}}(S)$ is independent of $S \in \mathcal{S}(\mathcal{H})$, and accordingly denotes it simply by $\hat{\mathcal{H}}$. The members of $\hat{\mathcal{H}}$ are called *$\mathcal{H}$-harmonic* (or, in a later context, *$\mathcal{H}$-affine*) functions. Now write $X^* = \check{S}_{\mathcal{H}}(X)$. A function $f \in C(X^*)$ is said to be *$\mathcal{H}$-resolutive* if there exists (a necessarily unique) $h \in \hat{\mathcal{H}}$ such that $f = h_{X^*}$. Bauer's *abstract Dirichlet problem* is that of finding necessary and sufficient conditions for a given $f \in C(X^*)$, or for all elements of $C(X^*)$, to be $\mathcal{H}$-resolutive. Before tackling it, Bauer establishes some preliminary results about $\hat{\mathcal{H}}$ [R17], [S2], [P4].

For example, $\hat{\mathcal{H}}$ is a closed linear subspace of $C(X)$, the Choquet and Šilov boundaries relative to $\mathcal{H}$ and $\hat{\mathcal{H}}$ are the same, and the $\hat{\mathcal{H}}$-harmonic functions are precisely those that belong to $\hat{\mathcal{H}}$. Next, given $f \in C(X)$ and $x \in X$, let

$$\hat{f}(x) = \inf\{h(x) : f \leq h \in \mathcal{H}\},$$

and $\check{f} = -\hat{g}$, where $g = -f$. Then for $f \in C(X)$ the following are equivalent:

1) $f \in \hat{\mathcal{H}}$;

2) $\hat{f}(x) = \check{f}(x)$ for all $x \in X$;

3) for each $\varepsilon > 0$ there exist $h_1, h_2, \ldots, h_m, k_1, k_2, \ldots, k_n$ in $\mathcal{H}$ such that $\underline{h} \leq f \leq \overline{k}$ and $\overline{k} - \underline{h} < \varepsilon$, where

$$\underline{h} = \sup(h_1, \ldots, h_m), \quad \overline{k} = \inf(k_1, \ldots, k_n).$$

It follows that (a) if $\mathcal{H}$ is a lattice with respect to the usual order $\leq$ then $\hat{\mathcal{H}} = \overline{\mathcal{H}}$, and (b) we have $\hat{\mathcal{H}} = C(X)$ if and only if $\mathrm{Ch}_{\mathcal{H}}(X) = X$.

Now denote by $\mathcal{E}_{\mathcal{H}}$ the smallest closed inf-stable convex cone in $C(X)$ with vertex 0 that contains $\mathcal{H}$. Then [R17], [S2] $\mathcal{E}_{\mathcal{H}}$ and $\mathcal{H}$ have the same Choquet and Šilov boundaries and

$$\hat{\mathcal{H}} = \mathcal{E}_{\mathcal{H}} \cap (-\mathcal{E}_{\mathcal{H}}).$$

It follows that we have $\mathcal{H} = \hat{\mathcal{H}}$ if and only if there exists a closed inf-stable convex cone $\mathcal{E}$ in $C(X)$ with vertex 0 such that

$$\mathcal{H} = \mathcal{E} \cap (-\mathcal{E}).$$

For the abstract Dirichlet problem we need some notation. We write $\mathcal{M}_x^*(\mathcal{H})$ for $\mathcal{M}_x^Y(\mathcal{H})$ and $h^*$ for $h_Y$ when $Y = X^*$. A typical result states that for each $f \in C(X^*)$ the following are equivalent:

1) $f$ is $\mathcal{H}$-resolutive;

2) for all $x \in X$

$$\inf\{h(x) : h^* \geq f,\ h \in \mathcal{H}\} = \sup\{h(x) : h^* \leq f,\ h \in \mathcal{H}\};$$

3) whenever $x \in X$ and $\mu_1, \mu_2 \in \mathcal{M}_x^*(\mathcal{H})$ we have

$$\int f\, d\mu_1 = \int f\, d\mu_2.$$

Another theorem establishes that the following assertions are equivalent:

1) all functions in $C(X^*)$ are $\mathcal{H}$-resolutive;

2) $\mathcal{M}_x^*(\mathcal{H})$ is a singleton for each $x \in X$;

3) $\hat{\mathcal{H}}$ is a lattice with respect to the usual order $\leq$;

4) each element of $\mathcal{E}_{\mathcal{H}}$ has a greatest $\mathcal{H}$-harmonic lower bound.

When these conditions are satisfied $\mathrm{Ch}_{\mathcal{H}}(X) = X^*$ and, if $X^*$ is metrizable, each point of $\mathrm{Ch}_{\mathcal{H}}(X)$ is $\hat{\mathcal{H}}$-exposed.

Applications of the above theory to convexity theory, the classical Dirichlet problem, and Korovkin theory will be described below. Further work on the abstract Dirichlet problem and variants thereof may be found in [ALFa], [EFF], [MCD], [ROG].

## 5 Extreme points and the Choquet boundary ([L2], [R17], [R33], [S2], [S5], [S6], [S17], [P4])

Let $K$ be a non-empty compact convex subset of a real Hausdorff locally convex space $E$ and let $\mathcal{L}$ or $\mathcal{L}(K)$ denote the set of all restrictions to $K$ of functions of the form $\phi + \alpha$, where $\phi \in E'$ ($E'$ being the dual of $E$) and $\alpha$ is a real constant. Taking $X = K$ and $\mathcal{H} = \mathcal{L}$, we can apply the results of §4 to this setting. In particular we have $\hat{\mathcal{L}} = \mathcal{E}_{\mathcal{L}} \cap (-\mathcal{E}_{\mathcal{L}})$. Bauer [R17], [S2], [S5], [S6], [P4], [L2] shows also that $\mathcal{E}_{\mathcal{L}}$ is precisely the set of all concave functions belonging to $C(K)$. It follows that $\hat{\mathcal{L}}$ is precisely $A(K)$, the set of all real continuous affine functions on $K$. Furthermore, for $S \in \mathcal{S}(\mathcal{L})$ (see §3) and $x \in K$ Bauer identifies $\mathcal{M}_x^S(\mathcal{L})$ as the set of all probability Radon measures on $S$ that have $x$ as barycentre. This is another basic result of Choquet theory (see [BDL], [PHEb], [PHEc]), and we have the immediate corollaries that $\mathrm{Ch}_{\mathcal{L}}(K) = K_e$ and $\check{S}_{\mathcal{L}}(K) = \overline{K_e}$. The minimum principle for $\mathrm{Ch}_{\mathcal{L}}(K)$ is thus equivalent to the Krein–Milman theorem.

Now let us revert to the study of a separating linear subspace $\mathcal{H}$ of $C(X)$ that contains the constant functions, where $X$ denotes a non-empty compact Hausdorff space. Let $E$ be the Banach dual $\mathcal{H}'$ of $\mathcal{H}$ endowed with the weak* topology $\sigma(\mathcal{H}', \mathcal{H})$ and let $Y$ be the state space of $\mathcal{H}$. That is, $Y$ is the set of positive linear functionals $g$ on $\mathcal{H}$ such that $g(1) = 1$. Then $Y$ is a non-empty compact convex subset of $E$. For each $x \in X$ denote by $j(x)$ the functional on $\mathcal{H}$ defined by $j(x)(f) = f(x)$. Then $j(X) \subseteq Y$, $j$ maps $X$ homeomorphically onto $j(X)$, and [L2], [R17], [S2], [S5], [S6], [P4] it is shown that $j(\mathrm{Ch}_{\mathcal{H}}(X)) = Y_e$. Hence $Y$ is the closed convex hull of $j(X)$, and we have

$$\mathcal{H} = \{\phi \circ j : \phi \in E'\}.$$

Finally it is shown in [R17] that a probability Radon measure $\mu$ on $X^*$ belongs to $\mathcal{M}_x^*(\mathcal{H})$, where $x \in X$, if and only if the image measure $j(\mu)$ on $Y$ has barycentre $j(x)$. This construction is used later in Korovkin approximation theory [R33], [S17].

Further work on the geometric interpretation of the Choquet boundary may be found in [LUMa], [LUMb].

## 6 Simplexes and ordered vector spaces ([L2], [R12], [R17], [R21], [R35], [S2], [S24], [P4])

In [R12] Bauer considers an additive abelian group $G$ that is (partially) ordered by a subsemigroup $P$ such that $P \cap (-P) = \{0\}$. This means that for $a, b \in G$ we stipulate that $a \leq b$ if and only if $b - a \in P$. We say that $G$ has the *(Riesz) decomposition property* if for every triple $x, a, b$ of elements of $P$ such that $x \leq a + b$ we can find $u, v \in P$ such that $u \leq a$, $v \leq b$ and $x = u + v$. This is trivially true if $G$ is lattice-ordered by $P$. In the reverse direction Bauer finds sufficient conditions for an ordered abelian group with the decomposition property to be lattice-ordered. We state here two results of this kind. The semigroup $P$ is said to be *Dedekind-complete* if every downward filtering non-empty subset $Q$ of $P$ has a greatest lower bound in $P$. Bauer demonstrates that if (i) $G = P - P$, (ii) $G$ has the decomposition property, and (iii) $P$ is Dedekind-complete, then $G$ is a lattice-ordered group.

Now let $G$ be a real Hausdorff topological vector space and $P$ a convex cone in $G$ with vertex 0. In this setting it is proved that if (i) $G = P - P$, (ii) $G$ has the decomposition property, and (iii) the cone $P$ has a compact base, then $G$ is a vector lattice.

Further results about ordered groups with the decomposition property can be found in [FUCa].

Now let $K$ be a non-empty compact convex subset of a real Hausdorff locally convex space $E$ and let

$$\Gamma = \{a + \lambda K : a \in E,\ \lambda \in \mathbb{R}_+\}.$$

We say that $K$ is a *Choquet simplex* if every non-empty intersection of two elements of $\Gamma$ belong to $\Gamma$. Much has been written on Choquet simplexes (see e.g., [BCa], [BCb], [GOU], [CHO], [ALFb], [AE], [L2]). We note that the following are equivalent: (i) $K$ is a Choquet simplex; (ii) $A(K)$ has the decomposition property with respect to the usual order; (iii) $A(K)'$ has the decomposition property; (iv) $A(K)'$ is an AL-space. (Here $A(K)'$ is the ordered Banach dual of $A(K)$.) Note that the equivalence of (iii) and (iv) results from [R12]. If $K$ is metrizable then it is a Choquet simplex if and only if for each $x \in K$ there is a unique $\mu \in \mathcal{M}_x(\mathcal{L})$ such that $\mu(K_e) = 1$. Pursuing this subject, Bauer [L2], [R17], [S2], [P4] establishes the theorem that the following are equivalent:

1) $K$ is a Choquet simplex such that $K_e$ is a closed subset of $K$;

2) For each $x \in K$ the set $\mathcal{M}_x^S(\mathcal{L})$, where $S = \overline{K}_e$, is a singleton;

3) Each function in $C(\overline{K}_e)$ is the restriction to $\overline{K}_e$ of an element of $A(K)$;

4) Each concave $f \in C(K)$ has a greatest lower bound in $A(K)$;

5) $A(K)$ is a vector lattice.

A Choquet simplex that satisfies the conditions of this last theorem is usually called a *Bauer simplex.* For another characterization see [R21]. For a metrizable Bauer simplex $K$ it is shown that each point of $K_e$ is $A(K)$-exposed. This result was extended to metrizable Choquet simplexes in [BCa], [BCb], [DAV].

Now let $X$ be a non-empty metrizable compact space and let $\mathcal{H}$ be a linear subspace of $C(X)$ that contains the constants and separates the points of $X$. Let $Y$ be the state space of $\mathcal{H}$. Following [EK], Bauer [R35], [S24] calls $\mathcal{H}$ a *simplicial space* if $Y$ is a Choquet simplex. He calls $\mathcal{H}$ *weakly simplicial* if for each $x \in X$ there exists a unique $\mu \in \mathcal{M}_x^X(\mathcal{L})$ such that $\mu(Z) = 1$, where $Z = \mathrm{Ch}_{\mathcal{H}}(X)$). (Note that in [BH] the spaces of this second class are termed *simplicial.*) Bauer gives simple examples to show that a weakly simplicial space can fail to be simplicial. The main theorem of [R35] states that the space $\mathcal{H}$ is weakly simplicial if and only if $\hat{\mathcal{H}}$ is simplicial.

## 7 The classical Dirichlet problem ([L2], [R13], [R17], [S2], [S5], [S6], [S24], [P4])

In the above publications Bauer applies his work in convexity to classical potential theory. In [R13] he indicates how a boundary minimum theorem for superharmonic functions follows from the first of his topological minimum principles. In the other works listed above he looks at the classical Dirichlet problem from the standpoint of convexity theory.

Let $\Omega$ be a bounded open subset of $\mathbb{R}^n$, where $n \geq 2$, let $F$ denote the topological boundary of $\Omega$, and let $X = \overline{\Omega}$. We take $\mathcal{H}$ to be the set of all real continuous functions on $X$ that are harmonic in $\Omega$, and denote by $\mathcal{S}$ the set of all $u \in C(X)$ such that $u$ is superharmonic in $\Omega$. Then $\mathcal{S}$ is a closed inf-stable convex cone with vertex 0, and $\mathcal{H} = \mathcal{S} \cap (-\mathcal{S})$. It follows (see §4) that $\mathcal{H} = \hat{\mathcal{H}}$. For each $f \in C(F)$ let $\mathcal{O}_f$ denote the set of all superharmonic functions $v$ on $\Omega$ such that

$$\liminf_{\substack{x \to x_0 \\ x \in \Omega}} v(x) \geq f(x_0)$$

for all $x_0 \in F$, and let $\mathcal{U}_f = -\mathcal{O}_{-f}$. By Wiener's resolutivity theorem we have, for all $x \in \Omega$,

$$\inf\{v(x) : v \in \mathcal{O}_f\} = \sup\{u(x) : u \in \mathcal{U}_f\}.$$

Denoting the common value by $H_f(x)$, one proves that $H_f(\cdot)$ is a harmonic function on $\Omega$. A point $x_0 \in F$ is said to be *regular* if

$$\lim_{\substack{x \to x_0 \\ x \in \Omega}} H_f(x) = f(x_0)$$

for all $f \in C(F)$. Using a lemma of Keldych, Bauer proves that $\mathrm{Ch}_{\mathcal{H}}(X)$ is precisely the set of regular points of $F$. The Šilov boundary $\check{S}_{\mathcal{H}}(X)$ is therefore the closure of the set of regular points. When $\Omega$ is the interior of $X$, Bauer shows that $\check{S}_{\mathcal{H}}(X) = F$.

Now denote by $\mathcal{G}$ the set of restrictions to $X$ of functions harmonic in some neighbourhood of $X$. In [S5], [S6] Bauer identifies $\mathrm{Ch}_{\mathcal{G}}(X)$ as the set of all *stable* points of $X$ (points where the complement of $X$ is not thin). (These findings led Vincent-Smith [VIN] to devise a proof by Choquet theory of Deny's theorem that $\mathcal{G}$ is uniformly dense in $\mathcal{H}$ if and only if every regular point of $F$ is stable. He also generalizes this result to Brelot harmonic spaces.)

For a full account of this work of Bauer the best reference is [R17], where the reader will also find some brief indications about discrete potential theory. For the situation in axiomatic potential theory see [NET].

## 8 Boundaries for function algebras ([R16], [R17], [S5], [S6])

Let $X$ be a non-empty compact Hausdorff space, let $C(X;\mathbb{C})$ denote the normed algebra of all complex continuous functions on $X$, and let $\mathcal{A}$ be a closed subalgebra of $C(X;\mathbb{C})$ that contains the constants and separates the points of $X$. In the above papers Bauer considers the sets

$$\mathcal{H} = \{\mathrm{Re} f : f \in \mathcal{A}\}, \quad \mathcal{F} = \{-|f| : f \in \mathcal{A}\},$$
$$\mathcal{E} = \{-\log|f| : f \in \mathcal{A}\}.$$

He shows that

$$\mathrm{Ch}_{\mathcal{F}}(X) \subseteq \mathrm{Ch}_{\mathcal{H}}(X) \subseteq \mathrm{Ch}_{\mathcal{E}}(X),$$

and that these three sets all have the same closure. Associated minimum principles for $\mathcal{H}$, $\mathcal{F}$, $\mathcal{E}$, are deduced via the general results of [R16], [R17], and the common closure of the Choquet boundaries is shown to be the Šilov boundary for $\mathcal{H}$, $\mathcal{F}$, $\mathcal{E}$ in the sense of [R17], and of $\mathcal{A}$ in the classical sense. The boundary $\mathrm{Ch}_{\mathcal{H}}(X)$ has subsequently become known as Bishop's *strong boundary* for $\mathcal{A}$, or, when $X$ is metrizable, as the *minimal boundary* [BIS]. In the latter case Bishop has shown that each point of $\mathrm{Ch}_{\mathcal{H}}(X)$ is $\mathcal{F}$-exposed (see §3). The boundary $\mathrm{Ch}_{\mathcal{E}}(X)$ is now referred to as the *Jensen boundary*, because it has been shown to be the appropriate boundary for the Choquet theory of Jensen measures [EDW], [GAMa]. Gamelin [GAMb] has pointed out that the 'Swiss cheese' construction of [MCK] shows that we can have $\mathrm{Ch}_{\mathcal{H}}(X) \neq \mathrm{Ch}_{\mathcal{E}}(X)$.

## 9 Supermartingales and the Choquet boundary ([R18])

Let $X$ be a locally compact Hausdorff space and $\mathcal{E}$ a non-empty set of lower-bounded lower semicontinuous numerical functions on $X$. It is not assumed that $\mathcal{E}$ separates the points of $X$. Bauer [R18] extends to this setting his definition of, as well as his minimum principle for, the Choquet boundary $\mathrm{Ch}_{\mathcal{E}}(X)$.

Next, he assumes that $X \subseteq \mathbb{N}$, and considers a substochastic matrix $\Pi = (p_{ij})_{X\times X}$. The elements of $X$ are called *states*. A *supermartingale* on $X$ is by definition a real-valued function $u$ on $X$ that is lower-bounded and such that $\Pi u \leq u$, and $\mathcal{E}$ is now taken to be the set of all supermartingales on $X$. The aim now is to relate various Markov chain concepts to Choquet theory concepts.

Given $i \in X$, Bauer denotes by $K_i$ the intersection

$$\bigcap \{u^{-1}(u(i)) : u \in \mathcal{E}\}.$$

A state $i$ is said to be *stochastically extreme* if $i \in \mathrm{Ch}_{\mathcal{E}}(X)$ and $\sum_k p_{jk} = 1$ for all $j \in K_i$. The following are typical results. The irreducible-recurrent sets are precisely the $K_i$ such that $i$ is stochastically extreme. A state is stochastically extreme if and only if it is recurrent. Thus, for stochastic $\Pi$, the recurrent states are precisely those that belong to $\mathrm{Ch}_{\mathcal{E}}(X)$. If $X$ is finite and $i$ is transient then $K_i = \{i\}$.

Feller's notion of a *sojourn set* (*Absorptionsmenge*) [FEL] plays an important part in the argument.

## 10 Locally compact convex sets ([S4])

This is joint work with D. Hinrichsen, in which the authors generalize, to locally compact convex sets, a number of standard theorems concerning compact convex sets.

Suppose first that $X$ is a locally compact convex subset of a Hausdorff locally convex space $E$. The authors prove that if there exists a closed affine submanifold $F$ of $E$ such that $0 \notin F$ and $X \subseteq F$ then the closed cone with vertex 0 generated by $X$ is locally compact.

Suppose next that $L$ is a topological vector space with respect to each of two Hausdorff topologies $\tau$ and $\tau'$, that $Y$ is a convex subset of $L$ that is locally compact with respect to $\tau$, and that $\tau$ is finer on $Y$ than $\tau'$. Under these conditions it is shown that the topologies $\tau$ and $\tau'$ coincide on $Y$. It follows that each closed locally compact convex subset of a Hausdorff locally convex space $E$ is weakly complete (i.e. complete with respect to the uniformity associated with $\sigma(E, E')$).

Next it is proved that the projective limit of a sequence of locally compact Hausdorff spaces is a Baire space. In particular, the authors note that this applies to certain linear projective limits of sequences of locally compact convex sets.

In the final section, the authors recall Klee's extension in [KLE] of the Krein–Milman theorem to closed locally compact lineless convex subsets of a Hausdorff locally convex space (*lineless* means containing no whole line). He shows that every such set $C$ is the closed convex hull of $C_e \cup C_r$, where $C_r$ is the union of the extreme rays of $C$. In [S4] the authors obtain results in this vein for weakly complete lineless convex subsets of a Hausdorff locally convex space and for sequential linear projective limits of such sets.

## 11 Parts and the part metric ([R26], [R27], [R28], [S5], [S6])

The decomposition of a convex set into *parts*, which is the concern of the papers [R26], [R27], [R28], evolved from the notion of the *Gleason parts* [GLE] of the spectrum of a function algebra. (See also [S5], [S6], [BEAb], [BEAc].)

In [R26], a joint paper with H. S. Bear, the authors study the *part metric*, as defined in [BW]. Let $C$ be a non-empty lineless convex subset of a real vector space. Given $x, y \in C$, let

$$A(x, y) = \{t : t > 0, x + t(x - y) \in C, y + t(y - x) \in C\}$$

and let

$$d(x, y) = \begin{cases} \inf\{\log(1 + t^{-1}) : t \in A(x, y)\}, & \text{if } A(x, y) \neq \emptyset, \\ +\infty, & \text{if } A(x, y) = \emptyset. \end{cases}$$

Then $d\colon C \times C \to [0, \infty]$ is a pseudometric in the sense of Bourbaki [BOU], but it is usually simply called the *part metric* for $C$. An equivalence relation on $C$ is obtained by writing $x \sim y$ whenever $A(x, y) \neq \emptyset$, and the associated equivalence classes are termed the *parts* of $C$. Each part $\Delta$ of $C$ is convex and the restriction of $d$ to $\Delta \times \Delta$ is a metric in the usual sense. The parts of $C$ are open and closed in the $d$-topology of $C$ and are the connected components for that topology. If $C$ is an open lineless convex subset of a topological vector space then $C$ has just one part. If we also have $0 \in C$, then the *Minkowski norm* associated with $C$ is defined by $q(x) = \max(p(x), p(-x))$, where $p$ is the Minkowski functional. Under these conditions it is proved that the $q$-topology on $C$ coincides with the $d$-topology.

A real vector space $L$ is termed a *weak space* when it is endowed with a weak topology $\sigma(L, M)$, where $M$ is a linear subspace of the algebraic dual $L'$ of $L$. A subset of $L$ is then said to be *complete* if it is complete with respect to the uniform structure induced by $\sigma(L, M)$. Given a closed lineless convex subset $C$ of a Hausdorff weak space $L$, the authors obtain a formula that expresses the part metric for $C$ in terms of the functions $\log F$, $F \in C^+$, where $C^+$ is the set of all real continuous affine functions $F$ on $L$ such that $F \geq 0$ on $C$. If also $C$ is $\sigma(L, M)$-complete, it follows that it is $d$-complete and that each part is $d$-complete. For the case of convex cones see [R26].

Generalizing a selection theorem of Michael [MIC], the authors establish the existence of a continuous selection function for a lower semicontinuous mapping of a paracompact space into the non-empty closed convex subsets of a single part of a complete lineless convex subset of a Hausdorff weak space. This result is applied to the convex cone of non-negative Radon measures on a locally compact space $X$. The authors thereby obtain an existence theorem for the selection in a continuous fashion of mutually absolutely continuous representing measures for points of $X$ in a single part of the state space of a function space or function algebra.

In [R27] Bauer considers a convex lineless subset $C$ of a real vector space $L$. He assumes that $0 \in C$, that $C$ has only one part, and that $L$ is generated by $C$, and he proves that $L$ is a Banach space with respect to the Minkowski norm for $C$ if and only if $C$ is $d$-complete, where $d$ is the part metric. He applies this result to show that if $C_1$, $C_2$ are lineless single-part convex sets, each complete with respect to its part metric, then each affine surjective mapping of $C_1$ onto $C_2$ is uniformly continuous and open with respect to the part metrics. This theorem leads to an analogue of the selection theorem of [BG].

In [R28] Bauer studies criteria for the completeness (here called *internal completeness*) of a lineless convex set with respect to its part metric. He first derives some consequences of two such criteria established in [R26], [R27]. Next he generalizes to convex sets the concept of *normality* with respect to a seminorm, hitherto defined only for cones. A convex subset of a locally convex set is then termed *normal* if it is normal with respect to all continuous seminorms. The principal result of the paper is that a normal complete convex subset of a Hausdorff locally convex space is lineless and internally complete. Several corollaries follow. In particular: (a) every bounded open convex subset of a Banach space is internally complete, and (b) the closed unit ball of a Banach space is internally complete. The paper concludes with a fixed-point theorem for strict contractions of an internally complete lineless single-part convex set.

## 12 Korovkin approximation theory ([R29], [R30], [R32], [R33], [S11], [S12], [S13], [S17], [S18], [S19], [S23])

This subject began with Korovkin's approximation theorem for $C([0, 1])$. Suppose that $(L_n)$ is a sequence of positive linear operators in $C([0, 1])$ such that $(L_n p)$ converges uniformly on $[0, 1]$ to $p$ for all quadratic polynomial functions $p$. Then Korovkin's theorem states that $(L_n f)$ converges uniformly on $[0, 1]$ to $f$ for all $f \in C([0, 1])$. The proof of the Weierstrass approximation theorem via Bernstein polynomials is an easy corollary.

The desire to generalize this result led in the first instance to the following setup. Let $X$ be a compact Hausdorff space and let $\mathcal{H}$ be a linear subspace of $C(X)$ that contains the constants and separates the points of $X$. By an *$\mathcal{H}$-admissible* net of operators we shall mean a net $(T_i)$ of positive linear operators in $C(X)$ such that $(T_i h)$ converges uniformly on $X$ to $h$ for all $h \in \mathcal{H}$. We say that $f$ belongs to the *Korovkin closure* $\mathrm{Kor}(\mathcal{H})$ of $\mathcal{H}$ if $(T_i f)$ converges uniformly on $X$ to $f$ for every $\mathcal{H}$-admissible net $(T_i)$. In this setting Korovkin theory is concerned with; (a) characterizing $\mathrm{Kor}(\mathcal{H})$, and (b) finding necessary and sufficient conditions for $\mathrm{Kor}(\mathcal{H}) = C(X)$ to be true. Šaškin [ŠAŠa], [ŠAŠb], using results due to Bauer [R17], [S2] , [P4] solved problem (a) by showing that $\mathrm{Kor}(\mathcal{H}) = \hat{\mathcal{H}}$, where $\hat{\mathcal{H}}$ is the space of $\mathcal{H}$-affine functions defined in §3 above. Combining this result of Šaškin with Bauer's characterization (see §3) of those $\mathcal{H}$ for which $\hat{\mathcal{H}} = C(X)$ yields a solution to (b), namely: $\mathrm{Kor}(\mathcal{H}) = C(X)$ if and only if $\mathrm{Ch}_{\mathcal{H}}(X) = X$ [ŠAŠc], [WUL]. For the case where $X$ is a compact metric space, Bauer's elegant expository paper [S12] (also published in translation as [S11], [S19]) gives a very clear introduction to these matters, with illuminating examples. The generalizations and variants of these results to be found in the other papers of Bauer cited above are very extensive, and in the space available we can only give a rather broad outline account.

In the first part of [R29] we are given a topological space $X$, and a linear subspace $\mathcal{H}$ of the space $C(X)$ of all real continuous functions on $X$. A function $f \in \mathbb{R}^X$ is said to be *$\mathcal{H}$-bounded* if $h_1 \le f \le h_2$ for some $h_1, h_2$ in $\mathcal{H}$, and the space of $\mathcal{H}$-bounded elements of $C(X)$ is denoted by $\mathcal{H}_0$. We denote by $(L_i)$ a net of increasing but possibly nonlinear maps $L_i : \mathcal{H}_0 \to \mathbb{R}^X$. For $\mathcal{H}$-bounded functions we can define $\hat{f}$ and $\check{f}$ via the formulae of §4. The *$\mathcal{H}$-affine functions* in this setting are then the elements of

$$\hat{\mathcal{H}} = \{f \in \mathcal{H}_0 : \hat{f} = \check{f}\}.$$

A characterization of $\hat{\mathcal{H}}$ is given, and it is proved, for instance, that if $(L_i h)$ converges pointwise on $X$ to $h$ for all $h \in \mathcal{H}$ then $(L_i f)$ converges pointwise on $X$ to $f$ for all $f \in \hat{\mathcal{H}}$, and that this statement remains true if pointwise convergence is replaced throughout by uniform convergence on compact sets.

In the second part of [R29] it is assumed that the space $X$ is locally compact and Hausdorff and that $\mathcal{H}$ is an *adapted* subspace of $C(X)$ in the sense of [CHO]. For $x \in X$, $\mathcal{M}_x(\mathcal{H})$ is now defined as the set of all positive Radon measures $\mu$ on $X$ such that $\mathcal{H} \subseteq L^1(\mu)$ and $\int h \, d\mu = h(x)$ for all $h \in \mathcal{H}$. It is shown that a function $f \in \mathbb{R}^X$ is $\mathcal{H}$-affine if and only if $f \in \mathcal{H}_0$ and $\int f \, d\mu = f(x)$ for all $x \in X$ and $\mu \in \mathcal{M}_x(\mathcal{H})$. Following [MS] one defines the Choquet boundary $\mathrm{Ch}_{\mathcal{H}}(X)$ in this setting to be

$$\{x \in X : \mathcal{M}_x(\mathcal{H}) = \{\varepsilon_x\}\},$$

and it is proved that $\hat{\mathcal{H}} = \mathcal{H}_0$ if and only if $\mathrm{Ch}_{\mathcal{H}}(X) = X$. Further convergence properties are established. In particular we note the theorem that $\hat{\mathcal{H}}$ is precisely the class of all $f \in \mathcal{H}_0$ having the Korovkin property that for every net $L_i : \mathcal{H}_0 \to \mathcal{H}_0$ of positive linear maps such that $(L_i h)$ converges pointwise on $X$ to $h$ for all $h \in \mathcal{H}$ it follows that $(L_i h)$ converges pointwise on $X$ to $f$. Moreover, this characterization of $\hat{\mathcal{H}}$ remains true if pointwise convergence is replaced by locally uniform convergence.

The study of Korovkin theory for an adapted space $\mathcal{H}$ of real continuous functions on a locally compact Hausdorff space $X$ is continued in [R30]. Let $\mathcal{H}^b$ denote the vector lattice of all $\mathcal{H}$-bounded elements of $\mathbb{R}^X$. The *order topology* $T_0$ for $\mathcal{H}^b$ is the finest locally convex topology on $\mathcal{H}^b$ for which all order intervals are bounded. It is finer than the topology of locally uniform convergence. The aim in [R30] is to discover how far convergence and allied theorems for a net of increasing maps $L_i\colon \mathcal{H} \to \mathcal{H}^b$ remain true when the topology of pointwise, or of locally uniform, convergence is replaced by that of order convergence. After a careful analysis of the order topology, it is proved, for example, that $\hat{\mathcal{H}}$ is the $T_0$-closure of $\mathcal{H}$ in $\mathcal{H}_0$. This makes possible an extremely simple proof of the theorem of Mokobodzki and Sibony [MS], [SIB] that if the adapted space $\mathcal{H}$ is a vector sublattice of $\mathbb{R}^X$ and is linearly separating then $\mathcal{H}$ is $T_0$-dense in $\mathcal{H}_0$. The $T_0$-Korovkin closure of $\mathcal{H}$ is defined as the set of all $f \in \mathcal{H}_0$ such that $(L_i f)$ is $T_0$-convergent to $f$ for every net of positive linear maps $L_i\colon \mathcal{H}_0 \to \mathcal{H}_0$ such that $(L_i f)$ is $T_0$-convergent to $h$ for all $h \in \mathcal{H}$. A major theorem of [R30] states that the $T_0$-Korovkin closure of $\mathcal{H}$ is precisely $\hat{\mathcal{H}}$. In the final section of [R30] the convergence of nets of increasing maps $L_i\colon \mathcal{H}_0 \to \hat{\mathcal{H}}$ is studied. This leads to an extension of a result of [LIO], and to a link with the abstract Dirichlet problem of §4.

In [R32], a joint paper with K. Donner (summarized in [S18]), the object of study is Korovkin convergence in $C_0(X)$, the Banach space of real continuous functions vanishing at infinity on a locally compact Hausdorff space $X$. Broadly speaking, the results described above for the case when $X$ is compact remain true here provided certain adjustments to the basic definitions and hypotheses are made. For example, an $\mathcal{H}$-admissible net of operators in $C_0(X)$ is defined (for any linear subspace $\mathcal{H}$ of $C_0(X)$) as a *norm-bounded* net $(T_i)$ of operators in $C_0(X)$ such that $(T_i h)$ converges to $h$ uniformly on $X$ for all $h \in \mathcal{H}$. In place of $\mathcal{H}$-bounded functions the authors use *almost $\mathcal{H}$-bounded functions*, and for the envelope functions $\hat{f}$ and $\check{f}$, they use a modified definition adapted from the proof in [R31] of the Choquet–Deny theorem. The symbol $\mathcal{M}_x(\mathcal{H})$ now denotes, for all $x \in X$, the set of all *bounded* positive Radon measures $\mu$ on $X$ such that $\int h\, d\mu = h(x)$ for all $h \in \mathcal{H}$. These changes make it possible to prove theorems closely analogous to those for the case of compact $X$.

The paper [S13] is a short survey article that gives a brief summary of matters treated in [R33], [S12] together with a quick survey of *relative* Korovkin approximation based mainly on the work of Leha [LEH].

The article [R33], a joint paper with G. Leha and S. Papadopoulou (summarized in [S17]), seeks an explicit determination of $\mathrm{Kor}(\mathcal{H})$ when $\mathcal{H}$ is a linear space of real continuous functions on a compact Hausdorff space $X$ such that $\mathrm{Kor}(\mathcal{H}) \neq C(X)$. It is assumed that $\mathcal{H}$ is a closed subspace of $C(X)$ that contains constants and separates points. By using the construction explained above in §5, the authors transfer the problem to one concerning functions on the state space of $\mathcal{H}$. This allows convexity methods to be used, and, in particular, the authors establish a characterization, involving Alfsen's *affine boundary dependences* [ALFb], of those functions in $C(\check{S}_{\mathcal{H}}(X))$ that are the restrictions of elements of $\hat{\mathcal{H}}$. (See [BB], [MCD], [SAI] for alternative proofs.) It is shown that under certain conditions this result makes possible a characterization of $\mathrm{Kor}(\mathcal{H})$ by means of simple geometric extension properties. Finally the authors apply their results to find $\mathrm{Kor}(\mathcal{H})$ when $X$ is a compact convex subset of $\mathbb{R}^n$

and $\mathcal{H}$ is the closed linear subspace of $C(X)$ generated by 1, the coordinate functions, and a given function $u \in C(X)$.

The paper [S23], a joint work with K. Donner, is a continuation of [R32]. In the same setting as before they define $\mathrm{Kor}_M(\mathcal{H})$ to be, for each constant $M \geq 1$, the set of all $f \in C_0(X)$ such that $(T_i f)$ converges uniformly to $f$ for all $\mathcal{H}$-admissible nets $(T_i)$ such that $\|T_i\| \leq M$ for all $i$. Then for $M > 1$ it transpires that $\mathrm{Kor}_M(\mathcal{H}) = \mathrm{Kor}(\mathcal{H})$. The proof requires an ingenious modification of their earlier methods. The results for the case $M = 1$ are more delicate and occupy the authors for a substantial part of the paper.

For more on the subject of Korovkin theorems see [DON], [AC].

## 13 The Choquet–Deny theorem ([S3], [R31], [S18])

Let $X$ be a Hausdorff space and let $C(X)$, the space of all real continuous functions on $X$, be endowed with the topology of uniform convergence on compact subsets of $X$. Denote by $\mathcal{M}^+$ the set of all positive Radon measures of compact support on $X$. Suppose we are given a set $\mathcal{S}$ of ordered pairs $(\sigma, x) \subseteq \mathcal{M}^+ \times X$, where $\sigma(\{x\}) = 0$ for all $(\sigma, x) \in \mathcal{S}$, and a set $\mathcal{T} \subseteq \mathcal{M}^+$. Let $\mathcal{K}$ be the set of all $u \in C(X)$ such that

$$\int u\, d\tau \leq 0 \quad \text{for all } \tau \in \mathcal{T}, \tag{1}$$

$$\int u\, d\sigma \leq u(x) \quad \text{for all } (\sigma, x) \in \mathcal{S}. \tag{2}$$

Then $\mathcal{K}$ is a closed convex cone such that $\inf(u, v) \in \mathcal{K}$ for all $u, v \in \mathcal{K}$. The Choquet–Deny theorem states that every such cone is of this form. In [R31] a new proof of this theorem is given for the case in which $X$ is a compact Hausdorff space. Given a closed inf-stable cone $\mathcal{K}$ in $C(X)$, define the cone $\mathcal{K}^*$ as the closure of the set of all elements of $C(X)$ that are majorized by elements of $\mathcal{K}$. By a Hahn–Banach separation argument it is proved that: (i) there exists $\mathcal{T} \subseteq \mathcal{M}^+$ such that $\mathcal{K}^*$ is the set of all $u \in C(X)$ that satisfy (1) above, and (ii) there exists $\mathcal{S}$ as above such that $\mathcal{K}$ is the set of all $u \in \mathcal{K}^*$ that satisfy (2). This establishes the theorem.

This work was announced, with some related matter, in [S3] and briefly referred to in [S18]. For a simplification see [LEM].

## 14 Expository works ([L2], [L6], [S5], [S6], [S11], [S12], [S13], [S19], [S24])

References to the survey articles [S5], [S6], [S11], [S12], [S13], [S19], [S24] have been made at appropriate points above. In addition we note that [S5], [S6] summarize work by various authors on pseudoconvex boundaries in $\mathbb{C}^n$ and related matters.

The lecture notes [L2] give one of the first book-length accounts of Choquet boundary theory, the lecture notes of Phelps [PHEa] (later published as [PHEb]) being their only rival at that time. Bauer's notes were timely, because the core theory had acquired definitive shape rather recently. His account is comprehensive and very clearly written, with detailed summaries of the functional analysis and measure theory required. The chapter headings are: Topological vector spaces, Extreme points and concave functions, The existence theorem of Choquet, The uniqueness theorem of Choquet, Examples and counter-examples. The book is still today an excellent introduction, and it is a pity that it was not published in a form more generally available.

The lecture notes [L6] are also the record of an introductory course on Choquet theory broadly similar to [L2], but shorter, somewhat less comprehensive, and without the summaries of prerequisite material.

**Acknowledgement.** I am grateful to Ivan Netuka and Niels Jacob for their assistance with the preparation of this article.

## References

[ALFa] E. M. Alfsen, On the Dirichlet problem of the Choquet boundary, *Acta Math.* **120** (1968), 149–159.

[ALFb] E. M. Alfsen, *Compact convex sets and boundary integrals*, Ergeb. Math. Grenzgeb. 57, Springer-Verlag, 1971.

[AC] F. Altomare and M. Capiti, *Korovkin-type approximation theory and its applications*, de Gruyter Stud. Math. 17, Walter de Gruyter, 1994.

[AS] R. Arens and I. M. Singer, Function values as boundary integrals, *Proc. Amer. Math. Soc.* **5** (1954), 735–745.

[ART] E. Artin, A proof of the Krein–Milman theorem (a letter to M. Zorn), *Picayune Sentinel*, University of Indiana, 1950, (= *The collected papers of Emil Artin*, Addison-Wesley, 1965, 523–525).

[AE] L. Asimow and A. J. Ellis, *Convexity theory and its applications*, Academic Press, 1980.

[BG] R. G. Bartle and L. M. Graves, Mappings between function spaces, *Trans. Amer. Math. Soc.* **72** (1952), 400–413.

[BEAa] H. S. Bear, The Silov boundary for a linear space of continuous functions, *Amer. Math. Monthly* **68** (1961), 483–485.

[BEAb] H. S. Bear, A geometric characterization of Gleason parts, *Proc. Amer. Math. Soc.* **16** (1965), 407–412.

[BEAc] H. S. Bear, *Lectures on Gleason parts*, Lecture Notes in Math. 121, Springer-Verlag, 1970.

[BW] H. S. Bear and M. L. Weiss, An intrinsic metric for parts, *Proc. Amer. Math. Soc.* **18** (1967), 812–817.

[BIS] E. Bishop, A minimal boundary for function algebras, *Pacific J. Math.* **9** (1959), 629–642.

[BDL] E. Bishop and de Leeuw, The representation of linear functionals by measures on sets of extreme points, *Ann. Inst. Fourier* **9** (1959), 305–331.

[BH] J. Bliedtner and W. Hansen, Simplicial cones in potential theory, *Invent. Math.* **29** (1975), 83–110.

[BJ] J. Bliedtner and K. Janssen, A generalization of H. Bauer's minimum principle, *Arch. Math.* **25** (1974), 505–510.

[BB] N. Boboc and G. H. Bucur, Sur le prolongement des fonctions affines, *Math. Scand.* **26** (1970), 42–50.

[BCa] N. Boboc and A. Cornea, Cônes de fonctions continues sur un espace compact, *C. R. Math. Acad. Sci. Paris* **261** (1965), 2564–2567.

[BCb] N. Boboc and A. Cornea, Convex cones of lower semicontinuous functions on compact spaces, *Rev. Roumaine Math. Pures Appl.* **12** (1967), 471–525.

[BOU] N. Bourbaki, *Elements of Mathematics: General Topology*, Part 2, Addison-Wesley, 1966.

[BRA] L. Bragard, Programmation quasi-concave, *Bull. Soc. Roy. Sci. Liége* **39** (1970), 478–485.

[CHO] G. Choquet, *Lectures on Analysis*, Vol. II, W. A. Benjamin, 1969.

[DAV] E. B. Davies, A generalized theory of convexity, *Proc. London Math. Soc.* **17** (1967), 644–652.

[DON] K. Donner, *Extension of positive operators and Korovkin theorems*, Lecture Notes in Math. 904, Springer-Verlag, 1982.

[DUB] S. Dubuc, Le maximum d'une fonction quasi-convexe, *C. R. Acad. Sci. Paris* **266** (1968), A271–A272.

[EDW] D. A. Edwards, Choquet boundary theory for certain spaces of lower semicontinuous functions, in *Function Algebras* (ed. F. T. Birtel), Scott, Foresman & Co., 1966, 300–309.

[EFF] E. G. Effros, Structure in simplexes II, *J. Funct. Anal.* **1** (1967), 379–391.

[EK] E. G. Effros and J. L. Kazdan, Applications of Choquet simplexes to elliptic and parabolic boundary value problems, *J. Differential Equations* **8** (1970), 95–134.

[FEL] W. Feller, Boundaries induced by non-negative matrices, *Trans. Amer. Math. Soc.* **83** (1956), 19–54.

[FUCa] L. Fuchs, Riesz groups, *Ann. Scuola Norm. Sup. Pisa Cl. Sci.* **19** (1965), 1–14.

[FUCb] B. Fuchssteiner, Bemerkungen zu den Minimumsätzen von H. Bauer, *Arch. Math.* **22** (1971), 287–290.

[FUCc] B. Fuchssteiner, Extrempunkte und Minimumsätze bei Hüllenbildungen, *Arch. Math.* **22** (1971), 523–527.

[FH] B. Fuchssteiner and W. Hackenbroch, Maximumpunkte, *Arch. Math.* **23** (1972), 415–421.

[GAMa] T. W. Gamelin, Uniform algebras and Jensen measures, *London Math. Soc. Lecture Note Ser.* 32, Cambridge University Press, 1978.

[GAMb] T. W. Gamelin, personal communication.

[GLE] A. M. Gleason, Function Algebras, in *Seminar on analytic functions*, Vol. II, Institute for Advanced Study, Princeton, 1957, 213–226.

[GOU] M. Goullet de Rugy, Géométrie des simplexes, *Collection Analyse Fonctionnelle, Centre de Documentation Universitaire*, Paris, 1968.

[KEL] J. L. Kelley, Note on a theorem of Krein and Milman, *J. Osaka Inst. Sci. Tech.* Part I **3** (1951), 1–2.

[KLE] V. L. Klee, Extremal structure of convex sets, *Arch. Math.* **8** (1957), 234–240.

[KÖN] H. König, On certain applications of the Hahn–Banach and minimax theorems , *Arch. Math.* **21** (1970), 583–591.

[KYF] Ky Fan, On the Krein-Milman theorem, in *Convexity*, Proc. Sympos. Pure Math. VII, Amer. Math. Soc., 1963, 211–219.

[LEH] G. Leha, Relative Korovkin-Sätze und Ränder, *Math. Ann.* **229** (1977), 87–95.

[LEM] J. Lembcke, Note zu "Funktionenkegel und Integralungleichungen" von H. Bauer, *Bayer. Akad. Wiss. Math. Natur. Kl. Sitzungsber.* **1977** (1978), 135–138.

[LIO] G. Lion, Familles d'opérateurs et frontière en théorie du potentiel, *Ann. Inst. Fourier* **16** (2) (1966), 389–453.

[LUMa] G. Lumer, Points extrémaux associés; frontières de Šilov et Choquet; principe du minimum, *C. R. Math. Acad. Sci. Paris* **256** (1963), 858–861.

[LUMb] G. Lumer, Points extrémaux associés; frontières de Šilov et Choquet; applications aux cônes de fonctions semi-continues, *C. R. Math. Acad. Sci. Paris* **256** (1963), 1066–1068.

[MCD] J. N. McDonald, Maximal measures and abstract Dirichlet problems, *Oxford Quart. J. Math.* **22** (1971), 239–246.

[MCK] R. McKissick, A non trivial sup norm algebra, *Bull. Amer. Math. Soc.* **69** (1963), 391–395.

[MIC] E. Michael, Continuous selections I, *Ann. of Math.* **63** (1956), 361–382.

[MS] G. Mokobodzki and D. Sibony, Cônes adaptés de fonctions continues et théorie du potentiel, *Séminaire Choquet (Initiation à l'Analyse)*, 6$^{e}$ année (1966/67), 5.01–5.35.

[NET] I. Netuka, The work of Heinz Bauer in potential theory, *Heinz Bauer Selecta*, Walter de Gruyter, Berlin–New York 2003, 29–41 (in this volume).

[PHEa] R. R. Phelps, *Lectures on Choquet's theorem*, University of Washington, 1963.

[PHEb] R. R. Phelps, *Lectures on Choquet's theorem*, van Nostrand, 1966.

[PHEc] R. R. Phelps, *Lectures on Choquet's theorem* (2nd edition), Lecture Notes in Math. 1757, Springer-Verlag, 2001.

[ROG] M. Rogalski, Espaces de Banach ordonnés, simplexes, frontières de Šilov et problème de Dirichlet *Séminaire Choquet* (*Initiation à l'Analyse*), 5$^e$ année (1965/66), 12.01–12.62.

[SAI] B. Saint-Loup, Sur le problème de Dirichlet de la frontière de Choquet, *Séminaire Brelot-Choquet-Deny, Théorie du potentiel*, 12$^e$ année (1967/68), 7.01–7.11.

[ŠAŠa] Ju. A. Šaškin, Convergence of positive linear operators in the space of continuous functions, *Dokl. Akad. Nauk. SSSR* **131** (1960), 525–527 (translated as *Soviet Math. Dokl.* **1** (1960), 303–305).

[ŠAŠb] Ju. A. Šaškin, Korovkin systems in spaces of continuous functions (in Russian), *Izv. Akad. Nauk. SSSR Ser. Mat* **26** (1962), 495–512.

[ŠAŠc] Ju. A. Šaškin, The Milman-Choquet boundary and the theory of approximation (in Russian), *Funktsional. Anal. i Prilozhen.* **1** (2) (1967), 95–96.

[SIB] D. Sibony, *Cônes de fonctions et potentiel*, Lecture Notes, Montreal, McGill University, 1968.

[SIC] J. Siciak, On function families with boundary, *Pacific J. Math.* **12** (1962), 375–384.

[VIN] G. F. Vincent-Smith, Uniform approximation of harmonic functions, *Ann. Inst. Fourier* **19** (2) (1969), 339–353.

[WUL] D. E. Wulbert, Convergence of operators and Korovkin's theorem, *J. Approx. Theory* **1** (1968), 381–390.

Mathematical Institute, 24-29 St Giles', Oxford, OX1 3LB, England
E-mail: edwardsd@maths.ox.ac.uk

# The work of Heinz Bauer in potential theory

*Ivan Netuka*

Bauer's work in potential theory extends over the period from 1958 to 1996. It can be divided as follows: *

1. Monograph: [TM2];

2. Original papers: [R19], [R20], [R23], [R24], [R34], [R36];

3. A summer school course: [S9];

4. Comptes Rendus Notes: [P5], [P6];

5. Proceedings contributions: [S7], [S8], [S10], [S14], [S20], [S24], [R39], [S27];

6. Survey papers: [S15], [S16], [S21], [S22], [O12], [S25], [S26].

Two *Seminar on potential theory* volumes (see p. xii) edited by Bauer, also reviews of five books, two obituaries and a preface to Constantinescu–Cornea's book [CC] should be included in this account concerning potential theory. Below we also mention several papers on convexity theory in which questions of potential theory are treated.
In this commentary, we briefly describe how Bauer's interest in potential theory began; then we comment on his original papers and, under the heading *Survey papers*, we offer a brief account of papers falling under the items 3, 5 and 6.

## Historical context

During the period 1956–1960, two sources of motivation leading to Bauer's deep and fruitful interest in potential theory can be traced: integral representation for compact convex sets and the beginnings of the theory of harmonic spaces. Let us quote from [O12]: [1)]

*References containing numbers refer to the Bibliography of Heinz Bauer whereas letters in brackets are related to references placed at the end of this commentary. The passages quoted here from the paper [O12] written in German are numbered and the translation into English was kindly provided by S. D. Chatterji and N. Jacob. The English version can be found under the corresponding numbers at the end of this commentary.

*Die Geburtsstunde der Theorie der harmonischen Räume fällt in das Jahr 1957. In diesem Jahr verbrachte ich die vorlesungsfreien Monate März und April in Paris, um in der Nähe von Gustave Choquet meine bereits 1956 bei ihm begonnenen Untersuchungen zum Thema der Integraldarstellungen in konvexen kompakten Mengen fortzuführen. In diesem vorausgehenden Jahr 1956 hatte ich durch das Centre National de la Recherche Scientifique als Attaché de Recherches neun Monate bei Choquet gearbeitet. Als „parrain", also als Pate und damit als menschlich-wissenschaftlicher Betreuer, wurde mir Marcel Brelot, der große Meister der Potentialtheorie, zugeordnet. Ich hörte damals seine Vorlesung über Potentialtheorie und studierte seine neuesten Veröffentlichungen, die er mir in die Hand drückte. Ein damals wie heute nicht einfaches Unterfangen. Diese – von einer Vorlesung von D. A. Kappos während meines Studiums abgesehen – erste intensive Begegnung mit der Potentialtheorie blieb zunächst eine Begegnung des ersten Einstimmens in ein riesiges Gebiet, das ich ohne zunächst weitergehende Ambitionen kennenlernen wollte.*

By a fortunate coincidence, a dominant role is played in both the above-mentioned fields by the *minimum principle.* Its form for a function space $\mathcal{H}$ of continuous functions on a compact space $X$ appears already in [P4]. It makes it possible to conclude that the Shilov boundary of $X$ with respect to $\mathcal{H}$ coincides with the closure of the set of the so-called *$\mathcal{H}$-extreme points* or, in later terminology, of the *Choquet boundary* of $X$ with respect to $\mathcal{H}$. For example, the following situation is considered in [P4]: $X$ is the closure of a bounded open subset $U$ of $\mathbb{R}^n$, and $\mathcal{H}$ consists of all functions continuous on $\overline{U}$ and harmonic on $U$. Using a result of M. V. Keldysh from 1941, H. Bauer shows that the $\mathcal{H}$-extreme points are exactly the regular points for the Dirichlet problem on $U$.

The preliminary communication [P4] was developed in [S2] and [R17]. In survey papers [S5] and [S6], as well as in lecture notes [L2] focused on Choquet theory, potential-theoretic situations are of importance (the Dirichlet problem on an open or on a compact set, Harnack's metric, Martin-type representation).

Before describing Bauer's contribution to the theory of harmonic spaces, we will borrow a quotation from [O12]:[2)]

*In diesem Jahr 1957 hatte sich erstmals das von Brelot und Choquet – ab 1958 auch von Jacques Deny – geleitete Séminaire de Théorie du Potentiel neu etabliert. Dort hörte ich am 9. April 1957 einen Vortrag Brelots zum Thema „Une axiomatique générale du problème de Dirichlet dans les espaces localement compacts" (vgl.* [BS]). *In diesem Vortrag erblickten die harmonischen Räume das Licht der Welt. Brelot hatte mir zwar schon im Jahr vorher andeutungsweise von seinen Überlegungen zu einem axiomatischen Zugang zur Behandlung des Dirichlet'schen Problems für möglichst allgemeine (stets lineare) elliptische Differentialgleichungen (zweiter Ordnung), auch auf allgemeineren Räumen und mit abstrakten Rändern, wie z. B. auf Green'schen Räumen, und im Zusammenhang mit dem damals gerade erst richtig bekannt werdenden Martin-Rand erzählt (vgl.* [BM]). *Er hatte dabei auch einen früheren, in eine ähnliche Richtung gehenden Versuch von Tautz* [T] *vor Augen. Jetzt aber, an jenem 9. April 1957, stellte Brelot erstmals seine diesbezügliche Ideen in einer bereits weitgehend ausgereiften Form der „Öffentlichkeit" vor, bei der es sich damals um einen recht kleinen Kreis von Zuhörern im Institut Henri Poincaré handelte.*

*Es war kein Zufall, daß Brelot gerade jetzt seine Resultate und Ideen vorstellte. Im Februar des gleichen Jahres hatte nämlich J. L. Doob in zwei Vorträgen des Séminaire de Théorie du Potentiel über das Thema „Le problème de Dirichlet généralisé – Interprétation probabiliste" vorgetragen (vgl.* [DO]) *und dabei vor allem auf seinen Beitrag* [D] *zum dritten Berkeley-Symposium von 1955 Bezug genommen.*

Brelot had already thought about the axiomatization of potential theory during the second world war. However, the circumstances described above completed the crystallization process and led him to the creation of what are nowadays called *Brelot harmonic spaces*, making it possible to develop a very satisfactory abstract potential theory that includes elliptic partial differential equations of second order as a principal example.

Brelot considers a locally compact, non-compact, connected and locally connected topological space, and a sheaf of real-valued continuous functions (called *harmonic*) such that regular sets for the Dirichlet problem form a base for the topology and the so-called Brelot convergence property holds: *the supremum of an increasing sequence of harmonic functions on a domain is harmonic, provided that it is finite at some point.* Here is Bauer's comment from [O12] on the situation: 3)

*Aus diesen Forderungen an das Verhalten der harmonischen Funktionen gewinnt Brelot in seinem Pariser Vortrag* [BS] *das allgemeine Randminimumprinzip. Im Gegensatz hierzu sah sich Doob dazu gezwungen, dieses Randminimumprinzip selbst zu einer definierenden Eigenschaft, also zu einem Axiom, zu erheben. Auch die Mischung analytischer und probabilistischer Methoden bei Doob* [D] *hängt mit der Problematik des Randminimumprinzips zusammen. Trotzdem kann Doob nicht so weit vorstoßen wie Brelot. Beispielsweise fehlt bei Doob ein Analogon zu dem klassischen Resolutivitätssatz von Brelot, welcher die resolutiven Funktionen in der PWB-Methode als diejenigen kennzeichnet, welche bezüglich eines (und damit jedes) harmonischen Maßes integrierbar sind. Eine kanonische Verallgemeinerung dieses Resolutivitätssatzes konnte Brelot bereits in seinem Pariser Vortrag beweisen.*

## Bauer harmonic spaces

It was by no means clear whether it was feasible to build up a satisfactory abstract potential theory covering the case of partial differential equations of parabolic type where e.g. a minimum principle and the resolutivity theorem could be deduced in a natural way. As we learn from Bauer in [O12], his previous results from convexity turned out to be extremely useful: 4)

*In dieser Situation kam mir im Sommer 1959 die Idee, meine parallel laufenden Untersuchungen über Integraldarstellungen in konvexen kompakten Mengen nach geeigneter Modifikation zum Einsatz zu bringen. Kurz vor der Dresdner DMV-Tagung hatte ich während einer verregneten Ferienwoche im Salzkammergut zeigen können, daß nach unten halbstetige, konkave, reelle Funktionen auf einer konvexen kompakten Menge (in einem geeigneten lokal-konvexen Hausdorff-Raum) ihr Minimum stets in einem Extremalpunkt annehmen (vgl.* [R13]). *Jetzt konnte ich die damalige Schlußweise so modifizieren, daß ein „allgemeines Minimumprinzip" folgender Art entstand: Auf einem kompakten Raum Y sei eine Menge* $\mathcal{E}$ *nach unten halbstetiger reeller Funktionen definiert, welche die Punkte von Y trennt. Dann nimmt jede dieser Funktionen ihr Minimum in einem Punkt* $y \in Y$ *mit folgender Eigenschaft an: Die Einheitsmasse* $\varepsilon_y$ *ist das einzige Radonmaß* $\mu \geq 0$ *auf Y mit Gesamtmasse 1 derart, daß* $\int u\, d\mu \leq u(y)$ *für alle* $u \in \mathcal{E}$ *gilt (vgl.* [R16]).

This was the key result and two modifications of Brelot's axioms turned out to be sufficient in order to develop the desired abstract potential theory of elliptic-parabolic type. In the first place, the harmonic structure had to be rich enough, namely the harmonic or at least hyperharmonic functions were supposed to separate linearly the points of the space. Secondly, a new convergence property had to be considered, because the Brelot convergence property already fails for solutions of the heat equation. In [TM2], Bauer works with the Doob convergence property which requires that *the supremum of an increasing sequence of harmonic functions on an open set U is harmonic, provided that it is finite on a dense subset of U*. As a preliminary replacement to the Doob convergence property, Bauer considers in [P5] and [R19], [R20] two weaker convergence properties. One of them, the so-called Bauer convergence property, reads as follows: *the supremum of an upper-bounded increasing sequence of harmonic functions is harmonic.* In fact, a large portion of the theory can be developed using this weak convergence property.

The results on harmonic spaces were announced in [P5] and developed in [R19], [R20] and [R23]. A very readable exposition of Bauer's theory of harmonic spaces was published as [TM2] in 1966 on the basis of a course held in Hamburg during the summer semester 1965. Incidentally, the term *harmonic space* was first introduced in this course; see [O12]. It should be emphasized that [TM2] is not a mere reprint of [R19] and [R20]. It develops the theory from the very beginning, step by step, in a self-contained way, using systematically, as already mentioned, the Doob convergence property. Special attention is paid to the relationship to ellipticity and the Brelot convergence property. In addition a Harnack type inequality involving absorbing sets is proved, and fine topology and nuclearity in a harmonic space are studied. Thus several results, not previously published, are included. For instance, in Bauer's theory, there may exist a compact set contained in no regular set. Many new results for the potential theory of the heat equation are obtained.

The theory of harmonic spaces attracted the attention of mathematicians working in analysis as well as in probability theory, and the state of art in the early seventies is well described by Bauer in the following part of his preface to [CC] written in 1971; cf. [O2]: *There has been a considerable revival of interest in potential theory during the last 20 years. This is made evident by the appearance of new mathematical disciplines in that period which now-a-days are considered as parts of potential theory.*

*Examples of such disciplines are: the theory of Choquet capacities, of Dirichlet spaces, of martingales and Markov processes, of integral representation in convex compact sets as well as the theory of harmonic spaces. All these theories have roots in classical potential theory.*

*The theory of harmonic spaces, sometimes also called axiomatic theory of harmonic functions, plays a particular role among the above mentioned theories. On the one hand, this theory has particularly close connections with classical potential theory. Its main notion is that of a harmonic function and its main aim is the generalization and unification of classical results and methods for application to an extended class of elliptic and parabolic second order partial differential equations. On the other hand, the theory of harmonic spaces is closely related to the theory of Markov processes. In fact, all important notions and results of the theory have a probabilistic interpretation.*

In 1973, a new subsection of the section 31 Potential theory, namely D 05 called *Axiomatic potential theory* (not including probabilistic potential theory) was introduced into *Mathematical Reviews*. Up to the end of the year 2000, reviews of more than 700 items appeared there. The impact of Bauer, as well as of many other mathematicians, on the further development of abstract potential theory is documented in detail in the monographs by C. Constantinescu and A. Cornea [CC], by J. Bliedtner and W. Hansen [BH] and in an excellent survey paper [S21].

We first turn to Bauer's fundamental papers [P5] and [R19], [R20]. We do not intend to analyse here all the subtleties of various forms of separation and convergence properties. We confine ourselves to recalling the basic results obtained. In the *Comptes Rendus Note* [P5], the Bauer convergence property is introduced and the minimum principle for hyperharmonic functions (called *fonctions hyperprincipales*) and the resolutivity theorem for the Perron–Wiener–Brelot solution of the Dirichlet problem are announced. Another result: *Differences of positive harmonic functions form a conditionally complete Riesz space.* The existence of regular points is a consequence of the following assertion: *For a relatively compact nonempty open set $U$, every function continuous on $\overline{U}$ and hyperharmonic on $U$ attains its minimum at a regular point.*

A detailed exposition of Bauer's theory is presented in [R19]: historical context, axioms, examples, minimum principle, convergence properties, relations to Brelot's theory, the Perron–Wiener–Brelot solution, resolutivity theorem, negligible sets, regular points (their existence and the local character), regularity criteria. Also, the notion of a *semiregular set* is introduced and studied.

The results obtained generalize those established by M. Brelot and also those obtained by R.-M. Hervé in her thesis [HR]. To obtain these generalizations, Bauer is frequently obliged to modify substantially the methods of earlier authors. One example of the obstacles encountered is described in [O12]:[5)]

*Im Bereich meiner auch parabolische Situationen erfassenden Theorie ergab sich bei der Weiterentwicklung eine neue grundsätzliche Schwierigkeit, nämlich die Frage nach der Existenz überall positiver Potentiale, welche diese Eigenschaft auch nach gewissen Modifikationen (z. B. der berühmten Perron-Modifikation) nicht verlieren. In der elliptischen, d. h. Brelot'schen Theorie, bereitet dies keine große Schwierigkeit, da dort nichtnegative hyperharmonische Funktionen in einem Gebiet entweder identisch Null oder überall strikt positiv sind. Bei der Wärmeleitungsgleichung ist dies aber nicht mehr richtig, wie bereits die auf einem Halbraum verschwindende Fundamentallösung dieser Gleichung zeigt.*

*Wie schon einmal kam auch diesmal die rettende Idee aus der konvexen Analysis. Ebenfalls im Jahr 1961 veröffentlichte Michel Hervé* [HM] *seinen Beweis des Choquet'schen Darstellungssatzes im metrisierbaren Fall, indem er in sehr einfacher Weise durch Überlagerung einer im Raum der stetigen Funktionen totalen Folge stetiger konkaver Funktionen eine stetige, strikt konkave Funktion konstruiert. Diese Idee ließ sich nach geeigneten Modifikationen auch bei harmonischen Räumen verwenden und führte zum Begriff des strengen Potentials und damit zur Überwindung der angedeuteten neuen Schwierigkeiten. Damit war einerseits der Weg frei für die Entwicklung der Balayage-Theorie, des Kernstücks einer guten Potentialtheorie. Andererseits gelangte man mit diesen Hilfsmitteln später zur probabilistischen Interpretation der Potentialtheorie eines harmonischen Raumes durch die Konstruktion eines geeigneten Hunt-Prozesses, der der Theorie in natürlicher Weise assoziiert ist. Der Weg zu den strengen Potentialen und ihrer Einführung als technisches Hilfsmittel der Potentialtheorie gelang mir im Winter 1961/62 während meines Gastaufenthaltes an der University of Washington in Seattle.*

The paper [R20] is a natural continuation of [R19] and presents important new notions and results. In Chapter I, the theory of superharmonic functions is further developed, and the Doob convergence property is introduced and applied to a more detailed study of polar sets. Chapter II deals with approximation by means of differences of potentials and with a construction of strict potentials. The important notion of *absorbing set* (characteristic for parabolic potential theories and having no place in elliptic ones) is introduced and investigated. Previous results on the Dirichlet problem are completed in Chapter III. Here the role of linear point separation by means of positive superharmonic functions is also cleared up and balayage theory for compactly supported measures is studied. It is shown that decomposition and extension theorems established by R.-M. Hervé in the framework of Brelot's theory hold in Bauer harmonic spaces. The final section of the paper is devoted to an application of the results proved to polar sets.

The article [R23] is Bauer's third contribution to axiomatic potential theory. The central result shows that the weak thinness of a set at a point in its complement and ordinary thinness are equivalent. Bauer's original approach used Hunt's theory, and therefore the set in question had to be supposed analytic. In [R23], a proof of N. Boboc, C. Constantinescu and A. Cornea of the equality of the reduite and the balayage outside the set is presented and the equality of the two types of thinness for an arbitrary set is obtained as a corollary. Further, absorbing sets and polar sets are investigated and those Bauer harmonic spaces which satisfy Brelot's axioms are characterized. In particular, an analytic set is shown to be polar if all its compact subsets are polar. It is also proved that the complement of a polar set in a connected harmonic space is connected. The Brelot convergence theorem formulated in terms of the notion of semipolar sets is of a great importance in Bauer's theory, since the classes of *polar, totally thin* and *semipolar* sets are pairwise different. This is illustrated at the end of [R23] within the harmonic space associated with the heat equation. For an arbitrary set $A$, the set of points of $A$ not belonging to the base of $A$ is shown to be semipolar. In particular, the set of irregular points of an open relatively compact set is semipolar. Six conditions equivalent to the *axiom of polarity* are given; five of them were known in Brelot's theory to be equivalent to the axiom of polarity from [HR]. Results of [R23] were included in [TM2] and later extended by several authors; cf. [CC], [BH].

## The Cauchy–Dirichlet problem

In the theory of parabolic partial differential equations, one traditionally makes a distinction between initial problems and boundary-value problems. The aim of the paper [R24] is to show that this distinction is superficial and that the Cauchy problem and the Dirichlet problem can be considered as special cases of a general Dirichlet problem formulated for an arbitrary open set. In [TM2], the Dirichlet problem is treated for relatively compact open sets only, which does not cover the most classical case of the Cauchy problem for the heat equation on a half-space.

Bauer considers a strong harmonic space and an open set with non-empty boundary, and he develops for *potentially bounded* functions on the boundary a modification of the Perron–Wiener–Brelot method. He establishes the resolutivity theorem and studies the corresponding notion of regular points. Finally, the results obtained in the abstract framework are applied to the Cauchy problem for the heat equation on a half-space. The Dirichlet problem for general open sets is extensively studied e.g. in [CC] and [BH].

## Heat balls and Fulks measures

An analogue of the spherical mean-value property of classical harmonic functions for caloric functions (i.e., solutions of the heat equation) was introduced by W. Fulks in [F]. Ordinary Euclidean balls, which are bodies bounded by level sets of the fundamental solution of the Laplace equation, are replaced by the so-called *heat balls*. These are bodies bounded by level sets of the fundamental solution of the heat equation. Here already a peculiar feature appears: the centre of the heat ball lies on its boundary and is an irregular point of the heat ball. Fulks introduced a measure having a density with respect to surface measure on the boundary of heat ball in such a way that the corresponding mean value equation characterizes caloric functions.

N. Watson used the Fulks measure in [W] to define a heat equation analogue of classical hyperharmonic functions, called by him *hypertemperatures*. For this he simply reformulates the super-mean-value inequality in terms of Fulks measures.

The main result of [R34], announced in [P6], reads as follows: *The Fulks measure on a heat ball centered at a point $z$ is obtained by sweeping the Dirac measure at $z$ onto the complement of the heat ball.* This leads to a probabilistic interpretation of the Fulks measure and to the identification of Watson's hypertemperatures with hyperharmonic functions of the harmonic-space theory associated with the heat equation.

The exposition uses results from harmonic spaces, and the presentation is very accessible. An alternative approach to the main result on the identification of the Fulks measure with the balayaged measure can be found in [N]. However, the proof suggested there relies heavily on a simpliciality result for the heat equation.

## Fine boundary behaviour of harmonic functions

M. Brelot [B], [BR] studied the fine limit behaviour of bounded classical harmonic functions at an irregular point of a bounded domain in an Euclidean space. An extension for Brelot harmonic spaces obtained by E. Smyrnélis in [S] was not easy, because the methods used by Brelot were based on Green's functions and the Martin representation theory. Therefore very restrictive additional assumptions were necessary, which ruled out the possibility of treating solutions of the heat equation.

Bauer was able to find in [R36] an approach making it possible to study the boundary behaviour of harmonic functions in general harmonic spaces.

Given a relatively compact open set $U$ of a strong harmonic space with a polar irregular point $z$, the object is to investigate the fine limit behaviour of the Perron–Wiener–Brelot solution for a bounded resolutive function $f$ on the boundary. One of the main results of [R36] states that such a function is integrable with respect to the balayage $\mu_z$ of the Dirac measure $\varepsilon_z$ at $z$ and the fine limit of the solution of the Dirichlet problem is equal to the integral of $f$ with respect to $\mu_z$.

The second main result has to do with maximal sequences. A sequence of points of $U$ is said to be maximal for an irregular point $z$, if the sequence tends to $z$ and the corresponding harmonic measures tend to $\mu_z$. It is shown that, for a bounded harmonic function $h$ on $U$, the fine limit at $z$ exists and is equal to the usual limit of $h$ along any maximal sequence. Maximal sequences for the heat equation are also considered.

Further improvements were subsequently obtained. W. Hansen established in [H] similar results for finely harmonic functions using a more sophisticated method. In [R39], H. Bauer presents in an expository form important results of [R36] and, moreover, shows that the main hypotheses of [R36] can be loosened considerably. In particular, in order to obtain the results, the Doob convergence property can be replaced by the weaker Bauer convergence property. There is also an improvement of a result on maximal sequences for the heat equation.

For further generalizations of results obtained in [R36] and [R39], see [N]; cf. also the survey papers [S25] and [S26].

## Survey papers

The contribution [S8] appeared before the publication of the monograph [TM2]. It provides an overview of the theory of Bauer harmonic spaces, including the application to the Cauchy–Dirichlet problem on general open sets of a harmonic space; cf. [R24]. A connection with the theory of Markov processes is briefly mentioned.

This connection is the central theme of the paper [S7]. The construction of sub-Markov semigroups as well as the construction of Markov processes with given transition probabilities are discussed and applied to the axiomatic theory of harmonic functions. The paper is based on results of W. Hansen's dissertation. Only sketches of proofs of the main results are given.

On the other hand, the course [S9] on the interplay between harmonic spaces and Markov processes provides all proofs concerning the equivalence of fundamental potential theoretic and probabilistic notions including balayage, polar sets, harmonic measure, absorbing sets, thinness, totally thin and semipolar sets.

In 1974 Bauer was an invited speaker at the International Congress of Mathematicians in Vancouver. His contribution [S10] to the Proceedings starts as follows: *In 1964 Pierre Jacquinot opened a colloquium on potential theory in Orsay, France, by comparing potential theory with a road intersection in mathematics.*** *This was ten years ago. Meanwhile traffic has increased, and crossroads had to be converted into interchanges of highways – also in potential theory.* The first part of the contribution addresses three aspects of classical potential theory: superharmonic functions, Newtonian kernel and potentials, Brownian motion. The role of balayage is emphasized. The second part reflects two main aspects of potential theory of the early seventies: harmonic spaces and Markov processes. The last part is devoted to Fuglede's theory of finely harmonic functions, including an application to asymptotic paths for subharmonic functions.

The main theme of [S14] is a description of an elliptic harmonic structure on infinite-dimensional torus, based on results obtained by A. D. Bendikov and C. Berg.

In [S15], the following topics of potential theory are discussed: relations between harmonic spaces and Feller semigroups, potential theory associated to a convolution semigroup of measures on a locally compact Abelian group, harmonic structures on an infinite-dimensional torus, and a semigroup approach to differentiation theory.

The article [S20] offers an image of potential theory presented on the occasion of Euler's Colloquium, Berlin, 1983. Three fundamental problems of classical potential

**La théorie du potentiel est un véritable carrefour de la Mathématique.

theory are discussed, namely the equilibrium problem, the conductor problem and the Dirichlet problem. Balayage theory which plays a prominent role in modern potential theory, is treated from the point of view of energy and, in more detail, from that of superharmonic functions. Two sections are devoted to harmonic spaces and to the basic notions of the theory: thinness, semipolar and polar sets and balayage. Then an overview of new results on finely harmonic functions and finely holomorphic functions is presented. The probabilistic interpretation of fundamental potential-theoretic notions is included in the last section. A historical observation, surprising even for specialists in the field, is documented: the Laplace equation already appears for the first time in mathematics in L. Euler's work *Principia Motus Fluidorum* in the fifties of the 18$^{th}$ century.

The most complete, detailed and extremely well-written survey on harmonic spaces is [S21], an extended version of [S16] published in German. It covers: history and motivation, axioms, a lot of examples, strong harmonic spaces, Brelot spaces, potential theory on harmonic spaces, associated Markov processes, probabilistic interpretations. The final section considers the outlook for further development: harmonic spaces in the setting of C. Constantinescu and A. Cornea, balayage spaces of J. Bliedtner and W. Hansen, local operators of L. Stoica, $H$-cones of N. Boboc, Gh. Bucur and A. Cornea, F.-Y. Maeda's investigations on the Dirichlet integral on harmonic spaces.

Elliptic-parabolic differential operators and associated diffusions on Euclidean spaces and on manifolds form a central theme of [S22]. In particular, the martingale problem and the explosion phenomenon are discussed. Also, the outlook for stochastic Riemannian geometry is briefly considered.

The survey paper [S24] is, so to say, in the *middle of the road* between Bauer's two favourite subjects: infinite-dimensional convexity and potential theory. The first part recalling basic notions and results on simplicial function spaces and cones is followed by an excursion into the potential theory for the Laplace equation. The next section describes fundamental results of J. Bliedtner and W. Hansen on the simpliciality of cones in potential theory, on the characterization of the Choquet boundary and on the weak Dirichlet problem in a general context; cf. [BH]. The final section is devoted to results related to the Keldych theorem on uniqueness of the operator approach to the Dirichlet problem.

The article [O12] dedicated to Prof. L. Schmetterer gives a personal description of the origin of Bauer harmonic spaces.

Papers [S25], [S26] present an exposition of results obtained by H. Bauer in [R36] and [R39]. These surveys also summarize more general results from [N].

The last paper devoted to potential theory is [S27] and bears the subtitle *Introductory thoughts to the International Conference on Potential Theory, Kouty, Czech Republic, 1994* and is best described by the following quotation from [S27]: *The author's* Reflections *centered around the widely accepted fact that mathematics becomes particularly interesting and promising whenever different fields of mathematics interfere or approach each other with the chance to amalgamate. In the main part of the lecture, it was then the author's aim to point out* Results *in connection with two such amalgamation processes: amalgamation of potential theory with Riemannian geometry as well as amalgamation of nonlinear aspects of potential theory with the probabilistic theory of superprocesses.*

**Acknowledgement.** The author of this commentary is grateful to D. A. Edwards for improvements of the language level of the text. He was supported in part by the grant GAČR 201/03/0935, and in part by the Research Project MSM 1132 0007 from the Czech Ministry of Education.

## References

[BH] Bliedtner, J., and Hansen, W., *Potential Theory, An analytic and probabilistic approach to balayage*, Springer-Verlag, Berlin–Heidelberg–NewYork–Tokyo 1986.

[B] Brelot, M., Sur les ensembles effilés, *Bull. Sci. Math.* (2) **68** (1944), 12–36.

[BM] Brelot, M., La théorie moderne du potentiel, *Ann. Inst. Fourier (Grenoble)* **4** (1954), 113–133.

[BR] Brelot, M., On the behavior of harmonic functions in the neighborhood of an irregular boundary point, *J. Anal. Math.* **4** (1954–1956), 209–221.

[BS] Brelot, M., Une axiomatique générale du problème de Dirichlet dans les espaces localement compacts, *Séminaire de Théorie du Potentiel* (dirigé par M. Brelot, G. Choquet), 1re année, 1957 (16 p.), Institut H. Poincaré, Paris 1958.

[CC] Constantinescu, C., and Cornea, A., *Potential Theory on Harmonic Spaces*, Grundlehren Math. Wiss. 158, Springer-Verlag, Berlin–Heidelberg–New York 1972.

[D] Doob, J. L., Probability methods applied to the first boundary value problem, Proc. 3rd *Berkeley Sympos. Math. Stat. and Probab.*, 1954–1955, 1956, 49–80.

[DO] Doob, J. L., Le problème de Dirichlet généralisée. Interprétation probabiliste, *Séminaire de Théorie du Potentiel* (dirigé par M. Brelot, G. Choquet), 1re année, 1957 (1 p.), Institut H. Poincaré, Paris 1958.

[F] Fulks, W., A mean value theorem for the heat equation, *Proc. Amer. Math. Soc.* **17** (1966), 6–11.

[H] Hansen, W., Fine boundary limits of finely harmonic functions, *Math. Ann.* **274** (1986), 643–647.

[HM] Hervé, M., Sur les représentations intégrales à l'aide des points extrémaux dans un ensemble compact convexe métrisable, *C. R. Acad. Sci. Paris* **253** (1961), 366–368.

[HR] Hervé, R.- M., Recherches axiomatiques sur la théorie des fonctions surharmoniques et du potentiel, *Ann. Inst. Fourier (Grenoble)* **12** (1962), 415–571.

[N] Netuka, I., Fine behaviour of solutions of the Dirichlet problem near an irregular boundary point, *Bull. Sci. Math.* (2) **114** (1990), 1–22.

[S] Smyrnélis, E., Sur les limites fines des fonctions harmoniques et les suites maximales, *Bull. Sci. Math.* (2) **97** (1973), 161–175.

[T] Tautz, G., Zur Theorie der ersten Randwertaufgabe, *Math. Nachr.* **2** (1949), 279–303.

[W] Watson, N. A., A theory of subtemperatures in several variables, *Proc. London Math. Soc.* (3) **26** (1973), 385–416.

Mathematical Institute, Faculty of Mathematics and Physics, Charles University, Sokolovská 83, 186 75 Praha 8, Czech Republic
E-mail: `netuka@karlin.mff.cuni.cz`

1) The hour of birth of the theory of harmonic spaces falls in the year 1957. In this year I spent the lecture free months of March and April in order to continue near Gustave Choquet my studies on integral representations in compact convex sets which I had already begun with him in 1956. During the year 1956 I had worked with Choquet for nine months as Attaché de Recherche of the Centre National de la Recherche Scientifique. Marcel Brelot, the great master of potential theory, was assigned to me as "parrain", i.e. as godfather and thereby as scientific mentor. At that time I attended his lecture on potential theory and studied his latest publications which he passed on to me. This was not an easy task neither then nor now. Except for a lecture course of D. A. Kappos when I was a student, this was my first intensive encounter with potential theory, an encounter of preliminary familiarization of a gigantic field which I wished to learn without having any further ambitions in the beginning.

2) The Séminaire de Théorie du Potentiel directed by Brelot and Choquet – from 1958 on also by Jacques Deny – was newly founded in this year 1957. On 9th April 1957 I heard a lecture there by Brelot on "Une axiomatique générale du problème de Dirichlet dans les espaces localement compacts" (cf. [BS]). This lecture was the moment of birth of harmonic spaces. Already a year before Brelot had indeed indicated to me his thoughts on an axiomatic approach for treating the Dirichlet problem for the most general (always linear) elliptic differential equations (of the second order), even on more general spaces with abstract boundaries, for example, on Green spaces and Martin boundary which in this connexion was beginning to be known at that time (cf. [BS]). He had also in view an earlier attempt of Tautz [T] in a similar direction. But now, on that 9th April 1957, Brelot presented his ideas on this for the first time in an already fairly matured form to the "public" which at that time consisted of a rather small group of auditors in Institut Henri Poincaré.

That Brelot presented his results and ideas just then was no accident. In fact, in February of the same year, J. L. Doob had given two lectures on "Le problème de Dirichlet généralisé – interprétation probabiliste" (cf. [DO]) at the Séminaire de Théorie du Potentiel and these had specially referred to his contribution [D] to the third Berkeley-Symposium of 1955.

3) In his Paris lecture [BS] Brelot deduces the general boundary minimum principle from these requirements on the behaviour of harmonic functions. In contrast, Doob saw himself obliged to assume this very boundary minimum principle as a defining property, i.e. as an axiom. The mixture of analytic and probabilistic methods in Doob's work [D] also depends on the question of the boundary minimum principle. In spite of this, Doob cannot advance as far as Brelot. For example, Doob lacks an analogue of the classical resolutivity theorem of Brelot which characterizes the resolutive functions in the PWB-method as those which are integrable with respect to a (and hence every) harmonic measure. Brelot could already prove a canonical generalization of the resolutivity theorem in his Paris lecture.

4) In this situation I had the idea in 1959 summer, of bringing to bear with suitable modification my parallel studies on integral representations in convex compacts sets. Shortly before the DMV-meeting in Dresden, I could prove during a rained out vacation week in Salzkammergut, that lower semicontinuous, concave, real functions on a convex compact set (in a suitable locally convex Hausdorff space) always assume their minimum at an extreme point (cf. [R13]). Now I could modify that argument so as to set up a "general minimum principle" of the following type: On a compact space $Y$, let $\mathcal{S}$ be a set of lower semicontinuous real functions which separates the points of $Y$. Then each of these functions assumes its minimum at a point $y \in Y$ having the following property: the unit mass $\varepsilon_y$ is the only Radon measure $\mu \geq 0$ on $Y$ of total mass 1 such that $\int u\,d\mu \leq u(y)$ for all $u \in \mathcal{S}$ (cf. [R16]).

5) A new fundamental difficulty arose in the further development of my theory covering also the parabolic situation, namely, the question of the existence of everywhere positive potentials which do not lose this property even after certain modifications (for example, the famous Perron modification). In the elliptic, i.e. in Brelot's theory, this presents no great difficulty, since there non-negative hyperharmonic functions in a domain are either identically zero or everywhere strictly positive. For the heat equation this is no longer true as shown already by a fundamental solution of this equation which vanishes in a half-space.

As once before so this time also the saving idea came from convex analysis. Also, in the year 1962 Michel Hervé [HM] published his proof of Choquet's representation theorem in the metrizable case by constructing very simply a continuous strictly concave function by the superposition of a sequence of continuous concave functions which is total in the space of continuous functions. This idea could also be applied in harmonic spaces after appropriate modifications and led to the notion of strict potentials and thereby to overcoming the indicated new difficulties. Thus the way was free, on the one hand, for the development of the balayage theory, the central topic of a good potential theory. On the other hand, with these tools one obtained later a probabilistic interpretation of potential theory in harmonic spaces via the construction of a suitable Hunt process which is naturally associated with the theory. The way to strict potentials and their introduction as technical tool of potential theory was achieved in winter 1961/62 during my stay at the University of Washington in Seattle.

# Reguläre und singuläre Abbildungen eines distributiven Verbandes in einen vollständigen Vektorverband, welche der Funktionalgleichung $f(x \vee y) + f(x \wedge y) = f(x) + f(y)$ genügen [R3]

J. Reine Angew. Math. **194** (1955), 141–179
[Zbl. Math. 65.01705; MR 17,177d]

## Einleitung.

**I.** In der klassischen Maßtheorie spielt eine wichtige Rolle folgender Zerlegungssatz von H. Lebesgue:

*Jedes auf einem Booleschen $\sigma$-Mengenverband $V$* (Hausdorffschen $\sigma$-Körper) *von Teilmengen einer Grundmenge $e \in V$ definierte, reelle* (endliche, nicht-negative) *Maß $f$ läßt sich bezüglich eines jeden, ebenfalls auf $V$ erklärten Maßes $m$ auf genau eine Weise darstellen in der Form*

$$f(x) = r(x) + s(x), \quad x \in V,$$

*wobei $r$ und $s$* (nicht-negative) *Maße auf $V$ mit den folgenden, für sie charakteristischen Eigenschaften sind:*

**(A)** Es ist $r$ $m$-stetig (total-stetig bezüglich $m$), d. h. für Mengen $x \in V$ mit hinreichend kleinem Maß $m(x)$ wird $r(x)$ beliebig klein. — Es ist $s$ $m$-rein-unstetig, d. h. es existieren Mengen $x \in V$ derart, daß $m(x)$ und $s(e - x)$ gleichzeitig beliebig klein sind.

Als gleichwertig mit den Eigenschaften (A) von $r$ und $s$ erweisen sich die folgenden, welche also für die Zerlegungsteile $r, s$ auch kennzeichnend sind:

**(B)** Es ist $r$ $m$-regulär ($m$-nulltreu), d. h. für jede $m$-Nullmenge $x \in V$ gilt $r(x) = 0$. — Es ist $s$ $m$-singulär, d. h. es existiert eine $m$-Nullmenge $n_0 \in V$ derart, daß $s(x) = 0$ für jede zu $n_0$ fremde Menge $x \in V$ ist.

Nennt man eine auf $V$ total-additive Funktion $m$-stetig bzw. $m$-rein-unstetig, wenn sie Differenz zweier $m$-stetiger bzw. $m$-rein-unstetiger *Maße* auf $V$ ist, so läßt sich auch jede auf $V$ total-additive Funktion auf genau eine Weise als Summe einer $m$-stetigen und einer $m$-rein-unstetigen, total-additiven Funktion darstellen.

**II.** Die Begriffe „$M$-stetig" und „$M$-rein-unstetig" lassen sich auch für reelle, relativ beschränkte, lineare Funktionale über einem Vektorverband $E$ erklären (vgl. Satz 5. 33 dieser Arbeit), wenn hierbei $M$ ein *positives lineares Funktional* auf $E$ bedeutet. Dann gilt nach F. Riesz [9] [1]) der zum Lebesgueschen Zerlegungssatz analoge Satz:

[1]) Zahlen in eckigen Klammern beziehen sich auf das Literaturverzeichnis am Schluß der Arbeit.

*Jedes relativ beschränkte, lineare Funktional $F$ über $E$ läßt sich auf genau eine Weise darstellen als Summe eines $M$-stetigen und eines $M$-rein-unstetigen, relativ beschränkten, linearen Funktionals; diese Zerlegungsteile sind positive Funktionale, wenn $F$ positiv ist.*

Diesem Satz liegt nach F. Riesz [9] folgender, rein *verbandsalgebraische* Sachverhalt zugrunde: Alle auf $E$ definierten, reellen, relativ beschränkten, linearen Funktionale bilden einen *vollständigen Vektorverband* $\Lambda$. Für jedes Band $B$ in $\Lambda$ (vgl. Nr. 1. 2 dieser Arbeit) bilden die zu den Elementen von $B$ disjunkten Elemente von $\Lambda$ ein Band $B'$; es ist $\Lambda$ *direkte ordnungstreue Summe dieser beiden Bänder.* Wählt man für $B$ das von dem oben genannten Funktional $M$ in $\Lambda$ erzeugte Band, so besteht dieses bzw. $B'$ genau aus den $M$-stetigen bzw. $M$-rein-unstetigen Funktionalen aus $\Lambda$. Bei der direkten ordnungstreuen Zerlegung von $\Lambda$ in $B$ und $B'$ ist für jedes $F \in \Lambda$ die $B$-Komponente der $M$-stetige und die $B'$-Komponente der $M$-rein-unstetige Teil des Funktionals $F$.

**III.** Mit Hilfe des Zerlegungssatzes von Ziff. II konnte F. Riesz [9] wenigstens in einem Spezialfall auch dem Lebesgueschen Zerlegungssatz von Ziff. I eine algebraische Wendung geben: Im Sinne der Bezeichnungen von Ziff. I sei nämlich $V$ der Boolesche $\sigma$-Verband aller Borelschen Teilmengen eines Intervalls $J = [a, b]$ der Zahlengeraden und $m$ das Lebesguesche Maß auf $V$. Jedem Maß $g \mid V$ ist ein positives lineares Funktional $G$ über dem Vektorverband $E$ aller auf $J$ stetigen Funktionen $\varphi$ zugeordnet, nämlich $G(\varphi) = \int_a^b \varphi dg$; insbesondere entspricht dem Lebesgueschen Maß $m$ auf diese Weise das Funktional $M(\varphi) = \int_a^b \varphi dx = \int_a^b \varphi dm$. Ist dann $f = r + s$ die Lebesguesche Zerlegung eines beliebigen Maßes $f \mid V$ bezüglich $m$, so ist $R(\varphi) = \int_a^b \varphi dr$ der $M$-stetige und $S(\varphi) = \int_a^b \varphi ds$ der $M$-rein-unstetige Teil des Funktionals $F(\varphi) = \int_a^b \varphi df$ [2]).

**IV.** S. Bochner und R. S. Phillips [3] haben die verbandsalgebraische Natur des Lebesgueschen Zerlegungssatzes im allgemeinen Fall und ohne Heranziehung der linearen Funktionale aufgezeigt. Als algebraischer Kern erscheint dabei wieder jener *Bändersatz* aus Ziff. II, welcher von F. Riesz [9] für den dort auftretenden, speziellen vollständigen Vektorverband $\Lambda$ und von Bochner-Phillips [3] für beliebige vollständige Vektorverbände bewiesen wird. Nach Bochner-Phillips bilden alle auf einem Booleschen Mengenverband mit Einheit $e$ beschränkten, (endlich-)additiven, reellen Funktionen einen vollständigen Vektorverband $\Phi$. Bezüglich einer nicht-negativen, auf $V$ additiven Funktion $m$ erklärt man die Begriffe $m$-stetig und $m$-rein-unstetig für Funktionen $f \in \Phi$ wie in Ziff. I für total-additive Funktionen. Ein $f \in \Phi$ erweist sich genau dann als $m$-stetig bzw. $m$-rein-unstetig, wenn $f$ zu dem von $m$ in $\Phi$ erzeugten Band bzw. zu dem aus allen zu $m$ disjunkten Elementen von $\Phi$ bestehenden Band gehört. Der Lebesguesche Zerlegungssatz unter wesentlich abgeschwächten Voraussetzungen besagt dann: *Der vollständige Vektorverband $\Phi$ ist direkte ordnungstreue Summe dieser beiden Bänder.* Erfüllen $V$ und $m$ die einschneidenderen in Ziff. I genannten Voraussetzungen, so gehört jede auf $V$ total-additive Funktion $f$ zu $\Phi$ und ihre Zerlegungsteile bei der genannten direkten ordnungstreuen Zerlegung sind identisch mit den Zerlegungsteilen $r$ und $s$ von Ziff. I.

---

[2]) Nach F. Riesz läßt sich umgekehrt auch jedes positive lineare Funktional über $E$ als Integral nach einem geeigneten Maß auf $V$ darstellen; der Rieszsche Zerlegungssatz von Ziff. II erweist sich daher in dem hier behandelten Spezialfall als äquivalent mit dem Zerlegungssatz von Lebesgue. Bez. der Weiterführung der von F. Riesz angeregten funktional-theoretischen Interpretation des Lebesgueschen Zerlegungssatzes vgl. *J. Dieudonné*, Sur le théorème de Lebesgue-Nikodym. I.—V., insbes. I., Ann. of Math. **42** (1941), S. 547—555, und *M. H. Stone*, Notes on Integration. IV., Proc. Nat. Acad. Sci. USA **35** (1949), S. 50—58.

V. H. Hahn und A. Rosenthal beweisen in ihrem Buch „Set Functions" [5], S. 42, den folgenden Zerlegungssatz für total-additive Mengenfunktionen, der den Lebesgueschen Satz als Spezialfall mit umfaßt und ebenfalls als Beitrag zu seiner Algebraisierung gewertet werden kann: *Bezüglich eines $\sigma$-Ideals $S$ in einem Booleschen $\sigma$-Mengenverband $V$ läßt sich jede auf $V$ total-additive Funktion $f$ auf genau eine Weise darstellen in der Form $f(x) = f_R(x) + f_S(x)$, $x \in V$, wobei $f_R$ $S$-regulär und $f_S$ $S$-singulär ist.* Eine auf $V$ total-additive Funktion $g$ heißt dabei $S$-regulär bzw. $S$-singulär, wenn $g(s) = 0$ ist für jedes $s \in S$ bzw. wenn ein $s_0 \in S$ existiert mit $f(x) = 0$ für alle zu $s_0$ fremden Mengen $x \in V$. Wählt man für $S$ speziell das $\sigma$-Ideal aller Nullmengen eines Maßes $m \mid V$, so ist die Hahn-Rosenthalsche Zerlegung mit der von Lebesgue identisch.

**VI.** Die Ergebnisse von F. Riesz, S. Bochner und R. S. Phillips legen die Vermutung nahe, daß auch der Hahn-Rosenthalschen Zerlegung letzten Endes ein verbandsalgebraischer Sachverhalt zugrunde liegt und daß sie sich wesentlich von einschränkenden Voraussetzungen befreien läßt. Im § 4 der vorliegenden Arbeit wird der Hahn-Rosenthalsche Zerlegungssatz für (reelle) Bewertungen *von relativ beschränkter Variation eines distributiven Verbandes $V$* bewiesen. Derartige Bewertungen sind nach G. Birkhoff [2] (reelle) Funktionen $f \mid V$, welche der Funktionalgleichung

$$f(x \vee y) + f(x \wedge y) = f(x) + f(y)$$

genügen und für welche außerdem die für alle Einteilungen $x = x_0 \leqq x_1 \leqq \cdots \leqq x_n = y$ eines festen, aber beliebigen Intervalls $[x, y]$ in $V$ gebildeten Summen $\sum_{\nu=1}^{n} | f(x_\nu) - f(x_{\nu-1}) |$ beschränkt sind. Normiert man diese Bewertungen $f$ noch durch die Vorschrift $f(c) = 0$, wobei $c$ ein festes Element von $V$ ist, so bilden diese $f$ in ihrer Gesamtheit einen vollständigen Vektorverband $\Phi_c$ (§ 3). Die Hahn-Rosenthalsche Zerlegung erscheint dann als eine direkte ordnungstreue Zerlegung dieses Vektorverbandes in zwei komplementäre *Bänder*, in das Band der $S$-regulären und das Band der $S$-singulären Elemente. Als für die verallgemeinerte Theorie brauchbare Bezugsmengen $S$ bieten sich die $c$ als Element enthaltenden gesättigten Unterverbände von $V$ an. Die ursprüngliche Definition der $S$-regulären und $S$-singulären Funktionen von Hahn-Rosenthal zeigt sich als nicht verallgemeinerungsfähig, insofern man mit ihr nur unter zusätzlichen Voraussetzungen zu einer direkten Zerlegung von $\Phi_c$ gelangt. Als auch der allgemeineren Problemstellung adäquat erweist sich jedoch eine von Hahn-Rosenthal [5] angegebene Darstellung des $S$-singulären Teils $f_S$ von $f$. Unserem verbandsalgebraischen Standpunkt entsprechend wird $f_S$ in Nr. 4. 1 für eine Bewertung $f \geqq 0$ aus $\Phi_c$ als in $\Phi_c$ gebildetes Infimum über alle Bewertungen $h \geqq 0$ aus $\Phi_c$ definiert, welche mit $f$ auf $S$ übereinstimmen; für beliebiges $f \in \Phi_c$ setzt man $f_S = (f^+)_S - (f^-)_S$. Die Abbildung $f \to f_S$ von $\Phi_c$ in sich ist dann eine Projektion; *$S$-regulär* heißen die *Urbilder der Null*, *$S$-singulär* die *Fixelemente* bei dieser Projektion.

Der distributive Verband $V$ kann insbesondere ein *Vektorverband $E$* sein. Wählt man für $c$ den Nullvektor 0 von $E$, so enthält $\Phi_0$ die Menge $\Lambda$ aller relativ beschränkten, linearen Funktionale über $E$ als linearen Unterraum. Dieser Unterraum $\Lambda$ wird bei der Projektion $f \to f_S$ in sich übergeführt, wenn $S$ speziell ein gesättigter linearer Unterraum von $E$ ist. Die Abbildung $f \to f_S$ liefert nach Verengerung auf $\Lambda$ wieder eine Projektion des vollständigen Vektorverbandes $\Lambda$ (vgl. Ziff. II) in sich und damit eine direkte ordnungstreue Zerlegung vom Hahn-Rosenthalschen Typus für $\Lambda$.

**VIIa.** In den bisherigen Angaben war nur von *reellen* Bewertungen die Rede. Im § 4 wird aber die Hahn-Rosenthalsche Zerlegungstheorie gleich für verallgemeinerte

Bewertungen entwickelt, das sind Bewertungen mit Werten aus einem vollständigen Vektorverband. Wir nehmen also diesbezüglich den gleichen Standpunkt ein wie in einer früheren Note [1][3]).

**VIIb.** Im § 5 der Arbeit führen wir noch einen *zweiten* Zerlegungstypus für verallgemeinerte Bewertungen ein, den wir aus einem sofort klar werdenden Grunde nach Lebesgue benennen: Eine Bewertung $f \in \Phi_c$ heißt bezüglich einer anderen Bewertung $g \in \Phi_c$ *g-stetig* bzw. *g-rein-unstetig*, wenn $f$ in dem von $g$ in $\Phi_c$ erzeugten Band liegt bzw. wenn $f$ und $g$ in $\Phi_c$ disjunkt sind. Es ist dann $\Phi_c$ direkte ordnungstreue Summe des Bandes seiner $g$-stetigen und des Bandes seiner $g$-rein-unstetigen Elemente. Aus diesem Zerlegungssatz folgen durch *Spezialisierung* der Rieszsche Zerlegungssatz für Funktionale von Ziff. II und die von Bochner-Phillips (Ziff. IV) angegebene Verallgemeinerung des klassischen Lebesgueschen Zerlegungssatzes, insbesondere dieser selbst.

**VIII.** Die beiden Zerlegungstypen von Ziff. VIIa und VIIb lassen sich *vergleichen*, wenn man bei der Hahn-Rosenthalschen Zerlegung für $S$ das System aller Elemente $a \in V$ mit $g(c \wedge a) = g(c \vee a) = 0$ für eine feste Bewertung $g \geqq 0$ aus $\Phi_c$ wählt. Der *Unterschied* zwischen den beiden Zerlegungsarten läßt sich dann, grob gesprochen, so ausdrücken: Es werden als *regulär* bezeichnet bei der Lebesgue-Zerlegung die *bezüglich g total-stetigen* (oder, wie wir sagen, $g$-stetigen) Bewertungen, bei der Hahn-Rosenthalschen Zerlegung die *bezüglich g nulltreuen* Bewertungen aus $\Phi_c$. Beide Zerlegungen verallgemeinern den klassischen Lebesgueschen Satz von Ziff. I; die Zerlegungsteile werden das eine Mal im Sinne von Ziff. I. A, das andere Mal im Sinne von Ziff. I. B charakterisiert. In zwei Spezialfällen erweisen sich die beiden Zerlegungen als *identisch*: (1) in dem in Ziff. I besprochenen, klassischen Fall, (2) für den Fall *reeller*, *stetiger*, *linearer Funktionale* über einem *σ-vollständigen Vektorverband.*

Auf eine andere, weniger naheliegende Vergleichsmöglichkeit werden wir an anderer Stelle zu sprechen kommen. Dort werden — wenigstens für reelle, auf einem Booleschen Verband additive Funktionen sowie für reelle, lineare Funktionale über einem Vektorverband — die Begriffe *g-stetig* und *S-regulär* einem *allgemeinen Stetigkeitsbegriff* für Bewertungen untergeordnet. Dann erweisen sich die Lebesguesche und die Hahn-Rosenthalsche Zerlegung als Spezialfälle eines Satzes[4]) über die Zerlegbarkeit jeder Bewertung in einen „stetigen" und einen „rein-unstetigen" Teil.

## § 1. Hilfsmittel aus der Theorie der Vektorverbände.

Bereits in einer früheren Veröffentlichung (Nr. [1] des Literaturverzeichnisses) gaben wir eine Zusammenfassung einiger für die damaligen Zwecke notwendigen Ergebnisse aus der Theorie der Vektorverbände. Die dort eingeführten Bezeichnungen werden im folgenden beibehalten. Die folgende Übersicht streift kurz die Resultate von [1], soweit sie hier benötigt werden, und ergänzt sie in mehreren Punkten. Bezüglich näherer Einzelheiten sei auf G. Birkhoff [2], S. 238ff., und N. Bourbaki [4], S. 17ff. verwiesen.

**1. 1. Vektorverbände.** Einen Vektorraum $E = \{a, b, \ldots\}$ über einem linear geordneten Körper $K = \{\alpha, \beta, \ldots\}$ nennen wir einen Vektorverein (über $K$), wenn $E$ bezüglich einer Ordnungsrelation $\leqq$ zugleich auch ein Verein (d. h. *partly ordered set* im Sinne von G. Birkhoff [2]) ist und wenn die beiden folgenden Axiome erfüllt sind:

[3]) Unberücksichtigt bleibt die Möglichkeit, die Theorie für Bewertungen mit Werten aus einer vollständigen Verbandsgruppe zu entwickeln.

[4]) Vgl. *H. Bauer* [1], S. 106; es handelt sich i. w. um den dortigen Zerlegungssatz. Betr. Vorankündigung vgl. Fußnote [9]) dieser Arbeit.

($K_1$) *Aus* $a \leqq b$ *folgt* $a + c \leqq b + c$ *für* $a, b, c \in E$.

($K_2$) *Aus* $a \geqq 0$ *und* $\lambda \geqq 0$ *folgt* $\lambda a \geqq 0$ *für* $a \in E$, $\lambda \in \mathsf{K}$.

Der Vektorverein $E$ heißt Vektorverband (über K), wenn für *beliebige* Elemente $a, b \in E$ *Vereinigung* (*Supremum*), i. Z.

$$a \vee b = \sup(a, b) = (E) \sup(a, b),$$

und *Durchschnitt* (*Infimum*), i. Z.

$$a \wedge b = \inf(a, b) = (E) \inf(a, b),$$

in $E$ existieren; es ist dann $E$ ein *distributiver Verband*. Ferner bestehen folgende Gleichungen:

(1. 1) $\sup(a, b) + \inf(a, b) = a + b$; $a, b \in E$,

(1. 2) $\sup(a, b) + c = \sup(a + c, b + c)$; $a, b, c \in E$,

(1. 3) $\sup(\lambda a, \lambda b) = \lambda \sup(a, b)$; $a, b \in E$, $\lambda \in \mathsf{K}$ mit $\lambda \geqq 0$.

Zu Elementen $a, b$ und $x$ aus $E$ mit $a \geqq 0, b \geqq 0$ und $0 \leqq x \leqq a + b$ existieren stets Elemente $a', b' \in E$ mit $0 \leqq a' \leqq a$, $0 \leqq b' \leqq b$ und $x = a' + b'$.

Aus $a \geqq 0$, $b \geqq 0$ und $c \geqq 0$, $a, b, c \in E$ folgt

(1. 4) $\inf(a + b, c) \leqq \inf(a, c) + \inf(b, c)$.

In einem Vektorverband $E$ definiert man für jedes $a \in E$ die Elemente $a^+$, $a^-$ und $|a| =_E |a|$ aus $E$:

(1. 5) $a^+ = \sup(a, 0)$, $a^- = (-a)^+$,

(1. 6) $|a| = \sup(a, -a)$.

Dann gelten für jedes Element $a \in E$ die Gleichungen

(1. 7) $a = a^+ - a^-$,

(1. 8) $|a| = a^+ + a^-$.

Die Abbildungen $a \to a^+$, $a \to a^-$ und $a \to |a|$ von $E$ in sich besitzen folgende Eigenschaften:

(1. 9) $(a + b)^+ \leqq a^+ + b^+$, $(a + b)^- \leqq a^- + b^-$, $|a + b| \leqq |a| + |b|$; $a, b \in E$,

(1. 10) $(\lambda a)^+ = \lambda a^+$, $(\lambda a)^- = \lambda a^-$; $a \in E$, $\lambda \in \mathsf{K}$ mit $\lambda \geqq 0$,

(1. 11) $|\lambda a| = |\lambda| \, |a|$; $a \in E$, $\lambda \in \mathsf{K}$.

Die Verbandsstruktur von $E$ ist bekannt, wenn man die Abbildung $a \to a^+$ von $E$ in sich kennt:

(1. 12) $\sup(a, b) = b + (a - b)^+$, $\inf(a, b) = -\sup(-a, -b)$; $a, b \in E$.

Zwei Elemente $a$ und $b$ von $E$ heißen disjunkt, wenn

(1. 13a) $\inf(|a|, |b|) = 0$.

Wegen (1. 1) bzw. (1. 12) ist hiermit *gleichwertig* jede der beiden folgenden Gleichungen

(1. 13b) $\sup(|a|, |b|) = |a| + |b|$,

(1. 13c) $|a| = (|a| - |b|)^+$.

Vermöge (1. 7) ist *jedes* Element $a \in E$ von der Form $a = a' - a''$ mit $a' \geqq 0$ und $a'' \geqq 0$. Es gilt gleichzeitig $a' = a^+$ und $a'' = a^-$, wenn und nur wenn die Elemente $a'$ und $a''$ *disjunkt* sind.

Ein Vektorverband $E$ heißt **vollständig**, wenn für jede nicht leere, in $E$ nach oben beschränkte Teilmenge $H$ von $E$ die Vereinigung (Supremum), i. Z.

$$\bigvee_{h \in H} h = \sup H = (E) \sup H = \sup (h;\ h \in H),$$

existiert. Für jede nicht leere, nach unten beschränkte Teilmenge $H$ von $E$ existiert dann der Durchschnitt (Infimum)

$$\bigwedge_{h \in H} h = \inf H = (E) \inf H = \inf (h;\ h \in H).$$

*Ein Vektorverband $E$ ist dann und nur dann vollständig, wenn für jede nicht leere, nach oben beschränkte und in $E$ steigend gerichtete Menge $H$ von Elementen $h \geqq 0$ aus $E$ die Vereinigung existiert.* Dabei heißt eine Teilmenge $H$ von $E$ **steigend gerichtet**, wenn zu beliebigen Elementen $h', h'' \in H$ stets ein $h \in H$ existiert mit $h' \leqq h$ und $h'' \leqq h$.

Existieren in einem (nicht notwendig vollständigen) Vektorverband $E$ für Teilmengen $A$ und $B$ von $E$ die Elemente $\sup A$ und $\sup B$, so existieren auch die Elemente $\sup (A + B)$ und $\sup (\lambda A)$ für jedes $\lambda \geqq 0$ aus $\mathsf{K}$. Dabei ist

$$A + B = \{x;\ x = a + b,\ a \in A,\ b \in B\},\ \lambda A = \{x;\ x = \lambda a,\ a \in A\}.$$

Es gilt

$$\sup (A + B) = \sup A + \sup B, \tag{1.14}$$

$$\sup (\lambda A) = \lambda \sup A, \text{ falls } \lambda \geqq 0. \tag{1.15}$$

Ist $A$ in $E$ nach oben beschränkt, so $-A$ nach unten; existiert $\sup A$, so auch $\inf(-A)$ (und umgekehrt). Es gilt

$$\sup A = -\inf(-A). \tag{1.16}$$

Ein Vektorverband $E$ heißt **$\sigma$-vollständig**, wenn für jede nach oben beschränkte Folge $\{a_\nu\}_{\nu=1,2,\ldots}$ von Elementen aus $E$ die Vereinigung $\bigvee_{\nu=1}^{\infty} a_\nu$ existiert.

**1. 2. Lineare Unterräume.** Eine Teilmenge $L$ eines Vektorvereins $E$ heißt **linearer Unterraum** von $E$, wenn $L$ linearer Unterraum von $E$ *ohne* Rücksicht auf die Vereinsstruktur von $E$ ist. Bezüglich der durch $E$ in $L$ induzierten Strukturen ist $L$ selbst ein Vektorverein.

Ist $E$ ein Vektorverband, so heißt $L$ ein **eigentlicher linearer Unterraum** oder ein **Unter-Vektorverband** von $E$, wenn aus $a \in L$ stets $a^+ = (E) \sup (a, 0) \in L$ folgt. Dann ist $L$ selbst ein Vektorverband; für Elemente $a, b \in L$ gilt

$$(L) \sup (a, b) = (E) \sup (a, b) \quad \text{und} \quad (L) \inf (a, b) = (E) \inf (a, b).$$

Für einen eigentlichen linearen Unterraum $L$ eines Vektorverbandes $E$ sind folgende Aussagen *gleichwertig*:

(I) Für jede nicht leere, in $E$ nach oben beschränkte Teilmenge $H$ von $L$ existiert $(E) \sup H$ und ist ein Element von $L$.

(II) Für jede nicht leere, in $E$ nach oben beschränkte und in $E$ steigend gerichtete Menge $H$ von Elementen $h \geqq 0$ aus $L$ existiert $(E) \sup H$ und ist ein Element von $L$.

Ein linearer Unterraum $G$ eines Vektorverbandes $E$ heißt **gesättigt**, wenn aus $a \in G$, $b \in E$ und ${}_E|b| \leqq {}_E|a|$ folgt $b \in G$. Es ist $G$ dann auch ein *eigentlicher* linearer Unterraum von $E$.

Ein *gesättigter* linearer Unterraum $B$ eines *vollständigen* Vektorverbandes $E$ heißt ein **Band** in $E$, wenn für jede in $E$ nach oben beschränkte Teilmenge $H$ von $B$ auch $(E) \sup H$ ein Element von $B$ ist. Ein gesättigter linearer Unterraum $B$ von $E$ ist *dann und nur dann* ein Band in $E$, wenn für jede nicht leere, in $E$ nach oben beschränkte und steigend gerichtete Menge $H$ von Elementen $h \geqq 0$ aus $B$ gilt: $(E) \sup H$ ist Element von $B$. Jedes Band $B$ in $E$ ist selbst ein vollständiger Vektorverband; die in $E$ nach oben beschränkten Teilmengen von $B$ sind identisch mit den in $B$ nach oben beschränkten Teilmengen. Für jede solche Teilmenge $H$ von $B$ gilt: $(B) \sup H = (E) \sup H$; entsprechendes gilt für die nach unten beschränkten Teilmengen.

Ist $B$ ein Band in $E$ und $U$ ein linearer Unterraum von $E$, welcher mit jeder in $E$ nach oben beschränkten Teilmenge auch deren $(E)$ Supremum enthält, so ist der mengentheoretische Durchschnitt $B \cap U$ ein Band im vollständigen Vektorverband $U$.

Der mengentheoretische *Durchschnitt beliebig vieler Bänder* in einem vollständigen Vektorverband $E$ ist wieder ein Band in $E$. Zu jeder Teilmenge $M$ von $E$ existiert daher ein *kleinstes*, $M$ als Teilmenge enthaltendes Band; dieses heißt von $M$ erzeugt und wird mit $[M]$ bezeichnet. Besteht $M$ nur aus dem einen Element $m$, so schreiben wir $[m]$ an Stelle von $[\{m\}]$. Das Band $[M]$ gewinnt man aus $M$ folgendermaßen: Ist $M'$ die Menge der zu *allen* Elementen von $M$ disjunkten Elemente von $E$, so ist $[M] = (M')'$, d. h. gleich der Menge der zu allen Elementen von $M'$ disjunkten Elemente von $E$.

Als Folgerung hieraus ergibt sich: Es sei $E'$ ein eigentlicher linearer Unterraum eines vollständigen Vekorverbandes $E$; $E'$ enthalte mit jeder in $E$ nach oben beschränkten Teilmenge auch deren $(E)$ Supremum. Ist dann $M$ eine Teilmenge von $E'$ und $M_{E'}$ bzw. $M_E$ das von $M$ in $E'$ bzw. $E$ erzeugte Band, so gilt: $M_{E'} = M_E \cap E'$.

1.3. **Direkte ordnungstreue Zerlegungen.** Ein Vektorverein $E$ (über dem Körper $\mathsf{K}$) heißt direkte ordnungstreue Summe seiner linearen Unterräume $M$ und $N$, wenn jedes Element $x \in E$ *auf genau eine Weise* darstellbar ist in der Form

$$x = x_M + x_N$$

mit den Komponenten $x_M \in M$ und $x_N \in N$ und wenn $x \geqq 0$ stets $x_M \geqq 0$ und $x_N \geqq 0$ nach sich zieht. Man sagt: $M$ und $N$ liefern eine *direkte ordnungstreue Zerlegung* von $E$ oder $M$ und $N$ sind zueinander komplementäre, lineare Unterräume von $E$.

Der Begriff der direkten ordnungstreuen Zerlegung läßt sich noch etwas anders fassen: Eine *lineare* Abbildung $F$ von $E$ in sich, das ist eine Abbildung, bei welcher $F(\lambda x + \mu y) = \lambda F(x) + \mu F(y)$ für beliebiges Elemente $x, y \in E$ und $\lambda, \mu \in \mathsf{K}$ gilt, nennen wir eine Projektion(sabbildung) (von $E$ in sich), wenn $0 \leqq F(x) \leqq x$ ist für alle $x \geqq 0$ aus $E$ und $FF = F$, d. h. $F(F(x)) = F(x)$ gilt für jedes $x \in E$.

Ist $E$ direkte ordnungstreue Summe der Unterräume $M$ und $N$, so ist die Abbildung $x \to x_M$ eine *Projektion* von $E$ in sich; bei dieser Abbildung ist $M$ die Menge der *Fixelemente* und $N$ die *Urbildmenge des Nullvektors* 0 von $E$. — Ist umgekehrt $F$ eine Projektion von $E$ in sich, $M$ die Menge der Fixelemente von $E$ bezüglich $F$ und $N$ die Urbildmenge von 0, so sind $M$ und $N$ lineare Unterräume von $E$ und $E$ deren direkte ordnungstreue Summe. Für jedes $x \in E$ gilt $x_M = F(x)$ bei dieser Zerlegung.

Nunmehr sei $E$ ein *Vektorverband*, ferner sei $E$ direkte ordnungstreue Summe der linearen Unterräume $M$ und $N$. Dann sind $M$ und $N$ notwendig gesättigt. Die Elemente von $M$ und $N$ sind zueinander disjunkt. Für die Komponente $x_M$ (und analog für $x_N$) eines $x \in E$ bei dieser Zerlegung gilt:

$$(1.17) \qquad (x^+)_M = (x_M)^+, \quad (x^-)_M = (x_M)^-, \quad |x|_M = |x_M|.$$

Die gesättigten, linearen Unterräume $M$ und $N$ sind sogar Bänder in $E$, wenn $E$ ein vollständiger Vektorverband ist. In diesem Fall existiert aber auch umgekehrt zu *jedem* Band $M$ in $E$ ein *eindeutig bestimmtes komplementäres Band*; dieses ist die Menge $M'$ aller Elemente von $E'$, welche zu jedem Element von $M$ disjunkt sind. (Vgl. hierzu die in Nr. 1. 2 angegebene Konstruktion des von einer Teilmenge von $E$ erzeugten Bandes.) In einem *vollständigen* Vektorverband $E$ sind also alle Projektionsabbildungen von der Form $x \to x_M$, wobei $M$ ein *Band* in $E$ ist. Bei einem beliebigen Band $M$ bestimmt sich die *Komponente* $x_M$ für ein Element $x \geqq 0$ von $E$ nach der Formel

(1. 18) $$x_M = (E) \sup (h;\ 0 \leqq h \leqq x,\ h \in M).$$

Für ein Band $M = [g]$ in $E$, welches von einem Element $g \in E$ erzeugt wird, besteht das komplementäre Band $N = M'$ aus den zu $g$ disjunkten Elementen von $M$. Für die Komponenten $x_M$ und $x_N$ eines Elementes $x \geqq 0$ aus $E$ gelten die Formeln:

(1. 19) $x_M = (E) \sup (u_n;\ n = 1, 2, \ldots)$, wobei $u_n = (E) \inf (x, n \mid g \mid)$, $n = 1, 2, \ldots$,

(1. 20) $x_B = (E) \inf (v_n;\ n = 1, 2, \ldots)$, wobei $v_n = (x - n \mid g \mid)^+$, $n = 1, 2, \ldots$.

Schließlich erwähnen wir noch: Der Vektorverband $E$ sei direkte ordnungstreue Summe seiner linearen Unterräume $M$ und $N$. Ist dann $G$ ein *gesättigter* linearer Unterraum von $E$, so ist der Vektorverband $G$ direkte ordnungstreue Summe seiner linearen Unterräume $M^* = M \cap G$ und $N^* = N \cap G$.

## § 2. Quotienten von gesättigten Unterverbänden in einem Verband.

**2. 1. Gesättigte Unterverbände.** Wir geben zunächst die erforderliche

**Definition.** Eine nicht leere Teilmenge $G$ eines Vereins $V$ heißt gesättigt, wenn aus $a \leqq x \leqq b$, $a, b \in G$ und $x \in V$ stets $x \in G$ folgt. Ist $V$ sogar ein Verband und $G$ zugleich ein Unterverband von $V$, folgt also aus $a, b \in G$ stets $a \vee b \in G$ und $a \wedge b \in G$, so heißt $G$ gesättigter Unterverband von $V$.

**Beispiele** für gesättigte Unter*verbände* liefern:

1. Die abgeschlossenen Intervalle in einem Verband. $J \leqq V$ heißt dabei abgeschlossenes Intervall, wenn (eindeutig bestimmte) Elemente $a, b \in J$ existieren, so daß $J$ aus genau den Elementen $x \in V$ mit $a \leqq x \leqq b$ besteht, in Zeichen $J = [a, b]$.

2. Die Ideale $I$ in einem Verband $V$. Eine Teilmenge $I \leqq V$ heißt Ideal in $V$, wenn aus $x, y \in I$ stets $x \vee y \in I$ und aus $y \leqq x$, $x \in I$, $y \in V$ auch $y \in I$ folgt.

3. Die Filter $F$ in einem Verband $V$. Der Filter wird dual zum Ideal definiert; es heißt also $F \leqq V$ Filter, wenn aus $x, y \in F$ stets $x \wedge y \in F$ und aus $x \leqq y$, $x \in F$, $y \in V$ auch $y \in F$ folgt.

4. Die im Sinne von Nr. 1. 2 gesättigten linearen Unterräume eines Vektorverbandes $E$. Genauer: *Ein eigentlicher linearer Unterraum $G$ von $E$ ist dann und nur dann gesättigt (im Sinne dieser Nr. 2. 1), wenn aus $|y| \leqq |x|$, $x \in G$, $y \in E$ auch $y \in G$ folgt.*

Beweis. *Nur dann*: Aus $|y| \leqq |x|$ folgt $-|x| \leqq y \leqq |x|$. Mit $x$ sind aber auch $\pm |x|$ Elemente von $G$. — *Dann*: Aus $a \leqq x \leqq b$ folgt $|x| \leqq |a| + |b|$. Es ist $|a| + |b| = ||a| + |b||$ ein Element von $G$, wenn $a \in G$ und $b \in G$.

Der mengentheoretische Durchschnitt *beliebig vieler* gesättigter Unterverbände eines Verbandes $V$ ist, falls nicht leer, wieder ein gesättigter Unterverband von $V$. Da $V$ selbst ein gesättigter Unterverband von sich ist, existiert somit zu jeder nicht leeren Teilmenge $M$ von $V$ ein *kleinster*, gesättigter Unterverband $M'$ von $V$, welcher $M$ als Teilmenge enthält. Man nennt $M'$ den von $M$ in $V$ erzeugten gesättigten Unterverband.

Beispiel: Der von einem *Unterverband* $U$ eines Verbandes $V$ erzeugte *gesättigte* Unterverband $U'$ besteht aus der Menge aller Elemente $x \in V$, zu welchen $u', u'' \in U$ mit $u' \leqq x \leqq u''$ existieren.

**Satz.** *Es seien $P$ und $Q$ zwei Unterverbände eines distributiven Verbandes $V$. Der von $P \cup Q$ in $V$ erzeugte gesättigte Unterverband ist gleich der Menge $R$ aller Elemente $x \in V$, zu welchen Elemente $p', p'' \in P$ und $q', q'' \in Q$ existieren mit $p' \wedge q' \leqq x \leqq p'' \vee q''$.*

Beweis. Zunächst sind $P$ und $Q$ Teilmengen von $R$; für beliebige $p \in P$ und $q \in Q$ gilt nämlich $p \wedge q \leqq p \leqq p \vee q$ und $p \wedge q \leqq q \leqq p \vee q$. — Nunmehr seien $x_1$ und $x_2$ Elemente aus $R$; wir zeigen, daß ihre Vereinigung $x = x_1 \vee x_2$ ebenfalls zu $R$ gehört. (Durch duale Überlegungen beweist man die Abgeschlossenheit von $R$ gegenüber Durchschnittsbildung in $V$.) Gemäß Definition existieren Elemente $p'_\nu, p''_\nu \in P$ und $q'_\nu, q''_\nu \in Q$ mit $p'_\nu \wedge q'_\nu \leqq x_\nu \leqq p''_\nu \vee q''_\nu$, $\nu = 1, 2$; es gilt also

$$(p'_1 \wedge q'_1) \vee (p'_2 \wedge q'_2) \leqq x \leqq (p''_1 \vee q''_1) \vee (p''_2 \vee q''_2).$$

Durch Anwendung des distributiven Gesetzes folgt

$$(p'_1 \wedge q'_1) \vee (p'_2 \wedge q'_2) \geqq (p'_1 \vee p'_2) \wedge (q'_1 \wedge q'_2);$$

wir erhalten also

$$(p'_1 \vee p'_2) \wedge (q'_1 \wedge q'_2) \leqq x \leqq (p''_1 \vee p''_2) \vee (q''_1 \vee q''_2).$$

Da $P$ und $Q$ Unterverbände von $V$ sind, folgt $x \in R$. — Unmittelbar ersichtlich ist, daß $R$ gesättigt ist. Weiter folgt aus der Konstruktion, daß $R$ der kleinste, $P$ und $Q$ als Teilmengen enthaltende, gesättigte Unterverband von $V$ ist.

**2. 2. Quotientenbildung.** Wir geben zunächst die benötigte

**Definition.** *Es sei $G$ ein gesättigter Unterverband eines Verbandes $V$, $S$ eine nicht leere, sonst aber beliebige Teilmenge von $V$ und $c$ ein beliebiges Element von $G$. Wir bezeichnen als $c$-Quotienten von $G$ nach $S$, in Zeichen $(G : S)_c$, die Menge aller Elemente $y \in V$, für welche gleichzeitig $c \vee (y \wedge s)$ und $c \wedge (y \vee s)$ Elemente von $G$ sind bei beliebigem $s \in S$.*

Unter den in dieser Definition genannten Voraussetzungen beweisen wir über $c$-Quotienten folgende Eigenschaften[5]):

(2. 1) $\quad G < (G : S)_c;$

(2. 2) $\quad S \cap (G : S)_c = S \cap G;$

(2. 3) $\quad S < (G : (G : S)_c)_c;$

(2. 4) $\quad (G : V)_c = G;$

(2. 5) $\quad (G : G)_c = V;$

(2. 6) *Monotonie-Eigenschaft*: Aus $G' < G''$, $S'' \leq S'$ und $c \in G'$ folgt

$$(G' : S')_c \leq (G'' : S'')_c.$$

Beweis. (2. 1) Es sei $g$ ein Element von $G$. Aus $c \leqq c \vee (g \wedge s) \leqq c \vee g$ und $c \wedge g \leqq c \wedge (g \vee s) \leqq c$ folgt wegen der Sättigung von $G$: Für jedes $s \in S$ sind $c \vee (g \wedge s)$ und $c \wedge (g \vee s)$ Elemente von $G$. — (2. 2) Wegen (2. 1) braucht nur noch gezeigt zu werden, daß $S \cap (G : S)_c$ Teilmenge von $G$ ist: Aus $y \in S \cap (G : S)_c$ folgt $c \vee (y \wedge y) = c \vee y \in G$ und $c \wedge (y \vee y) = c \wedge y \in G$. Wegen $c \wedge y \leqq y \leqq c \vee y$ ist dann auch $y$ ein Element von $G$. — (2. 3) ist eine unmittelbare Folgerung aus der Definition des $c$-Quotienten. — (2. 4) folgt aus (2. 2), wenn man dort $S = V$ setzt. — (2. 5) folgt bei Beachtung von (2. 4) aus (2. 3), wenn man dort $S = V$ setzt. — (2. 6) Ist $y$ ein Element von $(G' : S')_c$, so bedeutet dies:

[5]) Wir erwähnen nur die im folgenden benötigten Eigenschaften. Es lassen sich im wesentlichen alle über Filter- und Idealquotienten (vgl. Satz 2. 22) bekannten Rechenregeln auf unseren Fall übertragen. Vgl. *J. Schmidt*, Beiträge zur Filtertheorie. I., Math. Nachr. 7 (1952), S. 363.

Für alle $s' \in S'$ gilt $c \vee (y \wedge s') \in G'$ und $c \wedge (y \vee s') \in G'$. Wegen $G' \leq G''$ und $S' \geq S''$ gilt dann aber auch $c \vee (y \wedge s'') \in G''$ und $c \wedge (y \vee s'') \in G''$ für alle $s'' \in S''$, also ist $y$ Element von $(G'' : S'')_c$.

Die für uns wichtigste Eigenschaft des $c$-Quotienten $(G : S)_c$ enthält der folgende Satz.

**Satz 2. 21.** *Es sei G ein gesättigter Unterverband eines Verbandes V und S eine nicht leere Teilmenge von V. Dann ist für jedes Element $c \in G$ der c-Quotient $(G : S)_c$ eine gesättigte Teilmenge von V. Ist der Verband V distributiv, so ist $(G : S)_c$ sogar ein gesättigter Unterverband von V.*

Beweis. Aus $y' \leqq y \leqq y''$ folgt

$$c \vee (y' \wedge s) \leqq c \vee (y \wedge s) \leqq c \vee (y'' \wedge s)$$

und

$$c \wedge (y' \vee s) \leqq c \wedge (y \vee s) \leqq c \wedge (y'' \vee s)$$

für ein beliebiges Element $s \in V$. Wählt man speziell $y', y'' \in (G : S)_c$ und $s \in S$, so ergibt sich, daß mit $G$ auch $(G : S)_c$ eine gesättigte Teilmenge von $V$ ist. — Nun sei $V$ distributiv. Es seien $y_\nu$, $\nu = 1, 2$, zwei Elemente aus $(G : S)_c$ und $y$ deren Vereinigung. Wegen der Distributivität von $V$ gelten dann für ein beliebiges Element $s$ die Gleichungen

$$c \vee (y \wedge s) = \bigvee_{\nu=1}^{2} [c \vee (y_\nu \wedge s)] \text{ und } c \wedge (y \vee s) = \bigvee_{\nu=1}^{2} [c \wedge (y_\nu \vee s)].$$

Wählt man für $s$ ein Element aus $S$, so folgt $y \in (G : S)_c$, weil $G$ Unterverband von $V$ ist. Analog zeigt man, daß auch der Durchschnitt der $y_\nu$ zu $(G : S)_c$ gehört.

**Korollar.** *Ist der Verband V distributiv und sind G und S gesättigte, c als Element enthaltende Unterverbände von V, so ist $(G : S)_c$ der größte, gesättigte, c als Element enthaltende Unterverband $D_c$ von V, für welchen der mengentheoretische Durchschnitt $S \frown D_c$ in G enthalten ist.*

Beweis. Nach (2. 2) und Satz 2. 21 ist $(G : S)_c$ ein Unterverband $D_c$ mit den angegebenen Eigenschaften. Es verbleibt somit noch zu zeigen: Ist $D_c$ ein gesättigter Unterverband von $V$ mit $c \in D_c$ und $S \frown D_c \leq G$, so gilt $D_c \leq (G : S)_c$. Hierzu sei $y$ ein beliebiges Element aus $D_c$. Für beliebiges $s \in S$ sind dann $c \vee (y \wedge s)$ und $c \wedge (y \vee s)$ Elemente von $S \frown D_c$; dies folgt aus den Relationen $c \leqq c \vee (y \wedge s) \leqq c \vee s$, $c \leqq c \vee (y \wedge s) \leqq c \vee y$, $c \wedge s \leqq c \wedge (y \vee s) \leqq c$ und $c \wedge y \leqq c \wedge (y \vee s) \leqq c$, weil $S$ und $D_c$ gesättigte Unterverbände von $V$ mit $c \in S \frown D_c$ sind. Wegen $S \frown D_c \leq G$ ist dann auch $y$ ein Element von $(G : S)_c$.

Von besonderem Interesse sind die Fälle, wo der gesättigte Unterverband $G$ ein *Ideal* bzw. *Filter* oder ein *gesättigter linearer Unterraum eines Vektorverbandes* ist.

**Satz 2. 22.** *In einem Verband V sei G ein Ideal (bzw. ein Filter) und S eine nicht leere Teilmenge von V. Bei beliebiger Wahl des Elementes $c \in G$ ist dann der c-Quotient $(G : S)_c$ gleich der Menge aller Elemente $y \in V$, für welche bei beliebigem $s \in S$ gilt: Es ist $y \wedge s$ (bzw. $y \vee s$) ein Element von G; insbesondere ist also $(G : S)_c$ unabhängig von der Wahl des Elementes c in G. Wenn der Verband V distributiv ist, so ist $(G : S)_c$ selbst ein Ideal (bzw. Filter) in V.*

Beweis. Aus Dualitätsgründen können wir uns auf den Fall beschränken, wo $G$ ein Ideal ist. Aus $y \in (G : S)_c$ folgt $y \wedge s \in G$ für alle Elemente $s \in S$ wegen $y \wedge s \leqq c \vee (y \wedge s)$. Gilt umgekehrt $y \wedge s \in G$ für alle $s \in S$, so ergibt sich $c \vee (y \wedge s) \in G$ und $c \wedge (y \vee s) \in G$, da $c$ ein Element von $G$ und $c \wedge (y \vee s) \leqq c$ ist; somit ist $y$ ein Element von $(G : S)_c$. — Nunmehr setzen wir den Verband $V$ als distributiv voraus. Um nachzuweisen, daß $(G : S)_c$ selbst ein Ideal in $V$ ist, genügt es wegen Satz 2. 21 zu zeigen, daß aus $y' \leqq y$,

$y' \in V$ und $y \in (G:S)_c$ auch $y' \in (G:S)_c$ folgt. Dies ergibt sich aber aus $y' \wedge s \leqq y \wedge s$ und der bereits gegebenen Kennzeichnung von $(G:S)_c$.

Wenn $G$ ein Ideal (bzw. ein Filter) ist, schreiben wir nur noch kurz $G:S$ an Stelle von $(G:S)_c$, $c \in G$, und nennen $G:S$ den Idealquotienten (bzw. Filterquotienten) von $G$ nach $S$.

**Satz 2. 23.** *In einem Vektorverband $E$ über einem linear geordneten Körper $\mathsf{K}$ seien $G$ und $S$ zwei gesättigte lineare Unterräume. Der bezüglich $c = 0$ gebildete Quotient $(G:S)_0$ besteht dann genau aus den Elementen $y \in E$, für welche $(E)$ $\inf(|y|, |s|)$ bei beliebigem $s \in S$ ein Element von $G$ ist. $(G:S)_0$ ist selbst ein gesättigter linearer Unterraum von $E$.*

Beweis. Zunächst beweisen wir die in der Behauptung ausgesprochene Kennzeichnung der Elemente von $Q = (G:S)_0$: Nach der Definition in Nr. 2. 1 ist $y$ ein Element von $Q$, wenn gleichzeitig

$$\sup(0, \inf(y,s)) = \inf(y^+, s^+) \text{ und } \inf(0, \sup(y,s)) = \sup(-y^-, -s^-) = -\inf(y^-, s^-)$$

Elemente von $G$ für jedes $s \in S$ sind. Also gilt $y \in Q$ genau dann, wenn $\inf(y^+, s^+) \in G$ und $\inf(y^-, s^-) \in G$ ist für jedes $s \in S$. Da nach (1. 4)

$$\inf(|y|, |s|) \leqq \inf(y^+, |s|) + \inf(y^-, |s|) = \inf(y^+, |s|^+) + \inf(y^-, (-|s|)^-)$$

und mit $s \in S$ auch $\pm |s| \in S$ ist, gehört $\inf(|y|, |s|)$ zu $G$ für jedes $s \in S$. Ist umgekehrt $\inf(|y|, |s|)$ für jedes $s \in S$ ein Element von $G$, so folgt $y \in Q$ aus den für alle $s \in S$ gültigen Ungleichungen

$$0 \leqq \inf(y^+, s^+) \leqq \inf(|y|, s^+) = \inf(|y|, |s^+|)$$

und

$$0 \leqq \inf(y^-, s^-) \leqq \inf(|y|, s^-) = \inf(|y|, |s^-|),$$

sowie aus der Sättigung von $G$.

Nunmehr zeigen wir, daß $Q$ ein linearer Unterraum von $E$ ist: Mit $y$ und $z$ ist auch $y + z$ ein Element von $Q$, da nach (1. 4) und (1. 9)

$$\inf(|y + z|, |s|) \leqq \inf(|y|, |s|) + \inf(|z|, |s|)$$

gilt. Nach (1. 11) und der zu (1. 15) dualen Formel ist $\inf(|\lambda y|, |s|) = |\lambda| \inf\left(|y|, \frac{1}{\lambda} s\right)$ für jedes $\lambda \neq 0$ aus $\mathsf{K}$; mit $y$ gehört also auch $\lambda y$ zu $Q$ für beliebiges $\lambda \in \mathsf{K}$ (für $\lambda = 0$ trivialerweise). Schließlich ist $Q$ ein gesättigter linearer Unterraum, weil $|y'| \leqq |y|$ die Ungleichung $\inf(|y'|, |s|) \leqq \inf(|y|, |s|)$ für jedes $s \in S$ nach sich zieht.

## § 3. Bewertungen eines Verbandes.

**3. 1. Allgemeines.** *Für das Folgende seien fest vorgegeben ein Verband $V$ und ein vollständiger Vektorverband $W$ über einem linear geordneten Körper* $\mathsf{K}$. Die Menge aller eindeutigen Abbildungen von $V$ in $W$ bezeichnen wir mit $W^V$.

Eine Abbildung $f \in W^V$ heißt eine Bewertung von $V$ in $W$, wenn $f$ der Funktionalgleichung

$$f(x \vee y) + f(x \wedge y) = f(x) + f(y) \tag{3. 1}$$

genügt. Ist $W$ der vollständige Vektorverband $R$ der reellen Zahlen, so sprechen wir auch von einer *reellen Bewertung* von $V$.

Nunmehr stellen wir die wichtigsten und für alles weitere grundlegenden Ergebnisse über die Eigenschaften der Menge aller Bewertungen von $V$ in $W$ zusammen. Nähere Einzelheiten und Beweise für hier unbewiesene Behauptungen finden sich in [1].

*Die Menge der Bewertungen von $V$ in $W$, welche der Normierungsbedingung*

(3. 2) $$f(c) = 0$$

*für ein festes Element $c \in V$ genügen, bezeichnen wir mit $\Psi_c = \Psi_c(W, V)$.* Die Menge $\Psi_c$ betrachten wir stets als mit den beiden folgenden Strukturen versehen:

1. **Struktur eines Vektorraumes** über dem Skalarenkörper K von $W$.

*Definition von $f + g$, $f, g \in \Psi_c$: $(f + g)(x) = f(x) + g(x)$ für alle $x \in V$.*

*Definition von $\lambda f$, $f \in \Psi_c$, $\lambda \in$ K: $(\lambda f)(x) = \lambda f(x)$ für alle $x \in V$.* Der Nullvektor von $\Psi_c$ ist die Bewertung $n$ mit $n(x) = 0$ für alle $x \in V$; wir bezeichnen sie mit 0.

2. **Struktur eines Vereins.**

*Definition der Relation $f \leqq g$ oder $g \geqq f$ für $f, g \in \Psi_c$: Aus $x \leqq y$, $x, y \in V$, folgt stets $g(x) - f(x) \leqq g(y) - f(y)$.*

Bemerkung. *$g \geqq 0$ bedeutet also für ein $g \in \Psi_c$: Aus $x \leqq y$ folgt $g(x) \leqq g(y)$ (Monotonie); dann ist insbesondere $g(x) \geqq 0$ für alle $x \geqq c$ und $g(x) \leqq 0$ für alle $x \leqq c$, $x \in V$.*

Bezüglich der Strukturen 1 und 2 ist $\Psi_c$ ein Vektorverein über K.

Eine Bewertung $f$ von $V$ in $W$ heißt von relativ beschränkter Variation, wenn zu jedem abgeschlossenen Intervall $[a, b]$ in $V$ ein Element $s \in W$ existiert derart, daß

$$\sum_{\nu=1}^{n} {}_W | f(x_\nu) - f(x_{\nu-1}) | \leqq s$$

für *jede Einteilung*

$$a = x_0 \leqq x_1 \leqq \cdots \leqq x_n = b$$

des Intervalls $[a, b]$ gilt. *Die Menge der Elemente $f \in \Psi_c$, welche von relativ beschränkter Variation sind, bezeichnen wir mit $\Phi_c = \Phi_c(W, V)$.*

Bemerkung. *$\Phi_c$ enthält als Elemente insbesondere alle $f \in \Psi_c$ mit $f \geqq 0$, d. h. es sind alle $f \in \Psi_c$ mit $f \geqq 0$ von relativ beschränkter Variation.*

**Satz 3. 11.** *Es ist $\Phi_c(W, V)$ ein im Sinne von* Nr. 2. 1 *gesättigter, linearer Unterraum von $\Psi_c(W, V)$.*

Beweis. (1) $\Phi_c$ ist linearer Unterraum von $\Psi_c$: Setzen wir $h = \alpha f + \beta g$, $f, g \in \Phi_c$, $\alpha, \beta \in$ K, so gilt für jede Einteilung $x = x_0 \leqq x_1 \leqq \cdots \leqq x_n = y$ eines jeden Intervalls $[x, y]$ in $V$

$$\sum_{\nu=1}^{n} | h(x_\nu) - h(x_{\nu-1}) | \leqq | \alpha | \sum_{\nu=1}^{n} | f(x_\nu) - f(x_{\nu-1}) | + | \beta | \sum_{\nu=1}^{n} | g(x_\nu) - g(x_{\nu-1}) | .$$

Mit $f$ und $g$ ist also auch $h$ von relativ beschränkter Variation.

(2) $\Phi_c$ ist gesättigte Teilmenge von $\Psi_c$: Für die Elemente $f, g, h \in \Psi_c$ bestehe die Relation $f \leqq g \leqq h$. Für die in (1) genannte beliebige Einteilung des beliebigen Intervalls $[x, y]$ gilt dann

$$f(x_\nu) - f(x_{\nu-1}) \leqq g(x_\nu) - g(x_{\nu-1}) \leqq h(x_\nu) - h(x_{\nu-1}), \quad \nu = 1, 2, \ldots, n.$$

Hieraus folgt

$$\sum_{\nu=1}^{n} (W) \sup (g(x_\nu) - g(x_{\nu-1}); 0) \leqq \sum_{\nu=1}^{n} | h(x_\nu) - h(x_{\nu-1}) |$$

und

$$\sum_{\nu=1}^{n} (W) \sup (- g(x_\nu) + g(x_{\nu-1}); 0) \leqq \sum_{\nu=1}^{n} | f(x_\nu) - f(x_{\nu-1}) |,$$

mithin ist

$$\sum_{\nu=1}^{n} | g(x_\nu) - g(x_{\nu-1}) | \leqq \sum_{\nu=1}^{n} | f(x_\nu) - f(x_{\nu-1}) | + \sum_{\nu=1}^{n} | h(x_\nu) - h(x_{\nu-1}) | .$$

Zusammen mit $f$ und $h$ ist damit auch $g$ von relativ beschränkter Variation.

$\Phi_c$ ist unter allen linearen Unterräumen von $\Psi_c$ wie folgt ausgezeichnet: *$\Phi_c$ ist der größte aller linearen Unterräume von $\Psi_c$, welcher bezüglich der induzierten Strukturen sogar ein Vektorverband ist.* Darüber hinaus gilt:

**Satz** 3. 12. *Der Vektorverein $\Phi_c(W, V)$ ist ein vollständiger Vektorverband.*

Ein Beweis dieses für die weiteren Untersuchungen entscheidenden Satzes findet sich in [1], S. 101.

Wie in jedem abstrakten Vektorverband setzen wir auch für jede Bewertung $f \in \Phi_c$

(3. 3) $f^+ = (\Phi_c) \sup (f, 0),\quad f^- = (-f)^+$ und $|f| = (\Phi_c) \sup (f, -f)$.

Dann gilt

(3. 4) $$f = f^+ - f^- \text{ und } |f| = f^+ + f^-.$$

Die *Konstruktion der Bewertung $f^+$* mittels der Funktionswerte $f(x)$ verläuft folgendermaßen:

Es sei $[x, y]$ ein beliebiges Intervall in $V$ und $\mathfrak{t}$ eine beliebige Einteilung $x = x_0 \leqq x_1 \leqq \cdots \leqq x_n = y$ von $[x, y]$. Dann setzen wir

(3. 5) $$P(x, y; \mathfrak{t}) = \sum_{\nu=1}^{n} (W) \sup (f(x_\nu) - f(x_{\nu-1}); 0),$$

(3. 6) $$p(x, y) \quad = (W) \sup (P(x, y; \mathfrak{t});\ \mathfrak{t} \text{ beliebige Einteilung von } [x, y]);$$

es gilt

(3. 7) $$f^+(x) = p(c, c \vee x) - p(c \wedge x, c),\quad x \in V \text{ beliebig.}$$

Wenn $H$ eine nicht-leere, in $\Phi_c$ nach oben beschränkte und steigend gerichtete Menge von Elementen $h \in \Phi_c$ mit $h \geqq 0$ ist, so existiert nach Satz 3. 12 das Supremum $s = (\Phi_c) \sup H$. Dieses gewinnt man mittels der Funktionswerte der $h \in H$ wie folgt:

(3. 8) $$s(x) = (W) \lim_{h \in H} \operatorname{alg} h(x),\quad x \in V\ ^{6}).$$

Neben dem Begriff *Bewertung von relativ beschränkter Variation* benötigen wir noch den folgenden:

*Eine Bewertung $f$ von $V$ in $W$ heißt relativ beschränkt, wenn zu jedem Intervall $I = [a, b]$ in $V$ ein Element $s = s(I) \in W$ existiert mit ${}_W|f(x)| \leqq s$ für alle $x \in I$.*

**Satz** 3. 13. *Die relativ beschränkten Elemente von $\Psi_c = \Psi_c(W, V)$ bilden einen linearen Unterraum von $\Psi_c$, welcher den Vektorraum $\Phi_c(W, V)$ selbst wieder als linearen Unterraum enthält.*

Beweis. Wegen $|\alpha f(x) + \beta g(x)| \leqq |\alpha|\,|f(x)| + |\beta|\,|g(x)|$ ist mit $f, g \in \Psi_c$ auch $\alpha f + \beta g$, $\alpha, \beta \in \mathsf{K}$, relativ beschränkt. Für jedes Intervall $I = [a, b]$ und jedes $x \in I$ gilt

$$|f(x)| \leqq |f(a)| + |f(x) - f(a)| + |f(b) - f(x)|;$$

es ist also $f$ relativ beschränkt, wenn $f$ von relativ beschränkter Variation ist.

Schließlich beweisen wir noch einen Satz, der mehr den Charakter eines Hilfssatzes hat, sich aber im folgenden als nützlich erweisen wird.

**Satz** 3. 14. *Es sei $V$ ein distributiver Verband, $c$ ein Element von $V$ und $V'$ bzw. $V''$ der Unterverband aller Elemente $x \in V$ mit $x \geqq c$ bzw. $x \leqq c$. Mit $\Psi_c'$ bzw. $\Psi_c''$ sei die Menge aller $f \in \Psi_c(W, V)$ mit $f(x) = 0$ für alle $x \in V''$ bzw. $x \in V'$ bezeichnet. Es sind dann $\Psi_c'$ und $\Psi_c''$ lineare Unterräume von $\Psi_c(W, V)$ mit folgenden Eigenschaften:*

---

6) Es handelt sich um die sog. *Ordnungskonvergenz* (*o-Konvergenz*) im Sinne von *G. Birkhoff* [2], S. 59ff. Die Bezeichnung *algebraische Konvergenz* (*algebraischer Limes*) wird nach *Haupt-Aumann-Pauc* [6], S. 32 verwendet.

(1) *Für ein Element* $f \in \Psi_c(W, V')$ *bzw.* $f \in \Psi_c(W, V'')$ *und jedes* $x \in V$ *werde* $(\chi' f)(x) = f(c \vee x)$ *bzw.* $(\chi'' f)(x) = f(c \wedge x)$ *gesetzt. Dann ist* $\chi'$ *bzw.* $\chi''$ *ein Isomorphismus des Vektorvereins* $\Psi_c(W, V')$ *bzw.* $\Psi_c(W, V'')$ *auf den Vektorverein* $\Psi_c'$ *bzw.* $\Psi_c''$.

(2) *Der Vektorverein* $\Psi_c(W, V)$ *ist direkte ordnungstreue Summe seiner linearen Unterräume* $\Psi_c'$ *und* $\Psi_c''$.

(3) *Für jedes Paar von Bewertungen* $f \in \Psi_c(W, V')$ *und* $g \in \Psi_c(W, V'')$ *ist* $\chi' f + \chi'' g$ *die einzige Bewertung aus* $\Psi_c(W, V)$, *welche auf* $V'$ *mit* $f$ *und auf* $V''$ *mit* $g$ *wertverlaufsgleich ist.*

Beweis. Zu (1). Wir beschränken uns auf den $\chi'$ betreffenden Teil der Behauptung. Unter Benutzung des in $V$ gültigen distributiven Gesetzes bestätigt man, daß $\chi' f$ ein Element von $\Psi_c'$ ist für jedes $f \in \Psi_c(W, V')$. Aus $f' \in \Psi_c'$ folgt

$$f'(x) = f'(c \vee x) + f'(c \wedge x) = f'(c \vee x) \text{ für jedes } x \in V;$$

deshalb gilt $f' = \chi' f$, wenn $f \in \Psi_c(W, V')$ als Verengerung von $f'$ auf $V'$ erklärt, d. h. $f(x) = f'(x)$ für jedes $x \in V'$ gesetzt wird. Durch $\chi'$ wird also $\Psi_c(W, V')$ auf $\Psi_c'$ abgebildet. Die Abbildung $\chi'$ ist eineindeutig, weil $(\chi' f)(x) = f(x)$ für jedes $x \in V'$ gilt. Weiterhin ist $\chi'$ eine lineare Abbildung von $\Psi_c(W, V')$ auf $\Psi_c'$; sie ist sogar ein Isomorphismus des einen Vektorvereins auf den anderen, weil $\chi' f \geqq 0$ mit $f \geqq 0$ gleichwertig ist.

Zu (2). Es sei $f \in \Psi_c(W, V)$ beliebig. Wir definieren die Bewertungen $f' \in \Psi_c(W, V')$ und $f'' \in \Psi_c(W, V'')$ als Verengerungen von $f$ auf $V'$ bzw. $V''$. Dann gilt $f = \chi' f' + \chi'' f''$ mit $\chi' f' \in \Psi_c'$ und $\chi'' f'' \in \Psi_c''$. Diese Zerlegung ist direkt: Aus $f' + g' = f'' + g''$ mit $f', g' \in \Psi_c'$ und $f'', g'' \in \Psi_c''$ folgt nämlich $f'(x) = g'(x)$ für jedes $x \in V'$ und hieraus weiter $f'(x) = f'(c \vee x) = g'(c \vee x) = g'(x)$ für beliebiges $x \in V$. Es liegt eine direkte ordnungstreue Zerlegung vor, weil aus $f \geqq 0$ auch $f' \geqq 0$ und $f'' \geqq 0$ folgt, also $\chi' f' \geqq 0$ und $\chi'' f'' \geqq 0$ nach (1) ist.

Zu (3). Wegen $\chi' f \in \Psi_c'$ und $\chi'' g \in \Psi_c''$ ist $\chi' f + \chi'' g$ nach (2) ein Element von $\Psi_c(W, V)$; diese Bewertung setzt $f$ und $g$ auf ganz $V$ fort. Ist $h \in \Psi_c(W, V)$ irgendeine solche Fortsetzung, so gilt $h = h' + h''$ nach (2) mit $h' \in \Psi_c'$ und $h'' \in \Psi_c''$. Hieraus folgt $h'(x) = h(x) = f(x)$ für Elemente $x \in V'$ und $h''(x) = h(x) = g(x)$ für $x \in V''$. Nach (1) ist dann aber $h' = \chi' f$ und $h'' = \chi'' g$, also $h = \chi' f + \chi'' g$.

**3. 2. Additive Abbildungen eines Booleschen Verbandes in einen vollständigen Vektorverband.** *Der bislang beliebige Verband* $V$ *sei nunmehr* Boolesch, d. h. $V$ sei distributiv, besitze ein Nullelement $\emptyset$ und gestatte die Bildung des relativen Komplements $x - y$ für Elemente $x, y \in V$ mit $y \leqq x$. (Bei beliebigen $x, y \in V$ setzen wir $x - y = x - x \wedge y$.) *Die Existenz des Einselementes wird nicht gefordert*; man nennt deshalb $V$ auch einen *verallgemeinerten Booleschen Verband* (vgl. [1], S. 111). Wie bisher sei $W$ ein *vollständiger Vektorverband* über dem Körper K.

*Eine Abbildung von* $V$ *in* $W$ *heißt additiv, wenn aus* $x \wedge y = \emptyset$, $x, y \in V$, *die Gleichung*

$$f(x \vee y) = f(x) + f(y)$$

*folgt.*

Den Zusammenhang mit dem Bewertungsbegriff stellt der folgende Satz her:

**Satz 3. 21.** *Folgende drei Aussagen sind gleichwertig:* (1) *Es ist* $f$ *ein Element von* $\Psi_\emptyset(W, V)$. — (2) *Die Abbildung* $f$ *von* $V$ *in* $W$ *ist additiv.* — (3) *Die Abbildung* $f$ *von* $V$ *in* $W$ *ist subtraktiv, d. h. aus* $x \leqq y$; $x, y \in V$, *folgt stets* $f(y - x) = f(y) - f(x)$.

Beweis. (1) $\rightarrow$ (2): Klar, da $f(x \vee y) = 0$ ist für Elemente $x, y \in V$ mit $x \wedge y = \emptyset$. — (2) $\rightarrow$ (3): Für Elemente $x, y \in V$ mit $x \leqq y$ gilt $x \wedge (y - x) = \emptyset$ und $x \vee (y - x) = y$. Die Additivität von $f$ liefert daher $f(y) = f(x) + f(y - x)$. — (3) $\rightarrow$ (1): Für $x = y$ folgt aus der Subtraktivität von $f$ die Gleichung $f(\emptyset) = f(x) - f(x) = 0$. Aus der für beliebige Elemente $x, y \in V$ gültigen Gleichung $x \vee y - x = y - x \wedge y$ folgt

$$f(x \vee y) - f(x) = f(y) - f(x \wedge y),$$

also die Funktionalgleichung (3. 1).

**Korollar.** *Für additive Abbildungen $f, g \in \Psi_a(W, V)$ besteht die Relation $f \leqq g$ dann und nur dann, wenn $f(x) \leqq g(x)$ für jedes Element $x \in V$ gilt.*

Beweis. Aus $f \leqq g$ und $\emptyset \leqq x$ folgt $f(x) \leqq g(x)$ für jedes $x \in V$. Gilt umgekehrt $f(x) \leqq g(x)$ für jedes $x \in V$, so folgt aus $a \leqq b$, $a, b \in V$,

$$f(b - a) \leqq g(b - a), \text{ also } f(b) - f(a) \leqq g(b) - g(a) \text{ und somit } f \leqq g.$$

Für den Fall eines Booleschen Verbandes läßt sich also die Vereinsstruktur von $\Psi_a(W, V)$ mittels der Funktionswerte einfacher beschreiben als im allgemeinen Fall. Analoges gilt für die Kennzeichnung der Elemente von $\Phi_a = \Phi_a(W, V)$ und für die Verbandsstruktur von $\Phi_a$:

**Satz** 3. 22. *Eine additive Abbildung von $V$ in $W$ ist dann und nur dann von relativ beschränkter Variation, wenn sie relativ beschränkt ist; m. a. W.: es ist $\Phi_a = \Phi_a(W, V)$ gleich der Menge $\Pi$ aller relativ beschränkten, additiven Abbildungen von $V$ in $W$. Für jedes $f \in \Phi_a$ und jedes $x \in V$ gilt*

$$f^+(x) = (W) \sup (f(x');\ x' \leqq x, x' \in V). \tag{3. 9}$$

Beweis. Nach dem Satz 3. 13 ist die Menge $\Pi$ ein linearer Unterraum von $\Psi_a$, welcher $\Phi_a$ als Teilmenge enthält. Wir haben also noch zu zeigen, daß $\Pi$ Teilmenge von $\Phi_a$ ist. Hierzu sei $f$ ein Element von $\Pi$; dann ist für jedes $x \in V$ die Bildung von

$$f^*(x) = (W) \sup (f(x');\ x' \leqq x)$$

sinnvoll. $f^*$ ist ein Element von $\Psi_a$, d. h. eine additive Abbildung von $V$ in $W$. Für Elemente $x, y \in V$ mit $x \wedge y = \emptyset$ ist nämlich wegen der Distributivität von $V$ die Menge der Elemente $z \leqq x \vee y$ aus $V$ identisch mit der Menge der $z \in V$, welche sich in der Form $z = x' \vee y'$ mit $x' \leqq x$ und $y' \leqq y$ darstellen lassen; berücksichtigt man noch, daß dann auch $x' \wedge y' = \emptyset$ ist, so folgt

$$f^*(x \vee y) = (W) \sup (f(x') + f(y');\ x' \leqq x, y' \leqq y) = f^*(x) + f^*(y).$$

Nach der Bemerkung vor Satz 3. 11 ist $f^*$ sogar ein Element von $\Phi_a$; wegen $f(\emptyset) = 0$ ist nämlich $f^*(x) \geqq 0$ für alle $x \in V$, also $f^* \geqq 0$ nach dem Korollar zu 3. 21. Weiter gilt $f(x) \leqq f^*(x)$ für alle $x \in V$, also $f \leqq f^*$. Da $\Pi$ linearer Unterraum von $\Psi_a$ ist, gehört mit $f$ auch $-f$ zu $\Pi$; wir erhalten somit die Doppelungleichung $-(-f)^* \leqq f \leqq f^*$, wobei $f^*$ und $-(-f)^*$ Elemente von $\Phi_a$ sind. Nach 3. 11 folgt hieraus $f \in \Phi_a$ und damit die Gleichheit von $\Pi$ und $\Phi_a$.

Nun wenden wir uns dem Beweis der für jedes $f \in \Phi_a = \Pi$ gültigen Gleichung $f^+ = f^*$ zu. Für beliebiges $x \in V$ und jedes $y \in V$ mit $y \leqq x$ gilt nach (3. 5), (3. 6) und (3. 7)

$$f(y) \leqq (W) \sup (f(y) - f(\emptyset), 0) + (W) \sup (f(x) - f(y), 0) \leqq f^+(x).$$

Hieraus folgt $f^*(x) \leqq f^+(x)$ für jedes $x \in V$ oder die damit gleichwertige Relation $f \leqq f^+$. Bereits im ersten Teil des Beweises ergaben sich die Eigenschaften $0 \leqq f^*$ und $f \leqq f^*$ von $f^*$. Wegen $f^+ = (\Phi_a) \sup (f, 0)$ folgt hieraus $f^+ \leqq f^*$ und somit $f = f^+$.

**Korollar.** *Für das Infimum $h = (\Phi_a) \inf (f, g)$ zweier Elemente $f$ und $g$ aus $\Phi_a(W, V)$ gilt*

$$h(x) = (W) \inf (f(y) + g(x - y);\ y \leqq x, y \in V),\ x \in V. \tag{3. 10}$$

Beweis. Aus (1. 12) ergibt sich $h = g - (g - f)^+$. Hieraus und aus (3. 9) folgt die Behauptung.

Bemerkung. Nach 3. 22 sind die Elemente von $\Phi_8(W, V)$ relativ beschränkte, additive Abbildungen von $V$ in $W$. Mit jedem $f \in \Phi_8(W, V)$ sind daher auch $f^+$, $f^-$ und $|f|$ relativ beschränkte, additive Abbildungen von $V$ in $W$.

Abschließend erwähnen wir noch ein Ergebnis aus [1], S. 112.

Der (verallgemeinerte) Boolesche Verband $V$ sei *Boolescher Unterverband* eines (verallgemeinerten) *Booleschen $\sigma$-Verbandes $V'$. Eine Abbildung $f$ von $V$ in $W$ heißt total-additiv bezüglich $V'$ ($\sigma$-additiv bez. $V'$), wenn für jede Folge von Elementen $x_\nu \in V$ mit $x_\nu \wedge x_\varrho = \emptyset$, $\nu \neq \varrho$, $\nu, \varrho = 1, 2, \ldots$, deren in $V'$ gebildete Vereinigung $x$ ein Element von $V$ ist, die Reihe $\sum_{\nu=1}^{\infty} f(x_\nu)$ algebraisch konvergiert und $f(x) = \sum_{\nu=1}^{\infty} f(x_\nu)$ ist.*

Wenn $V$ selbst ein Boolescher $\sigma$-Verband und die Abbildung $f$ total-additiv bezüglich $V' = V$ ist, so sagen wir kurz: *$f$ ist auf $V$ total-additiv.* Ist $V$ speziell ein Boolescher Mengenverband mit der Trägermenge $G$, so soll total-additiv stets bedeuten: *total-additiv bezüglich des vollständigen Booleschen Mengenverbandes aller Teilmengen von $G$.*

Das angekündigte Ergebnis aus [1] lautet: *Ist $V$ Boolescher Unterverband eines Booleschen $\sigma$-Verbandes $V'$, so ist die Menge $\Phi^\sigma = \Phi^\sigma(W, V)$ aller bezüglich $V'$ total-additiven Abbildungen $f \in \Phi_8(W, V)$ ein Band im vollständigen Vektorverband $\Phi_8$.*

**3. 3. Lineare Abbildungen eines Vektorverbandes in einen vollständigen Vektorverband.** *Wir bezeichnen in dieser Nummer mit $E$ einen Vektorverband über dem Skalarenkörper* K *des vollständigen Vektorverbandes $W$.*

*Eine Abbildung $f$ von $E$ in $W$ heißt linear, wenn folgendes gilt:*

(3. 11) $\qquad f(x + y) = f(x) + f(y)$ *für beliebige* $x, y \in E$,

(3. 12) $\qquad f(\lambda x) \quad = \lambda f(x)$ *für beliebige* $x \in E$, $\lambda \in$ K.

*Die Menge aller linearen Abbildungen von $E$ in $W$ bezeichnen wir mit $\Lambda^* = \Lambda^*(W, E)$.* Den Zusammenhang mit den Bewertungen stellt der folgende Satz her:

**Satz** 3. 31. *Jede lineare Abbildung $f$ von $E$ in $W$ ist eine Bewertung von $E$ in $W$. Die Menge $\Lambda^*(W, E)$ ist ein linearer Unterraum von $\Psi_0(W, E)$, wenn 0 den Nullvektor von $E$ bedeutet.*

Beweis. Nach (1. 1) gilt für beliebige Elemente $x, y \in E$ die Gleichung $x \vee y + x \wedge y = x + y$. Aus der Eigenschaft (3. 11) allein folgt hieraus die Funktionalgleichung (3. 1). Ebenso folgt $f(0) = 0$ aus (3. 11) allein. Der Rest der Behauptung ist unmittelbar klar[7]).

Wie im reellen Fall (vgl. H. Nakano [7], S. 246) beweist man den folgenden Satz:

**Satz** 3. 32. *Eine Abbildung $f$ der Menge aller Elemente $x \geqq 0$ aus $E$ in $W$, welche die Gleichungen* (3. 11) *und* (3. 12) *nur für Elemente $x \geqq 0$, $y \geqq 0$ aus $E$ und $\lambda \geqq 0$ aus* K *erfüllt, läßt sich auf genau eine Weise zu einer linearen Abbildung von ganz $E$ in $W$ erweitern.*

Auf Grund des Satzes 3. 31 können alle für Bewertungen eingeführten Begriffe auch als für lineare Abbildungen erklärt betrachtet werden.

---

[7]) Der Beweis von Satz 3. 31 läßt die folgende Möglichkeit zu weiterer Verallgemeinerung verständlich erscheinen, von der wir jedoch keinen Gebrauch machen: Man kann in den folgenden Sätzen grundsätzlich die linearen Abbildungen von $E$ in $W$ durch Abbildungen einer abelschen Verbandsgruppe in eine vollständige Verbandsgruppe ersetzen, welche (3. 11) genügen. (Vgl. Fußnote [3]).) Für relativ beschränkte $f$ folgt im *reellen* Fall übrigens (3. 12) aus (3. 11). Vgl. hierzu *N. Bourbaki* [4], S. 33.

**Satz** 3. 33 **(1)** *Für lineare Abbildungen $f, g \in \Lambda^*(W, E)$ gilt $f \leqq g$ dann und nur dann, wenn $f(x) \leqq g(x)$ für jedes $x \in E$ mit $x \geqq 0$ gilt.* — **(2)** *Die lineare Abbildung $f \in \Lambda^*(W, E)$ ist dann und nur dann relativ beschränkt, wenn zu jedem $x \in E$ ein Element $s = s(x) \in W$ existiert, so daß $|f(y)| \leqq s$ für alle $y \in E$ mit $|y| \leqq |x|$ gilt.*

Beweis. Zu (1). *Nur dann*: Aus $f \leqq g$ und $x \geqq 0$ folgt $f(x) \leqq g(x)$. — *Dann*: Aus $x \leqq y$ folgt $y - x \geqq 0$, also $f(y-x) \leqq g(y-x)$. Wegen (3. 11) folgt hieraus $f(y) - f(x) \leqq g(y) - g(x)$, also $f \leqq g$. — Zu (2). *Nur dann*: Wegen (1. 6) sind die Relationen $|y| \leqq |x|$ und $-|x| \leqq y \leqq |x|$ gleichwertig. Hieraus ergibt sich die Behauptung. — *Dann*: Aus $a \leqq x \leqq b$ folgt $|x| \leqq |a| + |b|$, $a, b, x \in E$, und hieraus die Behauptung.

*Die Menge aller* **relativ beschränkten**, *linearen Abbildungen von $E$ in $W$ bezeichnen wir von nun an mit $\Lambda = \Lambda(W, E)$.*

**Satz** 3. 34. **(1)** *Eine lineare Abbildung von $E$ in $W$ ist dann und nur dann relativ beschränkt, wenn sie von relativ beschränkter Variation ist; m. a. W.: es gilt $\Lambda = \Lambda^* \cap \Phi_0$ $(\Phi_0 = \Phi_0(W, E))$.*

**(2)** *$\Lambda(W, E)$ ist ein* **eigentlicher** *linearer Unterraum von $\Phi_0(W, E)$. Für jedes $f \in \Lambda$ und jedes $x \geqq 0$ aus $E$ gilt*

$$(3.13) \qquad f^+(x) = (W) \sup (f(x');\ 0 \leqq x' \leqq x,\ x' \in E).$$

**(3)** *Für jede nicht leere und in $\Phi_0$ nach oben bzw. unten beschränkte Teilmenge $H$ von $\Lambda$ gilt $(\Phi_0)$ sup $H \in \Lambda$ bzw. $(\Phi_0)$* inf *$H \in \Lambda$.*

Beweis. Zunächst zeigen wir, daß für jede relativ beschränkte, lineare Abbildung $f$ von $E$ in $W$ durch

$$f^*(x) = (W) \sup (f(x');\ 0 \leqq x' \leqq x),\quad x \in E \text{ mit } x \geqq 0,$$

eine lineare Abbildung von $E$ in $W$ definiert wird: Es seien $x \geqq 0$ und $y \geqq 0$ Elemente von $E$; nach einer Bemerkung in Nr. 1. 1 (nach Formel (1. 3)) ist die Menge aller $z \in E$ mit $0 \leqq z \leqq x + y$ gleich der Menge der Elemente $z = x' + y'$ mit $0 \leqq x' \leqq x$ und $0 \leqq y' \leqq y$. Hieraus folgt

$$f^*(x+y) = f^*(x) + f^*(y).$$

Durch eine ähnliche Überlegung beweist man die Gültigkeit der Gleichung $f^*(\lambda x) = \lambda f^*(x)$ für alle $\lambda \geqq 0$ aus K und $x \geqq 0$ aus $E$. Nach Satz 3. 32 ist daher $f^*$ eindeutig zu einer linearen Abbildung von $E$ in $W$ erweiterbar, die wir wiederum mit $f^*$ bezeichnen. Für jedes $x \in E$ mit $x \geqq 0$ gilt $0 \leqq f^*(x)$ und $f(x) \leqq f^*(x)$, ersteres wegen $f(0) = 0$; nach Satz 3. 33, (1) bedeutet dies $0 \leqq f^*$ und $f \leqq f^*$. Nun erst wenden wir uns dem Beweis der drei Punkte der Behauptung zu:

Zu (1). *Nur dann*: Wie eben beschrieben, bilden wir zu $f \in \Lambda$ die lineare Abbildung $f^*$; diese ist wegen $f^* \geqq 0$ von relativ beschränkter Variation. Mit $f$ ist auch $-f$ relativ beschränkt und die zugehörige lineare Abbildung $(-f)^*$ von relativ beschränkter Variation. Nach dem ersten Teil des Beweises gilt $-(-f)^* \leqq f \leqq f^*$; nach Satz 3. 11 ist dann $f$ von relativ beschränkter Variation. — *Dann*: Folgt aus Satz 3. 13.

Zu (2). Genau so wie im Beweis von Satz 3. 22 beweist man nun die Gleichheit von $f^*$ und $f^+ = (\Phi_0) \sup (f, 0)$ für jedes $f \in \Lambda$ und damit auch die Gleichung (3. 13). Nach dem bisher Bewiesenen ist also mit jedem $f \in \Lambda$ auch $f^+ = (\Phi_0) \sup (f, 0) = f^*$ ein Element von $\Lambda$; gemäß der Definition in Nr. 1. 2 ist also der lineare Unterraum $\Lambda$ von $\Phi_0$ eigentlich.

Zu (3). Nach Nr. 1. 2, S. 146, genügt es, die Behauptung für eine Menge $H$ von Elementen $h \geqq 0$ aus $\Lambda$ zu beweisen, welche nicht leer, in $\Phi_0$ steigend gerichtet und in $\Phi_0$ nach oben beschränkt ist. Für $s = (\Phi_0) \sup H$ gilt dann nach (3. 8) die Gleichung

$$s(x) = (W) \lim_{h \in H} \operatorname{alg} h(x).$$

Hieraus ergibt sich: Es ist $s$ eine lineare Abbildung von $E$ in $W$ mit $s \geqq 0$; nach der Bemerkung vor Satz 3. 11 und nach Satz 3. 34 ist dann $s$ ein Element von $\Lambda$.

**1. Korollar.** *Der eigentliche lineare Unterraum $\Lambda(W, E)$ von $\Phi_0(W, E)$ ist bezüglich der in $\Phi_0$ erklärten Relation $\leqq$ ein vollständiger Vektorverband. Jede nicht leere in $\Phi_0$ nach oben beschränkte Teilmenge $H$ von $\Lambda$ ist auch in $\Lambda$ nach oben beschränkt und es gilt*

$$(\Lambda) \sup H = (\Phi_0) \sup H.$$

**2. Korollar.** *Für das in $\Lambda(W, E)$ gebildete Infimum zweier linearen Abbildungen $f$ und $g \in \Lambda$ gilt bei beliebigem $x \geqq 0$ aus $E$*

$$(3.\,14) \qquad h(x) = (W) \inf (f(y) + g(x - y);\ 0 \leqq y \leqq x,\ y \in E).$$

Beweis. Analog zum Beweis von (3. 10).

Nun noch eine abschließende Bemerkung über *stetige, lineare Abbildungen*: Der Vektorverband $E$ sei *eigentlicher* linearer Unterraum eines *$\sigma$-vollständigen* Vektorverbandes $E'$. Eine lineare Abbildung $f$ von $E$ in $W$ heißt stetig bezüglich $E'$, wenn für jede Folge von Elementen $x_\nu \in E$ mit $x_\nu \geqq x_{\nu+1} \geqq 0$, $\nu = 1, 2, \ldots$, deren bezüglich $E'$ gebildetes Infimum gleich 0 ist, der algebraische Limes der Folge der Bildelemente $f(x_\nu) \in W$ in $W$ existiert und gleich 0 ist. *Gleichwertig* hiermit ist die Forderung: Für jede in $E'$ beschränkte Folge von Elementen $x_\nu \in E$ mit $0 \leqq x_\nu \leqq x_{\nu+1}$, $\nu = 1, 2, \ldots$, deren in $E'$ gebildetes Supremum $x$ ein Element von $E$ ist, existiert $(W) \lim \operatorname{alg} f(x_\nu)$ und ist gleich $f(x)$. Wenn $E$ selbst ein $\sigma$-vollständiger Vektorverband und die lineare Abbildung $f$ von $E$ in $W$ stetig bezüglich $E' = E$ ist, so nennen wir $f$ stetig schlechthin. Ein in [1], S. 116 gewonnenes Ergebnis lautet nun[8]):

*Der Vektorverband $E$ sei ein eigentlicher linearer Unterraum des $\sigma$-vollständigen Vektorverbandes $E'$. Dann ist die Menge $\Lambda^\sigma = \Lambda^\sigma(W, E)$ aller bezüglich $E'$ stetigen Abbildungen $f \in \Lambda(W, E)$ ein Band im vollständigen Vektorverband $\Lambda$.*

## § 4. Reguläre und singuläre Bewertungen.

*In diesem Paragraphen soll wieder durchweg unter $W$ ein vollständiger Vektorverband über einem linear geordneten Körper $\mathsf{K}$ verstanden werden.*

[8]) Dort haben wir allerdings die Stetigkeit bezüglich $E'$ nur für relativ beschränkte, lineare Abbildungen $f$ von $E$ in $W$ definiert und zwar abweichend von obiger Definition: Ein $f \in \Lambda(W, E)$ hieß dort stetig bez. $E'$, wenn $|f|$ stetig bez. $E'$ im obigen Sinne war. Am Schluß von [1] hatten wir erwähnt, daß im reellen Fall die beiden Stetigkeitsdefinitionen für Elemente von $\Lambda$ *gleichwertig* sind; sie sind es sogar *im allgemeinen Fall*, wenn $W$ ein beliebiger vollständiger Vektorverband ist. *Beweis.* Es sei $x_\nu \in E$ mit $x_\nu \geqq x_{\nu+1} \geqq 0$, $\nu = 1, 2, \ldots$, und $(E') \inf x_\nu = 0$. Gilt dann für ein $f \in \Lambda$ die Gleichung $(W) \lim \operatorname{alg} |f|(x_\nu) = 0$, so ist auch $(W) \lim \operatorname{alg} f^+(x_\nu) = (W) \lim \operatorname{alg} f^-(x_\nu) = 0$ und damit $(W) \lim \operatorname{alg} f(x_\nu) = 0$. Man hat hierbei nur zu beachten, daß $0 \leqq f^+(x_\nu) \leqq |f|(x_\nu)$ und $0 \leqq f^-(x_\nu) \leqq |f|(x_\nu)$ ist für $\nu = 1, 2, \ldots$. — Umgekehrt: Es genügt zu zeigen, daß $f^+$ im Sinne des Textes stetig bez. $E'$ ist, wenn dies für $f \in \Lambda$ der Fall ist. Hierzu sei $x_\nu \in E$ mit $0 \leqq x_\nu \leqq x_{\nu+1}$, $\nu = 1, 2, \ldots$, und $(E') \sup x_\nu = x \in E$. Wegen $f^+ \geqq 0$ ist dann zunächst $(W) \lim \operatorname{alg} f^+(x_\nu) \leqq f^+(x)$. Weiter sei $y \in E$ mit $0 \leqq y \leqq x$; wir setzen $y_\nu = (E) \inf (y, x_\nu)$ für jedes $\nu$. Mit Hilfe des in $E'$ gültigen allgemeinen distributiven Gesetzes (vgl. *G. Birkhoff* [2], S. 231) folgt: $0 \leqq y_\nu \leqq y_{\nu+1}$, $\nu = 1, 2, \ldots$, und $(E') \sup y_\nu = y \in E$. Wir erhalten $f(y) = (W) \lim \operatorname{alg} f(y_\nu)$ und hieraus $f(y) \leqq (W) \lim \operatorname{alg} f^+(x_\nu)$ nach (3. 13). Nach (3. 13) folgt auch $f^+(x) \leqq (W) \lim \operatorname{alg} f^+(x_\nu)$ und damit die Behauptung.

**4. 1. Definition und Eigenschaften der Bewertung $f_S$.** Wir geben zunächst die

**Definition.** *Es sei $S$ eine beliebige Teilmenge eines Verbandes $V$. Für eine Bewertung $f \geqq 0$ aus $\Phi_c = \Phi_c(W, V)$ bezeichnen wir mit $f_S$ das in $\Phi_c$ gebildete Infimum über alle Elemente $h \geqq 0$ aus $\Phi_c$, für welche bei beliebigem $x \in S$ stets $h(x) = f(x)$ gilt. Für eine beliebige Bewertung $f \in \Phi_c$ setzen wir*

$$f_S = (f^+)_S - (f^-)_S. \tag{4. 1}$$

Wir bemerken, daß diese Definition sinnvoll ist, da für ein $f \geqq 0$ aus $\Phi_c$ zumindest $f$ selbst eine Bewertung $h$ mit den in der Definition geforderten Eigenschaften ist. Aus dieser Bemerkung und der Definition selbst folgt bereits der erste Teil des folgenden Satzes:

**Satz 4. 11.** *Für jede Bewertung $f \geqq 0$ aus $\Phi_c(W, V)$ gilt*

$$0 \leqq f_S \leqq f. \tag{4. 2}$$

*Mit $f \in \Phi_c$ ist stets auch $f_S$ ein Element von $\Phi_c$; es gilt*

$$(f_S)^+ = (f^+)_S, \quad (f_S)^- = (f^-)_S. \tag{4. 3}$$

Beweis. Eines Beweises bedarf nur noch (4. 3). Nach (4. 2) sind $(f^+)_S$ und $(f^-)_S$ Elemente $\geqq 0$ aus $\Phi_c$, sowie in $\Phi_c$ disjunkt, da $f^+$ und $f^-$ nach Nr. 1. 1 disjunkt sind. Vermöge der Gleichung (4. 1) ist also $f_S$ Differenz zweier disjunkter Elemente $\geqq 0$ aus $\Phi_c$. Hieraus folgt die Behauptung nach Nr. 1. 1.

*Wegen* (4. 3) *schreiben wir von nun an nur noch $f_S^+$ und $f_S^-$ an Stelle von $(f_S)^+$ und $(f_S)^-$.*

Für weitergehende Untersuchungen von $f_S$ benötigen wir einen Hilfssatz.

Hilfssatz. *Es sei $G$ ein gesättigter Unterverband eines Verbandes $V$ und $c$ ein Element von $G$. Für die Bewertung $f \in \Phi_c(W, V)$ sind dann folgende Aussagen gleichwertig:*

**(1)** *$f(x) = 0$ für jedes $x \in G$.*

**(2)** *$f^+(x) = f^-(x) = 0$ für jedes $x \in G$.*

**(3)** *$|f|(x) = 0$ für jedes $x \in G$.*

Beweis. (1) → (2) Wegen $f^- = (-f)^+$ genügt es, die $f^+$ betreffende Aussage zu beweisen; wegen (3. 1) und (3. 2) (angewandt auf $f^+$) genügt es weiter, nur Elemente $x \in G$ mit $x \geqq c$ oder $x \leqq c$ zu betrachten. Nach (3. 5), (3. 6) und (3. 7) gehen in die Berechnung von $f^+(x)$ für ein solches $x \in G$ nur Funktionswerte $f(y) \in W$ ein, welche Elementen $y \in V$ mit $c \leqq y \leqq x$ oder $x \leqq y \leqq c$ zugeordnet sind. Diese Elemente $y$ gehören aber sämtlich zu $G$; aus (3. 5) bis (3. 7) folgt daher $f^+(x) = 0$. — (2) → (3) Klar. — (3) → (1) Aus $-|f| \leqq f \leqq |f|$ folgt $-|f|(x) \leqq f(x) \leqq |f|(x)$ und damit $f(x) = 0$ für Elemente $x \geqq c$ aus $G$; entsprechend ergibt sich $f(x) = 0$ für Elemente $x \leqq c$ aus $G$. Nach (3. 1) und (3. 2) ist dann $f(x) = 0$ für alle $x \in G$.

**Satz 4. 12.** *Es sei $S$ ein gesättigter Unterverband des Verbandes $V$ und $c$ ein Element von $S$. Dann ist $f_S$ für jede Bewertung $f \geqq 0$ aus $\Phi_c(W, V)$ sogar das Minimum der Menge aller $h \geqq 0$ aus $\Phi_c$, für welche $h(x) = f(x)$ bei beliebigem $x \in S$ gilt. Für jede Bewertung $f \in \Phi_c$ ist*

$$f_S(x) = f(x), \quad \text{falls } x \in S. \tag{4. 4}$$

Beweis. Es sei $f \geqq 0$ und $H$ die Menge aller $h \geqq 0$ aus $\Phi_c$ mit $h(x) = f(x)$, $x \in S$. Es genügt, zu zeigen, daß $H$ in $\Phi_c$ fallend gerichtet ist; dann gilt nämlich nach (3. 8) für jedes $x \in V$ die Gleichung $f_S(x) = (W) \lim\limits_{h \in H} \operatorname{alg} h(x)$, aus welcher (4. 4) und damit $f_S \in H$ folgt. Der Rest der Behauptung ergibt sich hieraus mit Hilfe von (4. 1) und (4. 3). — Wir zeigen: Aus $h, k \in H$ folgt $q = (\Phi_c) \inf (h, k) \in H$. Aus (1. 12) ergibt sich die Gleichung $q = h - (h - k)^+$; hierbei ist $(h - k)(x) = 0$ für jedes $x \in S$ und damit nach dem eben bewiesenen Hilfssatz auch $(h-k)^+(x)=0$ für jedes $x \in S$. Wegen $q \geqq 0$ und $q(x)=h(x)=f(x)$ für jedes $x \in S$ ist $q \in H$, w. z. z. w.

**Satz 4. 13.** *Es sei V ein distributiver Verband, S ein gesättigter Unterverband von V und c ein Element von S. Dann gilt für jede Bewertung $f \geqq 0$ aus $\Phi_c(W, V)$:*

(4. 5) $f_S(x)=(W)\sup\big(f(x');\ c \leqq x' \leqq x,\ x' \in S\big)$, *falls* $x \geqq c,\ x \in V$,

(4. 6) $f_S(x)=(W)\inf\big(f(x');\ x \leqq x' \leqq c,\ x' \in S\big)$, *falls* $x \leqq c,\ x \in V$.

Beweis. (1) Es sei $V'$ die Menge aller $x \in V$ mit $x \geqq c$. Wir zeigen, daß durch die Gleichung

$$t'(x)=(W)\sup\big(f(x');\ c \leqq x' \leqq x,\ x' \in S\big),\quad x \in V',$$

eine Bewertung von $V'$, genauer ein Element $t' \geqq 0$ von $\Phi_c(W, V')$ definiert wird. Hierzu bezeichnen wir vorübergehend für ein $x \in V'$ mit $xS$ die Menge aller $x' \in S$ mit $c \leqq x' \leqq x$. Dann ist $(x \vee y)\,S$ für beliebige $x, y \in V'$ gleich der Menge aller $z \in V'$ mit $z=x' \vee y'$, $x' \in xS$, $y' \in yS$. Jedes solche $z$ ist nämlich ein Element von $(x \vee y)\,S$; aus $z \in (x \vee y)\,S$ folgt umgekehrt $z=x' \vee y'$ mit $x'=z \wedge x \in xS$ und $y'=z \wedge y \in yS$, weil $V$ distributiv und $S$ gesättigt ist. Analog, sogar ohne Benutzung des distributiven Gesetzes, beweist man, daß $(x \wedge y)\,S$, $x, y \in V'$, gleich der Menge aller $z \in V'$ ist mit $z=x' \wedge y'$, $x' \in xS$, $y' \in yS$. Bezüglich $t'$ zeigen wir zunächst, daß $t'$ ein Element von $\Psi_c(W, V')$ ist. Die Gültigkeit von $t'(c)=0$ ist unmittelbar ersichtlich, die Funktionalgleichung (3. 1) folgt so: Für beliebige $x, y \in V'$ ist

$$\begin{aligned} t'(x)+t'(y) &= (W)\sup\big(f(x');\ x' \in xS\big)+(W)\sup\big(f(y');\ y' \in yS\big) \\ &= (W)\sup\big(f(x' \vee y')+f(x' \wedge y');\ x' \in xS,\ y' \in yS\big). \end{aligned}$$

Nach dem über die Mengen $(x \vee y)\,S$ und $(x \wedge y)\,S$ Bewiesenen ergibt sich hieraus

$$\begin{aligned} t'(x)+t'(y) &\leqq (W)\sup\big(f(u)+f(v);\ u \in (x \vee y)\,S,\ \ v \in (x \wedge y)\,S\big) \\ &= t'(x \vee y)+t'(x \wedge y). \end{aligned}$$

Die hierzu duale Ungleichung folgt so: Für $x', x'' \in xS$ und $y', y'' \in yS$ mit $x, y \in V'$ setzen wir $x^*=x' \vee x''$ und $y^*=y' \vee y''$. Dann ist $x^* \in xS$ und $y^* \in yS$, sowie $x' \vee y' \leqq x^* \vee y^*$ und $x'' \wedge y'' \leqq x^* \wedge y^*$. Wegen $f \geqq 0$ folgt hieraus

$$f(x' \vee y')+f(x'' \wedge y'') \leqq f(x^* \vee y^*)+f(x^* \wedge y^*)=f(x^*)+f(y^*) \leqq t'(x)+t'(y)$$

und schließlich durch Übergang linkerhand zum $(W)$Supremum

$$t'(x \vee y)+t'(x \wedge y) \leqq t'(x)+t'(y).$$

Damit ist $t' \in \Psi_c(W, V')$ bewiesen. Da aus $c \leqq x \leqq y$ aber $t'(x) \leqq t'(y)$ folgt, gilt sogar $t' \geqq 0$ und somit $t' \in \Phi_c(W, V')$ nach der Bemerkung vor Satz 3. 11.

(2) Setzt man

$$t''(x)=(W)\inf\big(f(x'');\ x \leqq x'' \leqq c,\ x'' \in S\big)$$

für jedes $x$ aus dem Unterverband $V''$ aller $x \in V$ mit $x \leqq c$, so erweist sich durch analoge Betrachtungen $t''$ als Element von $\Phi_c(W, V'')$ mit $t'' \geqq 0$. Nach Satz 3. 14 setzt $t(x)=t'(c \vee x)+t''(c \wedge x)$, $x \in V$ die Bewertungen $t'$ und $t''$ auf ganz $V$ fort; es gilt $t \in \Phi_c(W, V)$ und $t \geqq 0$.

(3) Es ist $f_S \leqq t$, da $t \geqq 0$ und

$$t(x)=t'(c \vee x)+t''(c \wedge x)=f(c \vee x)+f(c \wedge x)=f(x)$$

für Elemente $x \in S$. Hieraus folgt $f_S(x) \leqq t'(x)$ für $x \in V'$ und $t''(x) \leqq f_S(x)$ für $x \in V''$. Wegen $f(x')=f_S(x')$ für $x' \in S$ gilt andererseits

$$t'(x)=(W)\sup\big(f_S(x');\ c \leqq x' \leqq x,\ x' \in S\big) \leqq f_S(x),\ x \in V'.$$

Also ist $t'(x) = f_S(x)$ für jedes $x \in V'$; ebenso beweist man $t''(x) = f_S(x)$ für $x \in V''$. Es folgt $t = f_S$ und damit die Behauptung.

Der dem Satz 4. 13 zugrundeliegende Fall ist für uns der allein wichtige; deshalb vereinbaren wir: *Wenn im folgenden $f_S$ für eine Bewertung $f \in \Phi_c(W, V)$ gebildet wird, so soll stets stillschweigend vorausgesetzt sein, daß der Verband V distributiv, S ein gesättigter Unterverband von V und c ein Element von S ist.*

**Korollar.** *Für eine Bewertung $f \in \Phi_c(W, V)$ gilt $f_S = 0$ dann und nur dann, wenn $f(x) = 0$ ist für jedes $x \in S$.*

Beweis. *Nur dann*: Folgt aus (4. 4). — *Dann*: Nach dem Hilfssatz dieser Nummer gilt mit $f(x) = 0$ auch $f^+(x) = f^-(x) = 0$ für alle $x \in S$; wir können daher o. B. d. A. $f \geqq 0$ voraussetzen. Dann aber folgt die Behauptung aus (4. 5), (4. 6) und (3. 1).

**Satz** 4. 14. *Für jede Bewertung $f \in \Phi_c(W, V)$ und je zwei gesättigte Unterverbände S und T des distributiven Verbandes V mit $c \in S \frown T$ gilt*

$$(4.\ 7) \qquad (f_S)_T = f_{S \frown T}.$$

Beweis. Zunächst sei $f \geqq 0$. Für jedes $x \in S \frown T$ gilt dann nach (4. 4)

$$(f_S)_T(x) = f_S(x) = f(x);$$

da zudem $(f_S)_T \geqq 0$ ist, folgt nach der Definition von $f_{S \frown T}$ die Ungleichung $f_{S \frown T} \leqq (f_S)_T$. Hiernach ist $f_{S \frown T}(x) \leqq (f_S)_T(x)$ für $x \in V$ mit $x \geqq c$; die duale Ungleichung ergibt sich bei gleichem $x \in V$ so: Aus $c \leqq y \leqq x$ mit $y \in T$ und aus $c \leqq y' \leqq y$ mit $y' \in S$ folgt $c \leqq y' \leqq x$ mit $y' \in S \frown T$; nach (4. 5) gilt daher $f(y') \leqq f_{S \frown T}(x)$ für alle $y' \in S$ mit $c \leqq y' \leqq y$ und somit auch $f_S(y) \leqq f_{S \frown T}(x)$ für alle $y \in T$ mit $c \leqq y \leqq x$. Übergang linkerhand zum (W)Supremum über alle diese $y \in T$ ergibt die gesuchte Ungleichung. Analog wie für Elemente $x \geqq c$ aus V beweist man die Gleichung $(f_S)_T(x) = f_{S \frown T}(x)$ für $x \in V$ mit $x \leqq c$. Nach Satz 4. 13 gilt dann $(f_S)_T = f_{S \frown T}$. Für beliebiges $f \in \Phi_c$ ergibt sich (4. 7) folgendermaßen:

$$(f_S)_T = (f_S)_T^+ - (f_S)_T^- = (f_S^+)_T - (f_S^-)_T = f_{S \frown T}^+ - f_{S \frown T}^- = f_{S \frown T}$$

(nach (4. 3) und dem Bisherigen).

Die für uns wichtigste Eigenschaft der Bewertung $f_S$ beweisen wir nunmehr.

**Satz** 4. 15. *Die Abbildung $f \to f_S$ von $\Phi_c(W, V)$ in sich ist eine Projektion* (vgl. Nr. 1. 3).

Beweis. (1) Die Linearität der Abbildung $f \to f_S$ ergibt sich aus folgender Zwischenbehauptung: Zu $x \in V$ mit $x \geqq c$ oder $x \leqq c$ existiert ein in V gerichtetes System $H(x)$ derart, daß für $H(x)$ und beliebiges $f \in \Phi_c$

$$f_S(x) = (W) \lim_{x' \in H(x)} \text{alg}\, f(x')$$

gilt. Hieraus folgt nämlich

$$(\alpha f + \beta g)_S(x) = \alpha f_S(x) + \beta g_S(x), \quad f, g \in \Phi_c, \quad \alpha, \beta \in \mathrm{K},$$

zunächst für $x \in V$ mit $x \geqq c$ oder $x \leqq c$, dann aber allgemein für beliebige $x \in V$ nach 4. 13. Wir definieren $H(x)$ für ein $x \geqq c$ als die in V steigend gerichtete Menge aller $x' \in S$ mit $c \leqq x' \leqq x$ und für ein $x \leqq c$ als die in V fallend gerichtete Menge aller $x' \in S$ mit $x \leqq x' \leqq c$. Die Formel (4. 5) bzw. (4. 6) schreibt sich für ein $f \geqq 0$ in der Form $f_S(x) = (W) \lim_{x' \in H(x)} \text{alg}\, f(x')$ mit $x \geqq c$ bzw. $x \leqq c$. Für beliebiges $f \in \Phi_c$ folgt daraus:

$$f_S(x) = f_S^+(x) - f_S^-(x) = (W) \lim_{x' \in H(x)} \text{alg}\, (f^+(x') - f^-(x')) = (W) \lim_{x' \in H(x)} \text{alg}\, f(x')$$

für $x \in V$ mit $x \geqq c$ oder $x \leqq c$.

(2) Die Abbildung $f \to f_S$ ist eine Projektion, weil $0 \leqq f_S \leqq f$ ist für ein $f \geqq 0$ nach (4. 2) und $(f_S)_S = f_S$ für jedes $f \in \Phi_c$ nach (4. 7).

Nach dem in Nr. 1. 3 über Projektionen Gesagten, insbesondere nach (1. 17) ergibt sich als Ergänzung zu (4. 3):

**Korollar.** *Für jede Bewertung* $f \in \Phi_c(W, V)$ *gilt*

(4. 8) $$|f_S| = |f|_S.$$

Für Anwendungen wichtig ist noch folgender Satz:

**Satz** 4. 16. *Es sei $E$ ein Vektorverband über dem Skalarenkörper* K *von $W$ und $S$ ein gesättigter linearer Unterraum von $E$. Mit jeder relativ beschränkten, linearen Abbildung $f$ von $E$ in $W$, d. h. mit jedem $f \in \Lambda(W, E)$ ist dann auch die Bewertung $f_S$ ein Element von $\Lambda(W, E)$.*

Beweis. Ohne Beschränkung der Allgemeinheit kann $f \geqq 0$ angenommen werden. Wegen Satz 3. 32 braucht nur

$$f_S(x + y) = f_S(x) + f_S(y) \text{ für Elemente } x \geqq 0, y \geqq 0 \text{ aus } E,$$
$$f_S(\lambda x) = \lambda f_S(x) \text{ für Elemente } x \geqq 0 \text{ aus } E \text{ und } \lambda \geqq 0 \text{ aus K}$$

gezeigt zu werden. Dies beweist man mittels der in (4. 5) gefundenen Darstellung von $f_S(x)$ für Elemente $x \geqq 0$ aus $E$ in Analogie zum Beweis von Satz 3. 34: Es ist nämlich die Menge der Elemente $z \in S$ mit $0 \leqq z \leqq x + y$, $x, y \geqq 0$, gleich der Menge aller $z \in E$ von der Form $z = x' + y'$ mit $x', y' \in S$, $0 \leqq x' \leqq x$ und $0 \leqq y' \leqq y$. Ferner ist die Menge aller $z \in S$ mit $0 \leqq z \leqq \lambda x$, $x \geqq 0$, $\lambda \geqq 0$, gleich der Menge der Elemente $z \in E$ von der Form $z = \lambda x'$ mit $x' \in S$ und $0 \leqq x' \leqq x$.

**4. 2. Totale Nullteile.** Wir geben zunächst wieder die erforderliche

**Definition.** *Es sei $f \in \Psi_c(W, V)$ eine Bewertung eines beliebigen Verbandes $V$. Es heißt $a \in V$ c-totaler Nullteil von $f$, wenn $f(x) = 0$ für alle $x \in V$ mit $c \wedge a \leqq x \leqq c \vee a$. Die Menge der c-totalen Nullteile von $f$ bezeichnen wir mit $N_c(f)$.*

**Bemerkung.** *Wegen $c \wedge a \leqq a \leqq c \vee a$ gilt $f(a) = 0$ für jedes $a \in N_c(f)$.*

An Eigenschaften der Mengen $N_c(f)$ mit $f \in \Psi_c(W, V)$ benötigen wir:

**Satz 4. 21. (1)** *Für jede Bewertung $f \geqq 0$ aus $\Phi_c(W, V)$ ist $N_c(f)$ gleich der Menge aller $a \in V$ mit $f(c \wedge a) = f(c \vee a) = 0$.*

**(2)** $N_c(f) = N_c(|f|) = N_c(f^+) \cap N_c(f^-)$ *für jedes* $f \in \Phi_c(W, V)$.

**(3)** *Aus* $|g| \leqq |f|$, $f, g \in \Phi_c(W, V)$, *folgt* $N_c(f) \leqq N_c(g)$.

Beweis. (1) Klar, weil $f(c \wedge a) \leqq f(x) \leqq f(c \vee a)$ ist für jedes $x \in V$ mit $c \wedge a \leqq x \leqq c \vee a$. — (2) Für jedes $a \in V$ ist das Intervall $[c \wedge a, c \vee a]$ ein $c$ als Element enthaltender, gesättigter Unterverband von $V$; die Behauptung folgt daher aus dem Hilfssatz der Nr. 4. 1. — (3) Die Behauptung folgt aus (1) und (2), sowie aus den Ungleichungen

$$|f|(c \wedge a) \leqq |g|(c \wedge a) \leqq 0 \text{ und } 0 \leqq |g|(c \vee a) \leqq |f|(c \vee a), \; a \in V.$$

**Satz 4. 22.** *Der Verband $V$ sei distributiv. Für jedes $f \in \Phi_c(W, V)$ ist dann $N_c(f)$ ein $c$ als Element enthaltender, gesättigter Unterverband von $V$.*

Beweis. Wegen $N_c(f) = N_c(|f|)$ können wir $f \geqq 0$ annehmen. (1) Es ist $c \in N_c(f)$: Klar. — (2) Aus $x \leqq z \leqq y$, $x, y \in N_c(f)$, $z \in V$, folgt $z \in N_c(f)$: Wegen $c \vee x \leqq c \vee z \leqq c \vee y$ und $c \wedge x \leqq c \wedge z \leqq c \wedge y$ gilt

$$0 = f(c \vee x) \leqq f(c \vee z) \leqq f(c \vee y) = 0 \text{ und } 0 = f(c \wedge x) \leqq f(c \wedge z) \leqq f(c \wedge y) = 0;$$

es ist also $f(c \wedge z) = f(c \vee z) = 0$ und somit $z \in N_c(f)$ nach Satz 4. 21. — (3) $N_c(f)$ ist ein Unterverband von $V$, d. h. aus $x, y \in N_c(f)$ folgt $x \vee y \in N_c(f)$ und $x \wedge y \in N_c(f)$. Nämlich: Wegen

$$c \leqq c \vee (x \wedge y) \leqq c \vee x \quad \text{und} \quad c \wedge x \leqq c \wedge (x \vee y) \leqq c$$

gilt $f(c \vee (x \wedge y)) = f(c \wedge (x \vee y)) = 0$ nach (2). Hieraus folgt unter Benutzung von (3. 1) und des distributiven Gesetzes:

$$f(c \wedge x \wedge y) = f(c \wedge x) + f(c \wedge y) - f(c \vee (x \vee y)) = 0,$$
$$f(c \vee x \vee y) = f(c \vee x) + f(c \vee y) - f(c \wedge (x \wedge y)) = 0.$$

Somit gilt $f(c \vee (x \vee y)) = f(c \wedge (x \vee y)) = 0$ und $f(c \vee (x \wedge y)) = f(c \wedge (x \wedge y)) = 0$, d. h. $x \vee y \in N_c(f)$ und $x \wedge y \in N_c(f)$. — Aus (1) bis (3) folgt die Behauptung.

**Satz 4. 23.** *Es sei $V = E$ ein Vektorverband über dem Skalarenkörper* K *von $W$ und $f$ eine lineare Abbildung von $E$ in $W$. Die Menge $N_0(f)$ der totalen Nullteile bezüglich $c = 0$ ist dann gleich der Menge aller Elemente $a \in E$, für welche aus $|x| \leqq |a|$ stets $f(x) = 0$ folgt. $N_0(f)$ ist ein gesättigter linearer Unterraum von $E$.*

Beweis. Gemäß der Definition dieser Nr. ist $N_0(f)$ die Menge aller $a \in E$ mit $f(x) = 0$ für $-a^- \leqq x \leqq a^+$, $x \in E$. Da $-|a| \leqq -a^- \leqq a^+ \leqq |a|$ und die Relationen $|x| \leqq |a|$ und $-|a| \leqq x \leqq |a|$ gleichwertig sind, gilt $a \in N_0(f)$, falls $f(x) = 0$ für alle $x \in E$ mit $|x| \leqq |a|$. Aus $a \in N_0(f)$ ist umgekehrt $f(x) = 0$ für alle $x \in E$ mit $|x| \leqq |a|$ zu folgern; wegen $x^+ \leqq |x|$, $x^- \leqq |x|$ und $f(x) = f(x^+) - f(x^-)$ kann hierbei o. B. d. A. $x \geqq 0$ angenommen werden. Es gilt dann $0 \leqq x \leqq |a| = a^+ + a^-$ und somit $x = x' + x''$ mit $0 \leqq x' \leqq a^+$ und $0 \leqq x'' \leqq a^-$ nach Nr. 1. 1. Hieraus folgt $f(x') = f(x'') = 0$, also $f(x) = 0$.

Nun zeigen wir, daß $N_0(f)$ ein linearer Unterraum von $E$ ist; aus der bereits bewiesenen Kennzeichnung der Elemente von $N_0(f)$ folgt dann, daß dieser Unterraum auch gesättigt ist. Es seien $a$ und $b$ Elemente von $N_0(f)$; wegen $|a + b| \leqq |a| + |b|$ genügt es für den Beweis von $(a + b) \in N_0(f)$, die Gleichung $f(x) = 0$ aus $|x| \leqq |a| + |b|$ zu folgern. Genau so wie im ersten Teil des Beweises begründet man, daß man hierbei $x \geqq 0$ annehmen darf. Dann ist aber $x$ von der Form $x = a' + b'$ mit $0 \leqq a' \leqq |a|$ und $0 \leqq b' \leqq |b|$; es folgt $f(a') = f(b') = 0$ und hieraus $f(x) = 0$. Schließlich ist mit $a$ auch $\lambda a$ bei beliebigem $\lambda \in$ K ein Element von $N_0(f)$. Für $\lambda = 0$ ist dies klar. Für $\lambda \neq 0$ folgt aus $|x| \leqq |\lambda a|$ nach (1. 11) $\left|\frac{1}{\lambda} x\right| \leqq |a|$, es ist also $f\left(\frac{1}{\lambda} x\right) = \frac{1}{\lambda} f(x) = 0$ und somit $f(x) = 0$.

**4. 3. Definition und Kennzeichnung $\underline{S}$-regulärer und $\underline{S}$-singulärer Bewertungen.** In dieser Nummer werden alle zu betrachtenden Verbände $V$ als *distributiv* vorausgesetzt. Es bedeutet wieder $S$ einen *gesättigten Unterverband* von $V$, $c$ ein Element von $S$ und $N_c(f)$ den *gesättigten Unterverband der c-totalen Nullteile* einer Bewertung $f \in \Phi_c(W, V)$.

Als wichtigste Eigenschaft der Bewertung $f_S$ hatte sich in Satz 4. 15 ergeben, daß die Abbildung $f \to f_S$ von $\Phi_c = \Phi_c(W, V)$ in sich eine *Projektion* ist. Nach Nr. 1. 3 wird von jeder Projektion in natürlicher Weise eine direkte ordnungstreue Zerlegung des Vektorverbandes erzeugt, in welchem sie erklärt ist. Dem Studium der von der Projektion $f \to f_S$ erzeugten direkten ordnungstreuen Zerlegung von $\Phi_c$ wenden wir uns nun zu.

**Definition.** *Eine Bewertung $f \in \Phi_c(W, V)$ heißt S-regulär bzw. S-singulär, wenn $f_S = 0$ bzw. $f_S = f$ ist.*

Aus Nr. 1. 3 folgt nun sofort:

**Zerlegungssatz.** *Die Menge der S-regulären Elemente sowie die Menge der S-singulären Elemente von $\Phi_c(W, V)$ ist ein Band in $\Phi_c$. Der vollständige Vektorverband $\Phi_c$ ist* *direkte ordnungstreue Summe dieser Bänder. Für beliebiges $f \in \Phi_c$ ist bei dieser Zerlegung die S-reguläre Komponente gleich $f - f_S$ und die S-singuläre Komponente gleich $f_S$.*

Auf Grund dieses Satzes bezeichnen wir von nun an die Bewertung $f - f_S$ bzw. $f_S$ als den $S$-regulären bzw. $S$-singulären Teil der Bewertung $f \in \Phi_c$.

Ist $B$ ein *beliebiges* Band in $\Phi_c(W, V)$, so folgt entweder aus einer entsprechenden Bemerkung in Nr. 1. 3 oder aus (4. 2) und (4. 3), daß mit $f$ auch $f_S$ ein Element von $B$ ist; m. a. W.: *Es induziert die direkte ordnungstreue Zerlegung von $\Phi_c$ in S-reguläre und S-singuläre Elemente eine direkte ordnungstreue Zerlegung in B, wenn man B als vollständigen Vektorverband auffaßt.* Insbesondere kann $B$ ein Band sein, welches aus den bezüglich eines (im Sinne von [1], S. 103—104) allgemeinen Stetigkeitsbegriffes $\mathfrak{g}$ stetigen Elementen von $\Phi_c$ besteht[9]). Dies bedeutet: Wenn $f$ $\mathfrak{g}$-stetig ist, so sind auch der $S$-reguläre und der $S$-singuläre Teil und der $S$-singuläre Teil von $f$ $\mathfrak{g}$-stetig. Ist $V$ speziell ein Boolescher Unterverband eines Booleschen $\sigma$-Verbandes $V'$ und $c = \emptyset$ das Nullelement von $V$, so sind für ein *bezüglich $V'$ total-additives* $f \in \Phi_\emptyset(W, V)$ auch der $S$-reguläre und der $S$-singuläre Teil total-additiv bezüglich $V'$ (vgl. die Bemerkung am Schluß der Nr. 3. 2).

Nunmehr wollen wir die $S$-regulären und $S$-singulären Elemente von $\Phi_c(W, V)$ auf verschiedene Arten kennzeichnen. Hierzu führen wir noch folgende Bezeichnung ein: Wir setzen für jedes $f \in \Phi_c$

$$R = R_c(f) = (N_c(f) : S)_c,$$

wobei $(N_c(f) : S)_c$ der in Nr. 2. 2 definierte $c$-Quotient von $N_c(f)$ nach $S$ ist. Nach (2. 1) und Satz 2. 21 ist $R_c(f)$ ein $c$ als Element enthaltender, gesättigter Unterverband von $V$.

**Satz** 4. 31. *Für jede Bewertung $f \in \Phi_c(W, V)$ sind folgende Aussagen gleichwertig:*

**(1)** *$f$ ist S-regulär.* — **(2)** *$f(s) = 0$ für alle $s \in S$.* — **(3)** *$S \leq N_c(f)$.* — **(4)** *$R_c(f) = V$.* — **(5)** *$f = f_R$, d. h. $f$ ist R-singulär ($R = R_c(f)$).*

Beweis. (1) → (2). Aus $f_S = 0$ folgt, nach dem Korollar zu 4. 13, $f(s) = 0$ für alle $s \in S$. — (2) → (3). Klar, da mit jedem $a \in S$ das Intervall $[c \wedge a, c \vee a]$ eine Teilmenge von $S$ ist. — (3) → (4). Aus (2. 5) und (2. 6) folgt $R \geq (N_c(f) : N_c(f))_c = V$, also $R = V$. — (4) → (5). Klar, da definitionsgemäß $f_V = f$ ist. — (5) → (1). Aus $f = f_R$ folgt $f_S = (f_R)_S = f_{S \cap R}$ nach (4. 7). Es ist aber $S \cap R$ nach dem Korollar zu 2. 21 eine Teilmenge von $N_c(f)$, d. h. es gilt $f(x) = 0$ für jedes $x \in S \cap R$. Nach dem Korollar zu 4. 13 folgt hieraus $f_{S \cap R} = 0$, also auch $f_S = 0$.

**Satz** 4. 32. *Jede S-singuläre Bewertung $f \in \Phi_c(W, V)$ erfüllt jede der folgenden, unter sich gleichwertigen Bedingungen:*

**(1)** $f_R = 0$. — **(2)** $R_c(f) = N_c(f)$. — **(3)** *Zu jedem $x \in V$ mit $f(x) \neq 0$ existiert ein Element $x' \in S$, für welches $f(x') \neq 0$ und entweder $c \leqq x' \leqq c \vee x$ oder $c \wedge x \leqq x' \leqq c$ ist.*

Beweis. Jedes $S$-singuläre $f$ erfüllt (1): Aus $f = f_S$ folgt $f_R = f_{S \cap R}$ nach (4. 7); wegen $S \cap R \leq N(f)$ ist also $f_R = 0$. — (1) → (2). Nach (4. 4) gilt $f_R(x) = f(x)$ für jedes $x \in R$; wegen $f_R = 0$ ist also $f(x) = 0$ für jedes $x \in R$. Da $R$ ein $c$ als Element enthaltender, gesättigter Unterverband von $V$ ist, bedeutet dies: $R \leq N_c(f)$. Nach (2. 1) gilt aber auch $N_c(f) \leq R$, also (2). — (2) → (3). Es sei $x \in V$ mit $f(x) \neq 0$; dann ist $x$ nicht Element von $N_c(f)$, also auch nicht von $R$. Es gibt somit mindestens ein $s \in S$, für welches entweder $c \vee (x \wedge s)$ oder $c \wedge (x \vee s)$ nicht Element von $N_c(f)$ ist. Dies bedeutet: Es existiert zu diesem Element $s$ mindestens ein $x' \in V$, so daß $f(x') \neq 0$ und entweder $c \leqq x' \leqq c \vee (x \wedge s)$ oder $c \wedge (x \vee s) \leqq x' \leqq c$ ist. Dieses Element $x'$ besitzt alle verlangten Eigenschaften. — (3) → (1). Wir nehmen an, es sei $f_R \neq 0$. Nach dem Korollar zu 4. 13 existiert dann in $R$

[9]) Vgl. hierzu auch eine später an anderer Stelle erscheinende Ergänzung zu [1].

mindestens ein Element $x$ mit $f(x) \neq 0$ und zu diesem nach (3) ein $x' \in S$, für welches $f(x') \neq 0$ und entweder $c \leqq x' \leqq c \vee x$ oder $c \wedge x \leqq x' \leqq c$ ist. Wegen $x \in R$ folgt $x' \in R$ aus jeder dieser Doppelungleichungen; es ist also $x'$ Element des Durchschnitts $S \frown R$, also auch Element von $N_c(f)$. Dann muß aber $f(x') = 0$ sein. — Widerspruch.

Nach (2. 1) gilt für den gesättigten Unterverband $R_c(f)$ von $V$ stets

$$N_c(f) < R_c(f) < V.$$

Gleichheit von $R_c(f)$ und $V$ ist nach Satz 4. 31 kennzeichnend dafür, daß $f$ $S$-regulär ist, also $f_S = 0$ gilt. Der andere Extremfall, in welchem $R_c(f)$ und $N_c(f)$ zusammenfallen, ist nach 4. 32 zwar gleichbedeutend mit $f_R = 0$, kennzeichnet aber im allgemeinen nicht die $S$-singulären Elemente $f \in \Phi_c$. Dies zeigt folgendes

**Beispiel.** Es sei $e$ das halboffene Einheitsintervall $[0, 1) = \{\xi;\ 0 \leqq \xi < 1\}$ auf der Zahlengeraden $R$. Der Verband $V$ bestehe aus der leeren Menge $\emptyset$ und allen Mengen $m$ der Gestalt

$$m = [\alpha_1, \beta_1) \smile [\alpha_2, \beta_2) \smile \cdots \smile [\alpha_n, \beta_n),$$

wobei $0 \leqq \alpha_1 < \beta_1 < \alpha_2 < \cdots < \beta_n \leqq 1$, $n \geqq 1$. Es ist $V$ ein *Boolescher Mengenverband* mit $e$ als Einheit. Auf dem abgeschlossenen Einheitsintervall $[0, 1]$ erklären wir eine reelle Punktfunktion $\varphi$:

$$\varphi(\xi) = \begin{cases} \xi, & 0 \leqq \xi \leqq \frac{1}{2} \\ \frac{1}{2} + \xi, & \frac{1}{2} < \xi \leqq 1 \end{cases}.$$

Mittels $\varphi$ definieren wir auf $V$ eine reelle Funktion $f$: Wir setzen $f(\emptyset) = 0$ und für jede nicht leere Menge $m \in V$ obiger Gestalt

$$f(m) = \sum_{i=1}^{n} (\varphi(\beta_i) - \varphi(\alpha_i)).$$

Dann ist $f$ ein Element von $\Phi_\emptyset(R, V)$ mit $f \geqq 0$; $f$ ist sogar *total-additiv*, da $\varphi$ in $[0, 1]$ linksseitig (von unten) stetig ist (vgl. [1], S. 114). Der Wert 0 wird von $f$ nur für die leere Menge $\emptyset$ angenommen, d. h. es ist $N_\emptyset(f) = \{\emptyset\}$. Das System $S$ aller Mengen $m \in V$, welche den Punkt $\frac{1}{2}$ nicht als Element enthalten, ist ein *Ideal* in $V$ (also ein gesättigter Unterverband mit $\emptyset$ als Element). Es gilt $R_\emptyset(f) = N_\emptyset(f) : S = \{\emptyset\}$, da in $V$ nur die leere Menge die Eigenschaft besitzt, mit allen Mengen von $S$ einen leeren Durchschnitt zu haben (vgl. Satz 2. 22). Somit ist zwar $R_\emptyset(f) = N_\emptyset(f)$, aber $f$ ist *nicht S-singulär*; es gilt vielmehr $f_S(e) = 1$ (nach (4. 5)) und $f(e) = \frac{3}{2}$.

Trotzdem kommt der Bewertung $f_R$ eine Sonderstellung innerhalb unserer Theorie zu, der wir uns jetzt zuwenden wollen.

**Satz** 4. 33. *Für jedes $f \in \Phi_c(W, V)$ ist die Bewertung $f_R$ S-regulär. Für jedes $f \geqq 0$ besteht zwischen $f_R$ und dem S-regulären Teil $f - f_S$ von $f$ die Relation*

$$(4.9) \qquad 0 \leqq f_R \leqq f - f_S.$$

Beweis. Wegen $S \frown R < N_c(f)$ gilt $(f_R)_S = f_{S \frown R} = 0$, also ist $f_R$ $S$-regulär. — Es sei nun $f \geqq 0$ und somit nach (4. 2) auch $f - f_S \geqq 0$. Da bei festem $f$ die Abbildung $g \to g_R$, $R = R_c(f)$, nach 4. 15 eine Projektion von $\Phi_c$ in sich ist, erhalten wir: $0 \leqq (f - f_S)_R \leqq f - f_S$ und $(f - f_S)_R = f_R - f_{S \frown R} = f_R$, also insgesamt $0 \leqq f_R \leqq f - f_S$.

**Satz** 4. 34. *Für jedes Element $x$ aus dem von $S$ und $R_c(f)$ in $V$ erzeugten gesättigten Unterverband $(S, R)$ gilt*

$$(4.10) \qquad f(x) = f_S(x) + f_R(x)$$

*bei beliebigem $f \in \Phi_c(W, V)$.*

Beweis. Wegen (4. 3) kann o. B. d. A. $f \geqq 0$ angenommen werden. Wegen der Gültigkeit von (3. 1) für $f$, $f_S$ und $f_R$ genügt es ferner, die Gleichung (4. 10) für Elemente $x \in (S, R)$ mit $x \geqq c$ oder $x \leqq c$ zu beweisen; wir behandeln nur den Fall $x \geqq c$, da für $x \leqq c$ duale Überlegungen zum Ziel führen. Zu $x \in (S, R)$ mit $x \geqq c$ existieren nach dem Satz von Nr. 2. 1 Elemente $s' \in S$ und $r' \in R$ mit $c \leqq x \leqq s' \vee r'$. Indem man erforderlichenfalls $s'$ durch $c \vee s'$ und $r'$ durch $c \vee r'$ ersetzt, kann man $s' \geqq c$ und $r' \geqq c$ erreichen.

Es folgt $x = s \vee r$ mit $c \leqq s = x \wedge s' \in S$ und $c \leqq r = x \wedge r' \in R$; somit ist $c \vee (r \wedge s) = r \wedge s$ ein Element von $N_c(f)$, also $f(r \wedge s) = 0$. Für $f_S(x)$ ergibt sich mittels (3. 1)

$$f_S(x) = f_S(s) + f_S(r) - f_S(r \wedge s);$$

nach (4. 4) ist hierbei $f_S(s) = f(s)$ und $f_S(r \wedge s) = f(r \wedge s) = 0$, ferner nach (4. 5) $f_S(r) = (W) \sup(f(x');\ c \leqq x' \leqq r,\ x' \in S) = 0$, weil für jedes $x' \in S$ mit $c \leqq x' \leqq r$ auch $x' \in S \cap R$ gilt und somit $x' \in N_c(f)$ nach dem Korollar zu 2. 21. Wir erhalten insgesamt $f_S(x) = f(s)$. Nun ist $x = s \vee r$ auch ein Element des von $R$ und $(N_c(f) : R)_c$ in $V$ erzeugten gesättigten Unterverbandes, da $S$ nach (2. 3) eine Teilmenge von $(N_c(f) : R)_c$ ist. Wir können daher im bisherigen Ergebnis $S$ durch $R$ ersetzen bei gleichzeitiger Vertauschung der Rollen von $s$ und $r$; so folgt $f_R(x) = f(r)$. Wegen $f(r \wedge s) = 0$ und der Gültigkeit der Funktionalgleichung (3. 1) folgt die Behauptung: $f(x) = f(s) + f(r) - f(r \wedge s) = f_S(x) + f_R(x)$.

**Korollar.** *Für eine Bewertung $f \in \Phi_c(W, V)$ sei der distributive Verband $V$ gleich dem von $S$ und $R_c(f)$ erzeugten gesättigten Unterverband. Dann ist $f_R$ der $S$-reguläre Teil von $f$, d. h. es gilt*

(4. 11) $$f = f_R + f_S.$$

*In diesem Fall ist $f$ genau dann $S$-singulär, wenn $f_R = 0$ ist.*

Der zweite Teil dieses Korollars bringt eine Ergänzung zu Satz 4. 32.

**4. 4. Darstellung des $S$-singulären Teils einer Bewertung mit Extremwerten auf $S$.** Der $S$-singuläre Teil $f_S$ einer Bewertung $f \in \Phi_c(W, V)$ nimmt unter geeigneten Zusatzannahmen über $f$ eine sehr einfache Form an, die im folgenden angegeben werden soll. Hierzu knüpfen wir wieder an die in Nr. 4. 3 eingeführten Bezeichnungen an; *die dort einleitend formulierten Voraussetzungen* (Distributivität von $V$ usw.) *sollen bestehen bleiben.*

**Definition.** *Wir sagen, eine Abbildung $t$ von $V$ in $W$ nimmt auf einer Teilmenge $Q$ von $V$ Extremwerte an, wenn Elemente $q', q'' \in Q$ existieren derart, daß $t(q') \leqq t(q) \leqq t(q'')$ für alle $q \in Q$ gilt. Es heißt $q'$ bzw. $q''$ Minimal- bzw. Maximalstelle von $t$ in $Q$ und $t(q')$ bzw. $t(q'')$ Minimum bzw. Maximum von $t$ in $Q$.*

**Satz 4. 41.** *Für eine Bewertung $f \in \Phi_c(W, V)$ sind folgende Aussagen gleichwertig:*

**(1)** *Es nehmen $f^+$ und $f^-$ auf $S$ Extremwerte an.*

**(2)** *Der $S$-singuläre Teil $f_S$ von $f$ ist mit Hilfe zweier nur von $f$ abhängigen Elemente $s' \leqq c$ und $s'' \geqq c$ darstellbar in der Form*

(4. 12) $$f_S(x) = f(c \wedge (x \vee s')) + f(c \vee (x \wedge s'')), \quad x \in V.$$

**Zusatz.** *Ist* (1) (*oder* (2)) *für $f \in \Phi_c(W, V)$ erfüllt, so kann man in* (4. 12) *das Element* $s' \leqq c$ bzw. $s'' \geqq c$ *als gemeinsame Minimal- bzw. Maximalstelle von $f^+$ und $f^-$ in $S$ wählen.*

Beweis. (1) → (2). Es existieren Elemente $s'_+, s''_+ \in S$ und $s'_-, s''_- \in S$ derart, daß $f^+(s'_+) \leqq f^+(s) \leqq f^+(s''_+)$ und $f^-(s'_-) \leqq f^-(s) \leqq f^-(s''_-)$ für alle $s \in S$ gilt. Wir setzen $s' = s'_+ \wedge s'_- \wedge c$ und $s'' = s''_+ \vee s''_- \vee c$; dann sind $s'$ und $s''$ Elemente von $S$ mit $s' \leqq c \leqq s''$. Wegen $f^+ \geqq 0$ und $f^- \geqq 0$ ist ferner $s'$ bzw. $s''$ eine gemeinsame Minimal- bzw. Maximalstelle von $f^+$ und $f^-$ in $S$. Wir zeigen, daß für jedes Paar von Elementen $s', s'' \in S$ mit den zuletzt genannten Eigenschaften die Gleichung

$$f_S^+(x) = f^+(c \wedge (x \vee s')) + f^+(c \vee (x \wedge s''))$$

bei beliebigem $x \in V$ erfüllt ist. Wendet man dieses Ergebnis auf die Bewertung $-f$ an, so folgt

$$f_S^-(x) = f^-(c \wedge (x \vee s')) + f^-(c \vee (x \wedge s'')) \text{ für jedes } x \in V$$

und hieraus (4. 12) wegen $f_S = f_S^+ - f_S^-$ und $f = f^+ - f^-$. Schließlich ist damit auch gleich der Zusatz bewiesen.

Zur Abkürzung setzen wir $p = f^+$. Dann liefert die Funktionalgleichung (3. 1)

$$p(y) = p(y \vee s'') + p(y \wedge s'') - p(s'') = p(y \wedge s'')$$

für jedes $y \in S$. Für ein Element $x \geqq c$ aus $V$ berechnen wir $p_S(x)$ gemäß (4. 5):

$$\begin{aligned} p_S(x) &= (W) \sup \left(p(x');\ c \leqq x' \leqq x,\ x' \in S\right) \\ &= (W) \sup \left(p(x' \wedge s'');\ c \leqq x' \leqq x,\ x' \in S\right) \\ &= (W) \sup \left(p(z);\ c \leqq z \leqq x \wedge s'' = p(x \wedge s'')\right); \end{aligned}$$

das letzte Gleichheitszeichen gilt, weil $x \wedge s''$ wegen $c \leqq x \wedge s'' \leqq s''$ ein Element von $S$ ist. Für ein Element $x \leqq c$ aus $V$ zeigt man analog $p_S(x) = p(x \vee s')$. Aus (3. 1), $s' \leqq c \leqq s''$ und dem distributiven Gesetz folgt dann die gesuchte Gleichung:

$$\begin{aligned} p_S(x) &= p_S(c \wedge x) + p_S(c \vee x) \\ &= p\left((c \wedge x) \vee s'\right) + p\left((c \vee x) \wedge s''\right) \\ &= p\left(c \wedge (x \vee s')\right) + p\left(c \vee (x \wedge s'')\right), \quad x \in V \text{ beliebig.} \end{aligned}$$

(2) → (1). Wir beweisen etwas mehr, nämlich folgenden

**Hilfssatz.** *Es seien $g, h \in \Phi_c(W, V)$ Bewertungen, ferner $s'$ und $s''$ Elemente aus $S$ mit $s' \leqq c \leqq s''$. Für jedes $x \in V$ sei*

$$g(x) = h(c \wedge (x \vee s')) + h(c \vee (x \wedge s'')).$$

*Dann gilt*

$$g^+(x) = h^+(c \wedge (x \vee s')) + h^+(c \vee (x \wedge s''))$$

*und*

$$g^+(s') = h^+(s') \leqq g^+(x) \leqq h^+(s'') = g^+(s'') \text{ für jedes } x \in V.$$

Wendet man die letzte Doppelungleichung des Hilfssatzes auf $h = f$ und $g = f_S$ sowie auf $h = -f$ und $g = (-f)_S = -f_S$ an, so folgt (1) bei Beachtung von (4. 4).

Beweis des Hilfssatzes. Es sei $x \geqq c$ ein Element aus $V$ und $c = x_0 \leqq x_1 \leqq \cdots \leqq x_n = x$ eine beliebige Einteilung von $[c, x]$. Dann ist $c = y_0 \leqq y_1 \leqq \cdots \leqq y_n = x \wedge s''$ mit $y_\nu = x_\nu \wedge s''$, $\nu = 0, 1, \ldots, n$, eine Einteilung von $[c, x \wedge s'']$. Wegen $g(y) = h(y \wedge s'')$ für $y \in V$ mit $c \leqq y$ erhalten wir

$$\sum_{\nu=1}^{n} (W) \sup \left(g(x_\nu) - g(x_{\nu-1}), 0\right) = \sum_{\nu=1}^{n} (W) \sup \left(h(y_\nu) - h(y_{\nu-1}), 0\right)$$

und hieraus $g^+(x) \leqq h^+(x \wedge s'')$ nach (3. 5) bis (3. 7). Ist andererseits

$$c = y_0 \leqq y_1 \leqq \cdots \leqq y_n = x \wedge s''$$

eine beliebige Einteilung von $[c, x \wedge s'']$, so ist $c = y_0 \leqq y_1 \leqq \cdots \leqq y_n = x \wedge s'' \leqq x$ eine Einteilung von $[c, x]$ und es gilt

$$\sum_{\nu=1}^{n} (W) \sup \left(h(y_\nu) - h(y_{\nu-1}), 0\right) \doteq \sum_{\nu=1}^{n} (W) \sup \left(g(y_\nu) - g(y_{\nu-1}), 0\right) + (W) \sup \left(g(x) - g(x \wedge s''), 0\right);$$

wegen $c \leqq y_\nu \leqq s''$ ist nämlich $g(y_\nu) = h(y_\nu)$, $\nu = 0, 1, \ldots, n$, wegen $c \leqq x$ ferner $g(x) = g(x \wedge s'')$. Wie oben erhalten wir $h^+(x \wedge s'') \leqq g^+(x)$; insgesamt folgt also $g^+(x) = h^+(x \wedge s'')$ für jedes $x \geqq c$ aus $V$. Analog ergibt sich $g^+(x) = h^+(x \vee s')$ für jedes $x \leqq c$ aus $V$ und hieraus wegen (3. 1)

$$\begin{aligned} g^+(x) &= g^+(c \wedge x) + g^+(c \vee x) \\ &= h^+\left((c \wedge x) \vee s'\right) + h^+\left((c \vee x) \wedge s''\right) \\ &= h^+\left(c \wedge (x \vee s')\right) + h^+\left(c \vee (x \wedge s'')\right). \end{aligned}$$

Wegen $s' \leqq c \wedge (x \vee s') \leqq c$ und $c \leqq c \vee (x \wedge s'') \leqq s''$ folgt hieraus noch

$$g^+(s') = h^+(s') \leqq g^+(x) \leqq g^+(s'') = h^+(s'') \text{ für jedes } x \in V.$$

Damit ist der Hilfssatz bewiesen.

**Bemerkungen. (1)** Besitzt der distributive Verband $V$ ein Nullelement $\emptyset$ und wählt man $c = \emptyset$, so ist $S$ ein Ideal in $V$ und in Satz 4. 41 können die Bedingungen (1) und (2) durch die folgenden ersetzt werden:

(1a) *Es nehmen $f^+$ und $f^-$ auf $S$ einen Maximalwert an*[10]).

(2a) *Der S-singuläre Teil $f_S$ von $f$ ist mit Hilfe eines nur von $f$ abhängigen Elementes $s \in S$ darstellbar in der Form*

(4. 12a) $$f_S(x) = f(x \wedge s), \quad x \in V.$$

Der Zusatz zu 4. 41 kann wie folgt abgeändert werden:

*Ist* (1a) (*oder* (2a)) *erfüllt, so kann man $s$ in* (4. 12a) *als gemeinsame Maximalstelle von $f^+$ und $f^-$ in $S$ wählen.*

**(2)** Der in (1) behandelte Fall liegt insbesondere dann vor, wenn $V$ ein Boolescher Verband ist und wenn man additive Abbildungen $f \in \Phi_a = \Phi_a(W, V)$ von $V$ in $W$ betrachtet. Hier gilt noch:

*Es nehmen $f^+$ und $f^-$ auf $S$ einen Maximalwert an, wenn $f \in \Phi_a$ auf $S$ Extremwerte besitzt.*

Beweis. Aus $f(s') \leqq f(s) \leqq f(s'')$, $s \in S$ beliebig, $s', s'' \in S$ fest, folgt nach (3. 9)

$$f^+(s) \leqq f^+(s'') = f(s'') \text{ für jedes } s \in S;$$

wegen $f^- = (-f)^+$ ist entsprechend $f^-(s) \leqq f^-(s') = -f(s')$ für jedes $s \in S$. Das Element $s_0 = s' \vee s''$ ist dann gemeinsame Maximalstelle von $f^+$ und $f^-$ auf $S$.

**(3)** Ist $V$ ein Boolescher $\sigma$-Verband, $S$ ein $\sigma$-Ideal in $V$ und $W$ der vollständige Vektorverband $R$ der reellen Zahlen, so sind die gleichwertigen Bedingungen (1a) und (2a) von (2) für jedes $f \in \Phi_a(R, V)$, d. h. für jede reelle, relativ beschränkte und auf $V$ additive Funktion erfüllt[11]).

**Satz 4. 42.** *Für eine Bewertung $f \in \Phi_c(W, V)$ sind folgende Aussagen gleichwertig:*

**(1)** *Es ist $f$ S-singulär, $f^+$ und $f^-$ nehmen auf $S$ Extremwerte an.*

**(2)** *Es existieren Elemente $s', s'' \in S$ mit $s' \leqq c \leqq s''$ derart, daß*

$$f(x) = f(c \wedge (x \vee s')) + f(c \vee (x \wedge s''))$$

*für alle $x \in V$ gilt.*

**(3)** *Es existieren Elemente $s', s'' \in S$ mit $s' \leqq c \leqq s''$ derart, daß $f(x) = f(s')$ für alle $x \leqq s'$ und $f(x) = f(s'')$ für alle $x \geqq s''$ aus $V$ gilt.*

Beweis. (1) → (2). Folgt aus 4. 41. — (2) → (3). Klar. — (3) → (1). Für beliebiges $x \in V$ folgt aus (3. 1)

$$f(x) = f(x \wedge s'') + f(x \vee s'') - f(s'') = f(x \wedge s'')$$

wegen $x \vee s'' \geqq s''$. Analog beweist man $f(x) = f(x \vee s')$ für jedes $x \in V$. Hieraus und aus (3. 1) folgt

$$f(x) = f(c \wedge (x \vee s')) + f(c \vee (x \wedge s''))$$

für jedes $x \in V$. Wendet man den Hilfssatz dieser Nr. 4. 4 auf $g = h = f$ und auf $g = h = -f$ an, so folgen die für alle $s \in S$ gültigen Ungleichungen

$$f^+(s') \leqq f^+(s) \leqq f^+(s'') \text{ und } f^-(s') \leqq f^-(s) \leqq f^-(s'').$$

---

[10]) Das Nullelement von $V$ ist gemeinsame Mimimalstelle von $f^+$ und $f^-$ auf $S$.

[11]) Nach *C. E. Rickart* [8], S. 657; der dort für einen Booleschen $\sigma$-Mengenverband geführte Beweis läßt sich leicht auf den Fall eines beliebigen Booleschen $\sigma$-Verbandes $V$ übertragen. Vgl. diesbezüglich *Haupt-Aumann-Pauc* [6], Nr. 5. 2. 2. Wendet man das Rickartsche Ergebnis auf das $\sigma$-Ideal $S = V$ an, so folgt übrigens, daß die Funktionen $f \in \Phi_a(R, V)$ auf $V$ *global* beschränkt sind.

Hiernach nehmen $f^+$ und $f^-$ auf $S$ Extremwerte an; es ist $s'$ bzw. $s''$ gemeinsame Minimal- bzw. Maximalstelle von $f^+$ und $f^-$ auf $S$. Nach 4. 41 und dem hierzu gehörigen Zusatz folgt

$$f_S^+(x) = f^+(c \wedge (x \vee s')) + f^+(c \vee (x \wedge s'')),$$
$$f_S^-(x) = f^-(c \wedge (x \vee s'')) + f^-(c \vee (x \wedge s'))$$

bei beliebigem $x \in V$. Es ist also

$$f_S(x) = f(c \wedge (x \vee s')) + f(c \vee (x \wedge s''))$$

für jedes $x \in V$ und somit $f_S = f$, d. h. $f$ ist $S$-singulär.

**Bemerkungen. (1)** Wenn (1) in 4. 42 erfüllt ist, besitzen $f^+$ und $f^-$ auf $V$ die gleichen Extremwerte wie auf $S$. Dies folgt aus dem Hilfssatz dieser Nr. 4. 4, angewandt auf die Fälle $g = h = f$ und $g = h = -f$.

**(2)** Ist $c$ das Nullelement $\emptyset$ von $V$, so vereinfachen sich die Aussagen (1) bis (3) in Satz 4. 42 insofern, als dann $s' = c = \emptyset$ ist und die $s'$ betreffenden Aussagen und Forderungen gestrichen werden können. Der Hinweis auf den in Bemerkung (1) zu 4. 41 behandelten analogen Fall mag hier genügen.

**(3)** Ist $V$ ein Boolescher Verband und $c = \emptyset$, so ist die Aussage (3) in 4. 42 mit der folgenden gleichwertig: *Es existiert ein Element $s'' \in S$ derart, daß $f(x) = 0$ ist für alle $x \in V$ mit $x \wedge s'' = \emptyset$.*

**4. 5. Zusammenhänge mit der Theorie von Hahn-Rosenthal.** Aus unserer Zerlegungstheorie lassen sich durch Spezialisierung alle wesentlichen Ergebnisse der Hahn-Rosenthalschen Theorie in [5], S. 40—43 gewinnen. Wir befinden uns im Hahn-Rosenthalschen Fall, wenn folgende, unter (HR) zusammengefaßte Voraussetzungen erfüllt sind:

(HR) $V$ = *Boolescher $\sigma$-Verband*[12]),
$c$ = *Nullelement $\emptyset$ aus $V$*,
$S$ = *$\sigma$-Ideal in $V$*,
$W$ = *vollständiger Vektorverband $R$ der reellen Zahlen*,
$f$ = *reelle, auf $V$ total-additive Fuhktion*[13]).

Die über $V$ definierten, reellen, total-additiven Funktionen $f$ brauchen dabei nicht als relativ beschränkt vorausgesetzt zu werden, da nach einem Satz von H. Hahn[14]) jede solche Funktion auf $V$ ein *absolutes Maximum und Minimum* annimmt, insbesondere also auf $V$ *global beschränkt* ist.

Wir erwähnten eben, daß wir durch Spezialisierung der *wesentlichen* Ergebnisse von Hahn-Rosenthal aus unserer Theorie gewinnen können. Genauer gesagt, erhalten wir *alle* Ergebnisse von Hahn-Rosenthal, soweit sie sich auf *endliche* reelle Funktionen beziehen. In [5] werden nämlich einige Sätze für total-additive Funktionen bewiesen, welche auch die uneigentlichen reellen Zahlen $+\infty$ und $-\infty$ als Werte annehmen können. An entscheidenden Stellen wird aber auch in [5] die Endlichkeit der Funktionen gefordert; dort angeführte Beispiele[15]) zeigen gerade, daß diese Forderung unvermeidbar wird, wenn man nicht auf die Eindeutigkeit der Zerlegung in $S$-regulären und $S$-singulären Teil verzichten will.

Um die Zusammenhänge zwischen unserer Darstellung und der von Hahn-Rosenthal noch klarer hervortreten zu lassen, sind noch einige Bemerkungen nötig. Bei Hahn-

[12]) Genau genommen bedeutet $V$ in [5] stets einen Booleschen $\sigma$-*Mengen*verband.

[13]) Es wird also nur die im Band $\Phi^\sigma$ der total-additiven Funktionen $f \in \Phi_\theta$ induzierte, direkte ordnungstreue Zerlegung in $S$-regulären und $S$-singulären Teil untersucht. Vgl. eine diesbezügliche Bemerkung nach dem Zerlegungssatz in Nr. 4. 3.

[14]) Vgl. *Hahn-Rosenthal* [5], S. 17; der dort für Boolesche $\sigma$-Mengenverbände geführte Beweis bleibt auch für beliebige Boolesche $\sigma$-Verbände gültig. Vgl. diesbezüglich auch *Haupt-Aumann-Pauc* [6], Nr. 5. 2, Satz 4.

[15]) Vgl. *Hahn-Rosenthal* [5], S. 42.

Rosenthal ist der $S$-reguläre Teil der total-additiven Funktion $f$ stets gleich $f_R$ im Sinne der Bezeichnungen von Nr. 4. 3. Unter den relativ schwachen Voraussetzungen der Nr. 4. 3 konnten wir im Korollar zu Satz 4. 43 nur eine hinreichende Bedingung dafür angeben, daß dieser Fall eintritt. Diese Bedingung ist unter den Voraussetzungen (HR) erfüllt. Um dies zu zeigen, beweisen wir zunächst zwei Sätze allgemeinerer Natur:

**Hilfssatz.** *Es seien $J$ und $S$ Ideale in einem Booleschen Verband $V$. Es ist $V$ gleich dem von $S$ und dem Idealquotienten $R = J : S$ erzeugten Ideal*[16]*), wenn in $S$ ein modulo $J$ größtes Element $s_0$ existiert; das ist ein Element $s_0 \in S$, für welches bei beliebigem $s \in S$ das Komplement $s - s_0$ zu $J$ gehört.*

Beweis. Nach dem Satz der Nr. 2. 1 genügt es zu zeigen, daß jedes Element $x \in V$ die Gestalt $x = s \vee r$ mit $s \in S$ und $r \in R$ hat. Ist $s_0$ ein modulo $J$ größtes Element von $S$, so setzen wir $s = x \wedge s_0$ und $r = x - s_0$. Da $S$ ein Ideal ist, gehört $s$ zu $S$; gemäß Satz 2. 22 bleibt zu zeigen, daß $r \wedge s' \in J$ für jedes $s' \in S$ gilt. Dies folgt wegen $r \wedge s' = x \wedge s' - s_0$ und $x \wedge s' \in S$ aus der Maximaleigenschaft von $s_0$.

**Satz.** *Es sei $V$ ein Boolescher Verband und $S$ ein Ideal in $V$; die additive Abbildung $f \in \Phi_s(W, V)$ nehme auf $S$ Extremwerte an. Dann ist $V$ gleich dem von den Idealen $S$ und $R_s(f) = N_s(f) : S$ erzeugten Ideal*[17]*).*

Beweis. Nach der Bemerkung (2) zu Satz 4. 41 besitzen $f^+$ und $f^-$ auf $S$ eine gemeinsame Maximalstelle $s_0 \in S$; wegen $|f| = f^+ + f^-$ ist $s_0$ auch Maximalstelle von $|f|$ auf $S$. Gemäß dem vorausgeschickten Hilfssatz genügt es zu zeigen, daß $s_0$ modulo $N_s(f)$ ein größtes Element von $S$ ist; wegen $N_s(f) = N_s(|f|)$ (Satz 4. 21) ist dies gleichbedeutend mit $|f|(s - s_0) = 0$ für jedes $s \in S$. Dies folgt so: Es ist

$$|f|(s - s_0) = |f|(s \vee s_0) - |f|(s_0);$$

wegen $|f| \geqq 0$ ist diese Differenz einerseits $\geqq 0$, wegen der Maximaleigenschaft von $s_0$ bezüglich $|f|$ andererseits $\leqq 0$, also ist $|f|(s - s_0) = 0$.

Wir kehren jetzt wieder zurück zum Hahn-Rosenthalschen Fall: Unter den Voraussetzungen (HR) ist $S$ ein $\sigma$-Ideal, also selbst ein Boolescher $\sigma$-Verband. Die Verengerung von $f$ auf $S$ ist total-additiv, erfüllt also nach dem Satz von H. Hahn die Voraussetzungen des Satzes dieser Nr. 4. 5; nach dem Korollar zu Satz 4. 34 ist daher für jedes $f$ im Sinne von (HR) *der $S$-reguläre Teil gleich $f_R$*. Nach der Bemerkung (2) zu 4. 41 existiert ein Element $s_0 \in S$ derart, daß $f_S(x) = f(x \wedge s_0)$ für alle $x \in V$ gilt. Aus 4. 42 und der zugehörigen Bemerkung (3) ergibt sich die in Ziff. V der Einleitung als Definition verwendete Kennzeichnung der $S$-singulären Funktionen: *Bei Gültigkeit von* (HR) *ist $f$ dann und nur dann $S$-singulär, wenn ein Element $s_0 \in S$ existiert derart, daß $f(x) = 0$ für alle $x \in V$ mit $x \wedge s_0 = \emptyset$ gilt.* Aus (HR), 4. 31 und 4. 32 folgt schließlich noch: *Es ist $f$ dann und nur dann $S$-regulär bzw. $S$-singulär, wenn $R_s(f) = V$ bzw. $R_s(f) = N_s(f)$ ist*[18]*).*

[16]) Gleichbedeutend ist die Aussage: $V$ ist gleich dem von $S$ und $R$ erzeugten gesättigten Unterverband. In einem Verband mit Nullelement sind nämlich die das Nullelement enthaltenden gesättigten Unterverbände identisch mit den Idealen (vgl. Nr. 2. 1).

[17]) Das Beispiel in Nr. 4. 3 lehrt übrigens, daß man auf die Forderung der Existenz von Extremwerten von $f$ auf $S$ im allgemeinen *nicht* verzichten kann.

[18]) Durch diese beiden Gleichungen werden bei *Hahn-Rosenthal* [5], S. 40, die $S$-regulären und $S$-singulären total-additiven Funktionen definiert. Nach 4. 31 ist auch in unserer allgemeinen Theorie das Bestehen der Gleichung $R_c(f) = V$ charakteristisch für die $S$-regulären Elemente von $\Phi_c(W, V)$. Die so definierten $S$-regulären Bewertungen bilden ein Band in $\Phi_c$; das zugehörige komplementäre Band ist eindeutig bestimmt. Das Beispiel in Nr. 4. 3 zeigt daher jetzt noch folgendes: Will man versuchen, durch die Bedingung $R_c(f) = N_c(f)$ oder die hiermit gleichwertige Aussage $f_R = 0$ die $S$-singulären Bewertungen zu definieren, so erhält man im neuen Sinne $S$-singuläre Elemente, die im allgemeinen nicht zu dem zum Band der $S$-regulären Bewertungen komplementären Band gehören. Ein dem unseren entsprechender Zerlegungssatz kann daher bei Zugrundelegung dieser Definition für die $S$-singulären Bewertungen aus den in Nr. 1. 3 dargelegten, verbandsalgebraischen Gründen nicht gelten.

**4. 6. $S$-reguläre und $S$-singuläre lineare Abbildungen.** Wir bezeichnen mit $E$ einen *Vektorverband* über dem Skalarenkörper K von $W$; im übrigen behalten wir die in Nr. 3. 3 eingeführten Bezeichnungen bei. Dort hatten wir u. a. gezeigt (Satz 3. 34), daß der Raum $\Lambda(W, E)$ der relativ beschränkten, linearen Abbildungen von $E$ in $W$ ein linearer Unterraum von $\Phi_0(W, E)$ ist. Gesucht wird nun eine Übertragung der in 4. 3 für Bewertungen entwickelten Zerlegungstheorie auf lineare Abbildungen; dies soll genauer bedeuten: Gesucht wird eine analoge Zerlegungstheorie in $\Lambda$, die auf die Einbettung von $\Lambda$ in $\Phi_0$ keinen Bezug mehr nimmt.

Es ist zu erwarten, daß hierbei an die Menge $S$, die zunächst nur ein gesättigter Unterverband war, zusätzliche Forderungen gestellt werden müssen. *Wir setzen voraus, daß $S$ ein gesättigter, linearer Unterraum von $E$ ist.* Die in 4. 3 ausgesprochenen Sätze sind grundsätzlich auf Elemente von $\Lambda(W, E)$ anwendbar, da $E$ ein distributiver Verband ist. Man beachte, daß im folgenden stets $c = 0$ gesetzt wird.

Für eine lineare Abbildung $f \in \Lambda$ bezeichnen wir mit $f_S$ zunächst die gemäß 4. 1 gebildete *Bewertung* aus $\Phi_0(W, E)$. Nach Satz 4. 16 ist $f_S$ dann wieder ein Element von $\Lambda$ und kann ohne Bezugnahme auf $\Phi_0$ definiert werden: Für ein $f \geqq 0$ aus $\Lambda$ ist $f_S$ nach 4. 12 das Minimum aller $h \geqq 0$ aus $\Lambda$, für welche $h(x) = f(x)$ bei beliebigem $x \in S$ gilt. Für ein beliebiges $f \in \Lambda$ ist $f_S = (f^+)_S - (f^-)_S$. Es ergibt sich nun bereits der entscheidende Satz:

**Satz** 4. 61. *Für einen gesättigten, linearen Unterraum $S$ von $E$ ist die Abbildung $f \to f_S$ von $\Lambda(W, E)$ in sich eine Projektion. Der vollständige Vektorverband $\Lambda$ ist direkte ordnungstreue Summe des Bandes seiner $S$-regulären Elemente (das sind die Elemente mit $f_S = 0$) und des Bandes seiner $S$-singulären Elemente (das sind die Elemente mit $f_S = f$).*

Der Beweis folgt aus 4. 15, 4. 16 und den Bemerkungen in Nr. 1. 3.

In Satz 4. 23 hatte sich ergeben, daß die Menge $N_0(f)$ der totalen Nullteile bezüglich $c = 0$ einer linearen Abbildung $f \in \Lambda$ ein gesättigter, linearer Unterraum von $E$ ist. Dann ist aber nach 2. 23 auch der Quotient $R_0(f) = (N_0(f) : S)_0$ ein gesättigter, linearer Unterraum von $E$. Für jedes $f \in \Lambda$ gehört damit auch $f_R$ zu $\Lambda$. Dies hat zur Folge, daß alle Sätze der Nr. 4. 3 auch als Sätze über lineare Abbildungen $f \in \Lambda(W, E)$ gelesen werden können. Insbesondere liefern alle dort angegebenen Kennzeichnungen $S$-regulärer und $S$-singulärer Bewertungen entsprechende Kennzeichnungen für $S$-reguläre und $S$-singuläre lineare Abbildungen aus $\Lambda$[19]).

Bemerkung. Die im Korollar zu Satz 4. 34 formulierte hinreichende Bedingung dafür, daß $f_R$ der $S$-reguläre Teil von $f$ ist, kann für den Fall einer linearen Abbildung $f \in \Lambda(W, E)$ so ausgesprochen werden: Es ist $E$ der von den gesättigten linearen Unterräumen $N_0(f)$ und $R_0(f)$ erzeugte gesättigte Unterverband. *Diese Bedingung ist insbesondere dann erfüllt, wenn zu $S$ ein komplementärer linearer Unterraum existiert*; das ist nach Nr. 1. 3 ein gesättigter linearer Unterraum $T$ derart, daß $E$ direkte ordnungstreue Summe von $S$ und $T$ ist. Sind $x_S$ und $x_T$ die Komponenten eines beliebigen $x \in E$ bei dieser Zerlegung, so sind nach Nr. 1. 3 diese Komponenten zueinander disjunkt und es gilt (vgl. (1. 13b) und (1. 17))

$$\inf(-(x^-)_S, -(x^-)_T) = -x^- \leqq x \leqq x^+ = \sup((x^+)_S, (x^+)_T).$$

Nach dem Satz der Nr. 2. 1 ist dann $E$ der von $S$ und $T$ erzeugte gesättigte Unterverband. Da weiter jedes Element von $S$ zu jedem Element von $T$ disjunkt ist, gilt $T = (\{0\} : S)_0$ nach Satz 2. 23, wenn $\{0\}$ der nur aus dem Nullvektor 0 bestehende Unterraum von $E$ ist. Es ist daher $T$ eine Teilmenge von $R_0(f) = (N_0(f) : S)_0$ und die Bedingung von Satz 4. 34 erfüllt. Wie im Beweis von 4. 34 zeigt man, daß der $S$-reguläre Teil $f_R$ von $f$ gleich $f_T$ ist. Hieraus folgt noch: Es ist $f$ dann und nur dann $S$-regulär, wenn $T$ Teilmenge von $N_0(f)$ ist. Besitzt schließlich auch $N_0(f)$ einen komplementären linearen Unterraum $M$ in $E$, so ist die Bedingung $T \leqq N_0(f)$ gleichwertig mit der folgenden: $T \cap M = \{0\}$[20]).

[19]) Eine Interpretation der Sätze 4. 41 und 4. 42 für den Fall linearer Funktionale ist gegenstandslos, da $f^+$ und $f^-$ für ein $f \in \Lambda(W, E)$ dann und nur dann Extremwerte auf $S$ annehmen, wenn $f(s) = 0$ ist für jedes $s \in S$, d. h. wenn $f$ $S$-regulär ist. Der Beweis dieser Behauptung darf dem Leser überlassen bleiben.

[20]) Vgl. zu dieser Bemerkung *N. Bourbaki* [4], S. 38, Exercice 4.

## § 5. Zerlegungen von Lebesgueschen Typus.

**5. 1. $g$-stetige und $g$-rein-unstetige Bewertungen.** Im folgenden bezeichnen wir mit $W$ wieder einen *vollständigen Vektorverband über einem linear geordneten Körper* K. Ferner bedeute $V$ einen *Verband*, der zunächst *beliebig* sein kann, insbesondere also nicht distributiv zu sein braucht.

**Definition.** *Eine Bewertung $f \in \Phi_c(W, V)$ heißt stetig bzw. rein-unstetig bezüglich einer Bewertung $g \in \Phi_c$, kurz g-stetig bzw. g-rein-unstetig, wenn $f$ ein Element des von $g$ in $\Phi_c$ erzeugten Bandes $[g]$ bzw. wenn $f$ ein Element des zu $[g]$ komplementären Bandes ist, d. h. wenn $f$ und $g$ in $\Phi_c$ disjunkt sind.*

Aus dieser Definition folgt nach Nr. 1. 3 sofort:

**Lebesguescher[21]) Zerlegungssatz für Bewertungen.** *Der vollständige Vektorverband $\Phi_c(W, V)$ ist für jedes $g \in \Phi_c$ direkte ordnungstreue Summe des Bandes seiner g-stetigen und des Bandes seiner g-rein-unstetigen Elemente.*

Die Formeln (1. 19) und (1. 20), auf unseren Fall angewendet, besagen: Der vermöge der Lebesgue-Zerlegung eindeutig bestimmte *g-stetige Teil* einer Bewertung $f \geqq 0$ aus $\Phi_c(W, V)$ ist gleich

(5. 1) $\quad (\Phi_c) \sup (u_n; n = 1, 2, \ldots)$, wobei $u_n = (\Phi_c) \inf (f, n \mid g \mid)$, $n = 1, 2, \ldots$;

der *g-rein-unstetige Teil* von $f \geqq 0$ ist gleich

(5. 2) $\quad (\Phi_c) \inf (v_n; n = 1, 2, \ldots)$, wobei $v_n = (f - n \mid g \mid)^+$, $n = 1, 2, \ldots$.

Hierbei gilt

(5. 3) $$u_n \leqq u_{n+1} \text{ und } v_{n+1} \leqq v_n \text{ für jedes } n = 1, 2, \ldots.$$

Formel (1. 13c) besagt: *Es ist $f \in \Phi_c(W, V)$ dann und nur dann g-rein-unstetig, wenn*

(5. 4) $$|f| = (|f| - |g|)^+.$$

Bei den nun folgenden Kennzeichnungen $g$-stetiger und $g$-rein-unstetiger Bewertungen $f \in \Phi_c(W, V)$ werden wir stets $f \geqq 0$ und $g \geqq 0$ voraussetzen. Diese Annahme bedeutet keine Beschränkung der Allgemeinheit, weil das von $g$ in $\Phi_c$ erzeugte Band gleich dem von $|g|$ in $\Phi_c$ erzeugten Band ist und es sich bei der Lebesgue-Zerlegung um eine direkte ordnungstreue Zerlegung von $\Phi_c$ handelt.

In den beiden nächsten Sätzen empfiehlt sich folgende, abkürzende Schreibweise: *Wir setzen $f(I) = f(b) - f(a)$, wenn $f \in \Phi_c(W, V)$ und $I = [a, b]$ ein Intervall in $V$ ist.* Es gilt $f(I) \geqq 0$ für jedes Intervall $I$ genau dann, wenn $f \geqq 0$ ist.

**Satz** 5. 11. *Es sei $W$ der vollständige Vektorverband $R$ der reellen Zahlen; $f$ und $g$ seien Bewertungen aus $\Phi_c(W, V)$ mit $f \geqq 0$ und $g \geqq 0$. Es ist $f$ dann und nur dann g-stetig, wenn zu jedem reellen $\varepsilon > 0$ und zu jedem $x \in V$ mit $x > c$ oder $x < c$ ein reelles $\delta = \delta(\varepsilon, x) > 0$ existiert derart, daß aus*

$$\sum_{\nu=1}^{n} g(I_\nu) \leqq \delta \quad \textit{stets} \quad \sum_{\nu=1}^{n} f(I_\nu) \leqq \varepsilon$$

*folgt, wenn hierbei $I_\nu = [x_\nu, y_\nu]$, $\nu = 1, 2, \ldots, n$, Intervalle in $V$ sind mit*

$$c \leqq x_1 < y_1 < x_2 < \cdots < y_n \leqq x \quad \textit{bzw.} \quad x \leqq x_1 < y_1 < x_2 < \cdots < y_n \leqq c.$$

Beweis. *Vorbemerkung.* Wir benützen die in (5. 2) angegebene Darstellung des $g$-rein-unstetigen Teils von $f$ und die dortigen Bezeichnungen. Aus (5. 3) und (3. 8) folgt dann: *Es ist $f$ dann und nur dann g-stetig, wenn* $\lim v_n(x) = 0$ *ist für jedes $x \in V$.*

[21]) Eine Rechtfertigung für diese Bezeichnung gaben wir bereits in Ziffer VIIb der Einleitung.

*Nur dann*: $f$ ist $g$-stetig, also gilt $\lim v_n(x) = 0$ für jedes $x \in V$. Zu $x > c$ aus $V$ und jeder reellen Zahl $\varepsilon > 0$ existiert dann eine natürliche Zahl $m_0 = m_0(\varepsilon, x)$ mit $v_{m_0}(x) \leqq 2^{-1}\varepsilon$. Für jede Einteilung $c = z_0 \leqq z_1 \leqq \cdots \leqq z_m = x$ gilt also nach (3. 5) bis (3. 7)

$$\sum_{i=1}^{n} \sup\,(f(I_i') - m_0 g(I_i'), 0) \leqq 2^{-1}\varepsilon, \quad I_i' = [z_{i-1}, z_i],\ i = 1, 2, \ldots, m.$$

Setzen wir $\delta = (2m_0)^{-1}\varepsilon$, so hängt $\delta$ nur von $\varepsilon$ und $x$ ab. Ferner seien $I_\nu = [x_\nu, y_\nu]$, $\nu = 1, 2, \ldots, n$, Teilintervalle von $[c, x]$ mit der in der Behauptung formulierten Eigenschaft; schließlich gelte noch

$$\sum_{\nu=1}^{n} g(I_\nu) \leqq \delta.$$

Da die Elemente $x_\nu, y_\nu$ eine Einteilung von $[c, x]$ definieren, folgt nach dem soeben Festgestellten

$$\sum_{\nu=1}^{n} f(I_\nu) - m_0 \sum_{\nu=1}^{n} g(I_\nu) \leqq \sum_{\nu=1}^{n} \sup\,(f(I_\nu) - m_0 g(I_\nu), 0) \leqq 2^{-1}\varepsilon$$

und hieraus

$$\sum_{\nu=1}^{n} f(I_\nu) \leqq 2^{-1}\varepsilon + m_0\delta = \varepsilon,$$

was gezeigt werden sollte. Im Falle $x < c$ schließt man analog.

*Dann*: Es genügt zu zeigen, $\lim v_n(x) = 0$ für alle $x \in V$ mit $x > c$ oder $x < c$; nach Satz 3. 14 folgt dann $\lim v_n(x) = 0$ für jedes $x \in V$. Wir behandeln nur den Fall $x > c$; für ein $x < c$ schließt man analog. Nach Annahme existiert zu beliebigem $\varepsilon > 0$ und zu $x$ ein $\delta = \delta(\varepsilon, x)$ mit der in der Behauptung formulierten Eigenschaft; $m_0$ sei eine natürliche Zahl $\geqq \delta^{-1} f(x)$. Wir zeigen $v_{m_0}(x) \leqq \varepsilon$. Für jede Einteilung $\mathfrak{t}$ mit

$$c = z_0 < z_1 < \cdots < z_n = x$$

von $[c, x]$ setzen wir

$$P_0(x, \mathfrak{t}) = \sum_{\nu=1}^{n} \sup\,(f(I_\nu) - m_0 g(I_\nu), 0), \quad I_\nu = [z_{\nu-1}, z_\nu],\ \nu = 1, 2, \ldots, n.$$

Unter den Intervallen $I_\nu$ seien $I_\varkappa'$, $\varkappa = 1, 2, \ldots, k$, diejenigen mit $f(I_\varkappa') - m_0 g(I_\varkappa') \geqq 0$. Für $v_{m_0}(x) = 0$ ist nichts zu beweisen. Da $v_{m_0}(x)$ gleich dem über alle Einteilungen gebildeten Supremum der $P_0(x, \mathfrak{t})$ ist, können wir o. B. d. A. annehmen, daß $P_0(x, \mathfrak{t}) > 0$ ist, also in der Einteilung $\mathfrak{t}$ solche Intervalle $I_\varkappa'$ existieren. Wäre nun $\sum_{\varkappa=1}^{k} g(I_\varkappa') > \delta$, so würde folgen:

$$P_0(x, \mathfrak{t}) = \sum_{\varkappa=1}^{k} f(I_\varkappa') - m_0 \sum_{\varkappa=1}^{k} g(I_\varkappa') < \sum_{\varkappa=1}^{k} f(I_\varkappa') - m_0\delta \leqq \sum_{\varkappa=1}^{k} f(I_\varkappa') - f(x) \leqq 0,$$

was der Annahme $P_0(x, \mathfrak{t}) > 0$ widerspricht. Es ist also $\sum_{\varkappa=1}^{k} g(I_\varkappa') \leqq \delta$ und damit auch $\sum_{\varkappa=1}^{k} f(I_\varkappa') \leqq \varepsilon$. Wir erhalten

$$P_0(x, \mathfrak{t}) = \sum_{\varkappa=1}^{k} f(I_\varkappa') - m_0 \sum_{\varkappa=1}^{k} g(I_\varkappa') \leqq \sum_{\varkappa=1}^{k} f(I_\varkappa') \leqq \varepsilon;$$

nach (3. 6) und (3. 7) folgt $v_{m_0}(x) \leqq \varepsilon$ und nach (5. 3) hieraus $0 \leqq v_n(x) \leqq v_{m_0}(x) \leqq \varepsilon$ für jedes $n \geqq m_0$, also $\lim v_n(x) = 0$.

**Satz** 5. 12. *Unter den Voraussetzungen von Satz* 5. 11 *gilt: Es ist $f$ dann und nur dann $g$-rein unstetig, wenn zu jedem reellen $\varepsilon > 0$ und zu jedem $x \in V$ mit $x > c$ oder $x < c$, sowie mit $f(x) \neq 0$ und $g(x) \neq 0$ eine Einteilung $c = x_0 < x_1 < \cdots < x_n = x$ von $[c, x]$*

bzw. $x = x_0 < x_1 < \cdots < x_n = c$ *von* $[x, c]$ *derart existiert, daß die Menge* $N = \{1, 2, \ldots, n\}$ *Vereinigung zweier nicht leeren, fremden Teilmengen* $N'$ *und* $N''$ *ist mit*

$$\sum_{\nu \in N'} f(I_\nu) \leqq \varepsilon \text{ und } \sum_{\nu \in N''} g(I_\nu) \leqq \varepsilon,\ I_\nu = [x_{\nu-1}, x_\nu],\ \nu = 1, 2, \ldots, n.$$

Beweis. *Vorbemerkung.* Da $f$ und $(f - g)^+$ Elemente von $\Phi_c$ sind, folgt aus (5. 3) und Satz 3. 14: *$f$ ist dann und nur dann g-rein-unstetig, wenn $f(x) = (f - g)^+(x)$ für jedes $x \in V$ mit $x > c$ oder $x < c$ gilt.*

*Nur dann*: Wir behandeln nur den Fall $x > c$; für $x < c$ führt man den Beweis analog. Da $f(x)$ und $g(x)$ ungleich Null, also beide $> 0$ sind, kann o. B. d. A. $0 < \varepsilon < \text{Min}\,(f(x), g(x))$ angenommen werden. Aus $f(x) = (f - g)^+(x)$ folgt die Existenz einer Einteilung $c = x_0 < x_1 < \cdots < x_n = x$ von $[c, x]$ mit

$$f(x) - \varepsilon \leqq \sum_{\nu=1}^{n} \sup\,(f(I_\nu) - g(I_\nu), 0),\ I_\nu = [x_{\nu-1}, x_\nu],\ \nu = 1, 2, \ldots, n.$$

Es sei $N'$ bzw. $N''$ die Menge aller $\nu = 1, 2, \ldots, n$ mit $f(I_\nu) - g(I_\nu) < 0$ bzw. mit $f(I_\nu) - g(I_\nu) \geqq 0$. Wegen $0 < \varepsilon < \text{Min}\,(f(x), g(x))$ sind $N'$ und $N''$ nicht leer; die beiden Mengen sind fremd und ihre Vereinigung ist $N$. Es gilt

$$f(x) = \sum_{\nu \in N'} f(I_\nu) + \sum_{\nu \in N''} f(I_\nu) \text{ sowie } \sum_{\nu=1}^{n} \sup\,(f(I_\nu) - g(I_\nu), 0) = \sum_{\nu \in N''} f(I_\nu) - \sum_{\nu \in N''} g(I_\nu).$$

Hieraus folgt $\sum_{\nu \in N'} f(I_\nu) + \sum_{\nu \in N''} g(I_\nu) \leqq \varepsilon$ und damit wegen $f \geqq 0, g \geqq 0$ die Behauptung.

*Dann*: Aus der in (3. 5) bis (3. 7) angegebenen Konstruktion von $h^+$ für $h \in \Phi_c$ folgt, daß die Gleichung $f(x) = (f - g)^+(x)$ für $x \in V$ mit $x > c$ oder $x < c$ trivialerweise erfüllt ist, wenn entweder $f(x) = 0$ oder $g(x) = 0$ oder $f(x) = g(x) = 0$ ist. Es seien also $f(x)$ und $g(x)$ ungleich Null und somit, da wir uns wie bisher auf den Fall $x > c$ beschränken, beide $> 0$. Im Sinne der Behauptung gilt nach Wahl von $\varepsilon > 0$ für eine geeignete Einteilung $c = x_0 < x_1 < \cdots < x_n = x$

$$\sum_{\nu \in N'} f(I_\nu) \leqq 2^{-1}\varepsilon \text{ und } \sum_{\nu \in N''} g(I_\nu) \leqq 2^{-1}\varepsilon,$$

wobei $I_\nu = [x_{\nu-1}, x_\nu]$, $\nu = 1, 2, \ldots, n$. Es folgt

$$0 \leqq \sum_{\nu \in N'} f(I_\nu) + \sum_{\nu \in N''} g(I_\nu) = f(x) - \sum_{\nu \in N''} f(I_\nu) + \sum_{\nu \in N''} g(I_\nu) \leqq \varepsilon$$

und somit

$$f(x) - \varepsilon \leqq \sum_{\nu \in N''} (f(I_\nu) - g(I_\nu)) \leqq \sum_{\nu=1}^{n} \sup\,(f(I_\nu) - g(I_\nu), 0) \leqq (f - g)^+(x).$$

Da $\varepsilon > 0$ beliebig gewählt war, folgt $f(x) \leqq (f - g)^+(x)$. Andererseits ist $f - g \leqq f$, also $(f - g)^+ \leqq f$ und somit $(f - g)^+(x) \leqq f(x)$. Für $x > c$ ist damit die Gleichung $f(x) = (f - g)^+(x)$ bewiesen.

Beispiel. Es sei $V$ ein abgeschlossenes Intervall $[c, d]$ auf der Zahlengeraden $R$ mit den Endpunkten $c$ und $d$. Die Punktmenge $V$ ist ein Verband bezüglich der in $R$ erklärten Ordnungsrelation. Der vollständige Vektorverband $\Phi_c(R, V)$ besteht aus allen reellen Funktionen $f \mid V$ von (im üblichen Sinne) beschränkter Variation mit $f(c) = 0$ [22]. Wählt man für $g \in \Phi_c$ die reelle Funktion $g(x) = x - c$, $x \in V$, so folgt aus Satz 5. 11, daß die $g$-stetigen Elemente von $\Phi_c$ genau die bezüglich des elementaren Inhalts auf $R$ totalstetigen Funktionen aus $\Phi_c$ sind [23]).

**5. 2. Spezialfall: Reelle additive Funktionen über einem Booleschen Verband.** Ist $V$ ein *Boolescher Verband* und $c = \emptyset$ das Nullelement von $V$, so ergeben sich aus 5. 11 und 5. 12 die beiden folgenden Sätze:

---

[22]) Vgl. *H. Bauer* [1], S. 109—110.

[23]) Vgl. *Haupt-Aumann-Pauc* [6], Nr. 10. 2. 5.

**Satz** 5. 21. *Es seien f und g reelle, auf einem Booleschen Verband V additive, nichtnegative Funktionen, d. h. Elemente $f \geqq 0$ und $g \geqq 0$ von $\Phi_s(R, V)$. Es ist f dann und nur dann g-stetig, wenn zu jedem reellen $\varepsilon > 0$ und zu jedem $x \in V$ eine reelle Zahl $\delta = \delta(\varepsilon, x) > 0$ existiert derart, daß aus $g(y) \leqq \delta$ stets $f(y) \leqq \varepsilon$ folgt für jedes Element $y \leqq x$ aus V.*

Beweis. *Nur dann*: O. B. d. A. können wir annehmen, daß $x$ nicht das Nullelement von $V$ ist. Zu $x > \emptyset$ und $\varepsilon > 0$ bestimmen wir $\delta = \delta(\varepsilon, x)$ nach 5. 11. Für ein $y \in V$ mit $\emptyset < y \leqq x$ und $g(y) \leqq 0$ folgt aus 5. 11:

$$f(y) = f(y) - f(\emptyset) \leqq \varepsilon.$$

Für $y = \emptyset$ ist dies trivialerweise richtig.

*Dann*: Zu $x \in V$ mit $x > \emptyset$ und $\varepsilon > 0$ wählen wir $\delta = \delta(\varepsilon, x) > 0$, so daß $f(y) \leqq \varepsilon$ für alle $y \leqq x$ aus $V$ mit $g(y) \leqq \delta$ gilt. Sind dann $I_\nu = [x_\nu, y_\nu]$, $\nu = 1, 2, \ldots, n$, Intervalle in $V$ mit $\emptyset \leqq x_1 < y_1 < x_2 < \cdots < y_n \leqq x$, so sind die Elemente $z_\nu = y_\nu - x_\nu$ paarweise fremd in $V$, d. h. es ist $z_\nu \wedge z_\varrho = \emptyset$ für $\nu \neq \varrho$, und ihre Vereinigung $z \leqq x$. Aus $\sum_{\nu=1}^{n} g(I_\nu) \leqq \delta$ folgt daher $\sum_{\nu=1}^{n} f(I_\nu) \leqq \varepsilon$, da $h(z) = \sum_{\nu=1}^{n} h(I_\nu)$ ist für jede auf $V$ additive Funktion $h$. Nach 5. 11 ist daher $f$ $g$-stetig.

**Satz** 5. 22. *Unter den Voraussetzungen des Satzes 5. 21 gilt: Es ist f dann und nur dann g-rein-unstetig, wenn zu jedem $x \in V$ und zu jedem reellen $\varepsilon > 0$ ein Element $y = y(\varepsilon, x) \leqq x$ in V existiert, für welches gleichzeitig $g(y) \leqq \varepsilon$ und $f(x - y) \leqq \varepsilon$ gilt.*

Der Beweis dieses Satzes folgt entweder aus 5. 12 oder unmittelbar aus Formel (3. 10).

**Bemerkung.** Besitzt der Boolesche Verband $V$ eine *Einheit* $e$, so gilt unter den Voraussetzungen von 5. 21: *f ist dann und nur dann g-stetig, wenn zu jedem $\varepsilon > 0$ ein $\delta(\varepsilon) > 0$ existiert derart, daß aus $g(x) \leqq \delta(\varepsilon)$ stets $f(x) \leqq \varepsilon$ folgt für jedes $x \in V$. — f ist dann und nur dann g-rein-unstetig, wenn zu beliebigem $\varepsilon > 0$ ein Element $x = x(\varepsilon) \in V$ existiert mit $g(x) \leqq \varepsilon$ und $f(e - x) \leqq \varepsilon$*[24]).

Der folgende Satz hat mehr den Charakter eines Hilfssatzes; er wird sich aber in Nr. 5. 4 als nützlich erweisen.

**Satz** 5. 23. *Unter den Voraussetzungen des Satzes 5. 21 läßt sich der g-rein-unstetige Teil u von f wie folgt darstellen:*

(5. 5) $$u(x) = \inf\,(u(x, \eta);\ \eta > 0),\ x \in V,$$

*wenn für jedes $x \in V$ und jede reelle Zahl $\eta > 0$*

(5. 6) $$u(x, \eta) = \sup\,(f(y);\ y \leqq x,\ g(y) \leqq \eta)$$

*gesetzt wird*[25]).

Beweis. Aus (5. 2) und (3. 8) ergibt sich folgende Darstellung von $u$:

$$u(x) = \lim v_n(x) \text{ mit } v_n(x) = \sup\,(f(y) - ng(y);\ y \leqq x),\ n = 1, 2, \ldots .$$

[24]) Unter den der Bemerkung zugrunde liegenden Voraussetzungen wurde der Zerlegungssatz von Nr. 5. 1 ebenfalls auf verbandsalgebraischem Wege von *S. Bochner* und *R. S. Phillips* [3] bewiesen. Vgl. Ziffer IV der Einleitung. Vgl. ferner Fußnote[25]).

[25]) Aus der Bemerkung zu Satz 5. 22 folgt ein Zerlegungssatz für Intervallfunktionen von *Chr. Pauc*, Contributions à une théorie de la différentiation de fonctions d'intervalle sans hypothèse de Vitali, C. R. Acad. Sci. Paris **236** (1953), S. 1937—1939. Es wird dort angedeutet, daß der als „singulär im Sinne von Saks" bezeichnete Teil der Intervallfunktion in Analogie zu dem in 5. 23 angegebenen Verfahren konstruiert wird. Ein Beweis des Zerlegungssatzes von Herrn Pauc findet sich in *Haupt-Aumann-Pauc* [6], Nr. 10. 2. 8.

Ist $\delta$ eine positive reelle Zahl, so gilt

$$f(y) - ng(y) \geqq f(y) - \delta \quad \text{bzw.} \quad f(y) - ng(y) < f(y) - \delta$$

für alle $y \leqq x$ aus $V$ mit $g(y) \leqq n^{-1}\delta$ bzw. $g(y) > n^{-1}\delta$, $n = 1, 2, \ldots$. Hieraus folgt

$$v_n(x) = \sup\,(f(y) - ng(y);\; y \leqq x, g(y) \leqq n^{-1}\delta);$$

mithin ist $v_n(x) \leqq u(x, n^{-1}\delta) \leqq v_n(x) + 0$. Läßt man $\delta$ eine Nullfolge $\delta_n$ positiver reeller Zahlen durchlaufen, so folgt aus dieser Doppelungleichung die Behauptung, wenn man noch $\inf\,(u(x, \eta);\, \eta > 0) = \lim_{\eta \to 0} u(x, \eta)$ beachtet.

**5. 3. Spezialfall: Lineare Abbildungen eines Vektorverbandes in einen vollständigen Vektorverband.** Nunmehr sei $V = E$ ein *Vektorverband über dem* g l e i c h e n *Körper* K *wie ein vollständiger Vektorverband W.*

Die Menge $\Lambda(W, E)$ der relativ beschränkten linearen Abbildungen von $E$ in $W$ ist nach Satz 3. 34 ein linearer Unterraum des vollständigen Vektorverbandes $\Phi_0(W, E)$. Sind also $f$ und $g$ Elemente von $\Lambda$, so ist $f$ *in* $\Phi_0$ eindeutig zerlegbar in einen $g$-stetigen und einen $g$-rein-unstetigen Teil. Der folgende Satz besagt, daß mit $f$ und $g$ auch die zunächst in $\Phi_0$ gebildeten Zerlegungsteile wieder Elemente von $\Lambda$ sind; er gibt an, wie man diese Zerlegungsteile in $\Lambda$ ohne Zuhilfenahme des Einbettungsraumes $\Phi_0$ gewinnen kann. Diese Lebesgue-Zerlegung in $\Lambda$ hat den gleichen algebraischen Charakter wie die Lebesgue-Zerlegung in $\Phi_0$.

**Satz** 5. 31. *Für Elemente f und g von $\Lambda(W, E)$ gehört auch der in $\Phi_0(W, E)$ gebildete g-stetige und g-rein-unstetige Teil von f zu $\Lambda$. Es ist f dann und nur dann g-stetig bzw. g-rein-unstetig, wenn f zu dem von g in $\Lambda$ erzeugten Band gehört bzw. wenn f und g in $\Lambda$ disjunkt sind.*

B e w e i s. Für den $g$-stetigen Teil $s(f)$ von $f$ gilt im Falle $f \geqq 0$ gemäß (5. 1)

$$s(f) = (\Phi_0)\sup\,(u_n;\, n = 1, 2, \ldots) \text{ mit } u_n = (\Phi_0)\inf\,(f, n\,|\,g\,|),\; n = 1, 2, \ldots.$$

Bei beliebigem $f \in \Lambda$ gilt $s(f) = s(f^+) - s(f^-)$. Hieraus und aus 3. 34, insbesondere Beh. (3), folgt, daß mit $f$ und $g$ auch $s(f)$ ein Element von $\Lambda$ ist. Folglich ist $f \in \Lambda$ dann und nur dann $g$-stetig, wenn $f$ zum mengentheoretischen Durchschnitt von $\Lambda$ mit dem von $g \in \Lambda$ in $\Phi_0$ erzeugten Band gehört. Dieser Durchschnitt ist nach einer Bemerkung am Schluß der Nr. 1. 2 das von $g$ in $\Lambda$ erzeugte Band. Da $\Lambda$ nach 3. 34 ein eigentlicher linearer Unterraum von $\Phi_0$ ist, sind zwei Elemente von $\Lambda$ dann und nur dann disjunkt in $\Lambda$, wenn sie es in $\Phi_0$ sind. Hieraus ergibt sich die behauptete Kennzeichnung der $g$-rein-unstetigen Elemente von $\Lambda$.

**Korollar (Lebesguescher Zerlegungssatz für lineare Abbildungen)** [26].

*Der vollständige Vektorverband $\Lambda(W, E)$ ist für jedes $g \in \Lambda$ direkte ordnungstreue Summe des Bandes seiner g-stetigen und des Bandes seiner g-rein-unstetigen Elemente.*

Die folgenden drei Sätze beweist man beinahe wortwörtlich genau so wie die entsprechenden Sätze 5. 21 bis 5. 23 [27]).

**Satz** 5. 32. *Es sei E ein Vektorverband über dem Körper der reellen Zahlen; f und g seien reelle, lineare Funktionale* [28]) *über E mit $f \geqq 0$ und $g \geqq 0$. Es ist f dann und nur dann* g - s t e t i g, *wenn zu jedem reellen $\varepsilon > 0$ und zu jedem $x \geqq 0$ aus E eine reelle Zahl $\delta = \delta(\varepsilon, x) > 0$ existiert, für welche aus $0 \leqq y \leqq x$, $y \in E$, und $g(y) \leqq \delta$ stets $f(y) \leqq \varepsilon$ folgt.*

---

[26]) Dieser Satz, zusammen mit den in 5. 31 bis 5. 33 gegebenen Kennzeichnungen $g$-stetiger und $g$-rein-unstetiger linearer Abbildungen, wurde im reellen Fall zuerst von *F. Rieß* [9] bewiesen.

[27]) Beweise für 5. 32 und 5. 33 auch bei *N. Bourbaki* [4], S. 36—37.

[28]) Ist $W$ der vollständige Vektorverband $R$ der reellen Zahlen und $E$ ein Vektorverband über $R$, so nennt man eine lineare Abbildung von $E$ in $R$ auch ein *reelles, lineares Funktional* über $E$.

**Satz** 5. 33. *Unter den Voraussetzungen des Satzes* 5. 32 *gilt: Es ist $f$ dann und nur dann $g$-rein-unstetig, wenn zu jedem reellen $\varepsilon > 0$ und zu jedem $x \geqq 0$ aus $E$ ein Element $y = y(\varepsilon, x)$ aus $E$ mit $0 \leqq y \leqq x$ sowie mit $g(y) \leqq \varepsilon$ und $f(x - y) \leqq \varepsilon$ existiert.*

**Satz** 5. 34. *Unter den Voraussetzungen von Satz* 5. 32 *läßt sich der $g$-rein-unstetige Teil $u$ von $f$ wie folgt darstellen: Setzt man*

$$(5.7) \qquad u(x, \eta) = \sup\left(f(y); 0 \leq y \leq x, g(y) \leq \eta\right)$$

*für jedes $x \geqq 0$ aus $E$ und jedes reelle $\eta > 0$, so ist*

$$(5.8) \qquad u(x) = \inf\left(u(x, \eta); \eta > 0\right) \text{ für jedes } x \geqq 0 \text{ aus } E.$$

**5. 4. Vergleich der Zerlegung in $g$-stetigen und $g$-rein-unstetigen Teil mit der Zerlegung in $N_c(g)$-regulären und $N_c(g)$-singulären Teil.** Zunächst liege der in § 4 betrachtete allgemeine Fall vor: Es sei also $V$ ein *distributiver Verband*, $c$ ein Element aus $V$ und $W$ ein vollständiger Vektorverband über einem linear geordneten Körper K.

Weiter sei $g$ ein Element aus $\Phi_c(W, V)$, welches wir als fest vorgegeben betrachten; $N = N_c(g)$ bezeichne den gesättigten Unterverband aller $c$-totalen Nullteile von $g$. Dann ist $\Phi_c$ nach dem Zerlegungssatz der Nr. 4. 3 direkte ordnungstreue Summe seiner $N$-regulären und $N$-singulären Elemente. Zwischen dieser Zerlegung und der Lebesgueschen in $g$-stetigen und $g$-rein-unstetigen Teil besteht folgender Zusammenhang:

**Satz** 5. 41. *Jedes $g$-stetige Element von $\Phi_c(W, V)$ ist auch $N$-regulär. Jedes $N$-singuläre Element von $\Phi_c$ ist $g$-rein-unstetig. Für eine Bewertung $f \geqq 0$ aus $\Phi_c$ ist der $g$-stetige Teil $s \leqq f - f_N$ und der $g$-rein-unstetige Teil $u \geqq f_N$ mit $N = N_c(g)$.*

Beweis. Die Bewertung $g$ ist $N$-regulär nach 4. 31. Es ist also $g$ ein Element des Bandes aller $N$-regulären Elemente von $\Phi_c$ und damit das von $g$ in $\Phi_c$ erzeugte Band $[g]$ eine Teilmenge des Bandes der $N$-regulären Elemente. Das Band $[g]$ besteht aber aus allen $g$-stetigen Elementen von $\Phi_c$. Das zu $[g]$ komplementäre Band der $g$-rein-unstetigen Elemente enthält dann das Band der $N$-singulären Elemente als Teilmenge. — Ist $f \geqq 0$ ein Element von $\Phi_c$, so ist $f - f_N$ der $N$-reguläre und $f_N$ der $N$-singuläre Teil von $f$; $f_N$ ist dann auch $g$-rein-unstetig. Bezeichnen wir mit $s'$ den $g$-stetigen und mit $u'$ den $g$-rein-unstetigen Teil von $f - f_N$, so ist $u' + f_N$ $g$-rein-unstetig; ferner folgt $s' \geqq 0$ und $u' + f_N \geqq 0$ aus $f_N \geqq 0$ und $f - f_N \geqq 0$. Demnach ist $s = s'$ der $g$-stetige und $u = u' + f_N$ der $g$-rein-unstetige Teil von $f$. Hieraus folgt der Rest der Behauptung.

Das folgende Beispiel zeigt, daß im allgemeinen die beiden Zerlegungen *nicht* zusammenfallen:

**Beispiel.** Wie im Beispiel der Nr. 4. 3 sei $e$ das halboffene Einheitsintervall $[0, 1)$ auf der Zahlengeraden $R$. Der Verband $V$ sei ein Boolescher Mengenverband bestehend aus der leeren Menge $\emptyset$ und allen Mengen $m$ der Gestalt

$$m = [\alpha_1, \beta_1) \cup [\alpha_2, \beta_2) \cup \cdots \cup [\alpha_n, \beta_n)$$

mit $0 \leqq \alpha_1 < \beta_1 < \alpha_2 < \cdots < \beta_n \leqq 1$, $n \geqq 1$. Die Einheit von $V$ ist $e$. Auf dem abgeschlossenen Einheitsintervall $[0, 1]$ definieren wir reelle Punktfunktionen $\varphi$ und $\psi$ wie folgt:

$$\varphi(\xi) = \begin{cases} 0, & 0 \leqq \xi \leqq \frac{1}{2}; \\ 1, & \frac{1}{2} < \xi \leqq 1; \end{cases}$$

$$\psi(\xi) = \xi, \quad \xi \in [0, 1].$$

Setzen wir dann $f(\emptyset) = g(\emptyset) = 0$ und für jede Menge $m \in V$ obiger Gestalt

$$f(m) = \sum_{i=1}^{n} (\varphi(\beta_i) - \varphi(\alpha_i)) \quad \text{bzw.} \quad g(m) = \sum_{i=1}^{n} (\psi(\beta_i) - \psi(\alpha_i)),$$

so sind $f$ und $g$ Elemente aus $\Phi_\emptyset(R, V)$, d. h. nicht-negative, reelle, auf $V$ additive Funktionen. $f$ und $g$ sind auf $V$ sogar *total-additiv*, da $\varphi$ und $\psi$ in $[0, 1]$ linksseitig (von unten) stetig sind (vgl. [1], S. 114). Weiter ist die Funktion $f$ $g$-rein-unstetig; nach der Bemerkung zu 5. 22 genügt es nämlich, zu jedem reellen $\varepsilon > 0$ eine Menge $y \in V$ anzugeben mit $f(y) \leqq \varepsilon$ und $g(e - y) \leqq \varepsilon$. Diese Eigenschaft besitzt aber die Menge $y = [0, \frac{1}{2}) \cup [\frac{1}{2} + \varepsilon, 1)$. Die Funktion $f$ ist aber nicht $N$-singulär, sondern vielmehr $N$-regulär: Es besteht nämlich $N_\emptyset = N(g)$ nur aus der leeren Menge $\emptyset$, also ist $f_N = 0$.

In zwei Fällen, wenn nämlich

(1) *$V$ ein Boolescher $\sigma$-Verband, $W$ der vollständige Vektorverband $R$ der reellen Zahlen und die additiven Funktionen $f, g \in \Phi_{\bullet}(R, V)$ auf $V$ total-additiv sind,*

oder wenn

(2) *$V$ ein $\sigma$-vollständiger Vektorverband über dem Körper der reellen Zahlen, $W$ der vollständige Vektorverband $R$ der reellen Zahlen und $f$ und $g$ stetige, lineare Funktionale aus $\Lambda(R, V)$ sind,*

führen beide Zerlegungsarten zu den gleichen Zerlegungsteilen. Dies wird bereits durch die Formeln (5. 5) und (5. 6) bzw. (5. 7) und (5. 8) nahegelegt.

**Satz** 5. 42[29]). *Im vollständigen Vektorverband $\Phi^{\sigma} = \Phi^{\sigma}(R, V)$ aller über einem Booleschen $\sigma$-Verband $V$ total-additiven, reellen Funktionen ist für jedes $g \in \Phi^{\sigma}$ die direkte ordnungstreue Zerlegung von $\Phi^{\sigma}$ in das Band der $g$-stetigen und das Band der $g$-rein-unstetigen Elemente identisch mit der direkten ordnungstreuen Zerlegung von $\Phi^{\sigma}$ in das Band der $N$-regulären und das der $N$-singulären Elemente ($N = N_{\bullet}(g)$)*[30]).

Beweis[31]). Wir zeigen, daß für jedes $f \in \Phi^{\sigma}$ der $g$-rein-unstetige Teil $u$ von $f$ gleich $f_N$ ist; dabei können wir o. B. d. A. $f \geqq 0$ und $g \geqq 0$ annehmen. Nach 5. 41 ist $f_N \leqq u$; es verbleibt also, die Ungleichung $u(x) \leqq f(x)$ für jedes $x \in V$ zu beweisen. Nach (5. 5) und (5. 6) ist aber

$$u(x) = \inf\,(u(x, \eta);\, \eta > 0) \quad \text{mit} \quad u(x, \eta) = \sup\,(f(y);\, y \leqq x,\, g(y) \leqq \eta).$$

Ferner gilt nach (4. 5) (Satz 4. 13)

$$f_N(x) = \sup\,(f(y);\, y \leqq x,\, g(y) = 0),\, x \in V.$$

Für jede natürliche Zahl $n$ und jede positive Zahl $\varepsilon_n$ existiert ein Element $y_n \leqq x$ aus $V$ mit $g(y_n) \leqq 2^{-n}$ und $u(x) \leqq u(x, \varepsilon_n) \leqq f(y_n) + \varepsilon_n$. Da $V$ ein Boolescher $\sigma$-Verband ist, existieren die Elemente $y'_n = \bigvee_{\nu=n}^{\infty} y_\nu$, $n = 1, 2, \ldots$ und $y = \bigwedge_{n=1}^{\infty} y'_n$; es gilt $y \leqq y'_{n+1} \leqq y'_n \leqq x$, für jedes $n = 1, 2, \ldots$. Aus der Total-Additivität von $g$ folgt

$$g(y'_n) \leqq \sum_{\nu=n}^{\infty} g(y_\nu) \leqq 2^{-n+1}, \quad n = 1, 2, \ldots,$$

und hieraus weiter $g(y) = \lim g(y'_n) = 0$[32]). Also ist $f(y) \leqq f_N(x)$. Schließlich folgt aus $y_n \leqq y'_n$

$$u(x) \leqq f(y_n) + \varepsilon_n \leqq f(y'_n) + \varepsilon_n, \quad n = 1, 2, \ldots,$$

und aus der Total-Additivität von $f$ die Gleichung

$$f(y) = \lim f(y'_n).$$

Wählt man für die Folge der $\varepsilon_n$ speziell eine Nullfolge, so hat dies $u(x) \leqq f(y)$ und damit die gesuchte Ungleichung $u(x) \leqq f_N(x)$ zur Folge.

Der zweite angekündigte Satz lautet:

**Satz** 5. 43. *Es sei $E$ ein $\sigma$-vollständiger Vektorverband über dem Körper der reellen Zahlen. Im vollständigen Vektorverband $\Lambda^{\sigma} = \Lambda^{\sigma}(R, E)$ aller reellen, stetigen, linearen*

---

[29]) Es handelt sich i. w. um den Nachweis, daß beim klassischen Lebesgueschen Zerlegungssatz die Zerlegungsteile auf zwei äquivalente Arten charakterisiert werden können (vgl. Einl., Ziff. I, A und B).

[30]) Nach einer Anmerkung am Schluß der Nr. 3. 2 und nach dem in Nr. 4. 5 zitierten Satz von H. Hahn ist $\Phi^{\sigma}$ das Band aller total-additiven Funktionen aus $\Phi_{\bullet}(R, V)$, also ein vollständiger Vektorverband. Jede direkte ordnungstreue Zerlegung von $\Phi_{\bullet}$ induziert eine ebensolche in $\Phi^{\sigma}$. Vgl. diesbez. eine Bemerkung in Nr. 1. 3.

[31]) Der Vollständigkeit halber beweisen wir hier diesen an und für sich bekannten Satz, zumal der Beweisgedanke auch beim Beweis von 5. 43 benötigt wird.

[32]) Es handelt sich um eine Anwendung des „absteigenden Limessatzes"; vgl. *Haupt-Aumann-Pauc* [6], Nr. 5. 2.

*Funktionale über $E$*[33]) *ist für jedes Funktional $g \in \Lambda^\sigma$ die direkte ordnungstreue Zerlegung von $\Lambda^\sigma$ in $g$-stetige und $g$-rein-unstetige Elemente identisch mit der direkten ordnungstreuen Zerlegung in $N$-reguläre und $N$-singuläre Elemente $(N = N_0(g))$.*

Beweis. Es sei $u$ der $g$-rein-unstetige Teil eines Funktionals $f \in \Lambda^\sigma$; dann soll $u = f_N$ gezeigt werden. Hierzu kann o. B. d. A. $f \geqq 0$ und $g \geqq 0$ angenommen werden; es genügt ferner die Gleichung $u(x) = f_N(x)$ für jedes $x \geqq 0$ aus $E$ zu beweisen. Dies kann genau so geschehen wie im Beweis von 5. 42; man hat nur von den Formeln (5. 7), (5. 8) auszugehen, welche den dort verwendeten Formeln (5. 5) und (5. 6) entsprechen.

## Literaturverzeichnis.

[1] *H. Bauer*, Eine Rieszsche Bandzerlegung im Raum der Bewertungen eines Verbandes. S.-Ber. math.-naturw. Kl. Bayer. Akad. Wiss., München (1953), S. 89—117.

[2] *G. Birkhoff*, Lattice Theory, Rev. Ed., Amer. Math. Soc. Colloq. Publ., Vol. XXV, New York (1948).

[3] *S. Bochner* und *R. S. Phillips*, Additive set functions and vector lattices. Ann. of Math. **42** (1941), S. 316—324.

[4] *N. Bourbaki*, Éléments de Mathématique, Livre VI, Intégration, Chap. I—IV, Actual. Scient. et Ind. **1175**, Hermann et C^ie^, Paris (1952).

[5] *H. Hahn* und *A. Rosenthal*, Set Functions, Albuquerque, New Mexico (1948).

[6] *O. Haupt*, *G. Aumann* und *Chr. Pauc*, Differential- und Integralrechnung, Bd. III, 2. Aufl., Göschens Lehrbücherei (1955).

[7] *H. Nakano*, Modern Spectral Theory, Tokyo Math. Book Series, Vol. II, Tokyo (1950).

[8] *C. E. Rickart*, Decomposition of additive set functions, Duke Math. J. **10** (1943), S. 653—665.

[9] *F. Riesz*, Sur quelques notions fondamentales dans la théorie générale des opérations linéaires, Ann. of Math **41** (1940), S. 174—206*).

---

[33]) Ein über $E$ definiertes, reelles, *stetiges*, lineares Funktional ist stets auch *relativ beschränkt*. Vgl. *H. Nakano* [7], S. 254. Nach der Bemerkung am Schluß der Nr. 3. 3 ist dann $\Lambda^\sigma$ das Band aller in $E$ stetigen Funktionale aus $\Lambda(R, E)$, also für sich betrachtet ein vollständiger Vektorverband.

*) Nr. des Bandes und Erscheinungsjahr dieser Arbeit wurden in [1] versehentlich falsch zitiert.

---

Eingegangen 6. Dezember 1953.

# Über die Beziehungen einer abstrakten Theorie des Riemann-Integrals zur Theorie Radonscher Maße *)
[R9]

Math. Z. **65** (1956), 448–482
[Zbl. Math. 73.04003; MR 18,645c]

## Einleitung

Neben die abstrakte Integrationstheorie von Stone [*21*], welche die klassische Lebesguesche Theorie zum Vorbild hat, ist neuerdings eine abstrakte Theorie des Riemann-Integrals von Loomis [*14*] getreten. „Abstrakt" heißt hierbei, daß auf eine Topologie im Definitionsbereich der zu integrierenden Funktionen kein Bezug genommen wird. Ausgangspunkt der Untersuchungen von Loomis ist vielmehr, ähnlich wie bei Stone, ein positives, lineares Funktional $I$ über einem System $\mathfrak{R}$ reeller Funktionen, deren gemeinsamer Definitionsbereich eine Menge $E$ ist; hierbei wird $\mathfrak{R}$ als Vektorverband angenommen, welcher als Ersatz für die eventuell fehlende Konstante 1 mit jeder Funktion $g$ auch $\min(1, g)$ enthält. Im Gegensatz zur Stoneschen Theorie wird die Stetigkeit des Funktionals $I$ nicht gefordert. Ziel ist, wie bei allen Untersuchungen dieser Art, durch Einführung einer geeigneten additiven Mengenfunktion $\mu | \mathfrak{Q}$ und Definition eines $\mu$-Unterteilungsintegrals eine Darstellung von $I(f)$ als $\mu$-Integral für eine möglichst umfassende Klasse von Funktionen $f \in \mathfrak{R}$ zu finden. Die fehlende Stetigkeit von $I$ führt hierbei zu einem neuen Gesichtspunkt; es wird nämlich ein gewisses Prinzip der Sparsamkeit bei der Bestimmung der $\mu$-meßbaren Mengen verfolgt. Bei Loomis ist $\mathfrak{Q}$ das System aller Teilmengen $Q$ von $E$, deren charakteristische Funktionen von oben und unten her durch Funktionen aus $\mathfrak{R}$ beliebig genau $I$-einschließbar sind; die charakteristische Funktion $\chi_Q$ einer Menge $Q \in \mathfrak{Q}$ erhält dann bei jeder möglichen Fortsetzung von $I$ zu einem positiven, linearen Funktional über einem $\chi_Q$ als Element enthaltenden Vektorraum von reellen Funktionen auf $E$ den gleichen Funktionalwert, nämlich $\mu(Q)$ zugeordnet. Mit Hilfe von $\mu | \mathfrak{Q}$ gewinnt Loomis das Integral $\int f\, d\mu$ durch Bildung von Ober- und Untersummen nach dem Vorbild des klassischen Riemann-Integrals.

Wendet man die Loomissche Theorie speziell auf den „konkreten" Fall eines lokal kompakten Raumes $E$ an und wählt man für $\mathfrak{R}$ die Menge $\mathfrak{K}(E)$ der auf $E$ stetigen Funktionen mit kompaktem Träger, so ist $I$ ein Radonsches Maß $\varrho$ auf $E$. Es zeigt sich, daß in diesem Falle $\mathfrak{Q}$ das System der $\varrho$-quadrierbaren Mengen ist, d.h. der relativ kompakten Teilmengen von $E$, deren Begrenzung eine $\varrho$-Nullmenge ist. Ferner erweist sich das System der bezüglich $I$ im Sinne von Loomis Riemann-integrierbaren Funktionen als das System aller $\varrho$-fast überall stetigen Funktionen auf $E$ mit kompaktem

***) Von der Naturwissenschaftlichen Fakultät der Universität Erlangen als Habilitationsschrift angenommen.**

Träger. Man gewinnt auf diese Weise aus der Theorie von Loomis die von Haupt und Pauc [10], [19], [12b], S. 174ff., entwickelte Theorie des Riemann-Integrals in lokal kompakten Räumen.

Das Hauptergebnis der vorliegenden Arbeit ist gewissermaßen die Umkehrung dieser Feststellung. Zunächst läßt sich zeigen, daß es für die Loomissche Theorie keine Beschränkung der Allgemeinheit bedeutet, die Funktionen aus $\mathfrak{R}$ als beschränkt vorauszusetzen und anzunehmen, daß je zwei verschiedene Elemente von $E$ durch eine Funktion aus $\mathfrak{R}$ getrennt werden können und in jedem Element von $E$ mindestens eine Funktion aus $\mathfrak{R}$ nicht verschwindet. Unter diesen zusätzlichen Annahmen über $E$ und $\mathfrak{R}$ wird dann die Existenz eines durch $E$ und $\mathfrak{R}$ im wesentlichen eindeutig bestimmten, lokal kompakten Raumes $E'_{\mathfrak{R}}$ nachgewiesen, der $E$ als dichte Teilmenge enthält und folgende Eigenschaft besitzt: Zu jedem positiven, linearen Funktional $I|\mathfrak{R}$ existiert ein eindeutig bestimmtes, positives Radonsches Maß $\varrho'$ auf $E'_{\mathfrak{R}}$ derart, daß das Mengensystem $\mathfrak{Q}$ genau die Spur des Systems der $\varrho'$-quadrierbaren Teilmengen von $E'_{\mathfrak{R}}$ und ferner das System der bezüglich $I$ Riemann-integrierbaren Funktionen $f|E$ gleich dem System der auf $E$ verengerten, $\varrho'$-fast überall stetigen Funktionen $f'|E'_{\mathfrak{R}}$ mit kompaktem Träger ist. Für jede $\varrho'$-quadrierbare Menge $Q'$ und deren Spur $Q$ in $E$ ist $\mu(Q)=\varrho'(Q')$; für jede $\varrho'$-fast überall stetige Funktion $f'|E'_{\mathfrak{R}}$ mit kompaktem Träger und deren Verengerung $f|E$ ist $\int f\,d\mu=\int f'\,d\varrho'$.

Durch diesen Darstellungssatz für die nach Loomis Riemann-integrierbaren Funktionen wird das nicht notwendig stetige Funktional $I$ mit dem Radonschen Maß $\varrho'$, also einem stetigen Funktional über dem Vektorraum $\mathfrak{K}(E'_{\mathfrak{R}})$ der auf $E'_{\mathfrak{R}}$ stetigen Funktionen mit kompaktem Träger in Beziehung gesetzt. Der Klärung des hier vorliegenden Sachverhalts dient eine Bemessung des Stetigkeitsdefektes von $I$ durch die Zerlegung von $I$ in seinen sog. stetigen Teil $I_c$ und rein-unstetigen Teil $I_p$. Da aus der Stoneschen Theorie die Existenz einer Integraldarstellung für den stetigen Teil $I_c$ folgt, lassen sich mit Hilfe dieser Zerlegung Bedingungen herleiten, unter welchen für eine vorgegebene Funktion $f\in\mathfrak{R}$ der Funktionalwert $I(f)$ als Integral darstellbar ist. Hierdurch erscheinen bekannte Sätze über Integraldarstellungen in neuem Lichte.

Die Arbeit gliedert sich in vier Paragraphen. § 1 gibt einen kurzen Überblick über die Theorie von Loomis und ergänzt sie durch einige auf unser Ziel ausgerichtete Bemerkungen. § 2 enthält einen Beweis für die Existenz der Zerlegung $I=I_c+I_p$ und bringt diese in Zusammenhang mit den Begriffsbildungen der Loomisschen Theorie. Insbesondere wird gezeigt, daß der Zerlegung von $I$ die Zerlegung von $\mu$ in $\sigma$-additiven und rein-endlich-additiven Teil entspricht. In § 3 wird die Loomissche Theorie auf den konkreten Fall eines lokal kompakten Raumes angewandt als Vorstufe für den Beweis des Darstellungssatzes, der dann in § 4 erbracht wird. In einem Beispiel am Schluß der Arbeit wird die Stonesche Kompaktifizierung einer Booleschen Mengenalgebra als Raum $E'_{\mathfrak{R}}$ gedeutet[1]).

[1]) Eine Vorankündigung der wichtigsten Resultate dieser Arbeit (ohne Beweise) findet sich in [7].

## § 1. Riemann-Integrierbarkeit bezüglich eines linearen Funktionals. Abstrakter Fall[2])

**1.1.** Es sei $E$ eine Menge beliebiger Elemente $x, y, \ldots$ und $\mathfrak{R}$ eine Menge reeller, endlicher, auf $E$ definierter Funktionen mit folgenden Eigenschaften:

(1) *sind* $g, h \in \mathfrak{R}$ *und* $\alpha, \beta$ *reelle Zahlen, so ist* $\alpha g + \beta h \in \mathfrak{R}$;

(2) *ist* $g \in \mathfrak{R}$, *so ist* $|g| \in \mathfrak{R}$;

(3) *ist* $g \in \mathfrak{R}$, *so ist* $\min(1, g) \in \mathfrak{R}$[3]).

Zufolge (1) und (2) ist $\mathfrak{R}$ ein Vektorverband reeller Funktionen über dem Körper $\mathsf{P}$ der reellen Zahlen; insbesondere liegen mit je endlich vielen Funktionen $g_1, \ldots, g_n$ auch die Funktionen $\max(g_1, \ldots, g_n)$ und $\min(g_1, \ldots, g_n)$ in $\mathfrak{R}$. Zufolge (1) bis (3) liegt $\min(\alpha, g)$ in $\mathfrak{R}$ für jede Funktion $g \in \mathfrak{R}$ und jede reelle Zahl $\alpha \geq 0$.

An anderer Stelle [*17*] haben NÖBELING und Verf. kennzeichnende Eigenschaften derjenigen beschränkten, reellen Funktionen auf $E$ angegeben, welche durch Funktionen aus $\mathfrak{R}$ gleichmäßig approximiert werden können. Wir entnehmen dieser Arbeit das folgende für uns wichtige Resultat[4]).

**Lemma 1.** *Die Menge* $\mathfrak{A}$ *aller beschränkten, reellen, auf* $E$ *definierten Funktionen, die durch Funktionen aus* $\mathfrak{R}$ *gleichmäßig approximiert werden können, ist eine Algebra*[5]). *Zu jeder Funktion* $f \in \mathfrak{A}$ *existiert eine gegen* $f$ *auf* $E$ *gleichmäßig konvergente Folge* $\{g_n\}_{n=1,2,\ldots}$ *von (beschränkten) Funktionen aus* $\mathfrak{R}$ *mit folgender Eigenschaft:*

$$(4)\quad \begin{cases} 0 \leq g_1(x) \leq g_2(x) \leq \cdots \leq f(x) & \text{für jedes } x \in E \text{ mit } f(x) \geq 0; \\ 0 \geq g_1(x) \geq g_2(x) \geq \cdots \geq f(x) & \text{für jedes } x \in E \text{ mit } f(x) \leq 0. \end{cases}$$

**1.2.** Neben $E$ und $\mathfrak{R}$ sei gegeben ein auf $\mathfrak{R}$ definiertes, reelles, *positives*, lineares Funktional $I$.

Als (zweiseitige) *Vervollständigung* $\overline{\mathfrak{R}} = \mathfrak{V}(\mathfrak{R}, I)$ von $\mathfrak{R}$ bezüglich $I$ bezeichnen wir die Menge aller reellen Funktionen $f|E$, zu welchen bei beliebig vorgegebenem $\varepsilon > 0$ Funktionen $g, h \in \mathfrak{R}$ existieren mit

$$g \leq f \leq h \quad \text{und} \quad I(h - g) \leq \varepsilon\text{[6])}.$$

Dann besitzt $\overline{\mathfrak{R}}$ die Eigenschaften (1) bis (3) von $\mathfrak{R}$.

---

[2]) Die in den Nummern 1.2 bis 1.6 dieses Paragraphen erwähnten Begriffsbildungen finden sich bei LOOMIS [*14*].

[3]) Ist $F(u_1, \ldots, u_n)$ eine reelle Funktion in $n$ reellen Variablen $u_1, \ldots, u_n$ und sind $g_1, \ldots, g_n$ auf $E$ definierte, reelle Funktionen, so bezeichnet $F(g_1, \ldots, g_n)$ die durch die Zuordnung $x \to F(g_1(x), \ldots, g_n(x))$ auf $E$ definierte Funktion. Es gilt also z. B. $|g|(x) = |g(x)|$ für alle $x \in E$.

[4]) Da sich die Eigenschaften (1) bis (3) von $\mathfrak{R}$ auf die Menge $\mathfrak{R}^*$ aller beschränkten Funktionen aus $\mathfrak{R}$ übertragen, folgt dies aus der Bemerkung zum Korollar des Kriteriums 2 und dem Satz 1 der zitierten Arbeit [*17*], S. 58 und 59.

[5]) Für je zwei Funktionen $f_1, f_2 \in \mathfrak{A}$ liegt also jede Linearkombination $\alpha f_1 + \beta f_2$ ($\alpha, \beta \in \mathsf{P}$) und das Produkt $f_1 \cdot f_2$ in $\mathfrak{A}$. Darüber hinaus besitzt $\mathfrak{A}$ auch die Eigenschaften (2) und (3) von $\mathfrak{R}$.

[6]) Wir schreiben $g \leq f \leq h$ für „$g(x) \leq f(x) \leq h(x)$ für alle $x \in E$".

Vermöge der Definition von $\overline{\mathfrak{R}}$ kann das Funktional $I$ *auf genau eine Weise* zu einem positiven, linearen Funktional $\bar{I}\,|\,\overline{\mathfrak{R}}$ fortgesetzt werden. Für ein beliebiges $f \in \overline{\mathfrak{R}}$ gilt

$$(5) \qquad \bar{I}(f) = \sup\big(I(g)\,;\, g \leq f,\, g \in \mathfrak{R}\big) = \inf\big(I(h)\,;\, h \geq f,\, h \in \mathfrak{R}\big)\,.$$

$\overline{\mathfrak{R}}$ ist vollständig im Sinne der Gleichung $\mathfrak{V}(\overline{\mathfrak{R}}, \bar{I}) = \overline{\mathfrak{R}}$. Wir bezeichnen daher $\bar{I}\,|\,\overline{\mathfrak{R}}$ als die Vervollständigung des Funktionals $I\,|\,\mathfrak{R}$.

**1.3.** Für eine Teilmenge $A$ von $E$ bezeichne $\chi_A$ die charakteristische Funktion von $A$ bezüglich der Grundmenge $E$. Eine Teilmenge $Q$ von $E$ heißt *quadrierbar* bezüglich $I$ (kurz ($I$-)quadrierbar), wenn ihre charakteristische Funktion $\chi_Q$ ein Element von $\overline{\mathfrak{R}}$ ist. Das System aller $I$-quadrierbaren Mengen werde mit $\mathfrak{Q} = \mathfrak{Q}(\mathfrak{R}, I)$ bezeichnet; definitionsgemäß ist $\mathfrak{Q}(\overline{\mathfrak{R}}, \bar{I}) = \mathfrak{Q}(\mathfrak{R}, I)$. Aus den Eigenschaften (1) bis (3) von $\overline{\mathfrak{R}}$ (mit $\overline{\mathfrak{R}}$ an Stelle von $\mathfrak{R}$) folgt, daß $\mathfrak{Q}$ ein Boolescher Mengenverband ist, welcher im allgemeinen keine Einheit besitzt. Ist jedoch die auf $E$ konstante Funktion $f = 1$ ein Element von $\mathfrak{R}$, so gilt $E \in \mathfrak{Q}$.

Die durch die Gleichung

$$(6) \qquad \mu(Q) = \bar{I}(\chi_Q) \qquad (Q \in \mathfrak{Q})$$

auf $\mathfrak{Q}$ definierte, reelle, endliche Mengenfunktion $\mu$ ist ein *vollständiger Inhalt*, d.h. $\mu\,|\,\mathfrak{Q}$ ist (endlich-)additiv und nicht negativ und es gilt $M \in \mathfrak{Q}$ für eine Menge $M \subseteq E$, wenn zu beliebig vorgegebenem $\varepsilon > 0$ Mengen $P, Q \in \mathfrak{Q}$ mit $P \subseteq M \subseteq Q$ und $\mu(Q - P) \leq \varepsilon$ existieren.

**1.4.** Mit $\mathfrak{E}$ bezeichnen wir die Menge aller endlichen Linearkombinationen von charakteristischen Funktionen $I$-quadrierbarer Mengen. Es besteht also $\mathfrak{E}$ aus allen reellen Funktionen der Gestalt $f = \sum_{\nu=1}^{n} \lambda_\nu \chi_{Q_\nu}$ $(Q_\nu \in \mathfrak{Q},\ \lambda_\nu \in \mathsf{P})$. Auch $\mathfrak{E}$ besitzt die Eigenschaften (1) bis (3) von $\mathfrak{R}$ und ist überdies eine Algebra.

Wegen $\mathfrak{E} \subseteq \overline{\mathfrak{R}}$ ist das Funktional $\bar{I}\,|\,\mathfrak{E}$ als Verengerung von $\bar{I}\,|\,\overline{\mathfrak{R}}$ und folglich auch die Vervollständigung $\mathfrak{S} = \mathfrak{V}(\mathfrak{E}, \bar{I})$ von $\mathfrak{E}$ bezüglich $\bar{I}$ definiert. $\mathfrak{S}$ besitzt daher ebenfalls die Eigenschaften (1) bis (3) von $\mathfrak{R}$ und ist ebenso wie $\mathfrak{E}$ eine Algebra von Funktionen. Wegen $\mathfrak{E} \subseteq \mathfrak{S} \subseteq \overline{\mathfrak{R}}$ ist die Verengerung $\bar{I}\,|\,\mathfrak{S}$ von $\bar{I}\,|\,\overline{\mathfrak{R}}$ gleich der Vervollständigung des Funktionals $\bar{I}\,|\,\mathfrak{E}$.

Nach Loomis [*14*] heißen die Funktionen aus $\mathfrak{S}$ (eigentlich) *Riemann-integrierbar* bezüglich $I$; die einer Funktion $f \in \mathfrak{S}$ zugeordnete reelle Zahl $\bar{I}(f)$ heißt *Riemann-Integral* von $f$ bezüglich $I$. Da $\bar{I}(f)$ durch Bildung von Ober- und Untersummen nach dem Vorbild des klassischen Riemann-Integrals definiert werden kann, setzt man

$$(7) \qquad \bar{I}(f) = \int f\, d\mu \qquad (f \in \mathfrak{S})\,.$$

**1.5.** Eine nicht negative, endliche, reelle Funktion $f\,|\,E$ heißt ($I$-)*meßbar*, wenn die Urbildmenge

$$[f > \lambda] = \{x : f(x) > \lambda\}$$

$I$-quadrierbar ist für alle reellen Zahlen $\lambda > 0$, abgesehen von höchstens abzählbar unendlich vielen Ausnahmen. Eine beliebige, endliche, reelle Funktion $f|E$ heißt ($I$-)meßbar, wenn ihr Positivteil $f^+ = \max(f, 0)$ und ihr Negativteil $f^- = (-f)^+$ meßbar sind.

Da $\mathfrak{R}$ die Eigenschaft (3) besitzt gilt nach LOOMIS [*14*], S. 171:

**Satz 1.** *Die Funktionen aus $\overline{\mathfrak{R}}$ sind sämtlich $I$-meßbar. Insbesondere sind die Funktionen aus $\mathfrak{R}$ meßbar in bezug auf jedes auf $\mathfrak{R}$ definierte, positive, lineare Funktional.*

**Zusatz.** *Für eine nicht negative Funktion $f \in \overline{\mathfrak{R}}$ gilt auch $[f \geq \lambda] \in \mathfrak{Q}$ für alle $\lambda > 0$, abgesehen von höchstens abzählbar unendlich vielen Ausnahmewerten.*

Mit Hilfe des eingeführten Meßbarkeitsbegriffes können die Riemann-integrierbaren Funktionen wie folgt gekennzeichnet werden.

**Satz 2.** *Notwendig und hinreichend für die Riemann-Integrierbarkeit einer reellen Funktion $f|E$ bezüglich des Funktionals $I$ ist folgende Bedingung: $f$ ist beschränkt, $I$-meßbar und im Komplement einer $I$-quadrierbaren Menge gleich Null.*

Beweis. Es sei $f$ Riemann-integrierbar. Wegen $\mathfrak{S} \subseteq \overline{\mathfrak{R}}$ ist $f \in \overline{\mathfrak{R}}$ und folglich nach Satz 1 meßbar. Die Funktionen aus $\mathfrak{E}$ sind als „$\mathfrak{Q}$-Treppenfunktionen" beschränkt und im Komplement einer quadrierbaren Menge gleich Null. Die gleichen Eigenschaften besitzt dann auch $f$ als Element der Vervollständigung $\mathfrak{S}$ von $\mathfrak{E}$ bezüglich $\bar{I}$. Also ist die Bedingung erfüllt. Nach LOOMIS [*14*], S. 172 folgt aus dem Bestehen der Bedingung umgekehrt die Riemann-Integrierbarkeit von $f$.

In diesem Zusammenhang beweisen wir das

**Lemma 2.** *Eine reelle Funktion $f|E$ ist dann und nur dann im Komplement einer quadrierbaren Menge gleich Null, wenn eine nicht negative Funktion $h \in \mathfrak{R}$ derart existiert, daß $f(x) = 0$ ist für jedes $x \in E$ mit $h(x) < 1$.*

Beweis. Die Funktion $f$ sei im Komplement der quadrierbaren Menge $Q$ gleich Null. Zu $Q$ existiert eine Funktion $h \in \mathfrak{R}$ mit $\chi_Q \leq h$; diese leistet das Verlangte. — Umgekehrt existiere zu $f$ eine nicht negative Funktion $h \in \mathfrak{R}$ mit der genannten Eigenschaft. Nach Satz 1 ist $h$ meßbar. Daher existiert eine reelle Zahl $\lambda$ mit $0 < \lambda < 1$, für welche die Urbildmenge $Q = [h > \lambda]$ quadrierbar ist. Für ein beliebiges nicht in $Q$ gelegenes Element $x \in E$ ist $h(x) \leq \lambda < 1$ und folglich $f(x) = 0$. Also ist $f$ im Komplement von $Q$ gleich Null.

Mit Hilfe dieses Lemmas ergeben sich nun zwei Korollare zu Satz 2.

**Korollar 1.** *Notwendig und hinreichend dafür, daß die beschränkten Funktionen aus $\mathfrak{R}$ sämtlich bezüglich $I$ und damit bezüglich eines jeden auf $\mathfrak{R}$ definierten, positiven, linearen Funktionals Riemann-integrierbar sind, ist folgende Bedingung:*

(8) *Zu jeder Funktion $g \in \mathfrak{R}$ existiert eine nicht negative Funktion $h \in \mathfrak{R}$ derart, daß $g(x) = 0$ ist für jedes $x \in E$ mit $h(x) < 1$*[7]).

---

[7]) Die Bedingung (8) tritt in anderem Zusammenhang auch bei ARENS [*1*] auf.

Beweis. Wegen Satz 2 und Lemma 2 ist nur noch zu zeigen, daß die Bedingung (8) notwendig ist. Hierzu sei $g$ eine Funktion aus $\mathfrak{R}$. Wir setzen $g' = \min(1, g^+) - \min(1, g^-)$, wobei $g^+$ den Positiv- und $g^-$ den Negativteil von $g$ bezeichnet. Dann ist $g'$ eine beschränkte Funktion aus $\mathfrak{R}$. Nach Satz 2 und Lemma 2 existiert also eine nicht negative Funktion $h \in \mathfrak{R}$ derart, daß aus $h(x) < 1$ stets $g'(x) = 0$ folgt. Die Funktion $h$ leistet auch bezüglich $g$ das in (8) Verlangte: Für ein $x \in E$ ist entweder $g^+(x) = 0$ oder $g^-(x) = 0$; folglich zieht $g'(x) = 0$ die Gleichungen $g^+(x) = g^-(x) = g(x) = 0$ nach sich.

**Korollar 2.** *Eine beschränkte, durch Funktionen aus $\mathfrak{R}$ gleichmäßig approximierbare, reelle Funktion $f|E$ ist dann und nur dann Riemann-integrierbar bezüglich $I$ und damit bezüglich eines jeden auf $\mathfrak{R}$ definierten, positiven, linearen Funktionals, wenn eine nicht negative Funktion $h \in \mathfrak{R}$ existiert derart, daß $f(x) = 0$ ist für jedes $x \in E$ mit $h(x) < 1$.*

Beweis. Wegen der Sätze 1 und 2 sowie wegen Lemma 2 genügt es zu zeigen, daß die Funktion $f$ in $\overline{\mathfrak{R}}$ liegt, falls zu $f$ ein $h \in \mathfrak{R}$ mit den genannten Eigenschaften existiert. Hierzu kann ohne Beschränkung der Allgemeinheit $f \geq 0$ angenommen werden, da mit $f$ auch $f^+$ und $f^-$ beschränkt und durch Funktionen aus $\mathfrak{R}$ gleichmäßig approximierbar sind und da aus $h(x) < 1$ auch $f^+(x) = f^-(x) = 0$ folgt. Zu $f \geq 0$ und einer beliebigen reellen Zahl $\varepsilon > 0$ existiert aber nach Lemma 1 eine Funktion $g \in \mathfrak{R}$ mit $0 \leq g \leq f \leq g + \varepsilon$. Da aus $f(x) \neq 0$ für ein $x \in E$ folgt $h(x) \geq 1$, gilt dann sogar $g' \leq f \leq h'$, wobei $g' = g$ und $h' = g + \varepsilon h$ Funktionen aus $\mathfrak{R}$ sind. Wegen $I(h' - g') = \varepsilon I(h)$ liegt also $f$ in $\overline{\mathfrak{R}}$.

**1.6.** Eine nicht negative, endliche, reelle Funktion $f|E$ heißt *uneigentlich Riemann-integrierbar* (bezüglich des Funktionals $I$), wenn $f$ $I$-meßbar und die Menge der für alle $g \in \mathfrak{S}$ mit $g \leq f$ gebildeten Zahlen $\int g \, d\mu$ nach oben beschränkt ist. Als uneigentliches Riemann-Integral $\overset{(u)}{\int} f \, d\mu$ von $f$ wird die reelle Zahl

$$\overset{(u)}{\int} f \, d\mu = \sup\left(\int g \, d\mu;\ g \leq f,\ g \in \mathfrak{S}\right) \tag{9}$$

definiert. Eine beliebige reelle Funktion $f|E$ heißt uneigentlich Riemann-integrierbar, wenn ihr Positiv- und Negativteil uneigentlich Riemann-integrierbar sind. Man setzt

$$\overset{(u)}{\int} f \, d\mu = \overset{(u)}{\int} f^+ \, d\mu - \overset{(u)}{\int} f^- \, d\mu. \tag{10}$$

Aus Satz 2 folgt, daß die Menge $\mathfrak{S}$ der eigentlich Riemann-integrierbaren Funktionen ein *gesättigter*, linearer Unterraum von $\overline{\mathfrak{R}}$ ist, d.h. ein linearer Unterraum mit folgender Eigenschaft:

(11) *Aus $|g| \leq |f|$, wobei $f \in \mathfrak{S}$ und $g \in \overline{\mathfrak{R}}$ ist, folgt $g \in \mathfrak{S}$.*

Also sind die Begriffsbildungen von [5] anwendbar. Danach heißt ein auf $\overline{\mathfrak{R}}$ definiertes, positives, lineares Funktional $L$ *$\mathfrak{S}$-singulär*, wenn

$$L(f) = \sup\left(L(g);\ g \leq f,\ g \in \mathfrak{S}\right) \tag{12}$$

für jede nicht negative Funktion $f \in \overline{\mathfrak{R}}$ gilt.

**Satz 3.** *Die Funktionen aus $\overline{\mathfrak{R}}$ sind sämtlich uneigentlich Riemann-integrierbar bezüglich $I$. Das somit für alle $f \in \overline{\mathfrak{R}}$ definierte uneigentliche Riemann-Integral $\overset{(u)}{\int} f\,d\mu$ ist die einzig mögliche Fortsetzung des auf $\mathfrak{S}$ definierten, positiven, linearen Funktionals $g \to \int g\,d\mu$ zu einem $\mathfrak{S}$-singulären, positiven, linearen Funktional auf $\overline{\mathfrak{R}}$. Hierbei gilt für jedes $f \in \overline{\mathfrak{R}}$*

$$\left|\overset{(u)}{\int} f\,d\mu\right| \leq \bar{I}\,(|f|). \tag{13}$$

Beweis. Es sei $f$ eine nicht negative Funktion aus $\overline{\mathfrak{R}}$. Zufolge (7) gilt dann $\int g\,d\mu \leq I(f)$ für jedes $g \in \mathfrak{S}$ mit $g \leq f$. Da $f$ nach Satz 1 auch meßbar ist, so ist also $f$ uneigentlich Riemann-integrierbar. Gleiches gilt dann für eine beliebige Funktion $f \in \overline{\mathfrak{R}}$, da mit $f$ auch $f^+$ und $f^-$ Elemente von $\overline{\mathfrak{R}}$ sind.

Für nicht negative Funktionen $f', f'' \in \overline{\mathfrak{R}}$ folgt aus (9) $\overset{(u)}{\int} (f'+f'')\,d\mu = \overset{(u)}{\int} f'\,d\mu + \overset{(u)}{\int} f''\,d\mu$. Hierzu genügt es zu bemerken, daß zu jeder Funktion $g \in \mathfrak{S}$ mit $0 \leq g \leq f'+f''$ Funktionen $g', g'' \in \mathfrak{S}$ mit $g = g'+g''$ sowie $0 \leq g' \leq f'$ und $0 \leq g'' \leq f''$ existieren; wegen (11) leisten z.B. $g' = \min(g, f')$ und $g'' = \max(0, g-f')$ das Verlangte. Aus der Gültigkeit der Gleichung

$$\overset{(u)}{\int} (f'+f'')\,d\mu = \overset{(u)}{\int} f'\,d\mu + \overset{(u)}{\int} f''\,d\mu$$

und $\overset{(u)}{\int} f'\,d\mu \geq 0$ für beliebige $f', f'' \in \overline{\mathfrak{R}}$ mit $f' \geq 0$ und $f'' \geq 0$ sowie aus (10) folgt aber (vgl. [8], S. 34), daß $f \to \overset{(u)}{\int} f\,d\mu$ auf $\overline{\mathfrak{R}}$ ein positives, lineares Funktional ist. Daß dieses die einzig mögliche $\mathfrak{S}$-singuläre Fortsetzung des auf $\mathfrak{S}$ erklärten, eigentlichen Riemann-Integrals ist, folgt aus (9) und (10). Aus (9) folgt (13) zunächst für den Spezialfall einer Funktion $f \geq 0$ aus $\overline{\mathfrak{R}}$ und hieraus allgemein bei Beachtung von (10).

## § 2. Zerlegungssätze

**2.1.** Es sei $\mathfrak{M}$ ein Vektorverband von auf einer Menge $A$ definierten, endlichen, reellen Funktionen und $L$ ein auf $\mathfrak{M}$ definiertes, positives, lineares Funktional. Dann heißt $L$ bekanntlich *stetig*, wenn es der folgenden Bedingung genügt: Für jede nicht fallende Folge $\{f_n\}_{n=1,2,\ldots}$ von Funktionen aus $\mathfrak{M}$, deren obere Einhüllende $f = \sup f_n$ ein Element von $\mathfrak{M}$ ist, gilt $L(f) = \sup L(f_n)$. Dagegen heißt $L$ *rein-unstetig*, wenn $K = 0$ das einzige stetige, (positive) lineare Funktional auf $\mathfrak{M}$ ist mit $0 \leq K(f) \leq L(f)$ für jede Funktion $f \geq 0$ aus $\mathfrak{M}$.

Nach einem in [3], S. 116 von uns bewiesenen Zerlegungssatz existieren zu einem positiven, linearen Funktional $L|\mathfrak{M}$ genau ein stetiges, positives, lineares Funktional $L_c|\mathfrak{M}$ und genau ein rein-unstetiges, positives lineares Funktional $L_p|\mathfrak{M}$ derart, daß $L = L_c + L_p$ ist. Es heißt $L_c$ der *stetige* und $L_p$ der *rein-unstetige Teil* von $L$.

Drücken wir den Sachverhalt, daß $\{f_n\}_{n=1,2,\ldots}$ eine nicht fallende Folge reeller Funktionen auf $A$ mit der oberen Einhüllenden $f$ ist, durch das Zeichen $f_n \uparrow f$ aus und bezeichnen wir mit $\mathfrak{M}_+$ die Menge aller $f \in \mathfrak{M}$ mit $f \geq 0$, so gilt

**Lemma 3.** *Für jede Funktion* $f \in \mathfrak{M}_+$ *ist*

$$L_c(f) = \inf\left(\sup L(f_n);\ f_n \uparrow f,\ f_n \in \mathfrak{M}_+\right).$$

Beweis. Wir setzen $L^*(f) = \inf\left(\sup L(f_n);\ f_n \uparrow f,\ f_n \in \mathfrak{M}_+\right)$ für eine beliebige Funktion $f \in \mathfrak{M}_+$ und behaupten zunächst, daß $L^*$ auf genau eine Weise zu einem positiven, linearen Funktional auf $\mathfrak{M}$ fortgesetzt werden kann, welches wir ebenfalls mit $L^*$ bezeichnen. Hierfür ist notwendig und hinreichend, daß (a) $L^*(f+g) = L^*(f) + L^*(g)$ und (b) $L^*(f) \geqq 0$ ist für beliebige Funktionen $f, g \in \mathfrak{M}_+$. Während (b) unmittelbar klar ist, folgt (a) aus der Bemerkung, daß die Existenz einer Folge von Funktionen $h_n \in \mathfrak{M}_+$ mit $h_n \uparrow f+g$ äquivalent ist mit der Existenz zweier Folgen von Funktionen $f_n \in \mathfrak{M}_+$ und $g_n \in \mathfrak{M}_+$ mit $h_n = f_n + g_n$ sowie $f_n \uparrow f$ und $g_n \uparrow g$. Bei gegebener Folge $\{h_n\}$ leisten nämlich $f_n = \min(h_n, f)$ und $g_n = h_n - f_n = \max(0, h_n - f)$ das Verlangte; aus $f_n \uparrow f$ und $g_n \uparrow g$ folgt umgekehrt $f_n + g_n \uparrow f+g$.

Wir behaupten weiter, daß $L^* \mid \mathfrak{M}$ stetig ist. Es genügt hierfür zu zeigen, daß $L^*(f) \leqq \sup L^*(f_n)$ ist für jede Folge $\{f_n\}_{n=1,2,\ldots}$ von Funktionen aus $\mathfrak{M}_+$ mit $f_n \uparrow f$ und $f \in \mathfrak{M}_+$. Nach Definition von $L^*$ existiert zu beliebigem $\varepsilon > 0$ und für jedes $n = 1, 2, \ldots$ eine Folge $\{g_{nk}\}_{k=1,2,\ldots}$ von Funktionen aus $\mathfrak{M}_+$ mit $g_{nk} \uparrow f_n$ und

$$\sup_{k=1,2,\ldots} L(g_{nk}) \leqq L^*(f_n) + \frac{\varepsilon}{2^n}.$$

Wir setzen $f_{nk} = \max(g_{1k}, \ldots, g_{nk})$ und zeigen mittels vollständiger Induktion nach $n$ die Gültigkeit der Ungleichung

$$\sup_{k=1,2,\ldots} L(f_{nk}) \leqq L^*(f_n) + \varepsilon\left(1 - \frac{1}{2^n}\right) \qquad (n = 1, 2, \ldots).$$

Für $n = 1$ ist nichts zu zeigen. Der Schluß von $n$ auf $n+1$ verläuft so: Setzt man $h_{nk} = \min(f_{nk}, g_{n+1,k})$, so ist $f_{n+1,k} = \max(f_{nk}, g_{n+1,k}) = f_{nk} + g_{n+1,k} - h_{nk}$ und folglich

$$\begin{aligned} L(f_{n+1,k}) &= L(f_{nk}) + L(g_{n+1,k}) - L(h_{nk}) \\ &\leqq L^*(f_n) + \varepsilon\left(1 - \frac{1}{2^n}\right) + L^*(f_{n+1}) + \frac{\varepsilon}{2^{n+1}} - L(h_{nk}); \end{aligned}$$

wegen $h_{nk} \uparrow f_n$ ist $L^*(f_n) \leqq \sup\limits_{k=1,2,\ldots} L(h_{nk})$ und daher, wie behauptet,

$$\sup_{k=1,2,\ldots} L(f_{n+1,k}) \leqq L^*(f_{n+1}) + \varepsilon\left(1 - \frac{1}{2^{n+1}}\right).$$

Nach Konstruktion gilt $f_{nk} \leqq f_{n+1,k+1}$ und entsprechend $L(f_{nk}) \leqq L(f_{n+1,k+1})$ für beliebige $n, k = 1, 2, \ldots$. Es ist also $\{f_{nn}\}_{n=1,2,\ldots}$ eine nicht fallende Folge von Funktionen aus $\mathfrak{M}_+$ mit $\sup f_{nn} = \sup\limits_n \left(\sup\limits_k f_{nk}\right) = f$ und $\sup L(f_{nn}) = \sup\limits_n \left(\sup\limits_k L(f_{nk})\right)$[8]. Aus der für alle $n = 1, 2, \ldots$ bewiesenen Ungleichung folgt also nunmehr

$$L^*(f) \leqq \sup L(f_{nn}) \leqq \sup L^*(f_n) + \varepsilon$$

für jedes $\varepsilon > 0$ und hieraus $L^*(f) \leqq \sup L^*(f_n)$. Also ist $L^*$ stetig.

[8]) Vgl. Haupt-Aumann-Pauc [*12a*], S. 147, Nr. 5.4.4.1.

Nach Definition von $L^*$ ist $L - L^*$ ein positives, lineares Funktional auf $\mathfrak{M}$. Wir zeigen, daß $L - L^*$ rein-unstetig ist. Hierzu sei $K|\mathfrak{M}$ ein stetiges, positives, lineares Funktional mit $0 \leq K(f) \leq L(f) - L^*(f)$ für alle $f \in \mathfrak{M}_+$. Für eine beliebige Folge $\{f_n\}_{n=1,2,\ldots}$ von Funktionen aus $\mathfrak{M}_+$ mit $f_n \uparrow f$ und $f \in \mathfrak{M}_+$ gilt dann die Ungleichung $0 \leq K(f) \leq \sup L(f_n) - L^*(f)$ wegen der Stetigkeit von $K$ und $L^*$. Hieraus folgt $K(f) = 0$ nach Definition von $L^*$. Also ist $L - L^*$ rein-unstetig.

Aus den bewiesenen Eigenschaften von $L^*$ folgt $L^* = L_c$ und damit die Behauptung.

**2.2.** Wie in § 1 sei nun wieder ein System $\mathfrak{R}$ auf einer Menge $E$ definierter, reeller Funktionen mit den Eigenschaften (1) bis (3) und ein positives, lineares Funktional $I|\mathfrak{R}$ vorgegeben.

Bezeichnet $I_c$ den stetigen und $I_p$ den rein-unstetigen Teil von $I|\mathfrak{R}$, so ergeben sich die zweiseitigen Vervollständigungen

$$\overline{\mathfrak{R}}_c = \mathfrak{V}(\mathfrak{R}, I_c) \quad \text{und} \quad \overline{\mathfrak{R}}_p = \mathfrak{V}(\mathfrak{R}, I_p)$$

von $\mathfrak{R}$ bezüglich $I_c$ und $I_p$ sowie die Booleschen Mengenverbände

$$\mathfrak{Q}_c = \mathfrak{Q}(\mathfrak{R}, I_c) \quad \text{und} \quad \mathfrak{Q}_p = \mathfrak{Q}(\mathfrak{R}, I_p)$$

bestehend aus allen $I_c$- bzw. $I_p$-quadrierbaren Teilmengen von $E$. Schließlich ergeben sich noch die Systeme $\mathfrak{S}_c$ und $\mathfrak{S}_p$ der bezüglich $I_c$ bzw. $I_p$ Riemann-integrierbaren Funktionen. Wir behaupten die Gültigkeit der folgenden drei Gleichungen[9]):

$$\overline{\mathfrak{R}} = \overline{\mathfrak{R}}_c \cap \overline{\mathfrak{R}}_p; \tag{14}$$

$$\mathfrak{Q} = \mathfrak{Q}_c \cap \mathfrak{Q}_p; \tag{15}$$

$$\mathfrak{S} = \mathfrak{S}_c \cap \mathfrak{S}_p. \tag{16}$$

Zum Beweis von (14) sei $f \in \overline{\mathfrak{R}}$. Zu beliebigem $\varepsilon > 0$ existieren dann Funktionen $g, h \in \mathfrak{R}$ mit $g \leq f \leq h$ und $I(h-g) \leq \varepsilon$. Wegen $I_c(h-g) \leq I(h-g)$ und $I_p(h-g) \leq I(h-g)$ ist also $f \in \overline{\mathfrak{R}}_c \cap \overline{\mathfrak{R}}_p$. Umgekehrt sei $f \in \overline{\mathfrak{R}}_c \cap \overline{\mathfrak{R}}_p$. Dann existieren zu beliebigem $\varepsilon > 0$ Funktionen $g', h', g'', h'' \in \mathfrak{R}$ mit $g' \leq f \leq h'$ und $g'' \leq f \leq h''$ derart, daß $I_c(h'-g') \leq \varepsilon$ und $I_p(h''-g'') \leq \varepsilon$ ist. Setzen wir also $g = \max(g', g'')$ und $h = \min(h', h'')$, so sind $g$ und $h$ Funktionen aus $\mathfrak{R}$ mit $g \leq f \leq h$ und $h - g \leq \min(h'-g', h''-g'')$. Folglich ist $I(h-g) = I_c(h-g) + I_p(h-g) \leq 2\varepsilon$, also $f \in \overline{\mathfrak{R}}$. — Aus (14) folgt (15), da $Q \in \mathfrak{Q}$ gleichbedeutend mit $\chi_Q \in \overline{\mathfrak{R}}$ ist. — Eine reelle Funktion $f|E$, die sowohl im Komplement einer $I_c$-quadrierbaren Menge als auch im Komplement einer $I_p$-quadrierbaren Menge gleich Null ist, ist nach Lemma 2 auch im Komplement einer $I$-quadrierbaren Menge gleich Null. Beachtet man dies, so folgt (16) aus (15) und Satz 2.

[9]) Bei deren Beweis wird nur benützt, daß $I_c$ und $I_p$ positive, lineare Funktionale auf $\mathfrak{R}$ mit $I = I_c + I_p$ sind.

**2.3.** Nach 1.2 können die Funktionale $I_c|\mathfrak{R}$ und $I_p|\mathfrak{R}$ zu positiven, linearen Funktionalen $\overline{I}_c|\overline{\mathfrak{R}}_c$ und $\overline{I}_p|\overline{\mathfrak{R}}_p$ vervollständigt werden; wegen (14) ist $\overline{\mathfrak{R}}$ der Durchschnitt der Definitionsbereiche von $\overline{I}_c$ und $\overline{I}_p$. Nach 1.3 leiten sich aus $I_c$ und $I_p$ die vollständigen Inhalte $\mu_c|\mathfrak{Q}_c$ und $\mu_p|\mathfrak{Q}_p$ ab, welche wegen (15) insbesondere auf $\mathfrak{Q}$ definiert sind. Der folgende Satz enthält unter anderem die Aussage, daß $\mu_c|\mathfrak{Q}$ der $\sigma$-additive und $\mu_p|\mathfrak{Q}$ der rein-endlich-additive Teil von $\mu|\mathfrak{Q}$ ist.

Hierzu sei bemerkt, daß eine auf einem Booleschen Verband $\mathfrak{B}$ von Teilmengen einer Grundmenge $A$ definierte, nicht negative, additive Mengenfunktion $\varrho$ *$\sigma$-additiv* heißt, wenn $\sup \varrho(M_n) = \varrho(M)$ ist für jede nicht fallende Folge $\{M_n\}_{n=1,2,\ldots}$ von Mengen aus $\mathfrak{B}$ mit $M = \bigcup_{n=1}^{\infty} M_n \in \mathfrak{B}$. Dagegen heißt $\varrho$ *rein-endlich-additiv*, wenn $\tau = 0$ die einzige $\sigma$-additive Funktion $\tau|\mathfrak{B}$ ist mit $0 \leq \tau(M) \leq \varrho(M)$ für alle $M \in \mathfrak{B}$. Nach Yosida und Hewitt [*23*][10]) existieren zu jeder nicht negativen, additiven Funktion $\varrho|\mathfrak{B}$ genau eine nicht negative, $\sigma$-additive Funktion $\varrho_c|\mathfrak{B}$ und genau eine nicht negative, rein-endlich-additive Funktion $\varrho_p|\mathfrak{B}$ mit $\varrho = \varrho_c + \varrho_p$. Man nennt $\varrho_c$ den *$\sigma$-additiven* und $\varrho_p$ den *rein-endlich-additiven Teil* von $\varrho$. In Analogie zu der in Lemma 3 behaupteten Formel gilt nach Woodbury [*22*][11])

$$\varrho_c(M) = \inf\left(\sup \varrho(M_n);\ M_n \uparrow M,\ M_n \in \mathfrak{B}\right) \tag{17}$$

für alle $M \in \mathfrak{B}$. Hierbei steht $M_n \uparrow M$ für $M_1 \subseteqq M_2 \subseteqq \cdots \subseteqq M = \bigcup_{n=1}^{\infty} M_n$.

Der bereits angekündigte Satz lautet nun:

**Satz 4.** I. *Es ist $\overline{I}_c|\overline{\mathfrak{R}}$ der stetige und $\overline{I}_p|\overline{\mathfrak{R}}$ der rein-unstetige Teil des Funktionals $\overline{I}|\overline{\mathfrak{R}}$; also gilt auf $\overline{\mathfrak{R}}$*

$$\overline{I}_c = \overline{I}_c \quad \textit{und} \quad \overline{I}_p = \overline{I}_p. \tag{18}$$

II. *Es ist $\mu_c|\mathfrak{Q}$ der $\sigma$-additive und $\mu_p|\mathfrak{Q}$ der rein-endlich-additive Teil von $\mu|\mathfrak{Q}$.*

Beweis. *Zu* I. Wir zeigen zunächst, daß das Funktional $\overline{I}_c|\overline{\mathfrak{R}}_c$ stetig ist; insbesondere ist dann $\overline{I}_c|\overline{\mathfrak{R}}$ stetig. Hierzu sei $f \in \overline{\mathfrak{R}}_c$ und $\{f_n\}_{n=1,2,\ldots}$ eine Folge von Funktionen aus $\overline{\mathfrak{R}}_c$ mit $f_n \uparrow f$; ferner sei $\varepsilon > 0$ eine reelle Zahl. Dann existiert ein $g \in \mathfrak{R}$ mit $g \leq f$ und $\overline{I}_c(f) - \varepsilon \leq I_c(g)$, sowie für $n = 1, 2, \ldots$ ein $h_n \in \mathfrak{R}$ mit $f_n \leq h_n$ und $I_c(h_n) \leq \overline{I}_c(f_n) + \varepsilon/2^n$. Setzen wir $g_n = \max(h_1, \ldots, h_n)$ für jedes $n$, so ist $\{g_n\}$ eine nicht fallende Folge von Funktionen aus $\mathfrak{R}$ mit $f_n \leq g_n$ und $I_c(g_n) \leq \overline{I}_c(f_n) + \varepsilon(1 - 1/2^n)$; diese Ungleichung beweist man wie die hierzu analoge Ungleichung im Beweis von Lemma 3 durch vollständige Induktion nach $n$. Nun ist $g \leq f \leq \sup g_n$ und folglich $I_c(g) \leq \sup I_c(g_n)$ wegen der Stetigkeit von $I_c|\mathfrak{R}$. Hieraus folgt $\overline{I}_c(f) - \varepsilon \leq I_c(g) \leq \sup I_c(g_n) \leq \sup \overline{I}_c(f_n) + \varepsilon$; also ist $\overline{I}_c(f) \leq \sup \overline{I}_c(f_n)$. Wegen der Positivität von $\overline{I}_c|\overline{\mathfrak{R}}_c$ folgt hieraus die behauptete Stetigkeit.

[10]) Wie in [*3*], S. 113 von Verf. gezeigt wurde, ist die bei Yosida-Hewitt gemachte Annahme der Existenz der Einheit $A$ in V unnötig.

[11]) Ein Beweis dieser Formel unter allgemeineren Voraussetzungen findet sich in [*6*], S. 219.

Das Funktional $\overline{I_p}|\mathfrak{R}$ (und damit auch $\overline{I_p}|\overline{\mathfrak{R}}_p$) ist rein-unstetig. Es sei nämlich $K|\overline{\mathfrak{R}}$ ein stetiges, positives, lineares Funktional mit $0 \leq K(f) \leq \overline{I_p}(f)$ für alle $f \geq 0$ aus $\overline{\mathfrak{R}}$. Da $\overline{I_p}(g) = I_p(g)$ für alle $g \in \mathfrak{R}$ und $I_p|\mathfrak{R}$ rein-unstetig ist, folgt dann zunächst $K(g) = 0$ für alle $g \in \mathfrak{R}$ und hieraus $K(f) = 0$ für alle $f \in \overline{\mathfrak{R}}$. Man hat hierzu nur zu berücksichtigen, daß $\overline{\mathfrak{R}} \subseteqq \mathfrak{B}(\mathfrak{R}, K)$ ist.

Schließlich ist $\overline{I_c} + \overline{I_p}$ eine Fortsetzung von $I|\mathfrak{R}$ zu einem positiven, linearen Funktional auf $\overline{\mathfrak{R}}$. Da $\overline{I}|\overline{\mathfrak{R}}$ nach 1.2 die einzige derartige Fortsetzung ist, gilt $\overline{I} = \overline{I_c} + \overline{I_p}$ auf $\overline{\mathfrak{R}}$. Also ist $\overline{I_c} = \overline{I}_c$ und $\overline{I_p} = \overline{I}_p$ auf $\overline{\mathfrak{R}}$, wie behauptet wurde.

*Zu* II. Es sei $\{Q_n\}_{n=1,2,\ldots}$ eine Folge von Mengen aus $\mathfrak{Q}_c$ mit $Q_n \uparrow Q$ und $Q \in \mathfrak{Q}_c$. Dann gilt $\chi_{Q_n} \uparrow \chi_Q$ für die Folge der zugehörigen charakteristischen Funktionen. Zufolge der bereits bewiesenen Stetigkeit von $\overline{I_c}|\overline{\mathfrak{R}}_c$ ist also $\sup \mu_c(Q_n) = \sup \overline{I_c}(\chi_{Q_n}) = \overline{I_c}(\chi_Q) = \mu_c(Q)$. Folglich ist $\mu_c|\mathfrak{Q}_c$ und damit auch $\mu_c|\mathfrak{Q}$ $\sigma$-additiv.

Für den Nachweis, daß $\mu_p|\mathfrak{Q}$ (und damit übrigens auch $\mu_p|\mathfrak{Q}_p$) rein-endlich-additiv ist, sei $\varrho|\mathfrak{Q}$ eine $\sigma$-additive Funktion mit $0 \leq \varrho(Q) \leq \mu_p(Q)$ für alle $Q \in \mathfrak{Q}$. Für eine beliebige Funktion $f$ aus der Menge $\mathfrak{E}$ der $\mathfrak{Q}$-Treppenfunktionen und eine Darstellung von $f$ in der Form $f = \sum_{\nu=1}^{n} \lambda_\nu \chi_{Q_\nu}$ mit $Q_\nu \in \mathfrak{Q}$ und reellen Zahlen $\lambda_\nu$ setzen wir $R(f) = \sum_{\nu=1}^{n} \lambda_\nu \varrho(Q_\nu)$. Dann ist bekanntlich[12] $R|\mathfrak{E}$ eindeutig definiert und ein stetiges, positives, lineares Funktional. Es ist $0 \leq R(f) \leq \overline{I_p}(f)$ für alle $f \geq 0$ aus $\mathfrak{E}$. Aus dieser Ungleichung folgt $\mathfrak{S}_p \subseteqq \mathfrak{B}(\mathfrak{E}, R)$; also kann $R$ nach 1.2 auf genau eine Weise zu einem positiven, linearen Funktional $\overline{R}$ auf $\mathfrak{S}_p$ fortgesetzt werden. Nach dem zu I. Bewiesenen (angewandt auf $R|\mathfrak{E}$ an Stelle von $I_c|\mathfrak{R}$) ist $\overline{R}|\mathfrak{S}_p$ stetig; ferner gilt $0 \leq \overline{R}(f) \leq \overline{I_p}(f)$ für alle $f \geq 0$ aus $\mathfrak{S}_p$. Wie im Beweis des Satzes 3 zeigt man, daß $\overline{R}|\mathfrak{S}_p$ auf genau eine Weise zu einem $\mathfrak{S}_p$-singulären, positiven, linearen Funktional $I_\varrho|\overline{\mathfrak{R}}_p$ fortgesetzt werden kann. Für jede Funktion $f \geq 0$ aus $\overline{\mathfrak{R}}_p$ gilt dann wegen (12) $I_\varrho(f) = \sup(\overline{R}(g); g \leq f, g \in \mathfrak{S}_p)$; hieraus folgt $0 \leq I_\varrho(f) \leq \overline{I_p}(f)$ für jedes solche $f$. Da $\overline{I_p}|\overline{\mathfrak{R}}_p$ nach dem zu I. Bewiesenen rein-unstetig ist, folgt also $I_\varrho(f) = 0$ für alle $f \in \overline{\mathfrak{R}}_p$ und hieraus $\varrho(Q) = 0$ für alle $Q \in \mathfrak{Q}$, falls $I_\varrho$ ein stetiges Funktional ist. Um zu zeigen, daß $\mu_p|\mathfrak{Q}$ rein-endlich-additiv ist, haben wir also noch die Stetigkeit von $I_\varrho$ zu beweisen. Hierzu sei $\{f_n\}_{n=1,2,\ldots}$ eine Folge nicht negativer Funktionen aus $\overline{\mathfrak{R}}_p$ mit $f_n \uparrow f$ und $f \in \overline{\mathfrak{R}}_p$. Wegen der Positivität von $I_\varrho$ ist nur $I_\varrho(f) \leq \sup I_\varrho(f_n)$ zu zeigen. Setzen wir $g_n = \min(f_n, g)$ für eine beliebige Funktion $g \in \mathfrak{S}_p$ mit $0 \leq g \leq f$, so sind, da $\mathfrak{S}_p$ ein gesättigter, linearer Unterraum von $\overline{\mathfrak{R}}_p$ ist, alle $g_n$ Elemente von $\mathfrak{S}_p$; ferner gilt $g_n \uparrow g$. Wegen der Stetigkeit von $\overline{R}|\mathfrak{S}_p$ ist also $\overline{R}(g) = \sup \overline{R}(g_n)$. Es ist $\overline{R}(g_n) \leq I_\varrho(f_n)$ für jedes $n$ und somit $\overline{R}(g) \leq \sup I_\varrho(f_n)$. Da $g \in \mathfrak{S}_p$ beliebig mit $0 \leq g \leq f$ gewählt war, folgt hieraus die behauptete Ungleichung $I_\varrho(f) \leq \sup I_\varrho(f_n)$.

---

[12]) Vgl. etwa AUMANN [2], S. 365, Satz 9.1.3.1.

Aus der für alle $f \in \overline{\mathfrak{R}}$ gültigen Gleichung $\bar{I} = \bar{I}_c + \bar{I}_p$ folgt schließlich $\mu(Q) = \mu_c(Q) + \mu_p(Q)$ für alle $Q \in \mathfrak{Q}$. Also ist auf $\mathfrak{Q}$ $\mu_c$ der $\sigma$-additive und $\mu_p$ der rein-endlich-additive Teil von $\mu$.

**Korollar.** *Auf $\mathfrak{S}$ ist $f \to \int f\,d\mu_c$ der stetige und $f \to \int f\,d\mu_p$ der rein-unstetige Teil des auf $\mathfrak{S}$ definierten Funktionals $f \to \int f\,d\mu$.*

Beweis. Es ist $\int f\,d\mu = \int f\,d\mu_c + \int f\,d\mu_p$ für jede Funktion $f \in \mathfrak{S}$, da auf $\overline{\mathfrak{R}}$ die Gleichung $\bar{I} = \bar{I}_c + \bar{I}_p$ gilt. Das auf $\mathfrak{S}$ definierte, positive, lineare Funktional $f \to \int f\,d\mu_c$ ist als Verengerung des stetigen Funktionals $\bar{I}_c|\overline{\mathfrak{R}}$ selbst stetig. Um zu zeigen, daß das auf $\mathfrak{S}$ definierte, positive, lineare Funktional $f \to \int f\,d\mu_p$ rein-unstetig ist, sei $K|\mathfrak{S}$ ein stetiges, positives, lineares Funktional mit $0 \leq K(f) \leq \int f\,d\mu_p$ für alle $f \geq 0$ aus $\mathfrak{S}$. Setzen wir $\varkappa(Q) = K(\chi_Q)$ für jedes $Q \in \mathfrak{Q}$, so ist $\varkappa|\mathfrak{Q}$ $\sigma$-additiv und es gilt $0 \leq \varkappa(Q) \leq \mu_p(Q)$ für alle $Q \in \mathfrak{Q}$. Nach der Behauptung II von Satz 4 ist also $\varkappa = 0$. Dann ist aber $K = 0$ auf $\mathfrak{S}$, da aus der für alle $f \geq 0$ aus $\mathfrak{S}$ gültigen Ungleichung $0 \leq K(f) \leq \int f\,d\mu_p$ folgt $\mathfrak{S} \subseteqq \mathfrak{B}(\mathfrak{C}, K)$ und somit die Fortsetzung von $K|\mathfrak{C}$ zu einem positiven, linearen Funktional auf $\mathfrak{B}(\mathfrak{C}, K)$ eindeutig bestimmt ist. Also ist das Funktional $f \to \int f\,d\mu_p$ rein-unstetig.

Nach Satz 3 sind die Funktionen aus $\mathfrak{R}$ in bezug auf jedes auf $\mathfrak{R}$ definierte, positive, lineare Funktional uneigentlich Riemann-integrierbar. Insbesondere existiert also $\overset{(u)}{\int} f\,d\mu_c$ für jedes $f \in \mathfrak{R}$. Über dieses auf $\mathfrak{R}$ definierte Funktional behaupten wir:

**Satz 5.** I. *Für jede Funktion $f \in \mathfrak{R}$ gilt*

$$I_c(f) = \overset{(u)}{\int} f\,d\mu_c. \tag{19}$$

II. *Das auf $\mathfrak{R}$ definierte, positive, lineare Funktional $f \to \overset{(u)}{\int} f\,d\mu$ hat $f \to \overset{(u)}{\int} f\,d\mu_c$ zum stetigen und $f \to \overset{(u)}{\int} f\,d\mu_p$ zum rein-unstetigen Teil.*

Beweis. *Zu* I. Es genügt die Gleichung $I_c(f) = \overset{(u)}{\int} f\,d\mu_c$ für jede nicht negative Funktion $f \in \mathfrak{R}$ zu beweisen. Es sei also $f \in \mathfrak{R}$ nicht negativ und außerdem vorerst beschränkt: $0 \leq f(x) \leq K$ für alle $x \in E$. Dann existiert eine Folge $\{g_n\}_{n=1,2,\ldots}$ von Funktionen aus $\mathfrak{S}$ mit $g_n \uparrow f$. Nach Satz 1 gibt es nämlich eine Folge $\{\alpha_n\}_{n=1,2,\ldots}$ reeller Zahlen mit $\alpha_1 > \alpha_2 > \cdots > 0$ und $\lim \alpha_n = 0$ derart, daß $Q_n = [f > \alpha_n]$ für jedes $n$ $I$-quadrierbar ist. Die Folge der Funktionen $g_n = \min(f, K\chi_{Q_n})$ leistet dann das Verlangte (wegen (11) ist $g_n \in \mathfrak{S}$). Nach Satz 4, Beh. I ist $\bar{I}_c$ auf $\overline{\mathfrak{R}}$ stetig; folglich gilt $I_c(f) = \sup \bar{I}_c(g_n) = \sup \int g_n\,d\mu_c \leq \overset{(u)}{\int} f\,d\mu_c$. Da nach (13) andererseits $\overset{(u)}{\int} f\,d\mu_c \leq I_c(f)$ ist, ist damit die behauptete Gleichung bewiesen.

Nunmehr sei $f$ eine beliebige nicht negative Funktion aus $\mathfrak{R}$. Dann ist $f_n = \min(n, f)$, $n = 1, 2, \ldots$, eine Folge beschränkter, nicht negativer Funktionen aus $\mathfrak{R}$ mit $f_n \uparrow f$; also gilt $I_c(f) = \sup I_c(f_n) = \sup \overset{(u)}{\int} f_n\,d\mu_c \leq \overset{(u)}{\int} f\,d\mu_c$ nach dem bisher Bewiesenen. Wegen (13) gilt also auch in diesem Falle (19).

Aus dem Beweis von Gl. (19) entnehmen wir noch das für das Folgende nützliche Ergebnis: Für jede Funktion $f \geqq 0$ aus $\mathfrak{R}$ ist

(20a) $$\overset{(u)}{\int} f\,d\mu_c = \sup\left(\int g\,d\mu_c;\ g \leqq f,\ g \in \mathfrak{S}\right).$$

*Zu* II. Nach I. ist das auf $\mathfrak{R}$ definierte, positive, lineare Funktional $f \to \overset{(u)}{\int} f\,d\mu_c$ stetig. Das auf $\mathfrak{R}$ definierte, positive, lineare Funktional $f \to \overset{(u)}{\int} f\,d\mu_p$ ist rein-unstetig, da $0 \leqq \overset{(u)}{\int} f\,d\mu_p \leqq I_p(f)$ für jede Funktion $f \geqq 0$ aus $\mathfrak{R}$ gilt. Es verbleibt also der Beweis der Gleichung $\overset{(u)}{\int} f\,d\mu = \overset{(u)}{\int} f\,d\mu_c + \overset{(u)}{\int} f\,d\mu_p$ für jedes $f \geqq 0$ aus $\mathfrak{R}$. Hierzu zeigen wir, daß neben (20a) auch

(20b) $$\overset{(u)}{\int} f\,d\mu_p = \sup\left(\int g\,d\mu_p;\ g \leqq f, g \in \mathfrak{S}\right)$$

für alle $f \geqq 0$ aus $\mathfrak{R}$ gilt. Die behauptete Gleichung folgt dann nämlich so: Es bezeichne $\Phi$ die steigend gerichtete Menge aller Funktionen $g \in \mathfrak{S}$ mit $g \leqq f$; dann ist

$$\overset{(u)}{\int} f\,d\mu = \lim_\Phi \int g\,d\mu = \lim_\Phi \int g\,d\mu_c + \lim_\Phi \int g\,d\mu_p = \overset{(u)}{\int} f\,d\mu_c + \overset{(u)}{\int} f\,d\mu_p.$$

Wir haben also nur noch den Beweis von (20b) nachzutragen. Es sei hierzu $Q \in \mathfrak{Q}_c$ mit $\chi_Q \leqq f$. Nach Satz 1 ist dann die Menge $Q' = [f > \alpha]$ für ein geeignetes $\alpha$ mit $0 < \alpha < 1$ Element von $\mathfrak{Q}$. Nach (11) ist $g = \min(f, \chi_{Q'})$ eine Funktion aus $\mathfrak{S}$; für diese gilt $\chi_Q \leqq g \leqq f$. — Aus der Definition der Riemann-integrierbaren Funktionen folgt unmittelbar die Gleichung

$$\overset{(u)}{\int} f\,d\mu_p = \sup\left(\int g\,d\mu_p;\ g \leqq f, g \in \mathfrak{T}_p\right),$$

wenn $\mathfrak{T}_p$ die Menge der $\mathfrak{Q}_p$-Treppenfunktionen bezeichnet. Die Gl. (20b) ist daher bewiesen, wenn gezeigt ist, daß zu beliebigem $g \in \mathfrak{T}_p$ mit $0 \leqq g \leqq f$ stets ein $g' \in \mathfrak{S}$ mit $g \leqq g' \leqq f$ existiert. Es kann $g$ dargestellt werden in der Form $g = \sum_{\nu=1}^{n} \lambda_\nu \chi_{Q_\nu}$ mit paarweise fremden Mengen $Q_\nu \in \mathfrak{Q}_p$ und reellen Zahlen $\lambda_\nu > 0$. Wegen $\chi_{Q_\nu} \leqq \frac{1}{\lambda_\nu} f$ existiert, wie bereits bewiesen wurde, eine Funktion $g_\nu \in \mathfrak{S}$ mit $\chi_{Q_\nu} \leqq g_\nu \leqq \frac{1}{\lambda_\nu} f$. Setzen wir also $g' = \max(\lambda_1 g_1, \ldots, \lambda_n g_n)$, so ist $g'$ eine Funktion aus $\mathfrak{S}$, die das Verlangte leistet.

Damit ist der Satz 5 vollständig bewiesen.

Im folgenden Satz vergleichen wir die auf $\mathfrak{R}$ definierten Funktionale $I$ und $f \to \overset{(u)}{\int} f\,d\mu$ hinsichtlich der sich aus ihnen ableitenden Riemann-Integrale.

**Satz 6.** I. *Die auf $\mathfrak{R}$ definierten, positiven, linearen Funktionale $I$ und $f \to \overset{(u)}{\int} f\,d\mu$ führen zur gleichen Menge $\mathfrak{S}$ Riemann-integrierbarer Funktionen und zum gleichen Riemann-Integral über $\mathfrak{S}$.*

II. *Das auf $\mathfrak{R}$ definierte, positive, lineare Funktional $f \to D(f) = I(f) - \overset{(u)}{\int} f\,d\mu$ ist rein-unstetig. Bezüglich $D$ sind genau diejenigen Funktionen auf $E$ Riemann-*

*integrierbar, welche beschränkt und im Komplement einer I-quadrierbaren Menge gleich Null sind. Für jede solche Funktion ist das Riemann-Integral bezüglich D gleich Null.*

Beweis. *Zu* I. Wir setzen $L(f) = \overset{(u)}{\int} f\,d\mu$ für jedes $f \in \mathfrak{R}$. Dann sei $\mathfrak{Q}_L$ das System der $L$-quadrierbaren Mengen und $\mu_L | \mathfrak{Q}_L$ der sich aus $L$ ableitende Inhalt. Es genügt zu zeigen, daß $\mathfrak{Q} = \mathfrak{Q}_L$ und $\mu = \mu_L$ ist. Zunächst ist $\mathfrak{Q} \subseteqq \mathfrak{Q}_L$, da nach (13) $0 \leqq L(f) \leqq I(f)$ für alle $f \geqq 0$ aus $\mathfrak{R}$ gilt. Ferner ist $\mu(Q) = \mu_L(Q)$ für jede Menge $Q \in \mathfrak{Q}$, da einerseits $f \to \bar{I}(f)$ bzw. $f \to \overset{(u)}{\int} f\,d\mu$ die einzige Fortsetzung von $I$ bzw. $L$ zu einem positiven, linearen Funktional auf $\overline{\mathfrak{R}}$ und andererseits $\overset{(u)}{\int} f\,d\mu = \int f\,d\mu$ für alle $f \in \mathfrak{S}$, also insbesondere für $f = \chi_Q$ ist. Es sei $Q$ nun eine Menge aus $\mathfrak{Q}_L$; dann gibt es zu beliebigem $\varepsilon > 0$ Funktionen $g, h \in \mathfrak{R}$ mit $0 \leqq g \leqq \chi_Q \leqq h$ und $L(h-g) \leqq \varepsilon$. Zu $g$ existiert nach (9) ein $g' \in \mathfrak{S}$ mit $g' \leqq g$ und $L(g) - \int g'\,d\mu \leqq \varepsilon$. Ferner beweist man wie im Beweis der Behauptung II von Satz 5 die Existenz einer Funktion $h' \in \mathfrak{S}$ mit $\chi_Q \leqq h' \leqq h$. Also gilt $g' \leqq \chi_Q \leqq h'$ und $\int (h' - g')\,d\mu \leqq L(h) - \int g'\,d\mu \leqq L(h-g) + \varepsilon \leqq 2\varepsilon$. Hieraus folgt $Q \in \mathfrak{Q}$ wegen der Vollständigkeit des Funktionals $f \to \int f\,d\mu$ auf $\mathfrak{S}$. Also ist $\mathfrak{Q} = \mathfrak{Q}_L$ und $\mu = \mu_L$.

*Zu* II. Nach Satz 5 ist $D(f) = I_c(f) + I_p(f) - \overset{(u)}{\int} f\,d\mu_c - \overset{(u)}{\int} f\,d\mu_p = I_p(f) - \overset{(u)}{\int} f\,d\mu_p$; da also $0 \leqq D(f) \leqq I_p(f)$ für alle $f \geqq 0$ aus $\mathfrak{R}$ gilt, ist $D$ reinunstetig. Nach der letzten Ungleichung ist ferner die Vervollständigung $\bar{I} | \overline{\mathfrak{R}}$ von $I$ Verengerung der Vervollständigung $\bar{D} | \mathfrak{B}(\mathfrak{R}, D)$ von $D$, insbesondere ist $\overline{\mathfrak{R}} \subseteqq \mathfrak{B}(\mathfrak{R}, D)$. Da die Fortsetzung von $D$ zu einem linearen Funktional auf $\mathfrak{B}(\mathfrak{R}, D)$ nur auf genau eine Weise möglich ist, gilt daher $\bar{D}(f) = \bar{I}_p(f) - \overset{(u)}{\int} f\,d\mu_p$ für jedes $f \in \overline{\mathfrak{R}}$. Also ist $\bar{D}(f) = 0$ für jedes $f \in \mathfrak{S}$. Hieraus folgt, daß jede $I$-quadrierbare Menge $Q$ und mit ihr jede Teilmenge $Q'$ von $Q$ $D$-quadrierbar ist; bezeichnet $\delta$ den zu $D$ gehörigen Inhalt, so ist $\delta(Q') = \bar{D}(\chi_{Q'}) \leqq \bar{D}(\chi_Q) = 0$, also $\delta(Q') = 0$ für jede solche Menge $Q'$. Ist umgekehrt $Q'$ eine beliebige $D$-quadrierbare Menge, so existiert eine $I$-quadrierbare Menge $Q$ mit $Q' \subseteqq Q$ und es gilt folglich wieder $\delta(Q') = 0$. Zu $Q'$ existiert nämlich ein $h \in \mathfrak{R}$ mit $\chi_{Q'} \leqq h$ und hierzu eine reelle Zahl $\alpha$ mit $0 < \alpha < 1$ derart, daß $Q = [h > \alpha]$ $I$-quadrierbar ist; diese Menge leistet das Verlangte. Somit ist das System der $D$-quadrierbaren Mengen gleich dem System der Teilmengen $Q'$ beliebiger $I$-quadrierbarer Mengen $Q$; für jede $D$-quadrierbare Menge $Q'$ gilt $\delta(Q') = 0$. Hieraus folgt die Behauptung (z. B. nach Satz 2).

## § 3. Riemann-Integrierbarkeit bezüglich eines linearen Funktionals. Konkreter Fall

**3.1.** Die in den beiden vorhergehenden Paragraphen gewonnenen Ergebnisse wenden wir nun auf folgenden Spezialfall an:

Die Menge $E$ sei ein *lokal kompakter*[13]), topologischer *Raum*. Die Elemente von $\mathfrak{R}$ seien auf $E$ definierte, *stetige*, (endliche) *reelle Funktionen, welche*

[13]) Im folgenden bedeutet „kompakt“ stets genauer „bikompakt und Hausdorffsch“.

*im Unendlichen verschwinden,* d.h. für jede solche Funktion $g$ und jede reelle Zahl $\alpha > 0$ ist die Urbildmenge $[|g| \geq \alpha]$ kompakt. Hieraus folgt insbesondere die Beschränktheit einer jeden Funktion $g \in \mathfrak{R}$. Die Funktionenmenge $\mathfrak{R}$ besitze wiederum die Eigenschaften (1) bis (3) und genüge daneben noch folgenden zwei Zusatzforderungen:

(21) *Zu jedem Punkt $x \in E$ existiert eine Funktion $g \in \mathfrak{R}$ mit $g(x) \neq 0$.*

(22) *Zu je zwei verschiedenen Punkten $x, y \in E$ existiert eine Funktion $h \in \mathfrak{R}$ mit $h(x) \neq h(y)$.*

Beispielsweise genügt die Menge $\mathfrak{R} = \mathfrak{K}(E)$ aller stetigen, reellen Funktionen $g|E$ mit kompaktem Träger[14]) den Bedingungen (1) bis (3) sowie (21) und (22).

Aus den genannten Eigenschaften von $\mathfrak{R}$ folgt

**Lemma 4.** *Zu jeder stetigen, im Unendlichen verschwindenden, reellen Funktion $f|E$ existiert eine gleichmäßig gegen $f$ konvergente Folge $\{g_n\}_{n=1,2,\ldots}$ von Funktionen aus $\mathfrak{R}$ mit der Eigenschaft* (4).

Beweis. Die Menge $\mathfrak{A}$ der auf $E$ definierten, durch Funktionen aus $\mathfrak{R}$ gleichmäßig approximierbaren Funktionen besteht aus stetigen, im Unendlichen verschwindenden Funktionen und ist nach Lemma 1 eine Algebra. Wegen $\mathfrak{R} \subseteqq \mathfrak{A}$ besitzt $\mathfrak{A}$ die Eigenschaften (21) und (22) von $\mathfrak{R}$. Nach einem Approximationssatz von NAKANO *[15]*[15]) kann daher jede stetige, im Unendlichen verschwindende Funktion auf $E$ durch Funktionen aus $\mathfrak{A}$ gleichmäßig approximiert werden. Da $\mathfrak{A}$ bezüglich gleichmäßiger Konvergenz abgeschlossen ist, so ist also $\mathfrak{A}$ das System aller stetigen, im Unendlichen verschwindenden Funktionen auf $E$. Die Behauptung folgt daher aus Lemma 1.

**3.2.** Neben $E$ und $\mathfrak{R}$ sei wie bisher ein auf $\mathfrak{R}$ definiertes, positives, lineares Funktional $I$ vorgegeben. Gemäß § 1 leiten sich aus $I$ ab: Das System $\mathfrak{Q}$ der $I$-quadrierbaren Teilmengen von $E$, der Inhalt $\mu|\mathfrak{Q}$ und das System $\mathfrak{S}$ aller bezüglich $I$ Riemann-integrierbaren Funktionen.

Wir behaupten von $\mu|\mathfrak{Q}$ die folgenden vier Eigenschaften: (1) jede Menge $Q \in \mathfrak{Q}$ ist relativ kompakt, d.h. es ist $\overline{Q}$[16]) kompakt; (2) in $\mathfrak{Q}$ ist eine offene Basis von $E$ enthalten; (3) mit $Q$ sind auch $\overline{Q}$ und $\underline{Q}$ Elemente von $\mathfrak{Q}$ und es ist $\mu(Q) = \mu(\overline{Q}) = \mu(\underline{Q})$; (4) zu jeder Menge $Q \in \mathfrak{Q}$ und jeder reellen Zahl $\varepsilon > 0$ existiert eine offene Menge $U \in \mathfrak{Q}$ mit $Q \subseteqq U$ und $\mu(U - Q) \leq \varepsilon$.

Nach HAUPT und PAUC *[11]* sind bei Gültigkeit von (1) und (2) die beiden Eigenschaften (3) und (4) äquivalent mit den beiden folgenden: (3′) für jedes $Q \in \mathfrak{Q}$ ist $\overline{Q} - \underline{Q} \in \mathfrak{Q}$; (4′) zu jeder Menge $Q \in \mathfrak{Q}$ und zu jedem $\varepsilon > 0$ existiert eine offene Menge $U \in \mathfrak{Q}$ mit $\overline{Q} - \underline{Q} \subseteqq U$ und $\mu(U) \leq \varepsilon$. Ein Inhalt $\mu|\mathfrak{Q}$ mit den Eigenschaften (1), (2), (3′) und (4′) heißt nach HAUPT-PAUC *[11]* an

[14]) Als *Träger* einer reellen Funktion $g|E$ bezeichnet man die abgeschlossene Hülle der Menge $[g \neq 0]$.

[15]) Vgl. hierzu auch NÖBELING und BAUER *[17]*, S. 72.

[16]) Mit $\overline{A}$ bezeichnen wir die abgeschlossene Hülle, mit $\underline{A}$ den offenen Kern einer Menge $A \subseteqq E$.

(die Topologie von) $E$ *adaptiert*. Unsere Behauptung kann also auch so ausgesprochen werden:

**Satz 7.** *Der Inhalt* $\mu|\mathfrak{Q}$ *ist an* $E$ *adaptiert.*

Beweis. Wir zeigen, daß $\mu|\mathfrak{Q}$ die Eigenschaften (1) bis (4) besitzt. — *Zu* (1). Zu einer Menge $Q \in \mathfrak{Q}$ existiert eine Funktion $h \in \mathfrak{R}$ mit $\chi_Q \leq h$; folglich ist $Q \subseteq [h \geq 1]$. Da $h$ stetig ist und im Unendlichen verschwindet, ist $[h \geq 1]$ kompakt, also $Q$ relativ kompakt. — *Zu* (2). Es sei $V$ eine offene Menge in $E$ und $x_0$ ein Punkt von $V$. Zu zeigen ist die Existenz einer offenen Menge $U \in \mathfrak{Q}$ mit $x_0 \in U$ und $U \subseteq V$. Zu $x_0$ und $V$ existiert eine stetige Funktion $f|E$ mit $0 \leq f \leq 1$, $f(x_0) = 1$ und in $V$ enthaltenem kompakten Träger (vgl. [8], S. 49). Nach Lemma 4 gibt es zu $f$ eine Funktion $g \in \mathfrak{R}$ mit $0 \leq g \leq f \leq g + \frac{1}{2}$ und weiter nach Satz 1 eine reelle Zahl $\alpha$ mit $0 < \alpha < \frac{1}{2}$, für welche die offene Urbildmenge $U = [g > \alpha]$ ein Element von $\mathfrak{Q}$ ist. Aus $f(x_0) = 1$ folgt $g(x_0) \geq \frac{1}{2}$; also ist $x_0 \in U$. Für einen beliebigen Punkt $x \in U$ ist $f(x) \geq g(x) > \alpha > 0$; da $f$ im Komplement von $V$ gleich Null ist, gilt also $x \in V$. Folglich ist $U \subseteq V$. Die Menge $U$ leistet somit das Verlangte. — *Zu* (3). Es sei $Q$ eine Menge aus $\mathfrak{Q}$, $g$ und $h$ seien Funktionen aus $\mathfrak{R}$ mit $g \leq \chi_Q \leq h$. Wegen der Stetigkeit von $g$ und $h$ gilt dann auch $g \leq \chi_{\underline{Q}} \leq \chi_Q \leq \chi_{\overline{Q}} \leq h$. Hieraus und aus der Definition von $\mu$ folgt, daß $\overline{Q}$ und $\underline{Q}$ quadrierbare Mengen mit $\mu(\overline{Q}) = \mu(\underline{Q}) = \mu(Q)$ sind. — *Zu* (4). Es sei $Q$ eine quadrierbare Menge. Zu einer beliebig reellen Zahl $\eta$ mit $0 < \eta < 1$ existiert dann ein $h \in \mathfrak{R}$ mit $\chi_Q \leq h$ und $I(h) \leq \mu(Q) + \eta$. Nach Satz 1 gibt es eine Zahl $\eta'$ mit $0 < \eta' \leq \eta$ derart, daß die offene Urbildmenge $U = [h > 1 - \eta']$ quadrierbar ist. Wir behaupten, daß $Q \subseteq U$ und $\mu(U - Q) \leq \frac{\eta}{1-\eta}(1 + \mu(Q))$ ist; hieraus folgt dann (4). Aus $x \in Q$ folgt $h(x) \geq 1 > 1 - \eta'$ und hieraus $x \in U$; also ist $Q \subseteq U$. Nach Definition von $U$ ist $(1 - \eta')\chi_U \leq h$ und daher $(1 - \eta')\mu(U) \leq I(h)$. Hieraus ergibt sich $(1-\eta)\mu(U) \leq (1-\eta')\mu(U) \leq I(h) \leq \mu(Q) + \eta$ und weiter $(1-\eta)\mu(U - Q) = (1-\eta)(\mu(U) - \mu(Q)) \leq \mu(Q) + \eta - (1-\eta)\mu(Q) = \eta(1 + \mu(Q))$. Es gilt also $\mu(U - Q) \leq \frac{\eta}{1-\eta}(1 + \mu(Q))$, wie behauptet. — Damit ist der Satz bewiesen.

Allein aus der Adaption von $\mu|\mathfrak{Q}$ an $E$ ergeben sich nun unter anderem die im folgenden Korollar genannten Eigenschaften von $\mu$ und $\mathfrak{Q}$.

**Korollar.** I. *Zu jeder kompakten Menge* $K$ *und jeder offenen Menge* $U$ *mit* $K \subseteq U$ *existiert eine Menge* $Q \in \mathfrak{Q}$ *mit* $K \subseteq Q \subseteq U$.

II. *Zu jeder Menge* $Q \in \mathfrak{Q}$ *sowie zu beliebigem* $\varepsilon > 0$ *existieren eine kompakte Menge* $K \in \mathfrak{Q}$ *und eine offene Menge* $U \in \mathfrak{Q}$ *mit* $K \subseteq Q \subseteq U$ *und* $\mu(U - K) \leq \varepsilon$.

Beweis. Vgl. Haupt-Pauc [11], S. 192.

**3.3.** Nach 3.2 genügt $\mathfrak{Q}$ den bei Bourbaki [8], S. 164—165 genannten Bedingungen ($PC_I$), ($PC_{II}$) und $\mu|\mathfrak{Q}$ den Bedingungen ($PM_I$), ($PM_{II}$). Folglich existiert nach Bourbaki *genau ein* positives Radonsches Maß $\varrho$ auf $E$, in bezug auf welches alle Mengen $Q \in \mathfrak{Q}$ $\varrho$-integrierbar mit $\varrho(Q) = \mu(Q)$ sind.

Dieses eindeutig bestimmte Radonsche Maß läßt sich auch wie folgt kennzeichnen[17]):

**Satz 8.** *Jede auf $E$ stetige Funktion mit kompaktem Träger ist Riemann-integrierbar bezüglich $I$. Das somit auf $\mathfrak{K}(E)$ definierte lineare Funktional $f \to \int f\,d\mu$ ist das einzige positive Radonsche Maß $\varrho$ auf $E$, in bezug auf welches jede Menge $Q \in \mathfrak{Q}$ $\varrho$-integrierbar mit $\varrho(Q) = \mu(Q)$ ist. Allgemeiner ist jede bezüglich $I$ Riemann-integrierbare Funktion $f$ sowohl $\varrho$-integrierbar als auch bezüglich $\varrho$ Riemann-integrierbar*[18]) *und es ist*

$$\int f\,d\varrho = \int f\,d\mu; \tag{23}$$

*in dieser Gleichung kann $\int f\,d\varrho$ sowohl als $\varrho$-Integral als auch als Riemann-Integral bezüglich $\varrho$ gedeutet werden.*

Beweis. Es sei $f$ eine Funktion aus $\mathfrak{K}(E)$ und $K$ der kompakte Träger von $f$. Nach dem Korollar zu Satz 7 gibt es dann eine Menge $Q \in \mathfrak{Q}$ mit $K \subseteqq Q$. Nach Lemma 4 kann $f$ durch Funktionen aus $\mathfrak{R}$ gleichmäßig approximiert werden. Also folgt aus dem Korollar 2 zu Satz 2 und aus Lemma 2, daß $f$ Riemann-integrierbar bezüglich $I$ ist. Setzen wir $\int f\,d\varrho = \int f\,d\mu$ für jedes $f \in \mathfrak{K}(E)$, so ist offenbar $\varrho$ ein positives Radonsches Maß auf $E$. Daß jede Menge $Q \in \mathfrak{Q}$ $\varrho$-integrierbar mit $\varrho(Q) = \mu(Q)$ ist, ergibt sich aus der jetzt zu beweisenden, allgemeineren Aussage: Jede Funktion $f \in \mathfrak{S}$ ist $\varrho$-integrierbar und zugleich bezüglich $\varrho$ Riemann-integrierbar; es gilt (23). Hierzu kann ohne Beschränkung der Allgemeinheit $f \geqq 0$ vorausgesetzt werden. Da $f$ im Komplement einer $I$-quadrierbaren Menge gleich Null ist, existiert nach

[17]) Bezüglich der im folgenden verwendeten Begriffe aus der Theorie Radonscher Maße sei auf BOURBAKI [*8*] verwiesen. Ein positives Radonsches Maß $\varrho$ auf $E$ ist demnach ein positives, lineares Funktional auf der Menge $\mathfrak{K}(E)$ aller stetigen Funktionen $f|E$ mit kompaktem Träger, wobei $\int f\,d\varrho$ für $\varrho(f)$ steht und den Funktionswert von $\varrho$ an der Stelle $f \in \mathfrak{K}(E)$ bezeichnet. Nach BOURBAKI definiert man für jede nicht negative Funktion $q|E$ das $\varrho$-Oberintegral $\int^* q\,d\varrho$. Es erweist sich dann $N^*(f) = \int^* |f|\,d\varrho$ als Quasi-Norm im System aller Funktionen $f|E$. Eine Funktion $f|E$ heißt $\varrho$-integrierbar, wenn $N^*(f) < +\infty$ ist und eine Folge von Funktionen $f_n \in \mathfrak{K}(E)$ mit $\lim\limits_{n\to\infty} N^*(f - f_n) = 0$ existiert; man setzt $\int f\,d\varrho = N^*(f^+) - N^*(f^-)$. Eine Teilmenge $A$ von $E$, deren charakteristische Funktion $\chi_A$ $\varrho$-integrierbar ist, wird selbst als $\varrho$-integrierbar bezeichnet; man setzt $\varrho(A) = \int \chi_A\,d\varrho$. Nachträglich erweist sich dann $\int f\,d\varrho$ als Unterteilungsintegral nach dieser Mengenfunktion. — Wendet man dagegen auf das auf $\mathfrak{K}(E)$ definierte Funktional $f \to \int f\,d\varrho$ den sog. ersten (mit Folgen und nicht mit gerichteten Systemen arbeitenden) Erweiterungsprozeß von STONE [*21*] (vgl. auch [*12b*], S. 130ff.) an, so gewinnt man ein für alle nicht negativen Funktionen $q|E$ erklärtes „*schwaches*" $\varrho$-Oberintegral $\bar{\int} q\,d\varrho$. Aus diesem leiten sich analog eine Quasi-Norm $N(f) = \bar{\int} |f|\,d\varrho$ über dem System aller Funktionen $f|E$ und ein Integralbegriff ab. Jede im Sinne von STONE $\varrho$-integrierbare Funktion ist auch im Sinne von BOURBAKI $\varrho$-integrierbar und die zugehörigen Integrale sind einander gleich. Wenn im folgenden von einer Funktion oder Menge (d.h. also von deren charakteristischen Funktionen) die $\varrho$-Integrierbarkeit behauptet wird, so kann hierbei, ohne daß dies eigens betont wird, „$\varrho$-integrierbar" sowohl im Sinne von STONE als auch im Sinne von BOURBAKI verstanden werden. Die Beweise sind dementsprechend angelegt.

[18]) „Bezüglich $\varrho$ Riemann-integrierbar" soll genauer heißen „bezüglich des auf $\mathfrak{K}(E)$ definierten Funktionals $f \to \int f\,d\varrho$ Riemann-integrierbar".

Lemma 2 eine nicht negative Funktion $h_0 \in \mathfrak{R}$ derart, daß für jedes $x \in E$ aus $h_0(x) < 1$ folgt $f(x) = 0$. Nach Lemma 4, angewandt auf $\mathfrak{K}(E)$ an Stelle von $\mathfrak{R}$, folgt die Existenz einer Funktion $h_0' \in \mathfrak{K}(E)$ mit $0 \leq h_0' \leq 2h_0 \leq h_0' + 1$. Auch diese Funktion besitzt dann die Eigenschaft, daß aus $h_0'(x) < 1$ stets $f(x) = 0$ folgt. Da $f$ in der Vervollständigung $\overline{\mathfrak{R}}$ von $\mathfrak{R}$ bezüglich $I$ liegt, gibt es zu jeder reellen Zahl $\varepsilon > 0$ Funktionen $g, h \in \mathfrak{R}$ mit $0 \leq g \leq f \leq h$ und $I(h-g) \leq \varepsilon$. Mit $f$ besitzt auch $g$ einen kompakten Träger, d.h. es ist $g \in \mathfrak{K}(E)$. Durch erneute Anwendung von Lemma 4 ergibt sich die Existenz einer Funktion $h' \in \mathfrak{K}(E)$ mit $0 \leq h' \leq h \leq h' + \varepsilon$. Setzt man $g'' = g$ und $h'' = h' + \varepsilon h_0'$, so sind $g''$ und $h''$ Funktionen aus $\mathfrak{K}(E)$, nach Wahl von $h_0'$ gilt $g'' \leq f \leq h''$ und es ist $\int (h'' - g'')\, d\varrho = \int (h' - g)\, d\mu + \varepsilon \int h_0'\, d\mu \leq I(h-g) + 2\varepsilon I(h_0) \leq \varepsilon(1 + 2I(h_0))$. Also ist $f$ bezüglich $\varrho$ Riemann-integrierbar. Wegen $\int g''\, d\varrho = \int g''\, d\mu$ und $\int h''\, d\varrho = \int h''\, d\mu$ gilt (23) in dem Sinne, daß $\int f\, d\varrho$ das Riemann-Integral von $f$ bezüglich $\varrho$ bedeutet. Nun ist aber $\int (f - g'')\, d\varrho \leq \int (h'' - g'')\, d\varrho$ und $g'' \in \mathfrak{K}(E)$; also ist $f$ auch $\varrho$-integrierbar und in (23) kann $\int f\, d\varrho$ auch als $\varrho$-Integral gedeutet werden.

**Korollar.** *Die Funktionale $I|\mathfrak{R}$ und $\int f\, d\varrho|\mathfrak{K}(E)$ führen zum gleichen an $E$ adaptierten Inhalt, zur gleichen Menge $\mathfrak{S}$ Riemann-integrierbarer Funktionen und zum gleichen Riemann-Integral über $\mathfrak{S}$. Es ist $\mathfrak{S}$ gleich der Vervollständigung von $\mathfrak{K}(E)$ bezüglich $\varrho$.*

Beweis. Bezeichnet wie bisher $\mathfrak{S}$ die Menge der bezüglich $I$ Riemann-integrierbaren Funktionen, so ist nach Satz 8 $\mathfrak{K}(E) \subseteqq \mathfrak{S}$ und $\int f\, d\varrho = \int f\, d\mu$ für jede Funktion $f \in \mathfrak{K}(E)$. Da $\mathfrak{S}$ bezüglich des über $\mathfrak{S}$ definierten Funktionals $f \to \int f\, d\mu$ vollständig ist, ergibt sich hieraus $\overline{\mathfrak{K}(E)} \subseteqq \mathfrak{S}$, wenn $\overline{\mathfrak{K}(E)}$ die Vervollständigung von $\mathfrak{K}(E)$ bezüglich $\varrho$ bezeichnet. Es gilt sogar $\overline{\mathfrak{K}(E)} = \mathfrak{S}$, da nach Satz 8 jede Funktion aus $\mathfrak{S}$ auch bezüglich $\varrho$ Riemann-integrierbar, also jedenfalls ein Element von $\overline{\mathfrak{K}(E)}$ ist. Aus der Gleichung $\overline{\mathfrak{K}(E)} = \mathfrak{S}$ und Satz 8 folgt aber der Rest der Behauptung.

Bemerkung. Nach dem zuletzt bewiesenen Korollar bedeutet es keine Beschränkung der Allgemeinheit, die Theorie des Riemann-Integrals in lokal kompakten Räumen nur für Radonsche Maße zu entwickeln, wie dies Haupt und Pauc in [*10*], [*19*], [*12b*], S. 174ff., für kompakte und in einer in Vorbereitung befindlichen Arbeit auch für lokal kompakte Räume getan haben. Ausgangspunkt der Untersuchungen von Haupt-Pauc ist dabei allerdings ein an den Raum adaptierter Inhalt. Vermöge der Bemerkung am Beginn der Nr. 3.3 entsprechen sich aber adaptierte Inhalte und positive Radonsche Maße in umkehrbar eindeutiger Weise.

**3.4.** Da Riemann-Integrierbarkeit bezüglich $I$ dasselbe wie Riemann-Integrierbarkeit bezüglich $\varrho$ bedeutet, ist zu erwarten, daß sich die von Pauc [*19*] für den Fall eines kompakten Raumes angegebene Kennzeichnung der Riemann-integrierbaren Funktionen als fast überall stetige Funktionen auch auf den Fall eines lokal kompakten Raumes übertragen läßt.

**Satz 9.** *$I$-quadrierbar sind genau die relativ kompakten Teilmengen $Q$ von $E$, deren Begrenzung $\overline{Q} - \underline{Q}$ eine $\varrho$-Nullmenge ist. Riemann-integrierbar* *bezüglich I sind genau die beschränkten, $\varrho$-fast überall stetigen Funktionen auf E mit kompaktem Träger*[19]).

Beweis. Nach der Adaptionseigenschaft (1) von S. 462 ist eine $I$-quadrierbare Menge $Q$ stets relativ kompakt; nach der Adaptionseigenschaft (3) ist $\overline{Q}-\underline{Q}\in\mathfrak{Q}$ und $\mu(\overline{Q}-\underline{Q})=0$. Aus Satz 8 folgt daher, daß $\overline{Q}-\underline{Q}$ eine $\varrho$-Nullmenge ist. Ist umgekehrt $Q$ eine relativ kompakte Menge mit $\varrho(\overline{Q}-\underline{Q})=0$, so existiert zu jedem $\varepsilon>0$ eine offene Menge $U\in\mathfrak{Q}$ mit $\overline{Q}-\underline{Q}\subseteqq U$ und $\mu(U)=\varrho(U)\leqq\varepsilon$. Wegen der Vollständigkeit des Inhalts $\mu$ folgt hieraus nach HAUPT-PAUC [*11*], S. 193 die $I$-Quadrierbarkeit von $Q$. Damit ist die erste Hälfte der Behauptung bewiesen. — Nunmehr sei $f$ eine bezüglich $I$ Riemann-integrierbare Funktion. Dann ist jedenfalls $f$ beschränkt, ferner besitzt $f$ einen kompakten Träger, da $f$ im Komplement einer Menge $Q\in\mathfrak{Q}$ gleich Null und $Q$ relativ kompakt ist. Es bezeichne $w(x)$ die Schwankung von $f$ in einem Punkte $x\in E$; für ein $\gamma>0$ sei $F_\gamma$ die Urbildmenge $[w\geqq\gamma]$. Zu beliebigem $\varepsilon>0$ existieren paarweise fremde Mengen $Q_1,\ldots,Q_n\in\mathfrak{Q}$ sowie reelle Zahlen $\alpha_\nu,\beta_\nu$ $(\nu=1,\ldots,n)$ mit $\sum_{\nu=1}^{n}\alpha_\nu\chi_{Q_\nu}\leqq f\leqq\sum_{\nu=1}^{n}\beta_\nu\chi_{Q_\nu}$ und $\sum_{\gamma=1}^{n}(\beta_\nu-\alpha_\nu)\,\mu(Q_\nu)\leqq\varepsilon$. Es sei $J$ die Menge aller Indizes $\nu=1,\ldots,n$, für welche der offene Kern $\underline{Q_\nu}$ von $Q_\nu$ mit $F_\gamma$ einen nicht leeren Durchschnitt hat. Für jeden Punkt $x\in\underline{Q_\nu}\cap F_\gamma$ ist dann $\underline{Q_\nu}$ eine Umgebung von $x$ und daher $\beta_\nu-\alpha_\nu\geqq w(x)\geqq\gamma$; also gilt $\beta_\nu-\alpha_\nu\geqq\gamma$ für jedes $\nu\in J$. Hieraus folgt $\sum_{\nu\in J}\mu(Q_\nu)\leqq\frac{1}{\gamma}\sum_{\nu\in J}(\beta_\nu-\alpha_\nu)\,\mu(Q_\nu)\leqq\frac{\varepsilon}{\gamma}$.

Ferner gilt $F_\gamma\subseteqq\bigcup_{\nu=1}^{n}\overline{Q_\nu}$, da zu jedem Punkt $x\notin\bigcup_{\nu=1}^{n}\overline{Q_\nu}$ eine Umgebung $U$ von $x$ mit $U\cap\bigcup_{\nu=1}^{n}\overline{Q_\nu}=\emptyset$ existiert und somit $w(x)=0$ ist. Wir erhalten daher die Relation $F_\gamma-\bigcup_{\nu\in J}Q_\nu\subseteqq N$, wobei $N=\bigcup_{\nu=1}^{n}(\overline{Q_\nu}-\underline{Q_\nu})$ nach dem ersten Teil des Beweises $I$-quadrierbar mit $\mu(N)=\varrho(N)=0$ ist. Setzen wir $Q_0=\bigcup_{\nu\in J}Q_\nu$, so ist $F_\gamma\subseteqq Q_0\cup N$ und $\mu(Q_0\cup N)=\sum_{\nu\in J}\mu(Q_\nu)\leqq\frac{\varepsilon}{\gamma}$. Wegen der Vollständigkeit des Inhalts $\mu|\mathfrak{Q}$ ist also $F_\gamma\in\mathfrak{Q}$ und $\mu(F_\gamma)=\varrho(F_\gamma)=0$. Da für die Menge $F$ aller Unstetigkeitsstellen von $f$ die Gleichung $F=\bigcup_{n=1}^{\infty}F_{1/n}$ gilt, ferner stets $F_{1/n}\subseteqq F_{1/n+1}$ ist, so ist $F$ eine $\varrho$-Nullmenge, also $f$ $\varrho$-fast überall stetig. — Umgekehrt sei $f|E$ beschränkt, $\varrho$-fast überall stetig und im Komplement einer kompakten Menge $K$ gleich Null. Nach dem Korollar zu Satz 7 kann ohne Beschränkung der Allgemeinheit $K$ als $I$-quadrierbar vorausgesetzt werden. Für ein $\varepsilon>0$ sei $F_\varepsilon$ wie oben als Urbildmenge $[w\geqq\varepsilon]$ definiert. Da $w$ oberhalb stetig ist, so ist $F_\varepsilon$ abgeschlossen und wegen $F_\varepsilon\subseteqq K$ sogar kompakt. Somit ist $\varrho(F_\varepsilon)=0$ und damit auch die Begrenzung von $F_\varepsilon$ eine $\varrho$-Nullmenge,

[19]) In Übereinstimmung mit dem in [17]) Gesagten ist hierbei „$\varrho$-fast überall stetig" im Sinne der BOURBAKIschen Theorie zu verstehen und heißt: es gilt $\varrho^*(F)=\int^*\chi_F\,d\varrho=0$ für die Menge $F$ aller Unstetigkeitsstellen der betreffenden Funktion $f$. Der folgende Beweis zeigt, daß $F$ dann sowohl im Sinne von BOURBAKI als auch von STONE $\varrho$-integrierbar und $\varrho(F)=0$ ist.

da $f$ $\varrho$-fast überall stetig ist; nach dem bereits bewiesenen ersten Teil der Behauptung ist dann $F_\varepsilon$ sogar $I$-quadrierbar und nach Satz 8 $\mu(F_\varepsilon) = \varrho(F_\varepsilon) = 0$. Folglich existiert eine offene Menge $U_0 \in \mathfrak{Q}$ mit $\mu(U_0) \leq \varepsilon$ und $F_\varepsilon \subseteqq U_0$. Für jeden Punkt $x \in K - F_\varepsilon$ ist $w(x) < \varepsilon$; daher existiert eine Umgebung $U_x \in \mathfrak{Q}$ von $x$, in welcher $f$ um höchstens $\varepsilon$ schwankt. Wegen der Kompaktheit von $K$ gibt es endlich viele Punkte $x_1, \ldots, x_n \in K - F_\varepsilon$ derart, daß $K$ von den $I$-quadrierbaren, offenen Mengen $U_0, U_1 = U_{x_1}, \ldots, U_n = U_{x_n}$ überdeckt wird. Setzen wir $Q_0 = U_0 \cap K$, $Q_1 = U_1 \cap K - Q_0$ und allgemein $Q_\nu = U_\nu \cap K - \bigcup_{\varrho=0}^{\nu-1} Q_\varrho$, so sind die Mengen $Q_0, \ldots, Q_n$ paarweise fremd und $I$-quadrierbar, ferner ist $K = \bigcup_{\nu=0}^{n} Q_\nu$. Bezeichnet $\alpha_\nu$ das Infimum und $\beta_\nu$ das Supremum von $f$ auf $Q_\nu$, $\nu = 0, 1, \ldots, n$, so gilt $\sum_{\nu=0}^{n} \alpha_\nu \chi_{Q_\nu} \leq f \leq \sum_{\nu=0}^{n} \beta_\nu \chi_{Q_\nu}$ und $\sum_{\nu=0}^{n} (\beta_\nu - \alpha_\nu)\mu(Q_\nu) \leq 2M\varepsilon + \varepsilon \sum_{\nu=1}^{n} \mu(Q_\nu) \leq \varepsilon(2M + \mu(K))$; hierbei bedeutet $M$ eine obere Schranke für den Betrag von $f$ auf $E$. Also ist $f$ Riemann-integrierbar bezüglich $I$.

**3.5.** Bislang blieb eine Folgerung aus der Adaption des Inhalts $\mu|\mathfrak{Q}$ an $E$ unberücksichtigt. Nach Haupt-Pauc [*11*] gilt nämlich: *Ein an E adaptierter Inhalt ist stets σ-additiv.* Es ist also

$$\mu_c = \mu \quad \text{und} \quad \mu_p = 0 \tag{24}$$

der $\sigma$-additive bzw. rein-endlich-additive Teil von $\mu$. Vermöge der in 2.3 aufgezeigten Koppelung der Zerlegung $\mu = \mu_c + \mu_p$ mit der Zerlegung $I = I_c + I_p$ von $I$ in seinen stetigen Teil $I_c$ und rein-unstetigen Teil $I_p$ lassen sich nunmehr über die Zerlegungsteile $I_c$ und $I_p$ genauere Aussagen machen. Wir benützen hierzu die in § 2 eingeführten Bezeichnungen.

**Satz 10.** I. *Jede Funktion $f \in \mathfrak{R}$ ist $\varrho$-integrierbar und uneigentlich Riemann-integrierbar; es gilt*

$$I_c(f) = \int f\,d\varrho = \int^{(u)} f\,d\mu. \tag{25}$$

II. *Es ist $\mathfrak{Q}_c = \mathfrak{Q}$ und $\mathfrak{S}_c = \mathfrak{S}$.*

III. *Es ist $\mathfrak{Q}_p$ das System aller relativ kompakten Teilmengen von E und $\mathfrak{S}_p$ das System aller beschränkten Funktionen $f|E$ mit kompaktem Träger.*

Beweis. *Zu* I. Da $\mu_c = \mu$ ist, folgt aus Satz 5, Beh. I die Gültigkeit der Gleichung $I_c(f) = \int^{(u)} f\,d\mu$ für alle $f \in \mathfrak{R}$. Zu zeigen ist also noch, daß jede Funktion $f \in \mathfrak{R}$ $\varrho$-integrierbar und $\int f\,d\varrho = \int^{(u)} f\,d\mu$ ist; hierzu kann ohne Beschränkung der Allgemeinheit $f \geq 0$ vorausgesetzt werden. Nach dem Lemma 4 (angewandt auf $\mathfrak{K}(E)$ an Stelle von $\mathfrak{R}$) existiert eine nicht fallende Folge $\{g_n\}$ nicht negativer Funktionen aus $\mathfrak{K}(E)$ mit $\sup g_n = f$. Wegen $\int g_n\,d\varrho = \int g_n\,d\mu \leq \int^{(u)} f\,d\mu < +\infty$, $n = 1, 2, \ldots$, ist also $f$ $\varrho$-integrierbar und $\int f\,d\varrho = \sup \int g_n\,d\varrho \leq \int^{(u)} f\,d\mu$. Für eine beliebige Funktion $g \in \mathfrak{S}$ mit $g \leq f$ ist nach

(23) $\int g\,d\mu = \int g\,d\varrho \leq \int f\,d\varrho$ und folglich $\overset{(u)}{\int} f\,d\mu \leq \int f\,d\varrho$. Also ist $\int f\,d\varrho = \overset{(u)}{\int} f\,d\mu$. — *Zu* II. Die Behauptung folgt aus I. und Satz 6, Beh. I. — *Zu* III. Da jede $I$-quadrierbare Menge relativ kompakt und jede kompakte Menge in einer $I$-quadrierbaren enthalten ist, folgt die Behauptung aus I. und Satz 6, Beh. II.

Der Satz 10 zeigt, daß es zu jedem *stetigen*, positiven, linearen Funktional $I|\mathfrak{R}$ ein eindeutig bestimmtes, positives Radonsches Maß $\varrho$ auf $E$ gibt, in bezug auf welches $I(f) = \int f\,d\varrho$ für alle Funktionen $f \in \mathfrak{R}$ gilt. Eine erste Kennzeichnung der *rein-unstetigen* Funktionale auf $\mathfrak{R}$ enthält das folgende

**Korollar.** *Gleichwertig sind folgende drei Aussagen:*

I. *Das Funktional $I|\mathfrak{R}$ ist rein-unstetig.*

II. *Es ist $I(f) = 0$ für jede Funktion $f \in \mathfrak{R}$ mit kompaktem Träger.*

III. *Der Inhalt $\mu|\mathfrak{Q}$ ist identisch Null.*

Beweis. *Aus* I. *folgt* II. Eine Funktion $f \in \mathfrak{R}$ mit kompaktem Träger ist nach Satz 8 Riemann-integrierbar; folglich gilt $\int f\,d\mu = \int f\,d\varrho = I(f)$. Nach Satz 10 gilt aber auch $\int f\,d\varrho = I_c(f)$. Wegen $I_c = 0$ ist also $I(f) = 0$. — *Aus* II. *folgt* III. Für jede Menge $Q \in \mathfrak{Q}$ ist nach (5) und (6) $\mu(Q) = \sup(I(g);\ 0 \leq g \leq \chi_Q,\ g \in \mathfrak{R})$. Da $Q$ relativ kompakt ist, besitzt jede Funktion $g \in \mathfrak{R}$ mit $0 \leq g \leq \chi_Q$ einen kompakten Träger und es ist somit $I(g) = 0$. Hieraus folgt $\mu(Q) = 0$. — *Aus* III. *folgt* I., da nach der Behauptung I des Satzes 10 $I_c = 0$ ist.

Dieses Korollar gestattet nunmehr die Herleitung einer einfachen Formel für die Berechnung von $I_p$ und damit auch von $I_c$.

**Satz 11.** *Für jede Funktion $f \geq 0$ aus $\mathfrak{R}$ ist*

$$I_p(f) = \lim_{\alpha \to +0} I(\min(\alpha, f)). \tag{26}$$

*Das Funktional $I$ ist dann und nur dann stetig bzw. rein-unstetig, wenn*

$$\lim_{\alpha \to +0} I(\min(\alpha, f)) = 0 \tag{27}$$

*bzw.*

$$I(\min(1, f)) = I(f) \tag{28}$$

*ist für alle $f \geq 0$ aus $\mathfrak{R}$.*

Beweis. Es sei $f$ eine nicht negative Funktion aus $\mathfrak{R}$. Da $f$ im Unendlichen verschwindet, ist die Funktion $g_\alpha = \min(\alpha, f) - f$ für jede reelle Zahl $\alpha > 0$ ein Element von $\mathfrak{R} \cap \mathfrak{K}(E)$. Nach dem Korollar zu Satz 10 gilt also $I_p(g_\alpha) = 0$ und somit die Gleichung $I(\min(\alpha, f)) = I_c(\min(\alpha, f)) + I_p(f)$ für jedes $\alpha > 0$. Läßt man in dieser Gleichung $\alpha$ gegen $+0$ konvergieren, so folgt (26); wegen der Stetigkeit von $I_c$ ist nämlich $\lim_{\alpha \to +0} I_c(\min(\alpha, f)) = \lim_{n \to \infty} I_c\left(\min\left(\frac{1}{n}, f\right)\right) = I_c(0) = 0$. — Aus (26) folgt unmittelbar die Kennzeichnung der stetigen Funktionale durch Gl. (27). — Ist $I$ rein-unstetig, so gilt nach dem Bisherigen $I(g_1) = 0$, also (28). Aus (28) folgt umgekehrt die Gleichung $I(\min(\alpha, f)) = I(f)$ für jedes $\alpha > 0$ und hieraus $I = I_p$ nach (26).

Wir erhalten als Korollar folgende Aussage über die Möglichkeit einer Integraldarstellung von $I(f)$.

**Korollar.** *Es sei $f$ eine nicht negative Funktion aus $\mathfrak{R}$. Dann ist die Bedingung (27) notwendig und hinreichend für das Bestehen der Gleichung*

$$I(f) = \int f\,d\varrho = \overset{(u)}{\int} f\,d\mu.$$

*Setzen wir $G = [f > 0]$, so ist für das Bestehen dieser Gleichungen dagegen nur hinreichend die Bedingung*

$$\sup\big(I(g)\,;\, 0 \leq g \leq \chi_G,\, g \in \mathfrak{R}\big) < +\infty\text{ [20]}.$$

Beweis. Daß die Bedingung (27) notwendig und hinreichend ist, folgt aus (25) und (26) sowie aus $I = I_c + I_p$. — Es sei $\gamma(G) = \sup\big(I(g)\,;\, 0 \leq g \leq \chi_G,$ $g \in \mathfrak{R}\big)$. Wegen $0 \leq \frac{1}{\alpha}\min(\alpha, f) \leq \chi_G$ ist dann $0 \leq I\big(\min(\alpha, f)\big) \leq \alpha\gamma(G)$ für jedes $\alpha > 0$. Hieraus folgt die Bedingung (27).

**3.6.** *Beispiele rein-unstetiger, positiver, linearer Funktionale.*

1. Der lokal kompakte Raum $E$ sei das halboffene Einheitsintervall $(0, 1] = \{x\colon 0 < x \leq 1\}$ der Zahlengeraden, versehen mit der üblichen Topologie. $\mathfrak{R}$ bestehe aus allen stetigen Funktionen $f\,|\,[0, 1]$, aufgefaßt als nur auf $E$ definiert, mit $f(0) = 0$ und in $x = 0$ existierender rechtsseitiger Ableitung $f'(0)$. Für jedes $f \in \mathfrak{R}$ sei $I(f) = f'(0)$ [21].

2. $E$ sei die Halbgerade $[0, +\infty) = \{x\colon 0 \leq x < +\infty\}$ mit der üblichen Topologie. $\mathfrak{R}$ bestehe aus allen stetigen Funktionen $f\,|\,[0, +\infty)$, für welche $\lim_{x\to+\infty} x\,f(x)$ (eigentlich) existiert. Für jedes $f \in \mathfrak{R}$ sei $I(f) = \lim_{x\to+\infty} x\,f(x)$.

Bemerkung. Im Beispiel 2 ist $I(f) = 0 = \int f\,d\varrho = \overset{(u)}{\int} f\,d\mu$ für die spezielle, nicht negative Funktion $f(x) = (1 + x^2)^{-1}$ $(x \in E)$ aus $\mathfrak{R}$; ferner ist $G = [f > 0]$ der Gesamtraum $E$. Bezeichnet $q$ die Funktion $(1 + x)^{-1}$ aus $\mathfrak{R}$, so ist für jede reelle Zahl $\alpha > 0$ die Funktion $g_\alpha = \min(1, \alpha q)$ ein Element von $\mathfrak{R}$ und $0 \leq g_\alpha \leq 1$. Wegen $I(g_\alpha) = \alpha$ ist folglich $\sup\big(I(g)\,;\, 0 \leq g \leq \chi_G,\, g \in \mathfrak{R}\big) = +\infty$. Die im Korollar zu Satz 11 genannte hinreichende Bedingung ist also nicht notwendig.

## § 4. Zurückführung des abstrakten Falles auf den konkreten

**4.1.** Im letzten Paragraphen ergab sich unter anderem, daß unter den dort gemachten speziellen Annahmen über $E$ und $\mathfrak{R}$ zu jedem positiven, linearen Funktional $I\,|\,\mathfrak{R}$ ein Radonsches Maß $\varrho$ auf $E$ derart existiert, daß die Theorie des Riemann-Integrals bezüglich $I$ mit der bezüglich $\varrho$ identisch ist. Es bedeutet also keine Einschränkung der Allgemeinheit, im konkreten Fall eines lokal kompakten Raumes den Standpunkt von Haupt und Pauc einzunehmen. In den folgenden Betrachtungen wird dieses Ergebnis dahingehend verschärft, daß die Loomissche Theorie des Riemann-Integrals selbst

[20]) Vgl. auch Satz 16.

[21]) Nach Loomis [*14*], S. 174, wo das Beispiel aber nicht in unserem Sinne interpretiert wird.

im abstrakten Fall keinen wesentlichen Gewinn an Allgemeinheit gegenüber dem konkreten Fall bringt. Wir behaupten nämlich: Ist $E$ eine Menge und $\mathfrak{R}$ ein System von beschränkten Funktionen $g|E$ mit den Eigenschaften (1) bis (3) sowie (21) und (22), so existiert ein lokal kompakter Raum $E'_{\mathfrak{R}}$, der $E$ als dichte Teilmenge enthält, mit folgender Eigenschaft: Zu jedem positiven, linearen Funktional $I|\mathfrak{R}$ gibt es ein positives Radonsches Maß $\varrho'$ auf $E'_{\mathfrak{R}}$ derart, daß eine Funktion $f|E$ dann und nur dann Riemann-integrierbar bezüglich $I$ ist, wenn sie zu einer bezüglich $\varrho'$ Riemann-integrierbaren Funktion $f'|E'_{\mathfrak{R}}$ fortgesetzt werden kann; für jede solche Fortsetzung $f'$ von $f$ ist $\int f\,d\mu = \int f'\,d\varrho'$.

**4.2.** Es liege der abstrakte Fall des § 1 vor: $E$ sei eine Menge, $\mathfrak{R}$ ein System auf $E$ definierter, reeller Funktionen mit den Eigenschaften (1) bis (3) und $I|\mathfrak{R}$ ein positives, lineares Funktional. Wir behaupten zunächst, daß es im Hinblick auf die Loomissche Theorie des Riemann-Integrals keine wesentliche Einschränkung bedeutet, über $\mathfrak{R}$ folgende drei Zusatz-Voraussetzungen zu machen: 1. *Jede Funktion aus* $\mathfrak{R}$ *ist beschränkt*; 2. *es besitzt* $\mathfrak{R}$ *die Eigenschaft* (21); 3. *es besitzt* $\mathfrak{R}$ *die Eigenschaft* (22).

*Zu* 1. Die Teilmenge $\mathfrak{R}^*$ von $\mathfrak{R}$ aller beschränkten Funktionen aus $\mathfrak{R}$ genügt den Bedingungen (1) bis (3). Bezeichnet $I^*|\mathfrak{R}^*$ die Verengerung von $I$ auf $\mathfrak{R}$, so führen die Funktionale $I|\mathfrak{R}$ und $I^*|\mathfrak{R}^*$ zum gleichen System $\mathfrak{Q}$ quadrierbarer Mengen, zum gleichen Inhalt $\mu|\mathfrak{Q}$, zum gleichen System $\mathfrak{S}$ Riemann-integrierbarer Funktionen und zum gleichen Riemann-Integral über $\mathfrak{S}$. Zum Beweis sei $Q$ eine $I$-quadrierbare Menge und $\varepsilon > 0$ eine reelle Zahl. Dann existieren Funktionen $g, h \in \mathfrak{R}$ mit $0 \leq g \leq \chi_Q \leq h$ und $I(h-g) \leq \varepsilon$. Setzen wir $h^* = \min(1, h)$, so sind $g$ und $h^*$ Funktionen aus $\mathfrak{R}^*$ mit $0 \leq g \leq \chi_Q \leq h^* \leq h$ und $I^*(h^* - g) \leq I(h-g) \leq \varepsilon$. Also ist $Q$ auch $I^*$-quadrierbar. Trivial ist, daß umgekehrt eine $I^*$-quadrierbare Menge auch $I$-quadrierbar ist. Da in die Definition des Inhalts $\mu|\mathfrak{Q}$ nur Funktionen aus $\mathfrak{R}^*$ eingehen, führen $I$ und $I^*$ zum gleichen Inhalt $\mu$. Hieraus folgt aber der Rest der Behauptung.

*Zu* 2. Es bezeichne $E_0$ die Menge aller Elemente $x \in E$ mit $g(x) = 0$ für alle $g \in \mathfrak{R}$. Aus der Definition der Vervollständigung $\overline{\mathfrak{R}}$ von $\mathfrak{R}$ bezüglich $I$ folgt, daß jede Funktion $f \in \overline{\mathfrak{R}}$ auf $E_0$ gleich Null ist. Hieraus folgt insbesondere, daß jede quadrierbare Menge $Q$ zu $E_0$ fremd und jede Riemann-integrierbare Funktion auf $E_0$ gleich Null ist. Es genügt daher, die Funktionen $g \in \mathfrak{R}$ nur auf dem verengerten Definitionsbereich $E - E_0$ zu betrachten. (Den Trivialfall, in welchem $E - E_0$ die leere Menge ist, also $\mathfrak{R}$ nur aus der konstanten Funktion $g = 0$ besteht, können wir unberücksichtigt lassen.)

*Zu* 3. Es bezeichne $R$ die folgende Äquivalenzrelation in $E$: Zwei Elemente $x, y \in E$ heißen äquivalent, wenn $g(x) = g(y)$ ist für jede Funktion $g \in \mathfrak{R}$. Es sei $E^* = E/R$ die Menge der Äquivalenzklassen bezüglich $R$; ohne Beschränkung der Allgemeinheit können wir $E^*$ als eine Teilmenge von $E$ annehmen, die aus jeder Äquivalenzklasse genau einen Punkt als Repräsentanten enthält. Das System $\mathfrak{R}^*$ der auf $E^*$ verengerten Funktionen aus $\mathfrak{R}$

besitzt neben den Eigenschaften (1) bis (3) auch die Eigenschaft (22) auf $E^*$. Wir übertragen das Funktional $I$ auf $\mathfrak{R}^*$ und erhalten ein positives, lineares Funktional $I^*|\mathfrak{R}^*$. Dann gilt: Eine Funktion $f^*|E^*$ liegt dann und nur dann in der Vervollständigung von $\mathfrak{R}^*$ bezüglich $I^*$, wenn sie zu einer in der Vervollständigung $\overline{\mathfrak{R}}$ von $\mathfrak{R}$ bezüglich $I$ gelegenen Funktion $f|E$ fortgesetzt werden kann. Die Fortsetzung $f \in \overline{\mathfrak{R}}$ von $f^*$ kann so gewählt werden, daß $f$ auf jeder Äquivalenzklasse bezüglich $R$ konstant ist. Hieraus folgt, daß eine Menge $Q^* \subseteqq E^*$ dann und nur dann $I^*$-quadrierbar ist, wenn eine $I$-quadrierbare Menge $Q \subseteqq E$ mit $Q^* = Q \cap E$ existiert; bezeichnet $\mu^*$ den sich aus $I^*$ ableitenden Inhalt, so ist $\mu^*(Q^*) = \mu(Q)$. Es folgt weiter, daß eine Funktion $f^*|E^*$ dann und nur dann Riemann-integrierbar bezüglich $I^*$ ist, wenn sie zu einer bezüglich $I$ Riemann-integrierbaren Funktion $f|E$ fortgesetzt werden kann; es gilt dann $\int f^* d\mu^* = \int f\, d\mu$.

**4.3.** Von nun an genüge $\mathfrak{R}$ neben (1) bis (3) auch noch den in 4.2 genannten drei Zusatz-Voraussetzungen. Dann behaupten wir:

**Satz 12.** *Zu $E$ und $\mathfrak{R}$ existiert ein lokal kompakter Raum $E'_{\mathfrak{R}}$ mit folgenden Eigenschaften:*

(29) *$E$ ist eine dichte Teilmenge von $E'_{\mathfrak{R}}$;*

(30) *jede Funktion $g \in \mathfrak{R}$ kann (auf genau eine Weise) zu einer auf $E'_{\mathfrak{R}}$ stetigen, im Unendlichen verschwindenden Funktion $g'$ fortgesetzt werden;*

(31) *die Menge $\mathfrak{R}'$ der stetigen Fortsetzungen $g'$ aller Funktionen $g \in \mathfrak{R}$ besitzt auf $E'_{\mathfrak{R}}$ die Trennungseigenschaften* (21) *und* (22)[22].

Beweis. Nach dem Lemma 1 ist die Menge $\mathfrak{A}_0$ aller (beschränkten) Funktionen auf $E$, die durch Funktionen aus $\mathfrak{R}$ gleichmäßig approximiert werden können, eine hinsichtlich gleichmäßiger Konvergenz abgeschlossene Algebra. Die kleinste, $\mathfrak{A}_0$ umfassende Algebra $\mathfrak{A}$, welche auch die auf $E$ konstanten Funktionen enthält, besteht dann genau aus den Funktionen $f = \alpha + f_0$, wobei $\alpha$ eine reelle Zahl und $f_0 \in \mathfrak{A}_0$ ist. Wir behaupten, daß auch $\mathfrak{A}$ hinsichtlich gleichmäßiger Konvergenz abgeschlossen ist. Wenn $\mathfrak{A} = \mathfrak{A}_0$ ist, also $\mathfrak{A}_0$ die auf $E$ konstanten Funktionen enthält, ist dies trivial. Wir können daher annehmen, daß $f = 0$ die einzige konstante Funktion in $\mathfrak{A}_0$ ist. Es sei $\{f_n\}$ eine gegen eine Funktion $f|E$ gleichmäßig konvergente Folge von Funktionen aus $\mathfrak{A}$; jedes $f_n$ kann in der Form $f_n = \alpha_n + f_n^0$ mit $\alpha_n \in \mathrm{P}$ und $f_n^0 \in \mathfrak{A}_0$ dargestellt werden. Ist $\alpha$ ein Limeswert der Folge $\{\alpha_n\}$, so existiert eine gegen $\alpha$ konvergente Teilfolge $\{\alpha_{n_\nu}\}$; folglich konvergiert $\{f_{n_\nu}^0\}$ gleichmäßig gegen $f - \alpha$, d.h. es ist $f - \alpha \in \mathfrak{A}_0$. Für einen zweiten Limeswert $\beta$ von $\{\alpha_n\}$ folgt entsprechend $f - \beta \in \mathfrak{A}_0$. Also sind $\alpha, \beta$ endlich und die konstante Funktion $\alpha - \beta = (f - \beta) - (f - \alpha)$ liegt in $\mathfrak{A}_0$; gemäß unserer Annahme über $\mathfrak{A}_0$ ist also $\alpha = \beta$, d.h. es existiert der Grenzwert der Folge $\{\alpha_n\}$ und zwar ist $\alpha = \beta = \lim \alpha_n$. Dann konvergiert aber auch die Folge $\{f_n^0\}$ gleichmäßig auf $E$ gegen eine Grenzfunktion $f_0 \in \mathfrak{A}_0$ und es ist $f = \alpha + f_0 \in \mathfrak{A}$. Auf Grund der festgestellten

---

[22]) In (21) und (22) sind also $E$ und $\mathfrak{R}$ sinngemäß durch $E'_{\mathfrak{R}}$ und $\mathfrak{R}'$ zu ersetzen.

Eigenschaften von $\mathfrak{A}$ existiert nach GELFAND und ŠILOV [9][23]) ein kompakter Raum $E''$, welcher $E$ als dichte Teilmenge enthält und auf welchen die Funktionen aus $\mathfrak{A}$ und nur diese stetig fortgesetzt werden können. Es bezeichne $\mathfrak{R}''$ bzw. $\mathfrak{A}_0''$ bzw. $\mathfrak{A}''$ die Menge der auf $E''$ stetig fortgesetzten Funktionen aus $\mathfrak{R}$ bzw. $\mathfrak{A}_0$ bzw. $\mathfrak{A}$; dann ist $\mathfrak{A}''$ die Menge aller stetigen Funktionen auf $E''$. Nun sind zwei Fälle zu unterscheiden. Erstens. Es sei $\mathfrak{A}=\mathfrak{A}_0$, d.h. es enthalte $\mathfrak{A}_0$ die auf $E$ konstanten Funktionen. Dann ist $E''$ selbst ein Raum $E'_{\mathfrak{R}}$. Die Trennungseigenschaften (21) und (22) sind auf $E''$ erfüllt, da jede auf $E''$ stetige, reelle Funktion durch Funktionen aus $\mathfrak{R}''$ zunächst auf $E$ und wegen der Dichtigkeit von $E$ in $E''$ dann auf ganz $E''$ gleichmäßig approximiert werden kann. Zweitens. Es sei $\mathfrak{A}\neq\mathfrak{A}_0$, d.h. $\mathfrak{A}_0$ enthalte $f=0$ als einzige auf $E$ konstante Funktion. In diesem Falle ist $\mathfrak{A}_0''$ ein maximales Ideal in $\mathfrak{A}''$. Nach STONE [20][24]) existiert folglich ein eindeutig bestimmter Punkt $p_0''$ in $E''$ derart, daß $\mathfrak{A}_0''$ die Menge aller stetigen Funktionen $f''|E''$ mit $f''(p_0'')=0$ ist. Der Unterraum $E'=E''-(p_0'')$ von $E''$ ist dann ein Raum $E'_{\mathfrak{R}}$. Die Trennungseigenschaften (21) und (22) ergeben sich auf Grund der gleichen Bemerkung wie im bereits erledigten Falle. Damit ist der Satz 12 vollständig bewiesen[25]).

Bemerkung. Aus dem Beweis des Satzes 12 (und wegen Satz 13 natürlich auch aus den Eigenschaften (29) bis (31) von $E'_{\mathfrak{R}}$) folgt, daß $E'_{\mathfrak{R}}$ *dann und nur dann kompakt* ist, wenn die konstante Funktion $g=1$ durch Funktionen aus $\mathfrak{R}$ gleichmäßig approximiert werden kann, also wegen (3) in $\mathfrak{R}$ liegt.

Der Vollständigkeit halber beweisen wir

**Satz 13.** *Der Raum $E'_{\mathfrak{R}}$ ist durch die Eigenschaften* (29) *bis* (31) *bis auf Homöomorphien, welche E punktweise festlassen, eindeutig bestimmt.*

Beweis. Es seien $E_1'$ und $E_2'$ zwei Räume $E'_{\mathfrak{R}}$. Wir setzen $E_i''=E_i'$, falls $E_i'$ kompakt ist; andernfalls sei $E_i''$ die durch Adjunktion des unendlich fernen Punktes $w_i$ entstehende Alexandroffsche Kompaktifizierung von $E_i'$. In jedem Falle ist $E_i''$ ein kompakter Raum, in dem $E$ dicht liegt und auf den jede stetige, im Unendlichen verschwindende Funktion $f'|E_i'$ stetig fortgesetzt werden kann; insbesondere können die Funktionen aus $\mathfrak{R}$ auf $E_i''$ stetig fortgesetzt werden. Aus den Eigenschaften von $E_i'$ und aus Lemma 4 folgt dann, daß jede stetige Funktion $f''|E_i''$ eine Darstellung $f''=\alpha+f_0''$ besitzt, wobei $\alpha$ eine reelle Zahl ist und $f_0''$ durch auf $E_i''$ stetig fortgesetzte Funktionen aus $\mathfrak{R}$ gleichmäßig approximiert werden kann $(i=1, 2)$. Daher kann jede beschränkte Funktion $f|E$, die zu einer auf $E_1''$ stetigen Funktion fortgesetzt werden kann, auch zu einer auf $E_2''$ stetigen Funktion fortgesetzt werden und umgekehrt. Nach STONE [20][26]) folgt hieraus die Existenz einer Homöomorphie $\psi$ von $E_1''$ auf $E_2''$, welche $E$ punktweise festläßt. Da $E$ in $E_1''$ und $E_2''$ dicht liegt, gilt

---

[23]) Vgl. hierzu auch NÖBELING und BAUER [16], S. 34 und 35.

[24]) Vgl. hierzu S. 25 der in [23]) zitierten Arbeit.

[25]) Einen anderen, nur einfache topologische Hilfsmittel verwendenden Beweis hat Verf. in [17], S. 65—67 gegeben.

[26]) Vgl. auch [16], S. 28.

dann $f_1''=f_2''\circ\psi$ für jede Funktion $f|E$, die zu einer stetigen Funktion $f_1''|E_1''$ und zu einer stetigen Funktion $f_2''|E_2''$ fortgesetzt werden kann. Ist $E_1'$ nicht kompakt, so folgt hieraus $0=f_1''(w_1)=f_2''(\psi(w_1))$ für jedes $f\in\mathfrak{R}$; wegen (31) ist also auch $E_2'$ nicht kompakt, vielmehr gilt $w_2=\psi(w_1)$. Durch Vertauschung der Rollen von $E_1'$ und $E_2'$ folgt, daß umgekehrt mit $E_1'$ auch $E_2'$ kompakt ist. Damit ist gezeigt, daß $\psi$ in jedem Falle $E_1'$ auf $E_2'$ homöomorph abbildet. Auch die auf $E_1'$ verengerte Homöomorphie $\psi$ läßt $E$ punktweise fest.

Der Satz 13 berechtigt uns, fernerhin von *dem* Raum $E'_{\mathfrak{R}}$ zu sprechen. Ist (wie in § 3) $E$ selbst ein lokal kompakter Raum und $\mathfrak{R}$ ein System stetiger, im Unendlichen verschwindender Funktionen auf $E$ mit den Eigenschaften (1) bis (3) sowie (21) und (22), so ist $E'_{\mathfrak{R}}=E$.

Die aus Nr. 1.5 bekannte Bedingung (8) für $\mathfrak{R}$ erfährt jetzt eine einfache und neue Deutung.

**Satz 14.** *Die Funktionenmenge $\mathfrak{R}$ genügt (unter den bisherigen Annahmen) dann und nur dann der Bedingung* (8), *wenn die stetige Fortsetzung $g'|E'_{\mathfrak{R}}$ einer jeden Funktion $g\in\mathfrak{R}$ einen kompakten Träger besitzt.*

Beweis. Es genüge $\mathfrak{R}$ der Bedingung (8). Dann existiert zu beliebigem $g\in\mathfrak{R}$ eine nicht negative Funktion $h\in\mathfrak{R}$ derart, daß für jedes $x\in E$ aus $h(x)<1$ folgt $g(x)=0$. Bezeichnet $g'$ bzw. $h'$ die stetige Fortsetzung von $g$ bzw. $h$ auf $E'_{\mathfrak{R}}$, so folgt, da $E$ in $E'_{\mathfrak{R}}$ dicht liegt, sogar für jedes $x'\in E'_{\mathfrak{R}}$ aus $h'(x')<1$ stets $g'(x')=0$. Da $h'\geq 0$ ist und im Unendlichen verschwindet, ist $[h'\geq 1]$ eine kompakte Menge. $g'$ ist im Komplement dieser Menge gleich Null und besitzt daher einen kompakten Träger. — Umgekehrt sei der Träger $T(g')$ der stetigen Fortsetzung $g'|E'_{\mathfrak{R}}$ einer Funktion $g\in\mathfrak{R}$ kompakt. Dann gibt es eine nicht negative, stetige, im Unendlichen verschwindende Funktion $t'|E'_{\mathfrak{R}}$ mit $t'(x')=1$ für alle $x'\in T(g')$. Wegen (31) und Lemma 4 existiert zu $t'$ eine nicht negative Funktion $h\in\mathfrak{R}$ derart, daß für deren stetige Fortsetzung $h'|E'_{\mathfrak{R}}$ gilt $0\leq\frac{1}{2}h'\leq t'\leq\frac{1}{2}h'+\frac{1}{2}$. Diese Funktion $h$ leistet bezüglich $g$ das in (8) Verlangte: Aus $h(x)<1$ für ein $x\in E$ folgt $t'(x)<1$; also ist $x\notin T(g')$ und somit $g(x)=g'(x)=0$. Die Bedingung (8) ist also erfüllt.

**4.4.** Der Raum $E'_{\mathfrak{R}}$ wurde unabhängig von dem auf $\mathfrak{R}$ definierten Funktional $I$ definiert. Erst jetzt beziehen wir $I$ in unsere Betrachtungen mit ein. Für eine beliebige Funktion $g\in\mathfrak{R}$ bezeichne im folgenden $g'$ deren eindeutig bestimmte stetige Fortsetzung auf $E'_{\mathfrak{R}}$; $\mathfrak{R}'$ sei die Menge der Fortsetzungen $g'$ aller Funktionen $g\in\mathfrak{R}$. Wir übertragen $I$ auf $\mathfrak{R}'$ durch die Festsetzung

$$I'(g')=I(g)\qquad(g\in\mathfrak{R}).\tag{32}$$

Dann ist auch $I'|\mathfrak{R}'$ ein positives, lineares Funktional. Da $E$ in $E'_{\mathfrak{R}}$ dicht liegt, genügt auch $\mathfrak{R}'$ den Bedingungen (1) bis (3) für $\mathfrak{R}$; nach Satz 12 besitzt daher $\mathfrak{R}'$ bezüglich $E'_{\mathfrak{R}}$ alle Eigenschaften, die in § 3 $\mathfrak{R}$ bezüglich $E$ hatte. Also sind die Resultate des § 3 auf $E'_{\mathfrak{R}}$, $\mathfrak{R}'$ und $I'$ an Stelle von $E$, $\mathfrak{R}$ und $I$ anwendbar. Es bezeichne $\mu'|\mathfrak{Q}'$ den sich aus $I'$ ableitenden, an $E'_{\mathfrak{R}}$ adaptierten Inhalt und $\varrho'$ das eindeutig bestimmte positive Radonsche Maß auf $E'_{\mathfrak{R}}$, in bezug auf welches alle Mengen $Q' \in \mathfrak{Q}'$ $\varrho'$-integrierbar mit $\varrho'(Q') = \mu'(Q')$ sind. Dann ist $\mathfrak{Q}'$ nach dem Korollar zu Satz 8 auch das System der $\varrho'$-quadrierbaren Mengen. Nach dem gleichen Korollar ist die Menge $\mathfrak{S}'$ der bezüglich $I'$ Riemann-integrierbaren Funktionen gleich der Menge der bezüglich $\varrho'$ Riemann-integrierbaren Funktionen. Es gilt nun der einleitend bereits angekündigte

**Satz 15.** I. *Eine Menge* $Q \subseteqq E$ *ist dann und nur dann* $I$*-quadrierbar, wenn eine* $\varrho'$*-quadrierbare Menge* $Q' \subseteqq E'_{\mathfrak{R}}$ *existiert mit* $Q = Q' \cap E$. *Für je zwei Mengen* $Q \in \mathfrak{Q}$ *und* $Q' \in \mathfrak{Q}'$ *mit* $Q = Q' \cap E$ *ist* $\mu(Q) = \mu'(Q') = \varrho'(Q')$.

II. *Eine reelle Funktion* $f|E$ *ist dann und nur dann Riemann-integrierbar bezüglich* $I$, *wenn sie zu einer bezüglich* $\varrho'$ *Riemann-integrierbaren Funktion* $f'|E'_{\mathfrak{R}}$ *fortgesetzt werden kann. Für je zwei Funktionen* $f \in \mathfrak{S}$ *und* $f' \in \mathfrak{S}'$, *von denen die zweite eine Fortsetzung der ersten ist, gilt* $\int f\,d\mu = \int f'\,d\mu' = \int f'\,d\varrho'$.

Beweis. *Zu* I. Es sei $Q' \in \mathfrak{Q}'$; wir behaupten die $I$-Quadrierbarkeit der Menge $Q = Q' \cap E$ und die Gültigkeit der Gleichung $\mu(Q) = \mu'(Q')$ $\big(= \varrho'(Q')\big)$. Zu beliebigem $\varepsilon > 0$ existieren Funktionen $g', h' \in \mathfrak{R}'$ mit $g' \leqq \chi'_{Q'} \leqq h'$ [27]) und $I'(h' - g') \leqq \varepsilon$. Für die zu $\mathfrak{R}$ gehörigen Verengerungen $g$ und $h$ von $g'$ und $h'$ auf $E$ ist dann $g \leqq \chi_Q \leqq h$ und $I(h - g) \leqq \varepsilon$. Also ist $Q \in \mathfrak{Q}$; ferner gilt $\mu(Q) = \mu'(Q')$ nach (6) und (32). — Umgekehrt sei $Q \subseteqq E$ $I$-quadrierbar; zu zeigen ist die Existenz einer Menge $Q' \in \mathfrak{Q}'$ mit $Q = Q' \cap E$. Für Funktionen $g, h \in \mathfrak{R}$ folgt aus $g \leqq \chi_Q \leqq h$ stets $g' \leqq \chi'_{\overline{Q}} \leqq h'$ für die stetigen Fortsetzungen $g'$ und $h'$ von $g$ und $h$ auf $E'_{\mathfrak{R}}$ und die abgeschlossene Hülle $\overline{Q}$ von $Q$ in $E'_{\mathfrak{R}}$. Setzen wir $Q' = Q \cup (\overline{Q} - E)$, so gilt auch $g' \leqq \chi'_{Q'} \leqq h'$. Die Menge $Q'$ leistet daher das Verlangte.

*Zu* II. Es sei $f'$ eine Funktion aus $\mathfrak{S}'$ und $f$ die Verengerung von $f'$ auf $E$. Zu $\varepsilon > 0$ existieren dann Mengen $Q'_1, \ldots, Q'_n \in \mathfrak{Q}'$ und reelle Zahlen $\alpha_1, \beta_1, \ldots, \alpha_n, \beta_n$ derart, daß $g' = \sum_{\nu=1}^{n} \alpha_\nu \chi'_{Q'_\nu} \leqq f' \leqq \sum_{\nu=1}^{n} \beta_\nu \chi'_{Q'_\nu} = h'$ und $\sum_{\nu=1}^{n} (\beta_\nu - \alpha_\nu)\mu'(Q'_\nu) \leqq \varepsilon$ ist. Für die auf $E$ definierten Funktionen $g = \sum \alpha_\nu \chi_{Q_\nu}$ und $h = \sum \beta_\nu \chi_{Q_\nu}$ mit $Q_\nu = Q'_\nu \cap E$, $\nu = 1, \ldots, n$, gilt folglich $g \leqq f \leqq h$ sowie $\sum (\beta_\nu - \alpha_\nu)\mu(Q_\nu) \leqq \varepsilon$. Also ist $f \in \mathfrak{S}$. Da $\mu(Q_\nu) = \mu'(Q'_\nu)$ ist folgt ferner noch $\int f\,d\mu = \int f'\,d\mu'$ $(= \int f'\,d\varrho')$. Umgekehrt sei $f$ eine Funktion aus $\mathfrak{S}$; wir beweisen die Existenz einer in $\mathfrak{S}'$ gelegenen Fortsetzung $f'$ von $f$ auf $E'_{\mathfrak{R}}$. Da jede Funktion aus $\mathfrak{S}$ gleich der Differenz zweier nicht negativer Funktionen aus $\mathfrak{S}$ ist, können wir ohne Beschränkung der Allgemeinheit $f \geqq 0$ annehmen. Wie in 1.4 bezeichne $\mathfrak{E}$ die Menge aller $\mathfrak{Q}$-Treppenfunktionen; entsprechend sei $\mathfrak{E}'$ die Menge aller $\mathfrak{Q}'$-Treppenfunktionen (auf $E'$). Zu $f$ existieren zwei Folgen $\{g_n\}_{n=1,2,\ldots}$ und $\{h_n\}_{n=1,2,\ldots}$ von Funktionen aus $\mathfrak{E}$ derart, daß $0 \leqq g_1 \leqq \cdots \leqq g_n \leqq g_{n+1} \leqq \cdots \leqq f \leqq \cdots \leqq h_{n+1} \leqq h_n \leqq \cdots \leqq h_1$ und $\int (h_n - g_n)\,d\mu \leqq 1/n$ für alle $n = 1, 2, \ldots$ gilt. Es genügt zu zeigen, daß jede Funktion $g_n$ bzw. $h_n$ zu einer Funktion $g'_n$ bzw. $h'_n$ aus $\mathfrak{E}'$ fortgesetzt werden kann, und zwar so, daß $g'_n \leqq g'_{n+1} \leqq h'_{n+1} \leqq h'_n$

[27]) Im folgenden bezeichnet $\chi_A$ die auf $E$ definierte, charakteristische Funktion einer Menge $A \subseteqq E$ und $\chi'_A$ die auf $E'_{\mathfrak{R}}$ definierte, charakteristische Funktion einer Menge $A' \subseteqq E'_{\mathfrak{R}}$.

und $\int (h'_n - g'_n)\, d\mu' = \int (h_n - g_n)\, d\mu$ für jedes $n$ ist. Dann folgt nämlich $\sup g'_n(x') \leq \inf h'_n(x')$ für jeden Punkt $x' \in E'_{\mathfrak{R}}$. Um also zu einer in $\mathfrak{S}'$ gelegenen Fortsetzung $f'$ von $f$ auf $E'_{\mathfrak{R}}$ zu gelangen, braucht man für einen beliebigen Punkt $x' \in E'_{\mathfrak{R}} - E$ als Funktionswert $f'(x')$ nur eine reelle Zahl $\xi$ mit $\sup g'_n(x') \leq \xi \leq \inf h'_n(x')$ zu wählen. Die Existenz der Fortsetzungen $g'_n$ und $h'_n$ ergibt sich folgendermaßen. Jeder Menge $Q \in \mathfrak{Q}$ ordnen wir die Menge $Q' = Q \cup (\overline{Q} - E)$ zu, wobei wieder $\overline{Q}$ die abgeschlossene Hülle von $Q$ in $E'_{\mathfrak{R}}$ bedeutet; nach dem Beweis der Behauptung I ist $Q' \in \mathfrak{Q}'$, $Q' \cap E = Q$ und $\mu'(Q') = \mu(Q)$. Ist $g$ eine beliebige nicht negative Funktion aus $\mathfrak{E}$ und $g = \sum \lambda_\nu \chi_{Q_\nu}$ eine ihrer Darstellungen mit paarweise fremden Mengen $Q_1, \ldots, Q_n \in \mathfrak{Q}$, so setzen wir sie zu einer nicht negativen Funktion $g' | E'_{\mathfrak{R}}$ wie folgt fort: Für einen Punkt $x' \in E'_{\mathfrak{R}} - E$ setzen wir $g'(x') = 0$, falls $x'$ in keiner der Mengen $Q'_1, \ldots, Q'_n$ liegt; liegt dagegen $x'$ in $Q'_{\nu_1}, \ldots, Q'_{\nu_r}$, $\nu_1 < \cdots < \nu_r$, aber in keiner weiteren Menge $Q'_\nu$, so setzen wir $g'(x') = \max(\lambda_{\nu_1}, \ldots, \lambda_{\nu_r})$. Für den Beweis der Unabhängigkeit der so definierten Fortsetzung $g'$ von der gewählten Darstellung von $g$ seien $g = \sum_{\mu=1}^{m} \alpha_\mu \chi_{A_\mu}$ und $g = \sum_{\nu=1}^{n} \beta_\nu \chi_{B_\nu}$ zwei Darstellungen von $g$ mit paarweise fremden Mengen $A_1, \ldots, A_m \in \mathfrak{Q}$ bzw. $B_1, \ldots, B_n \in \mathfrak{Q}$; ferner sei $x'$ ein Punkt von $E'_{\mathfrak{R}} - E$. Wir beweisen zuerst folgende einfache Bemerkung. Liegt $x'$ in $B'_\nu$, ist $\beta_\nu \neq 0$ und sind $A_{\mu_1}, \ldots, A_{\mu_r}$ genau die zu $B_\nu$ nicht fremden Mengen unter den $A_1, \ldots, A_m$, so liegt $x'$ in einer der Mengen $A'_{\mu_1}, \ldots, A'_{\mu_r}$ und es ist $\beta_\nu = \alpha_{\mu_1} = \cdots = \alpha_{\mu_r}$. Wegen $x' \in B'_\nu$ ist nämlich $B_\nu \neq \emptyset$ und wegen $\beta_\nu \neq 0$ existiert folglich wenigstens ein $A_\mu$ mit $A_\mu \cap B_\nu \neq \emptyset$; da $g$ auf $A_\mu$ und $B_\nu$ konstant ist, so ist also $\alpha_\mu = \beta_\nu$ für jedes $\mu$ mit $A_\mu \cap B_\nu \neq \emptyset$. Um die Gleichheit der Funktionswerte $g'(x')$ nachzuweisen, die man ausgehend von der einen bzw. anderen Darstellung von $g$ erhält, genügt es, die beiden folgenden Fälle zu betrachten. 1. Fall. Es liegt $x'$ in keiner der Mengen $A'_1, \ldots, A'_m$. 2. Fall. Es liegt $x'$ in den Mengen $A'_{\mu_1}, \ldots, A'_{\mu_k}$ und $B'_{\nu_1}, \ldots, B'_{\nu_l}$, aber in keiner weiteren Menge unter den $A'_1, \ldots, A'_m$ bzw. $B'_1, \ldots, B'_n$. Im ersten Falle folgt aus der vorangestellten Bemerkung, daß $x'$ höchstens in Mengen $B'_\nu$ mit $\beta_\nu = 0$ liegt. Im zweiten Falle folgt aus der gleichen Bemerkung, daß jedes von Null verschiedene $\alpha_{\mu_1}, \ldots, \alpha_{\mu_k}$ unter den $\beta_{\nu_1}, \ldots, \beta_{\nu_l}$ und umgekehrt jedes von Null verschiedene $\beta_{\nu_1}, \ldots, \beta_{\nu_l}$ unter den $\alpha_{\mu_1}, \ldots, \alpha_{\mu_k}$ auftritt. Da $g$ und damit alle Koeffizienten $\alpha_\mu, \beta_\nu$, die zu nicht leeren Mengen $Q_\mu$, $Q_\nu$ gehören, nicht negativ sind, führen also in beiden Fällen beide Darstellungen von $g$ zum gleichen Funktionswert $g'(x')$. Weiter behaupten wir, daß $g'$ eine Funktion aus $\mathfrak{E}'$ ist. Hierzu sei $g = \sum_{\nu=1}^{n} \lambda_\nu \chi_{Q_\nu}$ eine Darstellung von $g$; die Mengen $Q_\nu \in \mathfrak{Q}$ seien wieder paarweise fremd und es sei $0 < \lambda_1 < \cdots < \lambda_n$. Wir setzen $Q' = \bigcup_{\nu=1}^{n} Q'_\nu$ und $\tilde{Q}'_\nu = Q' - Q'_\nu$, $\nu = 1, \ldots, n$; dann sind $Q'$ und $\tilde{Q}'_\nu$ Mengen aus $\mathfrak{Q}'$. Die Urbildmenge $U'_\nu = [g' = \lambda_\nu]$ ist definitionsgemäß für jedes $\nu = 1, \ldots, n$ gleich $Q'_\nu \cap \tilde{Q}'_{\nu+1} \cap \cdots \cap \tilde{Q}'_n$, also ein Element von $\mathfrak{Q}'$. Da $g'$ nur die Werte $0, \lambda_1, \ldots, \lambda_n$ annehmen kann, ist $g' = \sum_{\nu=1}^{n} \lambda_\nu \chi_{U'_\nu}$;

und somit $g' \in \mathfrak{E}'$. Wegen $U'_\nu \cap E = Q_\nu$ gilt nach Behauptung I ferner $\mu'(U'_\nu) = \mu(Q_\nu)$ für jedes $\nu$ und somit $\int g' d\mu' = \int g\, d\mu$. Schließlich zeigen wir, daß aus $0 \leq g \leq h$, $g, h \in \mathfrak{E}$, stets $g' \leq h'$ folgt. Hierzu wählen wir in $\mathfrak{Q}$ paarweise fremde Mengen $Q_1, \ldots, Q_n$, so daß $g = \sum_{\nu=1}^{n} \alpha_\nu \chi_{Q_\nu}$ und $h = \sum_{\nu=1}^{n} \beta_\nu \chi_{Q_\nu}$ ist. Dann gilt $\alpha_\nu \leq \beta_\nu$ für jedes $\nu$ mit $Q_\nu \neq \emptyset$ und aus der Definition von $g'$ und $h'$ folgt sofort $g' \leq h'$. Wendet man das beschriebene Verfahren der Fortsetzung nicht negativer Funktionen aus $\mathfrak{E}$ zu nicht negativen Funktionen aus $\mathfrak{E}'$ auf die Funktionen $g_n$ und $h_n$ an, so erhält man Fortsetzungen $g'_n, h'_n \in \mathfrak{E}'$ mit den gewünschten Eigenschaften. Also ist nunmehr auch die Behauptung II vollständig bewiesen.

**Korollar.** *Für jede Funktion $f \in \mathfrak{R}$ ist $\overset{(u)}{\int} f\, d\mu = \overset{(u)}{\int} f' d\mu' = \int f' d\varrho'$.*

Beweis. Nach Satz 10 ist $\overset{(u)}{\int} f' d\mu' = \int f' d\varrho'$. Es genügt also die Gleichung $\overset{(u)}{\int} f\, d\mu = \overset{(u)}{\int} f'\, d\mu'$ und diese wiederum nur für nicht negative Funktionen $f \in \mathfrak{R}$ zu beweisen. Für ein $f \geq 0$ aus $\mathfrak{R}$ sei $g$ eine Funktion aus $\mathfrak{S}$ mit $0 \leq g \leq f$. Ist $g' \in \mathfrak{S}'$ eine Fortsetzung von $g$ auf $E'_{\mathfrak{R}}$, so ist deren Positivteil $g'^{+} \in \mathfrak{S}'$ und eine nicht negative Fortsetzung von $g$ auf $E'_{\mathfrak{R}}$; nach (11) ist dann $g'^{*} = \min(g'^{+}, f')$ ein Element von $\mathfrak{S}'$ sowie eine Fortsetzung von $g$ auf $E'_{\mathfrak{R}}$ mit $0 \leq g'^{*} \leq f'$. Nach Satz 15 ist daher die Menge der auf $E$ verengerten Funktionen $g' \in \mathfrak{S}'$ mit $0 \leq g' \leq f'$ identisch mit der Menge aller Funktionen $g \in \mathfrak{S}$ mit $0 \leq g \leq f$. Aus der Definition des uneigentlichen Riemann-Integrals und Satz 15 folgt daher $\overset{(u)}{\int} f\, d\mu = \overset{(u)}{\int} f' d\mu'$.

**4.5.** Das in § 3 bewiesene Korollar zu Satz 11 über die Darstellbarkeit von $I(f)$ durch ein Integral läßt sich jetzt wie folgt verallgemeinern.

**Satz 16.** *Für eine nicht negative Funktion $f \in \mathfrak{R}$ bestehen die Gleichungen*

$$I(f) = \overset{(u)}{\int} f\, d\mu = \overset{(u)}{\int} f' d\mu' = \int f' d\varrho' \tag{33}$$

*dann und nur dann, wenn*

$$\lim_{\alpha \to +0} I\big(\min(\alpha, f)\big) = 0 \tag{34}$$

*ist. Die Bedingung (34) ist insbesondere dann erfüllt, wenn für die Menge $G$ aller Elemente $x \in E$ mit $f(x) > 0$*

$$\sup\big(I(g)\,;\, 0 \leq g \leq \chi_G,\, g \in \mathfrak{R}\big) < +\infty \tag{35}$$

*ist*[28]).

Beweis. Da $\min(\alpha, f')$ die Funktion $\min(\alpha, f)$ auf $E'_{\mathfrak{R}}$ stetig fortsetzt, folgt die Gleichwertigkeit von (33) und (34) aus den Korollaren zu den

[28]) Ist $f$ eine beliebige Funktion aus $\mathfrak{R}$, so ist die gleiche Bedingung (35) hinreichend für die Gültigkeit der Gln. (33), wenn hierbei $G$ als Menge aller $x \in E$ mit $|f(x)| > 0$ definiert wird.

Sätzen 11 und 15. Daß aus (35) die Gl. (34) folgt, wurde bereits im Beweis des Korollars zu Satz 11 gezeigt.

Bemerkungen. 1. Durch Satz 16 haben wir auf anderem Wege ein Resultat von LOOMIS [*14*], S. 176 gewonnen. Danach ist im abstrakten Fall des § 1 für eine beschränkte, nicht negative Funktion $f \in \mathfrak{R}$ die Bedingung (34) notwendig und hinreichend für das Bestehen der Gleichung $I(f) = \overset{(u)}{\int} f\,d\mu$.

2. Aus dem Beweis des Satzes 16 und des Korollars zu Satz 11 kann folgende Deutung der Bedingung (34) entnommen werden. Sie ist gleichbedeutend damit, daß der rein-unstetige Teil des Funktionals $I'|\mathfrak{R}'$ für die stetige Fortsetzung $f'$ von $f$ auf $E'_{\mathfrak{R}}$ gleich Null ist.

3. Die Bedingung (35) tritt in ähnlichem Zusammenhang auch bei HEWITT [*13*], S. 275 auf. Dort wird wie bei uns von einem positiven, linearen Funktional $I$ über einem (dort mit $\mathfrak{F}$ bezeichneten) System $\mathfrak{R}$ von auf einer Menge $E$ definierten, reellen Funktionen ausgegangen; von $\mathfrak{R}$ werden die Eigenschaften (1) bis (3) gefordert. Es wird das Ziel verfolgt, eine Darstellung von $I(f)$ durch ein Unterteilungsintegral zu finden, dem eine nach dem Vorbild der Theorie Radonscher Maße konstruierte, additive Mengenfunktion $\gamma^*$ zugrunde liegt. In Theorem 2.27 wird die Gültigkeit der Gleichung $I(f) = \int f\,d\gamma^*$ für jede *beschränkte* Funktion $f \in \mathfrak{R}$ behauptet, welche der Bedingung (35) mit $G = [|f| > 0]$ genügt. Nach einer brieflichen Mitteilung von Herrn Hewitt ist sein Darstellungssatz (jedenfalls in dieser Allgemeinheit) falsch. Unser Satz 16 zeigt, daß das Hewittsche Ergebnis jedenfalls dann richtig ist, wenn man $\int f\,d\gamma^*$ entweder durch $\overset{(u)}{\int} f\,d\mu$ oder durch $\int f'\,d\varrho'$ ersetzt. (Für die zweite Ersetzung müssen allerdings an $\mathfrak{R}$ die in 4.2 formulierten drei Zusatzforderungen gestellt werden, die aber letzten Endes keine Beschränkung der Allgemeinheit bedeuten.) Herr Hewitt hat übrigens selbst angeregt, eine Richtigstellung seines Satzes entweder durch Verwendung eines anderen Integralbegriffes oder durch geeignete Einführung idealer Punkte zu versuchen.

**4.6.** Abschließend wenden wir uns der Frage zu, in welcher Weise sich der stetige Teil $I_c$ des Funktionals $I$ durch das auf $E'_{\mathfrak{R}}$ definierte Radonsche Maß $\varrho'$ ausdrückt.

**Satz 17.** *Es sei $f$ eine nicht negative Funktion aus $\mathfrak{R}$ und $\tilde{f}$ diejenige Fortsetzung von $f$ auf $E'_{\mathfrak{R}}$, welche in allen Punkten von $E'_{\mathfrak{R}} - E$ gleich Null ist. Dann gilt für den stetigen Teil $I_c$ von $I$*

(36) $$I_c(f) = \overline{\int} \tilde{f}\,d\varrho' \quad [29].$$

Beweis. Zunächst sei an die Definition von $\overline{\int} \tilde{f}\,d\varrho' = N(\tilde{f})$ erinnert. Es ist

$$\overline{\int} \tilde{f}\,d\varrho' = \inf\left(\sup \int k'_n\,d\varrho';\ \{k'_n\} \in \mathfrak{k}'(\tilde{f})\right),$$

wenn $\mathfrak{k}'(\tilde{f})$ die Menge aller nicht fallenden Folgen $\{k'_n\}$ nicht negativer Funktionen $k'_n \in \mathfrak{K}(E'_{\mathfrak{R}})$ mit $\tilde{f} \leq \sup k'_n$ bezeichnet. Ist andererseits $\mathfrak{r}(f)$ bzw. $\mathfrak{r}'(\tilde{f})$

[29]) Bezüglich $\overline{\int} \tilde{f}\,d\varrho'$ vgl. Fußnote [17]).

die Menge aller nicht fallenden Folgen $\{f_n\}$ bzw. $\{f'_n\}$ nicht negativer Funktionen $f_n \in \mathfrak{R}$ bzw. $f'_n \in \mathfrak{R}'$ mit $f \leq \sup f_n$ bzw. $\tilde{f} \leq \sup f'_n$, so kann (nach einer einfachen Überlegung) die in Lemma 3 gewonnene Formel für $I_c(f)$ auch so geschrieben werden

$$I_c(f) = \inf(\sup I(f_n); \{f_n\} \in \mathfrak{r}(f)).$$

Nun ist offenbar für eine Folge $\{f_n\} \in \mathfrak{r}(f)$ die Folge $\{f'_n\}$ der stetigen Fortsetzungen $f'_n$ der Funktionen $f_n$ auf $E'_{\mathfrak{R}}$ ein Element von $\mathfrak{r}'(\tilde{f})$ und umgekehrt für eine Folge $\{f'_n\} \in \mathfrak{r}'(\tilde{f})$ die Folge $\{f_n\}$ der Verengerungen $f_n$ von $f'_n$ auf $E$ ein Element von $\mathfrak{r}(f)$. Wegen (32) kann daher die letzte Formel zu

$$I_c(f) = \inf(\sup I'(f'_n); \{f'_n\} \in \mathfrak{r}'(\tilde{f}))$$

umgeschrieben werden. Wir beweisen die Gleichheit der rechten Seite dieser Gleichung mit $\overline{\int} \tilde{f}\, d\varrho'$. Für jede Folge $\{f'_n\} \in \mathfrak{r}'(\tilde{f})$ gilt nach bekannten Eigenschaften des Oberintegrals $\overline{\int} \tilde{f}\, d\varrho' \leq \sup \overline{\int} f'_n\, d\varrho'$; da nach (25) $\overline{\int} f'_n\, d\varrho' = \int f'_n\, d\varrho' \leq I'(f'_n)$ ist und $\{f'_n\}$ in $\mathfrak{r}'(\tilde{f})$ beliebig gewählt war, folgt $\overline{\int} \tilde{f}\, d\varrho' \leq I_c(f)$. Ist umgekehrt $\{k'_n\}$ eine beliebige Folge aus $\mathfrak{k}'(\tilde{f})$, so ist die Verengerung $k_n$ einer jeden Funktion $k'_n$ auf $E$ nach Satz 15 ein Element von $\mathfrak{S}$ und $\int k_n\, d\mu = \int k'_n\, d\varrho'$. Aus $0 \leq k_1 \leq \cdots \leq k_n \leq k_{n+1} \leq \cdots$ und $f \leq \sup k_n$ folgt $I_c(f) \leq \sup \overline{I}_c(k_n) \leq \sup \overline{I}(k_n)$ nach Satz 4, Beh. I; wegen $\overline{I}(k_n) = \int k_n\, d\mu = \int k'_n\, d\varrho'$ ist also $I_c(f) \leq \sup \int k'_n\, d\varrho'$ und somit $I_c(f) \leq \overline{\int} \tilde{f}\, d\varrho'$. Damit ist (36) bewiesen.

Nach 3.4 ist der sich aus $I'|\mathfrak{R}'$ ableitende, an $E'_{\mathfrak{R}}$ adaptierte Inhalt $\mu'|\mathfrak{Q}'$ $\sigma$-additiv. Aus $\mu'$ leitet sich daher das äußere Maß $\overline{\mu}'$ und das innere Maß $\underline{\mu}'$ ab; $\overline{\mu}'$ und $\underline{\mu}'$ sind definiert auf dem System aller Teilmengen von $E'_{\mathfrak{R}}$, welche durch eine Folge von Mengen aus $\mathfrak{Q}'$ überdeckt werden können[30]). Mit Hilfe von $\overline{\mu}'$ und $\underline{\mu}'$ errechnet sich der $\sigma$-additive Teil $\mu_c$ und der reinendlich-additive Teil $\mu_p$ von $\mu|\mathfrak{Q}$ wie folgt.

**Satz 18.** *Für jede Menge $Q \in \mathfrak{Q}$ ist*

$$\text{(37)} \qquad \mu_c(Q) = \overline{\mu}'(Q) \quad \textit{und} \quad \mu_p(Q) = \underline{\mu}'(\overline{Q} - Q),$$

*wenn $\overline{Q}$ die abgeschlossene Hülle von $Q$ in $E'_{\mathfrak{R}}$ bezeichnet.*

Beweis. Zu $Q \in \mathfrak{Q}$ und $\varepsilon > 0$ existieren nach (15) Funktionen $g, h \in \mathfrak{R}$ mit $0 \leq g \leq \chi_Q \leq h$, $I_c(g) \leq \mu_c(Q) \leq I_c(h)$ und $I_c(h-g) \leq \varepsilon$. Bezeichnen $\tilde{g}$ und $\tilde{h}$ die in allen Punkten von $E'_{\mathfrak{R}} - E$ verschwindenden Fortsetzungen von $g$ und $h$ auf $E'_{\mathfrak{R}}$, so gilt $\tilde{g} \leq \chi'_Q \leq \tilde{h}$ und $\overline{\int} \tilde{g}\, d\varrho' \leq \overline{\int} \chi'_Q\, d\varrho' \leq \overline{\int} \tilde{h}\, d\varrho'$. Nach Satz 17 ist $\overline{\int} \tilde{g}\, d\varrho' = I_c(g)$ und $\overline{\int} \tilde{h}\, d\varrho' = I_c(h)$; hieraus folgt $|\overline{\int} \chi'_Q\, d\varrho' - \mu_c(Q)| \leq \varepsilon$. Da $\varepsilon > 0$ beliebig gewählt war, ist also $\mu_c(Q) = \overline{\int} \chi'_Q\, d\varrho'$. Der erste Teil der Behauptung folgt daher aus der Gleichung $\overline{\mu}'(Q) = \overline{\int} \chi'_Q\, d\varrho'$, deren Richtigkeit aus der Stoneschen Integrationstheorie entnommen werden kann. Der Vollständigkeit halber geben wir hierfür jedoch einen Beweis an. Es bezeichne $\mathfrak{m}'(Q)$ das System aller nicht fallenden Folgen $\{Q'_n\}$ von Mengen aus $\mathfrak{Q}'$ mit

[30]) Vgl. etwa HAUPT-AUMANN-PAUC [*12b*], S. 55.

$Q \subseteqq \bigcup_n Q'_n$; nach Satz 15 ist $\mathfrak{m}'(Q)$ nicht leer. Nach Definition des äußeren Maßes ist

$$\bar{\mu}'(Q) = \inf\left(\sup \mu'(Q'_n);\ \{Q'_n\} \in \mathfrak{m}'(Q)\right).$$

Für jede Folge $\{Q'_n\} \in \mathfrak{m}'(Q)$ gilt daher $\bar{\int} \chi'_Q \, d\varrho' \leqq \sup \bar{\int} \chi'_{Q'_n} \, d\varrho' = \sup \int \chi'_{Q'_n} \, d\varrho' = \sup \mu'(Q'_n)$; also ist $\bar{\int} \chi'_Q \, d\varrho' \leqq \bar{\mu}'(Q)$. Weiter existiert zu jeder reellen Zahl $\varepsilon$ mit $0 < \varepsilon < 1$ eine nicht fallende Folge $\{f'_n\}$ nicht negativer Funktionen $f'_n \in \mathfrak{K}(E'_{\mathfrak{R}})$ mit $\chi'_Q \leqq \sup f'_n$ und $\sup \int f'_n \, d\varrho' \leqq \bar{\int} \chi'_Q \, d\varrho' + \varepsilon$. Nach Satz 8 ist jedes $f'_n$ Riemann-integrierbar bezüglich $I'$, also insbesondere $I'$-meßbar; folglich existiert eine Folge $\{\alpha_n\}$ reeller Zahlen mit $\alpha_1 > \alpha_2 > \cdots > \alpha_n > \cdots > 1 - \varepsilon = \eta^{-1}$ derart, daß $Q'_n = [f'_n > \alpha_n]$ ein Element von $\mathfrak{Q}'$ ist. Nach Konstruktion gilt $\{Q'_n\} \in \mathfrak{m}'(Q)$ und $\mu'(Q'_n) \leqq \alpha_n^{-1} \int f'_n \, d\varrho' \leqq \eta \int f'_n \, d\varrho'$; also ist $\bar{\mu}'(Q) \leqq \sup \mu'(Q'_n) \leqq \eta \sup \int f'_n \, d\varrho' \leqq \frac{1}{1-\varepsilon} \left(\bar{\int} \chi'_Q \, d\varrho' + \varepsilon\right)$. Hieraus folgt die noch fehlende Ungleichung $\bar{\mu}'(Q) \leqq \bar{\int} \chi'_Q \, d\varrho'$ und somit die Gleichung $\bar{\mu}'(Q) = \bar{\int} \chi'_Q \, d\varrho'$. — Nach dem Beweis des Satzes 15 ist die Menge $Q' = Q \cup (\bar{Q} - E)$ für jedes $Q \in \mathfrak{Q}$ ein Element von $\mathfrak{Q}'$; also ist auch $\bar{Q} = \overline{Q'} \in \mathfrak{Q}'$ und $\mu(Q) = \mu'(Q') = \mu'(\bar{Q})$ nach Satz 15 und der Adaption von $\mu'$ an $E'_{\mathfrak{R}}$. Wegen $\bar{Q} \in \mathfrak{Q}'$ gilt $\mu'(\bar{Q}) = \bar{\mu}'(Q) + \underline{\mu}'(\bar{Q} - Q)$. Hieraus und aus $\mu(Q) = \mu_c(Q) + \mu_p(Q) = \bar{\mu}'(Q) + \mu_p(Q)$ folgt schließlich $\mu_p(Q) = \underline{\mu}'(\bar{Q} - Q)$.

Die Formel (37) verallgemeinert die erste Hälfte von (25). Genügen nämlich $E$ und $\mathfrak{R}$ den in 3.1 formulierten Bedingungen, so ist $E'_{\mathfrak{R}} = E$, $\tilde{f} = f$ und $\mu' = \mu$, ferner ist dann $f$ $\varrho$-integrierbar.

**Korollar.** *Die auf $E$ konstanten Funktionen seien Elemente von $\mathfrak{R}$. Dann sind folgende Aussagen gleichwertig:* (1a) $\mu | \mathfrak{Q}$ *ist $\sigma$-additiv;* (2a) *es ist* $\underline{\mu}'(E'_{\mathfrak{R}} - E) = 0$; (3a) *es ist* $\mu(E) = \bar{\mu}'(E)$. *Ferner sind gleichwertig die Aussagen:* (1b) $\mu | \mathfrak{Q}$ *ist rein-endlich-additiv;* (2b) *es ist* $\bar{\mu}'(E) = 0$; (3b) *es ist* $\mu(E) = \underline{\mu}'(E'_{\mathfrak{R}} - E)$.

Beweis. Die Behauptung folgt unmittelbar aus Satz 18. Man hat nur zu beachten, daß $E \in \mathfrak{Q}$ und $\bar{E} = E'_{\mathfrak{R}}$ ist.

Bemerkung. Überträgt man von $\mathfrak{R}$ auf $\mathfrak{R}'$ auch die Zerlegungsteile $I_c$ und $I_p$ von $I$ durch die zu (32) analogen Festsetzungen $(I_c)'(g') = I_c(g)$ und $(I_p)'(g') = I_p(g)$ $(g \in \mathfrak{R})$, so spiegelt sich die Zerlegung $I = I_c + I_p$ in der Zerlegung $I' = (I_c)' + (I_p)'$ wider. Im allgemeinen gilt nicht $(I_c)' = (I')_c$ und $(I_p)' = (I')_p$, sondern nur $(I_c)'(g') \leqq (I')_c(g')$ und $(I_p)'(g') \geqq (I')_p(g')$ für alle $g' \geqq 0$ aus $\mathfrak{R}'$. Hierzu genügt es zu bemerken, daß mit $I_c | \mathfrak{R}$ auch $(I_c)' | \mathfrak{R}'$ stetig ist. *Notwendig und hinreichend für das Bestehen der Gleichungen $(I_c)' = (I')_c$ und $(I_p)' = (I')_p$ ist vielmehr die $\sigma$-Additivität des Inhalts $\mu | \mathfrak{Q}$.*

Beweis. Für alle $f \in \mathfrak{R}$ ist $(I_c)'(f') = I_c(f) = \overset{(u)}{\int} f \, d\mu_c$ nach Satz 5 sowie $(I')_c(f') = \overset{(u)}{\int} f' \, d\mu' = \overset{(u)}{\int} f \, d\mu$ nach Satz 10 und dem Korollar zu Satz 15. Folglich gilt $(I_c)' = (I')_c$ und damit auch $(I_p)' = (I')_p$ dann und nur dann, wenn $\overset{(u)}{\int} f \, d\mu_c = \overset{(u)}{\int} f \, d\mu$ ist für alle $f \in \mathfrak{R}$. Nach Satz 6 ist hiermit $\mu_c = \mu$

gleichwertig. — Falls $\mu$ $\sigma$-additiv ist, ergibt sich aus Satz 10 als Darstellung für den stetigen Teil von $I$ die für alle $f \in \mathfrak{R}$ gültige Gleichung $I_c(f) = \int f' d\varrho'$.

**4.7. Beispiel.** *Die Stonesche Kompaktifizierung einer Booleschen Mengenalgebra als Beispiel für einen Raum* $E'_{\mathfrak{R}}$. Es sei $\mathfrak{B}$ eine Boolesche Algebra von Teilmengen einer Grundmenge $E$, die in $\mathfrak{B}$ als Element enthalten ist; die Menge $E$ sei durch $\mathfrak{B}$ separiert, d.h. zu je zwei verschiedenen Elementen $x, y \in E$ existiert ein $A \in \mathfrak{B}$ mit $x \in A$ und $y \in E - A$. Mit $\mathfrak{R}$ werde bezeichnet die Menge aller $\mathfrak{B}$-Treppenfunktionen, also aller Funktionen $g = \sum_{\nu=1}^{n} \lambda_\nu \chi_{A_\nu}$, wobei $\lambda_1, \ldots, \lambda_n$ reelle Zahlen und $A_1, \ldots, A_n$ Mengen aus $\mathfrak{B}$ sind. Dann besitzt $\mathfrak{R}$ die Eigenschaften (1) bis (3) sowie (21) und (22), ferner ist jede Funktion $g \in \mathfrak{R}$ beschränkt. Also sind die in 4.2 gemachten Voraussetzungen über $\mathfrak{R}$ erfüllt und es existiert der Raum $E'_{\mathfrak{R}}$. Dieser ist kompakt, da die konstante Funktion $g = 1$ in $\mathfrak{R}$ liegt.

Nach STONE [20] existiert zu jeder Booleschen Algebra $\mathfrak{W}$ ein kompakter, total unzusammenhängender Raum $U_{\mathfrak{W}}$ und ein Isomorphismus $\Phi$ von $\mathfrak{W}$ auf die Boolesche Mengenalgebra aller zugleich offenen und abgeschlossenen Teilmengen von $U_{\mathfrak{W}}$. Die Punkte von $U_{\mathfrak{W}}$ sind die Ultrafilter in $\mathfrak{W}$; für ein beliebiges Element $A \in \mathfrak{W}$ besteht $\Phi A$ aus allen Ultrafiltern in $\mathfrak{W}$, die $A$ als Element enthalten. Die Mengen $\Phi A$ mit $A \in \mathfrak{W}$ bilden eine offene (und abgeschlossene) Basis von $U_{\mathfrak{W}}$. Wendet man das Stonesche Verfahren auf die Boolesche Mengenalgebra $\mathfrak{B}$ an, so kann $E$ als dichte Teilmenge von $U_{\mathfrak{B}}$ aufgefaßt werden, indem man jeden Punkt $x \in E$ mit dem durch $x$ in $\mathfrak{B}$ bestimmten Haupt(ultra)filter identifiziert. Für jedes $A \in \mathfrak{B}$ ist dann $\Phi A$ die abgeschlossene Hülle von $A$ in $U_{\mathfrak{B}}$ und $A = E \cap \Phi A$[31]). Wir behaupten:

*Es existiert eine die Punkte von E festlassende Homöomorphie von* $E'_{\mathfrak{R}}$ *auf* $U_{\mathfrak{B}}$.

Beweis. Für jede Menge $A \in \mathfrak{B}$ ist $g = \chi_A$ als Element von $\mathfrak{R}$ zu einer auf $E'_{\mathfrak{R}}$ stetigen Funktion $g'$ fortsetzbar. Aus der Stetigkeit von $g'$ folgt $g'(x') = 1$ für alle $x' \in \bar{A}$ und $g'(y') = 0$ für alle $y' \in \overline{E - A}$; wegen $E'_{\mathfrak{R}} = \bar{E} = \bar{A} \cup \overline{E - A}$ ist daher $g' = \chi'_{\bar{A}}$. Hieraus folgt weiter, daß die Zuordnung $A \to \bar{A}$ ein Isomorphismus von $\mathfrak{B}$ auf die Boolesche Algebra $\bar{\mathfrak{B}}$ der abgeschlossenen Hüllen $\bar{A}$ von Mengen $A \in \mathfrak{B}$ ist und $A = E \cap \bar{A}$ für jede Menge $A \in \mathfrak{B}$ gilt. Die zugleich offenen und abgeschlossenen Mengen $\bar{A} \in \bar{\mathfrak{B}}$ bilden eine offene (und abgeschlossene) Basis von $E'_{\mathfrak{R}}$. Zu einem beliebigen Punkt $x' \in E'_{\mathfrak{R}}$ und einer beliebigen Umgebung $U'$ von $x'$ in $E'$ existiert nämlich eine stetige Funktion $f' | E'_{\mathfrak{R}}$ mit $f'(x') = 1$ und $f'(y') = 0$ für alle $y' \notin U'$. Nach Lemma 4 und (31) existiert zu $f'$ eine Funktion $g' \in \mathfrak{R}'$ mit $|f'(z') - g'(z')| \leq 1/3$ für alle $z' \in E'_{\mathfrak{R}}$. $g'$ hat die Gestalt $g' = \sum \alpha_\nu \chi'_{\bar{A}_\nu} (\alpha_\nu \in \mathsf{P},\ \bar{A}_\nu \in \bar{\mathfrak{B}})$. Wegen $f'(x') = 1$ ist $g'(x') \neq 0$; folglich liegt $x'$ in einer Menge $\bar{A}_{\nu_0}$ und es ist $|1 - \alpha_{\nu_0}| \leq 1/3$. Da dann nicht $|\alpha_{\nu_0}| \leq 1/3$ sein kann, gilt $\bar{A}_{\nu_0} \subseteqq U'$. Also ist $\bar{\mathfrak{B}}$ eine offene Basis von $E'_{\mathfrak{R}}$. Ordnet man jedem Punkt $x' \in E'_{\mathfrak{R}}$ als Bild $\psi(x')$ in $U_{\mathfrak{B}}$ den aus allen Mengen $A \in \mathfrak{B}$ mit $x' \in \bar{A}$ bestehenden Ultrafilter in $\mathfrak{B}$ zu, so ist

[31]) Vgl. PAUC [*18*], S. 33.

$\psi(\bar{A}) = \Phi A$ für jede Menge $A \in \mathfrak{B}$. Hieraus und den bewiesenen Eigenschaften von $E'_{\mathfrak{R}}$ folgt, daß $\psi$ die behauptete Homöomorphie von $E'_{\mathfrak{R}}$ auf $U_{\mathfrak{B}}$ ist, welche $E$ punktweise festläßt.

Wir führen dieses Beispiel weiter und geben eine *Kennzeichnung des σ-additiven und rein-endlich-additiven Teils einer additiven Mengenfunktion* an. Auf der Booleschen Mengenalgebra $\mathfrak{B}$ sei nämlich eine nicht negative, (endlich-)additive Funktion $\varphi$ definiert. Wir übertragen sie von $\mathfrak{B}$ auf die zu $\mathfrak{B}$ isomorphe Boolesche Algebra $\bar{\mathfrak{B}}$ der abgeschlossenen Hüllen $\bar{A}$ von Mengen $A \in \mathfrak{B}$ in $E'_{\mathfrak{R}}$, indem wir $\varphi'(\bar{A}) = \varphi(A)$ für jedes $A \in \mathfrak{B}$ bzw. $\bar{A} \in \bar{\mathfrak{B}}$ setzen. Dann ist $\varphi' | \bar{\mathfrak{B}}$ trivialerweise $\sigma$-additiv; aus $\varphi'$ leitet sich daher das auf dem System aller Teilmengen von $E'_{\mathfrak{R}}$ definierte äußere Maß $\overline{\varphi'}$ und innere Maß $\underline{\varphi'}$ ab. Wir behaupten:

*Bezeichnet $\varphi_c$ den σ-additiven und $\varphi_p$ den rein-endlich-additiven Teil von $\varphi | \mathfrak{B}$, so gilt*

$$\varphi_c(A) = \overline{\varphi'}(A) \qquad \text{und} \qquad \varphi_p(A) = \underline{\varphi'}(\bar{A} - A)$$

*für jede Menge $A \in \mathfrak{B}$*[32]).

Beweis. Der additiven Funktion $\varphi | \mathfrak{B}$ entspricht in umkehrbar eindeutiger Weise ein positives, lineares Funktional $I | \mathfrak{R}$: Für eine Funktion $g = \sum \alpha_\nu \chi_{A_\nu}$ aus $\mathfrak{R}$ ist $I(g) = \sum \alpha_\nu \varphi(A_\nu)$. Aus $I | \mathfrak{R}$ leitet sich das Funktional $I' | \mathfrak{R}'$ und aus diesem der an $E'_{\mathfrak{R}}$ adaptierte Inhalt $\mu' | \mathfrak{Q}'$ ab. Aus der Definition von $\mu' | \mathfrak{Q}'$ folgt, da jede Funktion $g' \in \mathfrak{R}'$ Linearkombination von charakteristischen Funktionen $\chi'_{\bar{A}}$ $(\bar{A} \in \bar{\mathfrak{B}})$ ist, daß $\mu' | \mathfrak{Q}'$ die kleinste Erweiterung von $\varphi' | \bar{\mathfrak{B}}$ zu einem vollständigen Inhalt ist. Dann ist aber $\overline{\varphi'} = \overline{\mu'}$ sowie $\underline{\varphi'} = \underline{\mu'}$. Die Behauptung folgt daher aus Satz 18.

Aus den bewiesenen Formeln für $\varphi_c$ und $\varphi_p$ folgen nun notwendige und hinreichende Bedingungen dafür, daß $\varphi$ $\sigma$-additiv bzw. rein-endlich-additiv ist; diese Bedingungen sind zu denen im Korollar zu Satz 18 formulierten analog. Insbesondere erweist sich $\varphi$ dann und nur dann als $\sigma$-additiv, wenn $\underline{\varphi'}(\bar{E} - E) = 0$ ist. Dieses Kriterium findet sich bereits bei PAUC [*18*], S. 35.

## Literatur

[*1*] ARENS, R.: Representation of functionals by integrals. Duke math. J. **17**, 499—506 (1950). — [*2*] AUMANN, G.: Reelle Funktionen. Berlin-Göttingen-Heidelberg: Springer 1954. — [*3*] BAUER, H.: Eine Rieszsche Bandzerlegung im Raum der Bewertungen eines Verbandes. Sitzgsber. math.-nat. Kl. Bayer. Akad. Wiss. München **1953**, 89—117 (1954). — [*4*] BAUER, H.: Caractérisation topologique de la partie complètement additive et de la partie purement additive d'une fonction additive d'ensemble. C. r. Acad. Sci. Paris **238**, 1771—1773 (1954). — [*5*] BAUER, H.: Reguläre und singuläre Abbildungen eines distributiven Verbandes in einen vollständigen Vektorverband, welche der Funktionalgleichung $f(x \vee y) + f(x \wedge y) = f(x) + f(y)$ genügen. J. reine angew. Math. **194**, 141—179 (1955). — [*6*] BAUER, H.: Darstellung additiver Funktionen auf Booleschen Algebren als Mengenfunktionen. Arch. der Math. **6**, 215—222 (1955). — [*7*] BAUER, H.: Zur Theorie des Riemann-Integrals in lokal kompakten Räumen. Sitzgsber. math.-nat. Kl. Bayer. Akad. Wiss. München **1955**, 187—208 (1956). — [*8*] BOURBAKI, N.:

32) Da $E'_{\mathfrak{R}} = U_{\mathfrak{B}}$ ist, erweist sich jetzt dieses in [*4*] unter anderem von uns angekündigte Resultat als Spezialfall von Satz 18.

Intégration, Chap. I—IV. Actual. Scient. et Ind. **1175**, Paris 1952. — [*9*] GELFAND, I., u. G. ŠILOV: Über verschiedene Methoden der Einführung der Topologie in die Menge der maximalen Ideale eines normierten Ringes. Rec. math. Moscou, n. ser. **9**, 25—38 (1941). — [*10*] HAUPT, O., et CHR. PAUC: Mesure et topologie adaptées. Espaces mesurés topologiques. C. r. Acad. Sci. Paris **230**, 711—712 (1950). — [*11*] HAUPT, O., u. CHR. PAUC: Bemerkungen über Inhalte und Maße in lokal bikompakten Räumen. Akad. Wiss. Lit. Mainz, Abh. math.-nat. Kl. **1955**, 189—218 (1956). — [*12a*] HAUPT-AUMANN-PAUC: Differential- und Integralrechnung, Bd. I, 2. Aufl. Göschens Lehrbücherei **24**, Berlin 1948. — [*12b*] HAUPT-AUMANN-PAUC: Differential- und Integralrechnung, Bd. III, 2. Aufl. Göschens Lehrbücherei **26**, Berlin 1955. — [*13*] HEWITT, E.: Integral representation of certain linear functionals. Arkiv för Matematik **2**, 269—282 (1951). — [*14*] LOOMIS, L. H.: Linear functionals and content. Amer. J. Math. **76**, 168—182 (1954).— [*15*] NAKANO, H.: On the product of relative spectra. Ann. of Math., II. s. **49**, 281—315 (1948). — [*16*] NÖBELING, G., u. H. BAUER: Über die Erweiterungen topologischer Räume. Math. Ann. **130**, 20—45 (1955). — [*17*] NÖBELING, G., u. H. BAUER: Allgemeine Approximationskriterien mit Anwendungen. J.-Ber. Deutsch. Math.-Verein. **58**, 54—72 (1955). — [*18*] PAUC, CHR.: Darstellungs- und Struktursätze für Boolesche Verbände und σ-Verbände. Arch. der Math. **1**, 29—41 (1948). — [*19*] PAUC, CHR.: Intégrale de partition et intégrale topologique. Familles dérivantes topologiques. C. r. Acad. Sci. Paris **230**, 810—811 (1950). — [*20*] STONE, M. H.: Applications of the theory of Boolean rings to general topology. Trans. Amer. math. Soc. **41**, 375—481 (1937). — [*21*] STONE, M. H.: Notes on integration, I—IV. Proc. Nat. Acad. Sci. U.S.A. **34**, 336—342, 447—455, 483—490 (1948); **35**, 50—58 (1949). — [*22*] WOODBURY, M. A.: A decomposition theorem for finitely additive set functions (Preliminary report). Bull. Amer. math. Soc. **56**, 171 (1950). — [*23*] YOSIDA, K., and E. HEWITT: Finitely additive measures. Trans. Amer. math. Soc. **72**, 46—66 (1952).

*Erlangen, Mathematisches Institut der Universität*

*(Eingegangen am 2. März 1956)*

# Sur l'équivalence des théories de l'intégration selon N. Bourbaki et selon M. H. Stone (1) [R10]

Bull. Soc. Math. France **85** (1957), 51–75
[Zbl. Math. 77.10102; MR 19,1167h]

## INTRODUCTION.

On sait qu'on fait une distinction entre les mesures de Radon et les mesures abstraites. Une mesure de Radon positive sur un espace localement compact $E$ est par définition une forme linéaire positive sur l'espace vectoriel $\mathcal{K}(E)$ des fonctions numériques continues dans $E$ et à support compact. Par contre, une mesure abstraite est définie comme suit : Soient $E$ un ensemble quelconque, $\mathcal{R}$ un espace vectoriel de fonctions numériques définies dans $E$ tel que, pour toute fonction $f$ de $\mathcal{R}$, la fonction $|f|$ appartienne à $\mathcal{R}$; on appelle mesure abstraite toute forme linéaire positive $\mu$ sur $\mathcal{R}$ satisfaisant à la condition suivante : si l'enveloppe supérieure d'une suite croissante $(f_n)$ de fonctions positives de $\mathcal{R}$ appartient à $\mathcal{R}$, on a l'égalité $\mu(\sup f_n) = \sup \mu(f_n)$. Toute mesure de Radon positive est, en ce sens, une mesure abstraite particulière. Tandis que la notion de mesure de Radon est essentiellement liée à l'existence d'une topologie sur $E$, la notion de mesure abstraite ne se réfère à aucune topologie. N. Bourbaki [4] (2) et M. H. Stone [9] ont développé presque simultanément une théorie de l'intégration, l'un principalement pour les mesures de Radon, l'autre pour les mesures abstraites. Les méthodes utilisées et les résultats obtenus dans ces deux théories sont, en principe, les mêmes. On est donc amené à penser que la seconde théorie est beaucoup plus générale que la première.

En fait, on sait démontrer que les deux théories sont équivalentes au sens

---

(1) Travail effectué avec le concours du Centre National de la Recherche Scientifique. J'exprime ici ma profonde gratitude à MM. G. Choquet et M. Brelot pour l'intérêt qu'ils ont porté à mes recherches, et pour les entretiens stimulants que j'ai eus avec eux.

(2) *Voir* la bibliographie à la fin du Mémoire.

suivant (³) : Soit $\mu$ une mesure abstraite, donc une forme linéaire positive sur un espace vectoriel $\mathcal{R}$ de fonctions, définies dans un ensemble $E$, avec les propriétés citées. D'après S. Kakutani [6], il existe toujours une mesure de Radon positive $\tilde{\mu}$ sur un espace localement compact $\tilde{E}$ telle que les espaces de Riesz $L^1(\mu)$ et $L^1(\tilde{\mu})$ définis respectivement au sens de la théorie de Stone et de Bourbaki soient isomorphes avec conservation de la norme, donc aussi isomorphes en tant qu'espaces de Banach. On appelle $\tilde{E}$ l'espace de représentation de Kakutani.

Rappelons brièvement la construction de $\tilde{E}$ dans un cas particulier. Soient $\Gamma$ une algèbre booléenne de sous-ensembles d'un ensemble $E$ (avec $E \in \Gamma$), $\mathcal{R}$ l'espace vectoriel $E(\Gamma)$ des combinaisons linéaires de fonctions caractéristiques d'ensembles de $\Gamma$, $\mu$ une mesure abstraite sur $\mathcal{R}$. [On peut considérer une mesure abstraite sur $\mathcal{R}$ comme une fonction numérique (finie) positive $\sigma$-additive sur $\Gamma$, et réciproquement.] Soit $\Psi$ l'algèbre booléenne achevée des sous-ensembles $\mu$-intégrables de $E$ modulo les ensembles $\mu$-négligeables. D'après Stone, il existe un espace compact totalement discontinu $\tilde{E}$ et un isomorphisme $\psi$ de $\Psi$ sur l'algèbre booléenne $\tilde{\Psi}$ des sous-ensembles de $\tilde{E}$, à la fois ouverts et fermés. Pour un élément $\tilde{X}$ de $\tilde{\Psi}$, on pose $\tilde{\mu}(\tilde{X}) = \mu(X)$, où l'ensemble $\mu$-intégrable $X$ est un représentant de l'élément $\overset{-1}{\psi}(\tilde{X})$ de $\Psi$. La fonction $\tilde{\mu}$ est définie sans ambiguïté dans $\tilde{\Psi}$; il n'existe sur $\tilde{E}$ qu'une seule mesure de Radon positive donnant à chaque ensemble $\tilde{X}$ de $\tilde{\Psi}$ la mesure $\tilde{\mu}(\tilde{X})$. Nous désignerons encore cette mesure par $\tilde{\mu}$. L'espace $\tilde{E}$ ainsi obtenu est l'espace de représentation de Kakutani; les espaces de Riesz $L^1(\mu)$ et $L_1(\tilde{\mu})$ sont isomorphes avec conservation de la norme. Dans ce cas, même les espaces de Banach $L^p_F(\mu)$ et $L^p_F(\tilde{\mu})$ sont isomorphes quels que soient le nombre réel $p$ $(1 \leqq p < +\infty)$ et l'espace de Banach $F$. Signalons aussi que $\tilde{E}$ est un espace compact stonien, ce qui signifie que l'adhérence d'un ensemble ouvert est un ensemble ouvert. Dans le cas général, l'espace de Kakutani est somme d'espaces compacts stoniens.

Ces quelques remarques sur $\tilde{E}$ et $\tilde{\mu}$ nous permettent de constater deux inconvénients de la méthode de représentation de Kakutani.

1° L'espace de représentation de Kakutani dépend de la mesure abstraite $\mu$; un changement de $\mu$ entraîne, en général, un changement de $\tilde{E}$, même si l'ensemble $E$ et l'espace de fonctions $\mathcal{R}$ restent fixes.

2° Si l'on applique la méthode de Kakutani au cas où $E$ est un espace localement compact et $\mu$ une mesure de Radon positive sur $E$, on n'obtient pas généralement $E$ comme espace de représentation; l'espace $\tilde{E}$, qui est

(³) *Voir* Bourbaki [4], p. 175, exercice 10.

somme d'espaces compacts stoniens, possède une structure topologique souvent plus compliquée que celle de $E$. Nous avons indiqué ailleurs [1] un troisième inconvénient de la méthode de Kakutani : la méthode de Kakutani ne fournit pas une représentation concrète de la théorie abstraite de l'intégrale de Riemann développée récemment par L. H. LOOMIS [7], à l'aide de la théorie de l'intégrale de Riemann dans des espaces localement compacts développée par O. HAUPT et C. PAUC [5].

Il se pose donc une question : existe-t-il une méthode qui fournisse une représentation de la théorie de l'intégration de Stone par celle de Bourbaki, sans avoir les inconvénients de la méthode de Kakutani?

Nous allons montrer dans ce travail qu'une seule hypothèse supplémentaire sur l'espace de fonctions $\mathcal{R}$ permet de répondre affirmativement à cette question. Il s'agit de l'hypothèse suivante : étant donné une fonction quelconque $f$ de $\mathcal{R}$, la fonction inf $(1, f)$ appartient à $\mathcal{R}$. Cette hypothèse assure l'existence d'un espace localement compact $E'$ tel qu'à chaque mesure abstraite $\mu$ sur $\mathcal{R}$ corresponde une mesure de Radon positive $\mu'$ sur $E'$ pour laquelle l'espace de Banach $L^p_F(\mu')$ est isomorphe à l'espace de Banach $L^p_F(\mu)$ quels que soient le nombre réel $1 \leqq p < +\infty$ et l'espace de Banach $F$. Si $E$ est un espace localement compact et $\mathcal{R}$ l'espace vectoriel $\mathcal{K}(E)$ des fonctions continues dans $E$ à support compact, $E'$ est identique à $E$; la mesure $\mu'$ correspondant à $\mu$ est identique à $\mu$. Dans un autre travail [2] nous avons montré que l'existence de l'espace $E'$ conduit aussi à une représentation concrète de la théorie abstraite de l'intégrale de Riemann développée par LOOMIS. La différence essentielle entre notre méthode de représentation et celle de Kakutani peut s'énoncer comme suit : tandis qu'on retrouve, dans l'ensemble des fonctions continues définies dans l'espace de Kakutani $\tilde{E}$, seulement une image très vague de l'espace $\mathcal{R}$, l'espace $\mathcal{R}$ a une image fidèle dans l'ensemble des fonctions continues définies dans notre espace de représentation $E'$.

Ce travail se divise en quatre paragraphes. Le paragraphe 1 contient quelques conséquences importantes des propriétés de l'espace de fonctions $\mathcal{R}$. Dans le paragraphe 2 nous définirons les espaces $\mathcal{L}^p_F(\mu)$ et $\mathcal{L}^p_F(\mu)$ pour une mesure abstraite $\mu$ sur $\mathcal{R}$. Ceci est nécessaire puisqu'on ne trouve dans la littérature qu'une définition de ces espaces utilisant des hypothèses sur $\mathcal{R}$ trop restrictives pour nous. En fait, nos hypothèses sur $\mathcal{R}$ suffisent à définir ces espaces. Ceci n'est pas très surprenant puisque, d'après STONE [9], notre hypothèse supplémentaire sur $\mathcal{R}$ entraîne la mesurabilité de la fonction constante 1, donc une assez grande richesse d'ensembles intégrables. Dans le paragraphe 3 nous définirons l'espace de représentation $E'$ dans le cas le plus important, où $\mathcal{R}$ sépare les points de l'ensemble $E$; nous montrerons l'existence d'un isomorphisme naturel de l'espace vectoriel des formes linéaires, relativement bornées sur $\mathcal{R}$ dans l'espace vectoriel des mesures de Radon sur $E'$. Enfin, le paragraphe 4 contient la démonstration de notre théorème de représentation.

## Paragraphe 1. — L'espace $\mathcal{R}$ des fonctions élémentaires.

1.1. **L'espace $\mathcal{R}$.** — Soient $E$ un ensemble quelconque, $\mathcal{R}$ un ensemble de fonctions numériques finies définies dans $E$ et satisfaisant aux trois conditions suivantes :

($R_1$) *Étant données deux fonctions quelconques $f$, $g$ de $\mathcal{R}$ et deux nombres réels quelconques $\alpha$, $\beta$, la fonction $\alpha f + \beta g$ appartient à $\mathcal{R}$.*

($R_2$) *Étant donnée une fonction quelconque $f$ de $\mathcal{R}$, la fonction $|f|$ appartient à $\mathcal{R}$.*

($R_3$) *Étant donnée une fonction quelconque $f$ de $\mathcal{R}$, la fonction* $\inf(1, f)$ *appartient à $\mathcal{R}$.*

Les conditions ($R_1$) et ($R_2$) entraînent que $\mathcal{R}$ est un *espace de Riesz*. Pour deux fonctions $f$, $g$ de $\mathcal{R}$, les fonctions $\sup(f, g)$ et $\inf(f, g)$ appartiennent à $\mathcal{R}$. Chaque fonction $f \in \mathcal{R}$ est la différence des deux fonctions positives $f^+ = \sup(f, 0)$, $f^- = (-f)^+$ de $\mathcal{R}$ : $f = f^+ - f^-$. La condition ($R_3$) est une conséquence de ($R_1$) et ($R_2$) si $\mathcal{R}$ contient la fonction constante 1. Il résulte des axiomes ($R_1$), ($R_2$) et ($R_3$) que la fonction $\inf(\alpha, f)$ appartient à $\mathcal{R}$ quels que soient le nombre $\alpha \geq 0$ et la fonction $f \in \mathcal{R}$.

Nous désignerons par $\mathcal{R}_+$ l'ensemble des fonctions positives appartenant à $\mathcal{R}$.

1.2. **L'espace $\mathcal{R}^0$.** — Étudions maintenant un certain sous-espace de $\mathcal{R}$ qui joue un rôle particulier dans nos considérations.

Définition. — *Nous désignerons par $\mathcal{R}^0$ l'ensemble des fonctions bornées $g$ de $\mathcal{R}$ ayant la propriété suivante : il existe une fonction $h$ de $\mathcal{R}_+$ (fonction dépendant de $g$) telle que $h(x) < 1$ entraîne $g(x) = 0$ pour tout $x \in E$.*

Nous désignerons par $\mathcal{R}_+^0$ l'ensemble des fonctions positives appartenant à $\mathcal{R}^0$.

Il résulte immédiatement de la définition de $\mathcal{R}^0$ que c'est un sous-espace vectoriel de $\mathcal{R}$ vérifiant les axiomes de $\mathcal{R}$. Pour toute fonction bornée $f \in \mathcal{R}$ et tout nombre réel $\alpha > 0$, la fonction $f_\alpha = f - \inf(\alpha, f)$ appartient à $\mathcal{R}^0$ ; en effet, $f_\alpha$ est borné et $f_\alpha(x) = 0$ résulte de $\alpha^{-1} f^+(x) < 1$. Si $\mathcal{R}$ contient la fonction constante 1, $\mathcal{R}^0$ est simplement l'ensemble des fonctions bornées de $\mathcal{R}$.

Démontrons maintenant que $\mathcal{R}^0$ est une « base d'approximation » des fonctions de $\mathcal{R}$ au sens suivant :

Lemme. — *Pour toute fonction $f$ de $\mathcal{R}_+$, il existe une suite croissante $(g_n)$ de fonctions de $\mathcal{R}_+^0$ simplement convergente vers $f$. Pour toute fonction bornée $f$ de $\mathcal{R}_+$, il existe une suite croissante $(g_n)$ de fonctions de $\mathcal{R}_+^0$ qui converge uniformément vers $f$ dans $E$.*

DÉMONSTRATION. — Soit d'abord $f$ une fonction bornée de $\mathcal{R}_+$. Pour chaque couple de nombres réels $\alpha$, $\beta$ tels que $0 < \alpha < \beta$, les fonctions $f_\alpha = f - \inf(\alpha, f)$ et $k = \inf\left(1, \frac{1}{\beta - \alpha} f_\alpha\right)$ appartiennent à $\mathcal{R}^0$. On a $0 \leq k \leq 1$, la relation $f(x) \leq \alpha$ entraîne $k(x) = 0$, et la relation $f(x) \geq \beta$ entraîne $k(x) = 1$. D'après G. NÖBELING et l'auteur [8], p. 56, il existe donc une suite croissante $(g_n)$ de fonctions de $\mathcal{R}_+^0$ qui converge uniformément vers $f$ dans $E$. Soit maintenant $f$ une fonction quelconque de $\mathcal{R}_+$. Les fonctions bornées positives $h_n = \inf(n, f)$, $n = 1, 2, \ldots$ forment une suite croissante qui converge simplement vers $f$. En vertu de la première partie de la démonstration, on peut associer à chaque $h_n$ une fonction $g'_n \in \mathcal{R}_+^0$ telle que $g'_n \leq h_n \leq g'_n + \frac{1}{n}$. La suite croissante des fonctions $g_n = \sup(g'_1, \ldots, g'_n)$ de $\mathcal{R}_+^0$ répond à la question : on a $g'_n \leq g_n \leq h_n \leq g_n + \frac{1}{n}$ pour chaque $n = 1, 2, \ldots$ donc $\lim g_n = \lim h_n = f$.

En modifiant légèrement la démonstration du lemme nous obtenons le corollaire suivant :

COROLLAIRE 1. — *Pour toute suite croissante $(f_n)$ de fonctions de $\mathcal{R}_+$, il existe une suite croissante $(g_n)$ de fonctions de $\mathcal{R}_+^0$ telle que*

$$(1) \qquad \sup_n f_n = \sup_n g_n \quad \textit{et} \quad g_n \leq f_n \quad (n = 1, 2, \ldots).$$

DÉMONSTRATION. — Soit $\bar{f} = \sup f_n$ l'enveloppe supérieure de la suite $(f_n)$. Les fonctions bornées $f'_n = \inf(n, f_n)$ de $\mathcal{R}_+$ forment une suite croissante dont l'enveloppe supérieure est encore $\bar{f}$. Selon le lemme, il existe pour chaque $f'_n$ une fonction $g'_n \in \mathcal{R}_+^0$ telle que $g'_n \leq f'_n \leq g'_n + \frac{1}{n}$. La suite des fonctions $g_n = \sup(g'_1, \ldots, g'_n)$ répond à la question.

COROLLAIRE 2. — *Pour tout ensemble $H$ de fonctions de $\mathcal{R}_+$, filtrant pour la relation $\leq$, il existe un ensemble $H^0$ de fonctions de $\mathcal{R}_+^0$, filtrant pour $\leq$, tel que $\sup_{f \in H} f = \sup_{g \in H^0} g$ et tel que toute fonction de $H^0$ soit majorée par au moins une fonction de $H$.*

DÉMONSTRATION. — Montrons que l'ensemble $H^0$ des fonctions $g \in \mathcal{R}_+^0$, qui sont majorées par au moins une fonction de $H$, possède les propriétés en question. Pour deux fonctions $g_1$, $g_2$ de $H^0$, la fonction $\sup(g_1, g_2)$ appartient à $H^0$ ; l'ensemble $H^0$ est donc filtrant. De la définition de $H^0$ résulte l'inégalité $\sup_{g \in H^0} g \leq \sup_{f \in H} f$. On a même l'égalité. En effet, soit $\sup_{f \in H} f(x) = \alpha$ pour un élément quelconque $x \in E$. Il existe alors une suite croissante $(f_n)$ de fonctions de $H$ telle que $\sup f_n(x) = \alpha$. D'après le premier corollaire, il existe une suite croissante $(g_n)$ de fonctions de $H^0$ telle que $\sup f_n = \sup g_n$. On a donc $\sup g_n(x) = \alpha$, d'où résulte l'énoncé.

REMARQUE. — Il résulte de la définition de $\mathcal{R}^0$ et du lemme que, pour toute fonction $g \in \mathcal{R}^0$, il existe une fonction $h$ de $\mathcal{R}^0_+$ (et pas seulement de $\mathcal{R}_+$) telle que $h(x) < 1$ entraîne $g(x) = 0$ $(x \in E)$. On a donc $(\mathcal{R}^0)^0 = \mathcal{R}^0$.

## Paragraphe 2. — Les espaces $\mathcal{L}^p_F$ et $L^p_F$.

**2.1. Définitions.** — Considérons maintenant une *mesure abstraite* $\mu$ sur $\mathcal{R}$. C'est une forme linéaire positive sur $\mathcal{R}$ satisfaisant à la condition suivante :

(M) *Si une suite croissante $(f_n)$ de fonctions de $\mathcal{R}$ a pour enveloppe supérieure une fonction de $\mathcal{R}$, on a* $\mu\left(\sup_n f_n\right) = \sup_n \mu(f_n)$.

Partant de $\mu$ on définit, d'après le premier procédé d'intégration de Stone ([4]), une intégrale supérieure $\bar{\mu}(f)$ pour toute fonction numérique positive $f$, finie ou non, définie dans $E$ : désignons par $\sigma(f)$ l'ensemble des suites croissantes $(f_n)$ de fonctions de $\mathcal{R}_+$ telles que $f \leqq \sup f_n$ ; on posera

$$(2) \qquad \begin{cases} \bar{\mu}(f) = +\infty, & \text{si } \sigma(f) \text{ est vide;} \\ \bar{\mu}(f) = \inf\limits_{(f_n) \in \sigma(f)} \left(\sup\limits_n \mu(f_n)\right), & \text{si } \sigma(f) \text{ n'est pas vide.} \end{cases}$$

Pour une fonction $f \in \mathcal{R}_+$, on a $\bar{\mu}(f) = \mu(f)$.

Soient $p$ un nombre réel tel que $1 \leqq p < +\infty$, $F$ un espace de Banach sur le corps $\mathbf{R}$ des nombres réels ; nous écrirons $|\mathbf{z}|$ pour la norme d'un vecteur $\mathbf{z} \in F$. Pour toute application $\mathbf{f}$ de $E$ dans $F$, on posera

$$(3) \qquad N_p(\mathbf{f}) = \bar{\mu}(|\mathbf{f}|^p)^{\frac{1}{p}}.$$

Pour deux applications $\mathbf{f}$, $\mathbf{g}$ de $E$ dans $F$ et tout nombre réel $\alpha \neq 0$, on a

$$N_p(\mathbf{f} + \mathbf{g}) \leqq N_p(\mathbf{f}) + N_p(\mathbf{g}), \qquad N_p(\alpha\mathbf{f}) = |\alpha| N_p(\mathbf{f}).$$

On désignera par $\mathcal{F}^p_F(E, \mathcal{R}, \mu)$ [ou simplement $\mathcal{F}^p_F(\mu)$, ou $\mathcal{F}^p_F$] l'ensemble des applications $\mathbf{f}$ de $E$ dans $F$ telles que $N_p(\mathbf{f}) < +\infty$. L'ensemble $\mathcal{F}^p_F$ est un sous-espace vectoriel de l'espace vectoriel $F^E$ de toutes les applications de $E$ dans $F$; et $N_p$ est une semi-norme sur $\mathcal{F}^p_F$. On dit que la topologie sur $\mathcal{F}^p_F$ définie par cette semi-norme est la topologie de la convergence en moyenne d'ordre $p$. Enfin, nous désignerons par $\mathcal{R}_F$ (resp. $\mathcal{R}^0_F$) l'ensemble des fonctions $\mathbf{f} \in F^E$ de la forme

$$\mathbf{f} = \sum_{i=1}^{n} \mathbf{a}_i f_i,$$

([4]) *Voir* l'exposé de la théorie de Stone par BOURBAKI [4], p. 114 et suivantes, où $\bar{\mu}$ est désigné par $\mu^{**}$.

où les $\mathbf{a}_i$ sont des vecteurs de $F$ et où les $f_i$ appartiennent à $\mathcal{R}$ (resp. $\mathcal{R}^0$). Dans le cas $F = \mathbf{R}$, nous écrirons $\mathcal{F}^p$ au lieu de $\mathcal{F}^p_{\mathbf{R}}$; on a $\mathcal{R}_{\mathbf{R}} = \mathcal{R}$ et $\mathcal{R}^0_{\mathbf{R}} = \mathcal{R}^0$.

Proposition 1. — *$\mathcal{R}^0_F$ est un sous-espace vectoriel de $\mathcal{F}^p_F$.*

Démonstration. — Pour toute fonction $\mathbf{f} = \sum \mathbf{a}_i f_i$ de $\mathcal{R}^0_F$, on a

$$N_p(\mathbf{f}) \leqq \sum N_p(\mathbf{a}_i f_i) = \sum |\mathbf{a}_i| N_p(f_i).$$

Il suffit donc de démontrer que $N_p(f) < +\infty$ pour toute fonction $f \in \mathcal{R}^0$. La fonction $f$ est bornée; il existe un $h \in \mathcal{R}_+$ tel que $h(x) < 1$ entraîne $f(x) = 0$. On a donc $|f|^p \leqq K.h$, où $K > 0$ est la borne supérieure de $|f|^p$. Il en résulte $\bar{\mu}(|f|^p) \leqq \bar{\mu}(Kh) = K\mu(h) < +\infty$, d'où la proposition.

Ce résultat nous conduit à la définition suivante :

Définition. — *On désigne par $\mathcal{L}^p_F(E, \mathcal{R}, \mu)$ [ou simplement par $\mathcal{L}^p_F(\mu)$, ou $\mathcal{L}^p_F$] l'adhérence, dans l'espace vectoriel $\mathcal{F}^p_F(E, \mathcal{R}, \mu)$ muni de la topologie de la convergence en moyenne d'ordre $p$, de l'espace $\mathcal{R}^0_F$. On désigne par $L^p_F(E, \mathcal{R}, \mu)$ [ou $L^p_F(\mu)$, ou $L^p_F$] l'espace séparé normé associé à $\mathcal{L}^p_F(E, \mathcal{R}, \mu)$.*

On appellera les fonctions de $\mathcal{L}^p_F$, les fonctions de puissance $p^{\text{ième}}$ intégrables au sens du *premier* procédé d'intégration de Stone. Nous écrirons $\mathcal{L}^p$ et $L^p$ au lieu de $\mathcal{L}^p_{\mathbf{R}}$ et $L^p_{\mathbf{R}}$.

**2.2. Propriétés des espaces $\mathcal{L}^p_F$.** — Bourbaki [4], p. 143, ne définit $\mathcal{L}^p_F$ que dans le cas où $\mathcal{R}$ est un sous-espace de $\mathcal{F}^p$. De $\mathcal{R} \subset \mathcal{F}^p$ résulte $\mathcal{R}_F \subset \mathcal{F}^p_F$ pour tout espace de Banach $F$; Bourbaki définit donc $\mathcal{L}^p_F$ comme l'adhérence de $\mathcal{R}_F$ dans $\mathcal{F}^p_F$. En général, les axiomes de $\mathcal{R}$ n'entraînent pas la relation $\mathcal{R} \subset \mathcal{F}^p$; toutefois, nous ramènerons notre définition de $\mathcal{L}^p_F$ à celle de Bourbaki.

Soit pour cela $\mu^0$ la restriction de $\mu$ à $\mathcal{R}^0$. La forme linéaire $\mu^0$ sur $\mathcal{R}^0$ est une mesure abstraite; d'autre part $\mathcal{R}^0$ vérifie les axiomes de $\mathcal{R}$. On déduit de $\mu^0$ l'intégrale supérieure $\bar{\mu}^0$, les semi-normes $N^0_p$, et les espaces $\mathcal{F}^p_F(\mu^0)$, $\mathcal{L}^p_F(\mu^0)$, $L^p_F(\mu^0)$.

Proposition 2. — *Pour toute fonction numérique positive $f$, finie ou non, on a $\bar{\mu}^0(f) = \bar{\mu}(f)$. Les espaces $\mathcal{F}^p_F(\mu^0)$, $\mathcal{L}^p_F(\mu^0)$, $L^p_F(\mu^0)$ sont respectivement identiques aux espaces $\mathcal{F}^p_F(\mu)$, $\mathcal{L}^p_F(\mu)$, $L^p_F(\mu)$.*

Démonstration. — L'égalité $\bar{\mu}^0(f) = \bar{\mu}(f)$, pour une fonction numérique $f \geqq 0$, est une conséquence immédiate du corollaire 1 du lemme du paragraphe 1 ; elle implique l'égalité $N^0_p(\mathbf{f}) = N_p(\mathbf{f})$ pour toute application $\mathbf{f}$ de $E$ dans $F$. Il y a donc identité des espaces $\mathcal{F}^p_F(\mu^0)$ et $\mathcal{F}^p_F(\mu)$. Le reste de la proposition résulte de la remarque de la fin du premier paragraphe d'après laquelle $(\mathcal{R}^0)^0 = \mathcal{R}^0$ et par conséquent $(\mathcal{R}^0)^0_F = \mathcal{R}^0_F$.

Remarque. — Puisqu'on a $\mathcal{R}^0 \subset \mathcal{F}^p(\mu)$ et $\mathcal{F}^p(\mu^0) = \mathcal{F}^p(\mu)$, l'espace des applications de $E$ dans $F$ de puissance $p^{\text{ième}}$ $\mu^0$-intégrables, existe au sens de Bourbaki. A cause de $(\mathcal{R}^0)^0 = \mathcal{R}^0$, cet espace est identique à notre espace $\mathcal{L}^p_F(\mu^0)$, donc, d'après la proposition 2, identique à notre espace $\mathcal{L}^p_F(\mu)$. Cette remarque nous évite de faire toute la théorie des espaces $\mathcal{L}^p_F(\mu)$. Ainsi notre espace $\mathcal{L}^p_F(\mu)$ possède toutes les propriétés énoncées par Bourbaki, même sans la condition $\mathcal{R} \subset \mathcal{F}^p(\mu)$. Par exemple, $L^p_F(\mu)$ est un espace de Banach.

Montrons maintenant que notre espace $\mathcal{L}^p_F(\mu)$ est identique à l'espace correspondant de Bourbaki s'il est défini, donc si $\mathcal{R} \subset \mathcal{F}^p(\mu)$.

Proposition 3. — *Si $\mathcal{R}$ est un sous-espace vectoriel de $\mathcal{F}^p(\mu)$, l'espace $\mathcal{L}^p_F(\mu)$ est l'adhérence de $\mathcal{R}_F$ dans $\mathcal{F}^p_F(\mu)$.*

Démonstration. — Remarquons d'abord que $\mathcal{R} \subset \mathcal{F}^p(\mu)$ entraîne $\mathcal{R}_F \subset \mathcal{F}^p_F(\mu)$. Il suffit de démontrer que $\mathcal{R}_F \subset \mathcal{L}^p_F(\mu)$. Traitons d'abord le cas $F = \mathbf{R}$. Soit $f$ une fonction de $\mathcal{R}_+$. D'après le lemme du paragraphe 1, il existe une suite croissante $(g_n)$ de fonctions de $\mathcal{R}^0_+$ simplement convergente vers $f$. De $\mathcal{R} \subset \mathcal{F}^p(\mu)$ résulte $\bar{\mu}(f^p) < +\infty$; notre définition de $\mathcal{L}^p(\mu)$ entraîne $g_n \in \mathcal{L}^p(\mu)$. Le théorème de Lebesgue ([5]), qui est applicable à $\mathcal{L}^p(\mu)$ d'après la remarque ci-dessus, montre donc que $f$ appartient à $\mathcal{L}^p(\mu)$. Puisque chaque fonction de $\mathcal{R}$ est la différence de deux fonctions de $\mathcal{R}_+$, nous avons démontré que $\mathcal{R} \subset \mathcal{L}^p(\mu)$. Supposons maintenant que $F$ soit un espace de Banach quelconque, et considérons une fonction $\mathbf{f} = \sum \mathbf{a}_i f_i$ de $\mathcal{R}_F (\mathbf{a}_i \in F,\ f_i \in \mathcal{R})$. De la première partie de notre démonstration résulte qu'il existe, pour tout $\varepsilon > 0$ et tout $i = 1, \ldots, n$, une fonction $g_i \in \mathcal{R}^0$ telle que $N_p(f_i - g_i) \leqq \varepsilon$. La fonction $\mathbf{g} = \sum \mathbf{a}_i g_i$ appartient à $\mathcal{R}^0_F$, et l'on a

$$N_p(\mathbf{f} - \mathbf{g}) \leqq \sum N_p(\mathbf{a}_i(f_i - g_i)) \leqq \varepsilon \sum |\mathbf{a}_i|,$$

ce qui montre que $\mathbf{f}$ appartient à $\mathcal{L}^p_F(\mu)$. Nous avons donc démontré que $\mathcal{R}_F \subset \mathcal{L}^p_F(\mu)$.

Le théorème suivant est bien connu pour les mesures de Radon; pour les mesures abstraites, Bourbaki [4], p. 143, ne le démontre que sous des hypothèses restrictives sur $\mathcal{R}_F$.

Théorème 1. — *Soient $p$ et $q$ deux nombres réels tels que $1 \leqq p < +\infty$,*

---

([5]) *Voir* Bourbaki [4], p. 140 et 143.

$1 \leqq q < +\infty$. *Pour qu'une application* $\mathbf{f}$ *de* $E$ *dans* $F$ *appartienne à* $\mathcal{L}_F^p$, *il faut et il suffit que l'application* $|\mathbf{f}|^{\frac{p}{q}-1}\mathbf{f}$ *appartienne à* $\mathcal{L}_F^q$.

Démonstration. — Rappelons que, pour tout nombre $\alpha > 0$, l'application $\mathbf{z} \to |\mathbf{z}|^{\alpha-1}\mathbf{z}$ est définie et continue dans le complémentaire de $\mathbf{0}$ dans $F$; elle se prolonge par continuité au point $\mathbf{0}$ en lui donnant la valeur zéro.

Nous démontrerons d'abord ceci : soient $\Phi$ une application continue de $F$ dans $F$ qui laisse le point $\mathbf{0}$ invariant [par exemple $\Phi(\mathbf{z}) = |\mathbf{z}|^{\alpha-1}\mathbf{z}$ pour un $\alpha > 0$], $\mathbf{f}$ une fonction de $\mathcal{R}_F^0$; alors $\mathbf{g} = \Phi \circ \mathbf{f}$ appartient à $\mathcal{L}_F^q$. La fonction $\mathbf{f}$ est de la forme $\mathbf{f} = \sum_{i=1}^{n} \mathbf{a}_i f_i$ ($\mathbf{a}_i \in F$, $f_i \in \mathcal{R}^0$). La remarque de la fin du paragraphe 1 montre qu'il existe une fonction $h_0 \in \mathcal{R}_+^0$ telle que, pour tout $x \in E$, la relation $h_0(x) < 1$ entraîne $f_1(x) = \ldots = f_n(x) = 0$, donc $\mathbf{f}(x) = 0$. L'ensemble $\mathcal{R}^0(h_0)$ des fonctions de $\mathcal{R}^0$, qui s'annulent sur l'ensemble des $x \in E$ tels que $h_0(x) < 1$, est un sous-espace vectoriel de $\mathcal{R}^0$ vérifiant les axiomes de $\mathcal{R}$. Les fonctions $f_1, \ldots, f_n$ appartiennent à $\mathcal{R}^0(h_0)$. Sur chaque sous-ensemble relativement compact de l'espace euclidien $\mathbf{R}^n$ à $n$ dimensions, l'application $(\xi_1, \ldots, \xi_n) \to \Phi\left(\sum \mathbf{a}_i \xi_i\right)$ est uniformément continue. Puisque les fonctions $f_1, \ldots, f_n$ sont bornées, il existe donc, pour tout $\varepsilon > 0$, un nombre $\delta > 0$, tel que

$$\sup_{1 \leqq i \leqq n} |f_i(y) - f_i(z)| < \delta \quad \text{entraîne} \quad |\mathbf{g}(y) - \mathbf{g}(z)| < \varepsilon,$$
$$\sup_{1 \leqq i \leqq n} |f_i(x)| < \delta \quad \text{entraîne} \quad |\mathbf{g}(x)| < \varepsilon$$

quels que soient $x, y, z \in E$. D'après un critère d'approximation de Nöbeling et de l'auteur [8], p. 61, la fonction $\mathbf{g}$ peut être approchée uniformément par des combinaisons linéaires de fonctions de $\mathcal{R}^0(h_0)$ à coefficients dans $F$. Pour tout nombre $\varepsilon > 0$, il existe donc une fonction $\mathbf{g}' \in \mathcal{R}_F^0$ telle que $|\mathbf{g} - \mathbf{g}'| \leqq \varepsilon h_0$. On a par conséquent $N_p(\mathbf{g} - \mathbf{g}') \leqq \varepsilon N_p(h_0)$; puisque $h_0$ est indépendant de $\varepsilon$, il en résulte que $\mathbf{g} \in \mathcal{L}_F^q$.

Pour le reste de la démonstration nous nous inspirerons de Bourbaki [4], p. 142. Soit $\mathbf{f}$ une fonction de $\mathcal{L}_F^p$. D'après la remarque précédant la proposition 3, il existe une suite $(\mathbf{f}_n)$ de fonctions de $\mathcal{R}_F^0$ telle que

$$\sum_{n=1}^{\infty} N_p(\mathbf{f}_n) < +\infty \quad \text{et} \quad \mathbf{f}(x) = \sum_{n=1}^{\infty} \mathbf{f}_n(x)$$

presque partout dans $E$ (par rapport à $\mu$). La première partie de notre démonstration $\left[\text{avec } \Phi(\mathbf{z}) = |\mathbf{z}|^{\frac{p}{q}-1}\mathbf{z}\right]$ montre que la fonction

$$\mathbf{g}_n = |\mathbf{f}_1 + \ldots + \mathbf{f}_n|^{\frac{p}{q}-1}(\mathbf{f}_1 + \ldots + \mathbf{f}_n)$$

appartient à $\mathcal{L}_F^q$. On a

$$|\mathbf{g}_n|^q = |\mathbf{f}_1 + \ldots + \mathbf{f}_n|^p \leq \left(\sum_{n=1}^{\infty} |\mathbf{f}_n|\right)^p = h^q,$$

où $h$ est la fonction numérique positive (pas nécessairement finie) $\left(\sum_{n=1}^{\infty} |\mathbf{f}_n|\right)^{\frac{p}{q}}$ ;

elle vérifie l'inégalité

$$N_q(h))^q = \left(N_p\left(\sum |\mathbf{f}_n|\right)\right)^p \leq \left(\sum N_p(\mathbf{f}_n)\right)^p < +\infty,$$

d'où résulte $N_q(h) < +\infty$. D'autre part, $\mathbf{g}_n(x)$ tend presque partout vers $\mathbf{g}(x) = |\mathbf{f}(x)|^{\frac{p}{q}-1}\mathbf{f}(x)$. Le théorème de Lebesgue montre donc que $\mathbf{g}$ appartient à $\mathcal{L}_F^q$. La réciproque est immédiate puisque $\mathbf{g} = |\mathbf{f}|^{\frac{p}{q}-1}\mathbf{f}$ entraîne $\mathbf{f} = |\mathbf{g}|^{\frac{q}{p}-1}\mathbf{g}$.

Remarque. — Il résulte du théorème 1 que $\mathcal{L}^p$ est l'ensemble des fonctions numériques $f$ telles que $|f|^{p-1}f$ appartient à $\mathcal{L}^1$. C'est la définition de l'espace $\mathcal{L}^p$ à l'aide de $\mathcal{L}^1$ donné par Stone [9]. Cependant Stone, dans sa théorie de l'intégration, considère comme éléments de $\mathcal{L}^1$ non seulement des fonctions définies dans $E$ tout entier mais aussi des fonctions définies presque partout dans $E$. Nous avons laissé de côté cette possibilité de généralisation des espaces $\mathcal{L}_F^p$; ce ne sont que les espaces $L_F^p$ qui jouent un rôle dans notre théorème de représentation.

**2.3. Les espaces $\mathcal{L}_F^{p*}$.** — Considérons une forme linéaire positive $\mu$ sur $\mathcal{R}$ satisfaisant à la condition suivante, plus restrictive que (M) :

(M*) *Si $H$ est un ensemble de fonctions de $\mathcal{R}$ filtrant pour la relation $\leq$, ayant pour enveloppe supérieure une fonction de $\mathcal{R}$, on a* $\mu\left(\sup_{f\in H} f\right) = \sup_{f\in H} \mu(f)$.

$\mu$ est donc une mesure abstraite particulière. On sait que toute mesure de Radon positive sur un espace localement compact, considérée comme forme linéaire sur l'espace vectoriel des fonctions continues à support compact, vérifie la condition (M*).

Dans le reste du paragraphe nous supposerons que $\mu$ vérifie (M*). On définit, d'après le second procédé d'intégration de Stone [9] ([6]), une deuxième intégrale supérieure $\mu^*(f)$ pour toute fonction numérique positive

([6]) *Voir* aussi Bourbaki [4], p. 114 et suivantes.

$f$, finie ou non, définie dans $E$ : désignons par $\sigma^*(f)$ l'ensemble des sous-ensembles $H$ de $\mathcal{R}_+$ filtrant par $\leqq$ tels que $f \leqq \sup_{h\in H} h$ ; on posera

$$(4) \qquad \begin{cases} \mu^*(f) = +\infty, & \text{si } \sigma^*(f) \text{ est vide;} \\ \mu^*(f) = \inf\limits_{H\in\sigma^*(f)} \left(\sup\limits_{h\in H} \mu(h)\right), & \text{si } \sigma^*(f) \text{ n'est pas vide.} \end{cases}$$

Pour toute fonction $f \in \mathcal{R}_+$, on a $\mu^*(f) = \mu(f)$.

Pour $1 \leqq p < +\infty$ et toute application $\mathbf{f}$ de $E$ dans un espace de Banach $F$, on posera

$$(5) \qquad N_p^*(\mathbf{f}) = (\mu^*(|\mathbf{f}|^p))^{\frac{1}{p}}.$$

En remplaçant, dans la définition des espaces $\mathcal{F}_F^p$, $\mathcal{L}_F^p$, $L_F^p$, la semi-norme $N_p$ par $N_p^*$, on obtient la définition des espaces $\mathcal{F}_F^{p*}$, $\mathcal{L}_F^{p*}$, $L_F^{p*}$. Nous écrirons $\mathcal{F}^{p*}$ au lieu de $\mathcal{F}_{\mathbf{R}}^{p*}$, etc. On appellera les fonctions de $\mathcal{L}_F^{p*}$, les fonctions de puissance $p^{\text{ième}}$ intégrable au sens du *second* procédé d'intégration de Stone.

Étudions maintenant les relations entre les espaces $\mathcal{L}_F^p$ et $\mathcal{L}_F^{p*}$ ; mais remarquons d'abord que, pour toute application $\mathbf{f}$ de $E$ dans $F$, on a

$$(6) \qquad N_p^*(\mathbf{f}) \leqq N_p(\mathbf{f}).$$

Proposition 4. — *L'espace $\mathcal{L}_F^p$ est un sous-espace vectoriel de $\mathcal{L}_F^{p*}$. Pour toute fonction $\mathbf{f}$ de $\mathcal{L}_F^p$, on a*

$$(7) \qquad N_p^*(\mathbf{f}) = N_p(\mathbf{f}).$$

Démonstration. — Soit $\mathbf{f}$ une fonction de $\mathcal{L}_F^p$. Il existe une suite $(\mathbf{g}_n)$ de fonctions de $\mathcal{R}_F^0$ telle que $\lim\limits_{n\to\infty} N_p(\mathbf{f} - \mathbf{g}_n) = 0$; de (6) résulte d'une part : $\lim N_p^*(\mathbf{f} - \mathbf{g}_n) = 0$, d'autre part : $N_p^*(\mathbf{f}) \leqq N_p(\mathbf{f}) < +\infty$. Donc $\mathbf{f}$ appartient à $\mathcal{L}_F^{p*}$, ce qui prouve que $\mathcal{L}_F^p$ est un sous-espace vectoriel de $\mathcal{L}_F^{p*}$ D'après le théorème 1, on a $|\mathbf{f}|^{p-1}\mathbf{f} \in \mathcal{L}_F^1$; la fonction $|\mathbf{f}|^p = \big||\mathbf{f}|^{p-1}\mathbf{f}\big|$ appartient ainsi ([7]) à $\mathcal{L}^1$. Il suffit donc de montrer l'égalité (7) pour $p = 1$ et $F = \mathbf{R}$. Soient $f$ une fonction de $\mathcal{L}^1$, $(g_n)$ une suite de fonctions de $\mathcal{R}^0$ telle que $\lim N_1(f - g_n) = 0$. De l'inégalité $|N_1(f) - N_1(g_n)| \leqq N_1(f - g_n)$, résulte $N_1(f) = \lim N_1(g_n) = \lim \bar{\mu}(|g_n|)$. D'après (6) on a aussi $\lim N_1^*(f - g_n) = 0$, d'où résulte $N_1^*(f) = \lim N_1^*(g_n) = \lim \mu^*(|g_n|)$. Puisque $|g_n|$ appartient à $\mathcal{R}^0$ :

$$\mu(|g_n|) = \bar{\mu}(|g_n|) = \mu^*(|g_n|),$$

donc $N_1^*(f) = N_1(f)$.

---

([7]) *Voir* Bourbaki, [4], p. 135 et 143.

PROPOSITION 5. — *Pour toute fonction* $\mathbf{f}^*$ *de* $\mathcal{L}_F^{p*}$, *il existe une fonction* $\mathbf{f}$ *de* $\mathcal{L}_F^p$ *telle que* $N_p^*(\mathbf{f}^* - \mathbf{f}) = 0$. *Pour toute fonction positive* $f^*$ *de* $\mathcal{L}^{p*}$, *il existe une fonction positive* $f$ *de* $\mathcal{L}^p$ *telle que* $N_p^*(f^* - f) = 0$.

DÉMONSTRATION. — Pour $\mathbf{f}^* \in \mathcal{L}_F^{p*}$, il existe une suite $(g_n)$ de fonctions de $\mathcal{R}_F^0$ telle que $\lim\limits_{n\to\infty} N_p^*(\mathbf{f}^* - \mathbf{g}_n) = 0$. Il en résulte : $\lim\limits_{m,n\to\infty} N_p^*(\mathbf{g}_n - \mathbf{g}_m) = 0$, donc, d'après la proposition 4, $\lim\limits_{m,n\to\infty} N_p(\mathbf{g}_n - \mathbf{g}_m) = 0$. Puisque l'espace $\mathcal{L}_F^p$ est complet, la suite $(\mathbf{g}_n)$ possède une limite $\mathbf{f}$ dans $\mathcal{L}_F^p$ qui répond à la question. En effet, on a

$$N_p^*(\mathbf{f}^* - \mathbf{f}) \leqq N_p^*(\mathbf{f}^* - \mathbf{g}_n) + N_p^*(\mathbf{f} - \mathbf{g}_n) = N_p^*(\mathbf{f}^* - \mathbf{g}_n) + N_p(\mathbf{f} - \mathbf{g}_n),$$

donc $N_p^*(\mathbf{f}^* - \mathbf{f}) = 0$.

Soient maintenant $f^* \geqq 0$ une fonction de $\mathcal{L}^{p*}$ et $(g_n)$ une suite de fonctions de $\mathcal{R}^0$ telle que $\lim N_p^*(f^* - g_n) = 0$. De $\big| f^* - |g_n| \big| \leqq |f^* - g_n|$ résulte $N_p^*(f^* - |g_n|) \leqq N_p^*(f^* - g_n)$. On peut donc supposer que toute fonction $g_n$ est positive. Comme ci-dessus on montre l'existence d'une fonction $f \in \mathcal{L}^p$ telle que $\lim N_p(f - g_n) = 0$. De $\big| |f| - g_n \big| \leqq |f - g_n|$ résulte alors qu'on peut supposer $f \geqq 0$. La démonstration générale montre que $N_p^*(f^* - f) = 0$.

Désignons par $\mathcal{N}_F^p$ l'adhérence de $0$ dans $\mathcal{L}_F^p$, par $\mathcal{N}_F^{p*}$ l'adhérence de $0$ dans $\mathcal{L}_F^{p*}$ ; alors $\mathcal{N}_F^p$ est l'ensemble des $\mathbf{f} \in \mathcal{L}_F^p$ tels que $N_p(\mathbf{f}) = 0$, et $\mathcal{N}_F^{p*}$ est l'ensemble des $\mathbf{f}^* \in \mathcal{L}_F^{p*}$ tels que $N_p^*(\mathbf{f}^*) = 0$. Les propositions 4 et 5 entraînent que chaque classe d'équivalence de $\mathcal{L}_F^p$ modulo $\mathcal{N}_F^p$ est contenue dans une et une seule classe d'équivalence de $\mathcal{L}_F^{p*}$ modulo $\mathcal{N}_F^{p*}$, et qu'inversement chaque classe d'équivalence de $\mathcal{L}_F^{p*}$ modulo $\mathcal{N}_F^{p*}$ contient une classe d'équivalence de $\mathcal{L}_F^{p*}$ modulo $\mathcal{N}_F^{p*}$. Cette correspondance biunivoque entre les classes d'équivalences de $\mathcal{L}_F^p$ et de $\mathcal{L}_F^{p*}$ établit un isomorphisme canonique entre l'espace vectoriel $L_F^p = \mathcal{L}_F^p/\mathcal{N}_F^p$ et l'espace vectoriel $L_F^{p*} = \mathcal{L}_F^{p*}/\mathcal{N}_F^{p*}$ qui conserve la norme. Dans le cas $F = \mathbf{R}$, les espaces $L^p$ et $L^{p*}$ sont des espaces de Riesz complètement réticulés pour la structure d'ordre déduite de la structure d'ordre naturelle de $\mathcal{L}^p$ et $\mathcal{L}^{p*}$ (8). La deuxième partie de la proposition 5 montre que l'isomorphisme canonique de $L^p$ sur $L^{p*}$ est aussi un isomorphisme de l'espace de Riesz $L^p$ sur l'espace de Riesz $L^{p*}$. Or, nous avons démontré :

THÉORÈME 2. — *Soit* $\mu$ *une mesure abstraite avec la propriété* (M*). *Il existe un isomorphisme de l'espace de Banach* $L_F^p$ *sur l'espace de Banach* $L_F^{p*}$ ; *pour* $F = \mathbf{R}$, *cet isomorphisme est aussi un isomorphisme de* $L^p$ *sur* $L^{p*}$ *considérés comme espaces de Riesz* (9).

(8) *Voir* BOURBAKI, [4], p. 138 et 143.

(9) Pour $p = 1$, $F = \mathbf{R}$, ce théorème est connu; *voir* BOURBAKI [4], p. 143.

Remarques. — 1° Comme nous l'avons déjà remarqué, chaque mesure de Radon positive sur un espace localement compact $E$, considérée comme forme linéaire sur $\mathcal{K}(E)$ (espace des fonctions continues à support compact) possède la propriété (M*). Pour une mesure de Radon positive sur $E$, les espaces $\mathcal{L}_F^{p*}$ et $L_F^{p*}$ existent donc toujours. Une application $\mathbf{f}$ de $E$ dans l'espace de Banach $F$ est de puissance $p^{\text{ième}}$ intégrable au sens de la théorie de Bourbaki si $\mathbf{f}$ appartient à $\mathcal{L}_F^{p*}$. Pour une mesure de Radon positive, les espaces $\mathcal{L}_F^p$ et $L_F^p$ définis dans le livre de Bourbaki sont donc, dans notre terminologie, les espaces $\mathcal{L}_F^{p*}$ et $L_F^{p*}$ (10).

2° Pour une mesure abstraite $\mu$ sur $\mathcal{R}$ avec la propriété (M*), les espaces $\mathcal{F}_F^{p*}$, $\mathcal{L}_F^{p*}$, etc. possèdent les mêmes propriétés que celles démontrées pour $\mathcal{F}_F^p$, $\mathcal{L}_F^p$, etc. dans les propositions 2 et 3, et dans le théorème 1. Les démonstrations sont presque les mêmes; il faut simplement remplacer $\bar{\mu}$, $N_p$, etc. par $\mu^*$, $N_p^*$, etc. et, dans la démonstration de la proposition 2, le corollaire 1 du lemme du paragraphe 1 par le corollaire 2 de ce lemme.

## Paragraphe 3. — L'espace $E'_{\mathcal{R}}$.

**3.1. Propriétés topologiques.** — Rappelons quelques résultats que nous avons obtenus dans [2] et qui nous permettront de démontrer le théorème de représentation du prochain paragraphe.

Nous supposerons, dans le reste de ce paragraphe, que $\mathcal{R}$ vérifie, outre $(R_1)$, $(R_2)$ et $(R_3)$, les axiomes suivants :

$(S_1)$ *Étant donné un élément quelconque $x$ de $E$, il existe une fonction $f$ de $\mathcal{R}$ telle que $f(x) \neq 0$.*

$(S_2)$ *Étant donné un couple quelconque d'éléments différents $x$, $y$ de $E$, il existe une fonction $f$ de $\mathcal{R}$ telle que $f(x) \neq f(y)$.*

Il résulte du lemme du paragraphe 1 que $\mathcal{R}^0$ vérifie aussi les axiomes $(S_1)$ et $(S_2)$ de $\mathcal{R}$ [ainsi que $(R_1)$, $(R_2)$, et $(R_3)$].

Théorème 3. — *Si $\mathcal{R}$ vérifie les axiomes* $(R_1)$, $(R_2)$, $(R_3)$ *et* $(S_1)$, $(S_2)$, *il existe un espace topologique $E'_{\mathcal{R}}$ ayant les propriétés suivantes :*

1° *$E'_{\mathcal{R}}$ est localement compact et contient $E$ comme sous-ensemble dense;*

2° *toute fonction $f$ de $\mathcal{R}$ peut se prolonger en une fonction continue $f'$ définie dans $E'_{\mathcal{R}}$ s'annulant à l'infini* (11);

---

(10) *Voir* Bourbaki [4], p. 135, proposition 10.

(11) Cela signifie que, pour tout $\alpha > 0$, l'ensemble des $x' \in E'_{\mathcal{R}}$, tels que $|f'(x')| \geqq \alpha$, est compact. Donc, pour un espace compact $E'_{\mathcal{R}}$, toute fonction continue dans $E'_{\mathcal{R}}$ s'annule à l'infini.

*3° pour chaque point $x'$ de $E'_{\mathcal{R}}$, il existe une fonction $f$ de $\mathcal{R}$ dont le prolongement $f'$ ne s'annule pas au point $x'$;*

*4° pour chaque couple de points différents $x'$, $y'$ de $E'_{\mathcal{R}}$, il existe une fonction $f$ de $\mathcal{R}$ telle qu'on ait $f'(x') \neq f'(y')$ pour le prolongement $f'$ de $f$.*

*Ces propriétés déterminent l'espace $E'_{\mathcal{R}}$ à une homéomorphie près laissant $E$ ponctuellement invariant.*

DÉMONSTRATION. — Remarquons d'abord que le prolongement $f'$ d'une fonction $f \in \mathcal{R}$ est fini, si et seulement si $f$ est bornée. Pour une fonction $f \in \mathcal{R}$ non bornée, l'ensemble des points $x' \in E'_{\mathcal{R}}$, tels que $|f'(x')| = +\infty$, est compact et disjoint de $E$, donc est un ensemble rare.

Nous avons démontré le théorème 3 dans [2], p. 471, dans le cas particulier où toute fonction de $\mathcal{R}$ est bornée. Montrons maintenant qu'il n'est pas nécessaire de faire cette hypothèse.

Soit $\mathcal{R}^*$ l'ensemble des fonctions bornées de $\mathcal{R}$; c'est un sous-espace vectoriel de $\mathcal{R}$ vérifiant les axiomes ($R_1$), ($R_2$), ($R_3$) et ($S_1$), ($S_2$) de $\mathcal{R}$. L'espace $E'_{\mathcal{R}^*}$ existe donc d'après [2]. Nous démontrerons que $E'_{\mathcal{R}^*}$ est aussi un espace $E'_{\mathcal{R}}$. Pour cela il suffit de montrer que chaque fonction $f \in \mathcal{R}$ peut se prolonger en une fonction continue $f'$, définie dans $E'_{\mathcal{R}^*}$, s'annulant à l'infini. Les fonctions $f_n = \inf(n, f)$, $n = 1, 2, \ldots$ appartiennent à $\mathcal{R}^*$; chacune d'elles possède donc un prolongement continu $f'_n$ sur $E'_{\mathcal{R}^*}$. Les fonctions $f'_n$ sont positives, s'annulent à l'infini, et forment une suite croissante. La fonction $f' = \sup f'_n$ répond à la question. En effet, la fonction $f'$, étant l'enveloppe supérieure d'une suite de fonctions continues, est semi-continue inférieurement, donc continue en chaque point $a' \in E'_{\mathcal{R}^*}$ où $f'(a') = +\infty$. Soit $a' \in E'_{\mathcal{R}^*}$ tel que $f'(a') = \alpha < +\infty$. Puisque, pour un nombre naturel fixe $n_0 > \alpha$, on a $f'_{n_0}(a') \leqq \alpha < n_0$, il existe un voisinage $V'$ de $a'$ tel que $f'_{n_0}(x') < n_0$ pour tout $x' \in V'$. Il s'ensuit que $f_{n_0}(x) = \inf(n_0, f(x)) < n_0$, donc $f(x) = f_{n_0}(x) = f_n(x) < n_0$ pour tout $n \geqq n_0$ et tout $x \in V = V' \cap E$; d'où résulte $f'_{n_0}(x') = f'_n(x')$, pour tout $n \geqq n_0$ et tout $x' \in V'$, puisque $V$ est dense dans l'espace $V'$. Par conséquent on a $f'(x') = f'_{n_0}(x')$ pour tout $x' \in V'$. Donc, $f'$ est continu en $a'$. Nous avons ainsi démontré que $f'$ est un prolongement continu de $f$. Notre démonstration de cette propriété montre que, pour tout $\alpha > 0$, l'ensemble des $x' \in E'_{\mathcal{R}^*}$, tels que $f'(x') \geqq \alpha$, est égal à l'ensemble des $x'$ tels que $f'_{n_0}(x') \geqq \alpha$, pourvu que $n_0 > \alpha$. Ce dernier ensemble étant compact, $f'$ s'annule à l'infini.

Nous avons démontré l'unicité de $E'_{\mathcal{R}}$ dans le cas $\mathcal{R} = \mathcal{R}^*$. A cause de l'axiome ($R_3$), on voit immédiatement qu'un espace $E'_{\mathcal{R}}$ est aussi un espace $E'_{\mathcal{R}^*}$. L'unicité de $E'_{\mathcal{R}}$ est donc prouvée aussi dans le cas général.

D'après le lemme du paragraphe 1, toute fonction de $\mathcal{R}^*$ peut être approchée uniformément par des fonctions de $\mathcal{R}^0$. Il en résulte que chaque espace $E'_{\mathcal{R}^0}$ est aussi un espace $E'_{\mathcal{R}^*}$. Nous avons ainsi démontré :

COROLLAIRE. — *Les trois espaces $E'_{\mathcal{R}}$, $E'_{\mathcal{R}^*}$, $E'_{\mathcal{R}^0}$ sont égaux, à des homéomorphies près laissant $E$ ponctuellement invariant.*

Rappelons encore deux résultats démontrés dans [2], p. 473 et 462, sous l'hypothèse que chaque fonction de $\mathcal{R}$ est bornée. Le corollaire du théorème 4 nous permet d'abandonner cette hypothèse ou de remplacer $\mathcal{R}$ par $\mathcal{R}^0$.

PROPOSITION 6. — *Pour qu'une fonction $f$ de $\mathcal{R}$ appartienne à $\mathcal{R}^0$, il faut et il suffit que le prolongement continu de $f$ sur $E'_{\mathcal{R}}$ soit fini et à support compact.*

PROPOSITION 7. — *Soit $f' \geqq 0$ une fonction numérique finie continue dans $E'_{\mathcal{R}}$ qui s'annule à l'infini. Il existe une suite croissante $(g_n)$ de fonctions de $\mathcal{R}^0_+$ telle que la suite $(g'_n)$ des prolongements continus des fonctions $g_n$ sur $E'_{\mathcal{R}}$ converge uniformément vers $f'$ dans $E'_{\mathcal{R}}$.*

3.2. **Représentation de formes linéaires sur $\mathcal{R}$ par des mesures de Radon sur $E'_{\mathcal{R}}$.** — Désignons, dans ce qui suit, par $f'$ le prolongement continu sur $E'_{\mathcal{R}}$ d'une fonction $f \in \mathcal{R}$, et par $\mathcal{R}'$ (resp. $\mathcal{R}^{0\prime}$) l'ensemble des prolongements continus $f'$ de toutes les fonctions $f$ de $\mathcal{R}$ (resp. $\mathcal{R}^0$). D'après la proposition 6, $\mathcal{R}^{0\prime}$ est un sous-espace vectoriel de l'espace vectoriel $\mathcal{K}(E'_{\mathcal{R}})$ des fonctions numériques finies continues dans $E'_{\mathcal{R}}$ et à support compact.

THÉORÈME 4. — *Soit $\mu$ une forme linéaire positive sur $\mathcal{R}$. L'application $g' \to \mu(g)$, où $g'$ est une fonction de $\mathcal{R}^{0\prime}$, $g$ sa restriction à $E$, est une forme linéaire positive $\mu'_0$ sur $\mathcal{R}^{0\prime}$. Elle peut se prolonger d'une manière et d'une seule en une forme linéaire positive $\mu'$ sur $\mathcal{K}(E'_{\mathcal{R}})$, donc en une mesure de Radon positive sur $E'_{\mathcal{R}}$* [12].

DÉMONSTRATION. — Montrons d'abord que, pour toute fonction $k' \in \mathcal{K}(E'_{\mathcal{R}})$ et tout nombre $\varepsilon > 0$, il existe deux fonctions $g'$, $h' \in \mathcal{R}^{0\prime}$ telles que $g' \leqq k' \leqq h'$ et $\mu'_0(h' - g') \leqq \varepsilon$. Comme la partie positive et la partie négative de $k'$ appartiennent à $\mathcal{K}(E'_{\mathcal{R}})$, on peut supposer que $k' \geqq 0$. Il existe une fonction $h'_0$ continue dans $E'_{\mathcal{R}}$ telle que $0 \leqq h'_0 \leqq 1$, $h'_0(x') = 1$ pour tout point $x'$ du support compact de $k'$. La proposition 7 et la propriété $(R_3)$ de $\mathcal{R}^0$ montrent qu'il existe même une telle fonction dans $\mathcal{R}^{0\prime}$. La proposition 7 assure de plus l'existence d'une fonction $g' \in \mathcal{R}^{0\prime}$ telle que $0 \leqq g' \leqq k' \leqq g' + \varepsilon$ ;

---

[12] Ce théorème se trouve en principe déjà dans [2], p. 464. En ce qui nous concerne, une démonstration indépendante des résultats de [2] est nécessaire.

de ceci et de la définition de $h'_0$ résulte l'inégalité $0 \leq g' \leq k' \leq g' + \varepsilon h'_0$. Les fonctions $g'$ et $h' = g' + \varepsilon h'_0$ appartiennent à $\mathcal{R}^{0'}$; en outre $g' \leq k' \leq h'$, et enfin $\mu'_0(h' - g') = \varepsilon \mu'_0(h'_0)$, d'où notre proposition, puisque $h'_0$ ne dépend pas du choix de $\varepsilon$. Pour toute fonction $k' \in \mathcal{K}(E'_{\mathcal{R}})$, on a donc l'égalité

$$\sup_{g' \leq k', g' \in \mathcal{R}^{0'}} \mu'_0(g') = \inf_{k' \leq h', h' \in \mathcal{R}^{0'}} \mu'_0(h').$$

Si l'on désigne par $\mu'(k')$ le nombre réel défini par cette égalité, on constate immédiatement que la fonction $\mu'$ ainsi définie est une forme linéaire positive sur $\mathcal{K}(E'_{\mathcal{R}})$ qui prolonge $\mu'_0$. L'égalité montre en outre que $\mu'$ est la seule forme linéaire positive sur $\mathcal{K}(E'_{\mathcal{R}})$ ayant cette propriété ([13]).

Étudions maintenant quelques conséquences du théorème 4. Par le procédé de l'énoncé de ce théorème, on fait correspondre à chaque forme linéaire positive $\mu$ sur $\mathcal{R}$ une certaine mesure de Radon positive $\mu'$ sur $E'_{\mathcal{R}}$; désignons $\mu'$ par $\Phi(\mu)$. Il est immédiat, en vertu du théorème 4, qu'on a, pour chaque couple de formes linéaires positives $\mu_1$, $\mu_2$ sur $\mathcal{R}$ et de nombres positifs $\alpha_1$, $\alpha_2$, l'égalité

$$\Phi(\alpha_1 \mu_1 + \alpha_2 \mu_2) = \alpha_1 \Phi(\mu_1) + \alpha_2 \Phi(\mu_2).$$

Il en résulte que $\Phi$ se prolonge d'une seule manière en une application linéaire de $\mathcal{J}(\mathcal{R})$ dans l'espace vectoriel $\mathcal{M}(E'_{\mathcal{R}})$ des mesures de Radon sur $E'_{\mathcal{R}}$, où $\mathcal{J}(\mathcal{R})$ désigne l'espace vectoriel des formes linéaires sur $\mathcal{R}$ telles que chacune d'elles soit la différence de deux formes linéaires positives nous désignerons aussi ce prolongement par $\Phi$. Les éléments de $\mathcal{J}(\mathcal{R})$ sont précisément les formes linéaires *relativement bornées* sur $\mathcal{R}$; on a $\mathcal{M}(E'_{\mathcal{R}}) = \mathcal{J}(\mathcal{K}(E'_{\mathcal{R}}))$ ([14]). Le cône des formes positives de $\mathcal{J}(\mathcal{R})$ définit une structure d'ordre par rapport à laquelle $\mathcal{J}(\mathcal{R})$ est un espace de Riesz complètement réticulé; en particulier, $\mathcal{M}(E'_{\mathcal{R}})$ est un espace de Riesz complètement réticulé. L'application $\Phi$ est positive, c'est-à-dire que la relation $\mu \geq 0 \; (\mu \in \mathcal{J}(\mathcal{R}))$ entraîne $\Phi(\mu) \geq 0$.

PROPOSITION 8. — 1° *Le sous-espace* $\overset{-1}{\Phi}(0)$ *de* $\mathcal{J}(\mathcal{R})$ *est l'ensemble des formes linéaires* $\mu$ *de* $\mathcal{J}(\mathcal{R})$ *telles qu'on ait*

$$(8) \qquad \mu(f) = \mu(\inf(1, f))$$

*pour toute fonction bornée* $f$ *de* $\mathcal{R}$.

---

([13]) En appliquant une notion introduite dans LOOMIS [7], p. 170, on peut dire que la mesure de Radon $\mu'$ est la restriction à $\mathcal{K}(E'_{\mathcal{R}})$ de la complétion bilatérale de la forme linéaire $\mu'_0$.

([14]) *Voir* BOURBAKI [4], p. 34 et 54.

2° $\overset{-1}{\Phi}(0)$ *est une bande dans* $\mathcal{J}(\mathcal{R})$. *La bande complémentaire* [*des formes étrangères à toutes les formes de* $\overset{-1}{\Phi}(0)$] *se compose de toutes les formes* $\mu$ *de* $\mathcal{J}(\mathcal{R})$ *qui vérifient, pour toute fonction positive* $f$ *de* $\mathcal{R}$, *les deux conditions suivantes :*

$$\lim_{n\to\infty} \mu\left(\inf\left(\frac{1}{n}, f\right)\right) = 0; \tag{9}$$

$$\lim_{n\to\infty} \mu(\inf(n, f)) = \mu(f) \quad (^{15}). \tag{10}$$

Démonstration. — Désignons par $\mathcal{R}^*$ l'ensemble des fonctions bornées de $\mathcal{R}$, par $\mathcal{B}_0$ l'ensemble des $\mu \in \mathcal{J}(\mathcal{R})$ vérifiant la condition (8) pour tout $f \in \mathcal{R}^*$, par $\mathcal{B}_1$ l'ensemble des $\mu \in \mathcal{J}(\mathcal{R})$ vérifiant les conditions (9) et (10) pour tout $f \in \mathcal{R}_+$. Les ensembles $\mathcal{B}_0$ et $\mathcal{B}_1$ sont deux sous-espaces vectoriels de $\mathcal{J}(\mathcal{R})$. Signalons encore une conséquence immédiate de (8) : si $\mu$ est une forme de $\mathcal{B}_0$, alors $\mu(f) = \mu(\inf(\alpha, f))$ pour tout $f \in \mathcal{R}^*$ et tout $\alpha > 0$.

La deuxième partie de la proposition sera démontrée si l'on sait que $\mathcal{J}(\mathcal{R})$ est somme directe ordonnée de $\mathcal{B}_0$ et $\mathcal{B}_1$ (16). Montrons d'abord que toute forme positive $\mu \in \mathcal{J}(\mathcal{R})$ admet une décomposition en somme d'une forme positive $\mu_0 \in \mathcal{B}_0$ et d'une forme positive $\mu_1 \in \mathcal{B}_1$. Pour toute fonction positive $f \in \mathcal{R}^*$, nous posons $\mu'_0(f) = \inf\limits_{n=1,2,\ldots} \mu\left(\inf\left(\frac{1}{n}, f\right)\right)$. On a $0 \leq \mu'_0(f) \leq \mu(f)$. Pour deux fonctions positives $f, g \in \mathcal{R}^*$, il résulte des inégalités

$$\inf\left(\frac{1}{2n}, f\right) + \inf\left(\frac{1}{2n}, g\right) \leq \inf\left(\frac{1}{n}, f+g\right) \leq \inf\left(\frac{1}{n}, f\right) + \inf\left(\frac{1}{n}, g\right)$$

que $\mu'_0(f+g) = \mu'_0(f) + \mu'_0(g)$. Ainsi il existe une forme linéaire positive sur $\mathcal{R}^*$ et une seule qui prolonge $\mu'_0$ ; nous la désignerons aussi par $\mu'_0$. Considérons la forme linéaire positive $\mu'_1$ sur $\mathcal{R}^*$ définie par $\mu'_1(f) = \mu(f) - \mu'_0(f)$ pour tout $f \in \mathcal{R}^*$. En répétant la conclusion ci-dessus, on montre l'existence d'une forme linéaire positive $\mu_1$ sur $\mathcal{R}$ vérifiant $\mu_1(f) = \sup\limits_{n=1,2,\ldots} \mu'_1(\inf(n, f))$ pour tout $f \in \mathcal{R}_+$. On voit immédiatement que $\mu_1$ prolonge $\mu'_1$ et que $0 \leq \mu_1(f) \leq \mu(f)$ pour tout $f \in \mathcal{R}_+$. Donc $\mu_0 = \mu - \mu_1$ est une forme linéaire positive sur $\mathcal{R}$ telle que $\mu = \mu_0 + \mu_1$. Montrons que $\mu_0$ et $\mu_1$ répondent à la question. Pour une fonction $f \geq 0$ de $\mathcal{R}^*$, on a

$$\begin{aligned}\mu_0(f) = \mu'_0(f) &= \inf \mu\left(\inf\left(\frac{1}{n}, f\right)\right) = \inf \mu\left(\frac{1}{n}, \inf(1, f)\right)\\ &= \mu'_0(\inf(1, f)) = \mu_0(\inf(1, f));\end{aligned}$$

(15) Pour la définition d'une bande, *voir* Bourbaki [4], p. 23 et suivantes. Comme dans chaque espace de Riesz, on pose $|\mu| = \sup(\mu, -\mu)$, $\mu^+ = \sup(\mu, 0)$, $\mu^- = (-\mu)^+$; il en résulte $|\mu| = \mu^+ + \mu^-$ et $\mu = \mu^+ - \mu^-$. Deux formes linéaires $\mu$, $\nu$ de $\mathcal{J}(\mathcal{R})$ sont dites étrangères si $\inf(|\mu|, |\nu|) = 0$.

(16) *Voir* Bourbaki [4], p. 28, exercice 6.

pour une fonction quelconque $f \in \mathcal{R}^*$, on a $\inf(1, f) = \inf(1, f^+) - f^-$, d'où résulte l'égalité (8) pour $\mu_0$. La forme positive $\mu_0$ appartient donc à $\mathcal{B}_0$. Soit maintenant $f$ une fonction de $\mathcal{R}_+$. Pour une telle fonction, la forme $\mu_1$ satisfait à (10) par sa définition. Il résulte de l'identité $\inf\left(\frac{1}{n}, f\right) = \inf\left(\frac{1}{n}, \inf(1, f)\right)$ que, pour la démonstration de (9), nous pouvons supposer la fonction $f \in \mathcal{R}_+$ bornée. On a alors l'égalité

$$\mu_1\left(\inf\left(\frac{1}{n}, f\right)\right) = \mu'_1\left(\inf\left(\frac{1}{n}, f\right)\right) = \mu\left(\inf\left(\frac{1}{n}, f\right)\right) - \mu'_0\left(\inf\left(\frac{1}{n}, f\right)\right)$$
$$= \mu\left(\inf\left(\frac{1}{n}, f\right)\right) - \mu'_0(f)$$

qui, avec la définition de $\mu'_0(f)$, entraîne $\lim \mu_1\left(\inf\left(\frac{1}{n}, f\right)\right) = 0$. Par conséquent, $\mu_1$ appartient à $\mathcal{B}_1$. Une forme linéaire quelconque $\mu \in \mathcal{J}(\mathcal{R})$ peut aussi se décomposer en somme d'une forme $\mu_0 \in \mathcal{B}_0$ et d'une forme $\mu_1 \in \mathcal{B}_1$, puisqu'elle est la différence de deux formes positives de $\mathcal{J}(\mathcal{R})$. Pour montrer l'unicité de cette décomposition, il suffit de prouver que l'égalité $0 = \mu_0 + \mu_1$ ($\mu_0 \in \mathcal{B}_0$, $\mu_1 \in \mathcal{B}_1$) entraîne $\mu_0 = \mu_1 = 0$. Pour toute fonction positive $f \in \mathcal{R}^*$ et tout $n = 1, 2, \ldots$ on a

$$0 = \mu_0\left(\inf\left(\frac{1}{n}, f\right)\right) + \mu_1\left(\inf\left(\frac{1}{n}, f\right)\right) = \mu_0(f) + \mu_1\left(\inf\left(\frac{1}{n}, f\right)\right);$$

de (9) résulte donc $\mu_0(f) = \mu_1(f) = 0$.

Ainsi les formes $\mu_0$ et $\mu_1$ sont nulles sur $\mathcal{R}^*$. La condition (10) montre alors que $\mu_1$ est identiquement nulle sur $\mathcal{R}$, donc que $\mu_0 = \mu_1 = 0$. L'espace $\mathcal{J}(\mathcal{R})$ est ainsi somme directe ordonnée de $\mathcal{B}_0$ et $\mathcal{B}_1$.

La démonstration de la première partie de notre proposition est maintenant facile. Soit $\mu$ une forme linéaire de $\mathcal{J}(\mathcal{R})$. Si $\mu$ appartient à $\overset{-1}{\Phi}(0)$, le théorème 4 affirme que $\mu(g) = 0$ pour tout $g \in \mathcal{R}^0$. Pour une fonction $f \in \mathcal{R}^*$, $g = f - \inf(1, f)$ appartient à $\mathcal{R}^0$; par conséquent l'égalité (8) est vraie. Supposons inversement que (8) soit valable pour tout $f \in \mathcal{R}^*$. Comme $\mathcal{B}_0$ est une bande, la forme $\mu^+$ appartient ainsi que $\mu$ à $\mathcal{B}_0$; on a donc $\mu^+(f) = \mu^+(\inf(\alpha, f))$ pour tout $f \in \mathcal{R}^*$ et $\alpha > 0$. Soit en particulier $f$ une fonction de $\mathcal{R}^0_+$; il existe alors une fonction $h \in \mathcal{R}_+$ telle que $h(x) < 1$ entraîne $f(x) = 0$ $(x \in E)$. Par suite on a $\inf(\alpha, f) \leqq \alpha h$, donc $0 \leqq \mu^+(f) = \mu^+(\inf(\alpha, f)) \leqq \alpha \mu^+(h)$ pour tout $\alpha > 0$. Ainsi $\mu^+(f)$ s'annule pour tout $f \in \mathcal{R}^0_+$, donc pour tout $f \in \mathcal{R}^0$. D'après la définition de $\Phi(\mu^+)$ ceci équivaut à $\Phi(\mu^+) = 0$. En appliquant ce résultat à $-\mu$, on obtient l'égalité $\Phi(\mu^-) = 0$. Enfin, l'égalité

$$\Phi(\mu) = \Phi(\mu^+) - \Phi(\mu^-)$$

montre que $\mu$ appartient à $\overset{-1}{\Phi}(0)$.

**Corollaire 1.** — *La restriction de l'application $\Phi$ à l'ensemble des mesures abstraites dans $\mathcal{J}(\mathcal{R})$ est biunivoque.*

**Démonstration.** — D'après la deuxième partie de la proposition 8, toute mesure abstraite de $\mathcal{J}(\mathcal{R})$ appartient à la bande complémentaire de $\overset{-1}{\Phi}(0)$. La restriction de $\Phi$ à cette bande est biunivoque.

Il résulte du lemme du paragraphe 1 et du théorème de Lebesgue que toute fonction $g' \in \mathcal{R}'$ appartient à $\mathcal{L}_1(\Phi(\mu))$ pour tout $\mu \geqq 0$ de $\mathcal{J}(\mathcal{R})$. Une mesure de Radon non bornée n'appartient donc pas en général à l'image $\Phi(\mathcal{J}(\mathcal{R}))$ de $\mathcal{J}(\mathcal{R})$. Un cas particulier, où $\Phi$ est une application sur $\mathfrak{M}(E'_{\mathcal{R}})$, est le suivant :

**Corollaire 2.** — *Dans le cas $\mathcal{R} = \mathcal{R}^0$, l'application $\Phi$ est un isomorphisme de l'espace de Riesz $\mathcal{J}(\mathcal{R})$ sur l'espace de Riesz $\mathfrak{M}(E'_{\mathcal{R}})$.*

**Démonstration.** — $\Phi$ est un isomorphisme, puisque $\Phi(\mu) = 0$ entraîne $\mu(g) = 0$ pour tout $g \in \mathcal{R}^0$, donc $\mu = 0$. $\Phi$ est un isomorphisme de $\mathcal{J}(\mathcal{R})$ sur $\mathfrak{M}(E'_{\mathcal{R}})$, puisque, d'après la proposition 6, le prolongement continu $g'$ sur $E'_{\mathcal{R}}$ de toute fonction $g \in \mathcal{R}$ appartient à $\mathcal{K}(E'_{\mathcal{R}})$. Toute mesure de Radon $\mu'$ sur $E'_{\mathcal{R}}$ définit donc une forme linéaire $g \to \mu'(g')$ sur $\mathcal{R}$ qui appartient à $\mathcal{J}(\mathcal{R})$ car $\mu'$ est différence de deux mesures de Radon positives; on voit immédiatement que $\Phi(\mu) = \mu'$ et que $\mu' \geqq 0$ entraîne $\Phi(\mu) \geqq 0$.

**Remarques.** — 1° La deuxième partie de la proposition 8 peut être considérée comme cas particulier d'un théorème général de décomposition de [3], p. 116, si l'on donne à la notion de continuité introduite dans [3] le sens exprimé par les égalités (9) et (10).

2° Dans le cas particulier où toute fonction de $\mathcal{R}$ est bornée, l'égalité (10) dégénère en une identité et peut être supprimée. Dans ce cas-là on peut démontrer la deuxième partie de la proposition 8 en partant de résultats de [2] : Les espaces $\mathcal{J}(\mathcal{R})$ et $\mathcal{J}(\mathcal{R}')$ sont canoniquement isomorphes puisque toute forme linéaire sur $\mathcal{R}$ peut être considérée comme forme linéaire sur $\mathcal{R}'$, et inversement. D'après [2], p. 468, cet isomorphisme fait correspondre à $\overset{-1}{\Phi}(0)$ la bande des formes linéaires purement discontinues de $\mathcal{J}(\mathcal{R}')$ et à la bande complémentaire de $\overset{-1}{\Phi}(0)$ la bande des formes linéaires continues de $\mathcal{J}(\mathcal{R}')$.

## Paragraphe 4. — **Le théorème de représentation.**

Comme dans le premier paragraphe, nous supposons que $E$ est un ensemble quelconque, $\mathcal{R}$ un ensemble de fonctions numériques finies définies dans $E$, vérifiant les axiomes $(R_1)$, $(R_2)$ et $(R_3)$, mais pas nécessairement $(S_1)$ et $(S_2)$.

Nous nous proposons de démontrer le théorème suivant :

THÉORÈME 5. — *Il existe un espace localement compact $E'$ et une application linéaire positive* $\mathbf{T}$ *de l'espace de Riesz* $\mathcal{J}(\mathcal{R})$ *des formes linéaires relativement bornées sur* $\mathcal{R}$ *dans l'espace de Riesz* $\mathcal{M}(E')$ *des mesures de Radon sur* $E'$ *avec les propriétés suivantes : Pour toute mesure abstraite* $\mu$ *sur* $\mathcal{R}$, *tout nombre réel* $1 \leq p < +\infty$ *et tout espace de Banach* $F$, *les espaces* $L_F^p(E, \mathcal{R}, \mu)$ *et* $L_F^{p\star}(E', \mathcal{K}(E'), \mathbf{T}(\mu))$ *sont isomorphes en tant qu'espaces de Banach et, dans le cas* $F = \mathbf{R}$, *le même isomorphisme s'applique aussi à leur structure d'espaces de Riesz. La restriction de* $\mathbf{T}$ *à l'ensemble des mesures abstraites sur* $\mathcal{R}$ *est biunivoque.*

DÉMONSTRATION. — La démonstration se fait en trois étapes dont la dernière est la plus importante.

I. Soient $E_0$ l'ensemble des éléments $x \in E$ tels que $f(x) = 0$ pour tout $f \in \mathcal{R}$, $E_1 = \complement E_0$ le complémentaire de $E_0$, et $\mathcal{R}_1$ l'ensemble des restrictions des fonctions de $\mathcal{R}$ à $E_1$. Si $E = E_0$, $\mathcal{R}$ ne contient que la fonction 0 et le théorème est trivial : il suffit de prendre pour $E'$ l'espace formé d'un seul point. Nous supposons donc $E_1 \neq \emptyset$. Il est clair que $\mathcal{R}_1$ vérifie les axiomes $(R_1)$, $(R_2)$, $(R_3)$ et $(S_1)$ (en remplaçant $\mathcal{R}$ par $\mathcal{R}_1$ et $E$ par $E_1$ dans ces axiomes). A chaque forme linéaire $\mu \in \mathcal{J}(\mathcal{R})$ correspond canoniquement une forme linéaire $\mu_1 \in \mathcal{J}(\mathcal{R}_1)$ définie de la manière suivante : toute fonction $f_1 \in \mathcal{R}_1$ est la restriction d'une fonction unique $f \in \mathcal{R}$; on a $\mu_1(f_1) = \mu(f)$. L'application $\mu \to \mu_1$ est *un isomorphisme* $\mathbf{T}_1$ *de l'espace de Riesz* $\mathcal{J}(\mathcal{R})$ *sur l'espace de Riesz* $\mathcal{J}(\mathcal{R}_1)$; $\mu$ *est une mesure abstraite sur* $\mathcal{R}$, *si et seulement si* $\mathbf{T}_1(\mu)$ *est une mesure abstraite sur* $\mathcal{R}_1$. Soient maintenant $\mu$ une mesure abstraite fixée sur $\mathcal{R}$, et $\mu_1 = \mathbf{T}_1(\mu)$ son image. Pour une fonction numérique positive définie dans $E$ telle que $\overline{\mu}(f) < +\infty$, il existe une suite croissante $(f_n)$ de fonctions de $\mathcal{R}_+$ telle que $f \leq \sup f_n$; on a donc $f(x) = 0$ pour tout $x \in E_0$. Il est immédiat qu'on a $\overline{\mu}_1(f_1) = \overline{\mu}(f)$ pour la restriction $f_1$ de $f$ à $E_1$. Inversement toute fonction numérique positive $f_1$ définie dans $E_1$ telle que $\overline{\mu}_1(f_1) < +\infty$ ne possède qu'un prolongement $f \geq 0$ sur $E$ tel que $\overline{\mu}(f) < +\infty$ : cet $f$ doit être nul en chaque $x \in E_0$. Enfin, remarquons que le sous-espace $(\mathcal{R}_1)^0$ de $\mathcal{R}_1$ est l'ensemble des restrictions des fonctions de $\mathcal{R}^0$ à $E_1$. Il résulte de toutes ces remarques que, pour tout nombre réel $1 \leq p < +\infty$ et tout espace de Banach $F$, la rectriction à $E_1$ des fonctions de $\mathcal{L}_F^p(\mu)$ définit un isomorphisme de l'espace vectoriel $\mathcal{L}_F^p(\mu)$ sur l'espace vectoriel $\mathcal{L}_F^p(\mu_1)$, qui conserve les semi-normes $N_p$ définies sur ces espaces. Pour deux fonctions numériques $f$, $g$ définies dans $E$, on a $f \leq g$, si et seulement si ses restrictions $f_1$, $g_1$ à $E_1$ vérifient $f_1 \leq g_1$. Nous avons donc montré que : *Pour toute mesure abstraite* $\mu$ *sur* $\mathcal{R}$, *il existe un isomorphisme de l'espace de Banach* $L_F^p(E, \mathcal{R}, \mu)$ *sur l'espace de Banach* $L_F^p(E_1, \mathcal{R}_1, \mathbf{T}_1(\mu))$; *au cas où* $F = \mathbf{R}$, *les deux espaces sont, par le même isomorphisme, isomorphes en tant qu'espaces de Riesz.*

II. Soit $Q$ la relation d'équivalence suivante dans $E_1$ : deux éléments $x$, $y \in E_1$ seront dits équivalents modulo $Q$, si et seulement si $f_1(x) = f_1(y)$ pour toute fonction $f_1 \in \mathcal{R}_1$. Désignons par $E_2$ l'ensemble quotient $E_1/Q$, par $\varphi$ l'application canonique de $E_1$ sur $E_2$, par $\mathcal{R}_2$ l'ensemble des fonctions numériques $f_2$ définies sur $E_2$ telles que $f_2 \circ \varphi$ appartienne à $\mathcal{R}_1$. Il est clair que $\mathcal{R}_2$ vérifie les axiomes $(R_1)$, $(R_2)$, $(R_3)$ et $(S_1)$, $(S_2)$ ( en remplaçant $\mathcal{R}$ par $\mathcal{R}_2$ et $E$ par $E_2$). A toute forme linéaire $\mu_1 \in \mathcal{J}(\mathcal{R}_1)$ correspond canoniquement une forme linéaire $\mu_2 \in \mathcal{J}(\mathcal{R}_2)$ définie comme suit : $\mu_2(f_2) = \mu_1(f_2 \circ \varphi)$ pour tout $f_2 \in \mathcal{R}_2$. L'application $\mu_1 \to \mu_2$ est un *isomorphisme* $\mathbf{T}_2$ *de l'espace de Riesz* $\mathcal{J}(\mathcal{R}_1)$ *sur l'espace de Riesz* $\mathcal{J}(\mathcal{R}_2)$; $\mu_1$ *est une mesure abstraite sur* $\mathcal{R}_1$, *si et seulement si* $\mathbf{T}_2(\mu_1)$ *est une mesure abstraite sur* $\mathcal{R}_2$. Soient maintenant $\mu_1$ une mesure abstraite fixée sur $\mathcal{R}_1$ et $\mu_2 = \mathbf{T}_2(\mu_1)$ son image. On voit immédiatement que $\overline{\mu}_2(f_2) = \overline{\mu}_1(f_2 \circ \varphi)$ pour toute fonction numérique positive $f_2$ définie dans $E_2$. Remarquons encore qu'une fonction $f_2$ définie dans $E_2$ appartient à $(\mathcal{R}_2)^0$, si et seulement si $f_2 \circ \varphi$ appartient à $(\mathcal{R}_1)^0$. Nous obtenons donc le résultat suivant : pour tout nombre $1 \leq p < +\infty$ et tout espace de Banach $F$, une application $\mathbf{f}_2$ de $E_2$ dans $F$ appartient à $\mathcal{L}_F^p(\mu_2)$, si et seulement si $\mathbf{f}_2 \circ \varphi$ appartient à $\mathcal{L}_F^p(\mu_1)$ ; les semi-normes correspondantes de $\mathbf{f}_2$ et $\mathbf{f}_2 \circ \varphi$ sont égales. L'application $\mathbf{f}_2 \to \mathbf{f}_2 \circ \varphi$ est donc un isomorphisme de l'espace vectoriel $\mathcal{L}_F^p(\mu_2)$ dans l'espace vectoriel $\mathcal{L}_F^p(\mu_1)$ conservant les semi-normes. En particulier l'image inverse $\overset{-1}{\varphi}(Z_2)$ d'un ensemble $\mu_2$-négligeable $Z_2$ de $E_2$ est $\mu_1$-négligeable. Inversement, l'image directe $\varphi(Z_1)$ d'un ensemble $\mu_1$-négligeable $Z_1$ de $E_1$ est $\mu_2$-négligeable. En effet, soit $(f_n)$ une suite croissante de fonctions positives de $\mathcal{R}_1$, telle que $\chi_{Z_1} \leq \sup f_n$, où $\chi_{Z_1}$ est la fonction caractéristique de $Z_1$ dans $E_1$. Si $Z'_1$ désigne le plus petit ensemble contenant $Z_1$ et saturé relativement à $Q$, on a aussi $\chi_{Z'_1} \leq \sup f_n$. On a donc $\overline{\mu}(\chi_{Z'_1}) = \overline{\mu}_1(\chi_{Z_1}) = 0$, d'où l'assertion. Ceci montre que, pour deux fonctions numériques $f_2$, $g_2$ définies dans $E_2$, on a, $\mu_2$-presque partout dans $E_2$, la relation $f_2(x) \leq g_2(x)$, si et seulement si, $\mu_1$-presque partout dans $E_1$, on a la relation $(f_2 \circ \varphi)(x) \leq (g_2 \circ \varphi)(x)$.

Montrons maintenant que, pour toute fonction $\mathbf{f}_1 \in \mathcal{L}_F^p(\mu_1)$, il existe une fonction $\mathbf{f}'_1 \in \mathcal{L}_F^p(\mu_1)$ constante dans chaque classe d'équivalence, telle que $N_p^1(\mathbf{f}_1 - \mathbf{f}'_1) = 0$, où $N_p^1$ désigne la semi-norme de $\mathcal{L}_F^p(\mu_1)$. Il existe [17] une suite $(\mathbf{g}_n)$ de fonctions de $(\mathcal{R}_1)_F^0$, une fonction numérique $g \geq 0$ définie dans $E_1$ avec une semi-norme $N_p^1(g)$ finie et un ensemble $\mu$-négligeable $Z_1$ de $E_1$ tels qu'on ait $\mathbf{f}_1(x) = \lim \mathbf{g}_n(x)$ et $|\mathbf{g}_n(x)| \leq g(x)$ pour tout $x \in \complement Z_1$. Mais nous savons que le plus petit ensemble $Z'_1$, contenant $Z_1$ et saturé relativement à $Q$, est aussi $\mu_1$-négligeable. La fonction $\mathbf{f}'_1$ égale à $\mathbf{f}_1$ dans $\complement Z'_1$, et à 0 dans $Z'_1$ est alors constante dans chaque classe d'équivalence; d'après le théorème de Lebesgue, elle appartient à $\mathcal{L}_F^p(\mu_1)$ et l'on a $N_p^1(\mathbf{f}_1 - \mathbf{f}'_1) = 0$.

---

(17) *Voir* la remarque qui suit la proposition 2, et BOURBAKI [4], p. 133-143.

De notre isomorphisme de $\mathcal{L}_F^p(\mu_2)$ dans $\mathcal{L}_F^p(\mu_1)$ on déduit ainsi un isomorphisme de $L_F^p(\mu_2)$ sur $L_F^p(\mu_1)$. Il résulte des remarques précédentes que : *Pour toute mesure abstraite $\mu_1$ sur $\mathcal{R}_1$, il existe un isomorphisme de l'espace de Banach $L_F^p(E_1, \mathcal{R}_1, \mu_1)$ sur l'espace de Banach $L_F^p(E_2, \mathcal{R}_2, \mathbf{T}_2(\mu_1))$; dans le cas où $F = \mathbf{R}$, les deux espaces sont, par le même isomorphisme, isomorphes en tant qu'espaces de Riesz.*

III. Le théorème 3 assure l'existence de l'espace $(E_2)'_{\mathcal{R}_2}$ que nous désignerons par $E'$. Montrons que $E'$ possède les propriétés énoncées dans le théorème 5. Soit $\mathbf{\Phi}$ l'application linéaire positive de $\mathcal{J}(\mathcal{R}_2)$ dans $\mathfrak{M}(E')$ établie au n° 3.2 et posons $\mathbf{T} = \mathbf{\Phi} \circ \mathbf{T}_2 \circ \mathbf{T}_1$. Les deux premières parties de notre démonstration et le corollaire 1 de la proposition 8 montrent que $\mathbf{T}$ est une application linéaire positive de $\mathcal{J}(\mathcal{R})$ dans $\mathfrak{M}(E')$ dont la restriction à l'ensemble des mesures abstraites sur $\mathcal{R}$ est biunivoque. Le théorème 2 assure, pour toute mesure de Radon positive $\mu'$ sur $E'$, l'isomorphie des espaces de Banach $L_F^p(E', \mu')$ et $L_F^{p^*}(E', \mu')$; si $F = \mathbf{R}$, l'isomorphie s'étend aussi à leur structure d'espace de Riesz. Il reste donc à démontrer que, pour toute mesure abstraite $\mu_2$ sur $\mathcal{R}_2$, les espaces de Banach $L_F^p(\mu_2)$ et $L_F^p(\mathbf{\Phi}(\mu_2))$ sont isomorphes et que, dans le cas où $F = \mathbf{R}$, les deux espaces sont, par le même isomorphisme, isomorphes en tant qu'espaces de Riesz. Mais ceci est une conséquence immédiate du théorème suivant :

Théorème 6. — *Soient $\mathcal{R}$ un ensemble de fonctions numériques finies définies sur un ensemble $E$, vérifiant les axiomes* $(R_1)$, $(R_2)$, $(R_3)$ *et* $(S_1)$, $(S_2)$, *$\mu$ une mesure abstraite définie sur $\mathcal{R}$, $p$ un nombre réel tel que $1 \leq p < +\infty$, $F$ un espace de Banach. Toute fonction $\mathbf{f}$ de $\mathcal{L}_F^p(E, \mathcal{R}, \mu)$ peut être prolongée en une fonction $\mathbf{f}'$ de $\mathcal{L}_F^p(E'_{\mathcal{R}}, \mathcal{K}(E'_{\mathcal{R}}), \mathbf{\Phi}(\mu))$; inversement, la restriction $\mathbf{f}$ de toute fonction $\mathbf{f}'$ de $\mathcal{L}_F^p(E'_{\mathcal{R}}, \mathcal{K}(E'_{\mathcal{R}}), \mathbf{\Phi}(\mu))$ à $E$, appartient à $\mathcal{L}_F^p(E, \mathcal{R}, \mu)$. On a*

$$N_p(\mathbf{f}) = N'_p(\mathbf{f}') \tag{11}$$

*si $N_p$ et $N'_p$ désignent les semi-normes déduites de $\mu$ et $\mathbf{\Phi}(\mu)$ respectivement (au sens du premier procédé d'intégration de Stone). Si $F = \mathbf{R}$, toute fonction positive $f$ de $\mathcal{L}^p(E, \mathcal{R}, \mu)$ peut être prolongée en une fonction positive $f'$ de $\mathcal{L}^p(E'_{\mathcal{R}}, \mathcal{K}(E'_{\mathcal{R}}), \mathbf{\Phi}(\mu))$.*

Démonstration. — Nous désignerons par $g'$ le prolongement continu d'une fonction $g \in \mathcal{R}^0$ sur $E'_{\mathcal{R}}$, par $\mathcal{R}^{0\prime}$ l'ensemble des fonctions $g'$, où $g$ parcourt $\mathcal{R}^0$. Soit $\mu'_0$ la forme linéaire positive sur $\mathcal{R}^{0\prime}$ définie par $\mu'_0(g') = \mu(g)$ pour tout $g \in \mathcal{R}^0$. La mesure de Radon $\mu' = \mathbf{\Phi}(\mu)$ est, d'après le théorème 4, le seul prolongement possible de $\mu'_0$ en une forme linéaire positive sur $\mathcal{K}(E'_{\mathcal{R}})$; et $\mu'_0$, $\mu'$ sont des mesures abstraites [même avec la propriété $(M^*)$]. Nous allons montrer que $\mu'_0$ définit la même semi-norme $N'_p$ que $\mu'$ et que

$\mathcal{L}_F^p(E'_{\mathcal{R}}, \mathcal{R}^{0'}, \mu'_0) = \mathcal{L}_F^p(E'_{\mathcal{R}}, \mathcal{K}(E'_{\mathcal{R}}), \mu')$. De la même façon que nous avons déduit le corollaire 1 du lemme du paragraphe 1, nous déduisons de la proposition 7 que, pour toute suite croissante $(f'_n)$ de fonctions positives de $\mathcal{K}(E'_{\mathcal{R}})$, il existe une suite croissante $(g'_n)$ de fonctions de $\mathcal{R}^{0'}$ telles que $0 \leq g'_n \leq f'_n (n = 1, 2, \ldots)$ et $\sup f'_n = \sup g'_n$. De cela et du fait que $\mathcal{R}^{0'} \subset \mathcal{K}(E'_{\mathcal{R}})$, il résulte que $\overline{\mu}'_0(f') = \overline{\mu}'(f')$ pour toute fonction numérique $f' \geq 0$ dans $E'_{\mathcal{R}}$. Donc, la mesure $\mu'_0$ définit la même semi-norme $N'_p$ que $\mu'$. L'égalité $\mathcal{L}_F^p(\mu'_0) = \mathcal{L}_F^p(\mu')$ est une conséquence immédiate du fait que toute combinaison linéaire $\mathbf{f}' = \sum \mathbf{a}_i f'_i$ de fonctions positives de $\mathcal{K}(E'_{\mathcal{R}})$ à coefficients dans $F$ appartient à $\mathcal{L}_F^p(\mu'_0)$. Soit, en effet, $h'_0$ une fonction de $\mathcal{K}(E'_{\mathcal{R}})$ telle que $0 \leq h'_0 \leq 1$ et $h'_0(x') = 1$ pour tout point $x'$ du support compact de $\mathbf{f}'$. La proposition 7 assure, pour un $\varepsilon > 0$ quelconque, l'existence d'une fonction $g'_i \in \mathcal{R}^{0'}$ telle que $0 \leq g'_i \leq f'_i \leq g'_i + \varepsilon$, d'où résulte $|f'_i - g'_i| \leq \varepsilon h'_0 (i = 1, \ldots, n)$. La fonction $\mathbf{g}' = \sum \mathbf{a}_i g'_i$ de $\mathcal{R}_F^{0'}$ vérifie donc $N'_p(\mathbf{f}' - \mathbf{g}') \leq \varepsilon N'_p(h'_0) \sum |\mathbf{a}_i| = \varepsilon Q$, où $Q = N'_p(h'_0) \sum |\mathbf{a}_i|$ ne dépend que de $\mathbf{f}'$. Par conséquent, $\mathbf{f}'$ appartient à $\mathcal{L}_F^p(\mu'_0)$. Ces résultats nous permettent de démontrer le théorème en question sous la forme modifiée suivante : on remplace $\mu' = \Phi(\mu)$ par la mesure abstraite $\mu'_0$ définie sur $\mathcal{R}^{0'}$, et l'on considère $N'_p$ comme la semi-norme déduite de $\mu'_0$.

Nous étudierons d'abord les relations entre $N_p$ et $N'_p$. Soient $f' \geq 0$ une fonction numérique définie dans $E'_{\mathcal{R}}$, et $f$ la restriction de $f'$ à $E$. Si $(g'_n)$ est une suite croissante de fonctions positives de $\mathcal{R}^{0'}$ telle que $f' \leq \sup g'_n$, les restrictions $g_n$ à $E$ forment une suite croissante de fonctions de $\mathcal{R}_+^0$ telle que $f \leq \sup g_n$. On en déduit $\overline{\mu}(f) \leq \overline{\mu}'_0(f')$. Inversement, si $f'$ s'annule en tous les points de $E'_{\mathcal{R}} \cap \complement E$, toute suite croissante $(g_n)$ de fonctions de $\mathcal{R}_+^0$ donne naissance à la suite croissante $(g'_n)$ de fonctions positives de $\mathcal{R}^{0'}$ vérifiant l'inégalité $f' \leq \sup g'_n$. D'après la proposition 2, il en résulte que $\mu(f) = \overline{\mu}'_0(f')$. Nous obtenons donc le résultat suivant : pour toute application $\mathbf{f}'$ de $E'_{\mathcal{R}}$ dans $F$ et sa restriction $\mathbf{f}$ à $E$, on a $N_p(\mathbf{f}) \leq N'_p(\mathbf{f}')$ ; si $\mathbf{f}'(x') = 0$ pour tout $x' \in E'_{\mathcal{R}} \cap \complement E$, on a même $N_p(\mathbf{f}) = N'_p(\mathbf{f}')$.

Soit maintenant $\mathbf{f}' \in \mathcal{L}_F^p(\mu'_0)$ et $\mathbf{f}$ la restriction de $\mathbf{f}'$ à $E$. Nous démontrerons que $\mathbf{f} \in \mathcal{L}_F^p(\mu)$ et que $N_p(\mathbf{f}) = N'_p(\mathbf{f}')$. On a $N_p(\mathbf{f}) \leq N'_p(\mathbf{f}') < +\infty$, donc $\mathbf{f} \in \mathcal{F}_F^p(\mu)$. Il existe une suite $(\mathbf{g}'_n)$ de fonctions de $\mathcal{R}_F^{0'}$ (combinaisons linéaires de fonctions de $\mathcal{R}^{0'}$ à coefficients dans $F$) telle que $\lim N'_p(\mathbf{f}' - \mathbf{g}'_n) = 0$. La restriction $\mathbf{g}_n$ de chaque $\mathbf{g}'_n$ à $E$ appartient à $\mathcal{R}_F^0$; on a $N_p(\mathbf{f} - \mathbf{g}_n) \leq N'_p(\mathbf{f}' - \mathbf{g}'_n)$, donc $\lim N_p(\mathbf{f} - \mathbf{g}_n) = 0$, d'où résulte que $\mathbf{f} \in \mathcal{L}_F^p(\mu)$. Comme dans la démonstration de la proposition 4, on montre qu'il suffit de prouver l'égalité $N_p(\mathbf{f}) = N'_p(\mathbf{f}')$ seulement pour $p = 1$ et $F = \mathbf{R}$. Soient donc $f' \in \mathcal{L}^1(\mu'_0)$,

$f$ la restriction de $f'$ à $E$, $(g'_n)$ une suite de fonctions de $\mathcal{R}^{0'}$ telle que $\lim N'_1(f'-g'_n)=0$, $(g_n)$ la suite des restrictions des $g'_n$ à $E$. On a

$$N'_1(f')=\lim N'_1(g'_n) \qquad \text{et} \qquad N_1(f)=\lim N_1(g_n);$$

de $|g'_n|\in\mathcal{R}^{0'}$ résulte $N'_n(g'_n)=\mu'_0(|g'_n|)=\mu(|g_n|)=N_1(g_n)$, d'où $N_1(f)=N'_1(f')$.

Inversement, soit $\mathbf{f}$ une fonction de $\mathcal{L}^p_F(\mu)$; nous montrerons l'existence d'un prolongement $\mathbf{f}'$ de $\mathbf{f}$ appartenant à $\mathcal{L}^p_F(\mu'_0)$. Il existe une suite $(\mathbf{g}_n)$ de fonctions de $\mathcal{R}^0_F$ telle que $\lim\limits_{n\to\infty} N_p(\mathbf{f}-\mathbf{g}_n)=0$, donc telle que $\lim\limits_{m,n\to\infty} N_p(\mathbf{g}_n-\mathbf{g}_m)=0$. Le prolongement continu $\mathbf{g}'_n$ de $\mathbf{g}_n$ sur $E'_{\mathcal{R}}$ appartient à $\mathcal{R}^{0'}_F$, donc à $\mathcal{L}^p_F(\mu'_0)$. Nous avons ainsi $N_p(\mathbf{g}_n-\mathbf{g}_m)=N'_p(\mathbf{g}'_n-\mathbf{g}'_m)$ d'après la partie du théorème déjà démontrée. L'espace $\mathcal{L}^p_F(\mu'_0)$ étant complet, il existe un $\mathbf{f}^*\in\mathcal{L}^p_F(\mu'_0)$ tel que $\lim N'_p(\mathbf{f}^*-\mathbf{g}'_n)=0$. Nous posons $\mathbf{f}'(x)=\mathbf{f}(x)$ pour tout $x\in E$, $\mathbf{f}'(x')=\mathbf{f}^*(x')$ pour tout $x'\in E'_{\mathcal{R}}\cap \complement E$; nous allons montrer que ce prolongement $\mathbf{f}'$ de $\mathbf{f}$ répond à la question. Soit $\mathbf{f}^*_E$ la restriction de $\mathbf{f}^*$ à $E$; on a $N_p(\mathbf{f}-\mathbf{f}^*_E)=N'_p(\mathbf{f}'-\mathbf{f}^*)$ puisque $\mathbf{f}'-\mathbf{f}^*$ s'annule sur $E'_{\mathcal{R}}\cap\complement E$. Nous savons en outre que, pour tout $n$, $N_p(\mathbf{f}^*_E-\mathbf{g}_n)\leqq N'(\mathbf{f}^*-\mathbf{g}'_n)$, d'où $\lim N_p(\mathbf{f}^*_E-\mathbf{g}_n)=0$. L'inégalité $N_p(\mathbf{f}-\mathbf{f}^*_E)\leqq N_p(\mathbf{f}-\mathbf{g}_n)+N_p(\mathbf{f}^*_E-\mathbf{g}_n)$ entraîne alors $N_p(\mathbf{f}-\mathbf{f}^*_E)=N'_p(\mathbf{f}'-\mathbf{f}^*)=0$. La fonction $\mathbf{f}'$ appartient donc, comme $\mathbf{f}^*$, à $L^p_F(\mu'_0)$. Si $F=\mathbf{R}$, si $f$ est une fonction positive de $\mathcal{L}^p(\mu)$, et si $f'$ est une fonction de $\mathcal{L}^p(\mu'_0)$ prolongeant $f$, la fonction positive $f''=\sup(f',0)$ appartient à $\mathcal{L}^p(\mu'_0)$ et prolonge $f$.

Ceci achève la démonstration du théorème 6 et celle du théorème 5.

Remarques. — 1° C'est une conséquence immédiate des théorèmes 2 et 5 que, pour toute mesure abstraite $\mu$ sur $\mathcal{R}$ ayant la propriété $(\mathrm{M}^*)$, les espaces de Banach $L^{p*}_F(E,\mathcal{R},\mu)$ et $L^{p*}_F(E',\mathcal{K}(E'),\mathbf{T}(\mu))$ sont isomorphes. En supposant que la mesure abstraite $\mu$ du théorème 6 vérifie l'axiome $(\mathrm{M}^*)$, on peut aussi montrer que toute fonction $\mathbf{f}\in\mathcal{L}^{p*}_F(E,\mathcal{R},\mu)$ peut être prolongée en une fonction $\mathbf{f}'\in\mathcal{L}^{p*}_F(E'_{\mathcal{R}},\mathcal{K}(E'_{\mathcal{R}}),\mathbf{\Phi}(\mu))$, qu'inversement la restriction $\mathbf{f}$ à $E$ de toute fonction $\mathbf{f}'\in\mathcal{L}^{p*}_F(E'_{\mathcal{R}},\mathcal{K}(E'_{\mathcal{R}})),\mathbf{\Phi}(\mu)$ appartient à $\mathcal{L}^{p*}_F(E,\mathcal{R},\mu)$, et que $N^*_p(\mathbf{f})=N^{*\prime}_p(\mathbf{f}')$, où $N^*_p$ et $N^{*\prime}_p$ désignent les semi-normes déduites de $\mu$ et $\mathbf{\Phi}(\mu)$ respectivement d'après le second procédé d'intégration de Stone. La démonstration reste en principe la même que pour le cas traité dans le théorème 6. Il faut seulement remplacer les suites croissantes de fonctions positives de $\mathcal{R}^{0'}$ ou de $\mathcal{K}(E'_{\mathcal{R}})$ par des ensembles filtrants de fonctions positives de $\mathcal{R}^{0'}$ ou de $\mathcal{K}(E'_{\mathcal{R}})$. Dans la première partie de la démonstration, on utilise la remarque suivante : Pour tout ensemble filtrant $H$ (pour $\leqq$) de fonctions positives de $\mathcal{K}(E'_{\mathcal{R}})$, il existe un ensemble filtrant $H^0$ de fonctions positives de $\mathcal{R}^{0'}$ tel que toute fonction de $H^0$ soit majorée par au moins une fonction de H, et que $\sup_{f' \in H} f' = \sup_{g' \in H^0} g'$. On déduit cette remarque de la proposition 7 comme on a déduit le corollaire 2 du lemme du paragraphe 1.

2° Soit $\Gamma$ une algèbre booléenne de sous-ensembles de l'ensemble $E$. Nous supposons que $E$ appartient à $\Gamma$ et que, pour tout couple $x$, $y$ d'éléments de $E$, il existe un ensemble $A \in \Gamma$ tel que $x \in A$ et $y \in \complement A$. L'ensemble $\mathcal{R}$ des combinaisons linéaires de fonctions caractéristiques d'ensembles de $\Gamma$ vérifie alors les axiomes $(R_1)$, $(R_2)$, $(R_3)$ et $(S_1)$, $(S_2)$ : l'espace $E'_{\mathcal{R}}$ existe et est identique à l'espace $E'$ du théorème 5. Nous avons démontré dans [2], p. 480, que $E'_{\mathcal{R}}$ est un espace compact contenant $E$ comme sous-ensemble dense, et que l'application $A \rightarrow \overline{A}$, où $\overline{A}$ est l'adhérence de $A \in \Gamma$ dans $E'_{\mathcal{R}}$, est un isomorphisme de $\Gamma$ sur l'algèbre booléenne des sous-ensembles de $E'_{\mathcal{R}}$, à la fois ouverts et fermés ([18]).

## BIBLIOGRAPHIE.


[1] H. Bauer, *Zur Theorie des Riemann-Integrals in lokal kompakten Räumen* (*Sitzungsber. math.-nat. Kl. Bayer. Akad. Wiss. München*, 1955, p. 187-208).

[2] H. Bauer, *Über die Beziehungen einer abstrakten Theorie des Riemann-Integrals zur Theorie Radonscher Masse* (*Math. Z.*, t. 65, 1956, p. 448-482).

[3] H. Bauer, *Eine Rieszsche Bandzerlegung im Raum der Bewertungen eines Verbandes* (*Sitzungsber. math.-nat. Kl. Bayer. Akad. Wiss. München*, 1953, p. 89-117).

[4] N. Bourbaki, *Intégration*, Chap. I-IV (*Actual. scient. et ind.*, n° 1175, Paris, Hermann, 1952).

[5] Haupt-Aumann-Pauc, *Differential und Integralrechnung*, Bd. 3, 2. Aufl., Göschens Lehrbücherei 26, Berlin, 1955.

[6] S. Kakutani, *Concrete representations of abstract (L)-spaces and the mean ergodic theorem* (*Ann. Math.*, t. 42, 1941, p. 523-537).

[7] L. H. Loomis, *Linear functionals and content* (*Amer. J. Math.*, t. 76, 1954, p. 168-182).

[8] G. Nöbeling et H. Bauer, *Allgemeine Approximationskriterien mit Anwendungen* (*J. ber. Deutsch. Math. Verein.*, t. 58, 1955, p. 54-72).

[9] M. H. Stone, *Notes on integration*, I-IV (*Proc. Nat. Acad. Sc.*, *U. S. A.*, t. 34, 1948, p. 336-342, 447-455, 483-490, et t. 35, 1949, p. 50-58).


([18]) *Voir* l'introduction de ce travail (p. 52).

Manuscrit reçu le 31 juillet 1956.

# Minimalstellen von Funktionen und Extremalpunkte [R13]

Arch. Math. **9** (1958), 389–393
[Zbl. Math. 82.32601; MR 20 #7202]

**1. Einleitung.** Der Begriff des Extremalpunktes einer konvexen Menge $X$ ist insofern rein interner Natur, als zu seiner Definition nur ganz in $X$ liegende offene Strecken benützt werden und damit die externe Eigenschaft von $X$, in einen Vektorraum $E$ eingebettet zu sein, nur dazu verwendet wird, um dem Begriff „Strecke" einen Sinn zu geben. Die übliche Formulierung des die Existenz von Extremalpunkten sichernden und für den Beweis des Theorems von Krein-Milman [6] entscheidenden Satzes betont dagegen die Einbettung von $X$ in $E$. Nach diesem Satz enthält nämlich jede abgeschlossene Stützhyperebene einer konvexen kompakten Teilmenge $X$ eines lokalkonvexen separierten Raumes $E$ mindestens einen Extremalpunkt von $X$. Dieser Satz kann aber auch folgendermaßen formuliert werden: Ist $f$ eine auf $E$ stetige Linearform und $f_X$ ihre Restriktion auf $X$, so nimmt $f_X$ sein Minimum auf $X$ in einem Extremalpunkt von $X$ an. Gibt man dem Satz diese Wendung, so tritt seine interne Natur klarer zu Tage. Noch deutlicher tritt diese hervor durch eine Verallgemeinerung des Satzes, deren Beweis ein Ziel dieser Mitteilung ist. Es soll gezeigt werden, daß sogar jede auf $X$ nach unten halbstetige, konkave, numerische Funktion $f$ mit Werten $-\infty < f(x) \leqq +\infty$ ihr Minimum auf $X$ in einem Extremalpunkt von $X$ annimmt.

Während der Beweis des zu verallgemeinernden Satzes bei N. Bourbaki [2], S. 83, nach Ideen von R. Godement [3] mit Hilfe sogenannter Stützmannigfaltigkeiten geführt wird, also mit die Einbettung von $X$ in $E$ betonenden Hilfsmitteln, zwingt uns die zu beweisende Verallgemeinerung zu einem Beweis mit einer rein internen Methode. Dabei zeigt sich, daß sowohl die Konvexität von $X$ als auch die Einbettung von $X$ in einen lokalkonvexen Raum eine recht untergeordnete Rolle spielen. Unser Satz erweist sich als Spezialfall eines sich auf beliebige kompakte Räume beziehenden Satzes, den man als ein allgemeines Minimumprinzip bezeichnen könnte.

Herrn G. Nöbeling verdanke ich eine in die endgültige Fassung dieses Satzes eingearbeitete Diskussionsbemerkung.

**2. Definitionen und Bezeichnungen.** Im folgenden bezeichnen wir mit $\mathbf{R}$ die Menge der reellen Zahlen, aufgefaßt als topologischer Raum (Zahlengerade) bzw. Körper. Mit $\overline{\mathbf{R}}$ bezeichnen wir den aus der Zahlengeraden durch Adjunktion von $-\infty$ und $+\infty$ entstehenden topologischen Raum. *Numerische* Funktion heißt jede Abbildung $f$ einer Menge $A$ in den Raum $\overline{\mathbf{R}}$. Wir nennen *Minimalstelle* von $f$ (auf $A$) jeden Punkt $x_0 \in A$ mit der Eigenschaft:

$$f(x_0) = \inf_{x \in A} f(x) .$$

Existiert eine solche Minimalstelle von $f$, so sagen wir auch: die Funktion $f$ *nimmt ihr Minimum auf $A$ (in $x_0$) an.*

Ist $\mathscr{F}$ eine Menge von auf $A$ definierten numerischen Funktionen und existiert zu je zwei verschiedenen Punkten $x$ und $y$ aus $A$ eine Funktion $f \in \mathscr{F}$ mit $f(x) \neq f(y)$, so sagen wir: *$\mathscr{F}$ trennt die Punkte von $A$.*

**3. Der allgemeine Satz.** Es sei $X$ ein (nicht leerer) kompakter topologischer Raum. In $X$ sei ausgezeichnet ein (eventuell leeres) System $\mathscr{S}$ von Teilmengen $S$ von $X$ mit folgender Eigenschaft:

**(S)** *Jede Menge $S \in \mathscr{S}$ enthält mindestens zwei Punkte.*

Betrachtet werden dann alle auf $X$ definierten, *nach unten halbstetigen*, numerischen Funktionen $f$, die der folgenden Bedingung genügen:

**(K)** *Es sei $S$ eine Menge aus $\mathscr{S}$. Nimmt dann die Restriktion von $f$ auf $S$ ihr Minimum auf $S$ an, so ist $f$ auf $S$ konstant.*

Die Menge dieser Funktionen $f$ werde mit $\mathscr{K}_{\mathscr{S}} = \mathscr{K}_{\mathscr{S}}(X)$ bezeichnet.

Der in der Einleitung angekündigte Satz lautet nun:

**Satz 1.** *Es sei $X$ ein nicht leerer kompakter Raum und $\mathscr{S}$ ein System von Teilmengen von $X$ mit der Eigenschaft* (S). *Wenn dann die Funktionenmenge $\mathscr{K}_{\mathscr{S}}(X)$ die Punkte von $X$ trennt, so besitzt jede Funktion aus $\mathscr{K}_{\mathscr{S}}(X)$ mindestens eine Minimalstelle, die in keiner der Mengen aus $\mathscr{S}$ liegt. Insbesondere ist das Komplement der Menge $\bigcup_{S \in \mathscr{S}} S$ nicht leer.*

Für den Beweis benötigen wir den folgenden Begriff: *Extremalmenge* von $X$ (bezüglich $\mathscr{S}$) heiße jede nicht leere kompakte Menge $K \subset X$ derart, daß für jede Menge $S \in \mathscr{S}$ mit $S \cap K \neq \varnothing$ gilt: $S \subset K$. Das System aller Extremalmengen von $X$ werde mit $\mathscr{E}$ bezeichnet.

Dann ist $\mathscr{E}$ nicht leer, da offenbar $X$ zu $\mathscr{E}$ gehört. Eine einpunktige Menge $K = \{x_0\}$ ist wegen der Eigenschaft (S) von $\mathscr{S}$ genau dann eine Extremalmenge, wenn $x_0$ in keiner der Mengen $S \in \mathscr{S}$ liegt. Weitere Elemente von $\mathscr{E}$ werden durch die Funktionen aus $\mathscr{K}_{\mathscr{S}}$ geliefert. Es sei nämlich $f$ eine beliebige Funktion aus $\mathscr{K}_{\mathscr{S}}$ und $K_f$ die Menge aller Minimalstellen von $f$ auf $X$. Wir behaupten, daß $K_f$ in $\mathscr{E}$ liegt. Setzen wir

$$\alpha = \inf_{x \in X} f(x)\,,$$

so ist dies trivial im Falle $\alpha = +\infty$. Dann ist nämlich $K_f = X$ ein Element von $\mathscr{E}$. Also kann $\alpha < +\infty$ angenommen werden. Wir bemerken zunächst, daß zufolge der Kompaktheit von $X$ und der Halbstetigkeit von $f$ nach unten die Menge $K_f$ nicht leer und kompakt ist (vgl. N. Bourbaki [1], S. 112). Zum Nachweis der Extremaleigenschaft von $K_f$ sei $S$ eine Menge aus $\mathscr{S}$ mit $S \cap K_f \neq \varnothing$. Es gibt also einen Punkt $x_0$ in $S$ mit $f(x_0) = \alpha$. Folglich nimmt die Restriktion von $f$ auf $S$ ihr Minimum auf $S$ in $x_0$ an und nach (K) ist $f(x) = \alpha$ in allen Punkten $x \in S$. Daher ist $S \subset K_f$: $K_f$ ist eine Extremalmenge von $X$.

Nunmehr sei $f$ eine beliebige, aber für das Folgende fest gewählte Funktion aus $\mathscr{K}_{\mathscr{S}}$ und $\alpha$ ihr Minimum auf $X$. $\mathscr{E}_f$ bezeichne das System aller Extremalmengen $K$ von $X$, die in der Menge $K_f$ aller Minimalstellen von $f$ enthalten sind. Wegen $K_f \in \mathscr{E}_f$ ist $\mathscr{E}_f$ nicht leer. Bezüglich der Relation $K_1 \supset K_2$ ist $\mathscr{E}_f$ eine geordnete Menge; wir behaupten, daß $\mathscr{E}_f$ bezüglich $\supset$ sogar induktiv geordnet ist. Es sei nämlich $\mathscr{F}$ eine totalgeordnete Teilmenge von $\mathscr{E}_f$. Dann genügt es zu zeigen, daß die Menge

$$K_0 = \bigcap_{K \in \mathscr{F}} K$$

in $\mathscr{E}_f$ liegt. Nun ist aber $\mathscr{F}$ eine Filterbasis nicht leerer kompakter Mengen in $X$; da $X$ kompakt ist, ist also $K_0$ nicht leer und kompakt. Ist schließlich $S$ eine Menge aus $\mathscr{S}$ mit $S \cap K_0 \neq \varnothing$, so ist $S$ in jeder der Extremalmengen $K \in \mathscr{F}$ und damit in $K_0$ enthalten. Also liegt $K_0$ in $\mathscr{E}_f$. Da somit $\mathscr{E}_f$ induktiv geordnet ist, existiert nach dem Zornschen Lemma ein minimales Element $M$ in $\mathscr{E}_f$. Die Behauptung unseres Satzes ist bewiesen, wenn gezeigt werden kann, daß $M$ aus genau einem Punkt $x_0$ besteht. Dann ist nämlich $x_0$ eine Minimalstelle von $f$, die in keiner der Mengen $S \in \mathscr{S}$ liegt.

Wir nehmen an, es gäbe in $M$ mindestens zwei verschiedene Punkte. Da $\mathscr{K}_{\mathscr{S}}$ die Punkte von $X$ trennt, gibt es dann eine Funktion $g$ in $\mathscr{K}_{\mathscr{S}}$, die auf $M$ nicht konstant ist. Es bezeichne $g_M$ die Restriktion von $g$ auf $M$. Dann ist $g_M$ auf dem kompakten Raum $M$ nach unten halbstetig, also die Menge $K'$ aller Minimalstellen von $g_M$ auf $M$ nicht leer und kompakt. $K'$ ist eine Extremalmenge von $X$. Es sei nämlich $S$ eine Menge aus $\mathscr{S}$ mit $S \cap K' \neq \varnothing$. Wegen $K' \subset M$ ist dann auch $S \cap M \neq \varnothing$, wegen $M \in \mathscr{E}_f$ also $S$ eine Teilmenge von $M$. Hieraus folgt weiter, daß die Restriktion von $g$ auf $S$ gleich der Restriktion von $g_M$ auf $S$ ist, erstere also wegen $S \cap K' \neq \varnothing$ ihr Minimum auf $S$ annimmt. Nach (K) ist daher $g$ auf $S$ konstant, also $S \subset K'$. Damit ist gezeigt, daß $K'$ in $\mathscr{E}_f$ liegt. Nun ist aber $g$ auf $M$ nicht konstant, also $K'$ eine echte, zu $\mathscr{E}_f$ gehörige Teilmenge von $M$. Dies widerspricht der Minimaleigenschaft von $M$; also muß unsere Annahme falsch sein: $M$ ist einpunktig.

Damit ist insbesondere gezeigt, daß nicht in der Vereinigung aller Mengen $S \in \mathscr{S}$ gelegene Punkte existieren. Unser Satz ist also bewiesen.

Nur zur Illustration sei erwähnt, daß Satz 1 in gewissem Sinne eine Teilaussage des Minimumprinzips der klassischen Potentialtheorie für superharmonische Funktionen verallgemeinert. Es sei nämlich $X$ ein kompakter Unterraum eines topologischen Raumes $Y$. Das System $\mathscr{S}$ habe die zusätzliche Eigenschaft, daß die Vereinigung aller Mengen aus $\mathscr{S}$ der offene Kern von $X$ in $Y$ ist. Unser Satz besagt dann, daß, Trennung der Punkte von $X$ durch die Funktionen aus $\mathscr{K}_{\mathscr{S}}(X)$ vorausgesetzt, jede Funktion aus $\mathscr{K}_{\mathscr{S}}$ eine auf dem Rand von $X$ gelegene Minimalstelle besitzt.

Bemerkung. Mit einer ähnlichen Methode haben J. L. Kelley [5], J. Hotta [4] und R. Arens*) die Existenz von Extremalpunkten konvexer kompakter Mengen in einem lokalkonvexen separierten Raum bewiesen.

**4. Konkave Funktionen und Extremalpunkte.** Es sei $E$ ein Vektorraum über dem Körper $\mathbf{R}$ der reellen Zahlen.

Sind $x$ und $y$ zwei *verschiedene* Punkte (Vektoren) aus $E$, so heißt bekanntlich die Menge $S$ aller Punkte $z = \lambda x + (1 - \lambda)\, y$ mit $0 < \lambda < 1$ die durch $x$ und $y$ begrenzte *offene Strecke* in $E$; in Zeichen: $S = (x, y)$. Sind $x$ und $y$ beliebige Punkte aus $E$, so

*) Vgl. diesbezüglich das Referat von V. L. Klee, Jr. über die Arbeit von J. L. Kelley [5] in Math. Reviews **13**, 249 (1952).

heißt die Menge $S$ aller Punkte $z = \lambda x + (1 - \lambda) y$ mit $0 \leqq \lambda \leqq 1$ die *abgeschlossene Strecke* mit den Endpunkten $x$ und $y$; in Zeichen: $S = [x, y]$.

Eine numerische Funktion $f$, deren Definitionsbereich eine Teilmenge $A$ von $E$ ist, nennen wir *konkav*, wenn $-\infty < f(x) \leqq +\infty$ ist in jedem Punkt $x \in A$ und wenn für jede abgeschlossene Strecke $S = [x, y] \subset A$ gilt:

$$\lambda f(x) + (1 - \lambda) f(y) \leqq f(\lambda x + (1 - \lambda) y) \quad \text{für alle } 0 \leqq \lambda \leqq 1 .$$

Einen Punkt $x_0$ einer Teilmenge $A$ von $E$ nennen wir *Extremalpunkt* von $A$, wenn es keine offene Strecke $S \subset A$ gibt mit $x_0 \in S$.

Wenn speziell $A$ konvex ist, stimmen diese Definitionen mit der üblichen Definition der konkaven Funktionen und der Extremalpunkte überein. Vgl. N. Bourbaki [2], S. 81 und S. 90.

Als Anwendung von Satz 1 beweisen wir nun:

**Satz 2.** *Es sei $X$ eine nicht leere Teilmenge eines Vektorraumes $E$ über $\mathbf{R}$. Die Menge $X$ sei mit einer Topologie versehen und bezüglich dieser ein kompakter Raum. $\mathcal{K}(X)$ bezeichne die Menge aller auf $X$ definierten, nach unten halbstetigen, konkaven, numerischen Funktionen. Wenn dann $\mathcal{K}(X)$ die Punkte von $X$ trennt, so gibt es in der Menge der Minimalstellen einer jeden Funktion aus $\mathcal{K}(X)$ mindestens einen Extremalpunkt von $X$. Insbesondere besitzt $X$ mindestens einen Extremalpunkt.*

Beweis. Es sei $\mathcal{S}$ das System aller offenen Strecken $S$ in $E$ mit $S \subset X$. Dann genügt $\mathcal{S}$ der Bedingung (S) der letzten Nummer. Der zu beweisende Satz ist daher eine Folgerung aus Satz 1, wenn gezeigt werden kann, daß $\mathcal{K}(X) \subset \mathcal{K}_{\mathcal{S}}(X)$ ist. Hierfür genügt es zu zeigen, daß jede auf $X$ definierte konkave Funktion $f$ der Bedingung (K) genügt. Es sei also $S = (x, y)$ eine offene Strecke aus $\mathcal{S}$ und $x_0 \in S$ eine Minimalstelle der Restriktion von $f$ auf $S$; zu zeigen ist: $f(z) = f(x_0)$ für alle $z \in S$. Dies ist trivial, wenn $f(x_0) = +\infty$ ist; wir können also $f(x_0) < +\infty$ annehmen. Es sei nun $x'$ ein beliebiger Punkt aus $(x, x_0)$ und $y'$ ein beliebiger Punkt aus $(x_0, y)$. Dann ist $[x', y'] \subset S \subset X$ und $x_0 \in (x', y')$; folglich gibt es eine reelle Zahl $\lambda$ mit $0 < \lambda < 1$ und $x_0 = \lambda x' + (1 - \lambda) y'$. Da $f$ konkav ist, gilt

$$\lambda f(x') + (1 - \lambda) f(y') \leqq f(x_0) .$$

Weil $f(x') \geqq f(x_0)$, $f(y') \geqq f(x_0)$ und $|f(x_0)| < +\infty$ ist, folgt hieraus $f(x') = f(y') = f(x_0)$. Da $x'$ in $(x, x_0)$ und $y'$ in $(x_0, y)$ beliebig gewählt war, ist also $f$ auf $S$ konstant, unser Satz daher bewiesen.

**Korollar.** *Es sei $E$ ein lokalkonvexer separierter topologischer Vektorraum (über $\mathbf{R}$) und $X$ eine nicht leere kompakte Teilmenge von $E$. Dann nimmt jede auf $X$ definierte, nach unten halbstetige, konkave, numerische Funktion ihr Minimum auf $X$ in einem Extremalpunkt von $X$ an. Insbesondere besitzt $X$ mindestens einen Extremalpunkt*).*

Beweis. Zu je zwei verschiedenen Punkten $x$ und $y$ aus $E$, also erst recht aus $X$, existiert nach dem Satz von Hahn-Banach eine auf $E$ stetige Linearform $u$ mit

*) *Zusatz bei der Korrektur:* Das Korollar bleibt gültig, wenn der Raum $E$ nicht separiert, wohl aber $X$ kompakt (und daher auch separiert) ist. Zum Beweis verwendet man am besten prop. 4 von N. Bourbaki [2], chap. II, § 3.

$u(x) \neq u(y)$. Die Restriktion von $u$ auf $X$ liegt in der Funktionenmenge $\mathscr{K}(X)$. Die Behauptung folgt daher aus Satz 2.

Bemerkung. In dem speziellen Fall, wo $E$ der mit der schwachen Topologie versehene Dualraum eines Banach-Raumes und $X$ eine konvexe kompakte Teilmenge von $E$ ist, findet sich das Korollar zu Satz 2 bereits bei P. C. ROSENBLOOM [7], S. 54. Der dort geführte Beweis stützt sich wesentlich auf die Bemerkung, wonach für eine konkave, nach unten halbstetige Funktion auf $X$ die Menge aller Minimalstellen eine konvexe Teilmenge von $X$ sein soll. Offenbar ist aber diese Bemerkung im allgemeinen falsch; der betreffende Satz bleibt also bei ROSENBLOOM unbewiesen.

**5. Schlußbemerkungen.** Der Satz 1 verliert seine Gültigkeit, wenn man die Forderung fallen läßt, daß die Funktionenmenge $\mathscr{K}_{\mathscr{S}}(X)$ die Punkte von $X$ trennt. In der Tat: Es sei $X$ ein kompakter, mindestens zweipunktiger Raum. Das System $\mathscr{S}$ bestehe aus genau einer Menge $S$, nämlich $S = X$. Dann enthält $\mathscr{K}_{\mathscr{S}}(X)$ nur die konstanten numerischen Funktionen auf $X$, trennt also die Punkte von $X$ nicht. Die Behauptung des Satzes 1 trifft hier nicht zu, da kein Punkt existiert, der in keiner der Mengen aus $\mathscr{S}$ liegt.

Auch die Eigenschaft (S) des Systems $\mathscr{S}$ ist wesentlich für die Gültigkeit von Satz 1. Wählt man nämlich für $\mathscr{S}$ das System aller einpunktigen Mengen eines kompakten Raumes $X$, so ist die Behauptung des Satzes 1 falsch, da es wiederum keinen Punkt gibt, der in keiner der Mengen aus $\mathscr{S}$ liegt.

Schließlich sei noch auf die folgende Verallgemeinerung des Satzes 1 hingewiesen. Es sei jetzt $X$ ein beliebiger nicht leerer topologischer Raum und $\mathscr{S}$ ein System von Teilmengen von $X$ mit der Eigenschaft (S). $\mathscr{K}^{\circ}_{\mathscr{S}}(X)$ bezeichne die Menge aller auf $X$ definierten, nach unten halbstetigen, numerischen Funktionen mit der Eigenschaft (K), für welche die Menge ihrer Minimalstellen auf $X$ nicht leer und kompakt ist. Trennt dann $\mathscr{K}^{\circ}_{\mathscr{S}}(X)$ die Punkte von $X$, so nimmt jede Funktion aus $\mathscr{K}^{\circ}_{\mathscr{S}}(X)$ ihr Minimum auf $X$ sogar in einem Punkt $x_0$ an, der zu keiner der Mengen $S \in \mathscr{S}$ gehört. Der Beweis hierfür ist im wesentlichen identisch mit dem des Satzes 1 und braucht daher nicht ausgeführt zu werden. Eine analoge Verallgemeinerung ist für den Satz 2 möglich.

**Literaturverzeichnis**

[1] N. BOURBAKI, Topologie générale, Chap. III—IV. Actual. sci. industr. **1143**, (1951).
[2] N. BOURBAKI, Espaces vectoriels topologiques, Chap. I—II. Actual. sci. industr. **1189**, (1953).
[3] R. GODEMENT, Les fonctions de type positif et la théorie des groupes. Trans. Amer. math. Soc. **63**, 1—84 (1948).
[4] J. HOTTA, A remark on regularly convex sets. Kōdai Math. Sem. Rep. **1951**, 37—40 (1951).
[5] J. L. KELLEY, Note on a theorem of Krein and Milman. J. Osaka Inst. Sci. Technol., Part I **3**, 1—2 (1951).
[6] M. KREIN — D. MILMAN, On the extreme points of regularly convex sets. Studia math. **9**, 133—138 (1940).
[7] P. C. ROSENBLOOM, Quelques classes de problèmes extrémaux. Bull. Soc. math. France **79**, 1—58 (1951).

Eingegangen am 20. 2. 1958

# Konservative Abbildungen lokal-kompakter Räume *
# [R14]

Math. Ann. **138** (1959), 398–427
[Zbl. Math. 86.04503; MR 22 #8319]

## Einleitung

In dieser Arbeit untersuchen wir eine Klasse maßtheoretisch interessanter, stetiger Abbildungen lokal-kompakter Räume. Genauer handelt es sich um folgendes: Gegeben sei eine stetige Abbildung $\varphi : X \to Y$ eines lokal-kompakten Raumes $X$ in einen lokal-kompakten Raum $Y$. Dann ordnet $\varphi$ gewissen positiven (Radonschen) Maßen $\mu$ auf $X$ ein positives Maß $\varphi(\mu)$ auf $Y$ als Bild zu; $\varphi(\mu)$ ist definiert durch die für alle stetigen, reellen Funktionen $f$ auf $Y$ mit kompaktem Träger gültige Gleichung

$$\int f \, d\varphi(\mu) = \int f \circ \varphi \, d\mu$$

unter der Voraussetzung, daß jede der Funktionen $f \circ \varphi$ $\mu$-integrierbar ist. Stets wird hierbei das Maß $\varphi(\mu)$ von der Menge $\varphi(X)$ getragen. Hat nun die Abbildung $\varphi$ die spezielle Eigenschaft, daß umgekehrt zu jedem Maß $\nu \geqq 0$ auf $Y$, welches von der Menge $\varphi(X)$ getragen wird, mindestens ein Maß $\mu \geqq 0$ auf $X$ existiert mit $\varphi(\mu) = \nu$, gehen also gewissermaßen bei Anwendung von $\varphi$ auf die „zulässigen" positiven Maße auf $X$ keine auf $\varphi(X)$ konzentrierten, positiven Maße auf $Y$ verloren, so nennen wir die Abbildung $\varphi$ *konservativ*. Es soll hier gezeigt werden, daß eine Reihe topologisch interessanter Abbildungen konservativ, also auch von Interesse für die Maßtheorie sind. Die hier gewonnenen Resultate gestatten Anwendungen in der Potentialtheorie und in der Theorie der Flächenmaße.

Die Arbeit gliedert sich in sechs Paragraphen. In § 1 stellen wir zunächst die für unsere Bedürfnisse wichtigsten Begriffe und Resultate aus der Integrationstheorie zusammen und ergänzen sie sodann durch auf unser eigentliches Ziel abgestimmte Zusätze. Hierunter fallen u. a. die ersten elementaren Eigenschaften der konservativen Abbildungen, welche dort zunächst etwas allgemeiner als soeben, nämlich ohne Voraussetzung der Stetigkeit, definiert werden.

In § 2 wird als wichtiges Beweishilfsmittel ein *Lokalisationsprinzip* bereitgestellt. Es zeigt sich nämlich, daß die $\varphi$-Darstellbarkeit eines Maßes $\nu \geqq 0$ auf $Y$, d. h. die Existenz eines Maßes $\mu \geqq 0$ auf $X$ mit $\varphi(\mu) = \nu$ eine lokale

*) Der Verfasser ist der Vereinsbank in Hamburg für Forschungsmittel, welche die Fertigstellung der vorliegenden Arbeit ermöglichten, sehr dankbar.

Eigenschaft ist. Grob gesagt ist für die $\varphi$-Darstellbarkeit von $\nu$ die „$\varphi$-Darstellbarkeit im Kleinen" notwendig und hinreichend.

Eine erste Klasse topologisch interessanter Abbildungen, die sämtlich konservativ sind, wird dann in § 3 aufgezeigt. Es sind dies die *eigentlichen* und *lokal-eigentlichen* Abbildungen lokal-kompakter Räume, deren Bedeutung man z. B. aus der Theorie der holomorphen Abbildungen komplexer Räume kennt (vgl. K. Stein [18] und R. Remmert [17]). Entscheidendes Beweishilfsmittel ist hierbei neben dem Lokalisationsprinzip ein von I. Namioka [14] und Verf. [3] unabhängig und etwa gleichzeitig angegebener Fortsetzungssatz für positive Linearformen. Ferner wird der rein topologische Satz bewiesen, daß eine stetige Abbildung $\varphi : X \to Y$ eines lokal-kompakten Raumes $X$ in einen lokal-kompakten Raum $Y$ stets zu einer eigentlichen Abbildung $\tilde{\varphi} : \tilde{X} \to Y$ eines lokal-kompakten Raumes $\tilde{X}$ in $Y$ und damit zu einer konservativen Abbildung fortgesetzt werden kann. Jede stetige Abbildung ist daher wenigstens die Restriktion eine konservativen Abbildung.

Unter Verwendung der in § 3 gewonnenen Ergebnisse wird in § 4 u. a. gezeigt, daß jede stetige Abbildung $\varphi : X \to Y$ eines *im Unendlichen abzählbaren* lokal-kompakten Raumes $X$ in einen lokal-kompakten Raum $Y$ konservativ ist.

In § 5 wird bewiesen, daß eine stetige Abbildung $\varphi : X \to Y$ stets dann konservativ ist, wenn sie einen *lokalen Schnitt* besitzt, d. h. wenn in mindestens einer Umgebung $V$ eines jeden Punktes $y \in \varphi(X)$ eine stetige Abbildung $\sigma : V \to X$ definiert ist, für welche $\varphi \circ \sigma$ die identische Abbildung ist. Diese Eigenschaft besitzen die Projektionsabbildungen einer umfangreichen Klasse von lokal-kompakten Faserräumen.

Der abschließende § 6 beschäftigt sich mit der *Faktorisierung* konservativer Abbildungen. Zunächst wird gezeigt, daß das Transitivitätsgesetz in seiner allgemeinen Form nicht gilt, daß also die lokal-kompakten Räume als Objekte und die konservativen Abbildungen als Morphismen *keine* Kategorie bilden. Wohl aber gilt ein abgeschwächtes Transitivitätsgesetz: das Produkt $\psi \circ \varphi$ zweier konservativer Abbildungen $\varphi : X \to Y$ und $\psi : Y \to Z$ ist selbst wieder konservativ, sofern $\varphi(X) = Y$ ist. Bemerkenswerterweise gilt hiervon eine Umkehrung: ist $\psi \circ \varphi$ konservativ und $\varphi(X) = Y$, so ist wenigstens die Abbildung $\psi$ (nicht notwendig jedoch $\varphi$) konservativ. Eine analoge Eigenschaft besitzen die holomorphen Abbildungen komplexer Räume (vgl. K. Stein [18], Satz 2; R. Remmert [17], S. 363). Gerade diese letzte Eigenschaft gestattet es, eine Theorie *konservativer Basen und Zerlegungen* zu entwickeln, die analog zur Theorie analytischer Basen und Zerlegungen von K. Stein [18] ist. Auf den Zusammenhang dieses Teils der Arbeit mit einem Faktorisierungssatz von G. T. Whyburn [19], p. 141, wird in [4] genauer eingegangen.

Es sei noch bemerkt, daß die unter [2] zitierte C. R.-Note einen Vorläufer dieser Arbeit darstellt. Ein Teil der hier bewiesenen Ergebnisse findet sich dort ohne Angabe von Beweisen und meist unter entbehrlichen, zusätzlichen Voraussetzungen.

## § 1. Darstellbare Maße und konservative Abbildungen

In diesem einleitenden Paragraphen sollen zunächst die wichtigsten Begriffe und Ergebnisse aus der Integrationstheorie Radonscher Maße zusammengestellt werden, die im folgenden benötigt werden. Wir benützen hierbei weitgehend die Terminologie von N. BOURBAKI [5], [6]; dort sind auch nähere Einzelheiten zu den folgenden Nummern 1—5 zu finden. Sodann definieren wir die für die weiteren Untersuchungen zentralen Begriffe „darstellbares Maß" und „konservative Abbildung". An einem Beispiel wird die Existenz nicht-konservativer Abbildungen nachgewiesen.

**1.1.** *Radonsche Maße.* — Es sei $T$ ein lokal-kompakter[1]) topologischer Raum. Dann bezeichnen wir mit $\mathscr{K}(T)$ die Menge aller auf $T$ stetigen, reellen Funktionen mit kompaktem Träger. $\mathscr{K}(T)$ ist hinsichtlich der üblichen Operationen und der natürlichen Ordnungsrelation $\leqq$ ein geordneter Vektorraum über dem Körper $\mathbf{R}$ der reellen Zahlen. $\mathscr{K}_+(T)$ bezeichne den konvexen Kegel aller Funktionen $f \geqq 0$ aus $\mathscr{K}(T)$.

*Positives* (Radonsches) *Maß* auf $T$ heißt jede positive Linearform $\mu$ auf $\mathscr{K}(T)$. Beispielsweise ist $f \to f(t)$ für jeden Punkt $t \in T$ eine solche Linearform, also ein positives Maß auf $T$. Dieses wird mit $\varepsilon_t$ bezeichnet und das Maß der *Einheitsmasse im Punkte t* genannt.

Die Menge aller positiven Maße auf $T$ bezeichnen wir mit $\mathscr{M}(T)$. Bezüglich der üblichen Operationen ist $\mathscr{M}(T)$ ein konvexer Kegel und als solcher in natürlicher Weise geordnet: $\mu \leqq \nu$ ist gleichbedeutend mit $\mu(f) \leqq \nu(f)$ für alle $f \in \mathscr{K}_+(T)$. Ferner trägt $\mathscr{M}(T)$ die sogenannte *vage Topologie*; sie ist nichts anderes als die Topologie der punktweisen Konvergenz auf $\mathscr{K}(T)$.

**1.2.** *Integrierbare und meßbare reelle Funktionen.* — Es sei $\mu$ ein positives Maß auf einem lokal-kompakten Raum $T$. Dann wird jeder numerischen[2]) Funktion $f \geqq 0$ auf $T$ wie folgt das Oberintegral $\int^* f d\mu$ zugeordnet: Für jede nach unten halbstetige Funktion $f \geqq 0$ ist

$$\int^* f d\mu = \sup_{h \leqq f,\, h \in \mathscr{K}_+(T)} \mu(h)\,. \tag{1.1a}$$

Für ein beliebiges $f \geqq 0$ ist

$$\int^* f d\mu = \inf_{g \in \mathscr{I}_+(f)} \int^* g d\mu\,, \tag{1.1b}$$

wenn hierbei $\mathscr{I}_+(f)$ die Menge aller nach unten halbstetigen, numerischen Funktionen $g$ auf $T$ mit $g \geqq f$ bezeichnet. Für jede numerische Funktion $f$ auf $T$ wird

$$N_1^\mu(f) = \int^* |f|\, d\mu \tag{1.2}$$

gesetzt. Offenbar ist $N_1^\mu \geqq 0$.

---

[1]) Wir verwenden den Kompaktheitsbegriff von N. BOURBAKI [7]. Lokal-kompakte Räume sind demnach auch stets Hausdorffsch (separiert).

[2]) Unter einer *numerischen* bzw. *reellen* Funktion $f$ auf $T$ verstehen wir eine Abbildung $f: T \to \bar{\mathbf{R}}$ bzw. $f: T \to \mathbf{R}$ von $T$ in die durch die Adjunktion von $\pm\infty$ kompaktifizierte Zahlengerade $\bar{\mathbf{R}}$ bzw. in $\mathbf{R}$ selbst.

Die Menge $\mathscr{F}^1(\mu)$ aller *reellen* Funktionen $f$ auf $T$ mit $N_1^\mu(f) < +\infty$ ist bezüglich der üblichen Operationen ein Vektorraum und $N_1^\mu$ ist auf $\mathscr{F}^1(\mu)$ eine Halbnorm; $N_1^\mu$ definiert somit in $\mathscr{F}^1(\mu)$ eine lokal-konvexe Topologie: die Topologie der *Konvergenz im Mittel* (bezüglich $\mu$). Es ist $\mathscr{K}(T) \subset \mathscr{F}^1(\mu)$ und $\mu$ stetig bezüglich dieser Topologie. Daher kann $\mu$ auf genau eine Weise fortgesetzt werden zu einer stetigen Linearform $\mu_0$ auf der abgeschlossenen Hülle $\mathscr{L}^1(\mu)$ von $\mathscr{K}(T)$ in $\mathscr{F}^1(\mu)$. Eine reelle *Funktion* $f$ auf $T$ wird ($\mu$-)*integrierbar* genannt, wenn sie in $\mathscr{L}^1(\mu)$ liegt; $\mu_0(f)$ wird das $\mu$-*Integral* von $f$ genannt und mit $\int f d\mu$ bezeichnet. Speziell ist also $\mu(f) = \mu_0(f) = \int f d\mu$ für jedes $f \in \mathscr{K}(T)$. Daher schreibt man für Funktionen $f \in \mathscr{K}(T)$ anstelle von $\mu(f)$ meist gleich von vornherein $\int f d\mu$. Für Funktionen $f \geqq 0$ aus $\mathscr{L}^1(\mu)$ ist

$$\int f d\mu = \int^* f d\mu = N_1^\mu(f)\,. \tag{1.3}$$

Daher ist $f \to \int f d\mu$ eine positive Linearform auf dem durch die natürliche Ordnungsrelation $\leqq$ geordneten Vektorraum $\mathscr{L}^1(\mu)$.

Für eine Teilmenge $A$ von $T$ bezeichnen wir mit $\chi_A$ stets deren charakteristische Funktion (bezüglich $T$). Die *Menge A* wird ($\mu$-)*integrierbar* genannt, wenn $\chi_A \in \mathscr{L}^1(\mu)$ ist; $\mu(A) = \int \chi_A d\mu$ heißt dann das Maß von $A$. Die Menge $A$ wird ($\mu$-)*meßbar* genannt, wenn für jede in $T$ kompakte Menge $K$ der Durchschnitt $A \cap K$ integrierbar ist. Integrierbar sind z. B. alle kompakten und alle relativ-kompakten offenen Mengen; meßbar sind z. B. alle abgeschlossenen und alle offenen Mengen. Das System aller meßbaren Mengen ist eine Boolesche $\sigma$-Mengenalgebra mit der Einheit $T$.

Eine Menge $A \subset T$ heißt ($\mu$-)*vernachlässigbar* bzw. ($\mu$-)*lokal-vernachlässigbar* wenn $A$ integrierbar und $\mu(A) = 0$ bzw. wenn $A$ meßbar und $\mu(A \cap K) = 0$ ist für jede in $T$ kompakte Menge $K$.

Eine numerische *Funktion f* auf $T$ heißt ($\mu$-)*meßbar*, wenn jede der Mengen $\{x : f(x) \geqq \alpha\}$, $\alpha \in \mathbf{R}$, meßbar ist. Eine reelle Funktion $f$ auf $T$ ist dann und nur dann integrierbar, wenn sie meßbar und $N_1^\mu(f) < +\infty$ ist.

**1.3.** *Im wesentlichen integrierbare Funktionen.* — Eine reelle Funktion $f$ auf $T$ heißt *im wesentlichen* ($\mu$-)*integrierbar*, wenn sie lokal-fast-überall gleich einer integrierbaren Funktion ist, d. h. wenn eine Funktion $f' \in \mathscr{L}^1(\mu)$ existiert, für welche $f(x) = f'(x)$ ist für alle $x \in T$ mit Ausnahme der Punkte $x$ einer lokal-vernachlässigbaren Menge. Integral einer solchen Funktion $f$ wird dann die Zahl $\int f' d\mu$ genannt, die, wie sich zeigt, unabhängig von der speziellen Wahl von $f'$ ist. Offenbar ist jede integrierbare Funktion auch im wesentlichen integrierbar und ihr Integral im neuen Sinne gleich dem im alten Sinne. Da somit keine Mißverständnisse entstehen können, wird auch für jede im wesentlichen integrierbare Funktion $f$ das Integral mit $\int f d\mu$ bezeichnet.

Die Menge $\overline{\mathscr{L}}^1(\mu)$ aller im wesentlichen integrierbaren Funktionen ist hinsichtlich der üblichen Operationen und der natürlichen Ordnungsrelation $\leqq$ abermals ein geordneter Vektorraum; es ist $\mathscr{L}^1(\mu) \subset \overline{\mathscr{L}}^1(\mu)$ und $f \to \int f d\mu$ eine positive Linearform auf $\overline{\mathscr{L}}^1(\mu)$.

Der Raum $\overline{\mathscr{L}}^1(\mu)$ kann aus $\mathscr{K}(T)$ auf analoge Weise gewonnen werden wie $\mathscr{L}^1(\mu)$ aus $\mathscr{K}(T)$. Man hat hierbei nur die Halbnorm $N_1^\mu$ durch eine Halbnorm $\overline{N}_1^\mu \leqq N_1^\mu$ zu ersetzen, die wie folgt definiert ist: Für jede numerische Funktion $f$ auf $T$ ist

$$\overline{N}_1^\mu(f) = \sup_{K \in \mathfrak{K}} \int^* |f| \chi_K d\mu = \sup_{K \in \mathfrak{K}} N_1^\mu(f \chi_K), \tag{1.4}$$

wenn hierbei $\mathfrak{K}$ das System aller kompakten Teilmengen von $T$ bezeichnet. Es gilt dann wieder der Satz: Eine reelle Funktion $f$ auf $T$ ist dann und nur dann im wesentlichen integrierbar, wenn sie meßbar und $\overline{N}_1^\mu(f) < +\infty$ ist. In Analogie zu (1.3) gilt für jede Funktion $f \geqq 0$ aus $\overline{\mathscr{L}}^1(\mu)$:

$$\int f d\mu = \overline{N}_1^\mu(f). \tag{1.5a}$$

Für eine beliebige Funktion $f \in \overline{\mathscr{L}}^1(\mu)$ gilt:

$$\left|\int f d\mu\right| \leqq \overline{N}_1^\mu(f) = \int |f| d\mu. \tag{1.5b}$$

Ferner ist $f \to \int f d\mu$ die einzige, bezüglich der durch $\overline{N}_1^\mu$ definierten Topologie stetige Linearform auf $\overline{\mathscr{L}}^1(\mu)$, welche auf dem Unterraum $\mathscr{K}(T)$ mit dem Maß $\mu$, d. h. mit der Linearform $f \to \mu(f)$ übereinstimmt.

Für auf $T$ *stetige*, reelle Funktionen $f$ fallen übrigens die Begriffe „integrierbar“ und „im wesentlichen integrierbar“ zusammen. Dies folgt aus N. BOURBAKI [6], § 2, prop. 2, wonach für solche Funktionen $\overline{N}_1^\mu(f) = N_1^\mu(f)$ ist.

Jede Menge $A \subset T$ mit $\chi_A \in \overline{\mathscr{L}}^1(\mu)$ wird im wesentlichen ($\mu$-)integrierbar und $\mu(A) = \int \chi_A d\mu$ das Maß von $A$ genannt. Es erweist sich $A$ genau dann als lokal-vernachlässigbar, wenn $A$ im wesentlichen integrierbar und $\mu(A) = 0$ ist.

1.4. *Massentilgung und induziertes Maß.* — Es sei wieder $\mu$ ein positives Maß auf einem lokal-kompakten Raum $T$. Man sagt, es wird $\mu$ von einer Menge $M \subset T$ *getragen* oder es ist $\mu$ auf $M$ *konzentriert*, wenn die Komplementärmenge $\complement M = T - M$ bezüglich $\mu$ lokal-vernachlässigbar ist.

Ist $S$ eine $\mu$-meßbare Teilmenge von $T$, so ist für jede Funktion $f \in \mathscr{K}(T)$ die Funktion $f\chi_S$ integrierbar und daher $f \to \int f \chi_S d\mu$ eine positive Linearform auf $\mathscr{K}(T)$, also ein positives Maß auf $T$. Wir bezeichnen dieses mit $\chi_S \mu$ und nennen es das aus $\mu$ durch *Tilgung der Masse außerhalb* $S$ entstehende Maß. Gerechtfertigt wird diese Benennung durch die Bemerkung, daß $\chi_S \mu$ auf $S$ konzentriert ist. Ein Maß $\nu \in \mathscr{M}(T)$ wird genau dann von einer Menge $M \subset T$ getragen, wenn $M$ meßbar und $\nu = \chi_M \nu$ ist.

Zur Frage der Integrierbarkeit bezüglich des Maßes $\chi_S \mu$ sei an das folgende Resultat erinnert. Eine reelle Funktion $f$ auf $T$ ist dann und nur dann im wesentlichen $(\chi_S \mu)$-integrierbar, wenn die Funktion $f \chi_S$ im wesentlichen $\mu$-integrierbar ist. Es gilt dann

$$\int f d(\chi_S \mu) = \int f \chi_S d\mu. \tag{1.6}$$

Das Maß $\chi_S \mu$ ist insbesondere definiert für jeden lokal-kompakten Unterraum $S$ von $T$, da jeder solche als Durchschnitt einer in $T$ abgeschlossenen und einer in $T$ offenen Menge dargestellt werden kann und somit meßbar ist. Dann aber ist $\chi_S \mu$ wohl zu unterscheiden von dem von $\mu$ in einem solchen lokal-kompakten Unterraum *induzierten* Maß $\mu_S$. Während $\chi_S \mu$ ein Maß auf $T$ ist, handelt es sich bei $\mu_S$ um ein positives Maß auf $S$, welches wie folgt definiert ist: Für jede Funktion $g \in \mathcal{K}(S)$ ist diejenige Fortsetzung $g'$ auf $T$ $\mu$-integrierbar, welche auf $T - S$ gleich Null ist. Es ist $\int g d\mu_S = \int g' d\mu$ für jedes $g \in \mathcal{K}(S)$. Wegen $g' = g' \chi_S$ induzieren $\mu$ und $\chi_S \mu$ dasselbe Maß $\mu_S$ in $S$. Zu einem beliebigen Maß $\nu \in \mathcal{M}(S)$ gibt es im allgemeinen kein Maß $\mu \in \mathcal{M}(T)$ mit $\nu = \mu_S$. Wir werden hierauf in der nächsten Nummer zurückkommen.

Schließlich führen wir noch folgende Bezeichnung ein: Für eine Teilmenge $A$ von $T$ bezeichnen wir mit $\mathcal{M}(T, A)$ die Menge aller auf $A$ konzentrierten Maße aus $\mathcal{M}(T)$. Demnach ist speziell $\mathcal{M}(T, T) = \mathcal{M}(T)$.

**1.5.** *Bilder von Maßen*[3]). — Es sei $\varphi : X \to Y$ eine beliebige Abbildung eines lokal-kompakten Raumes $X$ in einen lokal-kompakten Raum $Y$. Dann ordnet $\varphi$ gewissen Maßen $\mu \in \mathcal{M}(X)$ ein Maß $\varphi(\mu) \in \mathcal{M}(Y)$ als Bild zu. Und zwar ist das *Bildmaß* $\varphi(\mu)$ für genau diejenigen Maße $\mu \in \mathcal{M}(X)$ definiert, für welche die Funktion $f \circ \varphi$ für jedes $f \in \mathcal{K}(Y)$ im wesentlichen $\mu$-integrierbar ist. $\varphi(\mu)$ wird dann erklärt durch die Gleichung

$$\int f d\,\varphi(\mu) = \int f \circ \varphi \, d\mu \quad \text{für alle } f \in \mathcal{K}(Y)\,. \tag{1.7}$$

Man zeigt, daß $\varphi(\mu)$ dann und nur dann definiert ist, wenn die Abbildung $\varphi$ *$\mu$-eigentlich*, d. h. wenn das Urbild $\overset{-1}{\varphi}(K)$ einer jeden in $Y$ kompakten Menge $K$ im wesentlichen $\mu$-integrierbar ist.

Die Gleichung (1.7) gilt allgemeiner für jede im wesentlichen $\varphi(\mu)$-integrierbare reelle Funktion $f$ auf $Y$, sofern $\varphi(\mu)$ überhaupt im Sinne der obigen Definition existiert. Genauer ist eine reelle Funktion $f$ auf $Y$ dann und nur dann im wesentlichen $\varphi(\mu)$-integrierbar, wenn $f \circ \varphi$ im wesentlichen $\mu$-integrierbar ist.

Für eine $\mu$-eigentliche Abbildung $\varphi : X \to Y$ ($\mu \in \mathcal{M}(X)$) und eine beliebige Abbildung $\psi : Y \to Z$ von $Y$ in einen lokal-kompakten Raum $Z$ gilt das folgende *Transitivitätsgesetz*: Es ist $\psi$ dann und nur dann $\varphi(\mu)$-eigentlich, wenn die zusammengesetzte Abbildung $\psi \circ \varphi : X \to Z$ $\mu$-eigentlich ist. Es gilt dann: $(\psi \circ \varphi)(\mu) = \psi(\varphi(\mu))$.

Die gegen Ende der letzten Nummer offengelassene Frage kann jetzt folgendermaßen beantwortet werden: Es sei $S$ ein lokal-kompakter Unterraum eines lokal-kompakten Raumes $T$ und $j : S \to T$ die kanonische Injektion von $S$ in $T$. Ein Maß $\nu \in \mathcal{M}(S)$ wird dann und nur dann von einem Maß aus $\mathcal{M}(T)$ in $S$ induziert, wenn das Bildmaß $j(\nu)$ definiert, wenn also die Injektion $j$ $\nu$-eigentlich ist. Dann wird nämlich $\nu$ durch $j(\nu)$ in $S$ induziert.

Schließlich ist für uns noch von Interesse die Frage nach der algebraischen Struktur (in $\mathcal{M}(X)$) des Definitionsbereiches $\mathcal{E}(\varphi)$ der durch eine Abbildung $\varphi : X \to Y$ vermittelten Abbildung aus $\mathcal{M}(X)$ in $\mathcal{M}(Y)$. Wir beweisen hierzu:

Hilfssatz 1. *Es sei $\varphi : X \to Y$ eine Abbildung eines lokal-kompakten Raumes $X$ in einen lokal-kompakten Raum $Y$; $\mu$ und $\nu$ seien positive Maße auf $X$. Dann gilt:* (a) *mit $\varphi(\mu)$ und $\varphi(\nu)$ ist auch $\varphi(\mu + \nu)$ definiert;* (b) *mit*

[3]) Vgl. hierzu insbesondere N. Bourbaki [6], § 6 und § 7.

*$\varphi(\mu)$ ist $\varphi(\alpha\mu)$ für jede Zahl $\alpha \geqq 0$ definiert;* (c) *ist $\nu \leqq \mu$ und ist $\varphi(\mu)$ definiert, so ist auch $\varphi(\nu)$ definiert. In den einzelnen Fällen gilt ferner:*

$$\text{(1.8)}\qquad \begin{aligned} &\text{(a)} && \varphi(\mu + \nu) = \varphi(\mu) + \varphi(\nu)\ ; \\ &\text{(b)} && \varphi(\alpha\mu) = \alpha\varphi(\mu)\ ; \\ &\text{(c)} && \varphi(\nu) \leq \varphi(\mu)\ . \end{aligned}$$

Beweis. Zu (a): Die Behauptung ist offenbar bewiesen, wenn gezeigt werden kann, daß $\mathscr{D} = \overline{\mathscr{L}}^1(\mu) \cap \overline{\mathscr{L}}^1(\nu)$ eine Teilmenge von $\overline{\mathscr{L}}^1(\mu + \nu)$ und $\int f d(\mu + \nu) = \int f d\mu + \int f d\nu$ für jedes $f \in \mathscr{D}$ ist. Bekannt ist (vgl. M. MORSE u. W. TRANSUE [12], theorem 6.1) die Relation $\mathscr{L}^1(\mu) \cap \mathscr{L}^1(\nu) \subset \mathscr{L}^1(\mu + \nu)$, aus der unmittelbar folgt, daß jede zugleich $\mu$- und $\nu$-meßbare Menge oder reelle Funktion auch $(\mu + \nu)$-meßbar ist. Ferner ist bekannt (vgl. N. BOURBAKI [5], p. 113, prop. 15) die Gleichung $N_1^{\mu+\nu} = N_1^{\mu} + N_1^{\nu}$, aus der bei Beachtung von (1.4) leicht die entsprechende Gleichung $\overline{N}_1^{\mu+\nu} = \overline{N}_1^{\mu} + \overline{N}_1^{\nu}$ folgt. Nun ist nach Nr. 1.3 jede Funktion $f \in \mathscr{D}$ $\mu$- und $\nu$-meßbar, also nach dem ersten Ergebnis $(\mu + \nu)$-meßbar; ferner sind $\overline{N}_1^{\mu}(f)$ und $\overline{N}_1^{\nu}(f)$ endlich, woraus nach dem zweiten Ergebnis die Endlichkeit von $\overline{N}_1^{\mu+\nu}(f)$ folgt. Abermals nach Nr. 1.3 ist dann aber $f$ ein Element von $\overline{\mathscr{L}}^1(\mu + \nu)$. Für jede Funktion $f \in \mathscr{D}$ ist nach (1.5b) weiter

$$\left|\int f d\mu + \int f d\nu\right| \leqq \left|\int f d\mu\right| + \left|\int f d\nu\right| \leqq \overline{N}_1^{\mu}(f) + \overline{N}_1^{\nu}(f) = \overline{N}_1^{\mu+\nu}(f)\ ,$$

also die Linearform $f \to \int f d\mu + \int f d\nu$ auf $\mathscr{D}$ stetig bezüglich der durch $\overline{N}_1^{\mu+\nu}$ definierten Topologie. Auf dem linearen Unterarm $\mathscr{K}(X)$ von $\mathscr{D}$ ist diese Linearform gleich $\mu + \nu$. Hieraus folgt nach dem in Nr. 1.3 Gesagten die behauptete Gleichheit $\int f d(\mu + \nu) = \int f d\mu + \int f d\nu$.

Zu (b): Für $\alpha = 0$ ist die Behauptung trivial. Für $\alpha > 0$ geht man von der bei N. BOURBAKI (loc. cit.) bewiesenen Gleichheit $N_1^{\alpha\mu} = \alpha N_1^{\mu}$ aus und beweist der Reihe nach durch Zurückgehen auf die entsprechenden Definitionen mühelos: $\overline{N}_1^{\alpha\mu} = \alpha \overline{N}_1^{\mu}$, $\overline{\mathscr{L}}^1(\alpha\mu) = \overline{\mathscr{L}}^1(\mu)$ sowie $\int f d(\alpha\mu) = \alpha \int f d\mu$ für jedes $f \in \overline{\mathscr{L}}^1(\mu)$. Hieraus folgt dann die Behauptung.

Zu (c): Nach N. BOURBAKI (loc. cit.) ist $N_1^{\nu} \leqq N_1^{\mu}$, woraus $\overline{N}_1^{\nu} \leqq \overline{N}_1^{\mu}$ und weiter $\overline{\mathscr{L}}^1(\mu) \subset \overline{\mathscr{L}}^1(\nu)$ folgt. Mit $\varphi(\mu)$ ist daher auch $\varphi(\nu)$ definiert. Weiter ist $\mu$ darstellbar in der Form $\mu = \nu + \varrho$ mit $\varrho \in \mathscr{M}(X)$; wegen $\varrho \leqq \mu$ ist nach dem soeben Bewiesenen auch $\varphi(\varrho)$ definiert. Aus der Behauptung (a) folgt daher $\varphi(\mu) = \varphi(\nu) + \varphi(\varrho)$ und hieraus $\varphi(\nu) \leqq \varphi(\varrho)$.

Aus diesem Hilfssatz ergibt sich nun sofort die Antwort auf unsere Frage:

Korollar. *Die Menge $\mathscr{E} = \mathscr{E}(\varphi)$ aller Maße $\mu \in \mathscr{M}(X)$, für welche das Bildmaß $\varphi(\mu)$ definiert ist, ist ein konvexer, erblicher Unterkegel von $\mathscr{M}(X)$. D. h. $\mathscr{E}$ besitzt folgende Eigenschaften:*

$$\text{(1.9)}\qquad \begin{aligned} &\mu, \nu \in \mathscr{E} \Rightarrow \mu + \nu \in \mathscr{E}\ ; \\ &\mu \in \mathscr{E} \Rightarrow \alpha\mu \in \mathscr{E} \quad (\alpha \geqq 0, \alpha \in \mathbf{R})\ ; \\ &\mu \in \mathscr{E}, \nu \leqq \mu \Rightarrow \nu \in \mathscr{E} \quad (\nu \in \mathscr{M}(X))\ . \end{aligned}$$

**1.6.** *Konservative Abbildungen.* — Es sei wieder $\varphi : X \to Y$ eine Abbildung eines lokal-kompakten Raumes $X$ in einen lokal-kompakten Raum $Y$. Für die charakteristische Funktion $f = \chi_A$ einer Menge $A \subset Y$ folgt aus (1.7) und dem im Anschluß an diese Formel erwähnten Satz: Die Menge $A$ ist genau dann lokal-vernachlässigbar bezüglich des Bildmaßes $\varphi(\mu)$ eines Maßes $\mu \in \mathscr{E}(\varphi)$, wenn die Menge $\overset{-1}{\varphi}(A)$ lokal-vernachlässigbar bezüglich $\mu$ ist. Speziell für die Menge $A = Y - \varphi(X)$, deren Urbild leer ist, gilt also: $Y - \varphi(X)$ ist $\varphi(\mu)$-lokal-vernachlässigbar, d. h. es wird $\varphi(\mu)$ für jedes Maß $\mu \in \mathscr{E}(\varphi)$ von $\varphi(X)$ getragen. Somit definiert $\varphi$ genauer eine Abbildung von $\mathscr{E}(\varphi)$ in $\mathscr{M}(Y, \varphi(X))$.

Hauptanliegen der vorliegenden Untersuchung ist die Frage nach Abbildungen $\varphi$, welche eine Abbildung von $\mathscr{E}(\varphi)$ auf $\mathscr{M}(Y, \varphi(X))$ vermitteln. Wir definieren daher:

**Definition 1.** *Es sei $\varphi: X \to Y$ eine Abbildung eines lokal-kompakten Raumes $X$ in einen lokal-kompakten Raum $Y$. Ein Maß $\nu \in \mathscr{M}(Y)$ heiße darstellbar bezüglich $\varphi$ (kürzer: ($\varphi$-)darstellbar), wenn mindestens ein Maß $\mu \in \mathscr{M}(X)$ existiert, für welches $\varphi(\mu)$ definiert und $\varphi(\mu) = \nu$ ist.*

*Die Abbildung $\varphi$ heiße konservativ, wenn jedes auf $\varphi(X)$ konzentrierte, positive Maß auf $Y$ $\varphi$-darstellbar ist.*

Beispiel. Es sei $X$ ein lokal-kompakter Unterraum eines lokal-kompakten Raumes $Y$ und $j: X \to Y$ die kanonische *Injektion* von $X$ in $Y$. Dann ist die Abbildung $j$ konservativ, denn für jedes Maß $\nu \in \mathscr{M}(Y, X) = \mathscr{M}(Y, j(X))$ ist das von $\nu$ in $X$ induzierte Maß $\nu_X$ ein Maß aus $\mathscr{M}(X)$ mit $j(\nu_X) = \nu$. (Vgl. N. Bourbaki [6], § 7, prop. 3.)

In den folgenden Paragraphen werden wir wichtige Klassen *stetiger*, konservativer Abbildungen kennenlernen. Zunächst soll jedoch die Existenz stetiger, nicht-konservativer Abbildungen durch Angabe eines Beispiels nachgewiesen werden.

1.7. *Beispiel einer stetigen, nicht-konservativen Abbildung.* — Es sei $I = [0,1]$ die Menge aller reellen Zahlen $x$ mit $0 \leqq x \leqq 1$. Der Raum $X$ bzw. $Y$ entsteht, indem man der Menge $I$ die diskrete bzw. die übliche (euklidische) Topologie aufprägt. Beide Räume sind lokal-kompakt ($Y$ ist sogar kompakt); die identische Abbildung $\varphi$ von $I$ auf sich ist eine stetige Abbildung von $X$ *auf* $Y$. Wir behaupten:

*Ein Maß $\nu \in \mathscr{M}(Y)$ ist dann und nur dann $\varphi$-darstellbar, wenn es die Gestalt $\nu = \sum_{n=1}^{\infty} a_n \varepsilon_{x_n}$[4]) besitzt, wobei $(a_n)_{n=1,2,\ldots}$ eine Folge reeller Zahlen $\geqq 0$ mit $\sum_{n=1}^{\infty} a_n < +\infty$ und $(x_n)_{n=1,2,\ldots}$ eine Folge von Punkten aus $I$ ist.*

Beweis. Es sei $\nu \in \mathscr{M}(Y)$ ein $\varphi$-darstellbares Maß und $\mu$ ein Maß aus $\mathscr{M}(X)$ mit $\varphi(\mu) = \nu$. Da auf einem diskreten Raum jedes Maß diskret ist (N. Bourbaki [5], p. 51), gibt es eine reelle Funktion $\alpha \geqq 0$ auf $X$ derart, daß $\int g\,d\mu = \sum_{x \in I} \alpha(x)\,g(x)$ ist für jedes $g \in \mathscr{K}(X)$. Eine reelle Funktion $h$ auf $X$ ist

[4]) Die Konvergenz dieser Reihe ist im Sinne der vagen Topologie zu verstehen.

genau dann $\mu$-integrierbar, wenn die Familie $(\alpha(x)h(x))_{x\in I}$ in $\mathbf{R}$ absolut summierbar oder, was hiermit gleichwertig ist[5]), summierbar ist; das Integral errechnet sich dann zu $\int h d\mu = \sum\limits_{x\in I} \alpha(x)h(x)$ (nach [5], p.146). Wegen $\varphi(\mu) = \nu$ und der Stetigkeit von $\varphi$, die das Übereinstimmen der Begriffe „integrierbar" und „im wesentlichen integrierbar" für alle Funktionen $f \circ \varphi$ mit $f \in \mathcal{K}(Y)$ zur Folge hat, ist also für jedes $f \in \mathcal{K}(Y)$ die Familie $(\alpha(x)f(x))_{x\in I}$ summierbar und $\int f d\nu = \sum\limits_{x\in I} \alpha(x)f(x)$. Da die konstante Funktion $f = 1$ zu $\mathcal{K}(Y)$ gehört, folgt hieraus die Summierbarkeit von $(\alpha(x))_{x\in I}$. D. h. aber: es gibt eine Folge $(x_n)_{n=1,2,\ldots}$ von Punkten aus $I$ derart, daß $\alpha(x) = 0$ ist für alle $x \neq x_n$, $x \in I$, und daß die Reihe $\sum \alpha(x_n)$ konvergiert. Für jedes $f \in \mathcal{K}(Y)$ ist dann $\int f d\nu = \sum \alpha(x_n) f(x_n)$ und somit $\nu = \sum \alpha(x_n)\,\varepsilon_{x_n}$. Somit hat $\nu$ die behauptete Gestalt. — Umgekehrt ist jedes Maß $\nu = \sum a_n \varepsilon_{x_n}$ aus $\mathcal{M}(Y)$ mit $a_n \geqq 0$ und $\sum a_n < +\infty$ darstellbar. Es ist nämlich $\mu = \sum a_n \varepsilon_{x_n}$, wobei jetzt $\varepsilon_{x_n}$ als Maß auf $X$ zu interpretieren ist, ein Maß aus $\mathcal{M}(X)$ mit $\varphi(\mu) = \nu$.

Aus dem soeben Bewiesenen folgt nun sofort, daß die Abbildung $\varphi$ nicht konservativ sein kann. Ist nämlich $\nu = \sum a_n \varepsilon_{x_n}$ ein darstellbares Maß aus $\mathcal{M}(Y)$, so ist (abermals nach [5], p. 146) $\nu(\{x\}) = 0$ für alle $x \neq x_n$, $x \in I$, und $\nu(\{x_n\}) = a_n$ für jedes $n$. Ein diffuses Maß $\nu \neq 0$ aus $\mathcal{M}(Y)$, also beispielsweise das Lebesguesche Maß auf $Y$, kann somit nicht darstellbar sein.

## § 2. Ein Lokalisationsprinzip

Von nun an wird unser Interesse den *stetigen*, konservativen Abbildungen gelten. Es soll hier zunächst gezeigt werden, daß die Frage nach der Darstellbarkeit eines Maßes $\nu \in \mathcal{M}(Y)$ bezüglich einer stetigen Abbildung $\varphi: X \to Y$ ein lokales Problem ist. Zu diesem Zweck führen wir folgenden Begriff ein:

**Definition.** *Es sei $\varphi: X \to Y$ eine stetige Abbildung eines lokal-kompakten Raumes $X$ in einen lokal-kompakten Raum $Y$. Wir nennen ein Maß $\nu \in \mathcal{M}(Y)$ ($\varphi$-)lokal-darstellbar, wenn jeder Punkt $y \in \varphi(X)$ eine offene Umgebung $V$ in $Y$ besitzt mit folgender Eigenschaft: Das durch $\nu$ in $V$ induzierte Maß $\nu_V \in \mathcal{M}(V)$ ist $\varphi'_V$-darstellbar, wenn hierbei $\varphi'_V: U \to V$ diejenige Abbildung bezeichnet, welche auf $U = \overset{-1}{\varphi}(V)$ mit $\varphi$ übereinstimmt.*

Bezeichnen wir mit $j_U: U \to X$ bzw. $j_V: V \to Y$ die kanonische Injektion von $U$ in $X$ bzw. von $V$ in $Y$, so ist also $\varphi'_V$ die einzige Abbildung von $U$ in $V$, für welche das folgende Diagramm kommutativ ist:

$$
\text{(2.1)} \qquad \begin{array}{ccc} U & \xrightarrow{\ \varphi'_V\ } & V \\ {\scriptstyle j_U}\downarrow & & \downarrow{\scriptstyle j_V} \\ X & \xrightarrow[\ \varphi\ ]{} & Y \end{array}
$$

Zu der Definition ist weiter zu bemerken, daß $U$ und $V$ als offene Mengen in lokal-kompakten Räumen selbst wieder lokal-kompakte Räume sind. Ferner

[5]) Vgl. hierzu N. BOURBAKI [8], chap. III, § 7, théorème 3.

ist die Abbildung $\varphi'_V$ zu unterscheiden von der Restriktion $\varphi_U : U \to Y$ von $\varphi$ auf $U$. Es wäre sinnlos, von der $\varphi_U$-Darstellbarkeit von $\nu_V$ zu sprechen, da $\nu_V$ kein Maß auf $Y$ ist. Es gilt jedoch folgender

Hilfssatz 2. *Es sei $\varphi : X \to Y$ eine stetige Abbildung eines lokal-kompakten Raumes $X$ in einen lokal-kompakten Raum $Y$, $T$ ein lokal-kompakter Unterraum von $Y$, $S = \overset{-1}{\varphi}(T)$ und $\nu$ ein positives Maß auf $Y$. Dann sind folgende drei Aussagen gleichwertig:*

(a) *Das durch $\nu$ in $T$ induzierte Maß $\nu_T \in \mathscr{M}(T)$ ist darstellbar bezüglich der Abbildung $\varphi'_T : S \to T$, die auf $S$ mit $\varphi$ übereinstimmt.*

(b) *Das aus $\nu$ durch Tilgung der Masse außerhalb $T$ entstehende Maß $\tau = \chi_T \nu \in \mathscr{M}(Y)$ ist $\varphi$-darstellbar.*

(c) *Das Maß $\tau$ ist darstellbar bezüglich der Restriktion $\varphi_S : S \to Y$ von $\varphi$ auf $S$.*

Beweis. Die Aussagen (a) und (c) sind rein formal betrachtet sinnvoll, da $S$ ein lokal-kompakter Unterraum von $X$ ist. Dies folgt aus dem bekannten Satz, wonach ein Unterraum eines lokal-kompakten Raumes genau dann lokal-kompakt ist, wenn er Durchschnitt einer abgeschlossenen und einer offenen Menge ist. Wir bezeichnen mit $j_S : S \to X$ bzw. $j_T : T \to Y$ die kanonische Injektion von $S$ in $X$ bzw. $T$ in $Y$.

(a) $\Rightarrow$ (b). Es sei $\sigma' \in \mathscr{M}(S)$ ein Maß mit $\varphi'_T(\sigma') = \nu_T$. Wir zeigen zunächst, daß die Abbildung $j_S$ dann $\sigma'$-eigentlich, also das Bildmaß $j_S(\sigma')$ definiert ist. Es sei hierzu $K$ eine in $X$ kompakte Menge. Wegen der Stetigkeit von $\varphi$ ist dann $L = \varphi(K)$ kompakt in $Y$; da $\nu_T$ durch $\nu$ in $T$ induziert wird, ist folglich die Spur $L \cap T$ von $L$ in $T$ im wesentlichen $\nu_T$-integrierbar (vgl. Nr. 1.5). Ebenfalls nach Nr. 1.5 ist dann wegen $\varphi'_T(\sigma') = \nu_T$ auch die Menge $(\varphi'_T)^{-1}(L \cap T)$ $= \overset{-1}{\varphi}(L) \cap S$ im wesentlichen $\sigma'$-integrierbar. Nun ist $K \cap S \subset \overset{-1}{\varphi}(L) \cap S$ und in $S$ abgeschlossen, also $\sigma'$-meßbar. Da $\overset{-1}{\varphi}(L) \cap S$ im wesentlichen $\sigma'$-integrierbar ist, folgt dann endlich nach Nr. 1.3, daß auch $K \cap S$ im wesentlichen $\sigma'$-integrierbar ist, und zwar für jede in $X$ kompakte Menge $K$. Somit ist $\sigma = j_S(\sigma')$ definiert und ein Element von $\mathscr{M}(X)$. Nach dem Beispiel der Nr. 1.6 ist weiter $j_T(\nu_T) = \tau$. Aus $\varphi'_T(\sigma') = \nu_T$ und $j_T(\nu_T) = \tau$ folgt aber nach dem Transitivitätsgesetz für Bildmaße (Nr. 1.6) die Gleichung $(j_T \circ \varphi'_T)(\sigma') = \tau$. Zufolge der Kommutativität des Diagramms (2.1) (mit $S$ bzw. $T$ anstelle von $U$ bzw. $V$) ist $j_T \circ \varphi'_T = \varphi \circ j_S$ und somit $(\varphi \circ j_S)(\sigma') = \tau$. Weil wir festgestellt haben, daß $\sigma = j_S(\sigma')$ definiert ist, liefert eine erneute Anwendung des Transitivitätsgesetzes für Bildmaße die Aussage, daß $\varphi(\sigma)$ definiert und $\varphi(\sigma) = \tau$ ist. Somit ist das Maß $\tau$ $\varphi$-darstellbar.

(b) $\Rightarrow$ (c). Es sei $\sigma \in \mathscr{M}(X)$ ein Maß mit $\varphi(\sigma) = \tau$. Die Menge $Y - T$ ist $\tau$-lokal-vernachlässigbar; aus (1.7) (mit $\mu = \sigma$ und $f = \chi_{Y-T}$) folgt daher, daß $\overset{-1}{\varphi}(Y - T) = X - S$ $\sigma$-lokal-vernachlässigbar ist, also $\sigma$ von $S$ getragen wird. Nach Nr. 1.6 gilt daher $j_S(\sigma_S) = \sigma$ für das von $\sigma$ in $S$ induzierte Maß $\sigma_S$. Da außerdem $\varphi(\sigma) = \tau$ ist, liefert das Transitivitätsgesetz für Bildmaße: $(\varphi \circ j_S)(\sigma_S) = \tau$. Es ist aber $\varphi_S = \varphi \circ j_S$ und somit das Maß $\tau$ $\varphi_S$-darstellbar.

(c) $\Rightarrow$ (a). Es sei $\sigma' \in \mathscr{M}(S)$ ein Maß mit $\varphi_S(\sigma') = \tau$. Dann folgt hieraus $\varphi'_T(\sigma') = \nu_T$ und damit (a). In der Tat: für jede Funktion $t \in \mathscr{K}(T)$ und deren

Fortsetzung $t'$ auf $Y$, die auf $Y - T$ gleich Null ist, gilt nach Definition des induzierten Maßes $\nu_T$ sowie (1.6):

$$\int t d\nu_T = \int t' d\nu = \int t' \chi_T d\nu = \int t' d\tau .$$

Wegen $\varphi_S(\sigma') = \tau$ ist nach (1.7) weiter:

$$\int t' d\tau = \int t' \circ \varphi_S d\sigma' .$$

Für jedes $t \in \mathscr{K}(T)$ ist daher die auf $S$ definierte Funktion $t \circ \varphi'_T = t' \circ \varphi_S$ im wesentlichen $\sigma'$-integrierbar (wegen der Stetigkeit von $t \circ \varphi'_T$ übrigens sogar $\sigma'$-integrierbar, vgl. Nr. 1.3) und es gilt $\int t d\nu_T = \int t \circ \varphi'_T d\sigma'$. Somit ist $\varphi'_T(\sigma') = \nu_T$ und der Hilfssatz damit bewiesen.

Nunmehr sind wir in der Lage, das zu Beginn dieses Paragraphen angekündigte Lokalisationsprinzip zu formulieren und zu beweisen.

**Satz 1** *(Lokalisationsprinzip). Es sei $\varphi : X \to Y$ eine stetige Abbildung eines lokal-kompakten Raumes $X$ in einen lokal-kompakten Raum $Y$. Ein positives Maß $\nu$ auf $Y$ ist dann und nur dann $\varphi$-darstellbar, wenn $\nu$ auf $\varphi(X)$ konzentriert und $\varphi$-lokal-darstellbar ist.*

Beweis. Sei $\nu \in \mathscr{M}(Y)$ ein darstellbares Maß. Wie in Nr. 1.6 bemerkt wurde, ist dann $\nu$ auf $\varphi(X)$ konzentriert. Das Maß ist lokal-darstellbar, denn offenbar leistet für jeden Punkt $y \in \varphi(X)$ die offene Umgebung $V = Y$ das in Definition 2 Verlangte. Zu beweisen ist daher nur die Aussage „dann".

Also sei jetzt $\nu$ ein lokal-darstellbares Maß aus $\mathscr{M}(Y, \varphi(X))$. Für jede in $Y$ lokal-kompakte Menge $Z$ bezeichne $\nu'_Z = \chi_Z \nu$ das aus $\nu$ durch Tilgung der Masse außerhalb $Z$ entstehende Maß aus $\mathscr{M}(Y)$. Mit $\mathfrak{G}$ werde die Menge aller Paare $(G, \mu)$ bezeichnet, in welchen $G$ eine offene Teilmenge von $Y$ und $\mu$ ein Maß aus $\mathscr{M}(X)$ mit $\varphi(\mu) = \nu'_G$ ist. $\mathfrak{G}$ ist dann nicht leer, denn es ist z. B. $(\emptyset, 0) \in \mathfrak{G}$. In $\mathfrak{G}$ definieren wir wie folgt eine Relation $\prec$: Die Relation $(G_1, \mu_1) \prec (G_2, \mu_2)$ bestehe zwischen zwei Elementen von $\mathfrak{G}$ genau dann, wenn $G_1 \subset G_2$ und $\mu_1 \leqq \mu_2$ ist. Offenbar ist $\prec$ eine Ordnungsrelation in $\mathfrak{G}$.

Wir wollen zeigen, daß $\mathfrak{G}$ durch $\prec$ sogar induktiv geordnet ist. Hierzu sei $\mathfrak{F}$ eine totalgeordnete, nicht-leere Teilmenge von $\mathfrak{G}$. Es bezeichne $\mathfrak{F}_Y$ das System aller Mengen $G \subset Y$, zu denen ein $\mu \in \mathscr{M}(X)$ mit $(G, \mu) \in \mathfrak{F}$ existiert; $\mathfrak{F}_{\mathscr{M}}$ sei das System aller Maße $\mu \in \mathscr{M}(X)$, zu denen eine Menge $G$ mit $(G, \mu) \in \mathfrak{F}$ existiert. Dann ist die Menge

$$H = \bigcup_{G \in \mathfrak{F}_Y} G \tag{2.2}$$

offen in $Y$ und $\mathfrak{F}_{\mathscr{M}}$ eine totalgeordnete Teilmenge von $\mathscr{M}(X)$. Weiter ist $\mathfrak{F}_{\mathscr{M}}$ beschränkt, d. h.

$$\sup_{\mu \in \mathfrak{F}_{\mathscr{M}}} \left|\int g d\mu\right| < +\infty \quad \text{für jedes } g \in \mathscr{K}(X) .$$

Zu jedem $g \in \mathscr{K}(X)$ gibt es nämlich ein $f \in \mathscr{K}_+(Y)$ mit $|g| \leqq f \circ \varphi$. Man hat dazu nur zu bemerken, daß das Bild $\varphi(T_g)$ des Trägers $T_g$ von $g$ kompakt ist und somit eine Funktion $f' \in \mathscr{K}_+(Y)$ existiert mit $f'(y) = 1$ für alle $y \in \varphi(T_g)$. Dann leistet $f = \alpha f'$ das Verlangte für hinreichend großes $\alpha > 0$. Ist aber

$f \in \mathcal{K}_+(Y)$ derart gewählt, so folgt die behauptete Beschränktheit von $\mathfrak{F}_{\mathcal{M}}$ aus der Abschätzung:

$$|\int g\,d\mu| \leqq \int |g|\,d\mu \leqq \int f \circ \varphi\,d\mu = \int f\,d\nu'_G \leqq \int f\,d\nu \quad (\mu \in \mathfrak{F}_{\mathcal{M}}).$$

Da $\mathfrak{F}_{\mathcal{M}}$ eine beschränkte und totalgeordnete Teilmenge von $\mathcal{M}(X)$ ist, konvergiert die Filterbasis der Abschnitte von $\mathfrak{F}_{\mathcal{M}}$ gegen ein Maß $\eta \in \mathcal{M}(X)$ im Sinne der vagen Topologie. Nach dem Monotonieprinzip existiert nämlich $\lim\limits_{\mu \in \mathfrak{F}_{\mathcal{M}}} \int g\,d\mu$ für jede Funktion $g \in \mathcal{K}_+(X)$ und damit für jedes $g \in \mathcal{K}(X)$. Folglich ist

$$(2.3) \qquad \int g\,d\eta = \sup_{\mu \in \mathfrak{F}_{\mathcal{M}}} \int g\,d\mu = \lim_{\mu \in \mathfrak{F}_{\mathcal{M}}} \int g\,d\mu \quad \text{für jedes } g \in \mathcal{K}_+(X).$$

Wir behaupten, daß das Paar $(H, \eta)$ in $\mathfrak{G}$ liegt. Um zu zeigen, daß $\varphi(\eta)$ definiert ist, genügt es zu beweisen, daß die Funktion $f \circ \varphi$ für jedes $f \in \mathcal{K}_+(Y)$ $\eta$-integrierbar ist. Es sei hierzu $g \in \mathcal{K}_+(X)$ mit $0 \leqq g \leqq f \circ \varphi$; dann gilt

$$\int g\,d\mu \leqq \int f \circ \varphi\,d\mu = \int f\,d\nu'_G \leqq \int f\,d\nu'_H \quad \text{für jedes } (G, \mu) \in \mathfrak{F}$$

und folglich nach (2.3)

$$\int g\,d\eta = \sup_{\mu \in \mathfrak{F}_{\mathcal{M}}} \int g\,d\mu \leqq \int f\,d\nu'_H.$$

Bei Beachtung von (1.1 a) und (1.2) ergibt dies:

$$N_1^\eta(f \circ \varphi) \leqq \int f\,d\nu'_H \qquad (f \in \mathcal{K}_+(Y)).$$

Da $f \circ \varphi$ als stetige Funktion meßbar und nach der vorausgehenden Ungleichung $N_1^\eta(f \circ \varphi)$ endlich ist, so ist $f \circ \varphi$ $\eta$-integrierbar und

$$(2.4) \qquad \int f \circ \varphi\,d\eta \leqq \int f\,d\nu'_H \qquad (f \in \mathcal{K}_+(Y)).$$

Es ist $\varphi(\eta) = \nu'_H$ und damit $(H, \eta) \in \mathfrak{G}$, wenn gezeigt werden kann, daß in (2.4) das Gleichheitszeichen steht. Aus (2.3) folgt zunächst $\mu \leqq \eta$ für jedes $\mu \in \mathfrak{F}_{\mathcal{M}}$ und hieraus

$$\int f\,\chi_G\,d\nu = \int f\,d\nu'_G = \int f \circ \varphi\,d\mu \leqq \int f \circ \varphi\,d\eta$$

für jedes Paar $(G, \mu) \in \mathfrak{F}$. Also ist

$$(2.5) \qquad \int f\,\chi_G\,d\nu \leqq \int f \circ \varphi\,d\eta \qquad \text{für jedes } G \in \mathfrak{F}_Y \text{ und } f \in \mathcal{K}_+(Y).$$

Nun ist $(f\,\chi_G)_{G \in \mathfrak{F}_Y}$ eine bezüglich $\leqq$ totalgeordnete Familie von $\nu$-integrierbaren, nach unten halbstetigen, nicht-negativen Funktionen mit der oberen Einhüllenden $f\,\chi_H$; nach N. Bourbaki [5], p. 151, gilt daher

$$\int f\,\chi_H\,d\nu = \sup_{G \in \mathfrak{F}_Y} \int f\,\chi_G\,d\nu \qquad (f \in \mathcal{K}_+(Y)).$$

Zusammen mit (2.5) ergibt dies

$$(2.6) \qquad \int f\,d\nu'_H \leqq \int f \circ \varphi\,d\eta \qquad (f \in \mathcal{K}_+(Y)).$$

Aus (2.4) und (2.6) folgt aber, daß in (2.4) das Gleichheitszeichen steht, also $(H, \eta) \in \mathfrak{G}$ ist. Aus (2.2) und (2.3) folgt schließlich, daß $(H, \eta)$ eine obere Schranke von $\mathfrak{F}$ in $\mathfrak{G}$ ist. Somit ist $\mathfrak{G}$ induktiv geordnet.

Das Zornsche Lemma sichert nunmehr die Existenz mindestens eines maximalen Elementes $(G_0, \mu_0)$ in $\mathfrak{G}$. Wir behaupten, daß $\varphi(X) \subset G_0$ ist. Angenommen, dies ist nicht der Fall. Dann gibt es einen Punkt $y \in \varphi(X) \cap (Y - G_0)$ und zu diesem nach Voraussetzung eine offene Umgebung $V$ derart, daß das induzierte Maß $\nu_V$ darstellbar ist bezüglich der Abbildung $\varphi'_V : U \to V$, die auf $U = \overset{-1}{\varphi}(V)$ mit $\varphi$ übereinstimmt. Nach dem Hilfssatz 2 ist dann das Maß $\nu'_V$ $\varphi$-darstellbar, d. h. es existiert ein Maß $\mu \in \mathcal{M}(X)$ mit $(V, \mu) \in \mathfrak{G}$. Wir setzen $T = V \cap (Y - G_0)$; $T$ ist somit ein zu $G_0$ fremder, lokal-kompakter Unterraum von $Y$. Wegen der Stetigkeit von $\varphi$ ist auch $S = \overset{-1}{\varphi}(T)$ lokal-kompakt in $X$. Folglich sind die aus $\mu$ bzw. $\nu$ durch Tilgung der Masse außerhalb $S$ bzw. $T$ entstehenden Maße $\mu'_S$ bzw. $\nu'_T$ definiert. Wir wollen zeigen, daß $\varphi(\mu'_S) = \nu'_T$ ist: Für jede Funktion $f \in \mathcal{K}(Y)$ ist $f\chi_T$ $\nu'_V$-integrierbar, wegen $\varphi(\mu) = \nu'_V$ und (1.7) also $(f\chi_T)\circ\varphi$ im wesentlichen $\mu$-integrierbar und

$$\int (f\chi_T)\circ\varphi \, d\mu = \int f\chi_T d\nu'_V .$$

Da $\nu'_T = \chi_T \nu = \chi_T \chi_V \nu = \chi_T \nu'_V$ und $(f\chi_T)\circ\varphi = (f\circ\varphi)\chi_S$ ist, folgt bei Beachtung von (1.6):

$$\int f\circ\varphi d\mu'_S = \int f d\nu'_T \qquad (f \in \mathcal{K}(Y)), \tag{2.7}$$

d. h. es ist $\varphi(\mu'_S) = \nu'_T$. Nun ist aber $H_0 = G_0 \cup V = G_0 \cup T$ offen in $Y$ und $\eta_0 = \mu_0 + \mu'_S$ ein Maß aus $\mathcal{M}(X)$; nach dem Hilfssatz 1 ist $\varphi(\eta_0)$ definiert und

$$\varphi(\eta_0) = \nu'_{G_0} + \nu'_T = (\chi_{G_0} + \chi_T)\nu = \chi_{H_0}\nu = \nu'_{H_0} .$$

Also ist auch $(H_0, \eta_0)$ ein Element von $\mathfrak{G}$; offenbar ist $G_0 \subset H_0$ und $\mu_0 \leqq \eta_0$, also $(G_0, \mu_0) \prec (H_0, \eta_0)$. Es ist aber $y \in V \cap (Y - G_0)$ und somit $G_0 \neq H_0$, also auch $(G_0, \mu_0) \neq (H_0, \eta_0)$. Dies widerspricht der Maximaleigenschaft von $(G_0, \mu_0)$. Also muß unsere Annahme falsch sein: es ist vielmehr $\varphi(X) \subset G_0$.

Schließlich ist nach Voraussetzung $\nu$ auf $\varphi(X)$ und damit auch auf $G_0$ konzentriert, also $\nu'_{G_0} = \nu$. Aus $\varphi(\mu_0) = \nu'_{G_0}$ folgt daher $\varphi(\mu_0) = \nu$, d. h. es ist $\nu$ $\varphi$-darstellbar. Das war aber zu beweisen.

## § 3. Eigentliche und lokal-eigentliche Abbildungen

**3.1.** *Konservativität lokal-eigentlicher Abbildungen.* — Von einer wichtigen Klasse stetiger Abbildungen soll nun gezeigt werden, daß sie aus lauter konservativen Abbildungen besteht. Wir erinnern zunächst an die Definition dieser Abbildungen:

**Definition 3.** *Es sei $\varphi : X \to Y$ eine stetige Abbildung eines lokal-kompakten Raumes $X$ in einen lokal-kompakten Raum $Y$. Dann heißt $\varphi$ e i g e n t l i c h, wenn das Urbild $\overset{-1}{\varphi}(K)$ jeder in $Y$ kompakten Menge $K$ kompakt in $X$ ist. Die Abbildung $\varphi$ heißt l o k a l - e i g e n t l i c h, wenn jeder Punkt $y \in \varphi(X)$ eine kompakte Umgebung $W$ in $Y$ besitzt, deren Urbild $\overset{-1}{\varphi}(W)$ kompakt in $X$ ist.*

Der Begriff der eigentlichen Abbildung findet sich bei N. BOURBAKI [7], der der lokal-eigentlichen Abbildung bei R. REMMERT [17].

Jede eigentliche Abbildung ist offensichtlich lokal-eigentlich. Dagegen gilt: *Eine lokal-eigentliche stetige Abbildung $\varphi : X \to Y$ ist dann und nur dann eigentlich, wenn $\varphi(X)$ in $Y$ abgeschlossen ist.* Wenn nämlich $\varphi(X)$ in $Y$ abgeschlossen und $\varphi$ lokal-eigentlich ist, so zeigt man mit Hilfe des Borelschen Überdeckungssatzes leicht, daß $\varphi$ eigentlich ist. Die Umkehrung folgt aus einem bekannten Satz (N. BOURBAKI [7], p. 99, prop. 13), wonach jede eigentliche Abbildung abgeschlossen ist, also abgeschlossene Mengen in abgeschlossene

Mengen überführt. Für eine stetige Abbildung $\varphi$ von $X$ *auf* $Y$ sind somit die Begriffe „eigentlich" und „lokal-eigentlich" gleichbedeutend. Eine stetige Abbildung eines kompakten Raumes in einen lokal-kompakten Raum ist stets eigentlich.

Über diese beiden Abbildungstypen beweisen wir zunächst folgenden

Hilfssatz 3. *Für jede stetige Abbildung* $\varphi: X \to Y$ *eines lokal-kompakten Raumes* $X$ *in einen lokal-kompakten Raum* $Y$ *gilt:*

(a) *Die Abbildung* $\varphi$ *ist dann und nur dann eigentlich, wenn aus* $f \in \mathcal{K}(Y)$ *stets* $f \circ \varphi \in \mathcal{K}(X)$ *folgt.*

(b) *Die Abbildung* $\varphi$ *ist dann und nur dann lokal-eigentlich, wenn jeder Punkt* $y \in \varphi(X)$ *eine offene Umgebung* $V$ *in* $Y$ *besitzt, für welche die Abbildung* $\varphi'_V: U \to V$ *eigentlich ist, welche auf* $U = \overset{-1}{\varphi}(V)$ *mit* $\varphi$ *übereinstimmt.*

Beweis. Zu (a): Trivial ist, daß für eine eigentliche Abbildung die angegebene Bedingung erfüllt ist. Ist umgekehrt diese Bedingung erfüllt, so ist $\varphi$ eigentlich, da zu jeder kompakten Menge $K \subset Y$ eine Funktion $h \in \mathcal{K}(Y)$ existiert mit $0 \leqq h \leqq 1$ und $h(y) = 1$ für alle $y \in K$.

Zu (b): Es sei $\varphi$ lokal-eigentlich. Dann gibt es zu jedem $y \in \varphi(X)$ eine kompakte Umgebung $W$ von $y$ mit kompaktem Urbild $\overset{-1}{\varphi}(W)$. Der offene Kern $V$ von $W$ ist dann eine offene Umgebung von $y$ in $Y$, die das Verlangte leistet: Für jede in $V$ kompakte Menge $K$ ist aus Stetigkeitsgründen $(\varphi'_V)^{-1}(K) = \overset{-1}{\varphi}(K)$ abgeschlossen, also wegen $\overset{-1}{\varphi}(K) \subset \overset{-1}{\varphi}(W)$ kompakt in $X$ und damit auch in $U$. Ist umgekehrt $V$ eine offene Umgebung von $y \in \varphi(X)$, für welche die Abbildung $\varphi'_V$ eigentlich ist, so ist für jede kompakte Umgebung $W$ von $y$ mit $W \subset V$ deren Urbild $\overset{-1}{\varphi}(W) = (\varphi'_V)^{-1}(W)$ kompakt in $U$ und damit in $X$. Zufolge der lokalen Kompaktheit von $X$ existiert aber mindestens ein solches $W$.[5a])

---

[5a]) **Zusatz bei der Korrektur.** Aus Hilfssatz 3 ergibt sich die folgende Kennzeichnung der lokal-eigentlichen Abbildungen: *Eine stetige Abbildung* $\varphi: X \to Y$ *eines lokal-kompakten Raumes* $X$ *in einen lokal-kompakten Raum* $Y$ *ist dann und nur dann lokal-eigentlich, wenn sie eine Faktorisierung* $\varphi = j \circ \varphi^*$ *besitzt, in welcher* $\varphi^*: X \to Y^*$ *eine eigentliche Abbildung von* $X$ *auf einen lokal-kompakten Unterraum* $Y^*$ *von* $Y$ *und* $j: Y^* \to Y$ *die kanonische Injektion von* $Y^*$ *in* $Y$ *ist.*

Beweis. „*Nur dann*": Ist $\varphi: X \to Y$ lokal-eigentlich, so ist offenbar das Urbild jeder kompakten Teilmenge von $Y^* = \varphi(X)$ kompakt in $X$. Zu zeigen ist daher nur, daß (der separierte Raum) $Y^*$ lokal-kompakt ist. Nach Hilfssatz 3 existiert zu einem beliebigen Punkt $y \in Y^*$ eine in $Y$ offene Umgebung $V$, für welche die zugehörige Abbildung $\varphi'_V: U = \overset{-1}{\varphi}(V) \to V$ eigentlich ist. Da eigentliche Abbildungen abgeschlossen sind, so ist $V' = V \cap Y^* = \varphi'_V(U)$ abgeschlossen in $V$, also lokal-kompakt. Folglich besitzt $y$ in $V'$ eine kompakte Umgebung $W$. Es ist aber $W$ auch Umgebung von $y$ in $Y^*$. In der Tat: es gibt eine in $Y$ offene Menge $G$ mit $y \in G$ und $G \cap V' \subset W$; die Menge $G \cap V$ ist offen in $Y$ und es gilt $G \cap V \cap Y^* = G \cap V'$. Also besitzt jeder Punkt $y \in Y^*$ eine kompakte Umgebung in $Y^*$, d. h. $Y^*$ ist lokal-kompakt. — „*Dann*": Es sei $\varphi = j \circ \varphi^*$ eine Faktorisierung mit den genannten Eigenschaften. Da $Y^* = \varphi^*(X) = \varphi(X)$ lokal-kompakt ist, gibt es in $Y$ eine abgeschlossene Menge $F$ und eine offene Menge $G$ mit $Y^* = F \cap G$. Jeder Punkt $y \in Y^*$ besitzt in $Y$ eine kompakte Umgebung $V$ mit $V \subset G$. Für diese ist $V \cap Y^* = V \cap F$ kompakt in $Y^*$, also $\overset{-1}{\varphi}(V) = \varphi^{*-1}(V \cap Y^*)$ kompakt in $X$. Somit ist $\varphi$ lokal-eigentlich. — Aus dieser Kennzeichnung folgt übrigens, daß sich im Beweis des folgenden Satzes 2 das Lokalisationsprinzip (Satz 1) noch vermeiden ließe.

Ein Analogon zu der Aussage (a) des Hilfssatzes 3 haben wir in Nr. 1.5 kennengelernt, wonach für eine Abbildung $\varphi: X \to Y$ eines lokal-kompakten Raumes in einen ebensolchen Raum das Bild $\varphi(\mu)$ eines Maßes $\mu \in \mathcal{M}(X)$ genau dann definiert ist, wenn die Abbildung $\mu$-eigentlich ist. Hieraus oder aus Hilfssatz 3 folgt übrigens, daß für eine eigentliche (stetige) Abbildung $\varphi$ das Bild $\varphi(\mu)$ für *jedes* $\mu \in \mathcal{M}(X)$ definiert ist. Im Sinne der Bezeichnungen von 1.5 ist also $\mathcal{E}(\varphi) = \mathcal{M}(X)$ und es definiert $\varphi$ eine Abbildung von $\mathcal{M}(X)$ in $\mathcal{M}(Y, \varphi(X))$. Daß sogar eine Abbildung auf $\mathcal{M}(Y, \varphi(X))$ vorliegt, lehrt der folgende Satz, aus dem die Bedeutung der eigentlichen und allgemeiner der lokal-eigentlichen Abbildungen für unser Problem erhellt.

**Satz 2.** *Jede lokal-eigentliche Abbildung $\varphi: X \to Y$ eines lokal-kompakten Raumes $X$ in einen lokal-kompakten Raum $Y$ ist konservativ.*

Beweis. Es sei $\nu$ ein Maß aus $\mathcal{M}(Y, \varphi(X))$ und $y$ ein Punkt aus $\varphi(X)$. Nach der Behauptung (b) des Hilfssatzes 3 gibt es eine offene Umgebung $V$ von $y$ in $Y$, für welche die zugehörige Abbildung $\varphi'_V: U \to V$ mit $U = \overset{-1}{\varphi}(V)$ eigentlich ist. Da $\nu$ auf $\varphi(X)$ konzentriert ist, so ist das von $\nu$ in $V$ induzierte Maß $\nu_V$ auf $V \cap \varphi(X) = \varphi'_V(U)$ konzentriert ([6], p. 88, cor. 2). Aus dem Lokalisationsprinzip (Satz 1) folgt daher, daß es genügt, den Satz für eigentliche Abbildungen zu beweisen.

Die Abbildung $\varphi$ werde also von nun an als eigentlich vorausgesetzt. Nach wie vor sei $\nu$ ein Maß aus $\mathcal{M}(Y, \varphi(X))$. Der Beweis der Darstellbarkeit von $\nu$ beruht auf dem folgenden allgemeinen Satz über die Fortsetzbarkeit positiver stetiger Linearformen[6]):

*Es sei $E$ ein geordneter lokal-konvexer Vektorraum (über* **R***), $P$ der konvexe Kegel aller Elemente $g \geqq 0$ aus $E$, $E_0$ ein linearer Unterraum von $E$ und $\mu_0$ eine auf $E_0$ definierte Linearform. Eine $\mu_0$ auf ganz $E$ fortsetzende, positive, stetige Linearform $\mu$ existiert dann und nur dann, wenn die Menge $\mu_0(E_0 \cap (V + P))$ reeller Zahlen für mindestens eine Umgebung $V$ des Nullvektors $0 \in E$ nach unten beschränkt ist.* ($V + P$ bezeichnet hierbei die Menge aller Vektoren $v + p$ mit $v \in V$ und $p \in P$.)

In der nun folgenden Anwendung dieses Satzes ist $E$ der geordnete Vektorraum $\mathcal{K}(X)$, $P = \mathcal{K}_+(X)$, $E_0$ die Menge aller Funktionen $f \circ \varphi$ mit $f \in \mathcal{K}(Y)$, die nach Hilfssatz 3 sämtlich in $E$ liegen, und $\mu_0$ die auf $E_0$ definierte Linearform $\mu_0(f \circ \varphi) = \int f \, d\nu$. Es genügt zu zeigen, daß $\mu_0$ zu einer positiven Linearform $\mu$ auf $E$ fortgesetzt werden kann. Dann ist nämlich $\mu$ ein positives Maß auf $X$ mit $\varphi(\mu) = \nu$. Um den Fortsetzungssatz anwenden zu können, versehen wir noch $E$ mit einer lokal-konvexen Topologie $\mathcal{T}$ und zeigen, daß $\mu$ sogar $\mathcal{T}$-stetig gewählt werden kann.

$\mathcal{T}$ sei die wie folgt definierte, sog. starke Topologie[7]) auf $E$: Für jede kompakte Menge $K$ in $X$ sei $E_K$ die Menge aller Funktionen $g \in E$, die in $X - K$ gleich Null sind. $E_K$ ist ein linearer Unterraum von $E$; er werde mit

---

[6]) Vgl. H. BAUER [3]. Unabhängig vom Verf. und etwa gleichzeitig wurde der Satz auch von I. NAMIOKA [14] gefunden. Vgl. ferner G. AUMANN [1].

[7]) Vgl. N. BOURBAKI [5], p. 64, exercice 1.

der Topologie $\mathcal{T}_K$ der gleichmäßigen Konvergenz auf $X$ versehen. Die Topologie auf $E$ ist dann der induktive Limes der Topologien $\mathcal{T}_K$[8]). Da die Vereinigung aller Räume $E_K$ gleich $E$ ist, erhält man nach einer Bemerkung von N. Bourbaki [9], p. 62, ein Fundamentalsystem von Umgebungen des Nullvektors von $E$ bezüglich $\mathcal{T}$, indem man alle Familien $(V_K)$ betrachtet, in welchen jeder kompakten Menge $K \subset X$ eine symmetrische Umgebung $V_K$ des Nullvektors von $E_K$ bezüglich $\mathcal{T}_K$ zugeordnet ist, und die konvexe Hülle der Vereinigungsmenge aller $V_K$ einer jeden solchen Familie bildet.

Die Existenz einer Fortsetzung $\mu$ ergibt sich nun wie folgt. Jeder kompakten Menge $K \subset X$ wird eine Funktion $f_K \in \mathcal{K}(Y)$ zugeordnet, und zwar derart, daß $0 \leqq f_K \leqq 1$ und $f_K(y) = 1$ ist für alle $y \in \varphi(K)$. Die Kompaktheit von $\varphi(K)$ sichert die Existenz einer solchen Funktion. Mittels $f_K$ wird weiter jeder kompakten Menge $K \subset X$ eine Zahl $\alpha_K > 0$ zugeordnet: und zwar sei $\alpha_K = 1$, wenn $\int f_K d\nu = 0$ ist, und $\alpha_K = (\int f_K d\nu)^{-1}$, wenn $\int f_K d\nu > 0$ ist. Schließlich bezeichne $V_K$ die Menge aller $g \in E_K$ mit $|g(x)| \leqq \alpha_K$ für alle $x \in X$. Für jedes $K$ ist dann $V_K$ eine symmetrische Umgebung der Null in $E_K$ bezüglich $\mathcal{T}_K$ und folglich die konvexe Hülle $V$ der Vereinigungsmenge aller dieser $V_K$ eine Umgebung der Null in $E$ bezüglich $\mathcal{T}$. Nach dem allgemeinen Fortsetzungssatz genügt es zu zeigen, daß $\mu_0$ auf der Menge $E_0 \cap (V + P)$ nach unten beschränkt ist. Jede Funktion aus $E_0$ ist von der Form $f \circ \varphi$ mit $f \in \mathcal{K}(Y)$. Zu jeder Funktion $f \circ \varphi \in E_0 \cap (V + P)$ gibt es endlich viele kompakte Mengen $K_1, \ldots, K_n$ in $X$, Funktionen $v_1, \ldots, v_n$ aus $V_{K_1}, \ldots, V_{K_n}$, reelle Zahlen $\lambda_1 \geqq 0, \ldots, \lambda_n \geqq 0$ mit $\Sigma \lambda_i = 1$ und eine Funktion $p \in P$ derart, daß

$$f \circ \varphi = \sum_{i=1}^{n} \lambda_i v_i + p \tag{3.1}$$

ist. Wegen $v_i \in V_{K_i}$ ist $v_i \geqq -\alpha_{K_i}$ und $v_i(x) = 0$ für alle $x \in X - K_i$, also

$$v_i \geqq -\alpha_{K_i} f_{K_i} \circ \varphi \qquad (i = 1, \ldots, n)\,. \tag{3.2}$$

Wegen $p \geqq 0$ folgt daher aus (3.1) und (3.2):

$$f \circ \varphi \geqq -\sum_{i=1}^{n} \lambda_i \alpha_{K_i} f_{K_i} \circ \varphi\,. \tag{3.3}$$

Da das Maß $\nu$ auf $\varphi(X)$ konzentriert und somit die Menge $Y - \varphi(X)$ $\nu$-lokal-vernachlässigbar ist, folgt weiter:

$$f \geqq -\sum_{i=1}^{n} \lambda_i \alpha_{K_i} f_{K_i} \quad \text{lokal fast-überall bez. } \nu\,. \tag{3.4}$$

Integration bezüglich $\nu$ ergibt daher

$$\mu_0(f \circ \varphi) = \int f d\nu \geqq -\sum_{i=1}^{n} \lambda_i \alpha_{K_i} \int f_{K_i} d\nu \geqq -\sum_{i=1}^{n} \lambda_i = -1\,. \tag{3.5}$$

Also ist $\mu_0$ auf $E_0 \cap (V + P)$ nach unten durch die Zahl $-1$ beschränkt.

Damit ist Satz 2 bewiesen.

---

8) Vgl. N. Bourbaki [9], chap. II, § 2, Nr. 4.

Bemerkungen. 1. Nach einer Mitteilung von Herrn CHOQUET findet sich der Satz 2 für den Spezialfall einer eigentlichen Abbildung von $X$ auf $Y$, sowie unter einschränkenden Voraussetzungen über die lokal-kompakten Räume $X$ und $Y$ in einer noch unveröffentlichten Arbeit von G. CHOQUET und J. DENY über potentialtheoretische Fragen.

2. Für den Beweis des Satzes 2 hätte es genügt, den Vektorraum $E = \mathscr{K}(X)$ mit der feinsten lokal-konvexen Topologie zu versehen. Die von uns verwendete starke Topologie $\mathscr{T}$ ist aber in diesem Zusammenhang natürlicher, da bekanntlich die Radonschen Maße auf $X$ gerade die $\mathscr{T}$-stetigen Linearformen auf $E$ sind. (Vgl. das Zitat in [7].)

**3.2.** *Fortsetzung stetiger Abbildungen zu eigentlichen Abbildungen.* — Die Nr. 1.7 hat die Existenz stetiger, nicht-konservativer Abbildungen ergeben. Der folgende Satz 3 (zusammen mit Satz 2) zeigt, daß aber immerhin jede stetige Abbildung eines lokal-kompakten Raumes in einen ebensolchen Raum wenigstens die Restriktion einer stetigen konservativen, genauer sogar einer eigentlichen Abbildung ist. Wir behaupten nämlich:

**Satz 3.** *Es sei $\varphi: X \to Y$ eine stetige Abbildung eines lokal-kompakten Raumes $X$ in einen lokal-kompakten Raum $Y$. Dann existiert ein lokal-kompakter Raum $\tilde{X}$ mit folgenden Eigenschaften:*

($E_1$) *Der Raum $X$ ist ein Unterraum von $\tilde{X}$ und liegt dicht in $\tilde{X}$.*

($E_2$) *Die Abbildung $\varphi$ kann (auf genau eine Weise) zu einer stetigen Abbildung $\tilde{\varphi}: \tilde{X} \to Y$ fortgesetzt werden.*

($E_3$) *Die Abbildung $\tilde{\varphi}$ ist eigentlich.*

($E_4$) *Für je zwei Punkte $x_1, x_2 \in \tilde{X} - X$ mit $x_1 \neq x_2$ ist $\tilde{\varphi}(x_1) \neq \tilde{\varphi}(x_2)$.*

**Zusatz.** *Ein lokal-kompakter Raum $\tilde{X}$ ist durch die Eigenschaften ($E_1$) — ($E_4$) bis auf Homöomorphien eindeutig bestimmt, die den Raum $X$ punktweise festlassen.*

Der Zusatz berechtigt uns im folgenden von *dem* zu einer Abbildung $\varphi$ gehörigen Raum $\tilde{X}$ zu sprechen. Insbesondere ist die Identität der Räume $X$ und $\tilde{X}$ gleichbedeutend mit der Eigentlichkeit der Abbildung $\varphi$.

Da es sich bei Satz 3 und seinem Zusatz um einen rein topologischen Sachverhalt handelt, werden wir den Beweis hierfür erst in Nr. 3.3 erbringen und uns zunächst den maßtheoretischen Folgerungen zuwenden. An topologischen Eigenschaften von $\tilde{X}$ erwähnen wir nur, daß der Raum $X$ in $\tilde{X}$ *offen* ist. Es ist nämlich $X$ als lokal-kompakter Unterraum von $\tilde{X}$ von der Form $X = \tilde{F} \cap \tilde{G}$, wobei $\tilde{F}$ in $\tilde{X}$ abgeschlossen und $\tilde{G}$ in $\tilde{X}$ offen ist. Da $X$ in $\tilde{X}$ dicht liegt, muß $\tilde{F} = \tilde{X}$ und somit $X = \tilde{G}$ sein.

Zunächst bemerken wir, daß für eine stetige Abbildung $\varphi: X \to Y$ zwar nicht jedes Maß $\nu \in \mathscr{M}(Y, \varphi(X))$ $\varphi$-darstellbar zu sein braucht, aber aus Satz 2 folgt:

*Jedes auf $\varphi(X)$ konzentrierte positive Maß auf $Y$ ist darstellbar bezüglich der Abbildung $\tilde{\varphi}: \tilde{X} \to Y$.*

Weiter können wir eine neue Interpretation des bereits beobachteten Phänomens geben, wonach nur für gewisse Maße $\mu \in \mathcal{M}(X)$ das Bild $\varphi(\mu) \in \mathcal{M}(Y)$ definiert ist:

**Satz 4.** *Für jede stetige Abbildung $\varphi: X \to Y$ eines lokal-kompakten Raumes $X$ in einen lokal-kompakten Raum $Y$ und jedes Maß $\mu \in \mathcal{M}(X)$ sind folgende Aussagen gleichwertig:*

(a) *Das Bildmaß $\varphi(\mu)$ ist definiert.*

(b) *Das Maß $\mu$ wird durch ein Maß $\tilde{\mu} \in \mathcal{M}(\tilde{X})$ in $X$ induziert.*

Beweis. Im Hinblick auf das in Nr. 1.5 Gesagte läßt sich die Behauptung auch so aussprechen: Die Abbildung $\varphi$ ist dann und nur dann $\mu$-eigentlich, wenn die kanonische Injektion $j: X \to \tilde{X}$ $\mu$-eigentlich ist. Wir beweisen die Behauptung in dieser Formulierung.

Sei also zunächst $\varphi$ $\mu$-eigentlich. Dann ist für jede in $\tilde{X}$ kompakte Menge $\tilde{K}$ und deren kompaktes Bild $L = \tilde{\varphi}(\tilde{K})$ die Urbildmenge $\overset{-1}{\varphi}(L)$ im wesentlichen $\mu$-integrierbar. Nun ist aber $X \cap \tilde{K}$ abgeschlossen in $X$ und somit $\mu$-meßbar; ferner ist $X \cap \tilde{K} \subset \overset{-1}{\varphi}(L)$. Nach Nr. 1.3 ist daher mit $\overset{-1}{\varphi}(L)$ auch $X \cap \tilde{K} = \overset{-1}{j}(\tilde{K})$ im wesentlichen $\mu$-integrierbar, also die Injektion $j$ $\mu$-eigentlich. — Nun sei umgekehrt $j$ $\mu$-eigentlich. Die eigentliche Abbildung $\tilde{\varphi}$ ist für jedes Maß $\tilde{\nu} \in \mathcal{M}(\tilde{X})$ $\tilde{\nu}$-eigentlich. Nach dem Transitivitätsgesetz für Bildmaße ist daher auch $\varphi = \tilde{\varphi} \circ j$ $\mu$-eigentlich.

**Korollar.** *Ein Maß $\nu \in \mathcal{M}(Y)$ ist dann und nur dann $\varphi$-darstellbar, wenn ein auf $X$ konzentriertes Maß $\tilde{\mu} \in \mathcal{M}(\tilde{X})$ existiert mit $\tilde{\varphi}(\tilde{\mu}) = \nu$.*

Beweis. Gilt $\varphi(\mu) = \nu$ für ein $\mu \in \mathcal{M}(X)$, so folgt die Existenz eines solchen Maßes $\tilde{\mu}$ aus Satz 4 und dem Transitivitätsgesetz für Bildmaße. Dann ist nämlich $\tilde{\mu} = j(\mu)$ definiert für die kanonische Injektion $j: X \to \tilde{X}$ und ein Maß aus $\mathcal{M}(\tilde{X}, X)$; wegen $\varphi = \tilde{\varphi} \circ j$ ist ferner $\nu = \tilde{\varphi}(\tilde{\mu})$. — Ist umgekehrt $\tilde{\mu}$ ein Maß aus $\mathcal{M}(\tilde{X}, X)$ mit $\varphi(\tilde{\mu}) = \nu$ und $\mu$ das von $\tilde{\mu}$ in $X$ induzierte Maß, so ist abermals nach Satz 4 $\varphi(\mu)$ definiert und $\varphi(\mu) = \nu$ wegen $\varphi = \tilde{\varphi} \circ j$ und $j(\mu) = \tilde{\mu}$.

**3.3.** *Beweis des Satzes* 3. — Zur Vorbereitung der nachfolgenden Konstruktion des Raumes $\tilde{X}$ beweisen wir zunächst:

Hilfssatz 4. *Für eine stetige Abbildung $\varphi: X \to Y$ eines lokal-kompakten Raumes $X$ in einen lokal-kompakten Raum $Y$ sei $\tilde{X}$ ein lokal-kompakter Raum mit den Eigenschaften $(E_1)$—$(E_4)$. Dann wird der Unterraum $\tilde{X} - X$ von $\tilde{X}$ durch die stetige Fortsetzung $\tilde{\varphi}: \tilde{X} \to Y$ von $\varphi$ homöomorph auf den Unterraum $Y_0$ von $Y$ abgebildet, der aus allen Punkten $y \in Y$ besteht mit der Eigenschaft, daß $\overset{-1}{\varphi}(V)$ für keine Umgebung $V$ von $y$ in $Y$ kompakt in $X$ ist.*

Beweis. Wir setzen $Y_0' = \tilde{\varphi}(\tilde{X} - X)$ und bezeichnen mit $\psi: \tilde{X} - X \to Y_0'$ die durch $\tilde{\varphi}$ vermittelte stetige Abbildung von $\tilde{X} - X$ auf $Y_0'$. Der Unterraum $\tilde{X} - X$ von $\tilde{X}$ ist abgeschlossen, insbesondere also lokal-kompakt; die Abbildung $\tilde{\varphi}$ ist eigentlich und daher abgeschlossen. Also ist $Y_0'$ in $Y$ abgeschlossen und somit selbst lokal-kompakt. Jede in $Y_0'$ kompakte Menge $K$ ist kompakt in $Y$ und daher ist $\overset{-1}{\psi}(K) = (\tilde{X} - X) \cap \tilde{\varphi}^{-1}(K)$ kompakt in $\tilde{X} - X$. Die Abbildung $\psi$ ist also ebenfalls eigentlich und folglich abgeschlossen. Da ferner $\psi$ nach $(E_4)$ eineindeutig ist, wird der Raum $\tilde{X} - X$ durch $\psi$ homöomorph auf $Y_0'$ abgebildet.

Nach diesem ersten Beweisschritt fehlt nur noch der Beweis der Gleichheit $Y_0' = Y_0$. Wäre für eine Umgebung $V$ eines Punktes $y \in Y_0'$ die Urbildmenge $\overset{-1}{\varphi}(V)$ kompakt in $X$ und damit auch in $\tilde{X}$, so gäbe es wegen der Stetigkeit von $\tilde{\varphi}$ eine zu $\overset{-1}{\varphi}(V)$ fremde Umgebung $\tilde{U}$ von $x = \overset{-1}{\psi}(y)$ in $\tilde{X}$ mit $\tilde{\varphi}(\tilde{U}) \subset V$. Da $X$ in $\tilde{X}$ dicht liegt, gäbe es einen Punkt $x' \in X \cap \tilde{U}$. Für diesen wäre dann $\tilde{\varphi}(x') = \varphi(x') \in V$, also $x' \in \overset{-1}{\varphi}(V)$ im Widerspruch zu $\tilde{U} \cap \overset{-1}{\psi}(V) = \theta$. Es ist also $\overset{-1}{\varphi}(V)$ für keine Umgebung $V$ in $Y$ eines Punktes $y \in Y_0'$ kompakt in $X$, d. h. es ist $Y_0' \subset Y_0$. — Da $Y_0'$ in $Y$ abgeschlossen ist, gibt es zu jedem Punkt $y \in Y - Y_0'$ eine zu $Y_0'$ fremde, kompakte Umgebung $V$ von $y$ in $Y$. Wegen $V \cap Y_0' = \theta$ ist $\tilde{\varphi}^{-1}(V) \subset X$, also $\overset{-1}{\varphi}(V) = \tilde{\varphi}^{-1}(V)$ kompakt in $X$ und somit $y \notin Y_0$. Also ist auch $Y - Y_0' \subset Y - Y_0$ und damit die Gleichheit $Y_0' = Y_0$ bewiesen.

Nun erst erbringen wir den

Beweis des Satzes 3. Wir bezeichnen hierzu wieder mit $Y_0$ den in Hilfssatz 4 definierten Unterraum von $Y$ bezüglich der gegebenen Abbildung $\varphi$ und bemerken zunächst, daß $Y_0$ in $Y$ abgeschlossen ist. Zu jeder Umgebung $V$ von $y \in \overline{Y}_0$ in $Y$ existiert nämlich ein $y_0 \in V \cap Y_0$ derart, daß $V$ auch Umgebung von $y_0$ ist. Daher muß $y \in Y_0$ sein.

Es sei weiter $\psi: X_0 \to Y_0$ eine Homöomorphie eines zu $X$ punktfremden Raumes $X_0$ auf den Raum $Y_0$. Wir setzen

$$\tilde{X}_0 = X \cup X_0 \tag{3.6}$$

und definieren die folgende Abbildung $\tilde{\varphi}_0: \tilde{X}_0 \to Y$:

$$\tilde{\varphi}_0(x) = \begin{cases} \varphi(x), \text{ für jedes } x \in X\,; \\ \psi(x), \text{ für jedes } x \in X_0\,. \end{cases} \tag{3.7}$$

Ferner sei $\tilde{\mathfrak{B}}_0$ das System aller Mengen $\tilde{U} \subset \tilde{X}_0$ der folgenden Gestalt:

$$\tilde{U} = [U \cup \overset{-1}{\varphi}(V) \cup \overset{-1}{\psi}(V \cap Y_0)] \cap (\tilde{X}_0 - K)\,, \tag{3.8}$$

wobei $U$ bzw. $K$ eine in $X$ offene bzw. kompakte Menge und $V$ eine in $Y$ offene Menge ist.

Wir überlassen es dem Leser nachzuprüfen, daß $\tilde{\mathfrak{B}}_0$ mit je endlich vielen Mengen auch deren Durchschnitt und insbesondere auch die leere Menge enthält. Wir versehen $\tilde{X}_0$ mit der Topologie, deren System aller offenen Mengen das System $\tilde{\mathfrak{O}}_0$ aller Vereinigungen von Mengen aus $\tilde{\mathfrak{B}}_0$ ist. Aus der speziellen Gestalt der Mengen von $\tilde{\mathfrak{O}}_0$ folgt, daß $X$ ein offener Unterraum von $\tilde{X}_0$ ist. Ferner liegt $X$ in $\tilde{X}_0$ dicht. Hierzu genügt es zu zeigen, daß aus $\tilde{U} \cap X = \theta$ folgt $\tilde{U} = \theta$ für jede Menge $\tilde{U}$ der Gestalt (3.8). Nun ist

$$\tilde{U} \cap X = [U \cup \overset{-1}{\varphi}(V)] \cap (X - K)\,,$$

wegen $\tilde{U} \cap X = \theta$ also $\overset{-1}{\varphi}(V) \subset U \cup \overset{-1}{\varphi}(V) \subset K$. Hieraus folgt $V \cap Y_0 = \theta$, da jeder Punkt $y \in V$ eine kompakte Umgebung $W \subset V$ besitzt mit $\overset{-1}{\varphi}(W) \subset \overset{-1}{\varphi}(V) \subset K$, also mit kompaktem Urbild. Dann aber ist $U \cup \overset{-1}{\varphi}(V) \cup \overset{-1}{\psi}(V \cap Y_0) = U \cup \overset{-1}{\varphi}(V)$ fremd zu $K$ und somit tatsächlich $\tilde{U} = \theta$. Die Abbildung $\tilde{\varphi}_0 : \tilde{X}_0 \to Y$ ist eine stetige Fortsetzung von $\varphi$ auf $\tilde{X}_0$, denn nach (3.7) gilt für jede in $Y$ offene Menge $V$:

$$\tilde{\varphi}_0^{-1}(V) = \overset{-1}{\varphi}(V) \cup \overset{-1}{\psi}(V \cap Y_0)\,.$$

Schließlich ist der Raum $\tilde{X}_0$ separiert: Es seien hierzu $x_1$ und $x_2$ zwei verschiedene Punkte aus $\tilde{X}_0$. Sie besitzen fremde Umgebungen in $\tilde{X}_0$, falls sie beide in $X$ liegen, da $X$ separiert und offener Unterraum von $\tilde{X}_0$ ist, oder falls genau einer von ihnen in $X$ liegt, da $X$ lokal-kompakt und nach (3.8) die Komplemente (in $\tilde{X}_0$) kompakter Teilmengen von $X$ offen in $\tilde{X}_0$ sind. Liegen jedoch beide Punkte in $X_0$, so ist nach (3.7) $\tilde{\varphi}_0(x_1) \neq \tilde{\varphi}_0(x_2)$ und die Existenz fremder Umgebungen folgt aus der Separiertheit von $Y$ und der Stetigkeit von $\tilde{\varphi}_0$.

Somit ist $\tilde{X}_0$ ein Raum mit den gewünschten Eigenschaften, wenn gezeigt werden kann, daß das Urbild $\tilde{\varphi}_0^{-1}(L)$ einer jeden in $Y$ kompakten Menge $L$ kompakt in $\tilde{X}_0$ ist. Wegen der lokalen Kompaktheit von $Y$ und der Stetigkeit von $\tilde{\varphi}_0$ ist dann nämlich $\tilde{X}_0$ lokal-kompakt und $\tilde{\varphi}_0$ eigentlich. Also sei $L$ kompakt in $Y$. Dann ist

$$\tilde{\varphi}_0^{-1}(L) = \overset{-1}{\varphi}(L) \cup \overset{-1}{\psi}(L \cap Y_0)\,. \tag{3.9}$$

Es sei $\mathfrak{U}$ ein Ultrafilter in $\tilde{\varphi}_0^{-1}(L)$; zu zeigen ist dessen Konvergenz in $\tilde{\varphi}_0^{-1}(L)$. Es genügt sogar, nur die Konvergenz der Filterbasis $\mathfrak{U}$ in $\tilde{X}_0$ zu beweisen, da $\tilde{\varphi}_0^{-1}(L)$ wegen der Stetigkeit von $\tilde{\varphi}_0$ abgeschlossen ist. Aus (3.9) folgt, daß entweder $\overset{-1}{\psi}(L \cap Y_0)$ oder $\overset{-1}{\varphi}(L)$ ein Element von $\mathfrak{U}$ ist. Da $\psi$ eine Homöomorphie, $L$ kompakt und $Y_0$ in $Y$ abgeschlossen ist, so ist $\overset{-1}{\psi}(L \cap Y_0)$ kompakt in $\tilde{X}_0$. Also konvergiert im ersten Falle, wo diese Menge zu $\mathfrak{U}$ gehört, der Ultrafilter $\mathfrak{U}$ gegen einen Punkt von $\overset{-1}{\psi}(L \cap Y_0)$. Zu behandeln ist daher nur noch der zweite Fall: $\overset{-1}{\varphi}(L) \in \mathfrak{U}$. Hier induziert $\mathfrak{U}$ einen Ultrafilter $\mathfrak{U}'$ in $L' = \overset{-1}{\varphi}(L)$ mit $\mathfrak{U}' \subset \mathfrak{U}$; $\mathfrak{B} = \varphi(\mathfrak{U}')$ ist die Basis eines Ultrafilters in $L$. Da $L$ kompakt ist, konvergiert somit $\mathfrak{B}$ gegen einen Punkt $y \in L$. Wiederum sind nun zwei Fälle zu unterscheiden. 1. Fall: Es ist $y$ ein Punkt aus $Y_0$, also $y \in L \cap Y_0$. Konvergiert $\mathfrak{U}$ gegen einen Punkt aus $X$, so ist nichts mehr zu beweisen. Konvergiert aber $\mathfrak{U}$ nicht in $X$, so konvergiert $\mathfrak{U}$ in $\tilde{X}_0$ gegen $x = \overset{-1}{\psi}(y)$. In der Tat: Es sei

$$\tilde{U} = [U \cup \overset{-1}{\varphi}(V) \cup \overset{-1}{\psi}(V \cap Y_0)] \cap (\tilde{X}_0 - K)$$

eine Menge aus $\tilde{\mathfrak{B}}_0$ (d. h. $U$ offen in $X$, $K$ kompakt in $X$, $V$ offen in $Y$) mit $x \in \tilde{U}$, also mit $y \in V$. Da $\mathfrak{B}$ gegen $y$ konvergiert, gibt es eine Menge $W_1 \in \mathfrak{U}' \subset \mathfrak{U}$ mit $\varphi(W_1) \subset V$, also mit $W_1 \subset \overset{-1}{\varphi}(V)$. Da $\mathfrak{U}$ nicht in $X$ konvergiert, gibt es weiter zur kompakten Menge $K$ eine Menge $W_2 \in \mathfrak{U}$ mit $W_2 \subset \tilde{X}_0 - K$. Für die zu $\mathfrak{U}$ gehörige Menge $W = W_1 \cap W_2$ gilt dann offenbar: $W \subset \tilde{U}$. Nun bilden aber die

$x$ enthaltenden Mengen $\tilde{U} \in \tilde{\mathfrak{B}}_0$ ein fundamentales Umgebungssystem von $x$. Daher konvergiert $\mathfrak{U}$ gegen $x$. — 2. Fall: Es ist $y \notin Y_0$. Dann gibt es eine kompakte Umgebung $V$ von $y$ in $Y$ mit in $X$ und $\tilde{X}_0$ kompaktem Urbild $\overset{-1}{\varphi}(V)$. Da $\mathfrak{V}$ gegen $y$ konvergiert, gibt es eine Menge $W \in \mathfrak{U}'$ mit $\varphi(W) \subset V \cap L$, also mit $W \subset \overset{-1}{\varphi}(V \cap L)$. Also enthält $\mathfrak{U}'$ und damit $\mathfrak{U}$ die zufolge der Stetigkeit von $\varphi$ und der Kompaktheit von $\overset{-1}{\varphi}(V)$ kompakte Menge $\overset{-1}{\varphi}(V \cap L)$. Dann aber konvergiert $\mathfrak{U}$ gegen einen Punkt dieser Menge, womit unsere Fallunterscheidung zu Ende geführt ist.

Der konstruierte Raum $\tilde{X}_0$ ist also ein lokal-kompakter Raum mit den Eigenschaften $(E_1)$ — $(E_4)$.

Abschließend führen wir noch den

Beweis des Zusatzes. Wir bezeichnen wieder mit $\mathcal{K}(X)$ bzw. $\mathcal{K}(Y)$ den Vektorraum aller stetigen reellen Funktionen mit kompaktem Träger auf $X$ bzw. $Y$; ferner sei $\mathcal{A}$ die Menge aller Funktionen $g = k + f \circ \varphi$ mit $k \in \mathcal{K}(X)$ und $f \in \mathcal{K}(Y)$. Dann ist $\mathcal{A}$ bezüglich der üblichen Operationen eine Algebra über dem Körper $\mathbf{R}$. Ist nun $\tilde{X}$ ein lokal-kompakter Raum mit den Eigenschaften $(E_1)$ — $(E_4)$, so kann jede Funktion $g = k + f \circ \varphi$ aus $\mathcal{A}$ auf genau eine Weise zu einer stetigen reellen Funktion $\tilde{g}$ auf $\tilde{X}$ fortgesetzt werden: es ist $\tilde{g} = \tilde{k} + f \circ \tilde{\varphi}$, wenn hierbei $\tilde{k}$ diejenige Fortsetzung von $k$ bezeichnet, welche auf $\tilde{X} - X$ gleich Null ist. Da die Abbildung $\tilde{\varphi}$ eigentlich und $k$ ein Element von $\mathcal{K}(X)$ ist, gilt $\tilde{g} \in \mathcal{K}(\tilde{X})$ für alle $g \in \mathcal{A}$. Die Menge $\tilde{\mathcal{A}}$ aller Fortsetzungen $\tilde{g}$ von Funktionen $g \in \mathcal{A}$ ist dann ebenfalls eine Algebra. $\tilde{\mathcal{A}}$ trennt die Punkte von $\tilde{X}$, d. h. zu je zwei Punkten $x_1 \neq x_2$ aus $\tilde{X}$ gibt es ein $\tilde{g} \in \tilde{\mathcal{A}}$ mit $\tilde{g}(x_1) \neq \tilde{g}(x_2)$. In der Tat: liegen $x_1$ und $x_2$ in $\tilde{X} - X$, so gilt $\tilde{\varphi}(x_1) \neq \tilde{\varphi}(x_2)$ nach $(E_4)$; also gibt es ein $f \in \mathcal{K}(Y)$ mit $f(\tilde{\varphi}(x_1)) \neq f(\tilde{\varphi}(x_2))$; es ist aber $f \circ \tilde{\varphi} \in \tilde{\mathcal{A}}$. Liegen $x_1$ und $x_2$ nicht beide in $\tilde{X} - X$, so gilt $\tilde{k}(x_1) \neq \tilde{k}(x_2)$ für ein geeignetes $k \in \mathcal{K}(X)$. Eine analoge Überlegung zeigt, daß es zu jedem $x \in \tilde{X}$ mindestens ein $\tilde{g} \in \tilde{\mathcal{A}}$ gibt mit $\tilde{g}(x) \neq 0$. Nach der von H. Nakano [13] stammenden Variante des Approximationssatzes von Stone-Weierstrass (vgl. auch [16], S. 72) kann daher jede auf $\tilde{X}$ stetige, im Unendlichen verschwindende, reelle Funktion[9]) gleichmäßig auf $\tilde{X}$ durch Funktionen aus $\tilde{\mathcal{A}}$ approximiert werden.

Jetzt seien $\tilde{X}_1$ und $\tilde{X}_2$ zwei lokal-kompakte Räume mit den Eigenschaften $(E_1)$ — $(E_4)$. $\mathcal{C}_0(\tilde{X}_i)$ bezeichne die Algebra aller auf $\tilde{X}_i$ stetigen reellen Funktionen, die im Unendlichen verschwinden $(i = 1, 2)$. Der erste Teil unseres Beweises zeigt, daß eine auf $\tilde{X}_1$ definierte Funktion $\tilde{g}_1$ dann und nur dann in $\mathcal{C}_0(\tilde{X}_1)$ liegt, wenn ihre Restriktion $g$ auf $X$ zu einer Funktion $\tilde{g}_2 \in \mathcal{C}_0(\tilde{X}_2)$ fortgesetzt werden kann. Für jedes $\tilde{g}_1 \in \mathcal{C}_0(\tilde{X}_1)$ ist diese Funktion $\tilde{g}_2 \in \mathcal{C}_0(\tilde{X}_2)$ eindeutig bestimmt. Die Abbildung $\tilde{g}_1 \to \tilde{g}_2$ ist ein Isomorphismus $\Psi: \mathcal{C}_0(\tilde{X}_1) \to$

[9]) Bekanntlich heißt eine auf einem lokal-kompakten Raum $T$ definierte, stetige, reelle Funktion $f$ im Unendlichen verschwindend, wenn für jede reelle Zahl $\alpha > 0$ die Menge aller $x \in T$ mit $|f(x)| \geqq \alpha$ kompakt ist. Für kompaktes $T$ hat offenbar jede stetige, reelle Funktion diese Eigenschaft.

$\mathscr{C}_0(\tilde{X}_2)$ der Algebra $\mathscr{C}_0(\tilde{X}_1)$ auf die Algebra $\mathscr{C}_0(\tilde{X}_2)$. Nach bekannten Sätzen (vgl. etwa [15], Satz 2, und [7], p. 100, théorème 4) existiert dann eine Homöomorphie $\psi : \tilde{X}_2 \to \tilde{X}_1$ von $\tilde{X}_2$ auf $\tilde{X}_1$ derart, daß für alle $\tilde{g}_1 \in \mathscr{C}_0(\tilde{X}_1)$ gilt:

$$\Psi(\tilde{g}_1) = \tilde{g}_1 \circ \psi .$$

Für jede Funktion $k \in \mathscr{K}(X)$ und deren stetige Fortsetzungen $\tilde{k}_1$ bzw. $\tilde{k}_2$ auf $\tilde{X}_1$ bzw. $\tilde{X}_2$ ist daher $\tilde{k}_2 = \tilde{k}_1 \circ \psi$, also insbesondere $k(x) = \tilde{k}_1(\psi(x))$ für jedes $x \in X$. Da $\tilde{k}_1$ in $\tilde{X}_1 - X$ gleich Null ist, muß dann $\psi(x) \in X$ und, da $\mathscr{K}(X)$ die Punkte von $X$ trennt, schließlich $\psi(x) = x$ für jedes $x \in X$ sein. Also ist $\psi$ eine den Raum $X$ punktweise festlassende Homöomorphie von $\tilde{X}_2$ auf $\tilde{X}_1$.

## § 4. Im Unendlichen abzählbare Räume

Wir kehren für einen Augenblick zurück zu dem in Nr. 1.7 konstruierten Beispiel einer nicht-konservativen, stetigen Abbildung $\varphi : X \to Y$. Das dort bewiesene Resultat, durch welches die $\varphi$-darstellbaren Maße $\nu \in \mathscr{M}(Y)$ charakterisiert wurden, kann auch folgendermaßen ausgesprochen werden: In diesem Beispiel ist ein Maß $\nu \in \mathscr{M}(Y)$ dann und nur dann $\varphi$-darstellbar, wenn es von einer Menge $M \subset \varphi(X)$ getragen wird, deren Urbild $\overset{-1}{\varphi}(M)$ in der Vereinigung abzählbar vieler kompakter Teilmengen von $X$ enthalten ist. In der Tat werden hierdurch im Beispiel gerade die abzählbaren Teilmengen $M$ von $Y$ gekennzeichnet. Wir werden nun zunächst zeigen, daß die Aussage „dann" allgemein richtig ist.

**Satz 5.** *Es sei $\varphi : X \to Y$ eine stetige Abbildung eines lokal-kompakten Raumes $X$ in einen lokal-kompakten Raum $Y$. Dann ist jedes Maß $\nu \in \mathscr{M}(Y)$ $\varphi$-darstellbar, welches von einer Menge $M \subset \varphi(X)$ getragen wird, deren Urbild $\overset{-1}{\varphi}(M)$ in der Vereinigung abzählbarer vieler kompakter Teilmengen von $X$ enthalten ist.*

Beweis. Nach Voraussetzung gibt es eine Folge $(K_n)_{n=1,2,\ldots}$ kompakter Mengen in $X$ mit $\overset{-1}{\varphi}(M) \subset \bigcup K_n$. Ohne Beschränkung der Allgemeinheit kann $K_n \subset K_{n+1}$ für alle $n = 1, 2, \ldots$ angenommen werden. Jede der Bildmengen $L_n = \varphi(K_n)$ ist kompakt in $Y$; es gilt $L_n \subset L_{n+1}$ für alle $n$ und $M \subset L = \bigcup L_n$, letzteres wegen $M \subset \varphi(X)$. Die Menge $L$ ist als Vereinigung einer Folge kompakter und daher $\nu$-meßbarer Mengen selbst $\nu$-meßbar. Also wird $\nu$ auch von der Menge $L$ getragen. Für jedes $n$ bezeichne nun $\nu_n$ das aus $\nu$ durch Tilgung der Masse außerhalb $L_n$ entstehende Maß: $\nu_n = \chi_{L_n}\nu$. Da $\nu$ auf $L$ konzentriert ist und wegen $L_n \subset L_{n+1} \subset L$, gilt:

(4.1) $$\nu_n \leqq \nu_{n+1} \leqq \nu \qquad (n = 1, 2, \ldots) .$$

Ferner ist

(4.2) $$\lim \nu_n = \nu ,$$

d. h. die Folge $(\nu_n)$ konvergiert vage gegen $\nu$. In der Tat: für beliebiges $f \in \mathscr{K}_+(Y)$ ist jede der Funktionen $f_n = \chi_{L_n} f$ $\nu$-integrierbar; wegen $f_n \leqq f_{n+1} \leqq$ $\leqq f\chi_L = \sup f_n$ gilt somit nach dem Lebesgueschen Konvergenzsatz $\lim \int f_n d\nu$ $= \int f \chi_L d\nu = \int f d(\chi_L \nu)$. Hieraus folgt (4.2), da $\nu_n(f) = \int f_n d\nu$ und $\nu = \chi_L \nu$ ist.

Nach dieser Vorbereitung behaupten wir die Existenz einer Folge $(\mu_n)_{n=1,2,\ldots}$ von Maßen $\mu_n \in \mathcal{M}(X)$ mit folgenden drei Eigenschaften: (a) $\mu_n$ ist auf $K_n$ konzentriert; (b) $\varphi(\mu_n) = \nu_n$; (c) $\mu_n \leqq \mu_{n+1}$ $(n = 1, 2, \ldots)$. Hierzu bezeichne $j_n : K_n \to X$ bzw. $j'_n : L_n \to Y$ die kanonische Injektion von $K_n$ in $X$ bzw. $L_n$ in $Y$, $\varphi'_n : K_n \to L_n$ diejenige Abbildung, die auf $K_n$ mit $\varphi$ übereinstimmt, und $\nu_{L_n}$ das durch $\nu$ in $L_n$ induzierte Maß. Dann ist jede der Abbildungen $\varphi'_n$ eigentlich und bildet $K_n$ auf $L_n$ ab, ferner ist $\varphi \circ j_n = j'_n \circ \varphi'_n$ und $j'_n(\nu_{L_n}) = \nu_n$. Wegen der beiden ersten Eigenschaften von $\varphi'_1$ existiert nach Satz 2 zu $\nu_{L_1}$ ein Maß $\mu'_1 \in \mathcal{M}(K_1)$ mit $\varphi'_1(\mu'_1) = \nu_{L_1}$. Da die Abbildung $j_1 : K_1 \to X$ eigentlich ist, so ist $\mu_1 = j(\mu'_1)$ definiert und ein auf $K_1$ konzentriertes Maß aus $\mathcal{M}(X)$. Das Transitivitätsgesetz für Bildmaße ergibt dann bei Beachtung von $\varphi \circ j_1 = j'_1 \circ \varphi'_1$ und $j'_1(\nu_{L_1}) = \nu_1$ die Gleichung $\varphi(\mu_1) = \nu_1$. Ausgehend von $\mu_1$ konstruieren wir die weiteren Elemente der Folge $(\mu_n)$ durch vollständige Induktion. Wir nehmen hierzu an, daß $\mu_1, \ldots, \mu_n$ für ein $n \geqq 1$ bereits konstruiert sind. Nach (4.1) ist $\tau = \nu_{n+1} - \nu_n$ ein auf $L_{n+1}$ konzentriertes Maß aus $\mathcal{M}(Y)$; für das hierdurch in $L_{n+1}$ induzierte Maß $\tau_{L_{n+1}}$ gilt daher $j'_{n+1}(\tau_{L_{n+1}}) = \tau$. Wiederum nach Satz 2 gibt es ein Maß $\sigma' \in \mathcal{M}(K_{n+1})$ mit $\varphi'_{n+1}(\sigma') = \tau_{L_{n+1}}$. Eine Wiederholung der soeben angestellten Überlegungen ergibt, daß $\sigma = j_{n+1}(\sigma')$ ein Maß aus $\mathcal{M}(X, K_{n+1})$ ist mit $\varphi(\sigma) = \tau$. Setzen wir daher $\mu_{n+1} = \mu_n + \sigma$, so ist $\mu_{n+1}$ ein auf $K_{n+1}$ konzentriertes Maß aus $\mathcal{M}(X)$ mit $\mu_n \leqq \mu_{n+1}$; nach Hilfssatz 1 ist ferner $\varphi(\mu_{n+1}) = \varphi(\mu_n) + \varphi(\sigma) = \nu_n + \tau = \nu_{n+1}$. Damit ist die Existenz einer Folge $(\mu_n)$ mit den Eigenschaften (a)—(c) gesichert.

Wir wollen jetzt zeigen, daß die Folge $(\mu_n)$ vage konvergiert[10]). Wegen der Eigenschaft (c) genügt es zu beweisen, daß die Folge $(\int g\,d\mu_n)$ für jedes $g \in \mathcal{K}_+(X)$ nach oben beschränkt ist. Zu $g \in \mathcal{K}_+(X)$ gibt es aber, wie im Beweis des Satzes 1 gezeigt wurde, eine Funktion $f \in \mathcal{K}_+(Y)$ mit $0 \leqq g \leqq f \circ \varphi$. Aus (b) und (4.1) folgt daher

$$(4.3) \qquad \int g\,d\mu_n \leqq \int f \circ \varphi\, d\mu_n = \int f\,d\nu_n \leqq \int f\,d\nu \qquad (n = 1, 2, \ldots).$$

Damit ist die Existenz des vagen Limes $\mu = \lim \mu_n$ bewiesen. Wir behaupten weiter, daß $\varphi(\mu) = \nu$ ist. Dazu genügt es zu zeigen, daß für jedes $f \in \mathcal{K}_+(Y)$ die Funktion $f \circ \varphi$ $\mu$-integrierbar und $\int f \circ \varphi\, d\mu = \int f\,d\nu$ ist. Zunächst ergibt sich genauso wie soeben die Gültigkeit von (4.3) für jedes $g \in \mathcal{K}_+(X)$ mit $0 \leqq g \leqq f \circ \varphi$. Hieraus folgt $\int g\,d\mu \leqq \int f\,d\nu$ durch Grenzübergang. Also ist die Funktion $f \circ \varphi$ $\mu$-integrierbar und $\int f \circ \varphi\, d\mu \leqq \int f\,d\nu$ (vgl. Nr. 1.2). Wegen $\mu_n \leqq \mu$ und $\varphi(\mu_n) = \nu_n$ ist weiter $\int f\,d\nu_n = \int f \circ \varphi\, d\mu_n \leqq \int f \circ \varphi\, d\mu$, woraus bei Beachtung von (4.2) folgt: $\int f\,d\nu \leqq \int f \circ \varphi\, d\mu$ und damit $\int f \circ \varphi\, d\mu = \int f\,d\nu$. Wie behauptet ist also $\varphi(\mu) = \nu$.

Jetzt sei an folgende bekannte Begriffsbildung erinnert:

**Definition 4.** *Ein lokal-kompakter Raum $T$ heißt abzählbar im Unendlichen, wenn eine Folge $(K_n)_{n=1,2,\ldots}$ in $T$ kompakter Mengen existiert mit $T = \bigcup K_n$.*

[10]) Der Rest des Beweises ähnelt einem Abschnitt des Beweises von Satz 1.

Dann ergibt sich mühelos das folgende Korollar zu Satz 5:

**Korollar.** *Jede stetige Abbildung $\varphi: X \to Y$ eines im Unendlichen abzählbaren, lokal-kompakten Raumes $X$ in einen lokal-kompakten Raum $Y$ ist konservativ.*

Unter Verwendung des Lokalisationsprinzips kann dieses Korollar verschärft werden zum

**Satz 6.** *Es sei $\varphi: X \to Y$ eine stetige Abbildung eines lokal-kompakten Raumes $X$ in einen lokal-kompakten Raum $Y$. Besitzt dann jeder Punkt $y \in \varphi(X)$ eine offene Umgebung $V$ in $Y$, deren Urbild $\overset{-1}{\varphi}(V)$, aufgefaßt als Unterraum von $X$, im Unendlichen abzählbar ist, so ist $\varphi$ konservativ.*

Beweis. Die Behauptung folgt aus dem Korollar zu Satz 5 und aus Satz 1. Man hat nur wie zu Beginn des Beweises von Satz 2 zu bemerken, daß aus $\nu \in \mathcal{M}(Y, \varphi(X))$ folgt $\nu_V \in \mathcal{M}(V, \varphi(X) \cap V)$ für jede offene Menge $V \subset Y$.

Bemerkungen. 1. Ist $\varphi: X \to Y$ eine stetige Abbildung eines im Unendlichen abzählbaren, lokal-kompakten Raumes $X$ in einen lokal-kompakten Raum $Y$, so ist offenbar das Urbild $\overset{-1}{\varphi}(K)$ einer jeden in $Y$ kompakten Menge $K$ darstellbar als Vereinigung abzählbar vieler kompakter Mengen in $X$. Es wäre interessant zu wissen, ob diese Bedingung allein schon impliziert, daß die Abbildung konservativ ist. Dann wären nämlich Satz 2 und das Korollar zu Satz 5 Folgerungen aus einem allgemeineren Satz.

2. Auf Grund der engen Beziehungen zwischen den im Unendlichen abzählbaren, lokal-kompakten Räumen und den parakompakten, lokal-kompakten Räumen[11]), könnte man vermuten, daß jede stetige Abbildung eines parakompakten, lokal-kompakten Raumes $X$ in einen lokal-kompakten Raum $Y$ konservativ ist. Daß dies jedoch nicht der Fall ist, zeigt das Beispiel in Nr. 1.7; der dort betrachtete Raum $X$ ist parakompakt und lokal-kompakt.

## § 5. Abbildungen mit lokalem Schnitt

Zunächst seien $X$ und $Y$ beliebige topologische Räume und $\varphi: X \to Y$ eine stetige Abbildung. Dann erweist es sich häufig als zweckmäßig, das Tripel $(X, Y, \varphi)$ als einen verallgemeinerten Faserraum aufzufassen, dessen Fasern im allgemeinen zueinander nicht homöomorph sind[12]).

*Fasern der Abbildung* $\varphi$ werden hierbei die nicht-leeren Urbilder einpunktiger Teilmengen von $Y$, also alle Mengen $\overset{-1}{\varphi}(y)$ mit $y \in \varphi(X)$ genannt. Vermöge dieser Definition ist keine Faser leer, je zwei verschiedene Fasern sind zueinander fremd und jeder Punkt $x \in X$ liegt in genau einer Faser $F = \overset{-1}{\varphi}(\varphi(x))$, genannt die Faser durch $x$ oder über $\varphi(x)$. Die Fasern von $\varphi$ definieren also in $X$ eine Klasseneinteilung.

Der aus der Theorie der Faserräume geläufige Begriff des (lokalen) Schnittes läßt sich auf unsere allgemeinere Situation übertragen:

---

[11]) Vgl. N. Bourbaki [7], p. 107, théorème 5.

[12]) Vgl. z. B. die Arbeiten von R. Remmert [17] und K. Stein [18].

**Definition 5.** *Es sei $\varphi: X \to Y$ eine stetige Abbildung eines topologischen Raumes $X$ in einen topologischen Raum $Y$. (Globaler) Schnitt (bezüglich $\varphi$) heiße dann jede stetige Abbildung $\sigma: Y \to X$, für welche $\varphi \circ \sigma$ die identische Abbildung von $Y$ auf sich ist. — Wir werden sagen, die Abbildung $\varphi$ besitzt einen lokalen Schnitt, wenn es zu jedem Punkt $y \in \varphi(X)$ eine Umgebung $V$ von $y$ in $Y$ und eine stetige Abbildung $\sigma_V: V \to X$ gibt, für welche $\varphi \circ \sigma_V$ die identische Abbildung von $V$ auf sich ist.*

Aus der Existenz eines Schnittes $\sigma$ folgt, daß die Abbildung $\varphi$ den Raum $X$ auf $Y$ abbildet. Der Unterraum $S = \sigma(Y)$ ist dann zu $Y$ homöomorph (vermöge der Restriktion von $\varphi$ auf $S$). Ferner ist $S$ in $X$ abgeschlossen, falls der Raum $X$ separiert ist. Es sei nämlich $x \in \bar{S}$ ein Punkt der abgeschlossenen Hülle von $S$ in $X$, $\mathfrak{B}$ der Umgebungsfilter von $x$ in $X$ und $\mathfrak{F}$ die Spur von $\mathfrak{B}$ in $S$. Da $\mathfrak{F}$ gegen $x$ konvergiert und die Abbildungen $\varphi$ und $\sigma$ stetig sind, konvergiert $\varphi(\mathfrak{F})$ gegen $\varphi(x)$ und $\mathfrak{F} = \sigma(\varphi(\mathfrak{F}))$ gegen $\sigma(\varphi(x))$. Wegen der Separiertheit von $X$ ist der Limes von $\mathfrak{F}$ eindeutig bestimmt, also $x = \sigma(\varphi(x))$, also $x \in S$ und somit $S = \bar{S}$.

Aus der Existenz eines lokalen Schnittes folgt die Offenheit von $\varphi(X)$ in $Y$. Wird also $Y$ zusätzlich als lokal-kompakt vorausgesetzt, so ist auch $\varphi(X)$ lokal-kompakt.

Die Bedeutung dieser Begriffsbildungen für unser Problem erhellt aus dem

**Satz 7.** *Es sei $\varphi: X \to Y$ eine stetige Abbildung eines lokal-kompakten Raumes $X$ in einen lokal-kompakten Raum $Y$. Besitzt dann $\varphi$ einen lokalen Schnitt, so ist $\varphi$ konservativ.*

Beweis. Zu jedem Punkt $y \in \varphi(X)$ gibt es nach Voraussetzung eine Umgebung $V$ von $y$ in $Y$ und eine stetige Abbildung $\sigma_V: V \to X$, für welche $\varphi \circ \sigma_V$ die identische Abbildung von $V$ auf sich ist. Die Abbildung $\varphi'_V: U \to V$, die auf $U = \overset{-1}{\varphi}(V)$ mit $\varphi$ übereinstimmt, besitzt dann einen globalen Schnitt. Es ist nämlich $\sigma_V(V) \subset U$ und somit diejenige Abbildung $\sigma'_V: V \to U$, die auf $V$ mit $\sigma_V$ übereinstimmt, ein Schnitt bezüglich $\varphi'_V$. Da $V$ ohne Beschränkung der Allgemeinheit offen gewählt werden kann, lehrt das Lokalisationsprinzip (Satz 1), daß es genügt, unseren Satz unter der Voraussetzung zu beweisen, daß ein Schnitt $\sigma: Y \to X$ bezüglich $\varphi$ existiert. Dies werde von jetzt an vorausgesetzt.

Wie bereits im Anschluß an die Definition 5 bemerkt wurde, ist dann $S = \sigma(Y)$ in $X$ abgeschlossen; ferner ist die Restriktion von $\varphi$ auf $S$ eine Homöomorphie von $S$ auf $Y$. Für jede Menge $K \subset X$ ist aber $\overset{-1}{\sigma}(K) = \varphi(S \cap K)$ und somit kompakt in $Y$, falls $K$ kompakt in $X$ ist. Es ist daher $\sigma$ eine eigentliche Abbildung und deswegen $\mu = \sigma(\nu)$ für jedes Maß $\nu \in \mathscr{M}(Y) = \mathscr{M}(Y, \varphi(X))$ definiert. Da $\varphi \circ \sigma$ die identische Abbildung von $Y$ auf sich ist, folgt aus dem Transitivitätsgesetz für Bildmaße, daß mit $\mu = \sigma(\nu)$ und $\nu = (\varphi \circ \sigma)(\nu)$ auch $\varphi(\mu)$ definiert und gleich $\nu$ ist. Es ist daher jedes Maß aus $\mathscr{M}(Y)$ $\varphi$-darstellbar, also $\varphi$ konservativ.

Beispiel. Es sei $X$ ein lokal-kompakter, *allgemeiner Faserraum* im Sinne von H. Cartan [10] (vgl. insbesondere Exposé 6), $Y$ die dann notwendig lokal-kompakte Basis von $X$, $p: X \to Y$ die Projektionsabbildung und $F$ die Faser. Ist dann $X$ lokal-trivial im Sinne von Cartan (loc. cit.), d. h. ist $X$ in geeigneter Weise im Kleinen zu einem Produktraum $V \times F$ homöomorph, so besitzt $p$ einen lokalen Schnitt. Nach Satz 7 ist dann $p$ eine konservative Abbildung.

## § 6. Faktorisierung konservativer Abbildungen

**6.1.** *Ein Transitivitätsgesetz.* — Für die folgenden Betrachtungen wird es sich als zweckmäßig erweisen, zunächst zu untersuchen, ob eine durch Zusammensetzen konservativer Abbildungen entstehende Abbildung wieder konservativ ist.

Dies ist ohne Zusatzvoraussetzung nicht der Fall, denn es kann *jede* stetige Abbildung $\varphi: X \to Y$ eines lokal-kompakten Raumes $X$ in einen lokal-kompakten Raum $Y$ durch Zusammensetzen zweier konservativer stetiger Abbildungen gewonnen werden. In der Tat: es sei $\mathfrak{X}$ der zur Abbildung $\varphi$ gehörige lokal-kompakte Raum des Satzes 3, $j: X \to \mathfrak{X}$ die kanonische Injektion von $X$ in $\mathfrak{X}$ und $\tilde{\varphi}: \mathfrak{X} \to Y$ die stetige Fortsetzung von $\varphi$ auf $\mathfrak{X}$. Nach Nr. 1.6 ist die Abbildung $j$ konservativ, da $X$ in $\mathfrak{X}$ offen, also ein lokal-kompakter Unterraum von $\mathfrak{X}$ ist. Nach Satz 3 ist die Abbildung $\tilde{\varphi}$ eigentlich und als solche nach Satz 2 ebenfalls konservativ. Da $\tilde{\varphi}$ eine Fortsetzung von $\varphi$ ist, gilt aber $\varphi = \tilde{\varphi} \circ j$, d. h. die beliebige stetige Abbildung $\varphi$ läßt sich aus zwei konservativen, stetigen Abbildungen zusammensetzen. Besonders hervorgehoben sei die Tatsache, daß $j(X) = X$ in $\mathfrak{X}$ absolut meßbar, d. h. meßbar in bezug auf jedes positive Maß auf $\mathfrak{X}$ ist.

Immerhin gilt aber folgendes schwächere *Transitivitätsgesetz:*

**Satz 8.** *Es seien $X, Y, Z$ lokal-kompakte Räume sowie $\varphi: X \to Y$ und $\psi: Y \to Z$ stetige Abbildungen, von denen die erste den Raum $X$ auf den Raum $Y$ abbildet. Sind dann $\varphi$ und $\psi$ konservativ, so ist auch die zusammengesetzte Abbildung $\psi \circ \varphi$ konservativ. Ist umgekehrt $\psi \circ \varphi$ konservativ, so ist wenigstens $\psi$ konservativ.*

Beweis. Wir setzen zur Abkürzung $\tau = \psi \circ \varphi$ und bemerken, daß $\tau(X) = \psi(Y)$ ist. Dies folgt aus der Voraussetzung: $\varphi(X) = Y$. — Nun seien zunächst $\varphi$ und $\psi$ konservativ. Dann existiert zu jedem Maß $\varrho \in \mathcal{M}(Z, \tau(X)) = \mathcal{M}(Z, \psi(Y))$ ein Maß $\nu \in \mathcal{M}(Y)$ mit $\psi(\nu) = \varrho$ und zu $\nu$ wegen $\varphi(X) = Y$ ein Maß $\mu \in \mathcal{M}(X)$ mit $\varphi(\mu) = \nu$. Nach dem Transitivitätsgesetz der Nr. 1.5 ist dann $\tau(\mu) = \varrho$, also jedes $\varrho \in \mathcal{M}(Z, \tau(X))$ $\tau$-darstellbar; d. h. $\tau$ ist konservativ. — Umgekehrt sei jetzt $\tau$ konservativ. Nach einem bereits mehrfach wiederholten Schluß (vgl. Beweis von Satz 1) gibt es zu jedem $f \in \mathcal{K}(Y)$ ein $g \in \mathcal{K}_+(Z)$ mit $|f| \leqq g \circ \psi$, also mit $|f \circ \varphi| \leqq g \circ \tau$. Zu jedem Maß $\varrho \in \mathcal{M}(Z, \tau(X)) = \mathcal{M}(Z, \psi(Y))$ gibt es nach Voraussetzung ein Maß $\mu \in \mathcal{M}(X)$ mit $\tau(\mu) = \varrho$.

Da $f \circ \varphi$ auf $X$ stetig und $g \circ \tau$ $\mu$-integrierbar ist, folgt aus $|f \circ \varphi| \leqq g \circ \tau$, daß auch $f \circ \varphi$ $\mu$-integrierbar ist, und zwar für jedes $f \in \mathscr{K}(Y)$. Also ist neben $\tau(\mu)$ auch $\nu = \varphi(\mu)$ definiert. Nach dem Transitivitätsgesetz von Nr. 1.5 ist dann $\psi(\nu)$ definiert und $\psi(\nu) = \varrho$. Also ist die Abbildung $\psi$ konservativ.

Wir bemerken noch, daß unter den Voraussetzungen des Satzes 8 aus der Konservativität der Abbildung $\psi \circ \varphi$ *nicht* auf die Konservativität von $\varphi$ geschlossen werden kann. Dies zeigt folgendes Beispiel: Es sei $\varphi : X \to Y$ eine stetige, nicht-konservative Abbildung von $X$ auf $Y$ (etwa die aus Nr. 1.7) und $\psi$ die Abbildung von $Y$ auf den nur aus einem Punkt bestehenden Raum $Z$. Die Abbildung $\psi \circ \varphi$ bildet $X$ auf $Z$ ab und ist konservativ (etwa weil ein globaler Schnitt existiert).

**6.2.** *Konservative Basen und Zerlegungen*[13]). — Unter einer *Zerlegung* (oder Klasseneinteilung) $Z$ einer Menge $X$ verstehen wir wie üblich ein System nicht leerer, paarweise fremder Teilmengen von $X$, die $X$ überdecken. Die Zerlegungen von $X$ entsprechen in bekannter Weise umkehrbar eindeutig den Äquivalenzrelationen in $X$. Eine Zerlegung $Z$ eines topologischen Raumes $X$ heißt *einfach*, wenn alle Elemente von $Z$ zusammenhängende Teilmengen von $X$ sind.

Für eine stetige Abbildung $\varphi : X \to Y$ eines topologischen Raumes in einen zweiten ist das System der Fasern $\overset{-1}{\varphi}(\varphi(x))$, $x \in X$, dieser Abbildung eine Zerlegung von $X$, die wir mit $Z(\varphi)$ bezeichnen und welche die *durch $\varphi$ bestimmte Zerlegung* heißt. Löst man die einzelnen Fasern von $\varphi$ in deren Zusammenhangskomponenten auf, so entsteht eine einfache Zerlegung $Z'(\varphi)$ von $X$, welche die *durch $\varphi$ bestimmte einfache Zerlegung* heißt. Die Elemente von $Z'(\varphi)$ werden die *Niveaumengen* von $\varphi$ genannt.

Sind $X$, $Y_1$ und $Y_2$ topologische Räume sowie $\varphi_1 : X \to Y_1$ und $\varphi_2 : X \to Y_2$ stetige Abbildungen, so heißt $\varphi_2$ *von* $\varphi_1$ *abhängig*, wenn die Zerlegung $Z'(\varphi_1)$ feiner ist als $Z'(\varphi_2)$, d. h. wenn jedes Element von $Z'(\varphi_1)$ in einem Element von $Z'(\varphi_2)$ enthalten ist. Hiermit gleichwertig ist die Aussage, daß $\varphi_2$ auf jeder Niveaumenge von $\varphi_1$ konstant ist. Wenn sowohl $\varphi_2$ von $\varphi_1$ als auch $\varphi_1$ von $\varphi_2$ abhängig, also $Z'(\varphi_1) = Z'(\varphi_2)$ ist, so nennt man die Abbildungen $\varphi_1$ und $\varphi_2$ *verwandt.*

Die Frage nach allen von einer konservativen, stetigen Abbildung abhängigen, konservativen, stetigen Abbildungen führt zum Begriff der konservativen Basis.

**Definition 6.** *Es sei $\varphi_0 : X \to Y_0$ eine konservative, stetige Abbildung eines lokal-kompakten Raumes $X$ in einen lokal-kompakten Raum $Y_0$. Ein Paar $(Y^*, \Phi)$ bestehend aus einem lokal-kompakten Raum $Y^*$ und einer stetigen Abbildung $\Phi : X \to Y^*$ heiße eine* k o n s e r v a t i v e  B a s i s *von $\varphi_0$, wenn es die folgenden Eigenschaften besitzt:*

[13]) Vgl. hierzu K. STEIN [18], wo analoge Untersuchungen für holomorphe Abbildungen eines komplexen Raumes in einen zweiten angestellt werden. Wir benützen weitgehend die dort verwendete Terminologie.

a) *die Abbildung* $\Phi$ *ist konservativ und mit* $\varphi_0$ *verwandt, ferner ist* $\Phi(X) = Y^*$;

b) *zu jeder von* $\varphi_0$ *abhängigen, konservativen, stetigen Abbildung* $\varphi : X \to Y$ *von* $X$ *in einen lokal-kompakten Raum* $Y$ *gibt es eine konservative, stetige Abbildung* $\varphi^* : Y^* \to Y$ *derart, daß* $\varphi = \varphi^* \circ \Phi$ *ist.*

Ist $(Y^*, \Phi)$ eine konservative Basis von $\varphi_0 : X \to Y_0$, so ist für jede konservative, stetige Abbildung $\varphi^* : Y^* \to Y$ die Abbildung $\varphi^* \circ \Phi$ von $\varphi_0$ abhängig und nach Satz 8 konservativ. Also durchläuft $\varphi = \varphi^* \circ \Phi$ alle konservativen, von $\varphi_0$ abhängigen, stetigen Abbildungen von $X$ in lokal-kompakte Räume, wenn $\varphi^*$ alle konservativen, stetigen Abbildungen von $Y^*$ in lokal-kompakte Räume durchläuft.

Man überlegt sich leicht, daß eine konservative Basis „*bis auf Homöomorphien*" *eindeutig* bestimmt ist. Genauer heißt dies: Sind $(Y_1^*, \Phi_1)$ und $(Y_2^*, \Phi_2)$ konservative Basen von $\varphi_0$, so gibt es eine Homöomorphie $\psi^* : Y_1^* \to Y_2^*$ von $Y_1^*$ auf $Y_2^*$ mit $\Phi_2 = \psi^* \circ \Phi_1$. Ferner ist mit $(Y_1^*, \Phi_1)$ stets auch $(Y_2^*, \psi^* \circ \Phi_1)$ eine konservative Basis von $\varphi_0$, sofern $\psi^* : Y_1^* \to Y_2^*$ eine Homöomorphie von $Y_1^*$ auf einen Raum $Y_2^*$ ist.

Um die Existenz konservativer Basen beweisen zu können, benötigen wir noch folgenden Begriff:

**Definition 7.** *Eine Zerlegung* $Z$ *eines lokal-kompakten Raumes* $X$ *heiße konservativ, wenn der Quotientenraum* $X/Z$ *lokal-kompakt und die kanonische Abbildung* $\Phi : X \to X/Z$ *konservativ ist.*

Wir erinnern noch daran, daß gemäß der Definition des Quotientenraumes die Abbildung $\Phi$ stetig und $Z(\Phi) = Z$ ist.

Beispiele konservativer Zerlegungen liefern die eigentlichen Abbildungen. Es sei nämlich $\varphi : X \to Y$ eine eigentliche Abbildung eines lokal-kompakten Raumes $X$ in einen lokal-kompakten Raum $Y$. Dann ist die durch $\varphi$ bestimmte Zerlegung $Z(\varphi)$ von $X$ konservativ. Nach N. Bourbaki [7], p. 104, prop. 17, ist nämlich der Raum $X/Z(\varphi)$ lokal-kompakt und die kanonische Abbildung $\Phi : X \to X/Z(\varphi)$ eigentlich, nach Satz 2 also die Zerlegung $Z(\varphi)$ konservativ.

Die Frage nach der Existenz konservativer Basen beleuchtet der

Hilfssatz 5. *Es sei* $\varphi_0 : X \to Y_0$ *eine konservative, stetige Abbildung eines lokal-kompakten Raumes* $X$ *in einen lokal-kompakten Raum* $Y_0$, *deren zugehörige einfache Zerlegung* $Z'(\varphi_0)$ *konservativ ist. Bezeichnet dann* $Y^*$ *den Quotientenraum* $X/Z'(\varphi_0)$ *und* $\Phi : X \to Y^*$ *die kanonische Abbildung, so ist* $(Y^*, \Phi)$ *eine konservative Basis von* $\varphi_0$.

Beweis. Nach Voraussetzung ist $Z'(\varphi_0)$ und damit $\Phi$ konservativ; ferner ist $\Phi(X) = Y^*$ und $\Phi$ mit $\varphi_0$ verwandt. Jede stetige, von $\varphi_0$ abhängige Abbildung $\varphi : X \to Y$ von $X$ in einen lokal-kompakten Raum $Y$ besitzt genau eine Faktorisierung $\varphi = \varphi^* \circ \Phi$ mit stetigem $\varphi^* : Y^* \to Y$. Da nach Satz 8 mit $\varphi$ auch $\varphi^*$ konservativ ist, so ist also $(Y^*, \Phi)$ wie behauptet eine konservative Basis.

Nun hat K. STEIN [18] (Sätze 5 und 6) folgenden Satz bewiesen: *Es sei $\varphi : X \to Y$ eine stetige Abbildung eines lokal-kompakten Raumes $X$ in einen lokal-kompakten Raum $Y$, deren sämtliche Niveaumengen kompakt sind. Dann ist die durch $\varphi$ bestimmte einfache Zerlegung $Z'(\varphi)$ eigentlich, d. h. es ist $X/Z'(\varphi)$ lokal-kompakt und die kanonische Abbildung $\Phi : X \to X/Z'(\varphi)$ eigentlich.*

Hieraus, aus Satz 2 und dem Hilfssatz 5 folgt dann sofort

**Satz 9.** *Sind die Niveaumengen einer konservativen, stetigen Abbildung $\varphi_0 : X \to Y_0$ eines lokal-kompakten Raumes in einen zweiten sämtlich kompakt, so besitzt $\varphi_0$ eine konservative Basis.*

Bemerkung. Genügt die Abbildung $\varphi_0 : X \to Y_0$ den Voraussetzungen des Satzes 9 und ist $(Y^*, \Phi)$ die zugehörige konservative Basis, so besitzt $\varphi_0$ selbst eine Faktorisierung $\varphi_0 = \varphi_0^* \circ \Phi$ in die stetigen, konservativen Abbildungen $\Phi : X \to Y^*$ und $\varphi_0^* : Y^* \to Y_0$. Der in [4] verallgemeinerte Faktorisierungssatz von G. T. WHYBURN [19], p. 141, kennzeichnet diese Faktorisierung durch topologische Eigenschaften der Faktoren $\Phi$ und $\varphi_0^*$. Es erweist sich nämlich $\Phi$ bzw. $\varphi_0^*$ als *monotone* bzw. *lichte* Abbildung, d. h. die Fasern der Abbildung sind zusammenhängend bzw. total-unzusammenhängend; außerdem ist $\Phi$ eigentlich und eine Abbildung *auf* $Y^*$. Durch die genannten Eigenschaften der Faktoren sind diese und der Faktorraum $Y^*$ „bis auf Homöomorphien" eindeutig bestimmt.

**6.3.** *Bemerkung zur Theorie des Lebesgueschen Flächenmaßes.* — Nach einer mündlichen Mitteilung von Herrn W. FLEMING gestattet der Satz 2 dieser Arbeit die folgende Anwendung: Es sei $X$ eine im Sinne von L. CESARI [11] zulässige, kompakte Teilmenge der Ebene $\mathbf{R}^2$, also z. B. das von einer einfach geschlossenen Jordan-Kurve umschlossene Gebiet einschließlich seiner Randkurve; ferner sei $\varphi : X \to \mathbf{R}^3$ eine stetige Abbildung von $X$ in den $\mathbf{R}^3$. Die zugehörige parametrische Fläche $(\varphi, X)$ besitze endliches Lebesguesches Flächenmaß. Bezeichnet $\varphi = \varphi^* \circ \Phi$ die monoton-lichte Whyburnsche Faktorisierung und $Y^* = X/Z'(\varphi)$ den (hier kompakten, metrisierbaren) Whyburnschen Faktorisierungsraum, so existiert nach A. ROSENTHAL (vgl. [11], p. 403) genau ein positives (Radonsches) Maß $\mu^*$ auf $Y^*$ derart, daß für jede offene Menge $G^*$ in $Y^*$ gilt: $\mu^*(G^*)$ ist das Lebesguesche Flächenmaß der Teilfläche $(\varphi', \Phi^{-1}(G^*))$ von $(\varphi, X)$, wenn hierbei $\varphi'$ die Restriktion von $\varphi$ auf $\Phi^{-1}(G^*)$ bedeutet. In diesem Sinne wird bei L. CESARI [11] das Lebesguesche Flächenmaß als Radonsches Maß gedeutet. Da die Abbildung $\Phi$ eigentlich ist, folgt aber aus Satz 2, daß auf dem Raum $X$ selbst mindestens ein positives Radonsches Maß $\mu$ existiert mit $\Phi(\mu) = \mu^*$. Dann aber gilt für jede bezüglich $Z'(\varphi)$ gesättigte offene Menge $G$ in $X$, also jede offene Menge $G \subset X$ mit $\Phi^{-1}(\Phi(G)) = G$: es ist $\mu(G)$ das Lebesguesche Flächenmaß der durch die Restriktion $\varphi_G$ von $\varphi$ auf $G$ definierten Teilfläche $(\varphi_G, G)$ von $(\varphi, X)$.

## Literatur

[1] Aumann, G.: Über die Erweiterung von additiven monotonen Funktionen auf regulär geordneten Halbgruppen. Arch. der Math. **8**, 422—427 (1957). — [2] Bauer, H.: Sur l'existence de mesures avec une image donnée. C. R. Acad. Sci. (Paris) **246**, 1953—1955 (1958). — [3] Bauer, H.: Über die Fortsetzung positiver Linearformen. Sitz.-Ber. math.-naturw. Kl. Bayer. Akad. Wiss. München **1957**, 177—190 (1958). — [4] Bauer, H.: Verallgemeinerung eines Faktorisierungssatzes von G. T. Whyburn. Arch. der Math. **10**, 373—378 (1959). — [5] Bourbaki, N.: Intégration, Chap. I—IV. Actual. Scient. et Ind. **1175**, Paris 1952. — [6] Bourbaki, N.: Intégration, Chap. V. Actual. Sci. et Ind. **1244**, Paris 1956. — [7] Bourbaki, N.: Topologie générale, Chap. I—II. Actual. Sci. et Ind. **1142**, Paris 1951. — [8] Bourbaki, N.: Topologie générale, Chap. III—IV. Actual. Sci. et Ind. **1143**, Paris 1951. — [9] Bourbaki, N.: Espaces vectoriels topologiques, Chap. I—II. Actual. Sci. et Ind. **1189**, Paris 1953. — [10] Cartan, H.: Espaces fibrés et homotopie. Séminaire E. N. S. 1949/50, 2e édition. — [11] Cesari, L.: Surface Area. Ann. of Math. Studies **35**, Princeton 1956. — [12] Morse, M., and W. Transue: Semi-normed vector spaces with duals of integral type. J. d'Analyse Math. **4**, 149—186 (1954/55). — [13] Nakano, H.: On the product of relative spectra. Ann. of Math. **49**, 281—315 (1948). — [14] Namioka, I.: Partially ordered linear topological spaces. Memoirs Amer. math. Soc. **24**, Providence 1957. — [15] Nöbeling, G., u. H. Bauer: Über die Erweiterungen topologischer Räume. Math. Ann. **130**, 20—45 (1955). — [16] Nöbeling, G., u. H. Bauer: Allgemeine Approximationskriterien mit Anwendungen. J.-Ber. Dtsch. Math.-Verein. **58**, 54—72 (1955). — [17] Remmert, R.: Holomorphe und meromorphe Abbildungen komplexer Räume. Math. Ann. **133**, 328—374 (1957). — [18] Stein, K.: Analytische Zerlegungen komplexer Räume. Math. Ann. **132**, 63—93 (1956). — [19] Whyburn, G. T.: Analytic Topology. Amer. math. Soc. Colloq. Publ. **28**, New York 1942.

*(Eingegangen am 1. Juni 1959)*

# Minimalstellen von Funktionen und Extremalpunkte. II
[R16]

Arch. Math. **11** (1960), 200–205
[Zbl. Math. 98.08003; MR 24 #A251]

**1. Einleitung.** In einer Note gleichen Titels [2] wurde ein allgemeiner Satz über die Existenz ausgezeichneter Minimalstellen nach unten halbstetiger Funktionen auf einem kompakten Raum angegeben. Aus ihm ließ sich in einfacher Weise folgern, daß eine auf einer kompakten konvexen Teilmenge $X$ eines lokal-konvexen Vektorraumes definierte, konkave, nach unten halbstetige Funktion ihr Minimum auf $X$ in mindestens einem Extremalpunkt von $X$ annimmt.

Im Rahmen meiner Untersuchungen über das Dirichletsche Problem und Šilovsche Berandungen stieß ich auf einen Satz ([3], théorème 1[1])), der ebenfalls die Existenz ausgezeichneter Minimalstellen von Funktionen garantiert. Sein Beweis verwendet ähnliche Methoden und Schlüsse wie der des zentralen Satzes aus [2]; das Resultat selbst konnte aber nicht aus dem letzteren Satz gefolgert werden. Es lag daher nahe, nach einer Verallgemeinerung des Satzes aus [2] zu fragen, aus welcher auch der Satz aus der Theorie des Dirichletschen Problems folgt. Diese Verallgemeinerung und Folgerungen aus ihr sollen hier mitgeteilt werden.

Der neue Satz und seine Folgerungen haben inzwischen nützliche Anwendungen in einer Theorie der ersten Randwertaufgabe für gewisse elliptische und parabolische Differentialgleichungen gefunden[2]). Eine von diesen doch etwas spezieller gearteten Anwendungen unabhängige Darstellung schien mir schon im Hinblick auf den bereits vorliegenden ersten Teil dieser Note zweckmäßig zu sein.

Im folgenden übernehmen wir die im ersten Teil eingeführten Bezeichnungen; wir verweisen diesbezüglich besonders auf [2], Nr. 2.

**2. Existenz extremaler Minimalstellen.** Unser erstes Ziel ist der Beweis des einleitend angekündigten „allgemeinen Minimumprinzipes".

Wir betrachten hierzu einen nicht leeren *kompakten* (Hausdorffschen) Raum $X$ und eine Teilmenge $\mathfrak{T}$ der Produktmenge $X \times \mathfrak{P}(X)$ von $X$ mit seiner Potenzmenge $\mathfrak{P}(X)$. Die Elemente von $\mathfrak{T}$ sind also geordnete Paare $(x, T)$ von Punkten $x \in X$ und Mengen $T \subset X$.

$\mathfrak{T}$ besitze folgende zwei Eigenschaften:

**(T$_1$)** *Zu jedem Punkt $x \in X$ existiert mindestens eine Menge $T \in \mathfrak{P}(X)$ mit $(x, T) \in \mathfrak{T}$.*

**(T$_2$)** *Aus $(x, T) \in \mathfrak{T}$ folgt stets: $T \neq \emptyset$.*

---

[1]) Eine ausführlichere Darstellung von [3] wird demnächst in den Annales de l'Institut Fourier erscheinen.

[2]) *Zusatz bei der Korrektur:* Vgl. H. Bauer, Une axiomatique du problème de Dirichlet pour certaines équations aux dérivées partielles elliptiques et paraboliques. C. r. Acad. Sci. Paris **250**, 2672—2674 (1960).

Ein *Punkt* $x_0 \in X$ heiße $\mathfrak{T}$*-extremal*, wenn aus $(x_0, T) \in \mathfrak{T}$ folgt $T = \{x_0\}$. Wegen $(T_1)$ ist dies gleichbedeutend damit, daß $(x_0, \{x_0\})$ das einzige Element aus $\mathfrak{T}$ ist mit $x_0$ als erster Komponente.

Aus beweistechnischen Gründen werde eine *Teilmenge* $E$ von $X$ $\mathfrak{T}$*-extremal* genannt, wenn $E$ nicht leer und kompakt ist und weiter folgende Eigenschaft besitzt:

(E) *Aus* $x \in E$ *und* $(x, T) \in \mathfrak{T}$ *folgt stets:* $T \subset E$.

Offenbar ist der Raum $X$ $\mathfrak{T}$-extremal. Eine einpunktige Menge $E = \{x_0\}$ ist genau dann $\mathfrak{T}$-extremal, wenn der Punkt $x_0$ $\mathfrak{T}$-extremal ist. Dies folgt aus $(T_2)$.

Mit einer einfachen Überlegung, die übrigens in völliger Analogie zu einer Stelle im Beweis des Satzes 1 in [2] verläuft und daher dem Leser überlassen werden kann, beweist man folgenden

**Hilfssatz 1.** *Das System* $\mathfrak{E}$ *aller* $\mathfrak{T}$*-extremalen Teilmengen von* $X$ *ist hinsichtlich der Relation* $\supset$ *induktiv geordnet.*

Nach dem Zornschen Lemma ist daher in jeder Menge aus $\mathfrak{E}$ mindestens eine *minimale* $\mathfrak{T}$-extremale Menge enthalten. Das angekündigte Minimumprinzip gibt hinreichende Bedingungen dafür an, daß jede minimale Menge aus $\mathfrak{E}$ einpunktig ist.

Zu diesem Zweck betrachten wir auf $X$ definierte, *numerische* Funktionen $f$ mit der folgenden Eigenschaft:

(F) *Ist* $(x, T)$ *ein beliebiges Element von* $\mathfrak{T}$ *und gilt*

$$f(x) \leqq f(y) \quad \textit{für alle } y \in T,$$

*so gilt sogar*

$$f(x) = f(y) \quad \textit{für alle } y \in T.$$

Es bezeichne $\mathscr{F}_{\mathfrak{T}} = \mathscr{F}_{\mathfrak{T}}(X)$ die Menge aller auf $X$ definierten, *nach unten halbstetigen, numerischen* Funktionen mit dieser Eigenschaft (F). $\mathscr{F}_{\mathfrak{T}}$ enthält dann z. B. alle konstanten, numerischen Funktionen auf $X$, ist also nicht leer.

Der Zusammenhang zwischen $\mathscr{F}_{\mathfrak{T}}$ und den $\mathfrak{T}$-extremalen Mengen wird hergestellt durch

**Hilfssatz 2.** *Für jede Funktion* $f \in \mathscr{F}_{\mathfrak{T}}$ *ist die Menge* $E_f$ *ihrer Minimalstellen* $\mathfrak{T}$*-extremal.*

Beweis. Zunächst ist die Menge $E_f$ nicht leer und kompakt, da $f$ nach unten halbstetig und $X$ kompakt ist (vgl. [5], p. 112). Es sei weiter $(x, T)$ ein Element aus $\mathfrak{T}$ mit $x \in E_f$. Dann gilt $f(x) \leqq f(y)$ für alle $y \in X$, also insbesondere für alle $y \in T$. Wegen (F) ist daher $f(x) = f(y)$ für alle $y \in T$, also $T \subset E_f$.

Das allgemeine Minimumprinzip lautet nun:

**Satz 1.** *Es sei* $X$ *ein nicht leerer, kompakter Raum und* $\mathfrak{T}$ *eine Teilmenge von* $X \times \mathfrak{P}(X)$ *mit den Eigenschaften* $(T_1)$ *und* $(T_2)$. *Wenn dann die Funktionenmenge* $\mathscr{F}_{\mathfrak{T}}$ *wenigstens je zwei verschiedene, nicht* $\mathfrak{T}$*-extremale Punkte von* $X$ *trennt, so besitzt jede Funktion aus* $\mathscr{F}_{\mathfrak{T}}$ *mindestens eine* $\mathfrak{T}$*-extremale Minimalstelle.*

Beweis. Wegen der vorangestellten zwei Hilfssätze und der Bemerkung im Anschluß an Hilfssatz 1 ist nur noch zu zeigen, daß jede minimale $\mathfrak{T}$-extremale Menge $M$

einen $\mathfrak{T}$-extremalen Punkt enthält. Die Annahme des Gegenteiles führt aber wie folgt zu einem Widerspruch. Die minimale Menge $M \in \mathfrak{E}$ enthalte keinen $\mathfrak{T}$-extremalen Punkt. Nach der im Anschluß an die Eigenschaft (E) gemachten Bemerkung enthält dann $M$ mindestens zwei verschiedene (nicht $\mathfrak{T}$-extremale) Punkte. Da $\mathscr{F}_{\mathfrak{T}}$ nach Voraussetzung je zwei verschiedene, nicht $\mathfrak{T}$-extremale Punkte trennt, gibt es also eine auf $M$ nicht konstante Funktion $g \in \mathscr{F}_{\mathfrak{T}}$. Es bezeichne $g_M$ die Restriktion von $g$ auf $M$. Dann ist $g_M$ auf dem nicht leeren kompakten Raum $M$ nach unten halbstetig, also die Menge $M'$ aller Minimalstellen von $g_M$ nicht leer und kompakt. Weiter besitzt $M'$ die Eigenschaft (E). Es sei nämlich $(x, T) \in \mathfrak{T}$ und $x \in M'$. Da $M$ $\mathfrak{T}$-extremal und $M' \subset M$ ist, folgt zunächst: $T \subset M$. Also gilt $g(x) = g_M(x) \leqq g_M(y) = g(y)$ für alle $y \in M$, insbesondere also für alle $y \in T$. Wegen (F) ist daher $g$ auf $T$ konstant und gleich $g(x)$, also ist $T \subset M'$. Damit ist $M'$ eine $\mathfrak{T}$-extremale Teilmenge von $M$; $M'$ ist eine echte Teilmenge von $M$, da $g$ auf $M$ nicht konstant ist. Wie zu zeigen war, führt unsere Annahme daher zu einem Widerspruch, nämlich zu einem Widerspruch zur Minimaleigenschaft von $M$.

Bemerkungen. 1. Die im Satz 1 geforderte Trennungsbedingung ist speziell dann erfüllt, wenn $\mathscr{F}_{\mathfrak{T}}$ sogar je zwei verschiedene Punkte von $X$ trennt. Im folgenden wird dieser Fall immer vorliegen. Für Anwendungen in der Theorie der ersten Randwertaufgabe benötigen wir aber die verwendete schwächere Trennungsbedingung.

2. Zu einem speziellen System $\mathfrak{T}$ gelangt man folgendermaßen: Wie in [2] sei $\mathfrak{S}$ eine Teilmenge von $\mathfrak{P}(X)$ derart, daß jede Menge $S \in \mathfrak{S}$ mindestens *zweipunktig* ist. Aus $\mathfrak{S}$ konstruieren wir nun eine Menge $\mathfrak{T} \subset X \times \mathfrak{P}(X)$: Es sei $x$ ein beliebiger Punkt aus $X$ und $S_0$ die Vereinigung aller Mengen $S \in \mathfrak{S}$. Liegt $x$ nicht in $S_0$, so werde als einziges Paar $(x, T)$ dasjenige mit $T = \{x\}$ in $\mathfrak{T}$ aufgenommen; liegt jedoch $x$ in $S_0$, so werden alle Paare $(x, T)$ mit $T \in \mathfrak{S}$ und $x \in T$ in $\mathfrak{T}$ aufgenommen. Offenbar besitzt dann $\mathfrak{T}$ die Eigenschaften ($T_1$) und ($T_2$). Es bezeichne nun weiter, wie auch in [2], $\mathscr{K}_{\mathfrak{S}}$ die Menge aller auf $X$ nach unten halbstetigen, numerischen Funktionen $f$ mit folgender Eigenschaft: *Nimmt für irgendeine Menge $S \in \mathfrak{S}$ die Restriktion von $f$ auf $S$ ihr Minimum (auf $S$) an, so ist $f$ auf $S$ konstant.* Wie man sofort sieht, gilt dann: $\mathscr{K}_{\mathfrak{S}} = \mathscr{F}_{\mathfrak{T}}$. Trennt daher $\mathscr{K}_{\mathfrak{S}}$ die Punkte von $X$, so besitzt nach Satz 1 jede Funktion aus $\mathscr{K}_{\mathfrak{S}}$ eine $\mathfrak{T}$-extremale Minimalstelle. Dies ist aber gerade die Aussage von Satz 1 in [2], da nach Konstruktion von $\mathfrak{T}$ und wegen der Voraussetzung über $\mathfrak{S}$ genau die nicht in $S_0$ gelegenen Punkte von $X$ $\mathfrak{T}$-extremal sind.

**3. Durch Integralungleichungen festgelegte Minimalstellen.** Wir geben hier eine Anwendung des Satzes 1, die dessen Vorläufer in [2] nicht gestattet.

Es sei wieder $X$ ein nicht leerer, *kompakter* Raum. Ferner sei $\mathscr{E}$ eine Menge von auf $X$ *nach unten halbstetigen, numerischen* Funktionen $h$ mit

(1) $$h(x) > -\infty \quad \text{für alle } x \in X.$$

Jede Funktion aus $\mathscr{E}$ ist dann nach unten beschränkt.

Für jeden Punkt $x \in X$ bezeichne $\mathscr{M}_x = \mathscr{M}_x(\mathscr{E})$ die Menge aller Radonschen Maße $\mu \geqq 0$ auf $X$ mit folgenden zwei Eigenschaften:

($\mathbf{M}_1$) $$\int d\mu = 1;$$

($\mathbf{M}_2$) $$\int^* h\, d\mu \leqq h(x) \quad \text{für alle } h \in \mathscr{E}.$$

Hierbei bezeichnet $\int^* h\,d\mu$ das Oberintegral von $h$. Seine Definition sowie die wichtigsten Eigenschaften findet man bei N. BOURBAKI [6] (p. 172—173, exercices 5, 6). Wegen unserer speziellen Annahmen über die Funktionen $h \in \mathscr{E}$ ist $\int^* h\,d\mu = \int h\,d\mu$, wenn $h$ integrierbar ist, und sonst $\int^* h\,d\mu = +\infty$ (vgl. [6], p. 150).

Die Menge $\mathscr{M}_x$ ist niemals leer; stets ist das Maß $\varepsilon_x$ der *Einheitsmasse im Punkte* $x$ Element von $\mathscr{M}_x$.

Unter den über $X$ und $\mathscr{E}$ gemachten Voraussetzungen gilt dann:

**Satz 2.** *Trennt die Funktionenmenge $\mathscr{E}$ die Punkte von $X$, so besitzt jede Funktion aus $\mathscr{E}$ mindestens eine Minimalstelle $x \in X$ mit $\mathscr{M}_x = \{\varepsilon_x\}$.*

Beweis. Es sei $\mathfrak{T} \subset X \times \mathfrak{P}(X)$ die Menge aller Paare $(x, T_\mu)$, in welchen $x$ ein beliebiger Punkt aus $X$ und $T_\mu$ der Träger irgendeines Maßes $\mu \in \mathscr{M}_x$ ist. Dann besitzt $\mathfrak{T}$ die Eigenschaften $(\mathrm{T}_1)$ und $(\mathrm{T}_2)$, da $\mathscr{M}_x$ nicht leer und nach $(\mathrm{M}_1)$ nur Maße der Gesamtmasse 1 enthält. Nun ist ein Maß $\mu \geqq 0$ mit $\int d\mu = 1$ bekanntlich dann und nur dann gleich einem Maß $\varepsilon_x$ mit $x \in X$, wenn sein Träger $T_\mu$ einpunktig (und dann gleich $\{x\}$) ist. $\mathfrak{T}$-extremal sind daher genau die Punkte $x \in X$ mit $\mathscr{M}_x = \{\varepsilon_x\}$. Die Behauptung folgt daher unmittelbar aus Satz 1, wenn $\mathscr{E} \subset \mathscr{F}_{\mathfrak{T}}$ gilt. Dies aber ergibt sich aus dem nachstehenden Hilfssatz, der damit den Beweis von Satz 2 zu Ende führt.

**Hilfssatz 3.** *Es sei $h$ eine auf einem kompakten Raum $X$ definierte, nach unten halbstetige und nach unten beschränkte, numerische Funktion. Weiter sei $\mu$ ein positives Radonsches Maß auf $X$ mit*

$$\int d\mu = 1 \quad \textit{und} \quad \int^* h\,d\mu \leqq h(x). \tag{2}$$

*Gilt dann $h(x) \leqq h(y)$ für jeden Punkt $y$ des Trägers $T$ von $\mu$, so gilt sogar $h(x) = h(y)$ für alle $y \in T$.*

Beweis. Die Behauptung ist trivial im Falle: $\int^* h\,d\mu = +\infty$. Nach (2) ist dann nämlich $h(x) = +\infty$ und somit $h(x) = h(y)$ für alle $y \in T$. Also kann $\int^* h\,d\mu < +\infty$ und damit $h$ als integrierbar bezüglich $\mu$ angenommen werden (vgl. die im Anschluß an $(\mathrm{M}_2)$ gemachte Bemerkung). Wir führen folgende Bezeichnungen ein:

$$\beta = \inf_{y \in T} h(y); \tag{3}$$

$$T' = \{y \in T : h(y) = \beta\}. \tag{4}$$

Dann ist $T'$ eine nicht leere, kompakte Teilmenge von $T$ und $\beta$ eine reelle Zahl. Es sei weiter $K$ eine kompakte Menge mit $\emptyset \neq K \subset T$ und $K \cap T' = \emptyset$. Dann folgt aus (2) und einfachen Eigenschaften des Integrals:

$$h(x) \geqq \int h\,d\mu = \int_T h\,d\mu = \int_K h\,d\mu + \int_{T \cap \complement K} h\,d\mu \geqq \mu(K) \inf_{y \in K} h(y) + \beta\,\mu(T \cap \complement K). \tag{5}$$

Nun ist $K$ zu $T'$ fremd und somit $\inf\limits_{y \in K} h(y) > \beta$. Wäre daher $\mu(K) > 0$, so würde aus (2) und (5) folgen

$$h(x) > \beta\,\mu(K) + \beta\,\mu(T \cap \complement K) = \beta\,\mu(T) = \beta. \tag{6}$$

Dies aber kann nicht sein, denn nach Annahme ist $h(x) \leqq h(y)$ für alle $y \in T$, also auch $h(x) \leqq \beta$. Somit ist $\mu(K) = 0$ für jedes derartige $K$ und wegen $\mu(\complement T) = 0$ dann sogar für jede zu $T'$ fremde kompakte Menge $K \subset X$. Hieraus folgt $\mu(\complement T') = 0$, also $T \subset T'$ und somit $T = T'$. Also ist $h(y) = \beta \geqq h(x)$ für alle $y \in T$. Aus

$$h(x) \geqq \int h\,d\mu = \int_T h\,d\mu - \beta \int d\mu = \beta \tag{7}$$

folgt schließlich $h(x) = h(y)$ für alle $y \in T$.

Bemerkungen. 1. Der Satz 2 verliert seine Gültigkeit, wenn man zuläßt, daß die nach unten halbstetigen Funktionen $h \in \mathscr{E}$ auch $-\infty$ als Wert annehmen dürfen. Der kompakte Raum $X$ bestehe nämlich z. B. nur aus zwei verschiedenen Punkten $x$ und $y$. Die Menge $\mathscr{E}$ bestehe aus der einen (stetigen) Funktion $h: h(x) = -\infty$, $h(y) = 0$. Dann sind alle Voraussetzungen des Satzes 2 mit Ausnahme von (1) erfüllt. Für die (einzige) Minimalstelle $x$ von $h$ besteht jedoch $\mathscr{M}_x$ aus allen Maßen $\mu = \alpha\varepsilon_x + (1-\alpha)\varepsilon_y$ mit $0 < \alpha \leqq 1$. Weiter zeigt dieses Beispiel, daß auch bei Hilfssatz 3 nicht auf die Forderung (1) verzichtet werden kann.

2. Selbstverständlich gilt Satz 2 auch für die Menge $\hat{\mathscr{E}}$ aller nach unten halbstetigen und nach unten beschränkten Funktionen $h$ auf $X$ mit $\int^* h\,d\mu \leqq h(x)$ für alle $x \in X$ und $\mu \in \mathscr{M}_x(\mathscr{E})$. $\hat{\mathscr{E}}$ ist übrigens ein konvexer Kegel, der wegen $(M_1)$ die konstanten Funktionen auf $X$ enthält; ferner liegt mit je zwei Funktionen auch deren untere Einhüllende in $\hat{\mathscr{E}}$. Vgl. hierzu auch G. Choquet—J. Deny [7].

3. Die Punkte $x \in X$ mit $\mathscr{M}_x(\mathscr{E}) = \{\varepsilon_x\}$ werden in [3] *$\mathscr{E}$-extremal* genannt und in speziellen Fällen (vor allem im Zusammenhang mit dem Dirichletschen Problem) genau untersucht. Nach einem Vorschlag von E. Bishop und K. de Leeuw [4] heißt die Menge der $\mathscr{E}$-extremalen Punkte auch die *Choquetsche Berandung* von $X$ bezüglich $\mathscr{E}$. In [4] ist allerdings $\mathscr{E}$ stets ein Vektorraum stetiger reeller Funktionen.

**4. Folgerungen aus Satz 2.** Aus Satz 2 folgt beinahe unmittelbar der Satz 2 aus [2] und das zugehörige Korollar. Zur Vermeidung von Wiederholungen beziehen wir uns nur auf dieses. Es sei also $X$ eine kompakte konvexe Menge in einem lokal-konvexen Raum und $h$ eine auf $X$ nach unten halbstetige, konkave, numerische Funktion mit Werten $> -\infty$. Ist dann $x \in X$ nicht (geometrischer) Extremalpunkt von $X$, so existieren Punkte $a, b \in X$ und eine reelle Zahl $\lambda$ derart, daß gilt: $x = \lambda a + (1-\lambda) b$, $a \neq b$, $0 < \lambda < 1$. Dann ist aber $\mu = \lambda\varepsilon_a + (1-\lambda)\varepsilon_b$ ein von $\varepsilon_x$ verschiedenes positives Maß auf $X$ mit $\int^* h\,d\mu \leqq h(x)$ und $\int d\mu = 1$. Da die Menge dieser Funktionen $h$ nach [2], Nr. 4 den Voraussetzungen des Satzes 2 genügt, folgt aus diesem, daß jede derartige Funktion $h$ einen Extremalpunkt von $X$ als Minimalstelle besitzt. Dies behauptete aber gerade das Korollar von Satz 2 in [2].

Weiter geht Satz 2 über in das bereits in der Einleitung genannte théorème 1 aus [3], wenn $\mathscr{E}$ speziell ein Vektorraum $\mathscr{H}$ von stetigen reellen Funktionen auf $X$ ist, der die konstanten Funktionen als Elemente enthält.

Als weitere Folgerung beweisen wir noch den folgenden Satz, aus dem übrigens leicht die Existenz Šilovscher Berandungen bezüglich Funktionenalgebren gefolgert werden kann (vgl. [3]):

**Satz 3.** *Es sei $X$ ein nicht leerer, kompakter Raum und $\mathscr{A}$ eine die Punkte von $X$ trennende Unteralgebra der Algebra $\mathscr{C}(X, \mathbf{C})$ aller auf $X$ stetigen, komplexwertigen Funktionen (über dem Körper $\mathbf{C}$ der komplexen Zahlen). Für jeden Punkt $x \in X$ bezeichne $\mathscr{M}_x(\mathscr{A})$ die Menge aller Radonschen Maße $\mu \geqq 0$ auf $X$ mit $\int d\mu = 1$ und $\int f\, d\mu = f(x)$ für alle $f \in \mathscr{A}$. Dann besitzt der absolute Betrag einer jeden Funktion aus $\mathscr{A}$ mindestens eine Maximalstelle $x$ mit $\mathscr{M}_x(\mathscr{A}) = \{\varepsilon_x\}$.*

Beweis. Es sei $\mathscr{E}$ die Menge aller Funktionen $h = -|f|$ mit $f \in \mathscr{A}$. Dann ist jede Funktion aus $\mathscr{E}$ reellwertig und stetig. Da ferner aus $f(x) = \int f\, d\mu$ folgt $|f(x)| \leqq \int |f|\, d\mu$, gilt $\mathscr{M}_x(\mathscr{A}) \subset \mathscr{M}_x(\mathscr{E})$ für alle $x \in X$. Schließlich trennt mit $\mathscr{A}$ auch $\mathscr{E}$ die Punkte von $X$. Zu je zwei verschiedenen Punkten $x, y \in X$ gibt es nämlich ein $f \in \mathscr{A}$ mit $f(x) \neq f(y)$. Ist $|f(x)| \neq |f(y)|$, so trennt $h = -|f|$ die beiden Punkte. Ist aber $|f(x)| = |f(y)|$, so leistet das Entsprechende die Funktion $h = -|f + \alpha f^2|$ mit $\alpha = -1/f(x)$. Die Behauptung folgt daher jetzt unmittelbar aus Satz 2.

**Korollar.** *Unter den Voraussetzungen von Satz 3 besitzt auch der Realteil $\Re f$ einer jeden Funktion $f \in \mathscr{A}$ eine Maximalstelle (und eine Minimalstelle) $x$ mit $\mathscr{M}_x(\mathscr{A}) = \{\varepsilon_x\}$.*

Beweis. Es sei $\mathscr{H}$ der Vektorraum aller Funktionen $\Re f$ mit $f \in \mathscr{A}$. Dann ist $\mathscr{H}$ auch der Vektorraum aller Imaginärteile von Funktionen aus $\mathscr{A}$. Daher trennt $\mathscr{H}$ die Punkte von $X$. Für jedes $x \in X$ ist offenbar $\mathscr{M}_x(\mathscr{H}) = \mathscr{M}_x(\mathscr{A})$. Die Behauptung folgt daher wiederum aus Satz 2.

Bemerkung. Aus Satz 3 und der Isotonie der Funktion $x \to \log x$ auf der reellen Halbgeraden $x \geqq 0$ folgt, daß auch jede Funktion $\log |f|$ mit $f \in \mathscr{A}$ und, da $\mathscr{A}$ eine Algebra ist, dann auch jede Funktion $\sum_{\nu=1}^{n} \lambda_\nu \log |f_\nu|$ mit $f_\nu \in \mathscr{A}$ und rationalen Koeffizienten $\lambda_\nu > 0$ eine Maximalstelle $x$ mit $\mathscr{M}_x(\mathscr{A}) = \{\varepsilon_x\}$ besitzt. Es wäre interessant zu wissen, ob diese Eigenschaft verloren geht, wenn man beliebige positive reelle Zahlen als Koeffizienten $\lambda_\nu$ zuläßt. Man vergleiche in diesem Zusammenhang R. Arens—I. M. Singer [1], insbesondere theorem 7.1.

## Literaturverzeichnis


[1] R. Arens and I. M. Singer, Function values as boundary integrals. Proc. Amer. Math. Soc. **5**, 735—745 (1954).

[2] H. Bauer, Minimalstellen von Funktionen und Extremalpunkte. Arch. Math. **9**, 389—393 (1958).

[3] H. Bauer, Frontière de Šilov et problème de Dirichlet. Séminaire de Théorie du Potentiel dirigé par M. Brelot, G. Choquet et J. Deny. 3e année, Institut H. Poincaré, Paris 1959.

[4] E. Bishop and K. de Leeuw, The representations of linear functionals by measures on sets of extreme points. Ann. Inst. Fourier **9**, 305—331 (1959).

[5] N. Bourbaki, Topologie générale, Chap. III—IV. Actual. sci. industr. **1143**, Paris 1951.

[6] N. Bourbaki, Intégration, Chap. I—IV. Actual. sci. industr. **1175**, Paris 1952.

[7] G. Choquet et J. Deny, Ensembles semi-réticulés et ensembles réticulés de fonctions continues. J. Math. pur. appl., IX. Sér., **36**, 179—189 (1957).


Eingegangen am 23. 3. 1960

# Šilovscher Rand und Dirichletsches Problem [R17]

Ann. Inst. Fourier (Grenoble) **11** (1961), 89–136
[Zbl. Math. 98.06902; MR 25 #443]

## EINLEITUNG

Aus der Theorie der Banach-Algebren kennt man Begriff und Bedeutung des Šilovschen Randes : Ist $\mathcal{A}$ eine *Algebra* von stetigen, komplexwertigen Funktionen auf einem kompakten Raum X, welche die konstanten Funktionen enthält und die Punkte von X trennt, so existiert nach Šilov [25] unter allen abgeschlossenen Teilmengen S von X mit der Eigenschaft, daß jede der Funktionen $|f|$ mit $f \in \mathcal{A}$ eine in S gelegene Maximalstelle besitzt, eine kleinste Menge. Diese wird mit $\partial_{\mathcal{A}}X$ bezeichnet und heißt der Šilovsche Rand von X bezüglich $\mathcal{A}$. Ist z.B. X die kompakte Kreisscheibe $|z| \leqslant 1$ in der komplexen $z$-Ebene und $\mathcal{A}$ die Algebra aller auf X stetigen, in der offenen Kreisscheibe holomorphen Funktionen, so folgt aus dem Maximum-Prinzip der Funktionentheorie, daß $\partial_{\mathcal{A}}X$ der topologische Rand von X, also die Kreislinie $|z| = 1$ ist.

Eine analoge Situation liegt vor, wenn man die Algebra $\mathcal{A}$ ersetzt durch einen *Vektorraum* $\mathcal{H}$ von auf einem kompakten Raum X definierten, stetigen, reellwertigen Funktionen, welcher die konstanten reellen Funktionen enthält und die Punkte von X trennt. Dann existiert nach Milman [33] und Arens-Singer [1] eine kleinste abgeschlossene Menge $\partial_{\mathcal{H}}X$ von X mit der Eigenschaft, daß jede Funktion aus $\mathcal{H}$ eine in $\partial_{\mathcal{H}}X$ gelegene Maximalstelle besitzt. Man nennt auch hier $\partial_{\mathcal{H}}X$ den Šilovschen Rand von X bezüglich $\mathcal{H}$.

Der Ausgangspunkt der vorliegenden Arbeit war die Frage nach der Bedeutung dieser Begriffsbildung für die Theorie des *Dirichletschen Problems*, oder genauer: Es sei $\Omega$ eine offene, relativ kompakte Menge im $\mathbf{R}^n$. Welches ist der Šilovsche Rand der abgeschlossenen Hülle $\overline{\Omega}$ von $\Omega$ bezüglich des Vektorraumes $\mathcal{H}_\Omega$ aller in $\overline{\Omega}$ stetigen, in $\Omega$ harmonischen, reellen Funktionen? In § 6 dieser Arbeit wird gezeigt, daß der gesuchte Šilovsche Rand die abgeschlossene Hülle der Menge der sog. regulären Randpunkte von $\Omega$ ist.

Darüber hinaus hat aber diese spezielle Frage zu einer allgemeineren Fragestellung geführt, nämlich zu einem *abstrakten Dirichletschen Problem*, das sich vom klassischen grob gesagt dadurch unterscheidet, daß der Raum $\overline{\Omega}$ durch einen beliebigen kompakten Raum X, der Vektorraum $\mathcal{H}_\Omega$ durch einen Vektorraum $\mathcal{H}$ stetiger reeller Funktionen auf X und der euklidische Rand $\Omega^*$ von $\Omega$ durch den Šilovschen Rand $X^* = \partial_{\mathcal{H}}X$ ersetzt wird. Von $\mathcal{H}$ wird zunächst nur vorausgesetzt, daß die konstanten Funktionen zu $\mathcal{H}$ gehören und daß die Punkte von X durch $\mathcal{H}$ getrennt werden. Durch einen sich in natürlicher Weise anbietenden Prozeß wird sodann $\mathcal{H}$ zu einem Vektorraum $\hat{\mathcal{H}}$ stetiger reeller Funktionen erweitert, der mit $\mathcal{H}_\Omega$ eine wichtige Eigenschaft gemeinsam hat: Es gibt eine bezüglich gleichmäßiger Konvergenz abgeschlossene Menge $\mathfrak{S}$ stetiger reeller Funktionen auf X, welche mit je zwei Funktionen auch deren untere Einhüllende enthält; ferner gilt $\hat{\mathcal{H}} = \mathfrak{S} \cap (-\mathfrak{S})$. Bezüglich $\mathcal{H}_\Omega$ hat die Menge aller auf $\overline{\Omega}$ stetigen, in $\Omega$ superharmonischen Funktionen diese Eigenschaft. Daher ist erst $\hat{\mathcal{H}}$ das genaue Analogon zu $\mathcal{H}_\Omega$. Das abstrakte Dirichletsche Problem besteht dann in der Aufgabe, notwendige und hinreichende Bedingungen dafür anzugeben, daß jede auf $X^*$ stetige reelle Funktion zu einer Funktion aus $\hat{\mathcal{H}}$ fortgesetzt werden kann. Gewisse Ansatzpunkte zu einem abstrakten, wenn auch andersartigen Dirichletschen Problem finden sich bereits bei Arens-Singer [1].

Innerhalb dieser allgemeinen Theorie spielt eine in $X^*$ dicht liegende Punktmenge, nämlich die Menge der sog. *$\mathcal{H}$-extremalen Punkte* von X eine wichtige Rolle. Die regulären Randpunkte von $\Omega$ erweisen sich hinterher als identisch mit den $\mathcal{H}_\Omega$-extremalen Punkten von $\overline{\Omega}$. Daher werden in § 1 zunächst diese ausgezeichneten Punkte eingeführt und mit ihrer Hilfe ein neuer Beweis für die Existenz des Šilovschen Randes gegeben. In den Paragraphen 2 und 3 wird dann das abstrakte Dirichletsche Problem entwickelt.

Der Rest der Arbeit ist im wesentlichen Anwendungen gewidmet. Es zeigt sich nämlich, daß die allgemeine Theorie auch nützliche Anwendungen außerhalb des Fragenkreises des klassischen Dirichletschen Problems besitzt. In § 4 wird X als *konvexe* kompakte Menge in einem lokal-konvexen Raum E angenommen und für $\mathcal{H}$ der Vektorraum der auf X eingeschränkten, stetigen, *affin-linearen* Funktionen auf E gewählt. Dann erweist sich $\hat{\mathcal{H}}$ als der Vektorraum aller auf X stetigen, affin-linearen Funktionen. Die $\mathcal{H}$-extremalen Punkte fallen mit den (geometrischen) Extremalpunkten von X zusammen. Das abstrakte Dirichletsche Problem reduziert sich also hier auf die Frage, wann jede stetige reelle Funktion auf der abgeschlossenen Hülle aller Extremalpunkte zu einer in X stetigen, affin-linearen Funktion fortgesetzt werden kann. Hierfür erweist es sich als notwendig und hinreichend, daß X ein Simplex im Sinne von Choquet [21] mit abgeschlossener Extremalpunktmenge ist. Die allgemeine Theorie liefert Kennzeichnungen dieser Klasse von Simplexen durch innere Eigenschaften.

Diese Anwendung wurde derjenigen über das klassische Dirichletsche Problem vorangestellt, da sie mit der allgemeinen Theorie in innigem Zusammenhang steht. Man kann nämlich im Rahmen der allgemeinen Theorie den kompakten Raum X homöomorph in den schwach topologisierten, topologischen Dualraum E von $\hat{\mathcal{H}}$ einbetten und z.B. zeigen, daß das abstrakte Dirichletsche Problem genau dann lösbar ist, wenn die abgeschlossene konvexe Hülle des Bildes von X in E ein Simplex mit kompakter Extremalpunktmenge ist. Der § 5 dient der Klärung dieser Beziehungen zwischen der allgemeinen Theorie und der soeben beschriebenen speziellen Anwendung des § 4.

Erst der § 6 bringt dann die Anwendung auf das *klassische Dirichletsche Problem.* Die allgemeine Theorie liefert hierbei neuartige Einsichten in diesen Problemkreis. Hervorzuheben ist besonders die bereits erwähnte Tatsache, daß die regulären Randpunkte einer offenen, relativ kompakten Menge im $\mathbf{R}^n$ als *Extremalpunkte* gedeutet werden können.

Der § 7 bringt eine Anwendung auf das Dirichletsche Problem für *diskrete harmonische Funktionen*, die neuerdings wieder im Zusammenhang mit der Theorie der Markovschen Ketten an Bedeutung gewonnen haben [29]. Der abschließende § 8 weist im wesentlichen auf offene Fragen im Zusammenhang mit *Funktionenalgebren* hin.

Die wichtigsten Resultate dieser Arbeit wurden ohne Beweis in einer Note in den Comptes rendus de l'Académie des Sciences (Paris) [3] angekündigt. In gedrängter Form habe ich einen im Pariser Seminar über Potentialtheorie gehaltenen Vortrag über dieses Thema in [4] ausgearbeitet.

## BEZEICHNUNGEN

Im folgenden schließen wir uns weitgehend der Terminologie von N. Bourbaki an.

Mit $\mathbf{R}$ bzw. $\mathbf{C}$ soll stets der topologische Körper der *reellen* bzw. *komplexen Zahlen* (Zahlengerade bzw. komplexe Zahlenebene) bezeichnet werden. Jede Abbildung $f\colon A \to \mathbf{R}$ einer Menge A in die Zahlengerade heiße *reelle* oder reellwertige Funktion auf A. Jede Abbildung $f\colon A \to \overline{\mathbf{R}}$ von A in die durch Adjunktion von $\pm\infty$ kompaktifizierte Zahlengerade $\overline{\mathbf{R}}$ heiße *numerische Funktion* auf A.

Für jede Abbildung $f\colon A \to B$ und jede Menge $S \subset A$ bezeichne $f_S$ die *Restriktion* der Abbildung $f$ auf S. Eine Menge $\mathfrak{F}$ von Abbildungen einer Menge A in eine Menge B heißt *punktetrennend*, wenn zu je zwei Punkten $x, y \in A$ mit $x \neq y$ eine Abbildung $f \in \mathfrak{F}$ existiert mit $f(x) \neq f(y)$.

Ist X ein *kompakter* (und daher Hausdorffscher) Raum, so bezeichnen wir mit $\mathcal{C}(X) = \mathcal{C}(X, \mathbf{R})$ *bzw.* $\mathcal{C}(X, \mathbf{C})$ die Menge aller stetigen, reellen bzw. komplexen Funktionen auf X. Bezüglich der üblichen (punktweise definierten) Operationen ist $\mathcal{C}(X, \mathbf{R})$ bzw. $\mathcal{C}(X, \mathbf{C})$ eine *Algebra*, insbesondere also ein *Vektorraum* über $\mathbf{R}$ bzw. $\mathbf{C}$. Beide Algebren sollen stets mit der *Topologie der gleichmäßigen Konvergenz* versehen sein. Außerdem trägt $\mathcal{C}(X, \mathbf{R})$ die übliche, mit der Vektorraumstruktur verträgliche Ordnungsrelation $\leqslant$.

Für einen kompakten Raum X soll ferner $\mathcal{M}(X)$ stets die Menge aller *positiven (Radonschen) Maße* auf X bezeichnen. $\mathcal{M}(X)$ ist bezüglich der üblichen Operationen ein konvexer Kegel. Es sei stets mit der *vagen* Topologie versehen. Für einen beliebigen Punkt $x \in X$ bezeichne $\varepsilon_x$ das durch die *Einheitsmasse im Punkte x* definierte positive Maß auf X.

## § 1. — Extremalpunkte und Šilovscher Rand.

1. 1. *Definition des Šilovschen Randes.* — Es sei X ein *kompakter* topologischer Raum und $\mathfrak{E}$ eine nicht leere Menge von auf X definierten, nach unten halbstetigen, numerischen Funktionen. Dann nimmt bekanntlich jede Funktion $u \in \mathfrak{E}$ ihr (globales) *Minimum* auf X an, d.h. es gibt mindestens ein $x_0 \in X$ mit

$$u(x_0) = \inf_{x \in X} u(x).$$

Jeden derartigen Punkt $x_0$ nennen wir eine *Minimalstelle* von $u$.

Wir betrachten nun das System $\mathfrak{S} = \mathfrak{S}(\mathfrak{E})$ aller *kompakten* Teilmengen S von X mit der Eigenschaft, daß jede Funktion $u \in \mathfrak{E}$ mindestens eine in S gelegene Minimalstelle besitzt. Es gilt dann $X \in \mathfrak{S}$; also ist $\mathfrak{S}$ nicht leer.

Ferner ist $\mathfrak{S}$ *induktiv geordnet* bezüglich der Relation $\supset$. Es sei nämlich $\mathfrak{K}$ eine bezüglich $\supset$ total-geordnete Teilmenge von $\mathfrak{S}$. Dann genügt es zu zeigen, daß $T = \bigcap_{S \in \mathfrak{K}} S$ ein Element von $\mathfrak{S}$ ist. Zunächst ist T kompakt und nicht leer, da X kompakt und $\mathfrak{K}$ eine Filterbasis kompakter Teilmengen von X ist. Weiter sei $u \in \mathfrak{E}$ und $M_u$ die Menge aller Minimalstellen von $u$. Da $u$ nach unten halbstetig ist, so ist $M_u$ abgeschlossen. Aus $M_u \cap S \neq \emptyset$ für alle $S \in \mathfrak{S}$, also insbesondere für alle $S \in \mathfrak{K}$, folgt dann $M_u \cap T \neq \emptyset$, da $\{M_u \cap S : S \in \mathfrak{K}\}$ eine Filterbasis kompakter Teilmengen von X ist. Daher gilt: $T \in \mathfrak{S}$.

Nach dem Zornschen Lemma ist also in jeder Menge $S \in \mathfrak{S}$ mindestens ein minimales Element von $\mathfrak{S}$ enthalten. Die Existenz genau eines minimalen Elementes in $\mathfrak{S}$ ist daher gleichwertig mit der Existenz eines kleinsten Elementes in $\mathfrak{S}$.

Definition 1. — *Existiert im System* $\mathfrak{S}(\mathfrak{E})$ *eine kleinste Menge* $S_0$, *so heiße* $S_0$ *der Šilovsche Rand von* X *bezüglich* $\mathfrak{E}$; *in Zeichen:* $S_0 = \partial_{\mathfrak{E}} X$.

1. 2. *$\mathfrak{E}$-extremale Punkte.* — Im folgenden sollen hinreichende Bedingungen für die Existenz von $\partial_{\mathfrak{E}} X$ angegeben werden.

Die Anwendungen, auf die wir in den Paragraphen 4 und 6 zu sprechen kommen, werden zeigen, daß $\partial_{\mathcal{E}}X$ im allgemeinen « unnatürliche » Punkte enthält. Der folgende Beweis für die Existenz von $\partial_{\mathcal{E}}X$ (unter gewissen Zusatzvoraussetzungen) soll daher die Definition der « natürlichen Punkte von $\partial_{\mathcal{E}}X$ » zum Ausgangspunkt haben.

Hierzu bezeichne $\mathcal{M}_x = \mathcal{M}_x(\mathcal{E})$ für jeden Punkt $x \in X$ die Menge aller Maße $\mu \in \mathcal{M}(X)$ mit folgenden zwei Eigenschaften:

$$(1.1) \qquad \int d\mu = 1;$$

$$(1.2) \qquad \int^* u d\mu \leqslant u(x) \quad \textit{für alle} \quad u \in \mathcal{E}.$$

Hierbei bezeichnet $\int^* u d\mu$ das $\mu$-Oberintegral von $u$ (vgl. Bourbaki [14], pp. 172-173, exercices 5,6). Wenn eine Funktion $u \in \mathcal{E}$ nach unten beschränkt ist, wie das im folgenden fast immer der Fall sein wird, ist $\int^* u d\mu = \int u d\mu$ unter der Voraussetzung der Integrierbarkeit von $u$ und sonst $\int^* u d\mu = +\infty$ (vgl. [14], p. 150).

Die Menge $\mathcal{M}_x$ ist für kein $x \in X$ leer, da offenbar stets $\varepsilon_x$ zu $\mathcal{M}_x$ gehört.

Definition 2. — *Ein Punkt $x \in X$ heiße $\mathcal{E}$-extremal, wenn gilt:*

$$(1.3) \qquad \mathcal{M}_x(\mathcal{E}) = \{\varepsilon_x\}.$$

*Die Menge aller $\mathcal{E}$-extremalen Punkte von X werde mit $X_e = X_e(\mathcal{E})$ bezeichnet und heiße* [1] *der Choquetsche Rand von X bezüglich $\mathcal{E}$.*

Ohne Zusatzvoraussetzungen über $\mathcal{E}$ kann die Menge $X_e$ leer sein, wie die beiden folgenden Beispiele zeigen:

Beispiel 1. — Der Raum X enthalte mindestens zwei Punkte; $\mathcal{E}$ sei die Menge aller auf X konstanten reellen Funktionen.

Beispiel 2. — Der Raum X sei die mit der diskreten Topologie versehene zweipunktige Menge $\{a, b\}$. $\mathcal{E}$ enthalte

[1] Nach einem Vorschlag von Bishop und de Leeuw [8].

als einziges Element die folgende Funktion $u$: $u(a) = -\infty$, $u(b) = 0$.

Die durch diese Beispiele nahegelegten Zusatzvoraussetzungen genügen, um die Existenz $\mathcal{E}$-extremaler Punkte zu beweisen:

Satz 1. — *Die Menge $\mathcal{E}$ besitze folgende zusätzliche Eigenschaften:*

$$(1.4) \qquad \mathcal{E} \text{ trennt die Punkte von } X;$$

$$(1.5) \qquad \text{es ist } u(x) > -\infty \quad \text{für alle} \quad x \in X \quad \text{und} \quad u \in \mathcal{E}.$$

*Dann besitzt jede Funktion aus $\mathcal{E}$ mindestens eine $\mathcal{E}$-extremale Minimalstelle. Insbesondere ist der Choquetsche Rand $X_e(\mathcal{E})$ nicht leer.*

Der Beweis dieses Satzes wurde in [5] (Satz 2) als Folgerung aus einem allgemeinen Minimumprinzip erbracht.

1. 3. *$\mathcal{E}$-exponierte Punkte.* — Spezielle $\mathcal{E}$-extremale Punkte lernen wir im folgenden Beispiel kennen:

Beispiel 3. — Ein Punkt $x \in X$ heiße *$\mathcal{E}$-exponiert*, wenn eine Funktion $u \in \mathcal{E}$ existiert derart, daß gilt:

$$(1.6) \qquad -\infty < u(x) < u(y) \quad \text{für alle} \quad y \in X \quad \text{mit} \quad y \neq x.$$

Es sei $x$ ein solcher Punkt und $u$ eine Funktion aus $\mathcal{E}$ mit der Eigenschaft (1.6). Für jedes Maß $\mu \in \mathcal{M}_x(\mathcal{E})$ ist dann $u$ auf dem Träger $T_\mu$ von $\mu$ konstant gleich $u(x)$ (nach [5], Hilfssatz 3). Wegen (1.6) folgt hieraus: $T_\mu = \{x\}$; wegen (1.1) ist dann $\mu = \varepsilon_x$, also $\mathcal{M}_x = \{\varepsilon_x\}$. Daher ist jeder $\mathcal{E}$-exponierte Punkt auch $\mathcal{E}$-extremal. Die Umkehrung hiervon gilt jedoch nicht, wie schon Beispiel 4 zeigen wird [2].

Aus der Definition der $\mathcal{E}$-exponierten Punkte folgt ferner, daß jeder derartige Punkt in jeder Menge $S \in \mathfrak{S}(\mathcal{E})$ und damit auch in $\partial_{\mathcal{E}} X$ liegt, sofern der Šilovsche Rand existiert.

1. 4. *Existenz des Šilovschen Randes.* — Die bisherigen Voraussetzungen über $\mathcal{E}$ (einschließlich (1.4) und (1.5)) haben nach Satz 1 zur Folge, daß die abgeschlossene Hülle $\overline{X_e}$ des

[2] Vgl. auch die Bemerkung 3 in Nr. 4.3.

Choquetschen Randes zum System $\mathfrak{S}(\mathcal{E})$ gehört. Sie lassen aber nicht den Schluß zu, daß der Šilovsche Rand existiert; aus der Existenz von $\partial_{\mathcal{E}}X$ folgt ferner nicht die Gleichheit mit $\overline{X}_e$. Dies zeigen die beiden nächsten Beispiele:

Beispiel 4. — Der Raum X sei die diskret topologisierte, aus drei Punkten bestehende Menge $\{x_1, x_2, x_3\}$. Die (stetigen) reellen Funktionen auf X entsprechen dann eineindeutig den Vektoren des $\mathbf{R}^3$. Die Funktionenmenge $\mathcal{E}$ bestehe aus den durch den Vektoren (1, 0, 0), (0, 1, 0), (0, 0, 1) definierten Funktionen. Es ist offenbar: $X_e(\mathcal{E}) = X$. Der Šilovsche Rand existiert nicht, da alle zweipunktigen Teilmengen von X minimale Elemente von $\mathfrak{S}(\mathcal{E})$ sind.

Beispiel 5. — Der Raum X sei wie im Beispiel 4 definiert. Jedoch sei $\mathcal{E}$ jetzt die durch die Vektoren (1, 0, 0), (0, 1, 0), (0, 0, 1), (0, 1, 1) und (1, 0, 1) definierte Funktionenmenge. Wiederum ist $X_e(\mathcal{E}) = X$. Jedoch existiert jetzt der Šilovsche Rand: $\partial_{\mathcal{E}}X = \{x_1, x_2\}$.

Es gilt nun aber der folgende.

Satz 2. — *Die Menge $\mathcal{E}$ besitze neben* (1. 4) *und* (1. 5) *noch folgende Eigenschaft:*

$$u, v \in \mathcal{E} \Longrightarrow u + v \in \mathcal{E}. \tag{1. 7}$$

*Dann existiert der Šilovsche Rand $\partial_{\mathcal{E}}X$ und er ist gleich der abgeschlossenen Hülle des Choquetschen Randes $X_e(\mathcal{E})$.*

Beweis. Wir setzen $S_0 = \overline{X}_e$. Dann folgt aus Satz 1, daß $S_0$ zum System $\mathfrak{S}$ gehört. Es sei nun S ein beliebiges Element aus $\mathfrak{S}$; dann ist noch $S_0 \subset S$ oder die gleichwertige Relation $X_e \subset S$ zu beweisen. Wir betrachten hierzu einen beliebigen Punkt $x \in X$ und konstruieren ein spezielles Maß $\mu \in \mathcal{M}_x(\mathcal{E})$, welches von S getragen wird. Für jeden Punkt $x \in \complement S$ ist dann dieses Maß von $\varepsilon_x$ verschieden, also $\complement S \subset \complement X_e$ und somit $X_e \subset S$.

Zur Konstruktion von $\mu$ bezeichnen wir mit $\mathcal{E}_1$ die Menge aller Funktionen $u + \alpha$ mit $u \in \mathcal{E}$ und $\alpha \in \mathbf{R}$ einschließlich aller auf X konstanten reellen Funktionen. Dann überträgt sich die Eigenschaft (1.7) auf $\mathcal{E}_1$; ferner besitzt auch jede

Funktion aus $\mathcal{E}_1$ eine in S gelegene Minimalstelle, es ist also $S \in \mathfrak{S}(\mathcal{E}_1)$. Für jede Funktion $f \in \mathcal{C}(S)$ setzen wir

$$(1.8)\qquad p(f) = \inf_{f \leqslant v_S,\ v \in \mathcal{E}_1} v(x),$$

wobei wie auf p. 93 vereinbart, $v_S$ die Restriktion von $v$ auf S bezeichnet. Dann gilt für alle $f \in \mathcal{C}(S)$:

$$(1.9)\qquad |p(f)| < +\infty.$$

Es ist nämlich $p(f) < +\infty$, da z.B. konstante Funktionen $v \in \mathcal{E}_1$ existieren mit $f \leqslant v_S$; ist weiter $\alpha \in \mathbf{R}$ eine untere Schranke von $f$, so gilt $p(f) \geqslant \alpha$ wegen $S \in \mathfrak{S}(\mathcal{E}_1)$. Diese Überlegung zeigt ferner, daß

$$(1.10)\qquad p(\alpha) = \alpha$$

ist für jede auf S konstante reelle Funktion $\alpha$. Aus $\mathcal{E}_1 + \mathcal{E}_1 \subset \mathcal{E}_1$ folgt noch:

$$(1.11)\qquad p(f+g) \leqslant p(f) + p(g) \quad \text{für beliebige} \quad f,\ g \in \mathcal{C}(S).$$

Nun wird durch $\nu_0(\alpha) = p(\alpha) = \alpha$ ein Homomorphismus der additiven Gruppe aller auf S konstanten reellen Funktionen in die additive Gruppe $\mathbf{R}$ definiert. ($\nu_0$ ist natürlich sogar ein Isomorphismus auf $\mathbf{R}$.) Nach AUMANN [2] (Satz 2) kann dann $\nu_0$ fortgesetzt werden zu einem Homomorphismus $\nu : \mathcal{C}(S) \to \mathbf{R}$ der additiven Gruppe $\mathcal{C}(S)$ in $\mathbf{R}$, und zwar derart, daß

$$(1.12)\qquad \nu(f) \leqslant p(f) \quad \text{für alle} \quad f \in \mathcal{C}(S)$$

gilt (3). Dann ist $\nu$ ein positives Maß auf S. In der Tat: aus $f \in \mathcal{C}(S)$ und $f \geqslant 0$ folgt $-f \leqslant 0$ und somit $p(-f) \leqslant 0$. Nach (1.12) ist daher $-\nu(f) = \nu(-f) \leqslant p(-f) \leqslant 0$, also $\nu(f) \geqslant 0$. Nach einem bekannten Satz ([14], p. 35, prop. 1) ist dann aber $\nu$ eine positive Linearform auf dem Vektorraum $\mathcal{C}(S)$, also $\nu \in \mathcal{M}(S)$.

Das Bildmaß $\mu = j(\nu)$ von $\nu$ bei der kanonischen Injektion $j : S \to X$ leistet dann das Verlangte. Zunächst ist

(3) Wir verwenden den zitierten Satz von AUMANN nur in der folgenden speziellen Form, die sich leicht aus [2] ableiten läßt: *Es sei* H *eine abelsche (additiv geschriebene) Gruppe und* E *eine Untergruppe von* H; *ferner sei p eine reelle Funktion auf H mit:* $p(0) = 0$ *und* $p(f+g) \leqslant p(f) + p(g)$ *für alle* $f,\ g \in$ H. *Dann existiert zu jedem Homomorphismus* $\nu_0 : E \to \mathbf{R}$ *mit* $\nu_0(f) \leqslant p(f)$ *für alle* $f \in$ E *ein Homomorphismus* $\nu : H \to \mathbf{R}$ *mit folgenden Eigenschaften:* $\nu$ *setzt* $\nu_0$ *fort und es ist* $\nu(f) \leqslant p(f)$ *für alle* $f \in$ H.

definitionsgemäß $\mu$ ein positives Maß auf X mit $\mu(\complement S) = 0$ und $\int g\, d\mu = \int g_S\, d\nu$ für alle $g \in \mathcal{C}(X)$. Speziell ist also: $\int d\mu = \int d\nu = \nu_0(1) = 1$. Für jede Funktion $u \in \mathcal{E}$, welche gemäß (1. 5) nach unten beschränkt ist, gilt:

$$(1.\,13) \qquad \int^* u\, d\mu = \sup_{g \leqslant u, g \in \mathcal{C}(X)} \int g\, d\mu.$$

Für jedes $g \in \mathcal{C}(X)$ mit $g \leqslant u$ ist $g_S \leqslant u_S$ und nach (1. 12): $\int g d\mu = \int g_S d\nu = \nu(g_S) \leqslant p(g_S)$. Nun ist aber $\mathcal{E} \subset \mathcal{E}_1$, also auch $u \in \mathcal{E}_1$; aus (1. 8) folgt daher noch: $p(g_S) \leqslant u(x)$. Zusammen mit (1. 13) ergibt dies die für alle $u \in \mathcal{E}$ gültige Ungleichung $\int^* u d\mu \leqslant u(x)$. Damit ist $\mu$ wie behauptet ein auf S konzentriertes Maß aus $\mathcal{M}_x(\mathcal{E})$, unser Satz also bewiesen.

Bemerkung. — Der Satz 2 überschneidet sich hinsichtlich der Aussage über die Existenz des Šilovschen Randes mit einem Resultat von Arens-Singer [1] (Theorem 2. 4). Unter der Voraussetzung, daß $\mathcal{E}$ nur stetige Funktionen enthält und die Eigenschaften (1. 4), (1. 5) und (1. 7) besitzt, genügt nämlich die Menge $H = \{e^{-u} : u \in \mathcal{E}\}$ den Annahmen in [1]. Dort werden allerdings $\mathcal{E}$-extremale Punkte und ihr Zusammenhang mit $\partial_{\mathcal{E}} X$ nicht diskutiert.

## § 2. — $\mathcal{H}$-harmonische Funktionen.

**2. 1.** *Der Vektorraum $\mathcal{H}$.* — Die Situation, mit der wir uns hier und im folgenden beschäftigen werden, wird spezieller sein als die im Paragraphen 1.

Gegeben sei jetzt nämlich ein kompakter Raum X und eine Menge $\mathcal{H}$ von auf X definierten Funktionen mit folgenden Eigenschaften:

(2. 1) *$\mathcal{H}$ ist ein linearer Unterraum von $\mathcal{C}(X)$;*
(2. 2) *$\mathcal{H}$ enthält alle konstanten Funktionen aus $\mathcal{C}(X)$;*
(2. 3) *$\mathcal{H}$ trennt die Punkte von X.*

Dann ist für jedes $x \in X$ die Menge $\mathcal{M}_x(\mathcal{H})$ von Maßen definiert. Wegen der speziellen Eigenschaften von $\mathcal{H}$ vereinfachen sich die an $\mathcal{M}_x(\mathcal{H})$ gestellten Forderungen (1. 1) und (1. 2):

$\mathcal{M}_x(\mathcal{H})$ ist für jedes $x \in \mathrm{X}$ die Menge aller Maße $\mu \in \mathcal{M}(\mathrm{X})$ mit:

$$(2.4) \qquad \int h d\mu = h(x) \qquad \text{für alle } h \in \mathcal{H}.$$

Nach dem Satz 2 existiert der Šilovsche Rand $\partial_{\mathcal{H}}\,\mathrm{X}$ und ist gleich der abgeschlossenen Hülle des Choquetschen Randes $\mathrm{X}_e(\mathcal{H})$ aller $\mathcal{H}$-extremalen Punkte. Das System $\mathfrak{S}(\mathcal{H})$ besteht somit aus allen kompakten Mengen S mit $\partial_{\mathcal{H}}\mathrm{X} \subset \mathrm{S} \subset \mathrm{X}$. Wegen (2.1) besitzt jede Funktion aus $\mathcal{H}$ eine in einem beliebigen $\mathrm{S} \in \mathfrak{S}(\mathcal{H})$ gelegene Minimal- und Maximalstelle.

2.2. *Die Mengen* $\mathcal{M}_x^{\mathrm{S}}$. — Für eine beliebige Menge $\mathrm{S} \in \mathfrak{S}(\mathcal{H})$ verallgemeinern wir nun die Mengen $\mathcal{M}_x(\mathcal{H})$ wie folgt: Für jeden Punkt $x \in \mathrm{X}$ bezeichne $\mathcal{M}_x^{\mathrm{S}} = \mathcal{M}_x^{\mathrm{S}}(\mathcal{H})$ die Menge aller Maße $\mu \in \mathcal{M}(\mathrm{S})$ mit:

$$(2.5) \qquad \int h_{\mathrm{S}} d\mu = h(x) \qquad \text{für alle } h \in \mathcal{H}.$$

Offenbar ist $\mathcal{M}_x(\mathcal{H}) = \mathcal{M}_x^{\mathrm{X}}(\mathcal{H})$ für jedes $x \in \mathrm{X}$. Aus (2.2) folgt: $\int d\mu = 1$ für alle $\mu \in \mathcal{M}_x^{\mathrm{S}}(\mathcal{H})$. Aus $x \in \mathrm{S}$ folgt $\varepsilon_x \in \mathcal{M}_x^{\mathrm{S}}$ und für jeden $\mathcal{H}$-extremalen Punkt $x$ ist sogar $\mathcal{M}_x^{\mathrm{S}} = \{\varepsilon_x\}$. Wir werden später sehen, daß diese letzte Eigenschaft die $\mathcal{H}$-extremalen Punkte sogar kennzeichnet (Satz 14).

Daß auch für Punkte $x \in \complement \mathrm{S}$ die Menge $\mathcal{M}_x^{\mathrm{S}}$ nicht leer ist, folgt aus der folgenden Betrachtung:

Für jeden Punkt $x \in \mathrm{X}$ und jede Funktion $f \in \mathcal{C}(\mathrm{S})$ werde gesetzt:

$$(2.6) \qquad \overline{\mathrm{Q}}_x^{\mathrm{S}}(f) = \inf_{f \leqslant h_{\mathrm{S}},\, h \in \mathcal{H}} h(x);$$

$$(2.7) \qquad \underline{\mathrm{Q}}_x^{\mathrm{S}}(f) = -\overline{\mathrm{Q}}_x^{\mathrm{S}}(-f).$$

Durch elementare Schlüsse, die dem Leser überlassen werden können, beweist man dann:

Hilfssatz 1. — *Für alle* $x \in \mathrm{X}$ *gilt:*

$$(2.8) \qquad -\infty < \underline{\mathrm{Q}}_x^{\mathrm{S}}(f) \leqslant \overline{\mathrm{Q}}_x^{\mathrm{S}}(f) < +\infty \qquad (f \in \mathcal{C}(\mathrm{S}));$$

$$(2.9) \qquad f \leqslant g \Longrightarrow \overline{\mathrm{Q}}_x^{\mathrm{S}}(f) \leqslant \overline{\mathrm{Q}}_x^{\mathrm{S}}(g) \qquad (f,\ g \in \mathcal{C}(\mathrm{S}));$$

$$(2.10) \qquad \overline{\mathrm{Q}}_x^{\mathrm{S}}(f+g) \leqslant \overline{\mathrm{Q}}_x^{\mathrm{S}}(f) + \overline{\mathrm{Q}}_x^{\mathrm{S}}(g) \qquad (f,\ g \in \mathcal{C}(\mathrm{S}));$$

$$(2.11) \qquad \overline{\mathrm{Q}}_x^{\mathrm{S}}(\lambda f) = \lambda \overline{\mathrm{Q}}_x^{\mathrm{S}}(f) \qquad (f \in \mathcal{C}(\mathrm{S});\ \lambda \geqslant 0);$$

$$(2.12) \qquad \underline{\mathrm{Q}}_x^{\mathrm{S}}(h_{\mathrm{S}}) = \overline{\mathrm{Q}}_x^{\mathrm{S}}(h_{\mathrm{S}}) = h(x) \qquad (h \in \mathcal{H}).$$

Hieraus folgern wir nunmehr:

SATZ 3. — (*a*) *Bei beliebiger Wahl von* $x \in X$, $\mu \in \mathcal{M}_x^S$ *und* $f \in \mathcal{C}(S)$ *gilt:*

(2. 13) $$\underline{Q}_x^S(f) \leqslant \int f\, d\mu \leqslant \overline{Q}_x^S(f).$$

(*b*) *Sind umgekehrt* $x \in X$, $f_0 \in \mathcal{C}(S)$ *und* $\gamma \in R$ *derart gewählt, daß*

(2. 14) $$\underline{Q}_x^S(f_0) \leqslant \gamma \leqslant \overline{Q}_x^S(f_0)$$

*gilt, so gibt es mindestens ein Maß* $\mu \in \mathcal{M}_x^S$ *mit:* $\gamma = \int f_0\, d\mu$.

Beweis. *Zu* (*a*): Für jedes $h \in \mathcal{H}$ mit $f \leqslant h_S$ gilt nach (2. 5): $\int f\, d\mu \leqslant \int h_S\, d\mu = h(x)$; somit ist $\int f\, d\mu \leqslant \overline{Q}_x^S(f)$. Bei Beachtung von (2. 7) folgt hieraus die Behauptung.

*Zu* (*b*): Es sei $\mathcal{F}_0 = \{\lambda f_0 : \lambda \in R\}$ der von $f_0$ in $\mathcal{C}(S)$ erzeugte lineare Unterraum. Dann ist $\lambda f_0 \to \lambda\gamma$ eine auf $\mathcal{F}_0$ definierte Linearform $\mu_0$; im Falle $f_0 = 0$ hat man hierbei zu berücksichtigen, daß $\gamma = 0$ ist. Für jedes $f \in \mathcal{F}_0$ gilt: $\mu_0(f) \leqslant \overline{Q}_x^S(f)$ nach (2. 11) und (2. 14). Wegen der Eigenschaften (2. 10) und (2. 11) der Funktion $f \to \overline{Q}_x^S(f)$ existiert nach dem Satz von HAHN-BANACH (vgl. [24], p. 9) eine $\mu_0$ auf $\mathcal{C}(S)$ fortsetzende Linearform $\mu$ mit $\mu(f) \leqslant \overline{Q}_x^S(f)$, also mit

(2. 15) $$\underline{Q}_x^S(f) \leqslant \mu(f) \leqslant \overline{Q}_x^S(f) \quad \text{für alle} \quad f \in \mathcal{C}(S).$$

Aus $f \geqslant 0$ folgt gemäß Hilfssatz 1: $0 = \underline{Q}_x^S(0) \leqslant \underline{Q}_x^S(f)$, also $\mu(f) \geqslant 0$. Somit ist $\mu$ ein positives Maß auf $\mathcal{C}(S)$. Nach Konstruktion gilt: $\int f_0\, d\mu = \mu_0(f_0) = \gamma$; aus (2. 15) folgt schließlich bei Beachtung von (2. 12), daß $\mu$ zu $\mathcal{M}_x^S$ gehört. Damit besitzt $\mu$ alle gewünschten Eigenschaften.

Zusammen mit (2. 8) zeigt der Satz 3 insbesondere, daß keine der Mengen $\mathcal{M}_x^S$ leer ist.

2. 3. *$\mathcal{H}$-harmonische Funktionen.* — Bei gegebenem $S \in \mathfrak{S}(\mathcal{H})$ fragen wir nun weiter nach der Menge aller Funktionen $h \in \mathcal{C}(X)$, welche der Gleichung (2. 5) für alle $x \in X$ und $\mu \in \mathcal{M}_x^S$ genügen.

Satz 4. — *Bei beliebig gegebenem* $S \in \mathfrak{S}(\mathcal{H})$ *sind für jede Funktion* $f \in \mathcal{C}(X)$ *folgende Bedingungen gleichwertig:*

(*a*) $f(x) = \int f_S \, d\mu$ *für alle* $x \in X$ *und alle* $\mu \in \mathcal{M}_x^S(\mathcal{H})$.

(*b*) $f(x) = \underline{Q}_x^S(f_S) = \overline{Q}_x^S(f_S)$ *für alle* $x \in X$.

(*c*) *Zu jedem* $\varepsilon > 0$ *gibt es endlich viele Funktionen* $h'_1, \ldots, h'_m$; $h''_1, \ldots, h''_n$ *in* $\mathcal{H}$ *derart, daß für die Einhüllenden*

$$\underline{h} = \sup(h'_1, \ldots, h'_m) \text{ und } \overline{h} = \inf(h''_1, \ldots, h''_n)$$

*gilt:*

$$\underline{h} \leqslant f \leqslant \overline{h} \text{ und } \overline{h} - \underline{h} \leqslant \varepsilon.$$

Beweis. (*a*) $\Longrightarrow$ (*b*): Dies ist eine unmittelbare Folgerung aus Satz 3.

(*b*) $\Longrightarrow$ (*c*): Es seien $\varepsilon > 0$ und $x \in X$ beliebig gewählt. Aus den Definitionsgleichungen (2. 6) und (2. 7) sowie aus (*b*) folgt dann die Existenz von Funktionen $h_x^1, h_x^2 \in \mathcal{H}$ mit folgenden Eigenschaften:

(2. 16) $$(h_x^1)_S \leqslant f_S \leqslant (h_x^2)_S;$$

(2. 17) $$h_x^2(x) - h_x^1(x) < \varepsilon.$$

Aus (2. 16) folgt wegen der Isotonie der Funktion $\overline{Q}_y^S$ bei Beachtung von (2. 12) und (*b*): $h_x^1(y) \leqslant f(y) \leqslant h_x^2(y)$ für alle $y \in X$, also $h_x^2 \leqslant f \leqslant h_x^1$. Wegen der Stetigkeit aller Funktionen aus $\mathcal{H}$ folgt aus (2. 17) die Existenz einer offenen Umgebung $U_x$ von $x$ mit: $h_x^2(y) - h_x^1(y) < \varepsilon$ für alle $y \in U_x$. Da der Raum X kompakt ist, genügen endlich viele der Mengen $U_x$ zu seiner Überdeckung: $X = U_{x_1} \cup \ldots \cup U_{x_n}$ $(x_i \in X)$. Es ist dann klar, daß die Funktionen $h'_i = h_{x_i}^1$ und $h''_i = h_{x_i}^2$ $(i = 1, \ldots, n)$ das Verlangte leisten.

(*c*) $\Longrightarrow$ (*a*): Zu gegebenem $\varepsilon > 0$ wählen wir die Funktionen $h'_1, \ldots, h'_m$; $h''_1, \ldots, h''_n$ gemäß der Bedingung (*c*). Durch Integration einer jeden Ungleichung $h'_i \leqslant f \leqslant h''_j$ folgt dann $h'_i(x) \leqslant \int f_S d\mu \leqslant h''_j(x)$ $(i = 1, \ldots, m; j = 1, \ldots, n)$ und somit $\underline{h}(x) \leqslant \int f_S d\mu \leqslant \overline{h}(x)$ für jedes $x \in X$ und alle $\mu \in \mathcal{M}_x^S$. Wegen $\underline{h} \leqslant f \leqslant \overline{h}$ und $\overline{h} - \underline{h} \leqslant \varepsilon$ folgt weiter: $\left|\int f_S d\mu - f(x)\right| \leqslant \varepsilon$ für alle $\varepsilon > 0$, $x \in X$ und $\mu \in \mathcal{M}_x^S$. Also genügt $f$ der Bedingung (*a*).

Die Tatsache, daß die Menge S nicht in der Bedingung (*c*) auftritt, gibt Anlaß zur folgenden Definition:

Definition 3. — *Eine Funktion* $f \in \mathcal{C}(X)$ *heiße* $\mathcal{H}$-*harmonisch, wenn sie den äquivalenten Bedingungen* (a)-(c) *in Satz 4 für eine Menge* $S \in \mathfrak{S}(\mathcal{H})$ *(und dann auch für jede derartige Menge) genügt.*

Wir bezeichnen die Menge aller $\mathcal{H}$-harmonischen Funktionen mit $\hat{\mathcal{H}}$. Aus der Definition (z. B. durch die Bedingung (*a*) von Satz 4) folgt sofort, daß $\hat{\mathcal{H}}$ ein *abgeschlossener, linearer Unterraum* von $\mathcal{C}(X)$ ist. Nach (2. 5) gilt: $\mathcal{H} \subset \hat{\mathcal{H}}$. Der Übergang von $\mathcal{H}$ zu $\hat{\mathcal{H}}$ ist ein gewisser Prozess der Vervollständigung, der, wie sich sofort zeigen wird, auf $\hat{\mathcal{H}}$ angewendet zu nichts Neuem führt.

2. 4. *Beziehungen zwischen* $\mathcal{H}$ *und* $\hat{\mathcal{H}}$. — Soeben wurde gezeigt, daß auch $\hat{\mathcal{H}}$ die Eigenschaften (2. 1)-(2. 3) von $\mathcal{H}$ besitzt. Also sind definiert die Mengen: $X_e(\hat{\mathcal{H}})$, $\partial_{\hat{\mathcal{H}}}X$, $\mathfrak{S}(\hat{\mathcal{H}})$, $\mathfrak{M}_x^S(\hat{\mathcal{H}})$ und $\hat{\hat{\mathcal{H}}}$. Aus der Definition von $\hat{\mathcal{H}}$ durch die Bedingung (*a*) von Satz 4 für $S = X$ folgt: $\mathfrak{M}_x(\hat{\mathcal{H}}) = \mathfrak{M}_x(\mathcal{H})$ für alle $x \in X$. Hieraus ergibt sich:

(2. 18) $$X_e(\hat{\mathcal{H}}) = X_e(\mathcal{H}).$$

Nach Satz 2 ist daher

(2. 19) $$\partial_{\hat{\mathcal{H}}}X = \partial_{\mathcal{H}}X.$$

Dann aber ist auch

(2. 20) $$\mathfrak{S}(\hat{\mathcal{H}}) = \mathfrak{S}(\mathcal{H})$$

und aus der Definition von $\hat{\mathcal{H}}$ folgt:

(2. 21) $$\mathfrak{M}_x^S(\hat{\mathcal{H}}) = \mathfrak{M}_x^S(\mathcal{H}) \quad \text{für alle} \quad S \in \mathfrak{S}(\mathcal{H}) \quad \text{und} \quad x \in X.$$

Schließlich folgt hieraus

(2. 22) $$\hat{\hat{\mathcal{H}}} = \hat{\mathcal{H}}.$$

Die Gleichungen (2. 18)-(2. 22) rechtfertigen die Verwendung der kürzeren Bezeichnungen $X_e$, $\mathfrak{S}$, $\mathfrak{M}_x^S$ und $\mathfrak{M}_x$ bei gegebenem $\mathcal{H}$.

Die naheliegende Vermutung, daß $\hat{\mathcal{H}}$ gleich der abgeschlossenen Hülle von $\mathcal{H}$ in $\mathcal{C}(X)$ ist, wird durch das folgende Beispiel widerlegt.

Beispiel 6. — Es sei X das kompakte Intervall [0,1] auf der Zahlengeraden und $\mathcal{H}$ die Menge aller auf X eingeschränkten, reellen, quadratischen Polynome $x \to \alpha + \beta x + \gamma x^2 (\alpha, \beta, \gamma \in \mathbf{R})$. Dann besitzt $\mathcal{H}$ die Eigenschaften (2. 1)-(2. 3); ferner ist $\mathcal{H}$ in $\mathcal{C}(X)$ abgeschlossen. Es gilt jedoch: $\hat{\mathcal{H}} = \mathcal{C}(X)$. Man zeigt dies entweder direkt mit Hilfe der Definition von $\hat{\mathcal{H}}$ durch die Bedingung ($c$) von Satz 4 oder einfacher wie folgt: Für jedes $x_0 \in X$ liegt die durch $x \to (x - x_0)^2$ definierte Funktion $h$ in $\mathcal{H}$ und es ist $h(x_0) < h(x)$ für alle $x \neq x_0$. Also ist jeder Punkt aus X $\mathcal{H}$-exponiert, also nach 1.3 $\mathcal{H}$-extremal. Dann aber ergibt sich $\hat{\mathcal{H}} = \mathcal{C}(X)$ nach dem folgenden Satz.

Satz 5. — *Dann und nur dann ist jeder Punkt aus X $\mathcal{H}$-extremal, wenn $\hat{\mathcal{H}} = \mathcal{C}(X)$ ist.*

Beweis. Aus $\hat{\mathcal{H}} = \mathcal{C}(X)$ folgt $X = X_e(\hat{\mathcal{H}})$ unmittelbar, also $X = X_e(\mathcal{H})$ nach (2. 18). Umgekehrt folgt aus $X = X_e$ zunächst $\mathcal{M}_x = \{\varepsilon_x\}$ für jedes $x \in X$. Die Definition der $\mathcal{H}$-harmonischen Funktionen liefert dann: $\hat{\mathcal{H}} = \mathcal{C}(X)$.

Unter einer Zusatzvoraussetzung über $\mathcal{H}$, die uns noch in § 3 beschäftigen wird, gilt jedoch:

Satz 6. — *Wenn $\mathcal{H}$ bezüglich der üblichen Ordnungsrelation $\leqslant$ ein Verband ist, so ist $\hat{\mathcal{H}}$ die abgeschlossene Hülle von $\mathcal{H}$ in $\mathcal{C}(X)$* (4).

Beweis. Zu jeder Funktion $f \in \hat{\mathcal{H}}$ und jedem $\varepsilon > 0$ gibt es nach Satz 4 Funktionen $h'_1, \ldots, h'_m$; $h''_1, \ldots, h''_n$ in $\mathcal{H}$ mit:

$$\underline{h} = \sup\ (h'_1, \ldots, h'_m) \leqslant f \leqslant \overline{h} = \inf\ (h''_1, \ldots, h''_n)$$

und $\overline{h} - \underline{h} \leqslant \varepsilon$. Nach Voraussetzung besitzen $h'_1, \ldots, h'_m$ eine kleinste gemeinsame Majorante $\underline{h}_0$ in $\mathcal{H}$, für die offenbar gilt: $\underline{h} \leqslant \underline{h}_0 \leqslant \overline{h}$. Man erhält daher: $|f - \underline{h}_0| \leqslant \varepsilon$. Damit ist gezeigt, daß $f$ in der abgeschlossenen Hülle $\overline{\mathcal{H}}$ von $\mathcal{H}$ liegt; es ist also: $\hat{\mathcal{H}} \subset \overline{\mathcal{H}}$. Da andererseits $\hat{\mathcal{H}}$ in $\mathcal{C}(X)$ abgeschlossen ist und $\mathcal{H}$ als Teilmenge enthält, gilt sogar $\overline{\mathcal{H}} = \hat{\mathcal{H}}$.

(4) Die Voraussetzung über $\mathcal{H}$ besagt, daß zu je zwei Funktionen $h_1$, $h_2$ in $\mathcal{H}$ eine kleinste gemeinsame Majorante und eine größte gemeinsame Minorante in $\mathcal{H}$ existiert. $\mathcal{H}$ ist dann bezüglich der Relation $\leqslant$ sogar ein *Rieszscher Raum* (Vektorverband).

Beispiel 7. — Es sei X das kompakte Intervall [0,1] auf der Zahlengeraden und $\mathcal{H}$ die Menge aller auf X eingeschränkten, affin-linearen, reellen Funktionen $x \to \alpha + \beta x$. Dann besitzt $\mathcal{H}$ die Eigenschaften (2. 1)-(2. 3). $\mathcal{H}$ ist in $\mathcal{C}(X)$ abgeschlossen und ein Verband. Also gilt $\hat{\mathcal{H}} = \mathcal{H}$. Übrigens ist hier $X_e = \{0,1\}$.

2. 5. *Kennzeichnungen von $\hat{\mathcal{H}}$ mittels* inf-*stabiler Teilmengen.* — Zunächst vereinbaren wir folgende Sprechweise: Eine Teilmenge $\mathcal{S}$ von $\mathcal{C}(X)$ heiße inf-*stabil*, wenn sie mit je endlich vielen Funktionen auch deren untere Einhüllende enthält. Offenbar existiert zu $\mathcal{H}$ (allgemeiner zu jeder Teilmenge von $\mathcal{C}(X)$) eine *kleinste abgeschlossene,* inf-*stabile Teilmenge* von $\mathcal{C}(X)$, *die $\mathcal{H}$ enthält.* Diese Menge soll mit $\mathcal{E}_{\mathcal{H}}$ bezeichnet werden. Es gilt nun folgende Verallgemeinerung von Satz 4:

Satz 7. — *Für jede Funktion $f \in \mathcal{C}(X)$ sind folgende Bedingungen gleichwertig:*

(*a*) $f \in \mathcal{E}_{\mathcal{H}}$;

(*b*) $\int f\,d\mu \leqslant f(x)$ *für alle* $x \in X$ *und* $\mu \in \mathcal{M}_x(\mathcal{H})$;

(*c*) $\overline{Q}_x^X(f) = f(x)$ *für alle* $x \in X$;

(*d*) *Zu jedem* $\varepsilon > 0$ *gibt es endlich viele Funktionen* $h_1, \ldots, h_n \in \mathcal{H}$ *mit* $f \leqslant \inf(h_1, \ldots, h_n) \leqslant f + \varepsilon$.

Beweis. $(a) \Longrightarrow (b)$: Es sei $\mathcal{S}$ die Menge aller Funktionen $u \in \mathcal{C}(X)$ mit $\int u\,d\mu \leqslant u(x)$ für alle $x \in X$ und $\mu \in \mathcal{M}_x$. Dann prüft man sofort nach, daß $\mathcal{S}$ in $\mathcal{C}(X)$ abgeschlossen und inf-stabil ist. Da ferner $\mathcal{H} \subset \mathcal{S}$ ist, so folgt $\mathcal{E}_{\mathcal{H}} \subset \mathcal{S}$ und damit (*b*).

$(b) \Longrightarrow (c)$: Aus (*b*) folgt zunächst $\overline{Q}_x^X(f) \leqslant f(x)$ für jedes $x \in X$, da nach Satz 3 speziell ein Maß $\mu \in \mathcal{M}_x$ existiert mit $\overline{Q}_x^X(f) = \int f\,d\mu$. Nach der Definitionsgleichung (2. 6) (für $S = X$) ist aber auch $f(x) \leqslant \overline{Q}_x^X(f)$ für alle $x \in X$.

$(c) \Longrightarrow (d)$: Zu $\varepsilon > 0$ und zu jedem $x \in X$ gibt es nach (2. 6) eine Funktion $h_x \in \mathcal{H}$ mit $f \leqslant h_x$ und

$$h_x(x) < f(x) + \varepsilon = \overline{Q}_x^X(f) + \varepsilon.$$

Der Rest des Beweises verläuft dann analog zum Schritt « $(b) \Longrightarrow (c)$ » des Beweises von Satz 4.

$(d) \Longrightarrow (a)$: Da $\mathcal{E}_{\mathcal{H}}$ definitionsgemäß eine abgeschlossene, inf-stabile Teilmenge von $\mathcal{C}(X)$ ist mit $\mathcal{H} \subset \mathcal{E}_{\mathcal{H}}$, zeigt die Bedingung $(d)$, daß $f$ in $\mathcal{E}_{\mathcal{H}}$ liegt.

Nunmehr ergeben sich eine Reihe von Folgerungen:

KOROLLAR 1. — *Es gilt* $\hat{\mathcal{H}} = \mathcal{E}_{\mathcal{H}} \cap (-\mathcal{E}_{\mathcal{H}})$. *Dann und nur dann ist* $\mathcal{H} = \hat{\mathcal{H}}$, *wenn es eine in* $\mathcal{C}(X)$ *abgeschlossene,* inf-*stabile Menge* $\mathcal{E}$ *gibt mit* $\mathcal{H} = \mathcal{E} \cap (-\mathcal{E})$.

Beweis. Nach Satz 7 ist $\mathcal{E}_{\mathcal{H}} \cap (-\mathcal{E}_{\mathcal{H}})$ die Menge aller $f \in \mathcal{E}(X)$ mit $\int f d\mu = f(x)$ für alle $x \in X$ und $\mu \in \mathcal{M}_x$. Nach Definition 3 ist dies die Menge $\hat{\mathcal{H}}$. — Ist $\mathcal{H} = \hat{\mathcal{H}}$, so ist nach dem bereits Bewiesenen $\mathcal{E} = \mathcal{E}_{\mathcal{H}}$ eine in $\mathcal{C}(X)$ abgeschlossene, inf-stabile Menge mit $\mathcal{H} = \mathcal{E} \cap (-\mathcal{E})$. Ist $\mathcal{E}$ irgendeine Menge mit diesen Eigenschaften, so folgt $\mathcal{H} = \hat{\mathcal{H}}$ aus der Gleichheit $\hat{\mathcal{H}} = \mathcal{E}_{\mathcal{H}} \cap (-\mathcal{E}_{\mathcal{H}})$ und aus $\mathcal{E}_{\mathcal{H}} \subset \mathcal{E}$.

KOROLLAR 2. — *Die Menge* $\mathcal{E}_{\mathcal{H}}$ *ist ein konvexer Kegel in* $\mathcal{C}(X)$ *mit der Spitze* 0 ([5]).

Beweis. Zu zeigen ist: $\mathcal{E}_{\mathcal{H}} + \mathcal{E}_{\mathcal{H}} \subset \mathcal{E}_{\mathcal{H}}$ und $\lambda\mathcal{E}_{\mathcal{H}} \subset \mathcal{E}_{\mathcal{H}}$ für jede reelle Zahl $\lambda \geqslant 0$. Dies aber folgt aus Satz 7, $(b)$.

KOROLLAR 3. — *Jede Funktion aus* $\mathcal{E}_{\mathcal{H}}$ *besitzt eine* $\mathcal{H}$-*extremale Minimalstelle.*

Beweis. Wegen $\mathcal{H} \subset \mathcal{E}_{\mathcal{H}} \subset \mathcal{C}(X)$ ist Satz 1 auf $\mathcal{E} = \mathcal{E}_{\mathcal{H}}$ anwendbar. Jede Funktion aus $\mathcal{E}_{\mathcal{H}}$ besitzt also eine $\mathcal{E}_{\mathcal{H}}$-extremale Minimalstelle. Nach Satz 7 gilt $\mathcal{M}_x(\mathcal{E}_{\mathcal{H}}) = \mathcal{M}_x(\mathcal{H})$ für alle $x \in X$. Daher ist $X_e(\mathcal{H}) = X_e(\mathcal{E}_{\mathcal{H}})$, woraus die Behauptung folgt.

Wegen $\mathcal{H} \subset \mathcal{E}_{\mathcal{H}}$ folgt aus dem letzten Korollar noch, daß $\partial_{\mathcal{H}} X = \partial_{\mathcal{E}_{\mathcal{H}}} X$ ist.

## § 3. — Das abstrakte Dirichletsche Problem.

3. 1. $\mathcal{H}$-*harmonische Maße und* $\mathcal{H}$-*resolutive Funktionen.* — Für eine beliebige Menge $S \in \mathfrak{S}(\mathcal{H})$ folgt aus der Gleichung (2. 20), daß jede $\mathcal{H}$-harmonische Funktion $h$ eindeutig durch

([5]) Vgl. hierzu auch CHOQUET-DENY [23].

ihre Restriktion $h_S$ auf S bestimmt ist. Für jede Funktion $g \in \hat{\mathcal{H}}$ mit $g_S = h_S$ gilt nämlich $(g - h)_S = 0$. Hieraus folgt $g = h$, da $g - h$ sowohl eine Maximal- als auch eine Minimalstelle in S besitzt. Wir stoßen daher auf das folgende « *abstrakte*» *Dirichletsche Problem:* Man gebe notwendige und hinreichende Bedingungen dafür an, daß *eine* Funktion $f \in \mathcal{C}(S)$ oder *jede* Funktion aus $\mathcal{C}(S)$ zu einer $\mathcal{H}$-harmonischen Funktion auf X fortgesetzt werden kann.

Eine Antwort auf die erste Frage gibt folgender Satz:

Satz 8. — *Für jede Menge* $S \in \mathfrak{S}(\mathcal{H})$ *und jede Funktion* $f \in \mathcal{C}(S)$ *sind folgende Aussagen gleichwertig:*

(*a*) *f kann zu einer* $\mathcal{H}$*-harmonischen Funktion auf* X *fortgesetzt werden.*

(*b*) *Für jeden Punkt* $x \in X$ *und je zwei Maße* $\mu_1, \mu_2 \in \mathcal{M}_x^S(\mathcal{H})$ *gilt:*

$$\int f\, d\mu_1 = \int f\, d\mu_2.$$

(*c*) $\underline{Q}_x^S(f) = \overline{Q}_x^S(f)$ *für jedes* $x \in X$.

Beweis. (*a*) $\Longrightarrow$ (*b*): Es gibt eine Funktion $h \in \hat{\mathcal{H}}$ mit $f = h_S$. Daher gilt $h(x) = \int h_S\, d\mu = \int f\, d\mu$ für alle $x \in X$ und $\mu \in \mathcal{M}_x^S$. Hieraus folgt (*b*).

(*b*) $\Longrightarrow$ (*c*): Dies folgt aus Satz 3.

(*c*) $\Longrightarrow$ (*a*): Wir definieren eine reelle Funktion *h* auf X durch: $h(x) = \underline{Q}_x^S(f) = \overline{Q}_x^S(f)$. Nach (2. 6) und (2. 7) ist sie nach oben und unten halbstetig, also stetig. Nach Satz 3 gilt $h(x) = \int f\, d\mu$ für alle $x \in X$ und $\mu \in \mathcal{M}_x^S$. Da für jeden Punkt $x \in S$ das Maß $\varepsilon_x$ zu $\mathcal{M}_x^S$ gehört, ergibt sich: $h(x) = \int f\, d\varepsilon_x = f(x)$. Also ist *h* eine Fortsetzung von *f*. Die für alle $x \in X$ und $\mu \in \mathcal{M}_x^S$ gültige Gleichung $h(x) = \int f\, d\mu = \int h_S\, d\mu$ besagt dann, daß *h* $\mathcal{H}$-harmonisch ist.

Bei der Behandlung der zweiten Frage, die uns im folgenden ausschließlich beschäftigen wird, können wir uns auf den Fall $S = \partial_{\mathcal{H}} X$ beschränken. In der Tat: Kann jede Funktion aus $\mathcal{C}(S)$ zu einer $\mathcal{H}$-harmonischen Funktion fortgesetzt werden, so kann auch jede Funktion aus $\mathcal{C}(\partial_{\mathcal{H}} X)$ zunächst zu einer Funktion aus $\mathcal{C}(S)$ und diese zu einer Funktion aus $\hat{\mathcal{H}}$ fortgesetzt werden. Wegen $\partial_{\mathcal{H}} X \in \mathfrak{S}(\mathcal{H})$ kann also jede Funktion aus $\mathcal{C}(\partial_{\mathcal{H}}X)$ auf genau eine Weise zu einer Funktion aus $\mathcal{C}(S)$ fortgesetzt werden. Wegen der Normalität kompakter Räume ist dies nur möglich, wenn $S = \partial_{\mathcal{H}}X$ ist.

Wir setzen daher im folgenden zur Abkürzung:

$$(3.1) \quad X^* = \partial_{\mathcal{H}}X \quad \text{und} \quad \mathcal{M}_x^* = \mathcal{M}_x^{X^*}(\mathcal{H}) \quad \text{für jedes } x \in X.$$

Die Elemente von $\mathcal{M}_x^*$ sollen die *zum Punkt* $x \in X$ *gehörigen* $\mathcal{H}$*-harmonischen Maße* genannt werden. Ferner werde eine Funktion $f \in \mathcal{C}(X^*)$ als $\mathcal{H}$*-resolutiv* bezeichnet, wenn sie zu einer Funktion aus $\hat{\mathcal{H}}$ fortgesetzt werden kann.

Aus dem Satz 8 erhalten wir nun sofort eine erste Antwort auf unsere zweite Frage:

Satz 9. — *Jede der beiden folgenden Bedingungen ist notwendig und hinreichend dafür, daß jede Funktion aus* $\mathcal{C}(X^*)$ $\mathcal{H}$*-resolutiv ist:*

(*a*) *Zu jedem Punkt* $x \in X$ *gehört genau ein* $\mathcal{H}$*-harmonisches Maß.*

(*b*) *Für alle* $x \in X$ *und* $f \in \mathcal{C}(X^*)$ *gilt:* $\underline{Q}_x^{X^*}(f) = \overline{Q}_x^{X^*}(f)$.

3. 2. *Verbandsstruktur von* $\hat{\mathcal{H}}$. — Es soll jetzt ein Kriterium anderer Art für die $\mathcal{H}$-Resolutivität aller Funktionen aus $\mathcal{C}(X^*)$ angegeben werden.

Wir betrachten hierzu die Abbildung $h \to h_{X^*}$ von $\hat{\mathcal{H}}$ in $\mathcal{C}(X^*)$, welche jedem $h \in \hat{\mathcal{H}}$ seine Restriktion $h_{X^*}$ auf $X^*$ zuordnet. Es handelt sich offenbar um einen Monomorphismus bezüglich der Ordnungsstruktur: aus $g \leqslant h$ folgt $g_{X^*} \leqslant h_{X^*}$ und umgekehrt. Wenn speziell jede Funktion aus $\mathcal{C}(X^*)$ $\mathcal{H}$-resolutiv ist, liegt ein Isomorphismus der geordneten Mengen $\hat{\mathcal{H}}$ und $\mathcal{C}(X^*)$ vor. Mit $\mathcal{C}(X^*)$ ist dann also auch $\hat{\mathcal{H}}$ ein *Verband*, wegen der Verträglichkeit der Relation $\leqslant$ mit der Vektorraumstruktur also sogar ein *Rieszscher Raum.*

Das angekündigte Kriterium lautet nun:

Satz 10. — *Dann und nur dann ist jede Funktion aus* $\mathcal{C}(X^*)$ $\mathcal{H}$*-resolutiv, wenn* $\hat{\mathcal{H}}$ *bezüglich der üblichen Ordnungsrelation* $\leqslant$ *ein Verband ist.*

Zu beweisen ist nur noch, daß die angegebene Bedingung auch hinreichend ist. Der Beweis soll erst in den beiden nächsten Nummern erbracht werden. Hier sollen zunächst einige Folgerungen aus Satz 10 besprochen werden.

Im Verband $\hat{\mathcal{H}}$ bezeichnen wir die Verbandsoperationen mit Sup und Inf. Offenbar gilt für je zwei Funktionen $h_1, h_2 \in \hat{\mathcal{H}}$:

$$\text{(3. 2)} \quad \operatorname{Inf}(h_1, h_2) \leqslant \inf(h_1, h_2) \leqslant \sup(h_1, h_2) \leqslant \operatorname{Sup}(h_1, h_2).$$

Im allgemeinen ist $\operatorname{Sup}(h_1, h_2) \neq \sup(h_1, h_2)$ wie Beispiel 7 zeigt, welches zudem eine erste Illustration unseres Satzes 10 liefert.

Mit Hilfe der Menge $\mathcal{E}_{\mathcal{H}}$ aus Nr. 2. 5 kann der Satz 10 auch wie folgt formuliert werden:

Korollar 1. — *Dann und nur dann ist jede Funktion aus $\mathcal{C}(X^*)$ $\mathcal{H}$-resolutiv, wenn jede Funktion $u \in \mathcal{E}_{\mathcal{H}}$ eine größte $\mathcal{H}$-harmonische Minorante $h_u$ besitzt.*

Beweis. Es sei jede Funktion aus $\mathcal{C}(X^*)$ $\mathcal{H}$-resolutiv. Speziell ist dann für jedes $u \in \mathcal{E}_{\mathcal{H}}$ die Restriktion $u_{X^*}$ auf $X^*$ $\mathcal{H}$-resolutiv, also existiert ein $h_u \in \hat{\mathcal{H}}$ mit $h_u(x) = u(x)$ für alle $x \in X^*$. Wegen $\hat{\mathcal{H}} \subset \mathcal{E}_{\mathcal{H}}$ ist $u - h_u$ eine Funktion aus $\mathcal{E}_{\mathcal{H}}$. Aus $u(x) - h_u(x) = 0$ für alle $x \in X^*$ folgt $u - h_u \geqslant 0$ nach dem Korollar 3 zu Satz 7. Also ist $h_u$ eine $\mathcal{H}$-harmonische Minorante von $u$. Sie ist sogar die größte: Aus $h \in \hat{\mathcal{H}}$ und $h \leqslant u$ folgt $h(x) \leqslant h_u(x)$ für alle $x \in X^*$ und hieraus $h \leqslant h_u$.

Nunmehr werde umgekehrt die Existenz von $h_u$ für jedes $u \in \mathcal{E}_{\mathcal{H}}$ vorausgesetzt. Für jede Funktion $h \in \hat{\mathcal{H}}$ gehört $u = \inf(h, 0)$ zu $\mathcal{E}_{\mathcal{H}}$. Die zugehörige Funktion $h_u$ ist dann eine $\mathcal{H}$-harmonische Minorante von $h$ und 0. Für jede andere Minorante $h' \in \hat{\mathcal{H}}$ von $h$ und 0 gilt $h' \leqslant u$ und somit $h' \leqslant h_u$. Also ist $h_u$ die größte $\mathcal{H}$-harmonische Minorante von $h$ und 0: es existiert $\operatorname{Inf}(h, 0) = h_u$. Hieraus folgt, daß $\hat{\mathcal{H}}$ ein Verband ist. Die Behauptung folgt daher aus Satz 10.

Korollar 2. — *Dann und nur dann kann jede Funktion aus $\mathcal{C}(X^*)$ zu einer Funktion aus dem Vektorraum $\mathcal{H}$ selbst fortgesetzt werden, wenn $\mathcal{H}$ in $\mathcal{C}(X)$ abgeschlossen und ein Verband ist.*

Beweis. Kann jede Funktion aus $\mathcal{C}(X^*)$ zu einer Funktion aus $\mathcal{H}$ fortgesetzt werden, so ist offenbar $\mathcal{H} = \hat{\mathcal{H}}$ und jede

Funktion aus $\mathcal{C}(X^*)$ $\mathcal{H}$-resolutiv. Nach Nr. 2. 3 ist daher $\mathcal{H}$ in $\mathcal{C}(X)$ abgeschlossen, nach Satz 10 ein Verband. — Besitzt umgekehrt $\mathcal{H}$ diese Eigenschaften, so ist $\mathcal{H} = \hat{\mathcal{H}}$ nach Satz 6 und die Behauptung folgt abermals nach Satz 10.

Bemerkung. — Wenn $\hat{\mathcal{H}}$ ein Verband ist, so ist $\hat{\mathcal{H}}$ bezüglich der Norm $\|h\| = \sup_{x \in X} |h(x)|$ sogar ein M-Raum im Sinne von Kakutani [27], d.h. es gilt

$$\|\mathrm{Sup}\,(h_1,\ h_2)\| = \sup\,(\|h_1\|,\ \|h_2\|)$$

für je zwei Funktionen $h_i \in \hat{\mathcal{H}}$ mit $h_i \geqslant 0$. Nach dem Hauptresultat von [27] ist jeder M-Raum isomorph zum Rieszschen Raum aller stetigen reellen Funktionen auf einem kompakten Raum Y. Der Satz 10 zeigt, daß der zu $\hat{\mathcal{H}}$ gehörige Raum Y zu $X^*$ homöomorph ist.

3. 3. *Hilfssätze über Rieszsche Räume.* — Wir erinnern zunächst an einige bekannte Begriffsbildungen über Rieszsche Räume ([6]). Es sei hierzu R ein Rieszscher Raum; mit Sup und Inf sollen die Verbandsoperationen bezeichnet werden. Ein linearer Unterraum N von R heißt *gesättigt*, wenn jedes Element $f \in N$ von der Form $f = f_1 - f_2$ ist mit $f_i \in N$, $f_i \geqslant 0$ $(i = 1, 2)$ und wenn aus $0 \leqslant g \leqslant f$, $f \in N$, $g \in R$ stets $g \in N$ folgt. Mit diesen Forderungen gleichwertig ist die Bedingung: aus $f \in N$, $g \in R$ und $\mathrm{Sup}\,(g, -g) \leqslant \mathrm{Sup}(f, -f)$ folgt $g \in N$.

Zu jedem linearen Unterraum N von R gehört der Quotientenraum R/N und die kanonische Abbildung $\varphi : R \to R/N$. Ist N gesättigt und $R^+$ die Menge aller Elemente $\geqslant 0$ in R, so gibt es in R/N genau eine mit der Vektorraumstruktur verträgliche Ordnungsrelation, bezüglich welcher $\varphi(R^+)$ die Menge aller Elemente $\geqslant 0$ in R/N ist. R/N ist dann sogar ein Rieszscher Raum, und es gilt:

$$(3.\,3) \qquad \varphi(\mathrm{Sup}(h_1, h_2)) = \mathrm{Sup}(\varphi(h_1), \varphi(h_2)) \qquad (h_1, h_2) \in R.$$

Ein gesättigter linearer Unterraum N von R heißt *maximal*, wenn $N \neq R$ ist und für jeden gesättigten linearen Unterraum N' mit $N \subset N' \subset R$ gilt: $N = N'$ oder $N' = R$.

([6]) Vgl. hierzu und für das Folgende Bourbaki [14], chap. II, insbesondere p. 27, exercice 4, sowie [11], p. 23, exercice 4.

Zur Vorbereitung des Beweises von Satz 10 benötigen wir einige Hilfssätze:

HILFSSATZ 2. — *Es sei* N *ein maximaler gesättigter linearer Unterraum eines Rieszschen Raumes* R. *Dann ist* N *eine Hyperebene in* R *und es gibt eine positive Linearform* $\chi$ *auf* R *mit* $N = \chi^{-1}(0)$. *Jede solche positive Linearform* $\chi$ *genügt den folgenden Bedingungen:*

$$(3.4)\quad \chi(\mathrm{Sup}\,(h_1, h_2)) = \sup\,(\chi(h_1), \chi(h_2)) \quad (h_1, h_2 \in R);$$

$$(3.5)\quad \chi(\mathrm{Inf}\,(h_1, h_2)) = \inf\,(\chi(h_1), \chi(h_2)) \quad (h_1, h_2 \in R);$$

$$(3.6)\quad \mathrm{Inf}\,(h_1, h_2) = 0 \Longrightarrow \chi(h_1) = 0 \text{ oder } \chi(h_2) = 0 \;(h_1, h_2 \in R).$$

*Diese sind für jede positive Linearform* $\chi$ *auf* R *äquivalent.*

Beweis. Es sei $\varphi$: $R \to R/N$ die kanonische Abbildung von R auf den Rieszschen Raum R/N. Aus der Maximaleigenschaft von N folgt, daß $\{0\}$ der einzige maximale gesättigte lineare Unterraum von R/N ist. Daher existiert ein Isomorphismus $\psi: R/N \to \mathbf{R}$ auf den Rieszschen Raum **R** der reellen Zahlen ([7]). Dann ist $\chi_0 = \psi \circ \varphi$ eine positive Linearform auf R mit $\chi_0^{-1}(0) = \varphi^{-1}(0) = N$, welche nach (3. 3) die Eigenschaft (3. 4) besitzt. Insbesondere ist dann N eine Hyperebene in R. Jede andere positive Linearform $\chi$ auf R mit $\chi^{-1}(0) = N$ ist von der Form $\chi = \lambda\chi_0$ mit $\lambda > 0$, $\lambda \in \mathbf{R}$, und besitzt daher ebenfalls die Eigenschaft (3. 4). Schließlich rechnet man leicht nach, daß die Eigenschaften (3. 4)-(3. 6) für jede positive Linearform $\chi$ auf R äquivalent sind ([8]).

HILFSSATZ 3. — *Für jede positive Linearform* $\chi \neq 0$ *auf einem Rieszschen Raum* R, *welche die Eigenschaften* (3. 4)-(3. 6) *besitzt, ist* $N = \chi^{-1}(0)$ *ein maximaler gesättigter linearer Unterraum.*

Beweis. Für jedes Element $f \in R$ gilt bekanntlich $f = f^+ - f^-$ und $\mathrm{Inf}\,(f^+, f^-) = 0$, wenn man setzt: $f^+ = \mathrm{Sup}(f, 0)$ und $f^- = (-f)^+$. Aus (3. 6) folgt daher $\chi(f^+) = 0$ oder $\chi(f^-) = 0$. Wegen $\chi(f) = \chi(f^+) - \chi(f^-)$ folgt hieraus: $\chi(f^+) = \chi(f^-) = 0$ für jedes $f \in N$; also liegen mit $f$ auch $f^+$ und $f^-$ in N. Somit ist jedes $f \in N$ die Differenz von Elementen $\geqslant 0$ aus N. Aus

([7]) Vgl. BIRKHOFF [7], p. 239; gesättigte lineare Unterräume werden dort $l$-Ideale genannt. Vgl. ferner BOURBAKI [11], p. 23, exercice 4.

([8]) Vgl. auch KAKUTANI [27], p. 1001.

$0 \leqslant g \leqslant f$, $f \in N$, $g \in R$ und der Positivität von $\chi$ folgt: $\chi(g) = 0$, also $g \in N$. Somit ist N ein gesättigter linearer Unterraum von R. N ist maximal, da N als Hyperebene sogar ein maximales Element in der Menge aller echten linearen Unterräume von R ist.

Hilfssatz 4. — *Es sei* Y *ein kompakter Raum und* $\mathfrak{R}$ *ein die Punkte von* Y *trennender, die konstanten reellen Funktionen enthaltender, linearer Unterraum von* $\mathcal{C}(Y)$. *Ferner sei* $\mathfrak{R}$ *ein Verband bezüglich der üblichen Relation* $\leqslant$ *(und damit ein Rieszscher Raum). Dann existiert zu jedem maximalen gesättigten linearen Unterraum* $\mathfrak{N}$ *von* $\mathfrak{R}$ *genau ein Punkt* $y_0 \in Y$ *derart, daß* $\mathfrak{N}$ *die Menge aller* $f \in \mathfrak{R}$ *ist mit* $f(y_0) = 0$.

Beweis. Auf Grund der Voraussetzungen über $\mathfrak{R}$ gibt es zu je zwei verschiedenen Punkten $y_1, y_2 \in Y$ eine Funktion $f \in \mathfrak{R}$ mit $f(y_1) = 0$ und $f(y_2) \neq 0$. Daher kann zu $\mathfrak{N}$ höchstens ein Punkt $y_0$ mit den genannten Eigenschaften existieren. Den Existenzbeweis führen wir indirekt: Angenommen zu jedem $y \in Y$ gibt es eine Funktion $f_y \in \mathfrak{N}$ mit $f_y(y) \neq 0$. Da sich jede Funktion aus $\mathfrak{N}$ als Differenz nicht-negativer Funktionen aus $\mathfrak{N}$ darstellen läßt, kann o.B.d.A. $f_y \geqslant 0$ angenommen werden. Daher gibt es zu $y \in Y$ eine offene Umgebung $U_y$ von $y$ mit: $f_y(z) > 0$ für alle $z \in U_y$. Da Y kompakt ist, gibt es endlich viele Punkte $y_1, \ldots, y_n \in Y$ mit $Y = U_{y_1} \cup \cdots \cup U_{y_n}$. Die Funktion $f_0 = f_{y_1} + \cdots + f_{y_n}$ liegt in $\mathfrak{N}$ und ist auf Y strikt positiv. Zu jedem $f \in \mathfrak{R}$ mit $f \geqslant 0$ gibt es daher eine Zahl $\alpha > 0$ mit $0 \leqslant f \leqslant \alpha f_0$, woraus $f \in \mathfrak{N}$ folgt. Hieraus ergibt sich aber $\mathfrak{N} = \mathfrak{R}$, was falsch ist. Damit ist die Existenz eines $y_0 \in Y$ bewiesen mit $f(y_0) = 0$ für alle $f \in \mathfrak{N}$. Es ist $\mathfrak{N}$ sogar gleich der Menge $\mathfrak{N}'$ aller $f \in \mathfrak{R}$ mit $f(y_0) = 0$. Es ist nämlich $\mathfrak{N}$ eine Hyperebene in $\mathfrak{R}$ nach Hilfssatz 2; $\mathfrak{N}'$ ist eine Hyperebene in $\mathfrak{R}$ nach Definition. Aus $\mathfrak{N} \subset \mathfrak{N}'$ folgt daher $\mathfrak{N} = \mathfrak{N}'$.

Der Hilfssatz 4 kann unter Heranziehung der vorausgehenden Hilfssätze auch wie folgt ausgesprochen werden: *Zu jeder positiven Linearform* $\chi$ *auf* $\mathfrak{R}$ *mit* $\chi(1) = 1$ *und den gleichwertigen Eigenschaften* (3. 4)-(3. 6) *gibt es genau einen Punkt* $y_0 \in Y$ *mit* $\chi(f) = f(y_0)$ *für alle* $f \in \mathfrak{R}$.

Wir fragen daher jetzt nach allen Punkten $y \in Y$, für welche die Linearform $f \to f(y)$ auf $\mathfrak{R}$ die Eigenschaften (3. 4)-(3. 6) besitzt. Hiermit gleichbedeutend ist die Frage nach allen Punkten $y \in \mathrm{Y}$, für welche die Menge $\mathfrak{N}_y = \{f \in \mathfrak{R} : f(y) = 0\}$ ein (notwendig maximaler) gesättigter linearer Unterraum ist.

Hilfssatz 5. — *Unter den in Hilfssatz 4 über* Y *und* $\mathfrak{R}$ *gemachten Voraussetzungen gilt: Die Menge* $\mathrm{Y}_0$ *aller Punkte* $y \in \mathrm{Y}$, *für welche die Linearform* $f \to f(y)$ *auf* $\mathfrak{R}$ *die Eigenschaften* (3. 4)-(3. 6) *besitzt, ist gleich dem Šilovschen Rand* $\partial_{\mathfrak{R}}\mathrm{Y}$.

Beweis. Wir zeigen zunächst, daß $\mathrm{Y}_0$ abgeschlossen und damit kompakt ist. Hierzu sei $y \in \complement \mathrm{Y}_0$. Dann gibt es Funktionen $f_1, f_2 \in \mathfrak{R}$ mit $\mathrm{Inf}\,(f_1, f_2) = 0$ sowie mit $f_1(y) > 0$ und $f_2(y) > 0$. Aus Stetigkeitsgründen sind $f_1$ und $f_2$ sogar noch in einer Umgebung U von $y$ strikt positiv. Also ist U zu $\mathrm{Y}_0$ fremd. Damit ist die Offenheit von $\complement \mathrm{Y}_0$, also die Abgeschlossenheit von $\mathrm{Y}_0$ bewiesen.

Weiter besitzt jede Funktion $f_0 \in \mathfrak{R}$ eine in $\mathrm{Y}_0$ gelegene Minimalstelle. Hierzu sei $\alpha = \inf\limits_{y \in \mathrm{Y}} f_0(y)$ und M die Menge aller Minimalstellen von $f_0$. Wir betrachten die Menge $\mathfrak{N}_0$ aller Funktionen $f \in \mathfrak{R}$, die sich in der Form $f = f_1 - f_2$ darstellen lassen, wobei $f_i \in \mathfrak{R}$, $f_i \geqslant 0$ und $f_i(x) = 0$ ist für alle $x \in \mathrm{M}$ $(i = 1, 2)$. Offenbar ist $\mathfrak{N}_0$ ein gesättigter linearer Unterraum, der $f_0 - \alpha$ aber nicht die konstante Funktion 1 als Element enthält. Eine einfache Anwendung des Zornschen Lemmas ergibt dann die Existenz eines maximalen gesättigten linearen Unterraumes $\mathfrak{N}$ mit $\mathfrak{N}_0 \subset \mathfrak{N}$. Nach Hilfssatz 4 entspricht diesem genau ein Punkt $y_0 \in \mathrm{Y}$ mit: $\mathfrak{N} = \{f \in \mathfrak{R} : f(y_0) = 0\}$. Nach der diesem Hilfssatz vorangestellten Bemerkung liegt $y_0$ in $\mathrm{Y}_0$; wegen $f_0 - \alpha \in \mathfrak{N}_0 \subset \mathfrak{N}$ gilt $f_0(y_0) = \alpha$. Also ist $y_0$ eine in $\mathrm{Y}_0$ gelegene Minimalstelle von $f_0$.

Schließlich haben wir noch zu zeigen, daß $\mathrm{Y}_0$ in $\mathrm{Y}^* = \partial_{\mathfrak{R}}\mathrm{Y}$ enthalten und damit gleich $\mathrm{Y}^*$ ist. Hierzu betrachten wir die Abbildung $f \to f_{\mathrm{Y}^*}$ von $\mathfrak{R}$ auf den Vektorraum $\mathfrak{R}_{\mathrm{Y}^*}$ aller Restriktionen $f_{\mathrm{Y}^*}$ mit $f \in \mathfrak{R}$; diese ist ein Isomorphismus bezüglich der Vektorraumstruktur und der Ordnungsstruktur. Insbesondere ist also mit $\mathfrak{R}$ auch $\mathfrak{R}_{\mathrm{Y}^*}$ ein Verband. Ferner folgt, daß für jeden Punkt $y_0 \in \mathrm{Y}_0$ die auf $\mathfrak{R}_{\mathrm{Y}^*}$ definierte positive Linearform $f_{\mathrm{Y}^*} \to f(y_0) = \chi(f_{\mathrm{Y}^*})$ die Eigenschaften (3. 4)-(3. 6) besitzt. Aus dem Hilfssatz 4 ergibt sich daher die Existenz eines Punktes $z_0 \in Y^*$ mit $f_{Y^*}(z_0) = \chi(f_{Y^*})$, also mit $f(z_0) = f(y_0)$ für alle $f \in \mathfrak{R}$. Hieraus folgt $z_0 = y_0 \in Y^*$. Also gilt $Y_0 \subset Y^*$, was noch zu zeigen war.

3. 4. *Beweis des Satzes 10.* — Es werde nun $\hat{\mathcal{H}}$ als Verband vorausgesetzt. Zu zeigen ist die $\mathcal{H}$-Resolutivität aller Funktionen aus $\mathcal{C}(X^*)$.

Nach dem Hilfssatz 5 (angewandt auf $Y = X$ und $\mathfrak{R} = \hat{\mathcal{H}}$) ist $X^*$ die Menge aller Punkte $x \in X$, für welche die positive Linearform $h \to h(x)$ die Eigenschaften (3. 4)-(3. 6) besitzt. Nach (3. 4) gilt dann also:

$$[\mathrm{Sup}\,(g, h)]_{X^*} = \sup\,(g_{X^*}, h_{X^*}) \tag{3. 7}$$

für je zwei Funktionen $g, h \in \hat{\mathcal{H}}$. Daher genügt der lineare Unterraum $\hat{\mathcal{H}}_{X^*}$ von $\mathcal{C}(X^*)$ aller Restriktionen $h_{X^*}$ mit $h \in \hat{\mathcal{H}}$ den Voraussetzungen des Approximationssatzes von M. H. Stone. Nach diesem Satz ist $\hat{\mathcal{H}}_{X^*}$ in $\mathcal{C}(X^*)$ dicht. Nun ist aber $\hat{\mathcal{H}}$ in $\mathcal{C}(X)$ abgeschlossen; aus den fundamentalen Eigenschaften des Šilovschen Randes folgt dann auch die Abgeschlossenheit von $\hat{\mathcal{H}}_{X^*}$ in $\mathcal{C}(X^*)$. Also ist $\hat{\mathcal{H}}_{X^*} = \mathcal{C}(X^*)$ und damit der Beweis für Satz 10 erbracht.

3. 5. *Aussagen über den Choquetschen Rand.* — Abschliessend untersuchen wir die Frage, in welcher Weise die $\mathcal{H}$-Resolutivität jeder Funktion aus $\mathcal{C}(X^*)$ den Choquetschen Rand $X_e$ einschränkt.

Satz 11. — *Ist jede Funktion aus $\mathcal{C}(X^*)$ $\mathcal{H}$-resolutiv, so ist der Choquetsche Rand $X_e(\mathcal{H})$ gleich dem Šilovschen Rand $\partial_{\mathcal{H}}X$.*

Beweis. Zu zeigen ist offenbar nur: $X^* \subset X_e$. Nach Satz 9 gehört zu jedem Punkt $x \in X$ genau ein $\mathcal{H}$-harmonisches Maß $\mu_x$; speziell ist $\mu_x = \varepsilon_x$ für $x \in X^*$. Die durch $\varphi(x) = \mu_x$ definierte Abbildung $\varphi : X \to \mathcal{M}(X^*)$ ist stetig. Es sei nämlich $\mathfrak{F}$ ein gegen $x_0 \in X$ konvergenter Filter in X; dann gilt aus Stetigkeitsgründen: $h(x_0) = \lim_{\mathfrak{F}} h(x)$, d. h. $\int h_{X^*}\, d\mu_{x_0} = \lim_{\mathfrak{F}} \int h_{X^*}\, d\mu_x$ für alle $h \in \hat{\mathcal{H}}$. Da nach Voraussetzung jede Funktion $f \in \mathcal{C}(X^*)$ von der Form $f = h_{X^*}$ mit $h \in \hat{\mathcal{H}}$ ist, bedeutet dies: $\mu_{x_0} = \lim_{\mathfrak{F}} \mu_x$, also die Stetigkeit von $\varphi$. Es sei jetzt $x_0$ ein beliebiger Punkt aus

$X^*$ und $\mu \in \mathcal{M}(X)$ ein Maß mit $\int h\,d\mu = h(x_0)$ für alle $h \in \mathcal{H}$. Wegen der Stetigkeit von $\varphi$ existiert $\int \varphi\,d\mu$ und es ist $\mu_\varphi = \int \varphi\,d\mu = \int \mu_x\,d\mu(x)$ ein Maß aus $\mathcal{M}(X^*)$; es gilt

$$\int h_{X^*}\,d\mu_\varphi = \int \Big( \int h_{X^*}\,d\mu_x \Big)\,d\mu(x) = \int h(x)\,d\mu(x) = h(x_0)$$

für alle $h \in \mathcal{H}$. Also ist $\mu_\varphi$ das zu $x_0$ gehörige $\mathcal{H}$-harmonische Maß, d.h. $\mu_\varphi = \varepsilon_{x_0}$. Aus $\varepsilon_{x_0} = \int \mu_x\,d\mu(x)$ und $\int d\mu = 1$ folgt aber $\mu = \varepsilon_{x_0}$. In der Tat: für die in $X^*$ offene Menge $U = X^* \cap \complement\{x_0\}$ gilt ([15], p. 19, corollaire)

$$0 = \varepsilon_{x_0}(U) = \int^* \mu_x(U)\,d\mu(x),$$

woraus $\mu_x(U) = 0$ für $\mu$-fast alle $x \in X$ folgt. Nun ist aber $\mu_x(U) = 0$ und $\int d\mu = 1$ gleichbedeutend mit $\mu_x = \varepsilon_{x_0}$. Daher ist $\mu_x(U) > 0$ für alle $x \neq x_0$ aus X. Wir erhalten somit $\mu(\complement\{x_0\}) = 0$, was zusammen mit $\int d\mu = 1$ die Gleichheit $\mu = \varepsilon_{x_0}$ liefert. Also ist damit gezeigt, daß jeder Punkt $x_0 \in X^*$ $\mathcal{H}$-extremal ist.

Wenn der Raum X (oder auch nur $X^*$) metrisierbar ist, läßt sich dieses Resultat verschärfen zu:

Satz 12. — *Ist jede Funktion aus $\mathcal{C}(X^*)$ $\mathcal{H}$-resolutiv und ist $X^*$ metrisierbar, so ist jeder Punkt des Šilovschen Randes $X^*$ $\hat{\mathcal{H}}$-exponiert.*

Beweis. Es sei $x_0$ in $X^*$ beliebig gewählt. Nach Voraussetzung gibt es eine Funktion $f \in \mathcal{C}(X^*)$ mit $f(x) > f(x_0) = 0$ für alle $x \neq x_0$ aus $X^*$. Wir behaupten, daß dann für die Funktion $h \in \hat{\mathcal{H}}$ mit $h_{X^*} = f$ gilt: $h(x) > h(x_0) = f(x_0) = 0$ für alle $x \neq x_0$ aus X. Wegen $f \geqslant 0$ gilt zunächst $h \geqslant 0$. Es sei $x \in X$ mit $h(x) = 0$. Wir zeigen, daß dies $x = x_0$ zur Folge hat. Für das zu $x$ gehörige $\mathcal{H}$-harmonische Maß $\mu_x$ gilt: $\int f\,d\mu_x = 0 = f(x_0)$. Hieraus folgt ([5], Hilfssatz 3), daß $f$ auf dem Träger $T_{\mu_x}$ von $\mu_x$ gleich Null ist. Da $x_0$ die einzige Nullstelle von $f$ ist, impliziert dies $T_{\mu_x} \subset \{x_0\}$. Zusammen mit $\int d\mu_x = 1$ ergibt dies: $\mu_x = \varepsilon_{x_0}$. Da $\mathcal{H}$ die Punkte von X trennt, muß dann schließlich $x = x_0$ sein.

## § 4. — Anwendung: Kennzeichnung gewisser Simplexe von G. CHOQUET.

4. 1. *Lineare und konkave Funktionen.* — Der bislang beliebige kompakte Grundraum X sei in diesem Paragraphen eine *konvexe, kompakte* Teilmenge eines *lokal-konvexen* topologischen Vektorraumes E (über **R**) (9). Mit $\mathfrak{L} = \mathfrak{L}(X)$ bezeichnen wir die Menge aller auf X eingeschränkten, stetigen, *affin-linearen* Funktionen auf E. Affin-linear auf E heißt dabei jede Funktion $f = l + \alpha$ auf E, wobei $l$ eine Linearform auf E und $\alpha$ eine konstante reelle Funktion ist. Im folgenden übernehme $\mathfrak{L}$ die Rolle des Vektorraumes $\mathcal{H}$ der allgemeinen Theorie. Dies ist möglich, da $\mathfrak{L}$ die Eigenschaften (2. 1)-(2. 3) besitzt. Die Eigenschaft (2. 3) ergibt sich dabei wie folgt: Sind $x$, $y$ verschiedene Punkte aus X, so liegt $y$ nicht in der abgeschlossenen Hülle N von $\{x\}$ in E, da X Hausdorffsch ist. Aus einem bekannten Trennungssatz ([12], p. 73, prop. 4) folgt dann die Existenz einer stetigen Linearform auf E, die N und $\{x\}$, also erst recht $x$ und $y$ trennt.

Wir interpretieren nun die wichtigsten Begriffe der allgemeinen Theorie in unserem Spezialfall. Wir beginnen mit der in 2. 5 definierten Menge $\mathcal{E}_{\mathfrak{L}}$.

Hilfssatz 6. — *$\mathcal{E}_{\mathfrak{L}}$ ist die Menge aller auf* X *definierten, konkaven, stetigen, reellen Funktionen.*

Beweis (10). Die Menge $\mathcal{K}$ aller auf X konkaven, stetigen, reellen Funktionen ist eine abgeschlossene, inf-stabile Teilmenge von $\mathcal{C}(X)$ mit $\mathfrak{L} \subset \mathcal{K}$. Daher gilt $\mathcal{E}_{\mathfrak{L}} \subset \mathcal{K}$. Für eine beliebige Funktion $u \in \mathcal{K}$ zeigen wir, daß sie die untere Einhüllende aller $h \in \mathfrak{L}$ mit $h \geqslant u$ ist. Dann ist $\overline{Q}_x^X(u) = u(x)$ für alle $x \in X$,

(9) Es ist also X, aber nicht notwendig E Hausdorffsch. Jedoch könnte man im folgenden E auch als Hausdorffsch voraussetzen, ohne dabei die Allgemeinheit wesentlich einzuschränken. Es bezeichne nämlich N die abgeschlossene Hülle von $\{0\}$ in E und $E_0 = E/N$ den zu E assoziierten Hausdorffschen lokal-konvexen Raum. Dann bildet die kanonische Abbildung $\varphi : E \to E_0$ den Raum X homöomorph auf $\varphi(X)$ ab. Durchläuft $l$ alle stetigen Linearformen auf $E_0$, so durchläuft $l \circ \varphi$ alle stetigen Linearformen auf E.

(10) Vgl. auch Klee [30], p. 99.

also $u \in \mathcal{E}_{\mathfrak{L}}$ nach Satz 7. Somit ergibt sich die behauptete Gleichheit: $\mathcal{E}_{\mathfrak{L}} = \mathfrak{K}$.

Es sei $u \in \mathfrak{K}$, $x_0 \in X$ und $\gamma \in \mathbf{R}$ mit $u(x_0) < \gamma$. Zu zeigen ist die Existenz eines $h \in \mathfrak{L}$ mit $u \leqslant h$ und $h(x_0) < \gamma$. Hierzu betrachten wir im lokal-konvexen Raum $E_1 = E \times \mathbf{R}$ die Menge

$$A = \{(x, \xi) : x \in X, \ \xi \leqslant u(x)\}.$$

Dann sind A und $\bar{A}$ konvex in $E_1$, da $u$ konkav ist. Der Punkt $(x_0, \gamma)$ liegt nicht in $\bar{A}$. Zu jeder Zahl $\gamma'$ mit $u(x_0) < \gamma' < \gamma$ gibt es nämlich eine Umgebung V von $x_0$ in E mit $u(x) < \gamma'$ für alle $x \in V \cap X$; setzen wir noch $W = \{\lambda \in \mathbf{R} : \lambda > \gamma'\}$, so ist $V \times W$ eine zu A fremde Umgebung von $(x_0, \gamma)$ in $E_1$. Nach dem bereits zitierten Trennungssatz gibt es eine stetige Linearform $f_1$ auf $E_1$ mit:

$$(4.\ 1) \qquad \eta = \sup_{(x, \xi) \in A} f_1(x, \xi) < f_1(x_0, \gamma).$$

Nun ist offenbar: $f_1(x, \xi) = f(x) + \tau\xi$ für alle $(x, \xi) \in E_1$, wobei $f$ eine stetige Linearform auf E und $\tau = f_1(0, 1)$ ist. Wegen $(x_0, u(x_0)) \in A$ ist $f_1(x_0, u(x_0)) < f_1(x_0, \gamma)$, wegen $\gamma - u(x_0) > 0$ also $\tau > 0$. Bezeichnen wir mit $h$ die Restriktion von $\tau^{-1}(\eta - f)$ auf X, so ist $h \in \mathfrak{L}$. Aus (4. 1) folgt $u \leqslant h$ und $h(x_0) < \gamma$.

Aus diesem Hilfssatz und dem Korollar 1 zu Satz 7 folgt nun die Interpretation der Menge $\hat{\mathfrak{L}}$ der *$\mathfrak{L}$-harmonischen* Funktionen:

KOROLLAR. — *$\hat{\mathfrak{L}}$ ist die Menge aller auf* X *stetigen, affinlinearen Funktionen, d.h. die Menge der auf* X *stetigen, zugleich konkaven und konvexen Funktionen.*

Bemerkung. — Im allgemeinen ist $\hat{\mathfrak{L}} \neq \mathfrak{L}$; dies folgt aus einer Bemerkung von CHOQUET [21], p. 15.

4. 2. *Geometrische Extremalpunkte.* — Für jeden Punkt $x_0 \in X$ und jede Menge $S \in \mathfrak{S}(\mathfrak{L})$ wurde in Nr. 2. 5 die Menge $\mathcal{M}_{x_0}^S = \mathcal{M}_{x_0}^S(\mathfrak{L})$ von Maßen auf S definiert. Nach BOURBAKI [14], p. 87 und CHOQUET [22], p. 236 ergibt sich im vorliegenden Spezialfall

folgende Interpretation: $\mathfrak{M}^S_{x_0}$ ist die Menge aller Maße $\mu \geqslant 0$ auf S mit der Gesamtmasse 1 und mit $x_0$ als *Schwerpunkt:*

$$(4.\ 2) \qquad \int d\mu = 1 \quad \text{und} \quad x_0 = \int x\, d\mu.$$

Umgekehrt existiert hier zu jedem Maß $\mu \geqslant 0$ auf S der Gesamtmasse 1 genau ein $x_0 \in X$ mit $\mu \in \mathfrak{M}^S_{x_0}$.

Nun soll gezeigt werden, daß die $\mathfrak{L}$-extremalen Punkte von X mit den Extremalpunkten von X identisch sind. Ein Punkt $x_0 \in X$ heißt bekanntlich ein Extremalpunkt oder, wie wir hier genauer sagen wollen, ein *geometrischer Extremalpunkt* von X, wenn die Menge $X \cap \complement \{x_0\}$ konvex ist.

Hilfssatz 7. — *Für jeden Punkt $x_0 \in X$ und jede Menge $S \in \mathfrak{S}(\mathfrak{L})$ sind folgende Aussagen gleichwertig:*

(*a*) *$x_0$ ist ein geometrischer Extremalpunkt von X.*

(*b*) $\mathfrak{M}^S_{x_0}(\mathfrak{L}) = \{\varepsilon_{x_0}\}$.

Beweis. $(a) \Longrightarrow (b)$: Es genügt offenbar den Fall $S = X$ zu behandeln, wenn man Satz 3 beachtet, wonach $\mathfrak{M}^S_{x_0}$ nicht leer ist. Dann aber hat man weiter nur zu zeigen, daß der Träger $T_\mu$ eines Maßes $\mu \in \mathfrak{M}_{x_0}$ keinen von $x_0$ verschiedenen Punkt enthalten kann. Wegen $\int d\mu = 1$ folgt hieraus nämlich $\mu = \varepsilon_{x_0}$; also ist $\mathfrak{M}_{x_0} = \{\varepsilon_{x_0}\}$. Angenommen für ein Maß $\mu \in \mathfrak{M}_{x_0}$ existiert ein Punkt $x_1 \neq x_0$ im Träger $T_\mu$. Da $\mathfrak{L}$ die Punkte von X trennt, gibt es dann eine konvexe, kompakte Umgebung V von $x_1$ in X mit $x_0 \notin V$. Dann ist $c_1 = \mu(V) > 0$, da V Umgebung eines Punktes aus $T_\mu$ ist. Ferner ist $c_2 = 1 - \mu(V) > 0$, da V konvex und kompakt ist und aus $\mu(V) = 1$ somit $x_0 = \int x\, d\mu \in V$ folgen würde. Bezeichnet $\chi_A$ die charakteristische Funktion einer Menge $A \subset X$ bezüglich X, so sind $\mu_1 = c_1^{-1}(\chi_V \mu)$ und $\mu_2 = c_2^{-1}(1 - \chi_V)\mu$ positive Maße auf X der Gesamtmasse 1. Für die zugehörigen Schwerpunkte $x_i^*$ aus X gilt:

$$x_0 = c_1 x_1^* + c_2 x_2^*, \quad c_1 > 0,\ c_2 > 0,\ c_1 + c_2 = 1.$$

Hieraus folgt $x_1^* = x_2^* = x_0$, da $x_0$ ein geometrischer Extremalpunkt ist. Dies aber kann nicht sein, da $\mu_1$ von V getragen wird und somit $x_1^*$ zu V gehört. Der Punkt $x_0$ liegt aber nicht in V. Wir stoßen somit auf einen Widerspruch zu unserer Annahme: $T_\mu \cap \complement \{x\} \neq \emptyset$.

Non $(a) \Longrightarrow$ non $(b)$: Ist $x_0$ kein geometrischer Extremalpunkt, so gibt es verschiedene Punkte $x_1, x_2, \in X$ und ein $\lambda \in \mathbf{R}$ mit $x_0 = \lambda x_1 + (1-\lambda)x_2$ und $0 < \lambda < 1$. Nach Satz 3 gibt es Maße $\mu_i \in \mathcal{M}^S_{x_i}$ $(i = 1, 2)$. Das Maß $\mu = \lambda\mu_1 + (1-\lambda)\mu_2$ liegt dann in $\mathcal{M}^S_{x_0}$ und ist ungleich $\varepsilon_{x_0}$. In der Tat: aus $\mu = \varepsilon_{x_0}$ folgt, daß $\{x_0\}$ der Träger von $\mu_1$ und $\mu_2$, also $\mu_1 = \mu_2 = \varepsilon_{x_0}$ ist. Dies kann aber wegen $x_1 \neq x_2$ nicht sein.

Setzt man speziell $S = X$, so erhält man das angekündigte Resultat:

Korollar. — *Die $\mathfrak{L}$-extremalen Punkte von X sind identisch mit den geometrischen Extremalpunkten von X.*

Folglich ist $\partial_{\mathfrak{L}}X$ die abgeschlossene Hülle $\overline{X_e}$ in X der Menge $X_e = X_e(\mathfrak{L})$ aller geometrischen Extremalpunkte von X. Dies ist ein bekanntes Resultat (vgl. Milman [33], Arens-Singer [1]).

3. 4. *Simplexe von Choquet.* — Nach den Resultaten der beiden letzten Nummern spezialisiert sich unser allgemeines Dirichletsches Problem hier auf die Aufgabe, notwendige und hinreichende Bedingungen dafür anzugeben, daß jede Funktion aus $\mathcal{C}(\overline{X_e})$ zu einer in X stetigen, affin-linearen Funktion fortgesetzt werden kann. Im folgenden kennzeichnen wir die konvexen kompakten Mengen X mit dieser Eigenschaft.

Hierzu erinnern wir zunächst an die von Choquet [21] gegebene Definition eines Simplex in einem Vektorraum F (über R): *Simplex* in F heißt jede konvexe Menge S in F mit der Eigenschaft, daß für je zwei positiv-homothetische Bilder $S_i = a_i + \lambda_i S$ $(a_i \in F, \lambda_i \in \mathbf{R}, \lambda_i > 0)$ die folgende Alternative besteht: Entweder ist der Durchschnitt $S_1 \cap S_2$ die leere Menge oder selbst ein positiv-homothetisches Bild von S.

Die affine Natur dieser Begriffsbildung wird deutlich durch die folgende Behauptung, deren elementarer Beweis dem Leser überlassen werden kann. Im Vektorraum F sei V eine affin-lineare Mannigfaltigkeit und $x_0$ ein Punkt aus V; ferner sei S eine konvexe Teilmenge von V. Dann und nur dann ist S ein

Simplex in F, wenn der Durchschnitt je zweier Mengen $S_i = a_i + \lambda_i(S - x_0)$ mit $a_i \in V$ und $\lambda_i \geqslant 0 (i = 1, 2)$ entweder leer oder von der gleichen Gestalt $a + \lambda(S - x_0)$ mit $a \in V$ und $\lambda \geqslant 0$ ist.

Nunmehr behaupten wir:

Satz 13. — *Für jede konvexe, kompakte Teilmenge X eines lokal-konvexen Raumes E sind folgende Aussagen gleichwertig:*

(*a*) *Jede Funktion* $f \in \mathcal{C}(\overline{X_e})$ *ist* $\mathfrak{L}$*-resolutiv.*

(*b*) *Jeder Punkt aus* X *ist der Schwerpunkt genau eines Maßes aus* $\mathcal{M}(\overline{X_e})$ *mit Gesamtmasse* 1.

(*c*) *Die Menge* $\mathfrak{L}$ *aller auf* X *stetigen, affin-linearen, reellen Funktionen ist ein Verband bezüglich der üblichen* $\leqslant$ *-Relation.*

(*d*) *Zu jeder auf* X *stetigen, konkaven, reellen Funktion existiert eine größte stetige, affin-lineare Minorante.*

(*e*) X *ist ein Simplex und* $X_e$ *ist abgeschlossen in* X.

Beweis. Die Äquivalenz der Aussagen (*a*)-(*d*) folgt unmittelbar aus den allgemeinen Sätzen 9 und 10 sowie dem Korollar 1 zu Satz 10 bei Beachtung der Resultate von 4. 1. und 4. 2. Zu beweisen ist also noch die Äquivalenz von (*e*) mit z. B. (*b*).

(*e*) $\Longrightarrow$ (*b*): Dies folgt einerseits aus einem Eindeutigkeitssatz von Choquet [21], wonach sich jeder Punkt $x$ eines kompakten Simplex X in E nur auf höchstens eine Weise als Schwerpunkt eines auf $X_e$ konzentrierten Maßes $\nu \in \mathcal{M}(X)$ mit $\int d\nu = 1$ darstellen läßt. Da $X_e$ in X abgeschlossen und somit gleich $\partial_{\mathfrak{L}} X$ ist, sichert andererseits Satz 3 die Existenz mindestens eines solchen Maßes $\nu$.

(*b*) $\Longrightarrow$ (*e*): Wegen der bereits bewiesenen Äquivalenz von (*b*) und (*a*) besagt Satz 11, daß die Menge $X_e$ abgeschlossen in X ist. Daß X ein Simplex ist, sieht man folgendermaßen ein: Es sei F der Vektorraum aller Radonschen Maße auf dem kompakten Raum $X_e$ und V die affin-lineare Mannigfaltigkeit aller Maße $\mu \in F$ mit $\int d\mu = 1$. Jedem Maß $\mu \in V$ ordne man als Bild $\varphi(\mu)$ in E zu den Punkt $\varphi(\mu) = \int x \, d\mu^+ - \int x \, d\mu^- \in E$, wenn hierbei $\mu^+$ bzw. $\mu^-$ den Positiv- bzw. Negativteil von $\mu$

bezeichnet ([11]). Die so definierte Abbildung $\varphi : V \to E$ ist affin-linear; man hat hierbei nur zu beachten, daß

$$\varphi(\mu) = \int x \, d\mu_1 - \int x \, d\mu_2$$

ist für jedes Paar von Maßen $\mu_1 \geqslant 0$, $\mu_2 \geqslant 0$ auf $X_e$ mit $\mu = \mu_1 - \mu_2$. Aus (*b*) folgt die Eineindeutigkeit von $\varphi$, wie man sofort nachrechnet. Daher ist $\varphi$ ein affiner Isomorphismus von V auf die affin-lineare Mannigfaltigkeit $\varphi(V)$ in E. Es gilt offenbar $\varphi(Y) = X$, wenn Y die Menge aller positiven Maße aus V bezeichnet. Nun ist bekanntlich Y ein Simplex in F (vgl. Choquet [21], p. 14) ([12]). Aus der dem Satz vorangestellten Hilfsbehauptung und den festgestellten Eigenschaften von $\varphi$ folgt dann, daß X ein Simplex in E ist.

Bemerkungen. — 1. Die Aussage « $(e) \Longrightarrow (a)$ » findet sich bereits bei Choquet [21].

2. Es wäre interessant zu wissen, ob ein beliebiges kompaktes Simplex in einem lokal-konvexen Raum durch zu (*a*)-(*d*) analoge Eigenschaften gekennzeichnet werden kann.

3. Aus den Sätzen 12 und 13 folgt, daß in einem kompakten, metrisierbaren Simplex X in E mit in X abgeschlossenem Choquetschen Rand $X_e$ jeder geometrische Extremalpunkt auch $\hat{\mathfrak{L}}$-exponiert ist. Im Hinblick auf ein analoges Resultat von Bishop [9] über Funktionenalgebren wäre es interessant zu wissen, ob diese Eigenschaft jedes kompakte, metrisierbare Simplex besitzt. Eine beliebige konvexe, kompakte, metrisierbare Menge X in einem lokal-konvexen Raum E besitzt diese Eigenschaft im allgemeinen nicht, wie folgendes Gegenbeispiel zeigt: X = konvexe Hülle des Torus $T^2$ im $R^3$.

## § 5. — Ergänzungen zur allgemeinen Theorie.

5. 1. *Einbettung von* X *in den Raum* $\mathcal{H}'$. — Wir kehren zurück zu der in den Paragraphen 2 und 3 betrachteten allgemeinen Situation. Gegeben sei also ein beliebiger kompakter

([11]) Für ein Maß $\mu \geqslant 0$ auf $X_e$ ist $\int x \, d\mu$ definitionsgemäß gleich 0, wenn $\mu = 0$ ist, und gleich $\alpha x_1$, wenn $\alpha = \int d\mu > 0$ und $x_1 \in X$ der Schwerpunkt von $\alpha^{-1}\mu$ ist. Ist E ein Hausdorff-Raum, so ist dies der einzige Punkt $x_0 \in E$ mit $l(x_0) = \int l \, d\mu$ für jede stetige Linearform $l$ auf E.

([12]) Man hat nur zu beachten, daß F ein Rieszscher Raum und Y eine Basis des Kegels aller Elemente $\geqslant 0$ aus F ist.

Raum X und eine Menge $\mathcal{H}$ reeller Funktionen auf X mit den Eigenschaften (2.1.)-(2. 3). Wir versehen $\mathcal{H}$ mit der Topologie der gleichmäßigen Konvergenz auf X und bezeichnen mit $E = \mathcal{H}'$ den mit der *schwachen Topologie* ausgestatteten topologischen *Dualraum* von $\mathcal{H}$. Die Menge $Y = Y_{\mathcal{H}}$ aller positiven Linearformen I auf $\mathcal{H}$ mit $I(1) = 1$ ist dann eine *konvexe, kompakte* Teilmenge von E (vgl. Bourbaki [13], chap. IV, § 5). Für jeden Punkt $x \in X$ ist $h \to h(x)$ ein Element $I_x$ von Y; durch die Zuordnung $x \to I_x$ ist also eine kanonische Abbildung $\varphi : X \to Y$ definiert. Wir behaupten:

Hilfssatz 8. — *Die Abbildung* $\varphi : X \to Y$ *besitzt folgende Eigenschaften:*

(*a*) *Der Raum* X *wird durch* $\varphi$ *homöomorph auf* $\varphi(X)$ *abgebildet.*

(*b*) *Die Menge* $X_e$ *der* $\mathcal{H}$*-extremalen Punkte von* X *wird auf die Menge* $Y_e$ *der geometrischen Extremalpunkte von* Y *abgebildet:* $Y_e = \varphi(X_e)$.

(*c*) *Ein positives Maß* $\mu$ *auf* $X^* = \partial_{\mathcal{H}}X$ *ist dann und nur dann ein zu einem Punkte* $x_0 \in X$ *gehöriges* $\mathcal{H}$*-harmonische Maß wenn das Bildmaß* $\varphi(\mu)$ *die Gesamtmasse* 1 *und* $\varphi(x_0)$ *zum Schwerpunkt hat.*

Beweis. *Zu* (*a*): Da $\mathcal{H}$ die Punkte von X trennt, ist die Abbildung $\varphi$ eineindeutig. Aus der Konvergenz eines Filters $\mathfrak{F}$ in X gegen einen Punkt $x_0 \in X$ folgt $\lim_{\mathfrak{F}} h(x) = h(x_0)$ also $\lim_{\mathfrak{F}} I_x(h) = I_{x_0}(h)$ für alle $h \in \mathcal{H}$. Also gilt $\lim_{\mathfrak{F}} \varphi(x) = \varphi(x_0)$ nach Definition der Topologie von E, was die Stetigkeit von $\varphi$ beweist. Da X kompakt und E Hausdorffsch ist, so ist $\varphi$ sogar eine Homöomorphie von X auf $\varphi(X)$.

*Zu* (*b*): Für jedes $h \in \mathcal{H}$ ist $I \to I(h)$ eine stetige Linearform auf E; nach einem bekannten Satz (Bourbaki [13], p. 50, prop. 1) ist umgekehrt jede stetige Linearform von dieser Form. Hieraus folgt zunächst: $Y_e \subset \varphi(X)$. Die Existenz eines geometrischen Extremalpunktes $I_0 \in Y_e$, der nicht in $\varphi(X)$ liegt, führt nämlich wie folgt auf einen Widerspruch: Zunächst kann $I_0$ nicht in der abgeschlossenen konvexen Hülle Q von $\varphi(X)$ in E liegen. Sonst wäre nämlich $I_0$ ein geometrischer Extremalpunkt von Q, müßte also nach Bourbaki [12], p. 84, prop. 4 bereits in der kompakten Menge $\varphi(X)$ liegen. Folglich existiert nach einem bereits früher verwendeten Trennungssatz ([12],

p. 73, prop. 4) eine in E abgeschlossene Hyperebene, welche Q und $\{I_0\}$ streng trennt. Nach der einleitenden Bemerkung bedeutet dies die Existenz einer Funktion $h_0 \in \mathcal{H}$ mit $I(h_0) < I_0(h_0)$ für alle $I \in Q$, also speziell für alle $I \in \varphi(X)$. Somit gilt $h_0(x) < I_0(h_0)$ für alle $x \in X$. Setzen wir $\alpha = \sup_{x \in X} h_0(x)$, so ergibt sich: $\alpha < I_0(h_0)$. Andererseits folgt aber aus $h_0 \leqslant \alpha$ und $I_0 \in Y$ die Relation: $I_0(h_0) \leqslant I_0(\alpha) = \alpha$. Dies ist der angekündigte Widerspruch zu unserer Annahme. Damit ist gezeigt worden, daß $Y_e$ in $\varphi(X)$ enthalten ist. — Hieraus folgt nun (*b*). In der Tat: Vermöge der Abbildung $\mu \rightarrow \varphi(\mu)$ entsprechen sich umkehrbar eindeutig die Maße $\mu \geqslant 0$ auf X der Gesamtmasse 1 und die Maße $\nu \geqslant 0$ auf $\varphi(X)$ der Gesamtmasse 1. Nach der einleitenden Bemerkung durchläuft $I \rightarrow I(h) = f(I)$ alle stetigen Linearformen $f$ auf E, wenn $h$ alle Funktionen aus $\mathcal{H}$ durchläuft. Somit gilt für jedes Maß $\mu \geqslant 0$ der Gesamtmasse 1 auf X:

$$\int f\, d\varphi(\mu) = \int f \circ \varphi\, d\mu = \int f(\varphi(x))\, d\mu(x) = \int f(I_x)\, d\mu(x)$$
$$= \int I_x(h)\, d\mu(x) = \int h\, d\mu.$$

Aus $\int h\, d\mu = h(x_0) = I_{x_0}(h) = \varphi(x_0)\,(h)$ für alle $h \in \mathcal{H}$ folgt daher $\int f\, d\varphi(\mu) = f(\varphi(x_0))$ für alle stetigen Linearformen $f$ auf E und umgekehrt. Die Behauptung (*b*) ergibt sich daher jetzt aus Hilfssatz 7, wenn man dort $S = \varphi(X)$ setzt.

*Zu* (*c*): Man beweist dies durch eine Überlegung, die zu derjenigen völlig analog ist, die wir soeben im zweiten Teil des Beweises von (*b*) durchgeführt haben. Man hat nur zu beachten, daß $\varphi(X^*)$ nach (*a*) und (*b*) gleich der abgeschlossenen Hülle von $Y_e$ in Y ist.

Bemerkungen. — 1. Der in Hilfssatz 8 festgestellte Sachverhalt kann umgekehrt zu einem Beweis der Sätze 1 und 2 für $\mathcal{E} = \mathcal{H}$ verwendet werden. Vgl. hierzu Bishop-de Leeuw [8].

2. Aus der Behauptung (*b*) von Hilfssatz 8 und dem Satz von Krein-Milman folgt, daß Y gleich der abgeschlossenen konvexen Hülle Q von $\varphi(X)$ ist.

5. 2. *Ergänzungen zu* § 2 *und* § 3. — Nunmehr ergibt sich die Antwort auf eine bereits in Nr. 2. 1 angeschnittene Frage:

Satz 14. — *Ein Punkt $x_0 \in X$ ist genau dann $\mathcal{H}$-extremal, wenn $\varepsilon_{x_0}$ das einzige zu $x_0$ gehörige $\mathcal{H}$-harmonische Maß ist.*

Beweis. Nach Hilfssatz 8 ist $\varepsilon_{x_0}$ genau dann das einzige zu $x_0$ gehörige $\mathcal{H}$-harmonische Maß, wenn $\varepsilon_{x_0}$ das einzige positive Maß auf $\varphi(X^*)$ der Gesamtmasse 1 mit $y_0 = \varphi(x_0)$ als Schwerpunkt ist. Da nach dem gleichen Hilfssatz $Y_e \subset \varphi(X^*)$ gilt, ist dies nach Hilfssatz 7 (mit $S = \varphi(X^*)$) gleichbedeutend mit der Aussage, daß $\varphi(x_0)$ ein geometrischer Extremalpunkt von Y ist. Abermals nach Hilfssatz 8 ist hiermit gleichwertig die Aussage, daß $x_0$ $\mathcal{H}$-extremal ist.

Ersetzt man in 5. 1 den Vektorraum $\mathcal{H}$ durch $\hat{\mathcal{H}}$, so gelangt man zum Raum $\hat{E} = \hat{\mathcal{H}}'$, zur konvexen kompakten Menge $\hat{Y} = Y_{\hat{\mathcal{H}}}$ in $\hat{E}$ und zur kanonischen Abbildung $\hat{\varphi}: X \to \hat{Y}$. Wir behaupten:

Satz 15. — *Dann und nur dann ist jede Funktion aus $\mathcal{C}(X^*)$ $\mathcal{H}$-resolutiv, wenn $\hat{Y}$ ein Simplex in $\hat{E}$ und die Menge $\hat{Y}_e$ seiner geometrischen Extremalpunkte abgeschlossen ist.*

Beweis. Es sei jede Funktion aus $\mathcal{C}(X^*)$ $\mathcal{H}$-resolutiv. Dann ist $\hat{\mathcal{H}}$ nach Satz 10 ein Rieszscher Raum, also (vgl. [14], p. 35) der konvexe Kegel $\hat{P}$ aller positiven Linearformen $I \in \hat{E}$ ein Verband bezüglich der durch $\hat{P}$ in $\hat{E}$ definierten Ordnungsrelation. Durch die Gleichung $I(1) = 1$ wird in $\hat{E}$ eine abgeschlossene Hyperebene definiert, welche jede Erzeugende von $\hat{P}$ in genau einem Punkt $\neq 0$ trifft. Der Durchschnitt von $\hat{P}$ mit dieser Hyperebene ist $\hat{Y}$. Somit ist $\hat{Y}$ eine Basis von $\hat{P}$, also $\hat{Y}$ ein Simplex in $\hat{E}$ nach Choquet [21]. Nach Satz 11 ist $X_e$ in X, nach Hilfssatz 8 also $\hat{Y}_e = \hat{\varphi}(X_e)$ in $\hat{Y}$ abgeschlossen. (Bei der Anwendung von Hilfssatz 8 tritt $\hat{\mathcal{H}}$ an die Stelle von $\mathcal{H}$ und man hat (2. 18) zu beachten, wonach $X_e(\hat{\mathcal{H}}) = X_e(\mathcal{H})$ ist.)

Umgekehrt besitze $\hat{Y}$ die genannten Eigenschaften. Nach Satz 13 existiert dann zu jedem Punkt $y_0 \in \hat{Y}$ genau ein positives Maß auf $\hat{Y}_e$ der Gesamtmasse 1 mit $y_0$ als Schwerpunkt. Da mit $\hat{Y}_e$ auch $X_e = \hat{\varphi}^{-1}(\hat{Y}_e)$ abgeschlossen, also $X_e = X^*$ ist, folgt aus Hilfssatz 8, daß zu jedem Punkt $x_0 \in X$ nur genau ein $\mathcal{H}$-harmonisches Maß gehört. Dann ist aber jede Funktion aus $\mathcal{C}(X^*)$ $\mathcal{H}$-resolutiv nach Satz 9.

Bemerkung. — Ein direkter Beweis von Satz 14, der also nicht die kanonische Einbettung $\varphi: X \to Y$ verwendet, wäre wünschenswert. Übrigens folgt aus Satz 14 leicht ein neuer Beweis von Satz 11.

5. 3. *Beziehungen zwischen* $\hat{Y}$ *und* Y. — Über die Beziehungen zwischen $\hat{E}$, $\hat{Y}$, $\hat{\varphi}$ einerseits und E, Y, $\varphi$ andererseits behaupten wir abschließend:

Hilfssatz 9. — *Die Abbildung* $\Phi: \hat{E} \to E$, *welche jeder Linearform aus* $\hat{E}$ *ihre Restriktion auf* $\mathcal{H}$ *zuordnet, besitzt folgende Eigenschaften:*

(*a*) $\Phi$ *ist eine stetige, lineare Abbildung von* $\hat{E}$ *auf* E.

(*b*) $\Phi(\hat{Y}) = Y$ *und* $\varphi = \Phi \circ \hat{\varphi}$.

(*c*) $\Phi$ *bildet* $\hat{\varphi}(X)$ *homöomorph auf* $\varphi(X)$ *ab.*

Beweis. *Zu* (*a*): Daß $\Phi$ linear und stetig ist, rechnet man sofort nach. Aus dem Satz von Hahn-Banach folgt, daß $\Phi$ eine Abbildung auf E ist.

*Zu* (*b*): Da die konstante Funktion 1 in $\mathcal{H}$ liegt und ein innerer Punkt des Kegels aller nicht-negativen Funktionen aus $\hat{\mathcal{H}}$ (bezüglich der Topologie der gleichmäßigen Konvergenz) ist, folgt die Gleichheit $\Phi(\hat{Y}) = Y$ aus dem Fortsetzungssatz von M. G. Krein ([12], p. 75, prop. 6). Aus der Definition der Abbildungen $\varphi$, $\hat{\varphi}$, $\Phi$ folgt: $\varphi = \Phi \circ \hat{\varphi}$.

*Zu* (*c*): Nach Hilfssatz 8 ist $\varphi$ eine homöomorphe Abbildung von X auf $\varphi(X)$ und $\hat{\varphi}$ eine homöomorphe Abbildung von X auf $\hat{\varphi}(X)$. Hieraus und aus $\varphi = \Phi \circ \hat{\varphi}$ folgt dann (*c*).

Bemerkung. — Im allgemeinen wird $\hat{Y}$ durch $\Phi$ nicht eineindeutig auf Y abgebildet. Dies zeigt das Beispiel 6. Wegen $\hat{\mathcal{H}} = \mathcal{C}(X)$ sind dort das Lebesguesche Maß $\lambda$ auf $[0, 1]$ und das diskrete Maß $\mu = \frac{1}{6}\varepsilon_0 + \frac{4}{6}\varepsilon_{\frac{1}{2}} + \frac{1}{6}\varepsilon_1$ zwei verschiedene Elemente aus $\hat{Y}$ mit $\Phi(\lambda) = \Phi(\mu)$.

## § 6. — Anwendung: Das klassische Dirichletsche Problem.

6. 1. *Reguläre Punkte.* — Wir betrachten im euklidischen Raum $\mathbf{R}^N$ von $N \geqslant 2$ Dimensionen eine *relativ kompakte, offene* Menge $\Omega$; F sei der euklidische Rand von $\Omega$ im $\mathbf{R}^N$. Der der allgemeinen Theorie zugrunde liegende kompakte Grundraum X sei jetzt speziell die abgeschlossene Hülle von $\Omega$ im $\mathbf{R}^N$, also: $X = \overline{\Omega} = \Omega \cup F$. Der Funktionenraum $\mathcal{H}$ der allgemeinen Theorie sei jetzt die Menge aller Funktionen $h \in \mathcal{C}(X)$, deren Restriktion $h_\Omega$ auf $\Omega$ im üblichen Sinne harmonisch, also in $\Omega$ eine Lösung der *Laplaceschen Differentialgleichung* $\Delta h_\Omega = 0$ ist. Dann besitzt offenbar $\mathcal{H}$ die Eigenschaften (2. 1)-(2. 3); die allgemeine Theorie ist also anwendbar. Die Eigenschaft (2. 3) folgt aus der Bemerkung, daß die im $\mathbf{R}^N$ harmonischen Funktionen die Punkte des $\mathbf{R}^N$ trennen und eine in $\mathcal{H}$ gelegene Restriktion auf X besitzen.

Zunächst bestimmen wir den Raum $\hat{\mathcal{H}}$ aller $\mathcal{H}$-harmonischen Funktionen. Hierzu bezeichne $\mathcal{E}$ die Menge aller Funktionen $u \in \mathcal{C}(X)$, deren Restriktion auf $\Omega$ *superharmonisch* ist. Offenbar ist $\mathcal{E}$ in $\mathcal{C}(X)$ abgeschlossen und inf-stabil. Da ferner $\mathcal{H} = \mathcal{E} \cap (-\mathcal{E})$ ist, folgt aus dem Korollar 1 zu Satz 7:

$$\hat{\mathcal{H}} = \mathcal{H}. \tag{6. 1}$$

Also fällt hier $\hat{\mathcal{H}}$ mit $\mathcal{H}$ zusammen.

Zur Bestimmung aller $\mathcal{H}$-extremalen Punkte von X erinnern wir an die Definition der *verallgemeinerten Lösung* $H_f$ des Dirichletschen Problems für Randfunktionen $f \in \mathcal{C}(F)$ ([13]). Man betrachtet hierzu für jede stetige, reelle Funktion $f$ auf F die Menge $\mathfrak{D}_f$ aller in $\Omega$ definierten superharmonischen Funktionen $v$ mit der Eigenschaft:

$$\liminf_{x \to x_0,\, x \in \Omega} v(x) \geqslant f(x_0) \quad \text{für alle} \quad x_0 \in F. \tag{6. 2}$$

Man betrachtet ferner die Menge $\mathfrak{U}_f = -\mathfrak{D}_{-f}$ und zeigt, daß die untere Einhüllende von $\mathfrak{D}_f$ gleich der oberen Einhüllenden von $\mathfrak{U}_f$ ist. Die somit durch

$$H_f = \inf \mathfrak{D}_f = \sup \mathfrak{U}_f \tag{6. 3}$$

([13]) Wegen der Einzelheiten sei auf Brelot [16] verwiesen.

für alle $f \in \mathcal{C}(F)$ in $\Omega$ definierte Funktion $H_f$ ist harmonisch und heißt die verallgemeinerte Lösung des Dirichletschen Problems (nach PERRON-WIENER).

Ein Punkt $x_0 \in F$ wird *regulär* genannt, wenn gilt:

$$(6.4) \qquad \lim_{x \to x_0,\, x \in \Omega} H_f(x) = f(x_0) \qquad \text{für alle} \qquad f \in \mathcal{C}(F).$$

Besitzt für eine Funktion $f \in \mathcal{C}(F)$ das klassische Dirichletsche Problem eine Lösung $u_f$, ist also $u_f$ eine in $\mathcal{H}$ gelegene Fortsetzung von $f$, so stimmt $u_f$ auf $\Omega$ mit der verallgemeinerten Lösung $H_f$ überein. Daher besitzt das klassische Dirichletsche Problem für jede Funktion $f \in \mathcal{C}(F)$ genau dann eine Lösung, wenn alle Punkte aus F regulär sind. In diesem Falle folgt aus einer Bemerkung in 3. 1 und Satz 11, daß die regulären Punkte mit den $\mathcal{H}$-extremalen Punkten zusammenfallen. Wir behaupten dies nun allgemein:

SATZ 16. — *Die $\mathcal{H}$-extremalen Punkte von $X = \overline{\Omega}$ fallen mit den regulären Punkten des euklidischen Randes* F *von* $\Omega$ *zusammen.*

Beweis. Wir bemerken zunächst, daß der Choquetsche Rand $X_e = X_e(\mathcal{H})$ in F enthalten ist. Dies folgt entweder aus dem Randmaximum-Prinzip der Potentialtheorie oder aus der Bemerkung, daß zu jedem Punkt $x_0 \in \Omega$ ein Maß $\mu \neq \varepsilon_{x_0}$ auf X existiert mit $\mu \in \mathcal{M}_{x_0}(\mathcal{H})$. Wegen der Mittelwerteigenschaft der harmonischen Funktionen genügt es für $\mu$ die auf einer in $\Omega$ enthaltenen Vollkugel mit Mittelpunkt $x_0$ « gleichmäßig verteilte Masse 1 » zu wählen.

Es sei nun $x_0$ ein $\mathcal{H}$-extremaler Punkt von X. Wegen $X_e \subset F$ gilt dann $\mathcal{M}^F_{x_0}(\mathcal{H}) = \{\varepsilon_{x_0}\}$ nach 2. 2. Hieraus folgt nach Satz 3:

$$\underline{Q}^F_{x_0}(f) = \overline{Q}^F_{x_0}(f) \qquad \text{für jedes } f \in \mathcal{C}(F).$$

Zu jeder Funktion $f \in \mathcal{C}(F)$ und jeder Zahl $\varepsilon > 0$ gibt es daher Funktionen $h^1, h^2 \in \mathcal{H}$ mit folgenden Eigenschaften:

$$(6.5) \qquad h^1(x) \leqslant f(x) \leqslant h^2(x) \qquad \text{für alle } x \in F;$$
$$(6.6) \qquad h^2(x_0) - h^1(x_0) \leqslant \varepsilon.$$

Für die Restriktionen $h^i_\Omega$ von $h^i$ auf $\Omega$ ergibt sich nach (6. 5): $h^1_\Omega \in \mathfrak{U}_f$, $h^2_\Omega \in \mathfrak{O}_f$. Hieraus folgt nach (6. 3)

$$h^1_\Omega \leqslant H_f \leqslant h^2_\Omega$$

und somit

$$(6.\ 7)\qquad h^1(x_0) \leqslant \liminf_{x \to x_0,\ x \in \Omega} \mathrm{H}_f(x) \leqslant \limsup_{x \to x_0,\ x \in \Omega} \mathrm{H}_f(x) \leqslant h^2(x_0).$$

Nach (6. 5)-(6. 7) weichen also der obere und der untere Limes von $\mathrm{H}_f(x)$ für $x \to x_0$ $(x \in \Omega)$ von $f(x_0)$ um höchstens $\varepsilon$ *ab*. Da $\varepsilon > 0$ beliebig gewählt war, erhält man : $f(x_0) = \lim_{x \to x_0,\ x \in \Omega} \mathrm{H}_f(x)$ für alle $f \in \mathcal{C}(\mathrm{F})$. Also ist der Punkt $x_0$ regulär.

Umgekehrt sei $x_0$ ein regulärer Randpunkt von $\Omega$. Nach einem Lemma von Keldych [28] ist dann $x_0$ $\mathcal{H}$-exponiert, also $x_0$ auch $\mathcal{H}$-extremal nach 1. 3.

Aus den Sätzen 2 und 16 folgt nun, daß der Šilovsche Rand $\mathrm{X}^* = \partial_{\mathcal{H}}\mathrm{X}$ gleich der abgeschlossenen Hülle der Menge aller regulären Randpunkte von $\Omega$ ist. Im Anschluß an diese Bemerkung wäre es leicht möglich, $\mathrm{X}^*$ unter Heranziehung klassischer Resultate von Vasilesco [34] und Bouligand [10] mit Hilfe potentialtheoretischer Begriffsbildungen zu kennzeichnen. Wenn $\Omega$ speziell ein Gebiet ist, folgt z.B. aus den Resultaten der beiden zitierten Arbeiten, daß $\mathrm{X}^*$ die Menge aller Punkte $x \in \mathrm{F}$ ist, in welchen F *lokal von positiver Kapazität* ist. In der Terminologie von [34] ist daher $\mathrm{X}^*$ mit dem *reduzierten Teil* von F identisch. Auf diese feineren Gesichtspunkte soll jedoch hier nicht eingegangen werden.

Bemerkungen. — 1. Für das am Schluß des Beweises von Satz 16 verwendete Lemma von Keldych hat Brelot [19] kürzlich einen einfachen Beweis gegeben. In der gleichen Arbeit wird das Lemma in geeigneter Form auch in Greenschen Räumen und allgemeiner im Rahmen der von Brelot [17], [18] entwickelten Axiomatik für das Dirichletsche Problem bewiesen. Diese Resultate und die Tatsache, daß der erste Teil des Beweises von Satz 16 nur einfache Schlüsse verwendet, gestatten es, Satz 16 auf allgemeineren Räumen zu beweisen. Wir kommen hierauf sowie auf die Beziehungen dieses Satzes zu Resultaten von Matsushita [31], [32] an anderer Stelle zurück.

2. Die Aussage, daß jeder $\mathcal{H}$-extremale Punkt regulär ist, bleibt richtig, wenn man die Laplacesche Differentialgleichung durch die Wärmeleitungsgleichung oder allgemeinere parabolische Differentialgleichungen ersetzt (vgl. hierzu Bauer [6]).

6. 2. *Minimal begrenztes* $\Omega$. — Von besonderem Interesse ist der Fall, daß $X^*$ mit dem euklidischen Rand F zusammenfällt. Dann liefern nämlich die Resultate von § 3 neuartige Kriterien für die uneingeschränkte Lösbarkeit des klassischen Dirichletschen Problems. Die Formulierung der Sätze in diesem Spezialfall kann dem Leser überlassen werden.

Notwendig und hinreichend für die Gleichheit $F = X^*$ ist nach Satz 16 die Bedingung, daß die regulären Punkte in F dicht liegen. Ein anderes Kriterium folgt aus den oben zitierten Resultaten von Bouligand-Vasilesco. Hier soll durch eine einfache Schlußweise gezeigt werden, daß F mit $X^*$ sicher dann zusammenfällt, wenn $\Omega$ *minimal begrenzt* ist, d.h. wenn für das Innere $\mathring{X}$ von X gilt: $\mathring{X} = \Omega$.

Wir erinnern hierzu an die Definition des *Newtonschen Kerns* K auf $R^N \times R^N$:

$$(6.8) \qquad K(x, y) = \begin{cases} \|x - y\|^{2-N}, & N \geqslant 3; \\ -\log \|x - y\|, & N = 2. \end{cases}$$

$\|...\|$ bezeichnet dabei die euklidische Norm.

Hilfssatz 10. — *Es sei* A *eine kompakte Teilmenge des* $R^N$ *und* $(x_n)_{n=1,2,...}$ *eine gegen einen Punkt* $x_0 \in \complement A$ *konvergente Folge von Punkten des* $R^N$. *Bezeichnet dann* $f_i$ *die Restriktion auf* A *der Funktion* $x \to K(x, x_i)$ $(i = 0, 1, ...)$, *so konvergiert die Folge* $(f_n)_{n=1,2,...}$ *auf* A *gleichmäßig gegen* $f_0$.

Beweis. Wir erbringen den Beweis nur für $N = 2$. Analoge Überlegungen führen für $N \geqslant 3$ zum Ziel. Da $(x_n)$ gegen $x_0 \in \complement A$ konvergiert und $\complement A$ offen ist, kann o.B.d.A. angenommen werden, daß kein $x_n$ zu A gehört. Die Menge B aller Punkte $x_0, x_1, ...,$ ist dann kompakt und zu A fremd, also der Abstand $\delta$ von A und B positiv. Für jedes $x \in A$ und je zwei Zahlen $i, j = 0, 1, ...$ gilt:

$$f_i(x) - f_j(x) = \log \frac{\|x - x_j\|}{\|x - x_i\|} \leqslant \log\left(1 + \frac{\|x_i - x_j\|}{\|x - x_i\|}\right)$$
$$\leqslant \log(1 + \delta^{-1}\|x_i - x_j\|).$$

Hieraus folgt: $|f_n(x) - f_0(x)| \leqslant \log(1 + \delta^{-1}\|x_n - x_0\|)$ für alle $x \in A$ und $n = 1, 2, ...,$ also die gleichmäßige Konvergenz auf A von $(f_n)$ gegen $f_0$.

Satz 17. — *Für den Šilovschen Rand* $X^* = \partial_{\mathcal{H}}X$ *gilt stets:*

$$F \cap \complement \mathring{X} \subset X^*.$$

Beweis. Angenommen, es gibt einen Punkt

$$x_0 \in F \cap \complement \mathring{X} \cap \complement X^*.$$

Wegen $x_0 \in \complement \mathring{X}$ hat dann jede Umgebung von $x_0$ im $R^N$ mit $\complement X$ einen Punkt gemeinsam. Also existiert eine gegen $x_0$ konvergente Folge von Punkten $x_n \in \complement X$. Wir bezeichnen mit $h_i^*$ bzw. $h_i$ die Restriktion auf $X^*$ bzw. X der Funktion $x \to K(x, x_i)$ für $i = 0, 1, \ldots$. Wegen $x_n \notin X$ ist dann jede der Funktionen $h_n$ $(n = 1, 2, \ldots)$ ein Element von $\mathcal{H}$. Wegen $x_0 \notin X^*$ konvergiert die Folge $(h_n^*)$ auf $X^*$ gleichmäßig gegen $h_0^*$ gemäß Hilfssatz 10. Da $X^*$ der Šilovsche Rand $\partial_{\mathcal{H}}X$ ist, folgt hieraus die gleichmäßige Konvergenz von $(h_n)$ gegen $h_0$ auf X. Somit liegt auch $h_0$ in $\mathcal{H}$. Dies kann aber nicht sein, da $x_0$ in F liegt und $h_0(x_0) = K(x_0, x_0) = +\infty$ ist. Unsere Annahme führt also zu einem Widerspruch; es gilt vielmehr $F \cap \complement \mathring{X} \cap \complement X^* = \emptyset$ und damit die Behauptung.

Wie angekündigt folgt nun:

Korollar. — *Ist die Menge* $\Omega$ *minimal begrenzt, so gilt:* $X^* = F$.

Beweis. Es ist $\mathring{X} = \Omega$ und somit

$$F \cap \complement \mathring{X} = F \cap \complement \Omega = F \subset X^*.$$

Nach Satz 16 gilt außerdem: $X^* \subset F$.

## § 7. — Bemerkungen zum Dirichletschen Problem für diskrete harmonische Funktionen.

Im folgenden soll angedeutet werden, daß die in den ersten Paragraphen entwickelte Theorie selbst dann noch von Interesse ist, wenn der kompakte Raum X nur aus endlich vielen Punkten besteht.

Wir betrachten hierzu im $R^N (N \geqslant 1)$ eine endliche Menge X von Punkten, deren sämtliche Koordinaten ganzzahlig sind *(Gitterpunkte)*. Dann ist X ein diskreter kompakter Raum. Für jeden Punkt $x \in X$ betrachten wir die Menge $U_x$ seiner $2^N$ Nachbargitterpunkte $y$, welche von $x$ den (euklidischen) Abstand 1 haben. Punkte $x \in X$ mit $U_x \subset X$ heißen dann *innere Punkte* von X; die Menge aller inneren Punkte von X soll mit $X_i$ bezeichnet werden. Die Menge $X_r = X \cap \complement X_i$ heißt Menge der *Randpunkte* von X.

Jedem Punkt $x \in X_i$ ordnen wir das von $U_x$ getragene Maß $\mu_x$ auf X zu, welches jeden Punkt von $U_x$ mit der Masse $2^{-N}$ belegt. Dann sei $\mathcal{H}$ die Menge aller (stetigen) reellen Funktionen $h$ auf X, welche auf $X_i$ *diskret harmonisch* sind, d.h. welche der Bedingung genügen:

$$(7.\ 1) \qquad \int h\, d\mu_x = h(x) \qquad \text{für jedes} \qquad x \in X_i.$$

Dann besitzt $\mathcal{H}$ wieder die Eigenschaften (2. 1)-(2. 3). ($\mathcal{H}$ trennt die Punkte von X, da z.B. die Restriktion auf X jeder auf dem $\mathbf{R}^N$ affin-linearen Funktion in $\mathcal{H}$ liegt.)

Die Menge $\mathcal{S}$ aller auf X definierten, reellen Funktionen $u$, welche in $X_i$ *diskret superharmonisch* sind, d.h. für jedes $x \in X_i$ der Ungleichung $\int u\, d\mu_x \leqslant u(x)$ genügen, ist eine in $\mathcal{C}(X)$ abgeschlossene inf-stabile Menge mit $\mathcal{H} = \mathcal{S} \cap (-\mathcal{S})$. Daher gilt $\hat{\mathcal{H}} = \mathcal{H}$ nach dem Korollar 1 zu Satz 7.

Da für jeden inneren Punkt $x \in X_i$ offenbar $\mu_x \neq \varepsilon_x$ und nach (7. 1) ein Maß aus $\mathcal{M}_x(\mathcal{H})$ ist, können $\mathcal{H}$-extremale Punkte von X nur Randpunkte sein: $X_e(\mathcal{H}) \subset X_r$. Der Satz 1 liefert also sofort das bekannte Resultat, wonach jede Funktion aus $\mathcal{H}$ eine in $X_r$ gelegene Minimalstelle besitzt (vgl. Heilbronn [26], theorem 1).

Daß sogar $X_e = X_r = \partial_{\mathcal{H}} X$ gilt, folgt aus dem bekannten Satz ([26], theorem 2), wonach jede reelle Funktion $f$ auf $X_r$ zu einer Funktion aus $\mathcal{H}$ fortgesetzt werden kann. Dies kann übrigens auch sehr einfach und durchsichtig durch eine naheliegende Abwandlung der Methode von Perron-Wiener aus der Theorie des klassischen Dirichletschen Problems gezeigt werden (vgl. § 6). Man betrachte hierzu zu $f$ die Menge $\mathfrak{O}_f$ aller reellen Funktionen $v$ auf X, welche auf $X_r$ die Funktion $f$ majorisieren und in $X_i$ diskret superharmonisch sind. Entsprechend setzt man $\mathfrak{U}_f = -\mathfrak{O}_{-f}$. Dann prüft man ohne Mühe nach, daß $\sup \mathfrak{U}_f = \inf \mathfrak{O}_f$ die gesuchte Fortsetzung von $f$ zu einer in $X_i$ diskret harmonischen Funktion ist.

## § 8. — Bemerkungen über Funktionenalgebren.

8. 1. *Existenz des Šilovschen Randes.* — Im folgenden betrachten wir einen *kompakten* Raum X und eine *Unteralgebra* $\mathcal{A}$ der Algebra $\mathcal{C}(X, \mathbf{C})$ aller komplexwertigen stetigen Funktionen auf X. $\mathcal{A}$ *trenne die Punkte* von X und enthalte die *konstanten Funktionen* aus $\mathcal{C}(X, \mathbf{C})$ als Elemente.

Die Menge $\mathcal{H} = \mathfrak{R}(\mathcal{A})$ aller Realteile $\mathfrak{R}(f)$ von Funktionen $f \in \mathcal{A}$ besitzt dann die Eigenschaften (2. 1)-(2. 3). Für jeden Punkt $x \in X$ ist $\mathcal{M}_x(\mathcal{H})$ die Menge aller Maße $\mu \geqslant 0$ auf X mit

$$(8.1) \qquad f(x) = \int f d\mu \qquad \text{für alle } f \in \mathcal{A}.$$

Wir setzen daher sinngemäß:

$$\mathcal{M}_x(\mathcal{A}) = \mathcal{M}_x(\mathcal{H}) \qquad \text{und} \qquad X_e(\mathcal{A}) = X_e(\mathcal{H}).$$

$X_e(\mathcal{A})$ soll auch der Choquetsche Rand von X bezüglich $\mathcal{A}$ genannt werden; die Punkte von $X_e(\mathcal{A})$ sollen $\mathcal{A}$-extremal heißen.

Es soll nun gezeigt werden, daß aus Satz 2 ein einfacher Beweis für die Existenz des Šilovschen Randes $\partial_{\mathcal{A}}X$ von X bezüglich $\mathcal{A}$ folgt, welcher zugleich wieder den Šilovschen Rand mit dem Choquetschen Rand verknüpft. $\partial_{\mathcal{A}}X$ ist bekanntlich die kleinste abgeschlossene Teilmenge von X mit der Eigenschaft, daß der absolute Betrag $|f|$ einer jeden Funktion $f \in \mathcal{A}$ eine in $\partial_{\mathcal{A}} X$ gelegene Minimalstelle besitzt.

Satz 18. — *Für jedes* $f \in \mathcal{A}$ *besitzt die Funktion* $|f|$ *eine* $\mathcal{A}$*-extremale Maximalstelle. Der Šilovsche Rand* $\partial_{\mathcal{A}}X$ *existiert und ist gleich der abgeschlossenen Hülle des Choquetschen Randes* $X_e(\mathcal{A})$.

Beweis. Bezeichnet $\overline{\mathcal{A}}$ die abgeschlossene Hülle von $\mathcal{A}$ in $\mathcal{C}(X, \mathbf{C})$ bezüglich der Topologie der gleichmäßigen Konvergenz, so ist $\overline{\mathcal{A}}$ wieder eine Unteralgebra von $\mathcal{C}(X, \mathbf{C})$ und aus (8. 1) folgt: $\mathcal{M}_x(\mathcal{A}) = \mathcal{M}_x(\overline{\mathcal{A}})$ für alle $x \in X$. Besitzt ferner jede

Funktion $|f|$ mit $f \in \mathcal{A}$ eine Maximalstelle in einer abgeschlossenen Menge $M \subset X$, so gilt dasselbe für jede Funktion $|f|$ mit $f \in \overline{\mathcal{A}}$. Daher kann für den Beweis $\mathcal{A}$ als in $\mathcal{C}(X, \mathbf{C})$ abgeschlossen vorausgesetzt werden.

Zunächst sei bemerkt, daß mit $\mathcal{A}$ auch die Menge $\mathcal{E}_0$ aller Funktionen $-|f|$ mit $f \in \mathcal{A}$ die Punkte von X trennt (vgl. auch [5]). Zu je zwei verschiedenen Punkten $x$, $y \in X$ gibt es nämlich ein $f \in \mathcal{A}$ mit $f(x) \neq f(y)$. Im Falle $|f(x)| \neq |f(y)|$ ist nichts zu zeigen. Ist jedoch $|f(x)| = |f(y)|$, so trennt $-|f + \alpha f^2|$ die beiden Punkte für $\alpha^{-1} = -f(x)$. Dann aber trennt auch die Menge $\mathcal{E}$ aller Funktionen $-\log |f|$ mit $f \in \mathcal{A}$ die Punkte von X; da $\mathcal{A}$ eine Algebra ist, enthält $\mathcal{E}$ mit je zwei Funktionen deren Summe. $\mathcal{E}$ besitzt somit die Eigenschaften (1. 4)-(1. 6). Nach Satz 2 existiert also $\partial_{\mathcal{E}} X$ und es ist $\partial_{\mathcal{E}} X = \overline{X_e(\mathcal{E})}$. Wegen der Isotonie von $\xi \to \log \xi$ auf der reellen Halbgeraden $\xi \geqslant 0$ existiert dann auch $\partial_{\mathcal{A}} X$ und es ist $\partial_{\mathcal{A}} X = \partial_{\mathcal{E}} X = \overline{X_e(\mathcal{E})}$.

Der Rest der Behauptung folgt so: Zunächst ist offenbar $\mathcal{M}_x(\mathcal{A}) \subset \mathcal{M}_x(\mathcal{E}_0)$ für jedes $x \in X$, wie durch Übergang zu den Absolutbeträgen in (8. 1) folgt; also ist $X_e(\mathcal{E}_0) \subset X_e(\mathcal{A})$. Nach Satz 1 besitzt dann jede Funktion $|f|$ mit $f \in \mathcal{A}$ eine $\mathcal{E}_0$-extremale, also auch $\mathcal{A}$-extremale Maximalstelle. Hieraus folgt weiter: $\partial_{\mathcal{A}} X \subset \overline{X_e(\mathcal{A})}$. Aus der Abgeschlossenheit von $\mathcal{A}$ folgt: $e^{-f} \in \mathcal{A}$ und somit $\mathcal{R}(f) = -\log |e^{-f}| \in \mathcal{E}$ für jedes $f \in \mathcal{A}$. Also ist $\mathcal{H} \subset \mathcal{E}$, $\mathcal{M}_x(\mathcal{E}) \subset \mathcal{M}_x(\mathcal{H}) = \mathcal{M}_x(\mathcal{A})$ für alle $x \in X$ und somit: $X_e(\mathcal{A}) \subset X_e(\mathcal{E})$. Damit ist auch noch $\overline{X_e(\mathcal{A})} \subset \overline{X_e(\mathcal{E})} = \partial_{\mathcal{A}} X$ gezeigt.

Bemerkung. — Wenn $\mathcal{A}$ abgeschlossen und X metrisierbar ist, so existiert nach Bishop [9] zu jedem $\mathcal{A}$-extremalen Punkt $x_0$ sogar ein $f \in \mathcal{A}$ mit $|f(x)| < |f(x_0)|$ für alle $x \neq x_0$ aus X. Vgl. auch [8], p. 325.

8. 2. *Zwei Fragen.* — Die Unteralgebra $\mathcal{A}$ werde jetzt als abgeschlossen vorausgesetzt. Zu jedem Punkt $x \in X$ betrachten wir die Menge $\mathcal{M}_x^*$ aller zu $x$ gehörigen $\mathcal{H}$-harmonischen Maße. Wegen $\mathcal{H} = \mathcal{R}(\mathcal{A})$ sind dies alle Maße $\mu \geqslant 0$ auf $X^* = \partial_{\mathcal{A}} X$ mit $f(x) = \int f_{X^*} d\mu$ für alle $f \in A$. In § 3 wurde gezeigt, daß zu jedem Punkt aus X dann und nur dann genau ein zugehöriges $\mathcal{H}$-harmonisches Maß existiert, wenn alle Funktionen

aus $\mathcal{C}(X^*)$ $\mathcal{H}$-resolutiv sind. Es wäre interessant zu wissen, welche Eigenschaften der Algebra $\mathcal{A}$ selbst hiermit gleichwertig sind. Vielleicht geben die schönen Resultate von WERMER [35], [36] einen Hinweis darauf, von welcher Natur diese gesuchten Eigenschaften sein könnten.

Im Anschluß an die in § 6 gewonnene Kennzeichnung der regulären Randpunkte einer offenen, relativ kompakten Teilmenge des $\mathbf{R}^N$ drängt sich weiter folgende Frage auf. Im Raum $\mathbf{C}^N$ von $N \geqslant 2$ komplexen Dimensionen sei $\Omega$ eine offene, relativ kompakte Menge. Man betrachte auf $X = \overline{\Omega}$ die Algebra $\mathcal{A}$ aller stetigen komplexwertigen Funktionen, welche in $\Omega$ holomorph sind. Welches ist dann der Choquetsche Rand $X_e(\mathcal{A})$? Wegen Satz 18 wäre mit dieser Frage auch die Frage nach dem Šilovschen Rand $\partial_{\mathcal{A}}X$ beantwortet, welche bereits durch BREMERMANN [20] Teilantworten gefunden hat.

## LITERATUR


[1] R. ARENS and I. M. SINGER, Function values as boundary integrals, *Proc. Amer. Math. Soc.*, **5**, 1954, 735-745.

[2] G. AUMANN, Über die Erweiterung von additiven monotonen Funktionen auf regulär geordneten Halbgruppen, *Arch. der Math.*, **8**, 1957, 422-427.

[3] H. BAUER, Un problème de Dirichlet pour la frontière de Šilov d'un espace compact, *C. r. Acad. Sci. Paris*, **247**, 1958, 843-846.

[4] H. BAUER, Frontière de Šilov et problème de Dirichlet, *Séminaire de Théorie du potentiel*, **3**, 1958-1959, n° 7, 23 p. (Institut H. Poincaré, Paris).

[5] H. BAUER, Minimalstellen von Funktionen und Extremalpunkte II, *Arch. der Math.*, **11**, 1960, 200-205.

[6] H. BAUER, Une axiomatique du problème de Dirichlet pour certaines équations aux dérivées partielles elliptiques et paraboliques, *C. r. Acad. Sci. Paris*, **250**, 1960, 2672-2674.

[7] G. BIRKHOFF, Lattice theorie, *Amer. Math. Soc. Colloq. Publ.*, **25**, New York, 1948.

[8] E. BISHOP and K. DE LEEUW, The representations of linear functionals by measures on sets of extreme points, *Ann. Inst. Fourier*, **9**, 1959, 305-331.

[9] E. BISHOP, A minimal boundary for function algebras, *Pacific Journ. Math.*, **9**, 1959, 629-642.

[10] G. BOULIGAND, Fonctions harmoniques. Principes de Picard et de Dirichlet, *Mém. Sci. Math.*, **11**, Paris, 1926.

[11] N. Bourbaki, Algèbre, Chap. vi-vii, *Actual. sci. et industr.*, **1179**, Paris, 1952.

[12] N. Bourbaki, Espaces vectoriels topologiques, Chap. i-ii, *Actual. sci. et industr.*, **1189**, Paris, 1953.

[13] N. Bourbaki, Espaces vectoriels topologiques, Chap. iii-v, *Actual. sci. et industr.*, **1299**, Paris, 1955.

[14] N. Bourbaki, Intégration, Chap. i-iv, *Actual. sci. et industr.*, **1175**, Paris, 1952.

[15] N. Bourbaki, Intégration, Chap. v, *Actual. Sci. et industr.*, **1244**, Paris, 1956.

[16] M. Brelot, Familles de Perron et problème de Dirichlet, *Acta Szeged*, **9**, 1939, 133-153.

[17] M. Brelot, Une axiomatique générale du problème de Dirichlet dans les espaces localement compacts, *Séminaire de Théorie du potentiel*, **1**, 1956-1957, n° 6, 16 p. (Institut H. Poincaré, Paris).

[18] M. Brelot, Axiomatique des fonctions harmoniques et surharmoniques dans un espace localement compact, *Séminaire de Théorie du potentiel*, **2**, 1957-1958, n° 1, 40 p. (Institut H. Poincaré, Paris).

[19] M. Brelot, Sur un théorème de prolongement fonctionnel de Keldych concernant le problème de Dirichlet, *Journal d'Analyse Math.*, Jérusalem (erscheint demnächst).

[20] H. J. Bremermann, On a generalized Dirichlet problem for plurisubharmonic functions and pseudoconvex domains. Characterisation of Šilov boundaries, *Trans. Amer. Math. Soc.*, **91**, 1959, 246-276.

[21] G. Choquet, Existence et unicité des représentations intégrales au moyen des points extrémaux dans les cônes convexes, *Séminaire Bourbaki*, Déc. 1956, n° 139, 15 p. (Institut H. Poincaré, Paris).

[22] G. Choquet, Theory of capacities, *Ann. Inst. Fourier*, **5**, 1953, 131-295.

[23] G. Choquet et J. Deny, Ensembles semi-réticulés et ensembles réticulés de fonctions continues, *J. Math. pur. appl.*, **36** (9e série), 1957, 179-189.

[24] M. M. Day, Normed linear spaces, *Ergebnisse der Math.*, **21** (Neue Folge), Berlin-Göttingen-Heidelberg, 1958.

[25] I. Gelfand, D. Raikov and G. Šilov, Commutative normed rings, *Usp. Mat. Nauk SSSR*, N. S. **2**, 1946, 48-146.

[26] H. A. Heilbronn, On discrete harmonic functions, *Proc. Cambridge philos. Soc.*, **45**, 1949, 194-206.

[27] S. Kakutani, Concrete representation of abstract (M-)spaces, *Ann. of Math.*, **42**, 1941, 994-1024.

[28] M. V. Keldych, On the solubility and the stability of Dirichlet's problem (russisch), *Usp. Mat. Nauk SSSR*, **8**, 1941, 172-231.

[29] J. G. Kemeny and J. L. Snell, Semimartingales of Markov chains, *Ann. Math. Statistics*, **29**, 1958, 143-154.

[30] V. L. Klee jr., Extremal structure of convex sets II, *Math. Zeitschr.*, **69**, 1958, 90-104.

[31] S. Matsushita, Théorème de Krein-Milman et le balayage de mesures dans la théorie du potentiel I-III, *Proc. Japan. Acad.*, **31**, 1955, 643-647; **32**, 1956, 29-34; **32**, 1956, 125-130.

[32] S. Matsushita, On the foundation of balayage theory, *J. Inst. Polytech. Osaka City Univ.*, Ser. A, **9**, 1958, 59-86.
[33] D. P. Milman, Characteristics of extremal points of regularly compact sets (russisch), *Dokl. Akad. Nauk SSSR*, N. S. **57**, 1947, 119-122.
[34] F. Vasilesco, Sur les singularités des fonctions harmoniques, *J. Math. pur. appl.*, **9** (9e série), 1930, 81-111.
[35] J. Wermer, On algebras of continuous functions, *Proc. Amer. Math. Soc.*, **4**, 1953, 866-869.
[36] J. Wermer, Subalgebras of the algebra of all complex-valued continuous functions on the circle, *Amer. J. Math.*, **78**, 1956, 225-242.

(Reçu le 22 juillet 1960.)

# Axiomatische Behandlung des Dirichletschen Problems für elliptische und parabolische Differentialgleichungen
[R19]

Math. Ann. **146** (1962), 1–59
[Zbl. Math. 107.08003; MR 26 #1612]

## Einleitung

**1.** Über das Dirichletsche Problem für die Laplacesche Differentialgleichung $\Delta u = 0$ im $\mathbf{R}^n$ liegt bekanntlich eine hochentwickelte Theorie vor, deren wichtigste Resultate vor allem mit den Namen PERRON, WIENER, KELLOGG, EVANS und BRELOT verknüpft sind. Die Gegenbeispiele von ZAREMBA und LEBESGUE, wonach das Dirichletsche Problem selbst bei relativ-kompakten, einfach-zusammenhängenden Gebieten im $\mathbf{R}^3$ nicht für jede stetige reelle Randfunktion lösbar zu sein braucht, waren richtungsweisend für die Entwicklung der Theorie: In ihrem Zentrum steht die Untersuchung der *verallgemeinerten Lösung* $H_f$ des Dirichletschen Problems, welche ohne Einschränkung für jede stetige, reelle, auf dem Rand eines relativ-kompakten Gebietes im $R^n$ definierte Funktion $f$ existiert. Diese verallgemeinerte Lösung $H_f$ gewinnt man mittels einer nach PERRON-WIENER-BRELOT benannten Methode der Ober- und Unterfunktionen.

Die Existenz dieser Theorie hat naturgemäß zu der Frage geführt, ob die Beschränkung auf die spezielle *elliptische* Differentialgleichung zweiter Ordnung $\Delta u = 0$ notwendig ist. Als erster hat wohl TAUTZ [23] gezeigt, daß diese Beschränkung weitgehend unnötig ist. Weiter wurde die Theorie von ihrer Bindung an den euklidischen Raum $\mathbf{R}^n$ befreit und zur Behandlung des Dirichletschen Problems auf gewissen Riemannschen Flächen entwickelt (vgl. AHLFORS-SARIO [1]). Schließlich hat diese Entwicklung zu einer Darstellung der Theorie des Dirichletschen Problems in gewissen Mannigfaltigkeiten, den durch BRELOT-CHOQUET [12] definierten *Greenschen Räumen* geführt[1]).

Noch vor TAUTZ, nämlich bald nach dem Erscheinen der grundlegenden Arbeiten von PERRON und WIENER, hatte STERNBERG [22] bemerkt, daß die Methode der verallgemeinerten Lösungen im Falle spezieller 2-dimensionaler Gebiete auch bei der Wärmeleitungsgleichung, also bei einer *parabolischen* Differentialgleichung zweiter Ordnung angewendet werden kann. Abgesehen von einem wichtigen Beitrag von PETROWSKY [21] wurde diese Bemerkung jedoch nicht aufgegriffen und systematisch verfolgt.

---

[1]) Eine ausführliche Darstellung der Entwicklung des Dirichletschen Problems bis zum Jahre 1954 findet der Leser bei BRELOT [11].

Erst recht nicht verfolgt wurde die Frage, ob etwa die für allgemeinere elliptische Differentialgleichungen und für Greensche Räume gewonnenen Resultate über das Dirichletsche Problem Spezialfälle einer allgemeinen Theorie sind, durch welche eine systematische Behandlung auch parabolischer Differentialgleichungen erschlossen wird.

Kürzlich hat nun aber DOOB [14] unter Heranziehung wahrscheinlichkeitstheoretischer Methoden und Begriffsbildungen den Versuch gemacht, eine solche, natürlich axiomatische Theorie zu entwickeln; diese Theorie spielt sich in einem lokal-kompakten Raum mit abzählbarer Basis ab. Ein weiterer, sich allerdings auf den $\mathbf{R}^n$ beschränkender Versuch wurde mit anderen Methoden von KAMKE [19] unternommen. Beide Theorien müssen jedoch als unbefriedigend bezeichnet werden, da in ihnen einerseits wichtige Sätze der klassischen Theorie, wie z. B. das Randminimum-Prinzip, zu Axiomen erhoben wurden, andererseits andere zentrale Sätze, wie etwa der Resolutivitätssatz von BRELOT, nicht bewiesen werden konnten. Bei den Untersuchungen von KAMKE kommt als weiterer Nachteil hinzu, daß die Topologie des $\mathbf{R}^n$ durch einen unnatürlich erscheinenden Verfeinerungsprozeß der speziell zu studierenden Differentialgleichung angepaßt werden muß.

Diese Mängel beider Theorien haben BRELOT [7]—[9] dazu veranlaßt, eine bewußt nur auf den elliptischen Fall zugeschnittene Axiomatik zu entwickeln, in deren Rahmen unter weitgehender Verwendung der im klassischen Fall bereitgestellten Methoden große Teile der klassischen Theorie verallgemeinert werden können. Ansatzpunkte zu der Betrachtungsweise von BRELOT finden sich bereits bei TAUTZ [24]. Wichtige Beiträge zum Ausbau der Theorie von BRELOT verdankt man HERVÉ [13], [18].

2. Ziel dieser Arbeit ist die Darstellung einer neuen axiomatischen Theorie des Dirichletschen Problems, welche die Behandlung *sowohl* elliptischer *als auch* parabolischer Differentialgleichungen gestattet. Diese Theorie hat einen ebenso einfachen Ausgangspunkt wie die von BRELOT und gestattet es trotz ihrer größeren Allgemeinheit, die Resultate der klassischen Theorie weitgehend zu übertragen. Die Brelotsche Theorie ist in ihr als Spezialfall enthalten. Der Grundraum ist ein lokal-kompakter, durch geeignete Zusatzannahmen eingeschränkter Raum. Bereits für die Wärmeleitungsgleichung sind viele der gewonnenen Resultate neu.

Methodisch weicht die hier dargestellte Theorie an vielen Stellen von der klassischen Theorie ab. Dies rührt vor allem von einem bei DOOB, KAMKE und BRELOT nicht verwendeten *Trennungsaxiom* her, welches besagt, daß je zwei verschiedene Punkte des lokal-kompakten Grundraumes $X$ durch eine in $X$ harmonische oder sogar nur hyperharmonische Funktion getrennt werden können. Dieses Axiom zusammen mit anderen, bereits durch die Theorie von BRELOT mehr oder weniger nahegelegten Axiomen hat weitreichende Konsequenzen: Es garantiert u. a. die Gültigkeit des Randminimum-Prinzips für hyperharmonische Funktionen in sehr allgemeiner Form, sichert die Gültigkeit des Resolutivitätskriteriums und die Existenz regulärer Randpunkte.

In die Beweisführung greift mehrmals ein *maßtheoretisches Minimumprinzip* ein, auf dessen Bedeutung bereits an anderer Stelle [3], [4] hingewiesen wurde. Es spielt eine ähnliche Rolle wie der Satz von KREIN-MILMAN in der Theorie der konvexen Mengen, der übrigens ein Spezialfall dieses Prinzips ist [3].

Trotz aller methodischen und sonstigen Eigenheiten haben die Untersuchungen von BRELOT die vorliegende Theorie mehr beeinflußt als vielleicht durch die in den Text eingestreuten Zitate zum Ausdruck kommt.

**3.** Die Arbeit gliedert sich in 3 Kapitel von insgesamt 13 Paragraphen. Das Kapitel I (§ 1 bis § 6) dient der Formulierung der Axiome der Theorie und der Bereitstellung der wichtigsten Sätze über hyperharmonische Funktionen. Die Theorie geht aus von einer Abbildung, welche jeder nicht-leeren offenen Teilmenge $U$ eines lokal-kompakten Raumes $X$ einen Vektorraum $\mathscr{H}_U$ stetiger reeller Funktionen auf $U$ zuordnet. Im Spezialfall der Laplaceschen Differentialgleichung, den der Leser stets als Leitlinie vor Augen haben möge, ist $X = \mathbf{R}^n (n = 1, 2, \ldots)$ und $\mathscr{H}_U$ die Menge aller in $U$ harmonischen Funktionen. Daher sprechen wir auch im allgemeinen Fall $\mathscr{H}_U$ als den Vektorraum der in $U$ *harmonischen* Funktionen an. Es wird gefordert (Axiom I), daß $U \to \mathscr{H}_U$ ein Garbendatum stetiger reeller Funktionen auf $X$ ist. Relativ-kompakte offene Mengen $V$, für welche das Dirichletsche Problem uneingeschränkt (im klassischen Sinne) lösbar ist, werden *regulär* genannt. Axiom II fordert die Existenz einer Basis regulärer Mengen. Sodann werden die *hyperharmonischen* Funktionen definiert, welche im Fall der Differentialgleichung $\Delta u = 0$ die superharmonischen Funktionen leicht verallgemeinern und für unsere Zwecke handlicher sind. Das Trennungsaxiom $T$ besagt, daß die Menge der in $X$ hyperharmonischen Funktionen die Punkte von $X$ trennt, sofern man die vereinfachende Annahme macht, daß die konstante Funktion 1 in $X$ harmonisch ist. (Diese Annahme wird allerdings in der Arbeit durch eine Bedingung anderer Art abgeschwächt.) Als letztes Axiom tritt ein Konvergenzaxiom vom Harnackschen Typus hinzu, das in einer schwachen Form (Axiom $K_1$) und einer scharfen Form (Axiom $K_2$) eingeführt wird. Der Paragraph 6 zeigt die Anwendungsmöglichkeiten der Theorie auf und klärt die Beziehungen zur Theorie von BRELOT.

Im Kapitel II (§ 7 bis § 10) wird das Dirichletsche Problem nach der Methode von PERRON-WIENER-BRELOT behandelt und insbesondere das Analogon zum *Resolutivitätssatz* der klassischen Theorie bewiesen.

Das Kapitel III (§ 11 bis § 13) beschäftigt sich mit dem Randverhalten der verallgemeinerten Lösungen. Mit Hilfe des bereits erwähnten maßtheoretischen Minimumprinzips wird die Existenz *regulärer Randpunkte* nachgewiesen und das Randminimumprinzip verschärft. Sodann werden in Analogie zur klassischen Theorie Kriterien für die Regularität eines Randpunktes hergeleitet.

**4.** Einige Resultate dieser Arbeit wurden ohne Beweis in einer Note in den Comptes rendus de l'Académie des Sciences (Paris) [2] angekündigt.

## Vorbereitungen

In dieser Arbeit schließen wir uns weitgehend der Terminologie von BOURBAKI an. Die folgenden vorbereitenden Bemerkungen sollen der Entlastung des eigentlichen Textes dienen.

Mit $\mathbf{R}$ bezeichnen wir stets den topologischen Körper der reellen Zahlen (Zahlengerade). Jede Abbildung $f: A \to \mathbf{R}$ einer Menge $A$ in $\mathbf{R}$ heiße *reelle* (reellwertige) Funktion auf $A$; jede Abbildung von $A$ in die durch die Adjunktion von $\pm \infty$ kompaktifizierte Zahlengerade $\overline{\mathbf{R}}$ heiße *numerische Funktion* auf $A$. Ein Punkt $x_0 \in A$ heiße Minimalstelle einer numerischen Funktion $f$ auf $A$, wenn $f(x_0) = \inf f(A)$ ist.

Sind $A$ und $B$ Mengen, so bezeichne $\mathrm{Rest}_Q f$ die *Restriktion* einer Abbildung $f: A \to B$ auf eine Menge $Q \subset A$. Ist $\mathscr{F}$ eine Menge von Abbildungen von $A$ in $B$, so bezeichne $\mathrm{Rest}_Q \mathscr{F}$ die Menge aller Abbildungen $\mathrm{Rest}_Q f$ mit $f \in \mathscr{F}$. Wir wollen sagen, die Menge $\mathscr{F}$ von Abbildungen *trenne die Punkte* von $A$, wenn zu je zwei verschiedenen Punkten $x, y \in A$ eine Abbildung $f \in \mathscr{F}$ existiert mit $f(x) \neq f(y)$.

Für einen topologischen Raum $T$ bezeichne $\mathscr{C}(T)$ stets den *Vektorraum aller stetigen, reellen Funktionen* auf $T$. Für eine Menge $U \subset T$ bedeute $U^*$ immer den *Rand* der Menge $U$ in $T$.

Ist $T$ ein lokal-kompakter (d. h. stets auch Hausdorffscher) Raum und $K$ eine kompakte Teilmenge von $T$, so existiert bekanntlich zu jedem positiven Radonschen Maß $\mu$ auf $K$ genau ein positives Radonsches Maß $\mu'$ auf $T$, welches von $K$ getragen wird und das Maß $\mu$ in $K$ induziert. Wie üblich sollen daher $\mu$ und $\mu'$ miteinander identifiziert werden. Statt von einem Radonschen Maß werden wir meist nur von einem *Maß* sprechen. $\varepsilon_t$ bezeichne stets das durch die *Einheitsmasse im Punkte* $t \in T$ definierte positive Maß auf $T$.

Der Raum $T$ sei wieder lokal-kompakt; $\mu$ sei ein positives Maß auf $T$. Dann ist gemäß BOURBAKI [5a] (pp. 172—173, exercice 5) für jede numerische Funktion $f$ auf $T$ das Oberintegral $\int^* f\,d\mu$ und das Unterintegral $\int_* f\,d\mu$ definiert. Ist $T$ speziell kompakt, so ist die dort angegebene Definition offenbar mit der im folgenden ausschließlich verwendeten äquivalent:

Für jede nach unten beschränkte, nach unten halbstetige, numerische Funktion $f$ auf $T$ ist

$$\int^* f\,d\mu = \sup_{\varphi \in \mathscr{C}(T),\ \varphi \leqq f} \int \varphi\,d\mu\,. \tag{1}$$

Außerdem ist $f$ die obere Einhüllende aller Funktionen $\varphi \in \mathscr{C}(T)$ mit $\varphi \leqq f$.

Für eine beliebige numerische Funktion $f$ auf $T$ ist

$$\int^* f\,d\mu = \inf_{\psi \in \mathscr{J}_f} \int^* \varphi\,d\mu\,, \tag{2}$$

wenn hierbei $\mathscr{J}_f$ die Menge aller nach unten beschränkten, nach unten halbstetigen, numerischen Funktionen $\psi$ auf $T$ mit $\psi \geqq f$ bezeichnet. Schließlich ist für jede numerische Funktion $f$ auf $T$:

(3) $$\int_{\bullet} f d\mu = -\int^{\bullet}(-f) d\mu .$$

Bezüglich der wichtigsten Eigenschaften des Oberintegrals sei auf BOURBAKI, loc. cit., verwiesen. Im folgenden benötigen wir noch folgendes

Lemma. *Es sei $T$ ein kompakter Raum, $\mu$ ein positives Maß und $f$ eine numerische Funktion auf $T$ mit endlichem Oberintegral $\int^{\bullet} f d\mu$. Dann besitzen für jede Funktion $g \in \mathscr{C}(T)$ auch die Funktionen $\inf(f, g)$ und $\sup(f, g)$ ein endliches $\mu$-Oberintegral.*

Beweis. Wir setzen $f_i = \inf(f, g)$ und $f_s = \sup(f, g)$. Dann ist $f_i$ nach oben und $f_s$ nach unten beschränkt, also $\int^{\bullet} f_i d\mu < +\infty$ und $-\infty < \int^{\bullet} f_s d\mu$. Nach (2) existiert eine nach unten beschränkte, nach unten halbstetige Funktion $\psi$ auf $T$ mit $\psi \geqq f$ und $\int^{\bullet} \psi d\mu < +\infty$. Hieraus folgt die Integrierbarkeit von $\psi$ nach [5a], p. 150. Nun ist aber $f_s \leqq \sup(\psi, g)$ und $\sup(\psi, g)$ integrierbar, also $\int^{\bullet} f_s d\mu < +\infty$ und somit $\left|\int^{\bullet} f_s d\mu\right| < +\infty$. Folglich ist $\int^{\bullet} f_i d\mu + \int^{\bullet} f_s d\mu$ definiert und nach [5a], p. 173, $\int^{\bullet}(f_i + f_s) d\mu \leqq \int^{\bullet} f_i d\mu + \int^{\bullet} f_s d\mu$. Berücksichtigt man noch, daß $f + g = f_i + f_s$ und nach (1) und (2) ferner $\int^{\bullet}(f + g) d\mu = \int^{\bullet} f d\mu + \int g d\mu$ ist, so erhält man $\int^{\bullet} f_i d\mu > -\infty$ und damit den Rest der Behauptung.

## I. Die Axiome der Theorie

### § 1. *Verallgemeinerte harmonische und hyperharmonische Funktionen*

**1.1.** *Harmonische Funktionen und reguläre Mengen.* — Die folgenden Betrachtungen spielen sich in einem *lokal-kompakten* (Hausdorffschen) Raum $X$ ab. Das System aller nicht-leeren, offenen Teilmengen $U$ von $X$ werde mit $\mathfrak{U}$ bezeichnet.

Jeder Menge $U \in \mathfrak{U}$ sei eine Menge $\mathscr{H}_U$ stetiger, reeller Funktionen auf $U$ zugeordnet. Die auf $\mathfrak{U}$ definierte Abbildung $U \to \mathscr{H}_U$ genüge den folgenden Axiomen I und II:

**Axiom I.** *Die Abbildung $U \to \mathscr{H}_U$ ist ein Garbendatum stetiger reeller Funktionen auf $X$, d. h. sie besitzt folgende Eigenschaften:*

(1.1.) *$\mathscr{H}_U$ ist ein linearer Unterraum von $\mathscr{C}(U)$.*

(1.2.) $U_1 \subset U_2 \Rightarrow \mathrm{Rest}_{U_1} \mathscr{H}_{U_2} \subset \mathscr{H}_{U_1}$.

(1.3.) *Ist $h$ eine auf einer Menge $U \in \mathfrak{U}$ definierte reelle Funktion und existiert zu jedem $x \in U$ eine offene Umgebung $V_x \subset U$ mit $\mathrm{Rest}_{V_x} h \in \mathscr{H}_{V_x}$, so gilt: $h \in \mathscr{H}_U$.*

Die Elemente von $\mathscr{H}_U$ sollen *harmonische Funktionen* in $U$ genannt werden. Die Eigenschaft (1.2) legt sogar die Verwendung dieser Sprechweise in einem allgemeineren Fall nahe: Ist $h$ eine auf einer Menge $D \subset X$ definierte numerische Funktion und gilt $\mathrm{Rest}_U h \in \mathscr{H}_U$ für eine Menge $U \in \mathfrak{U}$ mit $U \subset D$, so werde auch die Funktion $h$ harmonisch in $U$ genannt.

Dem zweiten Axiom schicken wir eine Definition voraus:

**Definition 1.** *Eine Menge $V \in \mathfrak{U}$ heiße regulär, wenn sie relativ-kompakt ist und folgende Eigenschaften besitzt:*

(1.4) *Der Rand $V^*$ von $V$ ist nicht leer.*

(1.5) *Jede Funktion $f \in \mathscr{C}(V^*)$ kann auf genau eine Weise zu einer in $\overline{V}$ stetigen Funktion fortgesetzt werden, deren Restriktion $H_f$ auf $V$ harmonisch in $V$ ist.*

(1.6) *Aus $f \in \mathscr{C}(V^*)$ und $f \geqq 0$ folgt $H_f \geqq 0$.*

*Ist $V$ regulär und gilt $\overline{V} \subset U$ für ein $U \in \mathfrak{U}$, so heiße die Menge $V$ regulär in $U$.*

Nunmehr lautet das zweite Axiom:

**Axiom II.** *Das System aller regulären Teilmengen von $X$ ist eine Basis der Topologie von $X$.*

Da $X$ speziell ein regulärer Raum ist, folgt aus Axiom II, daß jeder Punkt $x$ einer Menge $U \in \mathfrak{U}$ ein Fundamentalsystem von in $U$ regulären Umgebungen besitzt. Hieraus und aus (1.3) folgt sofort eine im folgenden häufig verwendete Bemerkung: Eine auf einer Menge $U \in \mathfrak{U}$ definierte reelle Funktion ist harmonisch in $U$, wenn sie in jeder in $U$ regulären Menge $V$ harmonisch ist.

**1.2.** *Harmonische Maße.* — Wir ziehen noch eine Folgerung aus der Definition 1. Hierzu sei $V$ eine reguläre Menge. Da $\mathscr{H}_V$ ein linearer Unterraum von $\mathscr{C}(V)$ ist, folgt aus (1.5), daß $f \to H_f$ eine lineare Abbildung von $\mathscr{C}(V^*)$ in $\mathscr{H}_V$ ist; nach (1.6) ist diese Abbildung ferner positiv, also auch isoton. Für jeden Punkt $x \in V$ ist daher $f \to H_f(x)$ eine positive Linearform auf $\mathscr{C}(V^*)$, also ein (Radonsches) Maß $\mu_x^V \geqq 0$ auf $V^*$. Definitionsgemäß gilt:

$$H_f(x) = \int f \, d\mu_x^V \quad \text{für alle } f \in \mathscr{C}(V^*)\,. \tag{1.7}$$

$\mu_x^V$ soll das zur regulären Menge $V$ und zu $x \in V$ gehörige *harmonische Maß* genannt werden.

**1.3.** *Hyperharmonische Funktionen.* — Der Begriff der harmonischen Funktion soll jetzt verallgemeinert werden.

**Definition 2.** *Eine in einer Menge $U \in \mathfrak{U}$ definierte, numerische Funktion $u$ heiße hyperharmonisch in $U$, wenn sie folgende Eigenschaften besitzt:*

(1.8) $u(x) > -\infty$ *für alle $x \in U$.*

(1.9) *$u$ ist nach unten halbstetig.*

(1.10) *Für jede in $U$ reguläre Menge $V$ und jede Funktion $f \in \mathscr{C}(V^*)$ mit $f(z) \leqq u(z)$ für alle $z \in V^*$ gilt: $H_f(x) \leqq u(x)$ für alle $x \in V$.*

Die Menge aller in $U$ hyperharmonischen Funktionen werde mit $\mathscr{H}_U^*$ bezeichnet. *Hypo*harmonisch in $U$ heiße jede numerische Funktion $v$ auf $U$ mit $-v \in \mathscr{H}_U^*$. Beispielsweise ist die konstante Funktion $+\infty$ in jeder Menge $U \in \mathfrak{U}$ hyperharmonisch.

Der folgende Satz lehrt, daß die in $U$ harmonischen Funktionen die einzigen in $U$ zugleich hyper- und hypoharmonischen Funktionen sind.

**Satz 1.** *Für jede Menge* $U \in \mathfrak{U}$ *ist* $\mathscr{H}_U = \mathscr{H}_U^* \cap (-\mathscr{H}_U^*)$.

Beweis. Es sei $h \in \mathscr{H}_U$, $V$ eine in $U$ reguläre Menge und $f \in \mathscr{C}(V^*)$; ferner sei $f(z) \leqq h(z)$ für alle $z \in V^*$. Bezeichnet $g$ die Restriktion von $h$ auf $V^*$, so ist $H_g$ die Restriktion von $h$ auf $V$ nach (1.2) und (1.5). Aus $g - f \geqq 0$ folgt $H_g - H_f = H_{g-f} \geqq 0$ nach 1.2. Also ist $H_f(x) \leqq h(x)$ für alle $x \in V$. Somit ist $h$ in $U$ hyperharmonisch; wegen $-h \in \mathscr{H}_U$ ist $h$ dann auch hypoharmonisch. — Umgekehrt sei $h \in \mathscr{H}_U^* \cap (-\mathscr{H}_U^*)$. Nach Definition 2 ist dann $h$ reellwertig und stetig. Ist also $V$ eine in $U$ reguläre Menge und $f = \mathrm{Rest}_{V^*} h$, so liegt $f$ in $\mathscr{C}(V^*)$ und es gilt: $H_f(x) \leqq h(x)$ sowie $-H_f(x) = H_{-f}(x) \leqq -h(x)$ für alle $x \in V$. Also ist $h(x) = H_f(x)$ für alle $x \in V$ und damit $\mathrm{Rest}_V h \in \mathscr{H}_V$. Nach der Bemerkung am Schluß von 1.1. ist dann $h$ in $U$ harmonisch.

Mit Hilfe der in 1.2. definierten harmonischen Maße lassen sich die hyperharmonischen Funktionen wie folgt kennzeichnen:

**Satz 2.** *Es sei $u$ eine in $U \in \mathfrak{U}$ definierte, nach unten halbstetige, numerische Funktion mit $u(x) > -\infty$ für alle $x \in U$. Die Funktion $u$ ist genau dann hyperharmonisch in $U$, wenn für jede in $U$ reguläre Menge $V$ und jedes $x \in V$ gilt:*

$$\overset{*}{\int} u \, d\mu_x^V \leqq u(x) \,. \tag{1.11}$$

Beweis. Es sei $u \in \mathscr{H}_U^*$, $V$ eine in $U$ reguläre Menge und $f$ eine Funktion aus $\mathscr{C}(V^*)$ mit $f(z) \leqq u(z)$ für alle $z \in V^*$. Dann gilt nach Definition 2:

$$\int f \, d\mu_x^V = H_f(x) \leqq u(x) \text{ für alle } x \in V \,.$$

Hieraus folgt (1.11) nach Definition des Oberintegrals. — Umgekehrt sei die genannte Bedingung für $u$ erfüllt. Ist dann $V$ eine in $U$ reguläre Menge, $f \in \mathscr{C}(V^*)$ und $f(z) \leqq u(z)$ für alle $z \in V^*$, so erhält man: $H_f(x) = \int f \, d\mu_x^V \leqq$ $\leqq \overset{*}{\int} u \, d\mu_x^V \leqq u(x)$, also $H_f(x) \leqq u(x)$ für alle $x \in V$. Danach ist $u$ in $U$ hyperharmonisch.

Aus den Sätzen 1 und 2 folgt sofort:

**Korollar 1.** *Eine auf einer Menge $U \in \mathfrak{U}$ definierte, stetige reelle Funktion $h$ ist genau dann in $U$ harmonisch, wenn für jede in $U$ reguläre Menge $V$ und alle $x \in V$ gilt:*

$$\int h \, d\mu_x^V = h(x) \,. \tag{1.12}$$

Das in Satz 2 formulierte Kriterium liefert zusammen mit bekannten Rechenregeln über das Oberintegral noch folgende Eigenschaften der hyperharmonischen Funktionen.

**Korollar 2.** *Für jede Menge $U \in \mathfrak{U}$ gilt:*

(1.13) $u, v \in \mathscr{H}_U^* \Rightarrow u + v \in \mathscr{H}_U^*$.

(1.14) $u \in \mathscr{H}_U^*, \lambda > 0 \Rightarrow \lambda u \in \mathscr{H}_U^*$.

(1.15) $u, v \in \mathscr{H}_U^* \Rightarrow \inf(u, v) \in \mathscr{H}_U^*$.

(1.16) *Für jede aufsteigend filtrierende, nicht-leere Menge $\mathscr{F} \subset \mathscr{H}_U^*$ ist* $\sup \mathscr{F} \in \mathscr{H}_U^*$.

(1.17) $\mathscr{H}_U$ *ist in* $\mathscr{C}(U)$ *abgeschlossen bezüglich der Topologie der kompakten Konvergenz.*

(1.18) $U_1 \in \mathfrak{U},\ U_1 \subset U \Rightarrow \mathrm{Rest}_{U_1} \mathscr{H}_U^* \subset \mathscr{H}_{U_1}^*$.

**1.4.** ***h-harmonische Funktionen.*** — Abschließend besprechen wir einen Prozeß, der es gestattet, aus dem Garbendatum $U \to \mathscr{H}_U$ ein neues, ebenfalls dem Axiom II genügendes Garbendatum zu konstruieren.

Hierzu sei $h \in \mathscr{C}(X)$ eine Funktion mit $h(x) > 0$ für alle $x \in X$. Für jede Menge $U \in \mathfrak{U}$ bezeichne ${}^h\mathscr{H}_U$ die Menge aller in $U$ *h-harmonischen* Funktionen, d. h. aller Funktionen

$$x \to \frac{u(x)}{h(x)}$$

auf $U$ mit $u \in \mathscr{H}_U$. Dann ist offenbar $U \to {}^h\mathscr{H}_U$ wieder ein Garbendatum stetiger reeller Funktionen; es genügt also auch diese neue Abbildung dem Axiom I. Aber auch das Axiom II ist erfüllt: Für jede bezüglich $U \to \mathscr{H}_U$ reguläre Menge $V$ und jedes $f \in \mathscr{C}(V^*)$ ist nämlich $\frac{1}{h} H_{hf}$[2]) die einzige $h$-harmonische Funktion in $V$, welche stetig an die Randfunktion $f$ anschließt; ferner folgt $\frac{1}{h} H_{hf} \geqq 0$ aus $f \geqq 0$. Die Menge $V$ ist daher auch regulär bezüglich des neuen Garbendatums. Eine ganz ähnliche Überlegung zeigt sogar, daß die regulären Mengen des ursprünglichen Garbendatums mit denen des neu konstruierten Garbendatums *identisch* sind. Dann aber ist klar, daß auch das Garbendatum $U \to {}^h\mathscr{H}_U$ dem Axiom II genügt.

Jeder regulären Menge $V$ und jedem Punkt $x \in V$ ordnet das Garbendatum der $h$-harmonischen Funktionen somit ein harmonisches Maß gemäß 1.2 zu; dieses heiße das zu $V$ und $x$ gehörige *h-harmonische Maß* und werde mit ${}^h\mu_x^V$ bezeichnet. Nach der soeben angestellten Überlegung gilt:

$$\int f\,d\,{}^h\mu_x^V = \frac{1}{h(x)} H_{hf}(x) = \frac{1}{h(x)} \int h f\, d\,\mu_x^V$$

für alle $f \in \mathscr{C}(V^*)$. Daher ist

$$(1.19)\qquad {}^h\mu_x^V = \frac{1}{h(x)} (h\,\mu_x^V)\,.$$

Aus Satz 2 folgt dann, daß die Menge ${}^h\mathscr{H}_U^*$ der in $U \in \mathfrak{U}$ *h-hyperharmonischen* Funktionen (d. h. der bezüglich $U \to {}^h\mathscr{H}_U$ hyperharmonischen Funktionen) gleich der Menge aller Funktionen $x \to \frac{u(x)}{h(x)}$ mit $u \in \mathscr{H}_U^*$ ist.

Von besonderem Interesse ist der Fall, daß $h$ eine Funktion aus $\mathscr{H}_X$ ist. Dann enthält offenbar jeder der Vektorräume ${}^h\mathscr{H}_U$ die konstanten reellen Funktionen auf $U$.

---

[2]) Eigentlich müßte es genauer heißen: $\frac{1}{h_1} H_{h_2 f}$, wobei $h_1 = \mathrm{Rest}_V h$ und $h_2 = \mathrm{Rest}_{V^*} h$. Diese und ähnliche Oberflächlichkeiten in der Schreibweise werden wir uns auch im folgenden immer dann gestatten, wenn keine Irrtümer zu befürchten sind.

**1.5.** *Bemerkung.* — Wir merken noch an, daß man durch *Restriktion* des Garbendatums $U \to \mathscr{H}_U$ auf einen offenen Unterraum $Z$ von $X$ ein Garbendatum stetiger reeller Funktionen auf $Z$ gewinnt, welches offenbar wiederum dem Axiom II genügt.

## § 2. *Das Randminimum-Prinzip*

**2.1.** *$\mathfrak{V}$-hyperharmonische Funktionen.* — Unter geeigneten Zusatzannahmen soll nun ein Randminimum-Prinzip für hyperharmonische Funktionen bewiesen werden. Aus beweistechnischen Gründen (vgl. Satz 5) soll zunächst der Begriff der hyperharmonischen Funktion etwas allgemeiner gefaßt werden:

**Definition 3.** *Es sei $\mathfrak{V}$ eine Abbildung, welche jedem Punkt $x \in X$ ein Fundamentalsystem $\mathfrak{V}(x)$ von regulären Umgebungen von $x$ zuordnet. Eine auf einer Menge $U \in \mathfrak{U}$ definierte, numerische, nach unten halbstetige Funktion $u$ mit Werten $u(x) > -\infty (x \in U)$ heiße $\mathfrak{V}$-hyperharmonisch in $U$, wenn gilt:*

$$\text{(2.1)} \qquad \int^{*} u\, d\mu_x^V \leq u(x) \quad \textit{für alle } x \in U \textit{ und alle } V \in \mathfrak{V}(x) \textit{ mit } \overline{V} \subset U.$$

Der Satz 2 besagt u. a., daß jede in $U$ hyperharmonische Funktion dort auch $\mathfrak{V}$-hyperharmonisch ist, wenn hierbei $\mathfrak{V}$ eine beliebige Abbildung der in der Definition 3 genannten Art ist.

**2.2.** *Randminimum-Prinzip.* — Es erweist sich als zweckmäßig, das nun folgende Randminimum-Prinzip in sehr allgemeiner Form auszusprechen.

**Lemma 1.** *Es sei $Y$ ein kompakter Raum, der $X$ als dichten Unterraum enthält; ferner sei $\mathfrak{V}$ eine Abbildung mit den in Definition 3 genannten Eigenschaften und $U$ eine Menge aus $\mathfrak{U}$. Es werde vorausgesetzt, daß $\mathscr{H}_U$ die konstanten reellen Funktionen auf $U$ enthält und daß $\mathscr{H}_U^*$ die Punkte von $U$ trennt. Dann besitzt jede in der abgeschlossenen Hülle $\overline{U}$ von $U$ in $Y$ definierte, numerische, nach unten halbstetige Funktion $u$, welche in $U$ $\mathfrak{V}$-hyperharmonisch ist, eine im Rand $U^*$ von $U$ in $Y$ gelegene Minimalstelle. Insbesondere ist $U^*$ nicht leer.*

Wir stellen dem Beweis zwei Bemerkungen voran:

1. Da $X$ ein lokal-kompakter, in $Y$ dicht liegender Unterraum des kompakten Raumes $Y$ ist, so ist $X$ in $Y$ offen. Daher ist auch $U$ in $Y$ offen. Für relativ-kompakte Mengen $V \in \mathfrak{U}$ stimmen die abgeschlossene Hülle und der Rand bezüglich $X$ mit den entsprechenden bezüglich $Y$ gebildeten Mengen überein.

2. Im Hinblick auf die Bemerkung in 1.5 würde es genügen, Lemma 1 für $U = X$ zu formulieren und zu beweisen. Der nachfolgende Beweis würde hierdurch aber kaum vereinfacht werden. Er stützt sich übrigens wesentlich auf die Resultate aus [3], deren Kenntnis hier vorausgesetzt wird.

Beweis. Wir bezeichnen mit $\mathfrak{P}(\overline{U})$ die Potenzmenge von $\overline{U}$ und definieren wie folgt eine Menge $\mathfrak{T} \subset \overline{U} \times \mathfrak{P}(\overline{U})$. Für jeden Punkt $x \in U^*$ sei $(x, \{x\})$ das einzige Paar $(x, T) \in \mathfrak{T}$ mit $x$ als erster Komponente. Für jeden Punkt $x \in U$ dagegen nehmen wir in $\mathfrak{T}$ auf alle Paare $(x, T)$, in welchen $T$ der Träger eines Maßes $\mu_x^V$ mit $V \in \mathfrak{V}(x)$ und $\overline{V} \subset U$ ist. Für den kompakten Raum $\overline{U}$ besitzt dann $\mathfrak{T}$ die Eigenschaften $(T_1)$ und $(T_2)$ aus [3][3]). Die Eigenschaft $(T_1)$, wonach zu $x \in \overline{U}$ stets ein $T \in \mathfrak{P}(\overline{U})$ existiert mit $(x, T) \in \mathfrak{T}$, folgt unmittelbar aus der Definition von $\mathfrak{T}$. Zum Nachweis von $(T_2)$, wonach aus $(x, T) \in \mathfrak{T}$ stets $T \neq \emptyset$ folgt, hat man nur zu bemerken, daß nach Voraussetzung die konstante Funktion 1 in $\mathscr{H}_U$ liegt, nach dem Korollar 1 zu Satz 2 also $\int d\mu_x^V = 1$ ist für alle $x \in U$ und alle in $U$ regulären Umgebungen $V$ von $x$. Nach Definition von $\mathfrak{T}$ sind alle Punkte $x \in U^*$ $\mathfrak{T}$-extremal im Sinne von [3], d. h. aus $(x, T) \in \mathfrak{T}$ folgt $T = \{x\}$. Ein Punkt $x \in U$ ist dagegen nicht $\mathfrak{T}$-extremal, da $\mu_x^V$ für jedes $V \in \mathfrak{V}(x)$ mit $\overline{V} \subset U$ ein Maß auf $V^*$ ist, der zugehörige nicht-leere Träger also in $V^*$ enthalten ist. Daher ist $U^*$ die Menge aller $\mathfrak{T}$-extremalen Punkte von $U$.

Es sei nun $\mathscr{E}$ die Menge aller numerischen, nach unten halbstetigen Funktionen auf $\overline{U}$, welche in $U$ $\mathfrak{V}$-hyperharmonisch sind. Im Sinne der Bezeichnungen von [3] ist dann $\mathscr{E} \subset \mathscr{F}_{\mathfrak{T}}$, d. h. jede Funktion $u \in \mathscr{E}$ genügt der dortigen Bedingung $(F)$: Ist $(x, T) \in \mathfrak{T}$ und $u(x) \leqq u(y)$ für alle $y \in T$, so ist sogar $u(x) = u(y)$ für alle $y \in T$. Zu beweisen ist offenbar nur etwas, wenn $x$ in $U$ liegt. $T$ ist dann der Träger eines Maßes $\mu_x^V$ mit $V \in \mathfrak{V}(x)$ und $\overline{V} \subset U$. Da $u$ in $U$ $\mathfrak{V}$-hyperharmonisch ist, gilt: $\int^* u\, d\mu_x^V \leqq u(x)$. Hieraus und aus $u(x) \leqq u(y)$ für alle $y \in T$ folgt aber tatsächlich $u(x) = u(y)$ für alle $y \in T$ gemäß [3], Hilfssatz 3. Die Voraussetzungen dieses Hilfssatzes sind erfüllt, da $u$ auf $V^*$ nach unten beschränkt und $\int d\mu_x^V = 1$ ist. Somit genügt jedes $u \in \mathscr{E}$ der Bedingung $(F)$.

Weiter kann jede Funktion $v \in \mathscr{H}_U^*$ zu einer Funktion $\overline{v}$ aus $\mathscr{E}$ fortgesetzt werden. Man hat nur zu setzen:

$$v(x) = \liminf_{y \to x,\, y \in U} v(y) \qquad (x \in \overline{U}) .$$

Daher trennt mit $\mathscr{H}_U^*$ (vgl. die Voraussetzungen) auch $\mathscr{E}$, also auch $\mathscr{F}_{\mathfrak{T}}$ je zwei verschiedene Punkte aus $U$, d. h. je zwei nicht $\mathfrak{T}$-extremale Punkte aus $U$. Nach dem Satz 1 aus [3] besitzt dann jede Funktion aus $\mathscr{F}_{\mathfrak{T}}$, speziell also aus $\mathscr{E}$ eine $\mathfrak{T}$-extremale, d. h. hier in $U^*$ gelegene Minimalstelle. Da $\mathscr{E}$ nicht leer ist, muß folglich $U^* \neq \emptyset$ sein. Damit ist der Beweis erbracht.

Nunmehr ergeben sich eine Reihe von Folgerungen:

**Lemma 2.** *Unter den in Lemma 1 gemachten Voraussetzungen über $Y$, $\mathfrak{V}$, $U$, $\mathscr{H}_U$ und $\mathscr{H}_U^*$ gilt für jede in $U$ $\mathfrak{V}$-hyperharmonische Funktion $u$:*

$$\inf_{x \in U} u(x) = \inf_{z \in U^*} \left( \liminf_{x \to z,\, x \in U} u(x) \right) . \tag{2.2}$$

Beweis. Wir setzen zur Abkürzung: $\alpha = \inf_{x \in U} u(x)$ und $\overline{u}(z) = \liminf_{x \to z,\, x \in U} u(x)$ für alle $z \in \overline{U}$. Dann gilt offenbar $\overline{u}(z) \geqq \alpha$ für alle $z \in U^*$. Zu zeigen ist daher noch: $\inf_{z \in U^*} \overline{u}(z) \leqq \alpha$. Nun ist aber $\overline{u}$ eine auf $\overline{U}$ nach unten halbstetige, numerische Funktion, welche wegen der Halbstetigkeit von $u$ auf $U$ mit $u$ überein-

[3]) An die Stelle des in [3] betrachteten Grundraumes $X$ tritt jetzt natürlich sinngemäß der Raum $\overline{U}$.

stimmt; $\bar{u}$ ist also in $U$ $\mathfrak{V}$-hyperharmonisch. Nach Lemma 1 ist dann $\inf_{z \in U^*} \bar{u}(z) = \inf_{z \in \bar{U}} \bar{u}(z)$. Hieraus und aus $u = \mathrm{Rest}_U\, \bar{u}$ folgt die noch fehlende Ungleichung.

**Korollar.** *Über $Y$ und $\mathfrak{V}$ werde dasselbe vorausgesetzt wie in Lemma 1. Weiter sei $U$ eine Menge aus $\mathfrak{U}$ und $h$ eine in $X$ harmonische und überall positive Funktion derart, daß $\inf_{x \in U} h(x) > 0$ ist und ${}^h\mathscr{H}^*_U$ die Punkte von $U$ trennt. Dann ist der Rand $U^*$ von $U$ in $Y$ nicht leer und für jede in $U$ $\mathfrak{V}$-hyperharmonische Funktion $v$ gilt:*

$$(2.3) \qquad \lim_{x \to z,\, x \in U} \inf v(x) \geqq 0 \quad \text{für alle } z \in U^* \Rightarrow v \geqq 0 .$$

Beweis. Wir gehen vom Garbendatum $W \to \mathscr{H}_W$ über zum Garbendatum $W \to {}^h\mathscr{H}_W$ $(W \in \mathfrak{U})$ der $h$-harmonischen Funktionen. Da nach 1.4 beide Garbendaten die gleichen regulären Mengen besitzen, sind bezüglich des neuen Garbendatums alle Voraussetzungen von Lemma 1 und 2 erfüllt. Daher ist $U^*$ nicht leer und aus (2.2) folgt (2.3) zunächst einmal für jede bezüglich $W \to {}^h\mathscr{H}_W$ in $U$ $\mathfrak{V}$-hyperharmonische Funktion. Ist nun $v$ eine in $U$ $\mathfrak{V}$-hyperharmonische Funktion bezüglich $W \to \mathscr{H}_W$, so folgt aus (1.19), daß $w = \frac{v}{h}$ eine in $U$ $\mathfrak{V}$-hyperharmonische Funktion bezüglich $W \to {}^h\mathscr{H}_W$ ist. Daher kann die Implikation (2.3) auf $w$ angewendet werden. Es sei nun $\lim_{x \to z,\, x \in U} \inf v(x) \geqq 0$ für alle $z \in U^*$. Da $h$ auf $U$ eine positive untere Schranke besitzt, ist dann auch $\lim_{x \to z,\, x \in U} \inf w(x) \geqq 0$ für jedes $z \in U^*$. Wie soeben dargelegt, folgt hieraus $w \geqq 0$ und damit $v \geqq 0$.

## § 3. *Trennungsaxiome*

**3.1.** *Das Trennungsaxiom T.* — In diesem Paragraphen ergänzen wir die Axiome I und II durch das folgende, bereits durch die Betrachtungen des § 2 nahegelegte Trennungsaxiom:

**Axiom T.** *Es existiert eine in $X$ harmonische und überall positive Funktion $h$, für welche die Menge ${}^h\mathscr{H}^*_X$ der $h$-hyperharmonischen Funktionen die Punkte von $X$ trennt.*

Zunächst bemerken wir, daß die hier ausgesprochene Trennungsbedingung unabhängig ist von der speziellen Wahl der harmonischen Funktion $h$. Ist nämlich $g$ eine weitere in $X$ harmonische, überall positive Funktion, so trennt mit ${}^h\mathscr{H}^*_X$ auch ${}^g\mathscr{H}^*_X$ die Punkte von $X$. Es seien zum Beweis $x$ und $y$ verschiedene Punkte aus $X$ und $u$ eine Funktion aus $\mathscr{H}^*_X$ mit $\frac{u(x)}{h(x)} \neq \frac{u(y)}{h(y)}$. Ist nun $\frac{h(x)}{g(x)} = \frac{h(y)}{g(y)}$, so werden $x$ und $y$ durch die Funktion $\frac{u}{g} = \frac{u}{h} \cdot \frac{h}{g}$ getrennt; ist aber $\frac{h(x)}{g(x)} \neq \frac{h(y)}{g(y)}$, so ist $\frac{h}{g}$ eine $x$ und $y$ trennende, sogar $g$-harmonische Funktion.

Ein besonders einfacher Fall liegt demnach vor, wenn die konstante Funktion 1 in $X$ harmonisch ist. Das Axiom I ist dann äquivalent mit der Forderung, daß $\mathscr{H}^*_X$ die Punkte von $X$ trennt.

Diese Aussage bleibt in der einen Richtung auch im allgemeinen Fall richtig: Aus dem Axiom $T$ folgt stets, daß $\mathcal{H}_X^*$ die Punkte von $X$ trennt. Man hat nur zu bemerken, daß aus $\frac{u(x)}{h(x)} \neq \frac{u(y)}{h(y)}$ entweder $h(x) \neq h(y)$ oder $u(x) \neq u(y)$ folgt. Unter zusätzlichen Annahmen läßt sich umgekehrt das Axiom $T$ herleiten aus der Forderung, daß $\mathcal{H}_X^*$ die Punkte von $X$ trennt:

**Satz 3.** *Es sei $h$ eine in $X$ harmonische, überall positive Funktion. In jedem der beiden folgenden Fälle trennt dann ${}^h\mathcal{H}_X^*$ die Punkte von $X$:*

(1) *$\mathcal{H}_X^*$ trennt die Punkte von $X$ und die konstante Funktion* 1 *ist hyperharmonisch in $X$.*

(2) *Zu jedem geordneten Paar $(x, y)$ verschiedener Punkte aus $X$ gibt es eine Funktion $u \in \mathcal{H}_X^*$ mit $0 \leq u(x) < u(y)$.*

Beweis. *Zu* (1): Es seien $x$ und $y$ verschiedene Punkte aus $X$. Gilt dann $h(x) = h(y)$ und trennt $u \in \mathcal{H}_X^*$ diese Punkte, so werden $x$ und $y$ durch die $h$-hyperharmonische Funktion $\frac{u}{h}$ getrennt. Ist jedoch $h(x) \neq h(y)$, so trennt die Funktion $\frac{1}{h}$ die beiden Punkte. Wegen $1 \in \mathcal{H}_X^*$ ist aber $\frac{1}{h}$ $h$-hyperharmonisch in $X$.

*Zu* (2): Sind $x, y$ verschiedene Punkte aus $X$, so ist wieder nur der Fall $h(x) \neq h(y)$ zu behandeln. Ohne Beschränkung der Allgemeinheit kann $h(y) < h(x)$ angenommen werden. Nach Voraussetzung gibt es zu dem geordneten Paar $(x, y)$ ein $u \in \mathcal{H}_X^*$ mit $0 \leq u(x) < u(y)$. Dann aber trennt $\frac{u}{h}$ die beiden Punkte.

Wir merken noch an, daß sich bei Restriktion des Garbendatums $U \to \mathcal{H}_U$ auf eine offene, nicht-leere Teilmenge von $X$ die durch Axiom $T$ ausgedrückte Eigenschaft offenbar auf das neue Garbendatum vererbt.

**3.2.** *Folgerungen.* — Mit Hilfe des erweiterten Axiomensystems lassen sich jetzt über den Grundraum $X$ und die hyperharmonischen Funktionen weitergehende Aussagen machen.

**Satz 4.** *Keine Menge $U \in \mathfrak{U}$, insbesondere nicht der Gesamtraum $X$ ist kompakt. Die Alexandroffsche Einpunkt-Kompaktifizierung $X_\omega$ von $X$ ist zusammenhängend*[4]).

Beweis. Wegen der Erblichkeit der durch die Axiome I, II und $T$ geforderten Eigenschaften bei Restriktion des Garbendatums $W \to \mathcal{H}_W$ auf Mengen aus $\mathfrak{U}$ genügt es, die erste Behauptung für $U = X$ zu beweisen. Wäre aber $X$ kompakt, so wäre $X$ ein Raum $Y$, $W \to {}^h\mathcal{H}_W$ ein Garbendatum und $X$ eine Menge $U \in \mathfrak{U}$ mit den in Lemma 1 geforderten Eigenschaften. Der Rand von $X$ in $Y = X$ wäre dann aber leer im Widerspruch zur Aussage dieses Lemmas. Also ist der lokal-kompakte Raum $X$ nicht kompakt. Wir betrachten seine durch Adjunktion eines unendlich fernen Punktes $\omega$ entstehende Alexandroffsche Kompaktifizierung $X_\omega$. Nach dem bisherigen Resultat ist dann $U = \emptyset$

---

[4]) Auf diese zweite, aus der ersten folgenden Aussage hat mich Herr H.-J. KOWALSKY in einem Gespräch aufmerksam gemacht.

die einzige in $X_\omega$ zugleich offene und abgeschlossene Menge $U$, welche $\omega$ nicht enthält. Daher ist $X_\omega$ zusammenhängend.

Aus diesem Satz folgt jetzt natürlich, daß jede Menge $U \in \mathfrak{U}$ in jedem kompakten Raum $Y$, der $X$ als dichten Unterraum enthält, einen nicht leeren Rand besitzt. Hierdurch wird eine Teilaussage des Korollars zu Lemma 2 verallgemeinert.

Der bislang angestrebten Allgemeinheit verdanken wir den jetzt folgenden einfachen Beweis der Identität von $\mathfrak{B}$-hyperharmonischen und hyperharmonischen Funktionen. Diese läuft auf eine Kennzeichnung der hyperharmonischen Funktionen durch „lokale Eigenschaften" hinaus.

**Satz 5.** *Es sei $\mathfrak{B}$ eine Abbildung, welche jedem Punkt $x \in X$ ein Fundamentalsystem von regulären Umgebungen zuordnet. Dann sind in jeder offenen Teilmenge von $X$ die $\mathfrak{B}$-hyperharmonischen Funktionen mit den hyperharmonischen Funktionen identisch.*

*Für jede in einer relativ-kompakten Menge $U \in \mathfrak{U}$ hyperharmonischen Funktion $u$ gilt:*

(3.1) $$\liminf_{x \to z,\, x \in U} u(x) \geqq 0 \quad \textit{für alle } z \in U^* \Rightarrow u \geqq 0\,.$$

Beweis. Wir beweisen zunächst den zweiten Teil des Satzes, und zwar für eine in einer relativ-kompakten Menge $U \in \mathfrak{U}$ $\mathfrak{B}$-hyperharmonischen Funktion $u$. Es sei hierzu $Y = X_\omega$ die Alexandroffsche Einpunkt-Kompaktifizierung von $X$. Da $U$ relativ kompakt ist und $h$ in $X$ nur positive Werte hat, ist $\inf h(U) > > 0$. Das Korollar zu Lemma 2, angewandt auf die jetzt spezielle Kompaktifizierung $Y$, liefert dann gerade die in (3.1) für $u$ behauptete Implikation. Man hat hierbei nur die dem Beweis des Lemmas 1 vorangestellte Bemerkung zu beachten, wonach für die relativ-kompakte Menge $U$ der Rand $U^*$ in $X$ mit dem Rand in $Y$ zusammenfällt.

Nunmehr erst beweisen wir den ersten Teil des Satzes. Zu zeigen ist nur, daß jede in einer Menge $W \in \mathfrak{U}$ $\mathfrak{B}$-hyperharmonische Funktion $v$ auch hyperharmonisch in $W$ ist. Hierzu sei $V$ eine in $W$ reguläre Menge und $f \in \mathscr{C}(V^*)$ eine Funktion mit $f(z) \leqq v(z)$ für alle $z \in V^*$. Nun ist $H_f$ in $V$ harmonisch; die Definition 3 lehrt daher sofort, daß mit $v$ auch $v - H_f$ in $V$ $\mathfrak{B}$-hyperharmonisch ist. Für jeden Punkt $z \in V^*$ gilt:

$$\liminf_{x \to z,\, x \in V} (v(x) - H_f(x)) \geqq \liminf_{x \to z,\, x \in V} v(x) - f(z) \geqq v(z) - f(z) \geqq 0\,.$$

Da $V$ offen und relativ-kompakt ist, folgt hieraus nach dem im ersten Teil des Beweises erzielten Resultat: $v(x) - H_f(x) \geqq 0$ für alle $x \in V$. Nach Definition ist dann $v$ in $U$ hyperharmonisch.

**Korollar.** *Es sei $u$ eine auf einer Menge $U \in \mathfrak{U}$ definierte numerische Funktion. Existiert dann zu jedem Punkt $x \in U$ eine offene Umgebung $V_x \subset U$ mit* $\mathrm{Rest}_{V_x} u \in \mathscr{H}^*_{V_x}$, *so ist $u \in \mathscr{H}^*_U$.*

Beweis. Die Funktion $u$ ist offenbar in $U$ $\mathfrak{B}$-hyperharmonisch, wenn $\mathfrak{B}(x)$ für jedes $x \in U$ das System aller regulären Mengen $V$ mit $x \in V \subset \overline{V} \subset V_x$ und für alle anderen Punkte $x \in X$ ein beliebiges Fundamentalsystem regulärer Umgebungen von $x$ ist. Die Behauptung folgt daher aus Satz 5.

Die im Korollar angesprochene Eigenschaft der Abbildung $U \to \mathcal{H}_U^*$ ist eine Verallgemeinerung von (1.3); die Eigenschaft (1.2) wurde bereits durch (1.18) verallgemeinert.

**Satz 6.** *Es sei $u$ eine in einer Menge $U \in \mathfrak{U}$ hyperharmonische Funktion und $V$ eine in $U$ reguläre Menge. Dann ist auch folgende Funktion $u_V$ in $U$ hyperharmonisch:*

$$(3.2) \qquad u_V(x) = \begin{cases} u(x), & x \in U \cap \complement V; \\ \int^* u\, d\mu_x^V, & x \in V. \end{cases}$$

*Für sie gilt*

$$(3.3) \qquad u_V \leqq u .$$

Beweis. Das harmonische Maß $\mu_x^V$ ist zunächst nur für Punkte $x \in V$ definiert; wir setzen hier die Abbildung $x \to \mu_x^V$ auf $\bar{V}$ fort, indem wir definieren: $\mu_x^V = \varepsilon_x$ für alle $x \in V^*$. Nach (1.5) und (1.7) ist dann $x \to \int f\, d\mu_x^V$ für jedes $f \in \mathscr{C}(V^*)$ eine in $\bar{V}$ stetige, in $V$ harmonische Funktion. Nach Definition des Oberintegrals ist

$$\int^* u\, d\mu_x^V = \sup_{\substack{f \leqq u^* \\ f \in \mathscr{C}(V^*)}} \int f\, d\mu_x^V$$

für alle $x \in \bar{V}$; dabei wurde abkürzend $u^* = \mathrm{Rest}_{V^*} u$ gesetzt. Daher ist die Funktion $x \to \int^* u\, d\mu_x^V$ auf $\bar{V}$, also die Restriktion von $u_V$ auf $\bar{V}$ nach unten halbstetig und überall $> -\infty$. Unmittelbar aus der Definition folgt, daß auch die Restriktion von $u_V$ auf $U \cap \complement V$ diese Eigenschaften besitzt. Hieraus folgt bereits, daß $u_V$ in $U$ nach unten halbstetig und überall $> -\infty$ ist.

Die Menge aller Funktionen $f \in \mathscr{C}(V^*)$ mit $f \leqq u^*$ ist aufsteigend filtrierend; gleiches gilt dann auch für die Menge der zugehörigen Funktionen $H_f$ auf $V$. Nun ist

$$u_V(x) = \int^* u\, d\mu_x^V = \sup H_f(x)$$

für alle $x \in V$, wenn dabei $f$ die oben beschriebene Menge durchläuft; nach (1.16) ist daher $u_V$ in $V$ hyperharmonisch. Nach Definition ist außerdem $u_V$ in $\complement \bar{V}$ hyperharmonisch. Nach Satz 5 ist dann $t = u_V$ in ganz $U$ hyperharmonisch, wenn noch für jedes $z \in V^*$ und jede in $U$ reguläre Menge $W$ mit $z \in W$ gezeigt werden kann: $\int^* t\, d\mu_z^W \leqq t(z)$. Nun ist aber nach Satz 2 $t \leqq u$, woraus folgt: $\int^* t\, d\mu_z^W \leqq \int^* u\, d\mu_z^W \leqq u(z) = t(z)$. Dies ist die gesuchte Ungleichung. Der Satz ist damit bewiesen, da bereits soeben die Gültigkeit von $t = u_V \leqq u$ bemerkt wurde.

**3.3.** *Konvergenz harmonischer Maße.* — Wir untersuchen weiter das Verhalten der Funktion $V \to \mu_{x_0}^V$ für kleine reguläre Umgebungen $V$ eines Punktes $x_0$.

**Satz 7.** *Es bezeichne $\mathfrak{F}$ den Filter der Abschnitte des bezüglich der Relation $\supset$ filtrierenden Systems aller regulären Umgebungen $V$ eines Punktes $x_0 \in X$. Dann gilt (im Sinne der vagen Topologie im Raum der Maße):*

$$(3.4) \qquad \lim_{\mathfrak{F}} \mu_{x_0}^V = \varepsilon_{x_0} .$$

Beweis[5]). Wir beweisen zunächst eine in (3.4) enthaltene speziellere Aussage, nämlich: $\lim_{\mathfrak{F}} \int d\mu_{x_0}^V = 1$. Nach dem Axiom $T$ existiert eine in $X$ harmonische, überall positive Funktion $h$; ohne Beschränkung der Allgemeinheit kann angenommen werden, daß für alle Punkte $x$ einer hinreichend kleinen Umgebung $W$ von $x_0$ gilt: $h(x) \geqq 1$. Zu jedem $\varepsilon > 0$ gibt es eine offene Umgebung $U \subset W$ von $x_0$ mit $|h(x) - h(x_0)| \leqq \varepsilon$ für alle $x \in U$. Für jede in $U$ reguläre Umgebung $V$ von $x_0$ ergibt sich hieraus durch Integration:

$$h(x_0) \,|\, 1 - \int d\mu_{x_0}^V| = |\int (h - h(x_0))\, d\mu_{x_0}^V| \leqq \varepsilon \int d\mu_{x_0}^V .$$

Wegen der speziellen Annahmen über $h$ ist $\int d\mu_{x_0}^V \leqq \int h\, d\mu_{x_0}^V = h(x_0)$. Wir erhalten daher die Abschätzung

$$|1 - \int d\mu_{x_0}^V| \leqq \frac{\varepsilon}{h(x_0)} \int d\mu_{x_0}^V \leqq \varepsilon$$

und damit die gewünschte spezielle Aussage.

Zum Beweis von (3.4) sei jetzt $f$ eine Funktion aus $\mathscr{C}(X)$ (mit oder ohne kompakten Träger). Zu jedem $\varepsilon > 0$ gibt es eine offene Umgebung $U$ von $x_0$ mit $|f(x) - f(x_0)| \leqq \varepsilon$ für alle $x \in U$. Also ist

$$|\int f\, d\mu_{x_0}^V - f(x_0)| - |f(x_0) - f(x_0) \int d\mu_{x_0}^V| \leqq |\int (f - f(x_0))\, d\mu_{x_0}^V| \leqq \varepsilon \int d\mu_{x_0}^V$$

und somit

$$|\int f\, d\mu_{x_0}^V - f(x_0)| \leqq \varepsilon \int d\mu_{x_0}^V + |f(x_0)|\, |1 - \int d\mu_{x_0}^V|$$

für alle in $U$ regulären Umgebungen $V$ von $x_0$. Hieraus und aus dem ersten Teil des Beweises folgt $\lim_{\mathfrak{F}} \int f\, d\mu_{x_0}^V = f(x_0) = \int f\, d\varepsilon_{x_0}$ für alle $f \in \mathscr{C}(X)$ und damit die Behauptung.

**3.4.** *Das starke Trennungsaxiom $T'$.* — In mehreren der noch zu studierenden Anwendungsbeispielen ist sogar eine verschärfte Form des Axioms $T$ erfüllt, welche auch an wenigen Stellen der allgemeinen Theorie eine Rolle spielen wird. Es handelt sich um das folgende starke Trennungsaxiom:

**(T′)** *Es existiert eine in $X$ harmonische und überall positive Funktion $h$, mit folgender Eigenschaft: Zu jedem geordneten Paar $(x, y)$ verschiedener Punkte aus $X$ gibt es eine $h$-hyperharmonische Funktion $v \in {}^h\mathscr{H}_X^*$ mit $v(x) < v(y)$.*

Offenbar ist $(T')$ eine Verschärfung von $(T)$. Ganz ähnlich wie im Anschluß an das Axiom $T$ zeigt man, daß die in $(T')$ ausgesprochene Trennungsbedingung unabhängig ist von der speziellen Wahl der harmonischen Funktion $h$. Aus $(T')$ folgt, daß zu jedem geordneten Paar $(x, y)$ verschiedener Punkte aus $X$ eine hyperharmonische Funktion $u \in \mathscr{H}_X^*$ existiert mit $u(x) < u(y)$.

## *§ 4. Konvergenzaxiome*

**4.1.** *Schwaches und starkes Konvergenzaxiom.* — In diesem Paragraphen wird vom Axiom $T$ kein Gebrauch gemacht werden. Dagegen ergänzen wir hier die Axiome I und II durch eines der folgenden Konvergenzaxiome:

---

[5]) Der Beweis verwendet das Axiom $T$ nicht in voller Schärfe. Benötigt wird nur die Existenz einer in einer Umgebung von $x_0$ definierten harmonischen Funktion $h$ mit $h(x_0) \neq 0$.

**Axiom $K_1$** (Schwaches Konvergenzaxiom). *Für jede nicht-leere, aufsteigend filtrierende, gleichmäßig nach oben beschränkte Menge* $\mathcal{F}$ *von harmonischen Funktionen in einer Menge* $U \in \mathfrak{U}$ *ist auch die obere Einhüllende* $\sup \mathcal{F}$ *in* $U$ *harmonisch.*

Eine Verschärfung hiervon ist das

**Axiom $K_2$** (Starkes Konvergenzaxiom). *Für jede nicht-leere, aufsteigend filtrierende Menge* $\mathcal{F}$ *von harmonischen Funktionen in einer Menge* $U \in \mathfrak{U}$ *ist auch die obere Einhüllende* $\sup \mathcal{F}$ *in* $U$ *harmonisch, sofern diese in* $U$ *überall endlich ist.*

**4.2.** *Äquivalente Aussagen.* — Wir zeigen nun, durch welche äquivalente Aussagen diese Zusatzaxiome ersetzt werden können. Trivial ist die Bemerkung, daß $K_1$ und $K_2$ äquivalent sind zu den entsprechenden Aussagen für absteigend filtrierende Mengen.

**Satz 8.** *Genügt die Abbildung* $U \to \mathcal{H}_U$ *den Axiomen I und II, so ist das Axiom* $K_1$ *mit jedem der drei folgenden Axiome gleichwertig:*

**($K_1'$)** *Für jede reguläre Menge* $V$ *und jede beschränkte, reelle Funktion* $f$ *auf* $V^*$ *ist die Funktion*

$$x \to \int^* f \, d\mu_x^V \tag{4.1}$$

*in* $V$ *harmonisch.*

**($K_1''$)** *Für jede reguläre Menge* $V$ *und jede beschränkte, reelle Funktion* $f$ *auf* $V^*$, *welche* $\mu_x^V$*-integrierbar ist für jedes* $x \in V$, *ist die Funktion*

$$x \to \int f \, d\mu_x^V \tag{4.2}$$

*in* $V$ *harmonisch.*

**($K_1'''$)** *Für jede Menge* $U \in \mathfrak{U}$ *und jede nicht-leere, aufsteigend filtrierende, lokal-gleichmäßig nach oben beschränkte Menge* $\mathcal{F} \subset \mathcal{H}_U$ *ist* $\sup \mathcal{F} \in \mathcal{H}_U$.

Beweis. $(K_1) \Rightarrow (K_1')$: Es sei $V$ eine reguläre Menge und $f$ eine beschränkte, reelle Funktion auf $V^*$. Zunächst werde $f$ als nach unten halbstetig vorausgesetzt. Dann ist die Menge $\mathcal{F}^*$ aller Funktionen $\varphi \in \mathcal{C}(V^*)$ mit $\varphi \leqq f$ aufsteigend filtrierend. Ebenfalls aufsteigend filtrierend (und nicht-leer) ist die Menge $\mathcal{F}$ aller Funktionen $H_\varphi \in \mathcal{H}_V$ mit $\varphi \in \mathcal{F}^*$. Ferner ist $\mathcal{F}$ gleichmäßig nach oben beschränkt. Um dies einzusehen, genügt es, eine Funktion $\psi \in \mathcal{C}(V^*)$ mit $f \leqq \psi$ zu wählen; dann ist nämlich $H_\psi$ eine in $V$ beschränkte Majorante von $\mathcal{F}$. Nach $(K_1)$ ist daher $u = \sup \mathcal{F} = \sup_{\varphi \in \mathcal{F}^*} H_\varphi$ in $V$ harmonisch. Es ist aber $u(x) = \sup_{\varphi \in \mathcal{F}^*} \int \varphi \, d\mu_x^V = \int^* f \, d\mu_x^V$ für alle $x \in V$ und damit die Behauptung in diesem Spezialfall bewiesen[6]).

Nunmehr sei $f$ eine beliebige beschränkte Funktion. Dann ist die Menge $\mathcal{G}^*$ aller nach unten halbstetigen, beschränkten Funktionen $\psi$ auf $V^*$ mit $f \leqq \psi$ absteigend filtrierend. Diese Eigenschaft überträgt sich auf die (nichtleere) Menge $\mathcal{G}$ aller nach dem ersten Teil des Beweises in $V$ harmonischen Funktionen $u_\psi$ mit $\psi \in \mathcal{G}^* : u_\psi(x) = \int^* \psi \, d\mu_x^V (x \in V)$. Analog wie im ersten Teil des Beweises zeigt man, daß $\mathcal{G}$ in $V$ gleichmäßig nach unten beschränkt ist.

[6]) Unter den bisherigen Annahmen ist $f$ natürlich sogar $\mu_x^V$-integrierbar für alle $x \in V$.

Nach ($K_1$) ist daher $v = \inf \mathscr{G}$ in $V$ harmonisch. Beachtet man noch, daß für alle $x \in V$ gilt

$$v(x) = \inf_{\psi \in \mathscr{G}^*} u_\psi(x) = \inf_{\psi \in \mathscr{G}^*} \overset{*}{\int} \psi \, d\mu_x^V = \overset{*}{\int} f \, d\mu_x^V ,$$

so ist damit ($K_1'$) bewiesen.

($K_1'$) $\Rightarrow$ ($K_1''$): Hier ist nichts zu beweisen.

($K_1''$) $\Rightarrow$ ($K_1'''$): Es sei $\mathscr{F} \subset \mathscr{H}_U$ nicht leer, aufsteigend filtrierend und lokal-gleichmäßig nach oben beschränkt. Dann ist $\mathscr{F}$ auf jeder kompakten Teilmenge von $U$ gleichmäßig nach oben beschränkt. Speziell trifft dies zu auf die abgeschlossene Hülle $\bar{V}$ jeder in $U$ regulären Menge $V$. Für jedes $u \in \mathscr{F}$ sei $u^* = \mathrm{Rest}_{V^*} u$ und $f$ die obere Einhüllende aller $u^*$. Dann ist $f$ beschränkt und nach unten halbstetig, also integrierbar bezüglich jedes positiven Maßes auf $V^*$; nach ($K_1''$) ist somit $x \to \int f \, d\mu_x^V$ eine harmonische Funktion in $V$. Setzen wir $s = \sup \mathscr{F}$, so ist

$$s(x) = \sup_{u \in \mathscr{F}} u(x) = \sup_{u \in \mathscr{F}} \int u^* \, d\mu_x^V = \int f \, d\mu_x^V$$

für alle $x \in V$; folglich ist die Restriktion von $s$ auf $V$ harmonisch. Da dies für jede in $U$ reguläre Menge $V$ gilt, ist $s$ harmonisch in $U$.

($K_1'''$) $\Rightarrow$ ($K_1$): Hier ist wieder nichts zu beweisen.

Der analoge, das Axiom $K_2$ betreffende Satz lautet:

**Satz 9.** *Genügt die Abbildung $U \to \mathscr{H}_U$ den Axiomen* I *und* II, *so ist das Axiom* $K_2$ *mit jedem der beiden folgenden Axiome gleichwertig:*

**($K_2'$)** *Für jede reguläre Menge $V$ und jede numerische Funktion $f$ auf $V$, welche für jedes $x \in V$ ein endliches $\mu_x^V$-Oberintegral besitzt, ist die Funktion* (4.1) *in $V$ harmonisch.*

**($K_2''$)** *Für jede reguläre Menge $V$ und jede numerische Funktion $f$ auf $V$, welche $\mu_x^V$-integrierbar ist für jedes $x \in V$, ist die Funktion* (4.2) *in $V$ harmonisch.*

Beweis. ($K_2$) $\Rightarrow$ ($K_2'$): Zunächst werde $f$ zusätzlich als nach oben beschränkt vorausgesetzt. $\mathscr{F}^*$ sei die Menge aller beschränkten, nach unten halbstetigen Funktionen $\psi$ auf $V^*$ mit $\psi \geqq f$. Wir betrachten dann die Menge $\mathscr{F}$ der nach Satz 8 in $V$ harmonischen Funktionen $x \to \overset{*}{\int} \psi d\mu_x^V = \int \psi d\mu_x^V$ für $\psi \in \mathscr{F}^*$. Mit $\mathscr{F}^*$ ist auch $\mathscr{F}$ nicht leer sowie absteigend filtrierend. Ferner ist die Funktion (4.1) nach Voraussetzung endlich und nach Definition des Oberintegrals die untere Einhüllende von $\mathscr{F}$. Nach ($K_2$) ist daher diese Funktion in $V$ harmonisch.

Erst jetzt fordern wir von $f$ nur noch die Endlichkeit von $\overset{*}{\int} f d\mu_x^V$ für jedes $x \in V$. Setzen wir $f_n = \inf(n, f)$, so ist nach dem Lemma des Abschnittes „Vorbereitungen" $\overset{*}{\int} f_n d\mu_x^V$ endlich für jedes $x \in V$; ferner ist $f_n$ nach oben beschränkt. Nach dem bereits Bewiesenen definiert also $x \to \overset{*}{\int} f_n d\mu_x^V$ eine in $V$ harmonische Funktion $u_n$. Die Funktionenfolge $u_1, u_2 \ldots$ ist isoton und hat auf Grund bekannter Eigenschaften des Oberintegrals die Funktion (4.1) zur oberen Einhüllenden. Nach ($K_2$) ist daher diese Funktion in $V$ harmonisch.

$(K_2') \Rightarrow (K_2'')$: Hier ist nichts zu beweisen.

$(K_2'') \Rightarrow (K_2)$: Es sei $\mathscr{F}$ eine nicht-leere, aufsteigend filtrierende Menge von harmonischen Funktionen in einer Menge $U \in \mathfrak{U}$ mit endlicher oberer Einhüllenden $s = \sup \mathscr{F}$; weiter sei $V$ eine in $U$ reguläre Menge. Für jedes $u \in \mathscr{F}$ bezeichne $u^*$ die Restriktion von $u$ auf $V^*$ und $f$ die obere Einhüllende aller $u^*$. Aus der für alle $x \in V$ und $u \in \mathscr{F}$ gültigen Gleichung $u(x) = \int u^* d\mu_x^V$ folgt dann nach bekannten Eigenschaften des Oberintegrals $s(x) = \int^* f d\mu_x^V$ für alle $x \in V$. Die Funktion $f$ ist nach unten beschränkt und nach unten halbstetig; ferner ist $\int^* f d\mu_x^V = s(x) < +\infty$. Also ist $f$ $\mu_x^V$-integrierbar und $s(x) = \int f d\mu_x^V$ für alle $x \in V$. Nach $(K_2'')$ ist daher $s$ in $V$ harmonisch. Da $V$ eine beliebige in $U$ reguläre Menge war, ist dann $s$ sogar in $U$ harmonisch.

**4.3.** *Grundraum mit abzählbarer Basis.* — Besitzt der Raum $X$ eine abzählbare Basis, so lassen sich die Axiome $K_1$ und $K_2$ durch formal schwächere, jedoch gleichwertige Aussagen ersetzen. Hierzu erinnern wir zunächst an ein

Lemma von CHOQUET[7]). *Es sei $(f_i)_{i \in I}$ eine Familie von numerischen Funktionen auf einem topologischen Raum $\Omega$ mit abzählbarer Basis. Dann existiert eine abzählbare Teilmenge $I_0$ der Indexmenge $I$ mit folgender Eigenschaft: Ist $g$ eine auf $\Omega$ definierte, numerische, nach oben halbstetige Funktion mit $g \geqq f_i$ für alle $i \in I_0$, so gilt sogar $g \geqq f_i$ für alle $i \in I$.*

Nunmehr lautet das angekündigte Resultat:

**Satz 10.** *Besitzt der Raum $X$ eine abzählbare Basis, so ist jedes der Axiome* $K_1$ *und* $K_2$ *äquivalent zu einem entsprechend formulierten Axiom, welches man aus* $K_1$ *und* $K_2$ *dadurch erhält, daß man dort den Ausdruck „nicht-leere, aufsteigend filtrierende Menge" durch „isotone Folge" ersetzt.*

Beweis. Es sei $U \in \mathfrak{U}$ und $\mathscr{F} \subset \mathscr{H}_U$ eine nicht-leere, aufsteigend filtrierende Menge mit der oberen Einhüllenden $u = \sup \mathscr{F}$. Da mit $X$ auch $U$ eine abzählbare Basis besitzt, gibt es nach dem Lemma von Choquet eine abzählbare Menge $\mathscr{F}_0 \subset \mathscr{F}$ mit folgender Eigenschaft: Für jede nach oben halbstetige Funktion $g$ auf $U$ gilt $g \geqq u$, sobald nur $g \geqq u_0$ mit $u_0 = \sup \mathscr{F}_0$ erfüllt ist. Da $\mathscr{F}$ aufsteigend filtrierend ist, kann ohne Beschränkung der Allgemeinheit $\mathscr{F}_0$ als die Menge der Elemente einer isotonen Folge gewählt werden. Ist $u_0$ stetig, so kann speziell $g = u_0$ gewählt werden, und es folgt $u_0 \geqq u$, also $u = u_0$. Aus diesen Betrachtungen folgt leicht die Behauptung.

**4.4.** *Bemerkung über h-harmonische Funktionen.* — Für jede überall positive Funktion $h \in \mathscr{C}(X)$ *vererbt* sich die Gültigkeit jedes der Axiome $K_1$ und $K_2$ auf das Garbendatum $U \to {}^h\mathscr{H}_U$ der $h$-harmonischen Funktionen. Man hat nur zu bemerken, daß die Abbildung $u \to \frac{u}{h}$ ein Automorphismus des geordneten Vektorraumes $\mathscr{C}(U)$ bei beliebigem $U \in \mathfrak{U}$ ist.

### § 5. *Untere Einhüllende hyperharmonischer Funktionen*

**5.1.** *Gesättigte Mengen hyperharmonischer Funktionen.* — Erstmals studieren wir nun das Zusammenspiel aller bislang formulierten Axiome. Generell

[7]) Vgl. BRELOT [9], p. 3.

setzen wir hierbei die Gültigkeit der Axiome I, II und T voraus. Über die Axiome $K_1$ und $K_2$ wird von Fall zu Fall gesondert verfügt.

**Definition 4.** *Eine Menge $\mathscr{S}$ von hyperharmonischen Funktionen in einer Menge $U \in \mathfrak{U}$ heiße gesättigt, wenn $\mathscr{S}$ mit je zwei Funktionen $u$, $v$ auch deren untere Einhüllende* $\inf(u, v)$ *und mit jeder Funktion $u$ auch die in* (3.2) *definierte Funktion $u_V$ enthält, und zwar für jede in $U$ reguläre Menge $V$.*

Nach (1.15) und Satz 6 ist für jedes $U \in \mathfrak{U}$ die Menge $\mathscr{H}_U^*$ aller in $U$ hyperharmonischen Funktionen gesättigt. Jeder Durchschnitt von gesättigten Teilmengen von $\mathscr{H}_U^*$ ist selbst wieder gesättigt.

Die Bedeutung dieser bereits von Brelot [7] verwendeten Begriffsbildung erhellt aus dem folgenden Satz.

**Satz 11.** *Es sei $\mathscr{S}$ eine nicht-leere, gesättigte Menge von hyperharmonischen Funktionen in einer Menge $U \in \mathfrak{U}$. Dann ist die untere Einhüllende $u_0 = \inf \mathscr{S}$ in jedem der beiden folgenden Fälle in $U$ harmonisch:*

(a) *Es gilt das Axiom* $K_1$. *Ferner ist $\mathscr{S}$ lokal-gleichmäßig nach unten beschränkt und jede Funktion aus $\mathscr{S}$ lokal-beschränkt.*

(b) *Es gilt das Axiom* $K_2$. *Ferner sind alle Funktionen aus $\mathscr{S}$ sowie $u_0$ reellwertig.*

Beweis. Zunächst zeigen wir unabhängig von den in (a) und (b) formulierten Zusatzvoraussetzungen, daß für jede in $U$ reguläre Menge $V$ gilt

$$u_0 = \inf_{u \in \mathscr{S}} u_V .$$

Dies folgt aus (3.3), wonach für jedes $u \in \mathscr{H}_U^*$ gilt $u_V \leqq u$, sowie aus der Voraussetzung über $\mathscr{S}$, wonach mit $u$ auch $u_V$ in $\mathscr{S}$ liegt. Weiter bemerken wir, daß die Menge $\mathscr{S}_V$ aller Funktionen $u_V$ mit $u \in \mathscr{S}$ absteigend filtrierend ist. Dies folgt aus der entsprechenden Eigenschaft von $\mathscr{S}$ und aus der Isotonie der Abbildung $u \to u_V$. Nun erst wenden wir uns den einzelnen Behauptungen des Satzes zu.

*Zu* (a): Es sei $V$ in $U$ regulär. Nach Voraussetzung ist jede Funktion aus $\mathscr{S}$ lokal-beschränkt, also insbesondere beschränkt auf $V^*$. Nach ($K_1'$) ist daher jede der Funktionen aus $\mathscr{S}_V$ harmonisch in $V$. Wie einleitend festgestellt wurde, ist $\mathscr{S}_V$ absteigend filtrierend und natürlich nicht-leer. Beachtet man noch, daß $\mathscr{S}_V \subset \mathscr{S}$ und somit mit $\mathscr{S}$ auch $\mathscr{S}_V$ lokal-gleichmäßig nach unten beschränkt ist, so folgt aus $u_0 = \inf \mathscr{S}_V$ und ($K_1''$), daß $u_0$ in $V$ harmonisch ist. Da dies für jede in $U$ reguläre Menge $V$ gilt, ist $u_0$ harmonisch in $U$.

*Zu* (b): Es genügt, wieder zu zeigen, daß $u_0$ in jeder in $U$ regulären Menge $V$ harmonisch ist. Bezeichnet $u^*$ die Restriktion auf $V^*$ einer Funktion $u \in \mathscr{S}$, so gilt $u_V(x) = \overset{*}{\int} u^* \, d\mu_x^V \leqq u(x)$ für alle $x \in V$. Da $u$ nach unten halbstetig und überall $> -\infty$ ist, so ist $u$ nach unten beschränkt. Hieraus und aus der vorausgesetzten Endlichkeit von $u$ folgt die Endlichkeit des Oberintegrals $\overset{*}{\int} u^* \, d\mu_x^V$ für alle $x \in V$. Nach ($K_2'$) ist daher $u_V$ in $V$ harmonisch. Schließlich ist $u_0$ eine reelle Funktion; aus den einleitenden Bemerkungen und $K_2$ folgt daher die Harmonizität von $u_0$ in $V$.

5.2. *Harmonische Majoranten.* — Wir besprechen zwei Folgerungen aus diesem Satz.

**Satz 12.** *Es sei $\mathcal{T}$ eine nicht-leere Menge von in $U \in \mathfrak{U}$ definierten hypoharmonischen Funktionen derart, daß $\mathcal{T}$ mindestens eine reelle, hyperharmonische Majorante $u_0$ in $U$ besitzt. In jedem der beiden folgenden Fälle besitzt dann $\mathcal{T}$ sogar eine kleinste hyperharmonische Majorante in $U$, welche außerdem harmonisch ist:*

(a) *Es gilt das Axiom* $K_1$. *Ferner ist $u_0$ und mindestens ein $t \in \mathcal{T}$ lokalbeschränkt.*

(b) *Es gilt das Axiom* $K_2$. *Ferner ist mindestens ein $t \in \mathcal{T}$ reellwertig.*

Beweis. Es sei $\mathcal{S}$ die Menge aller hyperharmonischen Majoranten $u$ von $\mathcal{T}$ in $U$ mit $u \leqq u_0$. Dann ist $u_0 \in \mathcal{S}$, also $\mathcal{S} \neq \emptyset$. Ist $V$ eine beliebige in $U$ reguläre Menge, so liegt mit jeder Funktion $u \in \mathcal{S}$ auch die Funktion $u_V$ in $\mathcal{S}$. Zum Beweis sei $t \in \mathcal{T}$ beliebig gewählt und im Sinne der Definition (3.2) $t_V = -(-t)_V$ gesetzt. Dann ist $t_V(x) = t(x)$ für alle $x \in U \cap \complement V$ und $t_V(x) = \int^{*} t\, d\mu_x^V$ für alle $x \in V$; aus $t \leqq u$ folgt daher $t_V \leqq u_V$. Zusammen mit (3.3) (angewendet auf $u$ und $-t$) erhält man weiter $t \leqq t_V \leqq u_V \leqq u \leqq u_0$ für alle $t \in \mathcal{T}$ und damit $u_V \in \mathcal{S}$. Da außerdem $\mathcal{S}$ mit je zwei Funktionen auch deren untere Einhüllende enthält, ist also $\mathcal{S}$ gesättigt. Da für alle $u$ gilt $t \leqq u \leqq u_0$ und hierbei $t \in \mathcal{S}$ den in (a) und (b) genannten Zusatzbedingungen unterworfen werden kann, ist nach Satz 11 in jedem der beiden Fälle $q = \inf \mathcal{S}$ harmonisch in $U$ und natürlich zudem eine Majorante von $\mathcal{T}$. Schließlich ist $q$ die kleinste hyperharmonische Majorante von $\mathcal{T}$ in $U$, da für jede Majorante $u \in \mathcal{H}_U^*$ von $\mathcal{T}$ die Funktion $\inf(u, u_0)$ zu $\mathcal{S}$ gehört. Dann aber ist $q \leqq \inf(u, u_0) \leqq u$.

**Korollar.** *Es werde das Axiom $K_1$ vorausgesetzt. Dann ist für jede Menge $U \in \mathfrak{U}$ der Vektorraum $\mathcal{H}_U^\delta$ aller Differenzen $h_1 - h_2$ nichtnegativer Funktionen $h_1, h_2 \in \mathcal{H}_U$ ein vollständiger Rieszscher Raum*[8]) *(bezüglich der üblichen $\leqq$-Relation).*

Beweis. $\mathcal{H}_U^\delta$ ist ein Rieszscher Raum, wenn je zwei Funktionen $d_1, d_2 \in \mathcal{H}_U^\delta$ eine kleinste Majorante in $\mathcal{H}_U^\delta$ besitzen. Nach Definition von $\mathcal{H}_U^\delta$ ist die Funktion $d_i$ von der Form $d_i = h_i' - h_i''$ mit $h_i', h_i'' \in \mathcal{H}_U$ und $h_i' \geqq 0$, $h_i'' \geqq 0$ $(i = 1, 2)$. Folglich ist $h_1' + h_2'$ eine in $\mathcal{H}_U^\delta$ gelegene Majorante von $d_1$ und $d_2$. Setzen wir $\mathcal{T} = \{d_1, d_2\}$, so sind offenbar die Voraussetzungen des Satzes 12 (Fall (a)) erfüllt; nach diesem existiert unter allen in $\mathcal{H}_U$ gelegenen Majoranten von $\mathcal{T}$ eine kleinste, welche mit $d_0$ bezeichnet werden möge. Aus $d_1 \leqq d_0$ folgt $d_0 = (d_0 - d_1) + d_1 = (d_0 - d_1) + h_1' - h_1''$. Also liegt $d_0$ in $\mathcal{H}_U^\delta$ und $d_0$ ist die gesuchte kleinste aller in $\mathcal{H}_U^\delta$ gelegenen Majoranten von $d_1, d_2$.

Schließlich ist der Rieszsche Raum $\mathcal{H}_U^\delta$ auch vollständig. Hierzu genügt es, folgendes zu beweisen[9]): Jede nicht-leere, aufsteigend filtrierende Menge $\mathcal{F} \subset \mathcal{H}_U$ von in $U$ nicht-negativen harmonischen Funktionen mit mindestens einer harmonischen Majorante $h_0$ besitzt eine kleinste harmonische Majorante.

---

[8]) Im Französischen: espace de Riesz complètement réticulé; vgl. BOURBAKI [5a].

[9]) Vgl. [5a], p. 21.

Da $h_0$ stetig ist, so ist $\mathscr{F}$ lokal-gleichmäßig nach oben beschränkt. Nach ($K_1'''$) ist daher sup$\mathscr{F}$ die gesuchte kleinste harmonische Majorante von $\mathscr{F}$.

## *§ 6. Beispiele zur Theorie*

**6.1.** *Laplacesche Differentialgleichung und Wärmeleitungsgleichung.* Die beiden folgenden Beispiele sind die einfachsten und wichtigsten für die ganze Theorie. Das Beispiel der Laplaceschen Differentialgleichung dient als Leitlinie für den Gesamtaufbau der Theorie, deren Resultate in diesem Spezialfall klassische Sätze der Potentialtheorie sind. Das zweite Beispiel ist ein Prüfstein für die nötige Allgemeinheit der Theorie, deren Ziel es ist, sowohl elliptische als auch parabolische Differentialgleichungen zu erfassen.

(a) Der lokal-kompakte Raum $X$ der allgemeinen Theorie sei jetzt speziell der $n$-dimensionale euklidische Raum $\mathbf{R}^n (n = 1, 2, \ldots)$. Wir betrachten in ihm die *Laplacesche Differentialgleichung*

$$\Delta u = \sum_{i=1}^{n} \frac{\partial^2 u}{\partial x_i^2} = 0\,. \tag{6.1}$$

Für jede Menge $U \in \mathfrak{U}$ sei $\mathscr{H}_U$ die Menge aller im üblichen Sinne in $U$ harmonischen Funktionen, d. h. aller auf $U$ definierten, zweimal stetig-differenzierbaren, reellen Funktionen $u$ mit $\Delta u = 0$. Dann genügt die Abbildung $U \to \mathscr{H}_U$ offenbar dem Axiom I. Nach der Poissonschen Integralformel der Potentialtheorie ist jede offene Vollkugel im $\mathbf{R}^n$ regulär, also das Axiom II erfüllt. Die $n$ Koordinatenfunktionen $x = (x_1, \ldots, x_n) \to x_i$ und die konstante Funktion $h = 1$ sind im $\mathbf{R}^n$ harmonisch. Daher trennt $\mathscr{H}_X$ die Punkte von $X = \mathbf{R}^n$; es ist somit das *starke Trennungsaxiom* $T'$ erfüllt. Das *starke Konvergenzaxiom* $K_2$ folgt aus dem zweiten Harnackschen Konvergenzsatz der Potentialtheorie (bei Beachtung von Satz 10).

Unsere Theorie macht keinen Gebrauch von der folgenden verschärften Form des Harnackschen Satzes: Für jedes Gebiet $\Omega$ im $\mathbf{R}^n$ und jede nicht-leere, aufsteigend filtrierende Menge $\mathscr{F}$ ist entweder sup$\mathscr{F}$ in $\Omega$ harmonisch oder die in $\Omega$ konstante Funktion $+\infty$.

Wir erwähnen noch, daß für Dimensionen $n \geqq 3$ die Menge aller in $X$ hyperharmonischen Funktionen $u \geqq 0$ die Punkte von $X = \mathbf{R}^n$ trennt. Man betrachte hierzu den auf $X \times X$ definierten Newtonschen Kern $K$ und bemerke, daß für jeden Punkt $x \in X$ die Funktion $x' \to K(x, x')$ überall positiv und auf dem Gesamtraum hyperharmonisch ist (letzteres nach Satz 5).

(b) Es sei $X = \mathbf{R}^{n+1} (n = 1, 2, \ldots)$; für jede Menge $U \in \mathfrak{U}$ sei $\mathscr{H}_U$ die Menge aller in $U$ definierten, zweimal stetig-differenzierbaren, reellen Lösungen $u$ der *Wärmeleitungsgleichung*

$$\Delta u = \frac{\partial u}{\partial x_{n+1}}\,. \tag{6.2}$$

Dann genügt die Abbildung $U \to \mathscr{H}_U$ dem Axiom I. Das Axiom II ist erfüllt, da beispielsweise jedes $(n + 1)$-dimensionale, offene, gleichseitige Simplex regulär ist, dessen eine Seite in einer Hyperebene $x_{n+1} = t_0$ liegt und dessen

Spitze eine $(n+1)$-te Koordinate $t_1 > t_0$ besitzt (vgl. GEVREY [16], HARTMAN-WINTNER [17] und DOOB [14], [15]). Es ist das *starke Trennungsaxiom* T' erfüllt, da die konstante Funktion $h = 1$ die Gleichung (6.2) löst und $\mathscr{H}_X$ die Punkte von $X$ trennt. Letzteres folgt etwa so: Die ersten $n$ Koordinatenfunktionen $(x_1, \ldots, x_{n+1}) \to x_i (i = 1, \ldots, n)$ sind Lösungen von (6.2) und trennen je zwei Punkte, die sich in den ersten $n$ Koordinaten unterscheiden. Zwei sich nur in der letzten Koordinate $x_{n+1}$ unterscheidende Punkte werden durch die Lösungen

$$(6.3) \qquad (x_1, \ldots, x_{n1}) \to e^{-x_{n+1}} \prod_{i=1}^{n} \sin \frac{x_i - a_i}{\sqrt{n}}$$

von (6.2) getrennt. Man hat nur die Konstanten $a_i$ geeignet zu wählen. Schließlich ist auch das *starke Konvergenzaxiom* $K_2$ erfüllt (vgl. [16] sowie die Sätze 9 und 10).

Unsere Theorie macht von folgender Verschärfung des Axioms $K_2$ keinen Gebrauch, obwohl diese in beiden Spezialfällen (a) und (b) zur Verfügung steht: Es sei $U \in \mathfrak{U}$ und $\mathscr{F}$ eine nicht-leere, aufsteigend filtrierende Teilmenge von $\mathscr{H}_U$. Ist dann $\sup \mathscr{F}$ auf einer in $U$ dichten Teilmenge endlich, so ist $\sup \mathscr{F} \in \mathscr{H}_U$. (Vgl. hierzu auch das Axiom TM 4 bei DOOB [14].) Für weiterreichende Fragestellungen dürfte jedoch diese Verschärfung von $K_2$ von Bedeutung sein.

Ebenso wie im Beispiel (a) zeigen wir noch, daß auch hier die Menge aller in $X$ hyperharmonischen Funktionen $u \geqq 0$ die Punkte von $X = \mathbf{R}^{n+1}$ trennt. Wir betrachten hierzu den auf $X \times X$ definierten „parabolischen Kern“ $P$.

Dabei ist $P(x, y) = p(x - y)$ für $x, y \in X$ und $p$ auf $X$ wie folgt definiert:

$$(6.4) \qquad p(x) = \begin{cases} \dfrac{1}{x_{n+1}{}^{n/2}} e^{-\frac{\sum_{i=1}^{n} x_i^2}{4x_{n+1}}}, & \text{falls } x_{n+1} > 0; \\ 0, & \text{falls } x_{n+1} \leqq 0. \end{cases}$$

Es ist bekannt und wegen Satz 5 leicht zu beweisen (vgl. DOOB [15]), daß $x \to P(x, y)$ für jedes $y \in X$ eine auf $X$ hyperharmonische Funktion ist. Man prüft sofort nach, daß zu je zwei verschiedenen Punkten $x, y \in X$ ein Punkt $z \in X$ existiert mit $P(x, z) \neq P(y, z)$. Da $p \geqq 0$ ist, so ist damit die behauptete verschärfte Trennungsaussage bewiesen.

6.2. *Beziehungen zur Axiomatik von* M. BRELOT. — Nunmehr erhebt sich die Frage, ob die Beschränkung auf die in (a) und (b) behandelten speziellen elliptischen und parabolischen Differentialgleichungen notwendig ist. Die Antwort auf die die elliptischen Differentialgleichungen betreffende Frage geben wir dadurch, daß wir gleich allgemeiner nach den Beziehungen unserer Theorie zu der von BRELOT [7]—[9] fragen, welche, wie einleitend erwähnt, speziell auf den elliptischen Fall zugeschnitten wurde.

Die Axiomatik von BRELOT geht von folgenden Annahmen aus: Ein lokalkompakter Grundraum $X$ wird zusätzlich als *nicht-kompakt*, *zusammenhängend* und *lokal-zusammenhängend* vorausgesetzt. Wie in unserer Theorie wird eine Abbildung $U \to \mathscr{H}_U (U \in \mathfrak{U})$ vorgegeben, welche den Axiomen I und II genügt. Dann existiert sogar zu jedem Punkt $x \in X$ ein Fundamentalsystem regulärer Umgebungen $V$ von $x$, welche sämtlich *zusammenhängend* sind (vgl. [9], p. 62). Zu den Axiomen I und II tritt als zusätzliches Axiom das bereits in 6.1., (a) angedeutete, verschärfte Konvergenzaxiom:

($\mathrm{III}_B$) *Für jedes Gebiet $\Omega \in \mathfrak{U}$ und jede nicht-leere, aufsteigend filtrierende Menge $\mathscr{F} \subset \mathscr{H}_\Omega$ ist entweder $\sup \mathscr{F} \in \mathscr{H}_\Omega$ oder $\sup \mathscr{F}$ die konstante Funktion $+\infty$.*

Eine Folgerung hieraus ist unser starkes Konvergenzaxiom $K_2$.

In allen entscheidenden Teilen der Theorie von Brelot wird weiter die Existenz eines *Potentials* $u > 0$ auf $X$ gefordert. Das ist eine hyperharmonische Funktion $u \in \mathscr{H}_X^*$ mit $u(x) > 0$ für alle $x \in X$, welche in mindestens einem Punkt endlich und für welche die konstante Funktion 0 die größte harmonische Minorante ist. Mit Hilfe dieser Begriffsbildung läßt sich nun zeigen[10]):

*Es sei $X$ ein lokal-kompakter, nicht-kompakter, zusammenhängender, lokal-zusammenhängender Raum und $U \to \mathscr{H}_U (U \in \mathfrak{U})$ ein den Axiomen* I, II *und* $\mathrm{III}_B$ *genügendes Garbendatum stetiger reeller Funktionen. Ferner existiere ein Potential $> 0$ auf $X$ und eine überall positive, harmonische Funktion $h \in \mathscr{H}_X$. Dann ist das Trennungsaxiom $T$ erfüllt. Genauer trennt bereits die Menge aller stetigen, reellen, in $X$ $h$-hyperharmonischen Funktionen $v \geqq 0$ die Punkte von $X$.*

Beweis. Es seien $x$ und $y$ verschiedene Punkte aus $X$ und $V$ eine reguläre, zusammenhängende Umgebung von $x$, in welcher $y$ nicht liegt. Nach Brelot [9], p. 95, gibt es ein Potential $u > 0$ aus $\mathscr{C}(X)$, welches in $V$ nicht harmonisch ist. Die Funktion $v = \frac{u}{h}$ ist in $X$ $h$-hyperharmonisch, stetig, reellwertig und $> 0$. Im Hinblick auf die zu beweisende Behauptung kann also angenommen werden, daß $v(x) = v(y)$ ist. Dann betrachten wir die nach [9], p. 75, in $X$ hyperharmonische Funktion $u' = u_V$[11]). Mit $u$ liegt auch $u'$ in $\mathscr{C}(X)$; ferner ist $u \geqq u' \geqq 0$. Weiter ist $u'(y) = u(y)$ und $u'(x) < u(x)$. Letzteres folgt aus $u' \leqq u$ und der Eigenschaft von $u$, in $V$ nicht harmonisch zu sein. Wäre nämlich $u'(x) = u(x)$, so würde die Harmonizität von $u'$ in $V$ und das Axiom $\mathrm{III}_B$ die Gleichheit $u'(z) = u(z)$ für alle $z \in V$ und damit die Harmonizität von $u$ in $V$ zur Folge haben (vgl. [9], p. 71). Somit erhalten wir $\frac{u'(x)}{h(x)} < v(x) = v(y) = \frac{u'(y)}{h(y)}$. Die Punkte $x$ und $y$ werden also durch die in $X$ $h$-hyperharmonische Funktion $v' = \frac{u'}{h}$ getrennt; ferner ist $v' \in \mathscr{C}(X)$ und $v' \geqq 0$.

Im Zusammenhang mit diesem Resultat verweisen wir noch auf Brelot [9], p. 94, wonach aus den Axiomen der Brelotschen Theorie (einschließlich der Forderung, daß ein Potential $> 0$ existiert) folgt, daß zu jeder *relativ-kompakten*

---

10) Vgl. auch Brelot [9], p. 97.

11) Die Hyperharmonizität von $u_V$ folgt nicht aus unserem Satz 6, da dort das Axiom $T$ herangezogen wird, welches durch diesen Beweis erst sichergestellt werden soll.

Menge $U \in \mathfrak{U}$ eine harmonische Funktion $h \in \mathscr{H}_U$ mit positiver unterer Schranke existiert.

Zusammenfassend können wir also feststellen, daß das volle Axiomensystem von BRELOT die Axiome I, II, T und $K_2$ unserer Theorie impliziert, sofern eine in $X$ überall positive, harmonische Funktion $h$ existiert. Für die Restriktion des Garbendatums der harmonischen Funktionen auf eine relativkompakte Menge $U \in \mathfrak{U}$ ist diese Zusatzforderung von selbst erfüllt. Im übrigen wird auch bei BRELOT fast durchwegs die Existenz eines überall positiven $h \in \mathscr{H}_X$ gefordert.

Es erscheint zweifelhaft, ob aus den Axiomen von BRELOT sogar das starke Trennungsaxiom T′ folgt.

**6.3.** *Allgemeinere elliptische Differentialgleichungen.* — Es sei $X$ eine offene (und daher lokal-kompakte) Teilmenge des $\mathbf{R}^n (n = 1, 2, \ldots)$. Für jede Menge $U \in \mathfrak{U}$ sei $\mathscr{H}_U$ die Menge aller in $U$ definierten, zweimal stetig differenzierbaren, reellen Funktionen $u$, welche der Differentialgleichung

$$\sum_{i,j=1}^{n} a_{ij} \frac{\partial^2 u}{\partial x_i \partial x_j} + \sum_{i=1}^{n} b_i \frac{\partial u}{\partial x_i} + cu = 0 \tag{6.5}$$

genügen. Die Differentialgleichung sei elliptisch, d. h. für jedes $x \in X$ sei $\Sigma a_{ij}(x) \lambda_i \lambda_j$ eine positiv-definite quadratische Form. Die Koeffizienten $a_{ij}$, $b_i$, $c$ seien sämtlich Funktionen aus $\mathscr{C}(X)$; ferner seien die Funktionen $a_{ij}$ einmal stetig-differenzierbar. Die partiellen Ableitungen der $a_{ij}$ sowie die Funktionen $b_i$ und $c$ sollen der lokalen Lipschitz-Bedingung genügen. Ferner werde $c \leqq 0$ vorausgesetzt.

Nach bekannten Sätzen (vgl. [20], [9]) genügt dann die Abbildung $U \to \mathscr{H}_U$ den Axiomen I, II und $K_2$; das Konvergenzaxiom $K_2$ ist sogar in der verschärften Form $III_B$ erfüllt. Unsere Theorie ist daher anwendbar, sofern auch das Axiom T erfüllt ist.

Es ist hier nicht unser Ziel, die Frage nach der Gültigkeit von $(T)$ möglichst umfassend zu beantworten. Hierauf soll an anderer Stelle eingegangen werden[11a]). Hier sei nur ohne Beweis erwähnt, daß aus $c \leqq 0$ folgt, daß die konstante Funktion 1 in $X$ hyperharmonisch ist. Nach Satz 3 ist daher nur zu zeigen, daß $\mathscr{H}_X^*$ die Punkte von $X$ trennt. Dann steht unsere Theorie in jeder offenen Teilmenge von $X$ zur Verfügung, in welcher eine überall positive Lösung $h$ von (6.5) existiert.

Man kann zeigen, daß zu jedem geordneten Paar $(x, y)$ von Punkten aus eine Funktion $u \in \mathscr{H}_X$ existiert mit $0 < u(x) < u(y)$, wenn die Koeffizienten der Differentialgleichung (6.5) noch folgenden Zusatzbedingungen genügen: Für jedes $i = 1, \ldots, n$ sind $a_{ii}$ und $b_i$ nur Funktionen von $x_i$ ($i$-te Koordinate von $x$) allein; ferner ist $c$ für $n \geqq 2$ eine Konstante. In diesem Fall kann dann der zweite Teil von Satz 3 direkt angewendet werden; da obige Funktion $u$ in $X$ harmonisch ist, zeigt der Beweis dieses Satzes, daß dann sogar das Axiom T′ in jeder offenen Teilmenge von $X$ erfüllt, in welcher eine überall positive

11a) *Zusatz bei der Korrektur:* Vgl. auch R. M. HERVÉ, Recherches axiomatiques sur la théorie des fonctions surharmoniques et du potential (Thèse, Paris, 1961), insbesondere p. 146 ff.

Lösung $h$ von (6.5) existiert. Die oben genannten Zusatzbedingungen sind speziell dann erfüllt, wenn alle Koeffizienten in (6.5) konstant sind. Im Falle $c = 0$ ist $h = 1$ eine (überall positive) Lösung von (6.5).

Schließlich weisen wir darauf hin, daß unter zusätzlichen Annahmen über $X$ die Vorzeichenbedingung $c \leqq 0$ auch fallen gelassen werden kann. Man prüft nämlich leicht nach, daß $X = ]0, \pi[ \subset \mathbf{R}$ und die gewöhnliche Differentialgleichung

$$u'' + u = 0 \tag{6.6}$$

ein weiteres Anwendungsbeispiel für unsere Theorie liefern. Dabei ist $\mathscr{H}_X$ die Menge aller Funktionen $x \to \alpha \sin x + \beta \cos x$ $(\alpha, \beta \in \mathbf{R}, x \in X)$ und $\mathscr{H}_U = \mathrm{Rest}_U \mathscr{H}_X$ für alle $U \in \mathfrak{U}$. Jedes offene Intervall $V = ]a, b[$ mit $a, b \in X$ ist regulär; das harmonische Maß $\mu_x^V$ ergibt sich für alle $x \in V$ zu:

$$\mu_x^V = \frac{\sin(x - b)}{\sin(a - b)} \varepsilon_a + \frac{\sin(a - x)}{\sin(a - b)} \varepsilon_b \,. \tag{6.7}$$

Die Lösung $h(x) = \sin x$ ist überall positiv in $X$. Die $h$-harmonische Funktion $x \to \mathrm{ctg}\, x$ trennt je zwei Punkte von $X$; folglich ist neben I, II und $\mathrm{K}_2$ sogar das starke Trennungsaxiom T′ erfüllt. Übrigens ist hier stets $\int d\mu_x^V > 1$ und somit die konstante Funktion $-1$ hyperharmonisch in $X$.

**6.4.** *Greensche Räume.* — Die Axiome unserer Theorie sind erfüllt, wenn $X$ ein *Greenscher Raum* [12], also zum Beispiel eine hyperbolische Riemannsche Fläche, und $U \to \mathscr{H}_U$ das Garbendatum der wie üblich definierten harmonischen Funktionen ist. Die Axiome I, II und $\mathrm{K}_2$ — letzteres sogar in der verschärften Form $\mathrm{III}_B$ — folgen beinahe unmittelbar aus der Definition dieser Räume. Ebenso einfach ergibt sich das starke Trennungsaxiom T′. Zu jedem Punkt $x \in X$ existiert nämlich die stets hyperharmonische Greensche Funktion $G_x^X$ im Sinne von [12]; diese ist überall positiv und nimmt ihr Maximum nur im Punkte $x$ an. Ferner ist die konstante Funktion 1 in $X$ harmonisch.

Dieses Beispiel ordnet sich auch in die Axiomatik von Brelot ein; es steht somit das allgemeine Resultat aus 6.2. zur Verfügung. Dieses liefert jedoch nur das Trennungsaxiom T.

**6.5.** *Allgemeinere parabolische Differentialgleichungen.* — Auf Grund der in 6.3 erwähnten Resultate ist zu erwarten, daß die Axiome unserer Theorie erfüllt sind, wenn $X$ eine offene Teilmenge des $\mathbf{R}^{n+1}$ ist und das Garbendatum $U \to \mathscr{H}_U$ durch die Lösungen von parabolischen Differentialgleichungen der Gestalt

$$\sum_{i,j=1}^{n} a_{ij} \frac{\partial^2 u}{\partial x_i \partial x_j} + \sum_{i=1}^{n} b_i \frac{\partial u}{\partial x_i} + cu = \frac{\partial u}{\partial x_{n+1}} \tag{6.8}$$

definiert wird. Dabei sind an die Koeffizientenfunktionen die gleichen Forderungen wie in 6.3 zu stellen.

## II. Verallgemeinerte Lösung des Dirichletschen Problems

*In diesem Kapitel wird durchwegs die Gültigkeit der Axiome* **I**, **II** *und* **T** *vorausgesetzt.* Über die Konvergenzaxiome $\mathrm{K}_1$ und $\mathrm{K}_2$ wird von Fall zu Fall gesondert verfügt.

## § 7. *Die Methode von* PERRON-WIENER-BRELOT

**7.1.** *Annahmen über Grundraum und Rand.* — Wie bei den Untersuchungen des § 2 bezeichne im folgenden $Y$ einen *kompakten* Raum, welcher $X$ als *dichten Unterraum* enthält. Der Rand $X^* = \complement_Y X$ von $X$ in $Y$ ist dann nach Satz 4 nicht leer sowie natürlich kompakt.

Wir machen weiter folgende, das Axiom T verschärfende Annahme:

**(P)** *Es existiert eine in $X$ harmonische Funktion $h \in \mathscr{H}_X$ mit positiver unterer Schranke, also mit:* $\inf\limits_{x \in X} h(x) > 0$.

Nach der im Anschluß an Axiom T gemachten Bemerkung kann angenommen werden, daß $h$ in der Formulierung von Axiom T auftretende harmonische Funktion ist.

Schon jetzt sei darauf aufmerksam gemacht, daß die Restriktion des den Axiomen I, II, T genügenden Garbendatums auf einen relativ-kompakten, offenen Unterraum von $X$ offenbar stets der Bedingung P genügt. Wie bereits in § 4 bemerkt wurde, vererben sich auch die Konvergenzaxiome bei Restriktion.

**7.2.** *Ober- und Unterfunktionen.* — Für jede numerische Funktion $f$ auf $X^*$ bezeichne nun $\mathscr{O}_f$ die Menge aller zu $f$ gehörigen *Oberfunktionen*. Das sind alle Funktionen $u \in \mathscr{H}_X^*$ mit folgenden zwei Eigenschaften:

(7.1) $$\liminf_{x \to z,\, x \in X} u(x) \geqq f(z) \qquad \text{für alle } z \in X^*;$$

(7.2) $$\liminf_{x \to z,\, x \in X} u(x) > -\infty \qquad \text{für alle } z \in X^*\,^{12)}.$$

Entsprechend heiße $\mathscr{U}_f = -\mathscr{O}_{-f}$ die Menge der zu $f$ gehörigen *Unterfunktionen*. Keine der beiden Mengen $\mathscr{O}_f$ und $\mathscr{U}_f$ ist leer, da die konstante Funktion $+\infty$ stets in $\mathscr{O}_f$ liegt. Es liegt daher die Bildung der folgenden Einhüllenden nahe:

(7.3) $$\overline{H}_f = \inf \mathscr{O}_f\,,$$

(7.4) $$\underline{H}_f = \sup \mathscr{U}_f\,.$$

Offenbar ist

(7.5) $$\underline{H}_f = -\overline{H}_{-f}\,.$$

$\overline{H}_f$ und $\underline{H}_f$ sind gemäß ihrer Definition numerische, auf $X$ definierte Funktionen. Soll der Bezugsraum $X$ besonders hervorgehoben werden, so schreiben wir an Stelle von $\mathscr{O}_f$, $\mathscr{U}_f$, $\overline{H}_f$, $\underline{H}_f$ genauer $\mathscr{O}_f^X$, $\mathscr{U}_f^X$, $\overline{H}_f^X$, $\underline{H}_f^X$.

Aus diesen Definitionen ergeben sich mühelos die drei folgenden Eigenschaften; $f$ und $g$ seien hierbei numerische Funktionen auf $X^*$:

(7.6) $$f \leqq g \Rightarrow \overline{H}_f \leqq \overline{H}_g\,;$$

(7.7) $$\overline{H}_{\lambda f} = \lambda \overline{H}_f \qquad \text{für jede reelle Zahl } \lambda > 0;$$

(7.8) $$\overline{H}_{f+g} \leqq \overline{H}_f + \overline{H}_g\,.$$

12) Da $Y$ kompakt ist, ist diese Forderung (7.2) äquivalent mit der Beschränktheit von $u$ nach unten.

Die Ungleichung (7.8) gilt dabei bereits unter den folgenden Voraussetzungen: Erstens ist die Bildung von $\overline{H}_f + \overline{H}_g$ sinnvoll im Sinne der üblichen Konventionen über das Rechnen in $\overline{\mathbf{R}}$. Zweitens ist $(f+g)(z) = f(z) + g(z)$ immer dann, wenn die rechte Seite sinnvoll ist; für alle anderen $z \in X^*$ kann für $(f+g)(z)$ eine beliebige Zahl aus $\overline{\mathbf{R}}$ gewählt werden.

**7.3.** *Weitergehende Eigenschaften von* $\overline{H}_f$. — Etwas tiefer liegen die folgenden Eigenschaften:

**Satz 13.** *Es sei $f$ eine numerische, auf $X^*$ definierte Funktion. Dann gilt:*

(a) *$\mathcal{O}_f$ ist eine gesättigte Teilmenge von $\mathcal{H}_X^*$;*

(b) $\underline{H}_f \leqq \overline{H}_f$.

Beweis. *Zu* (a): Für je zwei Funktionen $u_1, u_2 \in \mathcal{O}_f$ und deren untere Einhüllende $w = \inf(u_1, u_2)$ gilt:

$$\liminf_{x \to z, x \in X} w(x) = \inf\left(\liminf_{x \to z, x \in X} u_1(x), \liminf_{x \to z, x \in X} u_2(x)\right)$$

für jeden Randpunkt $z \in X^*$. Hieraus folgt $w \in \mathcal{O}_f$ bei Beachtung von (7.1) und (7.2). Ist weiter $V$ eine reguläre Menge, so ist $\complement_Y V$ eine Umgebung in $Y$ eines jeden Punktes $z \in X^*$ zufolge der relativen Kompaktheit von $V$. Da außerdem für jedes $u \in \mathcal{H}_X^*$ die Funktionen $u$ und $u_V$ im Komplement von $V$ bezüglich $X$ übereinstimmen, erhält man

$$\liminf_{x \to z, x \in X} u(x) = \liminf_{x \to z, x \in X} u_V(x)$$

für alle $z \in X^*$. Hieraus folgt aber, daß $\mathcal{O}_f$ mit jedem $u$ auch die Funktion $u_V$ enthält. Also ist $\mathcal{O}_f$ gesättigt.

*Zu* (b): Die Funktion $u$ sei in $\mathcal{O}_f$, die Funktion $v$ in $\mathcal{U}_f$ beliebig gewählt. Dann ist $u - v$ eine in $X$ hyperharmonische Funktion mit

$$\liminf_{x \to z, x \in X} (u(x) - v(x)) \geqq 0$$

für alle $z \in X^*$. Es ist nämlich

$$\liminf_{x \to z, x \in X} (u(x) - v(x)) \geqq \liminf_{x \to z, x \in X} u(x) + \liminf_{x \to z, x \in X} (-v(x))$$

für alle $z \in X^*$. Wegen $-v \in \mathcal{O}_{-f}$ und (7.1) folgt hieraus die behauptete Ungleichung für jeden Randpunkt $z$ mit $|f(z)| < +\infty$. Ist aber $f(z) = \pm\infty$, so gelangt man vermöge (7.1) und (7.2) zum gleichen Resultat. Da wir die Eigenschaft $P$ gefordert haben, ist das Korollar zu Lemma 2 anwendbar. Nach der dortigen Implikation (2.3) erhält man $u - v \geqq 0$, also $v \leqq u$ und hieraus $\underline{H}_f \leqq \overline{H}_f$.

**Lemma 3.** *Es sei $(f_n)_{n=1,2,\ldots}$ eine isotone Folge numerischer Funktionen auf $X^*$ derart, daß die zugehörigen Funktionen $\overline{H}_{f_n}$ sämtlich in $X$ harmonisch sind. Dann gilt:*

$$\overline{H}_{\sup f_n} = \sup \overline{H}_{f_n}. \tag{7.9}$$

Beweis. Wir setzen $f = \sup f_n$. Aus der in (7.6) formulierten Isotonie der Abbildung $g \to \overline{H}_g$ folgt dann: $\sup \overline{H}_{f_n} \leqq \overline{H}_f$. Zum Beweis der dualen Ungleichungen sei $\varepsilon > 0$ beliebig gewählt und $\varepsilon_1, \varepsilon_2, \ldots$ eine Folge positiver reeller Zahlen mit $\varepsilon = \Sigma \varepsilon_n$; weiter sei $y$ ein beliebiger Punkt aus $X$. Da $\overline{H}_{f_n}$ harmonisch, also reellwertig ist, gibt es zu jeder natürlichen Zahl $n$ eine Funktion $u_n \in \mathcal{O}_{f_n}$ mit $u_n(y) - H_{f_n}(y) \leqq \varepsilon_n$. Nun ist $(\overline{H}_{f_n})$ eine isotone Folge harmonischer Funktionen in $X$, nach (1.16) also $\sup \overline{H}_{f_n}$ in $X$ hyperharmonisch. Da $u_n - \overline{H}_{f_n} \geqq 0$ und hyperharmonisch ist, so ist abermals nach (1.16) die Funktion $\sum_{n=1}^{\infty} (u_n - \overline{H}_{f_n})$ hyperharmonisch. Also ist auch

$$w = \sup \overline{H}_{f_n} + \Sigma (u_n - \overline{H}_{f_n})$$

hyperharmonisch in $X$. Nun gilt $w \geqq \overline{H}_{f_n} + (u_n - \overline{H}_{f_n}) = u_n$ und damit $w \in \mathcal{O}_{f_n}$ für alle $n = 1, 2, \ldots$; also ist $w$ eine zu $f$ gehörige Oberfunktion und somit $w \geqq \overline{H}_f$. Wir erhalten daher

$$\overline{H}_f(y) \leqq w(y) \leqq \sup \overline{H}_{f_n}(y) + \Sigma \varepsilon_n = \sup \overline{H}_{f_n}(y) + \varepsilon .$$

Da $\varepsilon > 0$ und $y \in X$ beliebig gewählt waren, erhält man die noch fehlende Ungleichung.

Die Bedeutung von Lemma 3 unterstreicht der folgende Satz.

**Satz 14.** *Es sei $f$ eine auf $X^*$ definierte, numerische Funktion. Dann ist die Funktion $\overline{H}_f$ in jedem der beiden folgenden Fälle harmonisch in $X$:*

(a) *$f$ ist beschränkt; es gilt das Konvergenzaxiom* $K_1$.

(b) *$\overline{H}_f$ ist überall endlich; es gilt das Konvergenzaxiom* $K_2$.

Beweis. *Zu* (a): Aus der Eigenschaft $P$ und der Beschränktheit von $f$ folgt die Existenz einer harmonischen Funktion $h_0 \in \mathcal{H}_X$ mit $\sup_{z \in X^*} |f(z)| \leqq \inf_{x \in X} h_0(x)$. Daher ist $h_0$ eine zu $f$ gehörige Oberfunktion. Wir betrachten die Menge $\mathcal{S}$ aller $u \in \mathcal{O}_f$ mit $u \leqq h_0$. Wegen $h_0 \in \mathcal{S}$ ist $\mathcal{S}$ nicht leer; ferner ist mit $\mathcal{O}_f$ offenbar auch $\mathcal{S}$ gesättigt. Für jede Funktion $u \in \mathcal{O}_f$ ist weiter $u' = \inf(u, h_0)$ ein Element aus $\mathcal{S}$ mit $u' \leqq u$. Hieraus und aus $\mathcal{S} \subset \mathcal{O}_f$ folgt $\overline{H}_f = \inf \mathcal{S}$. Nach Wahl von $h_0$ gilt auch $h_0 \in \mathcal{O}_{-f}$ und damit $-h_0 \in \mathcal{U}_f$. Bei Beachtung der Ungleichung $\underline{H}_f \leqq \overline{H}_f$ erhält man dann $-h_0 \leqq u \leqq h_0$ für alle $u \in \mathcal{S}$. Wegen der Stetigkeit von $h_0$ ist also die Menge $\mathcal{S}$ auf $X$ lokal-gleichmäßig beschränkt. Die Behauptung folgt daher aus Satz 11.

*Zu* (b): Zunächst werde von $f$ zusätzlich die Beschränktheit nach oben vorausgesetzt. Dann ergibt sich wie unter (a) die Existenz einer Funktion $h_0 \in \mathcal{H}_X$ mit $h_0 \in \mathcal{O}_f$. Wie dort kann dann die Menge $\mathcal{S}$ aller $u \in \mathcal{O}_f$ mit $u \leqq h_0$ betrachtet werden; $\mathcal{S}$ ist nicht leer und gesättigt; es gilt $\overline{H}_f = \inf \mathcal{S}$. Da alle Funktionen $u \in \mathcal{S}$ sowie $\overline{H}_f$ reellwertig sind, folgt die Behauptung abermals aus Satz 11.

Nunmehr kann auch die über $f$ gemachte Zusatzvoraussetzung fallen gelassen werden. Setzen wir $f_n = \inf(n, f)$ für jedes $n = 1, 2, \ldots$, so ist $(f_n)$ eine isotone Folge nach oben beschränkter Funktionen. Zu zeigen ist daher nur, daß jede der Funktionen $\overline{H}_{f_n}$ endlich ist. Nach dem ersten Teil des Beweises ist dann nämlich $\overline{H}_{f_n}$ in $X$ harmonisch. Wegen $f = \sup f_n$ und $\overline{H}_{f_n} \leqq \overline{H}_{f_{n+1}}$ folgt daher die Behauptung aus ($K_2$) und Lemma 3.

Für den Beweis der Endlichkeit von $\overline{H}_{f_n}$ setzen wir noch $g_n = \sup(n, f)$ für jede natürliche Zahl $n$. Für jede Funktion $u \in \mathcal{O}_f$ ist dann offenbar

$u + \alpha h \in O_{g_n}$ für jedes hinreichend große $\alpha > 0$ und daher mit $\overline{H}_f(x) < +\infty$ auch $\overline{H}_{g_n}(x) < +\infty$ für alle $x \in X$; außerdem ist $\overline{H}_{g_n} \geqq \overline{H}_f$ wegen $g_n \geqq f$. Aus der Endlichkeit von $\overline{H}_f$ folgt daher die Endlichkeit von $\overline{H}_{g_n}$. Nun ist aber $f \leqq f + n = f_n + g_n$, also nach (7.6) und (7.8) $\overline{H}_f \leqq \overline{H}_{f_n} + \overline{H}_{g_n}$. Ferner ist $\overline{H}_{f_n} \leqq \overline{H}_f$ wegen $f_n \leqq f$. Mit $\overline{H}_f$ und $\overline{H}_{g_n}$ ist daher auch $\overline{H}_{f_n}$ endlich. Damit ist der Satz bewiesen.

Wir bemerken noch, daß die in Satz 14 über $\overline{H}_f$ bewiesenen Aussagen sofort die entsprechenden Behauptungen über $\underline{H}_f$ nach sich ziehen. Dies folgt aus (7.5).

**7.4.** *Resolutive Funktionen.* — Zur weiteren Orientierung betrachten wir eine stetige, reelle Randfunktion $f \in \mathscr{C}(X^*)$, für welche das Dirichletsche Problem im klassischen Sinn lösbar ist; es existiere also eine harmonische Funktion $h_0 \in \mathscr{H}_X$ mit $\lim\limits_{x \to z,\, z \in X} h_0(x) = f(z)$ für alle $z \in X^*$. Dann liegt offenbar $h_0$ sowohl in $\mathscr{O}_f$ als auch in $\mathscr{U}_f$; also gilt $h_0 \leqq \underline{H}_f \leqq \overline{H}_f \leqq h_0$ und damit $h_0 = \underline{H}_f = \overline{H}_f$. (Insbesondere ist damit auch die eindeutige Bestimmtheit von $h_0$ bewiesen, welche aber auch sofort aus dem Korollar 2 zu Lemma 2 folgt.)

Diese Vorbetrachtung legt folgende Definition nahe:

**Definition 5.** *Eine numerische Funktion $f$ auf $X^*$ heiße* r e s o l u t i v, *wenn $\underline{H}_f = \overline{H}_f$ ist und $\overline{H}_f$ nur reelle Zahlen als Werte annimmt. Wir setzen dann $H_f = \underline{H}_f = \overline{H}_f$.*

*Ist $f$ resolutiv und $H_f$ in $X$ harmonisch, so heiße $f$ harmonisch-resolutiv und $H_f$ die* v e r a l l g e m e i n e r t e  L ö s u n g *des Dirichletschen Problems für die Randfunktion $f$.*

Der Satz 14 lehrt nun, daß jede beschränkte resolutive Funktion harmonisch-resolutiv ist, falls das Konvergenzaxiom $K_1$ vorausgesetzt wird. Fordert man das Axiom $K_2$, so ist sogar jede resolutive Funktion harmonisch-resolutiv.

Wir beweisen zunächst das folgende Lemma:

**Lemma 4.** *Die Menge $\mathscr{R}$ aller reellwertigen, resolutiven Randfunktionen ist ein Vektorraum (bezüglich der üblichen Operationen). Für jeden Punkt $x \in X$ ist $f \to H_f(x)$ eine positive Linearform auf $\mathscr{R}$. Ferner ist $\mathscr{R}$ abgeschlossen bezüglich der Topologie der gleichmäßigen Konvergenz auf $X^*$.*

B e w e i s. Für je zwei Funktionen $f, g \in \mathscr{R}$ gilt nach (7.8) und (7.5): $\overline{H}_{f+g} \leqq \leqq H_f + H_g$ und $\underline{H}_{f+g} \geqq H_f + H_g$. Wegen $\underline{H}_{f+g} \leqq \overline{H}_{f+g}$ ist daher $f + g \in \mathscr{R}$ und $H_{f+g} = H_f + H_g$.

Wir zeigen weiter, daß aus $f \in R$ und $\lambda \in \mathbf{R}$ folgt: $\lambda f \in \mathscr{R}$ sowie $H_{\lambda f} = \lambda H_f$. Der Fall $\lambda = 0$ erledigt sich sofort, da $\overline{H}_0 = \underline{H}_0 = 0$ ist (etwa nach der einleitenden Bemerkung zu 7.4). Ebenso einfach läßt sich der Fall $\lambda = -1$ behandeln; es ist nämlich $\overline{H}_{-f} = -\underline{H}_f = -H_f = -\overline{H}_f = \underline{H}_{-f}$. Somit kann ohne Beschränkung der Allgemeinheit $\lambda > 0$ angenommen werden. Dann aber folgt die Behauptung aus (7.5) und (7.7). Daß weiter jede der Linearformen $f \to H_f(x)$ auf $\mathscr{R}$ positiv ist, ergibt sich aus (7.6), wonach $f \geqq 0$ die Relation $\overline{H}_f \geqq 0$ nach sich zieht.

Zum Beweis der Abgeschlossenheit von $\mathfrak{R}$ sei $(f_n)_{n=1,2,\ldots}$ eine gegen eine reelle Funktion $f$ auf $X^*$ gleichmäßig konvergente Folge von Funktionen $f_n \in \mathfrak{R}$ und $\varepsilon > 0$ eine reelle Zahl. Dann gilt

$$f_n - \varepsilon \leqq f \leqq f_n + \varepsilon$$

und somit

$$H_{f_n} - \varepsilon \overline{H}_1 \leqq \underline{H}_f \leqq \overline{H}_f \leqq H_{f_n} + \varepsilon \overline{H}_1$$

für schließlich alle $n$. Aus der Annahme $P$ folgt (wie im Beweis von Satz 14), daß $\overline{H}_1$ überall endlich ist. Da ferner $\varepsilon > 0$ beliebig gewählt war, erhält man somit $\overline{H}_f(x) = \underline{H}_f(x)$ für alle $x \in X$ und die Endlichkeit von $\overline{H}_f$. Also ist $f$ resolutiv.

**7.5.** *Vernachlässigbare Randmengen.* — Das Lemma 4 läßt sich wesentlich verallgemeinern. Hierfür benötigen wir eine weitere Begriffsbildung:

**Definition 6.** *Eine Menge $A \subset X^*$ von Randpunkten heiße vernachlässigbar, wenn für die charakteristische Funktion $\chi_A$ von $A$ bezüglich $X^*$ gilt: $\overline{H}_{\chi_A} = 0$.*

Hiermit gleichwertig ist offenbar die Forderung, daß $\chi_A$ resolutiv und $H_{\chi_A} = 0$ ist. Für eine vernachlässigbare Menge $A$ ist also insbesondere $\chi_A$ harmonisch-resolutiv. Die leere Menge ist offenbar vernachlässigbar.

Ähnlich wie in der Maßtheorie wollen wir sagen „fast alle Punkte aus $X^*$ besitzen eine gewisse Eigenschaft $E$", wenn die Menge aller Punkte $z \in X^*$, denen diese Eigenschaft nicht zukommt, vernachlässigbar ist. Damit ist dann auch die Bedeutung von „fast überall" klar.

**Lemma 5.** (a) *Jede Teilmenge einer vernachlässigbaren Menge ist vernachlässigbar.*

(b) *Die Vereinigungsmenge endlich oder abzählbar unendlich vieler vernachlässigbarer Mengen ist vernachlässigbar.*

(c) *Sind für eine numerische Funktion $f$ auf $X^*$ die Funktionen $\overline{H}_f$ und $\underline{H}_f$ überall endlich, so ist $|f(z)| < +\infty$ für fast alle $z \in X^*$.*

(d) *Sind zwei numerische Funktionen $f_1$ und $f_2$ auf $X^*$ fast überall gleich, so ist $\overline{H}_{f_1} = \overline{H}_{f_2}$ und $\underline{H}_{f_1} = \underline{H}_{f_2}$.*

Beweis. *Zu* (a): Die Behauptung folgt aus der Isotonie der Abbildung $f \to \overline{H}_f$.

*Zu* (b): Zunächst ist die Vereinigung von je zwei (und damit von je endlich vielen) vernachlässigbaren Mengen $A$, $B$ vernachlässigbar. Aus $\chi_{A \cup B} \leqq \chi_A + \chi_B$ folgt nämlich $\overline{H}_{\chi_{A \cup B}} \leqq \overline{H}_{\chi_A} + \overline{H}_{\chi_B} = 0$.

Schließlich ist auch die Vereinigung $A$ jeder Folge $(A_n)_{n=1,2,\ldots}$ vernachlässigbarer Mengen vernachlässigbar. Nach dem soeben Bewiesenen kann nämlich $A_1 \subset A_2 \subset \ldots$ angenommen werden; nach Lemma 3 ist dann aber $\overline{H}_{\chi_A} = \sup H_{\chi_{A_n}} = 0$.

*Zu* (c): Es sei $A = \{z \in X^* : |f(z)| = +\infty\}$. Für jede reelle Zahl $\lambda > 0$ ist dann $\lambda_{\chi_A} \leqq f + (-f)$ sofern $[f + (-f)](z) = +\infty$ für alle $z \in A$ und sonst $[f + (-f)](z) = 0$ gesetzt wird. Aus (7.6)—(7.8) folgt daher: $\lambda \overline{H}_{\chi_A} \leqq \overline{H}_f + \overline{H}_{-f} = \overline{H}_f - \underline{H}_f$ für alle $\lambda > 0$. Da $\overline{H}_f - \underline{H}_f$ nach Annahme überall endlich ist, ergibt sich also $\overline{H}_{\chi_A} \leqq 0$ und damit $\overline{H}_{\chi_A} = 0$.

*Zu* (d): Nach Voraussetzung ist die Menge $A = \{z \in X^*: f_1(z) \neq f_2(z)\}$ vernachlässigbar; es werde $g = \sup_{n \in \mathbf{N}} n\chi_A$ gesetzt. Dann gilt $f_1 \leqq f_2 + g$, wenn man wie soeben $(f_2 + g)(z) = \infty$ für alle $z \in A$ mit $f_2(z) = -\infty$ vereinbart. Man erhält dann wieder $\overline{H}_{f_1} \leqq \overline{H}_{f_2} + \overline{H}_g = \overline{H}_{f_2} + \sup n\, \overline{H}_{\chi_A} = \overline{H}_{f_2}$ bei Beachtung von (7.8) und Lemma 3. Durch Vertauschung der Rollen von $f_1$ und $f_2$ folgt $\overline{H}_{f_1} = \overline{H}_{f_2}$. Der Rest der Behauptung ergibt sich hieraus und aus (7.5).

Die angekündigte Verallgemeinerung des Lemmas 4 ist im folgenden Korollar enthalten:

**Korollar.** *Für je zwei numerische, auf $X^*$ definierte Funktionen $f$ und $g$ gilt:*

(a) *Ist $f$ resolutiv und $f(z) = g(z)$ für fast alle $z \in X^*$, so ist auch $g$ resolutiv und $H_f = H_g$.*

(b) *Sind $f$ und $g$ resolutiv, so ist $(f + g)(z) = f(z) + g(z)$ fast überall definiert. Legt man den Wert von $f + g$ in allen anderen Punkten $z \in X^*$ willkürlich fest, so ist $f + g$ resolutiv und $H_{f+g} = H_f + H_g$.*

(c) *Ist $f$ resolutiv und $\lambda$ eine reelle Zahl, so ist $(\lambda f)(z) = \lambda f(z)$ fast überall definiert. Legt man den Wert von $\lambda f$ in allen anderen Punkten $z \in X^*$ willkürlich fest, so ist $\lambda f$ resolutiv und $H_{\lambda f} = \lambda H_f$.*

Beweis. *Zu* (a): Dies folgt aus Lemma 5, Behauptung (d).

*Zu* (b): Gemäß Lemma 5, Behauptungen (b) und (c) ist $f + g$ fast überall definiert. Der Rest der Behauptung folgt aus (7.8) und (7.5).

*Zu* (c): Wegen (7.5) und (7.7) bereitet nur der Fall $\lambda = 0$ Schwierigkeiten. Nach Lemma (5), (c) ist für $\lambda = 0$ die Funktion $\lambda f$ im Sinne der obigen Vereinbarungen fast überall gleich 0. Nach der Behauptung (a) des Korollars ist dann aber $\lambda f$ resolutiv und $H_{\lambda f} = H_0 = 0 = \lambda H_f$.

Die hier bewiesenen Eigenschaften der vernachlässigbaren Mengen und der resolutiven Randfunktionen legen es nahe, abstrakte Maße $\mu_x (x \in X)$ auf $X^*$ zu konstruieren derart, daß $\mu_x(A) = 0$ gilt für alle $x \in X$ und alle vernachlässigbaren Mengen $A$ und daß weiter $H_f(x) = \int f\, d\mu_x (x \in X)$ gilt für eine möglichst umfangreiche Klasse resolutiver Randfunktionen. Dies kann ähnlich wie bei Brelot [9] im Rahmen der dortigen Axiomatik geschehen. Wir beschränken uns hier auf den Fall, daß die Maße $\mu_x$ sogar Radonsche Maße sind. Hiervon wird der nächste Paragraph handeln.

**7.6.** *Verhalten von $\overline{H}_f$ bei Restriktion.* — Wir bemerken zunächst, daß sich die bislang an das Garbendatum $W \to \mathscr{H}_W (W \in \mathfrak{U})$ gestellten Forderungen auf dasjenige Garbendatum vererben, welches man hieraus durch Restriktion auf eine Menge $U \in \mathfrak{U}$ erhält. Der kompakte Raum $\overline{U}$ (= abgeschlossene Hülle von $U$ in $Y$) leistet dann bezüglich $U$ dasselbe wie der Raum $Y$ bezüglich $X$. Vollzieht man diesen Übergang von $X$ zu $U$ und von $Y$ zu $\overline{U}$, so erhält man für jede numerische Funktion $f$ auf dem Rand $U^*$ von $U$ in $Y$ (oder, was dasselbe bedeutet, von $U$ in $\overline{U}$) entsprechend definierte Ober- und Unterfunktionen sowie Einhüllenden $\overline{H}_f = \overline{H}_f^U$ und $\underline{H}_f = \underline{H}_f^U$.

Der folgende Satz behauptet ein Analogen zur Eigenschaft (1.2) der harmonischen Funktionen.

**Satz 15.** *Es sei $F$ eine auf $X^*$ definierte numerische Funktion, für welche $\overline{H}_F = \overline{H}_F^X$ harmonisch in $X$ ist. Für eine beliebige Menge $U \in \mathfrak{U}$ bezeichne $f$ die Restriktion auf $U^*$ derjenigen Funktion, welche auf $X$ gleich $\overline{H}_F$ und auf $X^*$ gleich $F$ ist. Dann ist $\overline{H}_f = \overline{H}_f^U$ die Restriktion von $\overline{H}_F$ auf $U$.*

Beweis. Ohne die Voraussetzung der Harmonizität von $\overline{H}_F$ gilt zunächst:

$$\overline{H}_f \leqq \operatorname{Rest}_U \overline{H}_F .$$

Hierfür genügt es offenbar zu zeigen, daß die Restriktion $v$ auf $U$ jeder zu $F$ gehörigen Oberfunktion $V$ eine zu $f$ gehörige Oberfunktion ist. Nach (1.18) ist $v$ hyperharmonisch in $U$; nachzuweisen sind also nur die den Aussagen (7.1) und (7.2) entsprechenden Eigenschaften. Für jeden Punkt $z \in U^*$ sind zwei Fälle zu unterscheiden. Entweder liegt $z$ in $X^*$. Dann ist $\liminf_{x\to z, z\in U} v(x) \geqq \liminf_{x\to z, x\in X} V(x)$ und die beiden Eigenschaften folgen aus $f(z) = F(z)$ und $V \in \mathcal{O}_F$. Oder aber $z$ liegt in $X$; dann ist $\liminf_{x\to z, z\in U} v(x) \geqq \liminf_{x\to z, x\in X} V(x) = V(z)$ und die beiden Eigenschaften folgen aus $V(z) > -\infty$ und $V(z) \geqq \overline{H}_F(z) = f(z)$.

Zum Beweis der dualen Ungleichung

$$\operatorname{Rest}_U \overline{H}_F \leqq \overline{H}_f$$

werde $\overline{H}_F$ harmonisch in $X$ vorausgesetzt. Dann ist für jede zu $f$ gehörige Oberfunktion $v$ die wie folgt definierte Funktion $V$ hyperharmonisch in $X$:

$$V(x) = \begin{cases} \inf(v(x), \overline{H}_F(x)), & x \in U; \\ \overline{H}_F(x), & x \in X \cap \complement U . \end{cases}$$

Zunächst ergibt sich unmittelbar die Ungleichung $V(x) > -\infty$ für alle $x \in X$. Weiter ist $V$ in allen Punkten $x_0 \in X$ nach unten halbstetig; zu zeigen ist dabei nur etwas in Punkten $x_0 \in U^* \cap X$. Es sei $\alpha$ eine reelle Zahl mit $\alpha < V(x_0) = \overline{H}_F(x_0)$. Da $\overline{H}_F$ stetig ist, gibt es eine Umgebung $Q_1$ von $x_0$ in $X$ mit $\overline{H}_F(x) \geqq \alpha$ für alle $x \in Q_1$. Da $v$ eine zu $f$ gehörige Oberfunktion ist, gilt weiter $\liminf_{x\to x_0, x\in U} v(x) \geqq f(x_0) = \overline{H}_F(x_0) > \alpha$. Also gibt es eine Umgebung $Q_2$ von $x_0$ in $X$ mit $v(x) \geqq \alpha$ für alle $x \in Q_2 \cap U$. Hieraus folgt $V(x) \geqq \alpha$ für alle $x \in Q_1 \cap Q_2$; also ist $V$ in $x_0$ nach unten halbstetig. Schließlich ist $V$ hyperharmonisch in $X$. Dies folgt mit Hilfe von Satz 5 aus der Bemerkung, daß für jede reguläre Umgebung $T$ eines Punktes $x_0 \in U^* \cap X$ gilt:

$$\int^* V\, d\mu_{x_0}^T \leqq \int^* \overline{H}_F\, d\mu_{x_0}^T = \overline{H}_F(x_0) = V(x_0) .$$

Nunmehr betrachten wir neben $v$ und neben dem zugehörigen $V$ eine beliebige zu $F$ gehörige Oberfunktion $W$ und setzen $W_1 = W + V - \overline{H}_F$. Dann ist jedenfalls nach dem Bisherigen $W_1$ in $X$ hyperharmonisch. $W_1$ ist aber sogar eine zu $F$ gehörige Oberfunktion. Nachzuweisen sind also die Bedingungen (7.1) und (7.2) für jeden Punkt $z \in X^*$. Diese sind sofort zu sehen, wenn $z$ nicht in $\overline{U}$ liegt. Dann ist nämlich $\complement_Y \overline{U}$ eine Umgebung von $z$ in $Y$ derart, daß im Durchschnitt dieser Umgebung mit $X$ die Funktionen $V$ und $\overline{H}_F$ übereinstimmen. Dies hat aber $\liminf_{x\to z,\, x\in X} W_1(x) = \liminf_{x\to z,\, x\in X} W(x)$ und damit die Bedingungen (7.1) und (7.2) zur Folge. Also kann angenommen werden, daß $z$ in $X^* \cap U^*$ liegt. Wir bezeichnen mit $A$ die Menge aller $x \in U$ mit $V(x) = v(x)$ und mit $B$ die Menge aller $x \in X$ mit $V(x) = \overline{H}_F(x)$; ferner bezeichne $\mathfrak{F}$ den Umgebungsfilter von $z$ in $Y$. Wegen $A \cup B = X$ und $W \geqq \overline{H}_F$ ergibt sich dann[13]):

$$\begin{aligned}
\liminf_{x\to z,\, x\in X} W_1(x) &= \sup_{T\in\mathfrak{F}} \inf W_1((T\cap A)\cup(T\cap B))\\
&= \sup_{T\in\mathfrak{F}} \inf(\inf W_1(T\cap A), \inf W_1(T\cap B))\\
&\geqq \sup_{T\in\mathfrak{F}} \inf(\inf v(T\cap A), \inf W(T\cap B))\\
&\geqq \sup_{T\in\mathfrak{F}} \inf(\inf v(T\cap U), \inf W(T\cap X))\\
&= \inf\left(\liminf_{x\to z,\, x\in U} v(x), \liminf_{x\to z,\, x\in X} W(x)\right).
\end{aligned}$$

Da $v$ eine zu $f$ und $W$ eine zu $F$ gehörige Oberfunktion sowie $f(z) = F(z)$ ist, genügt dann also $W_1$ auch in Punkten $z \in U^* \cap X^*$ den Bedingungen (7.1) und (7.2). Somit ist $W_1$ eine zu $F$ gehörige Oberfunktion, und es gilt: $W_1 = W + V - \overline{H}_F \geqq \overline{H}_F$. Da bei gegebenem $x \in X$ die Differenz $W(x) - \overline{H}_F(x)$ durch geeignete Wahl der Oberfunktion $W$ beliebig klein gemacht werden kann, folgt weiter $\overline{H}_F \leqq V$.

Wir haben also zu jeder zu $f$ gehörigen Oberfunktion $v$ eine Funktion $V$ konstruiert mit $\overline{H}_F \leqq V$. Da offenbar $\mathrm{Rest}_U V \leqq v$ ist, folgt hieraus die gewünschte Ungleichung $\mathrm{Rest}_U \overline{H}_F \leqq \overline{H}_f$.

Unter Beibehaltung der Bezeichnungen aus Satz 15 ergibt sich noch:

**Korollar.** *Ist die Funktion $F$ harmonisch-resolutiv, so ist es auch die Funktion $f$. Ferner ist $H_f$ gleich der Restriktion von $H_F$ auf $U$.*

Beweis. Nach Voraussetzung ist $\overline{H}_F = \underline{H}_F = -\overline{H}_{-F}$. Folglich ist $-f$ die Restriktion auf $U^*$ derjenigen Funktion, welche auf $X$ gleich $\overline{H}_{-F}$ und auf $X^*$ gleich $-F$ ist. Da weiter $\overline{H}_F = H_F$ in $X$ harmonisch ist, gilt nach Satz 15: $\overline{H}_f = \mathrm{Rest}_U H_F$ und $\overline{H}_{-f} = \mathrm{Rest}_U(-H_F)$. Hieraus folgt die Behauptung.

## *§ 8. Resolutive Ränder*

**8.1.** *Resolutivität des Randes.* — Auch in diesem Paragraphen stellen wir an den Raum $X$ die *Zusatzforderung P*. $Y$ sei wieder ein kompakter Raum, der $X$ als dichten Unterraum enthält.

**Definition 7.** *Der Rand $X^*$ von $X$ in $Y$ heiße resolutiv, wenn alle stetigen, reellen Randfunktionen $f \in \mathscr{C}(X^*)$ resolutiv sind.*

Beispiele für reguläre Ränder werden wir in den beiden folgenden Paragraphen besprechen.

Von nun an setzen wir weiter voraus, daß $Y$ der Forderung genügt:

**(R)** *Der Rand $X^*$ von $X$ in $Y$ ist resolutiv.*

---

[13]) In der folgenden Abschätzung muß beachtet werden, daß $\inf\emptyset = +\infty$ ist.

Diese liefert uns ein wichtiges Werkzeug zur weiteren Behandlung des Dirichletschen Problems. Wegen $\mathscr{C}(X^*) \subset \mathscr{R}$ folgt nämlich aus dem Lemma 4, daß die für jeden Punkt $x \in X$ definierte Abbildung $f \to H_f(x)$ von $\mathscr{C}(X^*)$ in $\mathbf{R}$ eine positive Linearform auf $\mathscr{C}(X^*)$, also ein positives (Radonsches) Maß auf dem kompakten Raum $X^*$ ist. Wir bezeichnen dieses Maß mit $\mu_x$ oder genauer mit $\mu_x^X$ und nennen es das zu $x \in X$ (und $Y$) gehörige *harmonische Maß*. Definitionsgemäß gilt somit für alle $f \in \mathscr{C}(X^*)$

$$H_f(x) = \int f \, d\mu_x \qquad (x \in X) . \tag{8.1}$$

8.2. *Integrierbarkeit resolutiver Randfunktionen.* — Die Bedeutung der harmonischen Maße $\mu_x$ für das Studium der resolutiven Randfunktionen erhellt bereits aus folgendem Satz:

**Satz 16.** Für jede numerische Funktion $f$ auf $X^*$ gilt:

$$\underline{H}_f(x) \leqq \int_* f \, d\mu_x \leqq \int^* f \, d\mu_x \leqq \overline{H}_f(x) \qquad (x \in X) . \tag{8.2}$$

Beweis. Wir zeigen zunächst: Es sei $u$ eine in $X$ hyperharmonische Funktion und

$$\psi(z) = \liminf_{x \to z, x \in X} u(x) > -\infty \quad \text{für alle } z \in X^* .$$

Dann gilt für die auf $X^*$ definierte Funktion $\psi$:

$$\int^* \psi \, d\mu_x \leqq u(x) \qquad (x \in X) . \tag{8.3}$$

Für jede Funktion $\varphi \in \mathscr{C}(X^*)$ mit $\varphi \leqq \psi$ ist nämlich $u \in \mathfrak{O}_\varphi$, also $H_\varphi = \overline{H}_\varphi \leqq u$ und somit $H_\varphi(x) = \int \varphi \, d\mu_x \leqq u(x)$ für alle $x \in X$. Nach Definition ist $\psi$ nach unten halbstetig und auf dem kompakten Raum $X^*$ nach unten beschränkt. Folglich ist

$$\int^* \psi \, d\mu_x = \sup_{\substack{\varphi \leqq \psi \\ \varphi \in \mathscr{C}(X^*)}} \int \varphi \, d\mu_x \leqq u(x)$$

für alle $x \in X$; dies ist das gewünschte Resultat.

Nach dieser Vorbereitung betrachten wir eine beliebige Oberfunktion $u \in \mathfrak{O}_f$ der numerischen Funktion $f$. Wie soeben ordnen wir $u$ die Randfunktion $\psi$ zu; nach (7.2) ist die Bedingung $\psi(z) > -\infty$ für alle $z \in X^*$ erfüllt. Also besteht die Relation (8.3). Nach (7.1) ist weiter $\psi \geqq f$ und somit $\int^* f \, d\mu_x \leqq \int^* \psi \, d\mu_x$, also

$$\int^* f \, d\mu_x \leqq u(x)$$

für alle $x \in X$ und $u \in \mathfrak{O}_f$. Dann aber folgt nach Definition von $\overline{H}_f$:

$$\int^* f \, d\mu_x \leqq \overline{H}_f(x) \qquad (x \in X) .$$

Wendet man dieses Resultat auf die Funktion $-f$ an, so erhält man den Rest der Behauptung.

Hieraus folgt nun sofort:

**Korollar.** *Jede resolutive numerische Funktion $f$ auf $X^*$ ist $\mu_x$-integrierbar für alle $x \in X$. Die Funktion $H_f$ läßt sich darstellen in der Form*

(8.4) $$H_f(x) = \int f \, d\mu_x \qquad (x \in X).$$

*Für jede vernachlässigbare Menge $A \subset X^*$ ist $\mu_x(A) = 0$ für alle $x \in X$.*

Beweis. Man hat nur zu beachten, daß $f$ genau dann $\mu_x$-integrierbar ist, wenn $\int^* f \, d\mu_x = \int_* f \, d\mu_x$ und $\left|\int^* f \, d\mu_x\right| < \infty$ ist. Für $f = \chi_A$ ergibt sich die Aussage über die vernachlässigbaren Mengen $A$.

**8.3.** *Integration nach den harmonischen Maßen.* — Das folgende Resultat zeigt, daß die Maße $\mu_x$ analoge Eigenschaften besitzen wie die den regulären Mengen $V$ zugeordneten harmonischen Maße $\mu_x^V$. Man vergleiche hierzu die Sätze 8 und 9 sowie deren Beweis.

**Satz 17.** *Es sei $f$ eine auf $X^*$ definierte, numerische Funktion. In jedem der beiden folgenden Fälle ist dann die Funktion*

(8.5) $$x \to \int^* f \, d\mu_x \qquad (x \in X).$$

*harmonisch in $X$:*

(a) *$f$ ist beschränkt; es gilt das Konvergenzaxiom $K_1$.*

(b) *$\int^* f \, d\mu_x$ ist endlich für alle $x \in X$; es gilt das Konvergenzaxiom $K_2$.*

Beweis. *Zu* (a): Zunächst werde die beschränkte Funktion $f$ als nach unten halbstetig vorausgesetzt. Dann ist die Menge $\mathcal{F}^*$ aller $\varphi \in \mathcal{C}(X)^*$ mit $\varphi \leqq f$ und damit auch die Menge $\mathcal{F}$ aller Funktionen $H_\varphi$ mit $\varphi \in \mathcal{F}^*$ nichtleer und aufsteigend filtrierend. Es gilt $H_\varphi \leqq \overline{H}_f$ für alle $\varphi \in \mathcal{F}^*$; nach Satz 13 sind $\overline{H}_f$ und alle $H_\varphi \in \mathcal{F}$ harmonische Funktionen in $X$. Daher ist die Menge $\mathcal{F} \subset \mathcal{H}_X$ außerdem lokal-gleichmäßig nach oben beschränkt. Nach ($K_1''$) ist also die obere Einhüllende $\sup \mathcal{F}$ harmonisch in $X$. Hieraus folgt die Behauptung, da für alle $x \in X$ gilt:

$$\sup_{\varphi \in \mathcal{F}^*} H_\varphi(x) = \sup_{\varphi \in \mathcal{F}^*} \int \varphi \, d\mu_x = \int^* f \, d\mu_x .$$

Nunmehr sei $f$ eine beliebige beschränkte Funktion auf $X$. Dann ist die Menge $\mathcal{G}^*$ aller nach unten halbstetigen, beschränkten Funktionen $\psi$ auf $X^*$ mit $\psi \geqq f$ absteigend filtrierend. Diese Eigenschaft überträgt sich auf die (nicht-leere) Menge $\mathcal{G}$ aller nach dem ersten Teil des Beweises in $X$ harmonischen Funktionen $u_\psi$ mit $\psi \in \mathcal{G}^*: u_\psi(x) = \int^* \psi \, d\mu_x (x \in X)$. Aus $f \leqq \psi$ und Satz 16 folgt $\underline{H}_f(x) \leqq \underline{H}_\psi(x) \leqq u(x)$ für alle $\psi \in \mathcal{G}^*$ und $x \in X$; nach Satz 14 ist $\underline{H}_f$ eine in $X$ harmonische Funktion. Daher ist die Menge $\mathcal{G} \subset \mathcal{H}_X$ außerdem lokal-gleichmäßig nach unten beschränkt. Nach ($K_1''$) ist also die untere Einhüllende $\inf \mathcal{G}$ harmonisch in $X$. Hieraus folgt die Behauptung, da für alle $x \in X$ gilt:

$$\inf_{\psi \in \mathcal{G}^*} u_\psi(x) = \inf_{\psi \in \mathcal{G}^*} \int^* \psi \, d\mu_x = \int^* f \, d\mu_x .$$

*Zu* (b): Der Beweis dieser Behauptung verläuft völlig analog zu dem Schritt „$(K_2) \Rightarrow (K_2')$“ im Beweis von Satz 9 und kann daher dem Leser überlassen werden.

**8.4.** *Resolutivitätskriterien.* — Die Analogie der Sätze 14 und 17 sowie die Ungleichungen (8.5) legen die Frage nahe, wenn die Funktion $\overline{H}_f$ mit der Funktion (8.5) übereinstimmt. Wir beweisen hierfür folgendes

**Lemma 6.** *Gilt für jede nach unten beschränkte und nach unten halbstetige Funktion $f$ auf $X^*$*

$$\overline{H}_f(x) = \int^* f\, d\mu_x \qquad (x \in X),$$

*so gilt für jede numerische Funktion $f$ auf $X^*$:*

$$\underline{H}_f(x) = \int_* f\, d\mu_x \text{ und } \overline{H}_f(x) = \int^* f\, d\mu_x \qquad (x \in X). \tag{8.6}$$

*Fordert man zusätzlich das Konvergenzaxiom $K_1$, so genügt es, die Ausgangsforderung nur an die beschränkten, nach unten halbstetigen Randfunktionen $f$ zu stellen.*

Beweis. Wegen (8.2) und (7.5) folgt (8.6) für jede numerische Randfunktion $f$ bereits aus der jetzt zu beweisenden Ungleichung: $\overline{H}_f(x) \leqq \int^* f\, d\mu_x (x \in X)$. Bezeichnen wir mit $\mathcal{J}_f$ die Menge aller nach unten beschränkten und nach unten halbstetigen Funktionen $\psi$ auf $X^*$ mit $\psi \geqq f$, so ist $\int^* f\, d\mu_x = \inf\limits_{\psi \in \mathcal{J}_f} \int^* \psi d\mu_x$, also nach Voraussetzung $\int^* f du_x = \inf\limits_{\psi \in \mathcal{J}_f} \overline{H}_\psi(x)$. Wegen $\psi \geqq f$ ist $\overline{H}_\psi \geqq \overline{H}_f$ für alle $\psi \in \mathcal{J}_f$ und somit $\overline{H}_f(x) \leqq \int^* f\, d\mu_x$ für alle $x \in X$. Dies ist die zu beweisende Ungleichung.

Nunmehr werde das Axiom $K_1$ vorausgesetzt. $f$ sei eine nach unten beschränkte, nach unten halbstetige Randfunktion. Jede der Funktionen $f_n = \inf(n, f)$, $n = 1, 2, \ldots$, ist dann beschränkt und nach unten halbstetig. Nach Annahme gilt somit $\overline{H}_{f_n}(x) = \int^* f_n d\mu_x$. Hieraus, aus Satz 14 und dem Lemma 3 folgt dann $\overline{H}_f(x) = \int^* f\, d\mu_x$ für alle $x \in X$. Diese Überlegung beweist den Rest der Behauptung.

Man erhält nun sofort:

**Korollar.** *Unter den Voraussetzungen des Lemmas 6 ist jede numerische Funktion $f$ auf $X^*$ resolutiv, welche für jedes $x \in X$ $\mu_x$-integrierbar ist. Besteht für eine Menge $A \subset X$ die Gleichheit $\mu_x(A) = 0$ für jedes $x \in X$, so ist $A$ vernachlässigbar.*

Offenbar wird in diesem Korollar die Aussage des Korollars zu Satz 16 umgekehrt. Wir stoßen daher hier auf die Verallgemeinerung des im Fall der klassischen harmonischen Funktionen von BRELOT [10] entdeckten Resolutivitätssatzes.

Die Voraussetzungen des Lemmas 6 sind insbesondere erfüllt, wenn $(K_1)$ gilt und der Rand $X^*$ eine abzählbare Basis besitzt. Diese letzte Forderung ist bekanntlich äquivalent mit der Metrisierbarkeit von $X^*$.

**Satz 18.** *Es gelte das Konvergenzaxiom* $K_1$; *der Rand* $X^*$ *besitze eine abzählbare Basis. Dann sind die Voraussetzungen des Lemmas* 6 *erfüllt und es gilt* (8.6) *für jede numerische Funktion* $f$ *auf* $X^*$.

Beweis. Die Randfunktion $f$ sei (nach unten) beschränkt und nach unten halbstetig. Da $X^*$ eine abzählbare Basis besitzt, existiert zu $f$ eine isotone Folge von Funktionen $\varphi_n \in \mathscr{C}(X^*)$ mit $f = \sup \varphi_n$ (vgl. [6], p. 38). Jede der Funktionen $H_{\varphi_n}$ ist nach Satz 14 harmonisch in $X$; nach Lemma 3 ist weiter $\overline{H}_f = \sup H_{\varphi_n}$. Also ist $\overline{H}_f(x) = \int^* f\, d\mu_x$ für alle $x \in X$. Dies ist aber gerade die Voraussetzung zu Lemma 6.

**Korollar** (Resolutivitätssatz). *Es gelte das Konvergenzaxiom* $K_1$; *der Rand* $X^*$ *besitze eine abzählbare Basis. Notwendig und hinreichend für die Resolutivität einer numerischen Funktion* $f$ *auf* $X^*$ *ist dann die* $\mu_x$*-Integrierbarkeit von* $f$ *für alle* $x \in X$.

## *§ 9. Spezialfall der regulären Mengen*

**9.1.** *Dirichletsches Problem für reguläre Mengen.* — In diesem Paragraphen soll mit $V$ stets eine reguläre Teilmenge von $X$ bezeichnet werden. Im Axiom T, das in diesem Kapitel generell vorausgesetzt wird, wird die Existenz einer in $X$ überall positiven, harmonischen Funktion $h \in \mathscr{H}_X$ gefordert. Da $V$ relativ kompakt ist, besitzt die in $V$ harmonische Funktion $h_0 = \mathrm{Rest}_V h$ eine positive untere Schranke. Bildet man also die Restriktion auf $V$ des Garbendatums $U \to \mathscr{H}_U$, so genügt der neue Grundraum $V$ der in § 7 vorausgesetzten Annahme $P$. Ferner bietet sich die abgeschlossene Hülle $\overline{V}$ von $V$ in $X$ als Kompaktifizierung $Y$ des neuen Grundraumes $V$ an. Also steht die in § 7 entwickelte Theorie voll zur Verfügung. Im folgenden soll das Dirichletsche Problem in diesem Spezialfall weiter verfolgt werden.

Aus der Definition der regulären Menge und der zu Beginn der Nr. 7.4 gemachten Bemerkung folgt, daß der Rand $V^*$ von $V$ in $X$ resolutiv ist, also sogar die Annahme $R$ des § 8 erfüllt ist. Es folgt weiter, daß für jede Funktion $f \in \mathscr{C}(V^*)$ die in Definition 1 auftretende harmonische Funktion $H_f$, welche das Dirichletsche Problem im klassischen Sinn für $f$ löst, auch die verallgemeinerte Lösung des Dirichletschen Problems für $f$ ist. Insbesondere sind also hier alle Funktionen aus $\mathscr{C}(V^*)$ harmonisch resolutiv. Ein Vergleich von (8.1) mit (1.7) zeigt, daß die harmonischen Maße $\mu_x^V (x \in X)$ des § 8 mit denen des § 1 übereinstimmen.

**9.2.** *Resolutivitätssatz.* — Wir zeigen zunächst, daß das Lemma 6 der allgemeinen Theorie anwendbar ist.

**Lemma 7.** *Es werde das Axiom* $K_1$ *vorausgesetzt;* $V$ *sei eine reguläre Menge. Dann gilt füt jede beschränkte, nach unten halbstetige Funktion* $f$ *auf* $V^*$: $\overline{H}_f(x) = \int^* f\, d\mu_x^V$ *für alle* $x \in V$.

Beweis. Die Funktion $f$ ist die obere Einhüllende der Menge $\mathscr{F}^*$ aller $\varphi \in \mathscr{C}(V^*)$ mit $\varphi \leq f$. Die Menge $\mathscr{F}$ aller Funktionen $H_\varphi$ mit $\varphi \in \mathscr{F}^*$ ist dann nicht leer, aufsteigend filtrierend und lokal-gleichmäßig nach oben beschränkt. Letztere Eigenschaft folgt aus Satz 14, wonach $\overline{H}_f$ eine harmonische Majorante von $\mathscr{F}$ ist. Die obere Einhüllende $u = \sup \mathscr{F}$ ist nach ($K_1'''$) harmonisch in $V$; ferner gilt $u(x) = \int^{*} f \, d\mu_x^V$ für alle $x \in V$. Aus $H_\varphi \leqq u$ und der Regularität von $V$ folgt

$$\varphi(z) = \lim_{x \to z, x \in V} H_\varphi(x) \leqq \liminf_{x \to z, x \in V} u(x)$$

für alle $\varphi \in \mathscr{F}^*$ und hieraus $f(z) \leqq \liminf_{x \to z, x \in V} u(x)$ für alle $z \in V^*$. Somit ist $u$ eine zu $f$ gehörige Oberfunktion, also $\overline{H}_f \leqq u$. Die hierzu duale Ungleichung behauptet aber Satz 16 allgemein.

Das Lemma 6 und die übrigen Resultate des letzten Paragraphen liefern daher jetzt sofort das folgende umfassendere Ergebnis:

**Satz 19** (Resolutivitätssatz für reguläre Mengen). *Es gelte das Axiom* $K_1$; *$V$ sei eine reguläre Menge. Dann gilt für jede numerische Funktion $f$ auf $V^*$:*

$$\underline{H}_f(X) = \int_{*} f \, d\mu_x^V \text{ und } \overline{H}_f(x) = \int^{*} f \, d\mu_x^V \qquad (x \in V) . \tag{9.1}$$

*$f$ ist genau dann resolutiv, wenn $f$ $\mu_x$-integrierbar ist für alle $x \in V$.*

**9.3.** *Randverhalten von $\overline{H}_f$.* — Nach Definition der regulären Mengen $V$ gilt für alle Funktionen $f \in \mathscr{C}(V^*)$ und alle $z \in V^*$

$$\lim_{x \to z, x \in V} H_f(x) = f(z) .$$

Wir behaupten folgende Verallgemeinerung:

**Satz 20.** *Es sei $f$ eine nach oben beschränkte, numerische Funktion auf dem Rande $V^*$ einer regulären Menge $V$. Dann gilt für alle $z_0 \in V^*$:*

$$\limsup_{x \to z_0, x \in V} \overline{H}_f(x) \leqq \limsup_{z \to z_0, z \in V^*} f(z) . \tag{9.2}$$

Beweis. Da $f$ nach oben beschränkt ist, gilt $\lambda = \limsup_{z \to z_0} f(z) < +\infty$. Zu jeder reellen Zahl $\lambda'$ mit $\lambda < \lambda'$ gibt es eine offene Umgebung $T$ von $z_0$ in $V^*$ mit $f(z) \leqq \lambda'$ für alle $z \in T$. Berücksichtigt man nochmals die einseitige Beschränktheit von $f$, so existiert wegen der vollständigen Regularität kompakter Räume eine Funktion $\varphi \in \mathscr{C}(V^*)$ mit $\varphi(z_0) = \lambda'$, $\varphi \geqq \lambda'$ und $\varphi(z) \geqq f(z)$ für alle $z \in V^* \cap \complement T$. Dann aber ist sogar $\varphi \geqq f$ und damit $\overline{H}_\varphi \geqq \overline{H}_f$. Hieraus folgt

$$\limsup_{x \to z_0, x \in V} \overline{H}_f(x) \leqq \lim_{x \to z_0} H_\varphi(x) = \varphi(x_0) = \lambda' .$$

Da $\lambda'$ eine beliebige reelle Zahl mit $\lambda' > \lambda$ war, ist damit der Satz bewiesen.

**Korollar.** *Es sei $V$ eine reguläre Menge und $f$ eine beschränkte reelle Funktion auf $V^*$. Für jeden Punkt $z_0 \in V^*$, in welchem $f$ stetig ist, gilt dann:*

$$\lim_{x \to z_0} \overline{H}_f(x) = f(z_0) .$$

Beweis. Nach Satz 20 ist

$$\liminf_{z \to z_0} f(z) \leqq \liminf_{x \to z_0} \underline{H}_f(x) \text{ und } \limsup_{x \to z_0} \overline{H}_f(x) \leqq \limsup_{z \to z_0} f(z) .$$

Hieraus und aus $\underline{H}_f \leqq \overline{H}_f$ folgt die Behauptung.

Diese Aussagen über das Randverhalten von $\overline{H}_f$ gestatten jetzt den Beweis des folgenden Satzes:

**Satz 21.** *Es sei $u$ eine in einer Menge $U \in \mathfrak{U}$ definierte, hyperharmonische Funktion, welche in einer offenen Menge $T \subset U$ stetig ist; ferner sei $V$ eine in $U$ reguläre Menge. Dann ist in jedem der beiden folgenden Fälle die Funktion $u_V$ stetig in $T \cup V$:*

(a) *Es gilt das Axiom* $\mathrm{K}_1$; *ferner ist $u$ lokal-beschränkt.*

(b) *Es gilt das Axiom* $\mathrm{K}_2$; *ferner ist $u$ reellwertig.*

Beweis. Aus jeder der Annahmen (a) und (b) folgt zunächst die Stetigkeit von $u_V$ in $V$. Ist nämlich $u$ lokal-beschränkt, so ist $u$ auf der kompakten Menge $V^*$ beschränkt und dann $u_V$ nach ($\mathrm{K}_1'$) stetig in $V$. Ist $u$ reellwertig, so gilt $\int^* u\,d\mu_x^V \leqq u(x) < +\infty$; da $u$ auf jeder kompakten Menge nach unten beschränkt ist, gilt dann sogar $\left|\int^* u\,d\mu_x^V\right| < +\infty$ für alle $x \in V$. In diesem Fall ist $u_V$ nach ($\mathrm{K}_2'$) stetig in $V$. Nun ist weiter $u_V(x) = u(x)$ für alle $x \in U \cap \complement V$. Daher ist in jedem der beiden Fälle $u_V$ stetig in der offenen Menge $V \cup (T \cap \complement V^*)$. Zu beweisen ist daher nur noch die Stetigkeit von $u_V$ in jedem Punkt $z_0 \in T \cap V^*$. Nach Voraussetzung ist die Restriktion $u^*$ von $u$ auf $V^*$ stetig in $z_0$; sie ist ferner nach unten beschränkt. Nach Satz 20 gilt also

$$\liminf_{x \to z_0,\, x \in V} \underline{H}_{u^*}(x) \geqq \liminf_{z \to z_0,\, z \in V^*} u^*(z) = u(z_0) = u_V(z_0)\,.$$

Außerdem ist $\mathrm{Rest}_V u$ eine zu $u^*$ gehörige Oberfunktion und somit $\overline{H}_{u^*}(x) \leqq u(x)$ für alle $x \in V$. Hieraus folgt bei Beachtung der Stetigkeit von $u$ in $z_0$:

$$\limsup_{x \to z,\, x \in V} \overline{H}_{u^*}(x) \leqq u_V(z_0)\,.$$

Wegen $\underline{H}_{u^*} \leqq \overline{H}_{u^*}$ erhalten wir also:

$$\lim_{\substack{x \to z_0 \\ x \in V}} \overline{H}_{u^*}(x) = u_V(z_0)\,.$$

Da in beiden Fällen das Axiom $\mathrm{K}_1$ gilt, ist nach Satz 19 und (3.2) $u_V(x) = \overline{H}_{u^*}(x)$ für alle $x \in V$ und somit

$$\lim_{\substack{x \to z_0 \\ x \in V}} u_V(x) = u_V(z_0)\,.$$

Ferner folgt aus der Stetigkeit von $u$ in $z_0$:

$$\lim_{\substack{x \to z_0 \\ x \in \complement V}} u_V(x) = \lim_{x \to z_0} u(x) = u(z_0) = u_V(z_0)\,.$$

Beide Gleichungen zusammen ergeben die Stetigkeit von $u_V$ in $z_0$.

Nach Satz 6 ist für jede hyperharmonische Funktion $u \in H_U^*$ und jede in $U$ reguläre Menge $V$ auch die Funktion $u_V$ hyperharmonisch in $U$. Der Prozeß $u \to u_V$ läßt sich daher iterieren: Sind $V_1, \ldots, V_n$ reguläre Mengen in $U$, so ist mit $u$ auch die Funktion

$$u_{V_1,\ldots,V_n} = (\ldots(u_{V_1})_{V_2}\ldots)_{V_n} \tag{9.3}$$

hyperharmonisch in $U$.

**Korollar.** *Es sei $u$ eine in $U \in \mathfrak{U}$ hyperharmonische Funktion; $V_1, \ldots, V_n$ seien in $U$ reguläre Mengen. In jedem der in Satz 21 formulierten Fälle* (a) *und* (b) *ist dann die Funktion $u_{V_1,\ldots,V_n}$ stetig in $V_1 \cup \cdots \cup V_n$. Es ist $u_{V_1,\ldots,V_n} \leqq u$.*

Beweis. Eine hyperharmonische Funktion ist auf ihrem lokalkompakten Definitionsbereich $U \in \mathfrak{U}$ stets nach unten lokal-beschränkt; nach (3.2) und (3.3) gilt ferner $u_{V_1,\ldots,V_n} \leqq u$. Mit $u$ ist daher auch die Funktion $u_{V_1,\ldots,V_n}$ lokal-beschränkt bzw. reellwertig. Nach dieser Vorbemerkung folgt die Behauptung aus Satz 21 durch vollständige Induktion. Der Fall $n = 1$ folgt aus diesem Satz, indem man $T = \emptyset$ setzt. Beim Induktionsschluß von $n$ auf $n + 1$ setzt man $T = V_1 \cup \ldots \cup V_n$ und beachtet die Gleichheit $(u_{V_1,\ldots,V_n})_{V_{n+1}} = u_{V_1,\ldots,V_{n+1}}$.

9.4. *Verschärfung der Trennungsaxiome.* — Nunmehr sind wir in der Lage, mit Hilfe des schwachen Konvergenzaxioms $K_1$ verschärfte Formen der Trennungsaxiome T und T' zu beweisen. Wie dort bezeichne $h$ eine überall positive, in $X$ harmonische Funktion.

**Lemma 8.** *Es gelte das Axiom* $K_1$. *Es sei $U \in \mathfrak{U}$ eine Menge, welche entweder relativ-kompakt oder (als lokal-kompakter Raum) abzählbar im Unendlichen ist. Existiert dann zu Punkten $x, y \in U$ eine in $X$ $h$-hyperharmonische Funktion $v \in {}^h\mathscr{H}_X^*$ mit $v(x) < v(y)$, so gibt es eine in $X$ reellwertige, in $U$ stetige Funktion $w \in {}^h\mathscr{H}_X^*$ mit $w(x) < w(y)$ und $w \leqq v$. Ferner ist* $\inf w(X) = \inf v(X)$.

Beweis. Aus den Bemerkungen der Nr. 1.4 folgt, daß für den Beweis ohne Beschränkung der Allgemeinheit $h = 1$ und damit $\mathscr{H}_W^* = {}^h\mathscr{H}_W^*$ für alle $W \in \mathfrak{U}$ angenommen werden kann. Ferner kann $v$ ohne Beschränkung der Allgemeinheit als nach oben beschränkt vorausgesetzt werden. Man hat nur gegebenenfalls $v$ durch $\inf(\gamma, v)$ zu ersetzen und dabei $v(x) < \gamma < v(y)$ zu wählen.

Wir behandeln zuerst den Fall einer relativ-kompakten Menge $U \in \mathfrak{U}$ und wählen eine reelle Zahl $\alpha$ mit $v(x) < \alpha < v(y)$. Zufolge der Halbstetigkeit von $v$ gibt es dann eine in $U$ reguläre Umgebung $V_1$ von $y$ mit $v(y') \geqq \alpha$ für alle $y' \in \overline{V}_1$. Da $U$ relativ-kompakt ist, gibt es nach Axiom II endlich viele weitere reguläre Mengen $V_2, \ldots, V_n$ derart, daß $y$ in keiner dieser Mengen liegt und $U$ durch $V_1, \ldots, V_n$ überdeckt wird. Die Funktion $w = v_{V_1,V_2,\ldots,V_n}$ ist dann nach dem Korollar zu Satz 21 stetig in $V_1 \cup \cdots \cup V_n$; ferner ist $w \leqq v$ und $w$ hyperharmonisch in $X$. Wegen der einleitend gemachten Beschränktheitsannahme über $v$ ist also mit $v$ auch $w$ reellwertig. Da $y$ nicht in $V_2, \ldots, V_n$ liegt, gilt: $w(y) = v_{V_1}(y) = \int^* v \, d\mu_y^{V_1} \geqq \alpha \int d\mu_y^{V_1} = \alpha$. (Es ist $\int d\mu_y^{V_1} = 1$, da nach Annahme $h = 1$ in $X$ harmonisch ist.) Wegen $w \leqq v$ gilt somit auch $w(x) \leqq v(x) < \alpha \leqq w(y)$. Aus der Konstruktion von $w$ folgt schließlich, daß jede untere Schranke von $v$ auch eine untere Schranke von $w$ ist. Dies zusammen mit $w \leqq v$ ergibt $\inf w(X) = \inf v(X)$. Daher leistet $w$ das Verlangte.

Jetzt werde $U$ als abzählbar im Unendlichen vorausgesetzt. $\alpha$ sei wieder eine reelle Zahl mit $v(x) < \alpha < v(y)$; $V_1$ sei wieder eine in $U$ reguläre Umgebung von $y$ mit $v(y') \geqq \alpha$ für alle $y' \in \overline{V}_1$. Da $U$ abzählbar im Unendlichen ist, existieren relativ-kompakte Mengen $U_n \in \mathfrak{U}$, $n = 1, 2, \ldots$ mit folgenden

Eigenschaften: $U_1 = V_1 \subset \overline{U}_1 \subset U_2 \subset \overline{U}_2 \ldots$ und $\bigcup U_n = U$. Nach Axiom II existieren dann endlich viele in $U_3$ reguläre Mengen $W_1^{(2)}, \ldots, W_{k_2}^{(2)}$, welche die kompakte Menge $\overline{U}_2 \cap \complement U_1$ überdecken und den Punkt $y$ nicht enthalten. Ebenso existieren für jedes $n \geqq 3$ endlich viele in $U_{n+1}$ reguläre Mengen $W_1^{(n)}, \ldots, W_{k_n}^{(n)}$, welche die Menge $\overline{U}_n \cap \complement U_{n-1}$ überdecken und fremd sind zu $U_{n-2}$. In $V_1, W_1^{(2)}, \ldots, W_{k_2}^{(2)}, W_1^{(3)}, \ldots, W_{k_3}^{(3)}, \ldots, W_1^{(n)}, \ldots, W_{k_n}^{(n)}, \ldots$ haben wir dann eine Folge $V_1, V_2, \ldots$ regulärer Mengen mit den folgenden Eigenschaften vor uns: Zu jedem Punkt $z \in X$ existiert eine Menge $V_n$ mit $z \in V_n$ derart, daß schließlich alle Mengen $V_i$ der Folge fremd sind zu $V_n$; nur die Menge $V_1$ enthält den Punkt $y$. Wir setzen nun $w_n = v_{V_1, \ldots, V_n}$, $n = 1, 2, \ldots$ und zeigen, daß die Funktion $w = \inf w_n$ das Verlangte leistet. Nach dem Korollar zu Satz 21 ist jedes $w_n$ eine in $V_1 \cup \cdots \cup V_n$ stetige, in $X$ reelle hyperharmonische Funktion. Ferner ist die Folge $(w_n)$ antiton. Wie soeben bemerkt, gibt es zu einem beliebigen Punkt $z \in X$ eine Menge $V_n$ mit $z \in V_n$ und eine natürliche Zahl $m \geqq n$ derart, daß $V_i \cap V_n = \emptyset$ ist für alle $i > m$. Folglich gilt nach (3.2): $w_i(x') = w_m(x')$ für alle $i > m$ und alle $x' \in V_n$. Wegen der Antitonie von $(w_n)_{n=1,2,\ldots}$ folgt hieraus $w(x') = w_m(x')$ für alle $x' \in V_n$. Wegen $m \geqq n$ ist daher weiter die Restriktion von $w$ auf $V_n$ reellwertig, stetig und hyperharmonisch in $V_n$. Somit ist $w$ reellwertig, stetig und (nach dem Korollar zu Satz 5) hyperharmonisch in $X$. Wie im ersten Teil unseres Beweises zeigt man, daß $w_1(y) \geqq \alpha$ und $w \leqq w_n \leqq v$ gilt. Da nun $V_1$ die einzige Menge der Folge $(V_n)$ ist, welche $y$ enthält, ist $w_n(y) = w_1(y)$ für alle $n$ und somit $w(y) \geqq \alpha > v(x) \geqq w(x)$. Wieder ist jede untere Schranke von $v$ eine untere Schranke eines jeden $w_n$ und damit von $w$. Wegen $w \leqq v$ ist daher $\inf w(X) = \inf v(X)$. Somit leistet $w$ das Verlangte.

Die angekündigten Verschärfungen der Trennungsaxiome lauten nun:

**Satz 22.** *Es gelte das Axiom* $K_1$. *$U$ sei eine entweder relativ-kompakte oder im Unendlichen abzählbare Menge aus $\mathfrak{U}$. Dann trennt bereits die Menge aller stetigen, reellen, in $U$ h-hyperharmonischen Funktionen die Punkte von $U$. Gilt sogar das Trennungsaxiom* T′, *so gibt es zu jedem geordneten Paar $(x, y)$ von Punkten aus $U$ eine stetige, reelle, in $U$ h-hyperharmonische Funktion $w$ mit $w(x) < w(y)$.*

Der Beweis folgt unmittelbar aus dem Lemma 8 und aus (1.18).

Dieser Satz nimmt offenbar eine besonders einfache Form an, wenn der Grundraum $X$ selbst abzählbar im Unendlichen ist.

**9.5.** *Eine Dichtigkeitsaussage.* — Als für das Folgende wichtige Folgerung aus dem Satz 22 beweisen wir noch:

**Satz 23.** *Es gelte das Axiom* $K_1$, *und es sei $U$ eine Menge aus $\mathfrak{U}$, welche entweder relativ kompakt oder (als lokal-kompakter Raum) abzählbar im Unendlichen ist. Dann ist die Menge $\mathscr{S}$ aller in $U$ stetigen, reellen, hyperharmonischen Funktionen total*[14]) *im Raum $\mathscr{C}(U)$ bezüglich der Topologie der kompakten Konvergenz.*

[14]) Dies heißt bekanntlich, daß der von $\mathscr{S}$ erzeugte lineare Unterraum $\mathscr{D}$ von $\mathscr{C}(U)$ dicht in $\mathscr{C}(U)$ liegt (bezüglich der kompakten Konvergenz). Nach dem folgenden Beweis ist $\mathscr{D} = \mathscr{S} - \mathscr{S}$.

Beweis. Aus den Eigenschaften (1.13) und (1.14) folgt, daß der von $\mathcal{S}$ in $\mathcal{C}(U)$ erzeugte lineare Unterraum $\mathcal{D}$ aus allen Funktionen $d = u_1 - u_2$ besteht mit $u_1, u_2 \in \mathcal{S}$. Zu zeigen ist, daß $\mathcal{D}$ in $\mathcal{C}(U)$ dicht liegt. Hierzu betrachten wir für eine in $X$ überall positive, harmonische Funktion $h \in \mathcal{H}_X$ die Menge $\mathcal{D}'$ aller Funktionen $x \to \frac{d(x)}{h(x)}$ $(x \in U)$ mit $d \in \mathcal{D}$. Offenbar ist auch $\mathcal{D}'$ ein linearer Unterraum von $\mathcal{C}(U)$; er besteht aus allen Funktionen $d' = v_1 - v_2$, wobei $v_1$ und $v_2$ in $U$ stetige, reelle, $h$-hyperharmonische Funktionen sind. $\mathcal{D}'$ enthält alle konstanten reellen Funktionen auf $U$. Weiter liegt mit jeder Funktion $d' = v_1 - v_2$ auch die Funktion $|d'|$ in $\mathcal{D}'$. Dies folgt aus

$$\begin{aligned}|d'| &= \sup(d', -d') = \sup(v_1 + v_2 - 2v_2, v_1 + v_2 - 2v_1)\\ &= v_1 + v_2 - 2\inf(v_1, v_2)\end{aligned}$$

bei Beachtung von (1.15). Das Axiom $K_1$ und die speziellen Annahmen über $U$ haben schließlich nach Satz 22 zur Folge, daß $\mathcal{D}'$ die Punkte von $U$ trennt. Somit genügt $\mathcal{D}'$ den Voraussetzungen des Approximationssatzes von M. H. STONE. Nach diesem Satz liegt $\mathcal{D}'$ dicht in $\mathcal{C}(U)$ bezüglich der Topologie der kompakten Konvergenz. Die Abbildung $\Phi: \mathcal{C}(U) \to \mathcal{C}(U)$, welche jeder Funktion $f \in \mathcal{C}(U)$ die Funktion $x \to f(x)\,h(x)$ $(x \in U)$ als Bild zuordnet, ist auf Grund der speziellen Eigenschaften von $h$ eine Homöomorphie von $\mathcal{C}(U)$ auf sich (bezüglich der Topologie der kompakten Konvergenz). Wegen $\Phi(\mathcal{D}') = \mathcal{D}$ liegt daher mit $\mathcal{D}'$ auch $\mathcal{D}$ in $\mathcal{C}(U)$ dicht.

## *§ 10. Spezialfall beliebiger relativ-kompakter offener Mengen*

**10.1.** *Resolutivität des Randes.* — An die Stelle der regulären Menge $V$ des letzten Paragraphen trete jetzt eine beliebige relativ-kompakte Menge $U \in \mathfrak{U}$. Dementsprechend übernimmt jetzt $U$ die Rolle des Grundraumes $X$ im § 7 und $\overline{U}$ die Rolle der dortigen Kompaktifizierung $Y$. Die dort entwickelte allgemeine Theorie steht auch jetzt wieder voll zur Verfügung, da sich ebenso wie in Nr. 9.1 ergibt, daß die Annahme $P$ aus § 7 zufolge der relativen Kompaktheit von $U$ erfüllt ist.

Der folgende Satz zeigt, daß sogar die Annahme $R$ des § 8 erfüllt ist.

**Satz 24.** *Es gelte das Axiom* $K_1$. *Dann ist der Rand* $U^*$ *jeder relativ-kompakten Menge* $U \in \mathfrak{U}$ *resolutiv.*

Beweis[15]). Wir zeigen zunächst, daß jede Funktion $f \in \mathcal{C}(U^*)$, welche zu einer in $U$ hyperharmonischen Funktion $u \in \mathcal{C}(\overline{U})$ fortgesetzt werden kann, resolutiv ist. In der Tat: Die Restriktion $u_0$ von $u$ auf $U$ ist offenbar eine zu $f$ gehörige Oberfunktion, also $\overline{H}_f \leqq u_0$. Aus dieser Ungleichung folgt

$$\limsup_{x \to z,\, x \in U} \overline{H}_f(x) \leqq \lim_{x \to z,\, x \in U} u_0(x) = \lim_{x \to z} u(x) = f(z)$$

für jeden Randpunkt $z \in U^*$. Da nach Satz 14 die Funktion $\overline{H}_f$ in $U$ harmonisch ist, so ist $\overline{H}_f$ nach der vorausgehenden Ungleichung eine zu $f$ gehörige Unter-

[15]) Vgl. hinsichtlich der Beweismethode auch HERVÉ [18].

funktion, also $\overline{H}_f \leqq \underline{H}_f$. Nach Satz 13 ist dann $f$ resolutiv und, wie sich soeben ergeben hat, sogar harmonisch-resolutiv.

Betrachten wir also die Menge $\mathscr{S}$ aller in einer relativ-kompakten, offenen Umgebung $U_1$ von $\overline{U}$ definierten, stetigen, reellen, hyperharmonischen Funktionen, so ist die Restriktion $u^*$ auf $U^*$ jeder Funktion $u \in \mathscr{S}$ resolutiv. Die Menge $\mathscr{S}^*$ aller Funktionen $u^*$ mit $u \in \mathscr{S}$ ist aber nach Satz 23 total in $\mathscr{C}(U^*)$ bezüglich der Topologie der gleichmäßigen Konvergenz. Da aber, wie soeben bemerkt, $\mathscr{S}^*$ aus lauter resolutiven Funktionen besteht, sind dann nach dem Lemma 4 alle Funktionen aus $\mathscr{C}(U^*)$ resolutiv. Das aber behauptet gerade unser Satz.

Es sei noch erwähnt, daß bei Doob [14], p. 78, die Frage nach der Gültigkeit des obigen Satzes in der Theorie der Wärmeleitungsgleichung als offenes Problem bezeichnet wird.

**10.2.** *Resolutivitätssatz.* — Nach dem soeben bewiesenen Satz stehen nun auch die Resultate des § 8 der allgemeinen Theorie zur Verfügung. Insbesondere ist jedem Punkt $x \in U$ das harmonische Maß $\mu_x = \mu_x^U$ auf dem Rand $U^*$ zugeordnet. Zusammenfassend können wir das folgende Ergebnis formulieren:

**Satz 25.** (Resolutivitätssatz). *Es gelte das Axiom* $\mathrm{K}_1$; *$U$ sei eine relativ-kompakte Menge aus $\mathfrak{U}$, deren Rand $U^*$ eine abzählbare Basis besitzt. Dann gilt für jede numerische Funktion $f$ auf $U^*$*

$$\underline{H}_f(x) = \int_{\bullet} f\,d\mu_x^U \quad \textit{und} \quad \overline{H}_f(x) = \int^{\bullet} f\,d\mu_x^U \qquad (x \in U)\,. \tag{10.1}$$

*$f$ ist genau dann resolutiv, wenn $f$ $\mu_x^U$-integrierbar ist für alle $x \in U$.*

Ergänzend zeigen wir noch:

**Lemma 9.** *Für jede relativ-kompakte Menge $U \in \mathfrak{U}$ und jede beschränkte reelle Funktion $f$ auf $U^*$ sind die Funktionen $\overline{H}_f$ und $\underline{H}_f$ in $U$ beschränkt.*

Beweis. Es sei $h$ eine nach Axiom T existierende, überall positive Funktion aus $\mathscr{H}_X$. Dann existiert eine reelle Zahl $\alpha > 0$ mit $|f(x)| \leqq \alpha h(x)$ für alle $x \in U^*$. Bezeichnet $h_U$ die Restriktion von $h$ auf $U$, so ist also $\alpha h_U$ eine zu $f$ gehörige Oberfunktion und $-\alpha h_U$ eine zu $f$ gehörige Unterfunktion. Hieraus folgt $-\alpha h_U \leqq \underline{H}_f \leqq \overline{H}_f \leqq \alpha h_U$ und damit die Behauptung, da $h$ auf jeder relativ-kompakten Menge beschränkt ist.

## III. Randverhalten der verallgemeinerten Lösungen

*Auch in diesem Kapitel setzen wir durchwegs die Gültigkeit der Axiome* **I**, **II** *und* **T** *voraus. Weiter wird generell das schwache Konvergenzaxiom* $\mathbf{K}_1$ *gefordert.* Diese Forderung erscheint durch die Resultate des Kapitels II gerechtfertigt.

### *§ 11. Reguläre Randpunkte*

**11.1.** *Definition der regulären Punkte.* — Die Frage nach dem Verhalten der verallgemeinerten Lösungen $H_f = H_f^U$ am Rande einer relativ-kompakten offenen Menge $U$ legt es nahe, die folgende Begriffsbildung aus der Theorie des Dirichletschen Problems für die Laplacesche Differentialgleichung zu übernehmen.

**Definition 8.** *Ein Randpunkt z einer relativ-kompakten Menge $U \in \mathfrak{U}$ heiße* r e g u l ä r*, wenn für die verallgemeinerte Lösung $H_f$ einer jeden Funktion $f \in \mathscr{C}(U^*)$ gilt:*

$$(11.1) \qquad \lim_{x \to z, x \in U} H_f(x) = f(z)\,.$$

In anderer Formulierung besagt diese Definition: Ein Punkt $z \in U^*$ ist dann und nur dann regulär, wenn im Sinne der vagen Topologie gilt:

$$\lim_{x \to z, x \in U} \mu_x^U = \varepsilon_z\,.$$

Aus der Regularität eines Punktes $z \in U^*$ folgt die Gültigkeit von (11.1) für eine umfassendere Menge resolutiver Randfunktionen $f$. Dies behauptet der folgende Satz, dessen Beweis beinahe wörtlich mit dem Beweis von Satz 20 und dessen Korollar übereinstimmt; dieser kann daher dem Leser überlassen werden.

**Satz 26.** *Es sei $z_0$ ein regulärer Randpunkt von U. Dann gilt für jede auf $U^*$ definierte, nach oben beschränkte, numerische Funktion f:*

$$(11.2) \qquad \limsup_{x \to z_0, x \in U} \overline{H}_f(x) \leqq \limsup_{z \to z_0, z \in U^*} f(z)\,.$$

*Ist f sogar beschränkt und in $z_0$ stetig, so gilt:*

$$(11.3) \qquad \lim_{x \to z_0, x \in U} \overline{H}_f(x) = f(z_0)\,.$$

Für die regulären Mengen ergibt sich folgende neue Kennzeichnung:

**Satz 27.** *Eine relativ-kompakte Menge $V \in \mathfrak{U}$ ist dann und nur dann regulär, wenn alle Randpunkte von V regulär sind.*

B e w e i s. Aus den Bemerkungen der Nr. 9.1 sowie aus der Definition der regulären Mengen folgt, daß alle Randpunkte einer regulären Menge regulär sind. Zu beweisen ist daher nur die Aussage „dann". Da der Rand $V^*$ nach Satz 23 resolutiv ist und $(K_1)$ gefordert wird, ist nach Satz 14 jede Funktion $f \in \mathscr{C}(V^*)$ harmonisch-resolutiv. Da alle Randpunkte von $V$ regulär sind, wird jede Funktion $f \in \mathscr{C}(V^*)$ durch die verallgemeinerte Lösung $H_f$ stetig und harmonisch in $V$ hinein fortgesetzt. Nach der in Nr. 7.4 einleitend gemachten Bemerkung ist $H_f$ die einzige Funktion mit dieser Eigenschaft. Aus $f \geqq 0$, $f \in \mathscr{C}(V^*)$ folgt ferner $H_f \geqq 0$. Also ist $V$ eine reguläre Menge.

**11.2.** *Ein Lemma.* — Ein Punkt $z \in U^*$ ist bereits dann regulär, wenn die Beziehung (11.1) für eine totale Teilmenge von Funktionen $f \in \mathscr{C}(U^*)$ nachgewiesen werden kann:

**Lemma 10.** *Für einen Randpunkt $z \in U^*$ sei $\mathscr{N}_z$ der lineare Unterraum von $\mathscr{C}(U^*)$, welcher aus allen Funktionen $g \in \mathscr{C}(U^*)$ mit $g(z) = 0$ besteht. Genügen dann alle Funktionen f einer Menge $\mathscr{T}$, welche bezüglich der Topologie der gleichmäßigen Konvergenz total in $\mathscr{C}(U^*)$ bzw. nur total in $\mathscr{N}_z$ ist, der Bedingung* (11.1), *so ist z regulär.*

B e w e i s. Wir zeigen zunächst, daß aus dem Bestehen von (11.1) für alle Funktionen einer beliebigen Menge $\mathscr{B} \subset \mathscr{C}(U^*)$ das Bestehen von (11.1) für alle Funktionen aus der abgeschlossenen Hülle $\overline{\mathscr{B}}$ von $\mathscr{B}$ in $\mathscr{C}(U^*)$ (bezüglich der Topologie der gleichmäßigen Konvergenz) folgt. Hierzu sei $h$ eine nach dem

Axiom T existierende, in $X$ überall positive, harmonische Funktion und $h^*$ deren Restriktion auf $U^*$. Ohne Beschränkung der Allgemeinheit kann $h^* \geqq 1$ angenommen werden. Zu jeder Zahl $\varepsilon > 0$ und zu jeder Funktion $g \in \overline{\mathscr{B}}$ existiert dann ein $f \in \mathscr{B}$ mit $|g - f| \leqq \varepsilon h^*$. Hieraus folgt wegen der Linearität und Isotonie der Abbildungen $t \to H_t(x)$ die Ungleichung

$$|H_g(x) - H_f(x)| \leqq \varepsilon H_{h^*}(x) = \varepsilon h(x)$$

für alle $x \in U$. Nun genügen die Funktionen $f$ und $h$ der Beziehung (11.1); da $\varepsilon > 0$ beliebig gewählt war, folgt daher aus der obigen Ungleichung: $\lim\limits_{x \to z} H_g(x) = g(z)$. Wie behauptet ist also (11.1) auch für jedes $g \in \overline{\mathscr{B}}$ erfüllt.

Weiter bemerken wir, daß aus dem Bestehen von (11.1) für alle $f \in \mathscr{B}$ das Bestehen für alle Funktionen des von $\mathscr{B}$ in $\mathscr{C}(U^*)$ erzeugten linearen Unterraumes folgt. Dies ist eine unmittelbare Folgerung aus der Linearität der Abbildung $f \to H_f$.

Nunmehr sei $\mathscr{T}$ eine Teilmenge von $\mathscr{C}(U^*)$ mit den im Lemma formulierten Eigenschaften. Aus dem bislang Bewiesenen ergibt sich, daß dann (11.1) für alle $f \in \mathscr{C}(U^*)$ bzw. für alle $f \in \mathscr{N}_z$ erfüllt ist, je nachdem, ob $\mathscr{T}$ in $\mathscr{C}(U^*)$ oder in $\mathscr{N}_z$ total ist. Das Lemma ist somit bewiesen, wenn aus der Gültigkeit von (11.1) für alle $f \in \mathscr{N}_z$ die Gültigkeit für alle $f \in \mathscr{C}(U^*)$ folgt. Dies aber ergibt sich aus der Bemerkung, daß zu $f \in \mathscr{C}(U^*)$ stets eine reelle Zahl $\alpha$ existiert mit $f - \alpha h^* \in \mathscr{N}_z$.

**11.3.** *Lokale Natur der regulären Punkte.* — Wir beweisen zunächst die folgende Erblichkeitseigenschaft:

**Satz 28.** *Es seien $U$ und $U_1$ relativ-kompakte Mengen aus $\mathfrak{U}$ mit $U_1 \subset U$; $z$ sei ein gemeinsamer Randpunkt von $U$ und $U_1$. Ist dann $z$ ein regulärer Randpunkt von $U$, so ist $z$ auch ein regulärer Randpunkt von $U_1$.*

Beweis. Es sei $U_0$ eine relativ-kompakte, offene Umgebung von $\overline{U}$ und $\mathscr{S}$ die Menge aller in $U_0$ stetigen, reellen, hyperharmonischen Funktionen. Für jedes $u \in \mathscr{S}$ bezeichne $u^*$ die Restriktion von $u$ auf $U^*$ und $u_1^*$ die Restriktion auf $U_1^*$. Nach dem Satz 23 ist die Menge $\mathscr{S}_1^*$ aller Funktionen $u_1^*$ mit $u \in \mathscr{S}$ eine totale Teilmenge von $\mathscr{C}(U_1^*)$. Nach Lemma 10 ist daher $z$ ein regulärer Randpunkt von $U_1$, wenn gilt:

$$\lim_{x \to z, x \in U_1} H_{u_1^*}^{U_1}(x) = u_1^*(z) \text{ für alle } u_1^* \in \mathscr{S}_1^* .$$

Dies aber ergibt sich wie folgt: Für jedes $u \in \mathscr{S}$ ist $\mathrm{Rest}_U u$ eine zu $u^*$ gehörige Oberfunktion und somit

$$H_{u^*}^U(x) \leqq u(x) \text{ für alle } x \in U .$$

Aus obiger Ungleichung folgt für jeden Randpunkt $t$ von $U_1$:

$$\limsup_{x \to t,\, x \in U} H_{u^*}^U(x) \leqq \limsup_{x \to t,\, x \in U_1} u(x) = \lim_{x \to t} u(x) = u(t) = u_1^*(t) .$$

Da $H_{u^*}^U$ nach Satz 14 harmonisch in $U$ ist, so ist also $\mathrm{Rest}_{U_1} H_{u^*}^U$ eine zu $u_1$ gehörige Unterfunktion; daher gilt:

$$H_{u^*}^U(x) \leqq H_{u_1^*}^{U_1}(x) \quad \text{für alle } x \in U_1 .$$

Die Wiederholung der einleitend gemachten Überlegung zeigt, daß $\mathrm{Rest}_{U_1} u$ eine zu $u_1^*$ gehörige Oberfunktion ist und somit gilt:

$$H_{u_1^*}^{U_1}(x) \leqq u(x) \quad \text{für alle } x \in U_1 .$$

Somit ergibt sich die Doppelungleichung:

$$H_{u^*}^{U}(x) \leqq H_{u_1^*}^{U_1}(x) \leqq u(x) \quad \text{für alle } x \in U_1 .$$

Da nach Voraussetzung $z$ ein regulärer Randpunkt von $U$ und jede Funktion $u \in \mathscr{S}$ stetig ist, gilt:

$$\lim_{\substack{x \to z \\ x \in U}} H_{u^*}^{U}(x) = u^*(z) = u(z) = \lim_{x \to z} u(x) .$$

Hieraus und aus der zuletzt hergeleiteten Doppelungleichung folgt dann die zu beweisende Aussage:

$$\lim_{\substack{x \to z \\ x \in U_1}} H_{u_1^*}^{U_1} = u_1^*(z) = u(z) \quad \text{für alle } u_1^* \in \mathscr{S}^* .$$

**Korollar.** *Der Durchschnitt zweier regulärer Mengen ist entweder leer oder selbst eine reguläre Menge.*

Beweis. Es seien $U$ und $V$ reguläre Mengen mit nicht-leerem Durchschnitt $U_1 = U \cap V$. Dann sind $U, V$ und $U_1$ relativ-kompakte Mengen aus mit $U_1 \subset U$ und $U_1 \subset V$; ferner gilt $U_1^* \subset (\overline{U} \cap V^*) \cup (\overline{V} \cap U^*)$. Daher ist jeder Randpunkt von $U_1$ entweder ein gemeinsamer Randpunkt von $U_1$ und $U$ oder von $U_1$ und $V$. Nach Satz 27 sind alle Randpunkte von $U$ und $V$ regulär. Mit Hilfe des Satzes 28 ergibt sich daher, daß jeder Randpunkt von $U_1$ regulär ist. Eine erneute Anwendung von Satz 27 liefert dann die Behauptung, wonach $U_1$ eine reguläre Menge ist.

Wir bemerken noch, daß die Vereinigung zweier regulärer Mengen nicht regulär zu sein braucht. Man betrachte etwa in der Ebene $\mathbf{R}^2$ das Garbendatum der üblichen harmonischen Funktionen. Dann ist jede offene Kreisscheibe ohne Mittelpunkt eine nicht reguläre Menge. Diese Menge läßt sich aber auf viele Arten als Vereinigung zweier regulärer Mengen darstellen. (Man tilge z. B. aus der offenen Kreisscheibe auf zwei verschiedene Arten einen Radius.)

Über die angekündigte „lokale Natur" der regulären Randpunkte gibt nun der folgende Satz Auskunft.

**Satz 29.** *Es sei $z$ ein Randpunkt einer relativ-kompakten Menge $U \in \mathfrak{U}$ und $V$ eine offene Umgebung von $z$. Dann gilt: $z$ ist dann und nur dann ein regulärer Randpunkt von $U$, wenn $z$ ein regulärer Randpunkt von $U \cap V$ ist.*

Beweis. Es ist $U_1 = U \cap V$ eine relativ-kompakte Menge aus $\mathfrak{U}$ mit $U_1 \subset U$. Ferner ist $z$ ein gemeinsamer Randpunkt von $U_1$ und $U$. Ist daher $z$ ein regulärer Randpunkt von $U$, so ist $z$ nach dem Satz 28 auch ein regulärer Randpunkt von $U_1$. — Umgekehrt werde jetzt $z$ als regulärer Randpunkt von $U_1$ vorausgesetzt. Für eine beliebige Funktion $f \in \mathscr{C}(U^*)$ bezeichne $\varphi$ die Restriktion auf $U_1^*$ derjenigen Funktion, welche auf $U^*$ gleich $f$ und auf $U$ gleich $H_f^U$ ist. Nach dem Korollar zu Satz 15 ist dann mit $f$ auch $\varphi$ harmonisch-resolutiv und $H_\varphi^{U_1} = \mathrm{Rest}_{U_1} H_f^U$. Berücksichtigt man, daß $U_1$ die Spur in $U$

einer Umgebung von $z$ in $X$ ist, so ergibt sich hieraus

$$\lim_{\substack{x\to z\\ x\in U}} H_f^U(x) = \lim_{\substack{x\to z\\ x\in U_1}} H_f^U(x) = \lim_{\substack{x\to z\\ x\in U_1}} H_\varphi^{U_1}(x),$$

sofern der rechte Grenzwert existiert. Dies aber sieht man wie folgt ein: Mit $f$ ist $H_f$ und damit auch die Funktion $\varphi$ beschränkt (Lemma 9). Ferner ist $V \cap U^*$ eine Umgebung von $z$ in $U_1^*$, auf welcher $\varphi$ mit $f$ übereinstimme. Folglich ist die beschränkte Funktion $\varphi$ stetig in $z$. Nach Satz 27 existiert dann der fragliche Grenzwert, und es gilt

$$\lim_{\substack{x\to z\\ x\in U_1}} H_\varphi^{U_1}(x) = \varphi(z) = f(z).$$

Man erhält somit für jede Funktion $f \in \mathscr{C}(U^*)$:

$$\lim_{\substack{x\to z\\ x\in U}} H_f^U(x) = f(z).$$

Definitionsgemäß ist also $z$ ein regulärer Randpunkt von $U$.

**11.4.** *Regularitätskriterien.* — Aus der Theorie des Dirichletschen Problems für die Laplacesche Differentialgleichung übernehmen wir auch die folgende Begriffsbildung:

**Definition 9.** *Es sei $U$ eine relativ-kompakte Menge aus $\mathfrak{U}$ und $z$ ein Randpunkt von $U$. Barriere für den Punkt $z$ heiße dann jede Funktion $w$ mit folgenden Eigenschaften: $w$ ist im Durchschnitt $U \cap W$ von $U$ mit einer offenen Umgebung $W$ von $z$ in $X$ definiert; $w$ ist in $U \cap W$ hyperharmonisch; $w(x) > 0$ für alle $x \in U \cap W$; es gilt:*

$$\lim_{\substack{x\to z\\ x\in U\cap W}} w(x) = 0. \tag{11.4}$$

**Satz 30.** *Jeder Randpunkt einer relativ-kompakten Menge $U \in \mathfrak{U}$, zu dem eine Barriere existiert, ist regulär.*

Beweis. Es sei $z_0 \in U^*$ ein Randpunkt von $U$ und $w$ eine Barriere für $z_0$. Sie sei definiert in $U \cap W$, wobei $W$ eine offene Umgebung von $z_0$ in $X$ ist. Nach dem Satz 20 genügt es zu zeigen, daß $z_0$ ein regulärer Randpunkt von $U \cap W$ ist. Wir können daher ohne Beschränkung der Allgemeinheit annehmen, daß $U = U \cap W$ und somit $w$ in ganz $U$ definiert ist.

Wir betrachten die Menge $\mathscr{T}$ aller Funktionen $f \geqq 0$ aus $\mathscr{C}(U^*)$, welche in einer (von $f$ abhängigen) Umgebung von $z_0$ in $U$ verschwinden. Aus der Normalität des kompakten Raumes $U^*$ folgt leicht, daß $\mathscr{T}$ eine totale Teilmenge von $\mathscr{N}_{z_0}$ ist (vgl. Lemma 10). Fortan sei $f$ eine beliebige Funktion aus $\mathscr{T}$. Gemäß Lemma 10 ist dann zu zeigen:

$$\lim_{\substack{x\to z\\ x\in U_0}} H_f^U(x) = 0.$$

Zu $f$ existiert eine reguläre Umgebung $V$ von $z_0$ mit $f(z) = 0$ für alle $z \in \overline{V} \cap U^*$. Wir setzen $U_1 = U \cap V$ und bezeichnen mit $\varphi$ die Restriktion auf $U_1^*$ derjenigen Funktion, welche auf $U^*$ mit $f$ und auf $U$ mit $H_f^U$ übereinstimmt. Nach dem Korollar zu Satz 15 ist dann mit $f$ auch $\varphi$ harmonisch-resolutiv und

$H_{\varphi}^{U_1} = \mathrm{Rest}_{U_1} H_f^U$. Hieraus folgt (wie auch im Beweis von Satz 29), daß

$$\lim_{\substack{x \to z_0 \\ x \in U_1}} H_{\varphi}^{U_1}(x) = \lim_{\substack{x \to z_0 \\ x \in U}} H_f^U(x)$$

ist. Zu zeigen ist also

$$\lim_{\substack{x \to z_0 \\ x \in U_1}} H_{\varphi}^{U_1}(x) = 0 . \tag{11.5}$$

Dies aber ergibt sich wie folgt:

Nach dem Lemma 9 ist $\varphi$ beschränkt, also $\gamma = \sup \varphi(U_1^*)$ eine reelle Zahl; außerdem ist $\varphi \geqq 0$, also auch $\gamma \geqq 0$. Die Menge $V^* \cap U$ ist offen in $V^*$; zu jedem $\varepsilon > 0$ gibt es daher eine kompakte Menge $K \subset V^* \cap U$ derart, daß für die in $V^*$ offene Menge $G = V^* \cap U \cap \complement K$ gilt: $\mu_{z_0}^V(G) \leqq \varepsilon$. Bezeichnen wir mit $\chi_G$ die charakteristische Funktion von $G$ bezüglich $V^*$ und setzen wir $g = \gamma \chi_G$, so ist $g$ beschränkt und nach unten halbstetig, also nach den Sätzen 14 und 19 harmonisch-resolutiv. Folglich ist die Funktion $q = H_g^V$ harmonisch in $V$; aus $g \geqq 0$ folgt $q \geqq 0$.

Die durch

$$w_1(x) = \lambda w(x) + q(x)$$

für alle $x \in U_1$ definierte Funktion $w_1$ ist nach dem Bisherigen hyperharmonisch in $U_1$ für jedes $\lambda > 0$. Wir behaupten, daß $w_1$ für alle hinreichend großen $\lambda$ eine zu $\varphi$ gehörige Oberfunktion ist. Da $\varphi$ reellwertig ist, muß also gezeigt werden:

$$\tau(z) = \liminf_{x \to z,\, x \in U_1} w_1(x) \geqq \varphi(z)$$

für alle $z \in U_1^*$. Ist $z \in U_1^*$ gegeben, so unterscheiden wir wegen der Relation $U_1^* \subset (\overline{V} \cap U^*) \cup (\overline{U} \cap V^*)$ folgende drei Fälle für den Beweis: 1. $z$ liegt in $\overline{V} \cap U^*$. — 2. $z$ liegt in $G$. — 3. $z$ liegt in $K$. (Man beachte hierzu: Wenn der Fall 1 nicht vorliegt, so muß $z \in U \cap V^*$ sein.) Da $\tau \geqq 0$ und $\varphi(z) = f(z) = 0$ ist für alle $z \subset U_1^* \cap U^*$, erledigt sich der erste Fall sofort. Im zweiten und dritten Fall, wo $z$ in $U \cap V^*$ liegt, gilt:

$$\tau(z) \geqq \lambda \liminf_{x \to z,\, x \in U_1} w(x) + \liminf_{x \to z,\, x \in U_1} q(x)$$

sowie

$$\liminf_{x \to z,\, x \in U_1} q(x) \geqq \liminf_{x \to z,\, x \in V} q(x) \geqq \liminf_{z' \to z,\, z' \in V^*} g(z') = g(z) .$$

Bei dieser letzten Abschätzung wird der Satz 26 und die Halbstetigkeit von $g$ ausgenützt. Zusammenfassend gilt also in den beiden letzten Fällen:

$$\tau(z) \geqq \lambda \liminf_{x \to z,\, x \in U_1} w(x) + g(z) .$$

Im zweiten Fall ist $g(z) = \gamma$ und damit $\tau(z) \geqq g(z) = \gamma \geqq \varphi(z)$. Im dritten Fall ist $K$ eine nicht-leere kompakte Teilmenge von $U$, also auf Grund der Eigenschaften einer Barriere

$$\eta = \inf w(K) > 0 .$$

Die bislang noch beliebige Zahl $\lambda > 0$ braucht in diesem Fall nur $\geqq \gamma\eta^{-1}$ gewählt zu werden. Dann ergibt sich nämlich wegen $g(z) = 0$:

$$\tau(z) \geqq \lambda \liminf_{x \to z, x \in U_1} w(x) \geqq \lambda \liminf_{x \to z, x \in U} w(x) = \lambda\, w(z) \geqq \gamma \geqq \varphi(z)\,.$$

Also ist $w_1$ für $\lambda \geqq \gamma\eta^{-1}$ eine zu $\varphi$ gehörige Oberfunktion. Für den Rest des Beweises sei $w_1$ eine zu einem solchen $\lambda$ gehörige Oberfunktion.

Dann ist $0 \leqq H_\varphi^{U_1} \leqq w_1$; die Beziehung (11.5) ist daher bewiesen, wenn noch gezeigt werden kann, daß

$$\sigma = \limsup_{x \to z_0, x \in U_1} w_1(x) \leqq \gamma\varepsilon$$

gilt. Dies aber folgt so: Zunächst ist

$$\sigma \leqq \lambda \limsup_{x \to z_0, x \in U_1} w(x) + \limsup_{x \to z_0, x \in U_1} q(x) \leqq \lambda \limsup_{x \to z_0, x \in U} w(x) + $$
$$+ \limsup_{x \to z_0, x \in V} q(x) = \lambda \lim_{x \to z_0} w(x) + \lim_{x \to z_0} q(x) = q(z_0)\,;$$

hierbei wurde die Eigenschaft (11.4) einer Barriere und die Stetigkeit von $q$ in $V$ ausgenützt. Jetzt aber hat man nur noch die Gleichung (8.4) heranzuziehen und erhält

$$q(z_0) = H_g^V(z_0) = \int g\, d\mu_{z_0}^V = \gamma\, \mu_{z_0}^V(G) \leqq \gamma\varepsilon\,.$$

Damit ist auch die letzte Hilfsbehauptung bewiesen.

Mit Hilfe des starken Trennungsaxioms T', das hierbei erstmals verwendet wird, kann auch die Umkehrung der Aussage des Satzes 27 bewiesen werden.

**Satz 31.** *Es werde das starke Trennungsaxiom* T' *vorausgesetzt. Dann existiert zu jedem regulären Randpunkt* $z_0$ *einer relativ-kompakten Menge* $U \in \mathfrak{U}$, *für welchen* $\{z_0\}$ *eine* $G_\delta$*-Menge in* $U^*$ *ist*[16]), *eine (sogar in ganz* $U$ *definierte) Barriere.*

Beweis. Die einpunktige Menge $\{z_0\}$ ist genau dann eine $G_\delta$-Menge in $U^*$, wenn eine Funktion $f \in \mathscr{C}(U^*)$ existiert mit $f(z) > f(z_0) = 0$ für alle $z \neq z_0$ aus $U^*$. Für jede solche Funktion ist dann $w = H_f$ eine in $U$ definierte Barriere für $z_0$. Zunächst ist nämlich $w$ nicht-negativ und nach Satz 14 harmonisch in $U$; ferner gilt $\lim_{\substack{x \to z_0 \\ x \in U}} w(x) = 0$. Zu zeigen ist daher noch, daß $w$ nur positive Werte annimmt. Angenommen, es existiert ein $x_0 \in U$ mit $w(x_0) = \int f\, d\mu_{x_0}^U = 0$. Dann wird auf Grund der speziellen Wahl von $f$ das harmonische Maß $\mu_{x_0}^U$ von $\{z_0\}$ getragen, d. h. es ist $\mu_{x_0}^U = \lambda\varepsilon_{z_0}$ für eine geeignete reelle Zahl $\lambda \geqq 0$. Es ist $\lambda > 0$, da für jede in $X$ überall positive und harmonische Funktion $h$ (vgl. Axiom T) gilt: $h(x_0) = \int h\, d\mu_{x_0}^U = \lambda h(z_0)$. Die Gleichung $\mu_{x_0}^U = \lambda\varepsilon_{z_0} (\lambda > 0)$ führt aber wie folgt zu einem Widerspruch: Es sei $v$ eine in $U$ nach unten beschränkte, $h$-hyperharmonische Funktion; zu $v$ gibt es dann eine Funktion $u \in \mathscr{H}_U^*$ mit $v(x) = \frac{u(z)}{h(z)}$ für alle $x \in U$. Wegen der relativen Kompaktheit von $U$ ist $h$ auf $U$ beschränkt; daher ist mit $v$ auch $u$ nach unten beschränkt. Setzt man also $u^*(z) = \liminf_{x \to z, x \in U} u(x)$ für alle $z \in U^*$, so ist $u$ eine zu $u^*$ gehörige Oberfunktion. Dann aber folgt aus Satz 16: $\lambda u^*(z_0) = \int^* u^*\, d\mu_{x_0}^U \leqq \overline{H}_{u^*}(x_0) \leqq$

[16]) Dies ist insbesondere dann der Fall, wenn $U^*$ eine abzählbare Basis besitzt.

$\leqq u(x_0)$. Zusammen mit $h(x_0) = \lambda h(z_0)$ ergibt dies bei Beachtung der Tatsache, daß $\lim\limits_{\substack{x \to z_0 \\ x \in U}} h(x) = h(z_0) > 0$ ist, die Ungleichung

$$\liminf_{x \to z_0, x \in U} v(x) \leqq v(x_0) . \tag{11.6}$$

Man stößt daher auf den gesuchten Widerspruch, wenn die Existenz einer in $U$ nach unten beschränkten, $h$-hyperharmonischen Funktion $v$ gezeigt werden kann, welche nicht der Ungleichung (11.6) genügt. Nach dem starken Trennungsaxiom T' gibt es aber eine Funktion $v_0 \in {}^h\mathscr{H}_X^*$ mit $v_0(x_0) < v_0(z_0)$. Die Restriktion $v$ von $v_0$ auf $U$ ist dann nach unten beschränkt und $h$-hyperharmonisch; ferner ist $v_0(z_0) = \liminf\limits_{x \to z_0, x \in X} v_0(x) \leqq \liminf\limits_{x \to z_0, x \in U} v(x)$ und somit (11.6) verletzt. Damit ist gezeigt, daß $w$ überall positiv ist. Der Satz ist also bewiesen.

## *§ 12. Existenz regulärer Randpunkte*

**12.1.** *Extremale Randpunkte.* — Wir verschärfen zunächst den Begriff des regulären Randpunktes:

**Definition 10.** *Es sei $U$ eine relativ-kompakte Menge aus $\mathfrak{U}$. Ein Punkt $z \in \overline{U}$ heiße* e x t r e m a l, *wenn $\mu = \varepsilon_z$ das einzige positive Maß $\mu$ auf $\overline{U}$ ist mit folgender Eigenschaft: es gilt*

$$\int u \, d\mu \leqq u(z) \tag{12.1}$$

*für jede Funktion $u \in \mathscr{C}(\overline{U})$, welche in $U$ hyperharmonisch ist.*

Der folgende Satz zeigt, daß tatsächlich eine Verschärfung des Begriffes des regulären Randpunktes vorliegt.

**Satz 32.** *Jeder extremale Punkt $z \in \overline{U}$ ist ein regulärer Randpunkt von $U$.*

B e w e i s. Im folgenden setzen wir (wie schon früher) abkürzend $\mu_x = \mu_x^U$ für alle $x \in U$. Mit $\mathscr{E}$ bezeichnen wir die Menge aller Funktionen $u \in \mathscr{C}(\overline{U})$, welche in $U$ hyperharmonisch sind. Bezeichnet $u^*$ bzw. $u_U$ die Restriktion einer beliebigen Funktion $u \in \mathscr{E}$ auf $U^*$ bzw. auf $U$, so ist $u_U$ eine zu $u^*$ gehörige Oberfunktion und daher

$$\int u \, d\mu_x = H_{u^*}(x) \leqq u(x) \tag{12.2}$$

für alle $x \in U$. Für jeden Punkt $x \in U$ ist $\mu_x$ ein von $U^*$ getragenes, positives Maß auf $\overline{U}$, also $\mu_x \neq \varepsilon_x$. Daher kann kein Punkt aus $U$ extremal sein.

Jeder extremale Punkt $z \in \overline{U}$ ist also ein Randpunkt von $U$. Wir betrachten den Filter $\mathfrak{F}$ der Spuren $U \cap V$ in $U$ aller Umgebungen $V$ von $z$ in $X$. Weiter sei $h$ eine überall positive Funktion aus $\mathscr{H}_X$; setzen wir dann

$$\alpha = \inf h(U^*) \text{ und } \beta = \sup h(\overline{U}) ,$$

so ist $0 < \alpha \leqq \beta < \infty$. Aus $\int h \, d\mu_x = h(x)$ folgt dann: $\int d\mu_x \leqq \beta\alpha^{-1}$ für alle $x \in U$. Das Bild von $U$ bei der Abbildung $x \to \mu_x$ ist daher eine bezüglich der vagen Topologie relativ-kompakte Menge $H$ positiver Maße auf $U^*$ (vgl. [5a], p. 62). Folglich existiert ein Maß $\mu \geqq 0$ auf $U^*$, welches ein Limeswert der Abbildung $x \to \mu_x (x \in U)$ nach dem Filter $\mathfrak{F}$ ist. Da alle Funktionen $u \in \mathscr{E}$ in $\overline{U}$ stetig sind und $\mathfrak{F}$ gegen $z$ konvergiert, folgt aus (12.2) die Ungleichung

$$\int u \, d\mu \leqq u(z)$$

für alle $u \in \mathscr{E}$. Da $z$ ein extremaler Punkt ist, muß also $\mu = \varepsilon_z$ sein. Es ist daher $\varepsilon_z$ der einzige Limeswert der Abbildung $x \to \mu_x$ bezüglich $\mathfrak{F}$. Da $H$ relativ-kompakt ist, folgt hieraus die vage Konvergenz von $\mu_x$ gegen $\varepsilon_z$ nach dem Filter $\mathfrak{F}$, also $\lim\limits_{x \to z, x \in U} \mu_x = \varepsilon_z$. Nach der der Definition 8 angefügten Bemerkung ist dann $z$ regulär.

Dieser Satz berechtigt uns dazu, die extremalen Punkte von $\overline{U}$ fortan genauer als *extremale Randpunkte* von $U$ anzusprechen.

**12.2.** *Existenz extremaler Punkte.* — Die Frage nach der Existenz regulärer Randpunkte kann mit Hilfe von Satz 32 auf die Frage nach der Existenz extremaler Punkte zurückgeführt werden. Diese verschärfte Frage beantwortet der nächste Satz positiv; $h$ sei dabei wie immer eine überall positive Funktion aus $\mathscr{H}_X$.

**Satz 33.** *Jede Funktion $v \in \mathscr{C}(\overline{U})$, welche in $U$ $h$-hyperharmonisch ist, besitzt mindestens eine extremale Minimalstelle.*

Beweis. Es bezeichne wieder $\mathscr{E}$ die Menge aller Funktionen $u \in \mathscr{C}(\overline{U})$, welche in $U$ hyperharmonisch sind. Dann ist ${}^h\mathscr{E} = \left\{\frac{u}{h} : u \in \mathscr{E}\right\}$ die Menge aller Funktionen $v \in \mathscr{C}(\overline{U})$, welche in $U$ $h$-hyperharmonisch sind. Der Satz 22, angewandt auf eine relativ-kompakte offene Umgebung von $\overline{U}$, liefert die Aussage, daß ${}^h\mathscr{E}$ die Punkte von $\overline{U}$ trennt. Weiter enthält ${}^h\mathscr{E}$ die konstanten reellen Funktionen auf $\overline{U}$. Der Satz 2 aus [3] besagt dann, daß jede Funktion $v \in {}^h\mathscr{E}$ mindestens eine Minimalstelle $z \in \overline{U}$ besitzt derart, daß $\nu = \varepsilon_z$ das einzige Maß $\nu \geqq 0$ auf $\overline{U}$ ist mit $\int v\, d\nu \leqq v(z)$ für alle $v \in {}^h\mathscr{E}$. Jeder solche Punkt $z$ ist aber ein extremaler Punkt von $\overline{U}$. Ist nämlich $\mu \geqq 0$ ein Maß auf $\overline{U}$ mit $\int u\, d\mu \leqq u(z)$ für alle $u \in \mathscr{E}$, so ist $\nu = \frac{1}{h(z)}(h\mu)$ ein positives Maß auf $\overline{U}$ mit $\int v\, d\nu \leqq v(z)$ für alle $v \in {}^h\mathscr{E}$, also $\varepsilon_z = \nu = \frac{1}{h(z)}(h\mu)$. Hieraus folgt aber $\mu = \varepsilon_z$.

Hieraus läßt sich nun folgende Verschärfung des in (3.1) formulierten Randminimum-Prinzips ableiten:

**Korollar.** *Es sei $u$ eine in $U$ hyperharmonische Funktion aus $\mathscr{C}(\overline{U})$. Gilt dann $u(z) \geqq 0$ für alle extremalen (oder sogar regulären) Randpunkte $z$ von $U$, so ist $u \geqq 0$.*

Beweis. Die Funktion $v = \frac{u}{h}$ liegt in $\mathscr{C}(\overline{U})$ und ist $h$-hyperharmonisch in $U$. Nach Voraussetzung gilt $v(z) \geqq 0$ für alle extremalen Randpunkte $z \in U^*$. Nach dem Satz 33 ist dann $v \geqq 0$, also $u \geqq 0$.

**12.3.** *Semi-reguläre Mengen.* — Setzt man das starke Trennungsaxiom T′ voraus, so läßt sich für eine reguläre Menge $U$ zeigen, daß alle Randpunkte extremal sind, also die regulären mit den extremalen Randpunkten zusammenfallen. Dieses Resultat bleibt auch noch für einen allgemeineren Typus relativ-kompakter Mengen $U \in \mathfrak{U}$ richtig.

**Definition 11.** *Eine Menge $U \in \mathfrak{U}$ heiße semi-regulär, wenn sie relativ-kompakt ist und wenn für jede Funktion $f \in \mathscr{C}(U^*)$ die verallgemeinerte Lösung $H_f$ stetig auf $\overline{U}$ fortgesetzt werden kann.*

Aus den Ausführungen der Nr. 9.1 folgt, daß jede reguläre Menge auch semi-regulär ist. Die Umkehrung hiervon gilt nicht, wie folgendes Beispiel zeigt: Man betrachte in der Ebene $\mathbf{R}^2$ das Garbendatum der üblichen harmonischen Funktionen; $U$ sei eine offene Kreisscheibe ohne Mittelpunkt. Dann ist $U$ nicht regulär. Wohl aber folgt aus einem klassischen Satz von H. A. SCHWARZ, wonach isolierte Singularitäten beschränkter harmonischer Funktionen hebbar sind, daß $U$ semi-regulär ist.

**Satz 34.** *Es sei $U$ eine semi-reguläre Menge. Dann ist die Menge $U_r^*$ aller regulären Randpunkte von $U$ kompakt. Jedes der harmonischen Maße $\mu_x^U$ $(x \in U)$ wird von $U_r^*$ getragen. Jede Funktion $\varphi \in \mathscr{C}(U_r^*)$ kann auf genau eine Weise zu einer Funktion $u_\varphi \in \mathscr{C}(\overline{U})$ fortgesetzt werden, welche in $U$ harmonisch ist.*

Beweis. Für jede Funktion $f \in \mathscr{C}(U^*)$ bezeichne $\hat{H}_f$ die stetige Fortsetzung von $H_f$ auf $\overline{U}$; wegen der Beschränktheit von $H_f$ ist dann $\hat{H}_f \in \mathscr{C}(\overline{U})$. Nach Definition der regulären Randpunkte gilt $\hat{H}_f(z) = f(z)$ für alle $z \in U_r^*$. Wegen der Stetigkeit von $H_f$ und $f$ folgt hieraus $\hat{H}_f(z) = f(z)$ für alle $z \in \overline{U_r^*}$. Dies aber bedeutet, daß jeder Punkt aus $\overline{U_r^*}$ regulär, also $U_r^*$ abgeschlossen und damit kompakt ist.

Es sei weiter $K$ eine kompakte Menge mit $K \subset U^*$ und $K \cap U_r^* = \emptyset$. Dann gibt es eine Funktion $f \in \mathscr{C}(U^*)$ mit $0 \leqq f \leqq 1$, $f(z) = 0$ für alle $z \in U_r^*$ und $f(z) = 1$ für alle $z \in K$. Dann gilt $\hat{H}_f(z) \geqq 0$ und $-\hat{H}_f(z) \geqq 0$ für alle $z \in U_r^*$; nach dem Korollar zu Satz 33 ist daher $\hat{H}_f = 0$. Für jeden Punkt $x \in U$ folgt hieraus $\mu_x^U(K) \leqq \int f\, d\mu_x^U = H_f(x) = \hat{H}_f(x) = 0$, also $\mu_x^U(K) = 0$. Gemäß der Wahl von $K$ ist dann $\mu_x^U(U^* \cap \complement U_r^*) = 0$; es wird also jedes harmonische Maß $\mu_x^U (x \in U)$ von $U_r^*$ getragen.

Da $U_r^*$ und $U^*$ kompakt sind, kann jede Funktion $\varphi \in \mathscr{C}(U_r^*)$ zu einer Funktion $f \in \mathscr{C}(U^*)$ fortgesetzt werden; dann ist $\hat{H}_f$ eine stetige Fortsetzung von $\varphi$ auf $\overline{U}$, welche in $U$ harmonisch ist. Sind $u_1$ und $u_2$ zwei derartige Fortsetzungen von $\varphi$, so folgt aus $u_1(z) - u_2(z) = 0$ für alle $z \in U_r^*$ nach dem Korollar zu Satz 33 deren Identität: $u_1 = u_2$. Also kann jedes $\varphi \in \mathscr{C}(U_r^*)$ nur auf genau eine Weise zu einer in $U$ harmonischen Funktion $u_\varphi \in \mathscr{C}(\overline{U})$ fortgesetzt werden.

Nunmehr ergibt sich die einleitend angekündigte Aussage.

**Satz 35.** *Es gelte das starke Trennungsaxiom* T′. *Dann fallen für jede semi-reguläre Menge $U$ die regulären Randpunkte mit den extremalen Randpunkten von $U$ zusammen.*

Beweis. Wir übernehmen die im letzten Satz eingeführten Bezeichnungen und betrachten bei gegebenem $x \in \overline{U}$ die Abbildung $\varphi \to u_\varphi(x)$ von $\mathscr{C}(U_r^*)$ in $\mathbf{R}$. Nach Satz 34 ist klar, daß eine Linearform auf $\mathscr{C}(U_r^*)$ vorliegt, welche nach dem Korollar zu Satz 33 positiv ist. Da $U_r^*$ kompakt ist, handelt es sich also um ein positives Radonsches Maß $\mu_x$ auf $U_r^*$ mit $u_\varphi(x) = \int \varphi\, d\mu_x$ für alle $\varphi \in \mathscr{C}(U_r^*)$. Da für jede Fortsetzung $f \in \mathscr{C}(U^*)$ von $\varphi$ gilt $u_\varphi = \hat{H}_f$, wo $\hat{H}_f$ die stetige Fortsetzung von $H_f$ auf $\overline{U}$ ist, und da weiter jedes zu $x \in U$ gehörige harmonische Maß von $U_r^*$ getragen wird, gilt: $\mu_x = \mu_x^U$ für alle $x \in U$. Da $u_\varphi$ in $\mathscr{C}(\overline{U})$ liegt, ist die auf $\overline{U}$ definierte Abbildung $x \to \mu_x$ stetig bezüglich der vagen Topologie. Ist schließlich $u$ eine Funktion aus $\mathscr{C}(\overline{U})$, welche in $U$ hyperharmonisch ist, so gilt

$$(12.3) \qquad \int u\, d\mu_x \leqq u(x) \quad \text{für alle } x \subset \overline{U}\,.$$

Aus Stetigkeitsgründen genügt es, diese Ungleichung für Punkte $x \in U$ zu beweisen. Für diese Punkte wurde aber die Ungleichung bereits in (12.2) bewiesen.

Nach diesen Vorbereitungen zeigen wir jetzt, daß jeder Punkt $z \in U_r^*$ ein extremaler Randpunkt ist. Die Umkehrung ist nach Satz 32 allgemein richtig. Wir betrachten ein Maß $\mu \geqq 0$ auf $\overline{U}$ mit der Eigenschaft (12.1). Wegen der Stetigkeit der Abbildung $x \to \mu_x$ ist $\nu = \int \mu_x d\mu(x)$ definiert; es gilt $\nu = \varepsilon_z$. Für jede Funktion $\varphi \in \mathscr{C}(U_r^*)$ ist nämlich $\int \varphi d\nu = \int (\int \varphi d\mu_x) d\mu(x)$ $= \int u_\varphi d\mu = u_\varphi(z) = \varphi(z)$. Aus $\varepsilon_z = \int \mu_x d\mu$ folgt für die in $U_r^*$ offene Menge $G = U_r^* \cap \complement\{z\}$ nach [5b] (p. 19, corollaire)

$$0 = \varepsilon_z(G) = \int^* \mu_x(G)\, d\mu(x)\,;$$

hieraus ergibt sich $\mu_x(G) = 0$ für $\mu$-fast alle $x \in \overline{U}$. Aus $\mu_x(G) = 0$ folgt $\mu_x = \alpha(x)\varepsilon_z$, wobei $\alpha(x) \geqq 0$ ein Zahlenfaktor ist. Wählt man gemäß dem Axiom T eine überall positive Funktion $h \in \mathscr{H}_X$, so ist $h(x) = \int h\, d\mu_x$ und der Faktor $\alpha(x)$ ergibt sich zu $\frac{h(x)}{h(z)}$. Für jede in $U$ hyperharmonische Funktion $u \in \mathscr{C}(\overline{U})$ erhalten wir schließlich bei Berücksichtigung von (12.3) $\alpha(x)\, u(z)$ $= \int u\, d\mu_x \leqq u(x)$ und somit $\frac{u(z)}{h(z)} \leqq \frac{u(x)}{h(x)}$. Da das Axiom T′ vorausgesetzt wurde, liefert der Satz 22, angewandt auf eine relativ-kompakte offene Umgebung von $\overline{U}$, die Aussage: $x = z$. Also haben wir damit gezeigt, daß $x = z$ ist für $\mu$-fast alle $x \in \overline{U}$. Dies bedeutet aber, daß $\mu$ von $\{z\}$ getragen wird; somit existiert eine reelle Zahl $\alpha \geqq 0$ mit $\mu = \alpha\varepsilon_z$. Wegen $h(z) = \int h\, d\mu$ ist $\alpha = 1$ und damit $\mu = \varepsilon_z$. Also ist $z$ ein extremaler Randpunkt von $U$.

**12.4.** *Kennzeichnung semi-regulärer Mengen.* — Aus Gründen der Vollständigkeit zeigen wir:

**Satz 36.** *Eine relativ-kompakte Menge $U \in \mathfrak{U}$ ist genau dann semiregulär, wenn eine kompakte Menge $S \subset U^*$ existiert mit folgenden zwei Eigenschaften:*

(a) *Jede Funktion $\varphi \in \mathscr{C}(S)$ kann zu einer in $U$ harmonischen Funktion $u_\varphi \in \mathscr{C}(\overline{U})$ fortgesetzt werden.*

(b) *Für jede nach unten beschränkte Funktion $u \in \mathscr{H}_U^*$ gilt:*

$$(12.4) \qquad \liminf_{x \to z,\, x \in U} u(x) \geqq 0 \quad \textit{für alle } z \in S \Rightarrow u \geqq 0\,.$$

*Die Menge $S$ ist dann eindeutig bestimmt und besteht aus allen regulären Randpunkten von $U$.*

Beweis. Es werde $U$ als semi-regulär vorausgesetzt. Dann besitzt die Menge $S = U_r^*$ aller regulären Randpunkte von $U$ die genannten Eigenschaften. Wegen Satz 34 ist nur noch die zweite Eigenschaft nachzuweisen. Wir betrachten hierzu eine in $U$ nach unten beschränkte $h$-hyperharmonische Funktion $v \in {}^h\mathscr{H}_U^*$ und setzen sie zu einer in $\overline{U}$ definierten Funktion $\bar{v}$ fort durch die Festsetzung: $\bar{v}(z) = \liminf_{x \to z,\, x \in U} v(x)$ für alle $z \in U^*$. Dann genügt es zu zeigen, daß jede so gewonnene Funktion $\bar{v}$ eine in $S$ gelegene Minimalstelle besitzt. Die Menge $\mathscr{V}$ aller derartigen Funktionen trennt die Punkte von $\overline{U}$ nach Satz 22. Nach [3], Satz 2 besitzt also jede Funktion $\bar{v} \in \mathscr{V}$ eine Minimalstelle $z$, für welche $\nu = \varepsilon_z$ das einzige Maß $\nu \geqq 0$ auf $\overline{U}$ ist mit $\int^* \bar{v}\, d\nu \leqq \bar{v}(z)$ für alle $\bar{v} \in \mathscr{V}$. Jeder derartige Punkt $z$ liegt aber in $S$. In der Tat: Nach dem Beweis von Satz 35 existiert zu jedem Punkt $x \in \overline{U}$ ein Maß $\mu_x \geqq 0$ auf $S$ derart, daß die Abbildung $x \to \mu_x$ stetig und $\mu_x = \mu_x^U$ ist für alle $x \in U$. Dann ist auch die Abbildung $x \to {}^h\mu_x = \frac{1}{h(x)}(h\,\mu_x)$ stetig und für jede nach unten beschränkte Funktion $v \in {}^h\mathscr{H}_U^*$ gilt: $\int^* \bar{v}\, d\,{}^h\mu_x \leqq v(x)$ für alle $x \in U$. Wegen der Stetigkeit von $x \to {}^h\mu_x$ folgt hieraus:

$$\int^* \bar{v}\, d\,{}^h\mu_x \leqq \bar{v}(x)$$

für alle $x \in \overline{U}$. Da ${}^h\mu_x$ von $S$ getragen wird, so ist ${}^h\mu_x \neq \varepsilon_x$ für alle $x \in \overline{U} \cap \complement S$. Jeder Punkt $z$ mit der oben genannten Eigenschaft muß also in $S$ liegen.

Umgekehrt sei $S$ eine Teilmenge von $U^*$ mit den genannten Eigenschaften. Dann folgt zunächst aus (b), daß die Fortsetzung $u_\varphi$ von $\varphi$ durch $\varphi$ eindeutig bestimmt ist. Hieraus ergibt sich die Linearität der Abbildung $\varphi \to u_\varphi$ von $\mathscr{C}(S)$ in $\mathscr{C}(\overline{U})$. Es sei nun $\varphi$ die Restriktion auf $S$ einer beliebigen Funktion $f \in \mathscr{C}(U^*)$. Dann gilt offenbar für jede Oberfunktion $u \in \mathscr{O}_f : u_\varphi(x) \leqq u(x)$ für alle $x \in U$ und somit $u_\varphi(x) \leqq H_f(x)$ für alle $x \in U$. Indem man $f$ durch $-f$ ersetzt, besagt dieses Resultat, daß $-u_\varphi(x) = u_{-\varphi}(x) \leqq H_{-f}(x) = -H_f(x)$ und somit $u_\varphi(x) = H_f(x)$ ist für alle $x \in U$. Dann aber wird $H_f$ durch $u_\varphi$ stetig auf $\overline{U}$ fortgesetzt. Also ist $U$ semi-regulär.

Die letzte Überlegung zeigt noch, daß $S$ in der Menge $U_r^*$ aller regulären Randpunkte von $U$ enthalten ist. Da aber nach dem ersten Teil des Beweises $U_r^*$ selbst eine Menge $S$ ist, muß $S = U_r^*$ sein.

Aus diesem Satz folgt nun, daß z. B. jedes achsenparallele offene Rechteck $U$ in der Ebene $\mathbf{R}^2$ bezüglich des durch die Lösungen der Wärmeleitungsgleichung $u_{x_1 x_1} = u_{x_2}$ definierten Garbendatums semi-regulär ist. $S$ ist hierbei das aus drei Strecken zusammengesetzte „untere" Randstück von $U^*$ (vgl. [15], [17]).

**12.5.** *Bemerkungen.* — 1. Vorläufer der in diesem Paragraphen verwendeten Beweismethoden finden sich bereits in [4]. Aus dem dortigen Satz 2 folgt insbesondere sofort die folgende Aussage: Für eine beliebige relativ-kompakte Menge $U \in \mathfrak{U}$ bezeichne ${}^h\mathscr{E}$ die Menge aller Funktionen aus $\mathscr{C}(\overline{U})$, welche in $U$ $h$-hyperharmonisch sind. Dann ist die abgeschlossene Hülle aller extremalen Randpunkte von $U$ der Šilovsche Rand von $\overline{U}$ bezüglich ${}^h\mathscr{E}$. Es sei an dieser Stelle noch vermerkt, daß die extremalen Randpunkte von $U$ bezüglich der Garbendaten $W \to \mathscr{H}_W$ und $W \to {}^h\mathscr{H}_W$ dieselben sind. Dies folgt aus der Definition 10.

2. Für semi-reguläre Mengen $U$ wird nach Satz 34 jedes der harmonischen Maße $\mu_x^U$ von der Menge $U_r^*$ aller regulären Randpunkte getragen. Für das Garbendatum der üblichen harmonischen Funktionen im $\mathbf{R}^n$ besagt der Satz von KELLOGG-EVANS, daß dies für jede relativ-kompakte Menge $U \in \mathfrak{U}$ richtig

ist. DOOB [14] hat bemerkt, daß dieser Satz jedoch bereits falsch ist, wenn man im $\mathbf{R}^2$ von der Laplaceschen Differentialgleichung $\Delta u = 0$ zur Wärmeleitungsgleichung $u_{x_1 x_1} = u_{x_2}$ übergeht.

3. Der Satz 33 kann nicht dahingehend verallgemeinert werden, daß $v$ nur als in $U$ nach unten beschränkt, nach unten halbstetig und als in $U$ $h$-hyperharmonisch vorausgesetzt wird. Wenn der Rand von $U$ eine abzählbare Basis besitzt, ließe sich dann nämlich auch zeigen, daß jedes harmonische Maß $\mu_x^U$ von $U_r^*$ getragen wird. Dem widerspricht aber das soeben erwähnte Gegenbeispiel von DOOB.

4. Für die Laplacesche Differentialgleichung im $\mathbf{R}^n$ gilt nach [4], daß für jede relativ-kompakte Menge $U \in \mathfrak{U}$ die regulären Randpunkte mit dem extremalen Randpunkten von $U$ zusammenfallen. Es ist unbekannt, ob ein entsprechendes Resultat für die Wärmeleitungsgleichung gilt. Man vgl. hierzu auch [4], p. 127—128, wo auf die Beziehungen dieser Frage zu einem Satz von M. V. KELDYCH hingewiesen wird.

## § 13. *Weitergehende Regularitätskriterien*

**13.1.** *Reduktion und Extremisation.* — Es sei $E$ eine beliebige Teilmenge von $X$ und $\varphi \geqq 0$ eine numerische Funktion auf $X$. Dann bezeichne $R_\varphi^E$ die untere Einhüllende aller hyperharmonischen Funktionen $w \in \mathscr{H}_X^*$ mit

(13.1) $$w \geqq 0 \text{ und } \varphi(x) \leqq w(x) \quad \text{für alle } x \in E .$$

$R_\varphi^E$ hängt also genauer nur von der Restriktion von $\varphi$ auf $E$ ab. Weiter ergeben sich folgende Eigenschaften:

(13.2) $$F \subset E \Rightarrow R_\varphi^F \leqq R_\varphi^E ;$$

(13.3) $$R_{\lambda\varphi}^E = \lambda R_\varphi^E \qquad (\lambda \in \mathbf{R}, \lambda > 0) ;$$

(13.4) $$\psi \geqq 0, \psi(x) \leqq \varphi(x) \ (x \in E) \Rightarrow R_\psi^E \leqq R_\varphi^E .$$

Ausgehend von $R_\varphi^E$ gewinnt man die Funktion $\hat{R}_\varphi^E$ durch „Regularisation":

(13.5) $$\hat{R}_\varphi^E(x) = \liminf_{y \to x, y \in X} R_\varphi^E(y) .$$

Aus dieser Definition folgt sofort:

(13.6) $$0 \leqq \hat{R}_\varphi^E \leqq R^E .$$

Genauer ist $\hat{R}_\varphi^E$ die größte nach unten halbstetige, numerische Funktion $g$ auf $X$ mit $g \leqq R_\varphi^E$.

Ist nun speziell $\varphi$ eine nicht-negative, in $X$ hyperharmonische Funktion $u \in \mathscr{H}_X^*$, so heiße $R_u^E$ die *reduzierte* Funktion von $u$ bezüglich $E$ und $\hat{R}_u^E$ die *extremisierte* Funktion von $u$ bezüglich $\complement E$. In diesem Fall ergeben sich mühelos folgende weitere Eigenschaften:

(13.7) $$0 \leqq \hat{R}_u^E \leqq R_u^E \leqq u ;$$

(13.8) $$R_u^E(x) = u(x) \quad \text{für alle } x \in E ;$$

(13.9) $$\hat{R}_u^E(x) = u(x) \quad \text{für alle } x \in \mathring{E}\,^{17}) .$$

[17]) $\mathring{E}$ bezeichnet das Innere (= offener Kern) von $E$.

**Satz 37.** *Es sei $u \geqq 0$ eine in $X$ hyperharmonische Funktion und $E$ eine Teilmenge von $X$. Dann ist die reduzierte Funktion $R_u^E$ in jedem der beiden folgenden Fälle harmonisch in $\complement \overline{E}$:*

(a) *$u$ ist in $\complement \overline{E}$ lokal-beschränkt.*

(b) *Es gilt das Axiom* $K_2$; *$u$ ist in $\complement \overline{E}$ reellwertig.*

Beweis. Offenbar ist $R_u^E = \inf \mathscr{S}$, wenn hierbei $\mathscr{S}$ die (nicht-leere) Menge aller Funktionen $w \in \mathscr{H}_X^*$ ist mit $0 \leqq w \leqq u$ und $w(x) = u(x)$ für alle $x \in E$. Ferner enthält $\mathscr{S}$ mit je zwei Funktionen deren untere Einhüllende. Ist $V$ eine in $U = \complement \overline{E}$ reguläre Menge, so liegt nach Satz 6 mit jeder Funktion $w$ auch die Funktion $w_V$ in $\mathscr{S}$. Die Menge $\mathscr{S}_U$ aller Funktionen $\mathrm{Rest}_U w$ mit $w \in \mathscr{S}$ ist daher gesättigt. Die Behauptung folgt also aus Satz 11.

Mit Hilfe der Sätze 8 und 9 läßt sich übrigens zeigen, daß $\hat{R}_u^E$ in jedem der beiden folgenden Fälle hyperharmonisch ist: (a) $u$ ist lokal-beschränkt; (b) $u$ ist reellwertig; es gilt das Axiom $K_2$. Hierauf soll jedoch hier nicht eingegangen werden. (Vgl. auch BRELOT [9] sowie das nachfolgende Korollar zu Satz 38.)

**13.2.** *Zusammenhang mit dem Dirichletschen Problem.* — Der Zusammenhang mit dem Dirichletschen Problem wird durch folgenden Satz hergestellt:

**Satz 38.** *Es sei $U$ eine relativ-kompakte Menge aus $\mathfrak{U}$ und $u \geqq 0$ eine in $X$ hyperharmonische Funktion. Dann gilt für die Restriktion $f$ von $u$ auf $U^*$:*

$$\overline{H}_f(x) = R_u^{\complement U}(x) \quad \text{für alle } x \in U\,. \tag{13.10}$$

Beweis. Für jede Funktion $w \geqq 0$ aus $\mathscr{H}_X^*$ mit $w(x) \geqq u(x)$ für alle $x \in \complement U$ ist offenbar $\mathrm{Rest}_U w$ eine zu $f$ gehörige Oberfunktion. Also gilt $\overline{H}_f(x) \leqq w(x)$ für alle $x \in U$ und jedes derartige $w$. Dann aber ist $\overline{H}_f(x) \leqq R_u^{\complement U}(x)$ für alle $x \in U$. Zum Beweis der hierzu dualen Ungleichung sei $v$ eine beliebige zu $f$ gehörige Oberfunktion. Wir betrachten dann die wie folgt definierte Funktion $w$ auf $X$:

$$w(x) = \begin{cases} \inf(u(x), v(x)), & x \in U \\ u(x), & x \in \complement U \end{cases}.$$

Eine ähnliche Überlegung wie im Beweis von Satz 15 (beim Studium der dort definierten Funktion $V$) zeigt, daß $w$ in $X$ hyperharmonisch ist. Wegen $f \geqq 0$ ist $v \geqq 0$ und damit $w \geqq 0$; ferner stimmt $w$ auf $\complement U$ mit $u$ überein. Also gilt $w \geqq R_u^{\complement U}$. Hieraus folgt $R_u^{\complement U}(x) \leqq w(x) \leqq v(x)$ für alle $x \in U$ und $v \in \Phi_f$, also $R_u^{\complement U}(x) \leqq \overline{H}_f(x)$ für alle $x \in U$.

**Korollar.** *Es sei $u \geqq 0$ in $X$ hyperharmonisch. Dann gilt für jede reguläre Menge $V$:*

$$u_V = R_u^{\complement V} = \hat{R}_u^{\complement V}\,. \tag{13.11}$$

Beweis. Nach (3.2) und (13.8) ist $R_u^{\complement V}(x) = u(x) = u_V(x)$ für alle $x \in \complement V$. Nach den Sätzen 19 und 38 ist $u_V(x) = \int^* f\, d\mu_x^V = \overline{H}_f(x) = R_u^{\complement V}(x)$ für alle $x \in V$. Somit ist $u_V = R_u^{\complement V}$. Hieraus folgt der Rest der Behauptung, da $u_V$ nach Satz 6 nach unten halbstetig ist.

**13.3.** *Regularitätskriterien.* — Für jeden Randpunkt $z$ einer regulären Menge $V$ gilt nach (13.11) $u(z) = \hat{R}_u^{\complement V}(z)$; $z$ ist dabei automatisch ein regulärer

Randpunkt von $V$. Es soll nun gezeigt werden, daß durch Gleichungen der eben beschriebenen Art die regulären Randpunkte relativ-kompakter Mengen aus $\mathfrak{U}$ gekennzeichnet werden können.

**Satz 39.** *Notwendig für die Regularität eines Randpunktes $z$ einer relativ-kompakten Menge $U \in \mathfrak{U}$ ist die folgende Bedingung: Für jede in $X$ hyperharmonische Funktion $u \geqq 0$ und jede Umgebung $T$ von $z$ in $X$ gilt:*

$$(13.12) \qquad u(z) = \hat{R}_u^{T \cap \complement U}(z)\,.$$

Beweis. Nach (13.2) ist die Abbildung $T \to \hat{R}_u^{T \cap \complement U}$ isoton. Zu zeigen ist daher (13.12) nur für jede kompakte Umgebung $T$ von $z$. Zu $T$ wählen wir eine relativ-kompakte, offene Umgebung $W$ von $\overline{U} \cup T$ und setzen $Q = W \cap \cap (U \cup \complement T) = \complement_W(T \cap \complement U)$. Dann ist $Q$ offen und relativ-kompakt; es gilt $Q^* \subset W^* \cup (T \cap \complement U)^*$, wobei offenbar $W^*$ mit $(T \cap \complement U)^*$ keinen Punkt gemeinsam hat. Der Punkt $z$ liegt nicht in $Q$; wohl aber ist $z$ Randpunkt von $Q$, denn es gilt $V \cap Q = V \cap U$ für jede offene Umgebung $V$ von $z$ mit $V \subset T$. Für jede solche Umgebung $V$ ist $z$ nach Satz 29 ein regulärer Randpunkt von $V \cap Q = V \cap U$, also nach dem gleichen Satz ein regulärer Randpunkt von $Q$.

Nun sei $g$ eine wie folgt definierte Funktion auf $Q^*$:

$$g(x) = \begin{cases} 0, & \text{für } x \in W^*; \\ u(x), & \text{für } x \in (T \cap \complement U)^*. \end{cases}$$

Ist weiter $v \geqq 0$ eine Funktion aus $\mathscr{H}_X^*$ mit $v(x) \geqq u(x)$ für alle $x \in T \cap \complement U$, so rechnet man sofort nach, daß $\mathrm{Rest}_Q v$ eine zu $g$ gehörigen Oberfunktion ist. Somit gilt

$$R_u^{T \cap \complement U}(x) \geqq \overline{H}_g^Q(x) \quad \text{für alle } x \in Q\,.$$

Da $z$ ein regulärer Randpunkt von $Q$ ist, gilt nach Satz 27

$$\liminf_{x \to z,\, x \in Q} \underline{H}_g^Q(x) \geqq \liminf_{z' \to z,\, z' \in Q^*} g(z') \geqq u(z)\,.$$

Zusammen mit $\underline{H}_g^Q \leq \overline{H}_g^Q$ ergibt sich daher

$$\liminf_{x \to z,\, x \in Q} R_u^{T \cap \complement U}(x) \geqq u(z)\,.$$

Nun ist $W \cap \complement Q = T \cap \complement U$ die Spur in $\complement Q$ eine Umgebung von $z$; berücksichtigt man noch (13.8) für $E = T \cap \complement U$, so erhält man weiter

$$\liminf_{x \to z,\, x \in \complement Q} R_u^{T \cap \complement U}(x) = \liminf_{x \to z,\, x \in T \cap \complement U} u(x) \geqq u(z)\,.$$

Also ist $\hat{R}_u^{T \cap \complement U}(z) = \liminf_{x \to z,\, x \in X} R_u^{T \cap \complement U} \geqq u(z)$. Hieraus und aus (13.7) folgt die Behauptung[18]).

Unter einer Zusatzbedingung, die, wie wir zeigen werden, leicht erzwungen werden kann, gilt von Satz 39 auch die Umkehrung. Diese Umkehrung kann sogar auf zweierlei Art vorgenommen werden:

[18]) Wegen (13.7) ist also die Gleichung (13.12) äquivalent mit der Ungleichung $u(z) \leq \leq \hat{R}_u^{T \cap \complement U}(z)$. Die hier verwendeten Beweismethoden sind denen von Brelot [9] im elliptischen Fall verwendeten weitgehend analog.

**Satz 40.** *Die Menge aller in $X$ $h$-hyperharmonischen Funktionen $v \geqq 0$ trenne die Punkte von $X$. Jede der beiden folgenden Bedingungen ist dann hinreichend für die Regularität eines Randpunktes $z$ einer relativ-kompakten Menge $U \in \mathfrak{U}$:*

(A) *Für jede in $X$ hyperharmonische Funktion $u \geqq 0$ gilt:*

$$u(z) = \hat{R}_u^{\complement U}(z) . \tag{13.13}$$

(B) *Für eine in $X$ hyperharmonische, in $z$ stetige Funktion $u \geqq 0$ mit $0 < u(z) < +\infty$*[19]*) und jede Umgebung $T$ von $z$ in $X$ gilt:*

$$u(z) = \hat{R}_u^{T \cap \complement U}(z) . \tag{13.14}$$

Beweis. *Zu* (A): Es sei $U_1$ eine relativ-kompakte offene Umgebung von $\overline{U}$. Man betrachte die Menge $\mathcal{T}$ aller Funktionen $u \geqq 0$ aus $\mathcal{H}_X^*$, welche in $U_1$ stetig und reellwertig sind. Für jedes $u \in \mathcal{T}$ ist $u^* = \text{Rest}_{U^*} u$ als Funktion aus $\mathscr{C}(U^*)$ resolutiv, also nach Satz 38: $H_{u^*}(x) = R_u^{\complement U}(x)$ für alle $x \in U$. Bei Beachtung von (13.13) folgt hieraus:

$$\liminf_{x \to z,\, x \in U} H_{u^*}(x) \geqq \hat{R}_u^{\complement U}(z) = u(z) ;$$

nach (13.7) ist weiter $R_u^{\complement U} \leqq u$ und somit

$$\limsup_{x \to z,\, x \in U} H_{u^*}(x) \leqq \lim_{x \to z} u(x) = u(z) .$$

Beides zusammen ergibt

$$\lim_{\substack{x \to z \\ x \in U}} H_{u^*}(x) = u(z)$$

für alle $u \in \mathcal{T}$. Beachtet man Lemma 8, so liefert eine leichte Abwandlung der im Beweis von Satz 23 durchgeführten Schlüsse die Aussage, daß die Menge aller Funktionen $u^*$ mit $u \in \mathcal{T}$ in $\mathscr{C}(U^*)$ total ist. Dann aber ist $z$ regulär nach Lemma 10.

*Zu* (B): Wir führen den Beweis indirekt und setzen $z$ als nicht-regulär voraus. Nach (A) und (13.7) gibt es dann eine in $X$ hyperharmonische Funktion $q \geqq 0$ mit $\hat{R}_q^{\complement U}(z) < q(z)$. Wegen $0 < u(z) < +\infty$ existiert folglich eine reelle Zahl $\lambda > 0$ mit $\hat{R}_q^{\complement U}(z) < \lambda u(z) < q(z)$. Die Funktion $q - \lambda u$ ist wegen der speziellen Annahmen über $u$ im Punkte $z$ nach unten halbstetig. Folglich gibt es eine Umgebung $T$ von $z$ mit $\lambda u(x) < q(x)$ für alle $x \in T$. Aus den fundamentalen Eigenschaften der extremisierten Funktion folgt daher:

$$\lambda \hat{R}_u^{T \cap \complement U} = \hat{R}_{\lambda u}^{T \cap \complement U} \leqq \hat{R}_q^{T \cap \complement U} \leqq \hat{R}_q^{\complement U} .$$

Also ist $\lambda \hat{R}_u^{T \cap \complement U}(z) \leqq \hat{R}_q^{\complement U}(z) < \lambda u(z)$ und somit $\hat{R}_u^{T \cap \complement U}(z) < u(z)$. Die Bedingung (13.14) ist also nicht für jede Umgebung $T$ von $z$ erfüllt.

*Bemerkung.* Die in Satz 40 geforderte Verschärfung des Axioms T ist von selbst erfüllt, wenn man an die Stelle von $X$ eine relativ-kompakte, offene Umgebung $U_1$ von $\overline{U}$ als Grundraum treten läßt. Dies folgt aus dem Hinweis, daß die konstanten Funktionen $h$-harmonisch sind.

---

[19]) Beispielsweise ist $u = h$ eine derartige Funktion.

Übrigens wurde bei der Mehrzahl der im § 6 besprochenen Beispiele auf die Gültigkeit dieser verschärften Form des Axioms T im Grundraum $X$ bereits hingewiesen.

## Literatur


[1] Ahlfors, L. V., and L. Sario: Riemann Surfaces. Princeton Math. Series **26** (1960).

[2] Bauer, H.: Une axiomatique du problème de Dirichlet pour certaines équations aux dérivées partielles elliptiques et paraboliques. Compt. rend. Acad. Sci. (Paris) **250**, 2672—2674 (1960).

[3] Bauer, H.: Minimalstellen von Funktionen und Extremalpunkte, II. Arch. Math. **11**, 200—205 (1960).

[4] Bauer, H.: Šilovscher Rand und Dirichletsches Problem. Ann. Inst. Fourier **11**, 89—136 (1961).

[5a] Bourbaki, N.: Intégration, Chap. I—IV. Actual. sci. et ind. **1175**, Paris (1952).

[5b] Bourbaki, N.: Intégration, Chap. V. Actual. sci. et ind. **1244**, Paris (1956).

[6] Bourbaki, N.: Topologie générale, Chap. 9 (2e édition). Actual. sci. et ind. **1045**, Paris (1958).

[7] Brelot, M.: Une axiomatique générale du problème de Dirichlet dans les espaces localement compacts. Séminaire de Théorie du Potentiel **1**, No. 6, 16 p (1956—1957), Institut H. Poincaré.

[8] Brelot, M.: Axiomatique des fonctions harmoniques et surharmoniques dans un expace localement compact. Séminaire de Théorie du Potentiel **2**, No. 1, 40 p. (1957—1958), Institut H. Poincaré.

[9] Brelot, M.: Lectures on potentiel theory. Tata Inst. of Fund. Research, Bombay (1960).

[10] Brelot, M.: Familles de Perron et problème de Dirichlet. Acta Szeged **9**, 133—153 (1939).

[11] Brelot, M.: La théorie moderne du potentiel. Ann. Inst. Fourier **4**, 113—140 (1954).

[12] Brelot, M., et G. Choquet: Expaces et lignes de Green. Ann. Inst. Fourier **3**, 199—263 (1952).

[13] Brelot, M., et R. M. Hervé: Introduction de l'effilement dans une théorie axiomatique du potentiel. Compt. rend. Acad. Sci. (Paris) **247**, 1956—1959 (1958).

[14] Doob, J. L.: Probability methods applied to the first boundary value problem. Proc. 3rd Berkeley Symp. on Math. Stat. and Prob. 1954—1955, 49—80 (1956).

[15] Doob, J. L.: A probability approach to the heat equation. Trans. Am. Math. Soc. **80**, 216—280 (1955).

[16] Gevrey, M.: Sur les équations aux dérivées partielles du type parabolique. J. Math. pures. appl., VI Sér., **9**, 305—471 (1913) et **10**, 105—148 (1914).

[17] Hartman, P., and A. Wintner: On the solutions of the equation of heat conduction. Am. J. Math. **72**, 367—395 (1950).

[18] Hervé, R. M.: Développements sur une théorie axiomatique des fonctions surharmoniques. Compt. rend. Acad. Sci. (Paris) **248**, 179—181 (1959).

[19] Kamke, E.: Über die erste Randwertaufgabe bei der Laplace- und der Wärmeleitungs-Differentialgleichung. Jber. dtsch. Math.-Ver. **62**, 1—33 (1959).

[20] Miranda, C.: Equazioni alle derivate parziali di tipo ellitico. Ergeb. Math. **2** (neue Folge) (1955).

[21] Petrowsky, I.: Zur ersten Randwertaufgabe der Wärmeleitungsgleichung. Compositio Math. **1**, 383—419 (1935).

[22] Sternberg, W.: Über die Differentialgleichung der Wärmeleitung. Math. Ann. **101**, 394—398 (1929).

[23] Tautz, G.: Zur Theorie der elliptischen Differentialgleichungen. Math. Ann. **118**, 733—770 (1942).

[24] Tautz, G.: Zur Theorie der ersten Randwertaufgabe. Math. Nachr. **2**, 279—303 (1949).


*(Eingegangen am 6. Juli 1961)*

# Weiterführung einer axiomatischen Potentialtheorie ohne Kern (Existenz von Potentialen) * [R20]

Z. Wahrscheinlichkeitstheorie Verw. Gebiete **1** (1963), 197–229
[Zbl. Math. 216.10301; MR 27 #5926]

## Einleitung

**1.** In [*1*] wurde mit dem Aufbau einer axiomatischen Theorie harmonischer und hyperharmonischer Funktionen (Potentialtheorie ohne Kern) begonnen, welche sowohl auf *elliptische* als auch auf *parabolische* Differentialgleichungen angewendet werden kann. Dabei handelte es sich zunächst um die Grundlagen der Theorie und das Studium des Dirichletschen Problems mit Hilfe der verallgemeinerten Lösungen.

Diese Theorie unterscheidet sich von einer von Brelot [*3*] für den elliptischen Fall entwickelten Theorie hauptsächlich darin, daß das für die Theorie unentbehrliche Randminimum-Prinzip nicht auf dem üblichen Wege über ein Minimumprinzip und den lokalen Zusammenhang des Raumes gewonnen wurde. Dieser klassische Weg scheint wegen der so verschiedenen Minimumprinzipe für die Laplace-Gleichung und für die Wärmeleitungsgleichung in einer axiomatischen Theorie nicht gangbar zu sein. Diese Schwierigkeit wurde vielmehr in [*1*] durch die Einführung eines einfachen, in den Anwendungen meist leicht nachzuprüfenden Trennungsaxioms umgangen, aus welchem das Randminimum-Prinzip selbst auf nicht notwendig lokal-zusammenhängenden, wohl aber lokal-kompakten Grundräumen $X$ folgt. Dieses mit $T$ bezeichnete Axiom besagt, daß die Menge der hyperharmonischen Funktionen je zwei Punkte des lokal-kompakten Raumes $X$ trennt**. In [*1*] wurde bereits bemerkt, daß es für die dortigen Anwendungen im wesentlichen ausreicht, wenn man das Axiom $T$ durch das folgende schärfere Trennungsaxiom $T_+$ ersetzt: Die Menge aller auf $X$ nicht-negativen, hyperharmonischen Funktionen trennt je zwei Punkte von $X$.

Vergleicht man das in [*1*] Erreichte mit der sich bereits in einem reiferen Entwicklungsstadium befindlichen Brelotschen Theorie, so gewinnt man den Eindruck, daß nur eine neue Idee zu einer Weiterführung unserer Theorie führen kann. Man kommt dann nämlich zu Fragestellungen, wo in der Brelotschen Theorie ein neues Axiom herangezogen wird, nämlich die Existenz eines überall positiven Potentials auf $X$. Hinzu kommt, daß mit dem neuen Axiom laufend das für die elliptische Theorie charakteristische Minimumprinzip verwendet wird, wonach jede hyperharmonische Funktion in der Umgebung einer Minimalstelle konstant ist. Da dieses Prinzip in unserer Theorie nicht zur Verfügung steht, sieht man keine Möglichkeit, die Brelotschen Gedanken auch hier nutzbringend zu verwerten.

**2.** Nun ist aber bekannt, daß in der Brelotschen Theorie die Existenz eines überall positiven Potentials gleichwertig ist mit der Gültigkeit des Trennungs-

* Research supported in part by a grant form the National Science Foundation, U.S.A. (NSF-G 18975).

** Der Einfachheit halber soll hier in der Einleitung angenommen werden, daß die konstante Funktion 1 harmonisch auf $X$ ist.

axioms $T_+$. Es drängt sich daher die Frage auf, ob man nicht in unserer Theorie aus dem schärferen Trennungsaxiom $T_+$ auf die Existenz all jener Potentiale schließen kann, die man für die Weiterführung der Theorie braucht.

In dieser Arbeit soll gezeigt werden, daß dies möglich ist, wenn der Raum $X$ eine *abzählbare Basis* besitzt. Diese für die Anwendungen unerhebliche Voraussetzung soll von nun an hier in der Einleitung gemacht werden. Wir zeigen dann genauer, daß aus $T_+$ u. a. die Existenz eines stetigen reellen Potentials $p$ auf $X$ gefolgert werden kann, welches im folgenden Sinne ein *strenges Potential* ist: Für jede reguläre Menge $V$ und für jeden Punkt $x \in V$ gilt

$$\int p \, d\mu_x^V < p(x).$$

Ein solches Potential ist überall positiv und nirgends harmonisch. Diese ausgeprägten Eigenschaften erlauben es häufig, Schlüsse durchzuführen, bei denen in der elliptischen Theorie das bei uns fehlende Minimumprinzip herangezogen wird. Wir zeigen, daß die Existenz strenger Potentiale eine Reihe interessanter und weitreichender Folgen hat. Bei den Anwendungen, welche die Weiterführung der Theorie betreffen, legen wir weniger Wert auf Vollständigkeit als auf Vielseitigkeit.

**3.** Die Arbeit gliedert sich in 3 Kapitel von insgesamt 12 Paragraphen. Im Kapitel I (§ 1—§ 3) wird die Theorie der superharmonischen Funktionen weiterentwickelt und dabei das in [*1*] bereits angedeutete Konvergenzaxiom $K_D$ von DOOB [*5*] eingeführt und verwertet. Wirklicher Gebrauch wird von diesem stärkeren Konvergenzaxiom allerdings erst im letzten Paragraphen gemacht.

Im Kapitel II (§ 4—§ 7) handelt es sich um die Konstruktion strenger Potentiale und um die Approximation gewisser Funktionen durch Potentiale und Differenzen von solchen. In diesem Zusammenhang werden sog. *Absorptionsmengen* eingeführt, deren Auftreten in gewissem Sinne charakteristisch ist für eine Theorie, welche auch parabolische Differentialgleichungen erfassen will.

Das Kapitel III ist Anwendungen der neuen Resultate gewidmet. In § 8 und § 9 wird dabei zunächst die Theorie des Dirichletschen Problems vervollständigt. Insbesondere gelingt es dabei zu zeigen, daß ein in [*1*] an zwei Stellen herangezogenes schärferes Trennungsaxiom $T'$ entbehrlich für die Theorie ist. Nach Ideen von R. M. HERVÉ [*8*] wird dann in § 10 eine *Balayagetheorie* für Maße mit kompaktem Träger skizziert. In § 11 wird ein in der Brelotschen Theorie ebenfalls von R. M. HERVÉ [*8*] bewiesener Zerlegungs- und Fortsetzungssatz als in unserer Theorie gültig erkannt. Er gestattet u. a. in § 12 eine erschöpfende Behandlung der Eigenschaften *polarer Mengen*.

**4.** In der Einleitung zur Thèse von R. M. HERVÉ [*8*] wird die Meinung vertreten, daß man bei Aufgabe des lokalen Zusammenhangs und beim Versuch, auch parabolische Differentialgleichungen zu erfassen, zu einer von der klassischen Theorie wesentlich verschiedenen Theorie gelangt. Im Gegensatz hierzu zeigen unsere bisherigen Resultate, daß man zwar an einigen kritischen Stellen der neuen Theorie neuartige Beweismethoden zu erfinden hat, daß man aber letzten Endes zu ganz analogen Sätzen wie in der klassischen Theorie gelangt.* Diese neuen Beweismethoden sind so beschaffen, daß man an keiner Stelle eine Möglichkeit sieht,

* *Zusatz bei der Korrektur:* In diesem Zusammenhang vergleiche man auch M. BRELOT, Étude comparée de quelques axiomatiques des fonctions harmoniques et surharmoniques. Séminaire BRELOT-CHOQUET-DENY, 6e année, n° 1 b, 13 p. (1962), Institut H. Poincaré, Paris.

Schlüsse mit Hilfe des lokalen Zusammenhangs auszuführen. Dieser formale Grund erklärt das Fehlen des lokalen Zusammenhangs in unserer Theorie.

5. Für einen weiteren Ausbau der Theorie wäre es wichtig zu wissen, ob zu jedem den Axiomen I, II, $T_+$ und $K_D$ genügenden Garbendatum $U \to \mathcal{H}_U$ von harmonischen Funktionen auf einem lokal-kompakten Raum $X$ mit abzählbarer Basis ein starker *Markoffscher Prozeß* mit stetigen Trajektorien im Zustandsraum $X$ existiert, welcher mit dem Garbendatum $U \to \mathcal{H}_U$ in analoger Beziehung steht wie der Brownsche Bewegungsprozeß im $\boldsymbol{R}^n$ mit dem Garbendatum der Lösungen der Laplaceschen Differentialgleichung. Die Koppelung könnte darin bestehen, daß für jede relativ-kompakte offene Menge $U$ und jeden Punkt $x \in U$ die von $x$ ausgehenden Trajektorien die Menge $\complement U$ mit einer Verteilung treffen, welche gleich dem harmonischen Maße $\mu_x^U$ ist. Nach einem noch unveröffentlichten Resultat von R. M. Blumenthal, R. K. Getoor und H. P. McKean wäre hierdurch der Prozeß bereits bis auf eine zufällige stetige Zeittransformation eindeutig bestimmt.

## Vorbemerkung

Die Kenntnis der Arbeit [*1*] wird im folgenden vorausgesetzt. Die in [*1*] eingeführten Bezeichnungen und Verabredungen werden übernommen. So bezeichnet also im folgenden $X$ einen lokal-kompakten (Hausdorffschen) Raum, $\mathfrak{U}$ das System seiner nicht-leeren offenen Teilmengen und $U \to \mathcal{H}_U$ ($U \in \mathfrak{U}$) das Garbendatum der harmonischen Funktionen. *Stets wird vorausgesetzt, daß dieses Garbendatum den Axiomen* I, II, $T$ *und* $K_1$ *aus* [*1*] *genügt.* $h$ bezeichnet wieder eine gemäß $T$ existierende harmonische, positive Funktion auf $X$.

# I. Superharmonische Funktionen

## § 1. Doobsches Konvergenzaxiom

**1.1. Superharmonische Funktionen und Potentiale.** Ähnlich wie in der Theorie von Brelot definieren wir:

**Definition 1.** Eine hyperharmonische Funktion $u \in \mathcal{H}_U^*$ heiße *superharmonisch in* $U \in \mathfrak{U}$, wenn sie auf einer dichten Teilmenge von $U$ endliche Werte annimmt. *Sub*harmonisch in $U$ heiße jede Funktion $u$, für welche $-u$ superharmonisch in $U$ ist.

*Potential* in $U$ heiße jede in $U$ superharmonische Funktion $p \geqq 0$ mit der Eigenschaft, daß $q \leqq 0$ gilt für jede harmonische Minorante $q$ von $p$ in $U$.

Unmittelbar aus dieser Definition ergeben sich folgende Eigenschaften superharmonischer Funktionen:

1. Jede hyperharmonische Funktion in $U \in \mathfrak{U}$, welche durch eine in $U$ superharmonische Funktion majorisiert wird, ist selbst superharmonisch.

2. Jede nicht-negative hyperharmonische Funktion in $U \in \mathfrak{U}$, welche durch ein Potential majorisiert wird, ist selbst ein Potential in $U$.

3. Ist $u$ superharmonisch bzw. ein Potential in $U \in \mathfrak{U}$ und ist $\lambda > 0$ eine reelle Zahl, so ist auch $\lambda u$ superharmonisch bzw. ein Potential in $U$.

Für alle tieferliegenden Eigenschaften superharmonischer Funktionen benötigt man ein Konverganzaxiom, welches diesem neuen Begriff angepaßt ist.

**1.2. Das Konvergenzaxiom $K_D$.** Das dem Axiom TM 4 bei DOOB [5] entsprechende Konvergenzaxiom, auf dessen Gültigkeit bei den Anwendungsbeispielen in [1], § 6 bereits hingewiesen wurde, lautet:

**Axiom $K_D$** (Doobsches Konvergenzaxiom). *Für jede nicht-leere, aufsteigend filtrierende Menge $\mathcal{F}$ von harmonischen Funktionen in einer Menge $U \in \mathfrak{U}$ ist auch die obere Einhüllende* sup $\mathcal{F}$ *harmonisch in $U$, sofern diese auf einer in $U$ dichten Punktmenge endlich ist.*

Offenbar impliziert $K_D$ das Axiom $K_2$. Eine leichte Abwandlung der in den Beweisen der Sätze 8—10 in [1] durchgeführten Schlüsse liefert die folgenden Aussagen:

4. Unter alleiniger Voraussetzung der Axiome I und II ist das Axiom $K_D$ mit jedem der beiden folgenden Axiome gleichwertig:

($K_D'$) Für jede reguläre Menge $V$ und jede numerische Funktion $f$ auf $V^*$, welche für eine in $V$ dichte Menge von Punkten $x \in V$ ein endliches $\mu_x^V$-Oberintegral besitzt, ist die Funktion

$$x \to \int^* f \, d\mu_x^V \qquad (x \in V) \tag{1.1}$$

harmonisch in V.

($K_D''$) Für jede reguläre Menge $V$ und jede numerische Funktion $f$ auf $V^*$, welche $\mu_x^V$-meßbar für alle $x \in V$ und $\mu_x^V$-integrierbar für eine in $V$ dichte Menge von Punkten $x \in V$ ist, gilt: es ist $f$ $\mu_x^V$-integrierbar für alle $x \in V$ und

$$x \to \int f \, d\mu_x^V \qquad (x \in V) \tag{1.2}$$

eine in $V$ harmonische Funktion.

5. Besitzt der Raum $X$ eine abzählbare Basis (und setzt man nur die Axiome I, II voraus), so ist das Axiom $K_D$ einem entsprechend formulierten Axiom über isotone Folgen harmonischer Funktionen äquivalent.

Vereinbarung. Für den Rest dieses Paragraphen soll das Axiom $K_D$ gefordert werden.

Es ergibt sich dann:

**Satz 1.** *Eine in einer Menge $U \in \mathfrak{U}$ hyperharmonische Funktion $u$ ist genau dann superharmonisch, wenn sie $\mu_x^V$-integrierbar ist für alle in $U$ regulären Mengen $V$ und alle Punkte $x \in V$.*

*Beweis.* Es sei $u$ superharmonisch in $U$. Dann ist $-\infty < \int^* u \, d\mu_x^V \leqq u(x) < +\infty$ für eine in $V$ dichte Menge von Punkten $x \in V$. Nach $K_D'$ ist also die Funktion $x \to \int^* u \, d\mu_x^V$ endlich (und harmonisch) in $V$. Wegen der Halbstetigkeit von $u$ folgt hieraus die $\mu_x^V$-Integrierbarkeit von $u$ für alle $x \in V$. — Aus der uneingeschränkten $\mu_x^V$-Integrierbarkeit von $u$ folgt umgekehrt die Superharmonizität von $u$ (sogar ohne Verwendung von $K_D$), da dann auf dem Rand jeder in $U$ regulären Menge $V$ ein Punkt $y$ liegt mit $|u(y)| < +\infty$.

Der Beweis hat noch ergeben:

**Korollar 1.** *Für jede in $U \in \mathfrak{U}$ superharmonische Funktion $u$ und jede in $U$ reguläre Menge $V$ ist die Funktion $u_V$ in $V$ harmonisch (und superharmonisch in $U$).*

Ferner folgt aus obigem Satz:

**Korollar 2.** *Die Summe zweier in $U \in \mathfrak{U}$ superharmonischer Funktionen ist superharmonisch in $U$.*

**1.3. Verschärfung von Resultaten aus** [*1*]. Die folgenden Aussagen stellen Verschärfungen der Sätze 11, 12, 14, 17 und 37 aus [*1*] dar. Die Beweise ergeben sich durch leichte Abwandlungen der in [*1*] geführten Beweise unter Heranziehung von $K_D$ und Satz 1:

6. Es sei $\mathscr{S}$ eine nicht-leere, gesättigte Menge von superharmonischen Funktionen in einer Menge $U \in \mathfrak{U}$ und $u_0 = \inf \mathscr{S}$ die untere Einhüllende. Gilt dann $u_0(x) > -\infty$ für die Punkte $x$ einer in $U$ dichten Menge, so ist $u_0$ harmonisch in $U$.

7. Es sei $\mathscr{T}$ eine nicht-leere Menge subharmonischer Funktionen in einer Menge $U \in \mathfrak{U}$ derart, daß $\mathscr{T}$ mindestens eine superharmonische Majorante (in $U$) besitzt. Dann besitzt $\mathscr{T}$ sogar eine kleinste hyperharmonische Majorante, welche außerdem harmonisch in $U$ ist.

8. Es sei $Y$ eine Kompaktifizierung von $X$ mit den in [*1*], § 7 beschriebenen Eigenschaften. Ist dann für eine auf dem Rand $X^*$ von $X$ in $Y$ definierte, numerische Funktion $f$ die zugehörige Funktion $\overline{H}_f$ endlich auf einer dichten Teilmenge von $X$, so ist $\overline{H}_f$ harmonisch in $X$.

Wird der Rand $X^*$ zusätzlich als resolutiv vorausgesetzt, so gilt für jede numerische Funktion $f$ auf $X^*$: Die Funktion

$$x \to \int^* f \, d\mu_x \qquad (x \in X) \tag{1.3}$$

ist harmonisch in $X$, wenn sie auf einer dichten Teilmenge von $X$ endliche Werte annimmt.

9. Für jede superharmonische Funktion $u \geqq 0$ auf $X$ und jede Menge $E \subset X$ ist die reduzierte Funktion $R_u^E$ harmonisch in $\complement \overline{E}$.

Aus 7 erhält man:

10. Jede in $U \in \mathfrak{U}$ superharmonische Funktion $u \geqq 0$ ist die Summe eines Potentials in $U$ und einer harmonischen Funktion $m \geqq 0$ in $U$. Dabei ist $m$ die größte hypoharmonische Minorante von $u$.

Aus 9 folgert man wie bei Brelot [*3*], pp. 92—93:

11. Die Funktionen $u$ und $m$ seien wie in 10 definiert. Ferner sei $\mathfrak{K}$ eine aufsteigend filtrierende Menge relativ-kompakter Teilmengen von $X$ mit $X = \bigcup \mathfrak{K}$. Dann ist:

$$m = \inf_{E \in \mathfrak{K}} R_u^{\complement E} = \lim_{\mathfrak{K}} R_u^{\complement E}. \tag{1.4}$$

Wie in [*3*], p. 93 folgt hieraus schließlich:

12. Die Summe zweier Potentiale auf $U \in \mathfrak{U}$ ist ein Potential auf $U$.

Bemerkung. Hat man nur das Konvergenzaxiom $K_1$ (bzw. $K_2$) zur Verfügung, so gelten die Aussagen 10—12 wenigstens für lokal-beschränkte (bzw. reelle) superharmonische Funktionen und Potentiale.

## § 2. $S_{\mathfrak{B}}$-Funktionen

**2.1. Elementare Eigenschaften.** Wie in der Theorie von R. M. Hervé [*8*] definieren wir:

**Definition 2.** Es sei $\mathfrak{B}$ eine aus offenen, relativ-kompakten Mengen bestehende Basis von $X$. *$S_{\mathfrak{B}}$-Funktion* in einer Menge $U \in \mathfrak{U}$ heiße dann jede in $U$ definierte, numerische Funktion $v$ mit folgenden Eigenschaften:

a) $v$ ist nach unten lokal-beschränkt;

b) für jede Menge $B \in \mathfrak{B}$ mit $\overline{B} \subset U$ und jeden Punkt $x \in B$ gilt

$$\int^* v\, d\mu_x^B \leqq v(x). \tag{2.1}$$

Eine hyperharmonische Funktion ist offenbar eine $S_{\mathfrak{B}}$-Funktion für jede aus relativ-kompakten Mengen bestehende Basis von $X$. Solche Funktionen sollen auch *S-Funktionen* genannt werden.

Für nach unten halbstetige $S_{\mathfrak{B}}$-Funktionen gilt ein ähnliches Minimumprinzip wie für hyperharmonische Funktionen:

**Lemma 1.** *Es sei $Y$ ein kompakter Raum, der $X$ als dichten Unterraum enthält; ferner sei $U$ eine Menge aus $\mathfrak{U}$. Die konstante Funktion $h = 1$ sei harmonisch in $U$. Dann besitzt jede in der abgeschlossenen Hülle $\overline{U}$ von $U$ in $Y$ definierte, numerische, nach unten halbstetige Funktion $v$, welche in $U$ eine $S_{\mathfrak{B}}$-Funktion ist, eine im Rand $U^*$ von $U$ in $Y$ gelegene Minimalstelle.*

Der Beweis kann dem Leser überlassen werden, da er dem Beweis von Lemma 1 aus [*1*] analog ist. An die Stelle des dortigen Systems $\mathfrak{B}(x)$, $x \in U$, tritt jetzt das System aller Mengen $B \in \mathfrak{B}$ mit $x \in B \subset \overline{B} \subset U$.

Aus Lemma 1 folgert man wie in [*1*] den nächsten, dem Satz 5 in [*1*] entsprechenden Satz:

**Satz 2.** *Für jede Basis $\mathfrak{B}$ von $X$, welche aus offenen relativ-kompakten Mengen besteht, und für jede Menge $U \in \mathfrak{U}$ stimmen die nach unten halbstetigen $S_{\mathfrak{B}}$-Funktionen in $U$ mit den hyperharmonischen Funktionen in $U$ überein.*

Die nächsten Eigenschaften von $S_{\mathfrak{B}}$-Funktionen folgen unmittelbar aus Definition 2:

1. Mit $u$ und $v$ sind auch die beiden folgenden Funktionen $S_{\mathfrak{B}}$-Funktionen in $U \in \mathfrak{U}$: $u + v$ und $\lambda u$ ($\lambda > 0$, $\lambda \in \boldsymbol{R}$).

2. Die untere Einhüllende jeder lokal-gleichmäßig nach unten beschränkten Menge von $S_{\mathfrak{B}}$-Funktionen ist eine $S_{\mathfrak{B}}$-Funktion.

3. Die obere Einhüllende jeder isotonen Folge von $S_{\mathfrak{B}}$-Funktionen ist eine $S_{\mathfrak{B}}$-Funktion.

**2.2. Regularisierung von $S_{\mathfrak{B}}$-Funktionen.** Es gilt folgender Satz:

**Satz 3.** *Es gelte das Axiom $K_D$. Dann ist die untere Regularisierte* $\hat{v}$ einer jeden $S_{\mathfrak{B}}$-Funktion $v$ in $U \in \mathfrak{U}$, welche auf einer dichten Teilmenge von $U$ endliche Werte annimmt, in $U$ superharmonisch. Es gilt:*

$$\hat{v}(x) = \sup_{\substack{B \in \mathfrak{B} \\ x \in B \subset \overline{B} \subset U}} \int^* v\, d\mu_x^B \qquad (x \in U). \tag{2.2}$$

---

* Dies ist die größte nach unten halbstetige Funktion $\hat{v}$ auf $U$ mit $\hat{v} \leqq v$. Es gilt:

$$\hat{v}(x) = \liminf_{y \to x} v(y) \quad \text{für alle} \quad x \in U.$$

*Beweis*. Die Funktion $\hat{v}$ ist nach unten halbstetig; es gilt $\hat{v}(x) > -\infty$ für alle $x \in U$, da $v$ nach unten lokal-beschränkt ist. Für jedes $B \in \mathfrak{B}$ mit $\overline{B} \subset U$ und alle $x \in B$ besteht die Ungleichung $\int^* v d\mu_x^B \leqq v(x)$; vermöge der Endlichkeitsvoraussetzung ist also $|\int^* v d\mu_x^B| < +\infty$ für alle Punkte $x$ einer in $B$ dichten Teilmenge. Nach § 1, Eigenschaft 8 ist somit die Funktion

$$x \to \int^* v\, d\mu_x^B \qquad (x \in B)$$

harmonisch, also stetig in $B$. Da $v$ eine Majorante dieser Funktion ist, gilt dann sogar

$$\int^* v d\mu_x^B \leqq \hat{v}(x)$$

für alle $x \in B$. Wegen $\hat{v} \leqq v$ folgt hieraus, daß $\hat{v}$ eine nach unten halbstetige $S_{\mathfrak{B}}$-Funktion, nach Satz 2 also eine in $U$ hyperharmonische Funktion ist. Wegen $\hat{v} \leqq v$ ist $\hat{v}$ sogar superharmonisch.

Aus dem soeben Bewiesenen folgt die Ungleichung

$$\hat{v}(x) \geqq \sup_{\substack{B \in \mathfrak{B} \\ x \in B \subset \overline{B} \subset U}} \int^* v d\mu_x^B \qquad (x \in U).$$

Die behauptete Gleichheit ergibt sich wie folgt: Es sei $x$ ein Punkt aus $U$ und $\eta$ eine reelle Zahl $< \hat{v}(x)$. Dann existiert eine Umgebung $B \in \mathfrak{B}$ von $x$ mit $\overline{B} \subset U$ und $\inf \hat{v}(\overline{B}) \geqq \eta$. Folglich ist

$$\int^* v d\mu_x^B \geqq \eta \int d\mu_x^B.$$

Bei gegebenem $\varepsilon > 0$ kann $B$ weiterhin der Bedingung

$$\left|\int d\mu_x^B - 1\right| \leqq \varepsilon$$

unterworfen werden. Dies zeigt die im Beweis von Satz 7 aus [*1*] für ein spezielles System $\mathfrak{B}$ durchgeführte Überlegung. Somit erhält man:

$$\int^* v d\mu_x^B \geqq \eta + \eta\left(\int d\mu_x^B - 1\right) \geqq \eta - \varepsilon|\eta|$$

und hieraus

$$\sup_{\substack{B \in \mathfrak{B} \\ x \in B \subset \overline{B} \subset U}} \int^* v d\mu_x^B \geqq \eta - \varepsilon|\eta|$$

für alle $\eta < \hat{v}(x)$ und alle $\varepsilon > 0$. Daher besteht die Gleichheit (2.2).

Ergänzung. Für jeden Punkt $x \in U$ bezeichne $\mathfrak{F}(x)$ den Filter der Abschnitte des bezüglich der Relation $\supset$ filtrierenden Systems aller Umgebungen $B \in \mathfrak{B}$ von $x$ mit $\overline{B} \subset U$. Dann ist

$$\hat{v}(x) = \lim_{\mathfrak{F}(x)} \int^* v\, d\mu_x^B \qquad (x \in U). \tag{2.3}$$

Wegen (2.2) genügt es hierfür zu zeigen, daß aus $B, C \in \mathfrak{F}(x)$ und $\overline{C} \subset B$ folgt

$$\int^* v d\mu_x^B \leqq \int^* v d\mu_x^C.$$

Dies aber ergibt sich wie folgt: Die Funktion $w(y) = \int^* v d\mu_y^B$ $(y \in B)$ ist nach obigem Beweis harmonisch in $B$; somit gilt $w(x) = \int w d\mu_x^C$. Hieraus folgt die zu beweisende Ungleichung wegen $w(y) \leqq v(y)$ für alle $y \in B$.

**Korollar.** *Es gelte das Axiom $K_D$. Dann ist für jede superharmonische Funktion $u \geqq 0$ auf $X$ und jede Menge $E \subset X$ die extremisierte Funktion $\hat{R}_u^E$* superharmonisch in $X$.*

*Beweis.* Die Behauptung folgt aus Satz 3, da $\hat{R}_u^E$ die untere Regularisierte von $R_u^E$ ist. Diese letztere Funktion ist aber als untere Einhüllende von hyperharmonischen Funktionen $\geqq 0$ eine $S$-Funktion. Wegen $0 \leqq R_u^E \leqq u$ ist weiter $R_u^E$ auf einer dichten Teilmenge von $X$ endlich.

**2.3. $\mathfrak{B}$-vernachlässigbare Mengen.** Es bezeichne wieder $\mathfrak{B}$ eine aus offenen, relativ-kompakten Mengen bestehende Basis von $X$.

**Definition 3.** Eine Teilmenge $A$ einer Menge $U \in \mathfrak{U}$ heiße *$\mathfrak{B}$-vernachlässigbar in $U$*, wenn für alle Mengen $B \in \mathfrak{B}$ mit $\bar{B} \subset U$ gilt:

$$\mu_x^B(A \cap B^*) = 0 \qquad \text{für alle } x \in B. \tag{2.4}$$

Kommt eine Eigenschaft $E$ allen Punkten von $U$ mit Ausnahme der Punkte einer in $U$ $\mathfrak{B}$-vernachlässigbaren Menge zu, so wollen wir sagen: „*$\mathfrak{B}$-nahezu alle Punkte von $U$ besitzen diese Eigenschaft*" oder „*$E$ besteht $\mathfrak{B}$-nahezu überall in $U$*".

Wir stellen folgende Eigenschaften dieser Begriffsbildungen fest:

4. Jede Teilmenge einer in $U \in \mathfrak{U}$ $\mathfrak{B}$-vernachlässigbaren Menge ist $\mathfrak{B}$-vernachlässigbar in $U$.

5. Die Vereinigung abzählbar vieler in $U \in \mathfrak{U}$ $\mathfrak{B}$-vernachlässigbarer Mengen ist $\mathfrak{B}$-vernachlässigbar in $U$.

6. Eine Menge $A$ sei $\mathfrak{B}$-vernachlässigbar in $U \in \mathfrak{U}$. Dann ist $U \cap \complement A$ dicht in $U$.

In der Tat: Nach Axiom $T$ existiert eine in $X$ positive, harmonische Funktion $h$. Aus $h(x) = \int h\,d\mu_x^B$ folgt: $\int d\mu_x^B = \mu_x^B(B^*) > 0$ für alle $B \in \mathfrak{B}$ und $x \in B$. Somit ist $A \cap B^* \neq B^*$ für alle $B \in \mathfrak{B}$ mit $\bar{B} \subset U$. Also liegt auf dem Rand $B^*$ eines jeden derartigen $B \in \mathfrak{B}$ noch mindestens ein Punkt der Menge $U \cap \complement A$. Hieraus folgt die Behauptung, da $\mathfrak{B}$ eine Basis von $X$ ist.

7. Es gelte das Axiom $K_D$. $v_1$ und $v_2$ seien $S_{\mathfrak{B}}$-Funktionen in $U \in \mathfrak{U}$ derart, daß $v_1$ und $v_2$ auf je einer dichten Teilmenge von $U$ endlich sind und $\mathfrak{B}$-nahezu überall in $U$ die Gleichheit $v_1(x) = v_2(x)$ besteht. Dann ist $\hat{v}_1 = \hat{v}_2$.

Dies folgt unmittelbar aus Satz 3. Ein wichtiger Spezialfall hiervon ist:

8. Es gelte das Axiom $K_D$. Sind dann $v_1$ und $v_2$ superharmonische Funktionen in einer Menge $U \in \mathfrak{U}$ und gilt $v_1(x) = v_2(x)$ $\mathfrak{B}$-nahezu überall in $U$, so ist $v_1 = v_2$.

9. Es gelte das Axiom $K_D$. Eine Funktion $u \in \mathscr{H}_U^*$ ist genau dann superharmonisch in $U$, wenn $u(x)$ $\mathfrak{B}$-nahezu überall in $U$ endlich ist.

In der Tat: $u$ sei superharmonisch in $U$. Nach [1], Satz 16 gilt $\int^* u\,d\mu_x^B \leqq \bar{H}_u^B(x) \leqq u(x)$ für alle $B \in \mathfrak{B}$ mit $\bar{B} \subset U$ und alle $x \in B$. Aus § 1, Eigenschaft 8 folgt daher die $\mu_x^B$-Integrierbarkeit von $u$ für alle $x \in B$. Bezeichnet also $A$ die Menge aller $x \in U$ mit $u(x) = +\infty$, so ist $\mu_x^B(A \cap B^*) = 0$ für alle $x \in B$. D. h. aber $u(x)$ ist $\mathfrak{B}$-nahezu überall in $U$ endlich. — Die umgekehrte Aussage folgt aus der obigen Eigenschaft 6.

---

* Zur Definition von $\hat{R}_u^E$ siehe [1], § 13.

Bemerkung. Verfügt man nicht über das Doobsche Konvergenzaxiom, sondern nur über die Axiome $K_1$ bzw. $K_2$, so bleiben Satz 3 samt Ergänzung und Korollar sowie die Eigenschaften 7 und 8 wenigstens für lokal-beschränkte bzw. reelle superharmonische Funktionen bestehen. Die Beweise hierfür können analog geführt werden.

## § 3. Polare Mengen

**3.1. Definition und Beispiele.** Wie in der klassischen Theorie definieren wir:

**Definition 4.** Eine Menge $P \subset X$ heiße *polar* (in $X$), wenn eine superharmonische Funktion $u \geqq 0$ auf $X$ existiert mit

$$P \subset u^{-1}(+\infty). \tag{3.1}$$

Steht das Axiom $K_D$ zur Verfügung, so kann $u$ o.B.d.A. als Potential angenommen werden. Dies folgt aus § 1, Eigenschaft 10.

Zur Illustration bemerken wir, daß im Falle der Laplace-Gleichung auf dem Raum $X = \boldsymbol{R}^n$ $(n \geqq 3)$ jede einpunktige Menge polar ist. Für die Dimension $n = 2$ bleibt diese Aussage richtig, wenn $X$ eine offene relativ-kompakte Teilmenge des $\boldsymbol{R}^2$ ist.

Auch für das Garbendatum der Lösungen der Wärmeleitungsgleichung auf dem Raum $X = \boldsymbol{R}^{n+1}$ $(n \geqq 1)$ ist jede einpunktige Menge $\{x_0\}$ polar. Zum Beweise werde

$$x_k = (\xi_1, \ldots, \xi_n, \xi_{n+1} - k^{-4})$$

für $k = 1, 2, \ldots$ gesetzt; hierbei seien $\xi_1, \ldots, \xi_{n+1}$ die Koordinaten von $x_0$. Dann ist

$$\mu = \sum_{k=1}^{\infty} k^{-2} \varepsilon_{x_k}$$

ein positives Maß auf $X$, dessen Träger im Halbraum aller Punkte mit $(n+1)$-ter Koordinate $\leqq \xi_{n+1}$ enthalten ist. Bezeichnet also $P$ den auf $X \times X$ definierten parabolischen Kern (vgl. [*1*], S. 22), so ist nach Doob [*6*], p. 222 das dem Maß $\mu$ entsprechende Wärmepotential

$$x \to P^{\mu}(x) = \int^{*} P(x, y)\, d\mu(y)$$

auf $X$ superharmonisch und $\geqq 0$. Man rechnet leicht nach, daß $P^{\mu}(x_0) = +\infty$, also die Menge $\{x_0\}$ polar ist.

**3.2. Eigenschaften.** Wir benötigen die folgenden Eigenschaften polarer Mengen:

1. Jede Teilmenge einer polaren Menge ist polar.

2. Es gelte das Axiom $K_D$. Dann ist jede polare Menge $\mathfrak{B}$-vernachlässigbar in $X$ bezüglich jeder Basis $\mathfrak{B}$ offener relativ-kompakter Mengen.

3. Es gelte das Axiom $K_D$. Dann ist die Vereinigung je endlich vieler polarer Mengen polar.

Die Eigenschaft 1 folgt aus Definition 4. Eigenschaft 2 ist eine Folgerung aus § 2, Eigenschaft 9. Aus dem Korollar 2 zu Satz 1 folgt schließlich die letzte der drei Eigenschaften. Diese wird im folgenden Satz verallgemeinert.

**Satz 4.** *Es gelte das Axiom $K_D$; der Raum $X$ besitze eine abzählbare Basis. Dann ist die Vereinigung $P$ jeder Folge $P_1, P_2, \ldots$ polarer Mengen selbst wieder polar.*

*Beweis.* Es sei $V_1, V_2, \ldots$ eine abzählbare Basis von $X$. Aus dem Axiom II und dem Überdeckungssatz von Lindelöf folgt, daß die Mengen $V_n$ ohne Beschränkung der Allgemeinheit als regulär vorausgesetzt werden können. Für jedes $n = 1, 2, \ldots$ sei $x_n$ ein in $V_n$ beliebig gewählter Punkt und $\mu_n$ das zugehörige harmonische Maß $\mu_{x_n}^{V_n}$. Schließlich sei $\varepsilon_1, \varepsilon_2, \ldots$ eine Folge positiver reeller Zahlen mit $\sum \varepsilon_n < +\infty$.

Zu jeder natürlichen Zahl $n$ existiert nun eine in $X$ superharmonische Funktion $u_n \geqq 0$ mit $P_n \subset u_n^{-1}(+\infty)$. Nach Satz 1 ist $u_n$ $\mu_x^V$-integrierbar für jede reguläre Menge $V$ und alle $x \in V$. Somit kann ohne Beschränkung der Allgemeinheit angenommen werden, daß gilt:

$$\int u_n d\mu_i \leqq \varepsilon_n$$

für $i = 1, \ldots, n$ und alle $n = 1, 2, \ldots$. Man hat nur gegebenenfalls $u_n$ durch $\alpha u_n$ mit geeignetem $\alpha > 0$ zu ersetzen. Wir setzen nun

$$u = \sum_{n=1}^{\infty} u_n .$$

Dann ist $u$ hyperharmonisch und $\geqq 0$ auf $X$; ferner gilt $P \subset u^{-1}(+\infty)$. Zu zeigen ist also nur, daß $u$ superharmonisch ist. Nun gilt aber für jedes $j = 1, 2, \ldots$

$$\int^* u d\mu_j \leqq \int u_1 d\mu_j + \cdots + \int u_j d\mu_j + \sum_{n>j} \varepsilon_n .$$

Da alle Funktionen $u_n$ superharmonisch und $\sum \varepsilon_n < +\infty$ ist, so ist also

$$\int^* u d\mu_j < +\infty \qquad (j = 1, 2, \ldots).$$

Auf dem Rand jeder Menge $V_j$ liegt daher mindestens ein Punkt $y_j$ mit $|u(y_j)| < \infty$. Die Menge dieser Punkte liegt dicht in $X$. Also ist $u$ superharmonisch.

Wie bei BRELOT [3], pp. 125—126 beweist man die folgende Eigenschaft:

4. Es gelte das Axiom $K_D$. Es sei $P$ eine abgeschlossene polare Menge und $u$ eine in $\complement P$ definierte superharmonische bzw. harmonische Funktion. Dann und nur dann kann $u$ zu einer superharmonischen bzw. harmonischen Funktion $u_0$ auf $X$ fortgesetzt werden, wenn zu jedem Punkt $x \in P$ eine Umgebung $V$ existiert mit $\inf u(V \cap \complement P) > -\infty$ bzw. $\sup |u| (V \cap \complement P) < +\infty$. $u_0$ ist eindeutig bestimmt, falls die Fortsetzung möglich ist.

**3.3. Kennzeichnung polarer Mengen.** Genauso wie bei BRELOT [3], pp. 126 — 127 zeigt man:

5. Es sei $P$ eine polare und $E$ eine beliebige Teilmenge von $X$. Dann ist

$$\hat{R}_u^{E \cup P} = \hat{R}_u^E \tag{3.2}$$

für jede hyperharmonische Funktion $u \geqq 0$.

Indem man $E = \emptyset$ setzt, erhält man:

6. Für jede polare Menge $P \subset X$ und jede hyperharmonische Funktion $u \geqq 0$ auf $X$ gilt:

$$\hat{R}_u^P = 0 . \tag{3.3}$$

Umgekehrt gilt:

**Satz 5.** *Der Raum $X$ besitze eine abzählbare Basis. Existiert dann zu einer Menge $P \subset X$ eine in $X$ hyperharmonische Funktion $u \geqq 0$ mit $u(x) > 0$ für alle $x \in P$ sowie mit $\hat{R}_u^P = 0$, so ist $P$ polar.*

*Beweis.* Ohne Beschränkung der Allgemeinheit kann die Funktion $u$ als lokal-beschränkt vorausgesetzt werden; man hat nur $u$ durch $\inf(u, h)$ zu ersetzen. Nach Satz 3 und der Bemerkung am Schluß von § 2 gilt dann für jeden Punkt $x \in X$

$$0 = \hat{R}_u^P(x) = \sup \int^* R_u^P \, d\mu_x^B ,$$

wenn hierbei $B$ z.B. alle offenen relativ-kompakten Umgebungen von $x$ durchläuft. Wegen $R_u^P \geqq 0$ folgt hieraus und aus dem Vorhandensein einer abzählbaren Basis die Existenz einer in $X$ dicht liegenden Punktfolge $x_1, x_2, \ldots$ mit $R_u^P(x_n) = 0$ für alle $n$. Ist nun $(\varepsilon_n)_{n=1,2,\ldots}$ eine Folge positiver reeller Zahlen mit $\sum \varepsilon_n < +\infty$, so gibt es zu jeder natürlichen Zahl $n$ eine Funktion $v_n \in \mathscr{H}_X^*$ mit: $0 \leqq v_n \leqq u$, $v_n(x) = u(x)$ für alle $x \in P$ und $v_n(x_i) \leqq \varepsilon_n$ für $i = 1, \ldots, n$. Dabei hat man zu berücksichtigen, daß wegen $u \in \mathscr{H}_X^*$ die Funktion $R_u^P$ gleich der unteren Einhüllenden aller $w \in \mathscr{H}_X^*$ mit $0 \leqq w \leqq u$ und $w(x) = u(x)$ für alle $x \in P$ ist. Die Funktion $v = \sum_{n=1}^{\infty} v_n$ ist dann auf $X$ hyperharmonisch und $\geqq 0$; für jeden Punkt $x_j$ gilt

$$v(x_j) \leqq v_1(x_j) + \cdots + v_j(x_j) + \sum_{n>j} \varepsilon_n < +\infty ,$$

da mit $u$ auch jedes $v_n$ reellwertig ist. Somit ist $v$ sogar superharmonisch. Wegen $u(x) > 0$ für alle $x \in P$ ist schließlich noch $P \subset v^{-1}(+\infty)$. Daher ist die Menge $P$ polar.

## II. Existenz von Potentialen

### § 4. Absorptionsmengen

**4.1. Definition und Eigenschaften.** Wir beginnen mit der

**Definition 5.** Eine Menge $A \subset X$ heiße *Absorptionsmenge* (in $X$), wenn $A$ abgeschlossen ist und für jeden Punkt $x \in A$ sowie für jede reguläre Umgebung $V$ von $x$ das harmonische Maß $\mu_x^V$ von $A$ getragen wird.

Stets sind offenbar $A = \emptyset$ und $A = X$ Absorptionsmengen. Aus der Definition folgt, daß der Durchschnitt beliebig vieler und die Vereinigung endlich vieler Absorptionsmengen selbst wieder eine Absorptionsmenge ist. Weitere Beispiele für Absorptionsmengen liefert der folgende Satz:

**Satz 6.** *Für jede Menge $A \subset X$ sind folgende Aussagen gleichwertig:*

(a) *$A$ ist eine Absorptionsmenge.*

(b) *$A$ ist abgeschlossen. Es gibt eine Abbildung $\mathfrak{V}$, welche jedem Punkt $x \in X$ ein Fundamentalsystem $\mathfrak{V}(x)$ regulärer Umgebungen von $x$ derart zuordnet, daß für jeden Punkt $x \in A$ und jedes $V \in \mathfrak{V}(x)$ das harmonische Maß $\mu_x^V$ von $A$ getragen wird.*

(c) *Es gibt eine hyperharmonische Funktion $u \geqq 0$ auf $X$ mit $A = u^{-1}(0)$.*

*Beweis.* (a) $\Rightarrow$ (b): Dies folgt unmittelbar aus Definition 5.

(b) $\Rightarrow$ (c): Wir betrachten folgende, auf $X$ definierte Funktion $u$:

$$u(x) = \begin{cases} 0 \ , & x \in A; \\ h(x), & x \notin A. \end{cases}$$

Hierbei bezeichnet $h$ wieder eine auf $X$ harmonische und positive Funktion. Offensichtlich ist $u$ $\mathfrak{B}$-hyperharmonisch, also nach [*1*], Satz 5 hyperharmonisch auf $X$. Nach Konstruktion ist ferner $u \geqq 0$ und $A = u^{-1}(0)$.

(c) $\Rightarrow$ (a): $A$ ist abgeschlossen, da $u$ nach unten halbstetig ist. Für jeden Punkt $x \in A$ und jede reguläre Umgebung $V$ von $x$ ist $\int^* u d\mu_x^V = 0$ wegen $u \geqq 0$ und $\int^* u d\mu_x^V \leq u(x) = 0$. Somit wird $\mu_x^V$ von $V^* \cap A$ und damit von $A$ getragen.

Ähnlich wie im Beweis dieses Satzes ergeben sich folgende Eigenschaften:

1. Es sei $A$ eine Absorptionsmenge und $u$ eine in $X$ hyperharmonische Funktion mit $u(x) \geqq 0$ für alle $x \in \complement A$. Dann ist folgende Funktion $u^A$ auf $X$ hyperharmonisch:

$$u^A(x) = \begin{cases} u(x), & x \in \complement A; \\ 0 \ , & x \in A. \end{cases} \tag{4.1}$$

2. Es sei $A$ eine Absorptionsmenge, $x$ ein Punkt aus $A$ und $U$ eine offene relativkompakte Umgebung von $x$. Dann wird das harmonische Maß $\mu_x^U$ von $A$ getragen.

(Beim Beweis von 2 beachte man, daß nach [*1*], Satz 16 gilt $\int^* u d\mu_x^U \leqq u(x)$ für alle $u \in \mathscr{H}_X^*$.)

3. Schränkt man das Garbendatum der harmonischen Funktionen auf eine Menge $U_0 \in \mathfrak{U}$ ein, so ist für jede Absorptionsmenge $A$ in $X$ der Durchschnitt $A \cap U_0$ eine Absorptionsmenge in $U_0$ bezüglich des neuen Garbendatums.

4. Beim Übergang vom ursprünglichen Garbendatum zu dem der $h$-harmonischen Funktionen ändern sich die Absorptionsmengen nicht. Hierbei kann $h$ eine beliebige, überall positive Funktion aus $\mathscr{C}(X)$ sein.

(Dies folgt aus der in [*1*] angegebenen Formel (1.19).)

**4.2. Beispiele.** In der Theorie von BRELOT [*3*] sind (bei Zugrundelegung der Brelotschen Axiome 1—3 und im Falle der Existenz eines überall positiven Potentials $u_0$) $A = \emptyset$ und $A = X$ die einzigen Absorptionsmengen. Nach Eigenschaft 1 ist nämlich $u_0^A$ hyperharmonisch für jede Absorptionsmenge. Aus dem Minimumprinzip der Brelotschen Theorie folgt daher: $u_0^A(x) > 0$ für alle $x \in X$ oder $u_0^A(x) = 0$ für alle $x \in X$. Dies besagt aber gerade: $A = \emptyset$ oder $A = X$.

Die Existenz nicht-trivialer Absorptionsmengen in unserer Theorie zeigt das Beispiel der Wärmeleitungsgleichung. Es sei nämlich $X = \boldsymbol{R}^{n+1}$ $(n \geqq 1)$ und $U \to \mathscr{H}_U$ das Garbendatum der Lösungen von $\Delta u = \frac{\partial u}{\partial x_{n+1}}$. Dann ist für jede reelle Zahl $\xi$ die Menge

$$A_\xi = \{x \in X : x_{n+1} \leqq \xi\} \tag{4.2}$$

eine Absorptionsmenge, wie aus Satz 6 und den Betrachtungen in [*1*], Nr. 6.1 folgt. Setzt man noch $A_{-\infty} = \emptyset$ und $A_{+\infty} = X$, so durchläuft übrigens $A_\xi$, $-\infty \leqq \xi \leqq +\infty$, alle Absorptionsmengen in diesem Beispiel.

**4.3. Weitere Eigenschaften.** Nach der Aussage (c) des Satzes 6 kann man die Absorptionsmengen gewissermaßen als duales Gegenstück zu den polaren Mengen auffassen. Der Eigenschaft 6 polarer Mengen aus § 3 entspricht folgende Eigenschaft:

5. Für jede Absorptionsmenge $A \subset X$ und jede hyperharmonische Funktion $u \geqq 0$ auf $X$ gilt:

$$R_u^{\complement A}(x) = \hat{R}_u^{\complement A}(x) = 0 \text{ für alle } x \in A. \tag{4.3}$$

Es ist nämlich $R_u^{\complement A} = u^A$, woraus obige Behauptung folgt.

Umgekehrt gilt — man vergleiche die Analogie zu Satz 5:

6. Existiert zu einer abgeschlossenen Menge $A \subset X$ eine in $X$ hyperharmonische Funktion $u \geqq 0$ mit $u(x) > 0$ für alle $x \in \complement A$ sowie mit $\hat{R}_u^{\complement A}(x) = 0$ für alle $x \in A$, so ist $A$ eine Absorptionsmenge.

In der Tat: Wie im Beweis von Satz 5 zeigt man, daß $u$ als lokal-beschränkt angenommen werden kann. Dann ist $\hat{R}_u^{\complement A}$ hyperharmonisch nach der Bemerkung am Schluß von § 2. Da $\complement A$ offen ist, gilt $\hat{R}_u^{\complement A}(x) = u(x)$ für alle $x \in \complement A$ und somit $\hat{R}_u^{\complement A} = u^A$. Die Behauptung folgt daher aus Satz 6.

Abschließend zeigen wir noch:

**Satz 7.** *Die leere Menge ist die einzige kompakte Absorptionsmenge.*

*Beweis.* Nach Eigenschaft 4 kann zum Beweis angenommen werden, daß die konstante Funktion $h = 1$ auf $X$ harmonisch ist. Ist nun $A$ eine nicht-leere kompakte Absorptionsmenge, so betrachten wir die Menge $\mathscr{E}$ aller auf $A$ eingeschränkten Funktionen aus $\mathscr{H}_X^*$. Nach BAUER [2] existiert dann ein Punkt $x_0 \in A$ derart, daß $\mu = \varepsilon_{x_0}$ das einzige Maß $\mu \geqq 0$ auf $A$ ist mit $\int^* u\,d\mu \leqq u(x_0)$ für alle $u \in \mathscr{E}$. Für jede reguläre Umgebung $V$ von $x_0$ ist aber $\mu_{x_0}^V$ ein positives Maß auf $A$, weil $A$ eine Absorptionsmenge ist; es gilt weiter $\int^* u\,d\mu_{x_0}^V \leqq u(x_0)$ für alle $u \in \mathscr{E}$ und $\mu_{x_0}^V \neq \varepsilon_{x_0}$. Dies steht im Widerspruch zu der genannten Eigenschaft von $x_0$ und damit zu der Annahme, daß die kompakte Absorptionsmenge $A$ nicht-leer sei.

Aus den Sätzen 6 und 7 folgt nun, daß eine hyperharmonische Funktion $u \geqq 0$ entweder eine leere oder eine nicht-kompakte Menge von Nullstellen besitzt.

## § 5. Vorbereitende Approximationsaussagen

Wir beweisen eine Reihe von Hilfssätzen, die im nächsten Paragraphen benötigt werden.

**Lemma 2.** *Es sei $u \geqq 0$ eine lokal-beschränkte superharmonische Funktion auf $X$, welche in einer Umgebung einer kompakten Menge $K$ stetig ist. Dann existiert eine stetige superharmonische Funktion $u^*$ auf $X$ mit $0 \leqq u^* \leqq u$ und $u^*(x) = u(x)$ für alle $x \in K$.*

*Beweis.* Es sei $U$ eine offene relativ-kompakte Umgebung von $K$, in welcher $u$ stetig ist. Wir setzen

$$v = \hat{R}_u^U.$$

Damit ist $v$ nach der Schlußbemerkung des § 2 eine auf $X$ superharmonische Funktion mit folgenden Eigenschaften: $0 \leqq v \leqq u$; $v(x) = u(x)$ für alle $x \in U$; $v$ ist in

$\complement \overline{U}$ harmonisch und daher dort stetig (nach [1], Satz 37). Der kompakte Rand $U^*$ von $U$ kann durch endlich viele, zu $K$ fremde, reguläre Mengen $V_1, \ldots, V_n$ überdeckt werden. Dann aber leistet

$$u^* = v_{V_1, \ldots, V_n}$$

das Verlangte. Zunächst ist $u^*$ hyperharmonisch in $X$, $0 \leqq u^* \leqq v \leqq u$ sowie $u^*(x) = v(x) = u(x)$ für alle $x \in K$. Gemäß der Wahl von $U$ ist $T = U \cap \complement \overline{U}$ offen und $v$ in $T$ stetig. Nach [1], Satz 21 ist dann $u^*$ auf $T \cup V_1 \cup \ldots \cup V_n = X$ stetig. $u^*$ besitzt somit alle gewünschten Eigenschaften.

Am Ende von [1] trat erstmals eine verschärfte Form des Axioms $T$ auf, welche von nun an mit $T_+$ bezeichnet werden soll:

**Axiom $T_+$.** *Es existiert eine in $X$ harmonische und überall positive Funktion $h$, für welche die Menge aller auf $X$ nicht-negativen, $h$-hyperharmonischen Funktionen die Punkte von $X$ trennt.*

Es sei hier nochmals an die Schlußbemerkung von [1] erinnert, wonach das Axiom $T$ das Axiom $T_+$ in jeder relativ-kompakten offenen Teilmenge von $X$ (aufgefaßt als neuer Grundraum) nach sich zieht. Ferner wurde in den meisten der in [1], § 6 besprochenen Beispiele auf die Gültigkeit von $T_+$ hingewiesen.

Ähnlich wie im Anschluß an das Axiom $T$ kann auch hier bemerkt werden, daß die in $T_+$ ausgesprochene Trennungsbedingung unabhängig von der speziellen Wahl der positiven harmonischen Funktion $h$ ist.

**Lemma 3.** *Ist das Axiom $T_+$ erfüllt, so trennt sogar die Menge aller auf $X$ stetigen, reellen, nicht-negativen, $h$-superharmonischen Funktionen die Punkte von $X$.*

*Beweis.* Zu je zwei verschiedenen Punkten $x, y \in X$ gibt es eine $h$-hyperharmonische Funktion $v' \geqq 0$ auf $X$ mit $v'(x) \neq v'(y)$. Da die konstanten Funktionen $h$-harmonisch sind, so ist $v = \inf(v', \alpha)$ für geeignetes $\alpha > 0$ ($\alpha \in \boldsymbol{R}$) eine beschränkte $h$-superharmonische Funktion $\geqq 0$ mit $v(x) \neq v(y)$. Nach [1], Lemma 8 kann angenommen werden, daß $v$ in einer offenen Umgebung der kompakten Menge $K = \{x, y\}$ stetig ist. Nach Lemma 2, angewandt auf das Garbendatum der $h$-harmonischen Funktionen, gibt es eine stetige $h$-superharmonische Funktion $v^* \geqq 0$ auf $X$ mit: $0 \leqq v^* \leqq v$, $v^*(x) = v(x)$ und $v^*(y) = v(y)$. Mit $v$ trennt also auch $v^*$ die Punkte $x$ und $y$.

Bemerkung. Wenn der Raum $X$ abzählbar im Unendlichen ist, so folgt obiges Lemma auch aus [1], Lemma 8.

**Lemma 4.** *Es gelte das Axiom $T_+$. Ist dann $A$ eine Absorptionsmenge und $\mathscr{S}^A$ die Menge aller stetigen, reellen, superharmonischen Funktionen $u \geqq 0$ auf $X$ mit $A \subset u^{-1}(0)$, so kann jede Funktion aus $\mathscr{C}(X)$ auf jeder kompakten Menge $K \subset \complement A$ gleichmäßig durch Funktionen aus $\mathscr{S}^A - \mathscr{S}^A$ approximiert werden.*

*Beweis.* Es genügt den Beweis für den Fall $h = 1$ zu führen. Dann ist nämlich im allgemeinen Fall gezeigt, daß jede Funktion aus $\mathscr{C}(X)$ auf jeder kompakten Menge $K \subset \complement A$ gleichmäßig durch Funktionen aus ${}^h\mathscr{S}^A - {}^h\mathscr{S}^A$ approximiert werden kann, wenn ${}^h\mathscr{S}^A$ die Menge aller stetigen, reellen, $h$-superharmonischen Funktionen $v \geqq 0$ auf $X$ mit $A \subset v^{-1}(0)$ bezeichnet. Hieraus folgt aber sofort Lemma 3. Also werde von nun an $h = 1$ vorausgesetzt.

Offenbar ist $\mathscr{S}^A + \mathscr{S}^A \subset \mathscr{S}^A$ und $\lambda \mathscr{S}^A \subset \mathscr{S}^A$ für alle $\lambda \geqq 0$; folglich ist $\mathscr{D} = \mathscr{S}^A - \mathscr{S}^A$ ein linearer Unterraum von $\mathscr{C}(X)$. $\mathscr{S}^A$ enthält mit je zwei Funktionen deren untere Einhüllende; folglich enthält $\mathscr{D}$ mit jeder Funktion $d$ auch den Absolutbetrag $|d|$ dieser Funktion. Weiter trennt $\mathscr{S}^A$ und damit $\mathscr{D}$ je zwei Punkte $x \neq y$ aus $\complement A$. Nach Lemma 3 gibt es nämlich eine superharmonische Funktion $t \geqq 0$ aus $\mathscr{C}(X)$ mit $t(x) \neq t(y)$. Auf die Funktion $u = t^A$ * und die kompakte Menge $K = \{x, y\}$ kann Lemma 2 angewendet werden. Danach gibt es eine superharmonische Funktion $u^* \in \mathscr{C}(X)$ mit $0 \leqq u^* \leqq u$ sowie $u^*(x) = u(x) = t(x)$ und $u^*(y) = u(y) = t(y)$. Somit ist $u^* \in \mathscr{S}^A$ und $u^*(x) \neq u^*(y)$. Schließlich enthält $\mathscr{D}_K = \mathrm{Rest}_K\, \mathscr{D}$ für jede kompakte Menge $K \subset \complement A$ die konstanten Funktionen. Man hat hierzu nur Lemma 2 auf die Funktion $u = h^A$ anzuwenden, wobei $h = 1$ ist.

Für jede kompakte Menge $K \subset \complement A$ genügt also die Funktionenmenge $\mathscr{D}_K$ den Voraussetzungen des Approximationssatzes von M. H. Stone. Danach liegt $\mathscr{D}_K$ in $\mathscr{C}(K)$ dicht. Hieraus aber folgt die Behauptung.

## § 6. Potentiale und Absorptionsmengen

**6.1. Strenge Potentiale.** Entscheidend für die weiteren Betrachtungen ist die folgende Begriffsbildung:

**Definition 6.** Eine in einer Menge $U \in \mathfrak{U}$ definierte, hyperharmonische Funktion $u$ heiße *streng hyperharmonisch* in $U$, wenn für jede in $U$ reguläre Menge $V$ und jeden Punkt $x \in V$ gilt:

$$\int^* u\, d\mu_x^V < u(x). \tag{6.1}$$

Jedes Potential, welches streng hyperharmonisch ist, heiße ein *strenges Potential.*

Es ergeben sich sofort folgende Eigenschaften streng hyperharmonischer Funktionen:

1. Jede in $U \in \mathfrak{U}$ streng hyperharmonische Funktion $u$ ist $\mu_x^V$-integrierbar für jede in $U$ reguläre Menge $V$ und jedes $x \in V$. Insbesondere ist $u$ superharmonisch**. Wir werden daher $u$ auch *streng superharmonisch* nennen.

2. Es sei $u$ eine in $U \in \mathfrak{U}$ streng superharmonische und $g$ eine in $U$ harmonische Funktion. Dann ist auch $u + g$ streng superharmonisch in $U$.

3. Jede streng superharmonische und nicht-negative Funktion, insbesondere also jedes strenge Potential, ist in seinem Definitionsbereich überall positiv.

Diese drei Eigenschaften folgen unmittelbar aus (6.1)

4. Wenn $u$ streng superharmonisch bzw. ein strenges Potential ist, so ist $u/h$ streng $h$-superharmonisch bzw. ein strenges $h$-Potential und umgekehrt.

Dies folgt aus der in [*1*], Formel (1.19) angegebenen Gestalt der $h$-harmonischen Maße.

Etwas tiefer liegt die folgende Eigenschaft:

**Satz 8.** *Es gelte das Axiom $K_D$. Die Funktion $u$ sei streng superharmonisch in $U \in \mathfrak{U}$. Dann gilt für jede relativ-kompakte Menge $W \in \mathfrak{U}$ mit $\overline{W} \subset U$ und für jeden Punkt $x \in W$:*

$$\int u\, d\mu_x^W < u(x). \tag{6.2}$$

* Vgl. Formel (4.1).

** Vgl. hierzu den Beweis von Satz 1.

*Beweis.* Für gegebenes $x_0 \in W$ sei $V$ eine reguläre Menge mit $x_0 \in V \subset \overline{V} \subset W$. Da $u$ eine zu $\mathrm{Rest}_{W^*} u$ gehörige Oberfunktion ist, so gilt $\overline{H}_u^W \leqq \mathrm{Rest}_W u$; hieraus folgt nach § 1, Eigenschaft 8 die Harmonizität von $q = \overline{H}_u^W$ in $W$. Folglich erhält man:

$$q(x_0) = \int q \, d\mu_{x_0}^V \leqq \int u \, d\mu_{x_0}^V < u(x_0).$$

Andererseits ist nach [1], Satz 16

$$\int u \, d\mu_{x_0}^W \leqq \overline{H}_u^W(x_0) = q(x_0).$$

Beide Ungleichungen zusammen ergeben $\int^* u \, d\mu_{x_0}^W < u(x_0)$. Hieraus folgt die Endlichkeit des Oberintegrals $\int^* u \, d\mu_{x_0}^W$, also die $\mu_{x_0}^W$-Integrierbarkeit von $u$ für alle $x_0 \in W$ und damit (6.2).

Bemerkung. Verfügt man nur über das Axiom $K_1$ bzw. $K_2$, so bleibt die Behauptung des Satzes 8 richtig, wenn $u$ zusätzlich als lokal-beschränkt bzw. als reellwertig vorausgesetzt wird.

**6.2. Existenzsatz.** Im Falle der Wärmeleitungsgleichung ist die mit Hilfe des parabolischen Kerns $P$ (vgl. [1], S. 22) definierte Funktion $x \to P(x, y)$ für jeden Punkt $y \in X = \boldsymbol{R}^{n+1}$ $(n \geqq 1)$ ein Potential auf $X$. Für jede reelle Zahl $\xi$ und den zugehörigen Punkt

$$y_\xi = (0, \ldots, 0, \xi) \in X$$

ist dann $x \to P(x, y_\xi)$ ein Potential $p_\xi$ auf $X$, welches genau auf der in (4.2) definierten Absorptionsmenge $A_\xi$ verschwindet.

Der folgende Satz zeigt, daß dieser Bemerkung ein allgemeiner Sachverhalt zugrunde liegt:

**Satz 9.** *Der Raum $X$ besitze eine abzählbare Basis; es gelte das Axiom $T_+$. Dann existiert zu jeder Absorptionsmenge $A \subset X$ ein stetiges reelles Potential $p$ auf $X$ derart, daß die Restriktion von $p$ auf $\complement A$ ein strenges Potential auf $\complement A$ und $A = p^{-1}(0)$ ist.*

*Existiert eine beschränkte, überall positive, harmonische Funktion $h$ auf $X$, so kann $p$ zusätzlich noch beschränkt gewählt werden.*

*Beweis.* Wegen Eigenschaft 4 genügt es, den Beweis für den Fall zu führen, daß die konstante Funktion $h = 1$ harmonisch auf $X$ ist. Ohne Beschränkung der Allgemeinheit kann $A \neq X$ angenommen werden.

Es bezeichne $\mathscr{C}^A$ die Menge aller Funktionen $f \in \mathscr{C}(X)$ mit $A \subset f^{-1}(0)$. Jede stetige reelle Funktion auf einer kompakten Menge $K \subset \complement A$ kann zu einer Funktion aus $\mathscr{C}^A$ (sogar mit kompaktem Träger) fortgesetzt werden. Mit $X$ besitzt auch $\complement A$ eine abzählbare Basis; daher kann die offene Menge $\complement A$ durch eine Folge kompakter, metrisierbarer Mengen derart überdeckt werden, daß jede kompakte Teilmenge von $\complement A$ in einer dieser Mengen enthalten ist. Somit existiert eine abzählbare Teilmenge von $\mathscr{C}^A$, welche in $\mathscr{C}^A$ dicht liegt bezüglich der Topologie $\mathscr{T}^A$ der gleichmäßigen Konvergenz auf den kompakten Teilmengen von $\complement A$. Nach Lemma 4 kann jede Funktion aus $\mathscr{C}^A$ auf jeder kompakten Menge $K \subset \complement A$ gleichmäßig durch eine Differenz $u - v$ von nicht-negativen, superharmonischen Funktionen aus $\mathscr{C}^A$ approximiert werden. Da die Funktion $h = 1$ harmonisch ist, können $u$ und $v$ sogar als beschränkt angenommen werden. (Man ersetze gegebenenfalls $u$ durch $u' = \inf(u, \sup u(K))$ und $v$ durch $v' = \inf(v, \sup v(K))$.)

Berücksichtigt man dies, so gibt es also eine Folge $u_1, u_2, \ldots$ beschränkter, nicht-negativer, superharmonischer Funktionen aus $\mathscr{C}^A$, welche in $\mathscr{C}^A$ bezüglich der Topologie $\mathscr{T}^A$ total ist. Ohne Beschränkung der Allgemeinheit kann $0 \leqq u_n \leqq 1$ für alle $n = 1, 2, \ldots$ angenommen werden.

Man setze nun

$$u_0 = \sum_{n=1}^{\infty} 1/n^2 \, u_n .$$

Dann ist $u_0$ eine beschränkte, nicht-negative, superharmonische Funktion aus $\mathscr{C}^A$. Für jede reguläre Menge $V$ mit $\overline{V} \subset \complement A$ und jeden Punkt $x \in V$ gilt $\int u_0 \, d\mu_x^V < u_0(x)$. Aus $\int u_0 \, d\mu_x^V = u_0(x)$ würde nämlich $\int u_n \, d\mu_x^V = u_n(x)$ für alle $n = 1, 2, \ldots$ und hieraus bei Beachtung der Tatsache, daß die Folge $u_1, u_2, \ldots$ total in $\mathscr{C}^A$ ist, $\int f \, d\mu_x^V = f(x)$ für alle $f \in \mathscr{C}^A$ folgen. Da der Träger von $\mu_x^V$ eine kompakte Teilmenge von $\complement A$ ist, würde dies $\mu_x^V = \varepsilon_x$ bedeuten. Dies kann aber nicht sein, da der Träger von $\mu_x^V$ in $V^*$ enthalten ist. Somit ist die behauptete strenge Ungleichung bewiesen; $\mathrm{Rest}_{\complement A} u_0$ ist also streng superharmonisch in $\complement A$ *.

Weiter sei $m$ die nach [*1*], Satz 12 existierende größte harmonische Minorante von $u_0$ auf $X$ und $p = u_0 - m$. Dann ist $p$ ein Potential auf $X$; es ist beschränkt und liegt in $\mathscr{C}^A$ wegen $0 \leqq m \leqq u_0$. Aus den Eigenschaften 2 und 3 folgt, daß $p' = \mathrm{Rest}_{\complement A}\, p$ streng superharmonisch in $\complement A$ und $p'(x) > 0$ für alle $x \in \complement A$ ist. Somit gilt $A = p^{-1}(0)$. Zu zeigen ist daher noch, daß $p'$ ein Potential in $\complement A$ ist. Hierzu sei $t'$ die größte harmonische Minorante von $p'$ in $\complement A$. Wegen $0 \leqq t' \leqq p$ gilt dann

$$\lim_{\substack{x \to z \\ x \notin A}} t'(x) = 0$$

für jeden Randpunkt $z$ von $\complement A$. Folglich ist die Funktion

$$x \to t(x) = \begin{cases} 0 & , x \in A \\ t'(x) & x \in \complement A \end{cases}$$

auf $X$ stetig und außerdem offenbar harmonisch in $X$. Es gilt $0 \leqq t \leqq p$ und somit $t = 0$, da $p$ ein Potential auf $X$ ist. Dann aber ist $t' = 0$ und damit $p'$ ein (strenges) Potential auf $\complement A$.

**Korollar.** *Der Raum $X$ besitze eine abzählbare Basis; es gelte das Axiom $T_+$. Dann existiert ein stetiges, reelles, strenges Potential auf $X$.*

*Beweis.* Man wende Satz 9 auf die leere Absorptionsmenge $A = \emptyset$ an.

**6.3. Approximation durch Differenzen stetiger Potentiale.** Im folgenden bezeichnen wir mit $\mathscr{C}_0(X)$ die Menge aller im Unendlichen verschwindenden Funktionen $f \in \mathscr{C}(X)$. Für eine Teilmenge $A$ von $X$ bezeichne $\mathscr{C}_0^A(X)$ die Menge aller $f \in \mathscr{C}_0(X)$ mit $A \subset f^{-1}(0)$.

**Satz 10.** *Der Raum $X$ besitze eine abzählbare Basis; es gelte das Axiom $T_+$. Es sei $A$ eine Absorptionsmenge und $\mathscr{P}^A$ die Menge aller stetigen, reellen Potentiale $p$*

---

* Eine ähnliche Schlußweise hat M. Hervé [7] benützt, um die Existenz streng konvexer Funktionen auf konvexen, kompakten, metrisierbaren Teilmengen lokal-konvexer Räume zu beweisen.

*auf $X$ mit $A \subset p^{-1}(0)$; es werde $\mathcal{P}_\delta^A = \mathcal{P}^A - \mathcal{P}^A$ gesetzt. Dann liegt die Menge aller Funktionen $f \in \mathcal{P}_\delta^A$ mit kompaktem Träger dicht in $\mathcal{C}_0^A(X)$ bezüglich der Topologie der gleichmäßigen Konvergenz auf $X$.*

*Genauer gilt: zu jeder Funktion $f \geqq 0$ aus $\mathcal{C}_0^A(X)$ und zu jeder Zahl $\varepsilon > 0$ existiert eine Funktion $d \in \mathcal{P}_\delta^A$ mit kompaktem Träger und mit*

$$0 \leqq d \leqq f \leqq d + \varepsilon. \tag{6.3}$$

*Beweis.* Zunächst werde angenommen, daß die konstante Funktion $h = 1$ auf $X$ harmonisch ist. Ohne Beschränkung der Allgemeinheit kann $A \neq X$ vorausgesetzt werden.

Es bezeichne $\mathcal{V}$ die Menge aller Funktionen aus $\mathcal{P}_\delta^A$ mit kompaktem Träger. Nach § 1, Eigenschaften 3 und 12 (sowie der Bemerkung am Schluß von § 1) ist $\mathcal{V}$ ein Vektorraum beschränkter Funktionen auf $X$; aus § 1, Eigenschaft 2 folgt, daß $\mathcal{V}$ mit jeder Funktion auch deren Absolutbetrag enthält. Ferner enthält $\mathcal{P}_\delta^A$ und damit auch $\mathcal{V}$ mit jeder Funktion auch $\inf(1, d)$ als Element. Es sei nämlich $d = u - v$ mit $u, v \in \mathcal{P}^A$; dann ist

$$\inf(1, d) = \inf(1 + v, u) - v.$$

Wegen $h = 1 \in \mathcal{H}_X$ und § 1, Eigenschaft 2 folgt hieraus $\inf(1, d) \in \mathcal{P}_\delta^A$.

Es seien weiter $S$ und $T$ fremde Teilmengen von $\complement A$, von denen $S$ kompakt und $T$ abgeschlossen in $\complement A$ sei. Dann existiert eine Funktion $d \in \mathcal{V}$ mit $0 \leqq d \leqq 1$, $d(x) = 0$ für alle $x \in T$ und $d(x) = 1$ für alle $x \in S$. In der Tat: $S$ kann durch endlich viele reguläre Mengen $V_1, \ldots, V_n$ überdeckt werden, welche zu $A$ und $T$ fremd sind. Nach Satz 9 existiert ein Potential $p \in \mathcal{P}^A$, dessen Restriktion auf $\complement A$ ein strenges Potential liefert. Wir betrachten die Funktion

$$d' = np - p_{V_1} - \cdots - p_{V_n}.$$

Dann ist $d' \in \mathcal{V}$, $d'(x) = 0$ für alle $x \in T$ und $d'(x) > 0$ für alle $x \in S$. Setzt man also $\alpha = \inf d'(S)$, so ist $\alpha > 0$; $d = \alpha^{-1} \inf(\alpha, d)$ leistet dann das Verlangte. Aus NÖBELING-BAUER [9], Hilfssatz 3 folgt nun, daß zu jeder Funktion $f \geqq 0$ aus $\mathcal{C}_0^A(X)$ und zu $\varepsilon > 0$ ein $d \in \mathcal{V}$ existiert mit $0 \leqq d(x) \leqq f(x) \leqq d(x) + \varepsilon$ für alle $x \in \complement A$. Man hat hierzu nur zu bemerken, daß $\mathrm{Rest}_{\complement A} f$ im Unendlichen verschwindet. Gilt aber $0 \leqq d \leqq f \leqq d + \varepsilon$ auf $\complement A$, so gilt dieses System von Ungleichungen auf ganz $X$, da $d$ und $f$ Elemente von $\mathcal{C}_0^A(X)$ sind.

Im allgemeinen Fall gehen wir zum Garbendatum der $h$-harmonischen Funktionen über. Dann folgt aus dem bislang Bewiesenen, daß zu $f \geqq 0$ aus $\mathcal{C}_0^A(X)$ und $\varepsilon_1 > 0$ eine Funktion $d \in \mathcal{P}_\delta^A$ mit kompaktem Träger existiert derart, daß

$$0 \leqq d \leqq f \leqq d + \varepsilon_1 h$$

gilt. Da $f$ im Unendlichen verschwindet, gibt es zu gegebenem $\varepsilon_2 > 0$ eine kompakte Menge $K \subset X$ mit $f(x) \leqq \varepsilon_2$ für alle $x \in \complement K$. Dann aber ist

$$0 \leqq d \leqq f \leqq d + \max(\varepsilon_1 \sup h(K), \varepsilon_2).$$

Beachtet man, daß $K$ nur von $\varepsilon_2$ abhängt, so folgt hieraus die Behauptung im allgemeinen Fall.

Bemerkung. Für die Brelotsche Theorie wurde eine etwas schwächere Approximationsaussage von R. M. HERVÉ [8], p. 24 bewiesen.

## § 7. Approximation hyperharmonischer Funktionen durch stetige Potentiale

Wir beweisen zunächst folgendes Lemma:

**Lemma 5.** *Für jede lokal-beschränkte superharmonische Funktion $u$ auf $X$ und je endlich viele reguläre Mengen $V_1, \ldots, V_n$ ist die Funktion*

$$\inf(u_{V_1}, \ldots, u_{V_n}) \tag{7.1}$$

*stetig in $V_1 \cup \ldots \cup V_n$.*

*Beweis.* Nach [*1*], Sätze 14 und 19 ist $u_{V_i}$ eine in $V_i$ harmonische, also stetige Funktion ($i = 1, \ldots, n$). Hieraus folgt bereits die Richtigkeit der Behauptung für $n = 1$. Wir führen nun vollständige Induktion durch und zeigen, daß aus der Stetigkeit von

$$q' = \inf(u_{V_1}, \ldots, u_{V_n})$$

in $Q' = V_1 \cup \ldots \cup V_n$ die Stetigkeit von

$$q = \inf(u_{V_1}, \ldots, u_{V_{n+1}}) = \inf(q', u_{V_{n+1}})$$

in $Q = Q' \cup V_{n+1}$ folgt ($n = 1, 2, \ldots$).

Auf Grund der einleitenden Bemerkung ist klar, daß $q$ in $Q' \cap V_{n+1}$ stetig ist. Weiter ist die Restriktion von $q$ auf $Q' \cap \complement V_{n+1}$ und auf $V_{n+1} \cap \complement Q'$ stetig: für jeden Punkt $x$ der ersten bzw. zweiten Menge ist nämlich $q(x) = q'(x)$ bzw. $q(x) = u_{V_{n+1}}(x)$. Zu beweisen sind daher noch die beiden folgenden Behauptungen:

(a) $$\lim_{\substack{x \to z \\ x \in V_{n+1}}} q(x) = q(z) \qquad \text{für jedes } z \in Q' \cap V_{n+1}^*;$$

(b) $$\lim_{\substack{x \to z \\ x \in Q'}} q(x) = q(z) \qquad \text{für jedes } z \in V_{n+1} \cap Q'^* .$$

Zu (a): Nach [*1*], Satz 20 ist

$$\liminf_{x \to z, x \in V_{n+1}} u_{V_{n+1}}(x) \geqq \liminf_{y \to z, y \in V_{n+1}^*} u(y) \geqq u(z);$$

außerdem ist $q' \leqq u$. Wir erhalten somit:

$$\begin{aligned} \liminf_{x \to z, x \in V_{n+1}} q(x) &= \inf(\liminf_{x \to z, x \in V_{n+1}} q'(x), \quad \liminf_{x \to z, x \in V_{n+1}} u_{V_{n+1}}(x)) \\ &= \inf(q'(z), \liminf_{x \to z, x \in V_{n+1}} u_{V_{n+1}}(x)) \quad = q'(z) = q(z). \end{aligned}$$

Die gleiche Überlegung liefert: $\limsup_{x \to z, x \in V_{n+1}} q(x) = q(z)$. Beide Ergebnisse zusammen ergeben (a).

Zu (b): Beachtet man die Stetigkeit von $u_{V_{n+1}}$ in $V_{n+1}$, so liefert eine ähnliche Rechnung wie soeben die Gleichungen:

$$\liminf_{x \to z, x \in Q'} q(x) = \inf(\liminf_{x \to z, x \in Q'} q'(x), u_{V_{n+1}}(z))$$

$$\limsup_{x \to z, x \in Q'} q(x) = \inf(\limsup_{x \to z, x \in Q'} q'(x), u_{V_{n+1}}(z)) .$$

Wegen $q(z) = u_{V_{n+1}}(z)$ ist daher nur noch die folgende Ungleichung zu beweisen:

$$\liminf_{x \to z, x \in Q'} q'(x) \geqq u_{V_{n+1}}(z).$$

Diese aber ergibt sich durch folgende Überlegung: $z$ ist Randpunkt wenigstens einer der Mengen $V_1, \ldots, V_n$; ohne Beschränkung der Allgemeinheit kann $z \in V_1^* \cap \ldots \cap V_j^*$ und $z \notin \overline{V}_k$ für alle $k$ mit $j < k$ und $k \leqq n$ angenommen werden ($1 \leqq j \leqq n$). Es gilt:

$$\liminf_{x \to z, x \in Q'} q'(x) = \inf\,(\liminf_{x \to z, x \in Q'} u_{V_1}(x), \ldots, \liminf_{x \to z, x \in Q'} u_{V_n}(x)).$$

Für jedes $k = 1, \ldots, n$ ist aber

$$\liminf_{x \to z, x \in Q'} u_{V_k}(x) \geqq u(z) \geqq u_{V_{n+1}}(z).$$

Für $k > j$ ist dies klar, da $u_{V_k}$ in einer Umgebung von $z$ mit $u$ übereinstimmt. Für $1 \leqq k \leqq j$ hat man zu beachten, daß (abermals nach [*1*], Satz 20)

$$\liminf_{x \to z, x \in V_k} u_{V_k}(x) \geqq u(z)$$

und

$$\liminf_{x \to z, x \in Q' \cap \complement V_k} u_{V_k}(x) = \liminf_{x \to z, x \in Q' \cap \complement V_k} u(x) \geqq u(z)$$

ist. (Die letzte Abschätzung ist dabei natürlich nur für nicht leeres $Q' \cap \complement V_k$ in Betracht zu ziehen.)

Damit ist das Lemma bewiesen.

Nunmehr können wir die im Titel dieses Paragraphen angekündigte Approximationsaussage formulieren und beweisen:

**Satz 11.** *Der Raum $X$ besitze eine abzählbare Basis; es gelte das Axiom $T_+$. Dann ist jede in $X$ hyperharmonische Funktion $u \geqq 0$ die obere Einhüllende einer isotonen Folge stetiger reeller Potentiale.*

*Beweis**. Es genügt offenbar den Beweis unter der Annahme zu führen, daß die Funktion $h = 1$ harmonisch in $X$ ist.

Auf Grund der Abzählbarkeitsvoraussetzung über $X$ existiert eine Folge kompakter Mengen $K_0, K_1, \ldots$ derart, daß

$$K_0 = \emptyset \subset \mathring{K}_1 \subset K_1 \subset \ldots \subset K_n \subset \mathring{K}_{n+1} \subset \ldots$$

und $X = \bigcup_{n=1}^{\infty} K_n$ ist; ferner existiert eine isotone Folge von Funktionen $\varphi_n' \geqq 0$ aus $\mathscr{C}(X)$ ($n = 1, 2, \ldots$) mit $u$ als oberer Einhüllenden. Für jedes $n = 1, 2, \ldots$ wählen wir eine Funktion $\varphi_n'' \in \mathscr{C}(X)$ mit

$$0 \leqq \varphi_n'' \leqq 1 \quad \text{und} \quad \varphi_n''(x) = \begin{cases} 1, & x \in K_{n-1}; \\ 0, & x \notin K_n. \end{cases}$$

* Mit einer ähnlichen Idee hat BRELOT [*4*] den Beweis im Falle der klassischen superharmonischen Funktionen geführt.

Dann ist $\varphi_n = \varphi'_n \varphi''_n$ $(n = 1, 2, \ldots)$ eine isotone Folge von Funktionen aus $\mathscr{C}(X)$ mit $u = \sup \varphi_n$ und $\varphi_n(x) = 0$ für alle $x \notin K_n$.

Nach dem Korollar zu Satz 9 existiert ein stetiges reelles Potential $q$ auf $X$ mit $q(x) > 0$ für alle $x \in X$. Wir setzen

$$\psi_n = \inf(\varphi_n, nq) \quad \text{und} \quad q_n = \inf(u, nq) \qquad (n = 1, 2, \ldots).$$

Dann ist $(\psi_n)$ eine isotone Folge von Funktionen $\geqq 0$ aus $\mathscr{C}(X)$ mit $u$ als oberer Einhüllenden; jede der Funktionen $q_n \geqq 0$ ist lokal-beschränkt und superharmonisch auf $X$*. Wegen $\psi_n \leqq q_n$ und wegen $\psi_n(x) = 0$ für alle $x \notin K_n$ ist $\psi_n \leqq R_{q_n}^{K_n}$ und somit auch $\psi_n \leqq \hat{R}_{q_n}^{K_n}$. Jede der Funktionen $v_n = \hat{R}_{q_n}^{K_n}$ ist $\geqq 0$, lokal-beschränkt und superharmonisch (nach dem Korollar zu Satz 3 und Bemerkung am Schluß von § 2) sowie stetig in $\complement K_n$ (nach [*1*], Satz 37).

Setzen wir $\theta_n = \psi_n - \frac{1}{n}$, so ist $(\theta_n)_{n=1,2,\ldots}$ wiederum eine isotone Folge von Funktionen aus $\mathscr{C}(X)$ mit $u$ als oberer Einhüllenden; zusätzlich gilt die strenge Ungleichung

$$\theta_n(x) < v_n(x) \qquad (x \in X;\ n = 1, 2, \ldots).$$

Da $\theta_n$ stetig und $v_n$ nach unten halbstetig ist sowie wegen $h = 1 \in \mathscr{H}_X$, kann dann die kompakte Menge $K_n$ durch endlich viele reguläre Mengen $V_1^{(n)}, \ldots, V_{k_n}^{(n)}$ derart überdeckt werden, daß

$$\theta_n \leqq (v_n)_{V_k^{(n)}}$$

ist für alle $n = 1, 2, \ldots$ und alle $k$ mit $1 \leqq k \leqq n$. Wegen der Stetigkeit von $v_n$ in $\complement K_n$ sowie nach Lemma 5 ist dann

$$t_n = \inf\left((v_n)_{V_1^{(n)}}, \ldots, (v_n)_{V_{k_n}^{(n)}}\right)$$

eine auf $X$ stetige, reelle, superharmonische Funktion. Aus $0 \leqq t_n \leqq v_n \leqq q_n \leqq nq$ und § 1, Eigenschaften 2 und 3 folgt, daß $t_n$ sogar ein Potential auf $X$ ist. Es gilt

$$\theta_n \leqq t_n \leqq u \qquad (n = 1, 2, \ldots).$$

Setzen wir also

$$p_n = \inf_{m \geqq n} t_m,$$

so ist $p_1, p_2, \ldots$ eine isotone Folge von Funktionen mit der oberen Einhüllenden $u$. Aus der Stetigkeit aller $t_n$, den Eigenschaften der Folge $(\theta_n)$ und aus $\theta_n \leqq t_n \leqq u$ folgt die Stetigkeit einer jeden Funktion $p_n$. Dann aber ist $p_n$ offenbar sogar superharmonisch auf $X$. Endlich ist $0 \leqq p_n \leqq t_n$ und daher mit $t_n$ auch jedes $p_n$ ein Potential aus $\mathscr{C}(X)$. Die Folge dieser Potentiale $(p_n)$ hat damit bezüglich $u$ alle gewünschten Eigenschaften.

---

* Jede der Funktionen $q_n$ ist sogar ein Potential, welches aber im allgemeinen unstetig ist.

# III. Anwendungen

## § 8. Vermeidung des Trennungsaxioms $T'$

*Generelle Zusatzvoraussetzung:* In diesem Paragraphen setzen wir durchwegs voraus, daß der Raum $X$ eine *abzählbare Basis* besitzt.

**8.1. Existenz universeller Barrieren.** In [*1*] wurde an genau zwei Stellen das schärfere, vielleicht aber weniger natürliche Trennungsaxiom $T'$ herangezogen: Erstens beim Beweis von Satz 31, der die Existenz einer Barriere zu jedem regulären Randpunkt einer relativ-kompakten Menge $U \in \mathfrak{U}$ behauptet (deren Rand $U^*$ eine abzählbare Basis besitzt). Zweitens beim Beweis der Identität regulärer und extremaler Randpunkte von semi-regulären Mengen ([*1*], Satz 35).

Im folgenden zeigen wir, daß unsere bisherigen Ergebnisse den Beweis dieser Sätze allein mit dem Trennungsaxiom $T$ gestatten, sofern man über die Existenz einer abzählbaren Basis von $X$ verfügt. Der Satz 31 aus [*1*] erfährt dabei sogar noch eine Verschärfung.

**Satz 12.** *Es sei $U \in \mathfrak{U}$ relativ-kompakt. Dann existiert ein in einer offenen Umgebung von $U$ definiertes, stetiges, reelles, strenges Potential. Für jedes solche Potential $p$ ist die in $U$ definierte Funktion*

$$G_p^U(x) = p(x) - H_p^U(x) \qquad (x \in U) \tag{8.1}$$

*ein (stetiges, reelles) strenges Potential in $U$. $G_p^U$ ist eine universelle Barriere für die regulären Randprodukte von $U$, d. h. ein Punkt $z \in U^*$ ist genau dann regulär, wenn*

$$\lim_{x \to z,\, x \in U} G_p^U(x) = 0. \tag{8.2}$$

*Beweis.* Es sei $U_1$ eine offene relativ-kompakte Umgebung von $\overline{U}$. Das auf $U_1$ eingeschränkte Garbendatum der harmonischen Funktionen genügt dem Axiom $T_+$ (sowie den Axiomen I, II, $K_1$). Nach dem Korollar zu Satz 9 existiert also ein stetiges, reelles, strenges Potential in $U_1$. Im folgenden bezeichne $p$ irgendein stetiges, reelles, strenges Potential, welches in einer offenen Umgebung von $\overline{U}$ definiert ist.

Die zugehörige, in $U$ definierte Funktion $G_p^U$ ist dann stetig und reellwertig sowie aus dem folgenden Grunde ein Potential. Es sei $q$ die größte harmonische Minorante von $G_p^U$ (in $U$). Offenbar ist $q \geqq 0$; aus $q \leqq G_p^U$ folgt $q + H_p^U \leqq p$ in $U$. Hieraus folgt wegen der Stetigkeit von $p$ in einer Umgebung von $U$, daß $q + H_p^U$ eine zu $\mathrm{Rest}_{U^*} p$ gehörige (harmonische) Unterfunktion, also $q + H_p^U \leqq H_p^U$, also $q \leqq 0$ und damit $q = 0$ ist. Daher ist $G_p^U$ ein Potential. Dieses Potential ist streng nach § 6, Eigenschaft 2. Der Rest der Behauptung folgt nun aus [*1*], Satz 30 und der Definition der regulären Punkte.

**Korollar.** *Es sei $K$ eine kompakte Teilmenge einer regulären Menge $U$. Dann gibt es eine reguläre Menge $V$ mit $K \subset V \subset \overline{V} \subset U$.*

*Beweis.* Für ein in einer offenen Umgebung von $\overline{U}$ definiertes, stetiges, reelles, strenges Potential $p$ betrachten wir die Funktion $G_p^U$ in $U$. Nach Satz 12 ist dann

$G_p^U(x) > 0$ für alle $x \in U$ und $\lim\limits_{x \to z} G_p^U(x) = 0$ für alle $z \in U^*$. Folglich ist die Zahl

$$\beta_K = \inf_{x \in K} \frac{G_p^U(x)}{h(x)}$$

positiv und für jedes $\alpha$ mit $0 < \alpha < \beta_K$ ist

$$V = \{x \in U : G_p^U(x) - \alpha h(x) > 0\}$$

eine offene Teilmenge von $U$ mit $K \subset V \subset \overline{V} \subset U$. Die so konstruierte Menge $V$ ist regulär, da offenbar $x \to G_p^U(x) - \alpha h(x)$ eine Barriere für jeden Randpunkt von $V$ ist.

Bemerkung. Es wäre interessant zu wissen, ob dieses Korollar für jede kompakte Menge $K$ und *jede* offene Obermenge $U$ richtig ist. In der Brelotschen Theorie wurde diese allgemeinere Fassung des Korollars von R. M. Hervé [8] bewiesen.

**8.2. Identität regulärer und extremaler Randpunkte bei semi-regulären Mengen.** Die angekündigte Verschärfung von [1], Satz 35 lautet:

**Satz 13.** *Für jede semi-reguläre offene Menge $U$ fallen die regulären Randpunkte mit den extremalen Randpunkten zusammen.*

*Beweis.* Im Hinblick auf [1], Satz 32 ist nur zu zeigen, daß jeder reguläre Randpunkt $z_0$ von $U$ auch extremal ist. Wenn also $\mu \geqq 0$ ein Maß auf $\overline{U}$ ist mit $\int u\,d\mu \leqq u(z_0)$ für alle Funktionen $u \in \mathscr{C}(\overline{U})$, welche in $U$ superharmonisch sind, so ist $\mu = \varepsilon_{z_0}$ nachzuweisen.

Für jedes $f \in \mathscr{C}(U^*)$ kann gemäß der Definition semi-regulärer Mengen die verallgemeinerte Lösung $H_f^U$ zu einer Funktion $\hat{H}_f^U \in \mathscr{C}(\overline{U})$ fortgesetzt werden. Bezeichnet also $p$ ein Potential mit den in Satz 12 genannten Eigenschaften, so ist die folgende Funktion $G_p^U$ stetig und reell in $\overline{U}$ sowie superharmonisch in $U$:

$$\hat{G}_p^U(x) = p(x) - \hat{H}_p^U(x) \qquad (x \in \overline{U}).$$

Offenbar wird die Funktion $G_p^U$ aus (8.1) durch $\hat{G}_p^U$ stetig auf $\overline{U}$ fortgesetzt. Nach Satz 12 ist daher ein Punkt $z \in \overline{U}$ genau dann ein regulärer Randpunkt von $U$, wenn $\hat{G}_p^U(z) = 0$ ist. Da $z_0$ regulär ist, erhalten wir also

$$0 \leqq \int \hat{G}_p^U\,d\mu \leqq \hat{G}_p^U(z_0) = 0;$$

somit wird $\mu$ von der Menge $U_r^*$ aller regulären Randpunkte von $U$ getragen. Nach [1], Satz 34 ist $U_r^*$ kompakt* und jede Funktion $\varphi \in \mathscr{C}(U_r^*)$ kann auf genau eine Weise zu einer in $U$ harmonischen Funktion $u_\varphi \in \mathscr{C}(\overline{U})$ fortgesetzt werden. Folglich gilt weiter

$$\int u_\varphi\,d\mu = \int \varphi\,d\mu = \varphi(z_0)$$

für alle $\varphi \in \mathscr{C}(U_r^*)$. Hieraus aber ergibt sich die gewünschte Gleichheit $\mu = \varepsilon_{z_0}$.

* Dies folgt auch aus der soeben bewiesenen Gleichheit $U_r^* = \{z \in \overline{U} : \hat{G}_p^U(z) = 0\}$ und der Stetigkeit von $\hat{G}_p^U$.

## § 9. Kennzeichnung regulärer Randpunkte

**9.1. Dünne Mengen.** Aus der klassischen Potentialtheorie übernehmen wir die folgenden Begriffsbildungen:

**Definition 7.** Eine Menge $E \subset X$ heiße *dünn* in einem Punkte $x_0 \notin E$, wenn entweder $x_0$ nicht in $\overline{E}$ liegt oder wenn eine hyperharmonische Funktion $u \geqq 0$ auf $X$ existiert mit folgender Eigenschaft:

$$\lim_{x \to x_0, x \in E} \inf u(x) > u(x_0). \tag{9.1}$$

$E$ heiße dünn in einem Punkte $x_0 \in E$, wenn die Menge $\{x_0\}$ polar und $E \cap \complement\{x_0\}$ dünn im Punkte $x_0$ ist.

Als einfaches Beispiel für dünne Mengen erwähnen wir: Es sei $A$ eine nichtleere Absorptionsmenge. Dann ist $E = \complement A$ dünn in jedem Punkte $x_0 \in A$. Die Funktion $u = h^A$ besitzt nämlich die Eigenschaft (9.1) (vgl. § 4, Eigenschaft 1).

Wie in der Brelotschen Theorie [*3*], p. 137 zeigt man:

1. Es sei $u \geqq 0$ hyperharmonisch auf $X$ und stetig in einem Punkt $x_0 \in X$; ferner sei $0 < u(x_0) < +\infty$*. Eine Menge $E \subset X$ mit $x_0 \notin E$ ist genau dann dünn im Punkte $x_0$, wenn für mindestens eine Umgebung $V$ von $x_0$ gilt:

$$R_u^{E \cap V}(x_0) < u(x_0). \tag{9.2}$$

**9.2. Anwendung auf das Dirichletsche Problem.** Indem man Satz 39 aus [*1*] heranzieht, erhält man wie bei BRELOT [*3*], p. 138:

**Satz 14.** *Es sei $z$ ein Randpunkt einer relativ-kompakten Menge $U \in \mathfrak{U}$. Ist dann die Menge $\complement U$ dünn im Punkte $z$, so ist $z$ ein irregulärer Randpunkt.*

Es wäre wichtig zu wissen, welche Zusatzannahmen die Umkehrung der Aussage dieses Satzes gestatten. Bekanntlich gilt die Umkehrung im Falle der Laplace-Gleichung sowie in der Brelotschen Theorie unter der Annahme, daß der Grundraum eine abzählbare Basis besitzt und man über ein *Dominationsprinzip* verfügt.

Immerhin gestatten es uns die Resultate des § 6, die folgende abgeschwächte Umkehrung zu beweisen (vgl. auch R. M. HERVÉ [*8*], p. 21):

**Satz 15.** *Der Raum $X$ besitze eine abzählbare Basis; es gelte das Trennungsaxiom $T_+$. Ist dann für einen Randpunkt $z$ einer relativ-kompakten Menge $U \in \mathfrak{U}$ die Menge $\complement \overline{U}$ nicht dünn im Punkte $z$, so ist $z$ ein regulärer Randpunkt.*

*Beweis.* Die Menge $\complement \overline{U}$ ist nicht dünn im Punkte $z$, bedeutet: $z$ ist ein Randpunkt von $\overline{U}$, und es gilt

$$\liminf_{x \to z, x \notin \overline{U}} u(x) = u(z)$$

für alle $u \in \mathscr{H}_X^*$ mit $u \geqq 0$. Es sei nun $p \in \mathscr{C}(X)$ ein nach dem Korollar zu Satz 9 existierendes strenges Potential und $u_0 = \hat{R}_p^{\complement \overline{U}}$. Nach dem Korollar zu Satz 3 und der Schlußbemerkung des § 2 ist $u_0$ hyperharmonisch auf $X$; nach der einleitenden Bemerkung gilt also:

$$\liminf_{x \to z, x \notin \overline{U}} u_0(x) = u_0(z).$$

* Beispielsweise ist $u = h$ eine derartige Funktion.

Berücksichtigt man noch die für alle $x \in \complement \overline{U}$ bestehende Gleichheit $u_0(x) = R_p^{\complement \overline{U}}(x) = p(x)$, so ergibt sich

$$u_0(z) = \liminf_{x \to z, x \notin \overline{U}} p(x) = \lim_{x \to z} p(x) = p(z).$$

Wir betrachten nun folgende, in $U$ definierte Funktion $q$:

$$q(x) = p(x) - R_p^{\complement \overline{U}}(x) \qquad (x \in U).$$

Nach [1], Satz 37 ist $R_p^{\complement \overline{U}}$ in $U$ harmonisch; also ist $q$ nach § 6, Eigenschaft 2 in $U$ streng superharmonisch. Wegen $q \geqq 0$ gilt daher $q(x) > 0$ für alle $x \in U$ nach § 6, Eigenschaft 3. Somit muß nur noch gezeigt werden, daß

$$\lim_{x \to z, x \in U} q(x) = 0$$

ist. Dann ist nämlich $q$ eine Barriere für $z$ und die Behauptung folgt aus [1], Satz 30. Nun ist aber

$$\limsup_{x \to z, x \in U} q(x) \leqq p(z) - \liminf_{x \to z, x \in U} R_p^{\complement \overline{U}}(x)$$

$$\leqq p(z) - \liminf_{x \to z} R_p^{\complement \overline{U}}(x) = p(z) - u_0(z) = 0$$

und folglich tatsächlich $\lim_{x \to z, x \in U} q(x) = 0$.

**9.3. Regularität der Randpunkte von Absorptionsmengen.** Im Zusammenhang mit dem Dirichletschen Problem zeigen wir noch:

**Satz 16.** *Der Raum $X$ besitze eine abzählbare Basis. Es sei $A$ eine Absorptionsmenge und $U$ eine relativ-kompakte Menge aus $\mathfrak{U}$. Denn ist jeder in $A$ gelegene Randpunkt der Menge $U \cap \complement A$ regulär.*

*Beweis.* Es sei $U_1$ eine offene, relativ-kompakte Umgebung von $\overline{U}$. Die Restriktion auf $U_1$ des Garbendatums der harmonischen Funktionen genügt dann den Axiomen I, II, $K_1$ und $T_+$; ferner besitzt auch $U_1$ eine abzählbare Basis. Schließlich ist $A_1 = A \cap U_1$ eine Absorptionsmenge bezüglich $U_1$. Folglich existiert nach Satz 9 ein stetiges reelles Potential $p$ in $U_1$ mit $A_1 = p^{-1}(0)$. Für jeden in $A$ gelegenen Randpunkt von $U \cap \complement A$ ist daher die Restriktion von $p$ auf $U \cap \complement A$ eine Barriere für $z$. Daher ist $z$ regulär.

## § 10. Balayage für Maße mit kompaktem Träger

*Generelle Zusatzvoraussetzung:* In diesem Paragraphen wird ständig das *Trennungsaxiom $T_+$* und die Existenz einer *abzählbaren Basis* für den Raum $X$ vorausgesetzt.

**10.1. Vorbereitungen.** Der in § 6 gewonnene Approximationssatz (Satz 10) gestattet es, für jedes Maß $\mu \geqq 0$ auf $X$ mit kompaktem Träger Balayage-Maße zu definieren. Die Idee hierfür ist die gleiche wie in der durch R. M. Hervé [8] weitergeführten Brelotschen Theorie; sogar die einzelnen Beweisschritte können weitgehend übernommen werden. Wir werden uns daher häufig recht kurz fassen können. Entscheidend sind die folgenden Eigenschaften der reduzierten und extremisierten Funktionen von geeigneten hyperharmonischen Funktionen.

**Lemma 6.** *Eine isotone Folge nicht-negativer superharmonischer Funktionen $v_1, v_2, \ldots$ auf $X$ konvergiere gegen eine lokal-beschränkte* superharmonische Funktion $v$ auf $X$. Dann gilt für jede abgeschlossene Menge $E \subset X$:*

$$\lim_{n\to\infty} R^E_{v_n} = R^E_v; \tag{10.1}$$

$$\lim_{n\to\infty} \hat{R}^E_{v_n} = \hat{R}^E_v. \tag{10.2}$$

Der Beweis hierfür kann wie bei R. M. HERVÉ [*8*], p. 31 (unter Heranziehung der in [*1*] bereitgestellten Sätze) geführt werden.

**Lemma 7.** *Es seien $v_1$ und $v_2$ nicht-negative, lokal-beschränkte*, superharmonische Funktionen auf $X$. Dann gilt für jede abgeschlossene Menge $E \subset X$:*

$$R^E_{v_1+v_2} = R^E_{v_1} + R^E_{v_2}; \tag{10.3}$$

$$\hat{R}^E_{v_1+v_2} = \hat{R}^E_{v_1} + \hat{R}^E_{v_2}. \tag{10.4}$$

Den Beweis führt man wie bei R. M. HERVÉ [*8*], p. 32. Man hat nur zu bemerken, daß jedes $u \geqq 0$ aus $\mathscr{H}^*_X$ das Supremum einer isotonen Folge lokal-beschränkter Potentiale ist. Dies entnimmt man entweder dem Satz 11 oder der folgenden Überlegung: Es sei $p$ ein stetiges, reelles, strenges Potential auf $X$. Dann leistet die Folge $\inf(u, np)$, $n = 1, 2, \ldots$, das Verlangte.

**Lemma 8.** *Unter Beibehaltung aller übrigen Voraussetzungen gelten die Formeln* (10.1.)—(10.4) *auch für jede offene Menge $E \subset X$.*

Der Beweis wird wie bei R. M. HERVÉ [*8*], p. 32 geführt.

**Lemma 9.** *Es seien $v_1$ und $v_2$ nicht-negative superharmonische Funktionen aus $\mathscr{C}(X)$. Dann gelten die Formeln* (10.3) *und* (10.4) *für jede beliebige Menge $E \subset X$.*

Abermals wird der Beweis nach R. M. HERVÉ [*8*], pp. 32—33 geführt. Man verwendet dabei das folgende Lemma:

**Lemma 10.** *Es sei $v$ eine nicht-negative superharmonische Funktion aus $\mathscr{C}(X)$ und $E$ eine beliebige Teilmenge von $X$. Dann gilt:*

$$R^E_v = \inf_{\substack{U \text{ offen} \\ E \subset U}} R^U_v. \tag{10.5}$$

*Beweis.* Zu $\varepsilon > 0$ und $x_0 \in X$ existiert eine Funktion $u \geqq 0$ aus $\mathscr{H}^*_X$ mit $u(x) \geqq v(x)$ für alle $x \in E$ und $u(x_0) \leqq R^E_v(x_0) + \varepsilon$. Also ist $u(x) > v(x) - \varepsilon h(x)$ für alle $x \in E$ und aus Gründen der Stetigkeit besteht diese Ungleichung sogar für alle Punkte $x$ einer offenen Umgebung $U_0$ von $E$. (Wie immer ist dabei $h \in \mathscr{H}_X$ und $h(x) > 0$ für alle $x \in X$.) Die Funktion $u + \varepsilon h$ ist überall positiv und hyperharmonisch auf $X$ sowie $\geqq v$ auf $U_0$. Folglich ist $R^{U_0}_v \leqq u + \varepsilon h$ und man erhält

$$R^E_v \leqq \inf_{\substack{U \text{ offen} \\ E \subset U}} R^U_v \leqq R^{U_0}_v \leqq u + \varepsilon h.$$

---

* Diese Einschränkung kann weggelassen werden, wenn man über das Doobsche Konvergenzaxiom $K_D$ verfügt. Setzt man das Axiom $K_2$ voraus, so kann „lokal-beschränkt" zu „reell" abgeschwächt werden.

Insbesondere ist also

$$R_v^E(x_0) \leqq \inf R_v^U(x_0) \leqq R_v^E(x_0) + \varepsilon(1 + h(x_0)).$$

Hieraus folgt die Gleichheit (10.5) im Punkte $x_0$.

Schließlich werden wir noch benötigen:

**Lemma 11.** *Zu jedem Potential $p \geqq 0$ aus $\mathscr{C}(X)$ existiert eine Folge von Potentialen $p_n \in \mathscr{C}(X)$, $n = 1, 2, \ldots$, mit folgenden Eigenschaften:*

(10.6) $$0 \leqq p_{n+1} \leqq p_n \leqq p \qquad (n = 1, 2, \ldots);$$

(10.7) $$\lim_{n\to\infty} p_n = 0;$$

(10.8) $p - p_n$ *hat einen kompakten Träger* $(n = 1, 2, \ldots)$.

*Beweis.* Da $X$ eine abzählbare Basis besitzt, existiert eine Folge $U_1, U_2, \ldots$ offener, relativ-kompakter Mengen mit $\overline{U}_n \subset U_{n+1}$ $(n = 1, 2, \ldots)$ und $X = \bigcup U_n$. Jeder der kompakten Ränder $U_n^*$ kann durch endlich viele in $U_{n+1}$ reguläre Mengen $V_1^{(n)}, \ldots, V_{k_n}^{(n)}$ überdeckt werden. Für jedes $n$ setze man

$$p_n = (\hat{R}_p^{\complement U_n})_{V_1^{(n)}, \ldots, V_{k_n}^{(n)}}.$$

Nach § 1, Eigenschaft 2 ist jedes $p_n$ ein Potential mit $0 \leqq p_n \leqq p$. Nach [*1*], Nr. 13.1 ist $\hat{R}_p^{\complement U_n}$ stetig in der offenen Menge $\complement U_n^*$; nach [*1*], Satz 21 ist daher $p_n \in \mathscr{C}(X)$. Die Antitonie der Folge $(p_n)$ und damit (10.6) sieht man folgendermaßen ein: Aus $U_n \subset U_{n+1}$ folgt $\hat{R}_p^{\complement U_{n+1}} \leqq \hat{R}_p^{\complement U_n}$ und hieraus

$$(\hat{R}_p^{\complement U_{n+1}})_{V_1^{(n)}, \ldots, V_{k_n}^{(n)}} \leqq p_n.$$

Nun ist $R_p^{\complement U_{n+1}}$ und damit auch $\hat{R}_p^{\complement U_{n+1}}$ harmonisch in $U_{n+1}$ ([*1*], Satz 37); ferner ist jede der Mengen $V_j^{(n)}$ regulär in $U_{n+1}$. Folglich ist

$$(\hat{R}_p^{\complement U_{n+1}})_{V_1^{(n)}, \ldots, V_{k_n}^{(n)}} = \hat{R}_p^{\complement U_{n+1}}$$

und man erhält

$$p_{n+1} \leqq \hat{R}_p^{\complement U_{n+1}} \leqq p_n,$$

also die behauptete Antitonie.

Da $p$ ein Potential ist, gilt nach § 1, Eigenschaft 11 $\lim \hat{R}_p^{\complement U_n} = 0$. Hieraus folgt (10.7) wegen $0 \leqq p_n \leqq \hat{R}_p^{\complement U_n}$.

Schließlich ist

$$\hat{R}_p^{\complement U_n}(x) = p(x)$$

für alle $x \in \complement \overline{U}_n$. Da die Mengen $V_1^{(n)}, \ldots, V_{k_n}^{(n)}$ sämtlich regulär in $U_{n+1}$ sind, folgt hieraus

$$p_n(x) = p(x)$$

für alle $x \in \complement U_{n+1}$. Somit ist $p(x) - p_n(x) = 0$ in $\complement U_{n+1}$, also der Träger von $p - p_n$ enthalten in $\overline{U}_{n+1}$ und daher kompakt. Dies ist die letzte der zu beweisenden Eigenschaften.

**10.2. Definition der Maße $\mu^E$.** Ähnlich wie bei R. M. HERVÉ [8], p. 33 gilt nun:

**Satz 17.** *Zu jedem Maß $\mu \geqq 0$ auf $X$ mit kompaktem Träger und zu jeder Menge $E \subset X$ existiert genau ein Maß $\mu^E \geqq 0$ auf $X$ derart, daß für jedes Potential $p \in \mathscr{C}(X)$ gilt:*

$$\int \hat{R}_p^E \, d\mu = \int p \, d\mu^E . \tag{10.9}$$

*Ist $E$ offen oder abgeschlossen, so gilt sogar*

$$\int \hat{R}_v^E \, d\mu = \int v \, d\mu^E \tag{10.10}$$

*für jede lokal-beschränkte* superharmonische Funktion $v \geqq 0$ auf $X$.*

*Beweis.* Mit Hilfe von Satz 10 (für $A = \emptyset$) ergibt sich zunächst wie in [8], p. 33 die Existenz genau eines Maßes $\mu^E \geqq 0$ derart, daß

$$\int (\hat{R}_p^E - \hat{R}_{p'}^E) \, d\mu = \int (p - p') \, d\mu^E \tag{10.11}$$

gilt für alle Paare von Potentialen $p, p' \in \mathscr{C}(X)$, für welche $p - p'$ einen kompakten Träger besitzt. Zu jedem Potential $p \in \mathscr{C}(X)$ gibt es nun nach Lemma 11 eine Folge von Potentialen $p_n \in \mathscr{C}(X)$ mit den Eigenschaften (10.6)—(10.8). Nach (10.8) und (10.11) ist

$$\int (\hat{R}_p^E - \hat{R}_{p_n}^E) \, d\mu = \int (p - p_n) \, d\mu^E$$

für alle $n$. Aus (10.6) und (10.7) folgt

$$\lim_{n \to \infty} \int \hat{R}_{p_n}^E \, d\mu = \lim_{n \to \infty} \int p_n \, d\mu^E = 0 .$$

Wir erhalten somit (10.9).

Die Eindeutigkeit von $\mu^E$ ist beinahe selbstverständlich, da aus der Gültigkeit von (10.9) für alle Potentiale $p \in \mathscr{C}(X)$ die Gültigkeit von (10.11) folgt.

Der Rest der Behauptung ergibt sich schließlich aus Lemma 6 bzw. 8 sowie aus Satz 11.

**Definition 8.** Das zu jedem Maß $\mu \geqq 0$ mit kompaktem Träger und zu jeder Menge $E \subset X$ nach Satz 17 gehörige Maß $\mu^E$ heiße das aus $\mu$ durch *Balayage auf $E$* gewonnene Maß (oder das zu $\mu$ und $E$ gehörige Balayage-Maß).

Beispiel. Zu jeder relativ-kompakten Menge $U \in \mathfrak{U}$ und zu jedem Punkt $x \in U$ gehört das *harmonische Maß* $\mu_x^U$ auf $U^*$. Wir behaupten, daß dieses das zu $\varepsilon_x$ und $\complement U$ gehörige Balayage-Maß ist:

$$\mu_x^U = (\varepsilon_x)^{\complement U} . \tag{10.12}$$

In der Tat: für jedes Potential $p \in \mathscr{C}(X)$ gilt

$$\int p \, d\mu_x^U = H_p^U(x) = R_p^{\complement U}(x) = \hat{R}_p^{\complement U}(x) = \int \hat{R}_p^{\complement U} \, d\varepsilon_x .$$

Hierbei hat man die Sätze 37 und 38 aus [1] heranzuziehen.

**10.3. Eigenschaften der Maße $\mu^E$.** Zunächst beweisen wir die folgende Kennzeichnung polarer Mengen:

---

* Auch hier trifft die in der Fußnote von S. 222 formulierte Bemerkung zu.

**Satz 18.** *Eine Menge $P \subset X$ ist genau dann polar, wenn $\mu^P = 0$ ist für alle Maße $\mu \geqq 0$ auf $X$ mit kompaktem Träger**.

*Beweis.* Es sei $P$ eine polare Menge und $\mu \geqq 0$ ein Maß auf $X$ mit kompaktem Träger. Dann ist $\mu^P = 0$ nach § 2, Eigenschaft 6. — Für den Beweis der Umkehrung sei $p$ ein strenges Potential aus $\mathscr{C}(X)$, welches nach dem Korollar zu Satz 9 existiert. Es gilt $p(x) > 0$ für alle $x \in X$. Nach Annahme gilt für jedes Maß $\mu \geqq 0$ mit kompaktem Träger:

$$\int \hat{R}_p^P \, d\mu = \int p \, d\mu^P = 0 .$$

Spezialisiert auf die Maße $\mu = \varepsilon_x$ mit $x \in X$ liefert dies

$$\hat{R}_p^P(x) = \int \hat{R}_p^P \, d\varepsilon_x = 0$$

für alle $x \in X$. Dann aber ist $P$ polar nach Satz 5.

Die drei folgenden Eigenschaften der für Maße $\mu \geqq 0$ mit kompaktem Träger erklärten Abbildung $\mu \to \mu^E$ ($E \subset X$) ergeben sich schließlich wie bei R. M. Hervé [8], p. 34 und können daher ohne Beweis mitgeteilt werden:

1. Für jede Menge $E \subset X$ ist die Abbildung $\mu \to \mu^E$ auf ihrem Definitionsbereich linear, d. h.:

$$(\mu_1 + \mu_2)^E = \mu_1^E + \mu_2^E, \quad (\lambda\mu)^E = \lambda\mu^E \qquad (\lambda \geqq 0, \lambda \in \boldsymbol{R}).$$

2. Für jedes Maß $\mu \geqq 0$ mit kompaktem Träger $T_\mu$ und jede Menge $E \subset X$ wird $\mu^E$ von der Menge

$$(\overset{\circ}{E} \cap T_\mu) \cup E^*$$

getragen.

3. Es sei $(\mu_n)_{n=1,2,\ldots}$ eine isotone Folge positiver Maße, welche sämtlich von ein und derselben kompakten Menge getragen werden und vag gegen ein Maß $\mu$ konvergieren. Dann konvergiert $(\mu_n^E)$ vag gegen $\mu^E$.

## § 11. Übertragung eines Zerlegungs- und Fortsetzungssatzes von R. M. Hervé

*Generelle Zusatzvoraussetzung.* In diesem Paragraphen wird generell das *Doobsche Konvergenzaxiom $K_D$* vorausgesetzt.

**11.1. Spezifische Ordnung und Zerlegungssatz.** Wir bezeichnen mit $\mathscr{S}^+$ die Menge aller superharmonischen Funktionen $u \geqq 0$ auf $X$.

Wie bei Brelot [3] definiert man die sog. spezifische Ordnungsrelation $\prec$ in $\mathscr{S}^+$: Für Funktionen $u_1, u_2 \in \mathscr{S}^+$ sei $u_1 \prec u_2$ gleichbedeutend mit der Existenz eines $v \in \mathscr{S}^+$ mit

$$u_2 = u_1 + v . \tag{11.1}$$

Aus $u_1 \prec u_2$ folgt somit $u_1 \leqq u_2$. Berücksichtigt man dies, so ist leicht zu sehen, daß $\prec$ eine Ordnungsrelation in $\mathscr{S}^+$ ist.

---

* In der Brelotschen Theorie gilt eine schärfere Aussage; vgl. [8], p. 34. Übrigens lehrt der nachfolgende Beweis, daß $P$ bereits dann polar ist, wenn $(\varepsilon_x)^P = 0$ ist für alle Punkte $x \in X$.

Die zu $u_1, u_2 \in \mathcal{S}^+$ mit $u_1 \prec u_2$ gehörige Funktion $v \in \mathcal{S}^+$ mit $u_2 = u_1 + v$ ist übrigens eindeutig bestimmt. Es sei nämlich $v'$ eine zweite Funktion mit denselben Eigenschaften. Für jede reguläre Menge $V$ und jeden Punkt $x \in V$ gilt dann nach Satz 1

$$\int u_2\, d\mu_x^V = \int u_1\, d\mu_x^V + \int v\, d\mu_x^V = \int u_1\, d\mu_x^V + \int v'\, d\mu_x^V .$$

Hieraus folgt

$$\int v\, d\mu_x^V = \int v'\, d\mu_x^V$$

für alle derartigen $V$ und $x$. Nach Satz 3 ist daher $v = v'$.

Weiter übernehmen wir folgende Sprechweise von R. M. HERVÉ [8]: Es seien $u_1, u_2 \in \mathcal{S}^+$ und $U \in \mathfrak{U}$. Existiert dann eine superharmonische Funktion $w$ in $U$ (nicht notwendig $\geqq 0$) mit

(11.2) $$u_2(x) = u_1(x) + w(x) \qquad \text{für alle } x \in U,$$

so heiße $u_2$ eine *U-Majorante* von $u_1$.

Wie bei R. M. HERVÉ [8], p. 42 läßt sich dann der folgende *Zerlegungssatz* herleiten:

**Satz 19.** *Zu jeder Menge $U \in \mathfrak{U}$ und zu jeder Funktion $v \in \mathcal{S}^+$ existieren zwei Funktionen $v'_U$, $v''_U \in \mathcal{S}^+$, welche durch folgende zwei Eigenschaften gekennzeichnet sind:*

a) $$v = v'_U + v''_U;$$

b) *$v''_U$ ist hinsichtlich der spezifischen Ordnung in $\mathcal{S}^+$ die größte aller Minoranten von $v$, welche in $U$ harmonisch sind.*

*Außerdem ist $v'_U$ eine spezifische Minorante jeder U-Majorante von $v$; ferner ist $v'_U$ harmonisch in $\complement \overline{U}$.*

An dieser Stelle ließe sich die spezifische Ordnung in $\mathcal{S}^+$ noch weiter in Anlehnung an [3] und [8] verfolgen. Wir verzichten jedoch auf die Formulierung dieser Resultate.

**11.2. Fortsetzungssatz.** Da wir über die Approximationsaussage des Satzes 10 verfügen, bleiben (unter der Voraussetzung, daß man über das Axiom $T_+$ und die Existenz einer abzählbaren Basis von $X$ verfügt) die bei R. M. HERVÉ [8], pp.43 — 44 im Anschluß an den Zerlegungssatz durchgeführten Schlüsse auch in der vorliegenden Theorie richtig. Man erhält dann aus Satz 19 den folgenden *Fortsetzungssatz.* Die Durchführung der Beweise an Hand von [8] kann auch hier dem Leser überlassen werden.

**Satz 20.** *Es gelte das Axiom $T_+$; der Raum $X$ besitze eine abzählbare Basis. Die Funktion $u$ sei superharmonisch in einer Menge $U \in \mathfrak{U}$. Dann existieren zu jeder relativ-kompakten Menge $V \in \mathfrak{U}$ mit $\overline{V} \subset U$ zwei Potentiale $p$, $p'$ auf $X$ mit folgenden Eigenschaften:*

a) $$p(x) = u(x) + p'(x) \qquad \textit{für alle } x \in V;$$

b) *$p'$ ist ein Element von $\mathscr{C}(X)$ sowie harmonisch in $V$.*

*Wenn $u$ in $\mathscr{C}(U)$ liegt, so kann auch noch $p$ als Element von $\mathscr{C}(X)$ gewählt werden.*

## § 12. Weitere Eigenschaften polarer Mengen

*Generelle Zusatzvoraussetzung.* In diesem Paragraphen werden generell die *Axiome* $K_D$ und $T_+$ sowie die Existenz einer *abzählbaren Basis* von $X$ gefordert.

**12.1. Kennzeichnung polarer Mengen durch lokale Eigenschaften.** Wir definieren zunächst:

**Definition 9.** Eine Menge $P \subset X$ heiße *lokal*-polar, wenn jeder Punkt $x \in P$ eine offene Umgebung $U$ besitzt derart, daß $P \cap U$ polar in $U$ ist.

„Polar in $U$" heißt dabei natürlich: Es gibt eine in $U$ superharmonische Funktion $u \geqq 0$ mit $P \cap U \subset u^{-1}(+\infty)$.

Wir behaupten:

**Satz 21.** *Die lokal-polaren Mengen sind mit den polaren Mengen identisch.*

*Beweis.* Trivial ist, daß jede polare Menge auch lokal-polar ist. In umgekehrter Richtung zeigen wir zunächst: Es sei $E$ eine relativ-kompakte Menge und $U \in \mathfrak{U}$ eine Umgebung von $\overline{E}$. Ist dann $E$ polar in U, so ist $E$ auch polar in $X$. Es sei nämlich $V \in \mathfrak{U}$ relativ-kompakt mit $\overline{E} \subset V \subset \overline{V} \subset U$. Zu $E$ existiert eine in $U$ superharmonische Funktion $u \geqq 0$ mit $E \subset u^{-1}(+\infty)$. Nach Satz 20 existieren Potentiale $p$, $p'$ auf $X$ mit $p(x) = u(x) + p'(x)$ für alle $x \in V$. Folglich gilt $E \subset p^{-1}(+\infty)$ und $E$ ist polar in $X$.

Nunmehr erst beweisen wir die eigentliche Behauptung: Es sei $V_1, V_2, \ldots$ eine abzählbare Basis von $X$ bestehend aus lauter offenen, relativ-kompakten Mengen. Zu jedem Punkt $x$ der lokal-polaren Menge $P$ gibt es eine offene Umgebung $U$ von $x$ derart, daß $P \cap U$ polar in $U$ ist. Zu $x$ und $U$ gibt es eine Menge $V_n$ mit $x \in V_n \subset \overline{V}_n \subset U$. Dann ist $P \cap V_n$ polar in $U$ und $\overline{P \cap V_n} \subset \overline{V_n} \subset U$. Aus der einleitend bewiesenen Hilfsbehauptung folgt daher, daß $P \cap V_n$ polar in $X$ ist. Nun ist aber offenbar $P$ die Vereinigung aller derjenigen Mengen $P \cap V_n$, welche polar sind. Nach Satz 4 ist daher $P$ polar.

**Korollar 1.** *Eine Menge $P \subset X$ ist genau dann polar, wenn eine in $X$ superharmonische Funktion $u$ existiert mit $P \subset u^{-1}(+\infty)$*.*

*Beweis.* Es genügt zu zeigen, daß $P$ lokal-polar ist. Für einen beliebigen Punkt $x \in P$ sei hierzu $U$ eine relativ-kompakte offene Umgebung. Da $u$ auf $U$ nach unten beschränkt ist, gibt es eine Zahl $\alpha > 0$ derart, daß $v(x) = u(x) + \alpha h(x) \geqq 0$ ist für alle $x \in U$. Hierbei ist $h$ eine auf $X$ harmonische, überall positive Funktion. Offenbar ist $v$ auf $U$ superharmonisch und $P \cap U \subset v^{-1}(+\infty)$. Also ist $P \cap U$ polar in $U$. Dies beweist die lokale Polarität von $P$.

**Korollar 2.** *Jede in einer Menge $U \in \mathfrak{U}$ polare Menge $P \subset U$ ist auch polar in $X$.*

*Beweis.* $P$ ist lokal-polar, da $P = P \cap U$ polar in $U$ ist.

**12.2. Polare Randmengen.** Es sei $P$ eine polare Menge und $U \in \mathfrak{U}$ relativ-kompakt mit $P \subset U^*$. Dann ist das harmonische Maß

$$\mu_x^U(P) = 0 \tag{12.1}$$

für jeden Punkt $x \in U$.

* Im Gegensatz zur Definition 4 wird nicht $u \geqq 0$ gefordert.

In der Tat: zu $P$ gibt es eine superharmonische Funktion $u \geqq 0$ auf $X$ mit $P \subset u^{-1}(+\infty)$. Setzen wir also $v = \text{Rest}_U\, u$, so ist für jedes reelle $\lambda > 0$ die Funktion $\lambda v$ eine zur Randfunktion $\chi_P$ (charakteristische Funktion von $P$ bezüglich $U^*$) gehörige Oberfunktion. Folglich ist

$$\overline{H}^U_{\chi_P}(x) \leqq \lambda v(x)$$

für alle $x \in U$ und alle $\lambda > 0$; also ist $\overline{H}^U_{\chi_P}(x) = 0$ in den Punkten $x$ einer in $U$ dichten Menge. Nach [*1*], Satz 14 muß dann $\mu^U_x(P) = \overline{H}^U_{\chi_P}(x) = 0$ sein für alle $x \in U$. (Die einleitend gemachten Zusatzvoraussetzungen wurden dabei noch nicht verwendet.)

Weiter bemerken wir, daß eine polare Menge keine inneren Punkte besitzt. Dies folgt aus der Definition der superharmonischen Funktionen.

Wir behaupten nun umgekehrt (vgl. R. M. HERVÉ [*8*]):

**Satz 22.** *Es sei $K$ eine kompakte Menge ohne innere Punkte, $U \in \mathfrak{U}$ eine relativkompakte Umgebung von $K$ und $W = U \cap \complement K$. Gilt dann*

$$\mu^W_x(K) = 0$$

*für alle $x \in W$, so ist $K$ eine polare Menge.*

*Beweis.* Wir bemerken zunächst, daß $W$ offen und nicht-leer ist; letztere Eigenschaft folgt aus [*1*], Satz 4. Die Maße $\mu^W_x$ sind somit definiert. Mit $g$ soll die charakteristische Funktion von $K$ bezüglich $W^*$ bezeichnet werden.

Gemäß der Abzählbarkeitsvoraussetzung über $X$ existiert eine in $W$ dichte Punktfolge $(x_n)_{n=1,\,2,\dots}$. Weiter sei $(\varepsilon_n)_{n=1,\,2,\dots}$ eine Folge positiver reeller Zahlen mit $\sum \varepsilon_n < +\infty$. Es ist $H^W_{ng} = n H^W_g = 0$ in $W$ für alle natürlichen Zahlen $n$. Zu jedem solchen $n$ gibt es daher auf $W$ eine reelle (sogar beschränkte) superharmonische Funktion $w_n \geqq 0$ mit

$$\liminf_{x \to z,\, x \in W} w_n(x) \geqq n \qquad \text{für alle } z \in K$$

und

$$w_n(x_j) \leqq \varepsilon_n \qquad \text{für } j = 1, \dots, n.$$

Man hat nur zu beachten, daß die untere Einhüllende endlich vieler zu $ng$ gehöriger Oberfunktionen selbst wieder eine solche Oberfunktion ist. Wir setzen

$$w = \sum_{n=1}^{\infty} w_n .$$

Dann ist $w \geqq 0$ eine hyperharmonische Funktion in $W$ mit $w(x_n) < +\infty$ für alle $n$, also sogar eine superharmonische Funktion. Aus $w \geqq w_n$ für alle $n$ folgt

$$\lim_{x \to z,\, x \in W} w(x) = +\infty \qquad \text{für alle } z \in K .$$

Auf Grund der Kennzeichnung hyperharmonischer Funktionen durch lokale Eigenschaften ist dann folgende Funktion $u$ hyperharmonisch in $U$:

$$u(x) = \begin{cases} w(x), & x \in W; \\ +\infty, & x \in K. \end{cases}$$

Da $\mathring{K}$ leer und $w$ in $W$ superharmonisch ist, so ist auch $u$ superharmonisch in $U$. Wegen $K \subset u^{-1}(+\infty)$ ist also $K$ polar in $U$ und somit polar in $X$ (nach dem Korollar 2 zu Satz 21).

Bemerkung. Es wäre interessant zu wissen, ob man im obigen Satz die Voraussetzung $\mathring{K} = \emptyset$ entbehren kann. In der Brelotschen Theorie ist dies bekannt (vgl. [*8*], p. 45).

## Literatur

[*1*] BAUER, H.: Axiomatische Behandlung des Dirichletschen Problems für elliptische und parabolische Differentialgleichungen. Math. Ann. **146**, 1–59 (1962).

[*2*] — Minimalstellen von Funktionen und Extremalpunkte, II. Arch. der Math. **11**, 200–205 (1960).

[*3*] BRELOT, M.: Lectures on potential theory. Tata Inst. of Fund. Research, Bombay (1960).

[*4*] — Minorantes sousharmoniques, extrémales et capacités. J. Math. pur. appl. IX. Ser. **24**, 1–32 (1945).

[*5*] DOOB, J. L.: Probability methods applied to the first boundary value problem. Proc. 3rd Berkeley Symp. on Math. Stat. and Prob. 1954–1955, 49–80 (1956).

[*6*] — A probability approach to the heat equation. Trans. Amer. math. Soc. **80**, 216–280 (1955).

[*7*] HERVÉ, M.: Sur les représentation intégrales à l'aide des points extrémaux dans un ensemble compact convexe métrisable. C. r. Acad. Sci., Paris **253**, 366–368 (1961).

[*8*] HERVÉ, R. M.: Recherches axiomatiques sur la théorie des fonctions surharmoniques et du potentiel. Ann. Inst. Fourier **12**, 415–571 (1962).

[*9*] NÖBELING, G., und H. BAUER: Allgemeine Approximationskriterien mit Anwendungen. J.-Ber. Deutsch. Math.-Verein. **58**, 54–72 (1955).

*Zusatz bei der Korrektur*: Die in der Einleitung zitierte Arbeit von R. M. BLUMENTHAL, R. K. GETOOR und H. P. MCKEAN, JR. ist inzwischen erschienen: Markov processes with identical hitting distributions. Illinois J. of Math. **6**, 402–420 (1962).

*(Eingegangen am 16. Juli 1962)*

# Kennzeichnung kompakter Simplexe mit abgeschlossener Extremalpunktmenge [R21]

Arch. Math. **14** (1963), 415–421
[Zbl. Math. 196.42202; MR 29 #1352]

**1. Einleitung.** Für die CHOQUETschen Sätze über die Existenz und Eindeutigkeit von Integraldarstellungen der Punkte konvexer kompakter Mengen mittels der Extremalpunkte dieser Mengen wurden kürzlich neue durchsichtige Beweise von G. CHOQUET und P.-A. MEYER [3], [4], [7] angegeben. Hierdurch dürfte die Theorie ihre nunmehr wohl endgültige Form gefunden haben.

Der neue Zugang zur Theorie ermöglichte es CHOQUET [3], eine Kennzeichnung kompakter Simplexe anzugeben, welche analog ist zu einer früher vom Verf. [1] gewonnenen Kennzeichnung spezieller Simplexe, nämlich solcher mit abgeschlossener Extremalpunktmenge. Für die genaue Formulierung dieser Resultate bezeichne $X$ eine konvexe, kompakte Teilmenge eines Hausdorffschen, lokal-konvexen, topologischen Vektorraumes $E$ (über $\mathbf{R}$), $X_e$ die Menge aller Extremalpunkte von $X$, $\mathscr{A}$ die Menge aller reellen, stetigen, affin-linearen Funktionen auf $X$ und $\mathscr{A}_0$ die Menge aller numerischen Funktionen $l: X \to [-\infty, +\infty[$, welche nach oben halbstetig und affin-linear[1]) sind. Die erwähnte Kennzeichnung aus [1] besagt: *$X$ ist genau dann ein Simplex mit abgeschlossener Extremalpunktmenge $X_e$, wenn $\mathscr{A}$ ein Verband bezüglich der üblichen $\leqq$-Relation ist.* Da $\mathscr{A}$ ein Vektorraum ist, so ist hiermit gleichwertig die Forderung: Je zwei Funktionen aus $\mathscr{A}$ besitzen eine kleinste obere Schranke in $\mathscr{A}$. Das allgemeinere Resultat von CHOQUET [3] lautet dagegen: *$X$ ist genau dann ein Simplex, wenn je zwei Funktionen aus $\mathscr{A}_0$ eine kleinste obere Schranke in $\mathscr{A}_0$ besitzen.*

Die Analogie dieser beiden Aussagen ist unmittelbar ersichtlich. Jedoch fällt es auf Anhieb schwer, die gegenseitige Stellung der beiden kennzeichnenden Bedingungen zu durchschauen, da die eine Bedingung $\mathscr{A}$, die andere $\mathscr{A}_0$ verwendet. Ein erstes Ziel dieser Mitteilung soll der Klärung dieses Punktes dienen. Hierzu zeigen wir, daß $X$ genau dann ein Simplex mit abgeschlossenem $X_e$ ist, wenn jede nach oben halbstetige Funktion $f: \bar{X}_e \to [-\infty, +\infty[$ auf eine (und dann auch nur auf genau eine) Weise zu einer Funktion aus $\mathscr{A}_0$ fortgesetzt werden kann. Diese spezielle Eigenschaft des Halbverbandes $\mathscr{A}_0$ ist also für die Abgeschlossenheit von $X_e$ verantwortlich.

Das zitierte Resultat aus [1] erforderte einen ziemlich schwerfälligen Beweis. Ein zweites Ziel dieser Mitteilung ist daher die Herleitung der genannten Kennzeichnung

---

[1]) D. h. es gilt $l(\alpha x + (1 - \alpha) y) = \alpha l(x) + (1 - \alpha) l(y)$ für je zwei Punkte $x, y \in X$ und jedes reelle $\alpha$ mit $0 < \alpha < 1$.

kompakter Simplexe $X$ mit abgeschlossenem $X_e$ auf möglichst einfachem Wege. Dies wird durch die Heranziehung zweier einfach zu beweisender, für die Darstellung von Choquet-Meyer aber grundlegender Resultate ermöglicht. Auf diese Weise ergeben sich weitere in [1] bereits gewonnene Kennzeichnungen ganz zwangsläufig.

Im ersten Teil der Note werden einige Minimumprinzipe für halbstetige konvexe Funktionen hergeleitet, die anschließend benötigt werden. Diese Minimumprinzipe dürften auch ein gewisses Interesse für sich beanspruchen.

**2. Minimumprinzipe.** Die Bedeutung der bereits in der Einleitung eingeführten Bezeichnungen $E$, $X$, $X_e$, $\mathscr{A}$ und $\mathscr{A}_0$ soll auch im folgenden beibehalten werden. Eine numerische Funktion $k$ auf $X$ soll *konvex* heißen, wenn es sich um eine Abbildung $k : X \to [-\infty, +\infty[$ handelt mit

$$k(\alpha x + (1-\alpha) y) \leqq \alpha k(x) + (1-\alpha) k(y) \quad (x, y \in X; 0 < \alpha < 1).$$

Die Funktionen $-k$ mit konvexem $k$ heißen dann *konkav*.

**Lemma 1.** *Für je zwei numerische halbstetige*[2]) *Funktionen $k_1$ und $k_2$ auf $X$, von denen die erste konvex und die zweite konkav ist, gilt:*

$$(1) \qquad k_1(x) \leqq k_2(x) \text{ für alle } x \in X_e \Rightarrow k_1 \leqq k_2 \text{ auf } X.$$

Beweis. Es bezeichne $U_i$ bzw. $O_i$ den Fall, daß $k_i$ nach unten bzw. nach oben halbstetig ist ($i = 1, 2$). Dann führt folgende Fallunterscheidung zum Ziel:

$(U_1, U_2)$: Es ist $\inf k_2(X) > -\infty$. Für jedes $\alpha \in \mathbf{R}$ mit $\alpha \leqq \inf k_2(X)$ ist die Funktion $k_1' = \sup(\alpha, k_1)$ konvex und nach unten halbstetig; ferner ist $k_1'(x) \leqq k_2(x)$ für alle $x \in X_e$; aus $k_1' \leqq k_2$ auf $X$ folgt $k_1 \leqq k_2$. Somit kann $k_1$ als reell-wertig angenommen werden. Dann aber ist $k_1$ die obere Einhüllende aller $l \in \mathscr{A}$ mit $l \leqq k_1$ (vgl. Klee [6], p. 99). Für jedes solche $l$ ist $k_2 - l$ nach unten halbstetig, konkav und $\geqq 0$ auf $X_e$. Nach dem Minimumprinzip für nach unten halbstetige, konkave Funktionen (vgl. Bauer [2], p. 392) gilt dann $k_2 - l \geqq 0$ auf $X$ für jedes solche $l$, woraus $k_1 \leqq k_2$ folgt.

$(O_1, U_2)$: Da $k_2 - k_1$ nach unten halbstetig und konkav ist, folgt die Behauptung aus dem soeben bereits verwendeten Minimumprinzip für solche Funktionen.

$(U_1, O_2)$: Die Funktion $k = k_2 - k_1$ ist nach oben halbstetig, konkav und $\geqq 0$ auf $X_e$. Aus der die Konkavität kennzeichnenden Ungleichung folgt, daß dann $k(x) \geqq 0$ ist für alle $x$ aus der konvexen Hülle $c(X_e)$ von $X_e$. Nach dem Satz von Krein-Milman liegt $c(X_e)$ dicht in $X$; jede nicht-leere offene Teilmenge $G$ von $X$ hat also mit $c(X_e)$ einen nicht-leeren Durchschnitt. Für die Menge $G = \{x \in X : k(x) < 0\}$ ist $G \cap c(X_e) = \emptyset$. Wegen der Halbstetigkeit von $k$ nach oben ist $G$ offen. Folglich muß $G = \emptyset$, also $k \geqq 0$ und damit $k_1 \leqq k_2$ sein.

$(O_1, O_2)$: Vergleicht man $-k_2$ mit $-k_1$, so liegt der bereits behandelte Fall $(U_1, U_2)$ vor.

Damit ist das Lemma bewiesen.

---

[2]) Alle möglichen Kombinationen von Halbstetigkeit sind zugelassen: $k_1$ nach unten und $k_2$ nach unten halbstetig, $k_1$ nach unten und $k_2$ nach oben halbstetig, usw.

**Korollar 1.** *Eine numerische, halbstetige, konvexe Funktion $k$ auf $X$ ist nach oben beschränkt, wenn ihre Restriktion auf $X_e$ nach oben beschränkt ist.*

Beweis. Jede konstante reelle Funktion $\alpha$ auf $X$ liegt in $\mathscr{A}$. Aus $k(x) \leqq \alpha$ für alle $x \in X_e$ folgt daher $k \leqq \alpha$ nach Lemma 1.

**Korollar 2.** *Für je zwei reelle*[3]*), halbstetige, affin-lineare Funktionen $l_1$ und $l_2$ auf $X$ gilt:*

$$(2) \qquad l_1(x) = l_2(x) \text{ für alle } x \in X_e \Rightarrow l_1 = l_2 \text{ auf } X.$$

Beweis. Man wende Lemma 1 auf $k_1 = l_1$, $k_2 = l_2$ sowie $k_1 = l_2$, $k_2 = l_1$ an.

Wir wollen zeigen, daß die Aussage (2) des Korollars 2 auch für jedes Paar $l_1$, $l_2$ von Funktionen aus $\mathscr{A}_0$ zutrifft. Zur Vorbereitung des Beweises benötigen wir zunächst:

**Lemma 2.** *Jede Funktion $l \in \mathscr{A}_0$ ist die untere Einhüllende aller ihrer in $\mathscr{A}$ gelegenen Majoranten*[4]*).*

Beweis[5]). Man betrachte im Produktraum $E \times \mathbf{R}$ die offenbar abgeschlossene, konvexe Menge

$$L = \{(x, \lambda) : x \in X,\ \lambda \in \mathbf{R},\ l(x) \geqq \lambda\}.$$

Zu zeigen ist für jeden Punkt $x_0 \in X$ und jede reelle Zahl $\varrho > l(x_0)$ die Existenz eines $u \in \mathscr{A}$ mit $u \geqq l$ und $u(x_0) < \varrho$. Hierzu werde ein $\sigma \in \mathbf{R}$ mit $\varrho > \sigma \geqq l(x_0)$ gewählt und die abgeschlossene konvexe Hülle $Q$ von $L \cup \{(x_0, \sigma)\}$ betrachtet. Dann liegt $(x_0, \varrho)$ nicht in $Q$. In der Tat: Nach Wahl eines $\tau \in \mathbf{R}$ mit $\sigma > \tau$ betrachte man die konvexen Mengen

$$M = \{(x, \lambda) \in L : \lambda \geqq \tau\}$$

und

$$N = \{(x, \lambda) : x \in X, \lambda \leqq \sigma\}.$$

Dann ist $M$ kompakt, da $l$ nach oben beschränkt ist; folglich ist auch die konvexe Hülle $M_\sigma$ von $M \cup \{(x_0, \sigma)\}$ kompakt. Ferner ist $N$ und damit auch die Menge $M_\sigma \cup N$ abgeschlossen; sie enthält offenbar die konvexe Hülle von $L \cup \{(x_0, \sigma)\}$. Daher gilt $Q \subset M_\sigma \cup N$. Man rechnet sofort nach, daß $(x_0, \varrho)$ nicht in $M_\sigma \cup N$, also nicht in $Q$ liegen kann. Somit existiert eine durch eine stetige Linearform $F \neq 0$ definierte, abgeschlossene Hyperebene in $E \times \mathbf{R}$, welche $(x_0, \varrho)$ und $Q$ streng trennt:

$$F(x_0, \varrho) > \alpha > F(x, \lambda)$$

für alle $(x, \lambda) \in Q$ und ein $\alpha \in \mathbf{R}$. Die Linearform $F$ ist von der Form $F(x, \lambda) = f(x) + \omega\lambda$ $((x, \lambda) \in E \times \mathbf{R})$, wobei $f$ eine stetige Linearform auf $E$ und $\omega \in \mathbf{R}$ ist. Wegen

---

3) $l_1$ und $l_2$ haben also in $\mathbf{R}$ gelegene Werte.

4) Aussage und Beweis des Lemmas 2 bleiben für jede Funktion $l: X \to [-\infty, +\infty[$ richtig, welche der Konvexitätsungleichung

$$\alpha l(x) + (1 - \alpha) l(y) \leqq l(\alpha x + (1 - \alpha) y)$$

für alle $x, y \in X$ und alle $\alpha \in \mathbf{R}$ mit $0 < \alpha < 1$ genügt.

5) Der Beweis ist mehr oder weniger eine Routinearbeit, wenn $l$ reellwertig ist. (Dann kann im nachfolgenden Beweis nämlich $\sigma = l(x_0)$ gewählt werden, was $Q = L$ zur Folge hat.) Er erfordert etwas mehr Aufmerksamkeit, wenn, wie bei uns, $l$ auch $-\infty$ als Wert annehmen kann. Dieser Punkt bleibt bei Mokobodzki [8] ungeklärt.

$(x_0, \sigma) \in Q$ und $\sigma < \varrho$ folgt nunmehr, daß $\omega > 0$ ist und daß die in $\mathscr{A}$ gelegene Funktion

$$u = \omega^{-1}(\alpha - f)$$

das Verlangte leistet[6]).

Nun ergibt sich weiter:

**Lemma 3.** *Für jede numerische, halbstetige, konvexe Funktion $k$ auf $X$ und jede Funktion $l \in \mathscr{A}_0$ gilt:*

(3) $$k(x) \leqq l(x) \text{ für alle } x \in X_e \Rightarrow k \leqq l \text{ auf } X.$$

Beweis. Nach Lemma 2 ist $l$ die untere Einhüllende aller $u \in \mathscr{A}$ mit $u \geqq l$. Für jedes solche $u$ gilt $k(x) \leqq u(x)$ für alle $x \in X_e$, also $k \leqq u$ auf $X$ nach dem Lemma 1. Hieraus ergibt sich aber: $k \leqq l$ auf $X$.

**Korollar 1.** *Eine Funktion $l \in \mathscr{A}_0$ ist beschränkt, wenn ihre Restriktion auf $X_e$ beschränkt ist.*

Beweis. Die Beschränktheit von $l$ nach unten folgt aus Lemma 3, indem man dort für $k$ eine konstante reelle Funktion wählt, welche $l$ auf $X_e$ minorisiert. Nach dem Korollar 1 zu Lemma 1 ist $l$ ferner nach oben beschränkt.

Weiter folgt aus Lemma 3 unmittelbar das bereits angekündigte Resultat:

**Korollar 2.** *Für je zwei Funktionen $l_1, l_2 \in \mathscr{A}_0$ gilt:*

(4) $$l_1(x) = l_2(x) \text{ für alle } x \in X_e \Rightarrow l_1 = l_2 \text{ auf } X.$$

Offenbar besagen dieses Korollar und das Korollar 2 zu Lemma 1 in anderer Sprechweise folgendes: *Eine numerische Funktion $f: X_e \to [-\infty, +\infty[$ kann auf höchstens eine Weise zu einer Funktion aus $\mathscr{A}_0$ fortgesetzt werden. Eine reelle Funktion $f$ auf $X_e$ kann auf höchstens eine Weise zu einer reellen, halbstetigen, affin-linearen Funktion auf $X$ fortgesetzt werden.*

**3. Charakterisierung der Simplexe mit abgeschlossener Menge $X_e$.** Wir erinnern zunächst an zwei einfach zu beweisende, aber wichtige Resultate aus [3]. Hierzu bezeichne $\mathscr{S}$ eine Menge reeller, stetiger, konvexer Funktionen auf $X$. Es wird nur gefordert, daß $\mathscr{S}$ *total* ist im Vektorraum $\mathscr{C}(X)$ aller reellen stetigen Funktionen auf $X$ bezüglich der Topologie der gleichmäßigen Konvergenz. (Beispielsweise kann man für $\mathscr{S}$ die Menge aller oberen Einhüllenden $\sup(u_1, \ldots, u_n)$ von je endlich vielen Funktionen $u_1, \ldots, u_n \in \mathscr{A}$ wählen. $\mathscr{S}$ ist total in $\mathscr{C}(X)$, da $\mathscr{S} - \mathscr{S}$ den Voraussetzungen des Approximationssatzes von M. H. STONE genügt.) Ferner werde für jede reelle, stetige, konvexe Funktion $k$ auf $X$ gesetzt:

(5) $$\hat{k} = \inf_{\substack{u \in \mathscr{A} \\ u \geqq k}} u.$$

$\hat{k}$ ist die kleinste nach oben halbstetige, konkave Majorante von $k$.

[6]) Ist $l(x_0) = -\infty$ und fordert man von $F$ nur die strenge Trennung von $(x_0, \varrho)$ und $L$, so kann $\omega = 0$ sein.

Es gilt dann:

$$X_e = \bigcap_{k \in \mathscr{S}} \{x \in X : k(x) = \hat{k}(x)\}. \tag{6}$$

Ferner gilt:

$$X \textit{ ist Simplex} \Leftrightarrow \hat{k} \textit{ ist affin-linear für alle } k \in \mathscr{S}. \tag{7}$$

Man vergleiche hierzu CHOQUET [3], p. 2 und p. 6. Die Definition eines Simplexes findet man z. B. in [3] und [1].

Der einleitend angekündigte Satz lautet nun:

**Satz 1.** *Für jede konvexe, kompakte Teilmenge $X$ eines Hausdorffschen lokal-konvexen Raumes $E$ sind folgende Aussagen gleichwertig:*

(a) *$X$ ist ein Simplex; $X_e$ ist abgeschlossen.*

(b) *Jeder Punkt $x_0 \in X$ ist der Schwerpunkt genau eines von $\bar{X}_e$ getragenen Maßes $\mu \geqq 0$ auf $X$ mit Gesamtmasse 1.*

(c) *Jede reelle stetige Funktion $f$ auf $\bar{X}_e$ kann zu einer Funktion aus $\mathscr{A}$ fortgesetzt werden*[7]).

(d) *Jede nach oben halbstetige Funktion $f: \bar{X}_e \to [-\infty, +\infty[$ kann zu einer Funktion aus $\mathscr{A}_0$ fortgesetzt werden*[7]).

(e) *$\mathscr{A}$ ist bezüglich der üblichen $\leqq$-Relation ein Verband.*

Beweis. (a) $\Rightarrow$ (b): Dies folgt aus dem Eindeutigkeitssatz von CHOQUET.

(b) $\Rightarrow$ (c): Für jedes $x \in X$ bezeichne $\mu_x$ das einzige von $\bar{X}_e$ getragene, positive Maß der Gesamtmasse 1 auf $X$ mit $x$ als Schwerpunkt. Wegen der Kompaktheit der Menge aller Maße $\mu \geqq 0$ auf $X$ mit Gesamtmasse 1 (in der vagen Topologie) ergibt sich wie in [1], p. 114 die Stetigkeit der Abbildung $x \to \mu_x$. Für jede reelle stetige Funktion $f$ auf $X_e$ ist dann

$$x \to \int f \, d\mu_x$$

eine in $\mathscr{A}$ gelegene Fortsetzung von $f$.

(c) $\Rightarrow$ (d): Für gegebenes nach oben halbstetiges $f: \bar{X}_e \to [-\infty, +\infty[$ bezeichne $\Phi$ die nach unten filtrierende Menge aller reellen stetigen Majoranten $\varphi$ von $f$ auf $\bar{X}_e$. Zu jedem $\varphi \in \Phi$ gibt es nach (c) genau eine in $\mathscr{A}$ gelegene Fortsetzung $l_\varphi$. Mit $\Phi$ ist nach Lemma 1 auch die Menge aller $l_\varphi$ mit $\varphi \in \Phi$ nach unten filtrierend. Daher ist

$$l = \inf_{\varphi \in \Phi} l_\varphi$$

eine Funktion aus $\mathscr{A}_0$. Wegen $f = \inf \Phi$ leistet $l$ das Verlangte.

(d) $\Rightarrow$ (e): Für zwei Funktionen $u, v \in \mathscr{A}$ bezeichne $q$ die Restriktion von $\sup(u, v)$ auf $\bar{X}_e$. Dann kann $f = q$ bzw. $f' = -q$ zu einer Funktion $l$ bzw. $l'$ aus $\mathscr{A}_0$ fortgesetzt werden. Nach dem Korollar 1 zu Lemma 3 sind $l$ und $l'$ beschränkt, also reellwertig. Wegen $l(x) = -l'(x)$ für alle $x \in X_e$ ist dann $l = -l'$ nach dem Korollar 2 zu Lemma 1. Somit ist $l$ eine Funktion aus $\mathscr{A}$ mit $u \leqq l$ und $v \leqq l$ (nach Lemma 1). Für jede andere in $\mathscr{A}$ gelegene Majorante $l^*$ von $u$ und $v$ gilt $u(x) \leqq l^*(x)$ und $v(x) \leqq \leqq l^*(x)$, also $l(x) \leqq l^*(x)$ für alle $x \in X_e$, woraus abermals nach Lemma 1 $l \leqq l^*$

[7]) Die Fortsetzung ist nach dem Korollar zu Lemma 3 eindeutig bestimmt.

folgt. Somit ist $l$ die kleinste Majorante von $u$ und $v$ in $\mathscr{A}$. Dies zeigt, daß $\mathscr{A}$ ein Verband ist.

(e) $\Rightarrow$ (a): Es sei $\mathscr{S}$ die Menge aller Funktionen $k = \sup(u_1, \ldots, u_n)$ mit $u_1, \ldots, u_n \in \mathscr{A}$. Nach Annahme besitzen $u_1, \ldots, u_n \in \mathscr{A}$ eine kleinste Majorante in $\mathscr{A}$; diese werde mit $u_1 \vee \ldots \vee u_n$ bezeichnet. Dann gilt

$$\hat{k} = u_1 \vee \ldots \vee u_n$$

für alle $k = \sup(u_1, \ldots, u_n) \in \mathscr{S}$. In der Tat: es ist $\hat{k} \leqq u_1 \vee \ldots \vee u_n$, da $u_1 \vee \ldots \vee u_n$ eine Majorante von $k$ aus $\mathscr{A}$ ist. Jede Majorante von $k$ aus $\mathscr{A}$ ist aber eine gemeinsame Majorante von $u_1, \ldots, u_n$; daher gilt auch $u_1 \vee \ldots \vee u_n \leqq \hat{k}$. Damit ist aber $\hat{k}$ ein Element von $\mathscr{A}$ für alle $k \in \mathscr{S}$. Da $\mathscr{S}$ total in $\mathscr{C}(X)$ ist, muß $X$ nach (7) ein Simplex sein. Da $k$ und $\hat{k}$ stetig sind, ist $X_e$ nach (6) abgeschlossen.

Damit ist der Satz bewiesen.

Wir bemerken noch, daß für die Äquivalenz von (a) und (e) ein anderer Beweis von Edwards [5] gegeben wurde, der ebenfalls wesentlich einfacher als der ursprüngliche Beweis aus [1] ist.

Die im Beweisteil „(e) $\Rightarrow$ (a)" verwendete Schlußweise, die übrigens in ähnlicher Form auch beim Beweis der in der Einleitung zitierten Kennzeichnung allgemeiner Simplexe durch Choquet herangezogen wird, legt noch folgende Kennzeichnung nahe:

**Satz 2.** *Es sei $\mathscr{S}$ eine in $\mathscr{C}(X)$ totale Menge reeller, stetiger, konvexer Funktionen auf $X$. Dann gilt:*

(8) *$X$ ist Simplex mit abgeschlossener Menge $X_e \Leftrightarrow \hat{k} \in \mathscr{A}$ für alle $k \in \mathscr{S}$.*

Beweis. „$\Rightarrow$": Die Restriktion $f$ von $k \in \mathscr{S}$ auf $X_e$ kann nach Satz 1 auf genau eine Weise zu einer Funktion $l_f \in \mathscr{A}$ fortgesetzt werden. Offenbar ist $\hat{k} = l_f$.

„$\Leftarrow$": Dies folgt aus (6) und (7).

**4. Bemerkungen.** Das Lemma 1 kann als Beitrag zum folgenden, noch offenen Problem angesehen werden: Es sei $k$ eine numerische konkave Funktion auf $X$. Unter welchen zusätzlichen Annahmen gilt:

(9) $$k(x) \geqq 0 \text{ für alle } x \in X_e \Rightarrow k \geqq 0 \text{ auf } X.$$

Lemma 1 zeigt nämlich, daß (9) gilt, wenn sich $k$ als Summe einer nach oben und einer nach unten halbstetigen konkaven Funktion darstellen läßt.

Sicher ist (9) falsch, wenn man $k$ nur als Borelsche (konkave) Funktion voraussetzt. Man wähle nämlich für $E$ den mit der vagen Topologie versehenen Vektorraum aller Radonschen Maße auf dem kompakten Einheitsintervall [0, 1] und für $X$ die Menge aller Maße $\mu \geqq 0$ auf [0, 1] mit Gesamtmasse 1. Dann ist $X_e$ kanonisch isomorph zu [0, 1]. Die Funktion

$$k(\mu) = \begin{cases} -1, & \text{wenn } \mu \in X \text{ diffus;} \\ 0, & \text{wenn } \mu \in X \text{ nicht diffus,} \end{cases}$$

ist auf $X$ konkav, erfüllt aber (9) nicht. $k$ ist Borelsch, ja sogar von höchstens der dritten Baireschen Klasse. Bezeichnet nämlich $\mu_d$ den diffusen Teil eines $\mu \in X$, so ist die Funktion $\mu \to \int d\mu_d$ von der 2. Baireschen Klasse nach CHOQUET [3]. Es gilt aber

$$k(\mu) = \lim_{n \to \infty} (\int d\mu_d)^n - 1 \qquad (\mu \in X).$$

Resultate von CHOQUET [3] über allgemeine affin-lineare Funktionen legen die Vermutung nahe, daß (9) gilt, wenn $k$ in der ersten Baireschen Klasse liegt.

### Literaturverzeichnis


[1] H. BAUER, Šilovscher Rand und Dirichletsches Problem. Ann. Inst. Fourier **11**, 89—136 (1961).

[2] H. BAUER, Minimalstellen von Funktionen und Extremalpunkte. Arch. Math. **9**, 389—393 (1958).

[3] G. CHOQUET, Remarques à propos de la démonstration d'unicité de P.-A. Meyer. Séminaire BRELOT-CHOQUET-DENY (Théorie du Potentiel), 6e année, exposé 8, 13 p. (1962). (Institut H. Poincaré, Paris).

[4] G. CHOQUET et P.-A. MEYER, Existence et unicité des représentations intégrales dans les convexes compacts quelconques. Ann. Inst. Fourier **13**, 139—154 (1963).

[5] D. A. EDWARDS, On the representation of certain functionals by measures on the Choquet boundary. Ann. Inst. Fourier **13**, 111— 121 (1963).

[6] V. L. KLEE JR., Extremal structure of convex sets II. Math. Z. **69**, 90—104 (1958).

[7] P.-A. MEYER, Sur les démonstrations nouvelles du théorème de Choquet. Séminaire BRELOT-CHOQUET-DENY (Théorie du Potentiel), 6e année, exposé 7, 9 p. (1962). (Institut H. Poincaré, Paris.)

[8] G. MOKOBODZKI, Quelques propriétés des fonctions numériques convexes (s.c.i. ou s.c.s.) sur un ensemble convexe compact. Séminaire BRELOT-CHOQUET-DENY (Théorie du Potentiel), 6e année, exposé 9, 3 p. (1962). (Institut H. Poincaré, Paris.)


Eingegangen am 7. 11. 1962

# Propriétés fines des fonctions hyperharmoniques dans une théorie axiomatique du potentiel ([1]) [R23]

Ann. Inst. Fourier (Grenoble) **15** (1965), 137–154
[Zbl. Math. 127.05401; MR 32 #7772]

**Introduction.**

Ce travail représente la troisième contribution à une série d'articles concernant l'étude d'une théorie axiomatique du potentiel. Le point de départ de cette théorie est la notion d'un faisceau de fonctions « harmoniques ». Les applications concernent des équations aux dérivées partielles du second ordre de type elliptique et parabolique.

On s'intéresse ici surtout aux propriétés des fonctions hyperharmoniques qui, dans le cas classique, sont liées à la topologie fine et au théorème de convergence dû à H. Cartan. On aperçoit dans cette partie de la théorie pour la première fois des différences essentielles entre le cas elliptique, représenté par la théorie de Brelot [5], et le cas parabolique, englobé dans notre théorie. Les différences s'expriment par l'apparition des ensembles semi-polaires qui généralisent les ensembles polaires et coïncident avec eux dans le cas elliptique de Brelot.

Le résultat central concerne la notion d'effilement faible introduit par Brelot [6], [7]. On montrera que l'effilement faible d'un ensemble E en un point $x \notin$ E équivaut à l'effilement ordinaire de E en $x$. Notre première démonstration de ce résultat, valable seulement pour E analytique et exposée au Colloque, utilisait un théorème de Meyer [11] permettant l'application des méthodes probabilistes de la théorie de Hunt [10]. La démonstration donnée ici est basée sur un résultat de Boboc, Constantinescu et Cornea [4] et due à ces

([1]) Développement d'un exposé présenté au Colloque International du C.N.R.S. sur la Théorie du Potentiel. Ce travail a été effectué pendant un séjour de l'auteur à l'Université de Paris comme professeur d'échange au printemps 1964.

auteurs. Une conséquence de la première démonstration est maintenant le point de départ de la nouvelle présentation: On montre l'égalité de la réduite $R_u^E$ et de la balayée $\hat{R}_u^E$ dans le complémentaire de E.

L'étude comparée des deux notions d'effilement montre en particulier que, même pour $x \in E$, l'effilement faible de E en $x$ est la « bonne » notion d'effilement. On aperçoit ensuite d'après un théorème de Brelot [6] le théorème de convergence cherché dans cette théorie. Les applications de ce théorème concernent surtout des compléments à la théorie du problème de Dirichlet.

Dans une première partie du travail, on donnera un autre aspect de la notion d'un ensemble absorbant, introduite antérieurement. Cet aspect permettra une caractérisation de la théorie de Brelot en termes de la nôtre et une étude approfondie des ensembles polaires.

On utilisera dans tout l'article l'axiome de convergence de Doob. Mais une certaine partie du travail reste valable même avec l'axiome $K_1$ introduit dans [1].

Les notations seront les mêmes que dans les travaux antérieurs [1], [2]. En particulier, $\mathscr{C}(T)$ désignera l'espace vectoriel des fonctions réelles continues sur un espace topologique T.

## 1. Les axiomes de la théorie.

Rappelons brièvement les bases d'une théorie axiomatique des fonctions harmoniques et hyperharmoniques développée dans deux publications antérieures [1], [2].

On se donne un espace X *localement compact* et une application $U \to \mathscr{H}_U$ qui associe à tout ouvert $U \subset X$ un *sous-espace vectoriel* $\mathscr{H}_U$ de $\mathscr{C}(U)$. Les éléments de $\mathscr{H}_U$ sont appelés les *fonctions harmoniques* dans U. On définit ensuite les notions suivantes:

Un ouvert $V \subset X$ relativement compact avec frontière $V^* \neq \emptyset$ est dit *régulier* si toute fonction $f \in \mathscr{C}(V^*)$ se prolonge d'une seule manière en une fonction de $\mathscr{C}(\bar{V})$ dont la restriction $H_f^V$ à V est harmonique dans V et $\geqslant 0$ pour toute $f \geqslant 0$. Pour tout point $x \in V$, il existe alors une seule mesure de Radon $\mu_x^V \geqslant 0$ sur $V^*$, appelée la *mesure harmonique* associée à V et $x$, telle que l'on ait

$$(1) \qquad H_f^V(x) = \int f \, d\mu_x^V \qquad \text{pour toute } f \in \mathscr{C}(V^*).$$

Une fonction $u$: $U \to ]-\infty, +\infty]$ semi-continue inférieurement, définie dans un ouvert $U \subset X$ est dit *hyperharmonique* si

$$\int^* u \, d\mu_x^V \leqslant u(x) \tag{2}$$

quels que soient V régulier dans U (c'est-à-dire avec $\bar{V} \subset U$) et $x \in V$. On dénotera par $\mathscr{H}_U^*$ l'ensemble des fonctions hyperharmoniques dans U.

Les axiomes de la théorie s'énoncent alors de la manière suivante:

I) $U \to \mathscr{H}_U$ *est un faisceau (de fonctions continues réelles) sur* X. *Il existe une fonction* $h_0 \in \mathscr{H}_X$ *strictement positive partout.*

II) *Les ensembles réguliers forment une base de la topologie de* X.

III) *L'ensemble des fonctions* $\dfrac{u}{h_0}$ *avec* $u \in \mathscr{H}_X^*$ *et* $u \geqslant 0$ *sépare les points de* X.

IV) *Pour tout ouvert* $U \subset X$ *et tout ensemble non vide* $\mathscr{F} \subset \mathscr{H}_U$ *filtrant à droite dont l'enveloppe supérieure* sup $\mathscr{F}$ *est finie sur une partie partout dense dans* U, *on a* sup $\mathscr{F} \in \mathscr{H}_U$.

L'espace X est alors *non compact* et *localement connexe*; sa *compactification d'Alexandrov* est *connexe* [1], [3]. Les domaines réguliers forment une base de la topologie: Si V est un ensemble régulier, toute composante connexe W de V est aussi un ensemble régulier et on a

$$\mu_x^W = \mu_x^V \qquad \text{pour tout } x \in W. \tag{3}$$

Pour une démonstration, il suffit de considérer pour $f \in \mathscr{C}(W^*)$ un prolongement $\bar{f} \in \mathscr{C}(V^*)$ de $f$ et de remarquer que $H_f^W$ est la restriction de $H_{\bar{f}}^V$ à W.

Dans la suite on supposera toujours que $U \to \mathscr{H}_U$ satisfasse aux axiomes I, II, III, IV. Comme X est la somme topologique de ses composantes connexes, on pourrait même supposer que X soit connexe. On ne le fera pas sauf mention contraire.

## 2. Ensembles absorbants et supports surharmoniques.

Dans [2] nous avons introduit une notion dont l'apparition est typique pour une théorie axiomatique permettant aussi l'étude d'équations paraboliques: Un ensemble $A \subset X$ est dit *absorbant* si A est fermé et si, pour tout $x \in A$ et tout voisinage régulier V de $x$, la mesure harmonique $\mu_x^V$ est portée par A. Pour qu'un ensemble $A \subset X$ soit absorbant, il faut et il suffit que l'on ait $A = u^{-1}(0)$ pour au moins une fonction $u \geqslant 0$ de $\mathscr{H}_X^*$. Nous nous proposons de caractériser ces ensembles par une propriété « duale ».

Soit $u \in \mathscr{H}_X^*$ une fonction hyperharmonique dans X, non nécessairement $\geqslant 0$. Il existe un plus grand ensemble ouvert $G_u \subset X$ dans lequel on a $u(x) = +\infty$ $(x \in G_u)$; $G_u$ est l'ensemble des $x \in X$ tels que $u$ soit constante égale à $+\infty$ dans un voisinage de $x$. On appellera

$$A_u = \complement G_u$$

le *support surharmonique* de $u$. Il est caractérisé par l'égalité

$$A_u = \overline{\{x \in X : u(x) < +\infty\}}. \tag{4}$$

La restriction de $u$ à l'intérieur de $A_u$ est donc une fonction surharmonique au sens de [2].

THÉORÈME 1. — *Il y a identité entre les ensembles absorbants et les supports surharmoniques des fonctions de* $\mathscr{H}_X^*$.

*Démonstration.* — Pour un ensemble absorbant A, la fonction

$$u(x) = \begin{cases} +\infty, & x \in \complement A \\ 0, & x \in A \end{cases}$$

est hyperharmonique dans X (voir [2]). Le support surharmonique de cette fonction est A.

Soit inversement $A = A_u$ le support surharmonique d'une fonction $u \in \mathscr{H}_X^*$. Soit V un voisinage régulier d'un point $x \in X$ tel que le support $S\mu_x^V$ de la mesure harmonique $\mu_x^V$ possède une intersection non vide avec $G = \complement A$. La partie G étant ouverte, il en résulte, comme on le démontrera plus loin, que $S\mu_y^V \setminus A \neq \emptyset$ pour les points $y$ d'un voisinage de $x$. Dans ce même voisinage on a

$$u(y) \geqslant \int^* u \, d\mu_y^V = +\infty,$$

donc $u(y) = +\infty$. Un tel point $x$ n'appartient alors pas à A ce qui prouve que A est absorbant.

Il reste à démontrer que la relation $S\mu_x^V \cap G \neq \emptyset$ pour un ouvert G entraîne $S\mu_y^V \cap G \neq \emptyset$ pour les points $y$ d'un voisinage de $x$. Soit $f$ une fonction de $\mathscr{C}_+(V^*)$ telle que l'on ait $V^* \setminus G \subset f^{-1}(0)$ et $f(z) > 0$ pour au moins un point $z \in S\mu_x^V \cap G$. La solution $H_f^V$ du problème de Dirichlet est alors $>0$ au point $x$, donc strictement positive dans un voisinage W de $x$. En vertu de $H_f^V(y) = \int f \, d\mu_y^V > 0$

pour $y \in W$ on constate que $f$ ne peut pas être nulle sur $S\mu_y^W$. En utilisant la relation $V^* \setminus G \subset f^{-1}(0)$ il en résulte $S\mu_y^W \cap G \neq \emptyset$ pour tout $y \in W$, ce qui achève la démonstration.

COROLLAIRE. — *Une fonction* $u \in \mathscr{H}_X^*$ *égale à* $+\infty$ *en dehors d'un compact* $K \subset X$ *est la fonction constante* $+\infty$.

*Démonstration.* — Comme $A_u$ est contenu dans K, le support surharmonique $A_u$ de $u$ est compact. D'après [2; p. 209] l'ensemble vide est le seul ensemble absorbant compact. On a donc $A_u = \emptyset$ ce qui prouve que $u(x) = +\infty$ pour tout $x \in X$.

Nous donnerons deux applications du théorème 1 dans les deux paragraphes suivants.

*Remarque.* — Les résultats et démonstrations de ce paragraphe restent valables si on remplace les axiomes III et IV par les axiomes T et $K_1$ de [1].

### 3. Relations avec la théorie de M. Brelot.

Il est bien connu que l'axiomatique rappelée dans § 1 généralise la théorie de Brelot [5], [10]. Cette dernière utilise surtout un axiome de convergence plus fort que IV, à savoir

$IV_B$) *Pour tout domaine* $D \subset X$ *et tout ensemble non vide* $\mathscr{F} \subset \mathscr{H}_D$ *filtrant à droite, l'enveloppe* $\sup \mathscr{F}$ *est ou bien la fonction constante égale à* $+\infty$ *ou bien harmonique dans* D.

Le théorème 1 permet de clarifier les relations entre les deux théories :

THÉORÈME 2. — *Soit* $U \to \mathscr{H}_U$ *une application satisfaisant aux axiomes I, II, III. Les deux énoncés suivants sont alors équivalents* :

(*a*) *L'axiome IV est valable. Tout point* $x \in X$ *possède un système fondamental de voisinages réguliers* V *tels que le support* $S\mu_x^V$ *de* $\mu_x^V$ *soit toute la frontière* $V^*$ *de* V.

(*b*) *L'axiome* $IV_B$ *est valable et l'espace* X *est localement connexe.*

*Démonstration.* — (a) $\Rightarrow$ (b) : Nous avons déjà remarqué que X est localement connexe. Il suffira donc de montrer que, pour tout domaine D, les ensembles $\emptyset$ et D sont les seuls ensembles absorbants dans D (c'est à dire par rapport à la restriction de $U \to \mathscr{H}_U$ aux parties ouvertes de D). En effet, soit $\mathscr{F}$ non vide $\subset \mathscr{H}_D$ filtrant à droite et soit $u = \sup \mathscr{F}$ différente de la fonction constante $+\infty$.

Le support surharmonique de $u$ dans D sera alors non vide, donc égal à D. La fonction $u$ est par conséquent finie sur une partie dense dans D. L'axiome IV entraîne donc l'harmonicité de $u$ dans D.

Pour montrer que $\emptyset$ et D sont les seuls ensembles absorbants dans D, il suffira de supposer X connexe et de traiter le cas $D = X$. Considérons alors un ensemble absorbant $A \neq X$ et une composante connexe G de l'ouvert $\complement A$. L'ensemble G étant ouvert il suffira de montrer que G est aussi fermé. La connexion de X entraînera alors $G = X$ et par conséquent $A = \emptyset$. Supposons donc par l'absurde l'existence d'un point $x \in \bar{G} \setminus G$ qui appartient nécessairement à A. Soit V un voisinage régulier de $x$ tel que $S\mu_x^V = V^*$ et tel que G ne soit pas contenu dans V. L'ensemble A étant absorbant, on a $V^* = S\mu_x^V \subset A$ ce qui prouve que $V \cap G$ est à la fois ouvert et fermé relativement à G; l'intersection $V \cap G$ est aussi non vide car $x$ est adhérent à G. La connexion de G implique alors $V \cap G = G$ ce qui est une contradiction à notre hypothèse $G \not\subset V$. L'ensemble G est donc fermé.

(b) $\Rightarrow$ (a): C'est bien connu [5]. En effet, dans toute composante connexe de X les axiomes de la théorie de Brelot sont satisfaits.

Remarquons encore que la propriéte (*b*) *caractérise* dans toute composante connexe de X la théorie de Brelot avec les axiomes 1, 2, 3, l'existence d'un potentiel $>0$ et d'une fonction harmonique $>0$ (au sens des notations de [5]).

## 4. Quelques propriétés des ensembles polaires.

Une fonction $s \in \mathscr{H}_X^*$ est dite *surharmonique* dans X si $s$ est finie sur une partie partout dense dans X (donc, si $A_s = X$). Un ensemble $P \subset X$ est dit *polaire* s'il existe une fonction surharmonique $s \geqslant 0$ dans X telle que $P \subset s^{-1}(+\infty)$. L'étude des ensembles polaires dans cette théorie a été commencée dans [2].

Le lemme suivant prépare une autre application du théorème 1:

LEMME 1. — *L'ensemble vide est le seul ensemble* P *à la fois absorbant et polaire.*

*Démonstration.* — Soit $s$ une fonction surharmonique $\geqslant 0$ dans X telle que l'on ait $P \subset s^{-1}(+\infty)$. D'après [2], $s$ est $\mu_x^V$-intégrable pour tout ouvert regulier V et tout $x \in V$. Par conséquent $\mu_x^V$ ne peut pas être portée par P. L'ensemble P étant aussi absorbant, il en résulte $P = \emptyset$.

L'application annoncée concerne une propriété des ensembles polaires connue déjà dans la théorie de Brelot:

PROPOSITION 1. — *Le complémentaire* $\complement P$ *de tout ensemble polaire fermé* P *est connexe pourvu que l'espace* X *soit aussi connexe.*

*Démonstration.* — Supposons l'existence de deux ouverts $G_1, G_2$ non vides, disjoints tels que l'on ait $\complement P = G_1 \cup G_2$. L'intérieur de P étant vide, on a $\bar{G}_1 \cup \bar{G}_2 = X$. Nous avons supposé X connexe; il en résulte donc $\bar{G}_1 \cap \bar{G}_2 \neq \emptyset$. L'ensemble P étant polaire, il existe une fonction surharmonique $s \geqslant 0$ dans X telle que $P \subset s^{-1}(+\infty)$. Considérons les deux fonctions $s_i$ $(i = 1, 2)$ définies de la manière suivante:

$$s_i(x) = \begin{cases} s(x), & x \in P \cup G_i \\ +\infty, & x \in G_j \qquad (j \neq i, j = 1, 2). \end{cases}$$

Il s'agit de fonctions hyperharmoniques $\geqslant 0$ dans X. La fonction $s_i$ étant finie sur une partie dense dans $G_i$ et égale à $+\infty$ en dehors de $G_i$, son support surharmonique $A_{s_i}$ est égale à $\bar{G}_i$. D'après le théorème 1 les ensembles $\bar{G}_1$ et $\bar{G}_2$, donc aussi $\bar{G}_1 \cap \bar{G}_2$ sont absorbants. Mais comme $\bar{G}_1 \cap \bar{G}_2 \subset P$, l'intersection $\bar{G}_1 \cap \bar{G}_2$ est aussi polaire. Nous avons donc trouvé un ensemble absorbant et polaire $\bar{G}_1 \cap \bar{G}_2$ qui n'est pas vide. Le lemme 1 montre alors que notre hypothèse initiale est contradictoire.

Les résultats suivants seront utiles dans la suite. Rappelons d'abord que nous avons défini dans [1], pour toute partie A de X et toute fonction numérique $\varphi \geqslant 0$ dans X, la *réduite* $R_\varphi^A$ et la *balayée* $\hat{R}_\varphi^A$ de $\varphi$ sur A. Si l'espace X possède une base dénombrable, les ensembles polaires P sont caractérisés par la validité de $\hat{R}_u^P = 0$ pour toute fonction $u \geqslant 0$ de $\mathscr{H}_X^*$. La proposition suivante donne un résultat plus précis:

PROPOSITION 2. — *Supposons que l'espace* X *possède une base dénombrable. Pour tout ensemble polaire* P *et toute fonction numérique* $\varphi \geqslant 0$ *définie dans* X, *on a alors*

$$R_\varphi^P(x) = 0 \quad \textit{quel que soit } x \in \complement P. \tag{5}$$

*Démonstration.*— Pour $x \in \complement P$ donné, il suffira de montrer l'existence d'une fonction hyperharmonique $u \geqslant 0$ dans X telle que l'on ait $P \subset u^{-1}(+\infty)$ et $u(x) < +\infty$. Soit $s \geqslant 0$ une fonction surharmonique dans X telle que $P \subset s^{-1}(+\infty)$ et soit $(V_n)_{n=1,2,\ldots}$ une suite de voisinages réguliers de $x$ dont l'intersection est $\{x\}$. La fonction $s_{V_n}$ définie par

$$s_{V_n}(y) = \begin{cases} \int s\, d\mu_y^{V_n}, & y \in V_n \\ s(y), & y \in \complement V_n \end{cases}$$

est surharmonique dans X et finie en $x$ d'après [2; p. 200]. Pour terminer la démonstration, il suffit de définir $u = \sum_{n=1}^{\infty} \alpha_n s_{V_n}$, où $(\alpha_n)$ est une suite de nombres réels $>0$ telle que $u(x)$ soit finie. En effet, $u$ est hyperharmonique et $\geqslant 0$ dans X et, pour tout $y \in P$, il existe un voisinage $V_n$ ne contenant pas $y$. Il en résulte

$$u(y) = s(y) = s_{V_n}(y) = +\infty.$$

PROPOSITION 3. — *Supposons que l'espace* X *possède une base dénombrable. Pour tout ensemble polaire fermé* P, *il existe alors une fonction surharmonique* $s \geqslant 0$ *dans* X *telle que l'on ait* $P = s^{-1}(+\infty)$.

*Démonstration.* — Soit $v$ une fonction surharmonique dans X, finie et strictement positive, par exemple: $v = h_0$. D'après la proposition précédente on a $R_v^P(x) = 0$ quel que soit le point $x \in \complement P$. Pour un tel $x$ et tout $\varepsilon > 0$, il existe donc une fonction surharmonique $u \geqslant 0$ dans X avec les propriétés suivantes: $u \leqslant v$, $u(y) = v(y)$ pour tout $y \in P$, et $u(x) < \varepsilon$. On peut même supposer que $u$ est continue en $x$, il suffit de remplacer $u$ par $u_V$ (voir la démonstration précédente), où V est un voisinage régulier de $x$ disjoint de l'ensemble fermé P. Pour tout compact $K \subset \complement P$, il en résulte facilement l'existence d'une fonction surharmonique $u_K \geqslant 0$ dans X telle que $u_K \leqslant v$, $u_K(y) = v(y)$ pour tout $y \in P$, et $u_K(x) < \varepsilon$ pour tout $x \in K$.

Soit maintenant $(\varepsilon_n)_{n=1,2,\ldots}$ une suite de nombres $>0$ telle que $\sum_{n=1}^{\infty} \varepsilon_n < +\infty$, et soit $(K_n)_{n=1,2,\ldots}$ une suite croissante de compacts $\subset \complement P$ telle que $\bigcup_{n=1}^{\infty} K_n = \complement P$. Pour tout $n$, il existe une fonction surharmonique $u_n = u_{K_n} \geqslant 0$ dans X avec les propriétés suivantes: $u_n \leqslant v$, $u_n = v$ dans P, et $\sup u_n(K_n) \leqslant \varepsilon_n$. La fonction $s = \sum_{n=1}^{\infty} u_n$ répond alors à la question. Elle est hyperharmonique $\geqslant 0$ dans X, égale à $+\infty$ sur P et finie dans $\complement P$. En effet, pour tout $x \in \complement P$, il existe un $m = 1, 2, \ldots$ tel que $x$ appartienne à $K_m$; il en résulte $s(x) \leqslant (m-1)v(x) + \sum_{n=m}^{\infty} \varepsilon_n < +\infty$. L'intérieur de P étant vide, on voit que $s$ est même une fonction surharmonique.

Le résultat suivant utilise plusieurs propriétés des réduites et balayées démontrées pour $u \geqslant 0$ surharmonique par l'auteur (2) et pour $u \geqslant 0$ hyperharmonique par Boboc, Constantinescu et Cornea [4]:

Pour toute $u \geqslant 0$ de $\mathscr{H}_X^*$, la balayée $\hat{R}_u^E$ est *hyperharmonique* dans X quel que soit $E \subset X$. Pour toute $u \geqslant 0$ de $\mathscr{H}_X^*$, les fonctions $E \to R_u^E$ et $E \to \hat{R}_u^E$ définies sur l'ensemble $\mathfrak{P}(X)$ des parties de X sont *fortement sous-additives*, à savoir:

(6) $$R_u^{E\cup F} + R_u^{E\cap F} \leqslant R_u^E + R_u^F$$

(7) $$\hat{R}_u^{E\cup F} + \hat{R}_u^{E\cap F} \leqslant \hat{R}_u^E + \hat{R}_u^E.$$

Le résultat anoncé est bien connu dans la théorie de Brelot:

THÉORÈME 3. — *Pour toute fonction* $u \in \mathscr{H}_X^* \cap \mathscr{C}_+(X)$ *et tout* $x \in X$, *l'application*

$$K \to R_u^K(x)$$

*définie sur les compacts de X est une capacité forte de Choquet. L'application*

$$E \to R_u^E(x)$$

*est la capacité extérieure correspondante.*

*Démonstration.* – D'après le lemma 10 de [2], on a

(8) $$R_u^E = \inf_{\substack{E \subset U \\ U \text{ ouvert}}} R_u^U \qquad (E \subset X).$$

Il en résulte que $K \to R_u^K$ (K compact) est continue à droite, donc une capacité forte de Choquet d'après (6). Le reste de l'assertion résultera de (8) et de la formule suivante, valable pour tout ouvert U:

(9) $$R_u^U = \sup_{\substack{K \subset U \\ K \text{ compact}}} R_u^K.$$

Elle se démontre de la manière suivante: $(\hat{R}_u^K)_{K \text{ compact} \subset U}$ est une famille de fonctions hyperharmoniques, filtrante à droite, donc avec une enveloppe supérieure $u_0$ hyperharmonique dans X majorée par $\hat{R}_u^U = R_u^U$. Pour tout $x \in U$ et tout voisinage compact $K \subset U$ de $x$,

(2) Cf. l'exposé de ce travail présenté au Colloque International de la Théorie du Potentiel, Orsay (1964).

on a $\hat{R}_u^K(x) = u(x)$, ce qui entraîne $u_0(x) = u(x)$ quel que soit $x \in U$. On a donc $R_u^U \leqslant u_0$ et par conséquent l'égalité (9).

Le corollaire suivant généralise également un résultat de la théorie de Brelot [5]. Un ensemble $P \subset X$ est dit *intérieurement polaire* si toute partie compacte K de P est polaire.

COROLLAIRE. — *Si* X *est à base dénombrable, tout ensemble analytique* $P \subset X$ *intérieurement polaire est polaire.*

*Démonstration.* – Soit $u$ une fonction de $\mathscr{H}_X^* \cap \mathscr{C}_+(X)$. D'après le théorème 5 de [2], il suffit de montrer $\hat{R}_u^P = 0$. En appliquant le théorème de capacitabilité de Choquet [8] à la capacité du théorème 3, on trouve

$$R_u^P = \sup_{\substack{K \subset P \\ K \text{ compact}}} R_u^K.$$

P étant intérieurement polaire, il en résulte

(10) $$R_u^P(x) = 0 \qquad \text{pour tout } x \in \complement P.$$

En effet, pour tout compact $K \subset P$, on a $\hat{R}_u^K = 0$ et $R_u^K(x) = \hat{R}_u^K(x)$ quel que soit $x \in \complement K$.

Pour déduire $\hat{R}_u^P = 0$ de (10), il suffit de montrer que $\complement P$ soit partout dense dans X ou que l'on ait même $\mu_x^V(P) = 0$ pour tout ouvert régulier V et tout $x \in V$. En appliquant le théorème de capacitabilité à la mesure $\mu_x^V$, on obtient

$$\mu_x^V(P) = \sup_{\substack{K \subset P \\ K \text{ compact}}} \mu_x^V(K)$$

Tout tel compact K est polaire; on a donc $\mu_x^V(K) = 0$ d'après [2; p. 227]. Il en résulte finalement $\mu_x^V(P) = 0$.

## 5. Effilement faible et fort.

Dans tout ce paragraphe, on supposera l'espace X *à base dénombrable.*

Soit $\varphi$ une fonction de $\mathscr{C}(X)$ strictement positive sur X et soit $\mathfrak{V}(x)$ le filtre des voisinages d'un point $x \in X$. Pour tout ensemble $E \subset X$ et tout $x \notin E$, l'inégalité

(11) $$\inf_{V \in \mathfrak{V}(x)} R_\varphi^{E \cap V}(x) < \varphi(x)$$

*caractérise* alors l'effilement de l'ensemble E en $x$ (voir Brelot [6]).

Partant de cette remarque, Brelot [6], [7] a étudié deux autres notions d'effilement, à savoir :

Soit E une partie de X et $x$ un point de X. On dit que E est *faiblement effilé* en $x$ (même pour $x \in E$) si on a

$$\inf_{V \in \mathfrak{B}(x)} \hat{R}_\varphi^{E \cap V}(x) < \varphi(x). \tag{12}$$

Pour $x \notin E$, on dit que E est *fortement effilé* en $x$ si on a

$$\inf_{V \in \mathfrak{B}(x)} R_\varphi^{E \cap V}(x) = 0. \tag{13}$$

Les trois notions d'effilement sont indépendantes du choix de la fonction $\varphi$. En particulier on peut choisir $\varphi = 1$ ou $\varphi = h_0$.

Il est évident que, pour $x \notin E$, l'effilement fort entraîne l'effilement (ordinaire) qui, à son tour, entraîne l'effilement faible. Tout ensemble E est fortement effilé en tout $x \in \complement \bar{E}$ puisque $E \cap V = \emptyset$ pour $V \in \mathfrak{B}(x)$ assez petit. D'après la proposition 3, un ensemble polaire P est fortement effilé en tout $x \in \complement P$. Un tel ensemble est faiblement effilé en tout $x \in X$ selon [2, p. 206].

Étudions maintenant de plus près les relations entre les trois types d'effilement.

THÉORÈME 3. — *Soient* E *une partie de* X *et* $u \geqslant 0$ *une fonction de* $\mathscr{H}_X^*$, *on a alors*

$$\hat{R}_u^E(x) = R_u^E(x) \qquad \textit{quel que soit } x \in \complement E. \tag{14}$$

*Démonstration* [3]. — Soit $(V_n)_{n=1,2,\ldots}$ une suite décroissante de voisinages de $x \in E$ dont l'intersection est $x$. Le point $x$ n'étant pas adhérent à $E \setminus V_n$, on a

$$\hat{R}_u^{E \setminus V_n}(x) = R_u^{E \setminus V_n}(x). \tag{15}$$

En effet, c'est bien connu [1 ; p. 56] si $u$ est localement bornée. Dans le cas général, on applique ce résultat à la suite des fonctions $u_n = \inf(nh_0, u)$ de $\mathscr{H}_X^*$ $(n = 1, 2, \ldots)$. Grâce à un résultat de [4], on peut arriver à (15) par un passage à la limite. La suite $(E \setminus V_n)$ étant croissante et $\bigcup_{n=1}^{\infty} (E \setminus V_n)$ étant égale à E, on démontre (14) en appliquant un autre théorème de convergence de [4].

(3) Ce théorème a été énoncé au Colloque International de la Théorie du Potentiel (Orsay, 1964) pour E analytique et pour $x \in \complement E$ satisfaisant à une des hypothèses (16) ou (17). La démonstration utilisait la théorie des processus de Hunt. Le théorème dans le cas général et sa démonstration sont dus à MM. Boboc, Constantinescu et Cornea.

Une conséquence immédiate de ce théorème est le corollaire suivant :

COROLLAIRE. — *Pour qu'une partie* E *de* X *soit effilée en un point* $x \in \complement E$, *il faut et il suffit qu'elle soit faiblement effilée en* $x$.

L'effilement faible d'une partie $E \subset X$ en un point $x \in E$ entraîne donc en particulier l'effilement de $E \setminus \{x\}$ en $x$. Inversement, on a le résultat suivant :

PROPOSITION 4. — *Soit* $x$ *un point polaire* (4) *d'une partie* $E \subset X$. *L'effilement de* $E \setminus \{x\}$ *en* $x$ *entraîne alors l'effilement faible de* E *en* $x$.

*Démonstration.* — Il suffit de remarquer que la balayée $\hat{R}_u^A$ ne change pas si on remplace A par $A \cup P$ où P est un ensemble polaire [2 ; p. 206].

*Remarque.* — Dans la théorie de Brelot, on définit l'effilement d'une partie $E \subset X$ en un point $x \in E$ par les deux propriétés : le point $x$ est polaire et $E \setminus \{x\}$ est effilé en $x$. Dans ce sens l'effilement de E en $x \in E$ entraîne l'effilement faible de E en $x$. La réciproque est vraie dans la théorie de Brelot si on suppose l'axiome (D) de [5]. En effet, l'effilement faible de E en $x$ entraîne $\hat{R}_1^{\{x\}}(x) < 1$ ce qui montre [5 ; p. 135] que le point $x$ est polaire.

Nous allons voir que l'effilement faible est la bonne notion d'effilement dans notre théorie.

Passons maintenant à l'effilement fort :

THÉORÈME 4. — *Soit* $E \subset X$ *effilé en un point* $x \in \bar{E} \setminus E$. *Alors l'effilement est fort sous chacune des hypothèses suivantes :*

(16) *Le point* $x$ *est polaire.*

(17) *Tout potentiel localement borné sur* X *et harmonique dans* $\complement \{x\}$ *est continu en* $x$.

*Démonstration.* — En utilisant surtout le théorème de décomposition [2 ; p. 226], on montre, comme dans la théorie de Brelot [9 ; p. 46], l'existence d'une fonction $u \geqslant 0$ dans $\mathscr{H}_X^*$ telle que l'on ait

$$+\infty = \lim_{\substack{y \to x \\ y \in E}} u(y) > u(x).$$

En multipliant $u$ avec un nombre $\lambda > 0$, on peut supposer $u(x) \leqslant \varepsilon$ pour un $\varepsilon > 0$ donné. Il existe alors un voisinage V de $x$ tel que

(4) C'est-à-dire que l'ensemble $\{x\}$ est polaire.

la fonction constante 1 soit majorée par $u$ sur $E \cap V$. Par conséquent, on a

$$R_1^{E \cap V}(x) \leqslant u(x) \leqslant \varepsilon$$

d'où résulte l'assertion.

Remarquons que ce résultat s'applique immédiatement aux équations de Laplace et de la chaleur grâce à la condition (16).

## 6. Théorème de convergence de Brelot et conséquences.

Aussi dans tout ce paragraphe, on supposera l'espace X *à base dénombrable.*

DÉFINITION. — *On appellera un ensemble* $E \subset X$ *totalement effilé si* E *est faiblement effilé en tout point* $x$ *de l'espace* X. *Toute réunion dénombrable d'ensembles totalement effilés sera appelée un ensemble semi-polaire.*

D'après la proposition 2, tout ensemble polaire est totalement effilé. Evidemment tout ensemble totalement effilé est semi-polaire. Dans la théorie de Brelot munie de l'axiome (D), les trois types d'ensembles coïncident [5; p. 142]. On verra bientôt que ce n'est pas le cas dans notre théorie.

Toute réunion dénombrable d'ensembles semi-polaires et toute partie d'un ensemble semi-polaire sont semi-polaires.

Le théorème de convergence suivant est dû à Brelot [6] et valable sous des conditions beaucoup plus générales. Par exemple, il n'est pas nécessaire de supposer l'existence d'une base dénombrable. L'importance de ce théorème dans notre axiomatique vient du fait que nous avons constaté des relations étroites entre l'effilement faible et l'effilement ordinaire.

THÉORÈME 5 (M. Brelot). — *Pour toute famille* $(u_i)_{i \in I}$ *de fonctions hyperharmoniques* $\geqslant 0$ *dans* X, *l'ensemble*

$$\{x \in X : \widehat{\inf_{i \in I} u_i}(x) < \inf u_i(x)\} \tag{18}$$

*est semi-polaire.*

Les deux corollaires suivants sont immédiats:

COROLLAIRE 1. — *Pour tout ensemble* $E \subset X$ *et toute fonction numérique* $\varphi \geqslant 0$ *définie dans* X, *l'ensemble*

$$\{x \in X : \hat{R}_\varphi^E(x) < R_\varphi^E(x)\} \tag{19}$$

*est semi-polaire.*

Pour $\varphi \in \mathcal{H}_X^*$, $\varphi \geqslant 0$, l'ensemble (19) est une partie de E selon le théorème 3. Autrement dit :

COROLLAIRE 2.— *Pour tout ensemble* $E \subset X$ *et toute fonction* $u \geqslant 0$ *de* $\mathcal{H}_X^*$, *l'ensemble*

$$\{x \in E : \hat{R}_u^E(x) < u(x)\} \tag{20}$$

*est semi-polaire.*

COROLLAIRE 3.— *L'ensemble* $E_e$ *des points* $x$ *d'un ensemble* $E \subset X$, *où cet ensemble est faiblement effilé, est semi-polaire.*

*Démonstration.* — C'est une conséquence du corollaire 2. Il suffit de remarquer que l'on a

$$E_e = \bigcup_{B \in \mathcal{B}} \{x \in E \cap B : \hat{R}_{h_0}^{E \cap B}(x) < h_0(x)\}$$

où $\mathcal{B}$ désigne une base dénombrable de l'espace X.

Discutons enfin les conséquences de ces résultats pour le problème de Dirichlet pour un ouvert relativement compact U de X. D'après les théorèmes 39 et 40 de [1], on sait que l'ensemble $U_i^*$ des points irréguliers de $U^*$ est identique à l'ensemble des $z \in U^*$ où $\complement U$ est faiblement effilé. On a donc l'égalité

$$U_i^* = (\complement U)_e. \tag{21}$$

Cela conduit immédiatement au corollaire suivant :

COROLLAIRE 4. — *Pour tout ouvert relativement compact* U *de* X, *l'ensemble* $U_i^*$ *des points irréguliers de la frontière* $U^*$ *est semi-polaire.*

On sait de [1 ; p. 51] que l'on a toujours $U_i^* \neq U^*$.

## 7. Propriétés équivalentes au théorème fort de convergence.

Dans la théorie de Brelot, on montre à l'aide de l'axiome supplémentaire (D) un *théorème fort de convergence*, analogue au théorème de H. Cartan, à savoir :

(C) Pour toute famille $(u_i)_{i \in I}$ de fonctions hyperharmoniques $\geqslant 0$ dans X, l'ensemble des points $x \in X$, où $\widehat{\inf u_i}(x) < \inf u_i(x)$, est polaire.

M. Brelot et D. Sibony ont remarqué qu'un axiome supplémentaire analogue à l'axiome (D) permet la démonstration de (C) aussi dans la théorie exposée ici. Malheureusement, déjà pour l'équation de la chaleur, la propriété (C) n'est pas vraie comme le montre le théorème suivant.

THÉORÈME 6. — *Sous l'hypothèse que* X *possède une base dénombrable, la propriété* (C) *est équivalente à chacune des propriétés suivantes :*

($C_1$) *Pour tout compact* $K \subset X$ *et toute fonction* $u \geqslant 0$ *de* $\mathscr{H}^*_X$, *l'ensemble* $\{x \in K : \hat{R}^K_u(x) < u(x)\}$ *est polaire.*

($C_2$) *Pour tout compact* $K \subset X$ *non polaire et tout potentiel* $p \in \mathscr{C}(X)$, *on a* $\hat{R}^K_p(x) = p(x)$ *en au moins un point* $x \in K$.

($C_3$) *Pour tout compact* $K \subset X$ *non polaire d'intérieur vide et tout ouvert relativement compact* $U \supset K$, *il existe un point* $x_0 \in K$ *régulier par rapport à* $U \setminus K$.

($C_4$) *Pour tout ouvert relativement compact* $U \subset X$ *et tout* $x \in U$, *on a* $\mu^U_x(U^*_i) = 0$, *où* $U^*_i$ *désigne l'ensemble des points irréguliers de* $U^*$.

($C_5$) *Pour tout ouvert relativement compact* $U \subset X$, *l'ensemble* $U^*_i$ *est polaire.*

($C_6$) *Pour tout ouvert relativement compact* $U \subset X$ *et toute fonction* $u \in \mathscr{H}^*_U$ *bornée inférieurement, la relation* $\liminf_{x \to z,\, x \in U} u(x) \geqslant 0$ *en tout point régulier* $z \in U^*$ *entraîne* $u \geqslant 0$.

*Démonstration* (5). — Elle se fera d'après le schéma

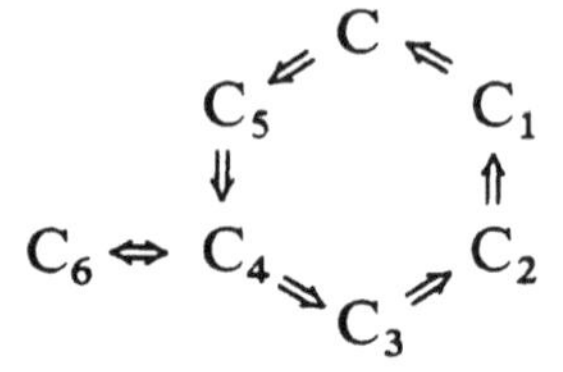


(C) ⇒ ($C_5$) : Il suffit de répéter des passages de Brelot [5], surtout dans les démonstrations des propositions 23, 25 et théorèmes 31, 32.

($C_5$) ⇒ ($C_4$) : C'est une conséquence du fait [2 ; p. 205] que tout ensemble polaire est négligeable.

($C_4$) ⇒ ($C_3$) : D'après le théorème 22 de [2], on a $\mu^W_x(K) > 0$ pour au moins au point $x \in W = U \setminus K$. Il en résulte que $K \setminus U^*_i$ est non vide.

($C_3$) ⇒ ($C_2$) : Si l'intérieur $\dot{K}$ de K est non vide, tout $x \in \dot{K}$ possède la propriété en question. Il suffit donc d'examiner le cas $\dot{K} = \emptyset$. On le fait avec le même raisonnement que dans [9 ; p. 115].

(5) Dans la théorie de Brelot, l'équivalence de (C), ($C_1$), . . . , ($C_5$) a été démontrée par Mme. Hervé [9].

$(C_2) \Rightarrow (C_1)$: Même démonstration que dans [9], lemme 27.3.

$(C_1) \Rightarrow (C)$: Même démonstration que dans [9], théorème 27.1.

$(C_4) \Rightarrow (C_6)$: Soit $\varphi$ la fonction indicatrice de $U_i^*$ par rapport à $U^*$. La fonction $u$ étant bornée inférieurement, il existe un $\lambda \in \mathbf{R}$ tel que l'on ait

$$\liminf_{x \to z} u(x) \geqslant \lambda\varphi(z) \qquad \text{pour tout } z \in U^*.$$

D'après [1; p. 37], il en résulte

$$u(x) \geqslant \bar{H}_{\lambda\varphi}(x) = \int^* \lambda\varphi \, d\mu_x^U = \lambda\mu_x^U(U_i^*) = 0$$

pour tout $x \in U$.

$(C_6) \Rightarrow (C_4)$: La démonstration se fait en plusieures étapes:

*a*) Pour toute $f \in \mathscr{C}(U^*)$, on a $H_f = H_{\tilde{f}}$ dans U, où la fonction $\tilde{f}$ est définie sur $U^*$ par $\tilde{f}(z) = \liminf_{x \to z} H_f(x)$ $(z \in U^*)$. En effet, pour toute $u \in \mathscr{H}_U^*$ bornée inférieurement et satisfaisant à $\liminf_{x \to z} u(x) \geqslant \tilde{f}(z)$ en tout $z \in U^*$, on a

$$\liminf_{x \to z} (u - H_f)(x) \geqslant \liminf_{x \to z} u(x) - f(z) \geqslant 0$$

en tout point régulier $z$ de $U^*$. D'après $(C_6)$, il en résulte $u \geqslant H_f$, d'où $\bar{H}_{\tilde{f}} \geqslant H_f$. La définition de $\tilde{f}$ donne l'inégalité duale. La fonction $\tilde{f}$ étant bornée et semi-continue inférieurement, on a aussi $\bar{H}_{\tilde{f}} = H_{\tilde{f}}$ d'après [1; p. 36].

*b*) Montrons maintenant que, pour tout $f \in \mathscr{C}(U^*)$ et $x_0 \in U$ donnés, la limite $\lim H_f(x)$ existe pour $\mu_{x_0}^U$ – presque tout $z \in U^*$. Soit $\tilde{f}$ définie comme dans *a*) et soit $\tilde{\tilde{f}} = -(\widetilde{-f})$. D'après *a*), on a $H_f = H_{\tilde{f}} = H_{\tilde{\tilde{f}}}$ ce qui entraîne $\int(\tilde{\tilde{f}} - \tilde{f})\, d\mu_{x_0}^U = 0$. Comme $\tilde{\tilde{f}} - \tilde{f} \geqslant 0$, il en résulte le résultat annoncé.

*c*) Pour tout $x \in U$, le support $S\mu_x^U$ est contenu dans l'adhérence de l'ensemble $U_r^*$ des points réguliers de $U^*$. En effet, pour toute $f \in \mathscr{C}(U^*)$ telle que $f(U_r^*) = \{0\}$, on a $H_f = 0$ d'après $(C_6)$.

*d*) L'espace X étant à base dénombrable, il existe une suite $(f_n)_{n=1,2,\ldots}$ dense dans $\mathscr{C}(U^*)$ pour la topologie de la convergence uniforme. Pour tout $x \in U$, l'ensemble $S_n(x)$ des points $z \in U^*$, pour lesquels $\lim_{y \to z} H_{f_n}(y)$ n'existe pas, est de $\mu_x^U$-mesure nulle d'après *b*). Pour $S(x) = \bigcup_{n=1}^{\infty} S_n(x)$, on a donc $\mu_x^U(S(x)) = 0$. Remarquons maintenant que, pour $f \in \mathscr{C}(U^*)$ et $\varepsilon > 0$, la relation $|f - f_n| \leqslant \varepsilon h_0$

entraîne $|H_f - H_{f_n}| \leqslant \varepsilon h_0$ dans U. Il en résulte aisément que

$$\lim_{y \to x} H_f(y)$$

existe pour tout $z \in U^* \setminus S(x)$ quelle que soit la fonction $f \in \mathscr{C}(U^*)$. L'ensemble des $z \in U^*$, où $\lim_{y \to z} H_f(y)$ existe pour toute $f \in \mathscr{C}(U^*)$, est donc égale à $U^* \setminus S(x)$. Il en résulte que $S = S(x)$ est indépendant de $x \in U$ et que l'on a $U_r^* \subset U^* \setminus S$. Pour toute $f \in \mathscr{C}(U^*)$ la fonction $\tilde{f}$ (resp. $\tilde{\tilde{f}}$) est semi-continue inférieurement (resp. supérieurement); la coïncidence de $\tilde{f}$ et $\tilde{\tilde{f}}$ avec $f$ sur $U_r^*$ entraîne donc $f(z) = \tilde{f}(z) = \tilde{\tilde{f}}(z)$ pour tout $z \in \overline{U_r^*} \setminus S$. Autrement dit, on a $\overline{U_r^*} \setminus S \subset U_r^* \subset U^* \setminus S$ ce qui prouve l'égalité $U_r^* = \overline{U_r^*} \setminus S$. Nous avons déjà vu que $\mu_x^U$ est portée par $\overline{U_r^*}$ et $U^* \setminus S$. La mesure $\mu_x^U$ est donc portée par $U_r^*$ pour tout $x \in U$, ce qui achève la démonstration.

Pour l'équation de la chaleur

$$\frac{\partial^2 u}{\partial \xi^2} = \frac{\partial u}{\partial t}$$

dans l'espace $X = \mathbf{R}^2$, le théorème fort de convergence ne peut pas être valable. En effet, considérons les rectangles

$$U = \{(\xi, t): 0 < \xi < 1, 0 < t < 1\}$$

et

$$V = \{(\xi, t): 0 < \xi < 1, 1 < t < 2\}.$$

On a

$$U_i^* = \{(\xi, t): 0 < \xi < 1, t = 1\}$$

et $\mu_x^V(U_i^*) > 0$ pour tout $x \subset V$. On voit donc que $U_i^*$ ne peut pas être polaire; la propriété $(C_5)$ n'a pas lieu.

-D'après le corollaire de la proposition 5, $U_i^*$ est un exemple d'un ensemble semi-polaire qui n'est pas polaire.

## BIBLIOGRAPHIE

[1] H. BAUER, Axiomatische Behandlung des Dirichletschen Problems für elliptische und parabolische Differentialgleichungen, *Math. Annalen*, 146 (1962), 1-59.

[2] H. BAUER, Weiterführung einer axiomatischen Potentialtheorie ohne Kern (Existenz von Potentialen), *Z. Wahrscheinlichkeitstheorie*, 1 (1963), 197-229.

[3] H. BOBOC, C. CONSTANTINESCU and A. CORNEA, Axiomatic theory of harmonic functions. — Non-negative superharmonic functions, *Annales Inst. Fourier*, 15 (1965), fasc. 1 (Colloque de Théorie du Potentiel).

[4] H. Boboc, C. Constantinescu and A. Cornea, Axiomatic theory of harmonic functions — Balayage, *Annales Inst. Fourier,* 15 (1965) (à l'impression).
[5] M. Brelot, Lectures on potential theory, *Tata Inst. of Fundamental Research, Bombay* (1960).
[6] M. Brelot, Quelques propriétés et applications nouvelles de l'effilement, *Séminaire de Théorie du Potentiel,* 6 (fasc. 1), no. 1c, 14 pp. (1961/62), Institut H. Poincaré, Paris.
[7] M. Brelot, Introduction axiomatique de l'effilement, *Annali di Matematica,* sér. 4, 57 (1962), 77-95.
[8] G. Choquet, Theory of capacities, *Annales Inst. Fourier,* 5 (1953/54), 131-295.
[9] R. M. Hervé, Recherches axiomatiques sur la théorie des fonctions surharmoniques et du potentiel, *Annales Inst. Fourier,* 12 (1962), 415-571.
[10] G. A. Hunt, Markoff processes and potentials I, *Illinois J. Math.,* 1 (1957), 44-93.
[11] P. A. Meyer, Brelot's axiomatic theory of the Dirichlet problem and Hunt's theory, *Annales Inst. Fourier,* 13, fasc. 2 (1963), 357-372.

# Zum Cauchyschen und Dirichletschen Problem bei elliptischen und parabolischen Differentialgleichungen [R24]

Math. Ann. **164** (1966), 142–153
[Zbl. Math. 142.38501; MR 34 #6135]

Herrn Gottfried Köthe zum 60. Geburtstag am 25.12.1965 gewidmet

## Einleitung

Den Anstoß zu der vorliegenden Arbeit gaben zwei Beobachtungen:

1. In der Theorie der parabolischen partiellen Differentialgleichungen zweiter Ordnung unterscheidet man in Fragestellung und Lösungsmethode zwischen dem Cauchyschen *Anfangswertproblem* und dem Dirichletschen *Randwertproblem*, obwohl diese Unterscheidung wenig gerechtfertigt erscheint. Betrachten wir zur Illustration etwa die Wärmeleitungsgleichung

$$\sum_{i=1}^{n} \frac{\partial^2 u}{\partial x_i^2} = \frac{\partial u}{\partial x_{n+1}} \tag{W}$$

im $\mathbf{R}^{n+1}$, so lautet das Cauchysche Problem wie folgt: Gegeben ist der offene Halbraum $L$ aller $x = (x_1, \ldots, x_{n+1}) \in \mathbf{R}^{n+1}$ mit $x_{n+1} > 0$ und eine auf dem Rand $L^*$ von $L$, also der Hyperebene mit der Gleichung $x_{n+1} = 0$ definierte, stetige, reelle Funktion $f$. Unter welchen Voraussetzungen über $f$ existiert dann eine in $L$ definierte Lösung $u$ von (W), welche $f$ stetig fortsetzt? Man sieht dieser Formulierung sofort an, daß es sich um die erste Randwertaufgabe für $L$, also um ein Dirichletsches Problem handelt. Bei diesem werden allerdings die Randwerte üblicherweise auf dem Rand $U^*$ einer relativ-kompakten offenen Teilmenge $U$ des $\mathbf{R}^{n+1}$ vorgegeben. Formal liegt aber dieselbe Situation vor.

2. Die einzige Rechtfertigung für die Unterscheidung beider Probleme scheint in der verschiedenen Form der zugehörigen Eindeutigkeitssätze zu liegen. Während beim Dirichletschen Problem für eine relativ-kompakte offene Menge $U \subset \mathbf{R}^{n+1}$ zu einer stetigen reellen Randfunktion höchstens eine Lösung der Randwertaufgabe existiert, ist die Lösung beim Cauchyschen Problem im allgemeinen mehrdeutig[1]). Die Eindeutigkeit kann hier aber durch Beschränktheitsbedingungen erzwungen werden. So existiert etwa zu jeder *beschränkten* stetigen Randfunktion $f$ auf $L^*$ genau eine *beschränkte* Lösung des Cauchyschen Problems[2]). Beim Dirichletschen Problem für $U$ ist eine Beschränktheitsbedingung dieser Art dagegen von selbst erfüllt: Jede stetige

[1]) Vgl. Friedman [5], p. 31 sowie Petrowski [6], S. 268.

[2]) Vgl. Petrowski [6], S. 267 ff.

reelle Funktion auf $U^*$ ist beschränkt; ebenso ist die nach der Perron-Wiener-Brelotschen Methode zugehörige verallgemeinerte Lösung in $U$ beschränkt.

Im folgenden soll dargelegt werden, daß die in 2. versuchte Rechtfertigung für die Unterscheidung beider Probleme oberflächlich ist. Es soll gezeigt werden, daß das Cauchysche und das Dirichletsche Problem Spezialfälle eines allgemeineren Dirichletschen Problems für beliebige offene Mengen $U \subset \mathbf{R}^{n+1}$ mit nicht leerem Rand $U^*$ (in $\mathbf{R}^{n+1}$) sind. Dieses allgemeinere Dirichletsche Problem unterscheidet sich dabei vom herkömmlichen durch die Bezugnahme auf ein vorgegebenes Potential $p$ auf $\mathbf{R}^{n+1}$, von dem nur gefordert werden muß, daß seine Restriktion auf den Abschluß $\overline{U}$ von $U$ eine stetige reelle Funktion mit Werten $>0$ ist. Das neue Problem lautet: Wann existiert zu jeder stetigen reellen Funktion $f$ auf $U^*$ mit $\sup\limits_{z \in U^*} \frac{|f(z)|}{p(z)} < +\infty$ eine, dann notwendigerweise eindeutig bestimmte Lösung $u$ von $(W)$, welche stetig an $f$ anschließt und der Bedingung $\sup\limits_{x \in U} \frac{|u(x)|}{p(x)} < +\infty$ genügt? Als notwendig und hinreichend wird sich die Regularität eines jeden Randpunktes $z \in U^*$ erweisen. Dabei heißt $z$ ein regulärer Randpunkt von $U$, wenn eine relativ-kompakte offene Umgebung $V$ von $z$ derart existiert, daß $z$ ein im üblichen Sinne regulärer Randpunkt von $U \cap V$ ist. Dieser Regularitätsbegriff ist somit von $p$ völlig unabhängig. Das Cauchysche Problem wird durch spezielle Wahl von $p$ und durch den Nachweis der Regularität aller Randpunkte von $L$ gelöst. Für relativ-kompaktes $U$ ist die Untersuchung von $p$ unabhängig und liefert die in [2] dargestellte Theorie des Dirichletschen Problems.

Methodisch gehen wir so vor, daß die Perron-Wiener-Brelotsche Methode geeignet modifiziert und auf $p$ bezogen wird. Entscheidend ist hierbei das folgende bekannte *Randminimum-Prinzip:* Ist $u$ eine Lösung von (W) in $U$ oder allgemeiner eine hyperharmonische Funktion in $U$ bezüglich (W), welche durch ein negatives Vielfaches von $p$ minorisiert wird, so gilt:

$$\liminf_{x \to z} u(x) \geqq 0 \quad \text{für alle } z \in U^* \ \Rightarrow \ u \geqq 0\,.$$

Dieses Randminimum-Prinzip gilt aber bei vielen elliptischen und parabolischen linearen Differentialgleichungen zweiter Ordnung, nämlich bei allen, welche der in [2] dargestellten axiomatischen Theorie harmonischer Funktionen genügen. Wir entwickeln daher die ganze Untersuchung gleich im Rahmen dieser Theorie, deren Kenntnis in Form der Darstellung [2] als bekannt vorausgesetzt wird. Die Arbeiten [2a]—[2c] sind Vorläufer dieser Darstellung.

## § 1. Eine Variante der Methode von Perron-Wiener-Brelot

Im folgenden sei $X$ ein *streng harmonischer Raum* im Sinne von [2] und $p$ ein zunächst beliebiges *Potential* auf $X$. Eine auf einer Menge $T \subset X$ definierte numerische Funktion $t$ heißt auf $T$ nach oben bzw. nach unten $p$-beschränkt, wenn eine reelle Zahl $\lambda \geqq 0$ existiert mit

$$t(x) \leqq \lambda p(x) \quad \text{bzw.} \quad t(x) \geqq -\lambda p(x) \quad \text{für alle} \quad x \in T\,.$$

Die Funktion $t$ heiße auf $T$ *p-beschränkt*, wenn sie auf $T$ nach oben und unten $p$-beschränkt ist.

Gegeben sei ferner eine offene Teilmenge $U$ von $X$ mit nicht-leerem Rand $U^*$ (bezüglich $X$)[3]). Dann kann die Methode von Perron-Wiener-Brelot — im folgenden kurz PWB-Methode genannt — wie folgt abgewandelt und auf das gegebene Potential $p$ bezogen werden. Sei $f$ eine auf dem Rand $U^*$ von $U$ definierte numerische Funktion. Zu $f$ gehörige *p-Oberfunktion* heiße jede in $U$ hyperharmonische Funktion $u$ mit folgenden drei Eigenschaften:

(1.1) $$\liminf_{x\to z, x\in U} u(x) \geqq f(z) \quad \text{für alle} \quad z \in U^*;$$

(1.2) $$\liminf_{x\to z, x\in U} u(x) > -\infty \quad \text{für alle} \quad z \in U^*;$$

(1.3) $u$ ist auf $U$ nach unten $p$-beschränkt.

Zu $f$ gehörige *p-Unterfunktion* heiße jede numerische Funktion $v$ auf $U$, für welche $-v$ eine zu $-f$ gehörige $p$-Oberfunktion ist. $\mathcal{O}_p(f)$ bzw. $\mathcal{U}_p(f)$ bezeichne die Menge aller zu $f$ gehörigen $p$-Ober- bzw. $p$-Unterfunktionen. Die konstante Funktion $+\infty$ liegt stets in $\mathcal{O}_p(f)$.

Zur Illustration sei sofort folgender *Spezialfall* hervorgehoben: $U$ werde zusätzlich als relativ-kompakt vorausgesetzt; ferner gelte $p(x) > 0$ für alle $x \in \overline{U}$. Dann ist bereits jede in $U$ hyperharmonische Funktion $u$, welche den Bedingungen (1.1) und (1.2) genügt, eine zu $f$ gehörige $p$-Oberfunktion. Es ist nämlich $u$ nach unten beschränkt und außerdem $\inf p(\overline{U}) > 0$. Also stimmen die zu $f$ gehörigen $p$-Oberfunktionen mit den im Sinne von [2] zu $f$ gehörigen Oberfunktionen überein. Die Betrachtung ist somit von der speziellen Wahl von $p$ unabhängig. Es liegt die PWB-Methode zur Lösung des Dirichletschen Problems in ihrer ursprünglichen Form vor.

Durch diese Illustration angeleitet soll jetzt die allgemeine Situation weiterverfolgt werden. Jeder auf $U^*$ definierten, numerischen Funktion $f$ werden die in $U$ definierten Funktionen $\overline{H}_f$ und $\underline{H}_f$ zugeordnet vermöge der Festsetzungen:

(1.4) $$\overline{H}_f = \inf \mathcal{O}_p(f);$$

(1.5) $$\underline{H}_f = \sup \mathcal{U}_p(f).$$

Sollen $U$ und $p$ besonders betont werden, so schreiben wir $\overline{H}_f^U$ oder ${}_p\overline{H}_f^U$ für $\overline{H}_f$ und entsprechend $\underline{H}_f^U$ oder ${}_p\underline{H}_f^U$ für $\underline{H}_f$. Nun gilt offenbar

(1.6) $$\underline{H}_f = -\overline{H}_{-f}$$

und außerdem ist

(1.7) $$\underline{H}_f \leqq \overline{H}_f.$$

Letzteres ergibt sich wie folgt: Seien $u \in \mathcal{O}_p(f)$ und $v \in \mathcal{U}_p(f)$; dann ist $u - v$ eine auf $U$ nach unten $p$-beschränkte, hyperharmonische Funktion. Aus (1.1) und (1.2) folgt außerdem

$$\liminf_{x\to z,\, x\in U} (u(x) - v(x)) \geqq 0 \quad \text{für alle} \quad z \in U^*.$$

[3]) Ist der Rand $U^*$ leer, so sind die folgenden Betrachtungen entweder leer oder trivial.

Nach dem bereits in der Einleitung zitierten Randminimum-Prinzip ist dann $u - v \geqq 0$, woraus (1.7) folgt.

Sind $f$ und $g$ auf $U^*$ definierte, numerische Funktionen, so ergeben sich aus obigen Definitionen sofort folgende Aussagen:

$$f \leqq g \Rightarrow \overline{H}_f \leqq \overline{H}_g ; \tag{1.8}$$

$$\overline{H}_{f+g} \leqq \overline{H}_f + \overline{H}_g ; \tag{1.9}$$

$$\overline{H}_{\lambda f} = \lambda \overline{H}_f \qquad (0 \leqq \lambda < +\infty) . \tag{1.10}$$

Bei (1.9) muß vorausgesetzt werden, daß beide Summen $f + g$ und $\overline{H}_f + \overline{H}_g$ überall in $U^*$ bzw. $U$ definiert sind.

In wichtigen Fällen erweisen sich die Funktionen $\overline{H}_f$ und $\underline{H}_f$ als in $U$ harmonisch. Wir zeigen diesbezüglich nur:

**Lemma 1.** *Ist die Randfunktion $f$ auf $U^*$ p-beschränkt, so sind $\overline{H}_f$ und $\underline{H}_f$ auf $U$ p-beschränkte und harmonische Funktionen.*

*Beweis.* Es gibt ein reelles $\lambda \geqq 0$ derart, daß $-\lambda p(z) \leqq f(z) \leqq \lambda p(z)$ für alle $z \in U^*$ gilt. Folglich liegt die Restriktion von $\lambda p$ auf $U$ in $\mathcal{O}_p(f)$ und die Restriktion von $-\lambda p$ auf $U$ in $\mathcal{U}_p(f)$. Hieraus folgt

$$-\lambda p(x) \leqq \underline{H}_f(x) \leqq \overline{H}_f(x) \leqq \lambda p(x)$$

für alle $x \in U$. Also sind $\overline{H}_f$ und $\underline{H}_f$ auf $U$ $p$-beschränkt. Es folgt weiter, daß $-\lambda p$ eine subharmonische Minorante von $\mathcal{O}_p(f)$ auf $U$ ist. Beachtet man noch, daß $\mathcal{O}_p(f)$ eine gesättigte Menge hyperharmonischer Funktionen mit mindestens einem superharmonischen Element (nämlich der Restriktion von $\lambda p$ auf $U$) ist, so folgt die Harmonizität von $\overline{H}_f$ aus [2]. Wegen (1.6) ist dann auch $\underline{H}_f$ in $U$ harmonisch.

Die beiden folgenden Lemmata beweist man analog wie die entsprechenden Behauptungen für die in [2] behandelte ursprüngliche Form der PWB-Methode. Die Beweise können daher hier übergangen werden.

**Lemma 2.** *Sei $(f_n)$ eine aufsteigende Folge numerischer Funktionen auf $U^*$ derart, daß jede der Funktionen $\overline{H}_{f_n}$ harmonisch in $U$ ist. Dann gilt*

$$\overline{H}_{\sup f_n} = \sup \overline{H}_{f_n} . \tag{1.11}$$

**Lemma 3.** *Sei $f$ eine auf $U^*$ p-beschränkte, numerische Funktion und sei $V$ eine offene Teilmenge von $U$ mit nicht-leerem Rand $V^*$. Dann gilt*

$$\overline{H}_g^V = \mathrm{Rest}_V \overline{H}_f^U \tag{1.12}$$

*für die folgende auf dem Rand $V^*$ von $V$ definierte Funktion $g$:*

$$g(z) = \begin{cases} f(z), \ z \in U^* \cap V^* \\ \overline{H}_f^U(z), z \in U \cap V^* \end{cases} . \tag{1.13}$$

Beim Beweis hat man Lemma 1 auszunützen, wonach $\overline{H}_f^U$ in $U$ harmonisch und $g$ auf $V^*$ $p$-beschränkt ist.

Nunmehr definieren wir in Analogie zur ursprünglichen Form der PWB-Methode:

**Definition 1.** *Eine auf $U^*$ definierte, numerische Funktion $f$ heißt p-resolutiv, wenn gilt*

$$\underline{H}_f = \overline{H}_f \quad \text{und} \quad |\overline{H}_f(x)| < +\infty \quad \text{für alle} \quad x \in U . \tag{1.14}$$

Wir setzen dann $H_f = \underline{H}_f = \overline{H}_f$ und nennen $H_f$ die zu $f$ gehörige *verallgemeinerte Lösung* des Dirichletschen Problems *bezüglich* $p$. Falls die Hervorhebung von $U$ und $p$ notwendig ist, schreiben wir $H_f^U$ oder ${}_pH_f^U$ für $H_f$.

Mit dieser Begriffsbildung ergibt sich sofort das folgende Korollar zu Lemma 3:

**Lemma 4.** *Sei $f$ eine auf $U^*$ p-beschränkte und p-resolutive Randfunktion und sei $V$ eine offene Teilmenge von $U$ mit nicht-leerem Rand $V^*$. Definiert man dann eine Funktion $g$ auf $V^*$ vermöge* (1.13), *so ist auch $g$ p-resolutiv und es gilt*

$$H_g^V = \mathrm{Rest}_V H_f^U . \tag{1.15}$$

*Beweis.* Es genügt, Lemma 3 auf die Funktionen $f$ und $-f$ anzuwenden.

Das nachfolgende, diesen Paragraphen abschließende Lemma gibt ein wichtiges Beispiel $p$-resolutiver Funktionen.

**Lemma 5.** *Sei $u \geqq 0$ eine auf $X$ hyperharmonische Funktion derart, daß ihre Restriktion $u^*$ auf $U^*$ p-beschränkt ist und für jeden Punkt $z \in U^*$ gilt:*

$$\lim_{\substack{x \to z \\ x \in U}} u(x) = u(z) \tag{1.16}$$

*Dann ist $u^*$ p-resolutiv.*

*Beweis.* Die Restriktion von $u$ auf $U$ ist eine zu $u^*$ gehörige $p$-Oberfunktion; folglich gilt

$$\overline{H}_{u^*}(x) \leqq u(x) \quad \text{für alle} \quad x \in U .$$

Nach Lemma 1 ist $\overline{H}_{u^*}$ harmonisch in $U$ und auf $U$ $p$-beschränkt, also wegen (1.16) eine zu $u^*$ gehörige $p$-Unterfunktion. Hieraus folgt $\overline{H}_{u^*} \leqq \underline{H}_{u^*}$ und somit die $p$-Resolutivität von $u^*$ wegen (1.7).

## § 2. *p*-Harmonische Maße

Für das Potential $p = 0$ ist offenbar $f = 0$ die einzige $p$-resolutive Randfunktion auf $U^*$. Zur Ausschaltung derartiger Trivialfälle soll nunmehr das Potential $p$ der folgenden Bedingung unterworfen werden:

($P_U$) *Die Restriktion von $p$ auf $\overline{U}$ ist eine stetige, reelle, überall positive Funktion.*

Diese Bedingung ist insbesondere erfüllt, wenn $p$ ein stetiges, reelles, strenges Potential auf $X$ ist, welches gemäß [2] stets existiert.

Für eine abgeschlossene Teilmenge $A$ von $\overline{U}$ bezeichnen wir mit $\mathscr{C}(A)$ den Vektorraum aller stetigen reellen Funktionen auf $A$ und mit $\mathscr{K}(A)$ den linearen Unterraum aller $f \in \mathscr{C}(A)$ mit kompaktem Träger. $\mathscr{C}_p(A)$ bezeichne denjenigen linearen Unterraum von $\mathscr{C}(A)$, welcher aus allen auf $A$ $p$-beschränkten Funktionen $f \in \mathscr{C}(A)$ besteht. Es gilt offenbar

$$\mathscr{K}(A) \subset \mathscr{C}_p(A) . \tag{2.1}$$

Für jede Funktion $f \in \mathscr{C}_p(A)$ werde gesetzt

(2.2) $$\|f\|_p = \sup_{x \in A} \left| \frac{f(x)}{p(x)} \right|$$

Dann ist $f \to \|f\|_p$ eine Norm auf $\mathscr{C}_p(A)$; die zugehörige Topologie heiße die der *p-gleichmäßigen Konvergenz* auf $A$.

Nunmehr ergibt sich eine erste Folgerung aus der Bedingung ($P_U$).

**Lemma 6.** *Es bezeichne $\mathscr{Q}_p$ die Menge aller auf $X$ p-beschränkten, reellen Potentiale, deren Restriktion auf $\overline{U}$ stetig ist. Für einen abgeschlossenen Unterraum $A$ von $\overline{U}$ bezeichne ferner $\mathscr{D}(A)$ die Menge aller Funktionen $f \in \mathscr{K}(A)$, welche eine Darstellung*

(2.3) $$f(x) = q_1(x) - q_2(x) \qquad (x \in A)$$

*mit Hilfe von Potentialen $q_1, q_2 \in \mathscr{Q}_p$ zulassen. Dann ist $\mathscr{D}(A)$ ein linearer Unterraum von $\mathscr{K}(A)$, welcher bezüglich der Topologie der p-gleichmäßigen Konvergenz auf $A$ in $\mathscr{K}(A)$ dicht liegt.*

*Beweis.* Für je zwei Potentiale $q_1, q_2 \in \mathscr{Q}_p$ liegt offenbar $\alpha_1 q_1 + \alpha_2 q_2$ für beliebige reelle Koeffizienten $\alpha_1 \geqq 0$, $\alpha_2 \geqq 0$ in $\mathscr{Q}_p$. Daher ist $\mathscr{D}(A)$ und damit auch die Menge $\mathscr{D}_p(A)$ aller auf $A$ definierten Funktionen $d = \frac{f}{p}$ mit $f \in \mathscr{D}(A)$ ein linearer Unterraum von $\mathscr{K}(A)$. Für jede Funktion $d \in \mathscr{D}_p(A)$ liegen die Funktionen $|d|$ und $\inf(1, d)$ in $\mathscr{D}_p(A)$. Ist nämlich $d = \frac{f}{p}$ und besitzt $f \in \mathscr{D}(A)$ die Darstellung (2.3), so gilt auf $A$

$$|d| = \frac{1}{p}\,[q_1 + q_2 - 2\inf(q_1, q_2)]$$

und

$$\inf(1, d) = \frac{1}{p}\,[\inf(p + q_2, q_1) - q_2]\,.$$

$\mathscr{D}_p(A)$ trennt außerdem je zwei Punkte $x_0 \neq y_0$ aus $A$ in dem Sinne, daß eine Funktion $d \in \mathscr{D}_p(A)$ existiert mit

$$d(x_0) > 0 \quad \text{und} \quad d(y_0) = 0\,.$$

Sei nämlich $V$ eine reguläre Umgebung von $x_0$, in welcher $y_0$ nicht liegt. Nach [2] existiert ein stetiges reelles Potential $q$ auf $X$ mit

$$\int q\, d\mu_{x_0}^V < q(x_0)\,.$$

Ohne Beschränkung der Allgemeinheit kann $q(x_0) \leqq p(x_0)$ und $q(y_0) \leqq q(y_0)$ angenommen werden, da $p(x_0) > 0$ und $p(y_0) > 0$ gilt. Setzt man nun wie üblich

$$q_V(x) = \begin{cases} q(x), & x \in \complement V \\ \int q\, d\mu_x^V, & x \in V\,, \end{cases}$$

so sind $q_1 = \inf(p, q)$ und $q_2 = \inf(p, q_V)$ Potentiale aus $\mathscr{Q}_p$, und die in $\mathscr{D}_p(A)$ gelegene Funktion $x \to \frac{q_1(x) - q_2(x)}{p(x)}$ leistet das Verlangte.

Nunmehr folgt die Behauptung aus dem Approximationssatz von M. H. Stone, wonach für jedes $k \in \mathscr{K}(A)$ die ebenfalls in $\mathscr{K}(A)$ gelegene Funktion $\frac{k}{p}$ gleichmäßig auf $A$ durch Funktionen aus $\mathscr{D}_p(A)$ approximiert werden kann.

Jetzt ergibt sich der erste entscheidende Satz.

**Satz 1.** *Jede Funktion $f \in \mathscr{K}(U^*)$ ist p-resolutiv. Für jeden Punkt $x \in U$ ist die auf $\mathscr{K}(U^*)$ definierte Abbildung*

$$f \to H_f^U(x) \tag{2.4}$$

*ein positives Radon-Maß $\mu_x^U$ auf $U^*$.*

*Beweis.* Es bezeichne $\mathscr{R}_p$ die Menge aller $p$-resolutiven Funktionen aus $\mathscr{C}_p(U^*)$. Aus den Eigenschaften (1.8)–(1.10) folgt wie bei der üblichen PWB-Methode, daß $\mathscr{R}_p$ ein linearer Unterraum von $\mathscr{C}_p(U^*)$ und $f \to H_f$ eine positive, lineare Abbildung von $\mathscr{R}_p$ in den Vektorraum $\mathscr{H}_U$ der in $U$ harmonischen Funktionen ist. Wegen Lemma 5 ist dann der Vektorraum $\mathscr{D}(U^*)$ aus Lemma 6 ein linearer Unterraum von $\mathscr{R}_p$. Die Behauptung folgt somit aus Lemma 6, falls $\mathscr{R}_p$ in $\mathscr{C}_p(U^*)$ abgeschlossen ist. Dies aber ergibt sich wie folgt: Sei $(f_n)$ eine gegen eine Funktion $f \in \mathscr{C}_p(U^*)$ auf $U^*$ $p$-gleichmäßig konvergente Folge aus $\mathscr{R}_p$. Bezeichnen wir mit $p^*$ die Restriktion von $p$ auf $U^*$, so gilt bei gegebenem $\varepsilon > 0$

$$f_n - \varepsilon p^* \leqq f \leqq f_n + \varepsilon p^*$$

für schließlich alle $n = 1, 2, \ldots$ . Da $p^*$ gemäß Lemma 5 $p$-resolutiv ist, folgt dann

$$H_{f_n}(x) - \varepsilon H_{p^*}(x) \leqq \underline{H}_f(x) \leqq \overline{H}_f(x) \leqq H_{f_n}(x) + \varepsilon H_{p^*}(x)$$

für schließlich alle $n$ gleichmäßig in $x \in U$. Folglich ist $\underline{H}_f = \overline{H}_f$ und somit $f$ wegen Lemma 1 $p$-resolutiv.

Das Maß $\mu_x^U$ heiße fortan das zu $x$ und $U$ gehörige *p-harmonische Maß*. Definitionsgemäß gilt

$$H_f^U(x) = \int f \, d\mu_x^U \quad \text{für alle} \quad f \in \mathscr{K}(U^*) \,. \tag{2.5}$$

**Korollar.** *Jede Funktion $f \in \mathscr{C}_p(U^*)$ ist p-resolutiv und $\mu_x^U$-integrierbar für jedes $x \in U$. Es gilt*

$$H_f^U(x) = \int f \, d\mu_x^U \quad \textit{für jedes} \quad x \in U \,. \tag{2.6}$$

*Beweis.* Sei $f \in \mathscr{C}_p(U^*)$ und $f \geqq 0$. Dann existiert eine Folge $(f_n)$ in $\mathscr{K}(U^*)$ mit $0 \leqq f_n \leqq f_{n+1}$ für alle $n = 1, 2, \ldots$ und $\sup f_n = f$. Aus (2.5) und den beiden ersten Lemmata folgt daher unter Verwendung des Konvergenzsatzes von B. LEVI die $\mu_x^U$-Integrierbarkeit von $f$ sowie

$$\overline{H}_f^U(x) = \int f \, d\mu_x^U \qquad (x \in U) \,.$$

Ferner gilt gemäß (1.8)

$$\underline{H}_{f_n}^U = H_{f_n} \leqq \underline{H}_f \quad \text{für alle} \quad n = 1, 2, \ldots,$$

woraus bei erneuter Heranziehung von Lemma 2 die Ungleichung $\overline{H}_f \leqq \underline{H}_f$ und damit die Behauptung für $f$ folgt. Da sich jede Funktion aus $\mathscr{C}_p(U^*)$ als Differenz nicht-negativer Funktionen aus $\mathscr{C}_p(U^*)$ darstellen läßt, ist dann auch der Fall eines beliebigen $f \in \mathscr{C}_p(U^*)$ erledigt.

Auf die Frage nach der Gültigkeit eines Analogons zum Resolutivitätssatz von BRELOT [3] soll hier nicht weiter eingegangen werden.

## § 3. Reguläre Randpunkte

Für jede Funktion $f \in \mathscr{C}_p(U^*)$ ist nach dem soeben bewiesenen Korollar die verallgemeinerte Lösung $H_f = H_f^U$ des Dirichletschen Problems bezüglich $p$ definiert und gemäß Lemma 1 in $U$ harmonisch. Zum Studium des Randverhaltens von $H_f$ definieren wir:

**Definition 2.** *Ein Randpunkt $z \in U^*$ heiße p-regulär, wenn für jede Funktion $f \in \mathscr{C}_p(U^*)$ gilt*

$$\lim_{\substack{x \to z \\ x \in U}} H_f(x) = f(z) . \tag{3.1}$$

Der Zusammenhang mit dem üblichen Regularitätsbegriff [2] wird durch die beiden nachfolgenden Sätze hergestellt.

**Satz 2.** *Für jede nicht-leere, relativ-kompakte, offene Menge $V \subset U$ stimmen die p-regulären Randpunkte von $V$ mit den regulären Randpunkten von $V$ überein.*

*Beweis.* Aus der Bedingung $(\mathrm{P}_U)$ folgt die Bedingung $(\mathrm{P}_V)$. Dann aber liegt der in § 1 besprochene Spezialfall vor, wonach die üblichen Begriffe der Resolutivität und der verallgemeinerten Lösung mit den hier studierten, auf $p$ bezogenen übereinstimmen. Beachtet man noch, daß $\mathscr{C}_p(V^*) = \mathscr{C}(V^*)$ wegen der Kompaktheit von $V^*$ und wegen $(\mathrm{P}_V)$ gilt, so folgt die Behauptung.

**Satz 3.** *Sei $z$ ein Randpunkt von $U$ und sei $W$ eine relativ-kompakte, offene Umgebung von $z$. Dann und nur dann ist $z$ ein p-regulärer Randpunkt von $U$, wenn $z$ ein regulärer Randpunkt von $U \cap W$ ist.*

*Beweis.* Zunächst werde $z$ als regulärer Randpunkt von $V = U \cap W$ angenommen. Da $V$ nicht-leer, relativ-kompakt und offen ist, so ist der Rand $V^*$ nach [2] nicht-leer. Definiert man daher für ein $f \in \mathscr{C}_p(U^*)$ die Funktion $g$ auf $V^*$ gemäß (1.13), so ist $g$ nach Lemma 4 $p$-resolutiv und nach Lemma 1 $p$-beschränkt. Da $V$ relativ-kompakt ist und $(\mathrm{P}_U)$ gilt, ist dann $g$ sogar resolutiv und beschränkt. Nach (1.15) gilt $H_g^V(x) = H_f^U(x)$ für alle $x \in V$. Beachtet man die Stetigkeit von $g$ in $z$, so folgt aus einer in [2] bewiesenen Eigenschaft regulärer Punkte:

$$\lim_{\substack{x \to z \\ x \in U}} H_f^U(x) = \lim_{\substack{x \to z \\ x \in V}} H_g^V(x) = g(z) = f(z) .$$

Somit ist $z$ ein $p$-regulärer Randpunkt von $U$.

Umgekehrt werde nun $z$ als $p$-regulärer Randpunkt von $U$ vorausgesetzt. Für jedes Potential $q$ aus der in Lemma 6 definierten Menge $\mathscr{Q}_p$ liegt die Restriktion $q^*$ bzw. $q^{**}$ auf $U^*$ bzw. $V^*$ in $\mathscr{C}_p(U^*)$ bzw. $\mathscr{C}_p(V^*) = \mathscr{C}(V^*)$, und es gilt

$$H_{q^*}^U(x) \leqq q(x) \quad \text{für alle} \quad x \in U$$

und

$$H_{q^{**}}^V(x) \leqq q(x) \quad \text{für alle} \quad x \in V .$$

Nach der ersten Ungleichung ist dann die Restriktion von $H_{q^*}^U$ auf $V$ eine zu $q^{**}$ gehörige $p$-Unterfunktion; man erhält somit die Doppel-Ungleichung

$$H_{q^*}^U(x) \leqq H_{q^{**}}^V(x) \leqq q(x) \quad \text{für alle} \quad x \in V . \tag{3.2}$$

Da $q$ in $\mathfrak{Q}_p$ liegt, gilt $\lim\limits_{\substack{x\to z\\ x\in V}} q(x) = q(z)$. Da $z$ ein $p$-regulärer Randpunkt von $U$ ist, gilt $\lim\limits_{\substack{x\to z\\ x\in V}} H^U_{q^*}(x) = q(z)$. Hieraus sowie aus (3.2) folgt dann

$$\lim_{\substack{x\to z\\ x\in V}} H^V_{q^{**}}(x) = q(z) = q^{**}(z) \tag{3.3}$$

für alle $q \in \mathfrak{Q}_p$. Beachtet man, daß die Topologie der $p$-gleichmäßigen Konvergenz mit der der gleichmäßigen Konvergenz in $\mathscr{C}_p(V^*) = \mathscr{C}(V^*)$ übereinstimmt, so besagt Lemma 6 für $A = V^*$, daß die Menge $\{q^{**} : q \in \mathfrak{Q}_p\}$ in $\mathscr{C}(V^*)$ total bezüglich der Topologie der gleichmäßigen Konvergenz auf $V^*$ ist. Hieraus und aus (3.3) folgt dann schließlich die Regularität von $z$ als Randpunkt von $V$ nach einem Resultat aus [2].

Der Satz 3 besagt insbesondere, daß unser Begriff der $p$-Regularität in Wirklichkeit von dem der Bedingung ($P_U$) genügenden Potential $p$ *unabhängig* ist. Fortan nennen wir daher die $p$-regulären Randpunkte von $U$ nur noch *regulär*. Jeder nicht reguläre Randpunkt von $U$ heiße *irregulär*.

Durch die Zurückführung des neuen Regularitätsbegriffes auf den bekannten, mittels der üblichen PWB-Methode eingeführten Regularitätsbegriff ergeben sich nunmehr leicht eine Reihe nützlicher Folgerungen:

**Korollar 1.** *Sei $U_1$ eine offene Teilmenge von $U$. Ist dann $z$ ein regulärer Randpunkt von $U$ und zugleich Randpunkt von $U_1$, so ist $z$ ein regulärer Randpunkt von $U_1$.*

*Beweis.* Dies folgt aus dem in [2] bewiesenen entsprechenden Resultat für relativ-kompaktes $U$ bei Beachtung von Satz 3.

**Korollar 2.** *Ein Randpunkt $z \in U^*$ ist genau dann irregulär, wenn $\complement U$ in $z$ dünn ist.*

*Beweis.* Da die Dünnheit von $\complement U$ in $z$ eine lokale Bedingung ist, folgt die Behauptung aus dem entsprechenden Satz im relativ-kompakten Fall mittels Satz 3.

**Korollar 3.** *Die Menge aller irregulären Randpunkte von $U$ ist semi-polar.*

*Beweis.* Nach [2] ist für jede Menge $E \subset X$ die Menge aller $z \in E$, in welchen $E$ dünn ist, semi-polar.

**Korollar 4.** *Ein Randpunkt $z$ von $U$ ist genau dann regulär, wenn eine zu $z$ gehörige Barriere existiert, wenn es also eine offene Umgebung $W$ von $z$ und eine in $U \cap W$ definierte hyperharmonische Funktion $w$ mit folgenden Eigenschaften gibt:*

$$w(x) > 0 \quad \textit{für alle} \quad x \in U \cap W; \tag{3.4}$$

$$\lim_{\substack{x\to z\\ x\in U\cap W}} w(x) = 0\,. \tag{3.5}$$

*Beweis.* Dies folgt wiederum aus dem entsprechenden Satz für relativ-kompaktes $U$, der in [2] behandelt wurde.

Ähnlich wie in [2] läßt sich schließlich die Existenz einer *universellen Barriere* beweisen:

**Satz 4.** *Es existiert ein stetiges, reelles, strenges Potential $G^U$ auf $U$ derart, daß genau diejenigen Randpunkte $z$ von $U$ regulär sind, für welche*

$$\lim_{x \to z} G^U(x) = 0 \tag{3.6}$$

gilt.

*Beweis.* Sei $p$ ein stetiges, reelles, strenges Potential auf $X$, welches nach [2] stets existiert. Das Potential $p$ genügt der Bedingung $(P_U)$; die Restriktion $p^*$ von $p$ auf $U^*$ liegt in $\mathscr{C}_p(U^*)$. Folglich ist die zu $p^*$ gehörige verallgemeinerte Lösung $H_{p^*} = {}_pH^U_{p^*}$ des Dirichletschen Problems bezüglich $p$ auf $U$ definiert und dort harmonisch. Setzt man $G^U(x) = p(x) - H_{p^*}(x)$ für alle $x \in U$, so ergibt sich wie in [2], daß $G^U$ das Verlangte leistet.

**Korollar.** *Die Menge aller regulären Randpunkte von $U$ ist eine $G_\delta$-Menge.*

*Beweis.* Dies folgert man aus Satz 4 wie in [2] im dort behandelten relativkompakten Fall.

Abschließend formulieren wir noch den besonders für die Anwendungen nützlichen

**Satz 5.** *Sind alle Randpunkte von $U$ regulär, so ist ${}_pH^U_f$ für jede Funktion $f \in \mathscr{C}_p(U^*)$ die einzige auf $U$ $p$-beschränkte harmonische Funktion $u$, welche $f$ stetig auf $\overline{U}$ fortsetzt.*

*Beweis.* Nur der Eindeutigkeitsbeweis ist noch zu führen. Jedes derartige $u$ ist aber eine zu $f$ gehörige $p$-Ober- und $p$-Unterfunktion. Also gilt $u \leqq {}_pH^U_f \leqq u$ und somit $u = {}_pH^U_f$.

**Korollar.** *Sei $A$ eine Absorptionsmenge in $X$ mit nicht-leerem Rand $A^*$ und sei $L = \complement A$. Genügt dann ein Potential $p$ auf $X$ der Bedingung $(P_L)$, so ist ${}_pH^L_f$ für jede Funktion $f \in \mathscr{C}_p(A^*) = \mathscr{C}_p(L^*)$ die einzige auf $L$ $p$-beschränkte harmonische Funktion, welche $f$ stetig auf $\overline{L}$ fortsetzt.*

*Beweis.* In [2] wurde gezeigt, daß jeder Punkt $z \in A^* = L^*$ ein regulärer Randpunkt von $L$ ist. Die Behauptung folgt daher aus Satz 5.

## § 4. Anwendungen auf das Cauchysche Problem

Wir begnügen uns hier mit einer Skizze der Anwendungsmöglichkeiten im Falle der Wärmeleitungsgleichung (W). Sei also im folgenden $X = \mathbf{R}^{n+1}$. Dann ist $X$ ein streng harmonischer Raum bezüglich des Garbendatums der (zweimal stetig nach $x_1, \ldots, x_n$ und einmal stetig nach $x_{n+1}$ differenzierbaren) Lösungen von (W). Setzt man

$$A_\tau = \{x \in \mathbf{R}^{n+1} : x_{n+1} \leqq \tau\} \qquad (-\infty \leqq \tau \leqq +\infty)\,, \tag{4.1}$$

so erhält man gemäß [2] gerade alle Absorptionsmengen von $X$. Es genügt offenbar, den Fall $\tau = 0$ zu betrachten. Dann ist $\complement A_0$ der in der Einleitung eingeführte offene Halbraum $L = \{x \in \mathbf{R}^{n+1} : x_{n+1} > 0\}$. Das Korollar zu Satz 5 liefert daher die Lösung des Cauchyschen Anfangwertproblems. Dabei kann man über das der Bedingung $(P_L)$ genügende Potential $p$ noch geeignet verfügen.

Nach [2] ist die Fundamentallösung der Wärmeleitungsgleichung

$$P(x) = \begin{cases} \left(\frac{1}{4\pi x_{n+1}}\right)^{n/2} e^{-\frac{|x|^2}{4x_{n+1}}}, & \text{falls } x_{n+1} > 0 \\ 0 & , \text{falls } x_{n+1} \leqq 0 \end{cases} \tag{4.2}$$

ein Potential auf $X$; dabei wird

$$|x|^2 = \sum_{i=1}^{n} x_i^2 \tag{4.3}$$

gesetzt. Nach ANGER [1], S. 45 existiert zu jedem Potential $p$ auf $X$ ein positives Radon-Maß $\mu$ auf $X$ mit

$$p(x) = \int P(x-y)\,\mu(dy) \qquad (x \in \mathbf{R}^{n+1}). \tag{4.4}$$

$(x, y) \to P(x-y)$ ist dabei der parabolische Kern. Nach DOOB [4] wird für ein positives Radon-Maß $\mu$ auf $X$ durch (4.4) umgekehrt ein Potential $p$ definiert, sofern die Funktion $x \to \int P(x-y)\,\mu(dy)$ auf einer dichten Teilmenge von $X$ endlich ist. Dies ist speziell dann der Fall, wenn $\mu$ von einer Absorptionsmenge $A_\tau$ mit $\tau \in \mathbf{R}$ getragen wird.

Wählt man für $\mu$ die Einheitsmasse im Punkte $x_\varepsilon = (0, \ldots, 0, -\varepsilon)$ für ein beliebig gegebenes reelles $\varepsilon > 0$, so erhält man das der Bedingung $(\mathrm{P}_L)$ genügende Potential $p_\varepsilon(x) = P(x + x_\varepsilon)$. Also besagt das Korollar zu Satz 5: *Jede stetige reelle Funktion $f$ auf $L^*$ mit*

$$|f(z)| \leqq \text{const.}\, e^{-\frac{|z|^2}{4\varepsilon}} \qquad (z \in L^*) \tag{4.5}$$

*kann auf genau eine Weise zu einer Lösung $u$ von* (W) *in $L$ stetig fortgesetzt werden, welche der Bedingung*

$$|u(x)| \leqq \text{const.}\,(x_{n+1} + \varepsilon)^{-n/2}\, e^{-\frac{|x|^2}{4(x_{n+1}+\varepsilon)}} \qquad (x \in L) \tag{4.6}$$

*genügt.*

Wählt man für $\mu$ dagegen das Produktmaß $\lambda \otimes \varepsilon_{-1}$, wobei $\lambda$ das Lebesguesche Maß auf $\mathbf{R}^n$ und $\varepsilon_{-1}$ die Einheitsmasse im Punkte $x_{n+1} = -1$ auf $\mathbf{R}$ bezeichnet, so liefert (4.4) ein Potential $p$ auf $X$ mit $p(x) = 1$ für alle $x \in X$ mit $x_{n+1} > -1$. Also genügt $p$ der Bedingung $(\mathrm{P}_L)$. Nach dem Korollar zu Satz 5 *existiert daher zu jeder beschränkten, stetigen, reellen Funktion $f$ auf $L^*$ genau eine beschränkte Lösung $u$ von* (W) *auf $L$, welche $f$ stetig fortsetzt.*

Selbstverständlich kann man nun auch Fortsetzungen einer Randfunktion $f$ auf $L^*$ nur in einen Streifen $\{x \in \mathbf{R}^{n+1} : 0 < x_{n+1} < t\}$ hinein behandeln. Hierzu hat man als Raum $X$ den Unterraum $\{x \in \mathbf{R}^{n+1} : x_{n+1} < t\}$ des $\mathbf{R}^{n+1}$ zu wählen. Als Beschränktheitsbedingungen bieten sich dann Bedingungen der Form $|f(z)| \leqq \text{const.}\, e^{h|z|^2}$ und $|u(x)| \leqq \text{const.}\, e^{h|x|^2}$ mit Konstanten $h > 0$ an. Solche sind in der Theorie der parabolischen Differentialgleichungen wohlbekannt (vgl. FRIEDMAN [5], p. 25).

Ausdehnungen dieser und ähnlicher Anwendungen auf allgemeinere parabolische und auch elliptische lineare Differentialgleichungen zweiter Ordnung sollen an anderer Stelle diskutiert werden.

## Literatur

[1] Anger, G.: Funktionalanalytische Betrachtungen bei Differentialgleichungen unter Verwendung der Methoden der Potentialtheorie I. Deutsche Akademie der Wissenschaften, Berlin (Manuskript).

[2] Bauer, H.: Harmonische Räume und ihre Potentialtheorie. Ausarbeitung einer im Sommersemester 1965 an der Universität Hamburg gehaltenen Vorlesung. (Erscheint demnächst in der Reihe Lecture Notes in Mathematics, Springer-Verlag.)

[2a]— Axiomatische Behandlung des Dirichletschen Problems für elliptische und parabolische Differentialgleichungen. Math. Ann. **146**, 1—59 (1962).

[2b]— Weiterführung einer axiomatischen Potentialtheorie ohne Kern (Existenz von Potentialen). Z. Wahrscheinlichkeitstheorie 1, 197—229 (1963).

[2c]— Propriétés fines des fonctions hyperharmoniques dans une théorie axiomatique du potentiel. Ann. Inst. Fourier (Grenoble) **15**/1, 137—154 (1965).

[3] Brelot, M.: Familles de Perron et problème de Dirichlet. Acta Szeged **9**, 133—153 (1939).

[4] Doob, J. L.: A probability approach to the heat equation. Trans. Am. Math. Soc. **80**, 216—280 (1955).

[5] Friedman, A.: Partial differential equations of parabolic type. Prentice-Hall, Inc., Englewood Cliffs, N.J. (1964).

[6] Petrowski, J. G.: Vorlesungen über partielle Differentialgleichungen. Leipzig: B. G. Teubner Verlagsges. 1955.

*(Eingegangen am 13. 8. 1965)*

# Mesures avec une image donnée [R25]

Rev. Roumaine Math. Pures Appl. **11** (1966), 747–752
[Zbl. Math. 145.05301; MR 34 #2814]

**Dans un travail antérieur [1], nous avons cherché à trouver quelques classes d'applications continues $\varphi : X \to Y$ d'un espace localement compact $X$ dans un autre tel espace $Y$ possédant la propriété suivante : Toute mesure de Radon $\nu \geqslant 0$ sur $Y$ concentrée sur $\varphi(X)$ est l'image par $\varphi$ au moins d'une mesure de Radon $\mu \geqslant 0$ sur $X$. Une telle application a été appelée *conservative*.**

Récemment, M. Hans Reiter a attiré notre attention sur un théorème de décomposition de HERZ [4] permettant d'écrire toute mesure bornée $\nu \geqslant 0$ sur $Y$ sous la forme $\nu_1 + \nu_2$, où $\nu_1$ est l'image par $\varphi$ d'une mesure $\mu \geqslant 0$ sur $X$ et où $\nu_2$ est une mesure positive telle que, pour tout compact $K$ dans $X$, on ait $\nu_2(\varphi(K)) = 0$. *) Malheureusement, la démonstration de Herz ne s'applique pas aux mesures non bornées. Cependant, on montrera dans cet article que l'assertion du théorème de Herz reste valable pour des mesures de Radon quelconques sur $Y$. On verra ensuite que les résultats essentiels du travail [1], démontrés par des méthodes différentes et plus compliquées, sont des conséquences simples du théorème généralisé de Herz. De plus, ce théorème permettra la démonstration de résultats nouveaux concernant la structure des images de mesures.

Dans ce qui suit, on utilisera partout la théorie de l'intégration de BOURBAKI [3], en particulier chap. 5, § 6 sur les images de mesures. Pour un espace localement compact $T$, on dénotera par $\mathfrak{M}(T)$ (resp. $\mathfrak{M}_+(T)$) l'ensemble des mesures de Radon (resp. positives) sur $T$. L'ensemble des mesures $\mu \in \mathfrak{M}_+(T)$ concentrées sur une partie $A$ de $T$ sera dénoté par $\mathfrak{M}_+(T, A)$.

---

*) Après avoir fini ce travail, M. Horst Leptin m'a communiqué une démonstration (non publiée) du théorème de Herz avec une méthode très proche de la nôtre.

## I. THÉORÈME DE DÉCOMPOSITION

Soit $\varphi : X \to Y$ une application continue d'un espace localement compact $X$ dans un espace localement compact $Y$. On dira qu'une mesure $\nu \in \mathfrak{M}_+ (Y)$ est *$\varphi$-représentable* si on a $\varphi(\mu) = \nu$ pour au moins une mesure $\mu \in \mathfrak{M}_+ (X)$. On dira qu'une mesure $\nu \in \mathfrak{M}_+ (Y)$ est *$\varphi$-diffuse* si, pour tout compact $K \subset X$, on a $\nu(\varphi(K)) = 0$. On appellera l'application *conservative* si toute mesure $\nu \in \mathfrak{M}_+ (Y, \varphi(X))$ est $\varphi$-représentable.

Remarquons tout de suite que les mesures positives *diffuses* sur un espace localement compact $Y$ coïncident avec les mesures positives $\varphi$-diffuses si on prend pour $X$ l'ensemble $Y$ muni de la topologie discrète et pour $\varphi$ l'application identique. Dans ce cas-là, les mesures $\varphi$-représentables sont les mesures positives *atomiques* sur $Y$.

La démonstration de notre théorème de décomposition sera basée sur les deux lemmes suivants :

LEMME 1. *Deux mesures $\varphi$-représentables $\nu, \nu' \in \mathfrak{M}_+ (Y)$ sont égales si, pour tout compact $K \subset X$, on a $\nu(\varphi(K)) = \nu'(\varphi(K))$.*

*Démonstration.* Il suffira de démontrer que, pour tout compact $L \subset Y$, on a

$$\nu(L) = \sup \nu(\varphi(K)),$$

où $K$ parcourt les compacts de $X$ tels que $\varphi(K) \subset L$. Comme $\nu = \varphi(\mu)$ pour une mesure $\mu \in \mathfrak{M}_+ (X)$, on a $\mu(\overset{-1}{\varphi}(L)) = \nu(L) < +\infty$. L'ensemble fermé $\overset{-1}{\varphi}(L)$ est donc $\mu$-intégrable. Par conséquent, pour tout nombre $\varepsilon > 0$ donné, il existe un compact $K \subset \overset{-1}{\varphi}(L)$ tel que $\mu(\overset{-1}{\varphi}(L) \setminus K) \leqslant \varepsilon$. La relation $K \subset \overset{-1}{\varphi}(L)$ entraîne $\varphi(K) \subset L$ et $K \subset \overset{-1}{\varphi}(\varphi(K)) \subset \overset{-1}{\varphi}(L)$. L'assertion résulte alors des inégalités

$$\nu(L \setminus \varphi(K)) = \mu(\overset{-1}{\varphi}(L) \setminus \overset{-1}{\varphi}(\varphi(K))) \leqslant \overset{-1}{\varphi}((L) \setminus K) \leqslant \varepsilon.$$

LEMME 2. *Toute application continue surjective $\psi : K \to L$ d'un espace compact $K$ sur un espace compact $L$ est conservative.*

*Démonstration.* Soit $\mathcal{C}(K)$ (resp. $\mathcal{C}(L)$) l'espace vectoriel des fonctions continues réelles sur $K$ (resp. $L$) muni de la topologie de la convergence uniforme. Pour toute $\nu \in \mathfrak{M}_+ (L)$, $g \circ \psi \to \int g d\nu$ est une forme linéaire positive $\mu_0$ sur le sous-espace vectoriel $\mathcal{C}_\psi = \{ g \circ \psi \mid g \in \mathcal{C}(L)\}$ de $\mathcal{C}(K)$. La fonction constante 1 appartient à $\mathcal{C}_\psi$; elle est un point intérieur du cône positif de $\mathcal{C}(K)$. D'après un théorème bien connu de M. KREIN (voir [2], p. 75), $\mu_0$ se prolonge alors en une forme linéaire positive $\mu$ sur $\mathcal{C}(K)$. L'image par $\psi$ de cette mesure $\mu$ est $\nu$.

Le théorème de décomposition suivant généralise celui de HERZ [4], p. 213.

THÉORÈME. *Soit* $\varphi : X \to Y$ *une application continue d'un espace localement compact* $X$ *dans un espace localement compact* $Y$. *Alors toute mesure* $\nu \in \mathfrak{M}_+(Y)$ *peut s'écrire d'une seule manière sous la forme* $\nu = \nu_1 + \nu_2$, *où* $\nu_1$ *est une mesure positive* $\varphi$*-représentable et* $\nu_2$ *une mesure positive* $\varphi$*-diffuse sur* $Y$.

*Démonstration.* L'unicité d'une telle décomposition est une conséquence immédiate du lemme 1. Montrons donc l'existence.

Soit $\mathfrak{M}_\nu$ l'ensemble des mesures $\mu \in \mathfrak{M}_+(X)$ pour lesquelles $\varphi(\mu)$ est défini et $\leqslant \nu$. En particulier, $\mu = 0$ appartient à $\mathfrak{M}_\nu$. On montrera que $\mathfrak{M}_\nu$ est inductivement ordonné pour l'ordre usuel de $\mathfrak{M}(X)$. Soit donc $\mathfrak{F}$ une partie non vide, totalement ordonnée de $\mathfrak{M}_\nu$. Pour toute fonction continue $f : X \to \mathbf{R}_+$ à support compact, il existe une fonction continue $g : Y \to \mathbf{R}_+$ à support compact telle que l'on ait $0 \leqslant f \leqslant g \circ \varphi$. La famille $(\int f d\mu)_{\mu \in \mathfrak{F}}$ est alors bornée supérieurement par le nombre $\int g d\nu$. Comme $\mu \to \int f d\mu$ est croissante sur $\mathfrak{F}$, la limite (prise pour l'ensemble filtrant $\mathfrak{F}$) $\lim_{\mathfrak{F}} \int f d\mu$ existe dans $\mathbf{R}$. Elle existe alors pour toute fonction continue $f : X \to \mathbf{R}$ à support compact. $f \to \lim_{\mathfrak{F}} \int f d\mu$ est une mesure $\mu_0 \in \mathfrak{M}_+(X)$ majorant $\mathfrak{F}$. On vérifie aisément que $\mu_0$ appartient à $\mathfrak{M}_\nu$.

D'après le lemme de Zorn, il existe un élément maximal $\mu$ dans $\mathfrak{M}_\nu$. Le théorème sera démontré si la mesure $\nu_2 = \nu - \varphi(\mu)$ est $\varphi$-diffuse. Raisonnons par l'absurde, en supposant l'existence d'un compact $K \subset X$ tel que $\nu_2(\varphi(K)) > 0$. Comme la restriction de $\varphi$ à $K$ nous fournit une application continue surjective $\psi : K \to \varphi(K)$, on peut appliquer le lemme 2. Il existe alors une mesure $\mu' \in \mathfrak{M}_+(K)$ telle que $\psi(\mu')$ soit la mesure $\nu'$ induite par $\nu_2$ dans $\varphi(K)$. On a $\varphi(\mu') = \nu'$ si on considère $\mu'$ resp. $\nu'$ comme mesure sur $X$ resp. $Y$. Il en résulte que la mesure $\mu + \mu'$ appartient à $\mathfrak{M}_\nu$. En effet, en utilisant quelques propriétés d'images de mesures démontrées dans [1], p. 404 on obtient $\varphi(\mu + \mu') = \varphi(\mu) + \varphi(\mu') = \varphi(\mu) + \chi_{\varphi(K)}(\nu - \varphi(\mu)) = (1 - \chi_{\varphi(K)})\varphi(\mu) + \chi_{\varphi(K)}\nu \leqslant \nu$, où $\chi_{\varphi(K)}$ est la fonction indicatrice de $\varphi(K)$ par rapport à $Y$. Comme l'hypothèse faite sur $K$ implique $\nu' \neq 0$, donc aussi $\mu' \neq 0$, on arrive à une contradiction avec la propriété maximale de $\mu$. Le théorème est donc démontré.

D'après [1], p. 404, l'ensemble $\mathfrak{R}_+^\varphi$ des mesures $\varphi$-représentables de $\mathfrak{M}_+(Y)$ est un sous-cône convexe de $\mathfrak{M}_+(Y)$. Evidemment, l'ensemble $\mathfrak{D}_+^\varphi$ des mesures $\varphi$-diffuses sur $Y$ est aussi un sous-cône convexe de $\mathfrak{M}_+(Y)$. Le théorème que l'on vient de démontrer implique que ces deux cônes $\mathfrak{R}_+^\varphi$ et $\mathfrak{D}_+^\varphi$ sont *héréditaires à gauche*. On obtient alors le corollaire suivant :

COROLLAIRE. *Soit* $\mathfrak{R}^\varphi$ *(resp.* $\mathfrak{D}^\varphi$*) l'ensemble des mesures* $\mu \in \mathfrak{M}(Y)$ *telles que la mesure* $|\mu|$ *appartienne à* $\mathfrak{R}^\varphi_+$ *(resp.* $\mathfrak{D}^\varphi_+$*). Alors* $\mathfrak{R}^\varphi$ *et* $\mathfrak{D}^\varphi$ *sont des bandes dans l'espace de Riesz complètement réticulé* $\mathfrak{M}(Y)$*. L'espace* $\mathfrak{M}(Y)$ *est somme directe ordonnée de* $\mathfrak{R}^\varphi$ *et* $\mathfrak{D}^\varphi$.

*Démonstration.* Comme $\mathfrak{R}^\varphi_+ = \mathfrak{R}^\varphi \cap \mathfrak{M}_+(Y)$ et $\mathfrak{D}^\varphi_+ = \mathfrak{D}^\varphi \cap \mathfrak{M}_+(Y)$ sont des cônes convexes, héréditaires à gauche, $\mathfrak{R}^\varphi$ et $\mathfrak{D}^\varphi$ sont des sous-espaces vectoriels de $\mathfrak{M}(Y)$. D'après le théorème, $\mathfrak{M}(Y)$ est somme directe ordonnée de $\mathfrak{R}^\varphi$ et $\mathfrak{D}^\varphi$. Il en résulte que $\mathfrak{R}^\varphi$ et $\mathfrak{D}^\varphi$ sont des bandes dans $\mathfrak{M}(Y)$.

La remarque précédant le lemme 1 montre que la décomposition de $\mathfrak{M}(Y)$ en mesures *atomiques* et mesures *diffuses* est un cas particulier de ce corollaire.

## II. PRINCIPE DE LOCALISATION — APPLICATIONS CONSERVATIVES

Dans ce qui suit, $\varphi : X \to Y$ sera de nouveau une application continue d'un espace localement compact $X$ dans un autre tel espace $Y$

Montrons d'abord que la propriété d'une mesure $\nu \in \mathfrak{M}_+(Y)$ d'être $\varphi$-représentable resp. $\varphi$-diffuse est d'une nature locale. Soit $V$ un sous-espace localement compact de $Y$. Son image inverse $\overset{-1}{\varphi}(V)$ est localement compacte dans $X$. On dénotera par $\varphi_V$ l'application $x \to \varphi(x)$ de $\overset{-1}{\varphi}(V)$ dans $V$ et par $\nu_V$ la mesure induite par $\nu$ dans $V$. Il est alors évident que $\nu_V$ est $\varphi_V$-diffuse si $\nu$ est $\varphi$-diffuse. Egalement, la mesure $\nu_V$ est $\varphi_V$-représentable si $\nu$ est $\varphi$-représentable. En effet, si $\mu$ est une mesure de $\mathfrak{M}_+(X)$ telle que $\varphi(\mu) = \nu$, on a $\varphi_V(\mu') = \nu_V$ pour la mesure $\mu'$ induite par $\mu$ dans $\overset{-1}{\varphi}(V)$. Toute mesure $\varphi$-représentable (resp. $\varphi$-diffuse) est donc aussi localement $\varphi$-représentable (resp. $\varphi$-diffuse) au sens suivant :

On dira qu'une mesure $\nu \in \mathfrak{M}_+(Y)$ est *localement $\varphi$-représentable* (resp. *$\varphi$-diffuse*) si tout point $y \in \varphi(X)$ possède dans $Y$ un voisinage localement compact $V$ tel que la mesure $\nu_V$ soit $\varphi_V$-représentable (resp. $\varphi_V$-diffuse). Il est évident que $V$ peut être choisi ou bien ouvert et relativement compact ou bien compact.

PROPOSITION 1. *Toute mesure localement $\varphi$-représentable (resp. $\varphi$-diffuse)* $\nu \in \mathfrak{M}_+(Y, \varphi(X))$ *est $\varphi$-représentable (resp. $\varphi$-diffuse).*

*Démonstration.* Soit $\nu \in \mathfrak{M}_+(Y, \varphi(X))$ localement $\varphi$-représentable et soit $\nu = \nu_1 + \nu_2$ la décomposition de $\nu$ selon le théorème. On sait alors que, pour tout ouvert $V$ de $Y$, $\nu_V = (\nu_1)_V + (\nu_2)_V$ est la décomposition de $\nu_V$ dans sa partie $\varphi_V$-représentable et sa partie $\varphi_V$-diffuse. Si donc $\nu_V$ est $\varphi_V$-représentable, on a $(\nu_2)_V = 0$. La mesure $\nu$ étant localement $\varphi$-représentable, les ouverts $V$ de $Y$ ayant la propriété

$(\nu_2)_V = 0$ recouvrent $\varphi(X)$. Comme $\nu$ est concentrée sur $\varphi(X)$, il en résulte $\nu_2 = 0$, ce qui prouve la première partie de la proposition.

Passons au cas d'une mesure localement $\varphi$-diffuse $\nu \in \mathfrak{M}_+(Y, \varphi(X))$. Pour tout compact $K \subset X$, la partie compacte $\varphi(K)$ peut être recouverte par un nombre fini de fermés $V_1, \ldots, V_n$ dans $Y$ tels que la mesure $\nu_{V_i}$ soit $\varphi_{V_i}$-diffuse. On a $K \subset \overset{-1}{\varphi}(V_1) \cup \ldots \cup \overset{-1}{\varphi}(V_n)$, d'où résulte $K = K_1 \cup \ldots \cup K_n$ pour les ensembles compacts $K_i = K \cap \overset{-1}{\varphi}(V_i)$ $(i = 1, \ldots, n)$. Toute mesure $\nu_{V_i}$ étant $\varphi_{V_i}$-diffuse, on a $\nu(\varphi(K_i)) = \nu_{V_i}(\varphi_{V_i}(K_i)) = 0$. Comme $\varphi(K) = \varphi(K_1) \cup \ldots \cup \varphi(K_n)$, il en résulte $\nu(\varphi(K)) = 0$, ce qui achève la démonstration.

Remarquons que l'assertion concernant les mesures localement $\varphi$-représentables est le « principe de localisation » démontré dans [1].

Les autres résultats principaux de [1] sont les corollaires 2, 3 de la proposition suivante. Le corollaire 1 répond à une question laissée ouverte dans [1], p. 421.

PROPOSITION 2. *Soit $\varphi : X \to Y$ une application continue d'un espace localement compact dans un autre ayant la propriété suivante : Pour tout point $y \in \varphi(X)$, il existe une suite $(K_n)_{n \in N}$ de compacts dans $X$ telle que $\bigcup_{n=1}^{\infty} \varphi(K_n)$ soit un voisinage de $y$ dans $\varphi(X)$. Alors $\varphi$ est conservative.*

*Démonstration.* Comme $Y$ est localement compact, la propriété particulière de l'application $\varphi$ peut être formulée de la manière suivante : Tout point $y \in \varphi(X)$ possède un voisinage compact $V$ dans $Y$ tel que l'on ait $V \cap \varphi(X) \subset \bigcup_{n=1}^{\infty} \varphi(K_n)$. Il en résulte $\bigcup_{n=1}^{\infty} \varphi(K'_n) = V \cap \varphi(X)$ pour la suite des compacts $K'_n = \overset{-1}{\varphi}(V) \cap K_n$ de $X$, $n = 1, 2, \ldots$. Comme $K'_n$ est une partie de $\overset{-1}{\varphi}(V)$, on voit que $\nu' = 0$ est la seule mesure $\varphi$-diffuse sur $Y$ concentrée sur $V \cap \varphi(X)$. Nous avons donc démontré que toute mesure $\nu \in \mathfrak{M}_+(Y, \varphi(X))$ est localement $\varphi$-représentable. La proposition 1 achève alors la démonstration.

COROLLAIRE 1. *Soit $\varphi : X \to Y$ une application continue d'un espace localement compact dans un autre telle que tout point $y \in \varphi(X)$ possède un voisinage $V$ dans $Y$ dont l'image inverse $\overset{-1}{\varphi}(V)$ est réunion dénombrable de compacts dans $X$. Alors $\varphi$ est conservative.*

COROLLAIRE 2. *Toute application continue, localement propre d'un espace localement compact dans un autre est conservative.* *)

---

*) On dit que $\varphi : X \to Y$ est *localement propre* si tout point $y \in \varphi(X)$ possède un voisinage $V$ dont l'image inverse $\overset{-1}{\varphi}(V)$ est compact. En particulier, toute application propre est localement propre.

Corollaire 3. *Toute application continue d'un espace localement compact dans un autre possédant des sections locales est conservative.* **)

En effet, tout point $y \in Y$ possède un voisinage ouvert $V$ tel que $\varphi$ possède une section $\sigma_V$ sur $V$. Pour tout voisinage compact $W$ de $y$, l'ensemble $K = \sigma_V(W)$ est compact et $\varphi(K) = W$.

*Reçu le 9 février 1966*

*Mathematisches Institut*
*der Universität Erlangen — Nürnberg*
*Erlangen, Allemagne*

BIBLIOGRAPHIE

1. H. Bauer, *Konservative Abbildungen lokal-kompakter Räume.* Math. Ann., 1959, **138**, 398—427.

2. N. Bourbaki, *Espaces vectoriels topologiques.* Chap. I—II. Actual. Sci. et Ind., 1953, **1189**.

3. — *Intégration,* Chap. I—IV et Chap. V. Actual. Sci. et Ind., 1952, **1175** et 1956, **1244**.

4. C. S. Herz, *The spectral theory of bounded functions.* Trans. Amer. Math. Soc., 1960, **94**, 181—232.

**) On dit que $\varphi : X \to Y$ possède des *sections locales* si, pour $y \in Y$, il existe un voisinage ouvert $V$ et une application continue $\sigma_V : V \to X$ (appelée section de $\varphi$ sur $V$) tel que $\varphi \circ \sigma_V$ soit l'application identique de $V$ sur lui-même.

# The part metric in convex sets [R26]

(with H. S. Bear)

Pacific J. Math. **30** (1969), 15–33
[Zbl. Math. 176.42801; MR 43 #859]

**Any convex set $C$ without lines in a linear space $L$ can be decomposed into disjoint convex subsets (called parts) in a way which generalizes the idea of Gleason parts for a function space or function algebra. A metric $d$ (called part metric) can be defined on $C$ in a purely geometric way such that the parts of $C$ are the components in the $d$-topology. This paper treats the connection between the convex structure of $C$ and the metric $d$. The situation is particularly interesting when $C$ is closed with respect to a weak Hausdorff topology on $L$ (defined by a duality between $L$ and another linear space). Then $C$ is characterized by the set $C^+$ of all continuous affine functions $F$ on $L$ satisfying $F(x) \geqq 0$ for all $x \in C$. This allows us to define $d$ in terms of the functions $\log F$, $F \in C^+$. Furthermore, $d$-completeness of $C$ can be derived from the completeness of $C$ in $L$. The "convexity" of the metric $d$ leads to the existence of a continuous selection function for lower semi-continuous mappings of a paracompact space into the nonempty $d$-closed convex subsets of one part of such a complete convex set $C$. We apply this result and the study of the part metric of the convex cone of positive Radon measures on a locally compact Hausdorff space to the problem of selecting in a continuous way mutually absolutely continuous representing measures for points in one part of a function space or function algebra.**

1. The part metric and convex structure. We consider a real linear space $L$, and a convex set $C$ in $L$ which contains no whole line. We do not necessarily assume that $L$ has a topology.

The closed segment from $x$ to $y$ is denoted $[x, y]$. If $x, y \in C$, we say that $[x, y]$ extends (in $C$) by $r(>0)$ if $x + r(x - y) \in C$ and $y + r(y - x) \in C$. We write $x \sim y$ if $[x, y]$ extends by some $r > 0$. It is shown in [1] that $\sim$ defines an equivalence relation in $C$.

The equivalence classes of $\sim$, called the *parts* of $C$, are clearly also convex. There is a metric $d$ on each part of $C$ defined by

$$d(x, y) = \inf \left\{\log\left(1 + \frac{1}{r}\right): [x, y] \text{ extends by } r\right\}.$$

If $[x, y]$ extends by $r$ (in $C$), then $x + r'(x - y)$ and $y + r'(y - x)$ are in the part $\Pi$ of $x$ and $y$ for all $r' < r$. It follows that one gets the same part metric on $\Pi$ if one replaces $C$ by $\Pi$ in the definition of $d(x, y)$.

If $x \not\sim y$, we write $d(x, y) = +\infty$. Then $d$ satisfies all axioms of a metric on $C$, except that it is not always finite. In spite of this, $d$ will be called the *part metric* on $C$. We could introduce $\min(1,d)$ or $d/1+d$ to obtain a bona fide metric on $C$ defining the same topology as $d$.

For each part $\Pi$ of $C$ and each $x \in \Pi$ we have

$$\Pi = \{y \in C: d(x, y) < \infty\} .$$

Therefore, the parts are open, and hence also closed.

The theorems of this sections establish connections between the part metric and the convex structure in $C$.

THEOREM 1. *Let $x, x', y, y' \in C$ and $0 \leqq \lambda \leqq 1$. If $[x, x']$ and $[y, y']$ extend by $r$, then $[\lambda x + (1-\lambda)y, \lambda x' + (1-\lambda)y']$ extends by $r$.*

*Proof.* We have the identity

$$\begin{aligned}&[\lambda x + (1-\lambda)y] + r\{[\lambda x + (1-\lambda)y] - [\lambda x' + (1-\lambda)y']\} \\ &\quad = \lambda[x + r(x - x')] + (1-\lambda)[y + r(y - y')] .\end{aligned}$$

The term on the right is a convex combination of points which are in $C$ by hypothesis. The extension beyond $\lambda x' + (1-\lambda)y'$ follows by a symmetric argument. Notice that we do not assume that $x \sim y$ or $x' \sim y'$.

The corollaries below and Theorem 2 are basic to the proof of the selection theorem of § 6. They are immediate consequences of Theorem 1.

COROLLARIES.

(i) $d(\lambda x + (1-\lambda)y, \lambda x' + (1-\lambda)y') \leqq \max[d(x, x'), d(y, y')]$

*for any points $x, x', y, y' \in C$ and $0 \leqq \lambda \leqq 1$.*

(ii) *If $S$ is a convex subset of $C$ (not necessarily in one part) and $d(x, S) < \varepsilon$, $d(y, S) < \varepsilon$, then $d(\lambda x + (1-\lambda)y, S) < \varepsilon$ (even if $x \not\sim y$).*

(iii) *$\{x: d(x, S) < \varepsilon\}$ is convex for any convex subset $S$ of $C$. In particular, $d$-balls are convex.*

LEMMA 1. *If $u$ is a nonnegative concave function on $C$ (in particular an affine function) and $\Pi$ is a part of $C$, then either $u(x) > 0$ for all $x \in \Pi$ or $u \equiv 0$ on $\Pi$.*

*Proof.* If $x, y \in \Pi$ and the segment $[x, y]$ extends by $r$, then

$$x_0 = x + r(x - y) \in C .$$

Hence

$$x = \frac{1}{1+r} x_0 + \frac{r}{1+r} y$$

is a convex combination of $x_0$ and $y$. Since $u$ is concave and $\geqq 0$, this implies

$$(1) \qquad u(x) \geqq \frac{1}{1+r} u(x_0) + \frac{r}{1+r} u(y) \geqq \frac{r}{1+r} u(y) .$$

Therefore it follows from $u(x) = 0$ and $u(y) \geqq 0$ that $u(y) = 0$.

THEOREM 2. *If $u$ is a nonnegative concave function on $C$, then* $\log u$ *is uniformly $d$-continuous on those parts where $u > 0$. Consequently, all lower bounded concave functions on $C$ are $d$-continuous.*

*Proof.* If $[x, y]$ extends by $r$, and $u > 0$ on the part containing $x$ and $y$, then (1) shows that

$$u(y)/u(x) \leqq 1 + \frac{1}{r} .$$

A symmetric argument shows that $u(x)/u(y) \leqq 1 + (1/r)$, and hence

$$\left(1 + \frac{1}{r}\right)^{-1} \leqq u(y)/u(x) \leqq 1 + \frac{1}{r} ,$$

$$|\log u(x) - \log u(y)| \leqq \log\left(1 + \frac{1}{r}\right) .$$

Since $d(x, y)$ is the infimum of the right side,

$$|\log u(x) - \log u(y)| \leqq d(x, y) .$$

This proves the uniform continuity of $\log u$ on that part. The remaining assertion is evident.

LEMMA 2. *If $x, y \in C$ and $\phi(\lambda) = \lambda x + (1 - \lambda)y$, then $\phi$ is $d$-continuous on $]0, 1[$. If $x \sim y$, then $\phi$ is even $d$-continuous on $[0, 1]$.*

*Proof.* The first assertion follows from the second, since for any two points $x \neq y$ in $C$ the open segment

$$]x, y[ = \{\lambda x + (1 - \lambda)y : 0 < \lambda < 1\}$$

lies in one part and since for $0 < \lambda_0 < 1$ the point $\phi(\lambda_0)$ is a convex combination of points $x_1 = \phi(\lambda_1)$ and $y_1 = \phi(\lambda_2)$ where $0<\lambda_1<\lambda_0<\lambda_2<1$.

Assume therefore that $x \sim y$, and denote $\phi(\lambda)$ by $z_\lambda$ for $0 \leqq \lambda \leqq 1$. We will show that $d(z_\lambda, y) \to 0$ as $\lambda \to 0$. The left-side continuity of

$\phi$ at a point $\lambda_0 \in ]0, 1[$ then follows from this by considering $\phi(\lambda_0)$ an end point, since

$$\lambda x + (1 - \lambda)y = \lambda' x + (1 - \lambda')\phi(\lambda_0)$$

for some $\lambda' = \lambda'(\lambda)$ which tends to zero as $\lambda \to \lambda_0$ from the left. A symmetric argument gives continuity from the right. The following identities can easily be checked:

$$y + r(y - x) = y + \frac{r}{\lambda}(y - z_\lambda)$$

$$x + r(x - y) = z_\lambda + \frac{1 - \lambda + r}{\lambda}(z_\lambda - y) .$$

Therefore, if $[x, y]$ extends by $r$, then $[z_\lambda, y]$ extends by the minimum of the numbers $r/\lambda$, $(1 - \lambda + r)/\lambda$, so $d(z_\lambda, y) \to 0$ as $\lambda \to 0$.

COROLLARY. *The parts of $C$ are connected, and are the components of $C$.*

We next study the continuity of convex combinations.

THEOREM 3. *For $x, y \in C$ and $0 \leqq \lambda \leqq 1$, denote by $\Psi(\lambda, x, y)$ the convex combination $\lambda x + (1 - \lambda)y$. Then for each part $\Pi$ of $C$, the mapping*

$$\Psi: [0, 1] \times \Pi \times \Pi \to \Pi$$

*is continuous. Moreover, the mapping*

$$\Psi: ]0, 1[ \times C \times C \to C$$

*is continuous.*

*Proof.* Let $x, y, x_0, y_0 \in C$, and $\lambda, \lambda_0 \in [0, 1]$. Then

$$\begin{aligned} &d(\lambda x + (1 - \lambda)y, \lambda_0 x_0 + (1 - \lambda_0)y_0) \\ &\quad \leqq d(\lambda x + (1 - \lambda)y, \lambda x_0 + (1 - \lambda)y_0 \\ &\qquad + d(\lambda x_0 + (1 - \lambda)y_0, \lambda_0 x_0 + (1 - \lambda_0)y_0) . \end{aligned}$$

The first term on the right is dominated by $\max[d(x, x_0)d(y, y_0)]$ by Corollary (i) of Theorem 1. The behavior of the second term was studied in Lemma 2.

The next result concerns the description of the $d$-topology in a special case.

Let $C$ be an open convex set in a linear topological space and assume $C$ contains no line. Assume for convenience that $0 \in C$. Let $p$ be the Minkowski functional for $C$: $p(x) = \inf\{r: x \in rC, r > 0\}$. Then $p$ is finite, subadditive and positive-homogeneous, and $C = \{x: p(x) < 1\}$. If we let $q(x) = \max[p(x), p(-x)]$, then $q$ is absolutely homogeneous, and subadditive, and hence a seminorm. Since $C$ contains no line, $q$ is a norm. We will call $q$ the *Minkowski norm* for $C$.

THEOREM 4. *If $C$ is an open convex set in a linear topological space, $0 \in C$, and $C$ contains no line, then $C$ has one part, and the part metric $d$ and the Minkowski norm define the same topology on $C$.*

*Proof*. Since $C$ is open, it is clear that $C$ is one part. Let $x_0 \in C$, so that $p(x_0) < 1$. If $q(x - x_0) < 1 - p(x_0)$, then

$$p(x) = p(x - x_0 + x_0) \leqq p(x - x_0) + p(x_0) \leqq q(x - x_0) + p(x_0) < 1 .$$

That is, $x \in C$ if $q(x - x_0) < 1 - p(x_0)$. Since

$$q[x_0 + r(x_0 - x) - x_0] = rq(x_0 - x) ,$$

$x_0 + r(x_0 - x) \in C$ if $rq(x_0 - x) < 1 - p(x_0)$, or $r < (1 - p(x_0))/q(x_0 - x)$. Also,

$$q[x + r(x - x_0) - x_0] = (1 + r)q(x - x_0) ,$$

so that $x + r(x - x_0) \in C$ if $(1 + r)q(x - x_0) < 1 - p(x_0)$, or

$$q(x - x_0) < (1 - p(x_0))/(1 + r) .$$

Combining these facts, we get that $[x, x_0]$ extends by $r$ if

$$1 + r < (1 - p(x_0))/q(x - x_0) \qquad (x \neq x_0) .$$

If $q(x - x_0) \to 0$, then $[x, x_0]$ extends by arbitrarily large $r$, and hence $d(x, x_0) \to 0$.

Now suppose that $[x, x_0]$ extends by $r$, so that $p(x_0 + r(x_0 - x)) > 1$ and $p(x + r(x - x_0)) < 1$. Since $r(x_0 - x) = x_0 + r(x_0 - x) - x_0$, we have

$$\begin{aligned} rp(x_0 - x) &< p(x_0 + r(x_0 - x)) + p(-x_0) \\ &< 1 + p(-x_0) . \end{aligned} \tag{2}$$

Similarly, interchanging $x$ and $x_0$ gives

$$rp(x - x_0) < 1 + p(-x) .$$

Subadditivity of $p$ also gives

$$p(-x) \leqq p(-x_0) + p(x_0 - x) .$$

Combining this last inequality and (2) gives

$$rp(x - x_0) < 1 + p(-x_0) + p(x_0 - x)$$
$$< 1 + p(-x_0) + \frac{1}{r}(1 + p(-x_0)) \tag{3}$$
$$= \left(1 + \frac{1}{r}\right)(1 + p(-x_0)) .$$

The inequalities (2) and (3) show that $p(x_0 - x)$ and $p(x - x_0)$ are both less than

$$\frac{1}{r}\left(1 + \frac{1}{r}\right)(1 + p(-x_0)) .$$

Consequently, $q(x - x_0) \to 0$ as $d(x, x_0) \to 0$ (i.e., $r \to \infty$).

COROLLARY. *A bounded open convex set $C$ in a normed linear space $E$ has one part, and the part metric defines the same topology on $C$ as the given norm.*

*Proof.* We may assume $0 \in C$. Then the Minkowski norm $q$ and the given norm are equivalent since $C$ contains some norm ball around 0, and some norm ball contains $C \cap (-C)$ which is the ball $\{x: q(x) < 1\}$. The result follows then from Theorem 4.

We remark that in the situation of the above corollary the part metric $d$ is in general not equivalent as a uniform structure to the norm metric. It suffices to assume that $E$ is the Euclidean space $R^n$. Then $C$ as an open set is not norm complete. But it will follow from Theorem 9 that $C$ as a part of the convex body $\bar{C}$ is $d$-complete.

Let us also point out that the above corollary together with Theorem 2 yields the well-known fact that a lower bounded concave function on a bounded open convex set in a normed linear space is continuous.

**2. The part metric in a cone.** In this section we give some additional properties of the equivalence relation $\sim$ and the part metric in the case that the convex set is a cone. We let $p$ denote a convex cone in the real linear space $L$ with vertex $0 \in P$, and assume that $P$ contains no line. We then have in particular no line in $P$ through 0, i.e., $P \cap (-P) = \{0\}$. Let $\leqq$ denote the partial order on $L$ induced by $P$: $x \leqq y$ means $y - x \in P$. Let us point out that $\{0\}$ is one part of $P$ since the vertex 0 is an extreme point of $P$.

THEOREM 5. *For any two points $x, y \in P$ the following statements hold:*

( i ) *$[x, y]$ extends by $r$ if and only if*

$$(4) \qquad \left(1 + \frac{1}{r}\right)x \geqq y \quad \text{and} \quad \left(1 + \frac{1}{r}\right)y \geqq x\,;$$

(ii) $[x, y]$ *extends by* $r$ *if and only if* $[px, py]$ *extends by* $r$ *for a given number* $p > 0$;

(iii) *if* $[x, y]$ *extends by* $r$ *and* $z \in p$, *then* $[x + z, y + z]$ *extends by* $r$;

(iv) $x \sim px$ *for all* $p > 0$;

(v) *if* $x \sim y$, *then* $x \sim x + y$.

*Proof.* Condition (4) of (i) is just a restatement of the definition of extension in terms of the partial order. Note that transitivity of $\sim$ is immediate from (4). Conditions (ii) and (iii) also follow easily from (4). Condition (iv) is clear and (v) follows from (iii) and (iv) as follows:

$$x + y \sim y + y = 2y \sim y \sim x\,.$$

From (iv) and (v) we obtain

COROLLARY. *Each part of* $P$ *is a convex subcone of* $P$.

THEOREM 6. *The part metric* $d$ *of* $P$ *has the following properties:*

(i) *if* $x \sim y$ *and* $x \neq 0$, $d(x, y) = \inf \{\log a : ax \geqq y$ *and* $ay \geqq x\}$;

(ii) $d(px, py) = d(x, y)$ *for all* $p > 0$, *all* $x, y \in P$;

(iii) $d(x + z, y + z) \leqq d(x, y)$ *for all* $x, y, z \in P$;

(iv) $d(rx, sx) = |\log r - \log s|$ *for* $x \in P, x \neq 0, r > 0, s > 0$.

*Proof.* Conditions (i), (ii), (iii) follows from (i), (ii) and (iii) of Theorem 5; (iv) follows from (i) by observing that $d(rx, sx) = d\{(r/s)x, x\}$ because of (ii).

THEOREM 7. *The mappings* $(\lambda, x) \rightarrow \lambda x$ *from* $]0, +\infty[\ \times p$ *into* $P$ *and* $(x, y) \rightarrow x + y$ *from* $P \times P$ *into* $P$ *are continuous (with respect to the part metric* $d$ *on* $P$). *Explicitly,*

$$(5) \qquad d(\lambda x, \lambda x_0) \leqq d(x, x_0) + |\log \lambda - \log \lambda_0| \qquad (x, x_0 \in P, x_0 \neq 0, \lambda > 0, \lambda_0 > 0);$$

$$(6) \qquad d(x + y, x_0 + y_0) \leqq \max [d(x, x_0), d(y, y_0)] \qquad (x, y, x_0, y_0 \in P)\,.$$

*Proof.* Using the triangle inequality and Theorem 6, (ii), (iv), we have

$$d(\lambda x, \lambda_0 x_0) \leqq d(\lambda x, \lambda x_0) + d(\lambda x_0, \lambda_0 x_0) = d(x, x_0) + |\log \lambda - \log \lambda_0|\,.$$

From Corollary (i) of Theorem 1 we have

$$d\left(\frac{1}{2}x + \frac{1}{2}y, \frac{1}{2}x_0 + \frac{1}{2}y_0\right) \leqq \max\,[d(x, x_0), d(y, y_0)]\,.$$

Multiplying both elements in the left term by 2 preserves distance (Theorem 6, (ii)) and gives (6). The continuity of $(\lambda, x) \to \lambda x$ at a point $(\lambda_0, 0)$, $\lambda_0 > 0$, is evident since $\{0\}$ is a part, and hence open in $P$.

**3. Completeness of part metric.** In this section we give some conditions which imply the completeness of the part metric. We begin with a functional analytic description of this metric.

Following Choquet [7], and Bourbaki [4] we will say that a linear space $L$ is a *weak space* if it is endowed with a weak topology $\sigma(L, M)$ derived from a duality between $L$ and some linear space $M$. Hence there is a bilinear form $(x, y) \to \langle x, y\rangle$ on $L \times M$, and $\sigma(L, M)$ is the weakest topology on $L$ making all linear functionals $x \to \langle x, y\rangle$, $y \in M$, continuous. Remember that $\sigma(L, M)$ is then a locally convex topology on $L$. We will always assume that the weak space $L$ is Hausdorff. This is equivalent to the fact that the linear forms $x \to \langle x, y\rangle$ with $y \in M$ separate the points of $L$.

A set $C$ in a weak space $L$ will be called complete if it is complete in the uniform structure induced in $C$ by the underlying topology $\sigma(L, M)$. Obviously, each complete set in a weak Hausdorff space $L$ is also closed.

Each closed convex subset $C$ of a weak Hausdorff space $L$ is an intersection of closed half-spaces. Therefore, we have

$$(7) \qquad C = \bigcap\nolimits_{F \in C^+} \{x \in L : F(x) \geqq 0\}$$

where $C^+$ denotes the set of all continuous affine functions on $L$ which are nonnegative on $C$. That is $C^+$ characterizes $C$ in the sense that $x \in C$ if and only if $F(x) \geqq 0$ for all $F \in C^+$. Remember [4; p. 91] that each continuous affine function $F$ on $L$ is of the form $F(x) = \langle x, y_F\rangle +$ constant with $y_F \in M$.

THEOREM 8. *If $C$ is a closed convex set containing no line in a weak Hausdorff space, then*

$$(8) \qquad d(x, y) = \sup\,\{|\log F(x) - \log F(y)| : F \in C^+, F(y) > 0\}^{1}$$

*for all points $x, y \in C$.*

*Proof.* Assume first that $x \sim y$. The condition $F(y) > 0$ is then equivalent to assuming that $F > 0$ on the part containing $y$ by Lemma 1.

[1] We make use of the convention: $\log 0 = -\infty$.

Since $C^+$ characterizes $C$, we have $x + r(x - y) = (1 + r)x - ry \in C$ if and only if

$$\left(1 + \frac{1}{r}\right)F(x) \geqq F(y)$$

for all $F \in C^+$. Similarly, $y + r(y - x) \in C$ if and only if $\{1+(1/r)\}F(y) \geqq F(x)$ for all $F \in C^+$. Hence $[x, y]$ extends by $r$ if and only if

$$\left(1 + \frac{1}{r}\right)^{-1} \leqq F(x)/F(y) \leqq 1 + \frac{1}{r}$$

for all $F \in C^+$ with $F(y) > 0$. Hence $[x, y]$ extends by $r$ if and only if

$$|\log F(x) - \log F(y)| \leqq \log\left(1 + \frac{1}{r}\right)$$

for all $F \in C^+$ with $F(y) > 0$. Therefore $d(x, y)$, which is the infimum of the right side, also is the supremum of the left side.

In the case $x \not\sim y$ we have to show that the supremum in (7) is infinite. We can assume specifically that the segment $[x, y]$ does not extend beyond $y$. Hence $y + \varepsilon(y - x) \notin C$ for each $\varepsilon > 0$. Therefore, for fixed $\varepsilon > 0$, there exists a function $F \in C^+$ such that

$$(1 + \varepsilon)F(y) - \varepsilon F(x) = F(y + \varepsilon(y - x)) < 0 .$$

From this follows that $F(x) > 0$ and

$$|\log F(x) - \log F(y)| = \log F(x) - \log F(y) > \log\left(1 + \frac{1}{\varepsilon}\right).$$

But $\log\{1 + (1/\varepsilon)\} \to +\infty$ as $\varepsilon \to 0$. If $F(y) = 0$, we can replace $F$ by $F + \alpha$, where $\alpha$ is a small positive number and draw the same conclusion.

COROLLARY ( i ). *If $C$ is a closed convex set containing no line in a weak Hausdorff space, then a fundamental system of d-neighborhoods of a point $x_0 \in C$ is given by the sets*

$$(9)\quad \begin{aligned} &\{x \in C: |F(x) - F(x_0)| < \varepsilon, \text{ all } F \in C^+, F(x_0) = 1\} \\ =&\{x \in C: |F(x) - F(x_0)| < \varepsilon, \text{ all } F \in C^+, 0 < F(x_0) \leqq 1\} , \end{aligned}$$

*where $\varepsilon$ ranges over the positive numbers.*

*Proof.* We first prove that the sets in (8) are the same for given $\varepsilon > 0$. Clearly, the second set is contained in the first one. Conversely, if $0 < F(x_0) = \alpha \leqq 1$ for $F \in C^+$, then

$$\frac{1}{\alpha} F(x_0) = 1 \quad \text{and} \quad |F(x) - F(x_0)| \leqq \left| \frac{1}{\alpha} F(x) - \frac{1}{\alpha} F(x_0) \right| .$$

By Theorem 8 we know that a sequence $\{x_n\}$ in $C$ is $d$-convergent to $x_0 \in C$ if and only if $\log F(x_0) \to \log F(x)$ uniformly for all $F \in C^+$ satisfying $F(x_0) > 0$. Since

$$\log F(x_n) - \log F(x_0) = \log [F(x_n)/F(x_0)]$$

does not change if $F$ is multiplied by a positive constant, $d(x_n, x_0) \to 0$ is equivalent to $F(x_n) \to F(x_0)$ uniformly for all $F \in C^+$ with $F(x_0) = 1$.

COROLLARY (ii). (cf. [10, Lemma 1.2]) *If $C$ is a closed convex set containing no line in a weak Hausdorff space, then the $d$-topology is the weakest topology such that for each $x_0 \in C$ the set $\{F \in C^+ : F(x_0) = 1\}$ is equicontinuous.*

In what follows we will assume that $C$ is a complete convex set without lines in a weak Hausdorff space $L$. Then a result of Choquet [7, Corollary 1, p. 1908] states that each continuous affine function on $L$ is the difference of two functions in $C^+$. This result will be fundamental for our proofs.

COROLLARY (iii). *Let $C$ be a complete convex set containing no line in a weak Hausdorff space $L$. Then the part metric topology is stronger than the topology of $C$.*

*Proof.* By the result of Choquet mentioned above the topology of $L$ is induced by the set $C^+$. Therefore, for any point $x_0 \in C$ exist arbitrary small $L$-neighborhoods $W$ of $x_0$ of the form

$$W = \{x \in C : |F_i(x) - F_i(x_0)| < \varepsilon, F_i \in C^+, i = 1, \cdots, n\} ,$$

where we may assume $F_i(x_0) > 0$ for all $i = 1, \cdots, n$. We find a smaller $d$-neighborhood of $x_0$ as follows: Let $F_i' = F_i/F_i(x_0)$ so that $F_i'(x_0) = 1$ and $|F_i'(x) - F_i'(x_0)| < \varepsilon'$ with $\varepsilon' = \min [\varepsilon/F_1(x_0), \cdots, \varepsilon/F_n(x_0)]$. Then

$$W_0 = \{x \in C : |F(x) - F(x_0)| < \varepsilon', F \in C^+, F(x_0) = 1\}$$

is a $d$-neighborhood of $x_0$, and clearly $W_0 \subset W$.

THEOREM 9. *If $C$ is a complete convex set containing no line in a weak Hausdorff space $L$, then $C$ is $d$-complete and each part of $C$ is $d$-complete.*

*Proof.* Again by the result of Choquet we can use the fact that the topology on $L$ is that induced by the set $C^+$ of all continuous affine functions on $L$ which are nonnegative on $C$. Let $\{x_n\}$ be a $d$-Cauchy sequence in $C$. Then all $x_n$ are in one part. Let $C_0^+$ be the functions $C^+$ which are strictly positive on this part. By Theorem 8, $\{\log F(x_n)\}$ is a Cauchy sequence uniformly for $F \in C_0^+$. For each fixed $F \in C^+$, $\{F(x_n)\}$ is therefore a Cauchy sequence. Since $C$ is complete there is $x_0 \in C$ such that $F(x_n) \to F(x_0)$ for each $F \in C^+$. Since $\{x_n\}$ is $d$-Cauchy, this convergence is uniform in those $F$ such that $F(x_0) = 1$. Therefore $d(x_n, x) \to 0$, and $C$ is $d$-complete. Then also each part is $d$-complete as a $d$-closed subset of $C$.

We next treat the special case where our convex set $C$ is a convex cone $P$ with vertex $0 \in P$. In this case Theorem 8 can be restated in a somewhat different form.

THEOREM 8′. *Let $P$ be a closed convex cone with vertex $0 \in P$ containing no line in a weak Hausdorff space $L$, and denote by $P_+$ the set of all continuous linear forms $f$ which are $\geqq 0$ on $P$. Then*

$$(8') \quad d(x, y) = \sup \{|\log f(x) - \log f(y)|: f \in P_+, f(x) > 0\ f(y) > 0\}$$

*for all points $x, y \in P$.*

*Proof.* $P_+$ characterizes $P$ in the sense

$$P = \bigcap_{f \in P_+} \{x \in L: f(x) \geqq 0\} .$$

This is well-known and can be deduced from (7) by remarking that each continuous affine function $F$ on $L$ satisfying $F \geqq 0$ on $P$ is of the form $F = f + \alpha$ where $f \in P_+$ and $\alpha \geqq 0$. But with this the proof of Theorem 8 can be repeated by replacing $C$ by $P$ and $C^+$ by $P_+$. Only the last sentence of the proof has to be dropped.

REMARK. If $x$ and $y$ are in the same part $\Pi$ of $P$ the condition $f(x) > 0$ or $f(y) > 0$ is equivalent to $f(y) > 0$ and even to $f > 0$ on $\Pi$. Therefore the sets

$$\{x \in \Pi: |f(x) - f(x_0)| < \varepsilon, \quad \text{all} \quad f \in P_+, 0 \leqq f(x_0) \leqq 1\}$$

for $\varepsilon > 0$ form a fundamental system of $d$-neighborhoods of $x_0 \in \Pi$ in $\Pi$ (cf. Theorem 8, Corollary (i)).

Let us point out that for a closed convex cone it is easy to check whether it contains a line:

LEMMA 3. *Let $P$ be a closed convex cone with vertex $0 \in P$ in a weak Hausdorff space. $P$ contains no line if and only if $P \cap (-P) = \{0\}$.*

*Proof.* The condition $P \cap (-P) = \{0\}$ is necessary since it says that $P$ does not contain a line through the origin. To prove its sufficiency, assume that $x + t(y - x) \in P$ for all real numbers $t$. Since $P_+$ characterizes $P$ this is equivalent to $f(x) + tf(y - x) \geqq 0$ for all $t$ and all $f \in P_+$. Therefore, $f(y-x) = 0$ for all $f \in P_+$; i.e., $y - x \in P \cap (-P)$ and hence $x = y$.

The next theorem gives a completeness criterion for the part metric in a cone. Remember that a convex subset $K$ of a convex cone $P$ with vertex $0 \in P$ is called a *section* or *base* for $P$ if each ray

$$\{rx: r > 0\}, x \in P, x \neq 0 ,$$

intersects $K$ exactly once. Hence $P = \{rx: x \in K, r \geqq 0\}$ if $K$ is a section of $P$.

THEOREM 10. *Let $P$ be a convex cone satisfying $P \cap (-P) = \{0\}$ in a weak Hausdorff space $L$. If $P$ has a compact section $K$ then $P$ and hence also $P \backslash \{0\}$ are d-complete.*

*Proof.* It is well-known [4, p. 113] that $P$ is complete in $L$. Therefore, by the above lemma $P$ does not contain any line. The result follows then from Theorem 9, since $P \backslash \{0\}$ is a part of $P$.

Let us remark that for a convex cone $P$ with a section $K$ and without lines the parts of $K$ are the intersections of the parts of $P$ with $K$. This results from the observation that by the definition of a section a segment $[x, y] \subset K$ extends by $r$ in $K$ if and only if it extends by $r$ in $P$. Hence also the part metric of $K$ is the restriction of the part metric in $P$.

Theorem 10 will be applied in particular to the following situation:

COROLLARY. *Let $B$ be a linear space of real continuous functions on a compact Hausdorff space containing the constant functions. Let $L = B'$ be the topological dual of $B$ viewed as a normed space with the norm of uniform convergence. Then $P = \{F \in B': F \geqq 0\}$ and $K = \{F \in P: F(1) = 1\}$ are d-complete.*

It suffices to remark that $K$ is compact in the weak topology $\sigma(L', L)$ and that $K$ is a section of the cone $P$.

**4. The space of Radon measures.** Let $\mathscr{M} = \mathscr{M}(X)$ be the space of all real Radon measures on a locally compact Hausdorff space $X$ with the *vague topology*.[2] This is the weak (Hausdorff) topology $\sigma(\mathscr{M}, \mathscr{K})$

[2] We follow Bourbaki [5], [6] in terminology and notation for integration theory.

induced by the linear space $\mathscr{K}=\mathscr{K}(X)$ of all continuous real-valued functions on $X$ with compact support. The convex cone $\mathscr{M}_+=\mathscr{M}_+(X)$ of all measures $\mu \geqq 0$ in $\mathscr{M}$ is then complete; it also satisfies $\mathscr{M}_+ \cap (-\mathscr{M}_+) = \{0\}$, and hence does contain any line by Lemma 3. Therefore by Theorem 9, $\mathscr{M}_+$ is complete with respect to the part metric $d$ on $\mathscr{M}_+$. In this section we will describe the parts of $\mathscr{M}_+$ and the metric $d$.

By Theorem 5 the equivalence relation $\sim$ in $\mathscr{M}_+$ is defined as follows: $\mu \sim \nu$ if and only if $\alpha\mu \geqq \nu$ and $\alpha\nu \geqq \mu$ for some number $\alpha > 0$ (necessarily $\alpha \geqq 1$ if $\mu \neq 0$ or $\nu \neq 0$). In particular the zero measure is a part by itself. Consequently [6; p. 51] two measures $\mu, \nu \in \mathscr{M}_+$ are equivalent if and only if they are mutually absolutely continuous, with (locally integrable) Radon-Nikodym derivatives which locally almost everywhere are bounded and bounded below by some number $\alpha > 0$. Hence the parts of $\mathscr{M}_+$ can be described as follows: for any fixed $\mu \in \mathscr{M}_+$, consider the set $P_\mu$ of all "functions" $g \in L^\infty(\mu)$ which are positive and bounded away from zero. Then the part $\Pi_\mu$ of $\mathscr{M}_+$ containing $\mu$ is the set

$$\Pi_\mu = \{g\mu \colon g \in P_\mu\} .$$

Therefore, we can transfer the $L^\infty(\mu)$-metric to $\Pi_\mu$ via the bijection $g \to g\mu$ of $P_\mu$ onto $\Pi_\mu$.

The following result describes the topology derived from $d$ in a part, thus generalizing and improving Theorem 3 of [2].

THEOREM 11. *The part metric $d$ in each part $\Pi_\mu$ of $\mathscr{M}_+(X)$ is topologically equivalent to the $L^\infty(\mu)$-metric on the corresponding functions of $P_\mu$ ($\mu \in \mathscr{M}_+(X)$).*

*Proof.* Let $g_0, g_1, g_2, \cdots$ be a sequence in $P_\mu$. We have to prove

$$\lim_{n\to\infty} \| g_n - g_0 \|_\infty = 0 \ \text{ if and only if } \ \lim_{n\to\infty} d(g_n\mu, g_0\mu) = 0 .$$

By a well-known result on $L^p$-spaces (cf. Bourbaki [5, pp. 211-213]), one has

$$\| g_n - g_0 \|_\infty = \sup \left| \int v(g_n - g_0) d\mu \right|$$

where the supremum is taken over all functions $v \in \mathscr{K}$ satisfying $\int |v|\, d\mu \leqq 1$. On the other hand, from Theorem 8′ and the remark following this theorem, we have $\lim_{n\to\infty} d(g_n\mu, g_0\mu) = 0$ if and only if

$$\lim_{n\to\infty} \int u(g_n - g_0) d\mu = 0$$

holds uniformly for all functions $u \in \mathscr{K}_+$ satisfying $\int u g_0 d\mu \leqq 1$. From this fact and the following two simple observations the theorem follows: (i) for each function $u \in \mathscr{K}_+$ satisfying $\int u g_0 d\mu \leqq 1$, one has

$$\int \alpha u d\mu \leqq \int u g_0 d\mu \leqq 1$$

where $\alpha > 0$ is locally almost everywhere a lower bound for $g_0$. (ii) For each function $v \in \mathscr{K}$ satisfying $\int | v | d\mu \leqq 1$ one has

$$\int | v | g_0 d\mu \leqq \| g_0 \|_\infty \int | v | d\mu \leqq \| g_0 \|_\infty \quad \text{and} \quad | v | \in \mathscr{K}_+ .$$

**5. A selection theorem for the part metric.** In this section we prove a generalization of the following theorem of E. Michael [11, Th. 1], [12, Th. 3.2″]:

THEOREM M. *If $T$ is a paracompact space and $\phi$ is a lower semi-continuous function from $T$ to the nonempty closed convex subsets of a Banach space $Y$ then there exists a continuous selection function $f: T \to Y$ such that $f(t) \in \phi(t)$ for all $t \in T$.*

In our version of Theorem $M$ we replace the Banach space $Y$ by a convex set with a complete metric which is convex in the sense of the following definition.

DEFINITION. *Let $Y$ be a convex set in a linear space. A metric $d$ on $Y$ will be called convex if*

( i ) *the mapping $(\lambda, x, y) \to \lambda x + (1 - \lambda) y$ of $[0, 1] \times Y \times Y$ into $Y$ is continuous, and if*

( ii ) *the set $\{y \in Y: d(y, S) < \varepsilon\}$ is convex for each $\varepsilon > 0$ and each convex subset $S$ of $Y$.*

Theorem M follows from our result by the remark that the norm metric on a Banach space $Y$ is convex and complete. From Theorem 1, Corollary (iii) and Theorem 3 it is clear that the part metric of a convex set $C$ without any line is convex on each part $\Pi$ of $C$.

We recall that a mapping $\phi$ from a topological space $T$ to the subsets of a topological space $Y$ is called *lower semi-continuous* (l.s.c.) if for any open set $U \subset Y$, the set of all $t \in T$ such that $\phi(t) \cap U \neq \varnothing$ is open. If $\phi$ is the inverse of a mapping $R$ from $Y$ onto $T$, then it is easy to see that $\phi$ is l.s.c. if and only if $R$ is an open mapping.

For the proof of our selection theorem the same kind of arguments can be used as in [11].

LEMMA 4. *Let $T$ be a paracompact space, $Y$ a convex set with a convex metric $d$, and let $\phi$ be a lower semi-continuous function from $T$ to the nonempty convex subsets of $Y$. Then for any $r > 0$, there is a $d$-continuous function $f: T \to Y$ such that*

$$d(f(t), \phi(t)) < r \text{ for all } t \in T .$$

*Proof.* For $y \in Y$, let $U_y = \{t \in T: d(y, \phi(t)) < r\}$; that is, $U_y$ is the open set of points $t \in T$ such that $\phi(t)$ intersects the open $r$-ball around $y$. The family $(U_y)_{y \in Y}$ covers $T$ since $t \in U_y$ if $y \in \phi(t) \neq \emptyset$. The space $T$ being paracompact, there exists an open locally finite refinement $(V_\alpha)_{\alpha \in A}$ of the covering $(U_y)$. Thus each $V_\alpha$ is contained in some $U_y$, and each $t \in T$ has a neighborhood which intersects only a finite number of $V_\alpha$. Let $(P_\alpha)_{\alpha \in A}$ be a partition of unity subordinate to $(V_\alpha)_{\alpha \in A}$; i.e., $p_\alpha$ is a nonnegative continuous function on $T$, $p_\alpha = 0$ off $V_\alpha$, and $\Sigma p_\alpha(t) = 1$ for all $t \in T$. For each $\alpha \in A$, pick $y_\alpha \in Y$ so that $V_\alpha \subset U_{y_\alpha}$. Let the function $f: T \to Y$ be defined by $f(t) = \sum_{\alpha \in A} p_\alpha(t) y_\alpha$, $t \in T$. For each $t_0 \in T$, there is a neighborhood $N_{t_0}$ which intersects only a finite number of $V_\alpha$, so that $p_\alpha = 0$ on $N_{t_0}$ for all but a finite number of $\alpha$. Thus in $N_{t_0}$ the sum $f(t)$ is a convex combination of a finite number of $y_\alpha \in Y$:

$$f(t) = p_{\alpha_1}(t) y_{\alpha_1} + \cdots + p_{\alpha_n}(t) y_{\alpha_n} \qquad (t \in N_{t_0}) .$$

Thus because of the convexity of the metric $f$ is $d$-continuous on each $N_{t_0}$, hence on $T$. Furthermore, $f(t) \in Y$ is a convex combination of points $y_\alpha$ such that $p_\alpha(t) \neq 0$, hence such that $t \in V_\alpha \subset U_{y_\alpha}$. That is all these $y_\alpha$ are within a distance $r$ of $\phi(t)$. Hence $d(f(t), \phi(t)) < r$ since each of the sets $\{y \in Y: d(y, \phi(t) < r\}$ is convex.

THEOREM 12. *Let $T$ be a paracompact space and $Y$ a convex set with a complete convex metric $d$. If $\phi$ is a lower semi-continuous mapping from $T$ to the nonempty closed convex subsets of $Y$, then there exists a continuous function $f: T \to Y$ such that $f(t) \in \phi(t)$ for all $t \in T$.*

*Proof.* Michael's proof [11, p. 235] for a Banach space works without change in view of the preceding lemma. We repeat the argument here for the reader's convenience.

We construct inductively a sequence $f_n: T \to Y$ of continuous mappings such that uniformly for all $t \in T$

$$(10) \qquad d(f_n(t), f_{n+1}(t)) < 1/2^{n-1}$$

$$(n = 1, 2, \cdots) .$$

$$(11) \qquad d(f_n(t), \phi(t)) < 1/2^n .$$

Then by condition (10) $(f_n)$ is uniformly Cauchy, and hence by the completeness of $Y$ converges to a continuous mapping $f: T \rightarrow Y$. The condition (11) insures $f(t) \in \phi(t)$ since each $\phi(t)$ is closed.

The preceding lemma proves the existence of a mapping $f_1: T \rightarrow Y$ such that $d(f_1(t), \phi(t)) < 1/2$ for all $t \in T$. Suppose $f_1, \cdots, f_n$ have been defined satisfying (10) and (11). Define for each $t \in T$

$$\phi_{n+1}(t) = \phi(t) \cap \{y \in Y: d(y, f_n(t) < 1/2^n\} .$$

Then $\phi_{n+1}(t) \neq \emptyset$ by (10). Since the metric $d$ is convex, $\phi_{n+1}(t)$ is the intersection of two convex sets and therefore convex. We show that $\phi_{n+1}$ is l.s.c., and then appeal to the lemma for the existence of $f_{n+1}$ such that $d(f_{n+1}(t), \phi_{n+1}(t)) < 1/2^{n+1}(t \in T)$. This implies then

$$d(f_n(t), f_{n+1}(t)) < 1/2^n + 1/2^{n+1} < 1/2^{n-1}$$

as required in (10).

To show that $\phi_{n+1}$ is l.s.c., suppose $\phi_{n+1}(t_0) \cap U \neq \emptyset$ for some open set $U \subset Y$ and some $t_0 \in T$. For each positive $\varepsilon < 1/2^n$, there is a point $y_0 \in \phi(t_0) \cap U$ such that $d(y_0, f_n(t_0)) < 1/2^n - \varepsilon$; i.e., $\phi(t_0) \cap U \cap S \neq \emptyset$ where $S$ is the open $(1/2^n - \varepsilon)$-ball around $f_n(t_0)$. Since $\phi$ is l.s.c., $\phi(t) \cap U \cap S \neq \emptyset$ for all $t$ of some neighborhood $N_1$ of $t_0$. Since $f_n$ is continuous, we also have $d(f_n(t), f_n(t_0)) < \varepsilon$ for all $t$ in some neighborhood $N_2$ of $t_0$. If $t \in N_1 \cap N_2$, then $\phi(t) \cap U \cap S \neq \emptyset$ and every point $y \in \phi(t) \cap U \cap S$ has the property that

$$d(y, f_n(t)) \leqq d(y, f_n(t_0)) + d(f_n(t), f_n(t_0)) < 1/2^n .$$

Hence $\phi_{n+1}(t) \cap U \neq \emptyset$ for all $t \in N_1 \cap N_2$, and $\phi_{n+1}$ is l.s.c.

COROLLARY (i). *Let $C$ be a complete convex set without lines in a weak Hausdorff space, and let $d$ be the part metric on $C$. If $\phi$ is a lower semi-continuous function on $T$ to the nonempty closed convex subsets of one part $\Pi$ of $C$, then there exists a continuous selection function $f: T \rightarrow \Pi$ for $\phi$.*

*Proof.* We mentioned already that $d$ is a convex metric on $\Pi$. By Theorem 9 $\Pi$ is $d$-complete.

COROLLARY (ii). *Let $\phi$ be a lower semi-continuous mapping from* *a paracompact space $T$ to the nonempty closed convex subsets of one part $\Pi$ of the cone $\mathscr{M}_+(X)$ of nonnegative Radon measures on a locally compact space $X$. Then there exists a continuous selection function for $\phi$. Let $f$ be any such function and $\mu \in \Pi$. Then $f$ has the form $f(t) = g_t\mu$, where $t \to g_t$ is a continuous mapping from $T$ into $L^\infty(\mu)$.*

This follows immediately from Corollary (i) and Theorem 11.

6. **Representing measures.** Let $X$ be a compact Hausdorff space, and $C_R(X)$ be the space of all continuous real functions on $X$. A function space $B$ is a separating subspace of $C_R(X)$, containing the constant functions. We summarize here some facts from [1].

Let $T_B = \{F \in B' : F \geqq 0, F(1) = 1\}$. Then $T_B$ is convex and compact in the weak topology $\sigma(B', B)$ on $B'$. The space $X$ can be embedded in $T_B$, and $B$ can be represented as the restrictions to $T_B$ of all weakly continuous affine functions on $B'$. Gleason parts can be defined in $X$ in terms of the Harnack-type of inequality: $x \sim y$ if and only if there is $\alpha > 1$ such that $\alpha^{-1} \leqq u(x)/u(y) \leqq \alpha$ for all $u > 0$ in $B$. The parts of this relation are the intersections of $X$, considered as a subset of $T_B$, with the geometrically defined parts of $T_B$. The distance in parts in $X$ defined by $d(x, y) = \sup\{|\log u(x) - \log u(y)| : u > 0\}$ is the same as the part metric in $T_B$.

For simplicity, by the boundary $\Gamma$ of $B$ we will mean the *Choquet boundary* if $X$ is metrizable (so that $\Gamma$ is a $G_\delta$), or otherwise the *Silov boundary* (so $\Gamma$ is closed). A representing measure $\mu$ for $x \in X$ is a positive Radon measure supported by $\Gamma$ such that $u(x) = \int_\Gamma u d\mu$ for all $u \in B$. Let $\mathscr{M}_+^1 = \mathscr{M}_+^1(X)$ be all positive probability measures on $X$ and let $\mathscr{M}_x$ be the representing measures for $x$. Clearly $\mathscr{M}_x \subset \mathscr{M}_+^1$ for each $x$, since $1 \in B$.

If $A$ is a complex function algebra on $X$, and $B = \operatorname{Re} A$, then the spectrum $S_A$ can be considered a subset of $T_B$. The Gleason parts of $S_A$ are the intersections of $S_A$ with the parts of $T_B$. The metric on $S_A$ from the norm of $A'$ gives the same topology as the part metric [1, 3]. The Silov and Choquet boundaries for $A$ are the same as for $B$, and a measure represents a point for $A$ if and only if it represents the point for $B$. Therefore we will confine our discussion to the case of a real function space $B$, but our results will also apply to complex function algebras.

Harkova [9] has given an example of a function algebra (hence also a function space) where the complement of the Silov = Choquet boundary is one part $\Delta$, and it is not possible to pick representing measures $\mu_x$ for all $x \in \Delta$ so that all $\mu_x$ are in one part of $\mathscr{M}_+^1$. It is shown in [9], however, that if the space of orthogonal real measures is finite dimensional, then it is possible to pick representing measures so that points in one part of $X$ will have measures in one part of $\mathscr{M}_+^1$. We give here a condition based on our selection theorem which insures that if measures $\mu_x$ for points in one part can be chosen in one part of $\mathscr{M}_+^1$, then we can pick them so that $x \to \mu_x$ is a continuous mapping with respect to the two part metrics.

**Lemma 5.** *The set $\mathscr{M}_x$ of positive Radon measures on $X$ supported by the Choquet or Silov boundary $\Gamma$ which represent $x$ is convex and closed in the part metric of $\mathscr{M}_+^1(X)$.*

*Proof.* The convexity is obvious. If $\mu_n \in \mathscr{M}_x$ and $d(\mu_n, \mu_0) \to 0$, then $\mu_n$, $\mu_0$ all are in one part of $\mathscr{M}_+^1$ (or $\mathscr{M}_x(X)$). Hence by Theorem 11 $\mu_n = g_n\mu$, $\mu_0 = g_0\mu$ for some $\mu$, where $g_n, g_0 \in L^\infty(\mu)$, and $\| g_n - g_0 \|_\infty \to 0$. It is clear that $\mu_0$ is supported by $\Gamma$. Since

$$\left|\int u(g_n - g_0)d\mu\right| \leqq \| u \| \, \| g_n - g_n \|_\infty \to 0 \,,$$

$\int ug_n d\mu = u(x) = \int ug_0 d\mu$ for all $u \in B$. Hence $\mu_0 \in \mathscr{M}_x$.

**Theorem 13.** *Let $B$ be a function space on $X$ with boundary $\Gamma$. Let $\Delta$ be a part of $X$ and suppose there is a part $\Pi_\mu \subset \mathscr{M}_+^1$ such that $\mathscr{M}_x \cap \Pi_\mu \neq \emptyset$ for all $x \in \Delta$. If the mapping $x \to \mathscr{M}_x \cap \Pi_\mu$ is* l.s.c. *with respect to the two part metrics then there is a continuous mapping $x \to \mu_x = g_x\mu$ on $\Delta$ into $\Pi_\mu$ such that $\mu_x \in \mathscr{M}_x$ for each $x$. Consequently, the mapping $x \to g_x$ is d-continuous from $\Delta$ into $L^\infty(\mu)$.*

*Proof.* Since each $\mathscr{M}_x$ is closed in $\mathscr{M}_+^1$, and each part $\Pi_\mu$ is $d$-closed, the sets $\mathscr{M}_+ \subset P_\mu$ are $d$-closed, and clearly also convex. The metric space $(\Delta, d)$ is paracompact. The part metric in $\mathscr{M}_+^1$ is complete, since $\mathscr{M}_+^1$ is compact in the vague topology. The theorem is now immediate from Theorem 12, Corollary (i).

We remark that the part metric in a part $\Delta$ of $X$ gives the relativization to $\Delta$ of the given compact topology of $X$ if the set

$$B^+(z_0) = \{u \mid \Delta : u \in B, u > 0, u(z_0) = 1\} \qquad (z_0 \in \Delta)$$

is equicontinuous [2, Th. 3], [10, Lemma 1]. This is the case for example for spaces $B$ of functions continuous on the closure of a bounded domain in $R^n$ and harmonic on the interior. In such a case the continuity of the mapping $x \to \mu_x = g_x\mu$ with respect to the part metric is the same as continuity with respect to the given Euclidean topology in $X$ and the part metric topology in $\mathscr{M}_+^1$ (or the $L^\infty(\mu)$ topology.)

*Added in proof.* The part metric is similar to the metric of Cayley—Klein (cf. W. BLASCHKE, Projektive Geometrie, Basel (1954), p. 80). A. C. Thompson has recently used the part metric in the form given here to simplify some proofs of Birkhoff and Samelson (On certain contraction mappings in a partially ordered vector space, P.A.M.S. vol. 14 (1963), 438-443).

## References

1. H. S. Bear, *A geometric characterization of Gleason parts*, Proc. Amer. Math. Soc. **16** (1965), 407-412.
2. H. S. Bear and Bertram Walsh, *Integral kernel for one-part function spaces*, Pacific J. Math. **23** (1967), 209-215.
3. H. S. Bear and M. L. Weiss, *An intrinsic metric for parts*, Proc. Amer. Math. Soc. **18** (1967), 812-817.
4. N. Bourbaki, *Espaces vectoriels topologiques*, Chap. 1-2 (2nd edition), Act. Sci. et Ind. 1189, Paris, 1966.
5. ———, *Intégration*, Chap. 1-4, Act. Sci. et Ind. 1175, Paris, 1952.
6. ———, *Intégration*, Chap. 5, Act. Sci et Ind. 1244, Paris, 1956.
7. G. Choquet, *Ensembles et cônes convexes faiblement complets*, C. R. Académie Sciences **254** (1962), 1908-1910; 2123-2125.
8. A. M. Gleason, *Function algebras*, Seminar on analytic functions, vol. 2, Institute for Advanced Study, Princeton, N. J., 1957.
9. N. V. Harkova, *Generalized Poisson formula* (Russian), Vestnik Moskov. Univ., Ser. I Mat. Meh. **22** (1967), 25-30.
10. J. Köhn, *Die Harnacksche Metrik in der Theorie der harmonischen Funktionen*, Math. Zeit. **91** (1966), 50-64.
11. E. Michael, *Selected selection theorems*, Amer. Math. Monthly **43** (1956), 233-238.
12. ———, *Continuous selections I*, Ann. of Math. **63** (1956), 361-381.

Received November 13, 1968. This is a revised form of a paper submitted by H. S. Bear on March 19, 1968. This author was supported in part by Grant GP7681 of the National Science Foundation.

Universität Erlangen-Nürnberg
New Mexico State University
and
University of Hawaii

# An open mapping theorem for convex sets with only one part [R27]

Aequationes Math. **4** (1970), 332–337
[Zbl. Math. 198.15702; MR 42 #8226]

Consider a real linear space $L$, and a convex set $C$ in $L$ which contains no line. Bear [3] has introduced an equivalence relation $\sim$ on $C$: two points $x, y \in C$ are called equivalent if the closed segment $[x, y]$ extends in $C$ by some number $r > 0$, i.e. $x + r(x-y) \in C$ and $y + r(y-x) \in C$. The corresponding equivalence classes are called the parts of $C$. A 'metric' with values in $[0, +\infty]$ is defined on $C$ by

$$d(x, y) = \begin{cases} \inf\left\{\log\left(1 + \frac{1}{r}\right) : [x, y] \text{ extends by } r\right\}, \text{ if } x \sim y \\ +\infty, \text{ if } x \nsim y . \end{cases} \tag{1}$$

$d$ is called the part metric on $C$.

By a result of [2], the parts are exactly the connected components of $C$ in the $d$-topology. Parts are always convex.*)

In this paper we are interested in the structure of a single part, hence of a convex set $C$ having only one part.

## 1. An Embedding Theorem

In what follows $C$ will denote a convex set without lines having only one part. We will assume that the origin 0 of $L$ is a point of $C$.

LEMMA 1. *The linear subspace generated by $C$ (in $L$) is equal to*

$$E = \bigcup_{t>0} tC. \tag{2}$$

*$C$ is an absorbing set in $E$.*

*Proof.* By definition, $tE \subset E$ for all $t \geqslant 0$ since $0 \in C$. We also have $-E \subset E$, and hence $tE \subset E$ for all $t \in \mathbf{R}$. To see this, consider $y = tx \in E$, $x \in C$, $t > 0$. Since $C$ has only one part, the segment $[x, 0]$ extends by some $r > 0$, hence in particular $-rx \in C$ and so $-y = (t/r)(-rx)$ is in $E$. From the convexity of $C$ it follows that $(s+t)\,C = sC + tC$ for all $s > 0$, $t > 0$. Hence $E + E \subset C$, and $E$ is a linear space (obviously generated by $C$). To see that $C$ is absorbing, notice that for each $x \in E$ there is a $t_1 > 0$ such that $x \in t_1 C$.

*) The above parts are closely related to the so-called Gleason parts of a function algebra or function space [3]. A slight modification of the metric $d$ is known to geometers as the projective metric of Cayley–Klein [2].

*Received February 12, 1969, and in revised form, April 16, 1969*

Since $C$ is convex and $0 \in C$, we then have $x \in tC$ for all $t \geqslant t_1$. Applying this to $-x$, gives the existence of $t_2 > 0$ such that $-x \in tC$ for all $t \geqslant t_2$. Hence $x \in tC$ for all $t$ satisfying $|t| \geqslant \max(t_1, t_2)$.

Introduce now the *Minkowski functional* $p$ on $E$ corresponding to $C$:

$$p(x) = \inf\{t : x \in tC, t > 0\} \quad (x \in E). \tag{3}$$

Then by the above Lemma, $p$ is finite, as well as sub-additive and positively homogeneous. We also define

$$q(x) = \max(p(x), p(-x)) \quad \text{for} \quad x \in E. \tag{4}$$

Then $q$ is a seminorm and, since $C$ contains no line, $q$ is a norm on $E$. As in [2], $q$ will be called the *Minkowski norm* for $C$ on $E$.

The following two results will be important in what follows:

THEOREM 1. *Under the initial assumptions on C, we have*

$$C = \{x \in E : p(x) < 1\}. \tag{5}$$

*The Minkowski norm q for C induces in C the d-topology.*

*Proof.* Clearly [4, p. 62],

$$\{x \in E : p(x) < 1\} \subset C \subset \{x \in E : p(x) \leqslant 1\}.$$

Since $C$ has only one part and since $[0, x]$ does not extend in $C$ by some $r > 0$ if $p(x) = 1$ for some $x \in C$, it follows that $C = \{x \in E : p(x) < 1\}$. We have $|p(x) - p(y)| \leqslant q(x - y)$ for all $x, y, \in E$. Therefore, $p$ is $q$-continuous and it follows that $C$ is an open set in the normed space $(E, q)$. The remaining part of the Theorem follows then from [2], Theorem 4.

THEOREM 2. *C is d-complete if and only if E is a Banach space with respect to the norm q.*

*Proof.* Assume $C$ to be $d$-complete, and let $(x_n)_{n=1,2,\ldots}$ be a $q$-Cauchy sequence in $E$. Then the sequence $(q(x_n))$ is convergent and hence bounded. After replacing the original sequence by $(\lambda x_n)$ for some $\lambda > 0$, we may assume

$$p(x_n) \leqslant q(x_n) \leqslant M < 1 \quad (n = 1, 2, \ldots). \tag{6}$$

Then by (5) we have $x_n \in C$ for all $n$. In the proof of [2], Theorem 4 it is shown that a segment $[x, y]$, $x, y, \in C$, $x \neq y$, extends by all $r > 0$ satisfying

$$1 + r < (1 - p(y))/q(x - y)$$

From this and (1) it follows that

$$d(x, y) \leqslant \log\left[1 + \frac{q(x-y)}{1 - p(y) - q(x-y)}\right],$$

an inequality which obviously also holds if $x = y \in C$. From this and (6) we obtain

$$\begin{aligned} d(x_m, x_n) &\leqslant \log\left[1 + \frac{q(x_m - x_n)}{1 - p(x_n) - q(x_m - x_n)}\right] \\ &\leqslant \log\left[1 + \frac{q(x_m - x_n)}{1 - M - q(x_m - x_n)}\right] \end{aligned}$$

for all $m, n$. Since $q(x_m - x_n) \to 0$ for $m, n \to +\infty$, this implies $d(x_m, x_n) \to 0$ for $m$, $n \to +\infty$. Hence, $(x_n)$ is a $d$-Cauchy sequence in $C$. By the $d$-completeness of $C$ it is $d$-convergent to a point $x \in C$. But then by Theorem 1, $x$ is also a $q$-limit of the sequence $(x_n)$.

Assume now that $E$ is a Banach space. For any $d$-Cauchy sequence $(x_n)_{n=1,2,\ldots}$ in $C$ we have by the definition of $d$: For every $\varepsilon > 0$ there exists a natural number $N_\varepsilon$ such that for every $m, n \geqslant N_\varepsilon$ the segment $[x_m, x_n]$ extends in $C$ by some $r \geqslant r_\varepsilon$ where $\log(1 + 1/r_\varepsilon) = \varepsilon$. Then $[x_m, x_n]$ also extends in $C$ by $r_\varepsilon$.

In particular, $[x_m, x_{N_1}]$ extends by $r_1$ for all $m \geqslant N_1$. From

$$(1 + r_1)\, x_{N_1} - r_1 x_m = x_{N_1} + r_1 (x_{N_1} - x_m) \in C$$

and Theorem 1 follows

$$r_1 p(-x_m) - (1 + r_1)\, p(-x_{N_1}) \leqslant p((1 + r_1)\, x_{N_1} - r_1 x_m) < 1 \quad (m \geqslant N_1).$$

Therefore, the sequence $(p(-x_n))_{n=1,2,\ldots}$ is bounded above by some number $M > 0$. From the proof of [2], Theorem 4, we know that

$$\begin{aligned} q(x_m - x_n) &\leqslant \frac{1}{r_\varepsilon}\left(1 + \frac{1}{r_\varepsilon}\right)(1 + p(-x_n)) \\ &\leqslant \frac{1}{r_\varepsilon}\left(1 + \frac{1}{r_\varepsilon}\right)(1 + M) \end{aligned}$$

for all $m, n \geqslant N_\varepsilon$. Hence $(x_n)$ is a $q$-Cauchy sequence in $E$ and by assumption convergent in $E$ to a point $x_0$ in the $q$-closure $\bar{C}$ of $C$. We will prove that $x_0$ is in $C$ and is the $d$-limit of $(x_n)$. For $m, n \geqslant N_\varepsilon$ we have

$$x_m + r_\varepsilon (x_m - x_n) \in C, \quad x_n + r_\varepsilon (x_n - x_m) \in C.$$

For $n \to +\infty$ we obtain

$$x_m + r_\varepsilon(x_m - x_0) \in \bar{C}, \quad x_0 + r_\varepsilon(x_0 - x_m) \in \bar{C}$$

for all $m \geqslant N_\varepsilon$. Since $x_m$ is in the open convex set $C$ and $x_0^* = x_0 + r_\varepsilon(x_0 - x_m)$ is in $\bar{C}$, it follows from [4, p. 54] that each point of the open segment $]x_m, x_0^*[$, hence in particular $x_0$ is in $C$. Since $q(x_n - x_0) \to 0$ we then also have $d(x_n, x_0) \to 0$ by Theorem 1.

Let us recall that [2] gives sufficient conditions for $C$ to be $d$-complete.

## 2. The Open Mapping Theorem

Consider now two convex sets $C_1$, $C_2$ containing no line in corresponding linear spaces $L_1$, $L_2$. Denote by $d_1$, $d_2$ the corresponding part metrics. A mapping $f: C_1 \to C_2$ is called *affine* if it preserves convex combinations.

LEMMA 2. *Each affine mapping $f: C_1 \to C_2$ satisfies*

$$d_2(f(x), f(y)) \leqslant d_1(x, y) \quad \text{for all } x, y \in C_1. \tag{7}$$

*Hence $f$ is uniformly continuous (with respect to $d_1$, $d_2$) and maps each part of $C_1$ into one part of $C_2$.*

*Proof.* If the segment $[x, y]$, $x, y \in C_1$, extends by $r$, then we have $x' = x + r(x - y) \in C$ and $y' = y + r(y - x) \in C$. Since

$$x = \frac{1}{1+r} x' + \frac{r}{1+r} y \quad \text{and} \quad y = \frac{1}{1+r} y' + \frac{r}{1+r} x$$

are convex combinations, they are preserved by $f$. This shows that also $[f(x), f(y)]$ extends by $r$. From this and (1) we get (7). The rest is immediate.

COROLLARY. *If $f: C_1 \to C_2$ is a surjective affine mapping and if $C_1$ has only one part, then $C_2$ has only one part.*

*Remark.* If $C_1$ is mapped by a translation $t(x) = x + a$ onto the convex set $C_2 = t(C)$, then $t$ is affine and obviously a $d_1$-$d_2$-isometry of $C_1$ onto $C_2$.

We now can prove an analogue of Banach's classical open mapping theorem (cf. [4], [5]).

THEOREM 3. *Let $f: C_1 \to C_2$ be a surjective affine mapping. Assume that $C_1$ (and hence $C_2$) has only one part. Then the mapping $f$ is*

(i) *almost-open if $C_2$ is $d_2$-complete;*

(ii) *open if $C_1$ is $d_1$-complete and $C_2$ is $d_2$-complete.*

Remember that a mapping $g: T_1 \to T_2$ of one topological space into another is called *almost-open* if for each neighborhood $U$ of a point $x \in T_1$ the closure $\overline{f(U)}$ is a neighborhood of $f(x)$.

*Proof.* By the above Remark, translations are isometries for the part metrics. It suffices therefore to prove the Theorem under the condition that both $C_1$ and $C_2$ contain the origin of $L_1$ and $L_2$, respectively, and that $f$ satisfies $f(0)=0$. Denote by $E_i$ the linear subspace generated by $C_i$, and by $q_i$ the Minkowski norm for $C_i$ on $E_i$ $(i=1, 2)$. Any affine mapping $f: C_1 \to C_2$ has then a unique linear extension $F: E_1 \to E_2$ which is surjective if $f$ is. This follows from (2) by a simple calculation. By Lemma 2, $f$ is $d_1$-$d_2$-continuous, and hence, by Theorem 1, $q_1$-$q_2$-continuous. So $F$, viewed as a linear mapping of the normed linear space $(E_1, q_1)$ into the normed linear space $(E_2, q_2)$, is continuous at the origin and hence continuous. Our result follows therefore from Theorems 1 and 2 and by the classical analogues of statements (i), (ii). These analogues state that a continuous surjective linear mapping $F$ of the normed space $E_1$ onto the normed space $E_2$ is almost open if $E_2$ is a Banach-space, and open if $E_1$ and $E_2$ are Banach spaces.

We now obtain an analogue of a well-known selection theorem of Bartle and Graves [1].

COROLLARY 1. *Let $f: C_1 \to C_2$ be a surjective affine mapping and assume that the convex sets $C_1$, $C_2$ have only one part and are complete in the part metric. Then for each continuous mapping $\Psi: T \to C_2$ of a paracompact space $T$ in $C_2$ there exists a continuous mapping $\phi: T \to C_1$ satisfying*

$$\Psi = f \circ \phi. \tag{8}$$

*Proof.* By Theorem 3, $f$ is open and hence $y \to f^{-1}(y)$ is a lower semi-continuous mapping of $C_2$ into the non-empty closed convex subsets of $C_1$. Then also $t \to f^{-1}(\Psi(t))$ is a lower semi-continuous mapping of $T$ into the non-empty closed convex subsets of $C_1$. The result then follows from the selection theorem proved in [2], Theorem 12, Corollary (i).

In [2] the parts of the convex set $M_+^1(X)$ of all positive Radon measures of total mass 1 on a compact space $X$ were determined: The part $\Pi_\mu$ of a measure $\mu \in M_+^1(X)$ consistst of all measures $g\mu \in M_+^1(X)$ where $g$ is a function in $L^\infty(\mu)$ positive and bounded away from zero. With this result we obtain the following Corollary.

COROLLARY 2. *Let $X$ be a compact convex set in a locally convex linear Hausdorff space. Assume that for a given part $\Delta$ of $X$ there exists a measure $\mu \in M_+^1(X)$ such that each point of $\Delta$ is the barycenter of a measure in $\Pi_\mu$. Then there exists a continuous mapping $x \to g_x$ of $\Delta$ (with the part metric) into $L^\infty(\mu)$ such that $g_x\mu$ has $x$ as barycenter.*

*Proof.* We apply Corollary 1 to the following situation: $C_1 = \Pi_\mu$, $C_2 = T = \Delta$, $f(v)$ = barycenter of $v$, $\Psi$ = identity. Then the result follows since by [2], Theorem 9 $\Pi_\mu$ and $\Delta$ are complete in the part metric, and since by [2], Theorem 10 the part metric on $\Pi_\mu$ is topologically equivalent to the $L^\infty(\mu)$-metric.

From [2], chapter 6 it is clear how this Corollary can be modified in order to obtain a continuous selection of mutually absolutely continuous representing measures for points in one part of a function space or function algebra. As in [2], Theorem 13 these measures can also be concentrated on boundaries.

REFERENCES

[1] BARTLE, R. G. and GRAVES, L. M., *Mappings between Function Spaces*, Trans. Amer. Math. Soc. *72*, 400–413 (1952).

[2] BAUER, H. and BEAR, H. S., *The Part Metric in Convex Sets*, to appear in Pacific J. Math.

[3] BEAR, H. S., *A Geometric Characterization of Gleason Parts*, Proc. Amer. Math. Soc. *16*, 407–412 (1952).

[4] BOURBAKI, N., *Espaces vectoriels topologiques*, Chap. 1–2, 2nd edition ([Act. Sci. et Ind. 1189] Paris 1966).

[5] HUSAIN, T., *The Open Mapping Theorem and Closed Graph Theorem in Topological Vector Spaces* (Oxford 1965).

*Universität Erlangen-Nürnberg*
*New Mexico State University*

# Theorems of Korovkin type for adapted spaces [R29]

Ann. Inst. Fourier (Grenoble) **23** (1973), 245–260
[Zbl. Math. 262.31005; MR 50 #10643]

Dedicated to Professor Karl Stein, Munich, on the occasion of his 60th birthday

## Introduction.

Korovkin's theorem [7, 8], sometimes also called Bohman-Korovkin theorem, arose from the study of the role of Bernstein polynomials in the proof of the Weierstrass approximation theorem. For sequences $(L_n)_{n\in\mathbf{N}}$ of positive linear maps $L_n : \mathscr{C}([0, 1]) \to \mathscr{C}([0, 1])$ the theorem of Korovkin states : $(L_n f)$ converges uniformly on $[0, 1]$ to $f$ for all $f \in \mathscr{C}([0, 1])$ provided that $(L_n f)$ converges uniformly to $f$ for the three functions $f(x) = 1, = x, = x^2$ defined on $[0, 1]$.

Several generalizations of this theorem have been obtained by different authors in recent years. The most typical ones are due to Šaškin [10, 11, 12] and Wulbert [15] where $[0, 1]$ is replaced by a compact (Hausdorff) space X and where the role of the three functions $id^\nu$, $\nu = 0, 1, 2$, is taken over by a subset $\mathscr{F}$ of $\mathscr{C}(X)$. One always assumes that the constant function 1 is in $\mathscr{F}$ and that $\mathscr{F}$ separates the points of X. The problem is to find conditions under which for every sequence or net $(L_i)_{i\in I}$ of positive linear maps $L_i$ : $\mathscr{C}(X) \to \mathscr{C}(X)$, uniform convergence of $(L_i f)$ to $f$ for all $f$ in $\mathscr{F}$ or, equivalently, for all functions $f$ in the linear subspace $\mathscr{H}$ of $\mathscr{C}(X)$ spanned by $\mathscr{F}$ implies uniform convergence of $(L_i f)$ to $f$ for all $f \in \mathscr{C}(X)$. It follows from the work of Šaškin and Wulbert that the condition " X is the Choquet boundary of X with respect to $\mathscr{H}$ " is both necessary and sufficient. Wulbert shows in addition that the operators $L_i : \mathscr{C}(X) \to \mathscr{C}(X)$ may be assumed to be only linear and of norm $\leqslant 1$. In the classical situation of Korovkin the above condition can be easily verified.

For positive linear maps $L_i$ of $\mathscr{C}(X)$ into the space $\mathscr{B}(X)$ of all bounded real-valued functions on X, also the more general problem of characterizing the Korovkin closure $\mathscr{H}^k$ of $\mathscr{H}$ in $\mathscr{C}(X)$ was treated by Baskakov [1] and Šaškin [10]. The Korovkin closure $\mathscr{H}^k$ is by definition the set of all functions $f$ in $\mathscr{C}(X)$ such that $(L_i f)$ converges uniformly to $f$ for all nets of positive linear maps $L_i : \mathscr{C}(X) \to \mathscr{B}(X)$ provided that $(L_i h)$ converges uniformly to $h$ for all $h$ in $\mathscr{H}$. Šaškin [10, 12] observed that this problem is closely related to the author's work [2] on an abstract Dirichlet problem and that $\mathscr{H}^k$ coincides with the space $\hat{\mathscr{H}}$ of all $\mathscr{H}$-harmonic functions introduced in [2].

The present paper has mainly two intentions. First, in all known proofs of Korovkin-type theorems linearity of the operators $L_i$ is used in an essential way. We will show that the methods developed in [2] lead in a very simple and natural way to generalizations of Korovkin's theorem for increasing, possibly non-linear maps $L_i : \mathscr{C}(X) \to \mathscr{B}(X)$. These generalizations reproduce with new and extremely easy proofs all the previously known results for positive linear maps. Secondly, we will show that all these generalizations can be obtained for locally compact (Hausdorff) spaces. The key to this second direction of study is Choquet's notion of an adapted vector space of continuous real-valued functions. In view of the results of Mokobodzki-Sibony [9] on adapted cones of continuous functions and their fine boundary, this second possibility of extending Korovkin's theorem seems to be quite natural.

The paper is split up into three sections. § 1 shows that Korovkin-type theorems can be obtained with our methods on arbitrary topological spaces without any use of measure theory. § 2 treats the special case of adapted spaces and shows how a limited amount of measure theory connects our results with older ones. § 3 is concerned with characterizations of the Korovkin closure. Special emphasis is given to the different role of increasing (or positive linear) maps of $\mathscr{C}(X)$ into $\mathscr{B}(X)$ and of maps with values in $\mathscr{C}(X)$ for compact spaces X and, with certain previsions, for locally compact spaces X. Some of our results seem to be new even for compact spaces and positive linear maps.

For the reader's convenience the paper is written in a self-contained manner. The author is grateful to Dr. A. Lupaş, Cluj, for having attracted his attention to the work of Šaškin.

## 1. $\mathscr{H}$-affine functions and Korovkin-type theorems.

In this section we denote by X a topological space and by $\mathscr{H}$ a linear subspace of the linear space $\mathscr{C}(X) = \mathscr{C}(X, \mathbf{R})$ of all real-valued continuous functions on X.

A function $f\colon X \to \mathbf{R}$ will be called *$\mathscr{H}$-bounded* if $f$ is minorized and majorized by functions $h_1, h_2 \in \mathscr{H}$, i.e. $h_1 \leqslant f \leqslant h_2$. The set $\mathscr{H}^b$ of all $\mathscr{H}$-bounded functions on X is a linear subspace of the linear space $\mathbf{R}^X$ of all real-valued functions on X. A linear subspace of $\mathscr{H}^b$ is

$$\mathscr{H}_0 := \mathscr{H}^b \cap \mathscr{C}(X). \qquad (1.1)$$

For every $\mathscr{H}$-bounded function $f$ we define the two envelopes

$$\hat{f} := \inf \{h \in \mathscr{H} | h \geqslant f\} \qquad (1.2)$$

and

$$\check{f} := \sup \{h \in \mathscr{H} | h \leqslant f\} = -\widehat{-f}. \qquad (1.3)$$

Then $\hat{f}$ and $\check{f}$ are upper resp. lower semi-continuous $\mathscr{H}$- bounded functions satisfying $\check{f} \leqslant f \leqslant \hat{f}$. For $f, g \in \mathscr{H}^b$ we have

$$\widehat{f+g} \leqslant \hat{f} + \hat{g} \qquad \text{and} \qquad \widehat{\lambda f} = \lambda \hat{f},\ \lambda \in \mathbf{R}_+. \qquad (1.4)$$

Hence the set

$$\hat{\mathscr{H}} := \{f \in \mathscr{H}^b | \check{f} = \hat{f}\} = \{f \in \mathscr{H}_0 | \check{f} = \hat{f}\} \qquad (1.5)$$

is a linear subspace of $\mathscr{H}_0$ satisfying

$$\mathscr{H} \subset \hat{\mathscr{H}} \subset \mathscr{H}_0 \subset \mathscr{H}^b. \tag{1.6}$$

The functions in $\hat{\mathscr{H}}$ will be called $\mathscr{H}$-*affine* (or, as in [2], $\mathscr{H}$-harmonic) functions on X. They can be characterized as follows:

**1.1** Lemma. — *A function* $f: X \to \mathbf{R}$ *is* $\mathscr{H}$-*affine if and only if for every quasi-compact set* $K \subset X$ *and for every* $\varepsilon > 0$ *there exist finitely many functions* $h'_1, \ldots, h'_n$ *and* $h''_1, \ldots, h''_n$ *in* $\mathscr{H}$, $n \in \mathbf{N}$, *satisfying*

$$h'_j \leqslant f \leqslant h''_j \quad (j = 1, \ldots, n) \tag{1.7}$$

*and*

$$\overline{h}(x) - \underline{h}(x) < \varepsilon \quad \textit{for all} \ \ x \in K, \tag{1.8}$$

*where*

$$\overline{h} := \inf(h''_1, \ldots, h''_n) \quad \textit{and} \quad \underline{h} := \sup(h'_1, \ldots, h'_n). \tag{1.9}$$

*Proof.* — For $f \in \hat{\mathscr{H}}$ the existence of $h'_1, \ldots, h'_n$ and $h''_1, \ldots, h''_n$ can be proved as in [2; p. 102]. Conversely, for a singleton $K = \{x\}$ the conditions (1.7) and (1.8) imply the existence of functions $h', h'' \in \mathscr{H}$ such that $h' \leqslant f \leqslant h''$ and $h''(x) - h'(x) < \varepsilon$ for given $\varepsilon > 0$. Therefore,

$$\hat{f}(x) - \check{f}(x) < \varepsilon,$$

i.e. $f$ is $\mathscr{H}$-affine.

We will now study the behavior of $\mathscr{H}$-affine functions in settings similar to the classical Korovkin theorem. In the remaining part of this section $(L_\iota)_{\iota \in I}$ will denote a *net* of *increasing* maps $L_\iota : H_0 \to \mathbf{R}^X$. Hence $L_\iota$ maps each $f \in \mathscr{H}_0$ into a real-valued function $L_\iota f$ on X such that for $f_1$, $f_2 \in \mathscr{H}_0$ the relation $f_1 \leqslant f_2$ implies $L_\iota f_1 \leqslant L_\iota f_2$. I is the directed index set of the net. More generally, I can be an arbitrary index set with a given filter in it.

**1.2** Proposition. — *Let* $x_0$ *be a point in* X. *If*

$$\lim_{\iota \in I} L_\iota h(x_0) = h(x_0) \quad \textit{for all} \ \ h \in \mathscr{H},$$

*then*

$$\check{f}(x_0) \leqslant \liminf_{i \in I} L_i f(x_0) \leqslant \limsup_{i \in I} L_i f(x_0) \leqslant \hat{f}(x_0)$$

*for all* $f \in \mathscr{H}_0$.

*Proof.* — Consider functions $g, h \in \mathscr{H}$ satisfying $g \leqslant f \leqslant h$. Since every $L_i$ is increasing we obtain

$$L_i g(x_0) \leqslant L_i f(x_0) \leqslant L_i h(x_0)$$

for all $i \in I$. This implies

$$g(x_0) = \lim L_i g(x_0) \leqslant \liminf L_i f(x_0) \leqslant \limsup L_i f(x_0) \leqslant \lim L_i h(x_0) = h(x_0);$$

from this and the definition of $\hat{f}$ and $\check{f}$ the result follows.

The applications which we intend to give suggest the following terminology: The net $(L_i)_{i \in I}$ will be said to *converge* pointwise on X (resp. uniformly on the quasi-compact subsets of X) (1) *to the identity on a subset* $\mathscr{P}$ *of* $\mathscr{H}_0$ if

$$\lim_{i \in I} L_i p(x) = p(x) \tag{1.10}$$

for all $p \in \mathscr{P}$ and all $x \in X$ (resp. if for all $p \in \mathscr{P}$ the convergence (1.10) is uniform on every quasi-compact subset K of X).

With the definition of $\mathscr{H}$-affine functions we thus obtain:

**1.3** Corollary. — *The net* $(L_i)_{i \in I}$ *converges pointwise on* X *to the identity on* $\hat{\mathscr{H}}$ *if it converges pointwise to the identity on* $\mathscr{H}$.

We now replace pointwise convergence by uniform convergence on quasi-compact subsets of X.

**1.4** Proposition. — *Let* K *be a quasi-compact subset of* X, *and suppose that* $(L_i h)_{i \in I}$ *converges uniformly on* K *to* $h$ *for all* $h \in \mathscr{H}$. *Then* $(L_i f)_{i \in I}$ *converges uniformly on* K *to* $f$ *for all functions* $f \in \mathscr{H}_0$ *satisfying* $\check{f}(x) = \hat{f}(x)$ *for all* $x \in K$.

(1) For a locally compact (Hausdorff) space X, uniform convergence on the compact subsets of X will be also called " locally uniform convergence on X ".

*Proof.* — Let $f$ be a function in $\mathscr{H}_0$ satisfying $\check{f} = \hat{f}$ on K. As in [2; p. 102] one proves for every $\varepsilon > 0$ the existence of functions $h'_1, \ldots, h'_n$ and $h''_1, \ldots, h''_n$ in $\mathscr{H}$, $n \in \mathbf{N}$, such that $h'_j \leqslant f \leqslant h''_j$, $j = 1, \ldots, n$, and

$$\bar{h}(x) - \underline{h}(x) < \varepsilon$$

for all $x \in K$. Here $\underline{h}$ and $\bar{h}$ are defined as in (1.9). By assumption there exists a section $S = \{i \in I | i \geqslant i_0\}$ of I such that for all $i \in S$, all $j = 1, \ldots, k$ and for all $x \in K$ we have

$$|L_i h'_j(x) - h'_j(x)| \leqslant \varepsilon$$

and

$$|L_i h''_j(x) - h''_j(x)| \leqslant \varepsilon.$$

Furthermore, $L_i h'_j \leqslant L_i f \leqslant L_i h''_j$ for all $i \in I$ and $j = 1, \ldots, n$ since all $L_i$ are increasing maps. From this we conclude

$$h'_j(x) - \varepsilon \leqslant L_i h'_j(x) \leqslant L_i f(x) \leqslant L_i h''_j(x) \leqslant h''_j(x) + \varepsilon$$

for $i \in S, j = 1, \ldots, n$, and $x \in K$; hence

$$\underline{h}(x) - \varepsilon \leqslant L_i f(x) \leqslant \bar{h}(x) + \varepsilon$$

holds on K for all $i \in S$. Together with $\underline{h} \leqslant f \leqslant \bar{h}$ this leads to

$$|L_i f(x) - f(x)| \leqslant \bar{h}(x) - \underline{h}(x) + \varepsilon < 2\varepsilon$$

for all $x \in K$ and $i \in S$. This proves the uniform convergence of $(L_i f)$ on K to $f$.

1.5 Corollary. — *The net* $(L_i)_{i \in I}$ *converges uniformly on the quasi-compact subsets of* X *to the identity on* $\hat{\mathscr{H}}$ *if it converges uniformly on the quasi-compact subset of* X *to the identity on* $\mathscr{H}$.

## 2. Case of an adapted space $\mathscr{H}$.

From now on X will be a *locally compact* (Hausdorff) space. Furthermore, the linear subspace $\mathscr{H}$ of $\mathscr{C}(X)$ will be assumed to be *adapted* in the sence of Choquet [3; p. 274].

This means by definition that $\mathscr{H}$ has the following three properties:

$$\mathscr{H} = \mathscr{H}^+ - \mathscr{H}^+; \tag{2.1}$$

for all $x \in X$ there is an $f \in \mathscr{H}^+$ such that $f(x) > 0$; (2.2)

$$\mathscr{H}^+ \subset o(\mathscr{H}^+). \tag{2.3}$$

Recall that $\mathscr{H}^+ = \{h \in \mathscr{H} | h \geqslant 0\}$ and that a real-valued function $f$ on X is said to be *dominated* by a real-valued function $g \geqslant 0$ on X if for any $\varepsilon > 0$ there exists a compact set $K \subset X$ such that $|f(x)| \leqslant \varepsilon g(x)$ holds on $X \setminus K$. We will then write $f \in o(g)$. By $o(\mathscr{H}^+)$ we denote the set $\bigcup_{h \in \mathscr{H}^+} o(h)$. Therefore (2.3) states that every function in $\mathscr{H}^+$ is dominated by some other function in $\mathscr{H}^+$. Observe that $o(\mathscr{H}^+)$ is a linear subspace of $\mathbf{R}^X$. Therefore (2.1) and (2.3) imply $\mathscr{H} \subset o(\mathscr{H}^+)$.

Under these assumptions we have

$$\mathscr{H}^b = \{f \in \mathbf{R}^X | \ |f| \leqslant h \text{ for some } h \in \mathscr{H}^+\}; \tag{2.4}$$

$$\mathscr{H}^b \subset o(\mathscr{H}^+); \tag{2.5}$$

$$\mathscr{H}_0 = o(\mathscr{H}^+) \cap \mathscr{C}(X); \tag{2.6}$$

$$\mathscr{C}_c(X) \subset \mathscr{H}_0, \tag{2.7}$$

where $\mathscr{C}_c(X)$ is the linear space of all $f \in \mathscr{C}(X)$ with *compact support*. — Obviously (2.4) follows from (2.1) since $\mathscr{H}^+$ is a convex cone. (2.5) is a consequence of (2.3) and (2.4). Since $\mathscr{H}_0 = \mathscr{H}^b \cap \mathscr{C}(X)$ it suffices to prove that

$$o(\mathscr{H}^+) \cap \mathscr{C}(X) \subset \mathscr{H}_0$$

in order to obtain (2.6). Every $f \in o(\mathscr{H}^+) \cap \mathscr{C}(C)$ is dominated by some $h \in \mathscr{H}^+$. Therefore, $|f(x)| \leqslant h(x)$ holds on the complement of a compact set $K \subset X$. From (2.2) follows the existence of a function $g \in \mathscr{H}^+$ which is strictly positive on K. The continuous function $f$ is bounded on K, hence $|f(x)| \leqslant \lambda g(x)$ on K for some $\lambda > 0$. But then $|f| \leqslant \lambda g + h$ on X, that is $f \in \mathscr{H}_0$. (2.7) is an immediate consequence of (2.6) since every $f \in \mathscr{C}_c(X)$ is dominated by even every $h \in \mathscr{H}^+$.

**2.1** *Remark.* — All our assumptions are satisfied when X is compact and $\mathscr{H}$ is a linear subspace of $\mathscr{C}(X)$ containing the constant functions. In this case $\mathscr{H}_0 = \mathscr{C}(X)$ and $\mathscr{H}^b$ is the space of all bounded real-valued functions on X. Some of the results in [2] are now extended to the more general situation of this section.

For a point $x \in X$ a positive Radon measure $\mu$ on X will be called an *$\mathscr{H}$-representing measure* if all functions in $\mathscr{H}$ are $\mu$-integrable and if $\int h\, d\mu = h(x)$ holds for all $h \in \mathscr{H}$. The set of all $\mathscr{H}$-representing measures for $x$ will be denoted by $\mathscr{M}_x = \mathscr{M}_x(\mathscr{H})$. For $\mu \in \mathscr{M}_x$ all functions in $\mathscr{H}_0$ are $\mu$-integrable by (2.4).

**2.2** Lemma. — *For all $x \in X$ and all $f \in \mathscr{H}_0$ the set*

$$\left\{\int f\, d\mu \,|\, \mu \in M_x\right\}$$

*is the compact interval* $[\check{f}(x), \hat{f}(x)]$.

*Proof* (compare with [2; p. 101] and [9; p. 5-18]). — Obviously $\int f\, d\mu \in [\check{f}(x), \hat{f}(x)]$ for all $\mu \in \mathscr{M}_x$. Conversely, let $\gamma$ be a number in $[\check{f}(x), \hat{f}(x)]$. On $\mathscr{H}_0$ the map $g \longmapsto \hat{g}(x)$ is a sublinear functional $p$. This functional majorizes the linear form $\lambda f \longmapsto \lambda\gamma$ defined on the linear subspace $\mathbf{R}f$ of $\mathscr{H}_0$ generated by $f$. The Hahn-Banach theorem hence implies the existence of a linear form T on $\mathscr{H}_0$ satisfying $T \leqslant p$ and $T(f) = \gamma$. T is positive because of the first condition: $g \in \mathscr{H}_0$, $g \leqslant 0$ implies $T(g) \leqslant p(g) = g(x) \leqslant 0$. It follows from (2.6) that together with $\mathscr{H}$ also $\mathscr{H}_0$ is an adapted linear subspace of $\mathscr{C}(X)$. Therefore, by Choquet [3; p. 276] there exists a positive Radon measure $\mu$ on X such that $\mathscr{H}_0 \subset \mathscr{L}^1(\mu)$ and $T(g) = \int g\, d\mu$ for all $g \in \mathscr{H}_0$. Then $\mu \in \mathscr{M}_x$ and $\int f\, d\mu = \gamma$ since

$$\int h\, d\mu = T(h) \leqslant p(h) = \hat{h}(x) = h(x)$$

for all $h \in \mathscr{H}$ implies $\int h\, d\mu = h(x)$ for all $h \in \mathscr{H}$.

**2.3** Corollary 1. — *A function $f\colon X \to \mathbf{R}$ is $\mathscr{H}$-affine if and only if $f \in \mathscr{H}_0$ and $\int f\, d\mu = f(x)$ for all $\mu \in \mathscr{M}_x$.*

Indeed, for $x \in X$ and $f \in \mathscr{H}_0$, $\check{f}(x) = \hat{f}(x)$ is equivalent to $\{f(x)\} = \left\{\int f\,d\mu : \mu \in M_x\right\}$ by the preceding lemma.

We are now able to connect propositions 1.2 and 1.4 with the notion of the Choquet boundary. Recall that

$$\partial_{\mathscr{H}}X := \{x \in X : \mathscr{M}_x = \{\varepsilon_x\}\} \tag{2.8}$$

is the Choquet boundary of $X$ with respect to $\mathscr{H}$.
From lemma 2.2. we then deduce:

**2.4** Corollary 2. — *A point $x \in X$ lies in the Choquet boundary $\partial_{\mathscr{H}}X$ if and only if $\check{f}(x) = \hat{f}(x)$ — or equivalently, $\hat{f}(x) = f(x)$ — holds for all $f \in \mathscr{H}_0$.*

*Proof.* — $\check{f}(x) = \hat{f}(x)$ for all $f \in \mathscr{H}_0$ is equivalent to

$$\int f\,d\mu = f(x)$$

for all $\mu \in \mathscr{M}_x$. Since, by (2.7), $\mathscr{C}_c(X) \subset \mathscr{H}_0$ this means that $\mathscr{M}_x$ reduces to $\{\varepsilon_x\}$. The rest of the statement follows from $-\check{f} = \widehat{-f}$ and $\check{f} \leqslant f \leqslant \hat{f}$.
In particular, we have (cf. [2]):

**2.5** Corollary 3. — $\partial_{\mathscr{H}}X = X$ *if and only if* $\hat{\mathscr{H}} = \mathscr{H}_0$.

**2.6** *Remark.* — By a result of Mokobodzki-Sibony [9; p. 5-26] $\partial_{\mathscr{H}}X$ is non-empty if $X$ is countable at infinity and if in addition $\mathscr{H}$ is linearly separating, i.e. for every pair of points $x \neq y$ of $X$ there exist functions $h_1, h_2 \in \mathscr{H}$ satisfying

$$h_1(x)\,h_2(y) \neq h_1(y)\,h_2(x).$$

Furthermore, $h \geqslant 0$ on $\partial_{\mathscr{H}}X$ implies $h \geqslant 0$ (on $X$) for all $h \in \mathscr{H}$.

**2.7** Theorem. — *Let $(L_i)_{i \in I}$ be a net of increasing maps $L_i : \mathscr{H}_0 \to \mathbf{R}^X$, and let $K$ be a compact subset of $\partial_{\mathscr{H}}X$. Assume that $(L_i h)_{i \in I}$ converges uniformly on $K$ to $h$ for all $h \in \mathscr{H}$. Then $(L_i f)_{i \in I}$ converges uniformly on $K$ to $f$ for all $f \in \mathscr{H}_0$.*

This is an immediate consequence of proposition 1.4 and corollary 2.4. By choosing a singleton for K it follows in particular, that pointwise convergence on $\partial_{\mathscr{H}}X$ of $(L_i h)$ to $h$ for all $h \in \mathscr{H}$ implies pointwise convergence on $\partial_{\mathscr{H}}X$ of $(L_i f)$ to $f$ for all $f \in \mathscr{H}_0$.

2.8 Corollary. — *Let $(L_i)_{i \in I}$ be a net of increasing maps $L_i : \mathscr{H}_0 \to \mathbf{R}^X$ and assume that $\partial_{\mathscr{H}}X = X$. Then pointwise (resp. locally uniform) convergence on X of the net $(L_i)$ to the identity on $\mathscr{H}$ implies pointwise (resp. locally uniform) convergence on X of the net $(L_i)$ to the identity on $\mathscr{H}_0$.*

We finally prove a Korovkin-type result which in its formulation is closest to the formulation of the classical Korovkin theorem. To simplify the statement of the result we define for a non-empty set $\mathscr{F} \subset \mathscr{C}(X)$

$$\mathscr{F}^\nu := \{f^\nu | f \in \mathscr{F}\} \quad \text{for } \nu = 0, 1, 2.$$

Hence $\mathscr{F}^0$ only contains the constant function 1, $\mathscr{F}^1 = \mathscr{F}$, and $\mathscr{F}^2$ is the set of all squares of functions in $\mathscr{F}$.

2.9 Proposition. — *Let $\mathscr{F}$ be a subset of $\mathscr{C}(X)$ separating the points of X and satisfying $\mathscr{F}^0 \cup \mathscr{F}^1 \subset o(\mathscr{F}^+)$. Consider a net $(L_i)_{i \in I}$ of positive linear maps of $\tilde{\mathscr{F}}$ into $\mathbf{R}^X$ where $\tilde{\mathscr{F}}$ is the linear space of all functions $f \in \mathscr{C}(X)$ such that $|f|$ is majorized by a linear combination of functions in $\mathscr{F}^0 \cup \mathscr{F}^1 \cup \mathscr{F}^2$. Assume that $(L_i f)$ converges on X pointwise* (resp. *locally uniformly*) *to $f$ for all functions $f \in \mathscr{F}^0 \cup \mathscr{F}^1 \cup \mathscr{F}^2$. Then $(L_i f)$ converges on X pointwise* (resp. locally uniformly) *to $f$ for all functions $f \in \tilde{\mathscr{F}}$, in particular for all bounded functions $f \in \mathscr{C}(X)$.*

*Proof.* — Consider the linear subspace $\mathscr{H}$ of $\mathscr{C}(X)$ generated by $\mathscr{F}^0 \cup \mathscr{F}^1 \cup \mathscr{F}^2$. Then $\mathscr{H}$ is adapted : By assumption, for every $f \in \mathscr{F}$ there exists a function $g \in \mathscr{F}^+$ such that $|f(x)| \leq g(x)$ holds in the complement $X \setminus K$ of a compact set $K \subset X$. Let $\alpha > 0$ be an upper bound for $f$ on K. Then $|f| \leq \alpha + g$ on X, and the equality

$$f = (f + \alpha + g) - (\alpha + g)$$

shows that $f$ is the difference of non-negative functions in the linear span of $\mathscr{F}^0 \cup \mathscr{F}^1$, hence of functions in $\mathscr{H}^+$. From the definition of $\mathscr{H}$ then follows $\mathscr{H} = \mathscr{H}^+ - \mathscr{H}^+$. Furthermore, $\mathscr{F} \subset o(\mathscr{F}^+)$ implies $\mathscr{F}^2 \subset o(\mathscr{F}^2) \subset o(\mathscr{H}^+)$. Since

$$\mathscr{F}^0 \cup \mathscr{F}^1 \subset o(\mathscr{F}^+) \subset o(\mathscr{H}^+),$$

we thus have $\mathscr{H} \subset o(\mathscr{H}^+)$. Property (2.2) of an adapted space follows trivially from $1 \in \mathscr{H}$.

The definition of $\tilde{\mathscr{F}}$ now shows that $\tilde{\mathscr{F}}$ is the space $\mathscr{H}_0$ derived from the adapted space $\mathscr{H}$. Because of the linearity of the maps $L_i : \mathscr{H}_0 \rightarrow \mathbf{R}^X$, our assumption implies pointwise (resp. locally uniform) convergence of $(L_i f)$ to $f$ for all functions $f$ in $\mathscr{H}$. Thus the statement of proposition 2.9 is a consequence of corollary 2.8 if $\partial_{\mathscr{H}} X = X$ holds. Therefore we consider $x \in X$ and $\mu \in \mathscr{M}_x$. For every $f \in \mathscr{F}$ the function $(f - f(x))^2$ is in $\mathscr{H}$ and vanishes at $x$. Hence

$$\int (f - f(x))^2 \, d\mu = 0$$

which implies that the support $S\mu$ of $\mu$ is contained in $\{f = f(x)\} := \{x' \in X | f(x') = f(x)\}$, i.e.

$$S\mu \subset \bigcup_{f \in \mathscr{F}} \{f = f(x)\}.$$

This intersection equals $\{x\}$ since $\mathscr{F}$ separates points. We have $\int d\mu = 1$ because of $1 \in \mathscr{H}$. Hence $\mu$ must be the unit mass $\varepsilon_x$ at $x$. This proves $\partial_{\mathscr{H}} X = X$.

**2.10.** *Remarks.* — 1) Let X be the unit interval [0, 1] and let $\mathscr{F}$ consist of the single function $x \longmapsto x$. Then the above proposition yields to the classical Korovkin theorem [7, 8].

2) For a compact space X all our assumptions about $\mathscr{F} \subset \mathscr{C}(X)$ reduce itself to a single one: $\mathscr{F}$ has to separate points. Under the assumption that each $L_i$ is a positive linear map of $\mathscr{C}(X)$ into $\mathscr{C}(X)$ this particular form of proposition 2.9 can be found implicitely in Felbecker-Schempp [4] and explicitely in Grossman [6]. However, the proof is totally different.

3) In the situation of remark 2.1 and under the assumption that each $L_i$ is a positive linear map of $\mathscr{C}(X)$ into the linear space $\mathscr{B}(X)$ of bounded real-valued functions, theorem 2.7 and its corollary are contained in a result of Wulbert [15]. However, Wulbert also treats nets of linear maps $L_i$: $\mathscr{C}(X) \to \mathscr{B}(X)$ with norm $\|L_i\| \leqslant 1$.

4) Corollary 2.8 can be applied in the following situation: $X = \mathbf{R}^n$ and $\mathscr{H}$ = linear space of all polynomials on $\mathbf{R}^n$ with real coefficients. Then $\mathscr{H}$ is adapted [3; p. 285] and $\partial_{\mathscr{H}} X = X$. Indeed, for $x = (x_1, \ldots, x_n) \in \mathbf{R}^n$ consider the polynomial $y \longmapsto p(y) = \sum_{i=1}^{n} (y_i - x_i)^2$ on $\mathbf{R}^n$; for every $\mu \in \mathscr{M}_x$ we must have $0 = p(x) = \int p \, d\mu_x$ and hence $\mu = \varepsilon_x$ since $S\mu = \{x\}$ and $1 \in \mathscr{H}$. Obviously, $\mathscr{H}_0$ is here the space of all $f \in \mathscr{C}(\mathbf{R}^n)$ satisfying

$$|f(x)| \leqslant \alpha |x_1|^p \cdot \ldots \cdot |x_n|^p + \beta \quad \text{for all } x \in \mathbf{R}^n,$$

where $\alpha, \beta \geqslant 0$ and $p \in \mathbf{N}_0$ are suitably chosen.

5) In proposition 1.4 uniform convergence on K of $(L_i f)$ to $f$ is proved for all functions $f \in \mathscr{H}_0$ satisfying $\check{f}(x) = \hat{f}(x)$ for all $x \in K$. In view of lemma 2.2 the latter condition states that $\int f \, d\mu = f(x)$ holds for all $x \in K$ and all $\mu \in \mathscr{M}_x$. The set $\hat{\mathscr{H}}_K$ of these functions $f \in \mathscr{H}_0$ is a linear subspace of $\mathscr{H}_0$. In the terminology of Franchetti [5] which is developed in the framework of remark 2.1, $\hat{\mathscr{H}}_K$ would be the maximal linear subspace of $\mathscr{H}_0$ such that K is contained in the relative Choquet boundary of $\mathscr{H}$ with respect to $\hat{\mathscr{H}}_K$. Thus, Franchetti's main result (for positive linear operators) is a special case of proposition 1.4.

### 3. Characterization of the Korovkin closure.

Our assumptions about X and $\mathscr{H}$ will be the same as in preceding paragraph: X is a locally compact space and $\mathscr{H}$ is an adapted linear subspace of $\mathscr{C}(X)$. Corollary 1.3 states that every $\mathscr{H}$-affine function, i.e. every function $f \in \hat{\mathscr{H}}$, has the following *Korovkin property:* For every net $(L_i)_{i \in I}$

of increasing maps $L_i : \mathscr{H}_0 \to \mathbf{R}^X$ pointwise convergence on X to the identity on $\mathscr{H}$ implies that the net $(L_i f)_{i\in I}$ converges pointwise on X to $f$.

We will now prove that the $\mathscr{H}$-affine functions are the only functions in $\mathscr{H}_0$ having this Korovkin property. In view of corollary 2.3 this is an almost obvious observation which, in the situation of remark 2.1 and for nets of positive linear maps $L_i$, has been made already by Baskakov [1] and Šaškin [10]. In fact, the following stronger statement holds:

**3.1** Proposition. — *Consider the set* **P** *of all positive linear maps* $L : \mathscr{H}_0 \to \mathscr{H}^b$ *which leave invariant all function* $h \in \mathscr{H}$, i.e. $Lh = h$. *Then* $\hat{\mathscr{H}}$ *is the set of all functions* $f \in \mathscr{H}_0$ *which are left invariant by all* $L \in \mathbf{P}$.

*Proof.* — By corollary 1.3, a function $f \in \hat{\mathscr{H}}$ is left invariant even by all increasing maps $L : \mathscr{H}_0 \to \mathbf{R}^X$ which leave invariant all functions in $\mathscr{H}$. Conversely, let $f \in \mathscr{H}_0$ be a function which is left invariant by all $L \in \mathbf{P}$. Let us choose for each $x \in X$ an arbitrary measure $\mu_x$ in $\mathscr{M}_x$. For $q \in \mathscr{H}_0$ define $Lq$ as the function $x \longmapsto \int q\, d\mu_x$ defined on X. Then $Lq \in \mathscr{H}^b$ since $|q| \leqslant h$ for some $h \in H$ implies $|Lq(x)| = \left|\int q\, d\mu_x\right| \leqslant \int h\, d\mu_x = h(x)$ for all $x \in X$. Therefore, L is a positive linear map of $\mathscr{H}_0$ in $\mathscr{H}^b$ which by definition leaves invariant all $h \in \mathscr{H}$. Hence $Lf = f$, i.e. $\int f\, d\mu_x = f(x)$ for all $x$, and since $\mu_x$ was arbitrarily chosen in $\mathscr{M}_x$, for all $\mu_x \in \mathscr{M}_x$. Then by corollary 2.3 $f$ is $\mathscr{H}$-affine.

The preceding proposition states in other words that, in order to characterize $\mathscr{H}$-affine functions by the Korovkin-property, it suffices to consider only *constant* nets of positive linear maps of $\mathscr{H}_0$ into $\mathscr{H}^b$.

The following example shows that one cannot replace $\mathscr{H}^b$ by $\mathscr{H}_0$.

*Example* (cf. Scheffold [14]). — Consider $X = [0, 1]$ and

$$\mathscr{H} = \left\{ h \in \mathscr{C}([0, 1]) \,\middle|\, \int_0^1 h(x)\, dx = h(1) \right\}.$$

For each $x_0 \in [0, 1[$ there exist (e.g. piecewise linear) functions $h \in \mathscr{H}$ satisfying $h(x) > h(x_0) = 0$ for all $x \neq x_0$ in $[0, 1]$. This proves that $\partial_{\mathscr{H}} X = [0, 1[$ since, by definition, the Lebesgue measure on $[0, 1]$ is a representing measure for $x_0 = 1$.

Corollary 2.3 and the definition of $\mathscr{H}$ yield $\hat{\mathscr{H}} = \mathscr{H}$; furthermore, $\mathscr{H}_0 = \mathscr{C}(X)$. Now let $L: \mathscr{C}([0, 1]) \to \mathscr{C}([0, 1])$ be an arbitrary positive linear map such that $Lh = h$ for all $h \in \mathscr{H}$. Then $f \longmapsto Lf(x)$, $f \in \mathscr{C}([0, 1])$, is an $\mathscr{H}$-representing measure $\mu_x$ for every $x$, hence $\mu_x = \varepsilon_x$ for all $x \in [0, 1[$, and, by continuity, for all $x \in [0, 1]$. This proves that $Lf = f$ holds even for *all* $f \in \mathscr{C}([0, 1])$.

However, one can replace $\mathscr{H}^b$ by $\mathscr{H}_0$ if one continues to work with *nets* of increasing maps $L_i : \mathscr{H}_0 \to \mathscr{H}_0$.

3.2 DEFINITION. — *Korovkin closure of $\mathscr{H}$ will be called the set of all functions $f \in \mathscr{H}_0$ having the following "continuous Korovkin property": For every net $(L_i)_{i \in I}$ of positive linear maps $L_i : \mathscr{H}_0 \to \mathscr{H}_0$ which converges pointwise on X to the identity on $\mathscr{H}$, the net $(L_i f)_{i \in I}$ converges pointwise on X to $f$.*

The adapted space $\mathscr{H}$ will be called a *Korovkin space* if $\mathscr{H}_0$ is its Korovkin closure.

3.3 THEOREM. — *$\hat{\mathscr{H}}$ is the Korovkin closure of $\mathscr{H}$.*

*Proof.* — Consider a function $f \in \mathscr{H}_0$ having the continuous Korovkin property. In view of corollary 2.3, we have to prove $\int f\, d\mu_0 = f(x_0)$ for every point $x_0 \in X$ and every measure $\mu_0 \in \mathscr{M}_{x_0}$. We choose a neighborhood base $\mathfrak{V}$ of $x_0$ consisting of open relatively compact neighborhoods of $x_0$. For every $V \in \mathfrak{V}$ denote by $q_V$ a function in $\mathscr{C}_c(X)$ satisfying $0 \leqslant q_V \leqslant 1$, $q_V(x_0) = 1$, $q_V(x) = 0$ for all $x \in \complement V$. For $V \in \mathfrak{V}$ and $g \in \mathscr{H}_0$ we define

$$L_V g := \left(\int g\, d\mu_0\right) q_V + g - g q_V.$$

Then $L_V g$ is a function in $\mathscr{H}_0$ since $\mathscr{C}_c(X) \subset \mathscr{H}_0$. $L_V : \mathscr{H}_0 \to \mathscr{H}_0$ is a positive linear map such that the net $(L_V h)_{V \in \mathfrak{V}}$ converges pointwise to $h$ for every $h \in \mathscr{H}$. Therefore, by assumption, $\lim_{V \in \mathfrak{V}} L_V f(x_0) = f(x_0)$. But this proves

$$\int f \, d\mu_0 = f(x_0)$$

since $L_V f(x_0) = \int f \, d\mu_0$ for all $V \in \mathfrak{V}$. — Conversely, by corollary 1.3, all functions $f \in \hat{\mathscr{H}}$ are in the Korovkin closure of $\mathscr{H}$.

3.4 Corollary. — *$\mathscr{H}$ is a Korovkin space if and only if $\partial_{\mathscr{H}} X = X$.*

*Proof.* — This is an immediate consequence of corollary 3.5 and of the above theorem.

3.5 *Remarks.* — 1) Our proof of theorem 3.3 shows that it suffices to use sequences $(L_n)_{n \in \mathbf{N}}$ in the definition of the Korovkin closure if the space X is metrizable.

2) For compact spaces X and finitely generated spaces $\mathscr{H}$, corollary 3.4 was obtained by different methods by Šaškin [11, 12].

3) In the definition of the Korovkin closure and of Korovkin spaces one can replace pointwise convergence by locally uniform convergence on X without changing the results 3.3 and 3.4. This follows from corollary 1.5 and the observation that the net $(L_V)$ from the proof of theorem 3.3 converges even uniformly on X to the identity on $\mathscr{H}$.

## BIBLIOGRAPHY

[1] V. A. Baskakov, Some convergence conditions for linear positive operators, *Uspehi Mat. Nauk*, 16 (1961), 131-135 (Russian).

[2] H. Bauer, Šilovscher Rand und Dirichletsches Problem. *Ann. Inst. Fourier*, 11 (1961), 89-136.

[3] G. Choquet, Lectures on Analysis, Vol. II (Representation Theory), Benjamin, New York-Amsterdam, 1969.

[4] G. Felbecker and W. Schempp, A generalization of Bohman-Korovkin's theorem, *Math. Zeitschrift*, 122 (1971), 63-70.

[5] G. Franchetti, Convergenza di operatori in sottospazi dello spacio C(Q), *Boll. d. Un. Matem. Ital.*, Ser. IV, 3 (1970), 668-675.

[6] M. V. Grossman, Note on a generalized Bohman-Korovkin theorem (to appear in J. of Math. Anal. and Appl.).

[7] P. P. Korovkin, On convergence of linear positive operators in the space of continuous functions, *Doklady Akad. Nauk* SSSR (N.S.), 90 (1953), 961-964.

[8] P. P. Korovkin, Linear operators and approximation theory, *Hindustan Publ. Corp.*, Delhi, India, 1960.

[9] G. Mokobodzki et D. Sibony, Cônes adaptés de fonctions continues et théorie du potentiel. Séminaire Choquet, Initiation à l'Analyse, 6e année (1966/1967), Fasc. 1, 35 p., Institut H.-Poincaré, Paris, 1968.

[10] Yu A. Šaškin, On the convergence of linear positive operators in the space of continuous functions, *Dokl. Akad. Nauk* SSSR, 131 (1960), 525-527 (Russian).

[11] Yu. A. Šaškin, Korovkin systems in spaces of continuous functions, *Amer. Math. Soc. Transl.*, Ser. 2, 54 (1966), 125-144.

[12] Yu. A. Šaškin, The Milman-Choquet boundary and approximation theory, *Functional Anal. Appl.*, 1 (1967), 170-171.

[13] Yu. A. Šaškin, On the convergence of linear operators. Proc. Intern. Conference on Constructive Function Theory, Varna (1970), 119-125 (Russian).

[14] E. Scheffold, Über die punktweise Konvergenz von Operatoren in Banachräumen (Manuskript).

[15] D. E. Wulbert, Convergence of operators and Korovkin's theorem, *J. of Appr. Theory*, 1 (1968), 381-390.

Added in proof. — The results of this paper should also be compared with :

W. B. Arveson, An approximation theorem for function algebras (*Preprint, Univ. of Texas, Austin*).

H. Berens and G. G. Lorentz, Theorems of Korovkin type for positive linear operators on Banach lattices. Proc. Intern. Symp. on Approximation Theory (*Univ. of Texas, Austin*, Jan. 22-24, 1973) (to appear).

Manuscrit reçu le 24 janvier 1973,
accepté par M. Brelot,

# Convergence of monotone operators [R30]

Math. Z. **136** (1974), 315–330
[Zbl. Math. 269.31009; MR 50 #14184]

## Introduction

In a preceding publication [2] we have shown that convergence theorems generalizing the classical Korovkin theorem can be obtained for nets $(L_i)_{i\in I}$ of monotone, i.e. increasing operators in the following situation: Let $\mathscr{H}$ be an adapted linear space of continuous real-valued functions on a locally compact space $X$. Denote by $\mathscr{H}^b$ the set of all $\mathscr{H}$-bounded functions, i.e. of all real-valued functions $f$ on $X$ satisfying $|f|\leqq h$ for some $h\in\mathscr{H}$, and denote by $\mathscr{H}_0$ the linear space of all continuous $\mathscr{H}$-bounded functions. Each $L_i$ is an increasing map of $\mathscr{H}_0$ into $\mathbb{R}^X$, the space of real-valued functions on $X$. We have proved in [2] that pointwise (resp. locally uniform) convergence of $(L_i f)_{i\in I}$ to $f$ for all $f\in\mathscr{H}$ implies the same convergence behavior for all functions $f$ which are $\mathscr{H}$-affine in the sense of [2]. The set $\hat{\mathscr{H}}$ of these functions is a linear subspace of $\mathscr{H}_0$ which coincides with $\mathscr{H}_0$ if and only if the Choquet boundary $\partial_{\mathscr{H}} X$ of $X$ with respect to $\mathscr{H}$ equals $X$.

It seems natural to ask whether these and some other results remain valid for nets of increasing maps $L_i$: $\mathscr{H}_0\to\mathscr{H}^b$ if one replaces pointwise (resp. locally uniform) convergence by convergence with respect to the order topology on $\mathscr{H}^b$. This topology is in general finer than the topology of locally uniform convergence, i.e. the topology of uniform convergence on compact sets.

Korovkin's classical theorem has its origin in the proof of the Weierstrass approximation theorem by means of the Bernstein polynomials. Therefore it seems to be another natural question to decide whether there is a corresponding close connection between the above Korovkin-type results and the Stone-Weierstrass approximation theorem.

The first three chapters of this paper are devoted to these questions. After some preliminaries about order topology in Chapter 1 we show in Chapter 2 that the structure of the space $\hat{\mathscr{H}}$ of $\mathscr{H}$-affine functions leads in a simple way not only to Stone's approximation theorem on compact spaces (and hence to the Stone-Weierstrass theorem) but also to its generalization for adapted spaces on locally compact spaces, a generalization which was obtained by means of quite different methods by Mokobodzki-Sibony [7]. Chapter 3 treats the extensions of all Korovkin-type results of our earlier paper [2] to the case where locally uniform convergence is replaced by order convergence.

In the final Chapter 4 we give an application to the convergence of increasing operators $L_i$ which have values not only in $\mathscr{H}^b$ but in $\hat{\mathscr{H}}$. Such $\mathscr{H}$-affine operators appear in connection with sub-Markov semigroups and resolvents. One of the main results of the thesis of Lion [5] turns out to be valid for this bigger class of operators. Furthermore an interesting relation between the order convergence of such operators and the abstract Dirichlet problem of [1] is discussed.

## 1. Preliminaries on Order Topology

In this section we consider a set $X$ and a linear subspace $\mathscr{H}$ of the linear space $\mathbb{R}^X$ of all real-valued functions on $X$. We assume

$$\mathscr{H} = \mathscr{H}^+ - \mathscr{H}^+, \tag{1.1}$$

where $\mathscr{H}^+ = \{h \in \mathscr{H} \mid h \geqq 0\}$. As in [2] we denote by $\mathscr{H}^b$ the set of all $\mathscr{H}$-bounded real-valued functions on $X$:

$$\mathscr{H}^b = \{f \in \mathbb{R}^X \mid |f| \leqq h \text{ for some } h \in \mathscr{H}^+\}. \tag{1.2}$$

Obviously, $\mathscr{H}^b$ is a sub-vector lattice of the vector lattice $\mathbb{R}^X$ with respect to the usual (pointwise) order relation $\leqq$.

On $\mathscr{H}^b$ and, more generally, on sub-vector lattices $\mathscr{G}$ of $\mathbb{R}^X$ satisfying

$$\mathscr{H} \subset \mathscr{G} \subset \mathscr{H}^b \tag{1.3}$$

we shall describe the *order topology* $T_0(\mathscr{G})$. This, by definition, is the finest locally convex topology on $\mathscr{G}$ for which every order interval in $\mathscr{G}$

$$I_g^{\mathscr{G}} := \{g' \in \mathscr{G} \mid -g \leqq g' \leqq g\} \qquad (g \in \mathscr{G}^+) \tag{1.4}$$

is bounded. First we introduce two notations:

$\mathbf{E} = \mathbf{E}(\mathscr{H}^+)$ denotes the set of all strictly positive functions $\varepsilon\colon \mathscr{H}^+ \to ]0, +\infty[$ defined on $\mathscr{H}^+$. For every $\varepsilon \in \mathbf{E}$ we denote by $V_\varepsilon$ the set of all functions $f \in \mathscr{H}^b$ with the following property: there exist finitely many functions $h_1, \ldots, h_r \in \mathscr{H}^+$ and corresponding numbers $\lambda_1, \ldots, \lambda_r \in \mathbb{R}_+$ with sum $\sum_{i=1}^{r} \lambda_i = 1$ such that

$$|f| \leqq \sum_{i=1}^{r} \lambda_i\, \varepsilon(h_i)\, h_i. \tag{1.5}$$

**1.1. Lemma.** *The family $(V_\varepsilon \cap \mathscr{G})_{\varepsilon \in \mathbf{E}}$ is a fundamental system of convex and symmetric $T_0(\mathscr{G})$-neighborhoods of the constant function* 0.

*Proof.* Obviously, $(V_\varepsilon \cap \mathscr{G})$ is a filter base of convex and symmetric subsets of $\mathscr{G}$. For every function $h \in \mathscr{H}^+$ the order interval $I_h^{\mathscr{G}}$ is absorbed by $V_\varepsilon \cap \mathscr{G}$, $\varepsilon \in \mathbf{E}$, since

$$\varepsilon(h)\, I_h^{\mathscr{G}} = \{g' \in \mathscr{G} \,|\, |g'| \leqq \varepsilon(h) h\} \subset V_\varepsilon \cap \mathscr{G}.$$

This proves that every order interval $I_g^{\mathscr{G}}$ is absorbed by $V_\varepsilon \cap \mathscr{G}$ since all functions $g \in \mathscr{G}$ are $\mathscr{H}$-bounded. Hence $(V_\varepsilon \cap \mathscr{G})_{\varepsilon \in \mathbf{E}}$ is a filter base of convex, symmetric, absorbing subsets of $\mathscr{G}$ and, consequently, a fundamental system of neighborhoods of 0 for a locally convex topology on $\mathscr{G}$ for which all order intervals in $\mathscr{G}$ are bounded. Hence, every set $V_\varepsilon \cap \mathscr{G}$ is a $T_0(\mathscr{G})$-neighborhood of 0.

Consider now an arbitrary convex, symmetric $T_0(\mathscr{G})$-neighborhood $V$ of 0. For every $h \in \mathscr{H}^+$ the order interval $I_h^{\mathscr{G}}$ is absorbed by $V$. Hence there exists a real number $\varepsilon'(h) > 0$ such that $\varepsilon'(h)\, I_h^{\mathscr{G}} \subset V$. The function $h \mapsto \varepsilon(h) := \frac{1}{2}\varepsilon'(h)$, $h \in \mathscr{H}^+$, is then in $\mathbf{E}$ and we shall see that $V_\varepsilon \cap \mathscr{G} \subset V$: For arbitrary $f \in V_\varepsilon \cap \mathscr{G}$ we have

$$|f| \leqq \sum_{i=1}^{r} \lambda_i\, \varepsilon(h_i)\, h_i$$

for certain functions $h_1, \ldots, h_r \in \mathscr{H}^+$ and numbers $\lambda_1, \ldots, \lambda_r \in \mathbb{R}_+$ satisfying $\sum_{i=1}^{r} \lambda_i = 1$. Now $f = \frac{1}{2}(2f^+) + \frac{1}{2}(-2f^-)$ where $f^+$ is the positive and $f^-$ is the negative part of $f$. Since $\mathscr{G}$ is a sub-vector lattice of $\mathbb{R}^X$ and since $V$ is convex and symmetric we only have to prove that $V$ contains all functions $g' \in \mathscr{G}^+$ satisfying

$$g' \leqq \sum_{i=1}^{r} \lambda_i\, \varepsilon'(h_i)\, h_i.$$

But the Riesz decomposition lemma [4; p. 19] yields the existence of functions $g_1, \ldots, g_r \in \mathscr{G}^+$ such that

$$g' = \lambda_1 g_1' + \cdots + \lambda_r g_r'$$

and

$$g_i' \leqq \varepsilon'(h_i)\, h_i \qquad (i = 1, \ldots, r).$$

Hence $g_i' \in \varepsilon'(h_i)\, I_{h_i}^{\mathscr{G}} \subset V$ for all $i = 1, \ldots, r$ and $g' \in V$ by the convexity of $V$. This finishes the proof of the lemma.

**1.2. Corollary.** *The order topology $T_0 = T_0(\mathscr{H}^b)$ of $\mathscr{H}^b$ induces the order topology $T_0(\mathscr{G})$ on every sub-vector lattice $\mathscr{G}$ of $\mathbb{R}^X$ satisfying* (1.3).

This is an immediate consequence of Lemma 1.1. Because of this fact the order topology $T_0(\mathscr{G})$ on $\mathscr{G}$ will be denoted from now on just by $T_0$.

**1.3. Remarks.** 1) Suppose that $\mathscr{U}$ is a subset of $\mathscr{H}^+$ such that each function in $\mathscr{H}^+$ is majorized by a positive multiple of some function in $\mathscr{U}$. Suppose furthermore that $\mathbf{E}_{\mathscr{U}}$ denotes the set of all strictly positive, real-valued functions $\varepsilon$ defined on $\mathscr{U}$ and that, for functions $\varepsilon \in \mathbf{E}_{\mathscr{U}}$, the above definition of $V_\varepsilon$ is modified in such a way that the functions $h_1, \ldots, h_r$ appearing in (1.5) are required to be in $\mathcal{U}$. Then Lemma 1.1 and its proof remain valid.

Suppose, in particular, that all functions in $\mathcal{H}$ are bounded and that $\mathcal{H}^+$ contains a strictly positive function $u_0$ bounded away from zero (for example the constant function 1). Then it suffices to take $\mathcal{U}=\{u_0\}$. The order topology is then just the topology of uniform convergence on the space of all bounded, real-valued functions on $X$. Lemma 1.1 then reduces to the usual description of the topology of uniform convergence.

2) The above lemma has obvious generalizations to arbitrary vector lattices which take over the role of $\mathcal{H}^b$.

Lemma 1.1 will also be applied to the following situation: Consider a subset $A$ of $X$ and the set $\mathcal{H}_A := \text{rest}_A \mathcal{H}$ of the restrictions of all functions of $\mathcal{H}$ to $A$. Then, obviously, $\mathcal{H}_A = \mathcal{H}_A^+ - \mathcal{H}_A^+$ and the space $(\mathcal{H}^b)_A := \text{rest}_A \mathcal{H}^b$ coincides with $(\mathcal{H}_A)^b$. This allows us to speak of the order topology on $(\mathcal{H}^b)_A$. Convergence of a filter in $(\mathcal{H}^b)_A$ with respect to this topology will be called order convergence *on* $A$.

To every $\varepsilon \in \mathbf{E} = \mathbf{E}(\mathcal{H}^+)$ corresponds the set $V_\varepsilon^A$ of all functions $\text{rest}_A f$ with $f \in \mathcal{H}^b$ such that

$$|f(x)| \leqq \sum_{i=1}^{r} \lambda_i \, \varepsilon(h_i) \, h_i(x) \qquad \text{for all } x \in A.$$

Here again the functions $h_1, \ldots, h_r \in \mathcal{H}^+$ and the numbers $\lambda_1, \ldots, \lambda_r \in \mathbb{R}_+$ satisfying $\sum_{i=1}^{r} \lambda_i = 1$ depend on $f$. Instead of saying $\text{rest}_A f \in V_\varepsilon^A$ we shall also say that the function $f$ *is in* $V_\varepsilon$ *on* $A$.

**1.4. Lemma.** *The family* $(V_\varepsilon^A)_{\varepsilon \in \mathbf{E}}$ *is a fundamental system of convex and symmetric neighborhoods of the constant function* 0 *on* $A$ *with respect to the order topology on* $(\mathcal{H}^b)_A$.

The proof follows the lines of the proof of Lemma 1.1. The reader should observe that, for reasons of convenience, we still work with $\mathbf{E} = \mathbf{E}(\mathcal{H}^+)$ and not with the corresponding set $\mathbf{E}(\mathcal{H}_A^+)$. According to Lemma 1.1, the latter set leads to another fundamental system of neighborhoods of 0 with respect to the order topology on $(\mathcal{H}^b)_A$.

## 2. Order Topology on Adapted Spaces

As in [2] we shall work within the following framework: For a *locally compact* Hausdorff space $X$, $\mathcal{H}$ will be an adapted linear subspace of $\mathcal{C}(X)$. Then the space

$$\mathcal{H}_0 = \mathcal{H}^b \cap \mathcal{C}(X) = o(\mathcal{H}^+) \cap \mathcal{C}(X) \tag{2.1}$$

is again adapted and a sub-vector lattice of $\mathbb{R}^X$. Therefore, by Corollary 1.2, the order topology on $\mathcal{H}_0$ is induced by the order topology $T_0$ on $\mathcal{H}^b$. As in [2] we shall be interested in the linear space $\hat{\mathcal{H}}$ of all $\mathcal{H}$-*affine functions.* These are all functions $f \in \mathcal{H}^b$ for which the two envelopes

$$\hat{f} := \inf\{h \in \mathcal{H} \mid h \geqq f\} \tag{2.2}$$

and

$$\check{f} := \sup\{h \in \mathcal{H} \mid h \leqq f\} = -\widehat{-f} \tag{2.3}$$

coincide. We have

$$\mathcal{H} \subset \hat{\mathcal{H}} \subset \mathcal{H}_0 \subset \mathcal{H}^b. \tag{2.4}$$

Our first intention is the characterization of $\hat{\mathcal{H}}$ by means of the order topology. In order to make the presentation self-contained we first prove the following known [7; p. 61] extension of Dini's theorem to the locally compact case.

**2.1. Lemma.** *Let $\mathcal{G}$ be a subset of $\mathcal{H}_0$ which is filtering to the left and satiesfies*

$$\inf_{g \in \mathcal{G}} g(x) = 0 \quad \textit{for all } x \in X.$$

*Then the section filter of $\mathcal{G}$ is $T_0$-convergent to the constant function* 0.

*Proof.* Choose $\varepsilon \in \mathrm{E}$ and $g_0 \in \mathcal{G}$ arbitrarily. From (2.1) follows the existence of a function $h_0 \in \mathcal{H}^+$ satisfying $g_0 \in o(h_0)$ and hence the existence of a compact set $K \subset X$ such that

$$g_0(x) \leqq \tfrac{1}{2}\varepsilon(h_0)\, h_0(x) \qquad \text{for all } x \in X \setminus K.$$

Since $\mathcal{H}$ is adapted a function $h_1 \in \mathcal{H}^+$ can be found such that $h_1(x) \geqq 1$ on $K$. By Dini's theorem there exists $g_1 \in \mathcal{G}$ satisfying $g_1 \leqq g_0$ and

$$g_1(x) \leqq \tfrac{1}{2}\varepsilon(h_1) \leqq \tfrac{1}{2}\varepsilon(h_1)\, h_1(x) \qquad \text{for all } x \in K.$$

Therefore,

$$g_1 \leqq \tfrac{1}{2}\varepsilon(h_0)\, h_0 + \tfrac{1}{2}\varepsilon(h_1)\, h_1$$

holds on $X$, and hence $g_1 \in V_\varepsilon$. The result then follows from Lemma 1.1.

The announced characterization of $\mathcal{H}$-affine functions generalizes the one given in [2], Lemma 1.1. There we used the topology $T_c$ of uniform convergence on compact sets instead of the finer order topology $T_0$ on $\mathcal{H}^b$.

**2.2. Lemma.** *A function $f\colon X \to \mathbb{R}$ is $\mathcal{H}$-affine if and only if for every $T_0$-neighborhood $V$ of 0 in $\mathcal{H}^b$ there exist finitely many functions $h_1', \ldots, h_n'$ and $h_1'', \ldots, h_n''$ in $\mathcal{H}$, $n \in \mathbb{N}$, satisfying*

$$\underline{h} \leqq f \leqq \bar{h} \tag{2.5}$$

*and*

$$\bar{h} - \underline{h} \in V, \tag{2.6}$$

*where*

$$\bar{h} := \inf(h_1'', \ldots, h_n'') \quad \textit{and} \quad \underline{h} := \sup(h_1', \ldots, h_n'). \tag{2.7}$$

*Proof.* A function $f$ satisfying the above conditions is $\mathscr{H}$-affine. This follows from the definition of $\hat{\mathscr{H}}$ since $T_0$ is finer than the topology of pointwise convergence on $X$.

Conversely, let $f$ be an $\mathscr{H}$-affine function and choose $V = V_\varepsilon$ for some $\varepsilon \in \mathbf{E}$. The set $\mathscr{H}_f \subset \mathscr{H}_0$ of all functions

$$\inf(h_1, \ldots, h_n) - f$$

with $n \in \mathbb{N}$, $h_1, \ldots, h_n \in \mathscr{H}$ satisfying $h_i \geqq f$ for $i = 1, \ldots, n$ is filtering to the left and has $\hat{f} - f = 0$ as infimum. Hence by Lemma 2.1 there exist finitely many functions $h_1'', \ldots, h_n'' \in \mathscr{H}$ satisfying

$$\bar{h} \geqq f \quad \text{and} \quad \bar{h} - f \in V_{\frac{1}{2}\varepsilon}$$

where $\bar{h} = \inf(h_1'', \ldots, h_n'')$. The same result, applied to $-f$, yields the existence of finitely many functions $h_1', \ldots, h_m' \in \mathscr{H}$ with upper envelope

$$\underline{h} \leqq f \quad \text{and} \quad f - \underline{h} \in V_{\frac{1}{2}\varepsilon}.$$

Obviously, we can assume $m = n$. Then $h_1', \ldots, h_n'$ and $h_1'', \ldots, h_n''$ have all desired properties since

$$\bar{h} - \underline{h} \in V_{\frac{1}{2}\varepsilon} + V_{\frac{1}{2}\varepsilon} \subset V_\varepsilon.$$

**2.3. Corollary 1.** *$\hat{\mathscr{H}}$ is $T_0$-closed.*

*Proof.* Let $f$ be a function in the $T_0$-closure of $\hat{\mathscr{H}}$ in $\mathscr{H}^b$ (or, what amounts to the same, in $\mathscr{H}_0$). Given $\varepsilon \in \mathbf{E}$ there is a function $f_0 \in \hat{\mathscr{H}}$ such that $f - f_0 \in V_{\frac{1}{2}\varepsilon}$, hence satisfying

$$|f - f_0| \leqq \tfrac{1}{2} \sum_{i=1}^{r} \lambda_i \, \varepsilon(h_i) \, h_i$$

$\left(h_1, \ldots, h_r \in \mathscr{H}^+; \lambda_1, \ldots, \lambda_r \in \mathbb{R}_+, \sum_{i=1}^{r} \lambda_i = 1\right)$. By Lemma 2.2 there are functions $h_1', \ldots, h_n'; h_1'', \ldots, h_n'' \in \mathscr{H}$ such that the corresponding two envelopes $\bar{h}$ and $\underline{h}$ (in the sense of (2.7)) satisfy $\underline{h} \leqq f_0 \leqq \bar{h}$ and $\bar{h} - \underline{h} \in V_{\frac{1}{2}\varepsilon}$.

Consequently,

$$h_j' - \tfrac{1}{2} \sum_{i=1}^{r} \lambda_i \, \varepsilon(h_i) \, h_i \quad \text{resp.} \quad h_j'' + \tfrac{1}{2} \sum_{i=1}^{r} \lambda_i \, \varepsilon(h_i) \, h_i$$

$(j = 1, \ldots, n)$ are functions in $\mathscr{H}$ which minorize resp. majorize $f$. The corresponding envelopes $\bar{\bar{h}}$ and $\underline{\underline{h}}$ then satisfy $\underline{\underline{h}} \leqq f \leqq \bar{\bar{h}}$ and

$$\bar{\bar{h}} - \underline{\underline{h}} = \bar{h} - \underline{h} + \sum_{i=1}^{r} \lambda_i \, \varepsilon(h_i) \, h_i \in V_{\frac{1}{2}\varepsilon} + V_{\frac{1}{2}\varepsilon} \subset V_\varepsilon.$$

Hence, again by Lemma 2.2, $f$ is $\mathscr{H}$-affine.

**2.4. Corollary 2.** *Suppose that $\mathscr{H}$ is a lattice with respect to the order relation $\leqq$. Then $\hat{\mathscr{H}}$ is the $T_0$-closure of $\mathscr{H}$ in $\mathscr{H}^b$.*

*Proof.* Choose $f \in \hat{\mathscr{H}}$ and $\varepsilon \in \mathbf{E}$. Then $\underline{h} \leqq f \leqq \bar{h}$ and $\bar{h} - \underline{h} \in V_\varepsilon$ where again $\underline{h}$ resp. $\bar{h}$ is the upper resp. lower envelope of a finite number of functions $h_1', \ldots, h_n'$ resp. $h_1'', \ldots, h_n''$ in $\mathscr{H}$. Since $\mathscr{H}$ is a lattice there is a function $h_0 \in \mathscr{H}$ satisfying $\underline{h} \leqq h_0 \leqq \bar{h}$, for example, the lattice-theoretic infimum of $h_1'', \ldots, h_n''$ or the lattice-theoretic supremum of $h_1', \ldots, h_n'$. Hence, $|f - h_0| \leqq \bar{h} - \underline{h}$ which implies $f - h_0 \in V_\varepsilon$.

Consequently, $f$ is in the $T_0$-closure of $\mathscr{H}$. Together with Corollary 1 this proves the assertion.

**2.5. Remarks.** 1. The proof shows that Corollary 2 remains valid if $\mathscr{H}$ satisfies the Riesz decomposition property (instead of being a lattice). In fact, it is well known that $\mathscr{H}$ then enjoys the Riesz interpolation property: Given functions $h_1', \ldots, h_n'; h_1'', \ldots, h_n''$ in $\mathscr{H}$ ($n \in \mathbb{N}$) such that the upper resp. lower envelope of the first resp. last $n$ functions satisfies $\underline{h} \leqq \bar{h}$. Then there exists a function $h_0 \in \mathscr{H}$ satisfying $\underline{h} \leqq h_0 \leqq \bar{h}$.

2. For a compact space $X$ Corollary 2 was already obtained in [1; p. 104].

As an application we give a simple proof of the extension of Stone's approximation theorem to locally compact spaces. This extension was obtained by Mokobodzki and Sibony [7; p. 57] (see also [6]) by quite a different and more complicated method. For a compact space $X$ our proof reduces to an elegant functional analytic proof of Stone's theorem which could have been given already in [1] [1].

**2.6. Theorem.** *Suppose that the adapted space $\mathscr{H} \subset \mathscr{C}(X)$ is a sub-vector lattice of $\mathbb{R}^X$ and linearly separating* [2]. *Then $\mathscr{H}$ is $T_0$-dense in $\mathscr{H}_0$.*

*Proof.* By the preceding Corollary 2 the $T_0$-closure of $\mathscr{H}$ in $\mathscr{H}^b$ and hence also in $\mathscr{H}_0$ (which is $T_0$-closed in $\mathscr{H}^b$) equals $\hat{\mathscr{H}}$. We thus have to prove $\hat{\mathscr{H}} = \mathscr{H}_0$. By [2], Corollary 2.5 this is equivalent to $\partial_{\mathscr{H}} X = X$ where $\partial_{\mathscr{H}} X$ denotes the Choquet boundary. This, however, can be proved as follows: Consider $x \in X$ and an $\mathscr{H}$-representing measure $\mu$ for $x$ (in the sense of [2]). Choose $h_0 \in \mathscr{H}^+$ such that $h_0(x) = 1$. For every $h \in \mathscr{H}$ the function $|h - h(x) h_0|$ is in $\mathscr{H}$ since $\mathscr{H}$ is a sub-vector lattice of $\mathbb{R}^X$. We conclude that $\int |h - h(x) h_0| \, d\mu = 0$ and that $\mu$ is supported by each of the sets $\{h = h(x) h_0\}, h \in \mathscr{H}$. Because of linear separation the intersection of these sets equals $\{x\}$. Hence $\mu = \tau \varepsilon_x$ with $\tau \in \mathbb{R}_+$. But

$$1 = h_0(x) = \int h_0 \, d\mu = \tau \, h_0(x) = \tau$$

implies $\mu = \varepsilon_x$. This proves $x \in \partial_{\mathscr{H}} X$.

[1] This observation was made independently by J. Bliedtner (private communication).

[2] This is equivalent to the following property: For all $\lambda \in \mathbb{R}_+$ and all $x, x_0 \in X$ with $x \neq x_0$ there exists $h \in \mathscr{H}^+$ such that $h(x) \neq \lambda h(x_0)$.

## 3. Korovkin-Type Theorems for the Order Topology

The following results extend the main propositions of [2] in the sense that uniform convergence on compact sets is replaced by order convergence. We consider a net or, more generally, a family $(L_i)_{i\in I}$ of *increasing* maps $L_i\colon \mathscr{H}_0 \to \mathscr{H}^b$ which is filtered with respect to a given filter $\mathfrak{F}$ in the index set $I$. (In [2] we could assume that each $L_i$ maps $\mathscr{H}_0$ into $\mathbb{R}^X$ since uniform convergence on compact sets is already defined on $\mathbb{R}^X$.)

**3.1. Theorem.** *Let $A$ be a closed subset of $X$, and suppose that $(L_i h)_{i\in I}$ is order convergent on $A$ to $h$ for all $h\in\mathscr{H}$. Then $(L_i f)_{i\in I}$ is order convergent on $A$ to $f$ for all functions $f\in\mathscr{H}_0$ satisfying $\check{f}(x)=\hat{f}(x)$ on $A$.*

*Proof.* Let $f$ be a function in $\mathscr{H}_0$ satisfying $\check{f}(x)=\hat{f}(x)$ on $A$. Then, given $\varepsilon\in\mathbf{E}$, there exist functions $h'_1,\dots,h'_n$ and $h''_1,\dots,h''_n$ in $\mathscr{H}$, $n\in\mathbb{N}$, such that

$$\sup(h'_1,\dots,h'_n)=:\underline{h}\leqq f\leqq \bar{h}:=\inf(h''_1,\dots,h''_n)$$

(on the total space $X$) and such that $\bar{h}-\underline{h}$ is in $V_{\frac{1}{3}\varepsilon}$ on $A$. This can be seen as follows: Since $A$ is closed the space $\mathscr{H}^A$ of restrictions of the functions of $\mathscr{H}$ to $A$ is an adapted linear subspace of $\mathscr{C}(A)$. As in the proof of Lemma 2.2 we consider the set $\mathscr{H}_f$ of all functions

$$\inf(h_1,\dots,h_n)-f$$

with $n\in\mathbb{N}$ and $h_1,\dots,h_n\in\mathscr{H}$ satisfying $h_i\geqq f$, $i=1,\dots,n$. It is filtering to the left and has zero as infimum on the set $A$. From Lemmas 2.1 and 1.4 then follows the existence of a finite number of functions $h''_1,\dots,h''_n\in\mathscr{H}$ such that $\bar{h}:=\inf(h''_1,\dots,h''_n)\geqq f$ and $\bar{h}-f$ is in $V_{\frac{1}{6}\varepsilon}$ on $A$. We now continue as in the proof of Lemma 2.2.

Since every map $L_i$, $i\in I$, is increasing we obtain

$$L_i\underline{h}-\bar{h}\leqq L_i f-f\leqq L_i\bar{h}-\underline{h}\qquad (i\in I)$$

from $\underline{h}\leqq f\leqq\bar{h}$. Furthermore,

$$L_i\bar{h}-\bar{h}\leqq \inf_{1\leqq j\leqq n} L_i h''_j-\inf_{1\leqq j\leqq n} h''_j\leqq\sum_{j=1}^{n}|L_i h''_j-h''_j|$$

and, correspondingly,

$$L_i\underline{h}-\underline{h}\geqq-\sum_{j=1}^{n}|L_i h'_j-h'_j|.$$

We thus obtain

$$|L_i f-f|\leqq\sum_{j=1}^{n}|L_i h'_j-h'_j|+\sum_{j=1}^{n}|L_i h''_j-h''_j|+\bar{h}-\underline{h}.$$

According to the convergence behavior of $(L_i)_{i\in I}$ on $\mathscr{H}$, there exists a set $F$ in the filter $\mathfrak{F}$ such that, for all $i\in F$ and all $j=1,\dots,n$, the functions

$$L_i h'_j-h'_j\quad\text{and}\quad L_i h''_j-h''_j\qquad\text{are in } V_{\frac{1}{3n}\varepsilon}\text{ on } A.$$

The above equality then yields the fact that $L_i f - f$ is in $V_\varepsilon$ on $A$ for all $i \in F$.

**3.2. Corollary 1.** *Let $A$ be a closed subset of the Choquet boundary $\partial_{\mathscr{H}} X$, and assume that $(L_i h)_{i\in I}$ is order convergent on $A$ to $h$ for all $h \in \mathscr{H}$. Then $(L_i f)_{i\in I}$ is order convergent on $A$ to $f$ for all $f \in \mathscr{H}_0$.*

This is an immediate consequence of [2], Corollary 2.4.

**3.3. Corollary 2.** *Suppose that*

$$T_0 - \lim_{\mathfrak{F}} L_i h = h \quad \text{for all } h \in \mathscr{H}. \tag{3.1}$$

*Then*

$$T_0 - \lim_{\mathfrak{F}} L_i f = f \tag{3.2}$$

*holds for all $\mathscr{H}$-affine functions $f$.*

This follows from the definition of $\mathscr{H}$-affine functions for $A = X$.

In [2] we have seen that $\hat{\mathscr{H}}$ is the Korovkin closure of $\mathscr{H}$ with respect to pointwise convergence and uniform convergence on compact sets. We can now define the $T_0$-*Korovkin closure* as the set of all functions $f \in \mathscr{H}_0$ having the following property: For every net $(L_i)_{i\in I}$ of positive linear maps $L_i\colon \mathscr{H}_0 \to \mathscr{H}_0$ satisfying (3.1),

$$T_0 - \lim L_i f = f$$

holds. Formally we obtain a smaller $T_0$-Korovkin closure if the preceding definition is formulated with filtered families of increasing maps $L_i\colon \mathscr{H}_0 \to \mathscr{H}_0$. However, we know from Corollary 2 that all $\mathscr{H}$-affine functions are in the coarser $T_0$-Korovkin closure of $\mathscr{H}$. The following result then proves in particular that the two definitions are equivalent.

**3.4. Theorem.** *$\hat{\mathscr{H}}$ is the $T_0$-Korovkin closure of $\mathscr{H}$.*

*Proof.* We modify the proof of [2], Theorem 3.3. Given a function $f \in \mathscr{H}_0$ in the $T_0$-Korovkin closure of $\mathscr{H}$, a point $x_0 \in X$ and an $\mathscr{H}$-representing measure $\mu_0$ for $x_0$, we choose the same net $(L_V)_{V\in\mathfrak{B}}$ of positive linear maps $L_V\colon \mathscr{H}_0 \to \mathscr{H}_0$ as in [2]. Namely,

$$L_V g = \left(\int g \, d\mu_0\right) q_V + g - g q_V \quad (g \in \mathscr{H}_0),$$

where $\mathfrak{B}$ is a neighborhood base of $x_0$ consisting of open relatively compact sets, and where $q_V \in \mathscr{C}(X)$ satisfies $0 \leqq q_V \leqq 1$, $q_V(x_0) = 1$, $q_V(x) = 0$ for all $x \in X \setminus V$.

For functions $h \in \mathscr{H}$ we have

$$|L_V h - h| = |h - h(x_0)| \, q_V \quad (V \in \mathfrak{B}).$$

For a given $V_0 \in \mathfrak{B}$ there exists a function $h_{V_0} \in \mathscr{H}$ such that $q_{V_0} \leqq h_{V_0}$ since $q_{V_0}$ has compact support. Hence, for arbitrary $\varepsilon \in \mathbf{E}$,

$$|L_V h - h| \leqq \varepsilon(h_{V_0})\, h_{V_0}$$

and hence $L_V h - h \in V_\varepsilon$ holds for all sufficiently small $V \in \mathfrak{B}$ contained in $V_0$. This proves $T_0 - \lim L_V h = h$ for all $h \in \mathscr{H}$. Consequently, $T_0 - \lim L_V f = f$ and therefore, in particular, $\int f\, d\mu_0 = \lim L_V f(x_0) = f(x_0)$. From [2], Corollary 2.3, then $f \in \hat{\mathscr{H}}$ follows.

## 4. $\mathscr{H}$-Affine Operators and Dirichlet Problem

Beyond the assumptions made on $X$ and $\mathscr{H}$ so far we shall assume from now on that $X$ is *countable at infinity* and that $\mathscr{H}$ is *linearly separating*.

According to Mokobodzki-Sibony [6, 7] the usual results of the theory of integral representation in compact convex sets then remain valid. In particular, we shall make use of the following known facts:

1. The Choquet boundary $\partial_{\mathscr{H}} X$ is non-empty, and for all $h \in \mathscr{H}$

$$h(x) \geqq 0 \quad \text{on } \partial_{\mathscr{H}} X \Rightarrow h \geqq 0 \tag{4.1}$$

([7], III, Théorème 1).

2. The implication holds for all $h \in \hat{\mathscr{H}}$ since $\partial_{\hat{\mathscr{H}}} X = \partial_{\mathscr{H}} X$ ([2], Corollary 2.3).

3. On the set $\mathscr{M}^+_{\mathscr{H}}$ of all positive Radon measures $\mu$ on $X$ satisfying $\mathscr{H} \subset \mathscr{L}^1(\mu)$ the relation

$$\nu \prec \mu \Leftrightarrow \int f\, d\nu \geqq \int f\, d\mu \quad \text{for all } f \in \mathscr{S} \tag{4.2}$$

is an order relation. Here $\mathscr{S}$ denotes the convex cone of all functions

$$\inf(h_1, \ldots, h_n) \quad (n \in \mathbb{N};\ h_1, \ldots, h_n \in \mathscr{H}).$$

Every $\nu \in \mathscr{M}^+_{\mathscr{H}}$ is majorized by a maximal measure $\mu \in \mathscr{M}^+_{\mathscr{H}}$. A measure $\mu \in \mathscr{M}^+_{\mathscr{H}}$ is maximal (with respect to $\prec$) if and only if $\mu$ is supported by all sets $\{f = \hat{f}\}$ where $f \in \mathscr{H}_0$. ([7], Chapter III, B, in particular: Remarque, p. 51, and Proposition 7.)

The set $\mathscr{M}_x$ of all $\mathscr{H}$-representing measures is the set of $\mu \in \mathscr{M}^+_{\mathscr{H}}$ satisfying $\varepsilon_x \prec \mu$ $(x \in X)$.

Maximal measures, though in general not supported by $\partial_{\mathscr{H}} X$, have the following Lebesgue-type dominated convergence property:

**4.1. Propositon.** *Let $f_0 \in \mathscr{H}_0$ and let $(f_i)_{i \in I}$ be a family of functions in $\mathscr{H}_0$ such that*

$$\lim_{\mathfrak{F}} f_i(x) = f_0(x) \quad \textit{for all } x \in \partial_{\mathscr{H}} X$$

*with respect to a filter* $\mathfrak{F}$ in $I$ *with countable base. Assume the existence of a function* $u \in \mathscr{H}_0$ *such that*

$$|f_i(x)| \leqq u(x) \quad \textit{for all } x \in \partial_{\mathscr{H}} X \textit{ and all } i \in I.$$

*Then*

$$\lim_{\mathfrak{F}} \int f_i \, d\mu = \int f_0 \, d\mu \tag{4.3}$$

*for every maximal measure* $\mu \in \mathscr{M}_{\mathscr{H}}^+$.

*Proof.* For every function $f \in \mathscr{H}_0$ we denote by $f'$ its restriction to $\partial_{\mathscr{H}} X$. We claim that, due to the maximality of $\mu$, $\int f \, d\mu$ depends only on $f'$. Because of linearity, we have to prove: $f' = 0 \Rightarrow \int f \, d\mu = 0$. But $f' = 0$ implies $\hat{f} \geqq 0$ and $\widehat{-f} \geqq 0$ because of (4.1). Property 3 then yields

$$\int f \, d\mu = \int \hat{f} \, d\mu \geqq 0 \quad \text{and} \quad -\int f \, d\mu = \int \widehat{-f} \, d\mu \geqq 0, \quad \text{hence} \int f \, d\mu = 0.$$

Thus $f' \mapsto \int f \, d\mu$, $f \in \mathscr{H}_0$, defines a positive linear form $I$ on $\mathscr{H}_0' := \{f' \mid f \in \mathscr{H}_0\}$. Positivity again results from the fact that $f' \geqq 0$ implies $\hat{f} \geqq 0$ and hence $\int f \, d\mu = \int \hat{f} \, d\mu \geqq 0$. By a well-known result of Mokobodzki from the theory of integral representation which is valid also in the present context ([7; p. 55], Théorème 12), $I$ is an abstract integral in the sense: $\lim I(f_n') = 0$ for every decreasing sequence $(f_n')_{n \in \mathbb{N}}$ in $\mathscr{H}_0'$ with $\inf f_n'(x) = 0$ for all $x \in \partial_{\mathscr{H}} X$. Furthermore, $\mathscr{H}_0'$ is a sub-vector lattice of $\mathbb{R}^{\partial_{\mathscr{H}} X}$ satisfying Stone's condition: $\inf(1, f') \in \mathscr{H}_0'$ for all $f' \in \mathscr{H}_0'$. It suffices to observe that $\inf(1, f)$ is $\mathscr{H}$-bounded for every $\mathscr{H}$-bounded function $f$.

We can then apply the Daniell-Stone-Theorem [3; p. 197]. It yields the existence of a $\sigma$-additive measure $\mu' \geqq 0$ on the $\sigma$-algebra $\mathfrak{A}(\mathscr{H}_0')$ generated by $\mathscr{H}_0'$ on $\partial_{\mathscr{H}} X$ such that $\mathscr{H}_0' \subset \mathscr{L}^1(\mu')$ and $I(f') = \int f' \, d\mu'$ for all $f' \in \mathscr{H}_0'$. For this measure $\mu'$ Lebesgue's dominated convergence theorem is valid. Since the filter $\mathfrak{F}$ in $I$ has a countable base the result follows [4; pp. 140–143] [3].

We shall now be interested in a particular type of increasing maps of $\mathscr{H}_0$ into $\mathscr{H}^b$:

**4.2. Definition.** *Every increasing map* $L: \mathscr{H}_0 \to \hat{\mathscr{H}}$ *of* $\mathscr{H}_0$ *into the space of* $\mathscr{H}$*-affine functions will be called an* $\mathscr{H}$*-affine operator.*

In the sequel we shall study a filtered family $(L_i)_{i \in I}$ of $\mathscr{H}$*-affine operators* such that the following conditions are satisfied:

(A) the underlying filter $\mathfrak{F}$ in $I$ has a *countable base;*

(B) $\lim_{\mathfrak{F}} L_i h(x) = h(x)$ for all $h \in \mathscr{H}$ and $x \in \partial_{\mathscr{H}} X$;

(C) the family $(L_i)_{i \in I}$ is *uniformly* $\mathscr{H}$*-bounded*, i.e. for every $u \in \mathscr{H}$ there exists a function $v \in \mathscr{H}$ such that

$$|L_i u| \leqq v \quad \text{for all } i \in I.$$

[3] A similar proposition (for $X$ compact) appears in Phelps [8; p. 47]. There the proof is based on a result of Bishop-De Leeuw.

**4.3. Examples.** 1) Following Lion [5; p. 437] we consider a compact space $X$, an index set $I$ filtered by a filter $\mathfrak{F}$ with countable base, and a family $(A_i)_{i\in I}$ of positive linear maps $A_i$: $\mathscr{C}(X)\to\mathscr{C}(X)$ with the following properties:

(a) $\lim_{\mathfrak{F}} A_i A_{i_0} f = A_{i_0} f$ for all $f\in\mathscr{C}(X)$ and all $i_0\in I$ in the sense of uniform convergence on $X$;

(b) $A_i A_j 1 \leqq \alpha A_j 1$ for some constant $\alpha \geqq 0$ and all $i, j\in I$;

(c) the linear subspace $\mathscr{H}$ of $\mathscr{C}(X)$ generated by

$$\bigcup_{i\in I} A_i(\mathscr{C}(X))$$

separates points linearly[4]. At no point of $X$ all functions of $\mathscr{H}$ vanish.

Positivity of the linear maps $A_i$ implies $\mathscr{H} = \mathscr{H}^+ - \mathscr{H}^+$ hence adaptedness of $\mathscr{H}$. All $A_i$ are $\mathscr{H}$-affine operators since $A_i(\mathscr{C}(X))\subset\mathscr{H}$. $\mathscr{H}$ contains strictly positive functions because of (c) and the compactness of $X$. The order topology on $\mathscr{H}$ is then the topology of uniform convergence. Conditions (A), (B) and (C) are easy consequences of (a), (b), (c).

Let us prove that $\hat{\mathscr{H}}$ coincides with the uniform closure of $\mathscr{H}$: It follows from (a) that

$$\lim_{\mathfrak{F}} A_i h = h$$

holds for all $h\in\mathscr{H}$ in the sence of uniform convergence. Therefore by Corollary 3.3,

$$\lim_{\mathfrak{F}} A_i f = f$$

for all $f\in\hat{\mathscr{H}}$. This proves that every $f\in\hat{\mathscr{H}}$ is the uniform limit of the functions $A_i f\in\mathscr{H}$. Hence (Corollary 2.3) $\hat{\mathscr{H}}$ is the uniform closure of $\mathscr{H}$[5].

Families $(A_i)_{i\in I}$ with the above properties appear as sub-Markovian resolvents or as strongly continuous sub-Markov semigroups (cf. [5]). The above example can be reformulated in more general terms on locally compact spaces by means of results of Mokobodzki-Sibony [6, 7].

2) Returning to the general context of this section, let us assume that the restriction $f'$ of an arbitrary function $f\in\mathscr{H}_0$ to $\partial_{\mathscr{H}}X$ has an $\mathscr{H}$-affine *Dirichlet* solution. This means the existence of a (necessarily unique) function $Df\in\hat{\mathscr{H}}$ such that $Df(x)=f(x)$ for all $x\in\partial_{\mathscr{H}}X$. Then, for $I=\{1\}$, $L_1 = D$ defines a family $(L_i)_{i\in I}$ of $\mathscr{H}$-affine operators with properties (A), (B) and (C). We shall see that this example is of fundamental importance for our further study. For compact spaces $X$ this example has been studied extensively in [1].

---

[4] Condition (c) is not assumed in [5]. However, Lion reduces his study to this case by passing to a quotient space. Condition (b) is a weaker form of condition (iii) in [5; p. 437]. The assumption $A_i A_j = A_j A_i$, $i, j\in I$, made in [5] turns out to be irrelevant for the sequel.

[5] The space $\mathscr{H}$ of Lion [5; p. 437] therefore coincides with our space $\hat{\mathscr{H}}$. Lion's proposition on p. 438 is closely related to this result.

We now return to the study of a family $(L_i)_{i\in I}$ of $\mathscr{H}$-affine operators satisfying conditions (A), (B) and (C).

**4.4. Proposition.** *The family $(L_i)_{i\in I}$ has the following properties:*

(1) $\lim_{\mathfrak{F}} L_i h(x)=h(x)$ *for all* $h\in\mathscr{H}$ *and* $x\in X$.

(2) *The limit*

$$\tilde{f}(x):=\lim_{\mathfrak{F}} L_i f(x) \tag{4.4}$$

*exists for all* $f\in\mathscr{H}_0$ *and all* $x\in X$.

(3) *For every* $x\in X$, *there exists exactly one maximal measure* $\mu_x\in\mathscr{M}_x$. *The equality*

$$\tilde{f}(x)=\int f\,d\mu_x \tag{4.5}$$

*holds for all* $f\in\mathscr{H}_0$ *and* $x\in X$. *In particular,* $\tilde{f}=f$ *for* $f\in\hat{\mathscr{H}}$ *and* $\tilde{f}(x)=f(x)$ *for* $f\in\mathscr{H}_0$ *and all points* $x\in\partial_{\mathscr{H}}X$.

*Proof.* (1) Given $x\in X$ there exists a maximal measure $\mu\in\mathscr{M}_x$. We have

$$L_i h(x)=\int L_i h\,d\mu \qquad (i\in I)$$

for arbitrary functions $h\in\mathscr{H}$ since $L_i h\in\hat{\mathscr{H}}$ by assumption. Assumptions (B) and (C) in connection with Proposition 4.1 imply

$$\lim_{\mathfrak{F}} L_i h(x)=\lim_{\mathfrak{F}}\int L_i h\,d\mu=\int h\,d\mu=h(x).$$

(2) Consider $x\in X$, a maximal measure $\mu\in\mathscr{M}_x$ and a function $f\in\mathscr{H}_0$. We again have

$$L_i f(x)=\int L_i f\,d\mu \qquad (i\in I)$$

since $L_i$ is $\mathscr{H}$-affine. Property (B) in connection with the Korovkin-type Theorem 2.7 of [2] yields

$$\lim_{\mathfrak{F}} L_i f(y)=f(y) \qquad \text{for all } y\in\partial_{\mathscr{H}}X.$$

Since property (C) is also available and since $\mathscr{H}_0\subset\mathscr{H}^b$ we can again apply Proposition 4.1. This gives us

$$\lim_{\mathfrak{F}} L_i f(x)=\lim_{\mathfrak{F}}\int L_i f\,d\mu=\int f\,d\mu,$$

hence the existence of the limit (4.4).

(3) Given $x\in X$ we have proved $\tilde{f}(x)=\int f\,d\mu$ for all maximal measures $\mu\in\mathscr{M}_x$ and all $f\in\mathscr{H}_0$. In view of the fact that $\mathscr{H}_0$ contains all functions in $\mathscr{C}(X)$ with compact support this proves that $\mu$ is uniquely determined and that (4.5) holds.

In the proof of (2) we obtained already $\lim_{\mathfrak{F}} L_i f(y)=f(y)$ for all $y\in\partial_{\mathscr{H}}X$. Hence $\tilde{f}=f$ on $\partial_{\mathscr{H}}X$ for all $f\in\mathscr{H}_0$. Finally, from (1) and Corollary 1.3 of [2] we conclude $\tilde{f}=f$ for all $f\in\hat{\mathscr{H}}$.

We now come to the main problem of this chapter. Assume that property (B) is replaced by the stronger condition

($B_0$) $T_0-\lim L_i h = h$ for all $h \in \mathscr{H}$.

Under which assumptions can one conclude that also $(L_i f)_{i\in I}$ is order convergent to $\hat{f}$ along the filter $\mathfrak{F}$? This question makes sense since $\hat{f}$ is $\mathscr{H}$-bounded by property (C). The reader should observe that ($B_0$) is satisfied in the Examples 4.3.

**4.5. Theorem.** *Assume that a family $(L_i)_{i\in I}$ of $\mathscr{H}$-affine operators satisfies properties* (A), ($B_0$) *and* (C). *Then the following statements are equivalent:*

(I) $T_0-\lim_{\mathfrak{F}} L_i f = \hat{f}$ *for all* $f \in \mathscr{H}_0$.

(II) $\hat{f} \in \mathscr{C}(X)$ *for all* $f \in \mathscr{H}_0$.

(III) $\partial_{\mathscr{H}} X$ *is closed.*

*Proof.* (I)⇒(II). From $L_i(\mathscr{H}_0) \subset \hat{\mathscr{H}}$ and Corollary 2.3 follows the even stronger statement: $\hat{f} \in \hat{\mathscr{H}}$ for all functions $f \in \mathscr{H}_0$.

(II)⇒(III). It suffices to prove

$$\partial_{\mathscr{H}} X = \bigcap_{f \in \mathscr{H}_0} \{f = \hat{f}\}.$$

$\partial_{\mathscr{H}} X$ is a subset of this intersection by Proposition 4.4, property (3). Conversely, we have for points $x$ of this intersection by (4.5)

$$f(x) = \hat{f}(x) = \int f \, d\mu_x \qquad \text{for all } f \in \mathscr{H}_0.$$

Consequently, $\varepsilon_x = \mu_x$ is maximal and hence supported by

$$\partial_{\mathscr{H}} X = \bigcap_{f \in \mathscr{H}_0} \{f = \hat{f}\}.$$

(III)⇒(I). For $f \in \mathscr{H}_0$ and $x \in X$ we have

$$L_i f(x) - \hat{f}(x) = \int (L_i f - f) \, d\mu_x$$

because of $L_i f \in \hat{\mathscr{H}}$ and (4.5). Since property ($B_0$) holds and since the Choquet boundary is closed, we know from Corollary 3.2 that $(L_i f)_{i\in I}$ is order convergent to $f$ on $\partial_{\mathscr{H}} X$. Therefore, given $\varepsilon \in \mathbf{E}$, there exists a set $F$ in $\mathfrak{F}$ such that $L_i f - f$ is in $V_\varepsilon$ on $\partial_{\mathscr{H}} X$ for all $i \in F$. Consequently, for every $i \in F$ there is a finite number $h_1, \ldots, h_r$ of functions in $\mathscr{H}^+$ and of corresponding numbers $\lambda_1, \ldots, \lambda_r \in \mathbb{R}_+$ such that $\sum_{j=1}^{r} \lambda_j = 1$ and

$$|L_i f(y) - f(y)| \leq \sum_{j=1}^{r} \lambda_j \varepsilon(h_j) \, h_j(y) \qquad \text{for all } y \in \partial_{\mathscr{H}} X.$$

For an arbitrary point $x \in X$ and the corresponding maximal measure $\mu_x \in \mathcal{M}_x$ this implies

$$|L_i f(x) - \tilde{f}(x)| = |\int (L_i f - f)\, d\mu_x| \leq \sum_{j=1}^{r} \lambda_j \varepsilon(h_j) \int h_j\, d\mu_x = \sum_{j=1}^{r} \lambda_j \varepsilon(h_j)\, h_j(x)$$

since $\mu_x$ is supported by $\partial_{\mathscr{H}} X$.[6] This proves $L_i f - \tilde{f} \in V_\varepsilon$ for all $i \in F$.

We can now see that the preceding Example 2 is closely related to the present problem:

**4.6. Corollary.** *Under the assumptions of Theorem* 4.5 *the following condition is also equivalent to* (I), (II), (III):

(IV) *Every function* $f' \in \text{rest}_{\partial_{\mathscr{H}} X} \mathscr{H}_0$ *has an* $\mathscr{H}$*-affine Dirichlet solution.*

*Proof.* (I)⇒(IV). For $f \in \mathscr{H}_0$ the function $\tilde{f}$ is $\mathscr{H}$-affine and satisfies $\tilde{f}(y) = f(y)$ on $\partial_{\mathscr{H}} X$.

(IV)⇒(II). For every function $f \in \mathscr{H}_0$ we denote by $\tilde{\tilde{f}} := Df'$ the $\mathscr{H}$-affine Dirichlet solution of its restriction $f'$ to $\partial_{\mathscr{H}} X$. For $x \in X$ we then have

$$\tilde{f}(x) = \int f\, d\mu_x = \int \tilde{\tilde{f}}\, d\mu_x = \tilde{\tilde{f}}(x)$$

since $\mu_x$ is concentrated on any closed set $A \supset \partial_{\mathscr{H}} X$, hence in particular on $\{f = \tilde{\tilde{f}}\}$. This proves $\tilde{f} = \tilde{\tilde{f}}$ and hence continuity of $\tilde{f}$.

**4.7. Remarks.** 1. If, in addition, $X$ is metrizable, i.e. has a countable base, $\text{rest}_{\partial_{\mathscr{H}} X} \mathscr{H}_0$ is just the space of $\mathscr{H}$-bounded continuous functions on the (closed) Choquet boundary $\partial_{\mathscr{H}} X$.

2. Condition (IV) appears already in our second example of 4.3. This example states in other words that the equivalent conditions (I)–(IV) imply the existence of an even constant *sequence* $(L_n)_{n \in \mathbb{N}}$ of *positive linear* maps $L_n: \mathscr{H}_0 \to \mathscr{H}$ such that $(B_0)$ and (C) (and, trivially, (A)) are satisfied.

3. Theorem 4.5 applied to the first example of 4.3 yields one of the main results of Lion [5]. In fact, Theorem 4.5 corresponds to Lion's Théorème 5.

4. It follows from Proposition 4.4 that, under the conditions (A), (B), (C), $f \mapsto \tilde{f}$ is a positive linear operator of $\mathscr{H}_0$ into $\mathscr{H}^b$ with the following properties: (i) Every function in $\hat{\mathscr{H}}$ is left invariant. – (ii) For every $f \in \mathscr{H}_0$ there exists a family $(f_i)_{i \in I}$ of functions in $\hat{\mathscr{H}}$ such that $\lim_{\mathfrak{F}} f_i(x) = \tilde{f}(x)$ for all $x \in X$ with respect to a countably generated filter $\mathfrak{F}$ in $I$: it suffices to define $f_i := L_i f$. – (iii) The family $(f_i)_{i \in I}$ of (ii) is uniformly $\mathscr{H}$-bounded, i.e. there exists a function $v \in \mathscr{H}$ such that $|f_i| \leq v$ for all $i \in I$. Therefore

[6] More generally every maximal measure $\mu \in M_{\mathscr{H}}^{+}$ is supported by every universally capacitable set $A \supset \partial_{\mathscr{H}} X$ ([7; p. 56], Théorème 13).

$f \mapsto \tilde{f}$ is a strong Lion operator with respect to $\hat{\mathscr{H}}$ in the sence of a suitably modified definition of Rogalski [9]. (In [9] only compact spaces $X$ and sequences of functions in $\hat{\mathscr{H}}$ approximating $\tilde{f}$ are considered.) For compact spaces $X$ the equivalence of the conditions (II), (III) and (IV) can then also be deduced from [9], Théorème 35. In fact, in this case and under condition (I) the operator $f \mapsto \tilde{f}$ even is a strong Lion operator with respect to $\hat{\mathscr{H}}$ without modification of Rogalski's definition.

## Bibliography


1. Bauer, H.: Šilovscher Rand und Dirichletsches Problem. Ann. Inst. Fourier **11**, 89–136 (1961)
2. Bauer, H.: Theorems of Korovkin type for adapted spaces. Ann. Inst. Fourier **23**, Fasc. 4, 245–260 (1973)
3. Bauer, H.: Wahrscheinlichkeitstheorie und Grundzüge der Maßtheorie (2. Auflage). Berlin: W. de Gruyter 1974
4. Bourbaki, N.: Éléments de mathématique, Livre VI: Intégration. Chap. 1–4 (2ᵉ édition). Paris: Hermann 1965
5. Lion, G.: Familles d'opérateurs et frontière en théorie du potentiel. Ann. Inst. Fourier **16**, Fasc. 2, 389–453 (1966)
6. Mokobodzki, G., Sibony, D.: Cônes adaptés de fonctions continues et théorie du potentiel. In: Séminaire Choquet, Initiation à l'Analyse, 6ᵉ année (1966/67), Fasc. 1, 35 p. Paris: Institut H. Poincaré 1968
7. Sibony, D.: Cônes de fonctions et potentiels. Lecture Notes, Montreal: McGill University 1968
8. Phelps, R. R.: Lectures on Choquet's theorem. Math. Studies **7**, New York-Cincinnati-Toronto: Van Nostrand Reinhold 1966
9. Rogalski, M.: Opérateurs de Lion, projecteurs boréliens et simplexes analytiques. J. functional Analysis **2**, 458–488 (1968)


*(Received January 28, 1974)*

# Korovkin approximation in $\mathscr{C}_0(X)$ * [R32]

(with Klaus Donner)

Math. Ann. **236** (1978), 225–237

[Zbl. Math. 368.46028; MR 58 #12115]

Dedicated to Professor Georg Nöbeling on the occasion of his 70th birthday, November 12, 1977

### Introduction

Korovkin approximation in the space $\mathscr{C}(X)$ of all real-valued continuous functions on a compact space $X$ has found a highly satisfactory form (cf., [2, 4]) by means of the envelope technique, wellknown from the theory of integral representation in convex compact sets. The main features of this approach to Korovkin approximation can be expressed as follows:

Given a function $f \in \mathscr{C}(X)$ and a linear subspace $\mathscr{H}$ of $\mathscr{C}(X)$ containing strictly positive functions, the following five statements are equivalent:

i) $f$ is in the *Korovkin closure* of $\mathscr{H}$, i.e. $(T_\alpha f)$ converges uniformly to $f$ for every net $(T_\alpha)$ of positive linear maps $T_\alpha : \mathscr{C}(X) \to \mathscr{C}(X)$ such that $(T_\alpha h)$ converges uniformly to $h$ for all $h \in \mathscr{H}$.

ii) $f$ is in the *universal Korovkin closure* of $\mathscr{H}$, i.e., for every locally convex Riesz space $E$, every continuous Riesz (i.e. the lattice structure preserving) homomorphism $P : \mathscr{C}(X) \to E$ and for every net $(T_\alpha)$ of positive linear maps $T_\alpha : \mathscr{C}(X) \to E$, $(T_\alpha f)$ converges to $Pf$ if $(T_\alpha h)$ converges to $Ph$ for all $h \in \mathscr{H}$.

iii) For every $x \in X$ and every positive Radon measure $\mu$ on $X$, the equality $\int f d\mu = f(x)$ holds if the corresponding equalities $\int h d\mu = h(x)$ hold for all $h \in \mathscr{H}$.

iv) $f$ is *trapped* by $\mathscr{H}$, i.e. given $\varepsilon > 0$ there exist finitely many functions $h'_1, \ldots, h'_m$ and $h''_1, \ldots, h''_n$ in $\mathscr{H}$ such that the functions

$$f - \sup(h'_1, \ldots, h'_m) \quad \text{and} \quad f - \inf(h''_1, \ldots, h''_n)$$

have sup-norm $\leqq \varepsilon$.

v) $f$ is $\mathscr{H}$*-affine*, i.e. the two envelopes

$$\hat{f} = \inf\{h : h \in \mathscr{H}, h \geqq f\} \quad \text{and} \quad \check{f} = \sup\{h : h \in \mathscr{H}, h \leqq f\}$$

coincide.

The use of condition v) is crucial and typical for the envelope technique. It is the key to the proof of the equivalence of statements i)—v).

---

* The first author's contribution to this paper was promoted by a visit to the Institut des Hautes Etudes Scientifiqes in Bures-sur-Yvette, France

In recent publications, in particular in those by Kitto-Wulbert [7] and Wolff [9], Korovkin approximation in more general Riesz function spaces has been investigated. However, for the simple space $\mathscr{C}_0(X)$ of all real-valued continuous functions vanishing at infinity on a locally compact space $X$ only partial results are available. The equivalence of the appropriately modified conditions ii) and iv) is established in [9]. In [7] only the case $X=\mathbb{N}$, i.e. the space $c_0=\mathscr{C}_0(\mathbb{N})$, is studied. For a finite dimensional subspace $\mathscr{H}$ of $c_0$ the authors of [7] prove that all elements $f\in c_0$ are trapped by $\mathscr{H}$ if and only if the Korovkin closure of $\mathscr{H}$ is $c_0$ and if $\mathscr{H}$ "contains essentially positive members". Here the definition of the Korovkin closure has to be modified as in [7] and [9]: only equicontinuous nets $(T_\alpha)$ are allowed. [For $X$ compact, every net $(T_\alpha)$ of positive linear maps $T_\alpha:\mathscr{C}(X)\to\mathscr{C}(X)$ is obviously equicontinuous provided that $(T_\alpha h)$ converges uniformly to $h$ for a strictly positive function $h\in\mathscr{C}(X)$.]

It is the aim of the present paper to prove that, for the case of function spaces $\mathscr{C}_0(X)$, the envelope technique also can be applied. However, the definition of the envelopes $\hat{f}$ and $\check{f}$ has to be modified. This modification appears in a different context in a recent proof of the Choquet-Deny theorem by Bauer [3] and Lembcke [8]. After the corresponding modifications, conditions i)—v) are again equivalent. The proof of the equivalence uses in an essential way a Hahn-Banach-type theorem of Anger and Lembcke [1] where sublinear forms are replaced by hypolinear ones.

For the case of a compact space $X$ our results yield improvements of the existing theory when $\mathscr{H}$ does not contain strictly positive functions.

In the sequel we shall denote by $X$ an arbitrary locally compact space (always assumed Hausdorff) and by $\mathscr{H}$ an arbitrary linear subspace of $\mathscr{C}_0(X)$. We shall consider $\mathscr{C}_0(X)$ as a normed space with respect to the sup-norm $\|\ldots\|$.

## 1. Almost $\mathscr{H}$-bounded Functions

For every function $f\in\mathscr{C}_0(X)$ the following two conditions are equivalent:

($B_1$) For every $\varepsilon>0$ there exist functions $h_1, h_2\in\mathscr{H}$ satisfying

$$h_1-\varepsilon\leqq f\leqq h_2+\varepsilon.$$

($B_2$) For every $\varepsilon>0$ there exist functions $h_1, h_2\in\mathscr{H}$ and $u\in\mathscr{C}_0(X)$ satisfying $h_1-u\leqq f\leqq h_2+u$ and $\|u\|\leqq\varepsilon$.

Indeed, the implication $B_2\Rightarrow B_1$ being evident we only prove $B_1\Rightarrow B_2$: If $h_1$, $h_2\in\mathscr{H}$ satisfy $h_1-\varepsilon\leqq f\leqq h_2+\varepsilon$ it suffices to choose $u:=\sup(0,h_1-f,f-h_2)$. (We then even have $0\leqq u\leqq\varepsilon$.)

A function $f\in\mathscr{C}_0(X)$ satisfying the equivalent conditions ($B_1$) and ($B_2$) will be called *almost $\mathscr{H}$-bounded*. The set of all almost $\mathscr{H}$-bounded functions in $\mathscr{C}_0(X)$ will be denoted by $\mathscr{H}^*$.

Obviously we then have

**1.1. Lemma.** *$\mathscr{H}^*$ is a closed linear subspace of $\mathscr{C}_0(X)$ satisfying $\mathscr{H}\subset\mathscr{H}^*$.*

For later use we formulate conditions under which all functions in $\mathscr{C}_0(X)$ are almost $\mathscr{H}$-bounded:

**1.2. Lemma.** *The equality $\mathscr{H}^* = \mathscr{C}_0(X)$ holds if and only if the following condition is satisfied:*

(P) *For every $\varepsilon > 0$ and every compact set $K \subset X$ there exists a function $h_K \in \mathscr{H}$ such that $h_K \geqq -\varepsilon$ and $h_K(x) \geqq 1$ for all $x \in K$.*

*Proof.* Suppose $\mathscr{H}^* = \mathscr{C}_0(X)$. Choose a function $u_K \in \mathscr{C}_0(X)$ such that $0 \leqq u_K \leqq 2$ and $u_K(x) = 2$ for all $x \in K$. For $0 < \varepsilon < 1$ there exists a function $h_K \in \mathscr{H}$ satisfying $u_K \leqq h_K + \varepsilon$ since $u_K$ is almost $\mathscr{H}$-bounded. Obviously $h_K$ has all desired properties.— Conversely, it suffices to prove that every function $f \in \mathscr{C}_0(X)$ with norm $\leqq 1$ is almost $\mathscr{H}$-bounded. Given $\varepsilon > 0$ we have $f(x) < \varepsilon$ for all $x$ in the complement of a compact set $K$. We then obtain $h_K(x) \geqq 1 \geqq f(x) > f(x) - \varepsilon$ for all $x \in K$ and $h_K(x) + \varepsilon \geqq 0 > f(x) - \varepsilon$ for all $x \in X \setminus K$. Hence $f \leqq h_K + 2\varepsilon$. The same argument applied to $-f$ yields a function $h'_K$ such that $-f \leqq h'_K + 2\varepsilon$. Thus $f$ is almost $\mathscr{H}$-bounded.

**1.3. Corollary.** *For a discrete space $X$ condition (P) is equivalent to the following discrete version:*

($P_d$) *For every $\varepsilon > 0$ and every $x \in X$ there exists a function $h_x \in \mathscr{H}$ such that $h_x(x) \geqq 1$ and $h_x \geqq -\varepsilon$.*

*Proof.* Only ($P_d$)$\Rightarrow$(P) has to be proved. Let $\varepsilon > 0$ and a compact subset $K$ of $X$ be given. We can assume $K$ to be non empty, hence $K$ is a set with finitely many points $x_1, \ldots, x_n$. Choose $\gamma > 0$ such that $\gamma - \frac{n-1}{n} \cdot \varepsilon \geqq 1$ and for every $i = 1, \ldots, n$ a function $h_{x_i} \in \mathscr{H}$ such that

$$h_{x_i}(x_i) \geqq 1 \quad \text{and} \quad h_{x_i} \geqq -\frac{\varepsilon}{\gamma n}.$$

Then $h_K := \gamma(h_{x_1} + \ldots + h_{x_n})$ has the properties desired in (P).

*1.4. Remarks.* 1. Condition (P) holds and hence the equality $\mathscr{H}^* = \mathscr{C}_0(X)$ is satisfied if for every $x \in X$ there exists a function $h \in \mathscr{H}$ such that $h \geqq 0$ and $h(x) > 0$.

2. For $X = \mathbb{N}$ condition ($P_d$) appears in Kitto-Wulbert [7]. There a linear subspace $\mathscr{H}$ of $c_0 = \mathscr{C}_0(\mathbb{N})$ is said to contain "essentially positive members" if ($P_d$) holds. According to Lemma 1.2 and its Corollarly this just means $\mathscr{H}^* = c_0$.

3. For $X$ compact (P) is equivalent to the existence of a strictly positive function in $\mathscr{H}$.

4. Conditions ($B_1$) and ($B_2$) split into two equivalent super- and sub-versions: A function $f \in \mathscr{C}_0(X)$ is called *almost $\mathscr{H}$-majorized* (resp. $\mathscr{H}$-minorized) if for every $\varepsilon > 0$ there is a function $h \in \mathscr{H}$ majorizing $f - \varepsilon$ (resp. minorizing $f + \varepsilon$). Lemma 1.1 then generalizes as follows: The set $\mathscr{H}_u^*$ of all almost $\mathscr{H}$-majorized functions is a closed convex cone in $\mathscr{C}_0(X)$.

## 2. $\mathscr{H}$-affine Functions

We associate to each $f \in \mathscr{C}_0(X)$ the two enveloping numerical functions

$$\hat{f}: X \to ]-\infty, +\infty] \quad \text{and} \quad \check{f}: X \to [-\infty, +\infty[$$

as follows

$$\hat{f}(x) := \sup_{\varepsilon>0} \inf\{h(x) : h\in\mathscr{H}, f\leqq h+\varepsilon\}, \tag{2.1}$$

$$\check{f}(x) := -\widehat{-f}(x) = \inf_{\varepsilon>0} \sup\{h(x) : h\in\mathscr{H}, h-\varepsilon\leqq f\}. \tag{2.2}$$

Here we use the conventions $\inf\emptyset = +\infty$ and $\sup\emptyset = -\infty$. Clearly $f\equiv +\infty$ or $f\equiv -\infty$ if $f\in\mathscr{C}_0(X)\setminus\mathscr{H}^*$. More precisely $f\equiv +\infty$ whenever $f$ is not almost $\mathscr{H}$-majorized. But also for $f\in\mathscr{H}^*$ the function $f$ can be the constant $+\infty$. It suffices to consider $X = ]0,1]$, $\mathscr{H} = \{x\mapsto\alpha x : \alpha\in\mathbb{R}\}$ and $f(x) = \sqrt{x}$.

The map $f\mapsto\hat{f}$ defined on $\mathscr{C}_0(X)$ possesses the following elementary properties:

$$\check{f}\leqq f\leqq\hat{f}; \tag{2.3}$$

$$\check{h} = h = \hat{h} \quad (h\in\mathscr{H}); \tag{2.4}$$

$$f\leqq g \Rightarrow \hat{f}\leqq\hat{g}; \tag{2.5}$$

$$\widehat{f+g}\leqq\hat{f}+\hat{g}; \tag{2.6}$$

$$\widehat{\lambda f} = \lambda\hat{f} \quad (\lambda>0). \tag{2.7}$$

Consequently, $f\mapsto\hat{f}(x)$ is an *increasing hypolinear form* on $\mathscr{C}_0(X)$ for every $x\in X$.

Important is the following property (cf. also [8]):

**2.1. Lemma.** *$f\mapsto\hat{f}(x)$ is lower semicontinuous on $\mathscr{C}_0(X)$ for every $x\in X$.*

*Proof.* The set $\mathscr{H}_u^*$ of all almost $\mathscr{H}$-majorized functions in $\mathscr{C}_0(X)$ is closed (Remark 1.4, 4). For all functions $f\in\mathscr{C}_0(X)\setminus\mathscr{H}_u^*$ we have $\hat{f}(x) = +\infty$. Consider now a function $f\in\mathscr{C}_0(X)$ and a real number $\alpha$ such that $\hat{f}(x)>\alpha$. We have to prove the existence of a neighborhood $U_f$ of $f$ in $\mathscr{C}_0(X)$ such that $\hat{g}(x)>\alpha$ holds for all $g\in U_f$. This is evident for $f\notin\mathscr{H}_u^*$: it suffices to take $U_f = \mathscr{C}_0(X)\setminus\mathscr{H}_u^*$. For $f\in\mathscr{H}_u^*$ there exists an $\varepsilon>0$ satisfying

$$\alpha < \inf\{h(x) : h\in\mathscr{H}, f\leqq h+\varepsilon\}.$$

Then $U_f := \left\{g\in\mathscr{C}_0(X) : \|g-f\|\leqq\frac{\varepsilon}{2}\right\}$ has the desired property: Indeed, for every $g\in U_f$ and every $h\in\mathscr{H}$ satisfying $g\leqq h+\frac{\varepsilon}{2}$ one has $f\leqq h+\varepsilon$. Hence

$$\alpha < \inf\{h(x) : h\in\mathscr{H}, f\leqq h+\varepsilon\} \leqq \inf\left\{h(x) : h\in\mathscr{H}, g\leqq h+\frac{\varepsilon}{2}\right\} \leqq \hat{g}(x).$$

*2.2. Definition.* Every bounded positive Radon measure $\mu$ on $X$ satisfying

$$\int h\,d\mu = h(x) \quad \text{for all } h\in\mathscr{H} \tag{2.8}$$

will be called an *$\mathscr{H}$-representing measure for $x$*. The set of these measures will be denoted by $\mathscr{M}_x(\mathscr{H})$.

Evidently, the unit mass $\varepsilon_x$ at $x$ is in $\mathscr{M}_x(\mathscr{H})$ for every $x\in X$.

As a consequence of Lemma 2.1 we now obtain an important lemma:

**2.3. Lemma.** *For every function $f \in \mathscr{C}_0(X)$ and every $x \in X$ the following inclusions hold:*

$$]\check{f}(x), \hat{f}(x)[\subset \{\int f\,d\mu : \mu \in \mathscr{M}_x(\mathscr{H})\} \subset [\check{f}(x), \hat{f}(x)]. \tag{2.9}$$

*Proof.* Consider $\mu \in \mathscr{M}_x(\mathscr{H})$ and $f \in \mathscr{C}_0(X)$. We have $\int f\,d\mu \leqq \hat{f}(x) = +\infty$ if $f$ is not almost $\mathscr{H}$-majorized. If $f$ is almost $\mathscr{H}$-majorized, for every $\varepsilon > 0$, there exist $h \in \mathscr{H}$ such that $f \leqq h + \varepsilon$ and thus

$$\int f\,d\mu \leqq h(x) + \varepsilon \|\mu\|.$$

This implies

$$\int f\,d\mu \leqq \inf\{h(x) : h \in \mathscr{H}, f \leqq h + \varepsilon\} + \varepsilon\|\mu\| \leqq \hat{f}(x) + \varepsilon\|\mu\|$$

for all $\varepsilon > 0$, and hence $\int f\,d\mu \leqq \hat{f}(x)$. Replacing $f$ by $-f$ we obtain the inclusion on the right hand side.

Since $g \mapsto \hat{g}(x)$ is a lower semicontinuous hypolinear form on $\mathscr{C}_0(X)$, Proposition 3.2 of Anger-Lembcke [1] applies[1]. Consequently, for $f \in \mathscr{C}_0(X)$ and $\alpha \in ]\check{f}(x), \hat{f}(x)[$, we obtain the existence of a continuous linear form $\varphi$ on $\mathscr{C}_0(X)$ such that $\varphi(f) = \alpha$ and $\varphi(g) \leqq \hat{g}(x)$ for all $g \in \mathscr{C}_0(X)$. Since $g \leqq 0$ implies $g(x) \leqq \hat{0}(x) = 0$ the linear form $\varphi$ is positive. Hence there exists a uniquely determined bounded positive Radon measure $\mu$ on $X$ such that $\varphi(g) = \int g\,d\mu$ for all $g \in \mathscr{C}_0(X)$. Because of (2.4) $\mu$ is an $\mathscr{H}$-representing measure for $x$ such that $\int f\,d\mu = \varphi(f) = \alpha$.

*2.4. Definition.* Functions $f \in \mathscr{C}_0(X)$ satisfying $\check{f} = \hat{f}$ will be called *$\mathscr{H}$-affine*. The set of these functions will be denoted by $\hat{\mathscr{H}}$.

$\hat{\mathscr{H}}$ is a linear subspace of $\mathscr{C}_0(X)$ satisfying

$$\mathscr{H} \subset \hat{\mathscr{H}} \subset \mathscr{H}^*. \tag{2.9}$$

Indeed, it follows from (2.3), (2.4) and (2.6), (2.7) that $\hat{\mathscr{H}}$ is a linear subspace containing $\mathscr{H}$. $\hat{\mathscr{H}}$ is a subspace of $\mathscr{H}^*$ since for functions $f \in \mathscr{C}_0(X) \backslash \mathscr{H}^*$ we have $\hat{f} \equiv +\infty$ or $\check{f} \equiv -\infty$.

Lemma 2.3 will lead to the subsequent characterization of $\mathscr{H}$-affine functions, which will be crucial for the rest of the paper.

**2.5. Proposition.** *For every function $f \in \mathscr{C}_0(X)$ the following four conditions are equivalent:*

a) *$f$ is $\mathscr{H}$-affine.*

b) *For every $\varepsilon > 0$, there exist finitely many functions $h'_1, \ldots, h'_m$ and $h''_1, \ldots, h''_n$ in $H$ such that*

$$\underline{h} := \sup(h'_1, \ldots, h'_m) \quad \text{and} \quad \bar{h} := \inf(h''_1, \ldots, h''_n)$$

*satisfy*

$$\underline{h} - \varepsilon \leqq f \leqq \bar{h} + \varepsilon \quad \text{and} \quad \|\bar{h} - \underline{h}\| \leqq \varepsilon.$$

[1] Choose $E = \mathscr{C}_0(X)$, $C = \{0\}$, $P = \{\lambda f : \lambda \in \mathbb{R}\}$. One could also apply [1], Theorem 2.11.

c) *For every* $\varepsilon>0$, *there exist finitely many functions* $h'_1, \ldots, h'_m$ *and* $h''_1, \ldots, h''_n$ *in* $\mathscr{H}$ *as well as a function* $u \in \mathscr{C}_0(X)$ *of norm* $\|u\| \leqq \varepsilon$ *such that the above defined envelopes* $\underline{h}$ *and* $\bar{h}$ *satisfy*

$$\underline{h}-u \leqq f \leqq \bar{h}+u \quad \textit{and} \quad \|\bar{h}-\underline{h}+2u\| \leqq \varepsilon.^2$$

d) $\int f\, d\mu = f(x)$ *for all* $x \in X$ *and all* $\mu \in \mathscr{M}_x(\mathscr{H})$.

*Proof.* a)⇒b): For given $\varepsilon>0$ and $f \in \mathscr{H}$, we have $f(x)=\hat{f}(x)<f(x)+\frac{\varepsilon}{2}$ and consequently

$$\inf\left\{h(x): h \in \mathscr{H}, f \leqq h+\frac{\varepsilon}{2}\right\}<f(x)+\frac{\varepsilon}{2}.$$

Therefore, given $x \in X$, there exists a function $h_x \in \mathscr{H}$ such that

$$f \leqq h_x+\frac{\varepsilon}{2} \quad \text{and} \quad h_x(x)<f(x)+\frac{\varepsilon}{2},$$

hence the inequality

$$h_x(y)<f(y)+\frac{\varepsilon}{2}$$

holds for all points $y$ of a certain neighborhood $U_x$ of $x$. Since $f \in \hat{\mathscr{H}} \subset \mathscr{H}^*$ there exists furthermore $h_0 \in \mathscr{H}$ such that

$$f \leqq h_0+\frac{\varepsilon}{2}.$$

For $h_0-f \in \mathscr{C}_0(X)$ we have

$$h_0(y)-f(y)<\frac{\varepsilon}{2}$$

for all points $y$ in the complement $X \backslash K$ of a compact set $K$. Finitely many of the sets $U_x, x \in K$, say $U_{x_1}, \ldots, U_{x_n}$, suffice for covering $K$. The $n+1$ functions $h_0$, $h_i := h_{x_i}$, $i=1, \ldots, n$, of $\mathscr{H}$ then have a lower envelope $\bar{h}:=\inf(h_0, h_1, \ldots, h_n)$ which satisfies

$$f \leqq \bar{h}+\frac{\varepsilon}{2}<\bar{h}+\varepsilon \quad \text{and} \quad \bar{h} \leqq f+\frac{\varepsilon}{2},$$

hence $\|\bar{h}-f\| \leqq \frac{\varepsilon}{2}$. Replacing $f$ by $-f$ we obtain an upper envelope $\underline{h}$ of finitely many functions in $\mathscr{H}$ such that

$$\underline{h}-\varepsilon<f \quad \text{and} \quad \|f-\underline{h}\| \leqq \frac{\varepsilon}{2}$$

are satisfied. Hence $\underline{h}$ and $\bar{h}$ have all desired properties.

---

[2] In addition to $u \in \mathscr{C}_0(X)$ and $\|u\| \leqq \varepsilon$ one may assume $u \geqq 0$, as will be shown by the proof

b)⇒c): Given $\varepsilon>0$, there exist finitely many functions $h'_1, \ldots, h'_m$ and $h''_1, \ldots, h''_n$ in $\mathscr{H}$ such that the corresponding envelopes satisfy

$$\underline{h}-\frac{\varepsilon}{4} \leqq f \leqq \bar{h}+\frac{\varepsilon}{4} \quad \text{and} \quad \|\bar{h}-\underline{h}\| \leqq \frac{\varepsilon}{4}.$$

The function $u:=\sup(0, f-\bar{h}, \underline{h}-f)$ is in $\mathscr{C}_0(X)$ and satisfies $0 \leqq u \leqq \frac{\varepsilon}{4}$, hence $\|u\| \leqq \frac{\varepsilon}{4}$, and $\underline{h}-u \leqq f \leqq \bar{h}+u$ as well as

$$\|\bar{h}-\underline{h}+2u\| \leqq \|\bar{h}-\underline{h}\|+2\|u\|<\varepsilon.$$

c)⇒d): For $\varepsilon>0$ we choose $\underline{h}$, $\bar{h}$ and $u$ according to c). For $x \in X$ and a measure $\mu \in \mathscr{M}_x(\mathscr{H})$ we then have

$$\int \underline{h}\, d\mu-\int u\, d\mu \leqq \int f\, d\mu \leqq \int \bar{h}\, d\mu+\int u\, d\mu$$

and, by observing the definition of $\underline{h}$ and $\bar{h}$,

$$\underline{h}(x) \leqq \int \underline{h}\, d\mu \quad \text{and} \quad \int \bar{h}\, d\mu \leqq \bar{h}(x).$$

Therefore,

$$\underline{h}(x)-\varepsilon\|\mu\| \leqq \int f\, d\mu \leqq \bar{h}(x)+\varepsilon\|\mu\|.$$

According to c), we also have

$$\underline{h}(x)-\varepsilon \leqq f(x) \leqq \bar{h}(x)+\varepsilon.$$

Consequently,

$$\left|\int f\, d\mu-f(x)\right| \leqq \bar{h}(x)-\underline{h}(x)+2\varepsilon(\|\mu\|+1).$$

Finally, $\|\bar{h}-\underline{h}+2u\| \leqq \varepsilon$ and $\|u\| \leqq \varepsilon$ imply $\|\bar{h}-\underline{h}\| \leqq 3\varepsilon$ and hence $|\int f\, d\mu-f(x)| \leqq 5\varepsilon(\|\mu\|+1)$. Since $\varepsilon>0$ was arbitrary this implies $\int f\, d\mu=f(x)$.

d)⇒a): The set $\{\int f\, d\mu : \mu \in \mathscr{M}_x(\mathscr{H})\}$ is the singleton $\{f(x)\}$. Therefore Lemma 2.3 yields $\check{f}(x)=\hat{f}(x)=f(x)$ for all $x \in X$. So $f$ is $\mathscr{H}$-affine.

*2.6. Remark.* In the terminology of Kitto-Wulbert [7] the equivalence of a) and c) states that a function $f \in \mathscr{C}_0(X)$ is *$\mathscr{H}$-affine if and only if $f$ is trapped by $\mathscr{H}$*. It is easy to see that, for a compact space $X$, $f$ is trapped by $\mathscr{H}$ if and only if the condition v) of the introduction is satisfied.

## 3. Korovkin Closure

*3.1. Definition.* A net $(T_\alpha)_{\alpha \in A}$ of positive linear maps $T_\alpha : \mathscr{C}_0(X) \to \mathscr{C}_0(X)$ will be called *$\mathscr{H}$-admissible* if it is equicontinuous, i.e. if

$$\sup_{\alpha \in A} \|T_\alpha\|<+\infty, \tag{3.1}$$

and if

$$\lim_{\alpha \in A} \|T_\alpha h-h\|=0 \tag{3.2}$$

holds for all $h \in \mathscr{H}$.

A function $f\in\mathscr{C}_0(X)$ is said to be a *Korovkin function* with respect to $\mathscr{H}$ if

$$\lim_{\alpha\in A}\|T_\alpha f-f\|=0$$

for all $\mathscr{H}$-admissible nets $(T_\alpha)_{\alpha\in A}$. The set of all Korovkin functions with respect to $\mathscr{H}$ will be called the *Korovkin closure of* $\mathscr{H}$ [in $\mathscr{C}_0(X)$] and denoted by $\mathrm{Kor}(\mathscr{H})$.

If in the preceding definition nets are replaced by sequences $(T_n)_{n\in\mathbf{N}}$ we obtain the definition of the *sequential Korovkin closure* which will be denoted by $\mathrm{Kor}_s(\mathscr{H})$.

**3.2. Theorem.** *For every linear subspace* $\mathscr{H}$ *of* $\mathscr{C}_0(X)$ *we have*

$$\mathrm{Kor}(\mathscr{H})=\hat{\mathscr{H}}. \tag{3.3}$$

*Proof.* Let $f$ be an $\mathscr{H}$-affine function and $(T_\alpha)_{\alpha\in A}$ an $\mathscr{H}$-admissible net of positive linear operators $T_\alpha:\mathscr{C}_0(X)\to\mathscr{C}_0(X)$. According to Proposition 2.5, for given $\varepsilon>0$ there exist finite families $(h_i')_{i=1,\dots,m}$ and $(h_j'')_{j=1,\dots,n}$ in $\mathscr{H}$ and a function $u\in\mathscr{C}_0(X)$ such that

$$\|u\|\leqq\varepsilon, \underline{h}-u\leqq f\leqq\bar{h}+u \quad\text{and}\quad \|\bar{h}-\underline{h}+2u\|\leqq\varepsilon,$$

where $\underline{h}=\sup h_i'$ and $\bar{h}=\inf h_j''$. Because of the positivity of the operators $T_\alpha$ we have (for $\alpha\in A$)

$$\sup T_\alpha h_i'\leqq T_\alpha\,\underline{h} \quad\text{and}\quad T_\alpha\bar{h}\leqq\inf T_\alpha h_j'' \;;$$

by means of linearity and positivity we thus obtain

$$\sup T_\alpha h_i'-T_\alpha u\leqq T_\alpha f\leqq\inf T_\alpha h_j''+T_\alpha u.$$

Furthermore,

$$|\inf T_\alpha h_j''-\bar{h}|\leqq\Sigma'':=\sum_{j=1}^{n}|T_\alpha h_j''-h_j''|$$

and

$$|\sup T_\alpha h_i'-\underline{h}|\leqq\Sigma':=\sum_{i=1}^{m}|T_\alpha h_i'-h_i'|.$$

We are thus led to

$$\begin{aligned}T_\alpha f-f&\leqq\inf T_\alpha h_j''-\underline{h}+T_\alpha u+u\\&\leqq\bar{h}-\underline{h}+2u+\Sigma''+T_\alpha u-u\end{aligned}$$

and, by a corresponding calculation, to

$$f-T_\alpha f\leqq\bar{h}-\underline{h}+2u+\Sigma'+T_\alpha u-u.$$

Both inequalities yield

$$|T_\alpha f-f|\leqq\bar{h}-\underline{h}+2u+\Sigma'+\Sigma''+T_\alpha u-u.$$

After introducing $M:=\sup_{\alpha\in A}\|T_\alpha\|$, we are thus led to

$$\begin{aligned}\|T_\alpha f-f\|\leqq\|\bar{h}-\underline{h}+2u\|&+\sum_{i=1}^{m}\|T_\alpha h_i'-h_i'\|\\&+\sum_{j=1}^{n}\|T_\alpha h_j''-h_j''\|+(M+1)\|u\|.\end{aligned}$$

From this inequality the uniform convergence of $(T_\alpha f)$ to $f$ follows. Hence $\hat{\mathscr{H}} \subset \mathrm{Kor}(\mathscr{H})$.

For the inverse inclusion let $f \in \mathrm{Kor}(\mathscr{H})$, $x \in X$ and $\mu \in \mathscr{M}_x(\mathscr{H})$ be given. We consider the set $F(\mathscr{H})$ of all finite, non-empty subsets of $\mathscr{H}$ ordered by inclusion $\subset$. For every $A \in F(\mathscr{H})$ we choose a neighborhood $U_A$ of $x$ such that

$$|h(y)-h(x)| \leqq \frac{1}{|A|} \quad \text{for all } y \in U_A \text{ and all } h \in A,$$

where $|A|$ denotes the number of elements of $A$. Denote by $q_A$ a function in $\mathscr{C}_0(X)$ (for example with compact support) such that

$$0 \leqq q_A \leqq 1, \quad q_A(x)=1 \quad \text{and} \quad q_A(y)=0 \quad \text{for all } y \in X \setminus U_A,$$

and define the map $T_A : \mathscr{C}_0(X) \to \mathscr{C}_0(X)$ by putting

$$T_A g := \int g \, d\mu \cdot q_A + g(1-q_A), \qquad g \in \mathscr{C}_0(X).$$

It is easy to check $(T_A)_{A \in F(\mathscr{H})}$ is a net of positive linear maps such that

$$\|T_A\| \leqq 1 + \|\mu\|$$

and

$$|T_A h - h| = q_A |h - h(x)| \leqq \frac{1}{|A|}$$

for all $A \in F(\mathscr{H})$ and all $h \in A$. Consequently, the net $(T_A)$ is admissible which implies, in particular,

$$\lim_{A \in F(\mathscr{H})} T_A f(x) = f(x)$$

since $f$ is a Korovkin function. But $T_A f(x) = \int f \, d\mu$ for every $A$, and hence $\int f \, d\mu = f(x)$. Since $x \in X$ and $\mu \in \mathscr{M}_x(\mathscr{H})$ were arbitrary, Proposition 2.5 yields $f \in \hat{\mathscr{H}}$.

A slight modification of the above proof yields

**3.3. Corollary 1.** *For a separable linear subspace $\mathscr{H}$ of $\mathscr{C}_0(X)$ $\hat{\mathscr{H}}$ is the sequential Korovkin closure of $\mathscr{H}$:*

$$\mathrm{Kor}_s(\mathscr{H}) = \mathrm{Kor}(\mathscr{H}) = \hat{\mathscr{H}}. \tag{3.4}$$

*Proof.* Clearly $\mathrm{Kor}(\mathscr{H}) \subset \mathrm{Kor}_s(\mathscr{H})$. So we only have to prove that every function $f \in \mathrm{Kor}_s(\mathscr{H})$ is $\mathscr{H}$-affine. As above we prove $\int f \, d\mu = f(x)$ for all $x \in X$ and all $\mu \in \mathscr{M}_x(\mathscr{H})$. By assumption there is a dense sequence $(h_n)_{n \in \mathbb{N}}$ in $\mathscr{H}$. Put $A_n := \{h_1, \ldots, h_n\}$ and denote by $U_n$ a neighborhood of $x$, such that $|h_i(y) - h_i(x)| \leqq n^{-1}$ for all $y \in U_n$ and all $i = 1, \ldots, n$. The above proof then yields the sequence $T_n := T_{A_n}$, $n = 1, 2, \ldots$, of positive linear maps $T_n : \mathscr{C}_0(X) \to \mathscr{C}_0(X)$ such that

$$\sup \|T_n\| < +\infty \quad \text{and} \quad \lim_{n \to \infty} \|T_n h_i - h_i\| = 0$$

for all $i = 1, 2, \ldots$,. Since $(h_n)$ is dense in $\mathscr{H}$ the latter property implies $\lim_{n \to \infty} \|T_n h - h\| = 0$ for all $h \in \mathscr{H}$. Thus $(T_n)$ is $\mathscr{H}$-admissible and, as above, we conclude that $\int f \, d\mu = f(x)$.

**3.4. Corollary 2.** *If $X$ has a countable base the equalities*

$$\mathrm{Kor}_s(\mathcal{H}) = \mathrm{Kor}(\mathcal{H}) = \hat{\mathcal{H}}$$

*hold for an arbitrary linear subspace $\mathcal{H}$ of $\mathcal{C}_0(X)$.*

**3.5. Corollary 3.** *For every linear subspace $\mathcal{H}$ of $\mathcal{C}_0(X)$ the inclusion*

$$\mathrm{Kor}(\mathcal{H}) \subset \mathcal{H}^*$$

*holds.*

As usual the *Choquet boundary* $\partial_{\mathcal{H}} X$ of $X$ with respect to $\mathcal{H}$ is defined as the set of all $x \in X$ for which $\varepsilon_x$ is the only $\mathcal{H}$-representing measure: $\mathcal{M}_x(\mathcal{H}) = \{\varepsilon_x\}$.

**3.6. Corollary 4.** *A linear subspace $\mathcal{H}$ of $\mathcal{C}_0(X)$ is a Korovkin space in $\mathcal{C}_0(X)$, i.e. $\mathrm{Kor}(\mathcal{H}) = \mathcal{C}_0(X)$, if and only if $\partial_{\mathcal{H}} X = X$.*

*Proof.* The equality $\hat{\mathcal{H}} = \mathcal{C}_0(X)$ implies $\int f\, d\mu = f(x)$ for all $f \in \mathcal{C}_0(X)$, $x \in X$ and $\mu \in \mathcal{M}_x(\mathcal{H})$. Hence $\mathcal{M}_x(\mathcal{H}) = \{\varepsilon_x\}$ for $x \in X$. Conversely, $\partial_{\mathcal{H}} X = X$ implies $\hat{\mathcal{H}} = \mathcal{C}_0(X)$ according to Proposition 2.5. The result then follows since $\hat{\mathcal{H}} = \mathrm{Kor}(\mathcal{H})$.

We have presented the second part of the proof of Theorem 3.2 in such a way that by simply rewriting it one obtains the following more general result about the *universal Korovkin closure* $\mathrm{Kor}_u(\mathcal{H})$. This, by definition, is the set of all functions $f \in \mathcal{C}_0(X)$ having the following property: For every locally convex Riesz space $E$, every continuous Riesz homomorphism $P: \mathcal{C}_0(X) \to E$ and every equicontinuous net $(T_\alpha)_{\alpha \in A}$ of positive linear maps $T_\alpha: \mathcal{C}_0(X) \to E$, the net $(T_\alpha f)$ converges uniformly to $Pf$ provided that $(T_\alpha h)$ converges uniformly to $Ph$ for all $h \in \mathcal{H}$.

**3.7. Theorem.** *$\hat{\mathcal{H}}$ is the universal Korovkin closure of $\mathcal{H}$.*

*3.8. Remark.* In Kitto-Wulbert [7], Theorem 1.7 it is proved that a finite dimensional linear subspace $\mathcal{H}$ of $c_0 = \mathcal{C}_0(\mathbb{N})$ traps all functions in $c_0$ if and only if $\mathrm{Kor}(\mathcal{H}) = c_0$ *and* $\mathcal{H}$ contains essentially positive members. According to Remark 1.4, 2, to $\mathrm{Kor}(\mathcal{H}) = \hat{\mathcal{H}}$ and the characterization of $\mathcal{H}$-affine functions given in Proposition 2.5, the latter condition as well as the assumption $\dim \mathcal{H} < \infty$ are superfluous.

## 4. Applications

**1.** Let us consider first the case of a *compact* space $X$. One then has $\mathcal{C}_0(X) = \mathcal{C}(X)$. Accordingly $\mathcal{H}$ is an arbitrary linear subspace of $\mathcal{C}(X)$. The results of the preceding sections then reduce to the results about Korovkin approximation in $\mathcal{C}(X)$ known from [2] and [4]. This proves that the assumption about the existence of a strictly positive function in $\mathcal{H}$ is not necessary for the validity of these results.

For the major results, in particular for Corollary 3.6 (again for $X$ compact), this has also been observed by Donner [6] and Berens-Lorentz [5]. Their key lemma for proving this states that $\mathcal{M}_x(\mathcal{H}) = \{\varepsilon_x\}$ for some $x \in X$ implies the existence of a strictly positive function in $\mathcal{H}$ (cf. [5], Lemma 2). Lemma 2.3 of the present article proves that even a weaker assumption leads to this conclusion: it suffices to assume $\hat{f}(x) < +\infty$ for some strictly positive function $f \in \mathcal{C}(X)$ and some $x \in X$.

Let us now turn again to the case of a *locally compact* space $X$. In order to prove that a given linear subspace $\mathscr{H}$ of $\mathscr{C}_0(X)$ is a Korovkin space in $\mathscr{C}_0(X)$, Corollary 3.6 is most useful in a given concrete situation. In order to prove $\partial_{\mathscr{H}}X = X$ it suffices to show that, given an arbitrary point $x_0 \in X$, there exists a *peak function* for $x_0$ in $\mathscr{H}$. This, by definition, is a function $h \in \mathscr{H}$ such that $h(x) > h(x_0) = 0$ holds for all points $x \in X$ different from $x_0$. We shall give several examples:

**2.** For $X = ]0, 1]$, consider the linear subspace $\mathscr{H}$ of $\mathscr{C}_0(X)$ generated by the three functions $x \mapsto x$, $x^2$ and $x^3$. Then $\mathscr{H}$ is a Korovkin space in $\mathscr{C}_0(X)$. Indeed,

$$x \mapsto x(x - x_0)^2$$

is a peak function for $x_0 \in X$ in $\mathscr{H}$.

**3.** For $X = [1, +\infty[$, the linear space $\mathscr{H}$ generated by the three functions $x \mapsto \frac{1}{x}, \frac{1}{x^2}$, and $\frac{1}{x^3}$ is a Korovkin space in $\mathscr{C}_0(X)$. It suffices again to remark that

$$x \mapsto \frac{1}{x}\left(\frac{1}{x} - \frac{1}{x_0}\right)^2$$

is a peak function for $x_0 \in X$ in $\mathscr{H}$.

**4.** For $X = \mathbb{R}_+$, the linear space $\mathscr{H}$ generated by the three functions $x \mapsto e^{-\lambda_i x}$, $i = 1$, 2, 3, is a Korovkin space whenever $0 < \lambda_1 < \lambda_2 < \lambda_3$ are three different positive real numbers. Indeed, for an arbitrary point $x_0 \in \mathbb{R}_+ = X$,

$$h_{x_0}(x) := e^{-\lambda_1 x} + \alpha e^{-\lambda_2 x} + \beta e^{-\lambda_3 x} = e^{-\lambda_1 x} H(x)$$

is a peak function for $x_0$ in $\mathscr{H}$ provided that

$$H(x) > H(x_0) = 0$$

holds for all $x \in \mathbb{R}_+ \setminus \{x_0\}$ for a certain choice of $\alpha$, $\beta$. But the system

$$H(x_0) = H'(x_0) = 0$$

of linear equations for $\alpha$, $\beta$, where

$$H(x) = 1 + \alpha e^{-(\lambda_2 - \lambda_1)x} + \beta e^{-(\lambda_3 - \lambda_1)x}$$

has a unique solution leading to $H''(x) = (\lambda_2 - \lambda_1)(\lambda_3 - \lambda_1) > 0$ for all $x \in \mathbb{R}_+$. Hence $H(x) > H(x_0) = 0$ for all $x \neq x_0$ in $\mathbb{R}_+$.

As a consequence we now obtain the following application: Consider a net $(t_\alpha)_{\alpha \in A}$ of Lebesgue-integrable functions $t_\alpha \geqq 0$ on the half-line $\mathbb{R}_+$. Assume that

$$\sup_{\alpha \in A} \int_0^\infty t_\alpha(x)dx < +\infty. \tag{4.1}$$

Denote by $\tilde{t}_\alpha$ the *Laplace transform* of $t_\alpha$ and by $T_\alpha$ the operator

$$T_\alpha f(x) := \int_0^\infty f(x+y) t_\alpha(y) dy \quad (\alpha \in A), \tag{4.2}$$

defined, at least, for functions $f \in \mathscr{C}_0(\mathbb{R}_+)$. It follows from Lebesgue's dominated convergence theorem that each $T_\alpha$ is a positive linear map from $\mathscr{C}_0(\mathbb{R}_+)$ into $\mathscr{C}_0(\mathbb{R}_+)$ such that $\|T_\alpha\| = \int_0^\infty t_\alpha(x)dx$. We then have

**4.1. Proposition.** *If there exist three different real numbers* $0<\lambda_1<\lambda_2<\lambda_3$ *for which*

$$\lim_{\alpha \in A} \tilde{t}_\alpha(\lambda_i)=1 \qquad (i=1,2,3), \tag{4.3}$$

*then* $(T_\alpha f)$ *converges uniformly to* $f$ *on* $\mathbb{R}_+$ *for all functions* $f \in \mathscr{C}_0(\mathbb{R}_+)$.

*Proof.* In view of the preceding remarks we only have to assure that $(T_\alpha e_i)$ converges uniformly to $e_i$ for the three functions $e_i(x) := e^{-\lambda_i x}$, $i=1,2,3$. But this follows from (4.3) since

$$T_\alpha e_i(x) = e^{-\lambda_i x} \tilde{t}_\alpha(\lambda_i).$$

5. Applications 2 and 3 are closely related to the fact that the three functions $x \mapsto 1$, $x$, $x^2$ generate a Korovkin space in $\mathscr{C}([0,1])$, hence for the compact interval $[0,1]$. By multiplying the restrictions of these functions to $]0,1]$ by the function $x \mapsto x$, one arrives at the three functions considered in Application 2. (A similar device, after mapping $]0,1]$ onto $[1,+\infty[$ by means of $x \mapsto \frac{1}{x}$, leads to Application 3.) The following two results show that one could equally well multiply $x \mapsto 1$, $x$, $x^2$ by an arbitrary strictly positive function in $\mathscr{C}_0(]0,1])$.

Consider the one-point compactification $X_0 = X \cup \{\omega\}$ of a locally compact space $X$, and let $f$ be a strictly positive function in $\mathscr{C}_0(X)$. (Such a function exists if and only if $X$ is countable at infinity.) We denote by $\mathscr{H}_0$ a given linear subspace of $\mathscr{C}(X_0)$ and define

$$\mathscr{H}_f := \{f \cdot (h|X) : h \in \mathscr{H}_0\}.$$

Then $\mathscr{H}_f$ is a linear subspace of $\mathscr{C}_0(X)$. We claim that the restriction of an $\mathscr{H}_0$-affine function to $X$ is the product of an $\mathscr{H}_f$-affine function with $\frac{1}{f}$.

**4.2. Lemma.** *In the above situation one has*

$$\hat{\mathscr{H}}_0|X \subset \frac{1}{f}\hat{\mathscr{H}}_f. \tag{4.5}$$

*Proof.* Consider an $\mathscr{H}_f$-representing measure $\mu \in \mathscr{M}_x(\mathscr{H}_f)$ for $x \in X$. Then $\nu := \frac{f}{f(x)}\mu$ is a positive Radon measure on $X$, hence on $X_0$ such that

$$\int h\,d\nu = \frac{1}{f(x)} \int fh\,d\mu = h(x)$$

for all $h \in \mathscr{H}_0$. Consequently, $\mu \in \mathscr{M}_x(\mathscr{H}_0)$. For every function $g \in \hat{\mathscr{H}}_0$ we thus have $\int g\,d\nu = g(x)$, i.e.

$$\int fg\,d\mu = f(x)g(x).$$

This proves that $f\cdot(g|X)$ is $\mathscr{H}_f$-affine. Hence $g|X$ is in $\frac{1}{f}\hat{\mathscr{H}}_f$.

**4.3. Proposition.** *$\mathscr{H}_f$ is a Korovkin space in $\mathscr{C}_0(X)$ if $X\subset\partial_{\mathscr{H}_0}X_0$.*

*Proof.* We shall prove that $\partial_{\mathscr{H}_f}X=X$. The result then follows from Corollary 3.6. So let $x\in X$ and $\mu\in\mathscr{M}_x(\mathscr{H}_f)$ be given. As in the above proof one shows that $\nu:=\frac{f}{f(x)}\mu$ is a measure in $\mathscr{M}_x(\mathscr{H}_0)$, hence equal to $\varepsilon_x$. But $\nu=\varepsilon_x$ implies $\mu=\varepsilon_x$ since $\mu(\{x\})=1$ and $\mu(X\setminus\{x\})=0$ because of the strict positivity of $f$.

**4.4. Corollary.** *$\mathscr{H}_f$ is a Korovkin space in $\mathscr{C}_0(X)$ if $\mathscr{H}_0$ is a Korovkin space in $\mathscr{C}(X_0)$.*

*4.5. Remarks.* I. In (4.5) one does not have equality, in general. Indeed, $\mathscr{H}_0=\mathscr{C}(X_0)$ trivially is a Korovkin space in $\mathscr{C}(X_0)$. Hence $\hat{\mathscr{H}}_f=\mathscr{C}_0(X)$ according to the last Corollary. But $\frac{1}{f}\mathscr{C}_0(X)\neq\mathscr{C}(X_0)|X$, in general.

2. Proposition 4.3 is a stronger result than Corollary 4.4. It suffices to consider $X=]0,1]$, $X_0=[0,1]$ and

$$\mathscr{H}_0:=\{h\in\mathscr{C}([0,1]):\int_0^1 h(x)dx=h(0)\}.$$

Then $\partial_{\mathscr{H}_0}X_0=]0,1]=X$. Consequently, $\mathscr{H}_0$ is not a Korovkin space in $\mathscr{C}(X_0)$.

## References


1. Anger, B., Lembcke, J.: Hahn-Banach type theorems for hypolinear functionals. Math. Ann. **209**, 127—151 (1974)
2. Bauer, H.: Theorems of Korovkin type for adapted spaces. Ann. Inst. Fourier **23**, Fasc. 4, 245—260 (1973)
3. Bauer, H.: Funktionenkegel und Integralungleichungen. Sitz. Ber. math. naturw. Kl. Bayer. Akad. Wiss. München 1977, 53—61 (1978)
4. Berens, H., Lorentz, G.G.: Theorems of Korovkin type for positive linear operators on Banach lattices. Approximation Theory (ed. G. G. Lorentz). New York, London: Academic Press 1973
5. Berens, H., Lorentz, G.G.: Geometric theory of Korovkin sets. J. Approximation Theory **15**, 161—189 (1975)
6. Donner, K.: Korovkin theorems for positive linear operators. J. Approximation Theory **13**, 443—450 (1975)
7. Kitto, W., Wulbert, D.E.: Korovkin approximation in $L_p$-spaces. Pacific J. Math. **63**, 153—167 (1976)
8. Lembcke, J.: Note zu „Funktionenkegel und Integralungleichungen" von H. Bauer. Sitz. Ber. math.-naturw. Kl. Bayer. Akad. Wiss. München 1977, 139—142 (1978)
9. Wolff, M.: On Korovkin-type theorems in special function lattices. J. Approximation Theory (to appear)


Received November 12, 1977

# Approximation and abstract boundaries [S12]

Amer. Math. Monthly **85** (1978), 632–647
[Zbl. Math. 416.41023; MR 80f:41014]

polynomial approximation is usually presented as the classical representative of the many approximation theorems. It states that for every continuous real-valued function $f$ on a compact interval $[a,b]$ of the real line **R**, there exists a sequence $(p_n)$ of real polynomials (of one variable) which converges uniformly to $f$ on $[a,b]$. Among the many proofs for this theorem, we select one in which the approximating sequence $(p_n)$ is defined explicitly by means of the polynomials introduced by S. Bernstein.

It suffices to consider the unit interval $[0,1]$ in **R**. For the given continuous function $f$ and $n=1,2,\ldots$ the $n$th Bernstein polynomial is then defined as

$$p_n^f(x)=\sum_{k=0}^{n}\binom{n}{k}f\left(\frac{k}{n}\right)x^k(1-x)^{n-k}. \tag{1}$$

Deviating from the original proof of S. Bernstein [**8**], we test the statement of the Weierstrass approximation theorem for three simple functions: the constant function 1, the function $f(x)=x$ which will be denoted by the symbol id (=identity), and its square $\mathrm{id}^2$, i.e., the function $f(x)=x^2$. From the binomial theorem and some simple calculations, we obtain for every $n=1,2,\ldots$

$$\begin{aligned} p_n^1&=1;\\ p_n^{\mathrm{id}}&=\mathrm{id}; \\ p_n^{\mathrm{id}^2}&=\frac{1}{n}\mathrm{id}+\frac{n-1}{n}\mathrm{id}^2. \end{aligned} \tag{2}$$

These formulas prove that for each of the three functions $f=1$, id, $\mathrm{id}^2$ the sequence $(p_n^f)$ converges uniformly to $f$ on $[0,1]$.

According to a theorem of Korovkin [**15**], [**16**]—the so-called three function theorem—this simple test immediately finishes the proof of the uniform convergence of $(p_n^f)$ to $f$ for every continuous real-valued function on $[0,1]$. To make this clear, we have to give an appropriate interpretation to the polynomial sequence $(p_n^f)$ in the context of the linear space $\mathcal{C}([0,1])$ of all real-valued functions on $[0,1]$. For every natural number $n$ we attach to $f$ the polynomial $p_n^f$ and thus arrive at a map

$$B_n:\mathcal{C}([0,1])\to\mathcal{C}([0,1]) \tag{3}$$

of the space $\mathcal{C}([0,1])$ into itself. That is, at $f\in\mathcal{C}([0,1])$ the map $B_n$ takes on the value $B_nf=p_n^f$, the $n$th Bernstein polynomial for $f$. From the definition (1) of these polynomials, one can see immediately that $B_n$ is in fact linear, hence a *linear operator* of $\mathcal{C}([0,1])$ into itself which, in addition, is also *positive*. The latter property means that every non-negative function $f\in\mathcal{C}([0,1])$ is transformed into a function $B_nf$ which is again non-negative. After this interpretation of our original situation, we may apply Korovkin's three function theorem to turn our test into a proof of the Weierstrass approximation theorem. The theorem in question is, more precisely,

KOROVKIN'S FIRST THEOREM. *Let $[a,b]$ be a compact interval in* **R** *and let $(L_n)$ be a sequence of positive linear operators $L_n$ of $\mathcal{C}([a,b])$ into itself. Suppose that $(L_nf)$ converges uniformly to $f$ for the*

The author studied at Erlangen and at Nancy and received his doctorate under the direction of O. Haupt at Erlangen. He has taught at Hamburg and at Erlangen, where he now holds a professorship; he has also been a visiting professor at the University of Washington (Seattle), the Sorbonne, California Institute of Technology, New Mexico State University (Las Cruces), Bahia Blanca, and Aarhus. His main fields of interest are potential theory, probability and functional analysis.—*Editors*

*three test functions* $f=1$, id, $\mathrm{id}^2$. *Then* $(L_n f)$ *converges uniformly to* $f$ *on* $[a,b]$ *for all functions* $f \in \mathcal{C}([a,b])$.*

A quick proof can be given by means of a trick. Every $f \in \mathcal{C}([a,b])$ is bounded:

$$|f(x)| \leq \gamma \qquad \text{for all} \quad x \in [a,b]$$

where $\gamma$ is a certain positive number. Also, $f$ is uniformly continuous on $[a,b]$: given $\varepsilon > 0$ there exists a number $\delta > 0$ such that, for all $x,y \in [a,b]$,

$$|x-y| < \sqrt{\delta} \quad \Rightarrow \quad |f(x)-f(y)| \leq \varepsilon,$$

or, equivalently,

$$(x-y)^2 < \delta \quad \Rightarrow \quad |f(x)-f(y)| \leq \varepsilon.$$

The trick consists of putting $\sqrt{\delta}$ where one normally puts $\delta$. It leads for arbitrary points $x,y \in [a,b]$ to the inequality

$$|f(x)-f(y)| \leq \varepsilon + \alpha(x-y)^2,$$

with $\alpha = 2\gamma\delta^{-1}$, which one immediately derives by considering the two cases $(x-y)^2 < \delta$ and $(x-y)^2 \geq \delta$. Thus we have for all $y \in [a,b]$ the following inequality between functions:

$$|f - f(y)| \leq \varepsilon + \alpha(\mathrm{id} - y)^2.$$

Linearity and positivity of the operators $L_n$ then imply†

$$|L_n f - f(y) L_n 1| \leq \varepsilon L_n 1 + \alpha\big(L_n \mathrm{id}^2 - 2y L_n \mathrm{id} + y^2 L_n 1\big).$$

Evaluating the above expression at $x=y$, we obtain

$$|L_n f - f L_n 1| \leq \varepsilon L_n 1 + \alpha\big(L_n \mathrm{id}^2 - 2\,\mathrm{id}\, L_n \mathrm{id} + \mathrm{id}^2 L_n 1\big).$$

From the assumption on $(L_n f)$ for the three functions $f=1$, id, $\mathrm{id}^2$ and from the triangle inequality it now follows that $L_n f$ converges to $f$ uniformly. We even obtain explicit bounds for the supremum norm

$$\|L_n f - f\| = \sup_{x \in [a,b]} |L_n f(x) - f(x)|$$

provided that such bounds are known for the three test functions as is the case for the original operator sequence $(B_n)$ according to (2).

Quite naturally, Korovkin's theorem led to a quest for a deeper understanding of the role of the three test functions involved. The above proof, based on the trick involving uniform continuity, does not, for example, provide answers to a number of questions. Can one replace the three functions 1, id, $\mathrm{id}^2$ by other test functions? Can one get along with two test functions? Is there an analogous theorem in higher dimensions?

In what follows, a generalized version of Korovkin's First Theorem will be derived and analyzed in order to answer such questions in a natural way.

**1. Test space and corresponding affine functions.** The above-mentioned questions will be treated within a framework broad enough for interesting applications. The compact interval $[a,b]$ will be

*We denote by $L_n f$ the image of $f$ under the operator $L_n$ and by $L_n f(x)$ the value of $L_n f$ at the point $x$. However, $L_n(f(x))$ is the image of the constant function $f(x)$ under $L_n$, hence $L_n(f(x)) = f(x) L_n 1$ by linearity.

†All positive linear operators $L: \mathcal{C}([a,b]) \to \mathcal{C}([a,b])$ are *increasing* since $f \leq g$ implies $g - f \geq 0$ and hence $Lg - Lf = L(g-f) \geq 0$, i.e. $Lf \leq Lg$. In particular, the inequality $|Lf| \leq L|f|$ is derived from $-|f| \leq f \leq |f|$ where $|g|$ denotes the function $x \mapsto |g(x)|$ $(f, g \in \mathcal{C}([a,b]))$.

replaced by an arbitrary *compact metric space* $X$.[†] The set $\{1,\mathrm{id},\mathrm{id}^2\}$ of the former three test functions will be replaced by an arbitrary (finite or infinite) subset $\mathfrak{T}$ of the linear space $\mathcal{C}(X)$ of all continuous real-valued functions on $X$. (All functions in $\mathcal{C}(X)$ are bounded since $X$ is compact.) The only assumption on the set $\mathfrak{T}$ will be that the constant function 1 is in it:

$$1 \in \mathfrak{T}.^{\ddagger} \tag{4}$$

$\mathfrak{T}$ will be called the set of *test functions* or the *test set*.

A sequence $(L_n)_{n=1,2,\ldots}$ of positive linear operators

$$L_n : \mathcal{C}(X) \to \mathcal{C}(X)$$

will be called $\mathfrak{T}$-*admissible* if the sequence $(L_n t)$ converges uniformly to $t$ on $X$ for all test functions $t \in \mathfrak{T}$, i.e.,

$$\lim_{n\to\infty} \|L_n t - t\| = 0 \qquad \text{for all} \quad t \in \mathfrak{T}, \tag{5}$$

where

$$\|f\| = \sup_{x\in X} |f(x)|$$

denotes the so-called sup-norm of a bounded real-valued function on $X$. A function $f \in \mathcal{C}(X)$ will be called a *Korovkin function* (with respect to $\mathfrak{T}$) if the sequence $(L_n f)$ converges to $f$ uniformly on $X$ for *all* $\mathfrak{T}$-admissible sequences $(L_n)$ of positive linear operators of $\mathcal{C}(X)$ into $\mathcal{C}(X)$.

With the help of these definitions, the main problems to be studied in this paper can be formulated as follows:

PROBLEM 1. *Which functions $f \in \mathcal{C}(X)$ are Korovkin functions with respect to a given test set $\mathfrak{T}$?*

PROBLEM 2. *What conditions on $\mathfrak{T}$ imply that all functions $f \in \mathcal{C}(X)$ are Korovkin functions?*

The linearity of the operators $L_n$ immediately leads to the observation that each linear combination

$$h = \lambda_1 t_1 + \cdots + \lambda_k t_k \qquad (\lambda_1,\ldots,\lambda_k \in \mathbf{R})$$

of test functions $t_1,\ldots,t_k \in \mathfrak{T}$ is a Korovkin function for $\mathfrak{T}$. In fact, the triangle inequality yields

$$\|L_n h - h\| \leq |\lambda_1|\,\|L_n t_1 - t_1\| + \cdots + |\lambda_k|\,\|L_n t_k - t_k\| \tag{6}$$

for all $n$. Hence all functions from the linear hull $\mathrm{lin}\,\mathfrak{T}$ of $\mathfrak{T}$ in $\mathcal{C}(X)$ are Korovkin functions. In addition, we remark that "Korovkin function with respect to $\mathfrak{T}$" and "Korovkin function with respect to $\mathrm{lin}\,\mathfrak{T}$" are equivalent properties for a function $f \in \mathcal{C}(X)$. In what follows $\mathrm{lin}\,\mathfrak{T}$ will be called the *test space* corresponding to the test set $\mathfrak{T}$.

The type of answer one expects to Problem 2 is that $\mathfrak{T}$ or $\mathrm{lin}\,\mathfrak{T}$ should be "sufficiently large." Hence we should try to measure the "size" of $\mathfrak{T}$ or $\mathrm{lin}\,\mathfrak{T}$. In the case of Korovkin's theorem, $\mathfrak{T}$ equals $\{1,\mathrm{id},\mathrm{id}^2\}$; consequently, $\mathrm{lin}\,\mathfrak{T}$ is the set of all real polynomials $x \mapsto \alpha + \beta x + \gamma x^2$ of degree $\leq 2$. As a linear space of dimension 3, $\mathrm{lin}\,\mathfrak{T}$ is a small subspace of $\mathcal{C}([a,b])$. However, $\mathrm{lin}\,\mathfrak{T}$ is large in the sense that an arbitrary function $f \in \mathcal{C}([a,b])$ is the infimum of all functions $h \in \mathrm{lin}\,\mathfrak{T}$ satisfying $h \geq f$ and, respectively, the supremum of all functions in $\mathrm{lin}\,\mathfrak{T}$ below $f$. This is geometrically plausible. One just has to observe that the graph of a polynomial $x \mapsto \alpha + \beta x + \gamma x^2$ of degree 2 (i.e., with $\gamma \neq 0$) is a parabola with an axis of symmetry parallel to the $y$-axis:

[†]An inexperienced reader should choose for $X$ a closed bounded subset of $\mathbf{R}^p$, $p = 1,2,\ldots$. A major part of the following considerations even holds for arbitrary compact topological spaces.

[‡]A formally weaker assumption would be to demand the existence of a strictly positive test function $t_0 \in \mathfrak{T}$. However, the following transformations reduce this case to the one considered above. Choose $\mathfrak{T}^* = \{(t/t_0) \mid t \in \mathfrak{T}\}$ as a new set of test functions and replace every linear operator $L : \mathcal{C}(X) \to \mathcal{C}(X)$ by the linear operator $L^*$ defined by

$$L^* f = \frac{1}{t_0} L(t_0 f).$$

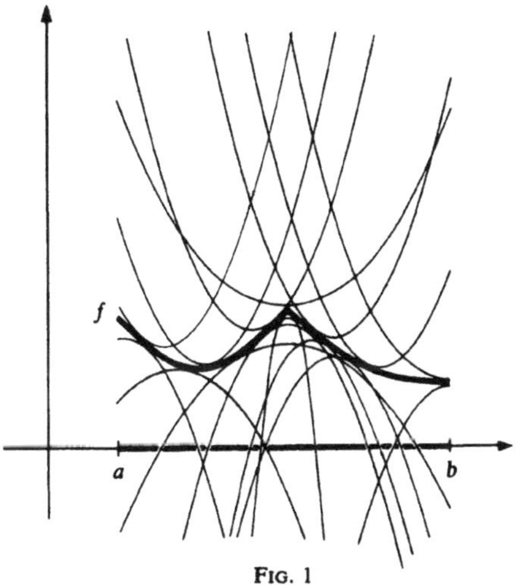


FIG. 1

As we will shortly see, it is exactly this behavior which hits the nerve of the problem and which leads to an adequate notion of $\mathfrak{T}$ being sufficiently large. In pursuing this claim, we define for each function $f \in \mathcal{C}(X)$ two, in general discontinuous, real-valued functions $f^*$ and $f_*$ on $X$ by putting

$$\begin{aligned} f^*(x) &= \inf\{h(x) \mid h \geq f, h \in \operatorname{lin} \mathfrak{T}\} \\ f_*(x) &= \sup\{h(x) \mid h \leq f, h \in \operatorname{lin} \mathfrak{T}\} \end{aligned} \qquad (x \in X). \tag{7}$$

This definition makes sense since $f$ is bounded on the compact metric space $X$ and since, as a consequence of (4), $\operatorname{lin} \mathfrak{T}$ contains the constant real functions as elements. We have

$$f_* \leq f \leq f^*. \tag{8}$$

The sup-norm $\|f^* - f_*\|$ of the non-negative function $f^* - f_*$ in some sense measures how well or how badly $f$ adjusts itself to the "shape" of the functions in the test space $\operatorname{lin} \mathfrak{T}$.

EXAMPLE 1. Consider the situation as in Korovkin's theorem: $X = [a,b]$, a compact interval in $\mathbf{R}$, and $\mathfrak{T} = \{1, \mathrm{id}, \mathrm{id}^2\}$. If one accepts the geometric argument preceding Figure 1, then one obtains $f_* = f = f^*$ for all $f \in \mathcal{C}([a,b])$. (A rigorous proof will be given later.)

EXAMPLE 2. Let again be $X = [a,b]$; but consider $\mathfrak{T} = \{1, \mathrm{id}\}$. The test space consists of all affine functions $x \mapsto \alpha + \beta x$. For all $f \in \mathcal{C}([a,b])$, $f^*$ is the smallest concave (necessarily continuous) function $\geq f$ and, correspondingly, $f_*$ the greatest convex (necessarily continuous) function $\leq f$ on $[a,b]$. Consequently, $f_* = f = f^*$ holds if and only if $f$ is affine, hence is in $\operatorname{lin} \mathfrak{T}$.

Inspired by this example, we introduce the following notions, again in the framework of the general situation. A function $f \in \mathcal{C}(X)$ will be called *$\mathfrak{T}$-affine*, if $f_* = f^* (= f)$ holds. The set of all $\mathfrak{T}$-affine-functions will be denoted by $\mathfrak{T}^*$. All functions of the test space are trivially $\mathfrak{T}$-affine; therefore we have

$$\mathfrak{T} \subset \operatorname{lin} \mathfrak{T} \subset \mathfrak{T}^* \subset \mathcal{C}(X). \tag{9}$$

It is easily checked that $\mathfrak{T}^*$ is a linear subspace of $\mathcal{C}(X)$; one just has to observe that for all functions $f, g \in \mathcal{C}(X)$ and every real number $\lambda \geq 0$ we have

$$(f+g)^* \leq f^* + g^*, \quad (\lambda f)^* = \lambda f^*, \quad (-f)^* = -f_*. \tag{10}$$

We now have to decide whether the new notion of $\mathfrak{T}$-affine function really is relevant for Problem

1. We start with a preparatory consideration. Let $f$ be a function in $\mathcal{C}(X)$ and let $(L_n)$ be a $\mathfrak{T}$-admissible sequence of positive linear operators of $\mathcal{C}(X)$ into itself. For $g,h \in \operatorname{lin}\mathfrak{T}$ satisfying $g \leq f \leq h$, we have $L_n g \leq L_n f \leq L_n h$ for all $n$. From this, we obtain

$$g(x) \leq \varliminf_{n\to\infty} L_n f(x) \leq \varlimsup_{n\to\infty} L_n f(x) \leq h(x)$$

keeping in mind that $(L_n g)$ resp. $(L_n h)$ converge uniformly and hence pointwise to $g$ resp. $h$. This being true for all such functions $g,h \in \lim\mathfrak{T}$, we deduce from (7) that

$$f_*(x) \leq \varliminf_{n\to\infty} L_n f(x) \leq \varlimsup_{n\to\infty} L_n f(x) \leq f^*(x). \tag{11}$$

Thus $(L_n f(x))$ converges to $f(x)$ for a given $x \in X$ if $f_*(x) = f^*(x)$. In particular we have proved that for a $\mathfrak{T}$-affine function $f$ the sequence $(L_n f)$ converges at least pointwise to $f$.

From this observation there is only a small step to the following partial answer to our Problem 1:

PROPOSITION 1. *Every $\mathfrak{T}$-affine function is a Korovkin function.*

*Proof.* We try to modify the proof given for the case of pointwise convergence. First we remark that, given $f \in \mathfrak{T}^*$ and $\varepsilon > 0$, there exist finitely many functions $h'_1, \ldots, h'_k$ and $h''_1, \ldots, h''_k$ in $\operatorname{lin}\mathfrak{T}$ such that we have

$$\underline{h} \leq f \leq \bar{h} \quad \text{and} \quad \bar{h} - \underline{h} < \varepsilon$$

for their corresponding envelopes

$$\underline{h} = \sup(h'_1, \ldots, h'_k) \quad \text{and} \quad \bar{h} = \inf(h''_1, \ldots, h''_k).$$

Indeed, it follows from $f_*(x) = f(x) = f^*(x)$ and (7) that for every $x \in X$ there exist functions $h'_x$ and $h''_x$ in $\operatorname{lin}\mathfrak{T}$ such that

$$h'_x \leq f \leq h''_x \quad \text{and} \quad h''_x(x) - h'_x(x) < \varepsilon.$$

Consequently, there is an open neighborhood $U_x$ of $x$ in $x$ such that

$$h''_x(y) - h'_x(y) < \varepsilon$$

holds for all points $y \in U_x$. Obviously, $(U_x)_{x\in X}$ is an open covering of $X$. By the definition of compactness finitely many of the $U_x$, say $U_{x_1}, \ldots, U_{x_k}$, suffice to cover $X$. Then the functions

$$h'_j = h'_{x_j} \quad \text{and} \quad h''_j = h''_{x_j} \qquad (j = 1, \ldots, k)$$

have all desired properties.

Now let $(L_n)$ be a $\mathfrak{T}$-admissible sequence of positive linear operators of $\mathcal{C}(X)$ into itself. Given $\varepsilon > 0$, there is a natural number $n_0$ such that

$$\|L_n h'_j - h'_j\| < \varepsilon \quad \text{and} \quad \|L_n h''_j - h''_j\| < \varepsilon$$

for all $n \geq n_0$ and all $j = 1, \ldots, k$. Since all maps $L_n$ are increasing, one has

$$L_n h'_j \leq L_n f \leq L_n h''_j$$

for all $n$ and $j = 1, \ldots, k$. Consequently, for all $n \geq n_0$, we have

$$h'_j - \varepsilon \leq L_n f \leq h''_j + \varepsilon \qquad (j = 1, \ldots, k)$$

and hence

$$\underline{h} - \varepsilon \leq L_n f \leq \bar{h} + \varepsilon.$$

By making use of the inequalities $\underline{h} \leq f \leq \bar{h}$, one finally obtains

$$|L_n f - f| \leq \bar{h} - \underline{h} + \varepsilon < 2\varepsilon \qquad \text{for all} \quad n \geq n_0$$

and from this uniform convergence of $(L_n f)$ to $f$. So each $\mathfrak{T}$-affine function $f$ is a Korovkin function.

REMARK. Contrary to the proof of Korovkin's First Theorem where we have used the linearity of the operators $L_n$ in an essential way, the above proof does not use linearity explicitly. One only uses the fact that the operators $L_n$ are increasing, which is a consequence of linearity and positivity. Therefore, if $\mathfrak{T}$ is a linear subspace of $\mathcal{C}(X)$, i.e., $\mathfrak{T} = \operatorname{lin}\mathfrak{T}$, Theorem 1 holds more generally for sequences $(L_n)$ of increasing, not necessarily linear, operators of $\mathcal{C}(X)$ into itself with the convergence behavior expressed by (5).

To summarize, we can state that Problem 2 has found a first answer in Theorem 1. Indeed, if all functions in $\mathcal{C}(X)$ are $\mathfrak{T}$-affine, then each function in $\mathcal{C}(X)$ is a Korovkin function. According to Example 1, this is the situation in Korovkin's First Theorem.

A necessary condition for the validity of $\mathfrak{T}^* = \mathcal{C}(X)$ can now be formulated: $\mathfrak{T}$ must *separate points*. That is, given an arbitrary pair $x_1, x_2$ of distinct points in $X$, there exists a test function $t \in \mathfrak{T}$ satisfying $t(x_1) \neq t(x_2)$. Indeed, there certainly is a function $f \in C(X)$ satisfying $f(x_1) \neq f(x_2)$, e.g., $x \mapsto d(x, x_2)$, where $d$ is the metric of $X$. As a consequence of $\mathfrak{T}^* = \mathcal{C}(X)$ and (7), then there also exists a function $h \in \operatorname{lin}\mathfrak{T}$ such that $h(x_1) \neq h(x_2)$. This finishes the proof since $h$ is a linear combination of test functions. The condition of point separation is certainly not sufficient for the validity of $\mathfrak{T}^* = \mathcal{C}(X)$ as can be seen from Example 2.

**2. Test set and Choquet boundary.** Whether the above first answer to Problem 2 is of practical use depends greatly on the difficulty of proving the equality $\mathfrak{T}^* = \mathcal{C}(X)$ in a given concrete case. We will now develop a method for attacking this problem. This method is based on the observation that, according to (10),

$$\lim L_n f(x) = f(x)$$

holds for all $f \in \mathcal{C}(X)$ and all $\mathfrak{T}$-admissible sequences $(L_n)$ at a given point $x \in X$ provided that

$$f_*(x) = f^*(x) \qquad \text{for all} \quad f \in \mathcal{C}(X).$$

This is the motivation for a deeper investigation of the set

$$\partial_{\mathfrak{T}} X = \{x \in X \mid f_*(x) = f^*(x) \qquad \text{for all} \quad f \in \mathcal{C}(X)\}. \tag{12}$$

In Example 1 this set equals $X = [a, b]$; in Example 2 it contains only the endpoints $a$ and $b$ of the interval as elements. In order to see this, one should observe that for the concave continuous function

$$f(x) = \begin{cases} x - a, & a \leqslant x \leqslant \dfrac{a+b}{2} \\ b - x, & \dfrac{a+b}{2} \leqslant x \leqslant b \end{cases}$$

we have $f^* = f$ and $f_* = 0$. Consequently $f^*(x) = f_*(x)$ only holds for $x = a$ and $x = b$.

In these two examples, it is evident that each function of the test space $\operatorname{lin}\mathfrak{T}$ attains its global maximum and minimum on $X$ at a point of $\partial_{\mathfrak{T}} X$. We mention without proof that this is true in general. (For a proof see [2], p. 96.) Because of the formal analogy to the boundary maximum principle of function and potential theory, $\partial_{\mathfrak{T}} X$ is called an *abstract boundary*; more precisely it is called the *Choquet boundary* of $X$ with respect to $\mathfrak{T}$.

By the very definition of this boundary the equality $\mathfrak{T}^* = \mathcal{C}(X)$ holds if and only if the Choquet boundary $\partial_{\mathfrak{T}} X$ equals $X$. This tautology will be transformed into a criterion for the validity of $\mathfrak{T}^* = \mathcal{C}(X)$ if we succeed in determining the Choquet boundary by using other tools. This can be done with the help of the notion of a measure and, in many important special cases, by means of the behavior of the zeros of functions in the test space.

As usual, a positive Radon measure $\mu$ on $X$ is by definition a positive linear form on $\mathcal{C}(X)$, i.e. a positive linear map $\mu : C(X) \to \mathbf{R}$. (According to the Riesz representation theorem $\mu$ can be viewed as a regular Borel measure $\mu_0 \geqslant 0$ on $X$. In fact there exists exactly one such measure $\mu_0$ such that $\mu(f) = \int f \, d\mu_0$ for all $f \in \mathcal{C}(X)$.) For every point $x \in X$, $f \mapsto f(x)$ is such a measure; it will be denoted by $\varepsilon_x$ and is called the *unit measure* (or *Dirac measure*) at $x$. In what follows, those measures which operate on the test set $\mathfrak{T}$ like a unit measure will be of particular importance. A positive Radon measure $\mu$ on $X$ will be called a *test measure* with respect to $\mathfrak{T}$ (or a $\mathfrak{T}$*-representing measure*) for a point $x \in X$ if

$$\mu(t) = t(x) \qquad \text{for all} \quad t \in \mathfrak{T}. \tag{13}$$

Obviously $\varepsilon_x$ is always such a test measure. In general, there are many test measures for a given point $x$. In Example 2 with $X = [0,1]$, the Lebesgue measure $\lambda$, which by definition is the linear form

$$\lambda(f) = \int_0^1 f(\xi)\,d\xi \qquad (f \in \mathcal{C}(X)),$$

as well as each of the measures $\frac{1}{2}\varepsilon_\alpha + \frac{1}{2}\varepsilon_{1-\alpha}, 0 \leq \alpha < \frac{1}{2}$, is a test measure for $x = \frac{1}{2}$ different from $\varepsilon_x$. However, it will soon turn out that the Choquet boundary $\partial_{\mathfrak{T}} X$ is exactly the set of those points $x \in X$ for which $\varepsilon_x$ is the only test measure. The reason for this is the following close connection between the envelopes of (7) and the test measures.

LEMMA. *For every point $x \in X$ and every function $f \in \mathcal{C}(X)$, the following two sets coincide:*

$$\left[ f_*(x), f^*(x) \right] = \{ \mu(f) \mid \mu \text{ test measure for } x \}. \tag{14}$$

*Proof.* Let $\mu$ be a test measure for $x$. Then $\mu(f)$ lies in the interval $[f_*(x), f^*(x)]$. In order to see this, consider functions $g, h \in \operatorname{lin} \mathfrak{T}$ satisfying $g \leq f \leq h$. Then according to (13)

$$g(x) = \mu(g) \leq \mu(f) \leq \mu(h) = h(x)$$

since $g$ and $h$ are linear combinations of test functions. Because of (7), this yields

$$f_*(x) \leq \mu(f) \leq f^*(x).$$

Conversely, let $\alpha$ be a real number in $[f_*(x), f^*(x)]$. On the linear subspace $\{\lambda f \mid \lambda \in \mathbf{R}\}$ of $\mathcal{C}(X)$ generated by $f$, the map

$$\lambda f \mapsto \lambda \alpha$$

defines a linear form $\mu_0$ which is majorized by the function $p(g) = g^*(x)$ defined on $\mathcal{C}(X)$. Indeed, for $\lambda \geq 0$, we have

$$\mu_0(\lambda f) = \lambda \alpha \leq \lambda f^*(x) = (\lambda f)^*(x) = p(\lambda f),$$

and for $\lambda < 0$

$$\mu_0(\lambda f) = \lambda \alpha \leq \lambda f_*(x) = -\lambda(-f)^*(x) = (\lambda f)^*(x) = p(\lambda f).$$

According to (10), $p$ is a sublinear form on $\mathcal{C}(X)$, that is, $p(g_1 + g_2) \leq p(g_1) + p(g_2)$ and $p(\lambda g) = \lambda p(g)$ for arbitrary functions $g_1, g_2, g \in \mathcal{C}(X)$ and arbitrary numbers $\lambda \geq 0$. Now the Hahn–Banach theorem ([13], p. 212) states that $\mu_0$ can be extended to a linear form $\mu$ defined on $\mathcal{C}(X)$ which still is majorized by $p$. $\mu$ is a positive linear form since $f \leq 0, f \in \mathcal{C}(X)$, implies

$$\mu(f) \leq p(f) = f^*(x) \leq 0.$$

(Observe that the constant function 0 is in $\operatorname{lin} \mathfrak{T}$ and majorizes $f$.) Consequently, $\mu$ is a positive Radon measure on $X$. For functions $h \in \operatorname{lin} \mathfrak{T}$ we have $h_* = h = h^*$ which implies $\mu(h) \leq h^*(x) = h(x)$ as well as $-\mu(h) = \mu(-h) \leq (-h)^*(x) = -h_*(x) = -h(x)$. So a test measure $\mu$ for $x$ has been found which by construction satisfies $\mu(f) = \mu_0(f) = \alpha$. This finishes the proof.

As already announced, we now obtain the crucial

PROPOSITION 2. *A point $x \in X$ lies in the Choquet boundary if and only if $\varepsilon_x$ is the only test measure for $x$.*

*Proof.* According to definition (12), $x \in \partial_{\mathfrak{T}} X$ is equivalent to the validity of $f_*(x) = f^*(x)$ for all $f \in \mathcal{C}(X)$. By the preceding Lemma, this means that an arbitrary test measure $\mu$ for $x$ satisfies $\mu(f) = f_*(x) = f^*(x) = f(x)$ for all $f \in \mathcal{C}(X)$. Equivalent to this is the statement that $\varepsilon_x$ is the only test measure for $x$.

It is this Proposition by which the tautology

$$\mathfrak{T}^* = \mathcal{C}(X) \Leftrightarrow \partial_{\mathfrak{T}} X = X$$

is transformed into a non-trivial statement, namely:

COROLLARY. *All functions in $\mathcal{C}(X)$ are $\mathfrak{T}$-affine if and only if $\varepsilon_x$ is the only test measure for an arbitrary point $x \in X$.*

Finally, we arrive at the decisive answer to our Problems 1 and 2.

THEOREM 1. *For a given test set $\mathfrak{T} \subset \mathcal{C}(X)$, the $\mathfrak{T}$-affine functions are the only Korovkin functions.*

THEOREM 2. *Let $\mathfrak{T} \subset \mathcal{C}(X)$ be a given test set. All functions in $\mathcal{C}(X)$ are Korovkin functions if and only if an arbitrary point $x \in X$ allows $\varepsilon_x$ as the only test measure.*

In view of the preceding Corollary, the second theorem is an immediate consequence of the first one. So we only have to prove Theorem 1.

*Proof.* Because of Proposition 1, we need only show that a Korovkin function $f \in \mathcal{C}(X)$ is necessarily $\mathfrak{T}$-affine. According to our Lemma, this is equivalent to the statement that $\mu(f) = f(x_0)$ for all points $x_0 \in X$ and all test measures $\mu$ of $x_0$. So let $x_0 \in X$ and let a test measure $\mu$ for $x_0$ be given. Denote by $d$ the metric of the space $X$. Let us consider for each natural number $n$ the closed set $A_n$ of all points $x \in X$ with distance $d(x, x_0) > (1/n)$ from $x_0$. It is well known that for non-empty $A_n$ the function $x \mapsto d(x, A_n)$ (which measures the distance between $x$ and $A_n$) is continuous and that $\alpha_n = d(x_0, A_n) > 0$. Putting

$$q_n(x) = \begin{cases} 1, & \text{if } A_n = \emptyset \\ \min\left(\dfrac{1}{\alpha_n} d(x, A_n), 1\right), & \text{if } A_n \neq \phi \end{cases}$$

we obtain a sequence $(q_n)$ of continuous real-valued functions on $X$ with the following properties.

$$0 \leqslant q_n \leqslant 1, \quad q_n(x_0) = 1, \quad q_n(x) = 0 \quad \text{on} \quad A_n \tag{15}$$

for all $n$. For $g \in \mathcal{C}(X)$

$$L_n g = \mu(g) q_n + (1 - q_n) g$$

defines a sequence $(L_n)$ of positive linear operators of $\mathcal{C}(X)$ into itself. It is $\mathfrak{T}$-admissible since for $t \in \mathfrak{T}$ and $x \in X$ we have

$$|L_n t(x) - t(x)| = |t(x_0) - t(x)| q_n(x)$$

and hence because of (15)

$$\|L_n t - t\| \leqslant \sup\{|t(x_0) - t(x)| \mid x \in X \setminus A_n\}.$$

But the right-hand side becomes arbitrarily small for $n$ sufficiently large since $t$ is continuous and $X \setminus A_n$ is the open ball with center $x_0$ and radius $1/n$. For the Korovkin function $f$ the sequence $(L_n f)$ converges uniformly to $f$, in particular

$$\lim L_n f(x_0) = f(x_0).$$

But according to our construction $L_n f(x_0) = \mu(f)$, and hence $\mu(f) = f(x_0)$. This proves that $f$ is $\mathfrak{T}$-affine.

Whether a point $x$ belongs to the Choquet boundary $\partial_{\mathfrak{T}} X$ can be decided in many examples by proving the existence of a function $h > 0$ in the test space $\operatorname{lin} \mathfrak{T}$, which has $x$ as *the only zero*, i.e., $0 = h(x) < h(x')$ for all $x' \in X$, $x' \neq x$.

PROPOSITION 3. *Suppose that a function $h \geqslant 0$ of $\operatorname{lin} \mathfrak{T}$ has $x \in X$ as the only zero. Then $\varepsilon_x$ is the only representing measure for $x$, i.e., $x$ lies in the Choquet boundary $\partial_{\mathfrak{T}} X$.*

*Proof.* The proof is almost trivial if we use some measure theory. Indeed, for every test measure $\mu$ for $x$ we derive from $\mu(h)=h(x)=0$ that the set of all $x'\in X$ satisfying $h(x')>0$, i.e., the complement of $\{x\}$ has measure zero. It then follows that $\mu=\varepsilon_x$ since the constant function 1 is a test function.

A direct proof can be given in the following way. Consider $f\in\mathcal{C}(X)$ such that $f(x)=0$. Then $f$ is bounded; consequently $f\leqslant\gamma$ for some real number $\gamma>0$. Also, $f$ is continuous at $x$; consequently, for a given number $\varepsilon>0$ there is an open neighborhood $V$ of $x$ such that $f(y)\leqslant\varepsilon$ holds for all $y\in V$. On the closed, hence compact, set $X\setminus V$ the function $h$ is positive and thus minorized by some number $\eta>0$. Therefore,

$$f\leqslant\varepsilon+\frac{\gamma}{\eta}h$$

and hence

$$\mu(f)\leqslant\varepsilon\mu(1)+\frac{\gamma}{\eta}\mu(h)=\varepsilon$$

for an arbitrary test measure $\mu$ for the point $x$. One only has to observe that 1 and $h$ are functions in the test space and that, accordingly, $\mu(1)=1$ and $\mu(h)=h(x)=0$. Since $\varepsilon>0$ was arbitrarily chosen, we conclude that $\mu(f)\leqslant 0$ for all $f\in\mathcal{C}(X)$ satisfying $f(x)=0$ and in fact $\mu(f)=0$, since these functions form a linear subspace. For an arbitrary $f\in\mathcal{C}(X)$ the function $f-f(x)$ is of the above type. So we have

$$0=\mu(f-f(x))=\mu(f)-f(x)\mu(1)$$

and thus $\mu(f)=f(x)$. This proves $\mu=\varepsilon_x$.

**3. Applications.** We now demonstrate the usefulness of our Theorems and of Proposition 3. First we return to

EXAMPLE 1. We have intentionally omitted, up to this point, a rigorous proof of the intuitive fact that all functions in $\mathcal{C}([a,b])$ are $\mathfrak{T}$-affine. This can now be easily shown. It suffices to observe that for each point $x_0\in[a,b]$ the function

$$x\mapsto(x-x_0)^2$$

lies in the test space lin $\mathfrak{T}$. It is non-negative and its only zero is $x_0$.

Theorem 2, together with Proposition 3, therefore leads to a new proof of Korovkin's first theorem which is totally different from the one given in the Introduction. In this connection, the reader is again referred to the Remark in Section 1.

EXAMPLE 3. Let now $X$ be the circle

$$\mathbf{T}=\{z\in\mathbf{C}\,|\,|z|=1\}$$

in the complex plane $\mathbf{C}$. By means of the map $x\mapsto e^{ix}$ we can identify $\mathbf{T}$ topologically with the interval $[0,2\pi]$ after having identified the endpoints 0 and $2\pi$. As test set we choose

$$\mathfrak{T}=\{1,\mathrm{Re},\mathrm{Im}\},$$

where Re [resp., Im] denotes the real [resp., imaginary] part of the identity mapping, i.e., $z\mapsto\mathrm{Re}\,z$ [resp., $z\mapsto\mathrm{Im}\,z$]. After the above identification, this amounts to considering $\{1,\cos,\sin\}$ as test set.

The test space lin $\mathfrak{T}$ then consists of all functions

$$x\mapsto\alpha+\beta\sin(x+\gamma)$$

with real coefficients $\alpha$, $\beta$, $\gamma$. For $x_0\in\mathbf{T}$

$$h(x)=1+\sin\left(x+\frac{3}{2}\pi-x_0\right)$$

is a function $\geqslant 0$ in lin $\mathfrak{T}$ with $x_0$ as its only zero in $\mathbf{T}$. So we arrive at another classical result of Korovkin as a consequence of Theorem 2:

KOROVKIN'S SECOND THEOREM. *For the circle* $\mathbf{T}$ *and the test set* $\mathcal{T}=\{1,\mathrm{Re},\mathrm{Im}\}$ *every function in* $\mathcal{C}(\mathbf{T})$ *is a Korovkin function.*

This has the following nice application. Associate to an arbitrary function $f\in\mathcal{C}(\mathbf{T},\mathbf{C})$, i.e., to a complex-valued continuous function $f$ on $\mathbf{T}$, its formal Fourier series

$$\sum_{n=-\infty}^{+\infty} c_n e^{inx}$$

with the coefficients

$$c_n=\frac{1}{2\pi}\int_0^{2\pi} f(t)e^{-int}\,dt \qquad (n=0,\pm 1,\ldots).$$

Define for every natural number $n$ an operator $L_n:\mathcal{C}(\mathbf{T},\mathbf{C})\to\mathcal{C}(\mathbf{T},\mathbf{C})$ as follows:

$$L_n f=\frac{s_0+\cdots+s_{n-1}}{n} \qquad (f\in\mathcal{C}(\mathbf{T},\mathbf{C}))$$

where $s_n$ is the $n$th partial sum

$$s_n(x)=\sum_{\nu=-n}^{+n} c_\nu e^{i\nu x} \qquad (x\in[0,2\pi])$$

of the Fourier series. Then $L_n$ obviously is linear. According to the classical representation formula of Fejér (c.f. [**13**, p. 292]),

$$L_n f(x)=\frac{1}{2\pi n}\int_0^{2\pi} f(x+t)\left[\frac{\sin\frac{n}{2}t}{\sin\frac{t}{2}}\right]^2 dt$$

for $x\in[0,2\pi]$ and $f\in\mathcal{C}(\mathbf{T},\mathbf{C})$. Thus non-negative functions are mapped into non-negative functions, so $L_n$ is a positive operator. After restricting $L_n$ to $\mathcal{C}(\mathbf{T})$ we end up with a sequence of positive linear operators of $\mathcal{C}(\mathbf{T})$ into itself. For the function $f_0=1$ all Fourier coefficients are zero except for $c_0=1$. For $f_1(x)=e^{ix}$ the coefficient $c_1=1$ is the only one different from zero. We thus obtain

$$L_n f_0=f_0$$
$$L_n f_1=\frac{n-1}{n}f_1$$

and hence after decomposition of $e^{ix}$ into real and imaginary parts

$$L_n\cos=\frac{n-1}{n}\cos,\qquad L_n\sin=\frac{n-1}{n}\sin$$

for all $n$. Consequently, the sequence $(L_n)$ is $\mathcal{T}$-admissible and Korovkin's Second Theorem yields uniform convergence of $(L_n f)$ to $f$ on $\mathbf{T}$ for all $f\in\mathcal{C}(\mathbf{T})$ and then even for all $f\in\mathcal{C}(\mathbf{T},\mathbf{C})$.

This is the classical theorem about the uniform Cesàro summability of Fourier series of continuous functions. From here one can easily go on to obtain Fejér's theorem about $L^1$-Cesàro summability of Fourier series of $L^1$-functions.

EXAMPLE 4. Now let $X$ be a compact subset of $\mathbf{R}^k$. We assume that for each point $x\in X$ there is at least one supporting hyperplane which intersects $X$ just in $x$, that is, there exists a linear form $l$ on $\mathbf{R}^k$ satisfying

$$l(x)<l(x') \qquad \text{for all } x'\in X, x'\neq x. \tag{16}$$

As a test set we choose

$$\mathcal{T}=\{1,p_1,\ldots,p_k\}$$

where $p_j:X\to\mathbf{R}$ is the $j$th coordinate function (restricted to $X$), which assigns to each $x\in X$ its $j$th coordinate $x_j$ $(j=1,\ldots,k)$.

Every linear form on $\mathbf{R}^k$ is a linear combination of the $k$ coordinate functions. Therefore, condition (16) just states that every point $x \in X$ is the only zero of a non-negative function of the test space lin $\mathfrak{T}$. So by Proposition 3, we can again apply Theorem 2 in order to obtain: *All functions in* $\mathcal{C}(X)$ *are Korovkin functions.*

This result contains Korovkin's Second Theorem as a special case, since through every point of the circle $\mathbf{T}$ in $\mathbf{C}$ $(=\mathbf{R}^2)$ there is exactly one supporting line, the tangent.

EXAMPLE 5. Let $X$ be an arbitrary metric space and let $\mathbf{F} \subset \mathbf{C}(X)$ be a set of real-valued continuous functions separating points (cf. the end of Section 1). If one takes as test set

$$\mathbf{T} = \{1\} \cup \mathbf{F} \cup \{f^2 \mid f \in \mathbf{F}\},$$

the set of powers $f^j$, $j = 0, 1, 2$, of functions $f$ in $\mathbf{F}$, then *all functions in* $\mathbf{C}(X)$ *are Korovkin functions.*

Indeed, consider an arbitrary test measure $\mu$ for $x \in X$. The function $(f - f(x))^2$ is for each $f \in \mathbf{F}$ in the test space; furthermore

$$\mu((f - f(x))^2) = 0.$$

By using a well-known measure theoretic argument (cf. [6, p. 225 and Theorem 2.5.2]) we obtain for the support $S_\mu$ of $\mu$

$$S_\mu \subset \bigcap_{f \in \mathbf{F}} \{x' \in X \mid f(x') = f(x)\}.$$

Point separation then yields

$$S_\mu \subset \{x\}$$

and finally $\mu = \varepsilon_x$ because of $\mu(1) = 1$. (A direct but longer proof can be given along the elementary ideas used in the proof of Proposition 3.)

In particular if $X$ is a compact subset of $\mathbf{R}^k$, $k = 1, 2, \ldots$, we can choose $\mathbf{F}$ to be the set $\{p_1, \ldots, p_k\}$ where $p_1, \ldots, p_k$ are again the coordinate functions. So $\mathbf{T}$ equals

$$\{1, p_1, \ldots, p_k, p_1^2, \ldots, p_k^2\}.$$

We can even reduce the number of test functions by considering

$$\mathbf{T} = \{1, p_1, \ldots, p_k, p_1^2 + \cdots + p_k^2\}$$

as a test set. In both cases it suffices to apply Proposition 3:

$$h = \sum_{j=1}^{k} (p_j - p_j(x))^2$$

is a function $\geqslant 0$ from $\operatorname{lin} \mathbf{T}$ with $x$ as the only zero.

Korovkin's First Theorem is a special case of this result: For $X = [a, b] \subset \mathbf{R}$ we have $p_1 = \mathrm{id}$ and are thus led to the test set $\{1, \mathrm{id}, \mathrm{id}^2\}$.

**4. Geometric interpretation.** We will now discuss the most important example of a Choquet boundary as well as its influence on the kind of problems we are treating.

Consider a *compact convex* set $C$ in $\mathbf{R}^k$. Then $C$ contains with each two of its points the segment joining them. As in Example 4, we introduce

$$\mathfrak{T}_C = \{1, p_1, \ldots, p_k\}$$

as test set. Then the test space lin $\mathfrak{T}_C$ is the space $\mathbf{A}(C)$ of all *affine* functions on $\mathbf{R}^k$ restricted to $C$, so all functions of the form

$$x = (x_1, \ldots, x_k) \mapsto \alpha_0 + \sum_{j=1}^{k} \alpha_j x_j$$

with real coefficients.

Let $a$ and $b$ be points of $C$ and let $x$ be a point of the segment joining $a$ and $b$. Then $x$ can be written in the form

$$x=\lambda a+(1-\lambda)b$$

where $\lambda$ is a real number in the unit interval. Thus

$$\mu=\lambda\varepsilon_a+(1-\lambda)\varepsilon_b$$

is a positive Radon measure on $C$ satisfying $\mu(t)=t(x)$ for all test functions $t\in\mathfrak{T}_C$, so it is a test measure for $x$. It equals $\varepsilon_x$ if and only if $x=a$ or $x=b$. Therefore, $x$ is not a point of the Choquet boundary $\partial_{\mathfrak{T}_C}C$ if $a\neq b$ and $0<\lambda<1$; that is, if $x$ is a point of a segment in $C$ but different from the two end points. Points of $C$ which can only be end points of segments in $C$ on which they are situated are called *extreme points* of $C$. These are exactly those points $x\in C$ for which $C\setminus\{x\}$ is convex. If we denote by $\operatorname{ex} C$ the set of all extreme points of $C$, we have just proved that $\partial_{\mathfrak{T}_C}C$ is a subset of $\operatorname{ex} C$. However, one even has equality:

PROPOSITION 4. *For every compact convex set $C\subset\mathbf{R}^k$ and the test set $\mathfrak{T}_C=\{1,p_1,\dots,p_k\}$ one has*

$$\partial_{\mathfrak{T}_C}C=\operatorname{ex} C. \tag{17}$$

We shall not give a proof of this result here, but rather refer the reader to Choquet [9, §29] or Phelps [19, §6] where Proposition 4 appears in much greater generality. An elementary introduction to the study of $\operatorname{ex} C$ can be found in Jacobs [14].

Let us now return to the general situation of a compact metric space $X$ and a test set $\mathfrak{T}\subset\mathcal{C}(X)$. In addition to condition (4), according to which the constant function 1 belongs to $\mathfrak{T}$, we now demand that $\mathfrak{T}$ is a finite set:

$$\mathfrak{T}=\{1,t_1,\dots,t_k\}, \tag{18}$$

and that $\mathfrak{T}$ separates points. (The case $\mathfrak{T}=\{1\}$ is uninteresting since $\partial_{\mathfrak{T}}X=X$ can hold only if $\mathfrak{T}$ is point separating, that is if $X$ is a singleton.) The test space $\operatorname{lin}\mathfrak{T}$ then is of finite dimension. Finiteness of $\mathfrak{T}$ allows the definition of a continuous map $\Phi: X\to\mathbf{R}^k$ by means of

$$\Phi(x)=(t_1(x),\dots,t_k(x)). \tag{19}$$

Consequently, $\Phi(X)$ is compact as a continuous image of $X$. Moreover, $\Phi$ is injective since $\mathfrak{T}$ separates the points of $X$. So, by a well-known theorem, $\Phi$ is a homeomorphism from $X$ onto the image $\Phi(X)$. Hence we have embedded $X$ homeomorphically in the euclidean $k$-space $\mathbf{R}^k$.

The relation between our new situation and the one considered in Proposition 4 becomes more evident by introducing the convex hull $C$ of the compact set $\Phi(X)$ in $\mathbf{R}^k$. This convex hull is also compact (cf. [24, Theorem 3.10, p. 40]), so we arrive at a compact convex set in $\mathbf{R}^k$ which is associated with $(X,\mathfrak{T})$ in a very natural way. In terms of this set $C$, a new and more lucid interpretation can be given to the Choquet boundary $\partial_{\mathfrak{T}}X$.

PROPOSITION 5. *For every test set $\mathfrak{T}=\{1,t_1,\dots,t_k\}$ separating the points of $X$, for the corresponding map $\Phi: X\to\mathbf{R}^k$ and the convex hull $C$ of $\Phi(X)$ one has*

$$\Phi(\partial_{\mathfrak{T}}X)=\operatorname{ex} C. \tag{20}$$

In other words: A point $x\in X$ belongs to the Choquet boundary $\partial_{\mathfrak{T}}X$ if and only if its image $\Phi(x)$ is an extreme point of $C$.

We restrict ourselves to a sketch of the proof: By means of $\Phi$ we can shift the test functions $t_0=1,t_1,\dots,t_k$ to $\Phi(X)$: the function $u_j=t_j\circ\Phi^{-1}$ on $\Phi(X)$ corresponds to $t_j$. Then $u_0$ is the constant function 1 and $u_j$ is the $j$th coordinate function on $\Phi(X)$, $j=1,\dots,k$, since for every point $y\in\Phi(X)$ and its pre-image $x=\Phi^{-1}(y)$ we have

$$u_j(y)=t_j(x)=p_j(\Phi(x))$$

because of (19). Thus the test set $\mathfrak{T}$ on $X$ is transferred into the test set $\mathfrak{T}_C$ of Proposition 4 restricted to $\Phi(X)$. The homeomorphism $\Phi: X\to\Phi(X)$ also transforms, in a natural way, every test measure for

$x$ (with respect to $\mathfrak{T}$) into a test measure for $\Phi(X)$ (with respect to $\mathfrak{T}_C$), which, however, is supported by $\Phi(X)$. Now Proposition 4 can be strengthened in such a way that $\mathrm{ex}\,C$ is the set of all points $y \in C$ for which $\varepsilon_y$ is the only test measure for $y$ (with respect to $\mathfrak{T}_C$) which is supported by $\Phi(X)$. Then Proposition 5 follows. Details can be found in [2] and [9].

The usefulness of Proposition 5 for our problems lies in the following consequence:

COROLLARY. *In the situation of Proposition 5 all functions in $\mathcal{C}(X)$ are Korovkin functions with respect to $\mathfrak{T}$ if and only if*

$$\Phi(X) = \mathrm{ex}\,C. \tag{21}$$

It suffices to observe that $\partial_{\mathfrak{T}}X = X$ is equivalent to $\Phi(\partial_{\mathfrak{T}}X) = \Phi(X)$, and hence to $\Phi(X) = \mathrm{ex}\,C$ because of (20). The result then follows from Theorem 2.

The equality (21) will enable us to decide in concrete cases by means of geometric arguments whether for a given test set all continuous functions are Korovkin functions.

EXAMPLE 6. Let $X$ be a compact interval $[a,b]$ in $\mathbf{R}$ and let $\mathfrak{T} = \{1,t\}$ be a test set with 2 elements. If every function in $\mathcal{C}([a,b])$ is a Korovkin function $\mathfrak{T}$, the function $t$ must separate points. The corresponding map $\Phi: X \to \mathbf{R}$ equals $t$. As a continuous image of a compact interval $\Phi(X) = t(X)$ is itself a compact interval $[c,d]$ in $\mathbf{R}$, hence convex and equal to its convex hull $C$. The extreme points of $[c,d]$ are the end points $c$ and $d$. Consequently, $\Phi(X) = \mathrm{ex}\,C$ holds if and only if $\Phi(X)$ and hence $X$ consists of just one point.

We have thus proved that on a compact interval $[a,b]$ which does not reduce to one point there is no test set $T = \{1,t\}$ of just 2 elements such that all functions in $\mathcal{C}(X)$ are Korovkin functions. This shows that the number of 3 test functions in Korovkin's First Theorem is *minimal*. It is this result which is called *Korovkin's Third Theorem.*

EXAMPLE 7. Again let $X$ be a compact interval $[a,b]$ in $\mathbf{R}$ such that $a<b$. Consider a test set

$$T = \{1, \mathrm{id}, u\}$$

which is derived from the test set in Korovkin's First Theorem by replacing the function $\mathrm{id}^2$ by a function $u \in \mathcal{C}([a,b])$ chosen arbitrarily for the moment. The set $\mathfrak{T}$ is point separating since the subset $\{\mathrm{id}\}$ already separates points, so we can use the map $\Phi$ from (19). We obtain

$$\Phi(X) = \{(x, u(x)) | x \in [a,b]\}$$

which is the graph of the function $u$. By passing from $\Phi(X)$ to its convex hull $C$, we arrive at a situation sketched in Figure 2.

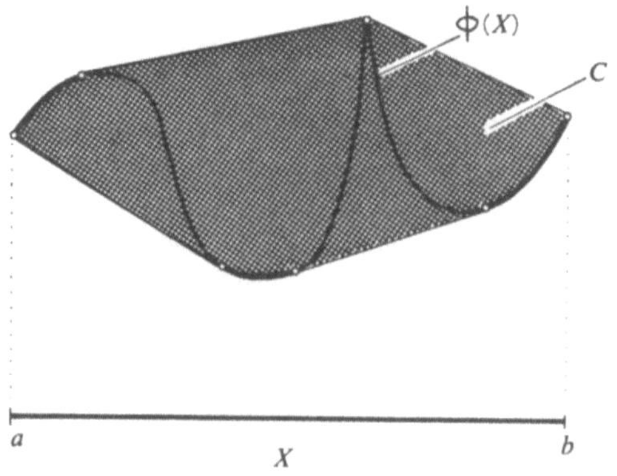


FIG. 2

Only the parts of the graph in boldfaced print consist of extreme points of $C$.

Now it is intuitively clear when the graph $\Phi(X)$ of $u$ coincides with the set $\operatorname{ex} C$ of all extreme points of $C$. This is the case if and only if $u$ is strictly concave or strictly convex. As usual $u \in \mathcal{C}([a,b])$ is called *strictly concave* if

$$\lambda u(x)+(1-\lambda)u(y)<u(\lambda x+(1-\lambda)y)$$

holds for all points $x,y$ satisfying $a \leqslant x<y \leqslant b$ and all $\lambda$ satisfying $0<\lambda<1$. We call $u$ *strictly convex* if $-u$ is strictly concave. We thus are led to

PROPOSITION 6. *Let $u$ be a continuous real-valued function on a compact interval $[a,b]$ in $\mathbf{R}$ such that $a<b$. All functions in $\mathcal{C}([a,b])$ are Korovkin functions with respect to the set $\mathfrak{T}=\{1,\mathrm{id},u\}$ if and only if $u$ is strictly concave or strictly convex.*

We leave the (not very difficult) proof to the reader and content ourselves with the geometric illustration.

This result clarifies the role of the function $\mathrm{id}^2$ appearing in Korovkin's First Theorem. It is strictly convex. By Proposition 6 it could be replaced by any other strictly convex function, for example, by the exponential function $x \mapsto e^x$. The proof for Korovkin's First Theorem given in the Introduction of course breaks down if we make this replacement.

EXAMPLE 8. Consider a compact convex set $C_0$ in $\mathbf{R}^k$ with a closed and hence compact set $X=\operatorname{ex} C_0$ of extreme points. According to (19) the map $\Phi: X \to \mathbf{R}^k$ is the identity map if we take

$$\mathfrak{T}=\{1,p_1,\ldots,p_k\}$$

as test set. Here $p_j$ again denotes the $j$th coordinate function restricted to $X$. The convex hull of $X=\Phi(X)$ (up to now denoted by $C$) coincides with $C_0$. This follows from a classical theorem of Minkowski or from the Krein-Milman theorem (cf. [9], [13], [19], [24]) according to which $C_0$ is the convex hull of the set of its extreme points. So we have $\Phi(X)=X=\operatorname{ex} C$. The Corollary following Proposition 4 then yields: *All functions in $\mathcal{C}(X)$ are Korovkin functions.*

This example contains Example 4 as a special case. If there exists a supporting hyperplane through a point $x \in X$, that is, if there exists a linear form $l$ on $\mathbf{R}^k$ such that

$$l(x)<l(x') \qquad \text{for all} \quad x' \in X,$$

then $X$ and consequently the convex hull $C$ of $X$ lie in the convex half space of all $x' \in X$ satisfying $l(x') \geqslant \alpha$ where $\alpha=l(x)$. The hyperplane

$$H=\{x' \in \mathbf{R}^k \mid l(x')=\alpha\}$$

therefore is a supporting hyperplane for $C$. Such a hyperplane always contains at least one extreme point since—as we have mentioned in Section 2 in connection with the introduction of the Choquet boundary—$l$ attains its minimum $\alpha$ on $C$ at a point of $\partial_{\mathfrak{T}_C} C=\operatorname{ex} C$. But in Example 4 $H$ hits $X$ only in the point $x$; the same is true if we replace $X$ by $C$. Therefore, $x$ must be an extreme point of $C$, and we have $X \subset \operatorname{ex} C$. We even have equality, since every point $c \in C$ is of the form $c=\sum_{j=1}^{r}\lambda_j x_j$ for distinct points $x_1,\ldots,x_r \in X$ and coefficients $\lambda_1 \geqslant 0,\ldots,\lambda_r \geqslant 0$ satisfying $\sum_{j=1}^{r}\lambda_j=1$. A simple calculation shows that $c$ can be an extreme point of $C$ only if $\lambda_j=0$ for all $j \in \{1,\ldots,r\}$ except one.

EXAMPLE 9. Our last example intends to show that Proposition 5 also leads in more complicated situations to a simple determination of the Choquet boundary. Consider the interval $X=[-1,+1]$ in $\mathbf{R}$ and let $T=\{1,\mathrm{id},\mathrm{id}^3\}$ be the test set. From Example 7 we know that $\Phi(X)$ is the graph of the function $x \mapsto x^3$, the well-known cubic parabola. Figure 2 has to be replaced in this case by Figure 3. The extreme points of $C=\operatorname{conv}\Phi(X)$ are all points $(x,x^3) \in \mathbf{R}^2$ where $x \in [-1,-\frac{1}{2}] \cup [\frac{1}{2},1]$. According to Proposition 5, $[-1,-\frac{1}{2}] \cup [\frac{1}{2},1]$ then is the Choquet boundary of $[-1,+1]$ with respect to $\mathfrak{T}$. A purely analytic proof based on Proposition 2 or even on formula (12) is much more complicated.

**5. Bibliographical comments.** Korovkin's famous three theorems first appeared in [15] and some-

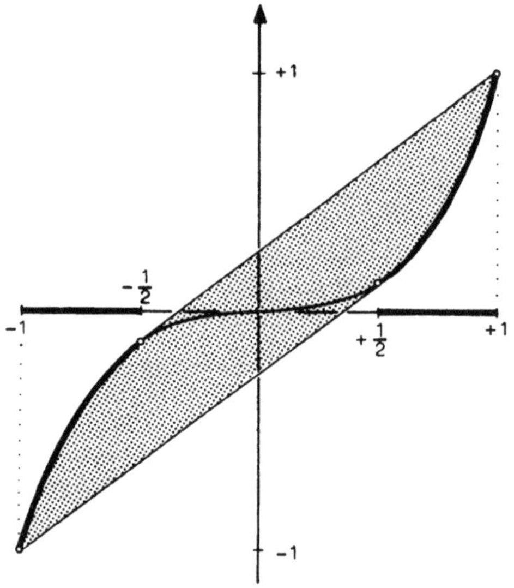


FIG. 3

what later in the book [16]. The tools of the present paper—in particular, the introduction of the space $\mathfrak{T}^*$ of $\mathfrak{T}$-affine functions (originally called $\mathfrak{T}$-harmonic functions) as well as properties of the Choquet boundary, including the geometric interpretation—appeared in the author's paper [2] in connection with potential-theoretic problems. Example 1 is already discussed there, however without reference to approximation theory. Šaškin [20] was the first to study conditions on a finite set of test functions on a compact metric space $X$ which imply that all functions in $\mathcal{C}(X)$ are Korovkin functions. The connection with the Choquet boundary arises implicitly there. Explicitly, this connection appears in Wulbert [27] where—similar to Šaškin [21]—sequences $(L_n)$ of (not necessarily positive) contraction maps $L_n : \mathcal{C}(X) \to \mathcal{C}(X)$ are also considered. The relevance of the space $\mathfrak{T}^*$ of $\mathfrak{T}$-affine functions was recognized by Šaškin [22] and Baskakov [1] though only in the form of an analogy for pointwise convergence. The methods of the present article have been developed in the author's paper [3] and independently at almost the same time by Berens–Lorentz [7]. In particular, Theorem 1 can be found there. Another proof of it appears in Šaškin [23]. Most of the examples of this article are part of the folklore in connection with Korovkin's theorems.

Further investigations on this subject can be found in [3], [4], [7], as well as in the work of Donner [10], [11], Grossman [12], Kutateladze [17], Leha [18] and Wolff [25], [26].

The present article is essentially a translation of the author's paper [5]. The author is grateful to R. Jamison for assistance with the translation.

## References

1. V. A. Baskakov, Some convergence conditions for positive linear operators, Uspehi Mat. Nauk., 16 (1961) 131–135 (Russian).

2. H. Bauer, Šilovscher Rand und Dirichletsches Problem, Ann. Inst. Fourier, 11 (1961) 89–136.

3. ———, Theorems of Korovkin type for adapted spaces, Ann. Inst. Fourier, 23, 4 (1973) 245–260.

4. ———, Convergence of monotone operators, Math. Z., 136 (1974) 315–330.

5. ———, Approximationssätze und abstrakte Ränder, Math.-Phys. Semesterber., 23 (1976) 141–173.

6. ———, Probability Theory and Elements of Measure Theory, Holt, Rinehart and Winston, New York, 1972.

7. H. Berens and G. G. Lorentz, Theorems of Korovkin type for positive linear operators on Banach lattices, Approximation Theory (Ed. G. G. Lorentz), Academic Press, New York, 1973.

8. S. Bernstein, Démonstration du théorème de Weierstrass fondée sur le calcul des probabilités. Comm. Soc. Math. Kharkow, (2) 13 (1912) 1–2.

9. G. Choquet, Lectures on Analysis, Vol. II, Representation Theory, Benjamin, New York, 1969.

10. K. Donner, Korovkin theorems for positive linear operators, J. Approximation Theory, 13 (1975) 443–450.

11. ———, Korovkin theorems and $P$-essential sets, to appear in J. Approximation Theory.

12. M. W. Grossman, Korovkin theorems for adapted spaces with respect to a positive operator, Math. Ann., 220 (1976) 253–262.

13. E. Hewitt and K. Stromberg, Real and Abstract Analysis, Springer-Verlag, New York, 1965.

14. K. Jacobs, Extremalpunkte konvexer Mengen, Selecta Mathematica III, Springer-Verlag, New York, 1971.

15. P. P. Korovkin, On convergence of linear positive operators in the space of continuous functions, Dokl. Akad. Nauk SSSR (N.S.), 90 (1953) 961–964 (Russian).

16. ———, Linear Operators and Approximation Theory, Hindustan Publ. Corp., Delhi, India, 1960.

17. S. S. Kutateladze, Choquet boundaries in $K$-spaces, Uspehi Mat. Nauk., 30 (1975) 107–146 (Russian). Translated in: Russian Math. Surveys, 30 (1975) 115–155.

18. G. Leha, Relative Korovkin-Sätze und Ränder, Math. Ann., 229 (1977) 87–95 and 233 (1978) 273–274.

19. R. R. Phelps, Lectures on Choquet's Theorem, Van Nostrand Math. Studies, 7, 1966.

20. Yu. A. Šaškin, Korovkin systems in spaces of continuous functions, Izv. Akad. Nauk SSSR, Ser. Mat., 26 (1962) 495–512 (Russian). Translated in: Amer. Math. Soc. Transl., Ser. 2, 54 (1966) 125–144.

21. ———, On the convergence of contractive operators, Mathematica (Cluj), 11 (34) (1969) 355–360 (Russian).

22. ———, On the convergence of linear operators, Konf. Konstruktiv. Teor. Funkciĭ, Varna 1970 (1972) 119–125 (Russian).

23. ———, Abstract harmonic functions and the problem of operator convergence. Mathematica (Cluj), 15 (38) (1973) 143–148 (Russian).

24. F. A. Valentine, Convex Sets, McGraw-Hill, New York, 1964.

25. M. Wolff, Uber die Korovkinhülle von Teilmengen in lokalkonvexen Vektorverbänden, Math. Ann., 213 (1975) 97–108.

26. ———, On Korovkin-type theorems in special function lattices, J. Approximation Theory (to appear).

27. D. E. Wulbert, Convergence of operators and Korovkin's theorem, J. Approximation Theory, 1 (1968) 381–390.

# Halbgruppen und Resolventen in der Potentialtheorie [1]
## [S15]

Enseign. Math. (2) **25** (1979), 9–22
[Zbl. Math. 416.31014; MR 80m:31012]

Der Vortrag verfolgte das Ziel, an Hand einiger neuer Resultate der Potentialtheorie die zunehmend deutlicher werdende zentrale Rolle von Halbgruppen und Resolventen in dieser Theorie aufzuzeigen. Der Verfasser hofft, damit auch einen Beitrag zu einer Standortbestimmung der Potentialtheorie zu liefern.

### 1. Ein Blick auf die klassische Theorie

Klassische Potentialtheorie heißt einerseits (lokaler Aspekt) das Studium der *Laplaceschen Differentialgleichung*

$$\Delta h = 0 \tag{1}$$

im $\mathbf{R}^n$, wobei wir uns auf den Fall $n \geqslant 3$ beschränken, und andererseits (globaler Aspekt) das Studium der *Newtonschen Kernfunktion*

$$N(x) = \frac{1}{|x|^{n-2}} \quad (x \in \mathbf{R}^n). \tag{2}$$ [2]

Aus ihr leiten sich die global definierten *Potentiale* $p$ ab. Dies sind alle Funktionen $p : \mathbf{R}^n \to [0, +\infty]$, welche nicht konstant gleich $+\infty$ sind und eine (notwendigerweise eindeutige) Darstellung

$$p = N * \mu \tag{3}$$

besitzen, wobei $\mu$ ein positives Radon-Maß auf $\mathbf{R}^n$ ist. Eine nicht-negative Funktion $u$: $\mathbf{R}^n \to [0, +\infty]$ heißt *hyperharmonisch* (auf $\mathbf{R}^n$), wenn sie von der Form

$$u = N * \mu + h$$

---

[1]) Ausarbeitung eines am 10. April 1978 an der ETH Zürich im Rahmen des International Symposium on Analysis gehaltenen Vortrages. Dieser Artikel wurde bereits in *Contributions to Analysis*, Monographie de l'*Ens. Math.* N° 27, Genève 1979, veröffentlicht.

[2]) $|x|$ bezeichnet die euklidische Norm des Vektors $x \in \mathbf{R}^n$.

ist, wobei jetzt $\mu$ ein beliebiges Radon-Maß $\geqslant 0$ auf $\mathbf{R}^n$ und $h$ eine positive harmonische Funktion auf $\mathbf{R}^n$, also eine (in diesem Fall konstante) Lösung $h \geqslant 0$ von (1) ist. Diese hyperharmonischen Funktionen $u \geqslant 0$ können als nach unten halbstetige Funktionen durch die übliche Mittelwerteigenschaft (Rieszscher Zerlegungssatz) oder durch die Gültigkeit von $\Delta u \leqslant 0$ im distributions-theoretischen Sinne gekennzeichnet werden.

Klassische Potentialtheorie muß aber auch unter dem Aspekt des Studiums der Brownschen Halbgruppe gesehen werden. Dieser (ebenfalls globale) Aspekt tritt historisch gesehen erst viel später in Erscheinung, nämlich durch die bahnbrechenden Arbeiten von KAKUTANI [21], [22]. Die *Brownsche Halbgruppe* (oder die Halbgruppe der Brownschen Bewegung) ist dabei die Faltungshalbgruppe $(\nu_t)_{t>0}$ der Wahrscheinlichkeitsmaße

(4) $$\nu_t = g_t \lambda^n,$$

wobei $\lambda^n$ das $n$-dimensionale Lebesgue-Maß und $g_t$ die Dichte

(5) $$g_t(x) = (4\pi t)^{-\frac{n}{2}} \exp\left(-\frac{|x|^2}{4t}\right)$$

ist. Wie üblich interpretieren wir $\nu_t$ als einen Kern, d.h. als den durch Faltung wirkenden Operator $P_t$:

$$P_t f = \nu_t * f.$$

Dieser operiert auf den beschränkten sowie auf den nicht-negativen Borel-meßbaren Funktionen linear und positiv. Somit erscheint die Faltungshalbgruppe $(\nu_t)_{t>0}$ als eine Halbgruppe $(P_t)_{t>0}$ von Kernen.

Bezüglich einer solchen Halbgruppe heißt eine nicht-negative Funktion $u\colon \mathbf{R}^n \to [0, +\infty]$ *exzessiv*, wenn sie Borel-meßbar ist und der Bedingung

(6) $$\sup_{t>0} P_t u = u$$

genügt. Die potentialtheoretische Bedeutung dieser Halbgruppe wird deutlich durch den fundamentalen Satz von DOOB [15] und HUNT [20]:

1.1. *Die exzessiven Funktionen bezüglich der Brownschen Halbgruppe fallen mit den nicht-negativen hyperharmonischen Funktionen zusammen.*

Der analytische Grund für diesen Zusammenhang ist die Gleichheit

(7) $$\int_0^\infty g_t(x)\,dt = c_n N(x) \qquad (x \in \mathbf{R}^n)$$

mit dem positiven Faktor

$$c_n = \frac{1}{4\pi^{n/2}} \Gamma\left(\frac{n-2}{2}\right).$$

Der sich durch Integration der Halbgruppe $(P_t)_{t>0}$ ergebende Kern

$$V = \int_0^\infty P_t \, dt \tag{8}$$

ist somit nichts anderes als der durch Faltung mit Hilfe des Maßes

$$\kappa = c_n N \lambda^n \tag{9}$$

wirkende Kern $Vf = \kappa * f$.

Damit sind es die drei Objekte $\Delta$, $N$ und $(P_t)_{t>0}$, die im Mittelpunkt der klassischen Potentialtheorie stehen. Jedes dieser Objekte erlaubt die Definition der (global definierten) hyperharmonischen Funktionen $\geqslant 0$. Aus der Kenntnis eines dieser Objekte folgt die der anderen, denn $N$ ist die Fundamentallösung von (1), interpretiert als der zu $\kappa$ gehörige Faltungskern $V$, berechnet sich $N$ gemäß (8) aus $(P_t)_{t>0}$, ferner ist $\Delta$ der infinitesimale Erzeuger von $(P_t)_{t>0}$, d.h. es gilt

$$\Delta f = \lim_{t \to 0} \frac{P_t f - f}{t} \tag{10}$$

für alle Funktionen $f \in C^2(\mathbf{R}^n)$ mit kompakten Träger. Hierzu vergleiche man BERG-FORST [7].

Der Satz 1.1 ist der Schlüssel zum wahrscheinlichkeitstheoretischen Verständnis potentialtheoretischer Begriffsbildungen. Hierauf soll aber hier nicht eingegangen werden. (Vgl. jedoch [2].)

## 2. HARMONISCHE RÄUME UND FELLERSCHE HALBGRUPPEN

Die weiteren Teile des Vortrages werden vor allem durch die Frage nach dem Zusammenhang zwischen lokaler Potentialtheorie und Halbgruppen von Kernen motiviert.

Als lokale Potentialtheorie verstehen wir dabei die Theorie der harmonischen Räume, die sich aus der Idee entwickelt hat, die Theorie der Laplace-Gleichung (1) auf allgemeinere elliptische und parabolische Differentialgleichungen etwa mit einer differenzierbaren Mannigfaltigkeit als Grundraum zu entwickeln. (Vgl. hierzu [1], BRELOT [10], [11], [12] und CONSTANTINESCU-CORNEA [13].)

Der Einfachheit halber sollen hier harmonische Räume nur im Sinne von Brelot [10] verstanden werden. Es sei demnach $X$ ein lokal-kompakter, (Hausdorff-) Raum und $\mathscr{H}$ ein Garbendatum von Vektorräumen stetiger reeller Funktionen. $(X,\mathscr{H})$ heißt dann ein *Brelot-Raum*, wenn $X$ keine isolierten Punkte besitzt und lokal-zusammenhängend ist, eine Basis regulärer Mengen existiert (per definitionem ist für diese das Dirichletsche Problem für alle stetigen Randwerte eindeutig und „positiv" lösbar) und das Brelotsche Konvergenzaxiom erfüllt ist. Wir werden nur *strenge* oder $\mathfrak{P}$-Brelot-Räume betrachten, also solche mit mindestens einem Potential $p \neq 0$. Die Definition dieser Begriffsbildungen findet man in [10], [11] und [1]. Für $n \geqslant 3$ liefert das Garbendatum $\mathscr{H}_\Delta$ der Lösungen von (1), also der klassischen harmonischen Lösungen, den strengen Brelot-Raum $(\mathbf{R}^n, \mathscr{H}_\Delta)$.

Ausgelöst durch die Untersuchung Markovscher Prozesse haben sich in zunehmendem Maße spezielle Halbgruppen von Kernen als wichtig erwiesen. Auf einem lokal-kompakten Raum $X$ mit abzählbarer Basis sei $(P_t)_{t>0}$ eine Halbgruppe von *Kernen.* Jedes $P_t$ ist also eine Funktion auf $X \times \mathfrak{B}_X$, wobei $\mathfrak{B}_X$ die $\sigma$-Algebra der Borelschen Mengen bezeichnet; für jedes $A_0 \in \mathfrak{B}_X$ bzw. $x_0 \in X$ werden dabei die Abbildungen $A \mapsto P_t(x_0, A)$ bzw. $x \mapsto P_t(x, A_0)$ als nicht-negative Maße auf $\mathfrak{B}_X$ bzw. als (nicht-negative) Borel-meßbare Funktionen vorausgesetzt. Durch Integration nach der zweiten Variablen wirkt jedes $F$ als linearer Operator auf geeigneten Räumen meßbarer Funktionen:

$$(11) \qquad P_t f(x) = \int P_t(x, dy)\, f(y)$$

(z.B. für Borel-meßbares $f \geqslant 0$). Die Halbgruppeneigenschaft besagt nichts anderes als

$$(12) \qquad P_{s+t} f = P_s P_t f$$

für alle Borel-meßbaren Funktionen $f \geqslant 0$ und für alle $s, t > 0$ oder — äquivalent hierzu —

$$(13) \qquad P_{s+t}(x, A) = \int P_s(x, dy)\, P_t(y, A)$$

für alle $x \in X$ und $A \in \mathfrak{B}_X$. Es müssen also die sogenannten Chapman-Kolmogorov-Gleichungen erfüllt sein.[1])

Eine solche Halbgruppe von Kernen heißt *Fellersch*, wenn jeder Kern $P_t$ sub-Markovsch, also $P_t 1 \leqslant 1$ erfüllt ist, wenn $P_t$ den Raum $C_0(X)$ der im Unendlichen verschwindenden, stetigen, reellen Funktionen auf $X$ in sich abbildet, und wenn außerdem

[1]) Vgl. hierzu [3] und Meyer [24].

$$\lim_{t\to 0} P_t f = f \qquad \textit{für alle } f \in C_0(X)$$

im Sinne der gleichmäßigen Konvergenz erfüllt ist. Einer solchen Fellerschen Halbgruppe sind wir in der klassischen Theorie bereits begegnet: die Brownsche Halbgruppe ist Fellersch. Jeder der Kerne $P_t$ ist dort sogar Markovsch, erfüllt also $P_t 1 = 1$.

Eines der Leitmotive für die Entwicklung der Theorie der harmonischen Räume war, über Jahre hinweg, die Frage nach der Existenz einer Halbgruppe von Kernen auf einem streng harmonischen Raum $(X, \mathscr{H})$ derart, daß, wie in der klassischen Theorie, die exzessiven Funktionen mit den hyperharmonischen Funktionen $\geqslant 0$ auf $X$ übereinstimmen. Hyperharmonisch heissen dabei die durch die Mittelwerteigenschaft (für reguläre Mengen) definierten nach unten halbstetigen Funktionen. Die Antwort auf diese Frage ist Ja. Sie wurde durch MEYER [23] für Brelotsche und später durch BOBOC-CONSTANTINESCU-CORNEA [9] und HANSEN [18], [19] für allgemeinere Typen harmonischer Räume gegeben:

2.1. *Auf einem streng harmonischen Raum* $(X, \mathscr{H})$ *existieren stets ein strikt positives, stetiges, reelles Potential* $q$ *und eine Fellersche Halbgruppe* $(Q_t)_{t>0}$ *derart, dass deren exzessive Funktionen mit den mit* $\frac{1}{q}$ *multiplizierten nicht-negativen hyperharmonischen Funktionen übereinstimmen.*

*Ist die konstante Funktion* 1 *hyperharmonisch, so fallen die nicht-negativen hyperharmonischen Funktionen mit den exzessiven Funktionen der neuen Halbgruppe*

$$P_t f = q\, Q_t\left(\frac{1}{q} f\right) \qquad (f \in C_0(X)) \tag{14}$$

*zusammen.*

Man nennt die Halbgruppe $(P_t)_{t>0}$ auf Grund ihrer Herkunft auch *quasi-Fellersch.*

## 3. DER SATZ VON G. FORST

Im Gegensatz zu der dem Resultat 2.1 zugrunde liegenden Fragestellung ist die umgekehrte Frage nach Eigenschaften einer Halbgruppe $(P_t)_{t>0}$ von Kernen auf einem Raum $X$, welche die Existenz eines Garbendatums $\mathscr{H}$ garantieren, so daß $(X, \mathscr{H})$ ein harmonischer Raum und die exzessiven Funktionen von $(P_t)_{t>0}$ mit den hyperharmonischen Funktionen $\geqslant 0$ zusammenfallen, neueren Datums. Es sind nur Teilantworten bekannt.

Jede Teilantwort ist aber von Interesse, da sie erhoffen läßt, dem Problem der Produktbildung harmonischer Räume näher zu treten.

Sind nämlich $(P_t)_{t>0}$ und $(P'_t)_{t>0}$ Halbgruppen von Kernen auf lokalkompakten Räumen $X$ bzw. $X'$, so liefert die Bildung der Produktmaße

$$Q_t((x,x'),.) = P_t(x,.) \otimes P'_t(x',.)$$

eine Halbgruppe $(Q_t)_{t>0}$ von Kernen auf $X \times X'$. Sind $(P_t)_{t>0}$ und $(P'_t)_{t>0}$ im Sinne von Satz 2.1 mit harmonischen Strukturen $\mathscr{H}$ auf $X$ bzw. $\mathscr{H}'$ auf $X'$ verknüpft und läßt sich mit Hilfe allgemeiner Sätze entscheiden, ob auch $(Q_t)_{t>0}$ zu einer harmonischen Struktur gehört, so liegt es nahe, diese als ein Produkt der harmonischen Räume $(X, \mathscr{H})$ und $(X', \mathscr{H}')$ zu interpretieren. Zu diesem Problem sind gerade in letzter Zeit Resultate von U. Schirmeier [28] erzielt worden. Hier geht es uns nur um den Hinweis auf die grundsätzliche Bedeutung dieser Fragestellung: Wir werden sehen, daß sie selbst unter stark einschränkenden Zusatzannahmen zu neuen, überraschenden Beispielen harmonischer Räume führt.

Es werde jetzt nämlich vorausgesetzt, daß der lokal-kompakte Grundraum eine lokal-kompakte abelsche Gruppe $G$ mit abzählbarer Basis ist; ferner sei die gegebene Halbgruppe $(P_t)_{t>0}$ Fellersch und translationsinvariant, also

$$P_t(f \circ \tau_{-a}) = \tau_a(P_t f)$$

für jede Translation $\tau_a(x) = x + a$ erfüllt ($t > 0$, $f$ Borel-meßbar $\geqslant 0$). Diese und allein diese Halbgruppen rühren von Faltungshalbgruppen $(\mu_t)_{t>0}$ von positiven Radon-Maßen auf $G$ her, d.h. es gilt

$$P_t f = \mu_t * f. \tag{15}$$

*Faltungshalbgruppe* heiße dabei, daß neben

$$\mu_{s+t} = \mu_s * \mu_t \qquad \textit{für } s, t > 0 \tag{16}$$

die Bedingungen

$$\mu_t(G) \leqslant 1 \qquad (t > 0) \tag{17}$$

und

$$\lim_{t \to 0} \mu_t = \varepsilon_0 \tag{18}$$

(in der vagen Topologie) erfüllt sind.[1])

[1]) $\varepsilon_0$ bezeichnet die Einheitsmasse in 0. Zu diesem und den folgenden Resultaten über Faltungshalbgruppen vom lokalen Typ vergleiche man Berg-Forst [7].

Eine solche Faltungshalbgruppe heißt vom *lokalen Typ*, wenn für ihren infinitesimalen Erzeuger $A$ mit Definitionsbereich $D_A \subset C_0(G)$ gilt: ist $f \in D_A$ Null in der Umgebung eines Punktes $x \in G$, so ist $f$ an der Stelle $x$ gleich Null. Im Fall einer *symmetrischen* Halbgruppe, wo also jedes der Maße $\mu_t$ bei der Spiegelung $x \mapsto -x$ in sich übergeht, ist dies äquivalent zum folgenden Verhalten der Fourier-Transformierten $\hat{\mu}_t$ von $\mu_t$: es gibt eine stetige quadratische Form $q$ auf der Charaktergruppe $\hat{G}$ und eine Konstante $c \geqslant 0$ mit

$$\hat{\mu}_t = e^{-t(c+q)} \quad \textit{für alle } t > 0. \tag{19}$$

Eine wichtige Beantwortung der zu Beginn dieses Abschnittes gestellten Frage gibt der folgende Satz von Forst [17]:

3.1. *Sei $G$ eine nicht diskrete, lokal-kompakte abelsche Gruppe mit abzählbarer Basis. Ferner sei $(\mu_t)_{t>0}$ eine Faltungshalbgruppe von Massen mit folgenden Eigenschaften:*

(i) *$(\mu_t)$ ist symmetrisch und vom lokalen Typ;*

(ii) *die Halbgruppe ist transient, d.h. es existiert das Integral*

$$\kappa(f) = \int_0^\infty \mu_t(f)\,dt$$

*für jede Funktion $f \in C_0(G)$ mit kompaktem Träger;*

(iii) *bezüglich des Haarschen Masses $\Theta$ von $G$ besitzt das Mass $\kappa$ eine nach unten halbstetige Dichte $N$, welche auf $G \setminus \{0\}$ stetig und reell ist.*

*Dann existiert ein translationsinvariantes, symmetrisches Garbendatum $\mathscr{H}$ auf $G$ derart, dass $(G, \mathscr{H})$ ein strenger Brelot-Raum und $N$ ein in $G \setminus \{0\}$ harmonisches Potential ist.*

Aus einem Resultat von Bliedtner [8] folgt ferner

3.2. *Die nicht-negativen hyperharmonischen Funktionen auf $G$ fallen mit den exzessiven Funktionen der gegebenen Faltungshalbgruppe zusammen.*

Speziell ordnet sich die klassische Potentialtheorie in das Ergebnis von Forst ein.

Kehren wir nun nochmals zu der eingangs erhobenen Frage nach der Bildung von Produkten harmonischer Räume zurück. Man sieht sofort, daß die Produkthalbgruppe $(\mu_t \otimes \mu_t')_{t>0}$ auf $G \times G'$ die Eigenschaften (i) und (ii) besitzt, wenn zwei vorgegebene Faltungshalbgruppen $(\mu_t)_{t>0}$

und $(\mu_t')_{t>0}$ auf den Gruppen $G$ bzw. $G'$ diese Eigenschaften besitzen. Schwierigkeiten bereitet die Eigenschaft (iii).

Wir werden aber selbst in einem relativ krassen Fall sehen, daß sich auch die Eigenschaft (iii) durch Zusatzbedingungen erhalten läßt. Es wird sich um den Fall eines unendlichen Produktes handeln.

## 4. Harmonische Strukturen auf $T^\infty$

Wir betrachten die Kreislinie $T$, also den eindimensionalen Torus. Auf der Charaktergruppe $\hat{T} = \mathbf{Z}$ sind $n \mapsto a\,n^2$ mit dem Normierungsfaktor $a \geqslant 0$ sämtliche quadratische Formen. Jeder dieser quadratischen Formen entspricht gemäß (19) eine symmetrische Faltungshalbgruppe $(\mu_{at})_{t>0}$ vom lokalen Typ auf $T$. Dabei hat $\mu_{at}$ eine stetige Dichte bezüglich des Haar-Maßes $\theta$ auf $T$ (mit zu 1 normierter Gesamtmasse):

$$\mu_{at} = g_{at}\,\theta\,. \tag{20}$$

Diese Dichte ist für $a = 1$ der Wärmekern

$$\begin{aligned} g_t(\theta) = \sum_{k\in\mathbf{Z}} e^{-tn^2} e^{ik\theta} &= 1 + 2\sum_{n=1}^{\infty} e^{-tn^2}\cos n\theta \\ &= \sqrt{\frac{\pi}{t}} \sum_{k\in\mathbf{Z}} \exp\left(-\frac{(\theta+2\pi k)^2}{4t}\right)\,. \end{aligned} \tag{21}$$

$(\mu_t)_{t>0}$ ist nichts anderes als die Faltungshalbgruppe der Brownschen Bewegung auf $T$.

Wir betrachten nun auf dem unendlich-dimensionalen Torus $T^\infty = \prod_{k=1}^{\infty} T_k$, wobei jedes $T_k$ gleich $T$ ist, für eine zunächst beliebig gegebene Folge $\mathscr{A} = (a_k)_{k\in\mathbf{N}}$ reeller Zahlen $a_k > 0$ die Faltungshalbgruppe $(\mu_t^{\mathscr{A}})_{t>0}$ mit

$$\mu_t^{\mathscr{A}} = \bigotimes_{k=1}^{\infty} \mu_{a_k t}\,. \tag{22}$$

Auf der Charaktergruppe $\hat{T^\infty} = \mathbf{Z}^{(\infty)}$, d.h. auf der direkten Summe von abzählbar unendlich vielen Kopien von $\mathbf{Z}$, ergibt sich die Fourier-Transformierte von $\mu_t^{\mathscr{A}}$ zu

$$\hat{\mu}_t^{\mathscr{A}}(n) = e^{-tq(n)} \tag{23}$$

mit $q(n) = \sum_{k=1}^{\infty} a_k n_k^2$, wobei $n \in \mathbf{Z}^{(\infty)}$ eine Folge $(n_k)$ ganzer Zahlen ist, von denen höchstens endlich viele $\neq 0$ sind. $q$ ist eine stetige quadratische Form auf $\mathbf{Z}^{(\infty)}$. Daher ist $(\mu_t^{\mathscr{A}})_{t>0}$ symmetrisch und vom lokalen Typ, also die Bedingung (i) des Forstschen Satzes erfüllt. Da $T^{\infty}$ kompakt, also die konstante Funktion 1 kompakten Träger besitzt und stets $\mu_t^{\mathscr{A}}(1) = 1$ gilt, liegt keine transiente Halbgruppe vor. Wir gehen daher für beliebiges $\lambda > 0$ zur Faltungshalbgruppe $(e^{-\lambda t} \mu_t^{\mathscr{A}})_{t>0}$ über, die (i) und (ii) erfüllt. Um (iii) erfüllen zu können, setzen wir

$$\rho_\lambda^{\mathscr{A}} = \int_0^{\infty} e^{-\lambda t} \mu_t^{\mathscr{A}} \, dt \,. \tag{24}$$

Nach einem bekannten Satz der harmonischen Analyse besitzt $\mu_t^{\mathscr{A}}$ genau dann eine *stetige* Dichte $g_t^{\mathscr{A}}$ bezüglich des Haar-Maßes $\Theta$ von $T^{\infty}$, wenn

$$\sum_{n \in \mathbf{Z}^{(\infty)}} \hat{\mu}_t^{\mathscr{A}}(n) < +\infty \,.$$

Dies aber erweist sich als zu

$$\sum_{k=1}^{\infty} e^{-ta_k} < +\infty \tag{25}$$

äquivalent. Unter der Voraussetzung der Gültigkeit von (25) für jedes $t > 0$ besitzt dann das Maß $\rho_\lambda^{\mathscr{A}}$ die Dichte

$$\tilde{\rho}_\lambda^{\mathscr{A}} = \int_0^{\infty} e^{-\lambda t} g_t^{\mathscr{A}} \, dt \,, \tag{26}$$

welche sich als nach unten halbstetig erweist. Stets ist $\tilde{\rho}_\lambda^{\mathscr{A}}(0) = +\infty$. Von Berg [5] wurde zunächst die Bedingung

$$\sum_{k=1}^{\infty} \frac{\sqrt{2^k}}{\sqrt{a_k}} < +\infty \,,$$

kurze Zeit darauf von Fuglede (unveröffentlicht)

$$\sum_{k=1}^{\infty} \frac{1}{\sqrt{a_k}} < +\infty$$

und schließlich 1977 erneut von Berg [6] die Bedingung

$$\sum_{k=1}^{\infty} \frac{1}{a_k} < +\infty \tag{27}$$

als hinreichend dafür erkannt, daß (25) für jedes $t > 0$ gilt und $\rho_\lambda^{\mathscr{A}}$ für jedes $\lambda > 0$ in $T^\infty \setminus \{0\}$ stetig und reell ist. Auf diese Weise erhielt BERG [5], [6] den Satz:

4.1 *Erfüllt eine Folge* $\mathscr{A} = (a_k)_{k \in \mathbb{N}}$ *positiver reeller Zahlen die Bedingung (27), so existiert auf* $T^\infty$ *für jedes* $\lambda > 0$ *ein translationsinvariantes, symmetrisches Garbendatum* $\mathscr{H}_\lambda^{\mathscr{A}}$*, für welches* $(T^\infty, \mathscr{H}_\lambda^{\mathscr{A}})$ *ein strenger Brelot-Raum und* $\tilde{\rho}_\lambda^{\mathscr{A}}$ *ein strikt positives, in* $T^\infty \setminus \{0\}$ *harmonisches Potential ist.*

Damit hat man mittels Produktbildung eine ganze Schar harmonischer Strukturen $\mathscr{H}_\lambda^{\mathscr{A}}$ auf dem unendlich-dimensionalen Torus konstruiert. Jede dieser Strukturen kann durch den linearen „*Differentialoperator*“

$$L_\lambda^{\mathscr{A}} = \sum_{k=1}^{\infty} a_k \frac{\partial^2}{\partial \theta_k^2} - \lambda \cdot id \tag{28}$$

beschrieben werden. Bezeichnet nämlich $\pi_p : T^\infty \to T^p$ die kanonische Projektion von $T^\infty$ auf den $p$-dimensionalen Torus $T^p$ der ersten $p = 1,2,\ldots$ Koordinaten, so liegen alle Funktionen $f \circ \pi_p$ mit $f \in C^2(T^p)$ im Definitionsbereich des infinitesimalen Erzeugers $L_\lambda^{\mathscr{A}}$ von $(\mu_t^{\mathscr{A}})_{t>0}$ und es gilt

$$L_\lambda^{\mathscr{A}}(f \circ \pi_p) = \sum_{k=1}^{p} a_k \frac{\partial^2 f}{\partial \theta_k^2} - \lambda f.$$

Mit distributionstheoretischen Methoden kann man, ausgehend von (28), das Garbendatum $\mathscr{H}_\lambda^{\mathscr{A}}$ direkt beschreiben: Es bezeichne hierzu $\mathscr{D}(\Omega)$ für offenes $\Omega \subset T^\infty$ die Menge aller Funktionen $f \circ \pi_p$ mit $p \in \mathbb{N}$ und $f \in C^\infty(T^p)$, deren Träger in $\Omega$ enthalten ist. Dann gilt

$$\mathscr{H}_\lambda^{\mathscr{A}}(\Omega) = \{ h \in C(\Omega) : \int_{T^p} h L_\lambda^{\mathscr{A}} g \, d\theta = 0 \ \text{ für alle } g \in \mathscr{D}(\Omega) \}.$$

Dabei kann man sogar $\lambda = 0$ zulassen. Allerdings ist der dann entstehende Brelot-Raum nicht mehr streng. Dieser Zugang findet sich ebenfalls bei BERG [5]. Analoge Resultate wurden von BENDIKOV [4] mit Methoden der Theorie der Markov-Prozesse erzielt.

## 5. BEZIEHUNGEN ZUR DIFFERENTIATIONSTHEORIE

Über die Theorie der harmonischen Räume hinaus werden der Potentialtheorie durch die Betrachtung von Halbgruppen von Kernen neue Dimensionen eröffnet. Dies soll nun noch kurz skizziert werden.

Eine der einfachsten Faltungshalbgruppen auf **R** ist die Halbgruppe $(\varepsilon_{-t})_{t>0}$ der Einheitsmassen in $-t$. Die zugehörigen Kerne $P_t$ operieren wie folgt

$$P_t f(x) = f(x+t).$$

Der durch Integration entstehende (Potential-) Kern $V$ hat also die Gestalt

$$Vf(x) = \int_0^\infty P_t f(x)\,dt = \int_x^\infty f(t)\,dt.$$

Die *supermedian* genannten Funktionen bezüglich einer Halbgruppe $(P_t)_{t>0}$, d.h. diejenigen Borel-meßbaren Funktionen $f \geqslant 0$ mit

$$P_t f \leqslant f \qquad \textit{für alle } t > 0,$$

sind im Falle dieser speziellen Halbgruppe gerade die monoton fallenden Funktionen $f \geqslant 0$ auf **R**. Für Lebesgue-integrierbares $f$ auf **R** ist $Vf$ absolut stetig und $\lim\limits_{x \to +\infty} Vf(x) = 0$. Setzen wir noch

$$D_t f = \frac{f - P_t f}{t} \qquad (t > 0)$$

und

$$Df = \limsup_{t \to 0} D_t f,$$

so erhält man aus den klassischen Differentiationssätzen der Lebesgueschen Theorie:

1. Für jede supermediane Funktion $u$ existiert der *reelle* Limes

(29) $$Du = \lim_{t \to 0} D_t u \qquad \textit{fast überall.}$$

2. Für jede Borel-meßbare Funktion $f \geqslant 0$ auf **R** mit $Vf(x) < +\infty$ für alle $x \in \mathbf{R}$ gilt

(30) $$DVf = f \qquad \textit{fast überall.}$$

Setzt man schließlich noch

$$D^* u = \sup_{t>0} D_t u,$$

so besagt das Maximallemma von Hardy-Littlewood:

3. Für jede supermediane Funktion $u$ gilt

(31) $$V1_{\{D^*u \geq \alpha\}} \leqslant \frac{u(x)}{\alpha} \qquad (\alpha > 0).$$

Dabei bezeichnet $1_A$ die Indikatorfunktion einer Menge $A$; die linke Seite der Ungleichung (31) ist das Lebesgue-Maß der Menge

$$\{y \in \mathbf{R} : D^* u(y) \geqslant \alpha\} \cap [x, +\infty[.$$

„Fast überall" heißt stets bis auf eine Borel-Menge $A$ vom Lebesgue-Maß Null. Hiermit äquivalent ist aber die Forderung

$$V 1_A(x) = 0 \qquad \textit{für alle } x \in \mathbf{R}. \tag{32}$$

Interpretiert man in den Aussagen 1 und 2 fast überall im Sinne von (32), so bekommen die Aussagen 1—3 einen Sinn für beliebige Halbgruppen $(P_t)_{t>0}$ von Kernen auf einem Meßraum.

Es ist höchst bemerkenswert, daß die Aussagen 1—3 nahezu ohne Zusatzbedingungen für *sub-Markovsche* Halbgruppen $(P_t)_{t>0}$ auf einem Meßraum von Mokobodzki [25], [26] bewiesen werden konnten. Man muß eigentlich nur voraussetzen, daß die $\sigma$-Algebra des Meßraumes von den exzessiven Funktionen der Halbgruppe erzeugt wird. Dies ist in unserem eingangs gewählten Beispiel der Fall.

## 6. Ausblick: Resolventen in der Potentialtheorie

Häufig — ein typisches Beispiel hierfür ist der Beweis des Satzes 2.1 — gelangt man zu einer Halbgruppe $(P_t)_{t>0}$ von Kernen nur auf dem Umweg über deren *Resolvente* $(V_\lambda)_{\lambda>0}$, wobei $V_\lambda$ den Kern,

$$V_\lambda = \int_0^\infty e^{-\lambda t} P_t \, dt, \tag{33}$$

also die Laplace-Transformierte von $(P_t)_{t>0}$ bezeichnet. Es ist der Satz von Hille-Yosida, der von einer Resolvente, d.h. genauer von einer der Resolventengleichung

$$V_\lambda - V_\mu + (\lambda - \mu) V_\lambda V_\mu = 0 \qquad (\lambda, \mu > 0) \tag{34}$$

genügenden Familie $(V_\lambda)_{\lambda>0}$ von Kernen zu einer zugehörigen, unter geeigneten Zusatzvoraussetzungen eindeutig bestimmten Halbgruppe $(P_t)_{t>0}$ von Kernen führt. (Vgl. Meyer [24].)

Außerdem ist es selbst bei gegebener Halbgruppe $(P_t)_{t>0}$ oft nur auf dem Umweg über die Resolvente möglich, gewisse Eigenschaften nachzuweisen. Ist beispielsweise $(P_t)_{t>0}$ eine Fellersche Halbgruppe auf einem lokal-kompakten Raum $X$ mit abzählbarer Basis, so läßt sich die Exzessivität einer Borel-meßbaren Funktion $u \geqslant 0$ auf $X$ durch die zur Definition äquivalente Bedingung

$$\sup_{\lambda > 0} \lambda V_\lambda u = u \tag{35}$$

für die Resolvente nachweisen.

Dies hat zur Folge, daß der explizite Gebrauch der Halbgruppe oft unnötig und Sätze gleich in der Sprache der Resolventen formuliert und bewiesen werden können. Häufig wird hierdurch größere Allgemeinheit erzielt, nämlich dann, wenn keine zugehörige Halbgruppe existiert. Ein Beispiel sind die Sätze von MOKOBODZKI des letzten Abschnitts, welche in [25], [26] gleich in der Sprache der Resolventen formuliert werden. Auch die Untersuchungen von CORNEA-LICEA [14] sind in diesem Licht zu sehen.

Resolventen treten auch bei *hyperbolischen* linearen Differentialgleichungen zweiter Ordnung auf — allerdings muß man dabei auf die Positivität der Kerne verzichten. RITTER [27] (vgl. auch [16]) hat nämlich gezeigt, daß für große Klassen solcher Differentialgleichungen der durch die Fundamentallösung definierte Kern in eine Resolvente reeller Kerne aufgelöst werden kann. Hierdurch dürfte es möglich werden, potentialtheoretische Methoden auch in das Gebiet der hyperbolischen Differentialgleichungen eindringen zu lassen. Das Gebiet der parabolischen Differentialgleichungen wurde potentialtheoretischen Methoden bereits durch den allgemeinen Begriff des harmonischen Raumes erschlossen (vgl. [1] und [13]).

## LITERATUR


[1] BAUER, H. *Harmonische Räume und ihre Potentialtheorie.* Lecture Notes in Math. 22, Springer-Verlag (1966).

[2] —— Aspects of modern potential theory. *Proc. Intern. Congress of Math., Vancouver 1974*, pp. 327-337 (1975).

[3] —— *Wahrscheinlichkeitstheorie und Grundzüge der Masstheorie. 3. Aufl.* Verlag W. de Gruyter, Berlin (1978).

[4] BENDIKOV, A. D. On functions that are harmonic for a certain class of diffusion processes on a group. *Theory Probability Appl. 20* (4), pp. 759-769 (1975).

[5] BERG, C. Potential theory on the infinite dimensional torus. *Inventiones math. 32* (1976), pp. 49-100.

[6] —— On Brownian and Poissonian convolution semigroups on the infinite dimensional torus. *Inventiones math. 38* (1977), pp. 227-235.

[7] BERG, C. and G. FORST. *Potential theory on locally compact abelian groups.* Ergebnisse der Math. 87, Springer-Verlag (1975).

[8] BLIEDTNER, J. Harmonische Gruppen und Huntsche Faltungskerne (in Seminar über Potentialtheorie). *Lecture Notes in Math. 69*, p.p. 69-102, Springer-Verlag (1968).

[9] Boboc, N., C. Constantinescu and A. Cornea. Semigroups of transitions on harmonic spaces. *Rev. Roumaine Math. Pures Appl. 12* (1967), pp. 763-805.

[10] Brelot, M. *Lectures on potential theory.* Tata Inst. of Fund. Research, Bombay (1960).

[11] —— *Axiomatique des fonctions harmoniques.* Les Presses de l'Université de Montréal (1966).

[12] —— Les étapes et les aspects multiples de la théorie du potentiel. *L'Enseignement Math. 18* (1972), pp. 1-36.

[13] Constantinescu, C. and A. Cornea. *Potential theory on harmonic spaces.* Grundlehren der Math. 158, Springer-Verlag (1972).

[14] Cornea, A. and G. Licea. *Order and potential, resolvent families of kernels.* Lecture Notes in Math. 494, Springer-Verlag (1975).

[15] Doob, J. L. Semimartingales and subharmonic functions. *Trans. Amer. Math. Soc. 77* (1954), pp. 86-121.

[16] Edwards, R. E., E. Hewitt and G. Ritter. Fourier multipliers for certain spaces of functions with compact supports. *Inventiones math. 40* (1977), pp. 37-57.

[17] Forst, G. Symmetric harmonic groups and translation invariant Dirichlet spaces. *Inventiones math. 18* (1972), pp. 143-182.

[18] Hansen, W. Konstruktion von Halbgruppen und Markoffschen Prozessen. *Inventiones math. 3* (1967), pp. 179-214.

[19] —— Hunt's theorem and axiomatic potential theory. *Inventiones math. 14* (1971), pp. 242-252.

[20] Hunt, G. A. Markoff processes and potentials I. *Illinois J. Math. 1* (1957), pp. 44-93.

[21] Kakutani, S. Two-dimensional Brownian motion and harmonic functions. *Proc. Imp. Acad. Tokyo 20* (1944), pp. 706-714.

[22] —— Markov processes and the Dirichlet problem. *Proc. Imp. Acad. Tokyo 21* (1945), pp. 227-233.

[23] Meyer, P. A. Brelot's axiomatic potential theory of the Dirichlet problem and Hunt's theory. *Ann. Inst. Fourier 13, fasc. 2* (1963), pp. 357-372.

[24] —— *Probabilités et potentiels.* Hermann, Paris (1966).

[25] Mokobodzki, G. Densité relative de deux potentiels comparables (in: Séminaire de Probabilités IV). *Lectures Notes in Math. 124,* pp. 170-194, Springer-Verlag (1970).

[26] —— Quelques propriétés remarquables des opérateurs positifs (in Séminaire de Probabilités IV). *Lecture Notes in Math. 124,* pp. 195-207, Springer-Verlag (1970).

[27] Ritter, G. Konstruktion von Operatoren und Kernen mit Hilfe von Absorptionsmengen. *Math. Ann. 216* (1975), pp. 51-66.

[28] Schirmeier, U. Produkte harmonischer Râume. *Sitz. Ber. math. nat. Kl. Bayer. Akad. Wiss. 1978* (1979), pp. 5-22.

*(Reçu le 3 octobre 1978)*

# Harmonic spaces – a survey [S21]

Conferenze del Seminario di Matematica dell'Università di Bari **197**, Bari 1984
[Zbl. Math. 572.31005; MR 86j:31001]

Conferenza tenuta l'8, 9 e 11.11.1983 (**)

**Introduction: Motivation and History.**

The main objects of classical potential theory are above all: the study of Newtonian potentials

$$N^{\mu}(x) = \int \| x - y \|^{2-p} \mu(\mathrm{d}y) \tag{1}$$

on $\mathbb{R}^p$, $p \geqslant 3$, where $\mu$ is a positive Radon measure on $\mathbb{R}^p$, and the study of harmonic functions which are defined as the solutions of the Laplace equation

$$\Delta h = 0.$$

The fundamental problems of the theory have already been formulated by C.F. Gauss. However, their definitive solutions have been obtained only during the first half of this century, more precisely during the period 1920 - 1950. An impressive report about the development of potential theory has been written by BRELOT [27], [32].

The existence of such a highly developed theory raised quite naturally the question

(**) Il manoscritto è pervenuto al Consiglio di Redazione il 25.5.1984.

to what extent the restriction to the particular kernel function

$$(2) \qquad x \mapsto \| x \|^{2-p}$$

or to the particular elliptic differential operator

$$(3) \qquad \Delta = \sum_{i=1}^{p} \frac{\partial^2}{\partial x_i^2}$$

is necessary. This question concerning the role of the Laplace operator stands at the beginning of the theory of harmonic spaces.

Early after the pioneering work of O. Perron and N. Wiener about the Dirichlet problem (cf. [27]) STERNBERG [65] observed in 1929 that their method of generalized solutions can also be applied to the heat equation (in the case of particular 2-dimensional domains), hence to the prototype of a parabolic differential equation. Apart from an important contribution by PETROWSKY [61], this observation had caused no further reaction for quite a while.

In 1949 TAUTZ [67] made the attempt to study the validity of certain potential-theoretic results in the case of elliptic differential equations – more general than the Laplace equation – by means of an axiomatic approach. But the notion of a harmonic space was born only in 1956 when BRELOT [29] pointed out that an axiomatic approach to the Dirichlet problem is possible for a large class of elliptic linear differential equations. More precisely, in [29] and in a series of succeeding publications, in particular in Brelot's lectures at the Tata Institute [30] (1), one finds the introduction of what is now called a Brelot space. By means of this notion Brelot succeeded in proving that the central theorems of classical potential theory remain valid for large classes of elliptic linear differential equations – even on manifolds.

The above mentioned observation of Sternberg motivated the attempt to include also parabolic differential equations. This was the main object of the author's first contribution [3] to this field. From this publication arose later on several very similar, but not fully equivalent definitions of the notion of a harmonic space with the aim to study the potential theory on such a space and to apply it to elliptic-parabolic differential equations. A certain final step in this development was done by CONSTANTINESCU and CORNEA [36].

(1) Part IV of [30] concerns harmonic spaces, and is essentially identical with [30a].

Two important historical remarks have to be added: First of all, Brelot had developed ideas about an axiomatic approach to potential theory already during World War II, hence long before the publication of his fundamental paper [29]. Secondly, DOOB [40] in 1956 published results about a unified approach to the Dirichlet problem for elliptic and parabolic differential equations by using a mixture of analytical and probabilistic methods. However, some of the highlights of the classical theory – like Brelot's resolutivity theorem – could not be obtained. Brelot, thanks to his earlier ideas, saw immediately the possibility to leave aside the probabilistic assumptions of Doob – however, by excluding the parabolic case. A few years later the author of the present paper succeeded to include the parabolic case – again without probabilistic assumptions. This was done in BAUER [3].

In conclusion, we can say that the theory of harmonic spaces is connected with probabilistic ideas since its existence. These connections with probability theory, in particular with the theory of Markov processes, turned out to be much closer than one had expected. Today, it is clear that the notion of a harmonic space, though being totally analytic, lies on the border line between analysis and probability theory.

In what follows we try to present the main ideas, results, methods and problems of the theory of harmonic spaces. We avoid to be as general as possible. So we leave aside the most general notion of a harmonic space. It is our intention to be as close as possible to interesting applications on the analytical side. At the same time we do not want to neglect the connections with the theory of Markov processes. Both subjects seem not to be explored totally; they offer the possibility of further fruitful research. Out point of view is exactly the one expressed by DOOB [41], p. xxii, in the introduction to his new monograph as follows: the theory of harmonic spaces or, more generally, abstract potential theory «is too important in its own right to be left to the probabilists».

The present survey is an improved version of the paper [11], originally published in German. At the same time the article develops lectures which the author gave in the Mathematical Seminar of the University of Bari during the period November 7 - 19, 1983. The author expresses his sincere thanks to his audience for its patience and interest in the subject. In particular, he thanks Professor Giovanni Aquaro for the kind invitation to publish these lectures in this series of «Conferenze del Seminario di Matematica della Università di Bari».

**§1. The notion of a harmonic space.**

A harmonic function is, by definition, a real-valued $C^2$-function defined on an open subset of $\mathbb{R}^p$ $p \geqslant 1$, which satisfies the Laplace equation

$$\Delta h = 0.$$

The following two results are of fundamental importance. They can be found in every introduction to potential theory (cf. BRELOT [28], DOOB [41], HELMS [48], KELLOGG [55], LANDKOF [57], DU PLESSIS [62]):

(A) For every open ball $B_r(x_0)$ in $\mathbb{R}^p$ (of centre $x_0$ and radius $r > 0$) the *Dirichlet problem is solvable without restrictions:* For every real-valued continuous function $f$ on the boundary $B_r^*(x_0)$ of the ball, hence on the sphere, there exists one and only one harmonic function $H_f$ on $B_r(x_0)$ which extends $f$ continuously. $H_f$ is given by the Poisson integral

$$\text{(4)} \qquad H_f(x) = \int P(x, y)\, f(y)\, \sigma_1(dy) \qquad (x \in B_r(x_0))$$

where $P$ is the Poisson kernel and $\sigma_1$ is the surface measure on $B_r^*(x_0)$ normalized to total mass 1. In dimension $p \geqslant 2$ one has

$$\text{(5)} \qquad P(x, z) = r^{p-2} \frac{r^2 - \| x - x_0 \|^2}{\| x - z \|^p}$$

$(x \in B_r(x_0),\ z \in B_r^*(x_0))$. Since $P$ is (in any dimension) strictly positive, $f \geqslant 0$ implies $H_f \geqslant 0$.

(B) Let $h_1, h_2, \ldots$ be an increasing sequence of harmonic functions on a *domain* $\Omega \subset \mathbb{R}^p$. Then the upper envelope sup $h_n$ is either harmonic or the constant function $+\infty$ on $\Omega$ (*Harnack's convergence theorem*).

The behavior of the solutions of the heat equation

$$\text{(6)} \qquad \square h = 0,$$

where

$$\text{(7)} \qquad \square = \Delta - \frac{\partial}{\partial x_{p+1}} = \sum_{i=1}^{p} \frac{\partial^2}{\partial x_i^2} - \frac{\partial}{\partial x_{p+1}}$$

is similar but not so well known. Here $h$ is defined on an open subset of $\mathbb{R}^{p+1}$, $p \geqslant 1$, which is a $C^2$-function with respect to $x_1, \ldots, x_p$, and a $C^1$-function with respect to $x_{p+1}$ and which satisfies (6) (cf. DOOB [40], [41], KAMKE [54] as well as BAUER [4], CONSTANTINESCU-CORNEA [36]) □:

(A′)˙ For every open ball $B_r(x_0)$ in $\mathbb{R}^{p+1}$ the *Dirichlet problem is solvable without restrictions:* For every real-valued continuous function $f$ on $B_r^*(x_0)$ there exists a uniquely determined solution $H_f$ of (6) on $B_r(x_0)$ which extends $f$ continuously. Furthermore, $f \geqslant 0$ implies $H_f \geqslant 0$.

The proof of property (A′) demands already quite a bit of potential theory. Computations are easier and we will lead to the same result when we replace the open ball $B_r(x_0)$ by the open cone

$$\Delta_r^a = \left\{ x \in \mathbb{R}^{p+1} : \left[ \sum_{i=1}^{p} (x_i - a_i)^2 \right]^{1/2} < a_{p+1} - x_{p+1}, \quad a_{p+1} - r < x_{p+1} < a_{p+1} \right\} \tag{8}$$

$(a = (a_1, \ldots, a_{p+1}) \in \mathbb{R}^{p+1}, r > 0)$. $\Delta_r^a$ is the convex hull of the $p$-dimensional open ball

$$\left\{ x \in \mathbb{R}^{p+1} : \sum_{i=1}^{p} (x_i - a_i)^2 < r^2, x_{p+1} = a_{p+1} - r \right\}$$

in the hyperplane $x_{p+1} = a_{p+1} - r$ and the vertex $a$ which has to be removed from the convex hull.

In both cases $H_f$ is given by a Poisson-type integral

$$H_f(x) = \int Q(x, z) f(z) \, \sigma_1(dz) \tag{9}$$

where integration refers to the surface measure $\sigma_1$ of $B_r(x_0)$ or $\Delta_r^a$, normalized to total mass 1. In contrast to (A) where the Poisson kernel $P$ was strictly positive, we have here $Q \geqslant 0$ and

$$Q(x, z) = 0 \tag{10}$$

for all points $(x, z) \in B_r(x_0) \times B_r^*(x_0)$ or $(x, z) \in \Delta_r^a \times (\Delta_r^a)^*$ which satisfy the condition $x_{n+1} < z_{n+1}$ (2).

(B′) Let $h_1, h_2, \ldots$ be a sequence of solutions of the heat equation defined on an open set $U$. Then $\sup h_n$ is again a solution of (6) provided that

$$\sup_{n=1,2,\ldots} h_n(x) < +\infty$$

holds on a dense subset of $U$.

This convergence theorem is analogous to the Harnack convergence theorem in classical potential theory. It is due to DOOB [40]. (See also [4], p. 21 and DOOB [41], p. 276).

It is the obvious analogy between (A) and (A′) as well as (B) and (B′) which leads to the definition of a harmonic space:

We consider a *locally compact space* $X$ which, in the above examples, equals $\mathbb{R}^p$ or $\mathbb{R}^{p+1}$, respectively. We imitate the solutions of the Laplace or heat equation by considering on $X$ a (complete) *sheaf* $\mathscr{H}$ of linear spaces of real-valued continuous functions. So, by definition, to every open subset $U \neq \phi$ of $X$ is associated a linear subspace $\mathscr{H}_U$ of $C(U)$, the linear space of all real-valued continuous functions on $U$, in such a way that the following two conditions are satisfied:

($G_1$) Restriction of a function $f \in \mathscr{H}_U$ to an open subset $V \neq \phi$ of $U$ yields a function $f | V \in \mathscr{H}_V$.

($G_2$) For every family $(U_i)_{i \in I}$ of open sets $U_i \neq \phi$ in $X$ and every function $f$ defined on $U = \bigcup_{i \in I} U_i$ one has $f \in \mathscr{H}_U$ provided that $f | U_i \in \mathscr{H}_{U_i}$ holds for all $i \in I$.

The elements of $\mathscr{H}_U$ will be called *harmonic functions* on $U$ (with respect to $\mathscr{H}$). So ($G_1$) and ($G_2$) state that the restriction of a harmonic function to a smaller open set is again harmonic and that a function, defined on an open set, is harmonic if it is locally harmonic. For the solution of the Laplace and the heat equation these properties are evident.

(2) $U^*$ denotes the boundary of a set $U$.

The sheaf $\mathscr{H}$ is called a *harmonic structure* on $X$ if the following three axioms are satisfied:

(I) $\mathscr{H}$ is *not degenerate*, i.e. for every point $x \in X$ there exists a harmonic function $h \in \mathscr{H}_U$ defined in an open neighborhood $U$ of $x$ such that $h(y) > 0$ for all $y \in U$.

(II) *Base Axiom:* The topology of $X$ has a base consisting of regular sets. Here a set $V \subset X$ is called *regular* (with respect to $\mathscr{H}$) if the Dirichlet problem is solvable without restrictions in the following sense: For every function $f \in C(V^*)$, i.e. real-valued continuous on the boundary of $V$, there exists a uniquely determined harmonic function $H_f \in \mathscr{H}_V$ which extends $f$ continuously. Furthermore, $f \geqslant 0$ implies $H_f \geqslant 0$.

(III) *Doob's Convergence Axiom* $K_D$: For every increasing sequence $h_1, h_2, \ldots$ of harmonic functions on an open set $U$, one has:

$$\{x \in U: \sup_{n \in \mathbb{N}} h_n(x) < +\infty\} \text{ dense in } U \Rightarrow \sup_{n \in \mathbb{N}} h_n \in \mathscr{H}_U.$$

For such a harmonic structure $\mathscr{H}$ on $X$, the pair $(X, \mathscr{H})$ is called a *harmonic space*. Other definitions of this notion will be discussed later.

A simple proof shows that a harmonic space is always *locally connected* (cf. [4], p. 17). Suppose now, conversely, that the given locally compact space $X$ is locally connected and that the sheaf $\mathscr{H}$ on $X$ satisfies the axioms I, II and

(III') *Brelot's Convergence Axiom* $K_B$: For every increasing sequence $h_1, h_2, \ldots$ of harmonic functions on a domain $\Omega \subset X$, one has:

$$\sup_{n \in \mathbb{N}} h_n(x) < +\infty \text{ for some } x \in \Omega \Rightarrow \sup_{n \in \mathbb{N}} h_n \in \mathscr{H}_U$$

Then, obviously, Doob's convergence axiom is satisfied, hence $\mathscr{H}$ is a harmonic structure. In that case, $\mathscr{H}$ is called a *Brelot harmonic structure* and $(X, \mathscr{H})$ is a *Brelot (harmonic) space.* Brelot's convergence axiom is identical with the Harnack convergence theorem (B) in classical potential theory.

More precisely we can say that $(\mathbb{R}^p, \mathscr{H}^\Delta)$ and $(\mathbb{R}^{p+1}, \mathscr{H}^\square)$ are harmonic spaces for every $p \geqslant 1$; the first space is even a Brelot space. Of course, $\mathscr{H}^\Delta$ and $\mathscr{H}^\square$ denote the sheaf of solutions of the Laplace and heat equation, respectively.

Also every Riemann surface is a Brelot harmonic space with respect to the usual harmonic functions on it. In all these cases axiom I is trivially satisfied since the constant functions are harmonic on the total space.

Every (Brelot) harmonic structure restricted to an open subset on the given (Brelot) harmonic space yields a (Brelot) harmonic structure.

Brelot harmonic spaces can be characterized among arbitrary harmonic spaces by means of the notion of ellipticity. We intend to return to this remark in § 3. Here we mention the following interpretation of the solutions $H_f$ of the Dirichlet problem for a regular set $V$ in a harmonic space:

For every point $x \in V$, $f \mapsto H_f(x)$ is a positive linear form on $C(V^*)$, hence a positive Radon measure $\mu_x^V$ on the boundary $V^*$ of $V$. $\mu_x^V$ is called the *harmonic measure* for $x$ (and $V$). By definition, we thus have

$$H_f(x) = \int f \, d\mu_x^V \qquad (f \in C(V^*),\ x \in V). \tag{11}$$

In the case of the harmonic space $(\mathbb{R}^p, \mathscr{H}^\Delta)$ and $V = B_r(x_0)$, one has

$$\mu_x^V = P(x, .)\sigma_1 \qquad (x \in V). \tag{12}$$

This is an immediate consequence from (4). Formula (9) can be interpreted similarly.

In the case of a Brelot space $(X, \mathscr{H})$ and a regular set $V$ in $X$ which is even connected and hence a *regular domain*, the support of the harmonic measure $\mu_x^V$ fills the whole boundary $V^*$, i.e.

$$\mathrm{Supp}(\mu_x^V) = V^* \qquad (x \in V). \tag{13}$$

This can be easily seen as follows: for every $f \in C(V^*)$ which is $\geqslant 0$ and $\neq 0$, we have to show

$$H_f(x) = \int f \, d\mu_x^V > 0$$

for a given point $x \in V$. Assume $H_f(x) = 0$ for such an $f$. We can apply Brelot's convergence axiom to the domain $\Omega = V$ and to the sequence of harmonic functions $h_n = nH_f$ on $V$. Since $h_n(x) = 0$ for all $n \in \mathbb{N}$, sup $h_n$ is harmonic on $V$ and, consequently,

$$\sup h_n(y) < +\infty \qquad \text{for all } y \in V.$$

This implies $H_f(y) = 0$ for all $y \in V$ and hence $f = 0$. We thus arrive at a contradiction to $f \neq 0$.

This argument shows that the harmonic space $(\mathbb{R}^{p+1}, \mathscr{H}^{\square})$ cannot be a Brelot space because of (10).

We conclude this paragraph by mentioning another convergence axiom which is very often used in the definition of a harmonic space. This is the *weak convergence axiom* $K_1$, sometimes named after the author (cf. [36]). It is only of technical importance as a first step towards Doob's axiom $K_D$. Under additional assumptions $K_1$ implies $K_D$ (cf. SCHIRMEIER [63] and BONY [26]).

## §2. Examples of harmonic spaces.

It seems now very natural to raise the following question:

Let $\Omega$ be a domain in $\mathbb{R}^p$, $L$ a linear differential operator of 2nd order defined on $\Omega$ and let $\mathscr{H}^L$ be the corresponding sheaf of $C^2$-functions $h$ which are solutions of the equation

$$Lh = 0.$$

When is $(\Omega, \mathscr{H}^L)$ a harmonic space?

2.1. We start with some answers which are due to BONY [25], [26]. Consider first order linear $C^\infty$-differential operators $X_1, \ldots, X_r, Y$ – or, equivalently, – $p$-dimensional $C^\infty$-vector fields on $\Omega$. Furthermore, consider a function $c \in C^\infty(\Omega)$. Suppose

(a): $L$ is of the form

$$L = \sum_{j=1}^{r} X_j^2 + Y + c. \tag{14}$$

Here, for first order linear $C^\infty$-differential operators $P$, $Q$, the product $PQ$ is defined by applying $P$ formally to $Q$. Suppose also

(b): $L$ is not degenerate in the sense that, given a point $x \in \Omega$, at least one of the vectors $X_1(x), \ldots, X_r(x)$ is $\neq 0$.

We denote by

$$[P, Q] = PQ - QP \tag{15}$$

the usual Lie-product of two $C^\infty$-vector fields $P$, $Q$. Then $[P, Q]$ is again a $C^\infty$-vector field.

Consider now a finite number $Y_1, \ldots, Y_s$ of $C^\infty$-vector fields. We denote by Lie $(Y_1, \ldots, Y_s)$ the $C^\infty(\Omega)$-module generated by $Y_1, \ldots, Y_s$ which is stable with respect to the Lie-product. Every vector field $Z \in \mathrm{Lie}(Y_1, \ldots, Y_s)$ is then a finite sum of terms

$$\varphi \cdot [Y_{i_1}, [Y_{i_2}, \ldots, [Y_{i_{k-1}}, Y_{i_k}] \ldots]]$$

where $\varphi \in C^\infty(\Omega)$ and $\{i_1, \ldots, i_k\} \subset \{1, \ldots s\}$ for arbitrary $k \in \mathbb{N}$. Each point $x \in \Omega$ leads to

$$\{Z(x) : Z \in \mathrm{Lie}\,(Y_1, \ldots, Y_s)\}$$

which is a linear space of dimension $\leq p$. We denote this dimension by

$$\mathrm{rank}_x\ \mathrm{Lie}(Y_1, \ldots, Y_s). \tag{16}$$

We then have

THEOREM 1 OF BONY. *Let the linear $C^\infty$-differential operator $L$ be of the form*

$$L = \sum_{j=1}^{r} X_j^2 + Y + c$$

*on a domain $\Omega \subset \mathbb{R}^p$, and assume that properties* (a) *and* (b) *hold. Then $(\Omega, \mathscr{H}^L)$ is a harmonic space if*

$$\mathrm{rank}_x\ \mathrm{Lie}(X_1, \ldots, X_r, Y) = p \qquad \textit{for all } x \in \Omega. \tag{17}$$

*$(\Omega, \mathscr{H}^L)$ is even a Brelot harmonic space if*

$$\mathrm{rank}_x\ \mathrm{Lie}(X_1, \ldots, X_r) = p \qquad \textit{for all } x \in \Omega. \tag{18}$$

Our standard examples $(\mathbb{R}^p, \mathscr{H}^\Delta)$ and $(\mathbb{R}^{p+1}, \mathscr{H}^\square)$ are special cases of this theorem. Assumption (b) is a consequence of condition (18), hence superfluous for the second part of the theorem.

The proof of Theorem 1 is based on a theorem of HÖRMANDER [52] according to which differential operators $L$ of the above type are *hypoelliptic*. So a distribution $u$ on an open subset of $\Omega$ is a $C^\infty$-function provided that $Lu$ is such a function. Therefore $\mathscr{H}^L$ can also be defined as the sheaf of continuous functions $u$ for which $Lu = 0$ holds in the sense of distribution theory. Harmonic functions are then even $C^\infty$-functions.

**2.2.** Bony's conditions in the above theorem are close to being necessary. This is demonstrated by the inverse problem treated by BONY [24], [26]. The problem is the following:

Let $\mathscr{H}$ be a harmonic structure on a domain $\Omega$ in $\mathbb{R}^p$. Is $\mathscr{H}$ necessary of the form $\mathscr{H}^L$? The answer is positive essentially when the sheaf $\mathscr{H}$ consists of $C^\infty$-functions. However, certain exceptional sets cannot be avoided. For simplicity's sake we assume that the constant function 1 is in $\mathscr{H}_\Omega$.

THEOREM 2 OF BONY. *Assume* $1 \in \mathscr{H}_\Omega$ *and assume that* $\mathscr{H}$ *contains sufficiently many* $C^\infty$*-functions in the following sense: For every harmonic function h defined on an open neighborhood of a compact set* $K \subset \Omega$ *there exists a sequence* $(h_n)$ *of harmonic* $C^\infty$*-functions defined on neighborhoods of K (depending on n) such that* $(h_n)$ *converges uniformly to h on K. Then there exists an open dense subset* $\Omega_0 \subset \Omega$ *and on it a differential operator*

$$L = \sum_{j=1}^{r} X_j^2 + Y$$

*with p-dimensional* $C^\infty$*-vector fields* $X_1, \dots, X_r$, $Y$ *such that*

$$\operatorname{rank}_x \operatorname{Lie}(X_1, \dots, X_r, Y) = p \qquad \textit{for all } x \in \Omega_0$$

*(or*

$$\operatorname{rank}_x \operatorname{Lie}(X_1, \dots, X_r) = p \qquad \textit{for all } x \in \Omega_0$$

*in the case of a Brelot harmonic structure) and such that the two sheafs* $\mathscr{H}$ *and* $\mathscr{H}^L$ *coincide after restriction to* $\Omega_0$.

The exceptional set $\Omega \setminus \Omega_0$ cannot be avoided, in general. This is shown by the following simple example of a Brelot space: $\Omega = \mathbb{R}$ and $\mathscr{H}_U$ is for every open interval $U \subset \mathbb{R}$ the set of all functions $x \mapsto \alpha x^3 + \beta$ restricted to $U (\alpha, \beta \in \mathbb{R})$. Here one can choose $\Omega_0 = \mathbb{R} \setminus \{0\}$, but not $\Omega_0 = \mathbb{R}$. In fact, a differential operator $L$ describing $\mathscr{H}$ is

$$L = \frac{d^2}{dx^2} - a(x) \frac{d}{dx}$$

where $a(x) = 2/x$ is uniquely determined only for $x \neq 0$. An arbitrary choice of the value $a(0)$ would lead to the solution

$$u(x)=\begin{cases}x^3, & x\leqslant 0\\ 2x^3, & x>0\end{cases}$$

of $Lu = 0$ defined on all of $\mathbb{R}$. However, $u$ is not in $\mathscr{H}_{\mathbb{R}}$.

It should also be pointed out, that the differential operator $L$ of Theorem 2 is essentially uniquely determined, namely up to a $C^\infty$-factor of proportionality (cf. BONY [26], p. 85).

Let us consider a particular case in which the above results can be improved. This is the case where $\Omega = \mathbb{R}^p$ and where the given harmonic structure is *translation-invariant:* So $h \in \mathscr{H}_U$ should imply the harmonicity of $x \mapsto h(x-t)$ on $t + U$ for every $t \in \mathbb{R}^p$. By a regularization procedure one can prove that $\mathscr{H}$ automatically contains sufficiently many $C^\infty$-functions. One can then choose $\Omega_0 = \mathbb{R}^p$ and $X_1, \ldots, X_r$, $Y$ as constant vector fields. In other words, $\mathscr{H} = \mathscr{H}^L$ where $L$ is a differential operator on $\mathbb{R}^p$ with constant coefficients (cf. [24], [25], [26]).

The situation becomes more complicated if the assumption about the existence of sufficiently many $C^\infty$-functions fails. Then unconventional examples of harmonic spaces enter into the picture. In that direction we content ourselves with a few hints.

2.3. Consider on a domain $\Omega \subset \mathbb{R}^p$ a differential operator in divergence form:

$$L = \sum_{i=1}^{p} \frac{\partial}{\partial x_i}\left(\sum_{j=1}^{p} a_{ij}\frac{\partial}{\partial x_j}\right) \tag{19}$$

where the coefficients $a_{ij}(x) = a_{ji}(x)$ are Borel-measurable functions on $\Omega$. We assume that $L$ is *uniformly elliptic* in the sense that, for some real number $\lambda \geqslant 1$,

$$\frac{1}{\lambda}\sum_i \xi_i^2 \leqslant \sum_{i,j} a_{ij}(x)\xi_i\xi_j \leqslant \lambda \sum_i \xi_i^2$$

holds for all $x \in \Omega$ and all $(\xi_1, \ldots, \xi_p) \in \mathbb{R}^p$. Then the local solutions of $Lu = 0$ (in the distribution-theoretic sense) form a Brelot harmonic structure $\mathscr{H}^L$. Here the harmonic functions are not differentiable, in general. For details see R.M. HERVÉ [49], [50] and BONY [26].

For other types of elliptic differentiable equations leading to Brelot harmonic structures see HERVÉ [49] - [51], BOBOC and MUSTAŢĂ [23], and BONY [26].

**2.4.** A remarkable class of Brelot harmonic structures has been discovered by BERG [14], [15] on the infinite dimensional torus $T^{\mathbb{N}}$ where $T$ is the unit circle $|z| = 1$ in $\mathbb{C}$. One considers a «differential operator»

$$L = \sum_{n=1}^{\infty} a_n \frac{\partial^2}{\partial \theta_n^2} - \lambda \tag{20}$$

on $T^{\mathbb{N}}$ with constant coefficients $a_1 > 0$, $a_2 > 0, \ldots,$ and $\lambda \geqslant 0$. $(\theta_n)_{n \in \mathbb{N}}$ denotes a point in $T^{\mathbb{N}}$. Under the assumption that

$$\sum_{n=1}^{\infty} \frac{1}{a_n} < +\infty, \tag{21}$$

the adequately defined solutions of the equation

$$Lu = 0$$

yield a Brelot harmonic structure $\mathscr{H}^L$. Though $T^{\mathbb{N}}$ is not a manifold in the ordinary sense, one imitates the distribution-theoretic definition of solutions. Details can be found in the original work of BERG [14], [15] as well as in BAUER [9], [10].

**2.5.** Following HANSEN [37] we consider in the plane $\mathbb{R}^2$ the compact subspace

$$X = [1, n] \times \{0\} \cup \bigcup_{i=1}^{n} \{i\} \times [-1, +1] \tag{22}$$

for a given natural number $n$:

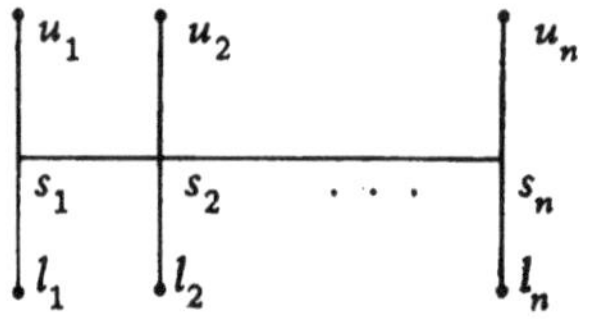


The points

$$s_i = (i, 0), \qquad u_i = (i, +1), \qquad l_i = (i, -1),$$

$i = 1, \ldots, n$, will be called support points, upper and lower comb points, respectively, whereas $X$ will be called *double comb of order n*. There exists a unique harmonic structure

$\mathscr{H}$ on $X$ such that for every open connected set $U \subset X$ of (euclidean) diameter $< 1$ the functions $h \in \mathscr{H}_U$ are defined as follows: $h$ is affine on $U$ if none of the points $s_i$, $u_i$, $l_i$ belongs to $U$; $h$ is constant on $U$ if an upper or lower comb point belongs to $U$; besides being continuous, $h$ is affine on each of the sets $U \cap [u_i, l_i]$, $U \cap [s_i, s_{i+1}]$ and $U \cap [s_{i-1}, s_i]$ if $s_i \in U$ for some $i$.

Consequently, harmonic functions on all of $X$ are those continuous functions which are affine on each segment $[s_i, s_{i+1}]$ and constant on each segment $[u_i, l_i]$. $\mathscr{H}_X$ is a linear space of dimension $n$. It is easy to see that only for $n = 1$ we have a Brelot space.

Finally, it should be pointed out that there are close connections between this example and the recent research of LUMER [58] about diffusion operators on networks.

## § 3. Strong harmonic spaces

Classical potential theory for the Laplace equation in $\mathbb{R}^p$ shows a sort of singular behavior in dimensions $p = 1$ and $p = 2$. In dimension $p = 2$ this can be explained in a somewhat superficial way by the fact that the Newtonian kernel has to be replaced by the logarithmic kernel $x \mapsto -\log \|x\|$ which attains also negative values. The peculiarities of the theory in dimensions $p \geqslant 3$ are reflected in the theory of harmonic spaces by the introduction of strong harmonic spaces. Positive superharmonic functions and, in particular, potentials are crucial for their definition.

Let again be $(X, \mathscr{H})$ a harmonic space. A *hyperharmonic function* on an open set $U \subset X$ is a lower semicontinuous function

$$u : U \to ]-\infty, +\infty]$$

which satisfies

$$\int u \, d\mu_x^V \leqslant u(x) \tag{23}$$

for every point $x \in U$ and all regular neighborhoods $V$ of $x$ such that $\bar{V} \subset U$. If, in addition,

$$u(x) < +\infty \tag{24}$$

holds on a *dense* subset of $U$, then $u$ is called *superharmonic* on $U$. We denote by $\mathscr{H}_U^*$ and $\mathscr{S}_U$ the sets of all hyperharmonic or superharmonic functions on $U$, respectively.

Obviously, one has

(25) $\mathscr{H}_U = \mathscr{H}_U^* \cap (-\mathscr{H}_U^*) = \mathscr{S}_U \cap (-\mathscr{S}_U)$

for every open set $U \neq \phi$.

Finally, a non-negative superharmonic function $p$ defined on all of $X$ is called a *potential* if $h = 0$ is the only harmonic function on $X$ satisfying $0 \leqslant h \leqslant p$. The set of all potentials will be denoted by $\mathscr{P}_X$.

The harmonic space $(X, \mathscr{H})$ is now called *strong* – or $\mathscr{P}$-harmonic in the terminology of [36] – if one of the following two equivalent conditions is fulfilled:

(S) The set $\mathscr{S}_X^+$ of all superharmonic functions $\geqslant 0$ on $X$ *separates* the *points* of $X$ *linearly:* for each pair of points $x_0 \neq y_0$ there exist functions $s, t \in \mathscr{S}_X^+$ which satisfy

(26) $s(x_0)\, t(y_0) \neq s(y_0)\, t(x_0).$

(P) The set $\mathscr{P}_X$ of potentials on $X$ separates the points of $X$ linearly.

Accordingly, $(X, \mathscr{H})$ is called *locally strong* if each point of $X$ has an open neighborhood $U$ such that the harmonic space $(U, \mathscr{H}|U)$ – obtained by restriction – is strong. Obviously, every strong harmonic space is also locally strong.

Many harmonic spaces are locally strong: All spaces $(\mathbb{R}^p, \mathscr{H}^\Delta)$, $p \geqslant 1$, are locally strong since the constant function 1 and the $p$ coordinate function are harmonic on $\mathbb{R}^p$. The harmonic space $(\mathbb{R}^{p+1}, \mathscr{H}^\square)$ of the heat equation is strong for every $p \geqslant 1$ since the 2p positive solutions of the heat equation

$$x \mapsto e^{\pm x_i} e^{x_{p+1}} \qquad (i = 1, \ldots, p)$$

together with the constant solution 1 separate the points of $\mathbb{R}^{p+1}$ linearly.

The harmonic space $(\mathbb{R}^p, \mathscr{H}^\Delta)$ is strong for all dimensions $p \geqslant 3$. This follows from the fact that the Newtonian kernel function (2), which is $+\infty$ at the origin, and consequently also the functions

$$p_1(x) = \frac{1}{\|x - y_0\|^{p-2}}, \quad p_2(x) = \frac{1}{\|x - z_0\|^{p-2}} \qquad (y_0, z_0 \in \mathbb{R}^p)$$

are potentials. For given points $x_0 \neq y_0$, one chooses a third point $z_0$ different from $x_0$ and $y_0$. Then, obviously,

$$p_1(x_0)\, p_2(y_0) < p_1(y_0)\, p_2(x_0) = +\infty.$$

One could also verify this fact by making use of the following result (cf. [36], p. 63 and p. 44):

3.1. *A locally strong harmonic space* $(X, \mathscr{H})$ *is strong if for every point* $x \in X$ *there exists a potential* $p$ *on* $X$ *which satisfies*

(27) $$p(x) > 0.$$

For Brelot spaces a simpler result holds (cf. [36], p. 67 and [30], p. 71):

3.2. *A connected Brelot harmonic space* $(X, \mathscr{H})$ *is strong if and only if there exists at least one potential* $p \neq 0$.

This is due to the consequences of the restrictive convergence axiom $K_B$. In particular, one has (cf. [30], p. 63, 66):

3.3. *Every Brelot harmonic space is locally strong.*

On strong harmonic spaces the important boundary minimum principle is valid. In fact, if $(X, \mathscr{H})$ is a strong harmonic space and if $U$ is an open subset, on which a harmonic function with strictly positive infimum exists, then one has [36], [4], [2]:

(MP) *A hyperharmonic function* $u$ *on* $U$ *is non-negative if, for all boundary points* $z$ *of* $U$ *in the one-point compactification of* $X$, *one has*

(28) $$\liminf_{\substack{y \to z \\ y \in U}} u(y) \geqslant 0.$$

In view of axiom $I$ we can conclude that, whenever the harmonic space $(X, \mathscr{H})$ is locally strong, (MP) holds for sufficiently small open neighborhoods of arbitrary points in $X$. An important consequence of this is the following *localized version* of the condition (23) for hyperharmonic functions: A lower semicontinuous function $u$ on an open set $U$ with values $> -\infty$ is hyperharmonic if

$$\int u \, d\mu_x^V \leqslant u(x)$$

holds for every point $x \in U$ and a fundamental system of regular neighborhoods $V$ of $x$ (such that $\bar{V} \subset U$).

Finally, Brelot harmonic spaces can now be characterized within the class of all harmonic spaces by means of the notion of ellipticity which is closely connected with property (13): A harmonic space $(X, \mathscr{H})$ is called *elliptic* if every point $x \in X$ has a fundamental system of regular neighborhoods $V$ such that

$$\operatorname{Supp} \mu_x^V = V^*$$

holds. The announced characterization is then the following (cf. [4], p. 40 and [36], p.66):

3.4. *A harmonic space is a Brelot space if and only if it is locally strong and elliptic.*

We have already seen that $(\mathbb{R}^p, \mathscr{H}^\Delta)$ for $p \geq 3$ and $(\mathbb{R}^{p+1}, \mathscr{H}^\square)$ for $p \geq 1$ are strong harmonic spaces. Let us discuss now also the other main examples of § 2.

First of all, every *Riemann surface* is a Brelot harmonic space. It is strong if and only if the surface is *hyperbolic* in the sense of AHLFORS-SARIO [1].

Every harmonic space appearing in Theorem 1 of BONY is locally strong. (Cf. [25], p. 19 and [26], p. 84).

The harmonic spaces from 2.3 are strong in the case of bounded domains $\Omega \subset \mathbb{R}^p$. For further related results the reader is referred to HERVÉ [49], p. 562.

The Brelot harmonic structures on the infinite dimensional torus discussed in 2.4 are strong if and only if $\lambda > 0$.

The double comb from 2.5 is a locally strong harmonic space for every $n \geq 1$, bur never a strong harmonic space.

We conclude this paragraph with the remark that the terminology of this survey sometimes is not in full agreement with the one used by CONSTANTINESCU and CORNEA [36]: The Brelot harmonic spaces of this survey coincide – even by definition – with the corresponding spaces in [36]. However, the locally strong harmonic spaces of this survey are called in [36] Bauer harmonic spaces in which Doob's convergence axiom holds. The definition of these spaces given in [36] – different from ours (but equivalent) – seems to be not quite satisfactory from the standpoint of applications.

## § 4. Potential theory of strong harmonic spaces.

*Strong harmonic* spaces offer a natural framework for potential-theoretic studies. In what follows we denote by $(X, \mathscr{H})$ such a space. We assume in addition that $X$ has a *countable base.* The existence of such a base is necessary for most of the results which will be discussed now. The assumption is satisfied in all important examples of applications.

Potential theory on $X$ has as main object the intensitive study of the balayage operator. Given is a hyperharmonic function $u \geqslant 0$ on $X$ and a subset $E$ of $X$. As already in classical potential theory one is interested in hyperharmonic functions $u \geqslant 0$ on $X$ which coincide on $E$ with $u$, and one tries to find a minimal function $v$ of this type. Accordingly, one defines the *reduced function* (or *réduite)* of $u$ with respect to $E$ by

$$(29) \qquad R_u^E(x) = \inf\{v(x) : v \in \mathscr{H}_X^*,\ v \geqslant 0,\ v = u \text{ on } E\}$$

for all $x \in X$. One then has

$$R_u^E(x) = u(x) \qquad (x \in E),$$

however $R_u^E$ is not lower semicontinuous and hence not hyperharmonic, in general. This is the reason why one passes from $R_u^E$ to $\hat{R}_u^E$ the *swept-out function* (or *balayée*) of $u$ with respect to $E$ which, by definition, is the greatest lower semicontinuous function $\leqslant R_u^E$:

$$(30) \qquad \hat{R}_u^E(x) = \liminf_{y \to x} R_u^E(y) \qquad (x \in X).$$

One has

$$(31) \qquad 0 \leqslant \hat{R}_u^E \leqslant R_u^E \leqslant u;$$

furthermore $\hat{R}_u^E$ turns out to be a hyperharmonic function. The justification for speaking about «sweeping» (in French: balayer) comes from the fact that, given a positive Radon measure $\mu$ on $X$ *with compact support,* there exists exactly one positive Radon measure $\mu^E$ on $X$ which satisfies

$$(32) \qquad \int \hat{R}_u^E \, d\mu = \int u \, d\mu^E \qquad \text{for all } u \in \mathscr{H}_X^*,\ u \geqslant 0.$$

$\mu^E$ is called the measure which is obtained by *sweeping $\mu$ onto $E$.*

For the particular choice $\mu = \epsilon_x$ (= unit mass at $x$), (32) yields

$$\hat{R}_u^E(x) = \int u \, \mathrm{d}\epsilon_x^E \qquad (x \in E) \tag{33}$$

From this one can conclude that $x \mapsto \epsilon_x^E(A)$ is Borel measurable for every Borel set $A \subset X$. Therefore,

$$B^E(x, A) = \epsilon_x^E(A) \qquad (x \in X,\ A \text{ Borel set} \subset X) \tag{34}$$

defines a *kernel* on $X$ in the usual measure theoretic sense (3).

This kernel $B^E$ permits to interpret sweeping by means of an operator. In fact, one has

$$\hat{R}_u^E = B^E u \qquad (u \in \mathscr{H}_X^*,\quad u \geqslant 0) \tag{35}$$

where

$$B^E u(x) = \int u(y)\, B^E(x, \mathrm{d}y)$$

according to the usual convention for kernels. In view of this interpretation $B^E$ is called the *balayage operator.*

It should be pointed out that examples of measures $\epsilon_x^E$ are already known to us. In fact, whenever $V$ is a regular set, the harmonic measure $\mu_x^V$ for a point $x \in V$ turns out to be

$$\mu_x^V = \epsilon_x^{\complement V}. \tag{36}$$

One can use (36) as a definition of the harmonic measures $\mu_x^V$ for an arbitrary open set $V$ and points $x \in V$. One arrives then at the treatment of the Dirichlet problem for arbitrary open, relatively compact sets $V$: For every function $f \in C(V^*)$, the function

$$x \mapsto \int f \, \mathrm{d}\mu_x^V = \int f \, \mathrm{d}\epsilon_x^{\complement V} \qquad (x \in V)$$

turns out to be the *generalized solution in the sense of Perron-Wiener-Brelot.* For arbitrary

(3) Cf. [13], pp. 323 - 326 and the explications given in the next paragraph.

open sets $V$ the same approach leads to the solution of the Cauchy problem,in particular for parabolic differential equations (cf. BAUER [6] and KÖHN and SIEVEKING [56a]). Boundary points $z$ of $V$ satisfying $\epsilon_z^{\complement V} \neq \epsilon_z$ are (in both cases) the irregular points of $V^*$ which are well-known in classical potential theory.

In general, a set $E \subset X$ is called *thin at a point* $x \in X$ if $\epsilon_x^E \neq \epsilon_x$. The set

(37) $$b(E) = \{x \in X : E \textit{ not } \text{thin at } x\}$$

is called the *base of* $E$. By means of $b(E)$ the general problem formulated at the beginning of this paragraph can be answered. The question was whether, given a set $E \subset X$ and a hyperharmonic function $u \geqslant 0$ on $X$, one can find a smallest possible hyperharmonic function $v \geqslant 0$ on $X$ which coincides with $u$ on $E$. The natural candidate for a solution of this problem is the reduced function $R_u^E$. It had to be replaced by the swept-out function $\hat{R}_u^E$ because of lack of semicontinuity.

Now one can prove that (cf. [4], p. 105 and [36], p. 129)

(38) $$\hat{R}_u^E(x) = R_u^E(x) \qquad \text{for all } x \in X \setminus E.$$

Furthermore, we know from (31) and (29) that

(39) $$\hat{R}_u^E(x) \leqslant R_u^E(x) = u(x) \qquad \text{for all } x \in E.$$

In general, one does not have $\hat{R}_u^E = R_u^E$ on $E$ since, otherwise, $\hat{R}_u^E$ would be lower semicontinuous. Therefore, one has to accept the fact that $\hat{R}_u^E(x) \neq u(x)$ – or, equivalently $\hat{R}_u^E(x) < u(x)$ – for some points $x$ of $E$. But it turns out that the set of points $x \in E$ where

$$\hat{R}_u^E(x) < u(x)$$

is small in a sense to be made precise now.

In view of (33), the definition of the base $b(E)$ can also be written in the form

(40) $$b(E) = \{x \in X : \hat{R}_u^E(x) = u(x) \text{ for all } u \in \mathscr{H}_X^*,\ u \geqslant 0\}.$$

Given a hyperharmonic function $u \geqslant 0$ on $X$, it then follows that all points $x \in E$ satisfying $\hat{R}_u^E(x) < u(x)$ are contained in $E \setminus b(E)$, hence in the set of points of $E$ where $E$ is thin. This set, however, is small in the sense that it is an *exceptional set* in the potential-theoretic sense. For the definition of these sets we need the following definition.

A set $P \subset X$ is called *totally thin* if it is thin at all points of $X$, hence if the base $b(P)$ is empty. Every union $\bigcup_{n=1}^{\infty} P_n$ of a *sequence* $P_1, P_2, \ldots,$ of totally thin sets is called a *semi-polar set*. In contrast to this definition, a set $P \subset X$ is called *polar* if there exists a superharmonic function $s \geqslant 0$ on $X$ such that

$$P \subset \{x \in X : s(x) = +\infty\}$$

Every polar set is also semi-polar but not conversely, in general.

In the classical harmonic spaces $(\mathbb{R}^p, \mathscr{H}^\Delta)$, $p \geqslant 3$, the semi-polar sets coincide with the polar ones. In general, this implication holds in strong harmonic spaces in which the so-called *domination axiom* of BRELOT [30] holds. According to a result of KÖHN and SIEVEKING [56] such spaces are necessarily elliptic and hence Brelot spaces. However, already for the harmonic spaces $(\mathbb{R}^{p+1}, \mathscr{H}^\square)$ of the heat equation there are totally thin sets which are not polar. A simple example of such a set is each «horizontal» hyperplane in $\mathbb{R}^{p+1}$ given by the equation $x_{p+1} = \text{const}$ (cf. [4], p. 108 - 109).

The importance of the semi-polar sets as the exceptional sets of potential theory becomes clear from a *convergence theorem* which generalizes the classical convergence theorem of H. CARTAN [34] in the case of the spaces $(\mathbb{R}^p, \mathscr{H}^\Delta)$, $p \geqslant 3$. One considers an arbitrary (not necessarily countable) family $(u_i)_{i \in I}$ of hyperharmonic functions $u_i \geqslant 0$ on $X$ with index set $I$. One defines $u_0$ to be the lower envelope of the family and $\hat{u}_0$ to be the greatest lower semicontinuous function $\leqslant u_0$, hence

$$\hat{u}_0(x) = \liminf_{y \to x} u(y) \qquad (x \in X)$$

Then the convergence theorem tells us that the two functions $\hat{u}_0$ and $u_0$ differ only in the points of a semi-polar set. (Cf. BAUER [5] and CONSTANTINESCU and CORNEA [36], p. 154).

The swept-out function $\hat{R}_u^E$ is a function of the type $u_0$. Consequently, the convergence theorem leads to a solution of the problem formulated at the beginning of this paragraph: for every hyperharmonic function $u \geqslant 0$ on $X$ and every set $E \subset X$, the two functions $\hat{R}_u^E$ and $R_u^E$ differ only on a semi-polar set. It follows from (38) that

$$\{x \in X : \hat{R}_u^E(x) < R_u^E(x)\} \tag{41}$$

is a *semi-polar subset of* $E$.

By means of a result of Constantinescu (cf. [36], p. 168), this result can be improved. There always exist continuous real-valued potentials $p$ of a particular type (so-called strict potentials) which allow the description of the base $b(E)$ by means of the single function $p$:

(42) $$b(E) = \{x \in X : \hat{R}_p^E(x) = p(x)\}.$$

This result improves the original definition (40). As a consequence, the base $b(E)$ of an arbitrary set $E \subset X$ turns out to be a $G_\delta$ *- set.* Furthermore, according to (41), *the set* $E \setminus b(E)$ *of all points* $x \in X$ *in which E is thin is always semi-polar.*

Far reaching, deep properties of the balayage operator have been discovered by means of the so-called *fine potential theory* of FUGLEDE [42] and by the investigations of BLIEDTNER and HANSEN [17] about simplicial function spaces derived from harmonic spaces. The reader is referred to the original publications as well as to the survey articles [8], [12] of the author.

For strong harmonic spaces which satisfy the above mentioned domination axiom, the importance of the base $b(E)$ can be made clear also by the behavior of swept-out measures: For every positive Radon measure $\mu$ with compact support and every set $E \subset X$, *the swept-out measure* $\mu^E$ *is supported by the base* $b(E)$. This result of HERVÉ [49], p. 532 generalizes the corresponding result of BRELOT [33] which gave the definitive form to the balayage problem in classical potential theory.

The corresponding result fails already for the heat equation as has been pointed out by DOOB [40]. In fact, for an open relatively compact set $V$ of the harmonic space ($\mathbb{R}^{p+1}$, $\mathscr{H}^{\square}$), the harmonic measure $\mu_x^V = \epsilon_x^{\complement V}$ of a point $x \in V$ is generally not supported by $b(\complement V)$ which is the set of regular boundary points of $V$. For further results illuminating this phenomenon the reader is referred to BLIEDTNER and HANSEN [17], [19].

## §5. Associated Markov processes.

The picture of the theory of harmonic spaces would be incomplete if we would not discuss the connections of this theory with the theory of Markov processes. Historically, the origin of these connections lies in the well-known interplay between classical potential theory and Brownian motion. Details about this interplay can be found in DOOB [41] as well as in the survey articles [8], [10] of the author and of BLIEDTNER [16].

We start with some preliminaries referring to an arbitrary locally compact space $X$ with countable base. Let us denote by $\mathscr{B}$ the $\sigma$-algebra of the Borel subsets of this space. In the last paragraph we have already made use of the notion of a *kernel* on $X$. By definition, this is a map

$$P : X \times \mathscr{B} \to [0, +\infty]$$

which as a function of the first variable $x \in X$ is $\mathscr{B}$-measurable and which as a function of the second variable $B \in \mathscr{B}$ is a positive measure on $\mathscr{B}$. In what follows only *sub-Markov* kernels will enter into the picture, hence kernels $P$ satisfying

$$P(x, X) \leqslant 1 \qquad \text{for all } x \in X \quad \square$$

Every kernel $P$ can be viewed as an operator on the Borel-measurable functions $f : X \to [0, +\infty]$ or, in the sub-Markov case, on the linear space of bounded, Borel-measurable, real-valued functions $f$ on $X$. This operator maps $f$ into the Borel function

$$Pf(x) = \int f(y)\, P(x, \mathrm{d}y)$$

which is non-negative or bounded, respectively. Consequently, two kernels $P$ and $Q$ on $X$ – viewed as operators – can be composed. The results is again a kernel $QP$ which equals

$$QP(x, B) = \int Q(x, \mathrm{d}y)\, P(y, B)$$

($x \in X$, $B \in \mathscr{B}$). The kernel $QP$ is sub-Markov in the case where both $P$ and $Q$ are sub-Markov kernels.

The analytical origin of a Markov process is a semigroup $(P_t)_{t>0}$ of sub-Markov kernels $P_t$. The semigroup property demands that

$$P_{s+t} = P_s P_t \qquad \text{for all } s > 0,\ t > 0,$$

or, equivalently,

$$P_{s+t}(x, B) = \int P_s(x, \mathrm{d}y)\, P_t(y, B)$$

for all $x \in X$, $B \in \mathscr{B}$, $s > 0$ and $t > 0$. In connection with such a *sub-Markov semigroup* one is interested – by reasons which will become clear in a minute – in the corresponding *excessive functions*. By definition, these are the Borel-measurable functions $u : X \to [0, +\infty]$ satisfying the condition

(43) $$\sup_{t>0} P_t u = u.$$

It should be observed that $t \mapsto P_t u$ is then necessarily decreasing and, consequently, (43) turns out to be equivalent to the two conditions

(44) $$P_t u \leqslant u \qquad \text{for all } t > 0, \text{ and}$$
$$\lim_{t \to 0} P_t u = u.$$

Among the *sub-Markov* semigroups $(P_t)_{t>0}$ which will be considered later there are two types of special general importance: the Feller and quasi-Feller semigroups. $(P_t)_{t>0}$ is called a *Feller semigroup* if each (sub-Markov) kernel $P_t$ maps the space $C_0(X)$ of continuous real functions on $X$ vanishing at infinity into itself and if

(45) $$\lim_{t>0} \| P_t f - f \| = 0 \qquad (f \in C_0(X))$$

holds with respect to the usual sup-norm.

If, in contrast to this, we have

(46) $$P_t(C_0(X)) \subset C(X) \qquad (t > 0)$$

as well as

(47) $$\lim_{t \to 0} P_t f = f$$

uniformly on the compact subsets of $X$ for all $f \in C_0(X)$ and

(48) $$\frac{p}{q} \in C_0(X)$$

for certain strictly positive, excessive functions $p$, $q \in C(X)$, then the sub-Markov semigroup $(P_t)_{t>0}$ is called *quasi-Feller*.

One justification for this naming is the fact that, starting out with an excessive function $q$ from property (48), the definition

(49) $$P_t^* f = \frac{1}{q} P_t(qf) \qquad (f \geqslant 0, \text{ Borel-measurable})$$

leads to a Feller semigroup $(P_t^*)_{t>0}$ (cf. HANSEN [45]).

All Feller semigroups $(P_t)_{t>0}$ for which the kernel

$$\int_0^\infty P_t \, dt$$

behaves nicely turn out to be also quasi-Feller (cf. [45], p. 246).

After these preparations we can formulate the central theorem which illuminates the connection of these notions with harmonic spaces. The theorem in this form is due to HANSEN [44]. Related results have been obtained by MEYER [60] and BOBOC, CONSTANTINESCU and CORNEA [22].

5.1. *Let $(X, \mathscr{H})$ be a strong harmonic space with countable base on which the constant function 1 is superharmonic. Then there exists a quasi-Feller semigroup $(P_t)_{t>0}$ on X such that the nonnegative hyperharmonic functions on X coincide with the excessive functions on X.*

The condition about the superharmonicity of the constant function 1 is necessary for the validity of this theorem since for quasi-Feller (and Feller) semigroups the constant function 1 is always excessive. Furthermore, the theorem allows certain improvements. According to them the function

$$p = \int_0^\infty P_t 1 \, dt$$

can be chosen arbitrarily in the class of all continuous real and strict potentials (cf. HANSEN [44], [45]; TAYLOR [68], and [7], [10]).

For all spaces $(\mathbb{R}^p, \mathscr{H}^\Delta)$, $p \geqslant 3$, of classical potential theory the semigroup $(P_t)_{t>0}$ of Brownian motion satisfies the conditions of the theorem. In particular, $(P_t)_{t>0}$ is at the same time a quasi-Feller and a Feller semigroup. For this case the theorem is due to DOOB [41] and HUNT [53] (cf. [10]).

Let us consider again a quasi-Feller semigroup $(P_t)_{t>0}$ on a locally compact space $X$ with countable base, not necessarily a harmonic space at the moment. The analytical properties of the semigroup are strong enough to interpret $(P_t)_{t>0}$ as the transition semigroup of a Markov process with good additional properties, more precisely as the semigroup of a standard or even a Hunt process. Without aiming at the very last precision this means the following: There is a family $(X_t)_{t\geqslant 0}$ of random variables

$$X_t : \Omega \to X$$

with values in $X$ which are defined on a suitable measurable space $(\Omega, \mathscr{A})$. For every point $x \in X$ there is a probability measure $P^x$ on $\mathscr{A}$ such that the stochastic process

$$(\Omega, \mathscr{A}, P^x, (X_t)_{t \geq 0})$$

satisfies the condition

$$P_t(x, B) = P^x\{X_t \in B\} \qquad (B \in \mathscr{B}), \tag{50}$$

in other words, such that the measure $P_t(x, .)$ is the distribution of $X_t$ with respect to $P^x$. [Hereby, the process $(X_t)_{t \geq 0}$ controlled by $(\Omega, \mathscr{A}, P^x)$ has only a finite «lifetime» since $P_t(x, X) < 1$, in general].

Furthermore, the strong Markov property and the usual regularity properties for the paths $t \mapsto X_t(\omega)$ should be present (cf. BLUMENTHAL - GETOOR [20] and WILLIAMS [70]). The total object that we have in mind is the Hunt process

$$\mathscr{X} = (\Omega, \mathscr{A}, (P^x)_{x \in X}, (X_t)_{t \geq 0}). \tag{51}$$

For the case of a harmonic space the situation simplifies itself since the quasi-Feller semigroups of 5.1 allow the interpretation by means of Hunt processes with *continuous paths*. The main result is then the following (cf. [60], [22], [44], [7]):

5.2. *Let $(X, \mathscr{H})$ be a strong harmonic space with countable base such that the constant function 1 is superharmonic. Then there exists a Hunt process $\mathscr{X}$ with continuous paths* (4) *such that the excessive functions of the transition semigroup $(P_t)_{t>0}$ defined by* (50) *coincide with the non-negative hyperharmonic functions on $X$.*

For the classical harmonic space $(\mathbb{R}^p, \mathscr{H}^\Delta)$, $p \geq 3$, Brownian motion has the above properties ([38], [41]). For the harmonic space of the heat equation $(\mathbb{R}^{p+1}, \mathscr{H}^\square)$ it suffices to choose the space-time modification which is Brownian motion parallel to the hyperplane $x_{p+1} = 0$ with constant drift vector $(0, 0, \ldots, 0, -1)$ (cf. [41]).

With the aid of the process $\mathscr{X}$ it is now possible to define the *hitting time* $T_E$ of a set $E \subset X$ as follows:

(4) This means that $t \mapsto X_t(\omega)$ is continuous on the interval $[0, \zeta(\omega)[$ where $\zeta$ is the «lifetime» of the process $(\omega \in \Omega)$.

(52) $$T_E(\omega) = \inf\{t > 0 : X_t(\omega) \in E\}.$$

According to a fundamental theorem of HUNT [53], $T_E$ is a stopping time for *Borel sets E* (more generally, for analytic sets). In particular, for a given Borel set $E$ the position of the process at the random time $T_E$ is a random variable, namely

$$\omega \mapsto X_{T_E(\omega)}(\omega),$$

defined on $\{T_E < +\infty\}$. [5] Consequently, the distribution $P_E$ of this random variable with respect to the probability measure $P^x$ is defined:

(53) $$P_E(x, B) = P^x(\{X_{T_E} \in B\} \cap \{T_E < +\infty\}) \qquad (x \in X,\ B \in \mathscr{B}).$$

$P_E$ turns out to be a kernel. By making essential use of results of HUNT [53] one can identify this kernel with the balayage operator $B^E$ of § 4. This result is the following theorem of central importance (cf. [7]):

5.3. *In the situation of Theorem 5.2., the kernels* $B^E$ *and* $P^E$ *coincide for every Borel set* $E \subset X$. *In particular, one has*

(54) $$\hat{R}_u^E = P_E u$$

*for every non-negative hyperharmonic function u on X.*

Hereby the last equality (54) results from (35). Theorem 5.3 is the key to the probabilistic interpretation of all potential-theoretic notions and results of importance. For the details the reader is referred to [7]. In this direction we content ourselves to the discussion of two results concerning the exceptional sets from paragraph 4.

5.4. *A Borel set* $E \subset X$ *is polar if and only if*

(55) $$P^x\{T_E < +\infty\} = 0 \qquad \textit{for all } x \in X.$$

Usually a Hunt Process is interpreted as the mathematical model for a particle moving at random in the space $X$ where $P^x$ is the random mechanism describing the movement of

(5) There are some technical difficulties to overcome which concerne the precise meaning of the measurability of $X_{T_E}$. For details the reader is referred to [20] and [35].

the particle when starting at the point $x$. With this interpretation in mind, one is led to the probabilistic meaning of (55): *The set E is not hit by the particle at times* $t > 0$ $P^x$*-almost surely whatever the starting point* $x \in X$ *is.*

5.5. *A Borel set* $E \subset X$ *is thin at a point* $x \in X$ *if and only if*

$$P^x\{T_E = 0\} = 0. \tag{56}$$

In prose: *When starting at x, the particle does not hit the set E at arbitrary small times* $t > 0$ ($P^x$*-almost surely*).

As a consequence, Borel sets which are semi-polar coincide with the semi-polar sets in the sense of the theory of Markov processes (cf. CHUNG [35], p. 110). The exceptional character of these sets can be seen from the following result:

5.6. *A Borel set* $E \subset X$ *is semi-polar if and only if, for every* $x \in X$, *the (random) set*

$$\{t \in \mathbb{R}_+ : X_t \in E\}$$

*is countable* $P^x$*-almost surely.*

In other words: *The particle starting from an arbitrary point* $x \in X$ *hits the Borel set E only a countable number of times* ($P^x$*-almost surely*) *if and only if E is semi-polar.* That semipolar sets have this property has been known for quite a while (cf. [20]). The converse is due to DELLACHERIE [37], p. 112 and, in a refined form, to HANSEN [47], p. 513.

## §6. Outlook

The notion of a harmonic function is fundamental for the definition of a harmonic space $(X, \mathscr{H})$. The harmonic functions on an open set $U \subset X$ form a linear space namely the space $\mathscr{H}_U$. For potential-theoretic problems the hyperharmonic functions are more important than the harmonic functions as we have seen in the preceding paragraphs. These hyperharmonic functions define the sheaf $U \to \mathscr{H}^*_U$ which is now a sheaf of convex cones of lower semicontinuous functions with values in $]-\infty, +\infty]$.

It is this observation which has led Constantinescu and Cornea to the definition of another, more general type of harmonic spaces, namely those which are defined by means of a *hyperharmonic structure* which is introduced by means of a sheaf $\mathscr{H}^*$ of convex cones of lower semicontinuous functions with values in $]-\infty, +\infty]$. These harmonic spaces defined by means of a hyperharmonic structure are introduced and studied in the monograph [36]. Furthermore, by reasons of mainly technical nature, not regular but resolutive sets have the priority in [36]. An open, relatively compact set is called resolutive if the Perron-Wiener-Brelot method leads to a generalized solution of the Dirichlet problem. Finally, the weak convergence axiom, which we mentioned at the end of § 1, is used in a systematic way.

This more general notion of a harmonic space has the advantage to cover a few more examples – e.g. the deterministic movement of a particle on the real line with constant velocity – and to facilitate certain constructions which, for harmonic spaces in the sense of this article, turn out to be difficult or even impossible without additional assumptions. These advantages are accompanied by disadvantages: The contact with problems from the theory of elliptic-parabolic differential equations is loosened. At some places the smoothness of the theory disappears though the principal results remain valid.

The general notion of a harmonic space in the sense of Constantinescu and Cornea has, however, considerable advantages as soon as one tries to characterize those standard or Hunt processes on a locally compact space $X$ with countable base $X$ for which the excessive functions are the non-negative hyperharmonic functions on $X$ with respect to an appropriate hyperharmonic structure on $X$. This problem is converse to the one studied in § 5 where the harmonic space was given. The problem has a beautiful solution which is due to BLIEDTNER and HANSEN [18], cf. also TAYLOR [69] and BLIEDTNER [16]. The key to the solution is the notion of a *balayage space* which is still more general than the harmonic spaces in the sense of Constantinescu and Cornea.

The standpoint of STOICA [66] is close to the original aims and motives of the theory of harmonic spaces. His point of departure is a *local operator* $L$ on a locally compact space $X$ with countable base, i.e. a homomorphism of a sheaf $\mathscr{D}$ of linear spaces of continuous real-valued functions in the full sheaf $\mathscr{C}$ of continuous real-valued functions on $X$. One assumes that $L$ satisfies a maximum principle so that $L$ is *locally dissipative*. In the applications $L$ is derived from an appropriate elliptic-parabolic linear differential operator on $\mathbb{R}^p$. Under additional assumptions it can be proved that $L$ is the restriction to $\mathscr{D}$ of the characteristic operator of a standard process with continuous paths. Necessary and sufficient conditions can be derived which assure that $Lh = 0$ defines a locally strong harmonic structure $\mathscr{H}$ on $X$.

Much more abstract is the standpoint of BOBOC, BUCUR, CORNEA [21] in their theory of *H-cones*. Large parts of balayage theory turn out to be valid in certain lattice cones. The great advantage of their point of view is the possibility of developing a duality theory in such a framework. In this «abstract» theory of *H-cones* the harmonic spaces yield the «concrete» examples.

In our survey we have left aside all those parts of potential theory which, in the classical case, have to do with the notions *energy* and *Dirichlet integral*. We content ourselves to refer the reader to the investigations of MAEDA [59] which connect the theory of harmonic spaces with the important area of *Dirichlet spaces* based on the fundamental work of A. BEURLING and J. DENY (cf. the book by FUKUSHIMA [43] and the literature mentioned there). The investigations of Maeda have been completed by an interesting results of SCHIRMEIER [64]. The connection between Dirichlet spaces and *H*-cones is discussed in [21].

**REFERENCES**


[1] L.V. AHLFORS, L. SARIO: *Riemann surfaces.* Princeton Univ. Press (1960), Princeton.

[2] Th. BARTH: *Sur le principe du minimum en théorie du potentiel.* C.R. Acad. Sc. Paris 296, Sér. I (1983), 733 - 736.

[3] H. BAUER: *Axiomatische Behandlung des Dirichletschen Problems für elliptische und parabolische Differentialgleichungen.* Math. Annalen 146 (1962), 1 - 59.

[4] H. BAUER: *Harmonische Räume und ihre Potentialtheorie.* Lecture Notes in Math. 22 (1966), Springer-Verlag.

[5] H. BAUER: *Propriétés fines des fonctions hyperharmoniques dans une théorie axiomatique du potentiel.* Ann. Inst. Fourier 15/1 (1965), 189 - 206.

[6] H. BAUER: *Zum Cauchyschen und Dirichletschen Problem bei elliptischen und parabolischen Differentialgleichungen.* Math. Annalen 164 (1966), 142 - 153.

[7] H. BAUER: *Harmonic spaces and associated Markov processes.* Published in «Potential Theory», Centro Internazionale Matematico Estivo (C.I.M.E.), Stresa 1969, Ed. Cremonese (1970), Roma.

[8] H. BAUER: *Aspects of modern potential theory.* Proc. Intern. Congress of Math., Vol. 1, Vancouver 1974 (1975), 41 - 51.

[9] H. BAUER: *Elliptische harmonische Strukturen.* Published in «Elliptische Differentialgleichungen», Vorträge der Tagung in Rostock 10-16/10/1977. Math. Ges. der DDR (1978), Rostock.

[10] H. BAUER: *Halbgruppen und Resolventen in der Potentialtheorie.* L'Enseignement math., 2ᵉ série, 25 (1979), 9 - 22.

[11] H. BAUER: *Harmonische Räume.* Jahrbuch Überblicke Mathematik 1981, 9 - 35. Bibliograph. Institut (1981), Mannheim.

[12] H. BAUER: *Zum heutigen Bild der Potentialtheorie.* Published in: Zum Werk Leonhard Eulers. Vorträge des Euler-Kolloquiums im Mai 1983 in Berlin (Herausgeg. von E. Knobloch, I.S. Louhivaara u. J. Winkler), pp. 3 - 20. Verlag Birkhäuser (1984), Basel-Boston-Stuttgart.

[13] H. BAUER: *Probability theory and elements of measure theory.* 2nd English edition. Academic Press (1981).

[14] C. BERG: *Potential theory on the infinite dimensional torus.* Inventiones math. 32 (1976), 49 - 100.

[15] C. BERG: *On Brownian and Poisson convolution semigroups on the infinite dimensional torus.* Inventiones math. 38 (1977), 227 - 235.

[16] J. BLIEDTNER: *Axiomatic foundation of potential theory.* Published in «Functional Analysis, Surveys and Recent Results II» (editors: K. D. Bierstedt, B. Fuchssteiner). North-Holland Publ. Comp. (1980).

[17] J. BLIEDTNER, W. HANSEN: *Simplicial cones in potential theory.* Inventiones math. 29 (1975), 83 - 110.

[18] J. BLIEDTNER, W. HANSEN: *Markov processes and harmonic spaces.* Z. Wahrscheinlichkeitstheorie verw. Gabiete 42 (1978), 309 - 325.

[19] J. BLIEDTNER, W. HANSEN: *The weak Dirichlet problem.* J. Reine Angew. Math. 348 (1984), 35 - 39.

[20] R.M. BLUMENTHAL, R.K. GETOOR: *Markov processes and potential theory.* Academic Press (1968).

[21] N. BOBOC, Ch. BUCUR, A. CORNEA: *Order and convexity in potential theory: H-cones.* Lecture Notes in Math. 853 (1981), Springer-Verlag.

[22] N. BOBOC, C. CONSTANTINESCU, A. CORNEA: *Semigroups of transition on harmonic spaces.* Rev. Roumaine Math. Pures Appl. 12 (1967), 763 - 805.

[23] N. BOBOC, P. MUSTAȚĂ: *Espaces harmoniques associés aux opérateurs différentiels linéaires du*

*second ordre de type elliptique*. Lecture Notes in Math. 68 (1968), Springer-Verlag.

[24] J.M. BONY: *Détermination des axiomatiques de théorie du potentiel dont les fonctions harmoniques sont différentiables*. Ann. Inst. Fourier 17/1 (1967), 353 - 382.

[25] J.M. BONY: *Principe du maximum et inégalités de Harnack pour des opérateurs elliptiques dégénérés*. Séminaire de Théorie du Potentiel (dirigé par M. Brelot, G. Choquet, J. Deny), 12$^{e}$ année: 1967/68, 20 pages (1969), Institut H. Poincaré, Paris.

[26] J.M. BONY: *Opérateurs elliptiques dégénérés associés aux axiomatiques de la théorie du potentiel*. Paru dans «Potential Theory», Centro Internazionale Matematico Estivo (C.I.M.E.), Stresa 1969. Ed. Cremonese (1970), Roma.

[27] M. BRELOT: *La théorie moderne du potentiel*. Ann. Inst. Fourier 4 (1954), 113 - 133.

[28] M. BRELOT: *Eléments de la théorie classique du potentiel*, 4$^{e}$ éd.. Centre de Documentation Universitaire (1969), Paris.

[29] M. BRELOT: *Une axiomatique générale du problème de Dirichlet dans les espaces localement compacts*. Séminaire de Théorie du Potentiel (dirigé par M. Brelot, G. Choquet), 1re année: 1957, 16 p. (1958), Institut H. Poincaré, Paris.

[30] M. BRELOT: *Lectures on potential theory*. Tata Inst. of Fund. Research (1960), Bombay.

[30a] M. BRELOT: *Axiomatique des fonctions harmoniques et surharmoniques dans un espace localement compact*. Séminaire de Théorie du Potentiel (dirigé par M. Brelot, G. Choquet), 2e année: 1958, 40 p. (1959). Institut H. Poincaré, Paris.

[31] M. BRELOT: *Axiomatique des fonctions harmoniques*. Les Presses de l'Université de Montréal (1966), Montréal.

[32] M. BRELOT:: *Les étapes et les aspects multiples de la théorie du potentiel*. L'Enseignement math., 2e série, 18 (1972), 1 - 36.

[33] M. BRELOT: *Minorantes sous-harmoniques, extrémales et capacités*. J. Math. Pures Appl. 24 (1945), 1 - 32.

[34] H. CARTAN: *Théorie du potentiel nowtonien: énergie, capacité, suites de potentiels*. Bull. Soc. Math. France 73 (1945), 74 - 106.

[35] K.L. CHUNG: *Lectures from Markov processes to Brownian motion*. Grundlehren d. math. Wiss. 249 (1982), Springer-Verlag.

[36] C. CONSTANTINESCU, A. CORNEA: *Potential theory on harmonic spaces*. Grundlehren d. math. Wiss. 158 (1972), Springer-Verlag.

[37] C. DELLACHERIE: *Ensembles aléatoires*, I. Séminaire de Probabilités, III, 97 - 114. Lecture Notes in Math. 88 (1969), Springer-Verlag.

[38] J.L. DOOB: *Semimartingales and subharmonic functions*. Trans. Amer. Math. Soc. 77 (1954), 86 - 121.

[39] J.L. DOOB: *A probability approach to the heat equation*. Trans. Amer. Math. Soc. 80 (1955), 216 - - 280.

[40] J.L. DOOB: *Probability methods applied to the first boundary value problem*. Proc. 3rd Berkeley Symp. on Math. Stat. and Prob. 1954 - 1955 (1956), 49 - 80.

[41] J.L. DOOB: *Classical potential theory and its probabilistic counterpart*. Grundlehren d. math. Wiss. 262 (1983), Springer-Verlag.

[42] B. FUGLEDE: *Finely harmonic functions*. Lectures Notes in Math. 289 (1972), Springer-Verlag.

[43] M. FUKUSHIMA: *Dirichlet forms and Markov processes*. North-Holland Publ. Comp. (1980).

[44] W. HANSEN: *Konstruktion von Halbgruppen und Markoffschen Prozessen*. Inventiones math. 3

(1967), 179 - 214.

[45] W. HANSEN: *Hunt's theorem and axiomatic potential theory*. Inventiones math. 14 (1971), 242 - 252.

[46] W. HANSEN: *Cohomology in harmonic spaces*. Published in «Seminar on Potential Theory II» (editor: H. Bauer), Lecture Notes in Math. 226 (1971), 63 - 101, Springer-Verlag.

[47] W. HANSEN: *Semi-polar sets and quasi-balayage*. Math. Ann. 257 (1981), 495 - 517.

[48] L.L. HELMS: *Introduction to potential theory*. Wiley-Interscience (1969).

[49] R.M. HERVÉ: *Recherches axiomatiques sur la théorie des fonctions surharmoniques et du potentiel*. Ann. Inst. Fourier 12 (1962), 415 - 571.

[50] R.M. HERVÉ: *Un principe du maximum pour les sous-solutions locales d'une équation uniformement elliptique de la forme*

$$Lu = -\sum_i \frac{\partial}{\partial x_i}\left(\sum_j a_{ij} \frac{\partial u}{\partial x_j}\right) = 0.$$

[51] R.M. et M. HERVÉ: *Les fonctions surharmoniques associées à un opérateur elliptique du second ordre à coefficients discontinus*. Ann. Inst. Fourier 19/1 (1969), 305 - 359.

[52] L. HÖRMANDER: *Hypoelliptic second order differential equations*. Acta Math., Uppsala, 119 (1967), 147 - 171.

[53] G.A. HUNT: *Markoff processes and potentials I - III*. Illinois J. Math. 1 (1957), 44 - 93, 316 - 369; 2 (1958), 151 - 213.

[54] E. KAMKE: *Über die erste Randwertaufgabe bei der Laplace - und der Wärmeleitungs - Differentialgleichung*. Jahresber. deutsche Math. Ver. 62 (1959), 1 - 33.

[55] O.D. KELLOG: *Foundations of potential theory*. Grundlehren d. math. Wiss. 31 (1929, 1967), Springer-Verlag.

[56] J. KÖHN, M. SIEVEKING: *Reguläre und extremale Randpunkte in der Potentialtheorie*. Rev. Roumaine Math. Pures et Appl. 12 (1967), 1489 - 1502.

[56a] J. KÖHN, M. SIEVEKING: *Zum Cauchyschen und Dirichletschen Problem*. Math. Ann. 177 (1968), 133 - 142.

[57] N.S. LANDKOF: *Foundations of modern potential theory*. Grundlehren d. math. Wiss. 180 (1972), Springer-Verlag (Translation from the Russian).

[58] G. LUMER: *Connecting of local operators and evolution equations on networks*. Published in «Potential Theory, Copenhagen 1979» (editors: C. Berg, G. Forst, B. Fuglede). Lecture Notes in Math. 787 (1980), 219 - 234.

[59] F.Y. MAEDA: *Dirichlet integrals on harmonic spaces*. Lecture Notes in Math. 803 (1980), Springer-Verlag.

[60] P.A. MEYER: *Brelot's axiomatic theory of the Dirichlet problem and Hunt's theory*. Ann. Inst. Fourier 13/2 (1963), 357 - 372.

[61] I. PETROWSKY: *Zur ersten Randwertaufgabe der Wärmeleitung*. Composito Math. 1 (1935), 383 - 419.

[62] N. DU PLESSIS: *An introduction to potential theory*. Oliver & Boyd (1970), Edinburgh.

[63] U. SCHIRMEIER: *Konvergenzeigenschaften in harmonischen Räumen*. Inventiones math. 55 (1979), 71 - 95.

[64] U. SCHIRMEIER: *Continuous measure representations on harmonic spaces*. Hiroshima Math. J. 13 (1983), 327 - 337.

[65] W. STERNBERG: *Über die Differentialgleichung der Wärmeleitung*. Math. Ann. 101 (1929), 394 - 398.

[66] L. STOICA: *Local operators and Markov processes*. Lecture Notes in Math. 816 (1980), Springer-Verlag.

[67] G. TAUTZ: *Zur Theorie der ersten Randwertaufgabe*. Math. Nachr. 2 (1949), 279 - 303.

[68] J.C. TAYLOR: *Strict potentials and Hunt Processes*. Inventiones math. 16 (1972), 249 - 259.

[69] J.C. TAYLOR: *The harmonic space associated with a «reasonable» standard process*. Math. Ann. 233 (1978), 89 - 96.

[70] D. WILLIAMS: *Diffusions, Markov processes, and martingales*. Vol. 1: Foundations. J. Wiley & Sons (1979).

# Heat balls and Fulks measures [R34]

Ann. Acad. Sci. Fenn. Ser. A I Math. **10** (1985), 67–82
[Zbl. Math. 592.35057; MR 87c:31010]

### Introduction

Euclidean spheres in $\boldsymbol{R}^p$ play an important role in classical potential theory, the theory of the Laplace equation. They enter into the picture as soon as one treats the Poisson formula which is the key to the definition of superharmonic functions. For the Laplace equation, the fundamental solution $N$ with pole at the origin is in every dimension $p \geqq 1$ a function of the euclidean norm of the point $x \in \boldsymbol{R}^p$. Hence spheres with center at the origin can be considered as the level surfaces of $N$.

The potential theory of the heat equation

$$\Delta h - \frac{\partial h}{\partial t} = 0$$

in $\boldsymbol{R}^{p+1} = \boldsymbol{R}^p \times \boldsymbol{R}$ is now well developed. In particular, there are formulas analogous to the Poisson formula. However, there is one crucial contrast to the Laplace case: In principle, $(p+1)$-dimensional balls could be used since all boundary points turn out to be regular. However, the proof of this fact demands already quite a bit of potential-theoretic machinery. So, for the definition of supercaloric functions — the analogue of superharmonic functions —, one usually uses either rectangular domains parallel to the coordinate axes where the inconvenience of having irregular boundary points is not serious or truncated rotational cones having a base parallel to the hyperplane $\boldsymbol{R}^p \times \{0\}$ and a vertex situated in positive time direction above this base. In both approaches the formula analogous to the Poisson formula is not easy to obtain since integral equations have to be solved. In any case, the final Poisson-type formula contains functions which cannot be written down explicitly.

In view of these difficulties it seems natural to proceed by analogy with the case of the Laplace equation: One considers

$$W(x,t) = \begin{cases} (-4\pi t)^{-p/2} e^{\|x\|^2/4t} & \text{when} \quad t < 0 \\ 0 & \text{when} \quad t \geqq 0 \end{cases}$$

— which is the fundamental solution of the adjoint heat equation with pole at the origin $\mathbf{0}$ — and one studies the level surfaces

$$(*) \qquad W(x,t) = d$$

where $d$ is a positive constant. This idea, due to Fulks [10], will be fundamental for us. In [10] Fulks considers the smooth convex surface $B_d$ defined by $(*)$ with inclusion of the limit point $\mathbf{0}$. It is the boundary of the relatively compact domain $\Omega_d$ in $\mathbf{R}^{p+1}$ defined by

$$W(x, t) > d.$$

We shall call $\Omega_d$ a *heat ball* and $B_d$ the corresponding *heat sphere*. In contrast to the case of the Laplace equation, the pole $\mathbf{0}$ of $W$ lies on the heat sphere $B_d$. The heat ball $\Omega_d$ is an egg-shaped domain which has $\mathbf{R}^p \times \{0\}$ as tangent hyperplane through $\mathbf{0}$ and which lies in the half-space of all points $(x, t)$ with negative time coordinate $t$. In [10] Fulks defines explicitly a continuous function $Q_d \geqq 0$ on $B_d$ which is strictly positive with the exception of one point. The measure $\mu_d$ which has $Q_d$ as density with respect to the surface measure of $B_d$ is a probability measure. This *Fulks measure* $\mu_d$ is connected with the solutions of the heat equation in a similar way as the normalized surface measure $\sigma_r$ of the sphere $\|x\| = r$ in $\mathbf{R}^p$ is connected with harmonic functions, however with one important difference: For the Laplace equation the usual mean value property says that integration of a harmonic function with respect to $\sigma_r$ yields the value of the function at the origin, hence at the center of the euclidean ball $\|x\| < r$. For solutions of the heat equation integration with respect to $\mu_d$ yields the value of the solution again at the origin $\mathbf{0}$ which is now a point of the heat sphere. According to [10], solutions of the heat equation can be characterized by mean value formulas where one integrates with respect to measures which are obtained from $\mu_d$ by translations. Also these translated measures will be called *Fulks measures*. Furthermore, Watson [18], [19] uses these measures for the definition and study of supertemperatures which correspond to superharmonic functions for the Laplace equation. However, no attempt seems to have been made to identify these supertemperatures with the supercaloric (or superparabolic) functions which are defined by means of the above mentioned rectangular domains or truncated cones, and hence in agreement with the general definition of superharmonic functions in the theory of harmonic spaces.

By making essential use of the harmonic space theory for the heat equation, we intend to show as the main result of this paper that every Fulks measure $\mu_d$ allows a simple potential-theoretic interpretation: $\mu_d$ is obtained from the unit mass at $\mathbf{0}$ by sweeping this mass out to the complement of $\Omega_d$, i.e.,

$$(**) \qquad \mu_d = \varepsilon_0^{\complement \Omega_d}.$$

In the proof we use a result of Boboc and Cornea [6] about cluster values of harmonic measures, a result which has been improved recently by Lukeš and Malý [16]. Our main result has several consequences which are interesting in themselves: The result implies that the origin $\mathbf{0}$ is an irregular boundary point of the heat ball $\Omega_d$ and that supertemperatures coincide with supercaloric functions. Furthermore, our result leads to a probabilistic interpretation of the Fulks measures: $\mu_d$ is the distribution of the first hit of the heat sphere $B_d$ (at times $t>0$) by the Brownian heat transfer starting at $\mathbf{0}$, where the Brownian heat transfer in $\mathbf{R}^{p+1}$ is the well-known space-time modification of $p$-dimensional Brownian motion. So, altogether, the measures $\mu_d$ find a very satisfactory and natural potential-theoretic characterization.

In order to make our presentation as smoothly accessible as possible, the paper is organized as follows: After the introduction of the Fulks measures we state our Main Theorem, namely the characterization of $\mu_d$. Then we immediately give applications. Finally, the proof of the Main Theorem is presented in the third and last paragraph of this paper.

In our presentation the harmonic space associated to the heat equation and the corresponding potential theory is used constantly and without further explanations. Standard references are [1], [3] and [7].

The main results of this paper have been announced in C.R. Acad. Sci. Paris, Sér. A.

## 1. Definition of the Fulks measures

We shall be concerned with the heat equation

$$\Delta h - \frac{\partial h}{\partial t} = 0 \tag{1.1}$$

in $\mathbf{R}^{p+1} = \mathbf{R}^p \times \mathbf{R}$ for $p=1, 2, \ldots$. The points of $\mathbf{R}^{p+1}$ will be denoted correspondingly by $(x, t) = (x_1, \ldots, x_p, t)$ where $x = (x_1, \ldots, x_p) \in \mathbf{R}^p$ and $t \in \mathbf{R}$. In particular, the Laplace operator refers to the first $p$ coordinates:

$$\Delta = \sum_{i=1}^{p} \frac{\partial^2}{\partial x_i^2}.$$

The euclidean norm of a point $x = (x_1, \ldots, x_p) \in \mathbf{R}^p$ will be denoted by

$$\|x\| = (x_1^2 + \ldots + x_p^2)^{1/2}.$$

Solutions of (1.1), more precisely functions $h\colon U \to \mathbf{R}$ defined on an open subset $U$ of $\mathbf{R}^{p+1}$ which are of class $C^2$ with respect to the space variable $x$ and of class $C^1$ with respect to the time variable $t$ and which satisfy (1.1), will be called *caloric functions* or *temperatures*.

We denote by $\tilde{W}$ the fundamental solution of (1.1) with pole at the origin $\mathbf{0} = (0, 0)$, i.e.,

$$\tilde{W}(x, t) = \begin{cases} \left(\dfrac{1}{4\pi t}\right)^{p/2} e^{-\|x\|^2/4t} & \text{when } t > 0 \\ 0 & \text{when } t \leqq 0. \end{cases} \tag{1.2}$$

Reflection at $\mathbf{0}$ yields the function $W$ from the introduction:

$$W(x, t) = \tilde{W}(-x, -t). \tag{1.3}$$

By translation we obtain for every point $z_0=(x_0, t_0)\in \boldsymbol{R}^{p+1}$ the function

$$W_{z_0}(x, t) = \tilde{W}(x_0-x, t_0-t) \tag{1.4}$$

which is the fundamental solution of the adjoint heat equation

$$\Delta h+\frac{\partial h}{\partial t} = 0 \tag{1.5}$$

with pole at $z_0$. In particular, $W=W_0$.

Given $z_0\in \boldsymbol{R}^{p+1}$ and a real number $c>0$ we denote by $\Omega(z_0, c)$ the set of all points $(x, t)\in \boldsymbol{R}^{p+1}$ satisfying

$$W_{z_0}(x, t) > \left(\frac{1}{4\pi c}\right)^{p/2}. \tag{1.6}$$

This set $\Omega(z_0, c)$ is a relatively compact domain, in fact a convex body. Its boundary

$$B(z_0, c) = \partial\Omega(z_0, c)$$

is a smooth surface given by the equation

$$W_{z_0}(x, t) = \left(\frac{1}{4\pi c}\right)^{p/2} \tag{1.7}$$

with inclusion of the point $z_0$. More precisely, let us define for $c>0$ the (continuous) function $F_c\colon [0, c]\to \boldsymbol{R}$ by

$$F_c(t) = \begin{cases} \left[-2pt\log\dfrac{t}{c}\right]^{1/2} & \text{when } 0<t\leqq c \\ 0 & \text{when } t=0; \end{cases} \tag{1.8}$$

then the surface $B(z_0, c)$ is given by the equation

$$\|x-x_0\| = F_c(t_0-t) \quad \text{with} \quad 0\leqq t_0-t\leqq c. \tag{1.7'}$$

From this it can be concluded that $B(z_0, c)$ is a ($p$-dimensional) $C^2$-manifold and that $B(z_0, c)\setminus\{z_0\}$ is even a $C^\infty$-manifold[1]). The domains $\Omega(z_0, c)$ will be called *heat balls* in $\boldsymbol{R}^{p+1}$; correspondingly, the manifolds $B(z_0, c)$ will be called *heat spheres*.

---

[1]) Denote by $\Psi_0$ the inverse of the strictly decreasing function $\Phi_0\colon [0, e^{-1}[\to]e^{-1/e}, 1]$ where $\Phi_0(\tau)=\tau^\tau$ for $\tau>0$ and $=1$ for $\tau=0$, and denote by $\Psi_1$ the inverse of the strictly increasing function $\Phi_1\colon ]e^{-1}, +\infty[\to]e^{-1/e}, +\infty[$ where $\Phi_1(\tau)=\tau^\tau$. Then one obtains two charts for neighborhoods of the points $z_0$ and $z_1=(x_0, t_0-c)$, respectively. In fact, it is easy to see that the equations

$$t = t_0-c\Psi_0\left(e^{-\frac{\|x-x_0\|^2}{2pc}}\right) \tag{a}$$

and

$$t = t_0-c\Psi_1\left(e^{-\frac{\|x-x_0\|^2}{2pc}}\right). \tag{b}$$

In what follows $\sigma=\sigma_{B(z_0,c)}$ will denote the surface measure of the heat sphere $B(z_0, c)$. We define a function $Q>0$ on $(\boldsymbol{R}^p\times]0, +\infty[)\cup\{0\}$ by means of

$$(1.9)\qquad Q(x,t)=\begin{cases}\|x\|^2[4\|x\|^2t^2+(\|x\|^2-2pt)^2]^{-1/2} & \text{when } t>0\\ 1 & \text{when } (x,t)=\mathbf{0}.\end{cases}$$

Furthermore, for $z_0=(x_0, t_0)\in\boldsymbol{R}^{p+1}$ we define a function

$$(1.10)\qquad q_{z_0}(x,t)=Q(x_0-x, t_0-t)$$

on $(\boldsymbol{R}^p\times]-\infty, t_0[)\cup\{z_0\}$. Its restriction to $B(z_0, c)$

$$(1.11)\qquad q_{z_0,c}=q_{z_0}|B(z_0,c)$$

turns out to be a *continuous* function which is strictly positive on $B(z_0, c)$ with the only exception of the point $z_1=(x_0, t_0-c)$ where $q_{z_0}(z_1)=0$. By taking $q_{z_0,c}$ as a density with respect to $\sigma$ we finally obtain the main object of our investigation.

**1.1** Definition. *The positive Radon measure*

$$(1.12)\qquad \mu_{z_0,c}=\left(\frac{1}{4\pi c}\right)^{p/2}q_{z_0,c}\sigma_{B(z_0,c)}$$

*on $B(z_0, c)$ is called the Fulks measure for the heat ball $\Omega(z_0, c)$.*

According to Fulks [10], $\mu_{z_0,c}$ is a probability measure. In fact, this follows from the following fundamental result of [10]:

(F) Every caloric function $h$ defined in a domain $D$ of $\boldsymbol{R}^{p+1}$ satisfies the following *mean value property:*

$$(1.13)\qquad \int h\,d\mu_{z_0,c}=h(z_0)$$

whenever $z_0\in D$ and $\bar{\Omega}(z_0, c)\subset D$.

---

in both cases with

$$\|x-x_0\|<\sqrt{2p\frac{c}{e}},$$

describe the neighborhoods

$$U_0=B(z_0,c)\cap\left\{(x,t)\colon t>t_0-\frac{c}{e}\right\}$$

and

$$U_1=B(z_0,c)\cap\left\{(x,t)\colon t<t_0-\frac{c}{e}\right\}$$

of $z_0$ and $z_1$, respectively. The right-hand side of (a) is a $C^1$-function, the right-hand side of (b) is a $C^\infty$-function of $x$. The value $t=t_0-c/e$ yields the "equator" of $B(z_0, c)$ which corresponds to the maximum of $F_c$:

$$F_c\left(\frac{c}{e}\right)=\sqrt{2p\frac{c}{e}}.$$

In [10] Fulks also shows the converse: A continuous function $h: D \to \boldsymbol{R}$ on a domain $D$ of $\boldsymbol{R}^{p+1}$ with the above mean value property is caloric. However, (F) as well as its converse will not be needed in the sequel. They will follow from our Main Result and its Addendum (see Corollary 2.3 and 2.6, Remark 2). Let us also remark that the Fulks measures have been rediscovered by Kuptsov [15] in an analytical presentation which, however, is less convenient for our purposes.

## 2. The main result and its consequences

We shall now give a potential-theoretic characterization of the Fulks measure $\mu_{z_0, c}$ which will be considered as a Radon measure on $\boldsymbol{R}^{p+1}$ concentrated on $B(z_0, c)$. Potential-theoretic notions for the heat equation should always be understood in the sense of harmonic spaces. More precisely, $\boldsymbol{R}^{p+1}$ is considered as the harmonic space with respect to the harmonic sheaf given by the caloric functions. It is a well-known fact [1] that this space is strongly harmonic or $\mathfrak{P}$-harmonic in the terminology of Constantinescu and Cornea [7].

We shall start with a simple observation:

**2.1** Proposition. *All points $z \neq z_0$ of the heat sphere $B(z_0, c)$ are regular boundary points of the heat ball $\Omega(z_0, c)$ (for arbitrary $z_0 \in \boldsymbol{R}^{p+1}$ and $c > 0$).*

*Proof.* This can be seen in many ways. A simple procedure is the following: The boundary point $z_1 = (x_0, t_0 - c)$ with the same space coordinate as $z_0$ is regular since, obviously,

$$v(x, t) = \|x - x_0\|^2 + 2p(t - t_0 + c)$$

is a caloric barrier for $z_1$. For all remaining boundary points $z = (\xi, \tau)$ in $B(z_0, c) \setminus \{z_0, z_1\}$ there exists a closed euclidean $(p+1)$-dimensional ball $K$ such that $K \cap \bar{\Omega}(z_0, c) = \{z\}$ and such that $\xi$ is not the space coordinate $\bar{x}$ of the center $(\bar{x}, \bar{t})$ of $K$. It then follows from Friedman [9], p. 69, property (b), that $z$ is regular. In fact for $\alpha > 0$ sufficiently large,

$$v(x, t) = R(z)^{-\alpha} - R(x, t)^{-\alpha},$$

where

$$R(x, t) = \|x - \bar{x}\|^2 + (t - \bar{t})^2,$$

is a barrier for $z$ (see also Effros and Kazdan [8], Lemma 2)[2]).

Our main result is now the following:

**2.2** Main Theorem. *For all $z_0 \in \boldsymbol{R}^{p+1}$ and $c > 0$, the Fulks measure $\mu_{z_0, c}$ is the unit mass at $z_0$ swept out to the complement of $\Omega = \Omega(z_0, c)$, i.e.*

$$\mu_{z_0, c} = \varepsilon_{z_0}^{\complement \Omega}. \tag{2.1}$$

[2]) It is an easy and well known consequence of the Hopf maximum principle that $\Delta v - \partial v / \partial t < 0$ for a $C^{2,1}$-function in a domain $D \subset \boldsymbol{R}^{p+1}$ implies that $v$ is supercaloric on $D$.

The proof will be given in the next paragraph. It will lead also to the following addendum:

**2.2′** Addendum. *In every heat ball* $\Omega=\Omega(z_0, c)$ *there exists a sequence* $(z_k)$ *of points in* $\Omega$ *such that the corresponding sequence* $(\mu_{z_k}^{\Omega})$ *of harmonic measures converges vaguely to the Fulks measure* $\mu_{z_0,c}$.

Let us remark that this Addendum can also be deduced from our Main Theorem and a lemma of Köhn and Sieveking [13] (see also [16], p. 356).

The Addendum 2.2′ leads directly to the mean value property (F) of Fulks:

**2.3** Corollary 1. *Let* $h$ *be a function which is caloric in an open set* $U\subset \boldsymbol{R}^{p+1}$. *Then the equality*

$$\int h\, d\mu_{z_0,c} = h(z_0)$$

*holds for all heat balls* $\Omega(z_0, c)$ *for which* $\bar{\Omega}(z_0, c)\subset U$.

*Proof.* Suppose that $(z_k)$ is a sequence in $\Omega=\Omega(z_0, c)$ with the properties stated in the Addendum. Then

$$\int h\, d\mu_{z_k}^{\Omega} = h(z_k) \quad \text{for all} \quad k\in \boldsymbol{N}.$$

From this the result follows for $k\to +\infty$ since $h$ is continuous on $U$.

An immediate consequence of the Main Theorem and Proposition 2.1 is

**2.4** Corollary 2. $z_0$ *is the only irregular boundary point of* $\Omega(z_0, c)$.

This result can be proved in several other ways. E.g. it follows from Proposition 2.1 and Watson [18], Theorem 9, p. 399. Another proof is indicated in the Remark following our Proposition 3.1.

We shall proeced now to more important consequences of the Main Theorem. As pointed out before, we consider $\boldsymbol{R}^{p+1}$ as a harmonic space with respect to the sheaf of caloric functions. The hyperharmonic (respectively superharmonic) functions for this harmonic space will be called *hypercaloric* (respectively *supercaloric*) functions. Consequently (see [1], p. 127), a hypercaloric function $u$ on an open set $U\subset \boldsymbol{R}^{p+1}$ satisfies the inequality

$$\int u\, d\mu_z^V \leqq u(z) \tag{2.2}$$

for every point $z\in U$ and every relatively compact, open neighborhood $V\subset \bar{V}\subset U$ of $z$. Here $\mu_z^V=\varepsilon_z^{\complement V}$ denotes again the harmonic measure. In particular, the approach of Gehring [11] — if extended to lower semicontinuous functions with values in $]-\infty, +\infty]$ — leads to an equivalent definition of hypercaloric functions. This is an immediate consequence of Köhn [12], p. 5 (Satz 3).

In contrast to the class of hypercaloric functions there is an analogous class of functions introduced by Watson [18] in his thesis and used in subsequent investigations, in particular in [19]. The definition makes essential use of the Fulks measures:

A function $u: U \to ]-\infty, +\infty]$ defined on an open subset $U$ of $\boldsymbol{R}^{p+1}$ is called a *hypertemperature* on $U$ if, besides being lower semicontinuous, it satisfies the inequality

$$\int u \, d\mu_{z_0, c} \leqq u(z_0) \tag{2.3}$$

for all choices of points $z_0 \in U$ and real numbers $c > 0$ for which the heat ball $\Omega(z_0, c)$ is contained together with its boundary $B(z_0, c)$ in $U$. We claim

**2.5** Theorem. *An extended real-valued function $u$ defined on an open subset $U$ of $\boldsymbol{R}^{p+1}$ is a hypertemperature on $U$ if and only if it is hypercaloric on $U$.*

*Proof.* Suppose that $u$ is a hypertemperature on $U$ and that $V$ is a regular set such that $V \subset \overline{V} \subset U$. It suffices to prove that for every function $f \in \mathscr{C}(\partial V)$, i.e. continuous real-valued function on the boundary $\partial V$ of $V$, satisfying $f(z) \leqq u(z)$ for $z \in \partial V$ one has

$$\int f \, d\mu_z^V \leqq u(z) \quad \text{for all} \quad z \in V. \tag{2.4}$$

But since $z \mapsto H_f^V(z) = \int f d\mu_z^V$ solves the Dirichlet problem for the boundary function $f$ (according to the definition of a regular set), $H_f^V$ is a temperature on $V$ and, consequently, the function

$$v(z) = u(z) - H_f^V(z) \quad (z \in V)$$

is a hypertemperature on $V$. For such functions the boundary minimum principle is valid (see [18], p. 397 [3])): So $v \geqq 0$ follows from

$$\liminf_{\substack{z' \to z \\ z' \in V}} v(z') = \liminf_{\substack{z' \to z \\ z' \in V}} u(z') - \lim_{\substack{z' \to z \\ z' \in V}} H_f^V(z') \geqq u(z) - f(z) \geqq 0$$

$(z \in \partial V)$. This proves (2.4). — Conversely, assume that $u$ is hypercaloric on $U$ and that $\overline{\Omega}(z_0, c) \subset U$ for some $z_0 \in U$ and $c > 0$. Since the closure $\overline{\Omega}$ of $\Omega = \Omega(z_0, c)$ is compact, $u$ is bounded from below on $\overline{\Omega}$: $u(z) > \gamma$ for all $z \in \overline{\Omega}$ and some $\gamma \in \boldsymbol{R}$. Consequently, there exists a relatively compact open set $U_0$ such that

$$\overline{\Omega} \subset U_0 \subset \overline{U}_0 \subset U \tag{2.5}$$

and

$$u(z) > \gamma \quad \text{for all} \quad z \in U_0. \tag{2.6}$$

Let us assume for a moment that $u$ is even bounded on $U_0$. Then, according to the extension theorem of [1], p. 159, there exists for a given open set $V$ satisfying $\overline{\Omega} \subset V \subset \overline{V} \subset U_0$ a potential $p \in \mathscr{C}(\boldsymbol{R}^{p+1})$ which is even caloric on $V$ as well as a potential $\tilde{u}$ on $\boldsymbol{R}^{p+1}$ such that

$$\tilde{u}(z) = u(z) + p(z) \quad \text{for all} \quad z \in V. \tag{2.7}$$

[3]) In order to be in formal agreement with [18], Theorem 8, it suffices to replace $u$ on $V$ by $\inf(u, \alpha)$ where $\alpha \in \boldsymbol{R}$ is an upper bound for $f$.

It follows from our Main Theorem that, whenever $w$ is a hypercaloric function $\geqq 0$ on $\boldsymbol{R}^{p+1}$, we have

$$\int w \, d\mu_{z_0, c} \leqq w(z_0). \tag{2.8}$$

(In fact, $\hat{R}_w^E \leqq w$ on $\boldsymbol{R}^{p+1}$ and

$$\hat{R}_w^E(z_0) = \int w \, d\varepsilon_{z_0}^E$$

according to the definition of the balayée $\hat{R}_w^E$ and the swept-out measure $\varepsilon_{z_0}^E$ for an arbitrary set $E \subset \boldsymbol{R}^{p+1}$, in particular for $E = \complement\Omega$.) The inequality (2.8) can be applied to $\tilde{u}$ which yields

$$\int \tilde{u} \, d\mu_{z_0, c} \leqq \tilde{u}(z_0). \tag{2.8'}$$

Since $p$ is caloric on $V$ we also have

$$\int p \, d\mu_{z_0, c} = p(z_0)$$

according to Corollary 2.3. Since the Fulks measure lives on $B(z_0, c) \subset V$, it then follows from (2.7) and (2.8′) that

$$\int \tilde{u} \, d\mu_{z_0, c} = \int u \, d\mu_{z_0, c} + p(z_0) \leqq \tilde{u}(z_0).$$

Consequently, we have

$$\int u \, d\mu_{z_0, c} \leqq u(z_0). \tag{2.9}$$

If $u$ is not bounded from above, we pass from $u$ to the function $u_n = \inf(u, n)$ which is hypercaloric for every $n \in \boldsymbol{N}$. We then have proved

$$\int u_n \, d\mu_{z_0, c} \leqq u_n(z_0)$$

for all $n \in \boldsymbol{N}$ and hence, by passing to the limit for $n \to +\infty$,

$$\int u \, d\mu_{z_0, c} \leqq u(z_0).$$

This, however, proves that $u$ is a hypertemperature on $U$.

**2.6** Remarks. 1) An immediate consequence of Theorem 2.5 is the result that the *supercaloric functions* on an open set $U$ *coincide with* the *supertemperatures* in the sense of Watson [18]. This follows immediately from [18], Lemma 5, by observing that finiteness of a function on a dense subset of an open set in $\boldsymbol{R}^{p+1}$ implies condition $(\delta_3)$ of [18].

2) Another consequence of Theorem 2.5 are the results of Fulks [10] mentioned at the end of our first paragraph: A function $h$ defined on an open set $U \subset \boldsymbol{R}^{p+1}$ is caloric if and only if the mean value property (1.13) holds for all heat balls $\Omega(z_0, c)$ for which $\bar{\Omega}(z_0, c) \subset U$.

We close this paragraph with a probabilistic interpretation of the Fulks measures. For this purpose we denote by $(X_t)_{t \geqq 0}$ or, more precisely by $(\Omega, \mathscr{F}, (P^z)_{z \in \boldsymbol{R}^{p+1}}, (X_t)_{t \geqq 0})$ the well-known diffusion process with $\boldsymbol{R}^{p+1}$ as state space which is associated to the heat equation (see Meyer [17], pp. 63—64). Hence $(X_t)_{t\geqq 0}$ is projected on the $p$-dimensional Brownian motion by the first canonical projection of $\boldsymbol{R}^{p+1}=\boldsymbol{R}^p\times\boldsymbol{R}$ and on the deterministic movement with constant velocity $-1$ by the second canonical projection. We will call $(X_t)_{t\geqq 0}$ the *Brownian heat transfer*.

**2.7** Proposition. *For every heat ball $\Omega=\Omega(z_0, c)$ the Fulks measure $\mu_{z_0,c}$ is the distribution of the first hit of the complement of $\Omega$ by the Brownian heat transfer starting at $z_0$. More precisely,*

$$\mu_{z_0,c} = P^{z_0}(X_{T_{\complement\Omega}}^{-1})$$

*where $T_{\complement\Omega}$ denotes the first hitting time of the complement of $\Omega$.*

*Proof.* It is well known that the excessive functions of the Brownian heat transfer (or, more precisely, of its transition semigroup) coincide with the hypercaloric functions $\geqq 0$ defined on $\boldsymbol{R}^{p+1}$. (A proof of this fact can be found in Bliedtner [5], p. 96, where the situation is even more general.) The arguments in the proof of [2], Theorem 7.1, can then be repeated. As a result one obtains the identification of the balayée $\hat{R}_u^A$ with the function

(2.10) $$P_A u(z) = E^z(1_{\{T_A<+\infty\}}u\circ X_{T_A}) \quad (z\in\boldsymbol{R}^{p+1})$$

for all hypercaloric functions $u\geqq 0$ on $\boldsymbol{R}^{p+1}$ and all Borel sets $A\subset\boldsymbol{R}^{p+1}$. For $A=\complement\Omega$ the result then follows since the kernel $P_{\complement\Omega}u(z)$ is just the integral of $u$ with respect to the hitting distribution $\tau=P^{z_0}(X_{T_{\complement\Omega}}^{-1})$ appearing in the Proposition [4]). So we have, according to our Main Theorem,

$$\int u\, d\mu_{z_0,c} = \hat{R}_u^{\complement\Omega}(z_0) = \int u\, d\tau$$

for all hypercaloric functions $u\geqq 0$ on $\boldsymbol{R}^{p+1}$. This implies $\tau=\mu_{z_0,c}$ (see [1], p. 113) and proves our result.

## 3. Proof of the Main Theorem

Let us put $\Omega=\Omega(z_0, c)$, $B=B(z_0, c)$, $q=q_{z_0}$ and

$$\mu_0 = \mu_{z_0,c} \quad (z_0 = (x_0, t_0),\ c > 0).$$

For $c'\in]0, c[$ the domains $\Omega(z_0, c')$ increase with $c'$ and have $\Omega$ as their union:

(3.1) $$\bigcup_{0<c'<c} \Omega(z_0, c') = \Omega(z_0, c).$$

Consequently, for a given point $(x, t)\in\Omega$ the number $c'\in]0, c[$ can be chosen in such a way that $(x, t)$ lies in $\Omega(z_0, c')$; it suffices to choose

(3.2) $$t_0-t < c' < c.$$

[4]) One has to use the obvious fact that $T_{\complement\Omega}<+\infty$ holds $P^{z_0}$-almost surely.

Since $c'$ will be fixed for a while, we put $\Omega'=\Omega(z_0, c')$ and $B'=B(z_0, c')$. We then have

$$\bar{\Omega}'\setminus\{z_0\} \subset \Omega. \tag{3.3}$$

Finally we introduce for an arbitrary set $A\subset \boldsymbol{R}^{p+1}$

$$A_t = A\cap(\boldsymbol{R}^p\times]-\infty, t[). \tag{3.4}$$

We intend to study the generalized solution $H_f=H_f^{\Omega}$ in $\Omega$ for a given boundary function $f\in\mathscr{C}(B)$. At the beginning we imitate the original proof of the property (F) in Fulks [10]. The domain $\Omega_t'$ is contained with its boundary in $\Omega$ according to (3.3). The boundary splits into three disjoint sets:

$$\partial\Omega_t' = B_t'\cup I'(t)\cup I_0'(t) \tag{3.5}$$

where

$$I'(t) = \{(\xi, t): \|\xi-x_0\| < F_{c'}(t_0-t)\} \tag{3.6}$$

and

$$I_0'(t) = \{(\xi, t): \|\xi-x_0\| = F_{c'}(t_0-t)\} \tag{3.7}$$

(and where $F_{c'}$ is defined in (1.7)). Since $u=H_f$ is a caloric function on $\Omega$ and hence caloric in a neighborhood of $\bar{\Omega}_t'$, an application of the Gauss divergence theorem [5]) yields as in [10]:

$$\int_{M'(t)} \left[\sum_{i=1}^p (uv_{\xi_i}-vu_{\xi_i})n_i+uvn_{p+1}\right] d\sigma' = 0. \tag{3.8}$$

Here $\sigma'$ is the surface measure of the smooth manifold

$$M'(t)=\partial\Omega_t'\setminus I_0'(t)$$

and $n_1, \ldots, n_{p+1}$ are the components of the outer normal field. Furthermore, $v$ is a solution of the adjoint heat equation in an open neighborhood of $\bar{\Omega}_t'$. We apply this formula (as in [10]) to two special functions $v$: to the constant function 1 and to the function $k=W_{z_0}$ of (1.4). We then obtain

$$\int_{M'(t)} \left[-\sum_{i=1}^p u_{\xi_i}n_i+un_{p+1}\right] d\sigma' = 0 \tag{3.9}$$

and

$$\int_{M'(t)} \left[\sum_{i=1}^p (uk_{\xi_i}-ku_{\xi_i})n_i+ukn_{p+1}\right] d\sigma' = 0, \tag{3.10}$$

respectively. We also observe that

$$n_1 = \ldots = n_p = 0, \quad n_{p+1} = 1 \quad \text{on} \quad I'(t), \tag{3.11}$$

$$k = d' \quad \text{on } B' \tag{3.12}$$

[5]) For an appropriate form of this theorem including "singular" boundary points (in the sense of differential geometry) like those in $I_0'(t)$ see König [14].

and

$$-\sum_{i=1}^{p} k_{\xi_i} n_i = d'q \quad \text{on} \quad B' \setminus \{z_0\}, \tag{3.13}$$

where

$$d' = \left(\frac{1}{4\pi c'}\right)^{p/2}.$$

We then obtain, in view of (3.5),

$$\int_{I'(t)} u\, d\lambda^p = \int_{B'_t} \left[\sum_{i=1}^{p} u_{\xi_i} n_i - u n_{p+1}\right] d\sigma' \tag{3.14}$$

and

$$\int_{I'(t)} uk\, d\lambda^p = d' \int_{B'_t} \left[\sum_{i=1}^{p} u_{\xi_i} n_i - u n_{p+1}\right] d\sigma' + d' \int_{B'_t} uq\, d\sigma', \tag{3.15}$$

where $\lambda^p$ is the Lebesgue measure on $\boldsymbol{R}^p \times \{t\}$ (so that integration with respect to $\lambda^p$ refers to the space variable). From these two equalities we finally deduce

$$\int_{I'(t)} u(k-d')\, d\lambda^p = d' \int_{B'_t} uq\, d\sigma'. \tag{3.16}$$

Obviously, the restriction of $u=H_f$ to $\Omega_t$ extends to $\bar{\Omega}_t$ continuously by means of $f$. This follows from Proposition 2.1 since $z_0 \notin \bar{\Omega}_t$. An elementary discussion of the surface measures $\sigma'$ for $c' \to c$ shows that we may pass to the limit for $c' \to c$. We then obtain from (3.16)

$$\int_{I(t)} H_f k\, d\lambda^p = d \int_{I(t)} H_f\, d\lambda^p + d \int_{B_t} fq\, d\sigma, \tag{3.17}$$

where $I(t)$ is defined in the same way as $I'(t)$ by replacing $c'$ by $c$ in (3.16), where

$$d = \left(\frac{1}{4\pi c}\right)^{p/2} \tag{3.18}$$

and where $\sigma$ denotes the surface measure of $B$.

Next we study the behaviour of (3.17) for $t < t_0$ tending to $t_0$. We observe that

$$\left|\int_{I(t)} H_f\, d\lambda^p\right| \leqq \|f\|\, \lambda^p(I(t))$$

since $|H_f| \leqq \|f\|$ on $\Omega$, where $\|f\|$ denotes the sup-norm on $\mathscr{C}(B)$. We then obtain

$$\lim_{t \to t_0 - 0} \int_{I(t)} H_f\, d\lambda^p = 0$$

since $I(t)$ is an open euclidean ball in $\boldsymbol{R}^p \times \{t\}$ with radius $F_c(t_0 - t)$. The second term on the right-hand side of (3.17) has

$$d \int_B fq\, d\sigma = \mu_0(f)$$

as a limit for $t \to t_0 - 0$. So we know that the limit of the left-hand side of (3.17) exists, namely

$$\lim_{t \to t_0 - 0} \int_{I(t)} H_f k\, d\lambda^p = \mu_0(f). \tag{3.19}$$

By changing the space variable $\xi$ in the integral over $I(t)$ by means of

$$\eta = \frac{1}{\sqrt{4(t_0-t)}}(\xi - x_0)$$

we obtain

$$\int_{I(t)} H_f k \, d\lambda^p = \pi^{-p/2} \int_{K(t)} H_f(x_0 + \sqrt{4(t_0-t)}\,\eta, t) e^{-\|\eta\|^2} d\eta \tag{3.20}$$

where $K(t)$ is the open ball in $\boldsymbol{R}^p$ with the origin as center and with radius

$$r(t) = \frac{F_c(t_0-t)}{\sqrt{4(t_0-t)}}.$$

Since

$$\lim_{t \to t_0 - 0} r(t) = +\infty,$$

we have

$$\lim_{t \to t_0 - 0} \pi^{-p/2} \int_{K(t)} e^{-\|\eta\|^2} d\eta = 1. \tag{3.21}$$

By applying a mean value argument [6]) to the integral (3.20), we obtain for each $t$ satisfying $0 < t_0 - t < c$ the existence of a point $\eta_t \in K(t)$, hence of a point

$$(\xi_t, t) = (x_0 + \sqrt{4(t_0-t)}\,\eta_t, t) \in I(t)$$

such that

$$\int_{I(t)} H_f k \, d\lambda^p = \pi^{-p/2} H_f(\xi_t, t) \int_{K(t)} e^{-\|\eta\|^2} d\eta.$$

Consequently, it follows from (3.19)—(3.21) that

$$\lim_{t \to t_0 - 0} H_f(\xi_t, t) = \mu_0(f).$$

Hereby, the point $(\xi_t, t) \in I(t)$ depends on the given boundary function $f \in \mathscr{C}(B)$.

So we have proved the following result:

**3.1** Proposition. *For every* $f \in \mathscr{C}(B)$ *and every* $t \in ]t_0 - c, t_0[$ *there exists a point* $\xi_t$ *— depending on* $f$ *— in the p-dimensional ball*

$$\|\xi - x_0\| < F_c(t_0 - t)$$

*such that*

$$\lim_{t \to t_0 - 0} H_f(\xi_t, t) = \mu_0(f). \tag{3.22}$$

**3.2** Remark. Corollary 2.4 follows already from this Proposition. In fact, $z_0$ cannot be regular since otherwise (3.22) would imply $f(z_0) = \mu_0(f)$ for all $f \in \mathscr{C}(B)$,

---

[6]) We are referring to the following fact which can be proved easily: Let $\mu$ be a positive bounded Radon measure on a connected, locally compact space $Y$ with support equal to $Y$. Then for every bounded function $g \in \mathscr{C}(Y)$ there exists a point $\eta \in Y$ such that

$$\int g \, d\mu = g(\eta) \|\mu\|.$$

and hence $\mu_0=\varepsilon_{z_0}$. However, $\mu_0=\mu_{z_0,c}$ has a density with respect to the surface measure $\sigma$ which implies $\mu_0(\{z_0\})=0$.

We continue with the proof of the Main Theorem and claim:

**3.3** Lemma. *Let $\Lambda$ denote the set of all measures*

$$\lambda = \alpha\varepsilon_{z_0}+(1-\alpha)\varepsilon_{z_0}^{\complement\Omega}, \quad 0 \leqq \alpha \leqq 1.$$

*Then for every function $f\in\mathscr{C}(B)$ there exists a measure $\lambda\in\Lambda$ such that $\lambda(f)=\mu_0(f)$.*

*Proof.* According to Proposition 3.1, for every $f\in\mathscr{C}(B)$ there exists a sequence $(z_n)$ in $\Omega$ converging to $z_0$ such that

$$\lim_{n\to\infty} \mu_{z_n}^{\Omega}(f) = \mu_0(f) \tag{3.23}$$

holds for the corresponding sequence of harmonic measures. From this sequence (of probability measures) we may extract a subsequence $(\mu_{z_{n_k}}^{\Omega})$ which converges vaguely to a probability (Radon) measure $\lambda$ on $B$ depending on $f$ (see [4], p. 243). So (3.23) implies $\lambda(f)=\mu_0(f)$. We can now apply — and this is crucial — a result of Boboc and Cornea [6] (for a simple proof see Lukeš and Malý [16], p. 362). According to it, the measure $\lambda$ is of the form

$$\lambda = \alpha\varepsilon_{z_0}+(1-\alpha)\varepsilon_{z_0}^{\complement\Omega\setminus\{z_0\}} \quad (0 \leqq \alpha \leqq 1).$$

But since all points in $\boldsymbol{R}^{p+1}$ are polar for the underlying harmonic structure (see [1], p. 79), we have

$$\varepsilon_{z_0}^{\complement\Omega\setminus\{z_0\}} = \varepsilon_{z_0}^{\complement\Omega}$$

(see [1], pp. 83 and 99). This proves $\lambda\in\Lambda$, and hence the Lemma.

The remaining part of the proof of the Main Theorem is now pure linear algebra. It suffices to apply the next lemma to the case $X=B$, $x_0=z_0$, $\nu=\varepsilon_{z_0}^{\complement\Omega}$. In fact, it follows from Lemmas 3.3 and 3.4 that $\mu_0$ is in $\Lambda$ which means

$$\mu_0 = \alpha\varepsilon_{z_0}+(1-\alpha)\varepsilon_{z_0}^{\complement\Omega} \quad \text{for some } \alpha\in[0, 1].$$

Since $\mu_0$ has a density with respect to $\sigma$, we have

$$0 = \mu_0(\{z_0\}) = \alpha+(1-\alpha)\varepsilon_{z_0}^{\complement\Omega}(\{z_0\}) \geqq \alpha$$

which implies $\alpha=0$ and hence $\mu_0=\varepsilon_{z_0}^{\complement\Omega}$, the statement of the Main Theorem.

**3.4** Lemma. *Let $X$ be a compact space, $x_0\in X$ and $\nu$ a positive Radon measure on $X$. Suppose that a Radon measure $\mu_0$ on $X$ has the property that for every $f\in\mathscr{C}(X)$ there exists a number $\alpha_f\in[0, 1]$ such that $\mu_0(f)=\lambda(f)$, where $\lambda=\alpha_f\varepsilon_{x_0}+(1-\alpha_f)\nu$. Then $\mu_0$ is of the form $\mu_0=\alpha\varepsilon_{x_0}+(1-\alpha)\nu$ for some $\alpha\in[0, 1]$.*

*Proof.* Let us denote by $\Lambda_\nu$ the set of all measures $\lambda=\alpha\varepsilon_{x_0}+(1-\alpha)\nu$ with $\alpha\in[0, 1]$ and by $H$ the set of all functions $h\in\mathscr{C}(X)$ satisfying $\nu(h)=h(x_0)$. Then obviously, for all $\lambda\in\Lambda_\nu$ and $h\in H$,

$$\lambda(h) = \nu(h) = h(x_0).$$

Our condition therefore implies

$$\mu_0(h) = \lambda(h) \quad \text{for all} \quad \lambda\in\Lambda_\nu \quad \text{and} \quad h\in H. \tag{3.24}$$

Consequently, in the case $H=\mathscr{C}(X)$, i.e. $\nu=\varepsilon_{x_0}$, the measure $\mu_0$ equals $\varepsilon_{x_0}$ and thus is of the desired form. In the remaining case where $H\neq\mathscr{C}(X)$ we choose $g\in\mathscr{C}(X)\setminus H$. The corresponding measure $\lambda_0=\alpha_g\varepsilon_{x_0}+(1-\alpha_g)\nu\in\Lambda_\nu$ satisfying $\lambda_0(g)=\mu_0(g)$ is then uniquely determined: the coefficient $\alpha_g$ equals

$$\alpha_g = \frac{\mu_0(g)-\nu(g)}{g(x_0)-\nu(g)}.$$

Since $H$ is a hyperplane in $\mathscr{C}(X)$ — defined by the linear form $\nu-\varepsilon_{x_0}\neq 0$ — every $f\in\mathscr{C}(X)$ is of the form $f=h+\tau g$ with some $h\in H$ and $\tau\in\boldsymbol{R}$. For the measure $\lambda_0$ we then have

$$\mu_0(g) = \lambda_0(g) \quad \text{and} \quad \mu_0(h) = \lambda_0(h),$$

the latter according to (3.24), and, consequently,

$$\mu_0(f) = \mu_0(h)+\tau\mu_0(g) = \lambda_0(h)+\tau\lambda_0(g) = \lambda_0(f).$$

Since $\lambda_0$ does not depend on the choice of $f$, this proves $\mu_0=\lambda_0$. Hence, also in this case, $\mu_0$ has the desired form.

As a corollary, we now obtain also a *proof of the Addendum* 2.2′: For every $f\in\mathscr{C}(B)$, the proof of Lemma 3.3 shows the existence of a sequence $(z_k)$ in $\Omega$ converging to $z_0$ such that the corresponding sequence $(\mu_{z_k}^{\Omega})$ of harmonic measures converges to a measure $\lambda\in\Lambda$ which satisfies $\lambda(f)=\mu_0(f)$. Since $\mu_0\neq\varepsilon_{z_0}$, the function $f\in\mathscr{C}(B)$ may be chosen in such a way that $\mu_0(f)\neq f(z_0)$. We have $\lambda=\alpha\varepsilon_{z_0}+(1-\alpha)\mu_0$ for some $\alpha\in[0, 1]$ according to the Main Theorem. Consequently, $\mu_0(f)=\alpha f(z_0)+(1-\alpha)\mu_0(f)$ which implies $\alpha=0$ and thus $\lambda=\mu_0$.

**3.5** Remark. *The set* $\Lambda$ *of Lemma* 3.3 *coincides with the set of all cluster points of* $(\mu_z^{\Omega})$ *for* $z\to z_0$, $z\in\Omega$. This is an easy consequence of Lukeš and Malý [16], Theorem 16. In fact, we know from Corollary 2.4 that all boundary points of $\Omega$ are regular, with $z_0$ as the only exception. From this the above result follows from [16], Theorem 16, in the same way as Corollary 17 of [16] was derived from that Theorem.

## References

[1] Bauer, H.: Harmonische Räume und ihre Potentialtheorie. - Lecture Notes in Mathematics 22. Springer-Verlag, Berlin—Heidelberg—New York, 1966.

[2] Bauer, H.: Harmonic spaces and associated Markov processes. - Potential Theory, C.I.M.E., I Ciclo, Stresa, 1969. Edizioni Cremonese, Roma, 1970, 23—67.

[3] BAUER, H.: Harmonische Räume. - Jahrbuch Überblicke Mathematik 1981. Bibliographisches Institut, Mannheim—Wien—Zürich, 1981, 9—35.

[4] BAUER, H.: Probability theory and elements of measure theory, 2nd edition. - Academic Press. Inc., London—New York—Toronto—Sydney—San Francisco, 1981.

[5] BLIEDTNER, J.: Harmonische Gruppen und Huntsche Faltungskerne. - Seminar über Potentialtheorie (Edited by H. Bauer). Lecture Notes in Mathematics 69. Springer-Verlag, Berlin—Heidelberg—New York, 1968, 69—102.

[6] BOBOC, N., et A. CORNEA: Comportement des balayées des mesures ponctuelles. Comportement des solutions du problème de Dirichlet aux points irréguliers. - C. R. Acad. Sci. Paris Sér. A 264, 1967, 995—997.

[7] CONSTANTINESCU, C., and A. CORNEA: Potential theory on harmonic spaces. - Die Grundlehren der mathematischen Wissenschaften 158. Springer-Verlag, Berlin—Heidelberg—New York, 1972.

[8] EFFROS, E. G., and J. L. KAZDAN: On the Dirichlet problem for the heat equation. - Indiana Univ. Math. J. 20, 1971, 683—693.

[9] FRIEDMAN, A.: Partial differential equations of parabolic type. - Prentice Hall, Inc., Englewood Cliffs, N. J., 1964.

[10] FULKS, W.: A mean value theorem for the heat equation. - Proc. Amer. Math. Soc. 17, 1966, 6—11.

[11] GEHRING, F. W.: On solutions of the equation of heat conduction. - Michigan Math. J. 5, 1958, 191—202.

[12] KÖHN, J.: Harmonische Räume mit einer Basis semiregulärer Mengen. - Seminar über Potentialtheorie (Edited by H. Bauer). Lecture Notes in Mathematics 69. Springer-Verlag, Berlin—Heidelberg—New York, 1968, 1—12.

[13] KÖHN, J., und M. SIEVEKING: Reguläre und extremale Randpunkte in der Potentialtheorie. - Rev. Roumaine Math. Pures Appl. 12, 1967, 1489—1502.

[14] KÖNIG, H.: Ein einfacher Beweis des Integralsatzes von Gauß. - Jber. Deutsch. Math.-Verein. 66, 1964, 119—138.

[15] KUPTSOV, L. P.: Mean property for the heat-conduction equation. - Mat. Zametki 29, 1981, 211—223. [Russian; English translation: Math. Notes 29, 1981, 110—116.]

[16] LUKEŠ, J., and J. MALÝ: On the boundary behaviour of the Perron generalized solution. - Math. Ann. 257, 1981, 355—366.

[17] MEYER, P.-A.: Processus de Markov. - Lecture Notes in Mathematics 26. Springer-Verlag, Berlin—Heidelberg—New York, 1967.

[18] WATSON, N. A.: A theory of subtemperatures in several variables. - Proc. London Math. Soc. (3) 26, 1973, 385—417.

[19] WATSON, N. A.: Green functions, potentials, and the Dirichlet problem for the heat equation. - Proc. London Math. Soc (3) 33, 1976, 251—298. Corrigendum: ibid. (3) 37, 1978, 32—34.

Received 26 January 1984

# Simplicial function spaces and simplexes [R35]

Exposition. Math. **3** (1985), 165–168
[Zbl. Math. 564.46007; MR 87c:46009]

Dedicated to Professor H. G. Tillmann on the occasion of his 60th birthday

We consider a metrizable compact space $X$ and a *function space* $\mathscr{H}$ on it. By this we mean a linear subspace of $\mathscr{C}(X)$ – the Banach space of all real-valued continuous functions on $X$ with the usual sup-norm – such that $\mathscr{H}$ separates points and contains the constant function 1. As usual, a positive Radon measure $\mu$ on $X$ is called ($\mathscr{H}$-)*representing measure* for a point $x \in X$ if

$$\int h \, d\mu = h(x) \quad \text{for all} \quad h \in \mathscr{H}.$$

$\mathscr{M}_x(\mathscr{H})$ will denote the set of these measures. It is well known that the *Choquet boundary* $Ch_{\mathscr{H}}(X)$ of $X$ with respect to $\mathscr{H}$ is a $G_\delta$-set in $X$ and that, for every point $x \in X$, there exist measures $\mu \in \mathscr{M}_x(\mathscr{H})$ which are carried by $Ch_{\mathscr{H}}(X)$ (cf. [7], p. 180).

Usually, properties of $\mathscr{H}$ and representing measures are reflected in corresponding properties of the *state space* $S(\mathscr{H})$ of $\mathscr{H}$ which, by definition, is the set of all positive linear forms $\varphi$ on $\mathscr{H}$ satisfying $\varphi(1) = 1$. With respect to the weak*-topology, $S(\mathscr{H})$ is a metrizable, compact, convex set. (It is metrizable since $\mathscr{H}$ is separable in the topology of uniform convergence on $X$.)

It seems natural to call the function space $\mathscr{H}$ *simplicial* if its state space $S(\mathscr{H})$ is a (Choquet) simplex. This is the original definition of a simplicial function space given by Effros-Kazdan [9]. There exists also the notion of a simplicial function cone which is due to Bliedtner-Hansen [4]. Their definition, specialized to the case of a function space $\mathscr{H}$ on $X$, demands the following property:

**(WS)** For every point $x \in X$, there exists exactly one representing measure $\mu \in \mathscr{M}_x(\mathscr{H})$ which is carried by $Ch_{\mathscr{H}}(X)$.

If property (WS) holds, we will call the function space $\mathscr{H}$ *weakly simplicial*. It is obvious that every simplicial function space $\mathscr{H}$ is also weakly simplicial.

The fact that in [4] weakly simplicial function spaces are called simplicial, seems to be the origin of some confusion*). It seems to have spread the belief that the

---

*) In Sibony [12] simpliciality is only introduced for *inf-stable* cones of continuous functions. This additional assumption avoids such confusions in the case of function spaces. The same situation is considered in Boboc-Cornea [5] – however, without point separation

two notions of simpliciality are equivalent. However, the following simple example shows that this is not true.

*Example.* Let $X$ be the compact unit interval $[0, 1]$ in $\mathbb{R}$ and let $\mathscr{H}$ be generated by 1, $id$, $id^2$ where $id$ is the identity map. It is well known that $Ch_{\mathscr{H}}(X)=X$. So, trivially, $\mathscr{H}$ is weakly simplicial. However, the state space $S(\mathscr{H})$ can be identified with the convex hull of the graph of the function $id^2$ (cf. [3]). Consequently, $S(\mathscr{H})$ is not a simplex, and $\mathscr{H}$ is not simplicial.

A more sophisticated example would be function space $\mathscr{H}=\operatorname{Re} H^\infty$ on the maximal ideal space $X$ of the unit disc $D$ where $H^\infty$ denotes the Banach algebra of bounded holomorphic functions on $D$ interpreted as function algebra on $X$. For details see Köhn [10], p. 81.

It is our intention to point out that also weak simpliciality of $\mathscr{H}$ can be characterized by means of a different type of state space. In order to make this clear, we have to pass from $\mathscr{H}$ to the bigger function space $\hat{\mathscr{H}}$ of all *$\mathscr{H}$-affine functions* (cf. [3]), i.e. the space of all functions $f \in \mathscr{C}(X)$ satisfying

$$\int f \, d\mu = f(x) \quad \text{for all } x \in X \text{ and all } \mu \in \mathscr{M}_x(\mathscr{H}).$$

The result is then the following:

**Theorem.** *A function space $\mathscr{H}$ on a metrizable compact space $X$ is weakly simplicial if and only if the space $\hat{\mathscr{H}}$ of $\mathscr{H}$-affine functions is simplicial.*

In the preceding example $\hat{\mathscr{H}}=\mathscr{C}([0, 1])$ and, consequently, $S(\hat{\mathscr{H}})$ is the simplex $\mathscr{M}^1([0, 1])$ of all probability measures on $[0, 1]$ (cf. [3]). This case is also covered by a predecessor of the above Theorem in [2], p. 124.

**Proof** of the Theorem. The following result is known from [5], Theorem 3.2. Assume that $\mathscr{H}$ is weakly simplicial. Then for every pair of functions $s, t \in \mathscr{C}(X)$ satisfying $s \le t$ with $s$ $\mathscr{H}$-convex and $t$ $\mathscr{H}$-concave, i.e.

$$\int t \, d\mu \le t(x) \quad \text{for all } x \in X \text{ and } \mu \in \mathscr{M}_x(\mathscr{H})$$

(and $\int s \, d\mu \ge s(x)$ for $x \in X$ and $\mu \in \mathscr{M}_x(\mathscr{H})$, respectively), there exists a function $v \in \hat{\mathscr{H}}$ such that

$$s \le v \le t.$$

Consequently, $\hat{\mathscr{H}}$ enjoys the Riesz decomposition property: Given non-negative functions $u, u_1, u_2$ in $\hat{\mathscr{H}}$ such that

$$u \le u_1 + u_2,$$

there exist functions $u'_1, u'_2 \in \hat{\mathscr{H}}$ satisfying

$$0 \le u'_i \le u_i \quad (i=1, 2)$$

and

$$u = u'_1 + u'_2.$$

It suffices to apply the above property to $s=0\vee(u-u_2)$ and $t=u\wedge u_1$. The corresponding interpolation function $v\in\hat{\mathscr{H}}$ then leads to the solution $u'_1=v$ and $u'_2=u-v$ (cf. also [1], p. 84). From the Riesz decomposition property it follows that the cone $P$ of all positive linear form s on $\hat{\mathscr{H}}$ is a lattice with respect to its natural order or - equivalently - that the linear space $E=P-P$ of all relatively bounded linear forms on $\hat{\mathscr{H}}$ is a Riesz space (=vector lattice). (In order to see this it suffices to rewrite the proof presented e.g. in Bourbaki [6], p. 28, for the case of a Riesz space). By definition, we have

$$S(\hat{\mathscr{H}})=P\cap H,$$

where $H$ is the hyperplane in $E$ defined by the equation $\varphi(1)=1$. So $S(\hat{\mathscr{H}})$ is a base of the lattice cone $P$. Hence $S(\hat{\mathscr{H}})$ is a simplex.

Conversely, assume that $\hat{S}=S(\hat{\mathscr{H}})$ is a simplex. We have the usual $\hat{\mathscr{H}}$-embedding

$$j\colon X\to j(X)\subset\hat{S}$$

at our disposal which maps $x\in X$ onto the evaluation functional $\delta_x$. It is well known that

$$j(Ch_{\hat{\mathscr{H}}}(X))=ex\hat{S}$$

where $ex\hat{S}$ is the set of extreme points of $\hat{S}$. By means of $j$ the measures $\nu\in\mathscr{M}_x(\hat{\mathscr{H}})$ for $x\in X$ are in 1-1-correspondence with the probability measures on $j(X)$ having $j(x)$ as barycenter. Consequently, according to Choquet's unicity theorem, every point $x\in X$ allows only one $\hat{\mathscr{H}}$-representing measure, which is carried by $Ch_{\hat{\mathscr{H}}}(X)$. This proves that $\mathscr{H}$ is weakly simplicial since

$$\mathscr{M}_x(\hat{\mathscr{H}})=\mathscr{M}_x(\mathscr{H})\qquad(x\in X)$$

and consequently

$$Ch_{\hat{\mathscr{H}}}(X)=Ch_{\mathscr{H}}(X).$$

According to a result of [2], p. 106, the space $\hat{\mathscr{H}}$ *coincides with* $\mathscr{H}$ *if and only if* there exists an inf-stable closed subset $\mathscr{S}$ of $\mathscr{C}(X)$ such that

$$(*)\qquad \mathscr{H}=\mathscr{S}\cap(-\mathscr{S}).$$

This condition is satisfied in most situations close to potential theory, e.g. in Netuka [11]. So in the presence of this condition the space $\mathscr{H}$ is weakly simplicial if and only if it is simplicial. In this case a result of Boboc-Cornea [5] and Davies [8] yields for every Choquet point $x_0\in Ch_{\mathscr{H}}(X)$ the existence of a function $h\in\mathscr{H}$ which peaks at $x_0: 0=h(x_0)<h(x)$ for all $x\in X\setminus\{x_0\}$. Without condition $(*)$ the above Theorem assures the existence of at least an $\mathscr{H}$-affine function which peaks at a given Choquet point.

It would be interesting to know an example of a closed, weakly simplicial function space $\mathscr{H}$ such that for some Choquet points no peak function exists in $\mathscr{H}$.

**Added in proof:** The following example answers the question raised at the end of this Note. It modifies an example which Mrs. S. Papadopoulou kindly communicated to the author. The example itself is a modification of an earlier example which appears in the author's paper in J.-Ber. Deutsche Math.-Verein. **69**, 89–104 (1967) on p. 100 (see also [2], p. 121). Let $X \subset \mathbb{R}^2$ be the union of the two semi-circles

$$X_1 = \{(x, y) \in \mathbb{R}^2 : (x+1)^2 + y^2 = 1, x \leq -1\}$$

and

$$X_2 = \{(x, y) \in \mathbb{R}^2 : (x-1)^2 + y^2 = 1, x \geq 1\}$$

with the line segments between the points $a_1 = (-1, 1)$ and $a_2 = (1, 1)$ as well as between $b_1 = (-1, -1)$ and $b_2 = (1, -1)$. Let $\mathscr{H}$ be the set of all affine functions on $\mathbb{R}^2$ restricted to $X$. Obviously, $\mathscr{H}$ is a closed, weakly simplicial function space on $X$ with $Ch_{\mathscr{H}}(X) = X_1 \cup X_2$. No function in $\mathscr{H}$ peaks at one of the four points $a_1, a_2, b_1, b_2$.

## References

[1] E.M. Alfsen: Compact convex sets and boundary integrals. Ergebnisse der Math. 57, Springer Verlag, Berlin-Heidelberg-New York 1971.

[2] H. Bauer: Šilovscher Rand und Dirichletsches Problem. Ann. Inst. Fourier **11**, 89–136 (1981).

[3] H. Bauer: Approximation and abstract boundaries. Amer. Math. Monthly **85**, 632–647 (1978).

[4] J. Bliedtner and W. Hansen: Simplicial cones in potential theory. Inventiones math. **29**, 83–110 (1975).

[5] N. Boboc and A. Cornea: Convex cones of lower semicontinuous functions on compact spaces. Rev. Roum. Math. Pures App. **12**, 471–525 (1967).

[6] N. Bourbaki: Intégration, Chap. 1–4, 2ᵉ édition. Hermann, Paris 1965.

[7] G. Choquet: Lectures on analysis, Vol. II. W.A. Benjamin, New York-Amsterdam 1969.

[8] E.B. Davies: A generalized theory of convexity. Proc. London Math. Soc. **17**, 644–652 (1967).

[9] E.G. Effros and J.L. Kazdan: Applications of Choquet simplexes to elliptic and parabolic boundary value problems. J. of Differential Equations **8**, 95–134 (1970).

[10] J. Köhn: Barycenters of unique maximal measures. J. of Functional Anal. **6**, 76–82 (1970).

[11] I. Netuka: The Dirichlet problem for harmonic functions. Amer. Math. Monthly **87**, 621–628 (1980).

[12] D. Sibony: Cônes de fonctions et potentiels. Lecture Notes, McGill University, Montreal 1968.

Received 23. 3. 84

# Fine boundary limits of harmonic and caloric functions (*) [R36]

Bull. Sci. Math. (2) **109** (1985), 337–361
[Zbl. Math. 594.31017; MR 87k:31010]

RÉSUMÉ. — On démontre que les résultats classiques de M. Brelot concernant l'existence et la représentation des limites fines de certaines fonctions harmoniques bornées en un point-frontière irrégulier et la liaison de ces limites avec les suites maximales (introduites par Brelot) restent valable pour une large classe d'espaces harmoniques. En particulier, on obtient des applications à des fonctions caloriques, c'est-à-dire à des solutions de l'équation de la chaleur.

ABSTRACT. — It will be shown that the classical results of M. Brelot about the existence and the representation of fine limits of certain bounded harmonic functions at an irregular boundary point and the connection of these results with the so-called maximal sequences (introduced by Brelot) remain valid for a large class of harmonic spaces. In particular, applications to caloric functions, i. e., solutions of the heat equation, thus become possible.

## Introduction

In his fundamental paper of 1944, BRELOT [10] has studied the fine limit behavior of bounded harmonic functions at an irregular boundary point $z$ of the bounded domain $U \subset \mathbb{R}^n$ on which these functions are defined. Later on, BRELOT [11] continued these studies by making essential use of the Green function $G_{y_0}$ of $U$ with pole $y_0 \in U$ and the Martin representation of positive harmonic functions. He introduced the maximal sequences for the irregular boundary point $z$. According to the

---

(*) Texte présenté par P. MALLIAVIN, reçu le 27 mars 1985.

Classification matières AMS (MOS) 1980 : 31C05.

Vedettes matières : Harmonic functions, Caloric functions, Fine boundary limits.

Heinz BAUER, Mathematisches Institut, Universität Erlangen-Nürnberg, Bismarckstrasse 1 1/2, D-8520 Erlangen (Allemagne fédérale).

The author is grateful to Mrs. Ursula SCHIRMEIER for critical comments.

original definition, these are sequences $(x_n)$ in $U$ converging to $z$ such that $(G_{y_0}(x_n))$ converges to the upper limit of $G_{y_0}(x)$ for $x \to z$ (in the euclidean topology). Brelot proved that, for every bounded harmonic function $u$ on $U$, the sequence $(u(x_n))$ tends to the fine limit of $u$ at $z$ whenever the sequence $(x_n)$ is maximal.

This intimate connection between the Green function, maximal sequences and fine limits made it difficult to prove similar results in more general situations. SMYRNÉLIS [20] studied this problem for Brelot harmonic spaces along the ideas of BRELOT [11]. He was forced to make very restrictive assumptions. He had to assume in particular: proportionality of potentials with point support in the original and in the adjoint sheaf as well as a weak form of Axiom D, namely the existence of a base of completely determining domains. These assumptions rule out the possibility of treating also caloric functions – i.e., solutions of the heat equation – by this approach.

In what follows we intend to show that the classical results of BRELOT ([10], [11]) remain valid under much weaker assumptions. They can be proved on every strong harmonic space in the author's sense [2], [3], [4]. This allows applications to solutions of elliptic and parabolic linear differential equations, in particular to the heat equation. Our results seem to be new already for this equation. The progress became possible by an approach which recalls the old ideas of BRELOT [10] and which avoids Green functions and Martin representation by a combination of old and new techniques. Maximal sequences are introduced as sequences $(x_n)$ converging to the irregular boundary point $z$ of $U$ such that the sequence $(\mu^U_{x_n})$ of harmonic measures for $U$ tends to the swept-out measure $\varepsilon_z^{\complement U}$ in the vague (or weak) sense. In the context of [20], SMYRNÉLIS has observed the equivalence of this property with maximality without applying it.

For the reader's convenience we collect now our main assumptions and notations: $X$ will always denote a *strong harmonic space* in the sense of [2] (*cf.* also [3], [4]). This means in particular that $X$ has a countable base and that Doob's convergence axiom holds. This convergence axiom plays an important role in the proof of our Theorem 2.4 which is basic for the rest of our presentation. However, our results do not require the presence of a base of regular sets. They seem to be valid on every $\mathfrak{P}$-harmonic space with countable base for which Doob's convergence axiom holds. For reasons of simplicity and easily available references we do not intend to present our results in this more general framework.

We study the boundary behavior of functions which are defined on a set $U \subset X$ which is supposed to be *open* and *relatively compact*. Since our results are essentially of a local type, relative compactness of $U$ is not a serious restriction. The notations will be those introduced in [2], [3], [4]. In particular, $U^*$ will denote the boundary of $U$. $U^*_{\text{irr}}$ stands for the set of irregular boundary points. $H_f = H^U_f$ denotes the generalized solution of the Dirichlet problem for a resolutive boundary function $f$ on $U^*$. For other open relatively compact sets $W$ it will be denoted by $H^W_f$ with explicit appearance of $W$. The letter $f$ in $f$-lim, $f$-lim sup and $f$-lim inf refers to the fine topology on $X$.

## 1. Preliminaries

For the case of classical potential theory, the following result is due to Brelot [10] (*see* also Brelot [13], p. 84 and Doob [15], p. 191):

1.1. Proposition. — *For every point $z \in U^*_{\text{irr}}$ one has*

$$f-\lim_{x \to z,\ x \in U} \mu^U_x = \varepsilon^{\complement U}_z$$

(*in the sense of vague convergence*).

*Proof.* — Let $\mathcal{S}$ be the set of all real-valued, continuous, superharmonic functions $\geqslant 0$ defined on $X$. It is well-known [2], p. 124 that for the restriction $u^* = u \mid U^*$ of a function $u \in \mathcal{S}$ one has

$$(1.1) \qquad H_{u^*} = \hat{R}^{\complement U}_u \mid U.$$

Since the balayee $\hat{R}^{\complement U}_u$ is superharmonic $\geqslant 0$ and since $U \cup \{z\}$ is a fine neighborhood of $z$ this implies

$$f-\lim_{x \to z,\ x \in U} \hat{R}^{\complement U}_u(x) = \hat{R}^{\complement U}_u(z)$$

and, consequently,

$$(1.2) \qquad f-\lim_{x \to z,\ x \in U} H_{u^*}(x) = \hat{R}^{\complement U}_u(z) = \int u\, d\varepsilon^{\complement U}_z.$$

According to the Stone approximation theorem, $\mathcal{S}$- $\mathcal{S}$ restricted to $U^*$ is dense in $\mathscr{C}(U^*)$. From this it follows easily that

$$(1.3) \qquad f-\lim_{x \to z,\ x \in U} H_f\,(x) = \int f d\,\varepsilon^{\complement U}_z$$

holds for all $f \in \mathscr{C}(U^*)$. By the definition of vague convergence, this is the result in question.

From this one can easily deduce the following result:

1.2. Lemma. — *Let $v$ be a hyperharmonic function on $U$ such that*

$$\lambda = f - \lim_{x \to z,\, x \in U} v(x)$$

*exists (in $\overline{\mathbb{R}}$) for $z \in U^*_{\mathrm{irr}}$.*

*Then there exists a sequence $(x_n)$ in $U$ having the following three properties:*

$$\lim_{n \to \infty} x_n = z; \tag{1.4}$$

$$\lim_{n \to \infty} \mu^U_{x_n} = \varepsilon^{\complement U}_z; \tag{1.5}$$

$$\lim_{n \to \infty} v(x_n) = \lambda. \tag{1.6}$$

*Proof.* — Let us denote by $d$ a metric on $X$ compatible with the given topology of $X$, by $\chi$ a metric on the space $\mathscr{M}_+(U^*)$ of positive Radon measures on $U^*$ compatible with the vague topology, and by $\rho$ a metric on $\overline{\mathbb{R}}$ compatible with the topology of the extended real line. Consequently, for every natural number $n$,

$$V^f_n = \left\{ x \in U : \rho(v(x), \lambda) < \frac{1}{n} \right\},$$

$$V_n = \left\{ x \in X : d(x, z) < \frac{1}{n} \right\},$$

and

$$W_n = \left\{ \mu \in \mathscr{M}_+(U^*) : \chi(\mu, \varepsilon^{\complement U}_z) < \frac{1}{n} \right\},$$

is the trace of a fine neighborhood of $z$ in $U$, a neighborhood of $z$ in $X$, and a neighborhood of $\varepsilon^{\complement U}_z$ in $\mathscr{M}_+(U^*)$, respectively. Since $V_n \cap V^f_n$ is the trace of a fine neighborhood of $z$ in $U$, Proposition 1.1 implies the existence of a point $x_n \in V_n \cap V^f_n$ such that $\mu^U_{x_n} \in W_n$ for every $n \in \mathbb{N}$. The sequence $(x_n)_{n \in \mathbb{N}}$ thus obtained then has all the desired properties.

By choosing for $v$ e.g. the constant function 0, Lemma 1.2 yields the existence of a sequence $(x_n)$ in $U$ satisfying (1.4) and (1.5). This is a lemma due to Köhn-Sieveking [18] (*see* also Effros-Kazdan [16]).

1.3. *Remark.* — If the irregular point $z$ is *polar*, then for every hyperharmonic function $v \geqslant 0$ on $U$ the fine limit $f-\lim_{x\to z,\, x\in U} v(x)$ exists. This result of BERLOT [12], p. 140 is valid also under our assumptions. The original proof for Brelot spaces can be copied without essential changes. This result will not be used in the proofs of this paper.

It follows that the same result holds for every *lower bounded hyperharmonic function* $v$ on $U$. Because of the local character of the regularity and irregularity of boundary points, it suffices that $v$ is a lower bounded hyperharmonic function on the trace $U \cap V$ of an open neighborhood $V$ of $z$ in $U$.

The following result will soon be needed for the proof of Theorem 2.4:

1.4. LEMMA. — *Let $z \in U^*$ be an arbitrary boundary point of $U$ and let $(V_n)_{n=0,1,\dots}$ be a sequence of open relatively compact neighborhoods of $z$ (in $X$) satisfying $\bar{V}_{n+1} \subset V_n$ for all $n=0, 1, \dots$ and $\bigcap_{n=0}^{\infty} V_n = \{z\}$. Then, for every bounded Borel function $\geqslant 0$ on $U^*$ vanishing on $U^* \cap V_0$:*

$$(1.7) \qquad \int_{U^*\setminus V_0} f\,d\varepsilon_z^{\complement(U\cup V_n)} \downarrow \int f\,d\varepsilon_z^{\complement U\setminus\{z\}} \qquad (\text{as } n\to\infty).$$

*Proof.* — Obviously, we have:

$$\complement(U\cup V_n)\uparrow \complement U\setminus\{z\}.$$

According to [2], p. 117 this implies

$$\lim_{n\to\infty} \varepsilon_z^{\complement(U\cup V_n)} = \varepsilon_z^{\complement U\setminus\{z\}}$$

(in the sense of vague convergence). We put

$$\mu_n = \varepsilon_z^{\complement(U\cup V_n)} \qquad \text{and} \qquad \mu = \varepsilon_z^{\complement U\setminus\{z\}} \qquad (n=1, 2, \dots),$$

where $\mu_n$ and $\mu$ can be considered as positive Radon measures on the compact set $U^* \cup \bar{V}_1$. For Borel sets $B \subset U^* \setminus V_0$ we then have

$$(1.8) \qquad \mu(B) \leqslant \mu_{n+1}(B) \leqslant \mu_n(B) \qquad (n=1, 2, \dots).$$

Indeed, according to a result of CONSTANTINESCU [14], p. 165 the equality

$$\complement(U\cup V_{n+1}) = \complement(U\cup V_n) \cup [\complement U \cap V_n \cap \complement V_{n+1}],$$

implies $\mu_{n+1} \leqslant \mu_n + \varepsilon_z^{A_n}$, where

$$A_n = \complement U \cap V_n \cap \complement V_{n+1}.$$

So we have

$$\mu_{n+1}(B) \leqslant \mu_n(B) + \varepsilon_z^{A_n}(B).$$

It follows from $B \cap V_0 = \emptyset$ that $B \cap \bar{V}_1 = \emptyset$; consequently $\varepsilon_z^{A_n}(B) = 0$ since $\varepsilon_z^{A_n}$ is supported by $\bar{V}_n \subset \bar{V}_1$. This yields $\mu_{n+1}(B) \leqslant \mu_n(B)$ for all $n = 1, 2, \ldots$. Similarly we have $\mu(B) \leqslant \mu_{n+1}(B)$ since

$$\complement U \setminus \{z\} = \complement (U \cup V_n) \cup [\complement U \cap V_n \cap \complement \{z\}].$$

The result can now be obtained as follows: Vague convergence of the sequence $(\mu_n)_{n=1,2,\ldots}$ to $\mu$ on the compact space $X_0 = U^* \cup \bar{V}_1$ is equivalent to the validity of

(1.9) $$\limsup_{n \to \infty} \mu_n(F) \leqslant \mu(F),$$

for all compact subsets $F$ of $X_0$ together with

(1.10) $$\lim_{n \to \infty} \mu_n(X_0) = \mu(X_0),$$

according to a well-known Portemanteau Theorem (*cf.* [1] and [6], p. 12). For the compact subset $Y = U^* \setminus V_0$ of $X_0$, (1.8) and (1.9) imply $\lim \mu_n(Y) = \mu(Y)$. Consequently, the sequence $(\nu_n)$ of the measures $\nu_n$ induced by $\mu_n$ on $Y (n = 1, 2, \ldots)$ converges vaguely to the measure $\nu$ induced by $\mu$ on $Y$. To this situation we apply our next Lemma in order to finish the proof.

1.5. Lemma. — *Let $\nu$, $\nu_1$, $\nu_2$, ... be positive Radon measures on a compact metrizable space $Y$. Suppose that the sequence $(\nu_n)$ converges vaguely to $\nu$ and that*

(1.11) $$\nu(B) \leqslant \nu_{n+1}(B) \leqslant \nu_n(B) \qquad (n = 1, 2, \ldots),$$

*holds for all Borel sets $B \subset Y$. Then for all bounded Borel functions $f \geqslant 0$ on $Y$ the sequence $\left( \int f d\nu_n \right)$ converges decreasingly to $\int f d\nu$.*

*Proof.* — For every Borel set $B \subset Y$ we put

$$\nu_0(B) = \lim_{n \to \infty} \nu_n(B).$$

Then $\nu_0$ is a finite Borel measure on $Y$. In fact, for every sequence $(B_k)$ of Borel subsets of $Y$ satisfying $B_k \downarrow \emptyset$ the double sequence $(\nu_n(B_k))$ is decreasing in $n$ and $k$ so that:

$$\inf_k \nu_0(B_k) = \inf_k \inf_n \nu_n(B_k) = \inf_n \inf_k \nu_n(B_k) = 0.$$

According to the Portemanteau Theorem recalled in the last proof, one has $\nu(F)=\nu_0(F)$ for all compact sets $F \subset Y$. The usual uniqueness theorem (*cf.* [5], p. 23) then leads to $\nu=\nu_0$. So $\nu_n(B) \downarrow \nu(B)$ for all Borel sets and, consequently, for all Borel functions $g \geqslant 0$ on $Y$ attaining only finitely many values. From this the result follows since for the given function $f$ and an arbitrary number $\varepsilon>0$ there exist Borel functions $g$ of the above type satisfying $0 \leqslant f-g \leqslant \varepsilon$ (*cf.* [5], p. 56, Exercice 1).

## 2. Fine limits of generalized solutions

We start with a statement which in classical potential theory is due to Brelot ([10], [11]). For Brelot spaces it has been obtained by Smyrnélis [20].

2.1. Proposition. — *Let $f \geqslant 0$ be a bounded resolutive function on $U^*$ which vanishes in a neighborhood of $z \in U^*_{\text{irr}}$. Then the corresponding generalized solution $H_f$ of the Dirichlet problem behaves as follows:*

$$f-\lim_{x \to z,\ x \in U} H_f(x) = \limsup_{x \to z,\ x \in U} H_f(x). \tag{2.1}$$

*Proof.* — Let us observe first that the existence of the fine limit in (2.1) will follow from Remark 1.3 if the irregular point $z$ is polar. However, this assumption is not made here. We proceed as follows:

Let $V_0$ be an open neighborhood of $z$ in $X$ such that $f$ vanishes on $U^* \cap V_0$. We denote by $u$ the restriction of $H_f$ to $U \cap V_0$ and define a Borel function $v$ on $V_0$ by

$$v(x)=\begin{cases} u(x), & x \in U \cap V_0, \\ 0, & x \in V_0 \setminus U. \end{cases}$$

Then we shall see that $v$ is nearly hypoharmonic on $V_0$ which means

$$v(x) \leqslant \int v \, d\mu_x^W \tag{2.2}$$

for all regular sets $W$ in $V_0$ and all $x \in W$. (1) This inequality is evident for $x \in W \setminus V$ since $f \geqslant 0$ and, consequently, $v \geqslant 0$. In order to treat the

(1) One could avoid the use of regular sets by assuming $W$ to be resolutive in the sense of [14].

remaining case $x \in W \cap U$, we observe that—according to the restriction lemma of [2], p. 130—one has

$$H_v^W \mid W \cap U = H_g^{W \cap U},$$

where the function $g$ on $(W \cap U)^*$ equals $v$ on $(W \cap U)^*$ and $H_v^W$ on $(W \cap U)^* \cap W$. Consequently, $g \geqslant v \mid (W \cap U)^*$ because of $H_v^W \geqslant 0 = v$ on $(W \cap U)^* \cap W$ which is subset of $V_0 \setminus U$. Now the inequality (2.2) follows also for $x \in W \cap U$:

$$v(x) = u(x) = H_f(x) = H_v^{W \cap U}(x) \leqslant H_g^{W \cap U}(x) = H_v^W(x) = \int v d\mu_x^W,$$

whereby the third equality sign follows from another application of the restriction lemma by observing again that $f$ and $v$ vanish on $U^* \cap W \subset U^* \cap V_0$. So we can regularize the nearly hypoharmonic, bounded function $v$ from above in order to obtain on $V_0$ the subharmonic function

$$v_0(x) = \limsup_{y \to x,\ y \in V_0} v(y) \qquad (x \in V_0). \tag{2.3}$$

The fine continuity of $v_0$ and the irregularity of $z$ then yield

$$v_0(z) = f - \lim_{x \to z,\ x \in U \cap V_0} v_0(x) = f - \lim_{x \to z,\ x \in U \cap V_0} v(x)$$
$$= f - \lim_{x \to z,\ x \in U \cap V_0} u(x)$$

so that

$$f - \lim_{x \to z,\ x \in U} H_f(x) = v_0(z).$$

But from (2.3) we can also conclude:

$$v_0(z) = \limsup_{x \to z,\ x \in U \cap V_0} v(x) = \limsup_{x \to z,\, x \in U \cap V_0} u(x)$$
$$= \limsup_{x \to z,\ x \in U} H_f(x),$$

just by observing the definition of $v$ as well as $u \geqslant 0$.

2.2. *Remarks.* — (1) If $H_f(x) > 0$ holds in the trace $U \cap V$ of a neighborhood $V$ of $z$ it follows that

$$\limsup_{x \to z} H_f(x) > 0.$$

In fact, otherwise $H_f$ would be a barrier for $z$ which contradicts the irregularity of $z$.

(2) The local character of Proposition 2.1 becomes clear if $H_f$ is replaced by a bounded harmonic function $u \geqslant 0$ defined in the trace $U \cap V$ of a neighborhood $V$ of $z$ which is 0-*associated* in the sense of SMYRNÉLIS [20]. This means $u = H_f^{U \cap V}$ with a function $f \geqslant 0$ on $(U \cap V)^*$ vanishing in a neighborhood of $z$. The restriction lemma implies that this property remains unchanged if $u$ is restricted to a trace smaller than $U \cap V$.

For a polar point $z \in U^*_{\text{irr}}$ we know from Remark 1.3 that

$$f-\lim_{x \to z} H_f(x)$$

exists for all bounded resolutive functions $f$ on $U^*$. Obviously, this fine limit depends linearly on $f$. We shall identify this linear form with the integral $\int f d\varepsilon_z^{\complement U}$ for all bounded resolutive functions on $U^*$. This result, presented as Theorem 2.4, is due to BRELOT [10], [10a] for the case of classical potential theory.

2.3. LEMMA. — *Let $\tau$ be a bounded positive Radon measure on $U$ such that $U$ is the smallest absorbing subset in the space $U$ containing the support of $\tau$ with respect to $U$. Let $\alpha > 0$ be a positive real number and $V_0$ a neighborhood of $z$ in $X$. Denote by $\Phi$ the set of all functions $\varphi : U^* \to [0, \alpha]$ vanishing on $U^* \cap V_0$. Then, given $\varepsilon > 0$, there exist $\delta > 0$ and a neighborhood $W$ of $z$ such that the lower generalized solution $\underline{H}_\varphi$ of the Dirichlet problem behaves as follows: Whenever $\varphi \in \Phi$ and $\int \underline{H}_\varphi d\tau < \delta$, one has*

$$\underline{H}_\varphi(x) < \varepsilon \quad \textit{for all} \quad x \in U \cap W.$$

Proof. — Let $h$ be a strictly positive harmonic function defined in an open neighborhood of $\bar{U}$ such that $h(x) \geqslant 1$ on $\bar{U}$, let $V$ be an open neighborhood of $z$ in $X$ satisfying $\bar{V} \subset V_0$, and define $\eta = \varepsilon/2\alpha$. Then there exists a compact set $K \subset V^* \cap U$ such that:

$$\mu_z^V(G) < \frac{\eta}{\beta}, \tag{2.4}$$

where $G=V^*\cap U\setminus K$ and $\beta=\sup h(\bar{U})$. To every $\varphi\in\Phi$ we associate a function $\varphi^*$ on $(U\cap V)^*$ by putting

$$\varphi^*(x)=\begin{cases}\underline{H}_\varphi(x), & x\in K\\ 0, & x\in(U\cap V)^*\setminus K.\end{cases}$$

We then have

$$\underline{H}_\varphi\leqslant H^{U\cap V}_{\varphi^*}+\alpha H^V_{h_G}\quad\text{on } U\cap V, \tag{2.5}$$

where $h_G$ equals $h$ on $G$ and zero elsewhere on $\bar{V}$. In fact, this can be seen as follows: According to the restriction lemma $\underline{H}_\varphi\,|\,U\cap V=\underline{H}^{U\cap V}_g$, where $g$ equals $\underline{H}_\varphi$ on $(U\cap V)^*\cap U$ and $\varphi$ on $(U\cap V)^*\cap U^*$. Obviously,

$$g\leqslant\varphi^*+\alpha h_G\quad\text{on } (U\cap V)^*$$

since $g=\varphi^*$ on $K$, $g=\underline{H}_\varphi\leqslant\alpha\underline{H}_1\leqslant\alpha h$ on $G$, and $g=\varphi^*=\varphi=0$ on $(U\cap V)^*\cap U^*$. Then (2.5) follows from

$$\underline{H}_\varphi=\underline{H}^{U\cap V}_g\leqslant\underline{H}^{U\cap V}_{\varphi^*+\alpha h_G}=H^{U\cap V}_{\varphi^*}+\alpha H^{U\cap V}_{h_G}\quad\text{on } U\cap V$$

since $\varphi^*$ and $h_G$ are resolutive and since

$$H^{U\cap V}_{h_G}\leqslant H^V_{h_G}\quad\text{on } U\cap V$$

according to the restriction lemma. The function $H^V_{1_G}$ is harmonic, hence continuous in $z$. So (2.4) implies the existence of a neighborhood $W\subset\bar{W}\subset V$ of $z$ such that

$$H^V_{1_G}(x)=\mu^V_z(G)<\frac{\eta}{\beta}\quad\text{for all } x\in W$$

and, consequently,

$$H^V_{h_G}(x)\leqslant\beta H^V_{1_G}(x)<\eta\quad\text{for all } x\in W. \tag{2.6}$$

We now apply the Harnack inequality [2], p. 35 to the harmonic functions $\underline{H}_\varphi\geqslant 0$. According to it, there exists a constant $k>0$ depending on $K$ and $\tau$ such that

$$\underline{H}_\varphi(x)\leqslant k\int\underline{H}_\varphi\,d\tau\qquad(x\in K,\ \varphi\in\Phi).$$

This prove that

$$\varphi^*\leqslant\left(k\int\underline{H}_\varphi\,d\tau\right)1_K\quad\text{on } (U\cap V)^*$$

and

$$\underline{H}_{\varphi^*}^{U \cap V} \leqslant \left(k \int \underline{H}_{\varphi}\, d\tau\right) H_{1_K}^{U \cap V} \leqslant \beta k \int \underline{H}_{\varphi}\, d\tau$$

on $U \cap V$ for all $\varphi \in \Phi$ (since $1 \leqslant h \leqslant \beta$ on $\bar{U}$). From (2.5) and (2.6) the result then follows: It suffices to choose besides $W$ the number $\delta = \varepsilon/2\beta k$.

Now we are in the position to prove the announced theorem:

2.4. THEOREM. – *Assume that $z \in U^*_{\text{irr}}$ is polar. Then every bounded resolutive function $f$ on $U^*$ is $\varepsilon_z^{\mathfrak{C}U}$-integrable and $H_f$ has $\int f d\varepsilon_z^{\mathfrak{C}U}$ as a fine limit at $z$:*

$$f-\lim_{x \to z} H_f(x) = \int f d\varepsilon_z^{\mathfrak{C}U}. \tag{2.7}$$

*Proof.* – We proceed in several steps:

1. Suppose $f \in \mathscr{C}(U^*)$. Then the result has been established in Proposition 1.1.

2. Suppose that $f$ is lower semicontinuous, $\geqslant 0$ and that $f$ vanishes in a neighborhood of $z$ (besides being bounded). Then $f$ is automatically $\varepsilon_z^{\mathfrak{C}U}$-integrable (and resolutive). We approximate $f$ by a sequence $(\varphi_n)$ in $\mathscr{C}(U^*)$ such that $\varphi_n \uparrow f$. The first step then yields

$$\int \varphi_n d\varepsilon_z^{\mathfrak{C}U} = f-\lim_{x \to z} H_{\varphi_n}(x) \leqslant f-\liminf_{x \to z} H_f(x),$$

which implies for $n \to \infty$:

$$\int f d\varepsilon_z^{\mathfrak{C}U} \leqslant f-\liminf_{x \to z} H_f(x). \tag{2.8}$$

By assumption $f$ vanishes on $U^* \cap V_0$ where $V_0$ is a certain open relatively compact neighborhood of $z$ in $X$. For the sake of simplicity of notations we assume that $f$ is defined on $U^* \cup V_0$ such that $f(x)=0$ on all of $V_0$. We then consider open neighborhoods $V_n$ of $z$ for $n=1, 2, \ldots$ such that $\bar{V}_{n+1} \subset V_n$ for all $n=0, 1, \ldots$ and $\bigcap V_n = \{z\}$. The restriction $f_n$ of $f$ to $(U \cup V_n)^*$ is then bounded $\geqslant 0$ and lower semicontinuous, hence also resolutive ($n=1, 2, \ldots$). According to the restriction lemma, we have

$H_{g_n}=H_{f_n}^{U\cup V_n}|U$ where $g_n$, equal to $H_{f_n}^{U\cup V_n}$ on $U^*\cap V_n$ and to $f$ on $U^*\setminus V_n$, is again resolutive ([2], p. 131). $f=0$ on $\bar{V}_n\subset V_0$ implies $f\leqslant g_n$ and, consequently,

$$H_f(x)\leqslant H_{f_n}^{U\cup V_n}(x)\quad \text{for all } x\in U.$$

From this we obtain

$$f\text{-}\limsup_{x\to z} H_f(x)\leqslant f\text{-}\lim_{x\to z,\, x\in U} H_{f_n}^{U\cup V_n}(x)=\lim_{x\to z} H_{f_n}^{U\cup V_n}(x)$$

$$=H_{f_n}^{U\cup V_n}(z)=\int f_n\, d\varepsilon_z^{\complement(U\cup V_n)}=\int f d\varepsilon_z^{\complement(U\cup V_n)}$$

$(n=1, 2, \ldots)$. Hence Lemma 1.4 enables us to pass to the limit for $n\to\infty$:

$$f\text{-}\limsup_{x\to z} H_f(x)\leqslant \int f d\varepsilon_z^{\complement U\setminus\{z\}}.$$

The last measure equals $\varepsilon_z^{\complement U}$ due to the polarity of $z$. Together with (2.8) we thus derived (2.7) for such an $f$.

3. Suppose that $f$ is upper semicontinuous and that $f$ vanishes in a neighborhood $V$ of $z$ in $U^*$ (besides being bounded, and thus $\varepsilon_z^{\complement U}$-integrable and resolutive). Let us assume $V$ to be closed in $U^*$. Then the function

$$g=\alpha 1_{U^*\setminus V}-f,$$

where $\alpha>0$ denotes an upper bound for $f$, is bounded, lower semicontinuous, $\geqslant 0$ and vanishes on $V$. Step 2 then yields

$$f\text{-}\lim_{x\to z} H_g(x)=\int g d\varepsilon_z^{\complement U},$$

as well as

$$f\text{-}\lim_{x\to z} H_{1_{U^*\setminus V}}(x)=\varepsilon_z^{\complement U}(U^*\setminus V).$$

By linearity, this implies again (2.7).

4. Suppose that the given function $f$ vanishes in a neighborhood of $z$. In addition to that we may assume $f\geqslant 0$ by reasons of linearity. Let $(x_n)$ be a sequence of points dense in $U$. Then

$$\tau=\sum_{n=1}^{\infty} 2^{-n}\varepsilon_{x_n},$$

is a probability measure on $U$ with support (with respect to $U$) equal to $U$. As before we consider a positive harmonic function defined in an open neighborhood of $\bar{U}$ which satisfies $1 \leqslant h(x) \leqslant \beta < +\infty$ on $\bar{U}$. We then have

$$(2.9) \qquad \|\mu_x^H\| \leqslant \int h d\mu_x^U = h(x) \leqslant \beta \qquad (x \in U)$$

for the total mass of the harmonic measures $\mu_x^U$, so that

$$\tau^* = \sum_{n=1}^{\infty} 2^{-n} \mu_{x_n}^U$$

is a positive measure on $U^*$. According to BOURBAKI [9], p. 28 and the resolutivity theorem [2], p. 127 the given function $f$ is $\tau^*$-integrable. For a given number $\delta > 0$, this implies the existence of an upper semicontinuous function $\varphi_1$ and a lower semicontinuous function $\varphi_2$ on $U^*$ such that

$$0 \leqslant \varphi_1 \leqslant f \leqslant \varphi_2 \leqslant \alpha$$

and

$$\int (\varphi_2 - \varphi_1)\, d\tau^* < \delta,$$

where $\alpha > 0$ denotes an upper bound for $f$. In addition to that we may assume that $\varphi_1$ and $\varphi_2$ vanish in a neighborhood of $z$. Indeed, let $V$ be an open neighborhood of $z$ in $U^*$ such that $f$ vanishes on $\bar{V}$; it suffices to replace $\varphi_1$ and $\varphi_2$ by $\varphi_1 1_{U^* \setminus V}$ and $\varphi_2 1_{U^* \setminus \bar{V}}$, respectively. We then have:

$$\int (\varphi_2 - \varphi_1)\, d\tau^* = \sum_{n=1}^{\infty} 2^{-n} \int (\varphi_2 - \varphi_1)\, d\mu_{x_n}^U$$

$$= \sum_{n=1}^{\infty} 2^{-n} (H_{\varphi_2}(x_n) - H_{\varphi_1}(x_n)) = \int H_{\varphi_2 - \varphi_1}\, d\tau.$$

Given $\varepsilon > 0$, we choose the above $\delta > 0$ as well as a neighborhood $W$ of $z$ according to Lemma 2.3. We can then conclude that

$$(2.10) \qquad H_{\varphi_2 - \varphi_1}(x) < \varepsilon \quad \text{for all } x \in W \cap U.$$

From the preceeding step we know that

$$f - \lim_{x \to z} H_{\varphi_i}(x) = \int \varphi_i\, d\varepsilon_z^{\mathfrak{C}U} \qquad (i = 1, 2).$$

Hence we obtain from (2.10) for $x \to z$:

$$\int (\varphi_2 - \varphi_1)\, d\varepsilon_z^{\mathfrak{C}U} \leqslant \varepsilon.$$

Since $\varphi_1 \leqslant f \leqslant \varphi_2$ we conclude from $H_{\varphi_1} \leqslant H_f \leqslant H_{\varphi_2}$:

$$\int \varphi_1\, d\varepsilon_z^{\mathfrak{C}U} \leqslant f - \liminf_{x \to z} H_f(x) \leqslant f - \limsup_{x \to z} H_f(x) \leqslant \int \varphi_2\, d\varepsilon_z^{\mathfrak{C}U}$$

$\varepsilon > 0$ being arbitrary, these two inequalities lead to the $\varepsilon_z^{\mathfrak{C}U}$-integrability of $f$ as well as to the equality (2.7).

5. The final step will concern the general case. Let $V_n$ be a sequence of open neighborhoods of $z$ in $U^*$ which shrink decreasingly to $z$. We define two sequences $(f_n)$ and $(g_n)$ of bounded functions on $V^*$ as follows:

$$f_n(y) = g_n(y) = f(y) \qquad \text{for} \quad y \in U^* \setminus V_n,$$
$$f_n(y) = \inf f(U^*) \qquad \text{and} \qquad g_n(y) = \sup f(U^*) \quad \text{for } y \in U^* \cap V_n.$$

According to the resolutivity theorem [2], p. 127, the functions $f_n$ and $g_n$ are resolutive. From steps 1 and 4 we obtain by linearity the $\varepsilon_z^{\mathfrak{C}U}$-integrability of $f_n$, $g_n$ as well as

$$f - \lim_{x \to z} H_{f_n}(x) = \int f_n\, d\varepsilon_z^{\mathfrak{C}U}$$

and

$$f - \lim_{x \to z} H_{g_n}(x) = \int g_n\, d\varepsilon_z^{\mathfrak{C}U} \qquad (n \in \mathbb{N}).$$

Since $f_n \leqslant f \leqslant g_n$ this implies

$$\int f_n\, d\varepsilon_z^{\mathfrak{C}U} \leqslant f - \liminf_{x \to z} H_f(x) \leqslant f - \limsup_{x \to z} H_f(x) \leqslant \int g_n\, d\varepsilon_z^{\mathfrak{C}U}.$$

From this the claim follows by Lebesgue's dominated convergence theorem: both sequences are uniformly bounded; furthermore, $f_n \uparrow f$ and $g_n \downarrow g$ on $U^* \setminus \{z\}$ and hence $\varepsilon_z^{\mathfrak{C}U}$-almost everywhere on $U^*$. In fact, $\varepsilon_z^{\mathfrak{C}U}$ does not charge the polar point $z$ according to a result of Constantinescu presented in [14], p. 171. (A more elementary proof of this result can be given by modifying a technique of Brelot which in [12], p. 140 leads to the result mentioned in Remark 1.3.)

2.5. *Remark.* — The importance of Theorem 2.4 lies in the fact that the calculation of $f-\lim_{x \to z} H_f(x)$ becomes possible whenever the measure $\varepsilon_z^{\complement U}$ is known. This is the case for the heat equation when $U$ is the *heat ball* studied in [6] and $z$ its only irregular (necessarily polar) boundary point. According to the main result of [6], $\varepsilon_z^{\complement U}$ is then the *Fulks measure* of $U$.

## 3. Maximal sequences

The idea of maximal sequences is due to BRELOT [11] who has introduced this concept in the framework of classical potential theory. We shall start with a definition which is the most natural one and which will soon turn out to be equivalent to the original definition.

3.1. DEFINITION. — *A sequence* $(x_n)$ *of points* $x_n \in U$ *will be called maximal (with respect to a given boundary point* $z \in U^*_{\mathrm{irr}}$*) if* $\lim_{n \to \infty} x_n = z$ *(in the topology of* $X$*) and if*

$$\lim_{n \to \infty} \mu_{x_n}^U = \varepsilon_z^{\complement U}$$

*(in the sense of vague convergence).*

It follows from Lemma 1.2 that there always exist maximal sequences for a point $z \in U^*_{\mathrm{irr}}$. This, in fact, is the statement of the Köhn-Sieveking-Lemma mentioned above.

We proceed immediately to a characterization of maximal sequences in order to clarify the relation to Brelot's original definition.

3.2. THEOREM. — *A sequence* $(x_n)$ *in* $U$ *converging to* $z \in U^*_{\mathrm{irr}}$ *is maximal with respect to* $z$ *if and only if there exists a hyperharmonic function* $u \geqslant 0$ *on* $X$ *having the following three properties:*

(3.1) $$u \mid U^* \in \mathscr{C}(U^*);$$

(3.2) $$\hat{R}_u^{\complement U}(z) < u(z);$$

(3.3) $$\lim_{n \to \infty} \hat{R}_u^{\complement U}(x_n) = \hat{R}_u^{\complement U}(z).$$

Before giving the proof let us remark that, due to the thinness of $\complement U$ at $z$, condition (3.3) is equivalent to

(3.3′) $$\lim_{n \to \infty} \hat{R}_u^{\complement U}(x_n) = \liminf_{x \to y,\, x \in U} \hat{R}_u^{\complement U}(x).$$

In fact, since $\hat{R}_u^{\complement U}$ is hyperharmonic, one has

$$\liminf_{x \to z} \hat{R}_u^{\complement U}(x) = \hat{R}_u^{\complement U}(z)$$

and, consequently,

$$\liminf_{x \to z,\ x \in U} \hat{R}_u^{\complement U}(x) = \hat{R}_u^{\complement U}(z). \tag{3.4}$$

Otherwise,

$$\liminf_{x \to z,\ x \in U} \hat{R}_u^{\complement U}(x) > \hat{R}_u^{\complement U}(z)$$

would imply that $U$ is thin at $z$. Since also $\complement U$ is thin at $z$ this would have the consequence that $X = U \cup \complement U$ is thin at $z$. This, however, is impossible. The equivalence of (3.3) and (3.3′) thus follows from (3.4).

*Proof* of the Theorem. – If the sequence $(x_n)$ is maximal we choose a strict potential $p \in \mathscr{C}(X)$ in the sense of [14], p. 166. Properties (3.1)-(3.3) are then evident for $u = p$. Conversely, let us assume that these properties hold for some hyperharmonic function $u \geqslant 0$ on $X$. We know from (2.9) that the total masses of the harmonic measures $\mu_x^U$, $x \in U$, are uniformly bounded. Consequently, there exists a subsequence $(x_{n_k})$ of the original sequence $(x_n)$ such that $(\mu_{x_{n_k}}^U)$ converges vaguely to some measure $\lambda \in \mathscr{M}_+(U^*)$. We know from a result of Boboc-Cornea [8] – for a simple proof see Lukeš-Malý [19], p. 363 and Hansen [17], p. 439 – that $\lambda$ is of the form $\lambda = \alpha\varepsilon_z + (1-\alpha)\varepsilon_z^{\complement U}$ for some $\alpha \in [0, 1]$. Since $u \,|\, U^*$ is in $\mathscr{C}(U^*)$, we have

$$\lim_{k \to \infty} \hat{R}_u^{\complement U}(x_{n_k}) = \lim_{k \to \infty} \int u \, d\mu_{x_{n_k}}^U = \int u \, d\lambda.$$

Therefore, according to (3.3)

$$\int u \, d\lambda = \hat{R}_u^{\complement U}(z) = \int u \, d\varepsilon_z^{\complement U}$$

and hence

$$\alpha\left[u(z) - \int u \, d\varepsilon_z^{\complement U}\right] = 0.$$

(3.2) then yields $\alpha = 0$, i.e. $\lambda = \varepsilon_z^{\complement U}$. So $\varepsilon_z^{\complement U}$ is the only vague cluster value of the sequence $(\mu_{x_n}^U)$ and, consequently, $\lim \mu_{x_n}^U = \varepsilon_z^{\complement U}$. Therefore the sequence is maximal.

Let us add that, according to what we have proved, a sequence $(x_n)$ in $U$ converging to $z \in U^*_{\text{irr}}$ is *maximal if and only if*

$$(3.5) \qquad \lim_{n \to \infty} \hat{R}_p^{\mathfrak{C}U}(x_n) = \hat{R}_p^{\mathfrak{C}U}(z),$$

*holds for some strict potential* $p \in \mathscr{C}(X)$ (and then for all such $p$).

3.4. *Example.* – Let $U$ be a bounded domain in $\mathbb{R}^p$, $p \geqslant 2$, and let $X$ be a bounded domain in $\mathbb{R}^p$ such that $\bar{U} \subset X$. (For $p \geqslant 3$ we could also take $X = \mathbb{R}^p$.) Then $X$ is a strong harmonic space with respect to harmonic functions in the classical sense, i.e. solutions of the Laplace equation. Denote by $G_{y_0}$ the Green function of $U$ with pole $y_0 \in U$. The original definition of Brelot [11] calls a sequence $(x_n)$ in $U$ converging to $z \in U^*_{\text{irr}}$ maximal if:

$$(3.6) \qquad \lim_{n \to \infty} G_{y_0}(x_n) = \limsup_{x \to z,\, x \in U} G_{y_0}(x),$$

for some (and then for all) $y_0 \in U$.

There is identity between the two notions of maximal sequences: For simplicity's sake we restrict ourselves to the case $p \geqslant 3$. Minor modifications allow the treatment of the case $p = 2$.

We have:

$$(3.7) \qquad G_{y_0} = N_{y_0} - \hat{R}_{N_{y_0}}^{\mathfrak{C}U} = N_{y_0} - H_{N_{y_0}} \quad (\text{on } U),$$

where $N_{y_0}(x) = \|x - y_0\|^{2-p}$ is the Newtonian kernel. Consequently, according to (3.4):

$$(3.8) \qquad \limsup_{x \to z,\, x \in U} G_{y_0}(x) = N_{y_0}(z) - \hat{R}_{N_{y_0}}^{\mathfrak{C}U}(z).$$

We now see that every maximal sequence for $z \in U^*_{\text{irr}}$ in Brelot's sense is maximal in our sense. In fact, for $u = N_{y_0}$ (3.1) is evident and (3.3) follows from (3.6) and (3.8); (3.8) yields also (3.2) since $\limsup_{x \to z,\, x \in U} G_{y_0}(x) > 0$ characterizes the irregularity of a boundary point $z$. The converse direction follows from (3.8) since:

$$\lim_{n \to \infty} G_{y_0}(x_n) = N_{y_0}(z) - \lim_{n \to \infty} \int N_{y_0}\, d\mu_{x_n}^U$$

$$= N_{y_0}(z) - \int N_{y_0}\, d\varepsilon_z^{\mathfrak{C}U} = N_{y_0}(z) - \hat{R}_{N_{y_0}}^{\mathfrak{C}U}(z),$$

according to the definition (3.7) of $G_{y_0}$, whenever $(x_n)$ is maximal for $z$ in our sense.

We return now again to the general situation. Maximality of a sequence $(x_n)$ in $U$ with respect to $z \in U^*_{\text{irr}}$ turns out to be a *local property:*

3.5. PROPOSITION. — *Let $(x_n)$ be a sequence in $U$ converging to a point $z \in U^*$, and let $V$ be an open neighborhood of $z$ with trace $W = V \cap U$ in $U$. Then the following two conditions are equivalent:*

(i) *$z \in U^*_{\text{irr}}$ and $(x_n)$ is maximal with respect to $z$ and $U$.*

(ii) *$z \in U^*_{\text{irr}}$ and $(x_n)$ is maximal with respect to $z$ and $W$ ([2]).*

*Proof.* — The equivalence of the conditions $z \in U^*_{\text{irr}}$ and $z \in W^*_{\text{irr}}$ is well-known [2], p. 132. The rest follows from a result of HANSEN [17], p. 440, according to which the conditions

$$\lim_{n \to \infty} \varepsilon_{x_n}^{\complement U} = \varepsilon_z^{\complement U} \qquad \text{and} \qquad \lim_{n \to \infty} \varepsilon_{x_n}^{\complement W} = \varepsilon_z^{\complement W}$$

are equivalent.

The importance of maximal sequences will become evident from the following Theorem 3.7. For the case of classical potential theory, this result is due to BRELOT [11]. We start with a lemma which completes Theorem 2.4.

3.6. LEMMA. — *Let $(x_n)$ be a maximal sequence in $U$ with respect to the irregular point $z \in U^*$ which is assumed to be polar. Then:*

$$\lim_{n \to \infty} H_f(x_n) = f - \lim_{x \to z} H_f(x) = \int f d\varepsilon_z^{\complement U}. \tag{3.11}$$

*for every bounded resolutive function $f$ on $U^*$.*

*Proof.* — 1. For $f \in \mathscr{C}(U^*)$ the result follows from Proposition 1.1.

2. Suppose that $f$ is lower semicontinuous $\geqslant 0$ and vanishes in a neighborhood of $z$. As in the proof of Theorem 2.4 we choose a sequence $(\varphi_n)$ in $\mathscr{C}(U^*)$ such that $\varphi_n \uparrow f$. From $H_{\varphi_n} \leqslant H_f$ it follows that

$$\int \varphi_n \, d\varepsilon_z^{\complement U} \leqslant \liminf_{k \to \infty} H_f(x_k) \quad \text{for all } n.$$

([2]) More precisely: finitely many $x_n$ have to be left away, namely those for which $x_n \in U \setminus W$.

Consequently,

$$\int f d\varepsilon_z^{\mathfrak{C}U} \leqslant \liminf_{k\to\infty} H_f(x_k). \tag{3.12}$$

We denote by $V_0$ an open neighborhood of $z$ in $X$ such that $f$ vanishes on $U^* \cap V_0$. We extend $f$ to $U^* \cup V_0$ by putting $f(x)=0$ for all $x \in V_0$. The new function $f$ is again lower semicontinuous $\geqslant 0$ on $U^* \cup V_0$. Exactly as in the corresponding part of the proof of Theorem 2.4 we introduce the sequences $(V_n)$ and $(f_n)$ which will lead to the inequalities $H_f \leqslant H_{f_n}^{U\cup V_n}$ valid on $U$. We then have

$$\limsup_{k\to\infty} H_f(x_k) \leqslant \limsup_{k\to\infty} H_{f_n}^{U\cup V_n}(x_k) = H_{f_n}^{U\cup V_n}(z) = \int f d\varepsilon_z^{\mathfrak{C}(U\cup V_n)},$$

for all $n=1, 2, \ldots$. As in the proof of Theorem 2.4 we conclude that

$$\limsup_{k\to\infty} H_f(x_k) \leqslant \int f d\varepsilon_z^{\mathfrak{C}U}. \tag{3.13}$$

The result then follows from (3.12) and (3.13) by applying Theorem 2.4.

3. Next we assume that $f$ vanishes in a neighborhood of $z$ (besides being bounded and resolutive). If $f \geqslant 0$ we consider the set $\mathscr{F}$ of all bounded lower semicontinuous functions $\varphi$ on $U^*$ which satisfy $\varphi \geqslant f$ and vanish on a given open neighborhood $V$ of $z$ in $U^*$ on which $f$ vanishes. We then have $H_f \leqslant H_\varphi$ on $U$ and, according to step 2,

$$\limsup_{n\to\infty} H_f(x_n) \leqslant \int \varphi \, d\varepsilon_z^{\mathfrak{C}U} \quad \text{for all } \varphi \in \mathscr{F}.$$

This implies

$$\limsup_{n\to\infty} H_f(x_n) \leqslant \int^* f d\varepsilon_z^{\mathfrak{C}U}. \tag{3.14}$$

This inequality remains valid if the assumption $f \geqslant 0$ is dropped. Indeed, we can add to $f$ a function $\psi \geqslant 0$ from $\mathscr{C}(U^*)$ vanishing in a neighborhood of $z$ (contained in $V$) such that $f+\psi \geqslant 0$. From step 1 we know

$$\lim_{n\to\infty} H_\psi(x_n) = \int \psi \, d\varepsilon_z^{\mathfrak{C}U};$$

furthermore, (3.14) holds for $f+\psi$ instead of $f$. By linearity we are then led to (3.14) for the function $f$. By replacing $f$ by $-f$ in (3.14) the result follows according to Theorem 2.4.

4. Now let $f$ be an arbitrary bounded resolutive function on $U^*$. We proceed as in step 5 of the proof of Theorem 2.4 and repeat the definition of the sequences $(f_n)$ and $(g_n)$ given there. Since each function $f_n$ and $g_n$ is constant in a neighborhood of $z$, we conclude from steps 1 and 3 that

$$\lim_{k \to \infty} H_{f_n}(x_k) = \int f_n \, d\varepsilon_z^{\complement U} \qquad \text{and} \qquad \lim_{k \to \infty} H_{g_n}(x_k) = \int g_n \, d\varepsilon_z^{\complement U}$$

for all $n \in \mathbb{N}$. Consequently, we obtain from $f_n \leqslant f \leqslant g_n$:

$$\int f_n \, d\varepsilon_z^{\complement U} \leqslant \liminf_{k \to \infty} H_f(x_k) \leqslant \limsup_{k \to \infty} H_f(x_k) \leqslant \int g_n \, d\varepsilon_z^{\complement U}$$

for all $n \in \mathbb{N}$. From this and Theorem 2.4 the result follows since $(f_n)$ and $(g_n)$ converge $\varepsilon_z^{\complement U}$-almost everywhere on $U^*$ to $f$ as the argument at the end of the proof of Theorem 2.4 shows.

The theorem, already announced, is the following:

3.7. Theorem. — *Let $u$ be a bounded harmonic function on $U$ and suppose that $z \in U_{\mathrm{irr}}^*$ is polar. Then $f\text{-}\lim_{x \to z} u(x)$ exists and*

$$\lim_{n \to \infty} u(x_n) = f\text{-}\lim_{x \to z} u(x), \tag{3.15}$$

*for every maximal sequence $(x_n)$ in $U$ for the boundary point $z$.*

*Proof.* [4]. — Since we have a strictly positive harmonic function at our disposition which is defined in a neighborhood of $\bar{U}$, we may assume $u \geqslant 0$ and then even $0 \leqslant u \leqslant 1$ on $U$. Let $V$ denote an open relatively compact neighborhood of $z$ in $X$ and let $W = U \cap V$ be its trace in $U$. We shall study the Dirichlet problem for the two functions $\varphi$, $\psi$ defined on $W^*$ as follows: $\varphi = \psi = u$ on $W^* \cap U$ and $\varphi = 0$, $\psi = 1$ on $W^* \cap U^*$. These are bounded semicontinuous (hence resolutive) functions $\geqslant 0$ on $W^*$ (which are constant in a neighborhood of $z$). We have

$$H_\varphi^W(x) \leqslant u(x) \leqslant H_\psi^W(x) \qquad (x \in W) \tag{3.16}$$

since $u$ is an upper (resp. lower) function for $\varphi$ (resp. $\psi$). Furthermore,

---

[3] Similar arguments have been used by Smyrnélis [20], p. 172.

$$H_{\psi}^{W} - H_{\varphi}^{W} = H_{1_{A_V}}^{W}, \tag{3.17}$$

where $A_V = W^* \cap U^*$. According to the local character (Lemma 3.5) of maximal sequences and to Lemma 3.6, the following limits exist:

$$\alpha_V = \lim_{n \to \infty} H_{\varphi}^{W}(x_n) = f - \lim_{x \to z} H_{\varphi}^{W}(x)$$

as well as

$$\beta_V = \lim_{n \to \infty} H_{\psi}^{W}(x_n) = f - \lim_{x \to z} H_{\psi}^{W}(x).$$

Therefore, (3.16) implies

$$\alpha_V \leqslant \liminf_{n \to \infty} u(x_n) \leqslant \limsup_{n \to \infty} u(x_n) \leqslant \beta_V$$

as well as

$$\alpha_V \leqslant f - \liminf_{x \to z} u(x) \leqslant f - \limsup_{x \to z} u(x) \leqslant \beta_V.$$

From these two series of inequalities the result will follow if $\beta_V - \alpha_V$ can be made arbitrarily small for appropriate choices of $V$. This can be seen as follows: According to the restriction lemma we have $H_{1_{A_V}}^{U} \mid W = H_{g}^{W}$ with a boundary function $g$ on $W^*$ which is $\geqslant 1_{A_V}$ on $W^*$. Consequently,

$$H_{1_{A_V}}^{U}(x) \geqslant H_{1_{A_V}}^{W}(x) \quad \text{for all } x \in W.$$

An application of Theorem 2.4 in connection with (3.17) yields:

$$\varepsilon_z^{\mathfrak{c}U}(A_V) = f - \lim_{x \to z} H_{1_{A_V}}^{U}(x) \geqslant \beta_V - \alpha_V.$$

But since the sets $A_V$ shrink to $\{z\}$ along with $V$, it suffices to prove that the measure $\varepsilon_z^{\mathfrak{c}U}$ does not charge the point $z$. This however is a fact which we have used already at the end of the proof of Theorem 2.4.

## 4. Maximal sequences for the heat equation

We close with an application which will show that the natural analogue of Brelot's original definition of maximal sequences in the case of classical potential theory—recalled in Example 3.4—is valid in the case of the heat equation:

$$\Delta u - \frac{\partial u}{\partial t} = 0, \tag{4.1}$$

on $\mathbb{R}^{p+1}=\mathbb{R}^p\times\mathbb{R}$, $p\geqslant 1$. More precisely, we consider the harmonic structure on $\mathbb{R}^{p+1}$ defined by the classical solutions of (4.1). $\mathbb{R}^{p+1}$ is then a strong harmonic space of which all points are polar. Harmonic and superharmonic functions on this harmonic space are usually called *caloric* and *supercaloric functions*, respectively.

The time (or second) coordinate of a point $x\in\mathbb{R}^p\times\mathbb{R}$ will be denoted by ord $x$. Given $\tau\in\mathbb{R}$, we denote by $A_\tau$ the open half space of all $x\in\mathbb{R}^{p+1}$ satisfying ord $x>\tau$. We denote by $W$ the fundamental solution of (4.1) with pole at the origin:

$$W(x)=\begin{cases}(4\pi\tau)^{-p/2}\exp\left[-\dfrac{\|\xi\|^2}{4\tau}\right], & \text{if } \tau>0,\\ 0, & \text{if } \tau\leqslant 0,\end{cases}$$

where $x=(\xi,\tau)$, $\tau=\text{ord } x$ and where $\|\xi\|$ stands for the euclidean norm of a point $\xi\in\mathbb{R}^p$. By translation we obtain the fundamental solution with pole $y_0\in\mathbb{R}^{p+1}$.

$$W_{y_0}(x)=W(x-y_0)\qquad (x\in\mathbb{R}^{p+1}).$$

In what follows we consider a fixed *bounded domain* $U$ in $\mathbb{R}^{p+1}$ and an *irregular boundary point* $z$ of $U$. The Green function of $U$ with pole at $y_0\in U$ is the function

$$G_{y_0}=W_{y_0}-H^U_{W_{y_0}},$$

defined on $U$.

4.1. Lemma. — *Whenever the pole* $y_0\in U$ *satisfies* ord $y_0<$ord $z$, *one has*

$$\limsup_{x\to z} G_{y_0}(x)=W_{y_0}(z)-\int W_{y_0}\,d\varepsilon_z^{\complement U}.$$

*Proof.* — We choose $\tau$ in the open interval $]$ord $y_0$, ord $z[$. Then $A_\tau$ is an open neighborhood of $z$ in $\mathbb{R}^{p+1}$. We put $V=U\cap A_\tau$ and define on $V^*$ the function

$$f(z')=\begin{cases}G_{y_0}(z'), & z'\in V^*\cap U,\\ 0, & z'\in V^*\cap U^*.\end{cases}$$

Obviously, $f\geqslant 0$ and $f(z')=0$ in a neighborhood of $z$ in $V^*$, namely, on

$U^* \cap A_\tau$. $f$ is also bounded and Borel measurable, hence resolutive. According to Proposition 2.1, this implies

(4.2) $$f-\lim_{x \to z,\ x \in V} H_f^V(x) = \limsup_{x \to z,\ x \in V} H_f^V(x).$$

With the abbreviation $u = W_{y_0} | U^*$ we introduce the function

$$v(z') = \begin{cases} H_u(z'), & z' \in V^* \cap U, \\ u(z'), & z' \in V^* \cap U^*, \end{cases}$$

defined on $V^*$. The restriction lemma yields that $v$ is resolutive and $H_v^V = H_u | V$. Since $f = W_{y_0} | V^* - v$, we conclude that

$$H_f^V(x) = H_{W_{y_0}}^V(x) - H_v^V(x) = G_{y_0}(x) \qquad (\text{for all } x \in V)$$

by observing that $W_{y_0}$ is caloric on $A_{\text{ord}\, y_0}$, hence on an open neighborhood of $\bar{V}$. Therefore, it follows from (4.2) that:

$$W_{y_0}(z) - f - \lim_{x \to z} H_u(x) = \limsup_{x \to z} G_{y_0}(x),$$

which, in view of Proposition 1.1, is the result in question.

The result generalizing Brelot's original definition of maximal sequences is now the following:

4.2. Proposition. — *Assume that $G_{y_0}$ is strictly positive in the trace $W \cap U$ of a neighborhood $W$ of $z$ in $\mathbb{R}^{p+1}$. Then a sequence $(x_n)$ in $U$ converging to $z$ is maximal with respect to $z$ if and only if*

$$\lim_{n \to \infty} G_{y_0}(x_n) = \limsup_{x \to z} G_{y_0}(x).$$

*Proof.* — According to the assumption we must have

(4.3) $$\limsup_{x \to z} G_{y_0}(x) > 0,$$

since otherwise $G_{y_0}$ would be a barrier for $z$. This implies ord $y_0 <$ ord $z$, the hypothesis of Lemma 4.1. In fact, there exist sequences $(x'_n)$ in $U$ converging to $z$ such that $\lim G_{y_0}(x'_n) = \limsup_{x \to z} G_y(x)$. Since

$$0 \leqslant G_{y_0}(x'_n) \leqslant W_{y_0}(x'_n) \qquad (n \in \mathbb{N})$$

and since (ord $x'_n$) converges to ord $z$, the assumption ord $z \leqslant$ ord $y_0$ would lead to $\lim G_{y_0}(x'_n) = 0$ which contradicts (4.3). Consequently, Lemma 4.1 leads to the inequality

$$\int W_{y_0}\, d\varepsilon_z^{\complement U} < W_{y_0}(z). \tag{4.4}$$

This proves that the supercaloric function $u = W_{y_0}$ satisfies condition (3.2) of Theorem 3.2. Since $u \mid U^*$ is continuous, condition (3.1) is also satisfied. So what remains is the verification of condition (3.3); the result then follows from Theorem 3.2. However, condition (3.3) is equivalent to

$$\lim_{n \to \infty} H_u(x_n) = \int u\, d\varepsilon_z^{\complement U},$$

hence to

$$\lim_{n \to \infty} G_{y_0}(x_n) = u(z) - \int u\, d\varepsilon_z^{\complement U}.$$

Consequently, another application of Lemma 4.1 yields the equivalence of (3.3) with the condition

$$\lim_{n \to \infty} G_{y_0}(x_n) = \limsup_{x \to z} G_{y_0}(x)$$

appearing in the Proposition.

4.3. *Remark.* – In concrete cases, the positivity hypothesis of Proposition 4.2 is easy to check. It amounts to a suitable choice of the pole $y_0 \in U$ depending on the given point $\hat{R}_u^{\complement U}$. One has to use the known fact (*cf.* WATSON [21] and DOOB [15], p. 300) that $G_{y_0}(x) > 0$ holds for a point $x \in U$ if and only if there exists a continuous path $\gamma : [0, 1] \to U$ in $U$ connecting $x = \gamma(0)$ with $y_0 = \gamma(1)$ in such a way that the function ord $\circ \gamma$ is strictly decreasing.

## REFERENCES


[1] BAUER (H.). – Zur Theorie des Riemann-Integrals in lokalkompakten Räumen, *Sitz.-Ber. math.-naturwis. Klasse Bayer. Akad. Wissenschaften München*, 1955, p. 187-208.

[2] BAUER (H.). – *Harmonische Räume und ihre Potentialtheorie*, Berlin, Heidelberg, New York, Springer-Verlag, 1966, *Lecture Notes in Mathematics*, Vol. 22.

[3] BAUER (H.). – Harmonische Räume. *Jahrbuch Überblicke Mathematik*, 1981, pp. 9-35, Mannheim, Wien, Zürich, Bibliographisches Institut, 1981.

[4] BAUER (H.). – *Harmonic spaces. A survey. Conferenze del Seminario di matematica dell'Università di Bari*, Vol. 197, 1984.

[5] BAUER (H.). – *Probability theory and elements of measure theory*, 2^e^ édition, London, New York, Toronto [etc.], Academic Press, 1981.

[6] BAUER (H.). – Heat balls and Fulks measures, *Ann. Acad. Scient. Fennicae*, serie A 1, Math., T. 10, 1985, pp. 317-332.

[7] BILLINGSLEY (P.). – *Convergence of probability measures*, London, Sydney, Toronto, John Wiley and Sons, 1968, *Wiley Series in Probability and mathematical Statistics.*

[8] BOBOC (N.) et CORNEA (A.). – Comportement des balayées des mesures ponctuelles. Comportement des solutions du problème de Dirichlet aux points irréguliers, *C. R. Acad. Sc.*, Paris, Série A, T. 264, 1967, pp. 995-997.

[9] BOURBAKI (N.). – *Éléments de mathématiques*, livre V, *Intégration*, chap. 5 : Intégration des mesures, 2^e^ édition, Hermann, Paris, 1967, *Actualités scientifiques et industrielles*, Vol. 1244.

[10] BRELOT (M.). – Sur les ensembles effilés, *Bull. Sc. math.*, 2^e^ série, Paris, T. 68, 1944, pp. 12-36.

[10a] BRELOT (M.). – Sur l'allure des fonctions harmoniques à la frontière. *C. R. Acad. Sc.*, Paris, T. 220, 1945, pp. 676-678.

[11] BRELOT (M.). – On the behavior of harmonic functions in the neighborhood of an irregular boundary point, *J. Anal. math.*, Jérusalem, T. 4, 1954-1956, pp. 209-221.

[12] BRELOT (M.). – *Lectures on potential theory*, 2^e^ édition, Bombay, Tata Institute of fundamental Research, 1967, *Tata Institute of fundamental Research, Lectures on Mathematics*, Vol. 19.

[13] BRELOT (M.). – *On topologies and boundaries in potential theory*, Berlin, Heidelberg, New York, Springer-Verlag, 1971, *Lecture Notes in Mathematics*, Vol. 175.

[14] CONSTANTINESCU (C.) and CORNEA (A.). – *Potential theory on harmonic spaces*, Berlin, Heidelberg, New York, Springer-Verlag, 1972, *Grundlehren der mathematischen Wissenschaften*, Vol. 158.

[15] DOOB (J.-L.). – *Classical potential theory and its probabilistic counterpart*, Berlin, Heidelberg, New York, Springer-Verlag, 1984, *Grundlehren der mathematischen Wissenschaften*, Vol. 262.

[16] EFFROS (E. G.) and KAZDAN (J.-L.). – Applications of Choquet simplexes to elliptic and parabolic boundary value problems, *J. of differential equations*, T. 8, 1970, pp. 95-134.

[17] HANSEN (W.). – Convergence of balayage measures, *Math. Annalen*, T. 264, 1983, pp. 437-446.

[18] KÖHN (J.) und SIEVEKING (M.). – Reguläre und extremale Randpunkte in der Potentialtheorie, *Revue roumaine de Math. pures et appl.*, T. 12, 1967, pp. 1489-1502.

[19] LUKEŠ (J.) and MALÝ (J.). – On the boundary behaviour of the Perron generalized solution, *Math. Annalen*, T. 257, 1981, pp. 355-366.

[20] SMYRNELIS (E. P.). – Sur les limites fines des fonctions harmoniques et les suites maximales, *Bull. Sc. math.*, Paris, T. 97, 1973, pp. 161-175.

[21] WATSON (N. A.). – Green functions, potentials, and the Dirichlet problem for the heat equation, *Proc. London math. Soc.*, 3rd series, T. 33, 1976, pp. 251-298.

# Simplices in potential theory [S24]

Aspects of Positivity in Functional Analysis, North-Holland Math. Stud. 122, Amsterdam 1986, 27–39
[Zbl. Math. 621.31008; MR 87k:31011]

## INTRODUCTION

Potential theory is perhaps the most prominent example of a branch of analysis in which positivity plays a fundamental role. Also the notion of a simplex - more precisely, of a Choquet simplex or an infinite-dimensional simplex - grew out of potential theory. In this survey lecture, a simplex will always be understood as a *metrizable* compact convex set $Y$ such that, for every point $y \in Y$, there exists exactly one probability measure $\mu$ on $Y$ which is carried by the set ex $Y$ of the extreme points of $Y$ and which has $y$ as its barycenter.

In 1941 R.S.MARTIN [24] discovered what is now called the Martin boundary of a (bounded) euclidean domain. By means of this boundary, one can prove an exact analogue to the classical Poisson integral formula. In today's language Martin proved that for every (bounded) domain $\Omega \subset \mathbb{R}^p$ and a given point $x_o \in \Omega$ the set $Y$ of positive harmonic functions $h$ defined on $\Omega$ and normalized by the condition $h(x_o) = 1$ is a simplex in the topology of uniform convergence on compact sets. To see this, the vision of a mathematician like H.Cartan was necessary: He formulated the fundamental existence problem of integral representation in compact convex sets. Ten years after Martin's discovery, R.GODEMENT - obviously influenced by Cartan - mentioned this problem in a paper on harmonic analysis [18; p.119] by saying: "... si l'on pourrait apporter une réponse affirmative au problème général, il en résulterait une économie de papier considérable ...". But it took five more years until CHOQUET

[11] with his powerful geometric intuition solved the existence as well as the corresponding unicity theorem by proving that a (metrizable) compact convex set is a simplex if and only if it can be viewed as the base of a cone which, in its intrinsic order, is a lattice. In the situation of the Martin boundary this cone is the lattice of all positive harmonic functions on $\Omega$. The base Y is given by the equation $h(x_o) = 1$.

It is our intention to present in this lecture some of the consequences which the results of Choquet and some of its refinements had on the development of potential theory itself. Potential theory in this lecture will be, more precisely, the potential theory of certain elliptic or parabolic differential equations including the classical case of the Laplace and the less classical case of the heat equation.

## §1. SIMPLICIAL FUNCTION CONES AND SPACES

We start with some functional-analytic background which will be useful later on.

We consider a metrizable compact space X and a convex cone $S \subset C(X)$ in the space $C(X)$ of all continuous real-valued functions on X. S is called a *function cone* if it satisfies the following two supplementary conditions:

(1.1) S contains a strictly positive function $s_o$.

(1.2) S separates linearly, i.e. for every two points $x_1 \neq x_2$ in X there are functions $s, t \in S$ such that $s(x_1)t(x_2) \neq s(x_2)t(x_1)$.

On the set $\mathcal{M}_+(X)$ of positive Radon measures on X the function cone S defines the Bishop - de Leeuw - Choquet order relation: $\mu \prec \nu$, for $\mu, \nu \in \mathcal{M}_+(X)$, means

(1.3) $$\mu(s_1 \wedge \ldots \wedge s_n) \geq \nu(s_1 \wedge \ldots \wedge s_n)$$

for every finite number $s_1, \ldots, s_n$ of functions in S. In the special case $\mu = \varepsilon_x$ = unit mass at $x \in X$, one has

(1.4) $\varepsilon_x \prec \nu \iff \nu(s) \le s(x)$ for all $s \in S$.

Measures $\nu \in \mathcal{M}_+(X)$ satisfying (1.4) are called *S-representing measures* for the point $x \in X$. $\mathcal{M}_x = \mathcal{M}_x(S)$ will denote the set of these measures.

The *Choquet boundary* of X with respect to S is defined as the set

(1.5) $\mathrm{Ch}_S(X) = \{x \in X : \mathcal{M}_x = \{\varepsilon_x\}\}$.

It is also the set of all $x \in X$ for which $\varepsilon_x$ is maximal in the order relation $\prec$. A measure $\mu \in \mathcal{M}_+(X)$ turns out to be *maximal* in the order relation $\prec$ *if and only if* it is *supported by* $\mathrm{Ch}_S(X)$. One should mention that the Choquet boundary is a $G_\delta$ - set.

The function cone S is called *simplicial* [7] *if, for every* $x \in X$ *there exists only one maximal measure* $\mu \in \mathcal{M}_x$. In particular, this notion is available when S is even a *function space*, i.e., if $S = -S$. The most important example of such a space is the following: Let Y be a metrizable compact convex set (in a Hausdorff locally convex space). Then Y is a simplex if and only if the function space aff(Y) of all continuous, real-valued, affine functions on Y is simplicial. The Choquet boundary of Y with respect to aff(Y) always coincides with ex Y.

For a function space H on (an arbitrary metrizable compact space) X the corresponding *state space* Y = state(H) is such a metrizable compact convex set [in the algebraic dual $H^*$ with the weak topology $\sigma(H^*,H)$]. It is the set of all $\varphi \in H^*$ normalized by $\varphi(s_o) = 1$. Simpliciality of H, in general, is not equivalent to the property that state(H) is a simplex. For a simple counter-example see [6; p.166]. However, the following is true: *A function space* H *on* X *is simplicial if and only if* state($\hat{H}$) *is a simplex*. Here $\hat{H}$ denotes the space of all H-*affine* functions:

$$\hat{H} = \{h \in C(X) : \mu(h) = h(x) \quad \forall\ x \in X,\ \forall\ \mu \in \mathcal{M}_x(H)\}.$$

It is the largest linear subspace of C(X) having the same

representing measures as H. As a consequence of a well-known theorem of BOBOC-CORNEA [9] and DAVIES [14], every point $x \in Ch_H(X)$ is an $\hat{H}$-peak point provided that the function space H is simplicial.

Function spaces H appearing in potential theory have the property that $\hat{H}$ coincides with H, in general. This is due to the following fact: *For a function space* H *on* X*, the space* $\hat{H}$ *coincides with* H *if and only if there exists an* inf-*stable closed set* $W \subset C(X)$ *such that* $H = W \cap (-W)$.

Additional informations about this subject, in particular proofs of the above results can be found in [6].

## §2. EXCURSION INTO POTENTIAL THEORY

We consider an open bounded set $U \subset \mathbb{R}^p$, $p \geq 1$. $\mathcal{H}_U$ will then denote the linear space $\mathcal{H}_U$ of all *harmonic functions*, i.e. all $C^2$-functions $u: U \to \mathbb{R}$ satisfying the Laplace equation $\Delta u = o$ on U or, equivalently, all $u \in C(U)$ satisfying the usual mean-value property

$$(2.1) \qquad \int u \, d\sigma_B^1 = u(x) \qquad (x \in U)$$

for all open balls B of center x such that $\bar{B} \subset U$. Here $\sigma_B^1$ is the surface measure on the sphere $B^*$ normalized to total mass 1.

We also consider the convex cone $\mathcal{S}_U$ of all *real-valued superharmonic* functions on U, i.e. all lower semi-continuous functions $u: U \to \mathbb{R}$ satisfying

$$(2.2) \qquad \int u \, d\sigma_B^1 \leq u(x) \qquad (x \in U)$$

for all open balls $B \subset \bar{B} \subset U$ of center x. On the compact space $\bar{U}$, we have then at our disposal: the function space

$$(2.3) \qquad H(U) = \{f \in C(\bar{U}) : f|U \in \mathcal{H}_U\}$$

and the function cone

$$(2.4) \qquad S(U) = \{f \in C(\bar{U}) : f|U \in \mathcal{S}_U\}.$$

Obviously, $H(U) = S(U) \cap [-S(U)]$. $H(U)$ coincides with the space of $H(U)$-affine functions since $S(U)$ is inf-stable and closed in $C(\overline{U})$.

It is well-known that, for every continuous boundary function $f \in C(U^*)$, the so-called generalized solution $H_f \in \mathcal{H}_U$ is defined. One can either use Perron's method of upper and lower functions or Wiener's exhaustion method. Both methods lead to the same object $H_f$. This classical fact is also a consequence of the following functional-analytic characterization of $H_f$ which is due to KELDYCH [19] (cf. also BRELOT [10]):

**2.1 KELDYCH-Theorem.** $f \to H_f$ *is the only positive linear map* $A : C(U^*) \to \mathcal{H}_U$ *which satisfies*

(2.5) $\quad A(h|U^*) = h|U \qquad$ *for all* $h \in H(U)$.

It has already been observed by KELDYCH [19],[10] that the result remains true when "positive linear map" is replaced by "increasing map" in the sense that $f \le g$ implies $Af \le Ag$.

The generalized solution leads to the well-known classification of boundary points: A boundary point $z \in U^*$ is called *regular* if

(2.6) $\quad \lim_{x \to z} H_f(x) = f(z) \qquad$ for all $f \in C(U^*)$;

otherwise, it is called irregular. Using an obvious notation, we thus have

(2.7) $\quad U^* = U^*_{reg} \cup U^*_{irr}$.

For the original proof of Theorem 2.1 the following Lemma is crucial (cf. also BRELOT [10]):

**2.2 KELDYCH-Lemma.** *Every* $z \in U^*_{reg}$ *is a peak point for the function space* $H(U)$.

From it follows a functional-analytic description of $U^*_{reg}$, a result obtained by the author [1],[2] in 1958:

**2.3 Corollary.** $U^*_{reg} = Ch_{H(U)}(\bar{U})$.

Another aspect of the generalized solution should be kept in mind. It has to do with one of the main techniques of potential theory, namely *balayage*: We assume $p \geq 3$ and consider a set $E \subset \mathbb{R}^p$ and a function $u \in \mathscr{S}^+_{\mathbb{R}^p}$, i.e. $u \geq 0$ and $u \in \mathscr{S}_{\mathbb{R}^p}$. Then the function $R^E_u$, defined on $\mathbb{R}^p$ by means of

$$(2.8) \qquad R^E_u(x) = \inf\{v(x) : v \in \mathscr{S}^+_{\mathbb{R}^p},\ v = u \text{ on } E\},$$

is called the *reduced function*. By passing to

$$(2.9) \qquad \hat{R}^E_u(x) = \liminf_{y \to x} R^E_u(y) \qquad (x \in \mathbb{R}^p)$$

one obtains another function $\hat{R}^E_u \in \mathscr{S}^+_{\mathbb{R}^p}$, the *swept-out function* (or *balayée*). For every Radon measure $\mu \geq 0$ on $\mathbb{R}^p$ with compact support, one then defines the *swept-out measure* $\mu^E \geq 0$ as follows: It is the only positive Radon measure on $\mathbb{R}^p$ which satisfies

$$(2.10) \qquad \int \hat{R}^E_u \, d\mu = \int u \, d\mu^E \qquad \text{for all } u \in \mathscr{S}^+_{\mathbb{R}^p}.$$

A simple example of a swept-out measure is the measure $\sigma^1_B$ for an open ball with center x. It is obtained by sweeping $\mu = \varepsilon_x$ to the set $E = \complement B$. So we have

$$\sigma^1_B = \varepsilon_x^{\complement B}.$$

Even more is true: The generalized solution $H_f$ for an open bounded set $U \subset \mathbb{R}^p$ permits (for $p \geq 3$) the representation

$$H_f(x) = \int f \, d\varepsilon_x^{\complement U} \qquad (x \in U).$$

for all $f \in C(U^*)$. Furthermore, again under the assumption $p \geq 3$, one has

$$U^*_{reg} = \{z \in U^* : \varepsilon_z^{\complement U} = \varepsilon_z\}$$

$$U^*_{irr} = \{z \in U^* : \varepsilon_z^{\complement U} \neq \varepsilon_z\}.$$

Usually this is expressed as follows: A point $z \in U^*$ is irregular if and only if the complement of U is thin at z. In general, a set $E \subset \mathbb{R}^p$ is called *thin at* z if $\varepsilon_z^E \neq \varepsilon_z$. For details the reader is referred to DOOB [15] or survey articles like [4],[5].

## §3. SIMPLICIAL CONES IN POTENTIAL THEORY

All what has been said in the last paragraph about the Dirichlet problem and balayage - with exception of the Keldych-Theorem, the Keldych-Lemma and its corollary - remains valid for the solutions of many linear, second order differential equations which are elliptic or parabolic. "Harmonic" then stands for "classical solution of the differential equation" under discussion. The generalized solutions are defined by Perron's method of upper and lower solutions.

More precisely, these results remain valid for all elliptic-parabolic differential equations whose solutions satisfy the axioms of a *harmonic space*. These spaces include, in particular, the *Brelot spaces* which - in a very precise sense - correspond to elliptic situations. All necessary details can be found in [3],[13] as well as in the survey articles [4],[5].

In what follows the space $\mathbb{R}^p$ together with the solutions of the Laplace equation - hence the classical harmonic functions - will be replaced by a *harmonic space* X. One can then speak again about harmonic and superharmonic functions. Also the objects $H(U)$, $S(U)$, $\mathscr{S}_U$, $\mathscr{S}_U^+$ for U open $\subset X$ as well as the generalized solution $H_f$ of the Dirichlet problem are defined. The dimension assumption $p \geq 3$, made for the treatment of balayage theory, is replaced by the corresponding assumption that $\mathscr{S}_X^+$ separates the points of X linearly. Such harmonic spaces are called *strong harmonic spaces* (or $\mathfrak{P}$-harmonic spaces with a countable base in the somewhat more general theory of CONSTANTINESCU - CORNEA [13]). Consequently, X should be from now on a strong harmonic space. The Laplace equation in $\mathbb{R}^p$ leads to such a space if and only if $p \geq 3$. The heat equation $\Delta u = \frac{\partial u}{\partial x_{p+1}}$ in

$\mathbb{R}^{p+1}$ yields for every $p \geq 1$ a strong harmonic space.

It was the Keldych-Lemma which early in the sixties led to the conjecture [2; p.121] and finally to the proof (see §1) that every extreme point of a metrizable simplex is a peak point for the space of continuous affine functions on it as well as to the conjecture that the space $H(U)$ is simplicial whenever $U$ is an open relatively compact set in a strong harmonic space $X$. This conjecture was proved for Brelot spaces satisfying an additional strong assumption - namely the axiom D which is not fulfilled for the harmonic space of the heat equation - by BOBOC-CORNEA [9]. They also proved the validity of Corollary 2.3. For the general case an example given by KÖHN-SIEVEKING [20] was a serious setback: They showed the existence of a bounded domain $U \subset \mathbb{R}^2$ such that, with respect to the heat equation, one has

$$U^*_{reg} \not\subset Ch_{S(U)}(\overline{U})$$

(whereas $Ch_{S(U)}(\overline{U}) \subset U^*_{reg}$ is quite easy to see). Their domain $U$ is the square $Q = ]0,1[ \times ]0,1[$ from which the segments

$$[\tfrac{1}{n}, 1[ \times \{1 - \tfrac{1}{n}\} \qquad (n = 2,3,\ldots)$$

have been deleted. Here $U^*_{reg} = Q^*$, whereas $Ch_{S(U)}(\overline{U})$ is $Q^*$ without the segment $]0,1[ \times \{1\}$ on the top of $Q$.

However, EFFROS and KAZDAN [16] succeeded in proving that for many bounded domains $U$ in $\mathbb{R}^{p+1}$ the function space $H(U)$ is simplicial for the heat equation.

The final break-through is due to BLIEDTNER-HANSEN [7]. They could prove the following:

**3.1 Theorem.** *Let $U$ be an open, relatively compact subset of a strong harmonic space. Then $S(U)$ and $H(U)$ are simplicial - the latter provided that linear point separation holds. Furthermore, $\mathcal{M}_x(S(U)) = \mathcal{M}_x(H(U))$ for all $x \in \overline{U}$; in particular,*

$Ch_{S(U)}(\overline{U}) = Ch_{H(U)}(\overline{U})$.

According to §1, this result implies in particular that *every point of* $Ch_{H(U)}(\overline{U})$ *is a peak point for* H(U), *provided that* H(U) *separates the points of* $\overline{U}$ *linearly.*

The fundamental paper [7] of Bliedtner and Hansen also gives a precise description of the Choquet boundary: There exists a largest subset of $\complement U$ which is *not* thin at any of its points. It is called the essential base of U, and denoted by $\beta(\complement U)$. The result then is:

**3.2 Theorem.** $Ch_{S(U)}(\overline{U}) = U^* \cap \beta(\complement U)$. *Furthermore,* $\varepsilon_x^{Ch_{S(U)}(\overline{U})}$ *is the (only) maximal representing measure for a point* $x \in \overline{U} \setminus Ch_{S(U)}(\overline{U})$.

There is some very recent development which may lead to new progress in the theory of general simplices. According to a result of CHOQUET [12] a metrizable compact convex set Y is a simplex if and only if for every compact set $K \subset \mathrm{ex}\, Y$ one has

$$\mathrm{aff}(Y)|K = C(K).$$

In other words, every function $f \in C(K)$ allows at least one extension to a function in aff(Y).

For the simplicial cone S(U) - or, in the case of linear point separation, for the simplicial space H(U) - more is true. In fact, the following theorem of BLIEDTNER-HANSEN [8] permits a new approach to the simpliciality of S(U):

**3.3 Theorem.** *For every compact set* $K \subset Ch_{S(U)}(\overline{U})$ *there exists a continuous kernel* $D_K$ *on* $\overline{U}$ *such that* $D_K$ *solves the "weak Dirichlet problem" in the following sense:* $D_K f \in H(U)$ *for all* $f \in C(K)$; *furthermore,* $D_K f(x) = f(x)$ *for all* $x \in K$.

(In addition to that $D_K$ can be constructed in such a way that all measures $D_K(x,\cdot)$, $x \in \overline{U}$, are supported by a fixed compact subset of $Ch_{S(U)}(\overline{U})$.)

It seems to be unknown whether the above mentioned weak Dirichlet problem for a metrizable compact convex set Y allows a "universal" solution by means of an appropriate kernel $D_K$ for K compact $\subset \text{ex}\, Y$.

## §4. KELDYCH SETS

In §2 we have pointed out the crucial role of the Keldych-Lemma for the proof of the Keldych-Theorem. In §3 we have seen that the Keldych-Lemma fails to be true for the heat equation. So it is natural to ask whether the analogue to the Keldych-Theorem is true for the heat equation or for more general harmonic spaces.

Let us consider an open, relatively compact set U in a strong harmonic space X. An operator

$$A : C(U^*) \to \mathscr{H}_U$$

will be called Keldych operator (or, K-operator) if it verifies property (2.5) and if it is positive linear (or, increasing). LUKEŠ [21] was the first to observe that for the heat equation and the Köhn-Sieveking domain $U \subset \mathbb{R}^2$ there exists a Keldych operator different from the one given by the generalized solution. The full truth behind this example was made evident for Keldych operators by LUKEŠ [21] and for K-operators, by means of a different method, by H. and U.SCHIRMEIER [25]:

**4.1 Theorem.** *Let* U *be an open, relatively compact subset of a strong harmonic space* X. *If* H(U) *separates the points of* $\overline{U}$ *linearly, then the following conditions are equivalent:*

(i) *the generalized solution yields the only Keldych operator;*

(ii) *the generalized solution yields the only K-operator;*

(iii) $\varepsilon_x^{\complement U}(U^*_{irr}) = 0$ *for all* $x \in U$.

Under the above assumptions, one can conclude that $U^*_{reg}$ and $Ch_{H(U)}(\overline{U})$ coincide if there is only one Keldych operator

[22], [25]. In classical potential theory, condition (iii) is a well-known fact.

Surprisingly there is the following weakened form of the Keldych-Theorem which is true without the additional assumption (iii). It is due to LUKEŠ-NETUKA [22] and H. and U. SCHIRMEIER [24]:

**4.2 Theorem.** *Let* U *be an open, relatively compact subset of a strong harmonic space* X *such that* H(U) *separates the point of* $\overline{U}$ *linearly. Consider an arbitrary exhaustion* $(U_n)_{n\in\mathbb{N}}$ *of* U *by open sets* $U_n$ *(i.e.* $\overline{U}_n \subset U_{n+1}$ *for all* n *and* $\bigcup U_n = U$*) and let* $A_n$ *be an arbitrary* K*-operator for* $U_n$, $n \in \mathbb{N}$. *Then, for every* $f \in C(U^*)$ *and every* $F \in C(\overline{U})$ *extending* f*, the generalized solution* $H_f$ *for* U *equals*

$$H_f(x) = \lim_{n\to\infty} A_n(F|U_n^*)(x) \qquad (x \in U).$$

We close with a remark: Consider a convex compact set Y. A map $\sigma: Y \to \mathcal{M}_+^1(Y)$ is called an affine selection if $\sigma$ is affine and if $\sigma(y)$ has y as barycenter for all $y \in Y$. FAKHOURY [17] has observed that Y *is a simplex if and only if there exists an affine selection.* Recently, Mrs. S. PAPADOPOULOU has given a very lucid (still unpublished) proof of the equivalence of (i) and (iii) in Theorem 4.1 along the ideas which are behind Fakhoury's result.

REFERENCES

[1] H. BAUER: Un problème de Dirichlet pour la frontière de Šilov d'un espace compact. C. r. Acad. Sci. Paris 247 (1958), 843-846.

[2] ——— : Šilovscher Rand und Dirichletsches Problem. Ann. Inst. Fourier 11 (1961), 89-136.

[3] ——— : Harmonische Räume und ihre Potentialtheorie. Lecture Notes in Math. 22, Springer-Verlag, Berlin - Heidelberg - New York, 1966.

[4] ——— : Harmonic Spaces - A Survey. Conferenze del Seminario di Matematica dell'Università di Bari, 30p. Gius. Laterza & Figli S.p.A., Bari, 1984.

[5] ——— : Zum heutigen Bild der Potentialtheorie. In: "Zum Werk Leonhard Eulers" (Eds.: E.Knobloch, I.S.Louhivaara, J.Winkler), 3-20. Birkhäuser Verlag, Basel - Boston - Stuttgart, 1984.

[6] ——— : Simplicial function spaces and simplexes. Expo. Math. 3 (1985), 165-168.

[7] J.BLIEDTNER and W.HANSEN: Simplicial cones in potential theory. Inventiones math. 29 (1975), 83-110.

[8] ——— : The weak Dirichlet problem. J. reine angew. Math. 348 (1984), 34-39.

[9] N.BOBOC and A.CORNEA: Convex cones of lower semicontinuous functions on compact spaces. Rev. Roum. Math. Pures Appl. 12 (1967), 471-525.

[10] M.BRELOT: Sur un théorème de prolongement fonctionnel de Keldych concernant le problème de Dirichlet. J. Analyse Math. 8 (1960/61), 273-288.

[11] G.CHOQUET: Existence et unicité des représentations intégrales. Sém. Bourbaki, 15p., décembre 1956.

[12] ——— : Lectures on Analysis, Vol II (Representation Theory). W.A.Benjamin, New York - Amsterdam, 1969.

[13] C.CONSTANTINCESCU and A.CORNEA: Potential Theory on Harmonic Spaces. Grundlehren d. math. Wiss. 158, Springer-Verlag, Berlin - Heidelberg - New York, 1972.

[14] E.B.DAVIES: A generalized theory of convexity. Proc. London Math. Soc. 17 (1967), 644-652.

[15] J.L.DOOB: Classical Potential Theory and its Probabilistic Counterpart. Grundlehren d. math. Wiss. 262 , Springer-Verlag, New York - Berlin - Heidelberg - Tokyo, 1984.

[16] E.G.EFFROS and J.L.KAZDAN: Applications of Choquet simplexes to elliptic and parabolic boundary value problems. J. of Differential Equations 8 (1970), 95-134.

[17] H.FAKHOURY: Une caractérisation des simplexes compacts et des cônes réticulés. Applications. Initiation à l'Analyse (Séminaire Choquet), 9e année, 1969/70, Fasc. 1, 12p. Institut H.Poincaré, Paris, 1970.

[18] R.GODEMENT: Théorie des représentations unitaires. Ann. of Math. (2), 53 (1951), 68-124.

[19] M.V.KELDYCH: On the solvability and stability of the Dirichlet problem (Russian). Usp. Mat. Nauk. SSSR 8 (1941), 171-231.

[20] J.KÖHN und M.SIEVEKING: Reguläre und extremale Randpunkte in der Potentialtheorie. Rev. Roum. Math. Pures Appl. 12 (1967), 1489-1502.

[21] J.LUKEŠ: Théorème de Keldych dans la théorie axiomatique de Bauer des fonctions harmoniques. Czech. Math. J. 24 (99) (1974), 114-125.

[22] ——— : Functional approach to the Brelot-Keldych theorem. Czech. Math. J. 27 (102) (1977), 609-616.

[23] J.LUKEŠ and I.NETUKA: The Wiener type solution of the Dirichlet problem in potential theory. Math. Ann. 224 (1976), 173-178.

[24] R.S.MARTIN: Minimal positive harmonic functions. Trans. Amer. Math. Soc. 49 (1941), 137-172.

[25] H. und U.SCHIRMEIER: Einige Bemerkungen über den Satz von Keldych. Math. Ann. 236 (1978), 245-254.

# Fine boundary limits and maximal sequences [R39]

Potential Theory and Its Related Fields, RIMS Kokyuroku 610, Kyoto 1987, 85–102
[Zbl. Math. 672.31014]

Dedicated to my teacher Professor Otto Haupt on the occasion of his 100th birthday

<u>Introduction.</u> A classical result of Brelot [4] concerns the limit behavior of bounded harmonic functions at an irregular boundary point. The result states that, given a bounded domain $\Omega$ in $R^n$ and an irregular boundary point z of $\Omega$, the fine limit

$$f - \lim_{\substack{x \to z \\ x \in \Omega}} v(x)$$

exists for every lower bounded, superharmonic function v on $\Omega$. In particular, this result applies to bounded harmonic functions on $\Omega$. Many years later Brelot [6] improved this result by introducing the so-called maximal sequences. These are sequences $(x_n)$ in $\Omega$ converging to z (in the ordinary sense) in such a way that the Green function $G_{y_o}$ of $\Omega$ with a pole $y_o \in \Omega$ converges along $(x_n)$ to

$$\limsup_{\substack{x \to z \\ x \in U}} G_{y_o}(x).$$

Brelot proved that, for a bounded harmonic function u on $\Omega$, the sequence $(u(x_n))$ converges for every maximal sequence and that one has

$$\lim_{n\to\infty} u(x_n) = f - \lim_{\substack{x\to z\\ x\in\Omega}} u(x).$$

Recently [3] I could prove that these results remain valid in very general situations, in particular for the heat equation. In the case of the heat equation irregular boundary points appear more naturally than for the Laplace equation. Even smooth domains may have irregular boundary points for the heat equation. A typical example is the "center" of the heat ball which we will discuss shortly.

The purpose of this paper is two-fold: We intend to present the main results of [3] in an expository form. At the same time the main hypotheses of [3] will be loosened considerably. We will show that it is possible to replace Doob's convergence axiom, predominant in [3], by the weak convergence axiom which is the standard convergence property for harmonic spaces in the sense of [8]. This improvement also follows from the more sophisticated results of Hansen [11] concerning finely harmonic functions. However, our techniques remain essentially the same as in [3] . Finally we will improve some results of [3] concerning maximal sequences for the heat equation.

Throughout this paper we will work within the following framework: X is a $\mathfrak{P}$-harmonic space (in the sense of Constantinescu-Cornea [8]) with a <u>countable base</u>. We study harmonic (or superharmonic) functions on an open and relatively compact set $U \subset X$ with topological boundary $U^* = \partial U$. $U^*_{irr}$ stands for the set of irregular boundary points. $H_f = H_f^U$ denotes the generalized solution of the Dirichlet problem for a resolutive boundary function f on U*. Limits (including upper and lower ones) with respect to the fine topology on X will be denoted by f-lim, f-lim sup, f-lim inf, respectively. All other undefined notations will be those of [3] and [8].

## 1. Maximal Sequences

In what follows z will denote a point in $U^*_{irr}$. When z is also polar, Brelot's classical result holds in the form that

$$\underset{\substack{x \to z \\ x \in U}}{\text{f-lim}}\ v(x)$$

exists for every hyperharmonic function $\geq 0$ on U. Brelot's proof given in [7], p.140, remains valid without essential modification. However, we start in a different way and will not need the above result.

<u>Proposition 1.</u> One has

$$\underset{\substack{x \to z \\ x \in U}}{\text{f-lim}}\ \mu_x^U = \varepsilon_z^{\complement U}$$

in the sense of vague convergence, i.e.

$$\underset{\substack{x \to z \\ x \in U}}{\text{f-lim}}\ H_f(x) = \int f d\varepsilon_z^{\complement U}$$

for all $f \in \mathcal{C}(U^*)$.

The proof relies on the fact that

$$H_{u*}(x) = \hat{R}_u^{\complement U}(x) \qquad \text{for all} \quad x \in U$$

whenever $u \geq o$ is a real-valued continuous superharmonic function on X and that $\hat{R}_u^{\complement U}$ is finely continuous as a superharmonic function.

From this we easily obtain the well-known Köhn-Sieveking-Lemma [10], [12]:

Corollary. There exist sequences $(x_n)$ in U such that

$$\text{(M)} \qquad \lim x_n = z \qquad \text{and} \qquad \lim \mu_{x_n}^U = \varepsilon_z^{\complement U}.$$

It will soon become evident that in the case of classical potential theory these sequences are just those which Brelot [6] had called maximal. Therefore we define:

Definition. A sequence $(x_n)$ is called maximal with respect to the point $z \in U_{irr}^*$ if the above condition (M) is satisfied.

Consequently, the Köhn-Sieveking-Lemma states the existence of maximal sequences for z. The following theorem characterizes these sequences.

Theorem 1. A sequence $(x_n)$ in U converging to $z \in U_{irr}^*$ is maximal with respect to z if and only if there exists a hyperharmonic function $u \geq o$ on X having the following two properties:

(1.1) $\hat{R}_u^{\complement U}(z) < u(z)$;

(1.2) $\lim_{n\to\infty} \hat{R}_u^{\complement U}(x_n) = \hat{R}_u^{\ U}(z)$.

The last condition is equivalent to

(1.2') $\lim_{n\to\infty} \hat{R}_u^{\complement U}(x_n) = \liminf_{\substack{x\to z\\ x\in U}} \hat{R}_u^{\complement U}(x)$.

The proof given in [3] remains essentially unchanged. However, two observations have to be made:

1. The total masses $\|\mu_x^U\|$ of the harmonic measures $\mu_x^U$, $x \in U$, are upper bounded. This can be seen by choosing a strictly positive potential $p \in \mathcal{C}(X)$, e.g. a strict potential $p \in \mathcal{C}(X)$. Then there are constants $\alpha, \beta \in \mathbb{R}$ such that $o < \alpha \leq p(x) \leq \beta$ for all $x \in \overline{U}$ due to the compactness of $\overline{U}$. Integration then yields

$$\alpha\|\mu_x^U\| \leq \int p d\mu_x^U < p(x) \leq \beta,$$

and hence

$$\|\mu_x^U\| < \frac{\beta}{\alpha} \qquad (x \in U).$$

(In [3] we gave a proof which uses the existence of a strictly positive harmonic function on an open neighborhood of $\overline{U}$. It may not exist under our weaker assumption.)

2. In [3] we used an additional condition, namely $u|U^* \in \mathcal{C}(U^*)$, which together with (1.1) and (1.2) characterized maximal sequences. However, as W.Hansen kindly has pointed out in a private communication, this additional condition can be avoided by means of the following argument: As in [3], consider a subsequence

$(x_{n_k})$ of $(x_n)$ such that $(\mu^U_{x_{n_k}})$ converges vaguely to a positive Radon measure $\lambda$ on $U^*$. As explained in [3], $\lambda$ has the form

$$\lambda = \alpha\varepsilon_z + (1-\alpha)\varepsilon_z^{\complement U}$$

for some $\alpha \in [0,1]$, and $\alpha = 0$ has to be proved. We have

$$\hat{R}_u^{\complement U}(z) = \lim_{k\to\infty} \hat{R}_u^{\complement U}(x_{n_k}) \geq \lim_{k\to\infty} \hat{R}_p^{\complement U}(x_{n_k}) = \int p\,d\lambda$$

$$= \alpha p(z) + (1-\alpha)\hat{R}_p^{\complement U}(z)$$

for every potential $p \leq u$ in $\mathcal{C}(X)$. This implies

$$\hat{R}_u^{\complement U}(z) \geq \alpha u(z) + (1-\alpha)\hat{R}_u^{\complement U}(z)$$

since u is the limit of some increasing sequence of such potentials p. The claim $\alpha = 0$ then follows from condition (1.1).

As an example, let U be a bounded domain in $\mathbb{R}^p$, $p \geq 3$. (The case $p = 2$ demands some simple modifications.) Consider the $\mathfrak{P}$-harmonic space $X = \mathbb{R}^p$ with respect to the sheaf of solutions of the Laplace equation. Choose a pole $y_o \in U$ and choose for u in the above theorem $u = N_{y_o}$ the Newtonian kernel

$$N_{y_o}(x) = \|x-y_o\|^{2-p} \qquad (x \in \mathbb{R}^p).$$

Then it is an immediate consequence of the theorem that a sequence $(x_n)$ in U converging to a point $z \in U^*_{irr}$ is maximal if and only if

$$\lim_{n\to\infty} G_{y_o}(x_n) = \limsup_{\substack{x\to z\\ x\in U}} G_{y_o}(x)$$

where $G_{y_o} = N_{y_o} - H_{N_{y_o}}$ is the Green function of U with pole $y_o \in U$. This, however, is Brelot's original definition of a maximal sequence given in [5].

## 2. Fine boundary limits of generalized solutions

Proposition 1 tells us that the generalized solution $H_f$ has a fine limit at $z \in U^*_{irr}$ for every $f \in \mathcal{C}(U^*)$. Furthermore, this limit is identified with the integral $\int f d\varepsilon_z^{\complement U}$ with respect to the swept-out measure $\varepsilon_z^{\complement U}$. This behavior of $H_f$ remains unchanged if we replace the continuous boundary function f by an arbitrary bounded resolutive boundary function f as we will see now. In order to obtain this result we have to assume that the point $z \in U^*_{irr}$ is polar.

<u>Theorem 2.</u> Under the assumption that the point $z \in U^*_{irr}$ is polar, every bounded resolutive function f on U* is $\varepsilon_z^{\complement U}$-integrable and $H_f$ has $\int f d\varepsilon_z^{\complement U}$ as a fine limit at z:

$$(2.1) \qquad \text{f-}\lim_{x \to z} H_f(x) = \int f d\varepsilon_z^{\complement U}.$$

The proof of this theorem given in [3] made use of Doob's convergence axiom in an essential way in order to derive the $\varepsilon_z^{\complement U}$-integrability of f. However, only mild modifications of the original proof have to be made in order to prove the theorem under the present assumptions.

P r o o f. As in [3], pp. 347-348, we can see that (2.1) holds when the bounded function f is upper or lower semicontinuous on $U^*$ and vanishes in a neighborhood of z.

Suppose now that f (besides being bounded resolutive) vanishes in a neighborhood of z. Then we observe that for every bounded lower semicontinuous function $\varphi \geq f$ on $U^*$ there exists a lower semicontinuous function $\varphi'$ on $U^*$ which also vanishes in a neighborhood of z and satisfies $\varphi \geq \varphi' \geq f$. In fact, we choose a closed neighborhood V of z in $U^*$ such that f vanishes in a neighborhood V' of V in $U^*$. Then $\varphi' = \varphi 1_{\complement V}$ has all desired properties since $\varphi(y) \geq f(y) = 0$ for all $y \in V'$. Consequently, $\int^* f d\varepsilon_z^{\complement U}$ is the infimum of the integrals $\int \varphi d\varepsilon_z^{\complement U}$ where $\varphi \geq f$ is bounded, lower semicontinuous and vanishes in a neighborhood of z. Since (2.1) holds for this type of functions we obtain - by applying this argument to f and -f -

$$\int_* f d\varepsilon_z^{\complement U} \leq \text{f-lim inf}\, H_f(x) \leq \text{f-lim sup}\, H_f(x) \leq \int^* f d\varepsilon_z^{\complement U}.$$

This, however, proves that (2.1) is valid for all bounded Borel functions f on $U^*$ which vanish in a neighborhood of z. This observation enables us to treat now also our bounded resolutive function f which vanishes in a neighborhood of z. Resolutivity of f is equivalent to the $\mu_x^U$-integrability of f for all $x \in U$. Hence, given $x \in U$, there exist bounded Borel functions $g_x$ and $h_x$ on $U^*$ vanishing in a (fixed) neighborhood of z such that

(2.2) $\qquad g_x \leq f \leq h_x$

and

(2.3) $\qquad \int g_x d\mu_x^U = \int f d\mu_x^U = \int h d\mu_x^U.$

Now we can use a trick which was also used by Hansen [11]:
We choose a countable dense subset $D \subset U$ and choose the functions $g_x, h_x$ for each $x \in D$ as before. Then

$$g := \sup_{x \in D} g_x \qquad \text{and} \qquad h = \inf_{x \in D} h_x$$

are again Borel functions on $U^*$ vanishing in a neighborhood of z such that

(2.4) $\quad g \leq f \leq h$

and

(2.5) $$\int g d\mu_x^U = \int f d\mu_x^U = \int h d\mu_x^U \qquad (x \in D).$$

But

$$x \to \int g d\mu_x^U = H_g(x) \quad \text{and} \quad x \to \int h d\mu_x^U = H_h(x)$$

are harmonic, hence continuous functions. Therefore (2.5) implies the equality

$$H_g(x) = H_h(x) \qquad \text{for all } x \in U.$$

Finally we can pass to the fine limit which yields

$$\int g d\varepsilon_z^{\complement U} = \underset{x \to z}{\text{f-lim}}\, H_g(x) = \underset{x \to z}{\text{f-lim}}\, H_h(x) = \int h d\varepsilon_z^{\complement U}.$$

This together with (2.4) proves that f is $\varepsilon_z^{\complement U}$-integrable and that (2.1) holds.

The rest of the proof given in [3], for an arbitrary bounded resolutive function f, remains unchanged.

It is not clear what can be said about unbounded resolutive functions (with or without Doob's convergence axiom). Even for classical potential theory this seems to be an open question. We also do not know whether Theorem 2 holds in full generality without assuming the polarity of the point z. However, the following weaker version of the theorem holds without the polarity of z. This observation, based on a preliminary version of [3], is due to Mrs. Ursula Schirmeier:

Remark. Assume that f is a bounded resolutive function on $U^*$ which is continuous at $z \in U^*_{irr}$. Then f is $\varepsilon_z^{\complement U}$-integrable and (2.1) holds.

P r o o f. As in the proof of Theorem 2.4 of [3] one can see that

$$(2.2) \qquad \int f d\varepsilon_z^{\complement U} \le \text{f-}\liminf_{x\to z} H_f(x) \le \text{f-}\limsup_{x\to z} H_f(x) \le \int f d\varepsilon_z^{\complement U\setminus\{z\}}$$

holds for all bounded lower semicontinuous functions $f \ge o$ which vanish in a neighborhood of z. We choose a sequence $(\varphi_n)$ of functions $\ge o$ in $\mathcal{C}(U^*)$ such that $\varphi_n \uparrow f$ on $U^*$. Then

$$\text{f-}\limsup_{x\to z} H_f(x) = \text{f-}\limsup_{x\to z} H_{f-\varphi_n}(x) + \text{f-}\lim_{x\to z} H_{\varphi_n}(x)$$

$$= \text{f-}\limsup_{x\to z} H_{f-\varphi_n}(x) + \int \varphi_n d\varepsilon_z^{\complement U}$$

since Proposition 1 holds without the polarity assumption about z. (2.2) can be applied to each function $f-\varphi_n$, $n \in \mathbb{N}$. So we have

$$\text{f-}\limsup_{x\to z} H_f(x) \le \int (f-\varphi_n) d\varepsilon_z^{\complement U\setminus\{z\}} + \int \varphi_n d\varepsilon_z^{\complement U}$$

for every $n \in \mathbb{N}$. For $n \to \infty$ this gives

$$\text{f-}\limsup_{x \to z} H_f(x) \leq \int f d\varepsilon_z^{\complement U}.$$

Consequently, (2.1) holds for bounded lower semicontinuous functions $f \geq o$ vanishing in a neighborhood of z.

Then we continue with the argument of step 3 in [3], p.348, and continue as above in the proof of Theorem 2 in order to obtain (2.1) for all bounded resolutive functions f which vanish in a neighborhood of z. From this we can proceed as follows:

We may assume $f(z) = o$ by passing from f to $f-f(z)$. Then we put

$$f_n = \sup(f, \tfrac{1}{n}) + \inf(f, -\tfrac{1}{n}) \qquad (n \in \mathbb{N}).$$

The continuity of f at z (where $f(z) = o$) implies that $\{|f| < \frac{1}{n}\}$ is a neighborhood of z on which $f_n$ vanishes. According to the result obtained so far, we have

$$\text{f-}\lim_{x \to z} H_{f_n}(x) = \int f_n d\varepsilon_z^{\complement U} \qquad (n \in \mathbb{N}).$$

From this the final result follows since

$$|f_n - f| < \frac{1}{n} \qquad \text{on } U^*,$$

so that $(f_n)$ converges uniformly to f on U*.

Let us also point out again that in classical potential theory Theorem 2 is due to Brelot [4],[5]. However, when Brelot studied the maximal sequences in [6] he did not not recognize the key rôle of his result.

This key rôle becomes evident from the following result:

<u>Lemma 1.</u> Assume that the point $z \in U^*_{irr}$ is polar and that $(x_n)$

is a maximal sequence in U with respect to z. Then, for every bounded resolutive function f on U*, the generalized solution $H_f$ of the Dirichlet problem has a limit along the sequence $(x_n)$, namely

$$(2.6) \qquad \lim_{n\to\infty} H_f(x_n) = \int f d\varepsilon_z^{\complement U}$$

Let us point out immediately that it is crucial to formulate (2.6) in this form and to combine (2.6) only after having proved it with Theorem 1. The proof itself can be copied from [3]. It remains valid in the new framework and is similar in spirit to the proof of Theorem 1.

Next we could make the combination of Lemma 1 and of Theorem 1 and state (2.6) in the more appropriate form:

$$(2.7) \qquad \lim_{n\to\infty} H_f(x_n) = \text{f-}\lim_{x\to z} H_f(x)$$

for all bounded resolutive functions. However much more is true and the proof goes via (2.7):

<u>Theorem 3.</u> Assume that the point $z \in U^*_{irr}$ is polar and that $(x_n)$ is a maximal sequence in U with respect to z. Then for every bounded harmonic function u on U the following two limits exist and are equal:

$$(2.8) \qquad \lim_{n\to\infty} u(x_n) = \text{f-}\lim_{x\to z} u(x).$$

Again the proof can be copied from [3]. In the case of classical potential theory this theorem was proved by Brelot [6] with totally different methods.

## 3. Applications to the heat equation

We consider the harmonic space given by the solutions of the heat equation

(3.1) $$\Delta u - \frac{\partial u}{\partial t} = o$$

on $\mathbb{R}^{p+1} = \mathbb{R}^p \times \mathbb{R}$, $p \geq 1$. In order to avoid confusions, harmonic (superharmonic) functions with respect to this new harmonic structure will be called caloric (supercaloric) functions.

The fundamental solution W of (3.1) with pole at the origin is given by

(3.2) $$W(x) = \begin{cases} \left(\frac{1}{4\pi\tau}\right)^{p/2} \exp\left(-\frac{\|\xi\|^2}{4\tau}\right) & , \tau > o \\ o & , \tau \leq o \end{cases}$$

where $x = (\xi, \tau) \in \mathbb{R}^p \times \mathbb{R}$ and where $\|\xi\|$ is the euclidean norm of $\xi$. We denote by ord x the time coordinate $\tau$ of x. By translation we obtain the fundamental solution

$$W_{x_o}(x) = W(x - x_o) \qquad (x \in \mathbb{R}^{p+1})$$

with pole $x_o \in \mathbb{R}^{p+1}$.

In [2] we have studied the relatively compact domain

$$\Omega(z_o, c) = \{x \in \mathbb{R}^{p+1} : W(z_o - x) > \left(\frac{1}{4\pi c}\right)^{p/2}\}$$

where $z_o \in \mathbb{R}^{p+1}$ and $c > o$. $\Omega(x_o, c)$ is called the heat ball with "center" $z_o$ and "radius" c. The center $z_o$ turns out to be the only irregular boundary point of $\Omega(z_o,c)$. The corresponding swept-out measure $\varepsilon_{z_o}^{\complement\Omega(z_o,c)}$ is explicitely known in this case; it is the co-called Fulks measure as we have shown in [2]. Furthermore all singletons of $R^{p+1}$ are polar. Consequently, Theorem 2 can be applied to this situation: the fine limit of $H_f$ can be calculated by integration f with respect to the Fulks measure whenener f is a bounded resolutive function for $\Omega(z_o,c)$.

Let us now improve two results of [3] by means of a different approach. We consider a relatively compact open subset U of $R^{p+1}$, an irregular boundary point z of U and for $y_o \in U$ the corresponding Green function on U:

$$G_{y_o} = W_{y_o} - H_{W_{y_o}} . \tag{3.3}$$

Here $H_{W_{y_o}}$ is the generalized solution of the Dirichlet problem with $W_{y_o}|U^*$ as boundary function.

<u>Lemma 2.</u> For arbitrary $y_o \in U$ one has

$$\limsup_{x\to z} G_{y_o}(x) = \text{f-}\lim_{x\to z} G_{y_o}(x) = W_{y_o}(z) - \int W_{y_o}\, d\varepsilon_z^U . \tag{3.4}$$

P r o o f. We choose a relatively compact open neighborhood $V_o$ of z (in $R^{p+1}$) such that $y_o \notin \bar{V}_o$, put $V = U \cap V_o$ and define a function f on $\dot{V}^*$ as follows:

$$f(y) = \begin{cases} G_{y_o}(y), & y \in V^* \cap U \\ o, & y \in V^* \cap U^*. \end{cases} \tag{3.5}$$

Obviously, we have $f \geq o$ and $f(y) = o$ in a neighborhood of z, namely for all $y \in U^* \cap V_o$. Furthermore, f is a bounded Borel function. From [3], Proposition 2.1 we know that we have

$$\text{(3.6)} \qquad \underset{x\to z}{\text{f-lim}}\ H_f^V(x) = \limsup_{x\to z} H_f^V(x)$$

for the corresponding generalized solution $H_f^V$. [The proof given in [3] remains valid even under the general assumptions of the previous paragraphs.] We denote by u the restriction of $W_{y_o}$ to $U^*$ and consider on $V^*$ the function

$$\text{(3.7)} \qquad v(y) = \begin{cases} H_u(y), & y \in V^* \cap U \\ u(y), & y \in V^* \cap U^*. \end{cases}$$

Then by the well-known restriction lemma [1], p.130, we know that v is a resolutive function for V and that the corresponding generalized solution $H_v^V$ equals the restriction of $H_u$ to V. But we have

$$\text{(3.8)} \qquad f = W_{y_o} - v \qquad\qquad \text{on } V^*$$

as one can easily check. Consequently, for all $x \in V$

$$H_f^V(x) = H_{W_{y_o}}^V(x) - H_v^V(x) = H_{W_{y_o}}^V(x) - H_u(x)$$

$$= W_{y_o}(x) - H_u(x) = G_{y_o}(x),$$

since $W_{y_o}$ is caloric (in particular continuous) on $\complement\{y_o\}$, hence in a neighborhood of $\overline{V}$. An application of (3.6) now yields

$$\limsup_{x\to z} G_{y_o}(x) = \underset{x\to z}{\text{f-lim}}\ G_{y_o}(x) = \underset{x\to z}{\text{f-lim}}(W_{y_o}(x) - H_{W_{y_o}}(x))$$

$$= W_{y_o}(z) - \text{f-lim}\ H_{W_{y_o}}(x)$$

$$= W_{y_o}(z) - \int W_{y_o} d\varepsilon_z^{\complement U}$$

according to Theorem 2.

We are now able to give a simpler proof of [3], Proposition 4.2. It characterizes maximal sequences for the heat equation in a similar way as Brelot's original definition of maximal sequences for classical potential theory.

Proposition 2. Let z be again an irregular boundary point of U. Assume that the Green function $G_{y_o}$ of U is strictly positive in the trace $W \cap U$ of a neighborhood W of z in $\mathbb{R}^{p+1}$ for some choice of the pole $y_o \in U$. Then a sequence $(x_n)$ in U converging to z is maximal with respect to z if and only if

$$\lim_{n\to\infty} G_{y_o}(x_n) = \limsup_{x\to z} G_{y_o}(x). \tag{3.9}$$

P r o o f. The positivity assumption about $G_{y_o}$ implies

$$\limsup_{x\to z} G_{y_o}(x) > o$$

because otherwise $G_{y_o}$ restricted to $W \cap U$ would be a barrier for z. Hence z would be a regular boundary point of $W \cap U$ and hence also of U. So the preceding Lemma 2 yields the inequality

$$\hat{R}_{W_{y_o}}^{\complement U}(z) = \int W_{y_o} d\varepsilon_z^{\complement U} < W_{y_o}(z).$$

Consequently, the superharmonic function $u = W_{y_o}$ on $\mathbb{R}^{p+1}$ ($W_{y_o}$ is even a potential) satisfies condition (1.1) of Theorem 1. According to this theorem the sequence $(x_n)$ is maximal if and only if (1.2) holds, i.e. if and only if

$$\lim_{n\to\infty} H_{W_{y_o}}(x_n) = \int W_{y_o}\, d\varepsilon_z^{\complement U}.$$

Due to the continuity of $W_{y_o}$ in $\complement\{y_o\}$ this is equivalent to

$$\lim_{n\to\infty} G_{y_o}(x_n) = W_{y_o}(z) - \int W_{y_o}\, d\varepsilon_z^{\complement U}$$

$$= \limsup_{x\to z} G_{y_o}(x)$$

according to Lemma 2. So the result follows.

Finally we point out that the condition concerning $G_{y_o}$ in the preceding proposition is easy to verify by choosing the pole $y_o$ appropriately. It suffices to observe (cf. Doob [9], p.300) that $G_{y_o}(x) > o$ holds true for $x \in U$ if and only if there exists a continuous path $\gamma : [o,1] \to U$ in U connecting $x = \gamma(o)$ with $y_o = \gamma(1)$ in such a way that the function $t \to \mathrm{ord}(\gamma(t))$ is strictly decreasing.

## REFERENCES

[1] BAUER, H.: Harmonische Räume und ihre Potentialtheorie. Lecture Notes in Mathematics, Vol.22. Springer Verlag, Berlin-Heidelberg-New York, 1966.

[2] BAUER, H.: Heat balls and Fulks measures. Ann. Acad. Scient. Fennicae, Ser.A 1, Math., 10, 317 - 332 (1985).

[3] BAUER, H.: Fine boundary limits of harmonic and caloric functions. Bull. Sc. math., 2e série, 109, 337 - 361 (1985).

[4] BRELOT, M.: Sur les ensembles effilés. Bull. Sc. math., 2e série, 68, 12 - 36 (1944).

[5] BRELOT, M.: Sur l'allure des fonctions harmoniques à la frontière. C.R. Acad. Sc., Paris, 220, 676 - 678 (1945).

[6] BRELOT, M.: On the behavior of harmonic functions in the neighborhood of an irregular boundary point. J. Anal. math., 4, 209 - 221 (1954/56).

[7] BRELOT, M.: Lectures on potential theory, 2nd edition. Tata Institute of Fund. Research, Lectures on Math., Vol. 19 (1967).

[8] CONSTANTINESCU, C., and A. CORNEA: Potential theory on harmonic spaces. Grundlehren der math. Wiss., Vol.158, Springer-Verlag, Berlin-Heidelberg-New York, 1972.

[9] DOOB, J.: Classical potential theory and its probabilistic counterpart. Grundlehren der math. Wiss., Vol.262. Springer-Verlag, Berlin-Heidelberg-New York, 1984.

[10] EFFROS, E.G., and J.-L. KAZDAN: Applications of Choquet simplexes to elliptic and parabolic boundary value problems. J. of differential equations, 8, 95 - 134 (1970).

[11] HANSEN, W.: Fine boundary limits of finely harmonic functions. Math. Ann., 274, 643 - 647 (1986).

[12] KÖHN, J., und M. SIEVEKING: Reguläre und extremale Randpunkte in der Potentialtheorie. Revue Roumaine de Math. pures et appl., 12, 1489 - 1502 (1967).

# Behaviour of solutions of elliptic-parabolic differential equations at irregular boundary points [S26]

Recenti sviluppi in analisi matematica e sue applicazioni
Conferenze del Seminario di Matematica dell'Università di Bari **237–244**,
Bari 1991, 27–37
[Zbl. Math. 792.35012; MR 94d:31001]

Conferenza tenuta l'8 Novembre 1990 (**)

Dedicated to Professor Giovanni Aquaro on the occasion of his 70th birthday

## 1. SOME FACTS FROM CLASSICAL POTENTIAL THEORY

We recall here some well-known and fundamental facts from classical potential theory. (Details can be found in Helms [15], Kellogg [17] and Doob [12]). The underlying differential equation is the Laplace equation

$$(1) \qquad \Delta h = 0$$

which will be considered in a *bounded domain* $\Omega$ of $\mathbf{R}^d$ for $d \geq 1$. The classical solutions $h : \Omega \to \mathbf{R}$ of equation (1) are the *harmonic functions* on $\Omega$. By the Gauss-Koebe theorem, these are exactly the continuous functions $h$ on $\Omega$ satisfying the mean-value property

$$(2) \qquad \int_{S_a(x_0)} h d\sigma_1 = h(x_0)$$

for every closed ball $\overline{B}_a(x_0)$ of center $x_0$ and radius $a > 0$ contained in $\Omega$. Hereby, $S_a(x_0)$ denotes the corresponding sphere of center $x_0$ and radius $a$; $\sigma_1$ denotes the surface measure on $S_a(x_0)$, normalized to 1. This characterization leads to the larger class of all *hyperharmonic functions* on $\Omega$. These are the functions $u : \Omega \to ]-\infty, +\infty]$ which are lower semicontinuous and

---

(**) Il manoscritto è pervenuto alla Redazione il 6 Giugno 1991.

satisfy

$$(2') \qquad \int_{S_a(x_0)} u\,d\sigma_1 \leq u(x_0)$$

for all balls satisfying $\overline{B}_a(x_0) \subset \Omega$.

The Perron-Wiener-Brelot-method (PWB-method) associates to every (extended real-valued) boundary function

$$f : \Omega^* \to \overline{\mathbb{R}} = [-\infty, +\infty]$$

(on the boundary $\Omega^*$ of $\Omega$ ) the function

$$\overline{H}_f : \Omega \to \overline{\mathbb{R}}$$

which is defined as the pointwise infimum on $\Omega$ of all *upper functions* for $f$. These are all hyperharmonic functions $u$ on $\Omega$ satisfying

$$\liminf_{\substack{x \to z \\ x \in \Omega}} u(x) \geq f(z) \quad \text{and} \quad \liminf_{\substack{x \to z \\ x \in \Omega}} u(x) > -\infty$$

for all boundary points $z \in \Omega^*$. One defines the *lower function* $\underline{H}_f$ for $f$ by means of $\underline{H}_f :=$ $-\overline{H}_{-f}$ and ends up with the inequality

$$\underline{H}_f \leq \overline{H}_f \qquad (on\,\Omega)$$

as a consequence of the boundary minimum principle. In the case where $\underline{H}_f = \overline{H}_f$ holds on $\Omega$ and where the function

$$(3) \qquad H_f := \underline{H}_f = \overline{H}_f$$

is *finite* everywhere on $\Omega$, this function $H_f$ is harmonic on $\Omega$. It is called the *generalized solution* of the Dirichlet problem for $f$. In this case $f$ is called *resolutive.*

All functions $f$ from $\mathscr{C}(\Omega^*)$, i.e. real-valued and continuous on $\Omega^*$, turn out to be resolutive. Consequently, every point $x \in \Omega$ yields the positive linear form

$$f \mapsto H_f(x)$$

on $\mathscr{C}(\Omega^*)$ and hence - by definition or by the Riesz representation theorem - one obtains

(4) $$H_f(x) = \int f d\mu_x^\Omega \qquad (f \in \mathscr{C}(\Omega^*), x \in \Omega)$$

with a uniquely determined Radon measure $\mu_x^\Omega$ on $\Omega^*$, the *harmonic measure* associated to $x$ (and $\Omega$ ). Brelot's classical *resolutivity theorem* states that a function $f : \Omega^* \mapsto \overline{\mathbf{R}}$ is resolutive if and only if it is $\mu_x^\Omega$ *-integrable* for all $x \in \Omega$ (in Bourbaki's sense, i.e. integrable with respect to the completion of $\mu_x^\Omega$ considered as a Borel measure. In fact, integrability with respect to just one of these measures would be sufficient). Formula (4) will then hold for every resolutive boundary function $f$.

The importance of the PWB-method lies in the fact that it yields a natural distinction between "good" and "bad" boundary points. A point $z \in \Omega^*$ is called *regular* if

(5) $$\lim_{\substack{x \to z \\ x \in \Omega}} H_f(x) = f(z) \qquad \text{for all } f \in \mathscr{C}(\Omega^*)$$

or, equivalently,

(5') $$\lim_{\substack{x \to z \\ x \in \Omega}} \mu_x^\Omega = \varepsilon_z$$

holds in the sense of vague (or weak) convergence of Radon measures. Here $\varepsilon_z$ denotes the unit mass (or Dirac-measure) at $z$. Points $z \in \Omega^*$ for which (5) or (5') does not hold are called *irregular* boundary points.

## 2. IRREGULAR BOUNDARY POINTS FOR PARABOLIC EQUATIONS

The appearance of irregular boundary points in the setting of §1 is a somewhat pathological phenomenon in view of applications: Domains with a smooth boundary are known to have only regular boundary points. In that case the Dirichlet problem has for every boundary function $f \in \mathscr{C}(\Omega^*)$ a unique solution, namely $H_f$. A typical example of an irregular point (for the Laplace equation) is the Lebesgue spine in $\mathbf{R}^3$ where an "exponentially thin" thorn points towards the interior of the domain, the thorn itself beingh part of the boundary of the domain.

However, the situation changes drastically if the elliptic equation $\Delta u = 0$ is replaced by a parabolic differential equation. We give two examples.

**2.1.** Consider the *heat equation*

$$(6) \qquad \Delta u = \frac{\partial u}{\partial t}$$

on $\mathbf{R}^d \times \mathbf{R}$ for $d \geq 1$. Points $x \in \mathbf{R}^d \times \mathbf{R}$ will be written in the form $x = (\xi, t)$ with coordinates $\xi \in \mathbf{R}^d$ and $t \in \mathbf{R}$. The (classical) solutions of (6) will be called *caloric functions*. The theory of these functions, as far as the PWB-method is concerned, is formally the same as the theory of harmonic functions recalled in §1. This goes back to contributions of Sternberg [26] and Pini [22]. An up-to-date presentation can be found in Bauer [1] and Doob [12]. Heat conduction in a solid bar leads to the Dirichlet problem for a rectangle

$$(7) \qquad \Omega := ]0, L[\times]0, T[ \qquad (L > 0, T > 0)$$

in $\mathbf{R} \times \mathbf{R}$. All boundary points from the top of the rectangle, i.e. from $]0, L[\times\{T\}$ are irregular in this case.

But there is a more stricking example. For the Laplace equation (1) in $\mathbf{R}^d$ with $d \geq 3$, the (Newtonian) function

$$N(x) := |x|^{2-d} \qquad (x \in \mathbf{R}^d)$$

(with euclidean norm $|x|$) is up to a constant the fundamental solution with pole at the origin. It allows to write an euclidean ball $B_a(\mathbf{0})$ with center at the origin $\mathbf{0}$ in the form

$$B_a(\mathbf{0}) = \{x \in \mathbf{R}^d : N(x) > a^{2-d}\}.$$

For the heat equation there are corresponding objects: The *heat ball* with "center" at the origin $z_0 := (\mathbf{0}, 0)$ of $\mathbf{R}^d \times \mathbf{R}$ is defined by

$$(8) \qquad \Omega_c(z_0) := \{x \in \mathbf{R}^d \times \mathbf{R} : W(x) > (4\pi c)^{-d/2}\}$$

where $W$ denotes (up to a constant factor) the fundamental solution of the adjoint heat equation $\Delta u + \dfrac{\partial u}{\partial t} = 0$ with singularity at $z_0$. It is given by

$$(9) \qquad W(x) := \begin{cases} (4\pi|t|)^{-d/2} exp(-|\xi|^2/4|t|), & t < 0 \\ 0, & t \geq 0 \end{cases}$$

Hereby, $c > 0$ is an arbitrarily chosen real constant which imitates the "radius" of $\Omega_c(z_0)$. The "center" $z_0$ of $\Omega_c$ is a boundary point of $\Omega_c$, in fact the north pole of the convex body $\overline{\Omega}_c$ (cf. Bauer [2]). It is known that $z_0$ is the only singular point on the boundary of $\Omega_c$ (cf. Watson [27], Bauer [2]).

**2.2.** Another simple parabolic differential equation - for reasons of simplicity we study it here only for the case of two space variables, i.e. on $\mathbf{R}^2 \times \mathbf{R}$ - is the so-called *Kolmogorov equation*

$$\frac{\partial^2 u}{\partial \xi_1^2} + \xi_1 \frac{\partial u}{\partial \xi_2} = \frac{\partial u}{\partial t} \tag{10}$$

where $(\xi_1, \xi_2, t) \in \mathbf{R}^2 \times \mathbf{R}$. It appears in a short note of Kolmogorov in 1934 in connection with random movements (Kolmogorov [18]). The equation may be irritating at the first moment since it is degenerate in the sense that the quadratic form coming from second order terms is only semidefinit. But the potential theory of the solutions of this equation as far as the PWB-method is concerned is as good as the one for the Laplace equation and the heat equation. In fact, if one writes (10) in the form

$$X_1^2 + X_0 = 0 \tag{10'}$$

by means of the two vector fields

$$X_1 := \frac{\partial}{\partial \xi_1} \qquad \text{and} \qquad X_0 := \xi_1 \frac{\partial}{\partial \xi_2} - \frac{\partial}{\partial t}$$

the usual Lie bracket $[X_1, X_0] := X_1 X_0 - X_0 X_1$ turns out to be the vector field $\dfrac{\partial}{\partial \xi_2}$. Consequently, the dimension of the linear span of $X_0, X_1$ and $[X_0, X_1]$ equals 3 at all points of $\mathbf{R}^2 \times \mathbf{R}$ which, by Hörmander's theorem, implies that the parabolic operator $X_1^2 + X_0$ is hypoelliptic. As a consequence, one knowns that the PWB-method can also be applied to Kolmogorov's equation (10) and its solutions with formally the same results (cf. §3 and the references given there). By analogy with 2.1, we consider the fundamental solution $K : \mathbf{R}^2 \times \mathbf{R} \mapsto \mathbf{R}$ of the adjoint equation of (10). It is again explicitely known: it equals

$$\begin{aligned} K(\xi, t) &= \frac{\sqrt{3}}{2\pi t^2}\, exp\left[\frac{\xi_1^2}{t} - 3\frac{\xi_1(\xi_2 + t\xi_1)}{t^2} + 3\frac{(\xi_2 + t\xi_1)^2}{t^3}\right] \\ &= \frac{\sqrt{3}}{2\pi t^2}\, exp\left[\frac{\xi_1^2}{4t} + \frac{3}{t^3}\,(\xi_2 + \frac{t}{2}\,\xi_1)^2\right] \end{aligned}$$

for $t < 0$ and equal to zero for $t \geq 0$. Corresponding to the heat ball we define the ***Kolmogorov ball***

(11) $$K_r(z_0) := \{x \in \mathbf{R}^2 \times \mathbf{R} : K(x) > r\}$$

with "center" $z_0 = (0,0,0)$ at the origin for an arbitrary number $r > 0$. Again $z_0$ turns out to be an irregular boundary point of $K_r(z_0)$, while all other boundary points are regular (cf. Scornazzani [24] and Hoh, Jacob [16] for all details).

Many additional examples can be derived from the work of Kröger [20] and Garofalo, Lanconelli [14]. In conclusion, these examples document the fact that parabolic differential equations lead to irregular boundary points for very natural domains. It was this observation which has led the author to a deeper study of irregular boundary points (cf. Bauer [3], [4]). These will be discussed in the next paragraph. For the case of the Laplace equation they reduce to classical results of Brelot, published in [7] - [10] as well as to results of Smyrnélis [25].

## 3. BOUNDARY BEHAVIOUR AT IRREGULAR BOUNDARY POINTS

In order to cover by our results a large class of (linear) elliptic and parabolic differential equations on $\mathbf{R}^d$ or even on differentiable manifolds, we shall work in the framework of harmonic spaces. We refer the reader to the exposition of the theory of harmonic spaces in this series (cf. Bauer [1]).

In what follows $X$ will be a *strong harmonic space* with its sheaf $\mathscr{H}$ of harmonic functions. (We could even assume that $X$ is a $\mathscr{P}$-harmonic space (with countable base) in the sense of Constantinescu, Cornea [11]). Examples of strong harmonic spaces are given by the Laplace equation on $\mathbf{R}^d$ for $d \geq 3$ and on a bounded subdomain $\Omega_0$ of $\mathbf{R}^d$ for $d = 1, 2$, by the heat equation on $\mathbf{R}^d \times \mathbf{R}$ for $d \geq 1$, and by the Kolmogorov equation on $\mathbf{R}^2 \times \mathbf{R}$ (cf. Bauer [1], Scornazzani [24] and Hoh, Jacob [16]). Here $X$ equals $\mathbf{R}^d, \Omega_0, \mathbf{R}^d \times \mathbf{R}$ or $\mathbf{R}^2 \times \mathbf{R}$, respectively; "harmonic function" stands for "solution of the corresponding differential equation". Examples of harmonic spaces derived from very general elliptic and parabolic differential equation on $\mathbf{R}^d$ have been studied by Kröger [20]. It should be underligned that a strong harmonic space may be exotic. Examples are given in Bauer [1]; in particular the infinite dimensional torus $T^\infty$ has to be mentioned here. The corresponding main results of Berg [6], discussed also in [1], have been obtained independently, improved and completed by Bendikov [5].

The results which will be surveyed in what follows have originally been published in Bauer [3] and [4]. Improvements and new approaches are due to Netuka [21]. They will be mentioned later explicitely.

The notations introduced in the preceding paragraphs in connection with the Dirichlet problem and the PWB-method will be used also in the new setting. All what has been said about the PWB-method remains valid. Consequently, $\Omega$ will be a given *relatively compact domain* in $X$ with non-empty boundary $\Omega^*$. $H_f$ stands for the generalized solution corresponding to a resolutive boundary function $f : \Omega^* \to \overline{\mathbf{R}}$. By means of the harmonic measures $\mu_x^\Omega, x \in \Omega$, one has again formula (4). The set of all irregular boundary points of $\Omega$ will be denoted by $\Omega^*_{irr}$.

A classical theorem of Hermann Amandus Schwarz (cf. Kellogg [17], p. 270) states that a classical bounded harmonic function in a punctured open ball

$$B_a(x_0)\backslash\{x_0\}$$

in $\mathbf{R}^d$ can be (uniquely) extended to a harmonic function on all of $B_a(x_0)$. Hence, $x_0$ is a removable singularity. The discussion which follows should be seen in the light of this result. However, we will have to do not with the given (locally compact) topology of the harmonic space $X$, but with its *fine topology*. By definition, this is the topology for which the system of all neighborhoods $\mathscr{V}_f(x)$ of an arbitrary point $x$ - the so-called fine neighborhoods of $x$ - is given by

$$\mathscr{V}_f(x) := \{V : x \in V \subset X, CV \text{ thin at } x\} \tag{12}$$

*Thinness* of a set $E \subset X$ at a point $x \in X$ is defined by $\varepsilon_x^E \neq \varepsilon_x$ where $\varepsilon_x^E$ is the *swept measure* derived from the Dirac measure $\varepsilon_x$ (as described in Bauer [1]). The new topology is the coarsest one on $X$ which makes all non-negative hyperharmonic function on $X$ continuous.

A point $z \in \Omega^*$ is known to be *irregular if and only if* $C\Omega$ is thin at $z$ (cf. [1], p. 20). Consequently, $\Omega \cup \{z\}$ is a fine neighborhood of $z \in \Omega^*$ if the point $z$ is irregular. (The converse is at least true if the point $z$ is also *polar* - a condition satisfied in most applications (cf. [1]). In probabilistic terms this means for the Laplace equation that Brownian motion starting at $z \in \Omega^*_{irr}$ with probability one immediately enters into $\Omega$ before having a chance to hit the complement of $\Omega$. The same is true for the heat equation if Brownian motion is replaced by the space-time modification of Brownian motion (or the heat transfer in the sense of [2]). A corresponding statement holds in general for the Hunt process (with continuous paths) associated to the given strong harmonic space $X$ provided that the constant function 1 is harmonic (cf. Bauer [1], p. 26).

After these preparations we now come to a first result which, for the Laplace equation, goes back to a result of Brelot [7] (cf. also Bauer [3]). In spirit, it generalizes the above mentioned result of H. A. Schwarz concerning removable singularities.

LEMMA. *Let* $u$ *be a lower bounded hyperharmonic function on* $\Omega$. *Then the fine limit (i.e. limit with respect to the fine topology)*

$$(13) \qquad \mathrm{f} - \lim_{\substack{x \to z \\ x \in \Omega}} u(x)$$

*exists whenever* $z$ *is a polar irregular boundary point of* $\Omega$.

It now follows that for every lower bounded resolutive function $f$ on $\Omega^*$ the fine limit

$$\mathrm{f} - \lim_{\substack{x \to z \\ x \in \Omega}} H_f(x)$$

exists for such a point $z \in \Omega^*_{irr}$. These limits allow an integral representation. For the Laplace equation this goes again back to Brelot [7], [8]. For bounded resolutive functions it is one of the main results of Bauer [3], [4]. Netuka [21] generalized it, as formulated below, to one-sided bounded resolutive function with a differential and shorter proof.

THEOREM 1. *Let* $f$ *be a one-sided bounded resolutive function on* $\Omega^*$. *Then, for every polar point* $z \in \Omega^*_{irr}$, *the function* $f$ *is* $\varepsilon_z^{C\Omega}$ *-integrable and*

$$(14) \qquad \mathrm{f} - \lim_{\substack{x \to z \\ x \in \Omega}} H_f(x) = \int f d\varepsilon_z^{C\Omega}$$

The difficult part of the proof concerns the integrability of $f$ with respect to $\varepsilon_z^{C\Omega}$. Without any boundedness condition it does not follow from resolutivity (cf. [21]). The interest in this result comes from the fact that $\varepsilon_z^{C\Omega}$ is known explicitely in certain cases, e.g. when $z$ is the center, i.e. north pole, of the heat ball - here $\varepsilon_z^{C\Omega}$ turns out to be identical with the Fulks or Pini-Fulks measure (cf. Bauer [2] and Pini [23]) - or the Kolmogorov ball (Scornazzani [24] and Hoh, Jacob [16]).

Let us return to the preceding Lemma (in the case of the Laplace equation) and add to it a probabilistic statement. Let $(B_t)_{t \geq 0}$ be a Brownian motion starting at $z \in \Omega^*_{irr}$. Then there exists a number $\alpha \in ]-\infty, +\infty]$ such that

$$\lim_{t \to 0^+} u(B_t) = \alpha$$

holds almost surely, i.e. along almost every path (entering into $\Omega$ immediately). The number $\alpha$ is nothing else than the fine limit appearing in the Lemma (cf. Doob [12]). A similar statement holds in the framework of this paragraph.

We shall now present a purely analytic counterpart to this statement. For functions $f \in \mathscr{B}(\Omega^*)$ it follows from Theorem 1 that

(15) $$\text{f} - \lim_{\substack{x \to z \\ x \in \Omega}} \mu_x^{\Omega} = \varepsilon_z^{C\Omega}$$

or, because of (5') equivalently,

(15') $$\text{f} - \lim_{\substack{x \to z \\ x \in \Omega}} \varepsilon_x^{C\Omega} = \varepsilon_z^{C\Omega}$$

holds in the vague topology. It is an easy and since long known consequence of this fact that, given $z \in \Omega^*_{irr}$, there exist sequences $(x_n)_{n \in \mathbb{N}}$ in $\Omega$ converging to $z$ in the original topology such that

(16) $$\lim_{n \to \infty} \mu_{x_n}^{\Omega} = \varepsilon_z^{C\Omega}$$

in the sense of vague convergence (cf. Köhn, Sieveking [19] and Bauer [3]). Such sequences are called *maximal*. The reason for this terminology comes from the fact that, e.g. in the case of the Laplace equation, a sequence $(x_n)$ in $\Omega$ converging to $z$ is maximal if and only if

$$\lim_{n \to \infty} G_{x_0}(x_n) = \limsup_{\substack{x \to z \\ x \in \Omega}} G_{x_0}(x)$$

for the Green function $G_{x_0}$ of $\Omega$ where the pole $x_0$ may be chosen arbitrarily. In this form Brelot [9] had originally introduced these sequences. Similar characterizations hold for the heat and Kolmogorov equation.

The already announced interpretation of fine limits in terms of the given topology on $X$ is now the following:

THEOREM 2. *Let $h$ be a bounded harmonic function on $\Omega$. Then*

$$\text{f} - \lim_{\substack{x \to z \\ x \in \Omega}} h(x) = \lim_{n \to \infty} h(x_n)$$

*holds whenever $(x_n)$ is a maximal sequence in $\Omega$ for a polar point $z \in \Omega^*_{irr}$.*

Also this result was proved in Bauer [3], [4]. Netuka [21] indicated a different approach. For the Laplace equation the same result can be found in Brelot [9].

A striking theoretical application of this result has been given by Netuka [21]. It yields the following result.

THEOREM 3. *For every polar point* $z \in \Omega_{irr}^{*}$ *there exists a (universal) set* $T \subset \Omega$ *such that*

(i) $T$ *is thin at* $z$

(ii) *all sequences in* $\Omega \setminus T$ *converging to* $z$ *are maximal sequences for* $z$.

Consequently, one has the existence of

$$\lim_{\substack{x \to z \\ x \in \Omega \setminus T}} h(x) = \mathrm{f} - \lim_{\substack{x \to z \\ x \in \Omega}} h(x)$$

for all bounded harmonic functions $h$ on $\Omega$.

For the case of Brelot spaces, in particular for the Laplace equation, this result was obtained in Brelot [9].

It was clear since long that maximal sequences should be seen as sequences in $\Omega$ which converge to $z$ in a "non-tangential" way. The proof for the explicit description of $\varepsilon_z^{C\Omega}$ for the heat ball $\Omega = \Omega_c(z_0)$ from (8) (as the Pini-Fulks measure) given in [2] or the corresponding proof for the Kolmogorov ball given in [16] should be seen in this light. A result of Wittmann [28] makes this intuition rigorous in the case of strong Brelot spaces, i.e. in the elliptic case (cf. [1]).

## REFERENCES


[1] H. BAUER, *Harmonic Spaces - A Survey*, Conferenze del Seminario di Matematica dell'Università di Bari 197, 1984.

[2] H. BAUER, *Heat balls and Fulks measures*, Ann. Acad. Scient. Fennicae, Series A.I., Math., 10 (1985), 67-82.

[3] H. BAUER, *Fine boundary limits of harmonic and caloric functions*, Bull. Sc. Math., 2ème série, 109 (1985), 337-361.

[4] H. BAUER, *Fine boundary limits and maximal sequences*, In: Potential theory and its Related Fields. Proc. of a Symposium held in Kyoto 1986, RIMS Kokyoroku 610, Kyoto Univ., Kyoto, 85-102, 1987.

[5] A.D. BENDIKOV, *Markov processes and partial differential equations on a group: the space-homogeneous case*, Russian Math. Surveys 42:5 (1987), 49-94.

[6] C. BERG, *Potential theory on the infinite dimensional torus*, Inventiones math. 32 (1976), 49-100.

[7] M. BRELOT, *Sur les ensembles effilés*, Bull. Sc. math., 2ème série, 68 (1944), 12-36.

[8] M. BRELOT, *Sur l'allure des fonctions harmoniques à la frontière*, C.R. Acad. Sc., Paris, 220 (1945), 676-678.

[9] M. BRELOT, *On the behavior of harmonic functions in the neighborhood of an irregular boundary point*, J. Anal. math., Jerusalem, 4 (1954-1956), 209-221.

[10] M. BRELOT, *Sur l'allure à la frontière des familles de fonctions harmoniques ou surharmoniques*, Rendiconti di Mat., Ser. 2, 8 (1975), 447-456.

[11] C. CONSTANTINESCU, C. CORNEA, *Potential theory on harmonic spaces*, Grundlehren d. math. Wiss. 158, Springer-Verlag, Berlin - Heidelberg - New York, 1972.

[12] J.L. Doob, *Classical potential theory and its probabilistic counterpart*, Grundlehren der math. Wiss. 262 Springer-Verlag, New York - Berlin - Heidelberg - Tokyo 1984.

[13] N. Garofalo, E. Lanconelli, *Wiener's criterion for parabolic equations with variable coefficients and its consequences*, Trans. Amer. Math. Soc. 308 (1988), 811-836.

[14] N. Garofalo, E. Lanconelli, *Level sets of the fundamental solution and Harnack inequality for degenerate equations of Kolmogorov type*, Trans. Amer. Math. Soc. 321 (1990), 775-792.

[15] L.L. Helms, *Introduction to potential theory*, Interscience, New York, 1969.

[16] W. Hoh, N. Jacob, *On the potential theory of the Kolmogorov equation*, To appear in Math. Nachrichten.

[17] O.D. Kellogg, *Foundations of Potential Theory*, Grundlehren der math. Wiss. 31, Springer-Verlag, Berlin - Heidelberg - New York, 1967.

[18] A.N. Kolmogorov, *Zufällige Bewegungen (Zur Theorie der Brownschen Bewegung)*, Annals of Math. 35 (1934), 116-117.

[19] T. Köhn, M. Sieveking, *Reguläre und extremale Randpunkte in der Potentialtheorie*, Rev. Roumaine Math. Pures et Appl. 12 (1967), 1489-1502.

[20] P. Kröger, *Harmonic spaces associated with parabolic and elliptic differential equations*, Math. Ann. 285 (1989), 393-403.

[21] I. Netuka, *Fine behaviour of solutions of the Dirichlet problem near an irregular boundary point*, Bull. Sc. math., 2ème série, 114 (1990), 1-22.

[22] B. Pini, *Sulla soluzione generalizzata di Wiener per il primo problema di valori al contorno nel caso parabolico*, Rend. Sem. Mat. Padova 23 (1954), 422-434.

[23] B. Pini, *Maggioranti e minoranti delle soluzioni delle equazioni paraboliche*, Annali di Mat. pura ed appl., IV. Ser., 37 (1954), 249-264.

[24] V. Scornazzani, *Sul problema di Dirichlet per l'operatore di Kolmogorov*, Bollettino U.M.I., Analisi Funzionale e Applicazioni, Ser. V, 18 (1981), 43-62.

[25] E. Smyrnélis, *Sur les limites fines des fonctions harmoniques et les suites maximales*, Bull. Sc. math. 2ème série, 97 (1973), 161-175.

[26] W. Sternberg, *Über die Differentialgleichung der Wärmeleitung*, Math. Ann. 101 (1929), 394-398.

[27] N.A. Watson, *Green functions, potentials and the Dirichlet problem for the heat equation*, Proc. London math. Soc., 3rd series, 33 (1976), 251-298.

[28] R. Wittmann, *Nicht-tangentielles Randverhalten von harmonischen Funktionen in irregulären Randpunkten*, Archiv Math. 48 (1987), 526-529.

# Acknowledgements

The editors and the publisher wish to thank the following for granting permission to reprint in this volume the papers as listed below. Abbreviations in brackets refer to the Bibliography of Heinz Bauer.

Annales Academiæ Scientiarum Fennicæ (R34)

Annales de l'Institut Fourier (R17, R23, R29)

Birkhäuser Verlag (R13, R16, R21, R27)

Elsevier Science Publishers (R36)

Gius. Laterza & Figli S.p.A. (S21, S26)

L'Enseignement Mathématique (S15)

Mathematical Association of America (S12)

Pacific Journal of Mathematics (R26)

Revue Roumaine de Mathématiques Pures et Appliquées (R25)

Research Institute for Mathematical Sciences, Kyoto University (R39)

Societé Mathématique de France (R10)

Springer-Verlag (R9, R14, R19, R20, R24, R30, R32)

Urban & Fischer Verlag (R35)

# Bibliography

## Textbooks and Monographs


[TM1] *Wahrscheinlichkeitstheorie und Grundzüge der Maßtheorie, Band* I. Sammlung Göschen, Band 1216/1216a, Walter de Gruyter, Berlin 1964.

[TM2] *Harmonische Räume und ihre Potentialtheorie.* Lecture Notes in Math. 22, Springer-Verlag, Berlin–Heidelberg–New York 1966.

[TM3] *Wahrscheinlichkeitstheorie und Grundzüge der Maßtheorie.* Walter de Gruyter, Berlin–New York 1968, 2nd ed. 1973, 3rd ed. 1978.

[TM4] *Probability Theory and Elements of Measure Theory.* Holt, Rinehard & Winston Inc., New York, 1972, 2nd ed.: Academic Press, London–New York–Sydney–San Francisco, 1981 (translation of [TM3]).

[TM5] *Mehrdimensionale Integration.* Sammlung Göschen, Band 2121, Walter de Gruyter, Berlin-New York, 1976 (co-author: B. Anger).

[TM6] *Maße auf topologischen Räumen.* Fernuniversität-Gesamthochschule Hagen, 1984.

[TM7] *Maß- und Integrationstheorie.* Walter de Gruyter, Berlin–New York 1990, 2nd revised ed., 1992.

[TM8] *Wahrscheinlichkeitstheorie.* Walter de Gruyter, Berlin–New York 1991, 4th completely revised ed. of [TM3], 5th ed. 2002.

[TM9] *Probability Theory.* De Gruyter Stud. Math. 23, Walter de Gruyter, Berlin–New York 1996 (translation of [TM8].)

[TM10] *Measure and Integration Theory.* De Gruyter Stud. Math. 26, Walter de Gruyter, Berlin–New York 2001 (translation of [TM7].)


## Lecture Notes


[L1] *Markoffsche Prozesse.* Universität Hamburg, 1963.

[L2] *Konvexität in topologischen Vektorräumen.* Universität Hamburg, 1964.

[L3] *Differential- und Integralrechnung, Band* I. Universitätsbuchhandlung R. Merkel, Erlangen, 1966. 5th ed. 1974, last printing 1984.

[L4] *Differential- und Integralrechnung,* II. Universitätsbuchhandlung R. Merkel, Erlangen, 1967. 4th ed. 1974, last printing 1984.

[L5] *Differential- und Integralrechnung,* IV (co-author: B. Anger). Universitätsbuchhandlung R. Merkel, Erlangen, 1971.

[L6] *Representaciones sobre conjunctos convexos compactos.* Instituto de Matematica, Universidad Nacional del Sur, Bahia Blanca, Argentina, 1971.

## Research Papers

[R1] Eine Rieszsche Bandzerlegung im Raum der Bewertungen eines Verbandes. *Bayer. Akad. Wiss. Math.-Natur. Kl. Sitzungsber.* **1953**, München 1954, 89–117.

[R2] Darstellung additiver Funktionen auf Boolschen Algebren als Mengenfunktionen. *Arch. Math.* **6** (1955), 215–222.

[R3] Reguläre und singuläre Abbildungen eines distributiven Verbandes in einen vollständigen Vektorverband, welche der Funktionalgleichung $f(x \vee y) + f(x \wedge y) = f(x) + f(y)$ genügen. *J. Reine Angew. Math.* **194** (1955), 141–179. (Dissertation, Erlangen 1953.)

[R4] Zur Theorie des Riemann-Integrals in lokal kompakten Räumen. *Bayer. Akad. Wiss. Math.-Natur. Kl. Sitzungsber.* **1955**, München 1956, 187–208.

[R5] Über die Erweiterungen topologischer Räume. *Math. Ann.* **130** (1955), 20–45 (co-author: G. Nöbeling).

[R6] Ergänzung zu „Über die Erweiterungen topologischer Räume". *Math. Ann.* **132** (1957), 451 (co-author: G. Nöbeling).

[R7] Kennzeichnende Eigenschaften der Alexandroffschen Erweiterung $\alpha E$ eines topologischen Raumes. *Bull. Soc. Math. Grèce* **30** (1956), 47–53.

[R8] Allgemeine Approximationskriterien mit Anwendungen. *Jahresber. Deutsch. Math.-Verein.* **58** (1955), 54–72 (co-author: G. Nöbeling).

[R9] Über die Beziehungen einer abstrakten Theorie des Riemann-Integrals zur Theorie Radonscher Maße. *Math. Z.* **65** (1956), 448–482. (Habilitationsschrift, Erlangen 1956.)

[R10] Sur l'équivalence des théories de l'intégration selon N. Bourbaki et selon M. H. Stone. *Bull. Soc. Math. France* **85** (1957), 51–75.

[R11] Über die Fortsetzung positiver Linearformen. *Bayer. Akad. Wiss. Math.-Natur. Kl. Sitzungsber.* **1957**, München 1958, 177–190.

[R12] Geordnete Gruppen mit Zerlegungseigenschaft. *Bayer. Akad. Wiss. Math.-Natur. Kl. Sitzungsber.* **1958**, München 1958, 25–36.

[R13] Minimalstellen von Funktionen und Extremalpunkte. *Arch. Math.* **9** (1958), 389–393.

[R14] Konservative Abbildungen lokal-kompakter Räume. *Math. Ann.* **138** (1959), 398–427.

[R15] Verallgemeinerung eines Faktorisierungssatzes von G. T. Whyburn. *Arch. Math.* **10** (1959), 373–378.

[R16] Minimalstellen von Funktionen und Extremalpunkte. II. *Arch. Math.* **11** (1960), 200–205.

[R17] Šilovscher Rand und Dirichletsches Problem. *Ann. Inst. Fourier (Grenoble)* **11** (1961), 89–136.

[R18] Supermartingale und Choquet-Rand. *Arch. Math.* **12** (1961), 210–223.

[R19] Axiomatische Behandlung des Dirichletschen Problems für elliptische und parabolische Differentialgleichungen. *Math. Ann.* **146** (1962), 1–59.

[R20] Weiterführung einer axiomatischen Potentialtheorie ohne Kern (Existenz von Potentialen). *Z. Wahrscheinlichkeitstheorie Verw. Gebiete* **1** (1963), 197–229.

[R21] Kennzeichnung kompakter Simplexe mit abgeschlossener Extremalpunktmenge. *Arch. Math.* **14** (1963), 415–421.

[R22] Darstellung von Bilinearformen auf Funktionenalgebren durch Integrale. *Math. Z.* **85** (1964), 107–115.

[R23] Propriétés fines des fonctions hyperharmoniques dans une théorie axiomatique du potentiel. *Ann. Inst. Fourier (Grenoble)* **15** (1965), 137–154. Also in: *La Théorie du Potentiel*, Colloque Internationaux du Centre National de la Recherche Scientifique, Éditions du C.N.R.S., Paris 1965.

[R24] Zum Cauchyschen und Dirichletschen Problem bei elliptischen und parabolischen Differentialgleichungen. *Math. Ann.* **164** (1966), 142–153.

[R25] Mesures avec une image donnée. *Rev. Roumaine Math. Pures Appl.* **11** (1966), 747–752.

[R26] The part metric in convex sets. *Pacific J. Math.* **30** (1969), 15–33 (co-author: H. S. Bear).

[R27] An open mapping theorem for convex sets with only one part. *Aequationes Math.* **4** (1970), 332–337.

[R28] Intern vollständige konvexe Mengen. In: *Theory of sets and topology*, Deutscher Verlag der Wissenschaften, Berlin 1972, 27–38.

[R29] Theorems of Korovkin type for adapted spaces. *Ann. Inst. Fourier (Grenoble)* **23** (1973), 245–260.

[R30] Convergence of monotone operators. *Math. Z.* **136** (1974), 315–330.

[R31] Funktionenkegel und Integralungleichungen. *Bayer. Akad. Wiss. Math.-Natur. Kl. Sitzungsber.* **1977**, München 1978, 53–61.

[R32] Korovkin approximation in $\mathscr{C}_0(X)$. *Math. Ann.* **236** (1978), 225–237 (co-author: Klaus Donner).

[R33] Determination of Korovkin closures. *Math. Z.* **168** (1979), 263–274 (co-authors: Gottlieb Leha and Susanne Papadopoulou).

[R34] Heat balls and Fulks measures. *Ann. Acad. Sci. Fenn. Ser. A I Math.* **10** (1985), 67–82.

[R35] Simplicial function spaces and simplexes. *Exposition. Math.* **3** (1985), 165–168.

[R36] Fine boundary limits of harmonic and caloric functions. *Bull. Sci. Math.* (2) **109** (1985), 337–361.

[R37] A class of means and related inequalities. *Manuscripta Math.* **55** (1986), 199–211.

[R38] Mittelwerte und Funktionalgleichungen. *Bayer. Akad. Wiss. Math.-Natur. Kl. Sitzungsber.* **1986**, München 1987, 1–9.

[R39] Fine boundary limits and maximal sequences. In: *Potential theory and Its Related Fields*, RIMS Kokyuroku 610, Res. Inst. Math. Sci., Kyoto University, Kyoto 1987, 85–102.

## Surveys, Conference and Seminar Contributions

[S1] Topologische Kennzeichnung des total-additiven und rein-endlich-additiven Teils einer additiven Mengenfunktion. In: *Proc. Internat. Congress of Mathematicians*, Amsterdam 1954, Vol. II, Groningen–Amsterdam 1954, 83.

[S2] Frontière de Šilov et problème de Dirichlet. In: *Séminaire de Théorie du Potentiel* dirigé par M. Brelot, G. Choquet et J. Deny, 3e année, 1958/59, Exposé No.7, 23 p., Institut H. Poincaré, Paris 1960.

[S3] Cônes convexes semi-réticulés de fonctions continues. Théorèmes de représentation et de stabilité. In: *Séminaire de Théorie du Potentiel* dirigé par M. Brelot, G. Choquet et J. Deny, 8e année, 1963/64, Exposé No.5, 2 p., Institut H. Poincaré, Paris 1965.

[S4] Einige Eigenschaften lokal-kompakter konvexer Mengen und ihrer projektiven Limiten. In: *Proc. Coll. Convexity*, Copenhagen 1965, Københavns Universitets Matematiske Institut, 1967, 143–153 (co-author: D. Hinrichsen).

[S5] Aspects of linearity in the theory of function algebras. In: *Function Algebras* (edited by F. T. Birtel), Scott, Foresman and Company, Glenview, IL, 1966, 122–137.

[S6] Choquetscher Rand und Integraldarstellungen. *Jahresber. Deutsch. Math.-Verein.* **69** (1967), 89–104.

[S7] Construction of Markov processes in axiomatic potential theory. In: *Markov Processes and Potential Theory* (edited by Joshua Chover), J. Wiley & Sons, Inc., New York–London–Sydney 1967, 3–11.

[S8] Recent developments in axiomatic potential theory. In: *Symposium on Probability Methods in Analysis*, Lecture Notes in Math. 31, Springer-Verlag, Berlin–Heidelberg–New York 1967, 20–27.

[S9] Harmonic spaces and associated Markov processes. In: *Potential Theory* (coordinator: M. Brelot), Centro Internazionale Matematico Estivo (C.I.M.E.), 1969, Edizioni Cremonese, Roma 1970, 25–67.

[S10] Aspects of modern potential theory. In: *Proc. Internat. Congress of Mathematicians*, Vancouver 1974, Vol. I, Canad. Math. Congress, Montreal, Que., 1975, 41–51.

[S11] Approximationssätze und abstrakte Ränder. *Math.-Phys. Semesterber.* **23** (1976), 141–173.

[S12] Approximation and abstract boundaries. *Amer. Math. Monthly* **85** (1978), 632–647.

[S13] Korovkin approximation in function spaces. In: *Approximation Theory and Functional Analysis* (edited by J. B. Prolla), North-Holland Math. Stud. 35, North-Holland Publ. Co., Amsterdam–New York–Oxford 1979, 19–36.

[S14] Elliptische harmonische Strukturen. In: *Elliptische Differentialgleichungen*, Vorträge der Tagung in Rostock 10–16.10.1977, Math. Ges. der DDR, Univ. Rostock, Rostock 1978, 59–70.

[S15] Halbgruppen und Resolventen in der Potentialtheorie. *Enseign. Math.* (2) **25** (1979), 9–22. Also in: *Contributions to Analysis*. International Symposium on Analysis in Honour of Professor Albert Pfluger, Zürich 1978. Enseign. Math. 27, Université de Genève 1979, 21–34.

[S16] Harmonische Räume. In: *Jahrbuch Überblicke Mathematik* 1981 (edited by S. D. Chatterji, B. Fuchssteiner, U. Kulisch, D. Laugwitz, R. Liedl), Bibliogr. Institut, Mannheim 1981, 9–35.

[S17] Détermination de la fermeture de Korovkin. In: *Séminaire d'Initiation à l'Analyse* (G. Choquet, M. Rogalski, J. Saint-Raymond), 19ᵉ année, 1979/80, Exposé No. 1, 2 p., Univ. Pierre et Marie Curie, Paris.

[S18] Sur l'approximation de Korovkin dans $C_0(X)$ et sur un théorème de Choquet-Deny. In: *Séminaire d'Initiation à l'Analyse* (G. Choquet, M. Rogalski, J. Saint-Raymond), 19ᵉ année, 1979/80, Exposé No. 2, 2 p., Univ. Pierre et Marie Curie, Paris.

[S19] Aproximace a abstraktní hranice. *Pokroky Mat. Fyz. & Astr.* **26** (6) (1981), 305–326. (Translation of [S12] into Czech language.)

[S20] Zum heutigen Bild der Potentialtheorie. In: *Zum Werk Leonhard Eulers – Vorträge des Euler-Kolloqiums 1983 in Berlin* (edited by E. Knobloch, I. S. Louhivaara, J. Winkler), Birkhäuser Verlag, Basel–Boston–Stuttgart 1984, 3–20.

[S21] Harmonic spaces – a survey. *Conferenze del Seminario di Matematica dell' Università di Bari* **197**, 34p., Gius. Laterza & Figli S.p.A., Bari 1984.

[S22] Elliptic differential operators and diffusion processes. *Bull. Austral. Math. Soc.* **30** (1984), 219–237.

[S23] Korovkin closures in $\mathscr{C}_0(X)$. In: *Aspects of Mathematics and its Application* (edited by J. A. Barroso), North-Holland Math. Library 34, Elsevier Science Publishers B. V. (North-Holland), Amsterdam 1986, 145–167 (co-author: Klaus Donner).

[S24] Simplices in potential theory. In: *Aspects of Positivity in Functional Analysis* (edited by R. Nagel, U. Schlotterbeck, M. P. H. Wolff), North-Holland Math. Stud. 122, Elsevier Science Publishers B.V. (North-Holland), Amsterdam 1986, 27–39.

[S25] Dirichlet problem and irregular boundary points. *Arkhimedes* **4** (1991), 290–294.

[S26] Behaviour of solutions of elliptic-parabolic differential equations at irregular boundary points. In: *Atti del Convegno Internazionale su Recenti Sviluppi in Analisi Matematica e sue Applicazioni, Conferenze del Seminario di Matematica dell'Università di Bari* **237–244**. Gius. Laterza & Figli S.p.A., Bari 1991, 27–37.

[S27] Recollections, reflections, results – Introductory thoughts to the ICPT 1994. In: *Potential Theory – ICPT 94* (edited by J. Král, J. Lukeš, I. Netuka, J. Veselý), Walter de Gruyter, Berlin–New York 1996, 1–8.

## Preliminary Communications

[P1] Caractérisation topologique de la partie complètement additive et de la partie purement additive d'une fonction additive d'ensemble. *C. R. Acad. Sci. Paris* **238** (1954), 1771–1773.

[P2] Sur le prolongement des formes linéaires positives dans un espace vectoriel ordonné. *C. R. Acad. Sci. Paris* **244** (1957), 289–292.

[P3] Sur l'existence de mesures avec une image donnée. *C. R. Acad. Sci. Paris* **246** (1958), 1953–1955.

[P4] Un problème de Dirichlet pour la frontière de Šilov d'un espace compact. *C. R. Acad. Sci. Paris* **247** (1958), 843–846.

[P5] Une axiomatique du problème de Dirichlet pour certaines équations aux dérivées partielles elliptiques et paraboliques. *C. R. Acad. Sci. Paris* **250** (1960), 2672–2674.

[P6] Sur les mesures de Fulks dans la théorie de l'équation de la chaleur. *C. R. Acad. Sci. Paris Sér. I Math.* **298** (1984), 165–167.

## Other Publications

[O1] Siegfried Guber (Obituary). *Jahresber. Deutsch. Math.-Verein.* **72** (1970), 1–2.

[O2] Preface to: C. Constantinescu and A. Cornea, *Potential Theory on Harmonic Spaces*. Grundlehren Math. Wiss. 158, Springer-Verlag, Berlin–Heidelberg–New York 1972.

[O3] Denkschrift zum Mathematikunterricht an Gymnasien. *Der math. u. naturw. Unterricht* **29/7** (1976), 431–434.

[O4] Grußadresse des Vorsitzenden der Deutschen Mathematiker-Vereinigung. *Internat. Math. Nachrichten* (Österreichische Math. Ges.), 31. Jahrg., Nr. 117 (1977), 4–6.

[O5] Hans Richter (2.5.1912–3.12.1978). *Bayer. Akad. d. Wiss., Jahrbuch* **1979**, München 1979, 284–286.

[O6] Eberhard Hopf (17.4.1902–24.7.1983). *Bayer. Akad. d. Wiss., Jahrbuch* **1984**, München 1984, 254–256.

[O7] Radons Arbeiten zur Maß- und Integrationstheorie sowie zur Potentialtheorie / The work of Johann Radon on measure theory, integration theory and potential theory. (Deutsch mit englischer Übersetzung.) In: *Johann Radon, Gesammelte Abhandlungen / Collected Works* (Herausgegeben von / edited by P. M. Gruber, E. Hlawka, W. Nöbauer, L. Schmetterer), Vol. 1, Verlag Österr. Akad. d. Wiss., Wien, und Birkhäuser Verlag, Basel–Boston 1987, 29–44.

[O8] Otto Haupt – zum 100. Geburtstag. *Aequationes Math.* **32** (1987), 1–5.

[O9] Otto Haupt – Zu Person und Werk. In: *Otto und Edith Haupt-Stiftung*, Univ. Erlangen-Nürnberg 1988, 16–29.

[O10] Marcel Brelot (29.12.1903–3.8.1987). *Bayer. Akad. d. Wiss., Jahrbuch* **1987**, München 1988, 273–275.

[O11] Preface to: Werner Fenchel, *Elementary Geometry in Hyperbolic Spaces*. De Gruyter Stud. Math. 11, Walter de Gruyter, Berlin–New York 1989.

[O12] Zur Entstehungsgeschichte der Theorie der harmonischen Räume. *Österr. Zeitschrift f. Statistik und Informatik* **19** (1989), 175–183.

[O13] Werner Fenchel (3.5.1905–24.1.1988). *Bayer. Akad. d. Wiss., Jahrbuch* **1988**, München 1989, 213–214.

[O14] Otto Haupt (5.3.1887–10.11.1988). *Bayer. Akad. d. Wiss., Jahrbuch* **1988**, München 1989, 234–237.

[O15] Quelques souvenirs à propos de Louis Antoine. In: *Journée Louis Antoine 1988*, Publ. Inst. Rech. Math. Rennes (IRMAR), Rennes 1989, 85–90.

[O16] Laudation for Professor Hans Föllmer (GMÖOR-Preis). *Newsletter, Ges. für Mathematik, Ökonomie und Operations Research* (GMÖOR) 4/89 (1989), 2–4.

[O17] Otto Haupt – Zu Person und Werk. *Jahresber. Deutsch. Math.-Verein.* **92** (1990), 169–181.

[O18] Otto Haupt – Zu Person und Werk. *Sitzungsber. Phys.-Med. Soz. Erlangen* (N.F.) **2** (4) (1990), 33–43.

[O19] Laudatio auf Prof. Dr. Dr. h.c. mult. Edmund Hlawka. *Internat. Math. Nachrichten* (Österreichische Math. Gesellschaft), 46. Jahrg., Nr. 159 (1992), 1–6.

[O20] Dankesrede von Heinz Bauer bei der feierlichen Ehrenpromotion in der Aula der Karls-Universität am 15. Oktober 1992. In: *Prof. Dr. Heinz Bauer, doctor honoris causa Universitatis Carolinae*. Private printing 1992, 15–19 (Czech translation, 20–23).

[O21] Wilhelm Maak (13.8.1912–6.6.1992). *Bayer. Akad. d. Wiss., Jahrbuch* **1992**, München 1993, 239–241.

[O22] Christian Pauc – Ein französischer Mathematiker als Wegbereiter der deutsch-französischen Freundschaft. *Das neue Erlangen*, Heft 92, Verlag Univ.-Buchhandl. Rudolf Merkel, Erlangen 1994, 4–5 (co-authors: H. Aldebert, J. Boclé).

[O23] Christian Pauc – un mathématicien français précurseur de l'amitié franco-allemande. *Das neue Erlangen*, Heft 92, Verlag Univ.-Buchhandl. Rudolf Merkel, Erlangen 1994, 6–7. (French version of [O22]; co-authors: H. Aldebert, J. Boclé.)

[O24] Dankesrede bei der Verleihung der Ehrendoktorwürde durch die Technische Universität Dresden. In: *Ehrenpromotion Prof. Dr. Dr. h.c. Heinz Bauer* (edited by Rektor der TU Dresden), 1994, 15–23.

[O25] Dank an Dresden. *Wiss. Z. TU Dresden* **43**/6 (1994), 3–5.

[O26] 50 Jahre Mathematisches Forschungsinstitut Oberwolfach: Verantwortung und Herausforderung. *DMV Mitteilungen*, Nr. 2 (1995), 10–14.

[O27] Christian Pauc – Ein französischer Mathematiker als Wegbereiter der deutsch-französischen Freundschaft. *DMV Mitteilungen*, Nr. 2 (1995), 24–26. (Modified and supplemented version of [O22]; co-authors: H. Aldebert, J. Boclé.)

[O28] Comments on Hans Hahn's work in measure theory. In: *Hans Hahn, Gesammelte Abhandlungen / Collected Works*, Vol. 3 (Herausgegeben von / edited by L. Schmetterer, K. Sigmund), Springer-Verlag, Wien–New York 1997, 1–14.

(The bibliography does not include Bauer's book reviews or reviews for Zentralblatt or Mathematical Reviews.)

Bei Fragen zur Produktsicherheit wenden Sie sich bitte an:
If you have any questions regarding product safety,
please contact:

Walter de Gruyter GmbH
Genthiner Straße 13
10785 Berlin
productsafety@degruyterbrill.com